WITHDRAWN
FROM
UNIVERSITIES
AT
MEDWAY
LIBRARY

D1426526

ENGINEERING DESIGN

A Synthesis of Stress Analysis and Materials Engineering

ENGINEERING DESIGN

A Synthesis of Stress Analysis and Materials Engineering

SECOND EDITION

Joseph H. Faupel, Ph.D.

Engineering Department
E. I. du Pont de Nemours & Co.

Fellow, The American Society of Mechanical Engineers

Franklin E. Fisher, Ph.D.

Professor of Mechanical Engineering
Loyola Marymount University

A WILEY-INTERSCIENCE PUBLICATION

John Wiley & Sons New York · Chichester · Brisbane · Toronto

Library of Congress Cataloging in Publication Data:

Faupel, Joseph Herman, 1916–
Engineering design.

"A Wiley-Interscience publication."
Includes bibliographies and indexes.
1. Engineering design. I. Fisher, Franklin E., 1933– joint author. II. Title.

TA174. F3 1980 620.1′12 80–16727
ISBN 0–471–03381–2

Photoset by Thomson Press (I) Limited, New Delhi and Printed in the United States of America

10 9 8 7 6 5 4 3 2 1

To Mary

To my late father
William Alfred Fisher

PREFACE

In this second edition the philosophy and intent of the first edition have been maintained. The addition of new chapters and new material to existing chapters reflects the current literature and the growing influence of computer technology on problem solving. A new feature is the addition of problems at the end of each chapter to facilitate using the book as a text for a two-semester course.

Many friends and associates offered aid and encouragement for which we are grateful. In particular, we are indebted to students who worked on and contributed suggestions for many of the homework problems; to Dr. James Foxworthy, Dean of the College of Science and Engineering, and Dr. Joseph Callinan, Chairman of the Mechanical Engineering Department, both of Loyola Marymount University, and to the management of the Engineering Department of the Du Pont Company.

The influence of the late Dr. Joseph Marin set the tone for the first edition. The second edition has benefited from the added influence of Professor Robert M. Rivello, author of *Theory and Analysis of Flight Structures*, and Professor John W. Jackson, who taught a year-long course from the first edition.

Like the first edition, the second edition could not have come to fruition without the assistance of many persons who contributed by their discussions. We are particularly indebted to the authors whose works are quoted in the text and to our wives and families for their support. It is also a pleasure to acknowledge the editorial and secretarial assistance contributed by Roxana Letamendi and Carol McMillan.

JOSEPH H. FAUPEL

Wilmington, Delaware

FRANKLIN E. FISHER

Los Angeles, California

April 1980

PREFACE TO FIRST EDITION

Many books, particularly those written by university professors, evolve over a span of years and represent a distillation of knowledge into a compact exposition of a discipline for others to follow. On the other hand, books are frequently written in response to a need in a particular field, and they represent the author's views of the subject at the time of writing; this book had its beginning somewhere in between.

The material given here is an integrated treatment of stress analysis and materials engineering based on lectures given for several years in the graduate division of mechanical engineering at the University of Delaware. The contents represent a combination of these fields at a level somewhat beyond that given in elementary texts, yet not approaching the degree of sophistication found in advanced treatises. The reader will find here the conventional subjects treated in elementary texts, but, in many cases, with extensions. For example, some beam problems are solved by numerical integration, Maclaurin's series, Laplace transform, and others. In addition, such subjects as minimum weight analysis, ductility and brittleness of materials, analysis of composite, honeycomb and reinforced materials, designing with plastics, metal-working and limit analysis in the plastic range, prestressing for strength, strength under combined stress, dynamic behavior of materials, stability and buckling, thermal stress analysis, creep, stress rupture, fatigue and stress concentration are presented from the point of view of the practicing engineer.

By the very nature of the contents, the format of this book departs considerably from what might be considered standard, but no apology is made for this. My purpose has been to present practical information in a form useful to a diverse audience and not to sacrifice clarity or usefulness for consistency or form; discussions with teachers, engineers, and scientists, many of whom were my students, have confirmed this view. Although directed principally to practicing engineers, this book can also serve as a reference for students and as a text for an intermediate or graduate course in engineering design.

The preparation required the cooperation and assistance of many people and I am indebted to my many friends and associates for their kind encouragement and assistance. I am particularly indebted to all those whose works are quoted in the text and to my graduate-school-day professors, Dr. Joseph Marin and Dr. Maxwell Gensamer. It is also a pleasure to acknowledge the generous support of H. C. Vernon, Director of the Du Pont Company's Engineering Research Laboratory, and Miss Jennie Di Bartolomeo, who supplied valuable editorial and secretarial assistance.

J. H. Faupel

Wilmington, Delaware
August 1964

CONTENTS

ENGINEERING DESIGN

A Synthesis of Stress Analysis and Materials Engineering

chapter 1

MATERIALS AND PROPERTIES

1-1 THE NATURE AND PROPERTIES OF MATERIALS

Theoretical strength of materials

The deformation of materials can be considered on at least three levels of the division of matter. At the atomic and molecular levels, strength is associated with elemental forms of matter of the order of 10^{-8} in. held together by electronic forces. At the microscopic level, groups of elemental particles of the order of 10^{-8} to 10^{-4} in. are also held together by electronic forces, but these forces are reduced by defects and imperfections in the structure. At the macroscopic or phenomenological level, which is of more interest to the engineer, the structure is held together by some average force determined by the defects and the electronic forces. Theoretically, the strength of a material should be reflected by the forces at the atomic level. However, because of defects in the structure, the practical strength of materials is several orders of magnitude less than theory would predict.

An estimation of the theoretical strength of a material may be made as follows. Consider Fig. 1.1, which shows an ideal elemental form of matter consisting of two rows of atoms. The shear stress necessary to produce slip by a distance x will be a function of x. Assuming that the attractive and repulsive forces are balanced in the position shown, there will be zero stress required to maintain the arrangement. If particle C is moved to a position m–n between A and B (Fig. 1.2), a force results which is zero at m–n, positive to the left of m–n and negative to the right; by assuming a sinusoidal stress variation, Fig. 1.3 is obtained. This figure shows how stress varies in the lattice structure. The shear stress is zero at 0, $d/2$, and d. For intermediate locations

$$\tau = f(\tau) \sin 2\pi x/d \qquad 1.1$$

or from Fig. 1.3, for small values of x,

$$\tau = (\tau_c) \sin 2\pi x/d = \tau_c(2\pi x/d) \qquad 1.2$$

where τ_c is the peak stress developed (maximum theoretical shear strength of the material). By virtue of Hooke's law and since shear strain equals x/a, the

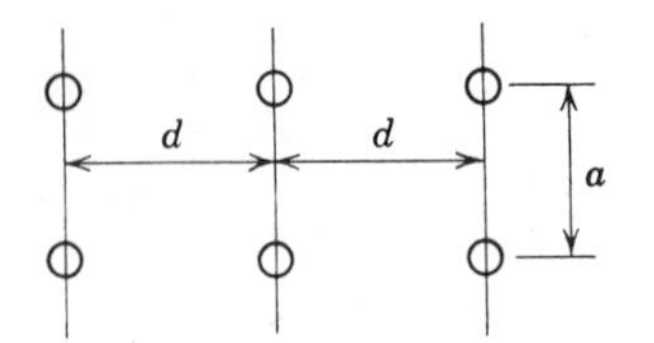

Fig. 1.1 Ideal elemental arrangement of atoms.

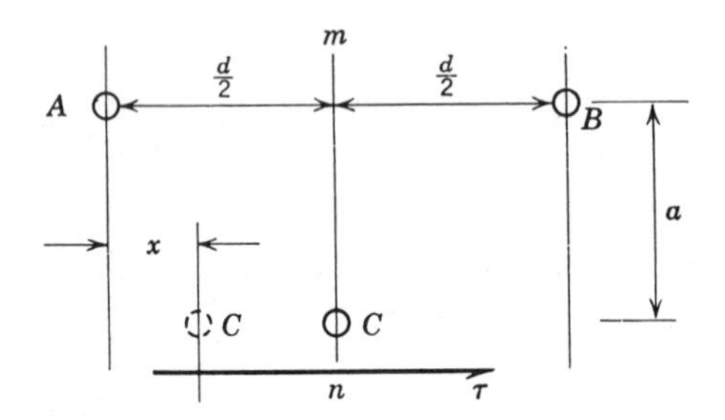

Fig. 1.2 Atoms displaced by applied load.

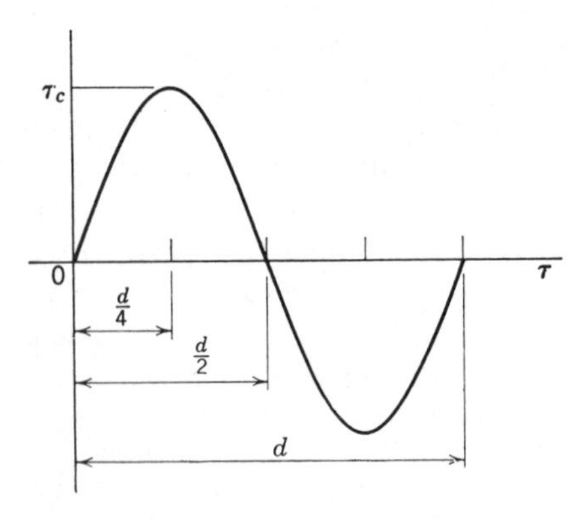

Fig. 1.3 Stress variation in lattice.

stress is given by the relationship

$$\tau = G(x/a) = \tau_c \sin 2\pi x/d \qquad 1.3$$

where G is the shear modulus of elasticity. From Eqs. 1.2 and 1.3 the theoretical maximum shear strength of a material is

$$\tau_c = G(d/2\pi a) \qquad 1.4$$

This means that for mild steel, for example, where $d = a$, the shear stress would be about $(11.5 \times 10^6)/2\pi$ or 1.8×10^6 psi. The observed value is about 3.0×10^4 psi, or about $\frac{1}{60}$ of the theoretical value. This large discrepancy is believed due to the presence of *dislocations* in the material structure (4).

Strain behavior

Most structural materials* exhibit one or more of the following types of strain: linear elastic, nonlinear elastic, viscoelastic, plastic, and anelastic. In linear

*Typical tensile stress-strain curves for some structural metals are shown in Figs. 1.4*B* and 1.4*C*.

elastic deformation, stress and strain are related by Hooke's law

$$\text{strain} = \text{stress/constant}$$

For uniaxial loading, Fig. 1.4*A*, this relationship becomes

$$\varepsilon = \sigma/E \qquad 1.5$$

where ε is strain, σ is the applied stress, and E is Young's modulus of elasticity. In nonlinear elastic behavior, the material behaves elastically in that no permanent residual deformation results when the applied load is removed from the

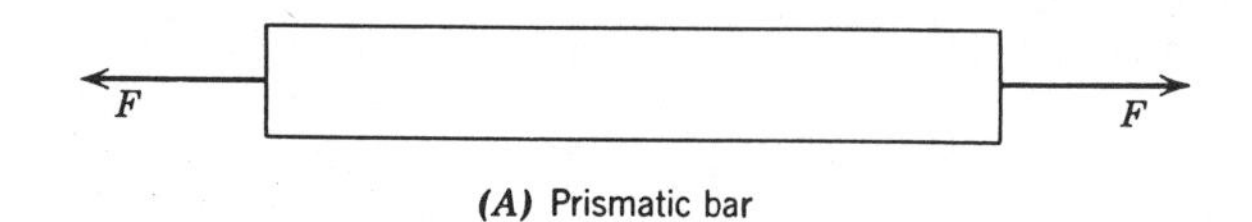

(A) Prismatic bar

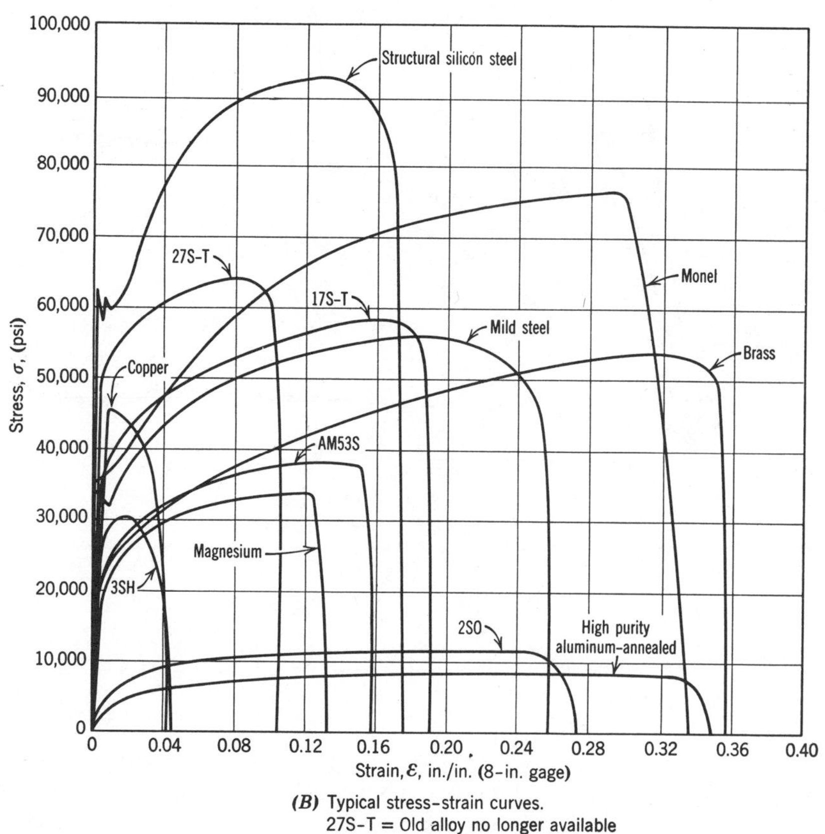

(B) Typical stress-strain curves.
27S-T = Old alloy no longer available
17S-T = 2017-T4
3SH = 3003-H14
2SO = 1100-0

Fig. 1.4 Tensile or uniaxial loading of a bar of material. [*After Templin and Sturm, (27), Courtesy of The Institute of Aeronautical Sciences.*]

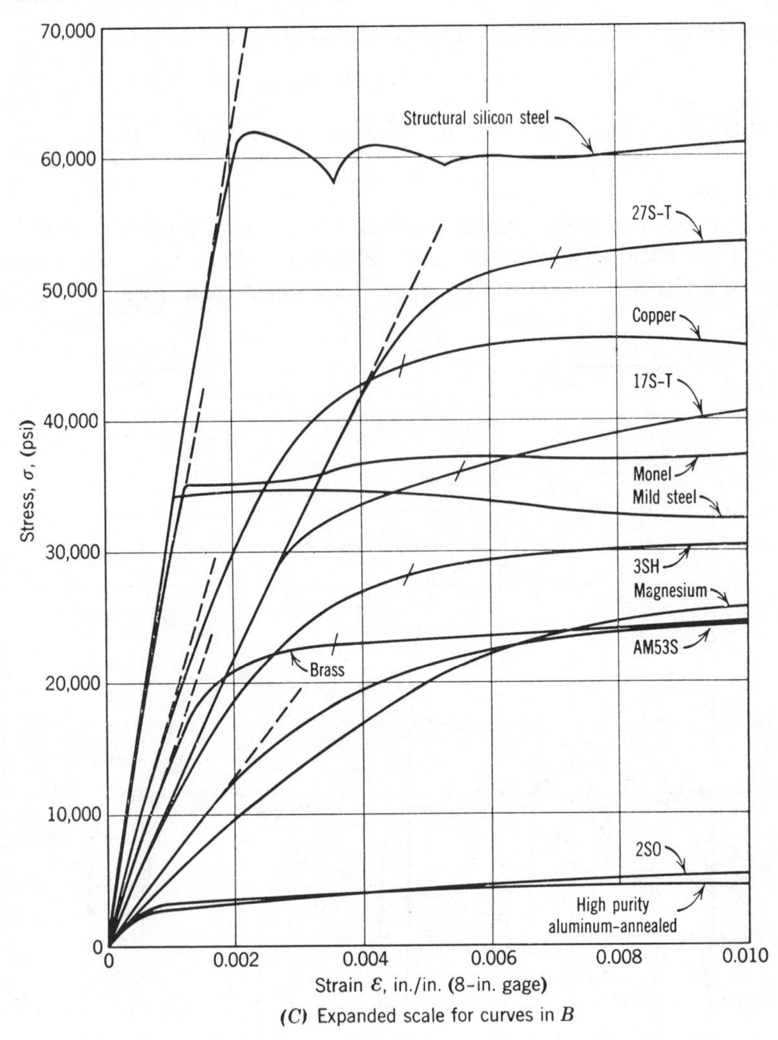

(*C*) Expanded scale for curves in *B*

Fig. 1.4 (Continued).

body. The relation between stress and strain is not linear as expressed by Eq. 1.5 but takes the general form

$$\varepsilon = (\sigma/\lambda)^n \tag{1.6}$$

where λ is a pseudoelastic modulus and n is a constant, both determined experimentally. This relationship is used in Chapter 15. In viscoelastic behavior there are strong resemblances to linear elasticity. In linear elastic strain the strain stops if the loading stops, whereas in viscoelastic deformation straining continues even though the loading stops and a residual strain remains when all the load is removed. Viscoelasticity is discussed in Chapter 5. In plastic deformation there is always a residual deformation on removal of the applied load. Anelastic behavior is characterized by noninstantaneous strain at small stress levels that is recoverable after a period of time (34). For practical structural applications the engineer will not ordinarily be concerned with anelasticity.

Simple static properties of materials

Load-deformation characteristics of materials are of great practical interest, and several common types of curves obtained in tension tests of materials have been observed. "Brittle" materials are characterized by a curve such as shown in Fig. 1.5*A*; the material deforms in accordance with Hooke's law (Eq. 1.5) to

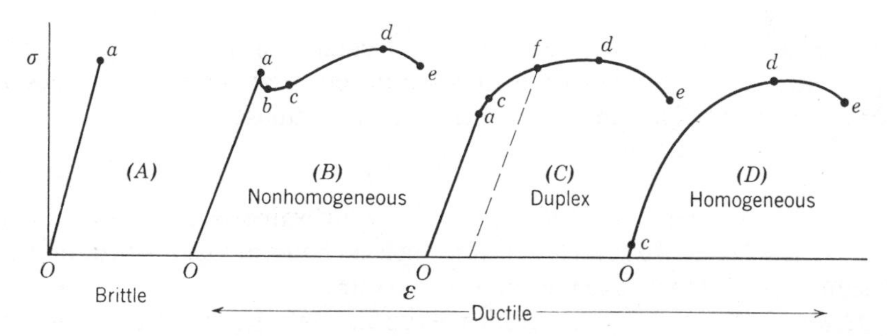

Fig. 1.5 Types of tension stress-strain curves.

fracture, point a. Materials such as some grades of tool steel, glass, and many ceramics, when tested in tension at room temperature, exhibit the characteristics shown in Fig. 1.5*A*. Most "ductile" materials exhibit behavior characterized by Figs. 1.5*B*, 1.5*C*, or 1.5*D*. *Nonhomogeneous* deformation (Fig. 1.5*B*) is most closely associated with low-carbon steel. Initially, the deformation is linearly elastic, line Oa; this is followed by a sudden drop in stress to b and then gradual stress increase through c to a maximum at d and finally fracture at e. In *duplex* deformation (Fig. 1.5*C*) there is an initial straight-line portion Oa, followed by gradual stress increase c and f, reaching a maximum at d and fracturing at e. In *homogeneous* deformation (Fig. 1.5*D*) there is no linear portion but continuously "smooth" stress buildup to d followed by fracture at e.

Modulus of elasticity. The modulus of elasticity E in Eq. 1.5 is a measure of the *inherent rigidity* of a material and is defined by the straight-line portion Oa of Fig. 1.5, and by OA in Fig. 1.6; thus, from Fig. 1.6,

$$E = \sigma/\varepsilon = HA/OH \qquad 1.7$$

Secant and tangent moduli. The *secant modulus*, frequently used in the theory of plasticity (Chapter 6), is a variable defined by Eq. 1.7 but represented by line OF in Fig. 1.6; that is,

$$\text{secant modulus} = E_s = \sigma/\varepsilon = CJ/OJ \qquad 1.8$$

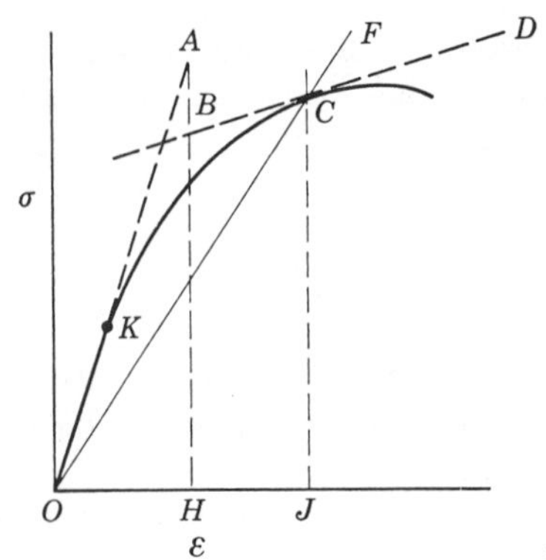

Fig. 1.6 Various moduli of elasticity.

The *tangent modulus* measures the rate of change of stress to strain and is represented by a tangent to the stress-strain curve at the point in question. For example, at point C in Fig. 1.6, the tangent modulus is

$$E_T = (d\sigma/d\varepsilon)_c \tag{1.9}$$

In the elastic range (OK, Fig. 1.6), all moduli values are identical; for the special case of stress-strain relationships such as shown in Fig. 1.5*D*, the modulus is defined as the tangent to the curve at zero stress.

Modulus of rupture. The modulus of rupture is the fracture stress in bending calculated by means of the formula $\sigma = M/Z$ discussed in Chapter 2.

Proportional limit. The proportional limit is defined as the terminus of the straight-line portion of the stress-strain curve; that is, it is the upper limit for which Hooke's law is valid (*a* in Fig. 1.5).

Elastic limit. The elastic limit defines a range of strain characterized by the absence of residual deformation on release of load (*c* in Fig. 1.5).

Upper yield point. An upper yield stress (*a*, Fig. 1.5*B*) is characteristic of mild steel, but it is believed to be a fictitious value obtained as a result of the deformation characteristics of steel and inertia effects in the testing machine.

Lower yield point. The lower yield point is also characteristic of low-carbon steel and is defined by *b* in Fig. 1.5*B*. Sometimes there is a "flat" at *b* (deformation at constant stress) as shown in Fig. 1.7. For behavior characterized by Fig. 1.7

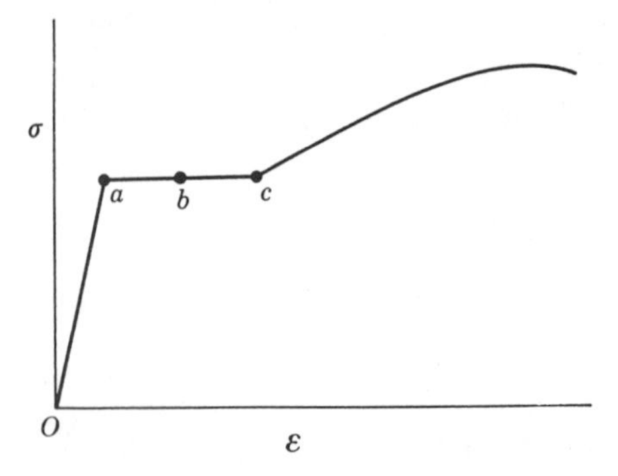

Fig. 1.7 Tensile deformation at constant stress.

the portion of the curve from a to c defines yielding and is frequently referred to as the *yield stress*, *lower yield stress*, or *fundamental yield stress*.

Offset yield strength. For materials characterized by the plots in Figs. 1.5*C* and 1.5*D*, there is no well-defined stress at which yielding occurs. Actually, the elastic limit defines this point, but this is frequently nearly impossible to discern. To overcome this difficulty the *offset yield strength* has been introduced and is defined as the stress corresponding to the intersection of the stress-strain curve with a parallel to the straight-line portion, with origin at some finite strain value (f in Fig. 1.5, for example). The strain offset is usually 0.01% or 0.20%. This stress has also been referred to as the *proof stress*.

Tensile strength. Tensile strength, also called the *ultimate strength*, is the greatest stress sustained per unit load by the material. It is characterized by the beginning of "necking down," or local instability, and is identified by d in Fig. 1.5.

Fracture strength. Fracture strength is, as its name implies, the stress at fracture (point e, Fig. 1.5). This value has little design significance, for it is, like the upper yield point, a fictitious value dependent on or influenced by structure, testing conditions, and the testing machine.

Working stress. The working or design stress is the yield or ultimate stress divided by a factor of safety.

Ductility. Ductility is a measure of the deformability of a material. In tension, the ductility is measured by *elongation* and *reduction of area* as

$$\varepsilon = \frac{L - L_0}{L_0} \qquad 1.10$$

$$q = \frac{A_0 - A}{A_0} \qquad 1.11$$

where ε is unit strain (in./in.), q is reduction of area, L is the final gauge length, L_0 is the initial gauge length, A is the final cross-sectional area, and A_0 is the initial cross-sectional area.

Shear modulus of elasticity. The shear modulus of elasticity, also called the *modulus of rigidity*, is the constant of proportionality between shear stress τ and shear strain γ in the elastic range of strain and has the same significance to shear as E, Young's modulus, has to tension stresses. Thus by definition

$$\gamma = \tau/G \qquad 1.12$$

It is shown in Chapter 3 that there is a relation between E and G that takes the form

$$G = \frac{E}{2(1 + \nu)} \qquad 1.13$$

where ν is Poission's ratio.

Poisson's ratio. Poisson's ratio ν is a measure of the unit strain of a material in the directions normal to the applied load. For example, in tension, if σ_x is the tensile stress and ε_x the strain, then

$$\varepsilon_x = \sigma_x/E \tag{1.14}$$

and the unit lateral contraction in the other two cartesian directions is

$$\varepsilon_y = \varepsilon_z = -\nu(\sigma_x/E) \tag{1.15}$$

Poisson's ratio, like modulus, is a constant depending on the material in the elastic range, and, as shown later, equal to 0.5 for all incompressible materials and materials operating in the plastic range.

Bulk modulus. Bulk modulus B, also called *modulus of volume expansion* and *modulus of volume elasticity*, is a measure of the elastic volume change in a material and is defined as

$$B = \frac{E}{3(1 - 2\nu)} \tag{1.16}$$

Compressibility. Compressibility C of a material (elastically) is associated with the reciprocal of bulk modulus; thus,

$$C = 1/B \tag{1.17}$$

Resilience. Resilience is defined as that property of a material by virtue of which it can release elastic strain energy as stress is removed. The measure of resilience most commonly used is the area enclosed by a stress-strain curve within the elastic limit. This is called the *modulus of elastic resilience U*.

RESILIENCE IN TENSION. A hypothetical tensile stress-strain curve is shown in Fig. 1.8. In this figure resilience would be considered as the area OAB. In terms of mechanical properties, if σ_y is the elastic limit stress and ε_y is the elastic limit strain, then

$$U = (OB)(AB)/2 = \varepsilon_y\sigma_y/2 \tag{1.18}$$

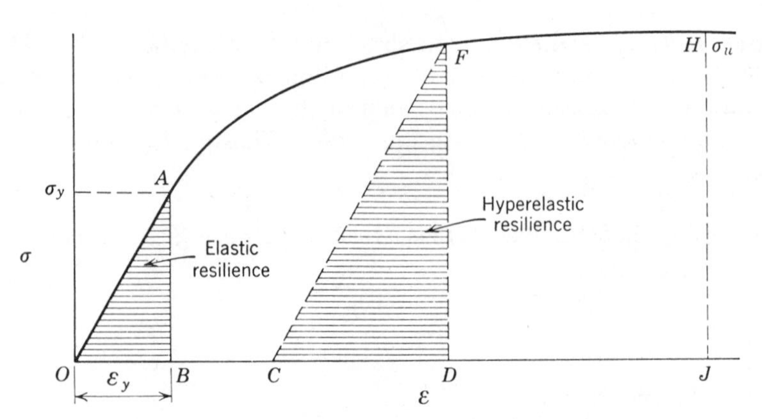

Fig. 1.8 Definition of areas of resilience.

or, since $\sigma_y = E\varepsilon_y$, Eq. 1.18 becomes

$$U = \sigma_y{}^2/2E \qquad 1.19$$

If load is released in the plastic range, say at F in Fig. 1.8 and recovery is linear (along FC), energy is released that is measured by the area CFD called *hyper-elastic resilience* U', which is defined by the relationship

$$U' = \sigma^2/2E \qquad 1.20$$

Examination of Eqs. 1.19 and 1.20 shows that maximum resilience is obtained by a combination of high strength and low modulus.

RESILIENCE IN BENDING. The modulus of elastic resilience in bending is most conveniently obtained by reference to the theory of elastic strain energy, which is discussed in Chapter 7.

It is shown in Chapter 7 that the elastic strain energy in bending is given by the relationship

$$U = \int_0^L \frac{M^2\,dx}{2EI} \qquad 1.21$$

where M is the bending moment, L the beam length, and I the moment of inertia of the cross section. Application of this equation reveals that for a simply supported beam with a central concentrated load the modulus of resilience per unit volume of beam is

$$u = \frac{U}{V} = \frac{2}{AL}\int_0^{L/2} \frac{F^2x^2\,dx}{2\ EI} = \frac{(\sigma_y)^2 I}{6Ec^2A} \qquad 1.22$$

where c is the distance from the centroid to the highest stressed fiber and σ_y, the elastic limit of the material in tension, is obtained by experiment and included analytically in Eq. 1.22 through application of the bending stress formula (Chapter 2), which for this case is

$$\sigma = \sigma_y = M/Z = Mc/I = (FL/4)(c/I) \qquad 1.23$$

Some results of resilience calculations are shown in Table 1.1 for beams of various cross section and loading. Note that the resilience is a function of the materials' properties: σ_y, E, the geometry of the cross section, and the manner of loading. For concentrated loads on beams of uniform cross section it does not matter whether the beam is in the form of a cantilever, a simply supported structure, or a fixed-end structure. For the constant strength cantilever beam shown in Fig. 1.46.

$$U = \int_0^L \frac{M^2\,dx}{2EI} = \frac{1}{2E}\int_0^L \frac{F^2x^2\,dx}{\frac{1}{12}b_0(x/L)h_0{}^3} = \frac{3F^2L^3}{Eb_0h_0{}^3} \qquad 1.24$$

Per unit volume,

$$u = \frac{3F^2L^3}{Eb_0h_0{}^3(b_0h_0L/2)} = \frac{6F^2L^2}{Eb_0{}^2h_0{}^4} \qquad 1.25$$

Table 1.1 Modulus of Elastic Resilience

Cross Section Form	Modulus of Elastic Resilience, in.-lb/in.³			
	Bending			Torsion
	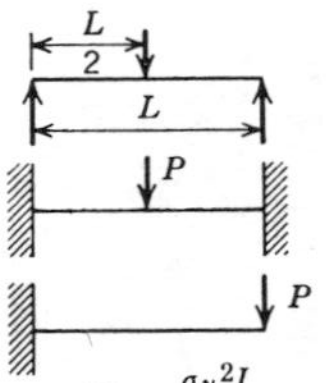 $\frac{L}{2}$, L, P, P; $U = \frac{\sigma_y{}^2 I}{6c^2EA}$	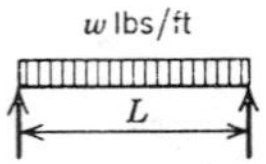 w lbs/ft, L; $U = \frac{4\sigma_y{}^2 I}{15c^2EA}$	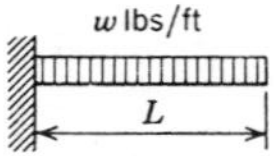 w lbs/ft, L; $U = \frac{\sigma_y{}^2 I}{10c^2EA}$	
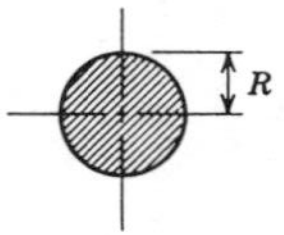 Solid rod (R)	$\frac{\sigma_y{}^2}{24E}$	$\frac{\sigma_y{}^2}{15E}$	$\frac{\sigma_y{}^2}{40E}$	$\frac{\tau^2}{4G}$

Table 1.1 (Continued).

Heavy-walled tube (R, R_0)	$\frac{\sigma_y^2}{24E}\left[1+\left(\frac{R_0}{R}\right)^2\right]$	$\frac{\sigma_y^2}{15E}\left[1+\left(\frac{R_0}{R}\right)^2\right]$	$\frac{\sigma_y^2}{40E}\left[1+\left(\frac{R_0}{R}\right)^2\right]$	$\frac{\tau^2}{4G}\left[1+\left(\frac{R_0}{R}\right)^2\right]$
Thin-walled tube (R to $t/2$)	$\frac{\sigma_y^2}{12E}$	$\frac{2\sigma_y^2}{15E}$	$\frac{\sigma_y^2}{20E}$	$\frac{\tau^2}{2G}$
Solid square (a)	$\frac{\sigma_y^2}{18E}$	$\frac{4\sigma_y^2}{45E}$	$\frac{\sigma_y^2}{30E}$	$\frac{\tau^2}{6.53G}$

Finally, from Eq. 1.23,

$$\sigma^2 = 36F^2L^2/b_0{}^2h_0{}^4$$

so that

$$u = \sigma^2/6E \tag{1.26}$$

Different values are obtained for beams with distributed loads in bending. In addition, note that the thin-walled tube in bending is twice as resilient as the solid rod. This is because the stress is uniform in the thin tube but varies from zero to a maximum at the outside surface of the solid rod. On the other hand, the resilience in pure tension (Eq. 1.19) is six times that of the thin-walled tube in bending and twelve times that of a solid cylinder in bending.

RESILIENCE IN TORSION. The modulus of elastic resilience in torsion is also obtained by application of the theory of elastic strain energy (Chapter 7). The general strain-energy relationship for torsion is

$$U = \int_0^L \frac{T^2\,dx}{2GJ} \tag{1.27}$$

where T is the torque, G is the shear modulus of elasticity, and J the polar moment of inertia of the cross section. Equation 1.27 has been solved for the geometries shown in Table 1.1 and the modulus of resilience calculated by application of the torsion formula for shear stress (Chapter 2)

$$\tau = Tr/J \tag{1.28}$$

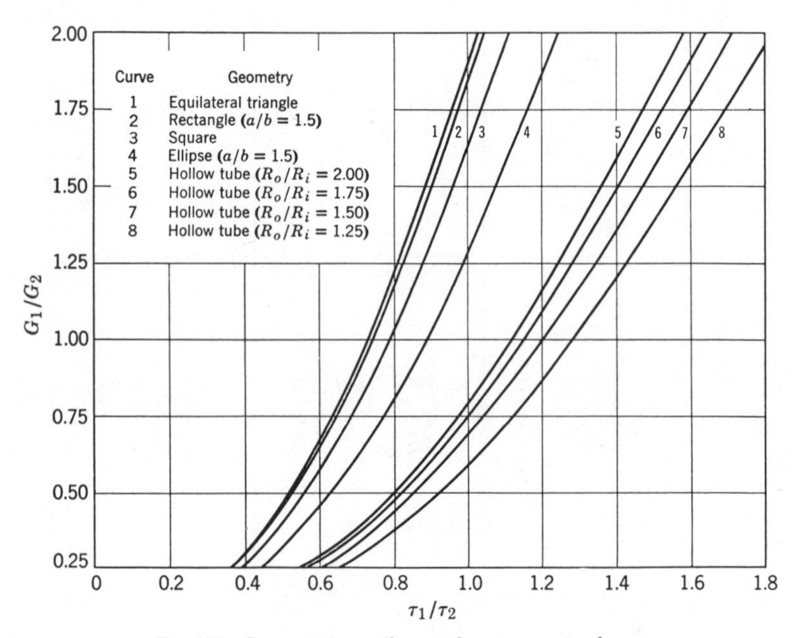

Fig. 1.9 Comparative resilience of various torsion bars.

for round rods and other equations developed in Chapter 3 for torsion of noncircular sections. Note that the thin-walled tube is the most resilient geometry; this is because the stress distribution is most uniform for this condition. In Fig. 1.9, curves are plotted showing the relation of modulus to shear strength for equivalent elastic resilience with a solid circular bar as the basis of comparison. For example, if the moduli of two bars are identical, then a hollow tube, say with outside to inside diameter ratio of 1.25, would require a shear strength of 0.78 times the shear strength of a solid circular bar.

Toughness. Toughness is a measure of the ability of a material to release energy in the plastic range; It is difficult to define precisely because it is a function of geometry, strain rate, temperature, and state of stress. In the simplest consideration, toughness may be considered as the area under a stress-strain curve from zero to the ultimate stress.

TOUGHNESS IN TENSION. An analytical expression for toughness in tension depends on the shape of the tensile stress-strain curve. To illustrate, three types of plots have been selected (Fig. 1.10). For these three plots the *modulus of toughness* U_T is defined by the following equations.*

For Fig. 1.10*A* $$U_T = (\sigma_y + \sigma_u)\varepsilon_u/2 \quad 1.29$$

For Fig. 1.10*B* $$U_T = (\sigma_y + 2\sigma_u)\varepsilon_u/3 \quad 1.30$$

For Fig. 1.10*C* $$U_T = 2\sigma_u\varepsilon_u/3 \quad 1.31$$

where σ_y is the elastic limit, σ_u the ultimate tensile stress, and ε_u the strain at the ultimate strength. For special purposes toughness of a material is sometimes measured by tension tests on notched specimens; this is discussed in the section on brittleness and ductility of materials.

TOUGHNESS IN BENDING. A measure of a material's toughness in bending is the area under the load-deflection curve up to fracture. Depending on the

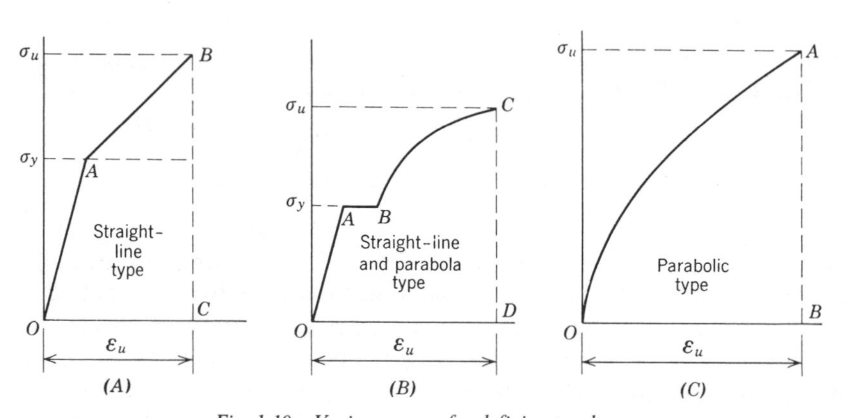

Fig. 1.10 Various curves for defining toughness.

*For simplicity the small area $OA\sigma_y$ is included.

geometry different curves will be obtained and analyses similar to those outlined for toughness in tension can be applied. An approximation of toughness in bending may be obtained by referring to Eq. 1.32, which is derived in Chapter 2.

$$\sigma = \frac{2M + k\,\partial M/\partial k}{bh\,\partial \varepsilon/\partial k} \qquad 1.32$$

where σ is an "equivalent" tensile stress, k the curvature of a beam in bending, ε the "equivalent" tensile strain in the most stressed fiber, and b and h the dimensions of the beam cross section. This equation applies only to beams of rectangular cross section and provides a way to transform a moment-strain curve into an equivalent tensile stress-strain curve. Having the stress-strain curve, the former equation for toughness in tension can be applied. Toughness in bending as measured by the foregoing method is ambiguous, since the stress in a beam is a function of the applied moment, and this can vary continuously along the length of the beam. Any σ–ε curve obtained, except for pure bending where the moment is constant, thus represents only a particular location. For this reason, toughness in bending is frequently determined by calculating the *modulus of rupture*, which is a fictitious value found by using the formula $\sigma = M/Z$ (truly applicable only in the elastic range) to calculate a fracture strength in the plastic range. If the material happens to be brittle, this method of calculation is more exact. Toughness in bending is also evaluated by impact tests on notched specimens. This is discussed later in the section on brittleness and ductility.

TOUGHNESS IN TORSION. A measure of toughness in torsion is the area under the torque-twist curve up to fracture. Methods are given in Chapter 2 for converting torque to shear stress so that stress-strain curves can be obtained. If it is assumed that the torsion curve is parabolic, the area under the curve up to the torsional ultimate strength (for a ductile material) is

$$\text{area} = U_T = 2\tau_u\theta_u/3$$

where τ_u is ultimate shear strength and θ_u the ultimate angle of twist. In terms of the ultimate torsional moment for a solid round rod,

$$U_T = \frac{(2)(3T)(\theta_u)}{(3)(2\pi R^3)} = \frac{T\theta_u}{\pi R^3} \qquad 1.33$$

where R is the radius of the rod. For brittle materials torsional toughness is better measured by an impact test devised by Luerssen and Greene (20). By applying this test to tool steels, optimum heat treatments can be spotted. For example, for a plain high-carbon, water-quenched tool steel, the torsional impact values as a function of tempering temperature are shown in Fig. 1.11. Note that the impact value "as quenched" is very low, but it increases rapidly as the tempering temperature is raised to 350 to 375°F; then it falls off again. From such a plot it is evident that for this particular grade of tool steel, toughness is at its highest value when the steel is tempered near 350°F.

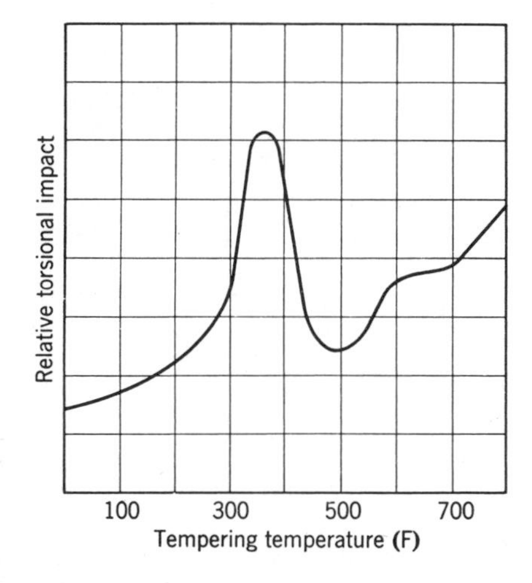

Fig. 1.11 Torsion impact data for a tool steel. (Courtesy of The Carpenter Steel Co.)

Hardness. Hardness as a mechanical property is also difficult to define. It is generally believed to be a measure of the resistance a material offers to indentation and scratching. In order to measure this property of materials, a number of tests have been devised (4, 23) that yield a numerical measure of hardness. These numbers are not material constants but are relative numerical indicators of how well a material resists indentation. Among the more common hardness tests for structural materials are the Brinell, Rockwell, and Vickers methods.

Brinell hardness is expressed as a number

$$B = \frac{2P}{\pi D(D - \sqrt{D^2 - d^2})} \tag{1.34}$$

where d is the diameter (in millimeters) of the impression left by a load P (in kilograms) applied to a flat surface by a steel ball of diameter D (usually 10 mm).

Rockwell hardness is also an indentation method similar to Brinell where steel balls or diamond points are pressed into materials by loads of varying magnitude; depending on the size of the indentor and the load used, Rockwell furnishes hardness data on various scales depending on the hardness and thickness of the material being tested. Rockwell C hardness is frequently used, as shown later, as an indicator of the tensile strength of steel. Nomographic methods have been developed for estimating yield strength from hardness data (37).

Vickers hardness is particularly suited for hard and thin materials where spot checks are required; it is similar to the Brinell test in that an indentation is used to obtain an area for a basic measurement. The indentor is a four-sided inverted pyramid with an apex angle of 136°. The Vickers hardness is

$$V = 1.854P/D \tag{1.35}$$

where D is the average length of the two diagonals of the impression in the plane of the surface (in millimeters) and P is the load (kilograms).

A measure of hardness is also obtained by comparing the scratching resistance

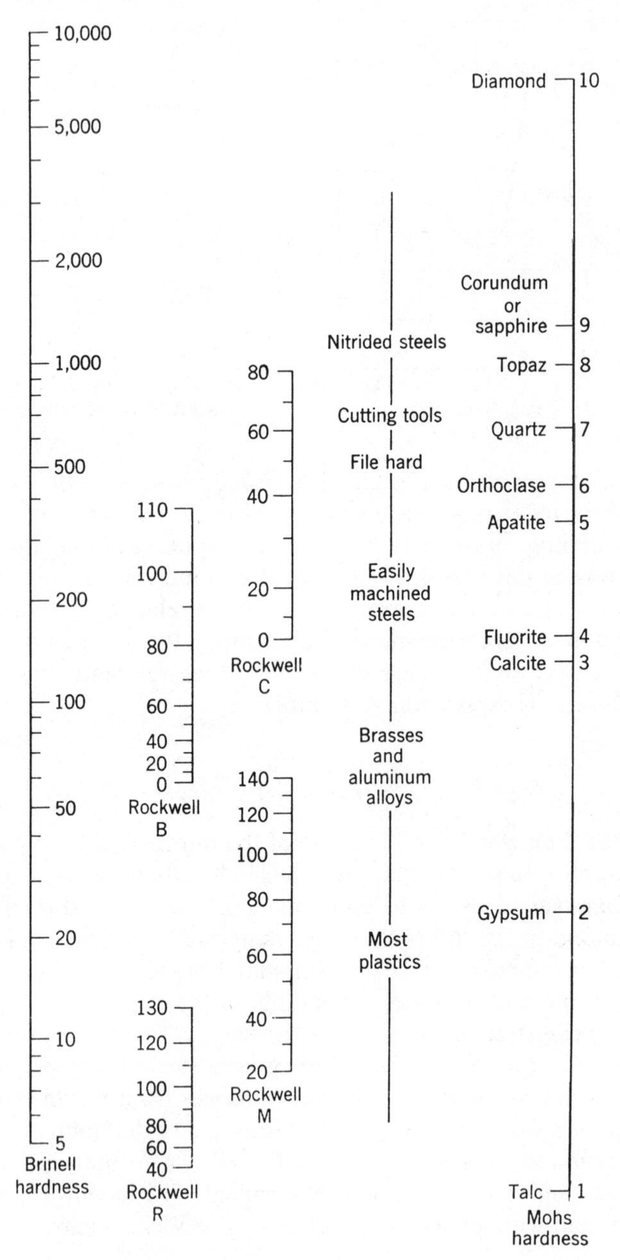

Fig. 1.12 Comparison of hardness scales. [*After Kinney, (16), Courtesy of John Wiley and Sons.*]

of the material with that of ten minerals arranged in order of increasing hardness from talc (1 on scale) to diamond (10 on scale). This system is called *Mohs' hardness test.*

Hardness is also measured on rubber products by various *durometers.* In these tests a blunt rod or ball is pressed into the surface of the material, the load required being recorded. There is no correlation of these readings with other properties, the test is used only as a control and not to obtain fundamental data (23).

The relationship among various hardness test results and strength of materials, where such a relationship exists, is shown in Figs. 1.12 and 1.13 and in Appendix E.

Coefficient of linear thermal expansion. The coefficient of linear thermal expansion of a material is the ratio of the change of length per unit length to the

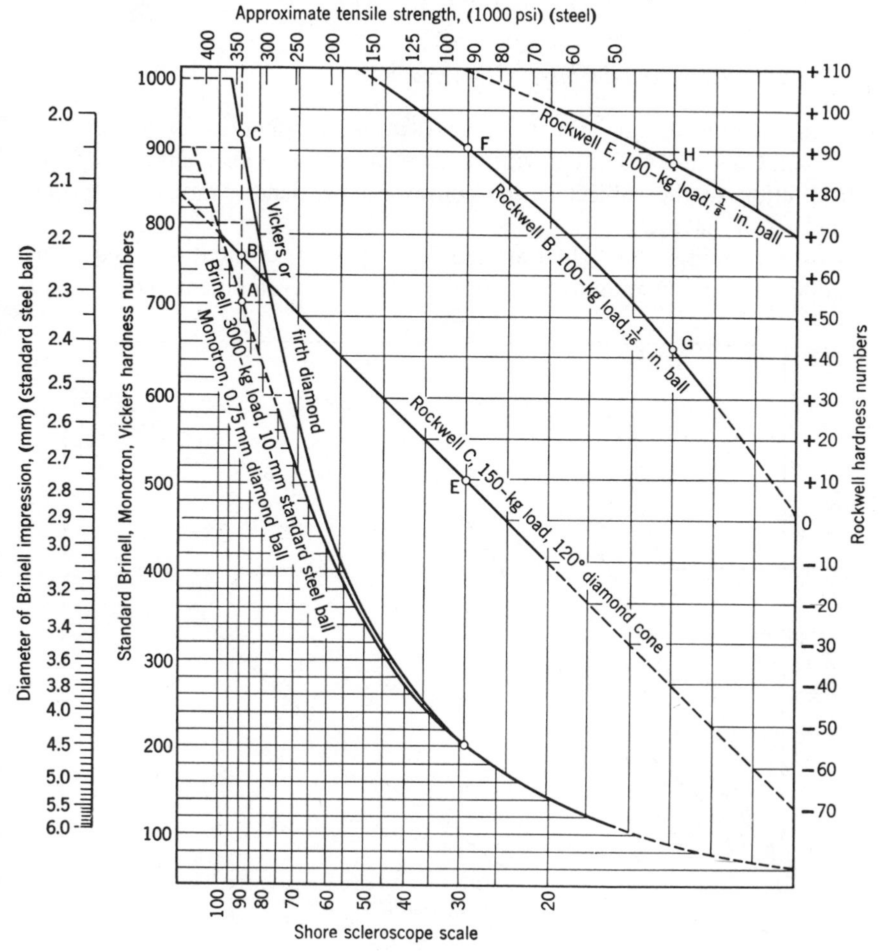

Fig. 1.13 Hardness-strength conversion plots. (Courtesy of The American Society for Metals.)

change in temperature. This coefficient varies depending on the temperature; mathematically

$$L = L_0(1 + \alpha\Delta T) \qquad 1.36$$

where L is a final length, L_0 is the initial length, ΔT the temperature differential, and α the coefficient of linear thermal expansion.

Coefficient of cubical thermal expansion. The coefficient of cubical thermal expansion is the ratio of the change of volume per unit volume to change in temperature. This coefficient also varies with the temperature. Mathematically,

$$V = V_0(1 + \alpha' \Delta T) \qquad 1.37$$

where V and V_0 are final and initial volumes, ΔT is the temperature change, and α' is the coefficient of cubical expansion. For homogeneous solids

$$\alpha' = 3\alpha \qquad 1.38$$

Brittleness and ductility of materials

The engineering design aspects of brittleness and ductility of materials are treated in Chapter 14. The descriptive material included here is presented as an introduction to the subject in the context of brittleness and ductility being properties of materials.

Most solid materials exhibit both elastic and plastic deformation when loaded; ceramics are no exception to this, but the extent of development of deformation

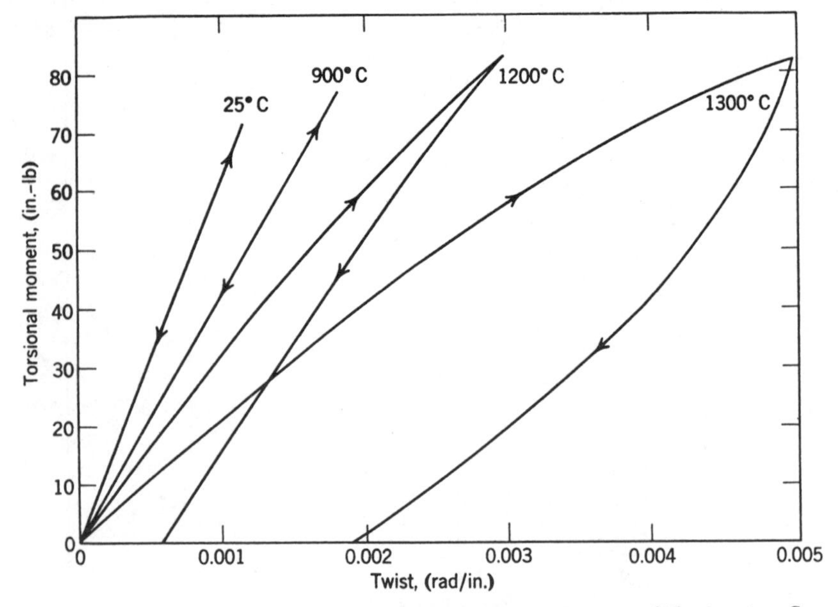

Fig. 1.14 Torsional properties of magnesia. [After Wygant (33), Courtesy of The American Ceramic Society.]

Table 1.2 Effect of Matrix-Grain Hardness on Ductility

Alloy	Ratio of Matrix-Grain Hardness	Bend Deflection for 1.25 in. Span (in.)
90W-10Ni	1.28	0.02
90W-6Ni-4Cu	0.61	0.05
90W-6Ni-4Fe	0.88	0.60

Data from Holtz, Reference 11.

with change in time, temperature, and stress may be quite different from metals. For example, when loaded at room temperature most ceramic materials exhibit elastic deformation to the point of failure as shown in Fig. 1.5*A*; torsional data for magnesium oxide are shown in Fig. 1.14. As temperature is increased, some ductility appears. At elevated temperature, creep of brittle materials also occurs, as shown in Chapter 15. Many materials like glass, tungsten, and ceramics have been considered as brittle materials exhibiting little or no ductility; however, advances in materials engineering have disclosed that, in some cases, brittleness is not necessarily an inherent property of a material but that it may be caused by trace impurities in the structure or sometimes dislocations. Tungsten and molybdenum, for example, have been rendered ductile (comparatively speaking) by a process called *liquid phase sintering*, which removes trace impurities interfering with plastic flow. Essentially, this process involves the heating of compacted powders at a high temperature in order to effect an aggregate of grains in an adhesive matrix; the more spherical the grains, the less likelihood of fracture. Furthermore, the closer the hardness of grains and matrix the less likelihood of fracture. Some results illustrating this technique are shown in Table 1.2. The following discussion is concerned with the behavior of materials like glass and ceramics that are ordinarily considered brittle; next are problems involving brittle failure of supposedly ductile materials.

Rationalization of brittle failure. Some insight into the problem of brittle fracture is gained through considering what takes place within the structure at the microscopic level. The brittleness of glass was first explained by Griffith (5) as being due to internal cracks generating excessively high stresses. In fact, Griffith suggested that the discrepancy between the theoretical and the measured strength of materials could be explained on this basis. Griffith using elastic energy theory derived an equation expressing the tensile stress to just cause a crack of a certain size to propagate at a high velocity to create a brittle fracture. Strictly speaking, the Griffith criterion does not define a rupture condition but merely the propagation of an already existing crack in the material. For convenience Griffith assumed that the existing crack was in the shape of a thin ellipse having a major axis of length L; if the load is applied normal to such a crack (Fig. 1.15), the strain energy per unit area is

$$U = -\pi L^2 \sigma^2 / 4E \qquad 1.39$$

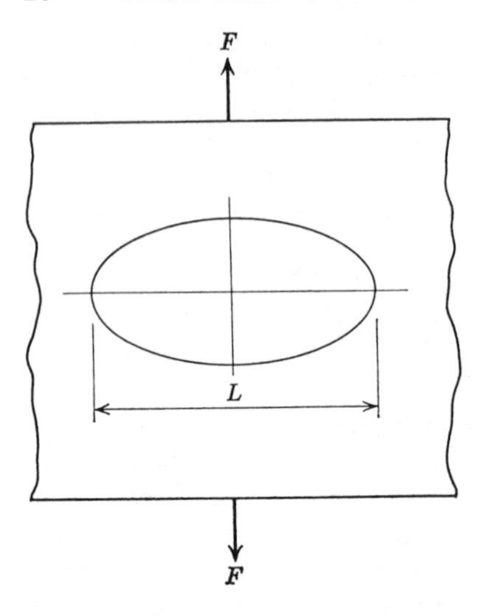

Fig. 1.15 Griffith crack in a material.

where σ is the applied stress and E is the modulus of elasticity of the material. The negative sign indicates released energy. The corresponding surface energy of the crack is

$$U_s = 2L\phi \tag{1.40}$$

where ϕ is the surface energy per unit area. For equilibrium the critical crack length L is

$$(d/dL)[2L\phi - (\pi L^2\sigma^2/4E)] = 0 \tag{1.41}$$

or

$$2\phi - (\pi L\sigma^2/2E) = 0 \tag{1.42}$$

from which the fracture stress σ is

$$\sigma = \sqrt{4\phi E/\pi L} \tag{1.43}$$

The value of ϕ (surface tension) for glass used by Griffith was about 0.003 lb/in. for the modulus about 10×10^6 psi, and for σ about 10,000 psi, which when put into Eq. 1.43 gives a critical crack length of about 0.0004 in. By mechanically introducing "cracks" in glass with a diamond cutter Griffith determined values of σ and found satisfactory agreement with predictions made by the use of Eq. 1.43. A few years after Griffith's theory was published Joffe found that rock salt exhibited its theoretical strength very closely if tested in hot water so that the "cracks" were "washed away." Several additional verifications of the Griffith theory have been reported over the years (30).

Related to the Griffith effect is the phenomenon of *notch sensitivity* of materials. On a gross scale, notch sensitivity may be loosely defined as the relative ability of a material to resist the stress-concentrating effect of a notch to produce brittle failure. Soft metals like copper and aluminium, for example, may be notched and yet exhibit a fairly high degree of ductility. Many steels, on the other hand, are embrittled by the presence of a notch, that is, they are notch sensitive. Another way of viewing notch sensitivity is to consider it as a measure of the degree to which the theoretical stress concentration factor is reached. This is further discussed in Chapter 13. Thus notch sensitivity may be symbolized as

$$q = \frac{k - 1}{k_t - 1} \qquad 1.44$$

where q is the notch sensitivity factor, k_t is the theoretical stress concentration factor for the state of stress involved (that is, tension, bending, torsion, or combined stress, design values for which are given in Chapter 13), and

$$k = \frac{\text{strength of unnotched material}}{\text{strength of notched material}}$$

where the "strength" refers to the state of stress and type of loading. For example, k might be the ratio of endurance limits in fatigue of unnotched and notched specimens. Unfortunately, q is a difficult factor to obtain and is usually either determined experimentally or allowed equal to one. The mechanism responsible for notch sensitivity in materials is not clearly known, but it is believed that in materials that are not particularly notch sensitive local deformation takes place at the leading edge of the advancing crack, spreading out the load and reducing the stress concentration factor. For example, consider notched specimens such as shown in Fig. 1.16 which are made from two steels A and B having the stress-strain properties shown. If, for example, the stress concentration factor for the particular notch geometry is 8, then application of σ stress to the specimen raises the stress in the material immediately to 8σ; for material A this stress is above the ultimate strength, and since the material is notch sensitive, the geometry of the notch does not change appreciably, and the specimen breaks in a brittle fashion. Material B, on the other hand, is much less notch sensitive, and as the stress starts to rise from σ to 8σ the material flows plastically, redistributing the load, changing the notch geometry, and possibly

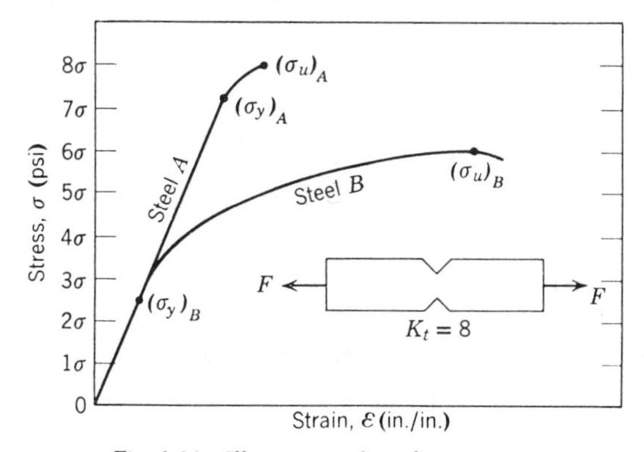

Fig. 1.16 Illustration of notch sensitivity.

reducing the effective stress concentration factor to 3 or 4, and the material follows its stress-strain curve as indicated. Often with non-notch-sensitive materials the initial stress concentration factor may be reduced to nearly unity, so that the only real effect of the notch is to decrease the cross-sectional area carrying the load.

Design with brittle materials. Obtaining realistic working stress (see Chapter 14) data from mechanical property tests is particularly difficult for brittle materials. Many of the brittle materials, like ceramics and cermets, are used at elevated temperature and are exposed to thermal shock; this problem is treated in Chapter 15. Consideration is given here to problems encountered at room temperature where most of the brittle materials will have stress-strain curves similar to those shown in Fig. 1.5*A*.

Tensile test data on brittle materials are not always available since most testing is done on bend specimens. From bend tests, plots of bending moment versus outer-fiber strain or deflection are obtained (Fig. 1.17) from which various properties like modulus, yield strength, and modulus of rupture can be deduced. For brittle materials only the portion OA of the plot is obtained. Such tests are frequently performed on center-loaded beams, but preferred practice is to use a condition of *pure bending* which can be accomplished by either of the means shown in Fig. 1.17. In pure bending the stress is constant between the supports, eliminating effects of variable stress and providing a larger volume of material to experience the maximum stress. In addition, the shear stress is zero. For the test employing the end moment the center deflection in the elastic range is*

$$\delta = ML^2/8EI \tag{1.45}$$

The more common method is to use symmetrical loading by transverse forces. Here the center deflection is

$$\delta = \frac{Wa(3L^2 - 4a^2)}{24EI} \tag{1.46}$$

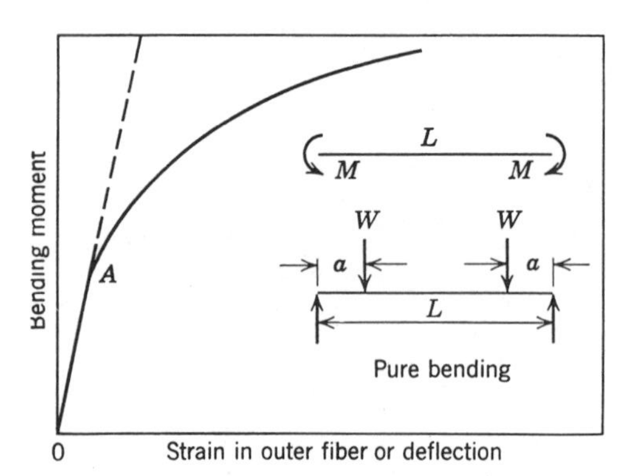

Fig. 1.17 Bending moment-deformation curve and two methods for pure bending.

*See Chapter 2 for derivations and definition of terms.

The deflection at the points of loading is

$$\Delta = \frac{Wa^2(3L - 4a)}{6EI} \qquad 1.47$$

and the stress is

$$\sigma = Mc/I = E\varepsilon \qquad 1.48$$

From Eq. 1.48 it is possible to interpret elastic bend test data on the basis of either strain or deflection. Since, for pure bending $M = Wa$, it is clear from Eqs. 1.46 and 1.48 that the fiber strain ε is

$$\varepsilon = \frac{24c\delta}{3L^2 - 4a^2} \qquad 1.49$$

The value of the modulus of elasticity E is also obtainable from either Eq. 1.46 or Eq. 1.48. At fracture, Eq. 1.48 defines the modulus of rupture, σ_{max}, which is a property commonly reported for brittle materials. This value divided by a factor of safety also provides the design-working stress value. For application involving combined stresses, design is based on the *maximum stress theory* (see Chapter 3), which states that failure takes place in a material subjected to combined stresses when the maximum principal stress equals the failure stress in tension (or at the stress defined by the modulus of rupture in bending). The maximum stress theory is applicable only to brittle materials and is a convenient design procedure when properly applied. In Chapter 3 a discussion is provided to show the limitations of this theory for other than brittle materials.

The following general rules will be found useful in designing with *brittle* materials.

1. Increase operating temperature if possible since brittleness tends to decrease with increase in temperature.
2. Apply loads slowly; this tends to increase ability of materials to absorb deformation.
3. Reduce or eliminate direct tensile loads.
4. Avoid or reduce stress concentration effects (see Chapter 13).
5. In structural applications, keep good alignment of parts.
6. If possible, use several small members rather than one large one since small parts have higher a fracture strength.
7. Be sure that data used are applicable. If bend test data are involved, make sure that the proper interpretation of the moment-deflection curve has been made; if thermal shock data are used, make sure that the tests were performed on section sizes of the same order as those used in the field application (see Chapter 15).

1-2 MATERIALS ENGINEERING

A discussion of the practical selection and use of materials to achieve maximum performance and safety at minimum cost is the objective of the following paragraphs.

Table 1.3 Typical Properties of Materials at 70° F

Material	Form and Condition	Yield Strength 0.2% Offset, 1000 psi	Tensile Strength, 1000 psi	Elongation in 2 in., %	Density, lb/in.^{3}	Coefficient of Expansion, in./in./°F × 10^{-6}	Modulus of Elasticity, psi × 10^{-6}
Carbon Steels							
SAE 1020	Annealed	40	60	35	0.284	6.7	30
	W-200°F[a]	80	104	6	0.284	6.7	30
	W-1000°F	62	90	22	0.284	6.7	30
SAE 1045	W-600°F	114	150	8	0.284	6.7	30
	W-1000°F	89	120	19	0.284	6.7	30
SAE 4340	0–400°F[b]	225	290	10	0.284	6.7	30
	0–1000°F	160	185	15	0.284	6.7	30
Stainless Steel							
Type 304	Annealed	30	85	50	0.286	9.6	29
Type 410	Heat-treated	115	150	15	0.277	6.1	28
17-7 PH	TH-1050	182	193	10	0.281	6.0	29
Nonferrous Metals							
Aluminum 1100-0	Annealed	5	13	45	0.098	12.9	10
Aluminum 2017-T4	Heat-treated	40	62	22	0.101	12.9	10
Beryllium copper	Heat-treated	130	175	5	0.297	9.2	18
Naval brass	Annealed	22	56	40	0.304	11.2	15
Admirality brass	Annealed	20	55	65	0.308	10.2	15
Nickel 200	Annealed	25	65	45	0.321	7.4	30
Magnesium AZ80A-T5	Heat-treated	40	55	7	0.064	14.0	6
Titanium 8-1-1	Heat-treated	150	160	15	0.170	4.7	17
Wood (W-Grain, 12% Moisture)							
Birch		8.3[c]	2.0[d]	–	0.026	1.1	2.1
White Oak		7.0[c]	1.9[d]	–	0.028	2.7	1.6

[a] Water quench, tempered at 200°F.
[b] Oil quench, tempered at 400°F.
[c] Compressive strength.
[d] Shear strength.

Comparative properties of materials

Various properties of materials at room temperature are listed in Table 1.3 High-temperature materials and their properties are listed separately in Chapter 15. Since the range of properties is large, it is difficult to select materials. In addition, many properties and ranges of applications of materials overlap as shown briefly in Tables 1.4 and 1.5.

A comparison of the energy properties of materials deserves special mention. For this purpose, it is convenient to compare the energy properties of some materials in tension and then to examine resilience as applied to springs. From the equations already given, a comparison can now be made among a few materials as shown in Table 1.6. This table shows, for example, that beryllium

Table 1.4 Comparative Properties at 70° F

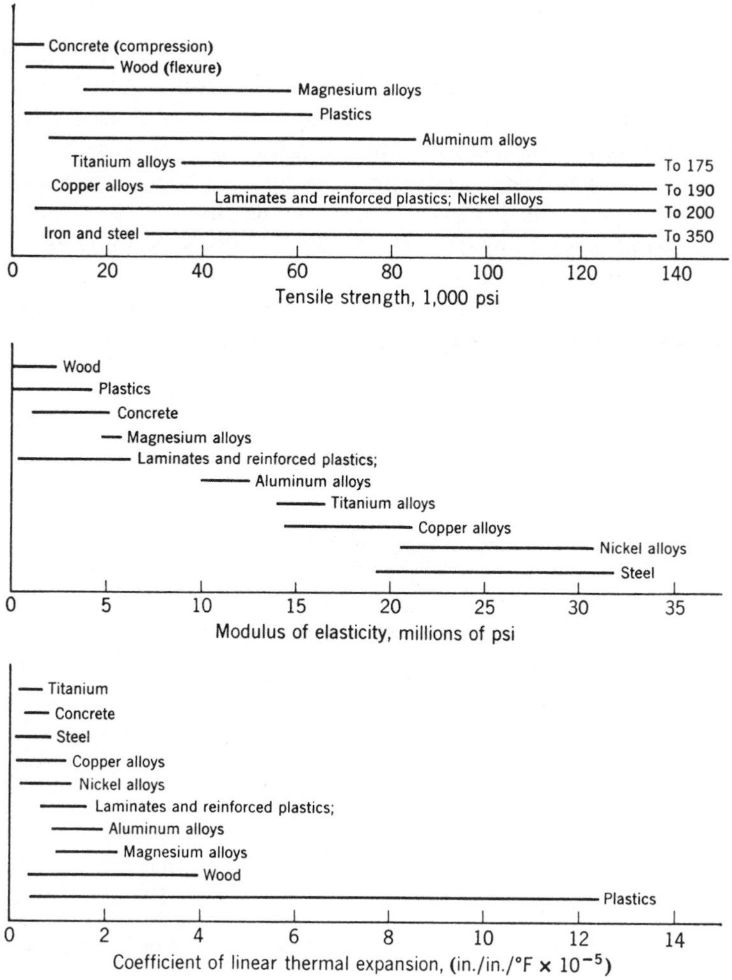

Concrete (compression)
Wood (flexure)
Magnesium alloys
Plastics
Aluminum alloys
Titanium alloys — To 175
Copper alloys — To 190
Laminates and reinforced plastics; Nickel alloys — To 200
Iron and steel — To 350
0 20 40 60 80 100 120 140
Tensile strength, 1,000 psi

Wood
Plastics
Concrete
Magnesium alloys
Laminates and reinforced plastics;
Aluminum alloys
Titanium alloys
Copper alloys
Nickel alloys
Steel
0 5 10 15 20 25 30 35
Modulus of elasticity, millions of psi

Titanium
Concrete
Steel
Copper alloys
Nickel alloys
Laminates and reinforced plastics;
Aluminum alloys
Magnesium alloys
Wood
Plastics
0 2 4 6 8 10 12 14
Coefficient of linear thermal expansion, (in./in./°F × 10^{-5})

Table 1.5 Maximum Operating Temperature °F of Some Common Structural Materials

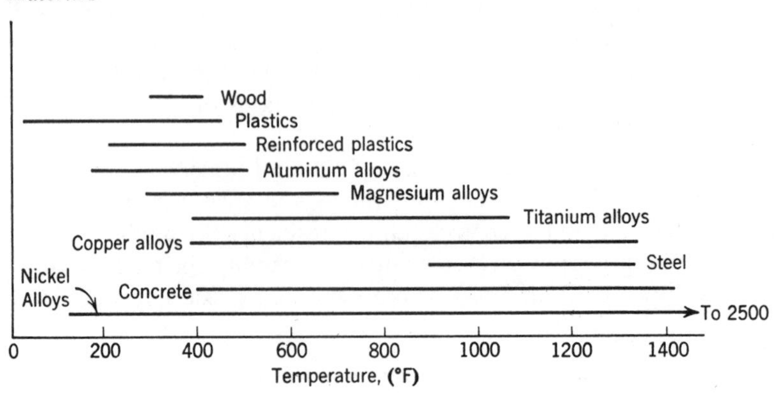

copper with its high resilience would be an excellent spring material and that cast iron would be a poor candidate for an application requiring toughness.

Springs of all types make use of the resilience properties of materials in conjunction with geometrical factors. Maier (21), for example, has shown that a tapered cross section is the most efficient shape for cantilever springs since it has maximum deflection, energy capacity, and volume efficiency (see also the discussion on beams of constant strength). Maier (22) also shows how to optimize a spring system by considering two factors, the form coefficient and the material. The form coefficient k is related to spring volume efficiency W/V as follows for E-type (tension) springs and G-type (shear) springs, respectively.

$$W/V = k(\sigma^2/2E) \tag{1.50}$$

and

$$W/V = k(\tau^2/2G) \tag{1.51}$$

Table 1.6 Energy Properties of Materials in Tension

Material	Yield Strength, psi × 10^{-3}	Ultimate Strength, psi × 10^{-3}	Modulus of Elastic Resilience, in.-lb/in.^{3}	Toughness, $\frac{2}{3}(\sigma_u \varepsilon_u)$ in.-lb/in.^{3}
SAE 1020 annealed	40	60	27	18,667
SAE 1020 heat treated	62	90	62	13,200
Type 304 stainless	30	85	15	28,300
17-7PH stainless	182	193	546	12,866
Cast iron		25		85
Ductile cast iron	58	73	67	7,300
Monel	100	110	200	3,667
Alcoa 2017	40	62	80	9,092
Red brass	60	75	120	2,000
Beryllium	130	175	470	5,832

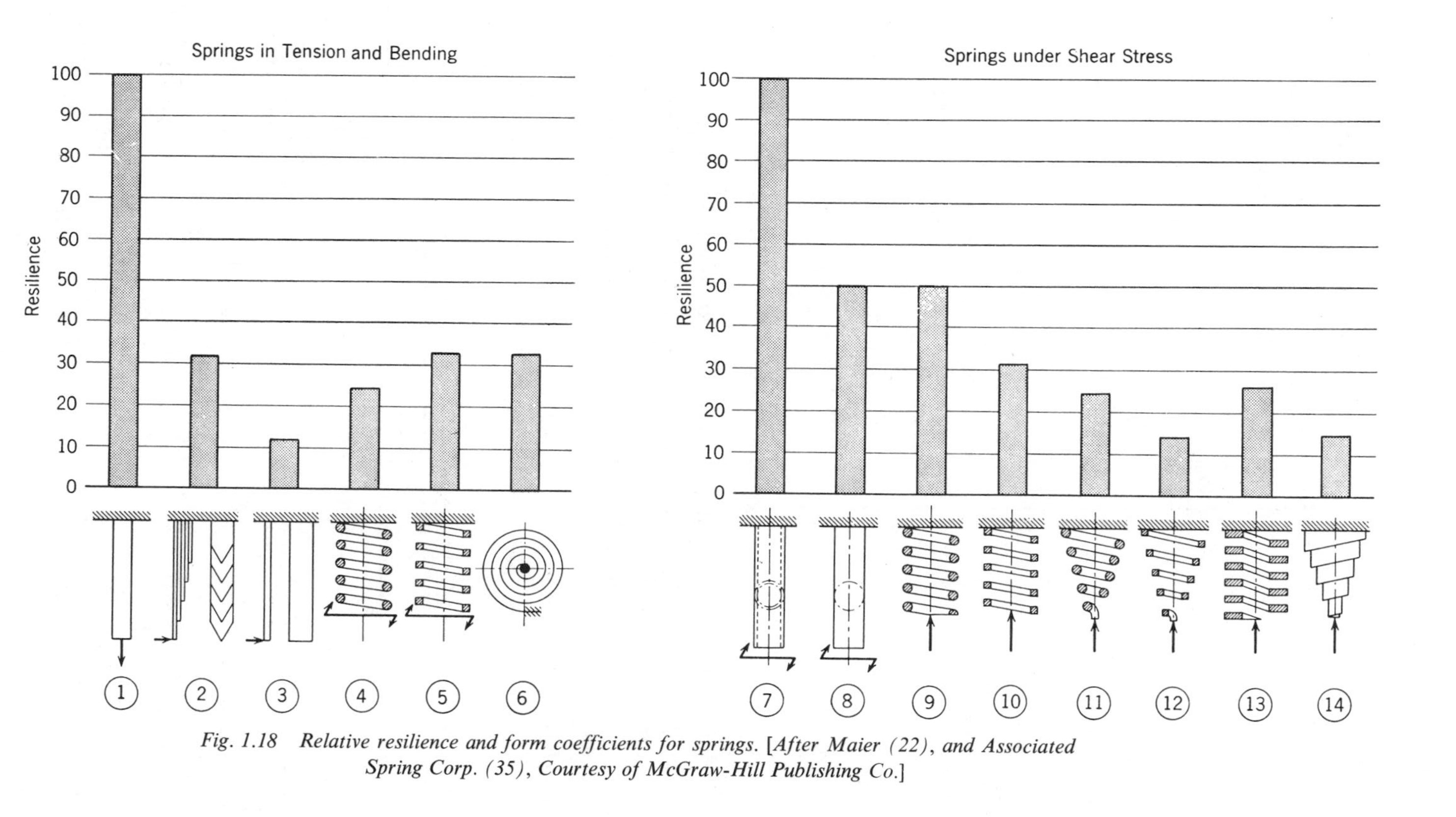

Fig. 1.18 Relative resilience and form coefficients for springs. [*After Maier* (22), *and Associated Spring Corp.* (35), *Courtesy of McGraw-Hill Publishing Co.*]

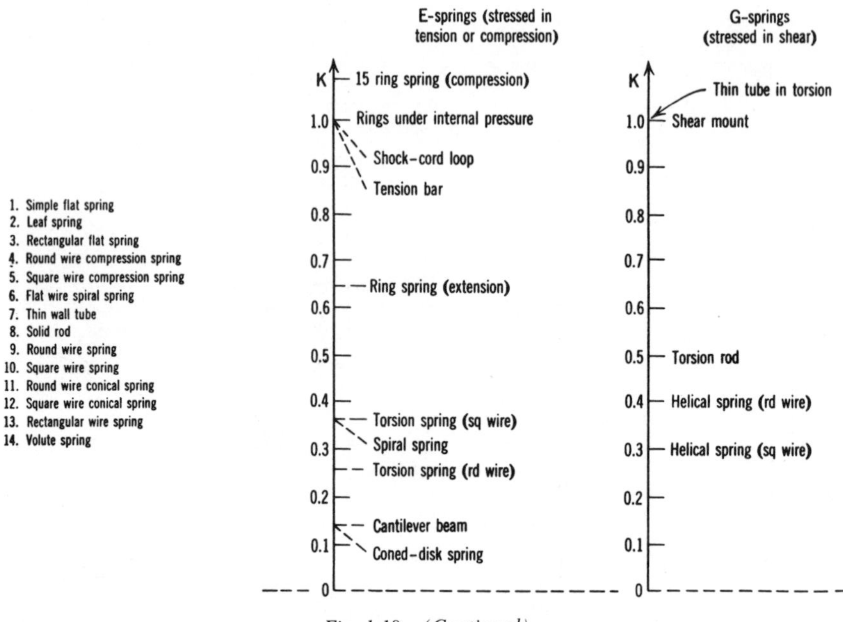

Fig. 1.18 (Continued).

where W is potential energy (in an end-loaded cantilever $W = P\delta/2$, where P is the end load and δ is the end deflection), V is volume of the spring material, and σ or τ is the maximum normal or shear stress acting. A survey of form coefficients and efficiencies for various springs and spring materials is shown in Fig. 1.18 and Table 1.7.

The elastic energy properties of some materials are compared further in Fig. 1.19. The data in this illustration have been arranged to show the relative

Table 1.7 Maximum Design Stress and Volume Efficiency of Spring Materials

	Tension			Shear		
Material	Modulus E, 10^6 psi	Max Design Stress, 1000 psi	Volume Efficiency, in.-lb/in.^{3}	Modulus G, 10^6 psi	Max Design Stress, 1000 psi	Volume Efficiency, in.-lb/in.^{3}
Music wire, ASTM A228-47	30	150–350	375–2042	11.5	90–180	352–1409
Oil-temp spring wire, ASTM A229-47	30	120–250	240–1042	11.5	80–130	278-735
Valve spring wire, ASTM A230-47	30	120–210	240–735	11.5	80–125	278–679
Chrome-silicon steel, SAE 9254	30	220–300	807–1500	11.5	130–160	735–1113
Hot-rolled bars, SAE 1085	30	115–150	220–375	11.5	75–110	245–526
Flat spring steel, SAE 1060	30	120–180	240–540	–	Not used	–
Clock-spring steel, SAE 1095	30	150–310	375–1602	–	Not used	–
Stainless steel wire, SAE 30316	28	60–260	64–1207	10	45–140	101–980
Phosphor-bronze wire, SAE 81	15	60–110	120–403	6.5	50–85	192–556
Shock cord, (rubber filaments)	0.002	2	1000	–	Not used	–
Parallel glass fiber, plastics	6	120–150	1200–1875	–	Not used	–

Data from Maier, Reference 22. (Courtesy McGraw-Hill Publishing Co.)

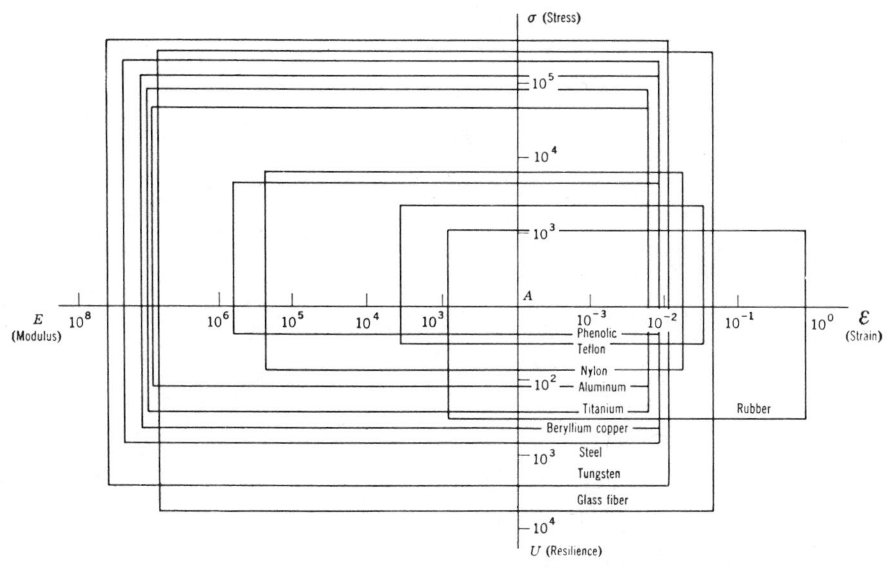

Fig. 1.19 Comparative elastic energy properties of some materials.

relation between modulus E, elastic limit σ, elastic limit strain ε, and modulus of resilience U so that materials can be selected on the basis of a limiting value for any of these properties. For example, if a modulus-stress ratio of unity is required, a material like rubber would be considered.

Effects of some variables on properties of materials

Up to this point the standard ambient properties of materials have been considered and comparison made among materials on the basis of strength, energy, and ductility. It is known that properties are significantly affected by time, temperature, strain rate, composition, state of stress, mechanical working, size and shape; information on all these factors is available in the literature. A brief account of the effects of these variables follows.

Time and temperature. The effect of combined time and temperature that leads to the creep and stress rupture properties of materials is treated in Chapters 5 and 15. In Chapter 15 are charts showing how strength-weight ratio and modulus of elasticity vary with temperature for several materials. The approximate variation in ultimate tensile strength with temperature for several materials is shown in Fig. 1.20. In general, as temperature increases, strength decreases; one notable exception to this generality is low-carbon steel, which goes through an apparent strengthening around 200 to 300°C (blue brittle range) and which should ordinarily be avoided in design. The advent of space vehicles that use cryogenic fluids for propellants has intensified design considerations for low-temperature applications; the properties of two hundred metallic and non-metallic materials have been assembled in Reference 45 of Chapter 15 for

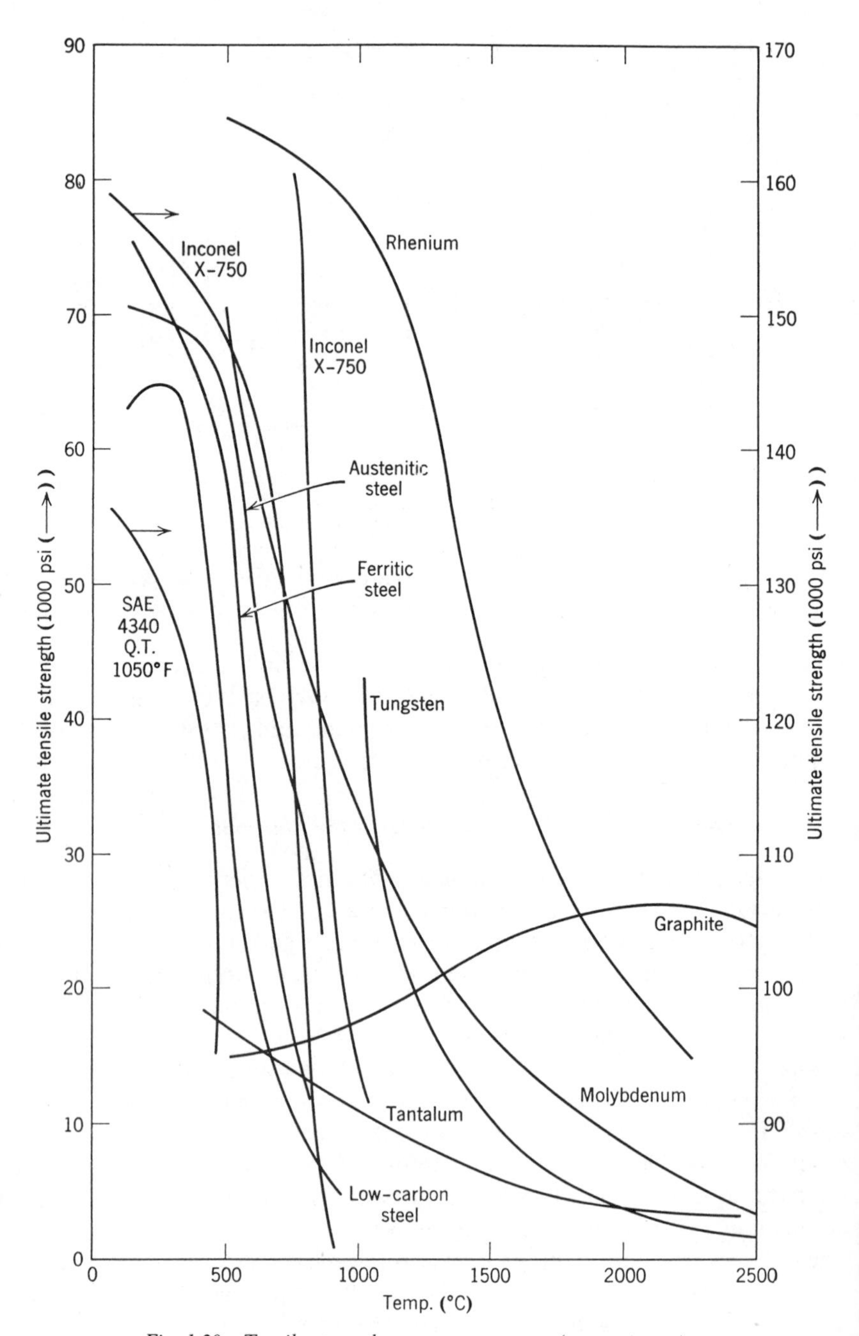

Fig. 1.20 Tensile strength versus temperature (approximate).

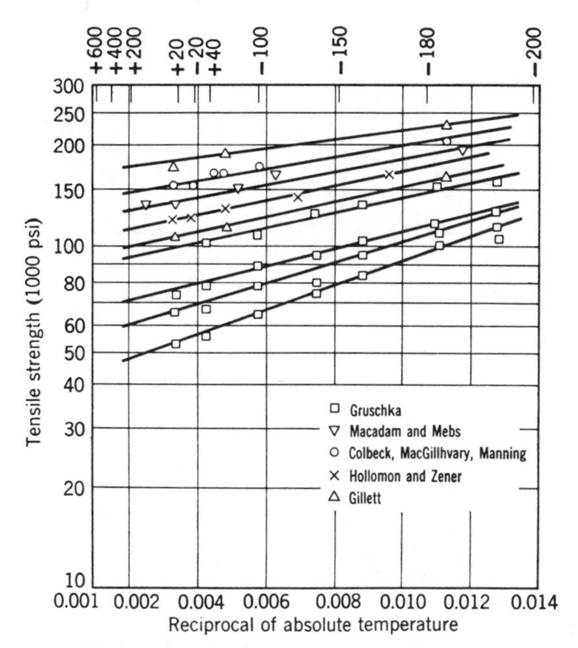

Fig. 1.21 Low temperature strength of some pearlitic steels. [*After Hollomon (10), Courtesy of The American Welding Society.*]

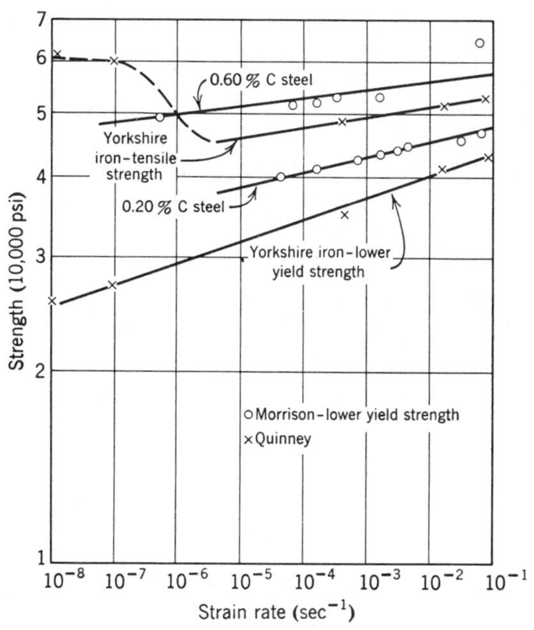

Fig. 1.22 Effect of strain rate on strength of steel. [*After Hollomon (10), Courtesy of The American Welding Society.*]

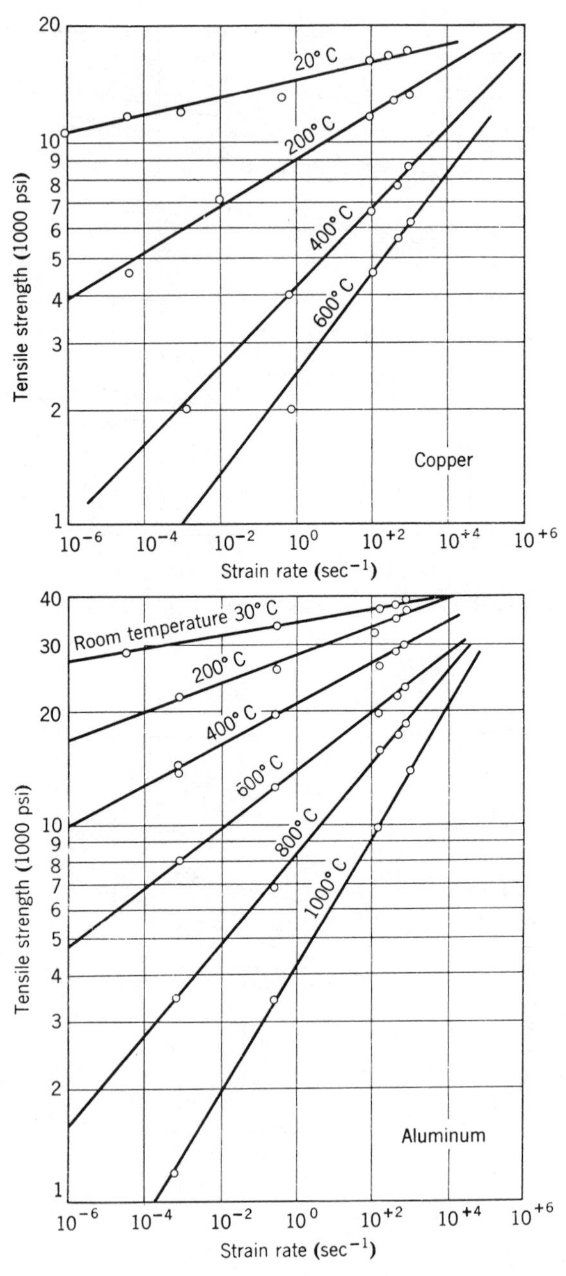

Fig. 1.23 Effect of strain rate on strength of copper and aluminum. [*After Hollomon (10), Courtesy of The American Welding Society.*]

temperatures between 4° and 300°K. This is also discussed in Chapter 15. Figure 1.21 illustrates the effect of low temperature on the tensile strength of steel.

Strain rate. In general, as strain rate increases, strength increases as shown in Figs. 1.22 and 1.23. Usually, the increased strength is accompanied by decreased ductility. The straight-line relationship indicated shows that at least approximately strain data fit the parabolic relationship.

$$\sigma = C(d\varepsilon/dt)^m \tag{1.52}$$

where σ is the stress at strain rate $(d\varepsilon/dt)$, C is the stress intercept at log strain

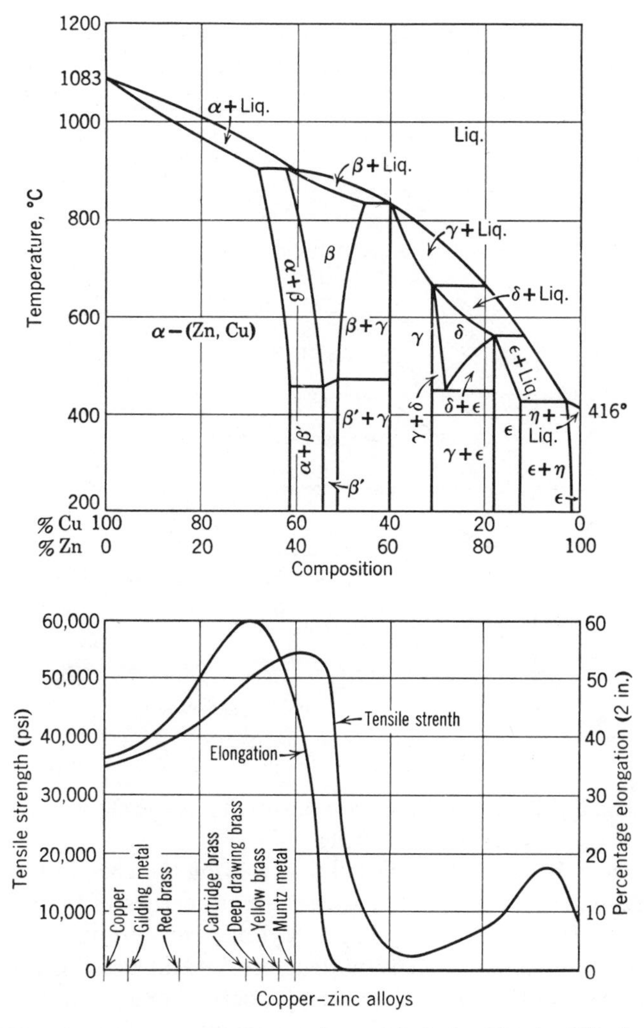

Fig. 1.24 Effect of composition on properties of materials. [After Murphy (25), Courtesy of The International Textbook Co.]

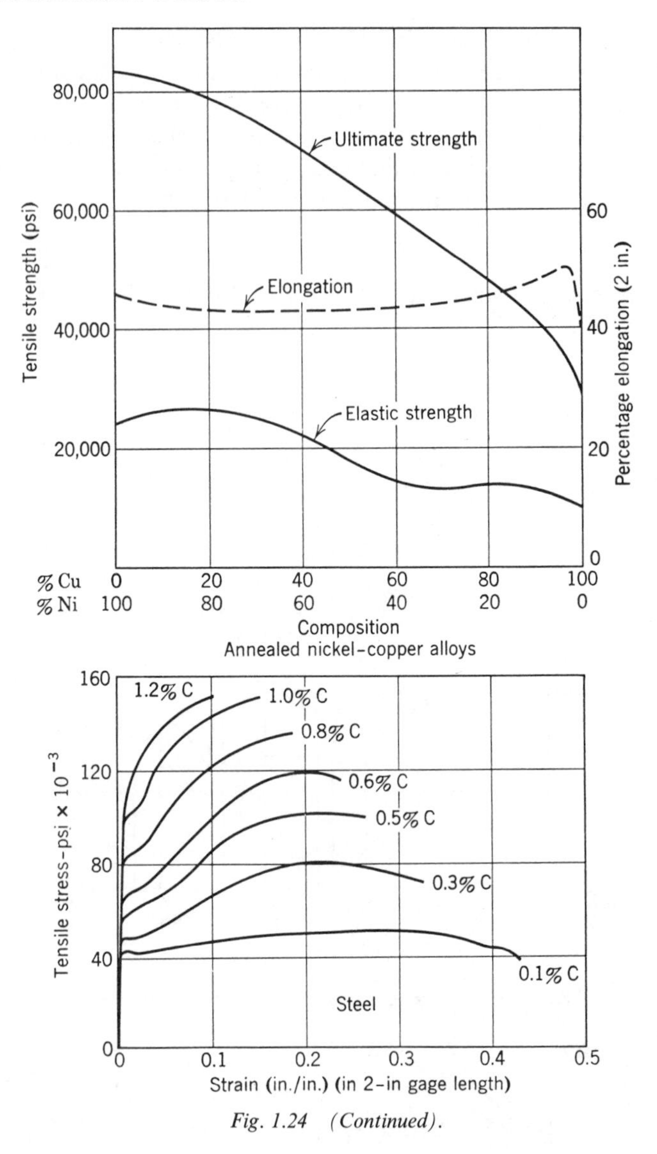

Fig. 1.24 (Continued).

rate 1.0, and *m*, a measure of strain rate sensitivity, is the slope of the line. This subject is pursued further in Chapter. 10

Composition. It would be expected that varying the chemical composition of a material would have an influence on the properties of the material. Examples of this effect are shown in Fig. 1.24. For steel, optimum properties are obtained with a tempered martensitic structure, and all steels with these structures have essentially the same mechanical properties for a given carbon content as shown in Fig. 1.25*A*; the effect of carbon content is shown in Fig. 1.25*B*.

Fabrication and working. Various methods used in fabricating materials

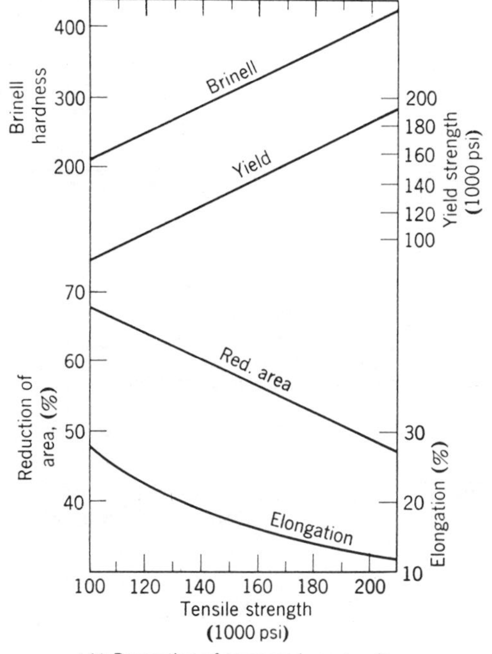

(A) Properties of tempered martensite

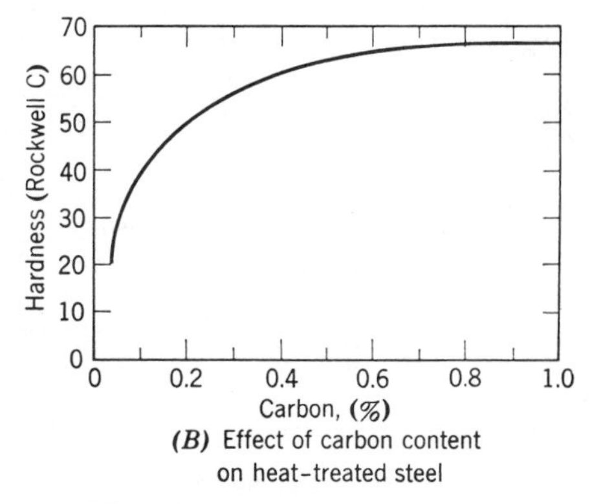

(B) Effect of carbon content on heat-treated steel

Fig. 1.25 Effect of composition on properties of steel.

may have an influence on mechanical properties by introducing varying amounts of cold working into the structure. Examples are shown in Figs. 1.26 and 1.27.

State of stress. It has already been shown that the state of stress can have a marked effect on the properties of a material. Recall, for instance, that the τ/σ ratio determines the relative brittleness or ductility of a material. It has also been indicated that materials show increased ductility when subjected to com-

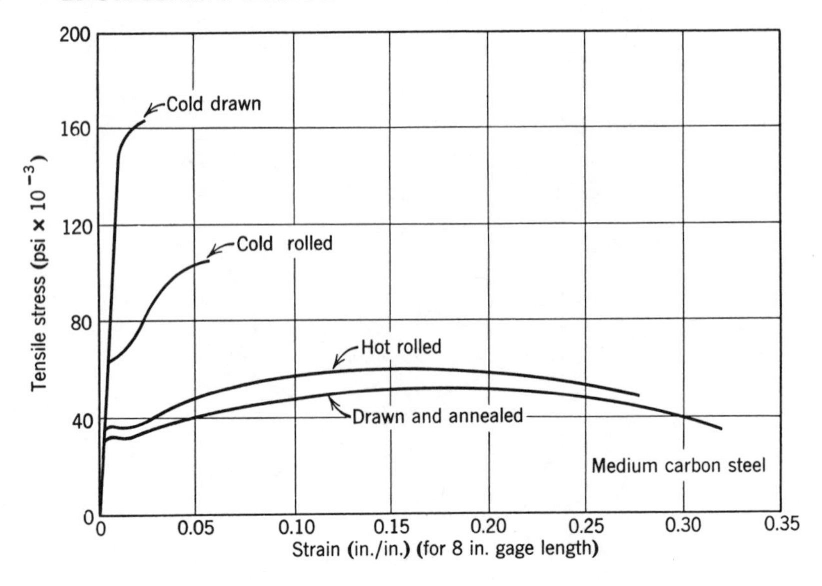

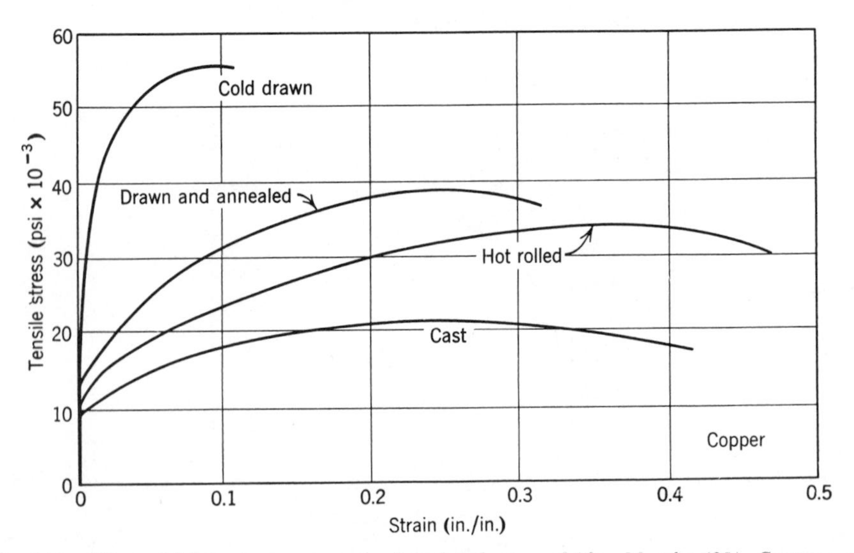

Fig. 1.26 Effect of fabrication on strength of steel and copper. [*After Murphy (25), Courtesy of The International Textbook Co.*]

pression loading. For example, Bridgman (2) found that 0.45% carbon steel in pure tension exhibited a tensile strength of about 60,000 psi, and when simultaneously subjected to a lateral pressure of 145,000 psi, it had a tensile strength of nearly 160,000 psi with a nearly 100% increase in ductility. More information on the effect of state of stress is in Chapter 3.

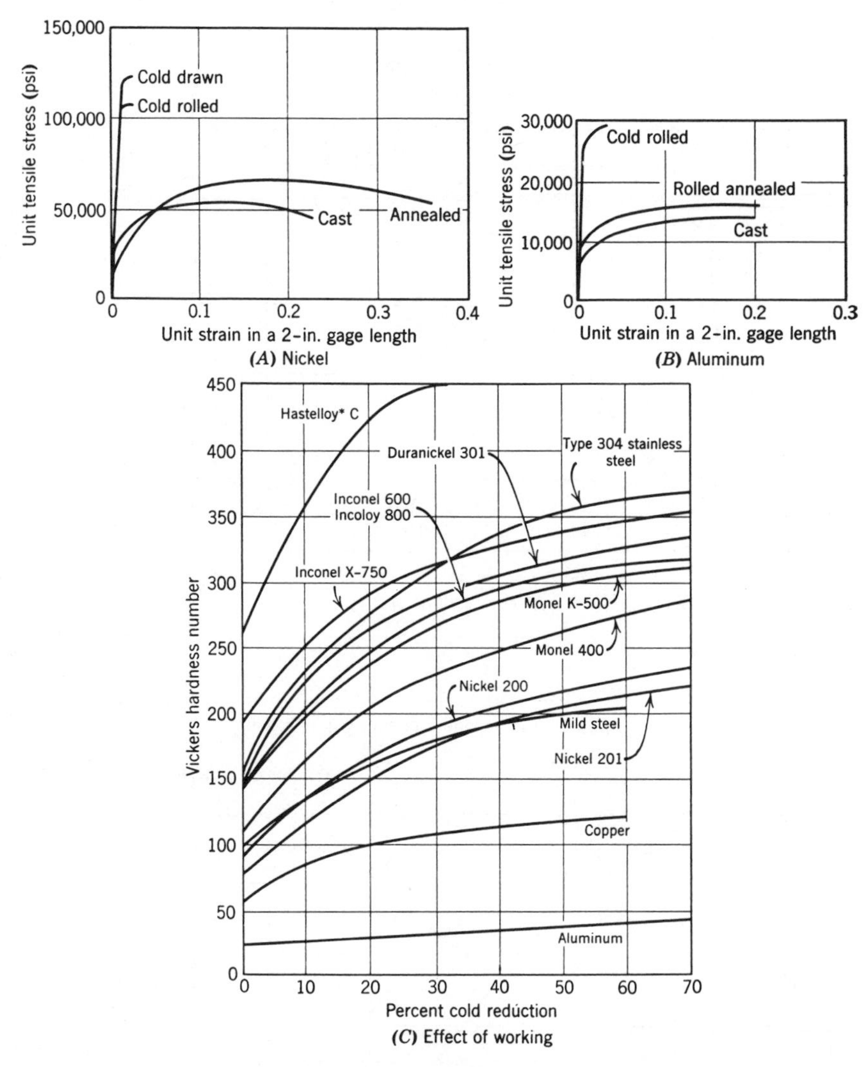

Fig. 1.27 Effect of working on properties. [*A and B (25), Courtesy of International Textbook Co.; C, Courtesy of International Nickel Co.*]

Size and shape. The effect of size and shape on mechanical properties of materials may take at least two forms; first are geometrical factors, which, through stress and strain gradients and section properties (moment of inertia), influence a material's reaction to imposed loads. This subject is considered in the next section. The other effect of size is related to the ability of a material to be properly heat treated or otherwise processed, depending on its size. In designing with steel, for example, one of the most important considerations is whether certain properties can be obtained in the required size and shape by

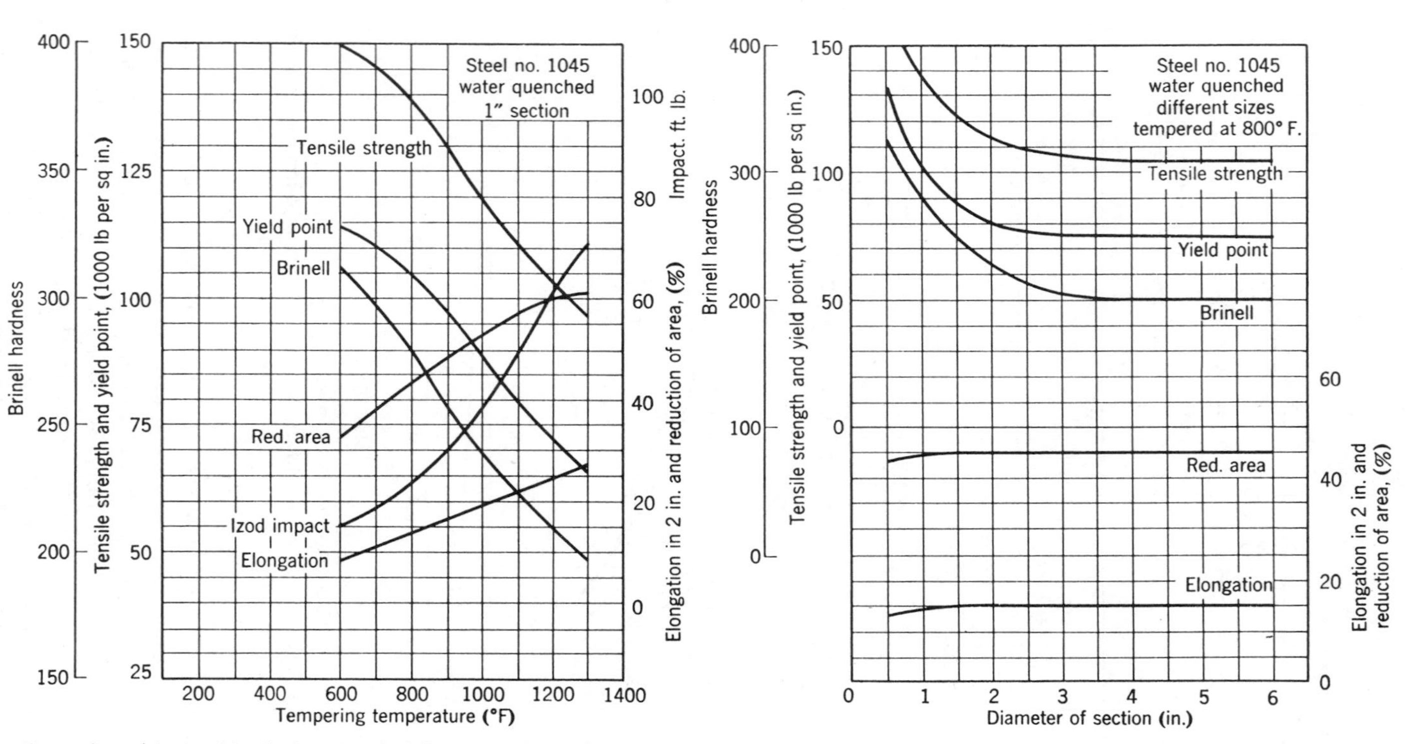

In sections ½ in. to 2 in. incl., quenched from 1475/1525°F; over 2 in. to 4 in. incl., from 1500/1550°F; over 4 in., from 1525/1575°F.

Fig. 1.28 Properties of water-quenched and tempered carbon steel SAE 1045 in different sizes. (Courtesy of International Nickel Co.)

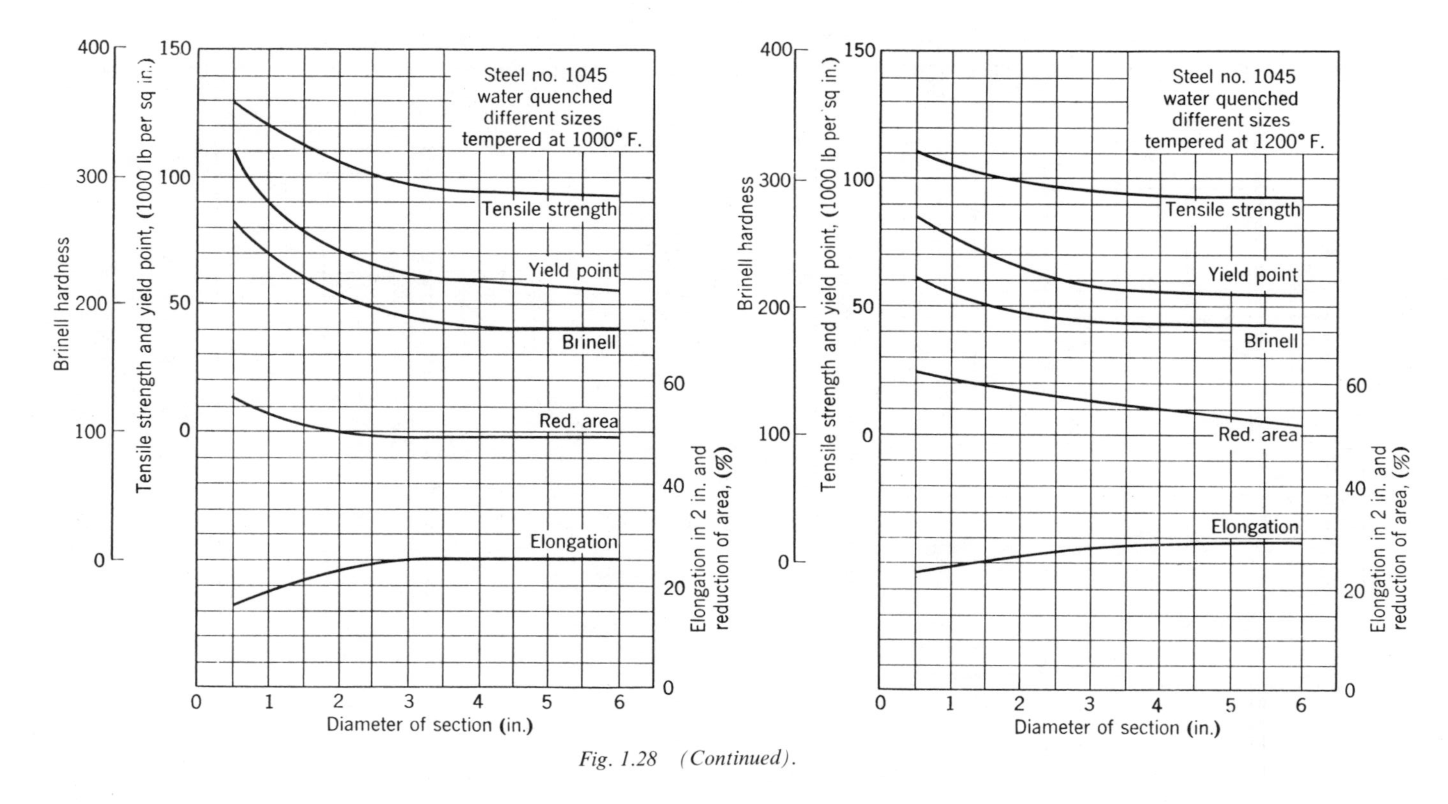

Fig. 1.28 (Continued).

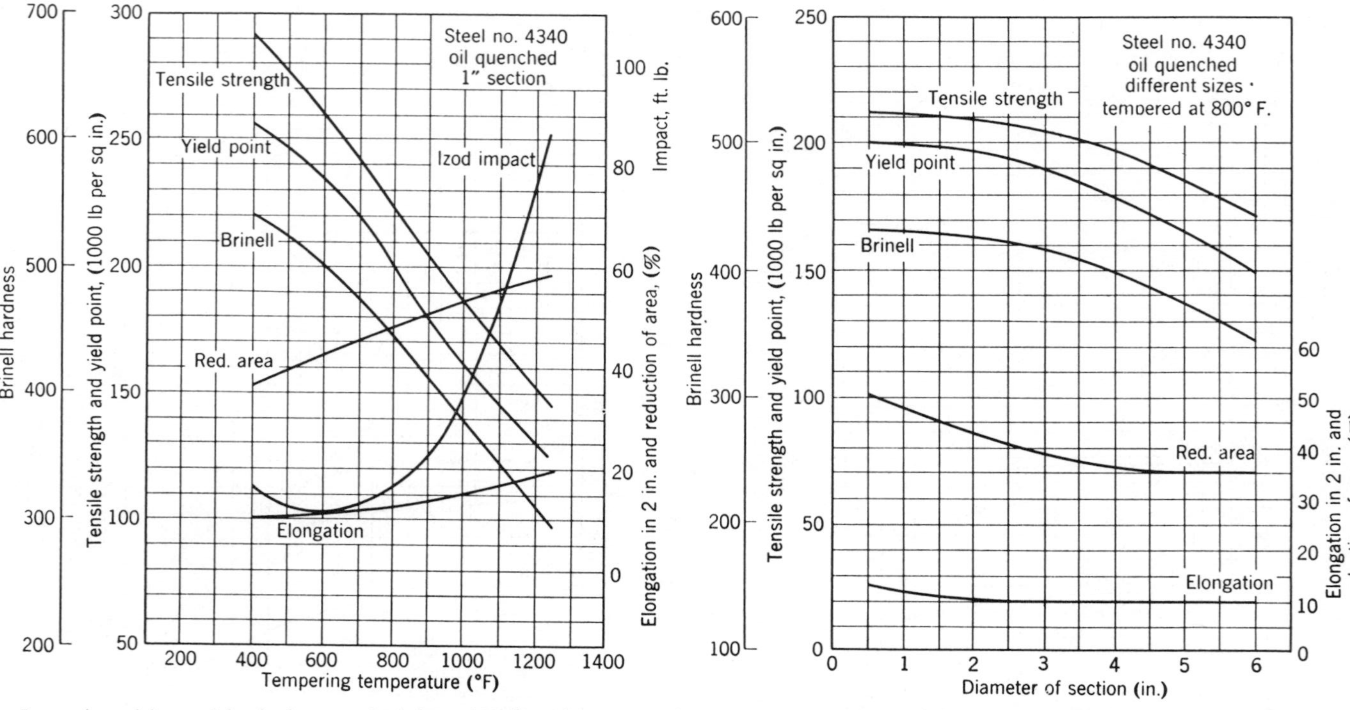

In sections ½ in. to 2 in. incl., quenched from 1500/1550°F; over 2 in. to 4 in. incl., from 1525/1575°F; over 4 in., from 1550/1600°F.

Fig. 1.29 Properties of oil quenched and tempered nickel-chromium-molybdenum steel SAE 4340 in different sizes. (Courtesy of International Nickel Co.)

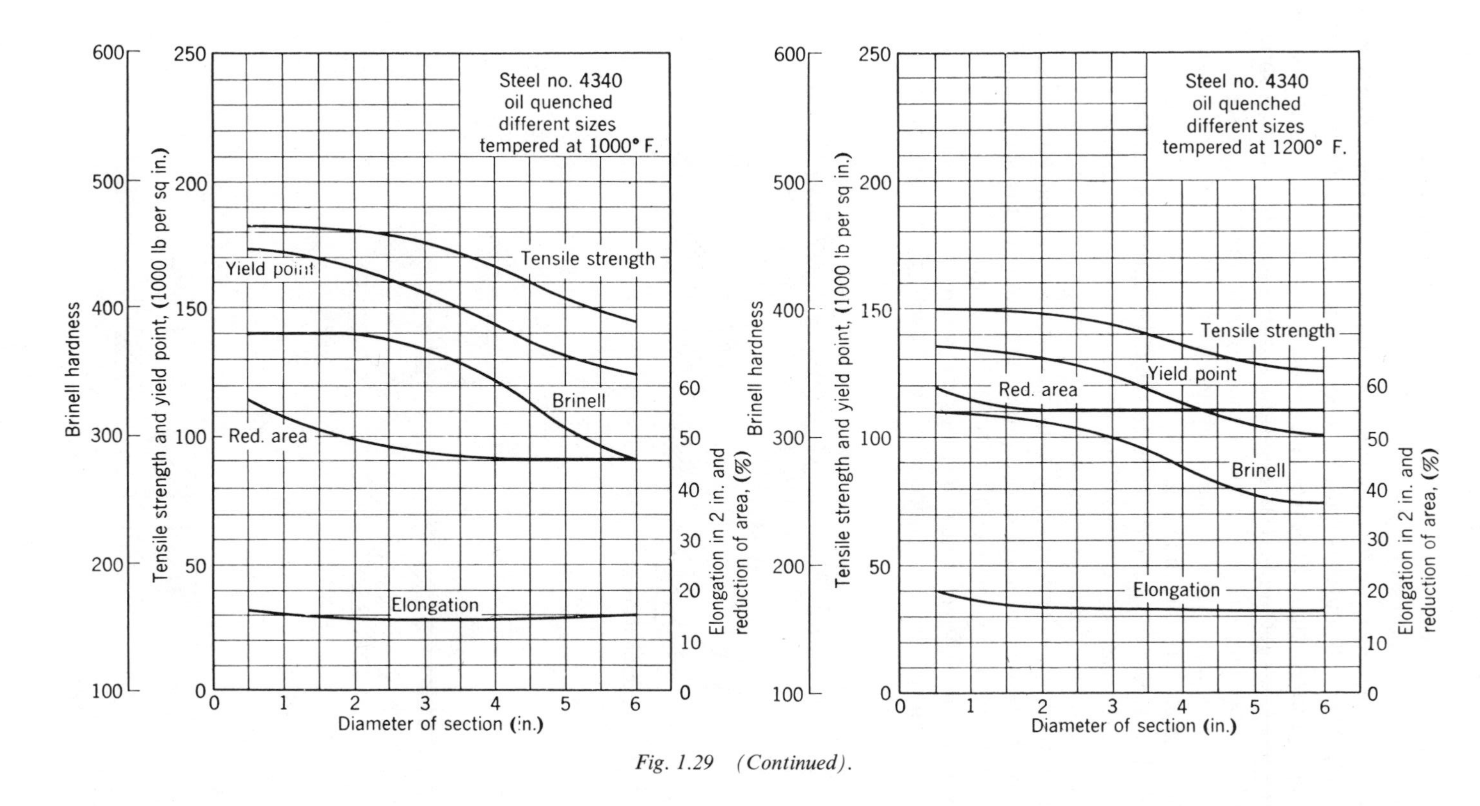

Fig. 1.29 (Continued).

commercial heat treatment. The variation in properties of some steels with heat treatment and section size is shown in Figs. 1.28 and 1.29. Hardness is a good indicator of the tensile strength of steel; but, for specific designs, charts such as shown in Figs. 1.28 and 1.29 must be used to insure that the ductility is not too low for a given strength. In designing steel parts requiring heat treatment, these factors must be kept in mind. In addition, abrupt changes in section should be avoided since thin parts will be cooled more rapidly than the heavier parts and, if holes are required, they should be symmetrically placed to avoid distortion. Some examples are shown in Fig. 1.30

Another aspect of size and shape as related to properties is shown in Fig. 1.31, which illustrates the variation of strain to fracture in tension tests of several metals as a function of gauge length. These data emphasize the fact that false impressions of the ductility of a material can be obtained if the gauge length used in the test is too small and representative only of the area of high flow instability and not of the over-all material behavior.

Environment. Environment includes temperature effects that have already been mentioned. The main concern in the following paragraphs is corrosion of

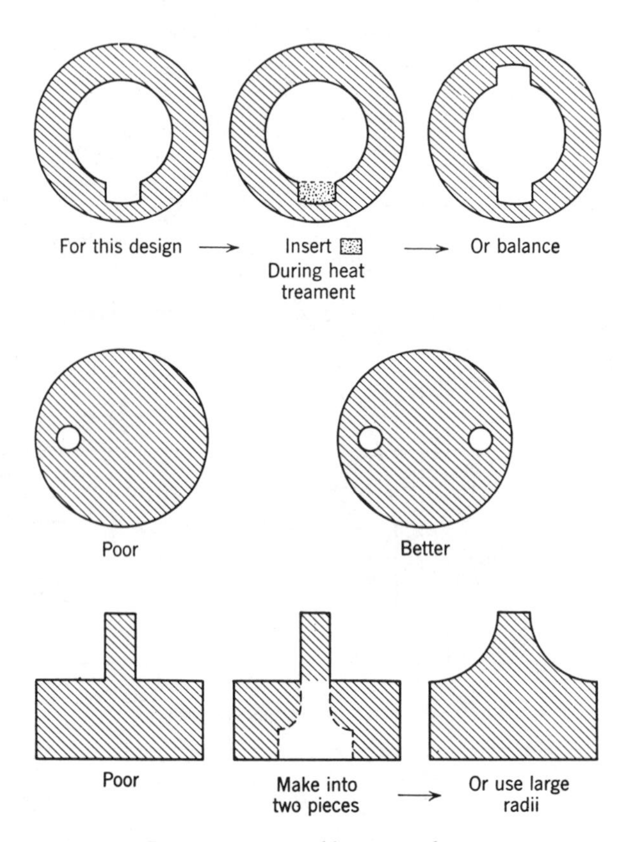

Fig. 1.30 Design of heat-treated parts.

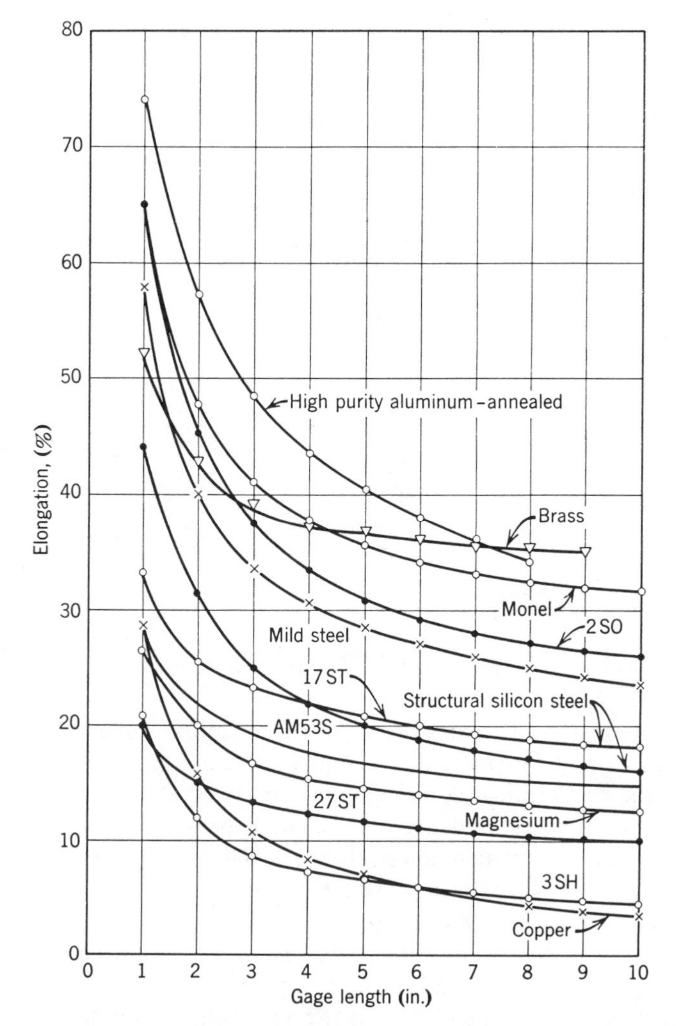

Fig. 1.31 Nominal strain to fracture as a function of initial gauge length. [After Templin and Sturm (27), Courtesy of The Institute of Aeronautical Sciences.] See Fig. 1.4 for alloy designation.

metals and its effect on design. Corrosion has been defined as the deterioration of materials through chemical or electrochemical attack. With this definition as a basis three important forms of corrosion are considered.

GALVANIC CORROSION. This is an electrochemical process that occurs when metals are exposed together in a common electrolyte. In this process the *less noble* metal corrodes. For example, if two bars of aluminum are held together by a steel bolt in a corrosive atmosphere, the aluminum, being less noble than steel, corrodes. To delay or prevent this attack a zinc washer may be inserted in the system. Zinc is less noble than either steel or aluminum and will corrode,

Table 1.8 Relative Corrodibility of Metals

Most Noble
Platinum
Gold
Silver
Titanium
Nickel alloys
Bronze
Copper
Brass
Tin
Lead
Steel
Cadmium
Aluminum
Zinc
Magnesium alloys
Least Noble

leaving the steel and aluminum intact. The relative corrodibility of some metals is given in Table 1.8.

STRESS CORROSION. The combined presence of stress and a corrosive atmosphere is particularly damaging to many metal systems and is a constant design problem. One of the most common forms of stress corrosion is the caustic embrittlement of boilers, pressure vessels, and process equipment. This type of failure is particularly prevalent in vessels exposed to caustics and stresses beyond the yield strength of the material. In the same process, corrosion products collect in crevices and cause additional stressing. It is often difficult to design around this type of failure since the embrittlement seems to be independent of alloy composition of mild steel. In general, stress corrosion is very difficult to handle since almost any metal can be affected. Some guidelines are available, however, that assist the designer in the selection of materials. Austenitic stainless steels seem to be prone to stress corrosion cracking in chloride atmospheres. Nitrogen in irons and steels tends to make these materials susceptible to cracking in nitrate-containing atmospheres. Copper alloys often crack in the presence of ammonia. Thus, in the area of corrosion, the designer should consider the following points:

1. Insulate dissimilar metals.
2. Avoid crevices and situations leading to condensation.
3. Avoid local hot spots.
4. Avoid stress concentration.

HYDROGEN ATTACK. Of special commercial interest is the effect of hydrogen on metals, particularly steels (38, 39, 40). One effect of hydrogen is chemical in nature, that is, hydrogen enters the metal, reacts with particular constituents, and induces loss in ductility. This effect is commonly known as hydrogen embrittlement. Hydrogen in the atomic form also enters materials and then forms molecular hydrogen. When this happens, high local pressure is generated, and the resulting effect is internal cracking and/or the formation of surface blisters and bulges. Generally these effects are more pronounced with increased strength of material, cold working, and the presence of notches. Typical data are shown in Tables 1.9 and 1.10.

Table 1.9 Effect of 5000 (10,000) Psig Hydrogen on Metals at Ambient Temperature (39, 40)

		Property in Hydrogen ÷ Property in Helium	
Material	Yield Strength 10^3 psi	Notched Tensile Strength	Notched Reduction in Area
A302B Steel	50	0.9 (0.8)	0.35 (0.3)
Inconel 718	198	0.7 (0.45)	0.40 (0.1)
Type 410 stainless	192	(0.22)	(0.27)
17–7 PH stainless	150	(0.23)	(0.40)
Type 430 stainless	45	(0.68)	(0.31)
Type 304L stainless	29	(0.87)	(0.48)
Type 316 stainless	31	(1.0)	(1.0)
Type 321 stainless	32	0.88	0.36
Type 347 stainless (cold-worked)	67	0.91	0.91

Principles of minimum weight analysis

The concept of minimum weight implies a structure having maximum strength combined with minimum weight. In this sense minimum weight is not necessarily the absolute minimum but the weight consistent with the geometry required to maintain integrity of the part. For example, in a column the minimum weight would be dictated by the geometry necessary to avoid buckling and not, in all cases, the weight required to only avoid plastic yielding. As shown later, relative costs, stiffness, and so on are also to be considered. The stress analysis formulas used in the following paragraphs are developed later. Minimum weight analyses and weight-reliability analyses, particularly for pressure vessels and aircraft, are given in (1, 7, 9, 12, and 26).

Tension (Fig. 1.4*A*). For simple tension the working stress is defined by the relationship

$$\sigma_w = F/A \qquad 1.53$$

where A is the cross-sectional area of the member. The weight of such a member is

$$W = \gamma V = \gamma AL \qquad 1.54$$

*Table 1.10 Operating Limits for Steels in Hydrogen Service (38)**

Steel	Temperature °F	Hydrogen Partial Pressure psia
Carbon	1000	100
	700	100
	600	150
	550	300
	500	1,000
	450	2,500
	430	13,000
0.5 Mo	1000	200
	975	200
	900	450
	775	1,000
	700	1,400
	630	2,500
	625	13,000

*(38) should be consulted for other steels at various pressures and temperatures.

where γ is the density of the material and V the volume. The *strength-weight ratio* can then be defined as

$$F/W = \sigma_w A/\gamma LA = \sigma_w/\gamma L \tag{1.55}$$

From this relationship it is clear that the same F/W value can be obtained from a variety of combinations of σ_w, γ, and L. Therefore, in order to arrive at an optimum selection among several candidate materials, the analysis must be continued. If the member in question is a solid round rod, the following basic equations can be set down:

$$\text{working stress} = \sigma_w = F/A = F/(\pi D^2/4) \tag{1.56}$$

$$\text{total strain} = \varepsilon = FL/AE = FL/(\pi D^2/4)E \tag{1.57}$$

From Eq 1.56 a comparison can now be made between two materials A and B relative to their ability to support a load F.

$$(\sigma_w)_A = 4F/\pi D_A{}^2 \quad \text{and} \quad (\sigma_w)_B = 4F/\pi D_B{}^2 \tag{1.58}$$

from which

$$D_B/D_A = (\sigma_w)_A/(\sigma_w)_B \tag{1.59}$$

Since weight equals volume times density,

$$W_A = V_A\gamma_A \quad \text{and} \quad W_B = V_B\gamma_B \tag{1.60}$$

Furthermore,

$$V_A = \pi D_A{}^2 L/4 \quad \text{and} \quad V_B = \pi D_B{}^2 L/4 \tag{1.61}$$

Therefore combining Eqs. 1.59 to 1.61 results in the relationship

$$W_B/W_A = (\sigma_A/\sigma_B)(\gamma_B/\gamma_A) \tag{1.62}$$

If material A costs $\$A$ per pound and material B costs $\$B$ per pound, the cost of the two materials is

$$C_A = W_A(\$A \text{ per pound}) \quad \text{and} \quad C_B = W_B(\$B \text{ per pound}) \tag{1.63}$$

which when combined with this equation gives

$$C_B/C_A = (\sigma_B/\sigma_A)(\gamma_B/\gamma_A)(\$B/\$A) \tag{1.64}$$

Similarly, starting with Eq. 1.57, a set of equations comparing the two materials A and B on the basis of equal rigidity can be made. The resulting relationships are as follows:

$$D_B/D_A = (E_A/E_B)^{1/2} \tag{1.65}$$

$$V_B/V_A = E_A/E_B \tag{1.66}$$

$$W_B/W_A = (E_A/E_B)(\gamma_B/\gamma_A) \tag{1.67}$$

$$C_B/C_A = (E_A/E_B)(\gamma_B/\gamma_A)(\$B/\$A) \tag{1.68}$$

Having the basic comparative equations, it is now possible to select the optimum material. Suppose the required structure has to support a load of 50,000 lb; space limitations restrict the length of the member to 2 ft, and the only materials to choose from are listed in Table 1.11. It will be assumed that the working stress

Table 1.11 Weight-Strength Cost Analysis of Materials

Material	Yield Strength, psi × 10^{-3} (σ_w)	Density, lb/in.^{3}, (γ)	σ_w/γ Ratio (× 10^{-3})	Modulus of Elasticity, psi × 10^{-6} (E)	Weight, 24-in. Bar,* (lb)	Cost (\$)†	1961 Cost/lb	1977 Cost/lb	Cost 1977 / Cost 1961
Monel (annealed)	35	0.319	110	26	10.90	90.80	0.90	8.33	9.26
Monel (cold-drawn)	100	0.319	314	26	3.82	44.31	3.14	11.60	3.69
Aluminum 1100–0	5	0.098	51	10	23.60	35.16	0.26	1.49	5.73
Aluminum 2017-T4	40	0.101	396	10	3.04	8.54	0.49	2.81	5.73
Copper (annealed)	10	0.322	31	10.4	23.70	54.04	0.52	2.28	4.38
Copper (cold-rolled)	48	0.322	149	10.4	8.10	18.47	0.52	2.28	4.38
Beryllium copper	130	0.297	437	18	2.77	29.25	1.44	10.56	7.33
Type 304 stainless	30	0.286	105	29	11.50	27.95	0.50	2.43	4.86
Type 410 stainless	115	0.277	416	28	4.92	10.82	0.76	2.20	2.89
SAE 1020 (annealed)	40	0.284	141	30	8.50	3.32	0.06	0.39	3.32
SAE 4340 (heat-treated)	150	0.284	528	30	2.24	2.76	0.56	1.23	2.76

*Based on round rod supporting a load of 50,000 lb.
†Approximately 1977 prices.

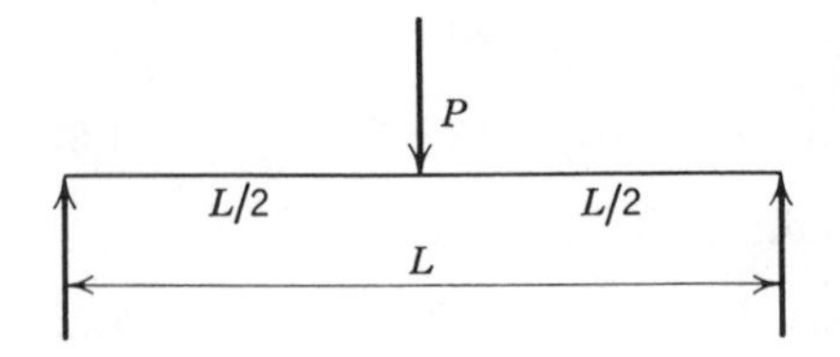

Fig. 1.32 Simply supported beam with center load.

is equal to the yield strength of the material. Examination of Table 1.11 reveals that the heat-treated 4340 steel would weigh the least of the materials available and would also be low enough in cost as compared with the other materials. Note that this material also has the highest materials strength-weight ratio. If there are limitations on deformation, resistance to corrosion, or volume, for example, some choice other than the 4340 might be satisfactory.

BENDING. In bending, the working stress is defined by the relationship (see Chapter 2)

$$\sigma_w = Mc/I \tag{1.69}$$

The weight of the part is given by Eq. 1.54 therefore, the general expression for *strength-weight ratio in bending* is given by the equation

$$M/W = (\sigma_w/\gamma)I/cLA \tag{1.70}$$

Since the bending strength-weight ratio depends on the geometry and type of loading, a specific value is obtained for each different condition; for example, for the beam shown in Fig. 1.32 the maximum bending moment is $PL/4$ so that Eq. 1.70 becomes

$$P/W = 4\sigma_w I/\gamma L^2 cA \tag{1.71}$$

Thus, based on Eq. 1.71, it is evident that for maximum strength-weight ratio the *materials strength-weight ratio* σ_w/γ, as well as the geometrical *section modulus*, I/c must be maximized. Note also in Eq. 1.71 that the cross-sectional area is a major factor assuming a constant value of length. The maximum value

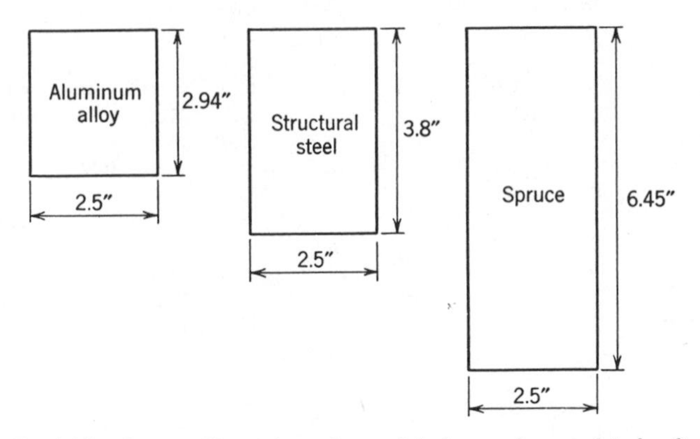

Fig. 1.33 Comparable sections of materials for equal strength in bending.

of I/cA occurs for thin-walled circular tubes; for such tubes the wall thickness is limited only by the material's ability to resist plastic flow or failure by buckling.

A comparison of relative weights, costs, stiffness, and volume for several materials may be desired. Equations expressing these comparisons similar to Eqs. 1.54, 1.62, 1.64, 1.65 to 1.68, for tension are easily derived and are not developed here.

As a numerical example, consider the relative performance of steel, wood, and aluminum for a structural application such as depicted in Fig. 1.32, a 30-ft long, simply supported beam with center load of 1000 Ib. The following information on the materials is given.

	Aluminium Alloy (A)	Structural Steel (B)	Wood (Spruce) (C)
Working stress σ (psi)	25,000	15,000	5000
Modulus of elasticity E (psi)	10×10^6	30×10^6	1.4×10^6
Density (lb/in.^{3})	0.10	0.28	0.014
Cost ($/lb)	2.81	0.39	—

By referring to the equations mentioned, the following comparisons for a beam of rectangular cross section can be made.

$$D_B/D_A = (25/15)^{1/2} = 1.292$$

$$D_C/D_B = (15/5)^{1/2} = 1.732$$

From the bending stress formula, $\sigma = Mc/I$, it is determined from the loading indicated that the section modulus I/c has to be

$$\frac{I}{c} = \frac{M}{\sigma} = \frac{(1000)(360)}{(4)(25{,}000)} = 3.6$$

Therefore taking a width of 2.5 in. the comparable rectangular cross sections would be those shown in Fig. 1.33. The metal members would not ordinarily be rectangular but equivalent I-beams as shown in Fig. 1.34. Thus, on an equal load-carrying basis, a 30-ft span would cost about $312 for aluminum, $176 for steel, and about the same for wood. However, on a stiffness basis,

$$D_B/D_A = (10/30)^{1/2} = 0.67$$

Therefore, for this geometry since

$$\text{deflection} = WL^3/48EI$$

and using the aluminum beam as the standard,

$$\text{deflection} = \frac{(1000)(360)^3}{48(10 \times 10^6)(7.18)} = 13.5 \text{ in.}$$

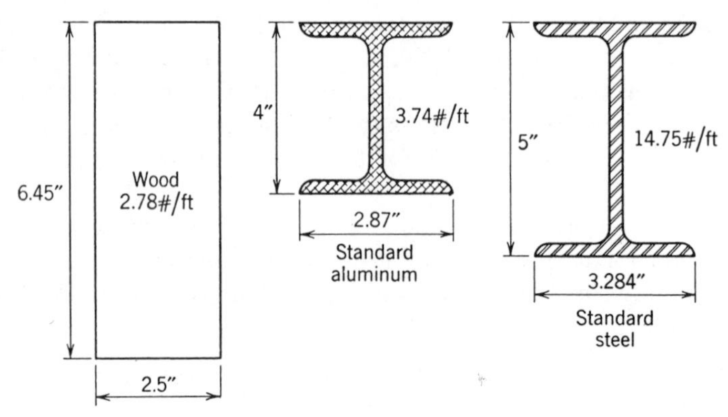

Fig. 1.34 Equivalent sections for equal strength.

For wood, for the same deflection, the required moment of inertia I is 51.6, a value supplied by a beam 3.5×5.5, approximately. For steel the required I-value is 2.4, represented by an I-beam approximately 3×2.5. Thus the sections shown in Fig. 1.35 would provide equal stiffness.

The preceding example shows how different materials may be compared on a size basis. An additional helpful comparison for strength and stiffness of some structural materials is given in Fig. 1.36. In this illustration the series of logarithmic scales are arranged to provide a means for finding the relative sheet, plate, or beam thickness required for various metals of equal stiffness or equal strength, respectively. Corresponding values for different materials may be read vertically in line within each of the two sections. For example, 0.049 in. aluminum sheet has the same stiffness as 0.035 in. steel or 0.058 in. magnesium. Similarly, an aluminum beam 4.9 in. deep corresponds in stiffness to 3.5 in. steel and 5.8 in. magnesium.

Torsion (Fig. 1.37). In torsion the working stress τ_w is expressed by different equations, depending on the shape of the cross section of the member (see Chapter 2). For circular members this relationship is

$$\tau_w = Tr/J \qquad 1.72$$

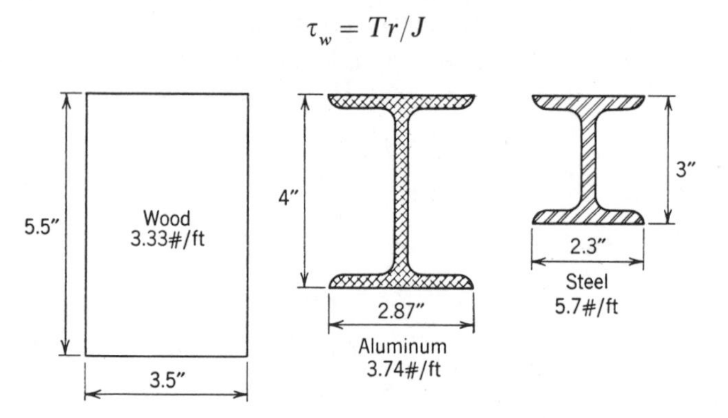

Fig. 1.35 Equivalent sections for equal stiffness.

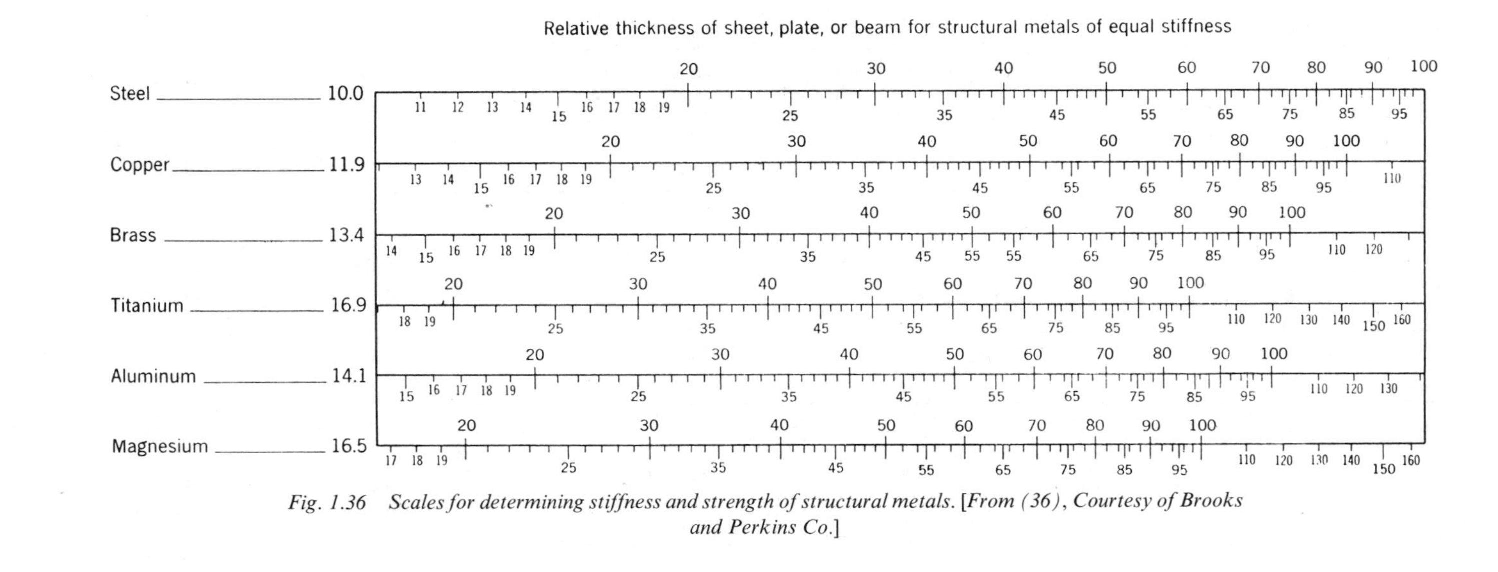

Fig. 1.36 *Scales for determining stiffness and strength of structural metals.* [*From* (36), *Courtesy of Brooks and Perkins Co.*]

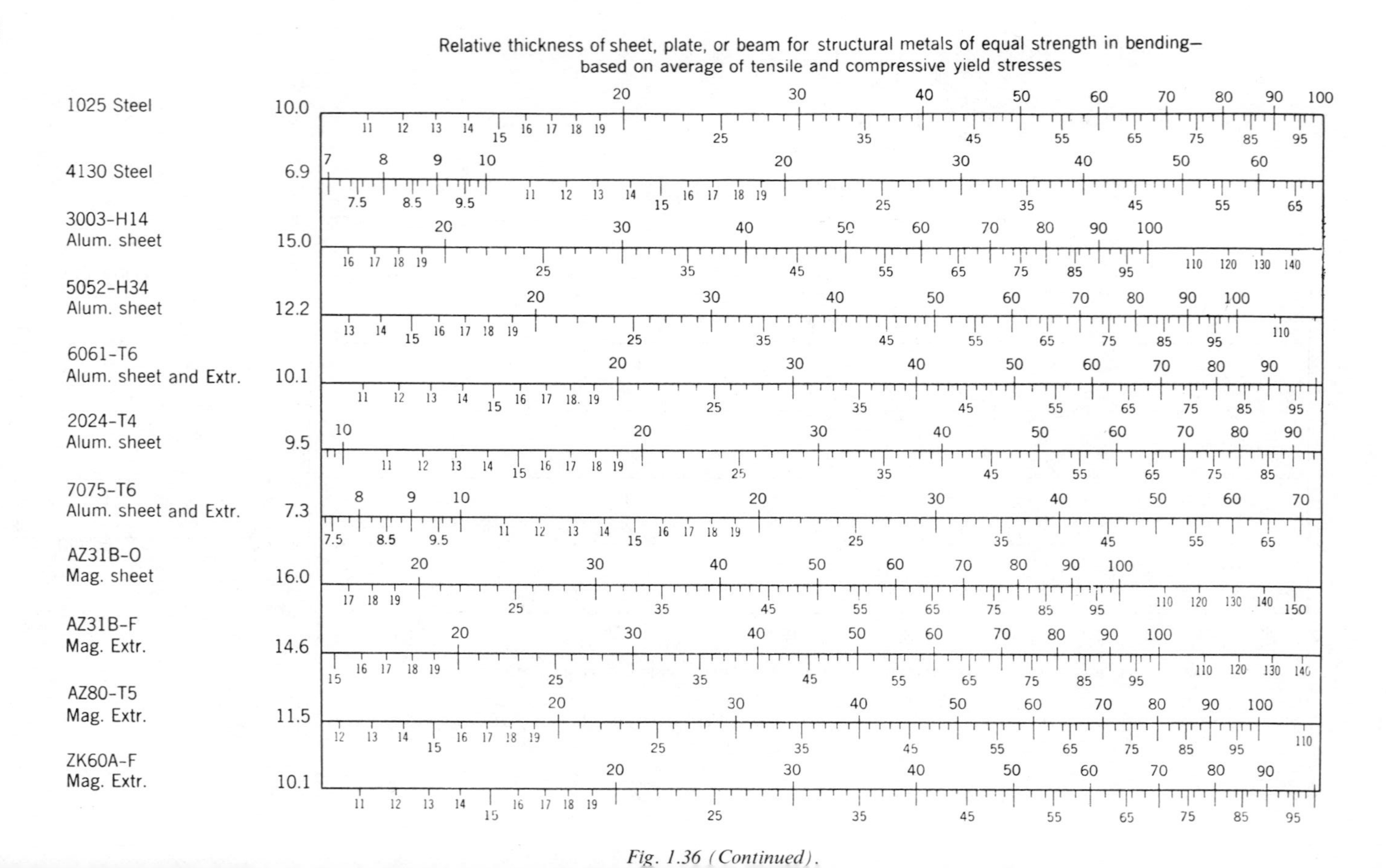

Fig. 1.36 (Continued).

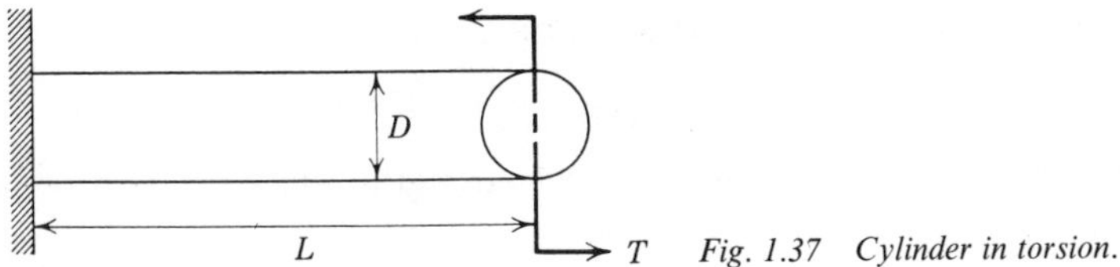

Fig. 1.37 Cylinder in torsion.

Weight is given by Eq. 1.54; therefore the strength-weight ratio in torsion is given by

$$T/W = \tau_w J/\gamma ArL \tag{1.73}$$

For a solid circular rod

$$\frac{T}{W} = \left(\frac{\tau_w}{\gamma L}\right)\left[\frac{\pi r^4}{(2r)(\pi r^2)}\right] = \left(\frac{\tau_w}{\gamma L}\right)\left(\frac{r}{2}\right) \tag{1.74}$$

For a thin-walled tube,

$$\frac{T}{W} = \left(\frac{\tau_w}{\gamma L}\right)\left[\frac{2\pi r^3 t}{(2\pi rt)(r)}\right] = \frac{\tau_w r}{\gamma L} \tag{1.75}$$

These equations indicate increase in T/W with increase in section radius, with the thin-walled tube being the more efficient. However, as shown in Chapter 9, the critical stress for torsional buckling decreases as the tube size r increases. This condition tends to offset the advantage gained by increasing r. The wall thickness is, in general, governed by the ability of the material to resist plastic flow. Furthermore, the *minimum* wall thickness of the tube governs. Therefore the most efficient shape in torsion is a thin-walled tube of *uniform* thickness since it conforms to maximizing the torsional section modulus as well as providing the minimum cross-sectional area. Again, as in tension, comparative equations can be set up to illustrate the relative strength, stiffness, and so on, for various materials.

Buckling of slender column (Fig. 1.38). In a column failure may occur by plastic flow of the material at stresses higher than the working stress level, or failure may occur by buckling. Because of these two considerations it may not always be possible to design a column having minimum weight. For the column shown in Fig. 1.38, the critical buckling load (Chapter 9) is given by the relationship

$$F_{cr} = \pi^2 EI/4L^2 \tag{1.76}$$

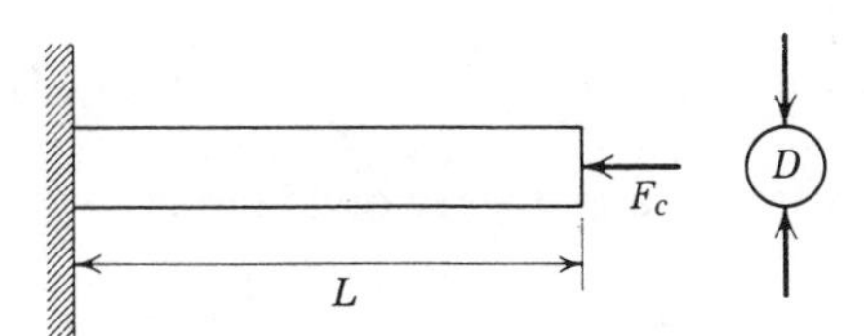

Fig. 1.38 End-loaded column.

or, alternatively, in terms of a critical stress,

$$\sigma_{cr} = F_{cr}/A = \pi^2 EI/4L^2 A \tag{1.77}$$

Weight is given by Eq. 1.54; therefore the strength-weight ratio for the member shown is

$$F_{cr}/W = \pi^2 EI/4\gamma L^3 A \tag{1.78}$$

If the working stress is approximated by the critical stress for bucking,

$$\sigma_{cr} = \sigma_w = \pi^2 EI/4L^2 A \tag{1.79}$$

Equation 1.73 can be written in the form

$$(F/A)\sigma_w = (\pi^2 EI/4L^2 A)(F/A) \tag{1.80}$$

from which

$$\sigma_w = (\pi^2 EIF/4L^2 A^2)^{1/2} \tag{1.81}$$

Equation 1.81 defines the working or optimum stress in terms of the loading and geometry involved. From Eq. 1.77 the optimum area is defined by

$$\sigma_w L^2 = (F/A)L^2 \tag{1.82}$$

or

$$A/L^2 = F/L^2\sigma_w \tag{1.83}$$

Therefore, from Eqs. 1.54 and 1.83,

$$(W/L^3)/(F/L^2) = \gamma/\sigma_w \tag{1.84}$$

which when substituted in Eq. 1.81 gives Eq. 1.78. A more detailed analysis of such problems is given by Gerard (3). Relative comparison among several materials can also be made by formulating equations similar to those previously given for tension.

Another important type of buckling problem is the thin-walled structures used in airframe and missile design. For weight saving the light metals are frequently used, but this is not always optimum. For example, steel can often be used effectively in place of aluminum since it has a higher modulus of elasticity and therefore resists buckling to a greater extent for the same geometry even though the part weighs three times as much.

Thin-walled cylinder under internal pressure (Fig. 1.39). In Chapter 3 it is shown that two principal stresses for the geometry shown are as follows:

$$\text{hoop stress} = \sigma_h = pd/2t \tag{1.85}$$

$$\text{axial stress} = \sigma_z = pd/4t \tag{1.86}$$

The working stress defined by the distortion energy theory of failure is

$$\sigma_w{}^2 = \sigma_h{}^2 + \sigma_z{}^2 - \sigma_z\sigma_h \tag{1.87}$$

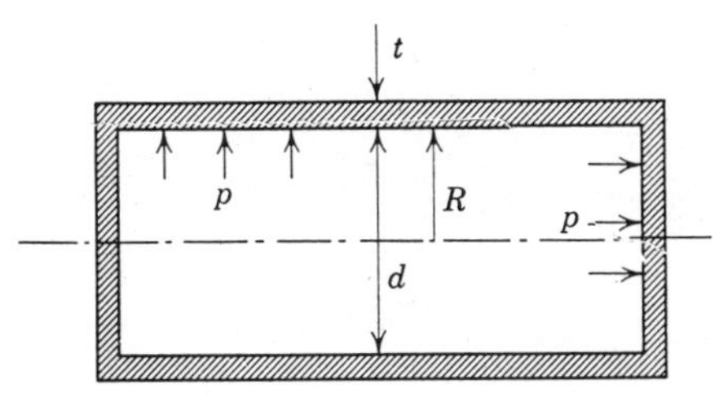

Fig. 1.39 Thin-walled cylinder under internal pressure.

Combining Eqs. 1.85 to 1.87 gives an expression for the working internal pressure*

$$p_w = (4\sigma_w/\sqrt{3})(t/d) \tag{1.88}$$

Weight is given by Eq. 1.54; therefore the strength-weight ratio for the vessel in question is

$$\frac{p}{W} = \left(\frac{8\sqrt{3}}{3\pi}\right)\left(\frac{\sigma_w}{\gamma}\right)\left(\frac{t}{d}\right)\left[\frac{1}{(td)(2L+d)}\right] \tag{1.89}$$

Multiple-sphere vessel. It is well known that a spherical vessel is a very efficient type of structure since it has a uniformly stressed membrane. In addition, this type of shell has the largest volume per surface area. However, for many applications the spherical vessel, as such, is not adaptable to other requirements, and so some departures from maximum efficiency have to be tolerated. One way to approach the ideal condition of a sphere is to use the multiple-sphere arrangement; this type of vessel gives a longitudinal dimension approaching that of a cylinder while retaining many of the advantages of the sphere (1). A typical multiple-sphere arrangement is shown in Fig. 1.40. In the vessel shown the bulkhead attachment A–A is critical; ideally it should be such that the uniform membrane stress in the shell is not disturbed. Usually A–A consists of a concentric ring for access to all segments of the vessel. Here

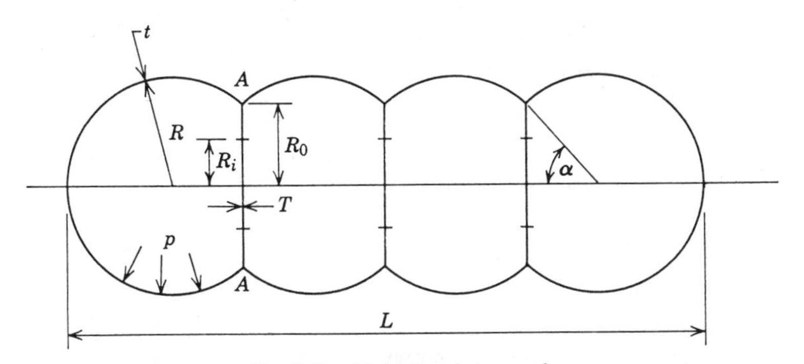

Fig. 1.40 Multiple-sphere vessel.

*Optimum design of pressure vessels is not always possible because of proof test requirements. For a discussion of this problem see (18).

the required thickness is

$$T = t\left\{\left(\frac{2E_{\text{vessel}}}{E_{AA}}\right)\left[\frac{1-\nu_{AA}+(1+\nu_{AA})(R_i/R_0)^2}{(1-\nu_{\text{vessel}})[1-(R_i/R_0)^2]}\right]\cos\alpha\right\} \qquad 1.90$$

The enclosed volume of a vessel with n intersections is

$$V = \frac{4\pi R^3}{3}\left[\frac{n\cos\alpha}{2}(3-\cos^2\alpha)+1\right] \qquad 1.91$$

and the weight is

$$W = 4\pi R^2 t\gamma(n\cos\alpha+1)$$
$$+\,2R^2 tn\pi\left(\frac{E_{\text{vessel}}}{E_{AA}}\right)\gamma_{AA}\sin^2\alpha\cos\alpha\left[\frac{1-\gamma_{AA}+(1+\gamma_{AA})(R_i/R_0)^2}{1-\nu_{\text{vessel}}}\right] \qquad 1.92$$

where γ is specific weight. From the preceding the W/V ratio can be calculated. For many such vessels a particular L/R ratio is required; however, from the geometry

$$L/R = 2(n\cos\alpha+1) \qquad 1.93$$

and this value can be selected to obtain the desired proportions. For a simple sphere

$$(W/V)_{\text{sphere}} = 3p\gamma/2\sigma \qquad 1.94$$

where σ is the uniform membrane stress

$$\sigma = pR/2t \qquad 1.95$$

To compare the multiple sphere with the single sphere, use is made of Eqs. 1.91 to 1.94.

$$\frac{(W/V)_{\text{intersecting}}}{(W/V)_{\text{single}}} = 1+\left\{1+\frac{E/\gamma}{2E_{AA}/\gamma_{AA}}\left[1-\nu_{AA}+(1+\nu_{AA})\left(\frac{R_i}{R_0}\right)^2\right]\sin\alpha\right\}$$
$$\times\frac{(L/2R)-1}{1+\frac{1}{2}(3-\cos^2\alpha)(L/2R-1)} \qquad 1.96$$

For A–A the stress σ_{AA} is

$$\sigma_{AA} = E_{AA}\left[\frac{2(1-\nu)\sigma}{1-\nu_{AA}+(1+\nu_{AA})E(R_i/R_0)^2}\right] \qquad 1.97$$

From Eq. 1.96 the optimum geometry is for a solid bulkhead.

Thick-walled cylinder under internal pressure. In the thick-walled cylindrical vessel, initial yielding (see Chapter 3) can be controlled somewhat by various prestressing techniques; there is also a limiting ratio of outside to inside diameter beyond which there is little gained in elastic range behavior by simply increasing the size of the vessel (see discussion in Chapter 3). However, burst pressure is not only a function of material properties but also of the wall thickness of the vessel so that theoretically there is no limit on this quantity and considerable saving can often be accomplished by a judicious choice of materials. For example, the burst pressure of a thick cylinder is

$$p_b = (2\sigma_y/\sqrt{3}) \ln R(2 - \sigma_y/\sigma_u) \qquad 1.98$$

where σ_y and σ_u are the yield (0.2% offset) and ultimate strengths of the material and R is the ratio of outside to inside diameter. Now, consider two candidate materials of widely different properties, low-carbon steel and heat-treated alloy steel, as shown in the following.

	Yield Strength (psi)	Ultimate Strength (psi)	Density (lb/in.^{3})	Cost ($/lb)
Low-carbon steel	30,000	80,000	0.28	0.39
Alloy steel	190,000	250,000	0.28	1.23

By using these values with Eq. 1.98, the following costs are obtained:

	Outside Diameter, D (in)		Weight/Foot (lb)		Cost/Foot ($)	
Inside Diameter, d (in)	Low-Carbon Steel	Alloy Steel	Low-Carbon Steel	Alloy Steel	Low-Carbon Steel	Alloy Steel
1	7.14	1.5	133	3.3	51.87	4.18
2	14.28	3.0	535	13.3	208.65	16.40
4	28.56	6.0	2140	53.3	834.60	65.60
8	57.12	12.0	8450	213.0	3295.50	262.40

It is obvious that for large cylinders, a more expensive steel having higher mechanical properties is cheaper in the long run. Added to the preceding is additional machining time for the low-carbon steel and higher transportation costs because of the increased size and weight.

1-3 ANALYSIS OF SIZE AND SHAPE EFFECTS

In this section several situations where size and shape of a part have an influence on its reaction to imposed loads are considered. Effects associated with section modulus have already been discussed as well as the effect of size on the ability of a material to react to heat treatment. In the following paragraphs, con-

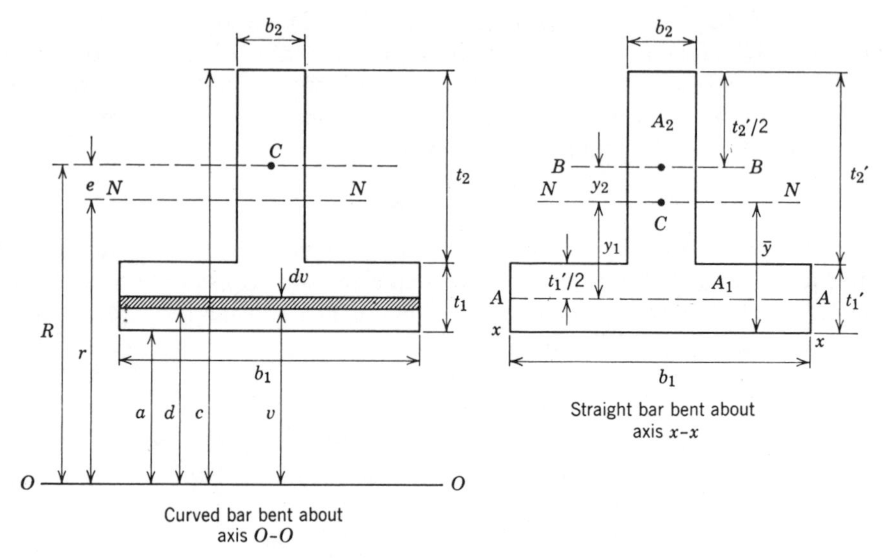

Fig. 1.41 Bending of T-sections.

sideration is given first to the bending of beams whose shape is such that either the centroidal and neutral axes do not coincide (curved bars) or one that does not occur at mid-depth (straight bars with only one axis of symmetry).

Bending of a T-section. Consider the bending of a bar having the section shown in Fig. 1.41 with first, a straight bar where the neutral and centroidal axes coincide and then an initially curved bar where this condition does not exist.

In the initially straight bar, the position of the neutral (centroidal) axis must be located. This is done by equating the moment of the cross-sectional area about axis x–x to the moment of the areas of the two component areas, $b_1 t'_1$ and $b_2 t'_2$. Thus (see Appendix A)

$$\bar{y} = \frac{1}{A}\left[\frac{A_1 t'_1}{2} + A_2\left(\frac{t'_2}{2} + t'_1\right)\right] \qquad 1.99$$

where $\bar{y}$ is the distance from the x–x line to the centroid of the area A (entire cross section), $A_1 = b_1 t'_1$, and $A_2 = b_2 t'_2$. The moment of inertia of the cross section can then be determined about N–N by adding the moments of inertia of the areas A_1 and A_2 about the neutral axis; that is,

$$(I_A)_{NN} = (I_{A_1})_{NN} + (I_{A_2})_{NN} \qquad 1.100$$

The inertias I_{A_1} and I_{A_2} about the section centroid are determined by the parallel-axis theorem (Appendix B) so that

$$(I_{A_1})_{NN} = (I_{A_1})_{AA} + A_1 y_1{}^2 \qquad 1.101$$

and

$$(I_{A_2})_{NN} = (I_{A_2})_{BB} + A_2 y_2{}^2 \qquad 1.102$$

Having these quantities the stresses at any location are calculated by application of the bending stress formula $\sigma = M/Z$.

If the beam is curved such that the centroidal radius of curvature is R, the centroidal and neutral axes do not coincide, and the distance e separating these axes must be determined. The development of the theory is given in Chapter 3. The result is that the distance r to the neutral axis is

$$r = \frac{A}{\int dA/v} \qquad 1.103$$

where A is the area of the cross section. For this particular case (29)

$$r = \frac{b_1 t_1 + b_2 t_2}{b_1 \int_a^d dv/v + b_2 \int_d^c dv/v}$$

or

$$r = \frac{b_1 t_1 + b_2 t_2}{b_1 \ln d/a + b^2 \ln c/d} \qquad 1.104$$

where ln is the symbol for natural logarithms. Having the value of r, e is determined from the relation

$$e = R - r \qquad 1.105$$

and the bending stresses from the relationship

$$\sigma = (My/Ae)(r - y) \qquad 1.106$$

The stresses in the outermost fibers thus become

$$\sigma_{max} = Mh_1/Aea \qquad 1.107$$

$$\sigma_{min} = -Mh_2/Aec \qquad 1.108$$

where h_1 and h_2 are the distances from the neutral axis to the fibers at the inner radius a and the center radius c, respectively. Analyses for other cross-sectional types are given in Chapter 3.

Variable cross section. Another important shape effect is illustrated by members having a variable cross section. Such members, when the cross section is optimumly controlled, are frequently called *shapes of constant strength* since the section along the length can be varied to produce a constant stress σ.

COMPRESSION. The attainment of a constant stress in compression is relatively simple since the cross-sectional area can be varied continuously to adjust to the increasing weight of material above a given section (28). In Fig. 1.42, consider a differential element *abcd*. The compressive stress on plane *cd* is greater than the stress on plane *ab* by the weight of the element *abcd*; thus, since a condition of constant stress is desired, the difference in the cross-sectional areas *ab* and *cd* must be such as to compensate for the additional compressive

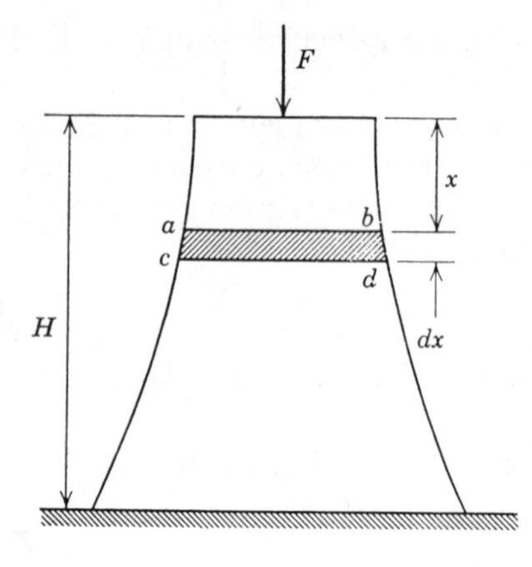

Fig. 1.42 Constant strength compression member. [After Timoshenko (28), Courtesy of D. Van Nostrand Co.]

force on *cd*. Thus, if the difference in the areas is dA and the compressive stress is σ, then

$$dA\sigma = \gamma A \, dx \tag{1.109}$$

where γ is the density of the material. Dividing through by $A\sigma$ and integrating gives

$$\int \frac{dA}{A} = \int \frac{\gamma \, dx}{\sigma} \tag{1.110}$$

the solution of which is

$$\ln A = \gamma x/\sigma + C \tag{1.111}$$

where C is a constant of integration. Rewriting Eq. 1.111 gives

$$A = e^C e^{\gamma x/\sigma} \tag{1.112}$$

where e is the base of natural logarithms. To find the value of C, note that at $x = 0$, $\sigma = F/A$, and $A = e^C$; therefore

$$A = (F/\sigma)e^{\gamma x/\sigma} \tag{1.113}$$

At the top,

$$A_{\text{top}} = F/\sigma \tag{1.114}$$

and at the bottom where $x = H_1$

$$A_{\text{bottom}} = (F/\sigma)e^{\gamma H/\sigma} \tag{1.115}$$

These equations define the form of the constant stress compression member in Fig. 1.42.

Torsion. For constant torsional stress as obtained in a thin tube, (Fig. 1.43) the shear stress is given by

$$\tau = T/2\pi R^2 t \tag{1.116}$$

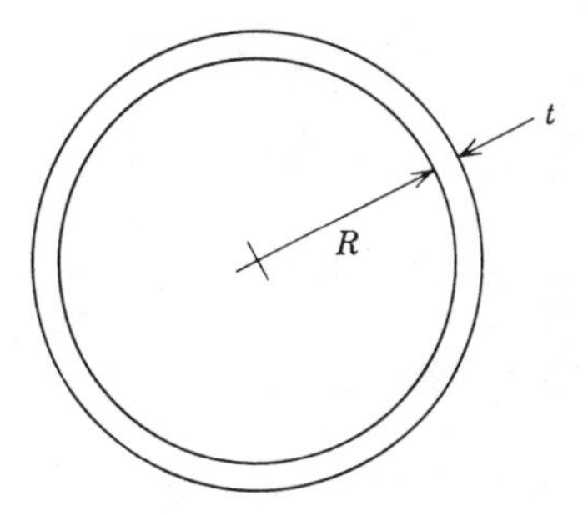

Fig. 1.43 Thin-walled tube.

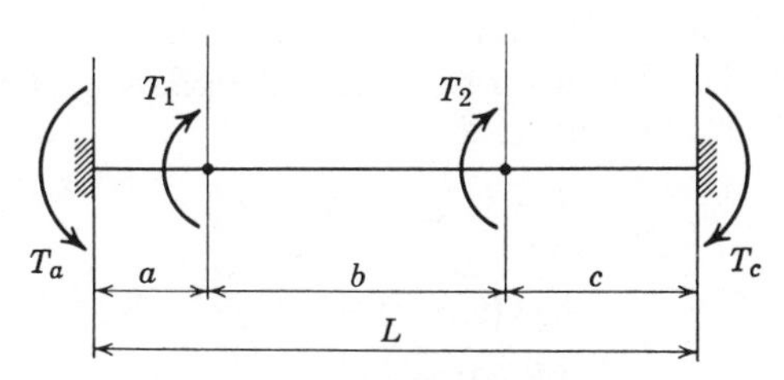

Fig. 1.44 Bar with variable torque.

where T is the torque. Suppose, however, that a constant maximum shear stress is required for the member shown in Fig. 1.44. For this case

$$T_a L = T_1(b + c) + T_2 c \tag{1.117}$$

Since $\theta = TL/GJ = \text{constant}$,

$$T_a = \frac{T_1(b + c) + T_2 c}{L} \tag{1.118}$$

and also

$$T_c L = T_2(a + b) + T_1 a$$

or

$$T_c = \frac{T_1 a + T_2(a + b)}{L} \tag{1.119}$$

For the center section,

$$T_b = T_1 - T_a = T_1 - \frac{T_1(b + c) + T_2 c}{L}$$

or

$$T_b = \frac{T_1 a - T_2 c}{L} \tag{1.120}$$

Here for a constant maximum shear stress, with T_1, T_2, a, b, and c fixed, the polar moment of inertia J must be varied for each section of the shaft. Thus

$$T_a R_A / J_A = T_b R_B / J_B = T_c R_C / J_C \tag{1.121}$$

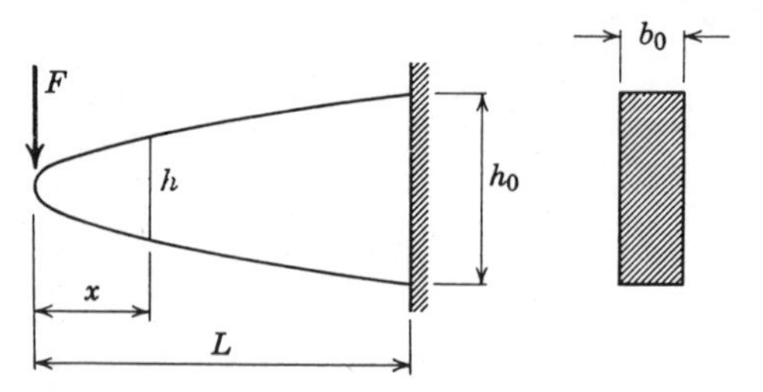

Fig. 1.45 Constant strength cantilever.

BENDING For constant strength in bending, the critical factor is the section modulus, which can be varied along the length of a beam. Consider the beam shown in Fig. 1.45. This beam is of constant thickness b_0, and varies in depth along the length. For constant stress

$$\sigma = M/Z = 6M/b_0h^2 \tag{1.122}$$

or

$$h = (6M/\sigma)^{1/2}(1/b_0)^{1/2} \tag{1.123}$$

For this particular case of end loading

$$M/Z = 6Fx/b_0h^2 = 6FL/b_0h_0{}^2 = \text{constant} \tag{1.124}$$

from which

$$h = h_0\sqrt{x/L} \tag{1.125}$$

which is parabolic. The deflection is (see Chapter 7)

$$\delta = \frac{\partial U}{\partial F} = \frac{\partial}{\partial F}\int_0^L \frac{M^2\,dx}{2EI} = \frac{\partial}{\partial F}\left[\int_0^L \frac{F^2x^2\,dx}{2E(b_0/12)(h_0\sqrt{x/L})^3}\right]$$

or

$$\delta = 8FL^3/Eb_0h_0{}^3 \tag{1.126}$$

Note that the deflection is twice that of a prismatic beam of constant depth; thus the above beam is of constant strength, but it is at the same time consider-

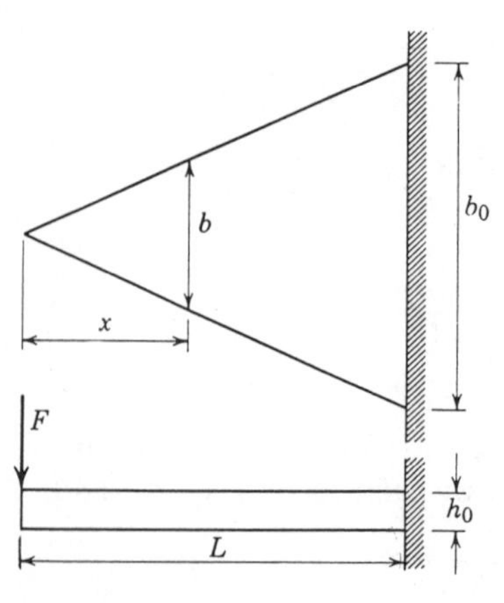

Fig. 1.46 Constant strength cantilever.

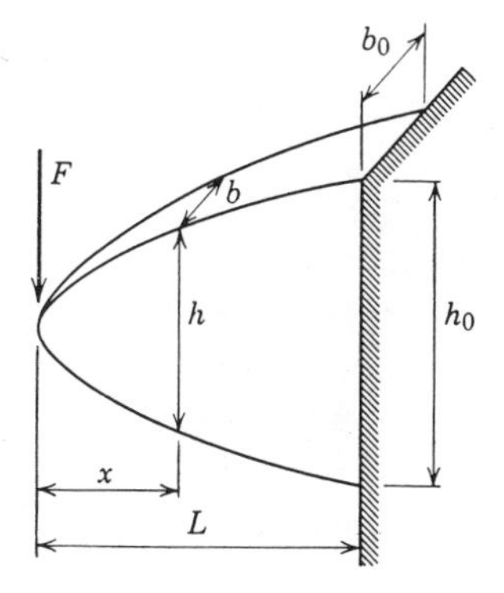

Fig. 1.47 Constant strength cantilever.

ably more flexible. The width could also have varied with the thickness kept constant as shown in Fig. 1.46. Here

$$b = (6M/\sigma)(1/h_0)^2 \qquad 1.127$$

and for end loading

$$b = b_0(x/L) \qquad 1.128$$

This is for a triangular-shaped plate. The optimum geometry is obtained when both width and depth vary simultaneously (Fig. 1.47); here

$$\sigma = M/Z = \text{constant} = Mh/2(\tfrac{1}{12}bh^3) = 6M/bh^2 \qquad 1.129$$

or

$$b = b_0(h_0/h)^3 \qquad 1.130$$

For additional analyses see (6 and 29).

Not all beams having a variable cross section are of constant strength. For the beam shown in Fig. 1.48 the maximum deflection is

$$y_{max} = (P/3EI_1)[L^3 - A^3 + (I_1/I_2)A^3] \qquad 1.131$$

and for a uniformly loaded beam of the same geometry

$$y_{max} = (wL/8EI_1)[L^3 + (A^4/L)(I_1/I_2 - 1)] \qquad 1.132$$

where w is the unit loading. Other examples are given in (17).

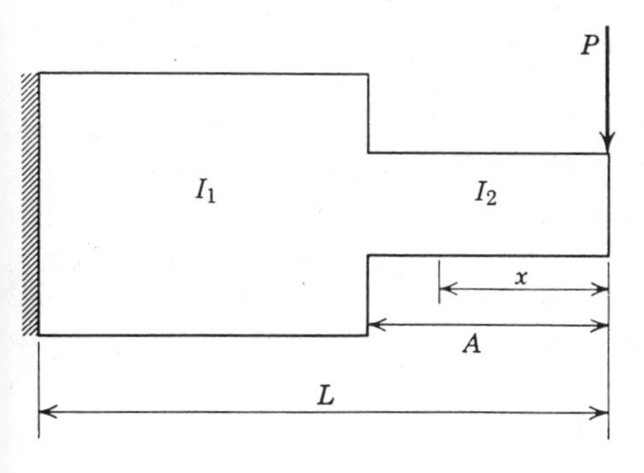

Fig. 1.48 Variable-section cantilever.

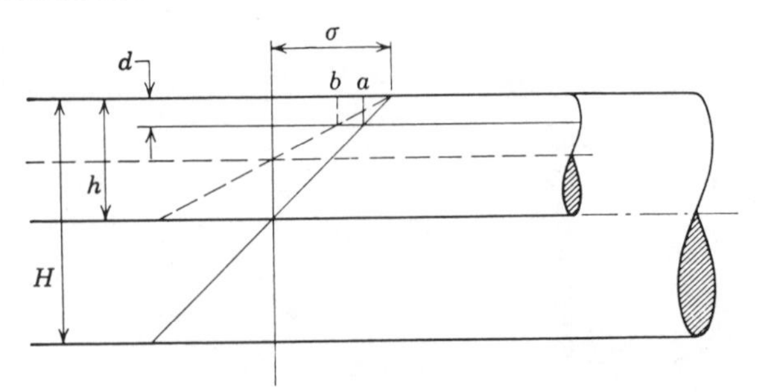

Fig. 1.49 Stress gradients in bars of different size.

Inherent nature of size effect. The foregoing effects, all applicable to the elastic range, have been explained on the basis of the section properties of the members. The effect of size and shape on the yield and ultimate strength of a material will now be considered.

There is no well-defined effect of size on the stress required for initial yielding, but size does have a marked effect on the propagation of plastic flow and on the fracture strength of materials. Early thoughts on this subject indicated that in nonuniform stress distribution, like bending and torsion, the outer fibers were somehow strengthened by the lower-stressed inner fibers, giving rise to increased yield stress for smaller parts. This was explained as follows: Take two bars, such as shown in Fig. 1.49, one of thickness H and the other of thickness h, and stress them to the same value of σ on the outside fibers; then, in the larger bar, a fiber distant d from the surface is stressed to an amount given by a, whereas for the smaller bar the stress on the same fiber is only b. Thus the lesser-stressed inner fibers "strengthen" the bar, and the effect is greater the smaller the bar. Although observations made years ago seem to support this theory, it is now known that gross misinterpretations were made because of the presence of upper yield points especially in steel, experimental inaccuracy, and residual stresses in the material. Similar claims of a "strengthening" effect have been claimed for the yielding of heavy-wall cylinders under pressure, but it is now known that cylinders exhibiting such an effect nearly always contain residual stresses, which when allowed for, cancel out the effect.

It is true that the safe maximum static load that can be applied to a ductile steel subjected to nonuniform distribution of stress is larger than the load that causes a stress in the most-stressed fiber equal to the yield point of the material found, for example, from the tension test. These most-stressed fibers yield when the stress in these fibers reaches the yield point of the material; however, since yielding occurs in a small portion of the member, readjustment of stress occurs and the understressed fibers are made to bear more of the load. In other words, structural damage to the member may not occur until the load reaches a value sufficiently large to cause yielding not only in the most-stressed fibers but in the fibers to a measurable depth in the member.

Size effects concerned with crack initiation and propagation are quite real,

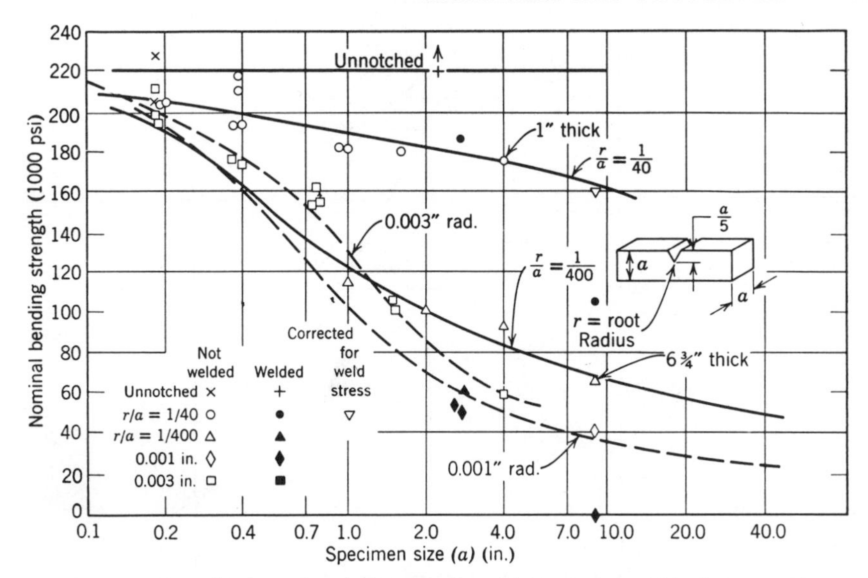

Fig. 1.50 Size effect in bending of notched Ni-Mo-V steel. [*After Lubahn (19), Courtesy of Syracuse University Research Institute.*]

however, and have been observed under many different circumstances. Although a precise explanation of the effect is not known, the appropriate mathematical theory given earlier can be used to determine size and shape parameters to be incorporated in a design schedule. However, for more assurance of success experiments should be conducted on each material in order to discern what, if any, size or shape effect may be present; this is especially true for applications or designs likely to contain large stress concentration sources such as sharp corners or inclusions. For example, it has been shown that for some high-strength steels (19) the ductility of V-notched bend specimens dropped from 25% for 0.2-in. thick bars to 3% for a 9-in. bar, whereas unnotched specimens retained 30% ductility throughout the same size range. Similarly, as shown in Fig. 1.50 the bend strength of similar material was reduced drastically when notches were present. In tension similar effects have been observed as shown in Figs. 1.51 and 1.52.

The nature of size effect in brittle fracture has received considerable attention, and predictions concerning the effect have been formularized, particularly by Weibull (31). According to Weibull's analysis

$$(\sigma_u)_1/(\sigma_u)_2 = (V_2/V_1)^{1/m} \qquad 1.133$$

where σ_u is ultimate strength in tension and the subscripts 1 and 2 indicate specimens of different volume V. The letter m is a material constant. Thus by conducting a few tests m can be determined and used to predict the size effect at any level for the material in question.

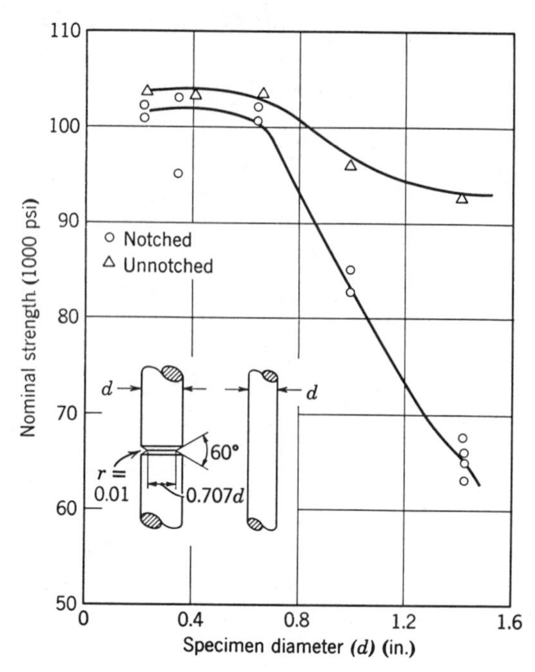

Fig. 1.51 Size effect in tension for aluminum. [*After Lubahn (19), Courtesy of Syracuse University Research Institute.*]

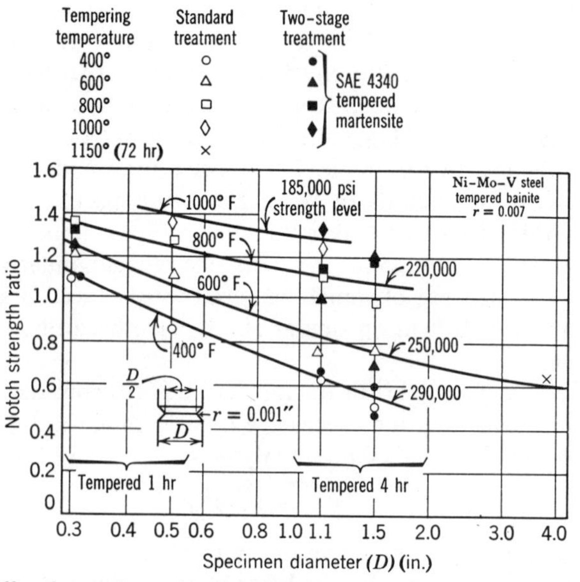

Fig. 1.52 Size effect for notch tensile tests of SAE 4340 steel. [*After Lubahn (19), Courtesy of Syracuse University Research Institute.*]

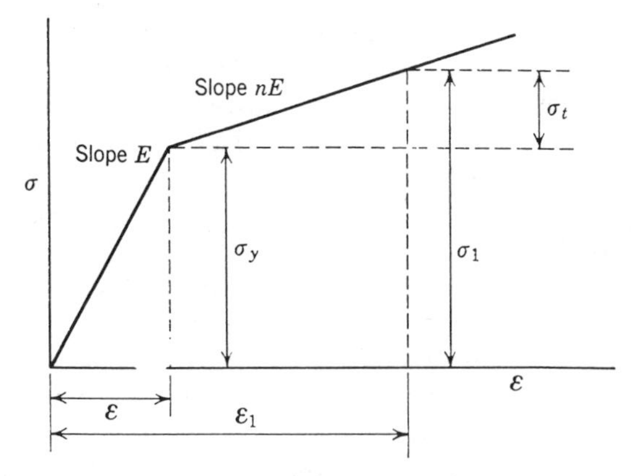

Fig. 1.53 Tensile stress-strain diagram.

Theoretical analysis of size effect in plastic bending. A relationship has been developed (24) between applied moment and strain for both elastic and plastic strains in a beam based on the assumptions that the strain distribution is linear at any section of the beam and that the material does not have an upper yield point. The stress distribution is found from the strain distribution by assuming that the stress in any fiber for a given value of strain can be read from the tensile test or compressive stress-strain diagram of the material; hence it is assumed that the stress gradient does not influence the stress at which yielding starts. Since the moment on any section can be found from a known stress distribution and the dimensions of the cross section of the beam, a theoretical relation between bending moment and strain in the extreme fibers may be determined for a beam of a given cross section.

A theoretical moment-strain diagram will be obtained first for a beam of rectangular cross section of material having the stress-strain diagram shown in Fig. 1.53.

In Fig. 1.54, σ_1 represents any stress greater than the yield stress σ_y, ε_1 repre-

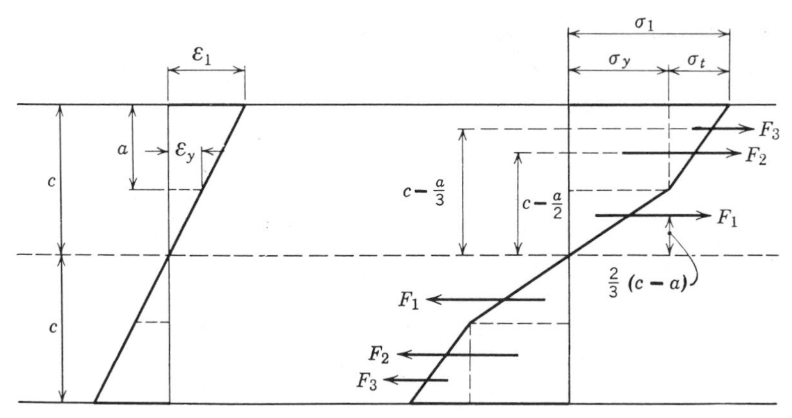

Fig. 1.54 Theoretical strain and stress distributions.

sents the corresponding strain greater than the yield strain ε_y, and σ_t represents the increment of stress above the yield stress ($\sigma_t = \sigma_1 - \sigma_y$). The beam is loaded in such a manner that the strain in the most-stressed fibers has a magnitude ε_1. Because the strain distribution is linear, the depth to which yielding should theoretically have progressed is found by locating the depth on the beam at which the strain has a magnitude ε_y. This depth is denoted by the letter a, and the corresponding stress σ_1 can be found from the stress-strain diagram. It is shown in Fig. 1.54 that the portion of the beam within the distance $(c - a)$ from the neutral surface remains elastic with a stress distribution of triangular shape. The resultant force in this elastic region is F_1 acting at a distance $\frac{2}{3}(c - a)$ from the neutral surface. The stress distribution in the yielded portion to depth a from the extreme fibers is trapezoidal. This trapezoid is considered to be composed of a rectangle $a\sigma_y$ and a triangle $\frac{1}{2}a\sigma_t$. The resultant force corresponding to the rectangle is F_2, located at a distance $(c - a/2)$ from the neutral axis. The resultant force corresponding to the remaining triangle is F_3, at a distance $(c - a/3)$ from the neutral axis. The moment about the neutral axis of all the forces on the section is equal to the bending moment M for the reaction considered. Thus

$$M = 2[\tfrac{2}{3}F_1(c - a) + F_2(c - a/2) + F_3(c - a/3)] \qquad 1.134$$

An examination of Figs. 1.54 and 1.55 shows that

$$\sigma_t = nE(\varepsilon_1 - \varepsilon_y) = n\sigma_y(\varepsilon_1/\varepsilon_y - 1) \qquad 1.135$$

and

$$\varepsilon_1/\varepsilon_y = c/(c - a) \qquad 1.136$$

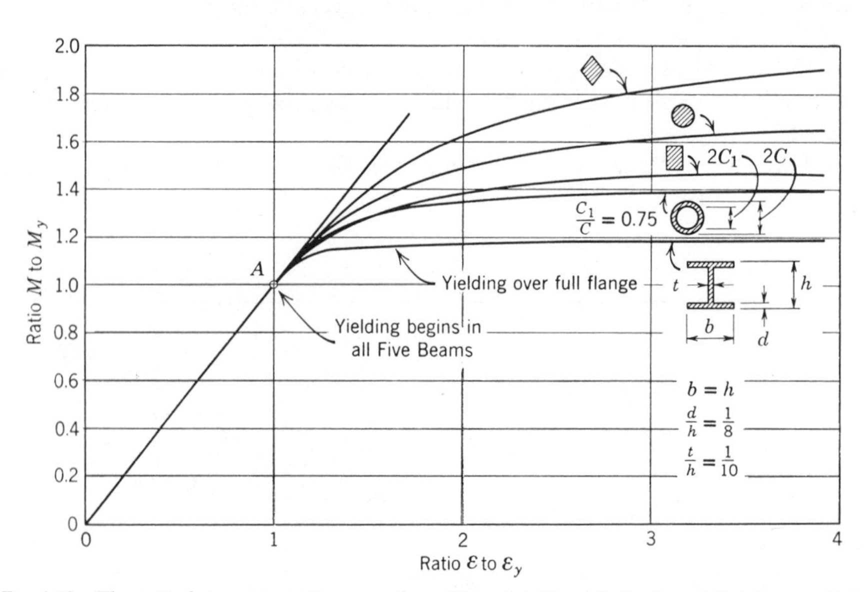

Fig. 1.55 Theoretical moment-strain curves for mild steel. [After Morkovin and Sidebottom (24), Courtesy of The University of Illinois Engineering Experiment Station.]

The magnitudes of the forces F_1, F_2, and F_3 are

$$F_1 = (\sigma_y/2)b(c - a) \tag{1.137}$$

$$F_2 = \sigma_y ba \tag{1.138}$$

$$F_3 = \sigma_t ba/2 \tag{1.139}$$

When yielding is about to take place in the extreme fibers, the stress distribution is linear and the stress in the extreme fiber is σ_y. At that instant the bending moment M at the given section in the beam has a magnitude M_y, which is

$$M_y = \frac{\sigma_y b(2c)^2}{6} \tag{1.140}$$

By substituting the values of F_1, F_2, and F_3 in Eq. 1.134, dividing through by Eq. 1.140, and eliminating σ_t, c, and a through Eqs. 1.135 and 1.136, a relation between bending moment and strain in the center fibers is obtained; thus,

$$\frac{M}{M_y} = \tfrac{3}{2}(1 - n) - \frac{1 - n}{2(1/y)^2} + n\left(\frac{\varepsilon_1}{\varepsilon_y}\right) \tag{1.141}$$

Equation 1.141 is simplified for mild steel since $n = 0$ for this material; thus, for this steel, in rectangular section,

$$\frac{M}{M_y} = \frac{3}{2} - \frac{1}{2}\left(\frac{\varepsilon_1}{\varepsilon_y}\right)^2 \tag{1.142}$$

or, also,

$$\frac{\varepsilon_1}{\varepsilon_y} = \frac{1}{\sqrt{(3 - 2M/M_y)}} \tag{1.143}$$

Equation 1.143, which is valid for mild steel in a beam of rectangular cross section, is illustrated in Fig. 1.55. The analysis for a circular bar is given in (24).

REFERENCES

1. W. Bert, *Space/Aero.*, **38** (5), 77–83 (October 1962).
2. P. W. Bridgman, *Studies in Large Plastic Flow and Fracture*, McGraw-Hill Book Co., New York (1952).
3. G. Gerard, *Minimum Weight Analysis of Compression Structures*, New York University Press, New York, N.Y. (1956).
4. J. E. Goldman (Ed.), *The Science of Engineering Materials*, John Wiley and Sons, Inc., New York (1957).
5. A. A. Griffith, *Phil. Trans. Royal Soc. (London)*, **221** (A), 163–198 (October 1920).
6. W. A. Gross, and J. P. Li, *J. Appl. Mech.*, **24** (1), 105–108 (March 1957).
7. H. H. Hilton, and M. Feigen, *J. Aerosp. Sci.*, **27** (9), 641–652 (September 1960).
8. G. A. Hoffman, *J. Appl. Mech.*, **29** (4), 662–668 (December 1962).
9. G. A. Hoffman, *J. Aerosp. Sci.*, **29** (12), 1471–1475 (December 1962).

10. J. H. Hollomon, *The Problem of Fracture*, American Welding Society, New York, N.Y. (1946).

11. F. C. Holtz, *Frontier*, **24** (1) (Winter), 20–25 (1963).

12. J. R. Hunt, *Mater. Methods*, **38** (3), 102–107 (September 1953).

13. G. R. Irwin, and J. A. Kies, *Weld. J. Research Supp.*, **33** (4), 193s–198s (April 1954).

14. G. R. Irwin, *Weld. J. Res. Suppl.*, **41**(11), 519s–528s (November 1962).

15. Z. B. Jastryzebski, *Nature and Properties of Engineering Materials*, John Wilëy and Sons. Inc., New York, N.Y. (1976).

16. G. F. Kinney, *Engineering Properties and Applications of Plastics,* John Wiley and Sons., Inc., New York (1957).

17. W. R. Leopold, *Prod. Eng.*, **27** (7), 215, 217, 219 (July 1956).

18. C. R. Lieberman, *Space/Aero.*, **36** (4), 56–60, November (1961).

19. J. D. Lubahn, *Proc. 1955 Sagamore Research Conf.*, 143–161, Syracuse University Research Institute, Syracuse (1955).

20. G. V. Luerssen, and O. V. Greene, *Proc. Am. Soc. Testing and Materials*, **33** (Part II), 315–333 (1933).

21. K. Maier, *Prod. Eng.*, **29** (7), 83–86 (February 17, 1958).

22. K. Maier, *Prod. Eng.*, **29** (46), 71–75 (November 10, 1958).

23. D. F. Minër, and J. B. Seastone, *Handbook of Engineering Materials*, John Wiley and Sons, Inc., New York, N.Y. (1955).

24. D. Morkovin, and O. Sidebottom, *The Effect of Non-Uniform Distribution of Stress on the Yield Strength of Steel, Univ. of Ill. Bull.*, **45** (26), Eng. Exp. Station Bull. Series No. 372, University of Illinois, Urbana, Ill (December 18, 1947).

25. G. Murphy, *Properties of Engineering Materials*, International Textbook Co., Scranton, Pa. (1957).

26. F. R. Shanley, *Weight-Strength Analyses of Aircraft Structures*, McGraw-Hill Book Co., New York, N.Y. (1952).

27. R. L. Templin, and R. G. Sturm, *J. Aero. Sci.*, **7** (7), 189–198 (July 1940).

28. S. Timoshenko, *Strength of Materials, Part 1*, D. Van Nostrand Co., Princeton, N.J. (1955).

29. S. Timoshenko, *Strength of Materials, Part 2*, D. Van Nostrand Co., Princeton, N.J. (1956).

30. S. Timoshenko, *History of the Strength of Materials*, McGraw-Hill Book Co., New York, N.Y. (1953).

31. W. Weibull, *Roy. Swed. Inst. Eng. Res. Proc.* (151), Stockholm (1939).

32. D. H. Winnie, and B. M. Wundt, *Trans. Am. Soc. Mech. Engrs.,* **80** (8), 1643–1658 (November 1958).

33. J. F. Wygant, *J. Am. Ceram. Soc.*, **34** (12), 374–380 (December 1951).

34. C. Zener, *Elasticity and Anelasticity of Metals*, University of Chicago Press, Chicago, Ill. (1948).

35. *Prod. Eng.*, **32** (17), 14 (April 24, 1961).

36. *The Magazine of Magnesium*, Brooks and Perkins Co., Detroit, Mich. (August 1951).

37. *Hardness Testing Handbook*, American Chain and Cable Co. (1969). See also, George, R. A. et al., *Metal Progress*, 30–35 (May 1960).

38. *Interpretative Report on Effect of Hydrogen in Pressure Vessel Steels*, Welding Research Council Bulletin No. 145, Welding Research Council, New York, N.Y. (October 1969).

39. *Hydrogen Environment Embrittlement*, NASA Report No. N72–27574, National Technical Information Service, Washington, D.C. (1972).

40. *How High Pressure Hydrogen Affects Stainless Steels*, Metal Progress Review Paper, A. W. Thompson, Metal Progress, 30–33 (July 1976).

PROBLEMS

1.1 Derive an expression for bending stress using Equation 1.6 in place of Hooke's law and solve for the case of a rectangular section.

1.2 Using Equation 1.12 in the form $\gamma = \tau^n/G$ derive an expression for shear stress for a bar of circular cross section subject to torsion.

1.3 For an I beam subjected to bending, derive an expression for strength-weight ratio M/W in terms of the ratio of flange thickness t to flange width b. Use a ratio of web thickness to distance between flange faces of 1/170. Plot the relationship between M/W and t/b.

1.4 A thin-wall tube of square cross section, $a \times a$, and wall thickness t is subjected to a twisting moment M_t. Derive a relationship between strength weight ratio M_t/W and the wall-thickness ratio t/a. Plot the relationship.

1.5 There is a choice between white oak and structural steel for a compression member of length L. If the compressive elastic strength of the steel is 35,000 psi, determine the most economical material. Use a factor of safety of 2 for the steel and 4 for the white oak. Assume that the cost of steel is $0.39/lb and that the cost of white oak is $400.00/1000 board feet.

chapter 2

TENSION, TORSION, AND BENDING

In this chapter consideration is given to the analyses of tension, compression, torsion of cylindrical sections, and bending of prismatic bars. In order to place more emphasis on their importance, beams on elastic foundations, curved bars, torsion of noncircular sections, and beams of constant strength are treated elsewhere. See also (10, 17–21, and 24).

2-1 DIRECT AXIAL LOADING

Tension

A typical tension member is shown in Fig. 2.1; it has a length L_0, cross-sectional area A_0, and is subjected to an axial load F. It is assumed that the force acts uniformly over the cross section so that the unit stress at any point is

$$\sigma_0 = F/A_0 \tag{2.1}$$

As a result of the applied force, the bar elongates an amount δ; in terms of unit strain ε_0 along the bar,

$$\varepsilon_0 = \delta/L_0 \tag{2.2}$$

The quantities σ_0 and ε_0 are called *engineering* stress and strain since they are based on the original dimensions of the bar. In the elastic range where Hooke's law applies (Eq. 1.5) $\sigma = E\varepsilon$, and from Eq. 2.1 it is seen that

$$\varepsilon = F/A_0E \tag{2.3}$$

or

$$\delta = FL_0/A_0E \tag{2.4}$$

Examples illustrating the use of the foregoing relationships follow.

1. A solid aluminum cylinder 3 in. in diameter and 6 in. long is subjected to a tensile load of 50,000 lb(Fig. 2.1). Determine the axial and lateral deformations and change in volume.

From Eq. 2.1, $\sigma = F/A = (50{,}000)/[\pi(3)^2/4] = 7080$ psi. The unit strain $\varepsilon = \sigma/E = 7080/(10 \times 10^6) = 0.00708$ in./in. Total strain $= \varepsilon L = 0.00708(6) =$

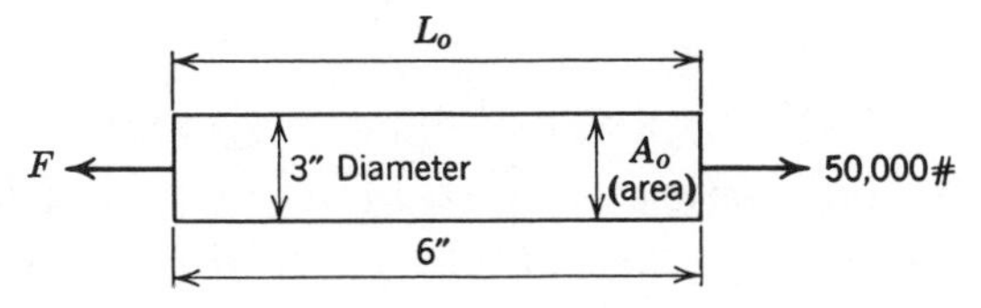

Fig. 2.1 Cylinder subjected to tension.

0.0425 in. The lateral strain is $-\nu\sigma/E = -(0.3)(7080)/(10 \times 10^6) = -0.00212$ in./in.

The decrease in diameter is thus $-0.00212(3) = -0.00636$ in. and the unit volume change is $\varepsilon - 2\nu\varepsilon = \varepsilon(1 - 2\nu)$ or $\Delta V = 0.00708(0.4) = 0.00283$ in.^{3} The total volume change is $V(\Delta V) = \pi(3)^2(6)(0.00283)/4 = 0.12$ in.^{3}

2. Compare the elongation of bars produced by their own weights for circular, conical, and pyramidal geometries (Fig. 2.2).

The weights of the three bars are

$$\text{circular} \quad W_a = (\pi D^2/4)(L\gamma_a)$$
$$\text{conical} \quad W_b = (\pi D^2/4)(\gamma_b)(L/3)$$
$$\text{pyramidal} \quad W_c = (D^2)(\gamma_c)(L/3)$$

The stress in the bars at any location is due to the weight of the bar below that location; therefore, for a distance x from the unsupported end of the bars,

$$\text{Circular} \quad (W_a)_x = (\pi D^2/4)(\gamma_a)(x)$$
$$\text{Conical} \quad (W_b)_x = (\pi D^2/4)(\gamma_b)(L/3)(x^3/L^3)$$
$$\text{Pyramidal} \quad (W_c)_x = (D^2)(\gamma_c)(L/3)(x^3/L^3)$$

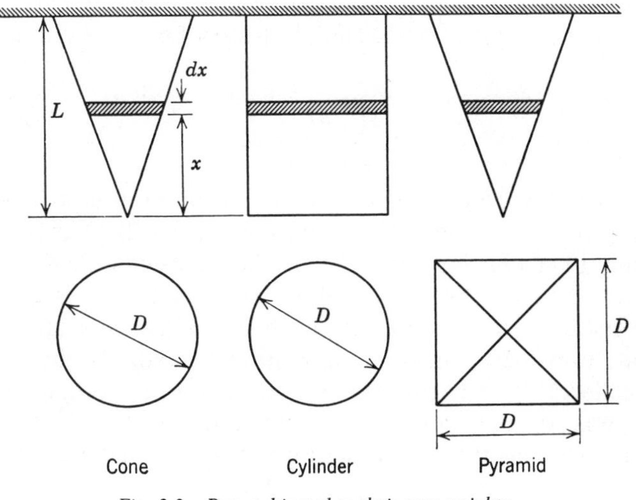

Fig. 2.2 Bars subjected to their own weights.

From Eq. 2.3 and Fig. 2.2,

$$\begin{aligned}
\text{circular} \quad & d\delta = (\pi D^2/4)(\gamma_a)(x)/(\pi D^2/4)E\,dx = \gamma_a x\,dx/E \\
\text{conical} \quad & d\delta = (\pi D^2/4)(\gamma_b)(L/3)(x^3/L^3)/(\pi D^2/4)(x^2/L^2)E\,dx = \gamma_b x\,dx/3E \\
\text{pyramidal} \quad & d\delta = (D^2)(\gamma_c)(L/3)(x^3/L^3)/(D^2)(x^2/L^2)E\,dx = \gamma_c x\,dx/3E
\end{aligned}$$

Total elongation is then $\delta = \int_0^L d\delta$ or

$$\text{Circular} \qquad \delta = (\gamma_a/E)\int_0^L x\,dx = \gamma_a L^2/2E$$

$$\text{Conical} \qquad \delta = (\gamma_b/3E)\int_0^L x\,dx = \gamma_b L^2/6E$$

$$\text{Pyramidal} \qquad \delta = (\gamma_c/3E)\int_0^L x\,dx = \gamma_c L^2/6E$$

Thus the elongation of the cone and pyramid is the same and equal to one-third that of the circular bar.

The length of the bar L_0 in the elastic range is considered to be essentially constant regardless of the load imposed on it; thus a more precise definition of Eq. 2.2 is

$$\varepsilon_0 = \int_{L_0}^{L} \frac{dL}{L_0} = \frac{L - L_0}{L_0} = \frac{\delta}{L_0} \tag{2.5}$$

However, for large deformations, particularly those occurring in the plastic range the quantity δ becomes large compared with L_0; therefore, to be precise, it is necessary to consider the change in length of the bar during stressing and to base deformation on the actual length of the bar at any particular instant rather than on the original length. Thus, instead of Eq. 2.5, strain can be defined as (11)

$$\varepsilon = \int_{L_0}^{L} \frac{dL}{L} = \ln \frac{L}{L_0} \tag{2.6}$$

The value of ε in Eq. 2.6 is called the *true strain* since it is based on a changing length rather than on a constant length as in Eq. 2.5. To illustrate, suppose a bar originally 10 in. long is elongated 1 in.; the extension is then 10%. If elongation is continued until the bar is 100 in. long and then it is elongated 1 in. more, this last elongation is only 1% of the actual gauge length; if, however, the calculation of the strain is by conventional standards, the last extension would be called 10% and set down as equivalent to the original extension of 1 in., which was 10% of the gauge length at that time (1, 5).

In the plastic range, the volume is assumed to remain constant; that is,

$$A_0 L_0 = AL \tag{2.7}$$

where A_0 and L_0 are original cross-sectional area and length, respectively, and A and L are final values. From Eq. 2.7, by rearrangement

$$A = A_0 L_0 / L \qquad 2.8$$

and if this value is placed in Eq. 2.1 with A_0 replaced by A to give the *true stress*,

$$\sigma = (F/A_0)(L_0/L) = (F/A_0)(L/L_0) \qquad 2.9$$

But, from Eq. 2.5,

$$L/L_0 = 1 + \varepsilon_0 \qquad 2.10$$

Therefore, putting L/L_0 from Eq. 2.10 into Eq. 2.9 gives

$$\sigma = (F/A_0)(1 + \varepsilon_0) \qquad 2.11$$

or, in other words, the true stress equals the *nominal* or *engineering stress* F/A_0 *multiplied by the nominal strain*, ε_0, *plus one.* Furthermore, combining Eqs. 2.6 and 2.10 results in another definition for true strain in terms of nominal strain; thus

$$\varepsilon = \ln L/L_0 = \ln (1 + \varepsilon_0) \qquad 2.12$$

If deformation is measured in terms of reduction of area q rather than on the basis of length change,

$$q = -\int_{A_0}^{A} dA/A = \ln A_0/A \qquad 2.13$$

Again, by virtue of the condition of constancy of volume, from Eqs. 2.6 and 2.13

$$\varepsilon = q \qquad 2.14$$

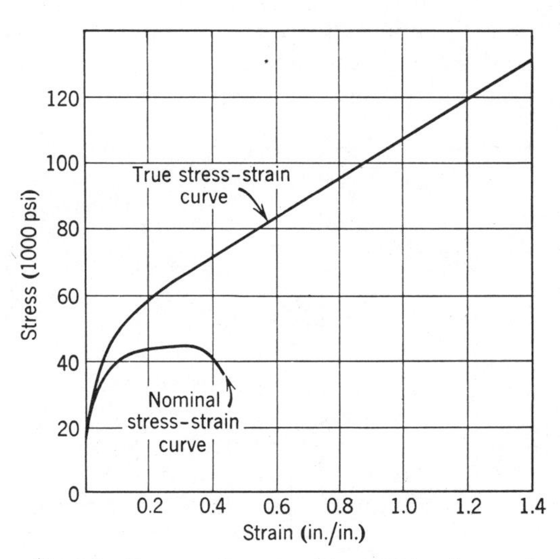

Fig. 2.3 Stress-strain curves for a 0.05% carbon steel.

A comparison of true and nominal stress-strain curves for a low-carbon steel is shown in Fig. 2.3. These plots show how a distorted picture of the plastic deformation process can occur when nominal stress and strain values are used. In the nominal stress-strain plot it is indicated that beyond the ultimate strength the stress decreases with increase in strain, which is a false indication of material behavior. The true stress-strain plot shows that as straining progresses more and more stress develops. This interpretation has particular significance for large deformations involved in metal forming operations (see Chapter 6).

Compression

Compression may be considered as negative tension, thus the analyses developed for tensile loading should also be applicable to structures under compression loads. There are three exceptions to this generality. First, in compression, instability failure by buckling may occur depending on the geometry, especially the length; this subject is considered in Chapter 9. Second, for ductile materials, there is no ultimate strength in compression since, as load increases, the cross-sectional area of the compressed part also increases. Third, for fabricated products, residual stresses in the material can have the effect of reducing the yield strength (effective) so that the basic compression stress-strain properties of the material are not identical with those observed in tensile tests. In addition, in many compression tests there is high frictional restraint between the ends of the test bar and the platens of the testing machine; this effect is most pronounced at low height-diameter ratios of the test piece. Generally speaking, for best results the H/D ratio should be between about 1.5 and 3.0.

Some examples of compression analyses follow.

1. Three cylinders rest on a rigid base (Fig. 2.4). The copper cylinder and aluminum cylinder are initially 4 in. high, and the steel cylinder is 4.0003 in. high. Calculate the load supported by each cylinder if a weight of 100,000 lb is lowered onto the three cylinders, the center of gravity being over the central axis of the steel cylinder. It is assumed that the load remains in a horizontal

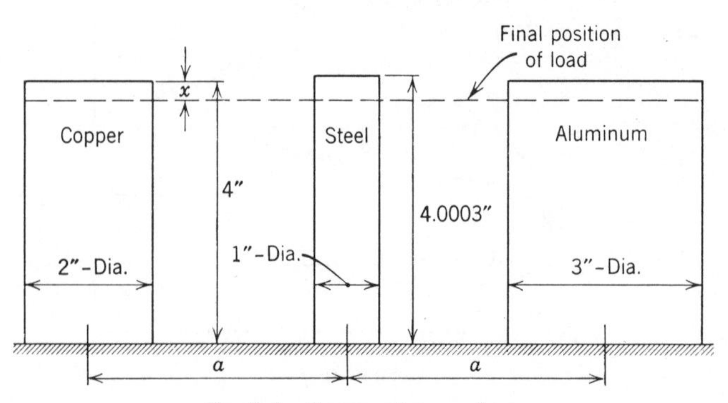

Fig. 2.4 Compression members.

plane. The following properties are given:

Modulus of elasticity of aluminum	$E_A = 10 \times 10^6$ psi
Modulus of elasticity of copper	$E_C = 15 \times 10^6$ psi
Modulus of elasticity of steel	$E_S = 30 \times 10^6$ psi
Yield strength of aluminum	$(\sigma_y)_A = 10{,}000$ psi
Yield strength of copper	$(\sigma_y)_C = 15{,}000$ psi
Yield strength of steel	$(\sigma_y)_S = 30{,}000$ psi

The load required to remove the 0.0003 in. differential between the steel and the other cylinders is obtained from the strain formula

$$\varepsilon_S = PL/AE$$

or

$$P_S = \varepsilon_S A_S E_S/L = [(0.0003)(0.785)(30 \times 10^6)]/4 = 1766 \text{ lb}$$

Since the load remains horizontal, the stress is the same for all cylinders after the 0.0003 in. differential has been removed; then

$$\varepsilon = P_C/A_C E_C = P_S/A_S E_S + \varepsilon_S A_S E_S/L = P_A/A_A E_A$$

$$P_C/(3.14)(15 \times 10^6) = P_S/(0.785)(30 \times 10^6) + [(0.0003)(0.785)(30 \times 10^6)]/4 = P_A/(7.05/10 \times 10^6)$$

Also

$$P_C + P_S + 1766 + P_A = 100{,}000$$

$$P_C/P_A = 47.1/70.5 = 0.67$$

and

$$\frac{P_S + 1766}{23.5 \times 10^6} = \frac{P_A}{70.5 \times 10^6}$$

$$\frac{P_S + 1966}{P_A} = \frac{23.5}{70.5} = 0.334$$

From this relationship

$$0.67 P_A + 0.334 P_A - 1766 + P_A = 100{,}000$$

or

$$P_A = 49{,}000 \text{ lb}$$

The area of the aluminum cylinder is 7.05 in.^{2}; therefore, the compressive stress is $49{,}000/7.05 = 6960$ psi, which is below the yield strength $(\sigma_y)_A$.

For the steel, the final load is

$$\frac{P_S + 1766}{23.5 \times 10^6} = \frac{49{,}000}{70.5 \times 10^6}$$

or

$$P_S = 15{,}532 + 1766 = 16{,}300 \text{ lb}$$

For the copper, $P_C = 100{,}000 - 65{,}300 = 34{,}700$ lb. The area of the copper is 3.14 in.^{2}; therefore the stress induced is $34{,}700/3.14 = 11{,}000$ psi; for the steel, the stress is $16{,}300/0.785 = 21{,}000$ psi. Thus all induced stresses are below the yield values for the various materials, and no modifications are required.

2-2 SHEAR AND TORSION

Shear is considered a process where parallel planes move relative to one another in the direction of the load. For the purposes of this book shear will be discussed under the categories of *direct shear*, as in rivets, *torsion or twisting* of bars, and *transverse shear* in beams. In this chapter the discussion is concerned mainly with torsion.

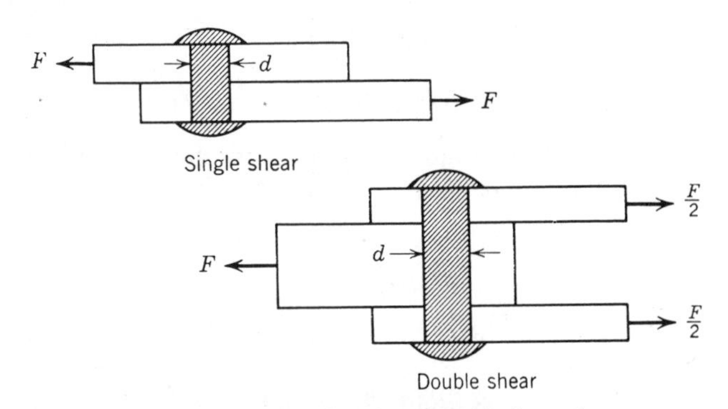

Fig. 2.5 Illustration of single and double direct shear.

Direct shear

In direct shear the shear stress is calculated by dividing the applied load by the cross-sectional area in shear. In Fig. 2.5 two types of direct shear are shown, single and double. In this example of direct shear the shear stress τ is calculated as follows, assuming the rivet diameter to be d.

$$\text{Single shear} \qquad \tau = F/A = F/(\pi d^2/4) = 4F/\pi d^2 \qquad 2.15$$

$$\text{Double shear} \qquad \tau = F/A = F/(2\pi d^2/4) = 2F/\pi d^2 \qquad 2.16$$

Another type of direct shear is shown in Fig. 2.6, which represents a punch and plate. For this case, if the punch diameter is d and the plate thickness is t, the shear stress τ is

$$\tau = F/A = F/\pi\, dt \qquad 2.17$$

Torsion

Torsion of solid circular rod. Figure 2.7 shows a rod having a radius R and length L, fixed at one end and subjected to a torque T at the other end,

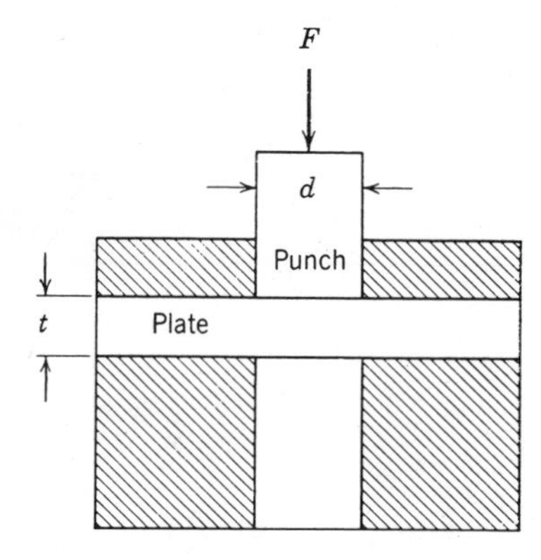

Fig. 2.6 Direct shear in punching operation.

which produces a twist θ in the bar. It is assumed that the shear stress produced at any cross section is a function of the distance to the axis of the bar and that it acts normal to the radius; that is,

$$\tau = Cr \tag{2.18}$$

where C is a constant. In the elastic range, from Hooke's law, the shearing strain γ is related to the shear stress by the torsional or shear modulus of elasticity G; thus

$$\gamma = \tau/G \tag{2.19}$$

Substituting Eq. 2.18 into Eq. 2.19 gives the relationship

$$\gamma = Cr/G \tag{2.20}$$

In order to evaluate the constant C note that the net force on the cross section

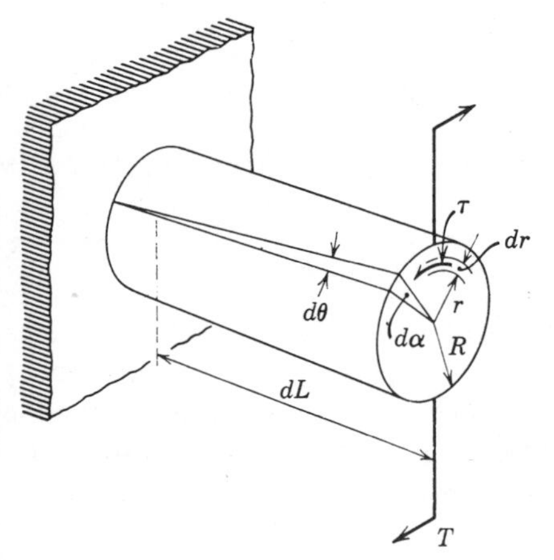

Fig. 2.7 Torsion of circular rod.

must balance the applied torque; that is,

$$T = \int \tau(2\pi r)(dr)r \qquad 2.21$$

or, using Eq. 2.18,

$$T = \int Cr(2\pi r)(dr)r$$

from which

$$T = 2\pi C \int_0^R r^3\, dr = \frac{\pi C R^4}{2} \qquad 2.22$$

Therefore, from Eq. 2.22

$$C = 2T/\pi R^4 \qquad 2.23$$

Now, substituting Eq. 2.23 into 2.18 gives a value for the shear stress τ in terms of the variable radius r; thus

$$\tau = 2Tr/\pi R^4 \qquad 2.24$$

the quantity $\pi R^4/2$ in Eq. 2.24, however, is the polar moment of inertia J; therefore Eq. 2.24 can be rewritten as

$$\tau = \frac{Tr}{J} \qquad 2.25$$

The shearing strain γ, for any section of the bar, from Eq. 2.19 is

$$\gamma = \tau/G = Cr/G = 2Tr/\pi R^4 G = Tr/GJ \qquad 2.26$$

In addition, since at any distance dL from the fixed end of the bar, $\gamma = r\, d\alpha/dL$, Eq. 2.20 shows that

$$d\alpha = (Cr/G)(dL/r)$$

which on integrating gives

$$\alpha = CL/G = TL/GJ \qquad 2.27$$

At the outside surface of the rod, $\theta = \alpha$; therefore, at the free end, the angle of twist is expressed as follows:

$$\text{angle of twist} = \theta = \alpha = TL/GJ \qquad 2.28$$

In some materials, particularly wood, twisting produces splits along the axis of the member, which is called *longitudinal shearing*. Such splits come about because, for equilibrium, there have to be longitudinal shear stresses along the length of the member to balance the transverse shear stresses; also, many materials, particularly wood, have high shear strength normal to the grain but are quite weak in shear along the grain.

Table 2.1 Torsion Properties of Various Sections

Section	Angle of Twist, θ, radians	Max. Shear Stress, τ_{max} psi	Modulus of Elastic Resilience U, in. lb/in.^{3}	Strength-Weight Ratio
Solid Circular Rod (R)	$\frac{2TL}{\pi G R^4}$	$\frac{2T}{\pi R^3}$	$\frac{\tau^2}{4G}$	$\frac{\tau R}{2L\gamma}$
Heavy-walled Tube (R_o, R_i)	$\frac{2TL}{\pi G(R_0^4 - R_i^4)}$	$\frac{2TR_0}{\pi(R_0^4 - R_i^4)}$	$\frac{\tau^2}{4G}\left(\frac{R_i^2}{R_0^2} + 1\right)$	$\frac{\tau^2}{2R_0\gamma L}(R_0^2 + R_i^2)$
Thin-walled Tube (R, $t/2$)	$\frac{TL}{\pi G(2tR^3)}$	$\frac{T}{2\pi R^2 t}$	$\frac{\tau^2}{2G}$	$\frac{\tau R}{\gamma L}$

Table 2.1 (Continued).

Any Thin-walled Tube ($t/2$)	$\frac{TSL}{4G\bar{A}^2t}$ $\bar{A}$ = Total area enclosed by midwall perimeter	$\frac{T}{2\bar{A}t}$	$\frac{\tau^2}{2G}$	$\frac{2\tau\bar{A}}{\gamma LS}$
Ellipse (b, a)	$\frac{TL(a^2+b^2)}{\pi Ga^3b^3}$	$\frac{2T}{\pi ab^2}$	$\frac{\tau^2}{8G}\left(\frac{b^2}{a^2}+1\right)$	$\frac{\tau b}{2\gamma L}$
Square (a, a)	$\frac{7.11TL}{a^4G}$	$\frac{4.81T}{a^3}$	$\frac{\tau^2}{6.53G}$	$\frac{\tau a}{4.81\gamma L}$
Rectangle (a, b)	$\frac{3.33TL(a^2+b^2)}{a^3b^3G}$	$\frac{T(3a+1.8b)}{a^2b^2}$	$\frac{\tau^2}{2G}\left[\frac{3.33(a^2+b^2)}{(3a+1.8b)^2}\right]$	$\frac{\tau ab}{(3a+1.8b)}\left(\frac{1}{\gamma L}\right)$
Equilateral Triangle ($2a/\sqrt{3}$, a)	$\frac{26TL}{a^4G}$	$\frac{13T}{a^3}$	$\frac{\tau^2}{7.5G}$	$\frac{\tau a}{7.5\gamma L}$

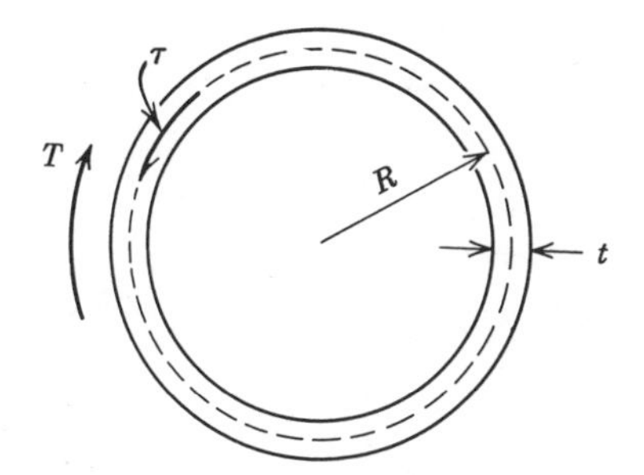

Fig. 2.8 Thin-walled tube in torsion.

Torsion of thick and thin-walled circular tubes. The foregoing analysis for a solid circular section can also be applied to a thick-walled tube by simply inserting the proper value of J in Eqs. 2.25 or 2.28 (see Table 2.1). For a thin-walled cylindrical tube, the analysis is modified slightly.

Consider the tube shown in Fig. 2.8. The external torque T is balanced by the internal resistance (shear force), $\tau(2\pi Rt)R$, or

$$T = 2\pi R^2 t\tau \tag{2.29}$$

where R is the radius to midwall. From Eq. 2.29, the shear stress τ is equal to

$$\tau = T/2\pi R^2 t \tag{2.30}$$

Since the polar moment of inertia for a thin circular section is $J = 2\pi R^3 t$, the shear stress can be written as

$$\tau = TR/J \tag{2.31}$$

The twist per unit length, from Eq. 2.28 is

$$\theta/L = T/GJ \tag{2.32}$$

Therefore for the thin circular section the angle of twist is obtained by substituting J for a thin tube into Eq. 2.32 which gives the result

$$\theta = TL/GJ = TL/2\pi GtR^3 \tag{2.33}$$

Combining Eqs. 2.30 and 2.33 gives the Batho torsion formula

$$\tau = 2GA\theta/S$$

where A is the area enclosed by the average perimeter S and θ is angle of twist per unit length.

Torsion of thin-walled noncircular tubes. Noncircular tubes are frequently used, particularly in aircraft applications. In the following analysis, in addition to the usual assumptions for solid bars, it is assumed that the wall thickness is constant and small compared to the other dimensions. It is also assumed that the stress is uniform through the wall thickness (that is, there are no abrupt changes in curvature) and that buckling will not occur. For the tube shown in Fig. 2.9, of arbitrary cross section, a torque T induces a shearing stress $d\tau$ to act

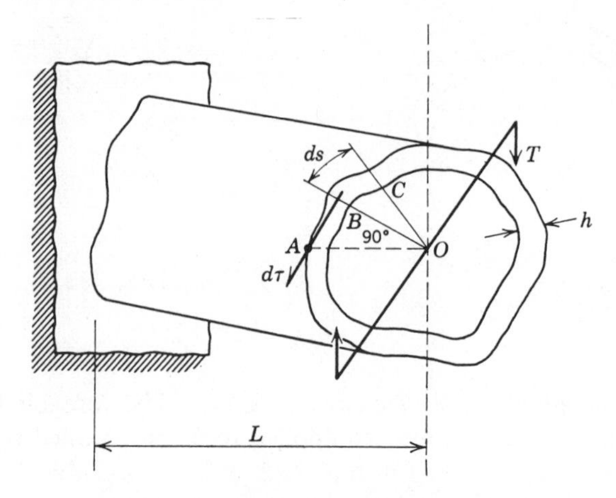

Fig. 2.9 Noncircular thin-walled tube in torsion.

on an element of length ds. If any point O is chosen and the perpendicular OA is drawn to the line of action of $d\tau$, the moment $d\tau$ about O is

$$dT = d\tau(\overline{OA})$$

or

$$dT = \tau h\, ds(\overline{OA}) = 2\tau h\, dA \tag{2.34}$$

where dA is the area of the element OBC. Integration of Eq. 2.34 gives the result

$$T = 2\tau h\bar{A}$$

or

$$\tau = T/2\bar{A}h \tag{2.35}$$

where $\bar{A}$ is the area enclosed by the center line of the wall, including the void. The angle of twist is calculated by the elastic strain-energy method (see Chapter 7) and is

$$\theta = TSL/4G(\bar{A})^2 h \tag{2.36}$$

where S is the length of the center line of the cross section.

The analysis of torsion for geometries other than those given, especially for tubes of variable wall thickness and for solid noncircular sections, is most conveniently handled by methods developed in the theory of elasticity (see Chapter 3). A summary of torsion equations for some common cross sections is given in Table 2.1. Equations for structural members like I beams may be found in (3 and 6).

Shear stress-shear strain relations. A typical form of torsional data for solid circular bars is shown in Fig. 2.10. From the data given various properties can be deduced. The *torsional proportional limit* is at A, and for all points on the curve from O to A the shear stress is given by Eq. 2.25 for a solid bar, Eq. 2.30 for a tube, and so on. Also point A is approximately the torsional yield stress,

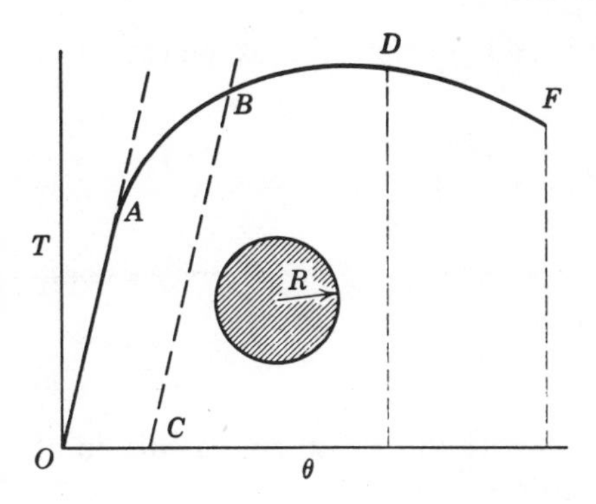

Fig. 2.10 Offset torsional yield strength B on torque (T)-twist (θ) diagram.

although, as in tension, it is common practice to define the yield strength in terms of an offset (B in Fig. 2.10). The *ultimate torsional strength* is given by D, Fig. 2.10, and the fracture stress by point F.

Torsion, unlike tension, gives rise to a nonuniform stress distribution that can be deduced from any of the stress equations in Table 2.1. Schematically, the stress distribution is shown in Fig. 2.11 for a solid circular bar. The shear stress τ is zero at the center of the bar and maximum at the outside surface. This means that, at some value of torque, a shear stress is reached that exceeds the torsional shear strength of the material; yielding then occurs. It will be recalled that in tension the stress distribution was uniform and, at yield, the entire cross section of the bar yielded. In torsion, only the outside (in a circular bar) yields first; then, as loading progresses, the plastic zone penetrates the bar until finally the entire cross section is in the plastic range. Because of this nonuniform behavior, it is more difficult to make calculations involving the torsional properties of materials. To transform Fig. 2.10 from a torque-angle of twist curve to a shear stress-shear strain curve consider only the outside fibers and proceed as follows. Up to point A there is no difficulty; shear stress is given by

$$\tau = Tr/J = 2T/\pi R^3 \tag{2.37}$$

and the corresponding shear strain is

$$\gamma = r/G \tag{2.38}$$

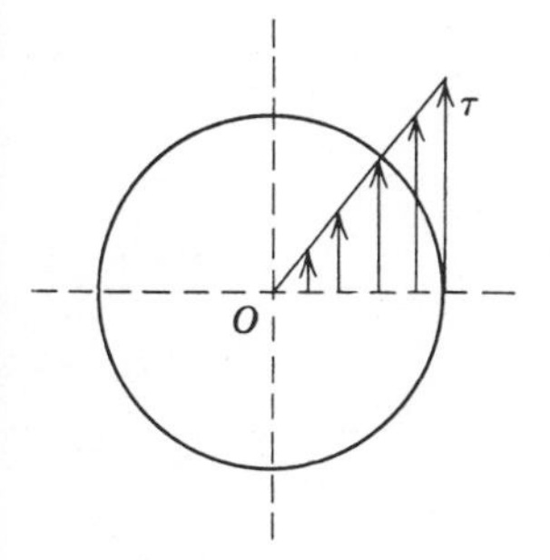

Fig. 2.11 Torsional stress distribution in a solid circular bar.

Beyond point A plasticity relationships derived in Chapter 6 must be used. For the outside fibers, the shear stress is

$$\tau = \left(\frac{1}{2\pi R^3}\right)\left(3T + \theta\frac{dT}{d\theta}\right) \tag{2.39}$$

where θ is the unit angle of twist and $dT/d\theta$ is the slope of the T versus θ curve at the particular θ in question. The shear strain is given by

$$\gamma = \theta R \tag{2.40}$$

For thin-walled circular tubes the average stress and strain are calculated at midwall and for this case (again in the plastic range)

$$\tau = \left(\frac{2T}{\pi D^2 t}\right)\left(\frac{1}{1 - 2t/D + 2t^2/D^2}\right) \tag{2.41}$$

and

$$\gamma = (\theta D/2)(1 - t/D) \tag{2.42}$$

where D is the outside diameter of the tube and t is the wall thickness.

The ultimate torsional strength determined by Eq. 2.39 can be applied to both ductile and brittle materials as shown in Fig. 2.12. For a brittle material the ultimate torsional strength of a solid circular bar is

$$\tau_u = (1/2\pi R^3)[3(\overline{BD}) + \overline{OD}(\overline{BC}/\overline{AC})] \tag{2.43}$$

and for a ductile circular bar it is seen that $dT/d\theta = 0$ at τ_u; therefore from Eq. 2.39

$$\tau_u = 3T/2\pi R^3 \tag{2.44}$$

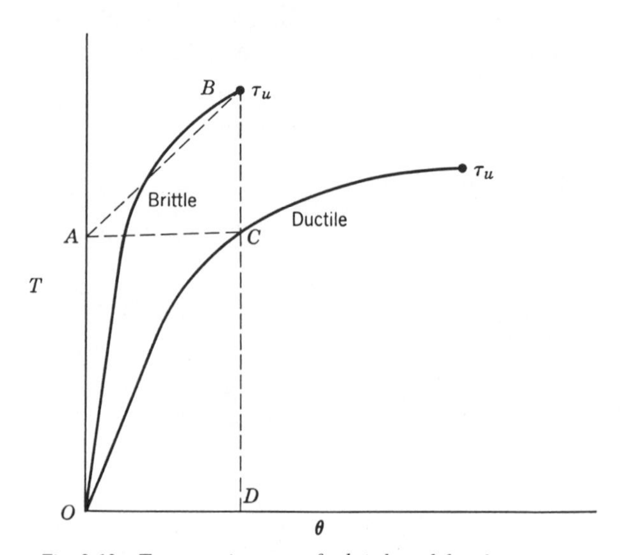

Fig. 2.12 Torque-twist curves for brittle and ductile materials.

Note how the value of τ given by Eq. 2.44, based on plasticity theory, compares with the elastic range value given by Eq. 2.37; that is,

$$\tau_u = (\text{Eq. 2.37})(\tfrac{3}{4}) = \text{Eq. 2.44}$$

or

$$\tau_u = (2T/\pi R^3)(\tfrac{3}{4}) = 3T/2\pi R^3 \qquad 2.45$$

which is 25% lower than given by a calculation of τ_u based on elastic theory.

In determining the torsional strength of materials, it is sometimes advisable, if high accuracy is needed, to use thin-walled tubes where the stress is constant across the cross section. A specimen of considerable shortness is required (approximately a length-diameter ratio of about 0.5) so that buckling will not occur. The subject of buckling is discussed in Chapter 9 where it is shown that for short tubes ($D/t > L^2/D^2 > 25t/D$) the critical buckling stress is

$$\tau_{cr} = \frac{1.26E(E_t/E)^{1/2}}{(D/t)^{5/4}(L/D)^{1/2}} \qquad 2.46$$

where E_t is the tangent modulus at a value $\sigma = 2\tau$ on a tension stress-strain curve of the material. For long tubes

$$\tau_{cr} = \frac{0.73E(E_t/E)^{1/2}}{(D/t)^{3/2}} \qquad 2.47$$

Comparative torsional properties. A comparison of torsional stiffness for various cross sections is given in Fig. 2.13. The plots in this figure were obtained by using the values for θ in Table 2.1, with a solid circular rod as the basis of comparison. For example, if the shear moduli are equal for various bars ($G_1 = G_2$), then Fig. 2.13 shows that the cross-sectional area of a thick-walled tube with an outside-to-inside diameter ratio of 1.25 is about 0.467 times the area of a solid bar for equal stiffness; if, at the same time, the modulus of the hollow bar is twice that of the solid bar, the area required for equal torsional stiffness is only one-third the area of the solid bar. It is thus clear that, using the relationships shown in Fig. 2.13, a variety of conditions can be produced. In any event the hollow section is the most efficient—it requires less material and is stiffer; in fact, under any condition, as a hollow section tends to become more and more circular it gets stiffer and stronger. A comparison of the resilience properties of torsion bars and simple springs was presented in Chapter 1. Stresses and deflections of springs are considered further in Chapter 3.

Some examples of torsion analyses are given.

1. Torsional equations for a thin-walled circular tube may be obtained by two routes Eqs. 2.28 and 2.29 or Eqs. 2.35 and 2.36. Compare the final results using both methods.

The tube is shown in Fig. 2.14. The outside diameter is $D + t$ and the internal diameter is $D - t$. From Eq. 2.29

$$\tau = TR/J = TR/2\pi t R^3 = (T/2\pi t)(2/D)^2 = 2T/\pi t D^2$$

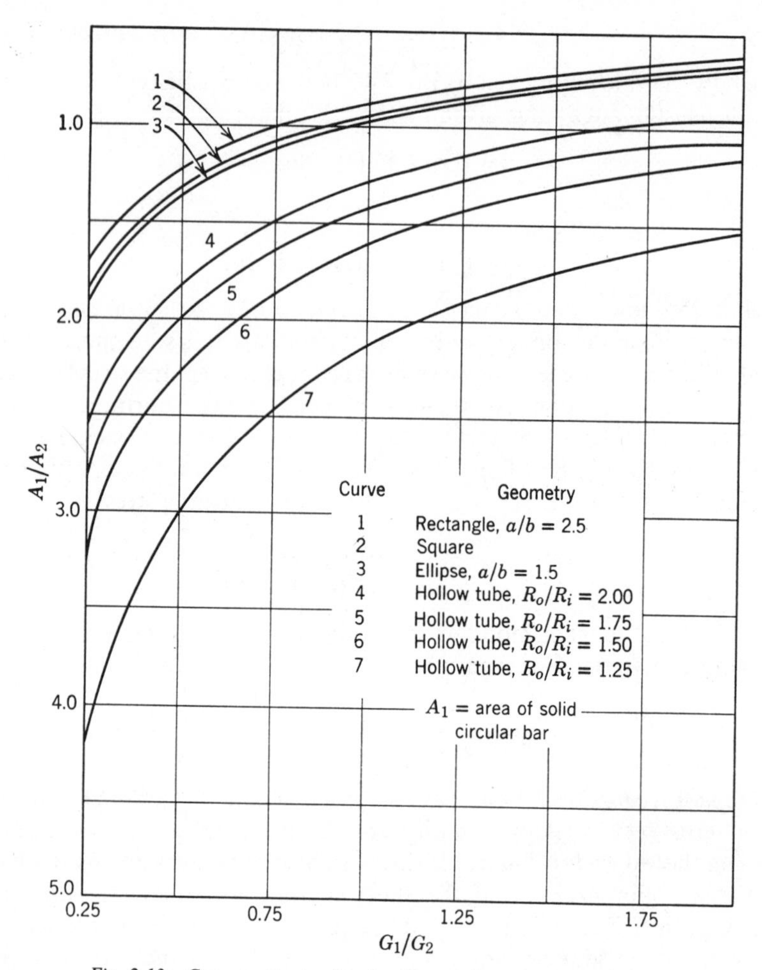

Fig. 2.13 Comparative torsional stiffness of various cross sections.

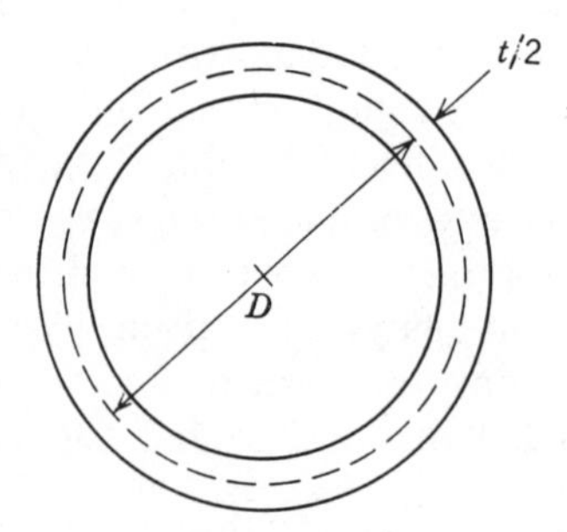

Fig. 2.14 Thin-walled tube section.

From Eq. 2.33

$$\theta = TL/2\pi GtR^3 = (TL/2\pi Gt)(2/D)^3 = 4TL/\pi GtD^3$$

For comparison, by Eqs. 2.35 and 2.36

$$\tau = T/2\bar{A}h = (T/2)(4/\pi D^2 t) = 2T/\pi t D^2$$

$$\theta = \frac{TSL}{4G(\bar{A})^2 h} = \frac{T(\pi D)L}{4G(\pi D^2/4)^2 t} = \frac{4TL}{\pi GtD^3}$$

Thus the same results are obtained by either method.

2. Calculate the increase in the maximum shearing stress and angle of twist caused by slitting lengthwise a torsion member 1 in. mean diameter and 0.10 in. wall thickness.

For an unslit tube by Eqs. 2.35 and 2.36,

$$\tau = T/2\bar{A}h = 20T/\pi$$
$$\theta = (TSL/4G)(\bar{A})^2 h = 40TL/G\pi$$

For the slotted tube (see Chapter 3)

$$\tau = 3T/ab^2 = 300T/\pi$$
$$\theta = 3TL/ab^3 G = 3000TL/G\pi$$

Therefore the shearing stress is increased by a factor of 15 by slitting the tube and the angle of twist is increased by a factor of 75.

3. Derive an equation for calculating the horsepower transmitted by a circular shaft and illustrate the equation by determining the greatest horsepower that can be transmitted at 100 revolutions per minute to a steel shaft 18 in. outside diameter and 12 in. internal diameter. The maximum shearing stress may not exceed 15,000 psi.

If a torque T is transmitted by a shaft rotating at N revolution per minute, the speed in radians per second is $2\pi N/60$. The work done on the shaft per second is thus

$$W = 2\pi NT/60$$

If torque is in inch-pounds, W is in inch-pounds/sec also. One horsepower equals 6600 in.-lb work/sec.; therefore, the horsepower transmitted H is

$$H = 2\pi NT/(60)(6600) = NT/63{,}000$$

By Eq. 2.25

$$\tau = \frac{TR}{J} = \frac{T(9)}{(\pi/2)(9^4 - 6^4)} = \frac{9T}{8260} = 15{,}000 \text{ psi}$$

from which

$$T = (8260)(15{,}000)/9 = 13.8 \times 10^6 \text{ in. lb}$$

From the horsepower equation,

$$H = NT/63{,}000 = (100)(13{,}800{,}000)/63{,}000 = 21{,}900$$

Now from Eq. 2.28, letting L be 100 ft, and G equal to 12×10^6 psi

$$\theta = \frac{TL}{GJ} = \frac{TL}{G(\pi/2)(9^4 - 6^4)} = \frac{(13.8 \times 10^6)(100)(12)}{(12 \times 10^6)(8260)} = 0.167 \text{ radian}$$

Since π radians equals 180°,

$$\theta = (180/\pi)(0.167) = 9.55°$$

4. Design a circular torsion bar with a spring rate of 2400 lb-in./rad and a total angle of twist of 0.20 rad. Use Type 304 stainless steel.

For this material the maximum design stress (proportional limit) is about 20,000 psi; therefore

$$T = (T/\theta)(\theta) = (2400)(0.20) = 480 \text{ lb-in.}$$

$$\theta = \frac{TL}{GJ}$$

or

$$L = \frac{(\theta)(G)(J)}{T} = \frac{(0.20)(10.5 \times 10^6)(\pi)(R^4)}{(480)(2)}$$

But $\tau = TR/J$; therefore

$$L = (0.20)(10.5 \times 10^6)(\pi)\frac{[(2)(480)/(20{,}000)(\pi)]^{4/3}}{(480)(2)}$$

or the required bar length $L = 28.6$ in.

Torsion of multicelled thin-walled noncircular tubes. Frequently multicell thin-wall tubes are used as structural members. The same basic assumptions for the single-cell tube apply for a multicell tube. Shown in Fig. 2.15 is a multicell that can be made of areas $\bar{A}_1, \bar{A}_2, \bar{A}_3, \ldots, \bar{A}_n$ enclosed by the center lines of the walls. Each area forms a cell that has a shear flow q_i, lb/in. acting. From Eq. 2.35

$$T = (\tau h)2\bar{A}$$

and further noting with h the wall thickness that

$$q = \tau h \qquad 2.48$$

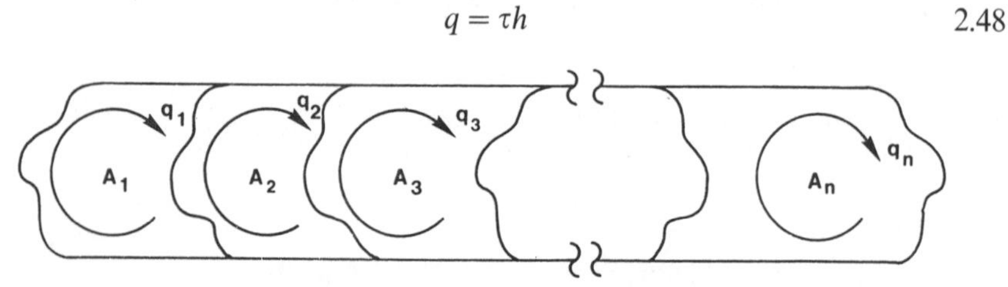

Fig. 2.15 Multicell thin-walled tube.

Therefore, extending Eq. 2.35 to multicells yields

$$T = 2 \sum_{i=1}^{i=n} q_i \bar{A}_i \tag{2.49}$$

For the angle of rotation for each cell Eq. 2.36 is used

$$\frac{\theta}{L} = \left(\frac{T}{2\bar{A}}\right) \frac{1}{2\bar{A}G} \frac{s}{h}$$

It is noted that for a single cell that

$$q = \frac{T}{2\bar{A}}$$

Then for the multicell tube

$$\theta_i = \frac{\theta}{L} = \frac{1}{2\bar{A}_i G} q_i \frac{s_i}{h_i} \tag{2.50}$$

Should h_i vary around the length of the center line wall s_i, a segment addition is performed. Since the angle of twist is the same for each cell in the tube and for the whole tube, the θ_i equations and Eq. 2.49 will yield a solution. After the solution for the q_i's, one θ_i is selected and a value for the twist is used in Eq. 2.32. A value for the torsional stiffness GJ can be found for the whole tube from

$$\theta_i = \frac{\theta}{L} = \frac{T}{GJ} \tag{2.51}$$

In cases where G is not a constant a more general treatment of the multicell tube is given in (16).

An example of torsion in a multicell tube is presented in Fig. 2.16. A three-cell aluminum tube fixed at one end and 100 in. long has a 125,000 in.-lb torque applied at the free end. Find the angle of twist, stresses in the walls, and J for the tube.

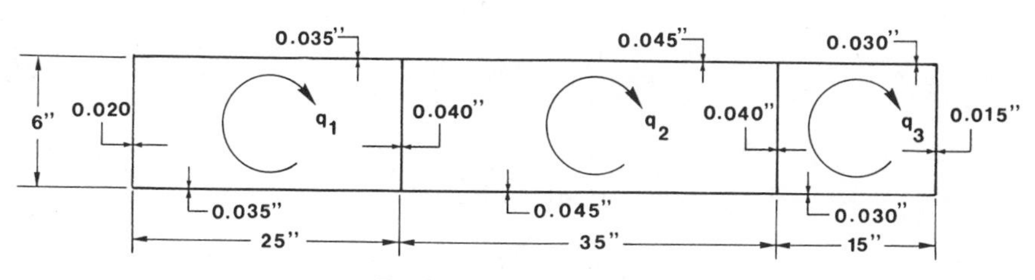

Fig. 2.16 Three-cell tube.

From Eq. 2.49 and 2.50

$$T = 2[q_1 \bar{A}_1 + q_2 \bar{A}_2 + q_3 \bar{A}_3]$$

$$\theta_1 = \frac{1}{2\bar{A}_1 G}\left[\left(\frac{6}{0.020} + \frac{50}{0.035}\right)q_1 + \frac{6}{0.040}(q_1 - q_2)\right]$$

$$\theta_2 = \frac{1}{2\bar{A}_2 G}\left[\frac{6}{0.040}(q_2 - q_1) + \frac{70}{0.045}q_2 + \frac{6}{0.040}(q_2 - q_3)\right]$$

$$\theta_3 = \frac{1}{2\bar{A}_3 G}\left[\frac{6}{0.040}(q_3 - q_2) + \left(\frac{30}{0.030} + \frac{6}{0.015}\right)q_3\right]$$

Using $\theta_1 = \theta_2, \theta_1 = \theta_3$, and values for $\bar{A}_1, \bar{A}_2, \bar{A}_3$, and G, the equation reduces to the following:

$$300\, q_1 + 420\, q_2 + 180\, q_3 = 125{,}000$$
$$1771.429\, q_1 - 1475.397\, q_2 + 107.143\, q_3 = 0$$
$$1878.571\, q_1 + 100\, q_2 - 2583.333\, q_3 = 0$$

A solution yields,

$$q_1 = 129.187 \text{ lb/in.}$$
$$q_2 = 162.387 \text{ lb/in.}$$
$$q_3 = 100.230 \text{ lb/in.}$$

When the q_i's are substituted into Eq. 2.50,

$$\theta_1 = \theta_2 = \theta_3 = 1.766 \times 10^{-4} \text{ rad/in.}$$

The total angle of twist from Eq. 2.51 is,

$$\theta = \theta_1 (100 \text{ in.}) = 1.766 \times 10^{-2} \text{ rad}$$

For the value of J solve from Eq. 2.51

$$J = \frac{T}{\theta_1 G} = 171.76 \text{ in.}^4$$

The stresses in the walls are found using Eq. 2.48.

$$\tau_1 = \frac{q_1}{0.020 \text{ in}} = 6459 \text{ lb/in.}^2$$

$$\tau_2 = \frac{q_1}{0.035} = 3691 \text{ lb/in.}^2$$

$$\tau_3 = \frac{q_1 - q_2}{0.040} = 830 \text{ lb/in.}^2$$

$$\tau_4 = \frac{q_2}{0.045} = 3608 \text{ lb/in.}^2$$

$$\tau_5 = \frac{q_2 - q_3}{0.040} = 1554 \text{ lb/in.}^2$$

$$\tau_6 = \frac{q_3}{0.030} = 3341 \text{ lb/in.}^2$$

$$\tau_7 = \frac{q_3}{0.015} = 6682 \text{ lb/in.}^2$$

2-3 BENDING OF PRISMATIC BARS

Bending of a bar is caused by transverse loads or by moments applied at right angles to the principal axes of the body.

In the following discussion a variety of beam-bending problems are examined; some of them are statically determinate but many others are not. Therefore it is convenient at the outset to define determinacy and to show how beam problems can be classified. It is not implied that only bending problems can be statically indeterminate; however, the methods used in bending are different from those applied to other cases. Therefore in the following discussion the concept of indeterminacy is confined to bending.

Rule of determinacy

In bending problems that are statically determinate the solution of forces and reactions can be accomplished by applying two of the basic laws of mechanics, namely

$$\sum F = 0 \tag{2.52}$$

and

$$\sum M = 0 \tag{2.53}$$

where Σ means a summation and F and M are force and moment, respectively.

In statically indeterminate bending problems, which are discussed later, the preceding laws, of themselves, are inadequate to allow for solution of all the

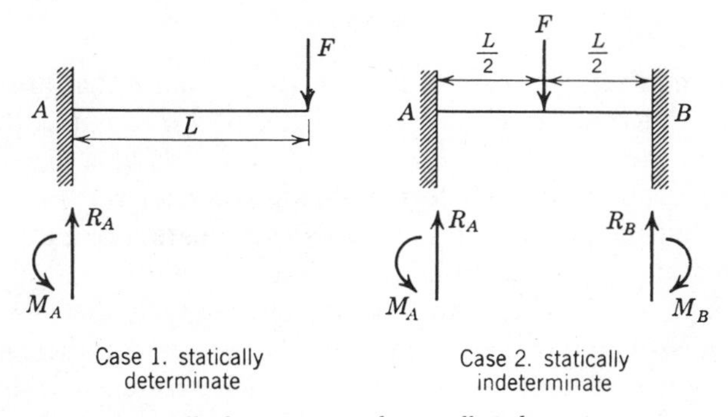

Fig. 2.17 Statically determinate and statically indeterminate structures.

Connection	Unknown Reactions
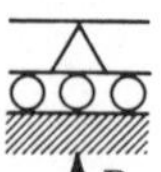 R_V	Roller Joint—Vertical reaction R_V
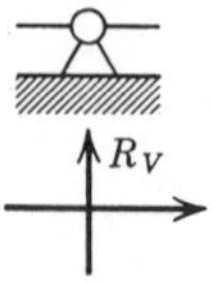 R_V, R_H	Hinge Joint—Vertical reaction R_V and horizontal reaction R_H
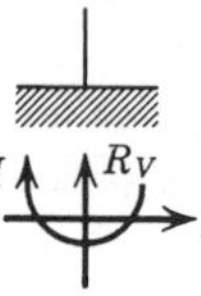 M, R_V, R_H	Fixed Joint— Vertical reaction R_V Horizontal reaction R_H Bending moment M

Fig. 2.18 Basic connections in structures.

forces and reactions. To illustrate, consider Fig. 2.17; case 1 is statically determinate and case 2 is statically indeterminate. In case 1, reaction R_A, by Eq. 2.52 is F; by application of Eq. 2.53, $M_A = FL$. In case 2, by symmetry, the reactions R_A and R_B are both equal to $F/2$; the moments M_A and M_B are also equal, but their value cannot be determined by Eq. 2.53 since

$$-M + FL/2 - RL + M = 0 \qquad \text{or} \qquad 0 = 0 \tag{2.54}$$

The method of solution for this type of problem is treated later; however, at this point consider briefly a method suggested by Bogusolvsky (2) on how to determine whether or not a bending problem is statically determinate, and, if it is indeterminate, to what degree. To begin with, consider the three basic types of connections as shown in Fig. 2.18. The number of independent equations available for solution of a particular problem n is found from Boguslovsky's equation,

$$n = 3 + \sum(p - 1) \tag{2.55}$$

where p is the number of separate parts of the structure that are formed on removal of one internal hinge. In case 1, Fig. 2.17, for example, there are no internal hinges; therefore $\Sigma(p-1) = 0$ and $n = 3$. End A is fixed; therefore there are three reactions and the degree of indeterminacy is $r - n$ or $3 - 3 = 0$, which means that the problem is statically determinate. For case 2, there are two fixed ends giving six reactions; there are no internal hinges. Therefore $\Sigma(p-1) = 0$ and $n = 3$. The degree of indeterminacy is then $6 - 3 = 3$ (for this case the horizontal forces are assumed to be zero so that there are actually only four unknown reactions and the degree of indeterminacy is thus 1). Now consider the structure shown in Fig. 2.19. In this structure, removal of hinge A

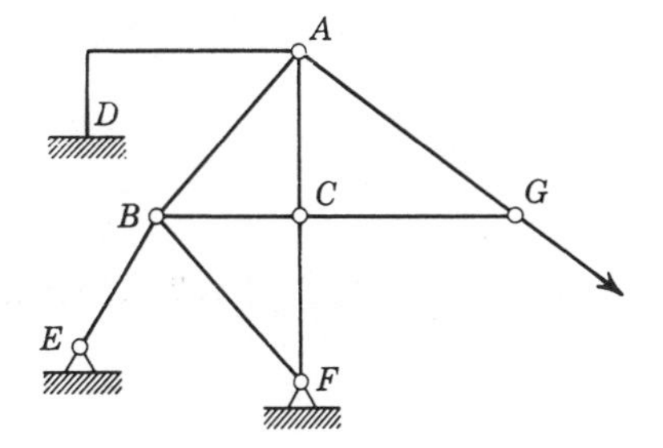

Fig. 2.19 Frame structure.

breaks the structure into two parts, AD and $EBAGCF$, giving $p - 1 = 2 - 1 = 1$. Removal of hinge B gives two parts, BE and $BFCGAD$, with $p - 1 = 1$. Removal of hinges C or G does not break up the structure into separate parts; therefore for the entire operation $\Sigma(p - 1) = 2$ and

$$n = 3 + \sum(p - 1) = 3 + 2 = 5 \tag{2.56}$$

Since the number of unknown reactions is seven, the degree of indeterminacy is $7 - 5 = 2$.

Shear and bending-moment diagrams

Shear and bending-moment diagrams for various types of loaded beams in bending have been of great practical assistance to engineers since, at a glance, stress distributions can be appraised and locations of high stress noted. In a word, such diagrams give a visual presentation of the inner reactions in a beam. The mathematical justification for such diagrams will now be given.

Consider an elemental length of a beam (Fig. 2.20) that has bending moments and shear forces acting on it. In the absence of any external forces between AB and CD, Eq. 2.52 shows that $S_1 = S$. Also, taking moments about any point, a gives

$$M + S\,dx = M + dM \tag{2.57}$$

from which

$$S = dM/dx \tag{2.58}$$

Equation 2.58 indicates that shear is the rate of change of the moment with respect to x. If a distributed load w acts on the differential element (Fig. 2.21),

$$S = S + dS + w\,dx \tag{2.59}$$

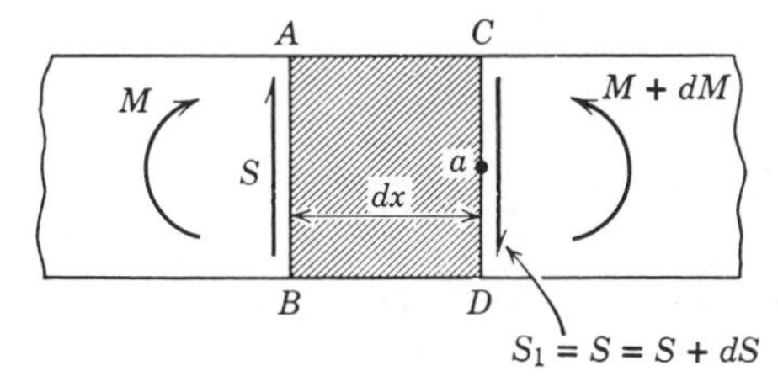

Fig. 2.20 Differential element in beam.

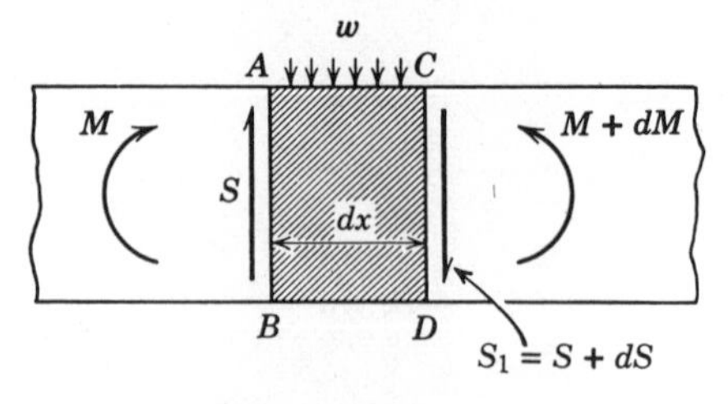

Fig. 2.21 *Differential element under distributed load w.*

which gives

$$dS = -w\,dx \tag{2.60}$$

or also

$$dS/dx = -w \tag{2.61}$$

indicating that a distributed load is the rate of change of shear. Again applying Eq. 2.53 it is seen that

$$S\,dx + M = M + dM - w\,dx(dx/2) \tag{2.62}$$

from which, neglecting $(dx)^2$ terms, gives

$$S = dM/dx \tag{2.63}$$

the same as Eq. 2.58. If a concentrated load acts (Fig. 2.22), equilibrium shows that

$$S = F + S + dS \tag{2.64}$$

or

$$dS = -F \tag{2.65}$$

and finally

$$S_1 = S - F \tag{2.66}$$

By application of Eq. 2.53

$$S - F/2 = dM/dx \tag{2.67}$$

Equation 2.67 indicates the condition over a differential element of length dx with the load at midlength $dx/2$. In actual problems, Eq. 2.67 is adjusted appropriately, depending on the location of the load F. The foregoing relations can now be used as the basis of a graphical representation of shear and bending moment in a beam, which portrays at once the total stress situation. For any system of concentrated loads the bending moment is a linear function of distance

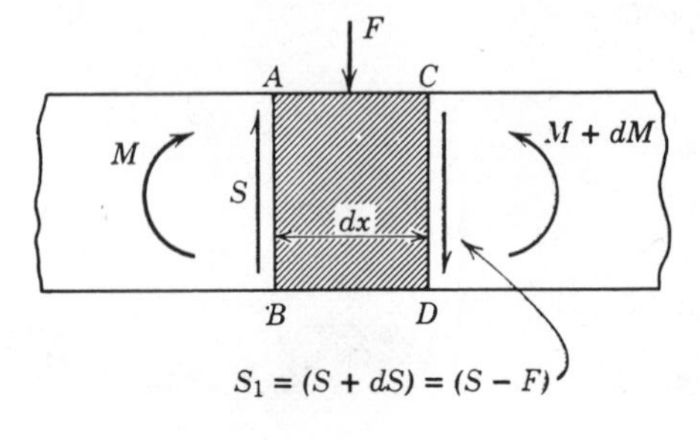

Fig. 2.22 *Differential element under concentrated load F.*

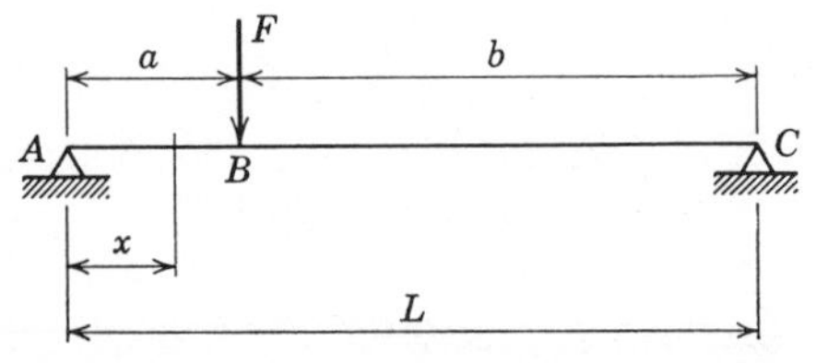

Fig. 2.23 Beam with off-center load.

along the beam. For example, in Fig. 2.23, the reaction at A is $F(b/L)$ and the moment at any point between A and B is $F(b/L)(x)$. At location B the moment is $F(b/L)(a)$. From B to C, $M = F(b/L)(x)$ minus $F(S - a)$; thus, graphically, the situation is as shown in Fig. 2.24. From Eq. 2.67 the shear at location A is $F(b/L)$, and since dM/dx is constant, this value of shear persists to location B. At location B the shear is decreased by the amount of the applied load; from here to location C the shear is constant at the value $-[(Fb/L) - F]$ since the slope dM/dx is constant as seen in the moment diagram. Thus for concentrated loads the moment diagram consists of a series of straight lines of various slopes, and the corresponding shear diagram consists of a series of rectangles.

In a distributed load the situation is somewhat different as shown in Fig. 2.25 Here the moment at any location x is

$$M_x = wLx/2 - wx^2/2 \qquad 2.68$$

This moment varies with x (a parabolic function) with a maximum value for this particular case of $wL^2/8$. The moment diagram is thus a parabola. The shear at A is equal to the end reaction $wL/2$; from A to C the shear decreases linearly, that is,

$$S = w(L/2 - x) \qquad 2.69$$

The slope of the moment diagram continuously changes reflecting the decreasing shear. At midspan, the slope is zero, and the shear is also zero at this point. A summary of some of the useful results of the foregoing analyses is given below.

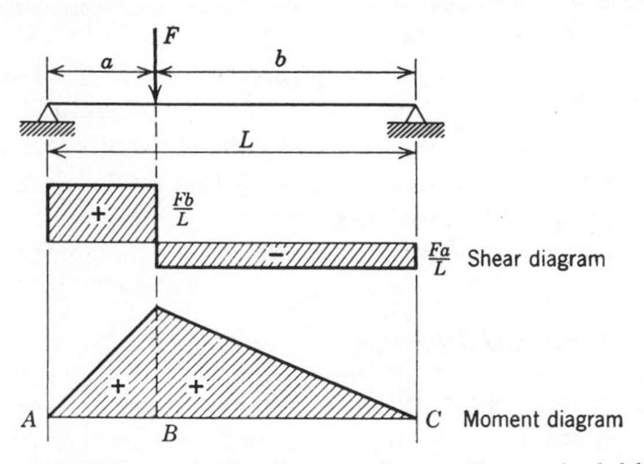

Fig. 2.24 Moment and shear diagrams for an off-center loaded beam.

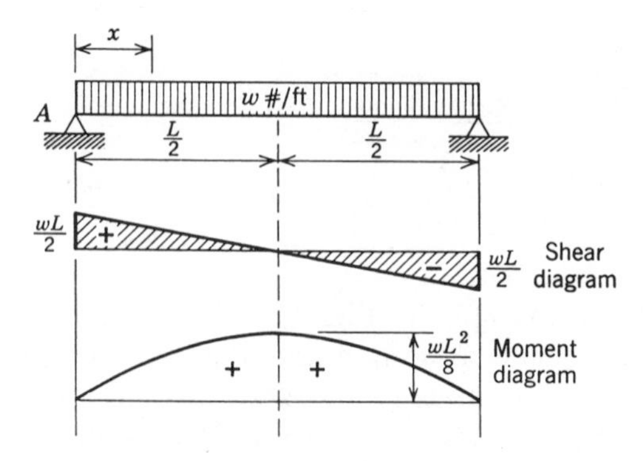

Fig. 2.25 Shear and moment diagrams for a uniformly loaded beam.

Shear

1. The shear force at the supported ends of a beam is equal to the reaction at these points. For a cantilever beam the shear at the fixed end equals the total load on the beam.
2. For beams with only concentrated loads the shear diagram consists only of horizontal and vertical straight lines. At a concentrated load the shear diagram changes abruptly in the vertical direction of magnitude equal to the load.
3. For beams with distributed loads the shear diagram consists of inclined straight lines.
4. The load intensity at any point is equal to the slope of the shear diagram at that point.

Moment

1. The bending moment is zero at the supports of a simply supported beam. For a cantilever beam the fixed-end moment is equal to the sum of the moments of the forces about the fixed end.
2. For beams with concentrated loads the moment diagram consists of inclined straight lines. For beams carrying distributed loads the moment diagram is composed of parabolic curves.
3. Where couples are applied, a discontinuity in the moment diagram occurs. The magnitude of the discontinuity equals the magnitude of the couple.
4. The shear at any point is equal to the slope of the moment diagram at that point. The change in bending moment between two points is equal to the area of the shear diagram between the two points.
5. Where the shear is zero, the bending moment is maximum or minimum.

Beam stresses and deflections

Algebraic sign convention. In order to keep the sign of moments and deflections consistent the system shown in Fig. 2.26 will be adopted. The bending moment M is positive when tensile stress is on the bottom fiber or the

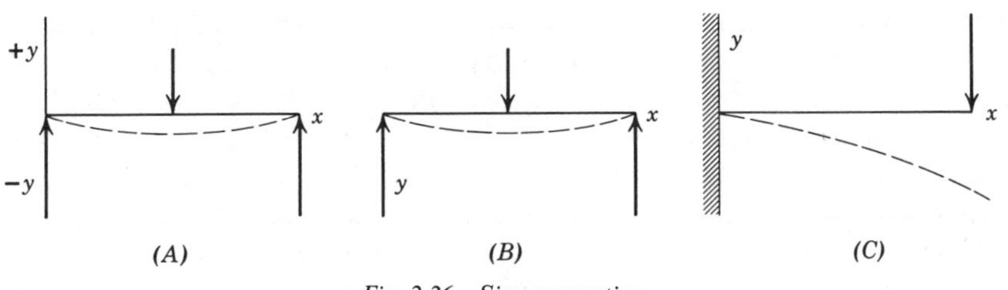

Fig. 2.26 Sign convention.

center of curvature is above the beam. M is negative when compression stress is on the bottom fiber or the center of curvature is below the beam. The quantity d^2y/dx^2 is positive when the x axis is positive to the right and the y axis is positive upward, or, when x increases, the slope changes from minus to plus. The quantity d^2y/dx^2 is negative when the y axis is positive downward or when the slope is always negative. In Fig. 2.26A x is plus to the right, y is plus upward, the center of curvature is above the beam, and the slope changes with increasing x from minus to plus. Therefore, for this case,

$$M = EId^2y/dx^2 \qquad 2.70$$

This formula will be derived presently.

General bending equations. The problem of finding stress and deflection for a general case of beam loading is now considered. To solve this problem it will be assumed that the stresses do not exceed the proportional limit, that the beam is initially straight, and that it is composed of homogeneous and isotropic material. Figure 2.27 shows the deflection of an arbitrarily loaded

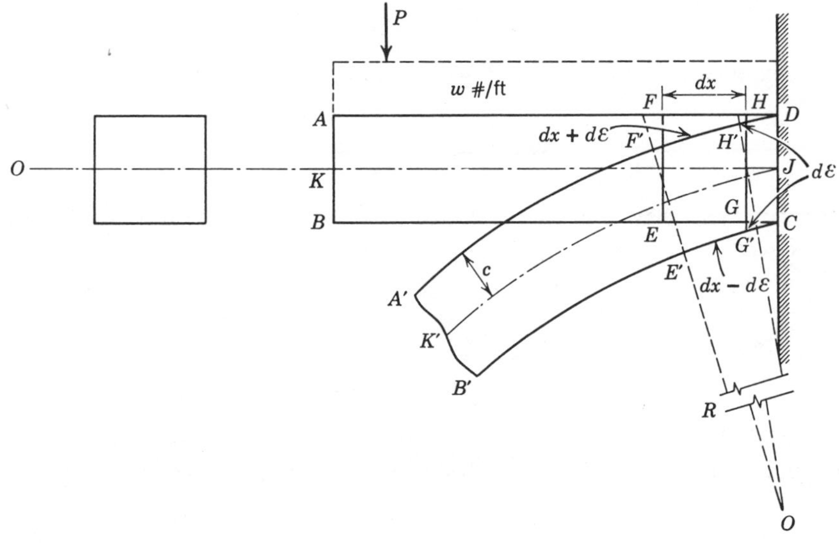

Fig. 2.27 Bending of a beam section.

beam. $ABCD$ represents the beam before loading, and $A'B'CD$ represents the beam in the deflected position. EF and GH represent initial gauge lines on the beam and JK the neutral axis. After deflection the gauge lines are rotated, and the neutral axis is described by line JK'. After deflection the original gauge lines EF and GH are rotated so that they intersect at O, forming lines OF' and OH'. If the gauge lines are initially dx distance apart, then after rotation there is no increase or decrease in length on the neutral axis. However, at the uppermost tension fiber, the material has elongated a distance $d\varepsilon$, and on the bottommost compression fiber there has been a decrease in length of $d\varepsilon$. Therefore $F'H'$ now has the value $(dx + d\varepsilon)$. Also let c be the distance from the neutral axis to the outermost fiber and R the radius of curvature. By similar triangles

$$d\varepsilon/c = dx/R \qquad 2.71$$

or

$$d\varepsilon/dx = c/R \qquad 2.72$$

But $\quad d\varepsilon/dx = \varepsilon \quad$ and $\quad \varepsilon = \sigma/E = Mc/IE \qquad 2.73$

From Eqs. 2.72 and 2.73

$$1/R = (d\varepsilon/dx)(1/c) = (1/c)(Mc/IE) = M/EI \qquad 2.74$$

Therefore $R = EI/M$, and the equation for calculating bending stress in a beam is

$$\sigma = Ec/R = Mc/I = M/Z \qquad 2.75$$

where M is the bending moment, I is the moment of interia of the cross section (see Appendix B), and c is the distance from the neutral axis to the plane where stress is to be calculated. The quantity I/c is sometimes designated Z, the section modulus. From Eq. 2.74

$$M = EI/R \qquad 2.76$$

The general equation for the radius of curvature is given by calculus (see any text) as

$$R = \frac{[1 + (dy/dx)^2]^{3/2}}{d^2y/dx^2} \qquad 2.77$$

Since, however, in the bending problem dy/dx is small,* the term $(dy/dx)^2$ in Eq. 2.77 can be neglected; thus

$$R = 1/d^2y/dx^2 \qquad 2.78$$

and, combining Eqs. 2.76 and 2.78.

$$M = EI(d^2y/dx^2) \qquad 2.79$$

Now recall Fig. 2.26. In Fig. 2.26A, x is plus to the right, y is plus upward, and the center of curvature is above the beam, and the slope changes with increasing

*For large elastic deflections this is not true; see Appendix C.

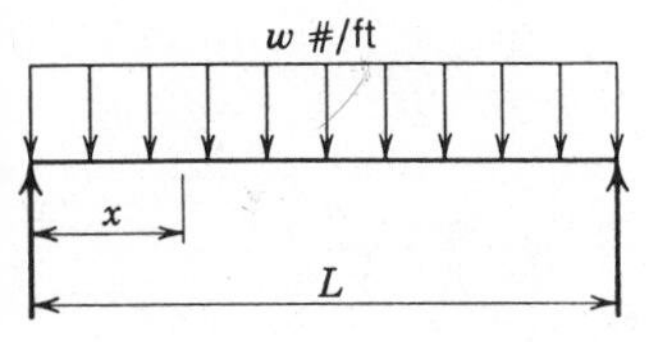

Fig. 2.28 Simply supported beam bent by its own weight.

x from minus to plus. Therefore, for this case

$$M = EI(d^2y/dx^2) \tag{2.80}$$

In Fig. 2.26B, x is plus to the right and y is plus downward; therefore

$$M = -EI(d^2y/dx^2) \tag{2.81}$$

In Fig. 2.26C the bottom fibers are in compression, and the slope is always negative; therefore

$$-M = -EI\, d^2y/dx^2 \qquad \text{or} \quad M = EI\, d^2y/dx^2 \tag{2.82}$$

Application of the deflection equation. The basic equations for determining stress and deflection in beams† subject to bending are given by 2.75 and 2.79. Some examples of the use of these equations are given in the following paragraphs. A few cases are summarized in Table 2.2.

BENDING OF A BEAM UNDER ITS OWN WEIGHT. Figure 2.28 shows a beam, simply supported and bent by its own weight. Let the weight of the beam be w lb/unit length; thus the reactions at the supports are $wL/2$, and consequently the bending moment at a section distant x from the left support is

$$M_x = (wL/2)x - (wx)(x/2) \tag{2.83}$$

From Eq. 2.79

$$d^2y/dx^2 = M/EI = (wx/2)(L - x)/EI \tag{2.84}$$

Double integration of Eq. 2.84 gives

$$y = (wx^3/12EI)(L - x/2) + C_1x + C_2 \tag{2.85}$$

The constants of integration C_1 and C_2 are evaluated from the boundary conditions; for $y = 0$, $x = 0$, and L; thus $C_2 = 0$ and

$$C_1 = -wL^3/24EI \tag{2.86}$$

Therefore putting C_1 and C_2 into Eq. 2.85 gives

$$y = (wx/24EI)(2Lx^2 - x^3 - L^3) \tag{2.87}$$

Equation 2.87 is the general deflection equation for the beam in question. The maximum deflection occurs at midspan, so that, from Eq. 2.87

$$y_{max} = -5wL^4/384EI \tag{2.88}$$

†See Chapter 7 for bending of rings, arches, and so on.

Table 2.2 Beam Formulas

Simple Beam—Concentrated Load at Center

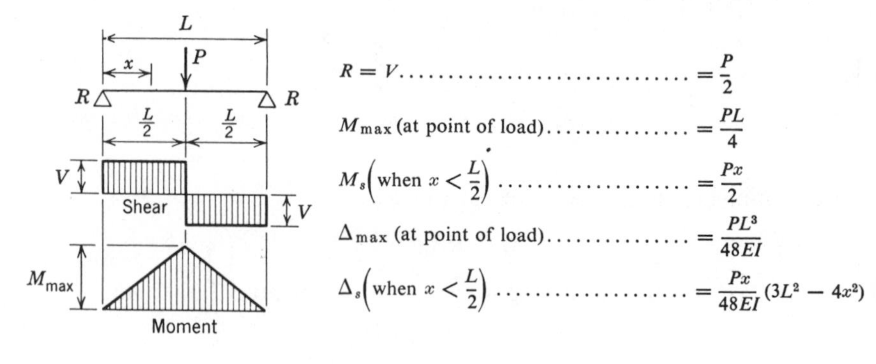

$R = V \ldots\ldots = \frac{P}{2}$

M_{max} (at point of load) $\ldots\ldots = \frac{PL}{4}$

$M_x\left(\text{when } x < \frac{L}{2}\right) \ldots\ldots = \frac{Px}{2}$

Δ_{max} (at point of load) $\ldots\ldots = \frac{PL^3}{48EI}$

$\Delta_x\left(\text{when } x < \frac{L}{2}\right) \ldots\ldots = \frac{Px}{48EI}(3L^2 - 4x^2)$

Simple Beam—Concentrated Load at Any Point

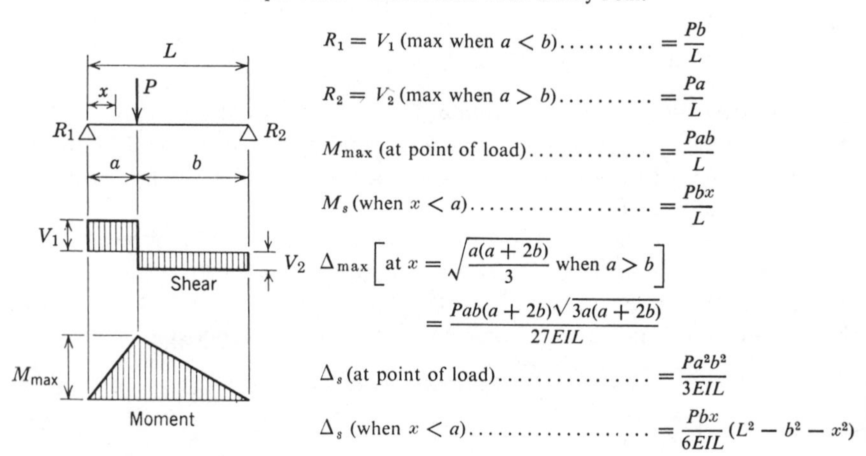

$R_1 = V_1$ (max when $a < b$) $\ldots\ldots = \frac{Pb}{L}$

$R_2 = V_2$ (max when $a > b$) $\ldots\ldots = \frac{Pa}{L}$

M_{max} (at point of load) $\ldots\ldots = \frac{Pab}{L}$

M_x (when $x < a$) $\ldots\ldots = \frac{Pbx}{L}$

$\Delta_{max}\left[\text{at } x = \sqrt{\frac{a(a+2b)}{3}} \text{ when } a > b\right]$

$= \frac{Pab(a+2b)\sqrt{3a(a+2b)}}{27EIL}$

Δ_x (at point of load) $\ldots\ldots = \frac{Pa^2b^2}{3EIL}$

Δ_x (when $x < a$) $\ldots\ldots = \frac{Pbx}{6EIL}(L^2 - b^2 - x^2)$

Simple Beam—Uniformly Distributed Load

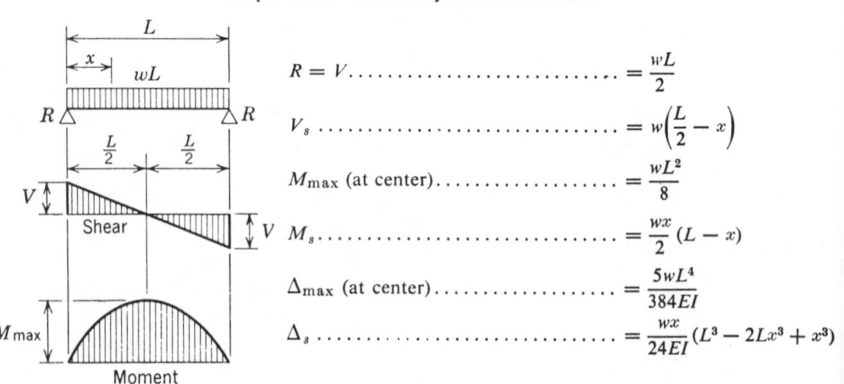

$R = V \ldots\ldots = \frac{wL}{2}$

$V_x \ldots\ldots = w\left(\frac{L}{2} - x\right)$

M_{max} (at center) $\ldots\ldots = \frac{wL^2}{8}$

$M_x \ldots\ldots = \frac{wx}{2}(L - x)$

Δ_{max} (at center) $\ldots\ldots = \frac{5wL^4}{384EI}$

$\Delta_x \ldots\ldots = \frac{wx}{24EI}(L^3 - 2Lx^2 + x^3)$

Table 2.2 (Continued).

Simple Beam—Load Increasing Uniformly to Center

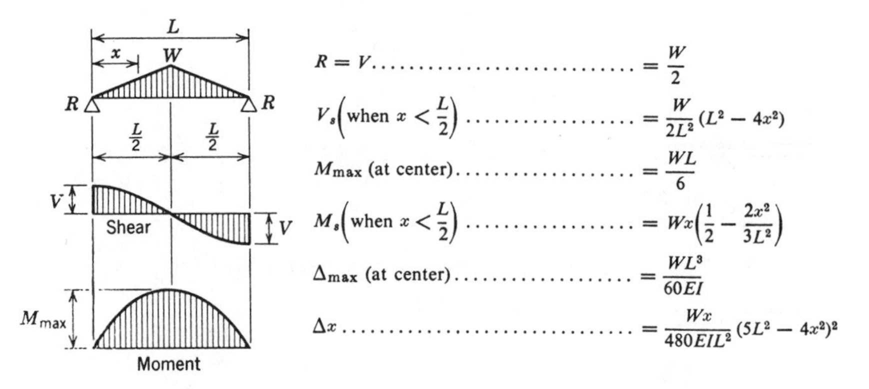

$R = V \ldots = \dfrac{W}{2}$

$V_x\left(\text{when } x < \dfrac{L}{2}\right) \ldots = \dfrac{W}{2L^2}(L^2 - 4x^2)$

$M_{\max}$ (at center) $\ldots = \dfrac{WL}{6}$

$M_x\left(\text{when } x < \dfrac{L}{2}\right) \ldots = Wx\left(\dfrac{1}{2} - \dfrac{2x^2}{3L^2}\right)$

$\Delta_{\max}$ (at center) $\ldots = \dfrac{WL^3}{60EI}$

$\Delta x \ldots = \dfrac{Wx}{480EIL^2}(5L^2 - 4x^2)^2$

Beam Fixed at Both Ends—Uniformly Distributed Loads

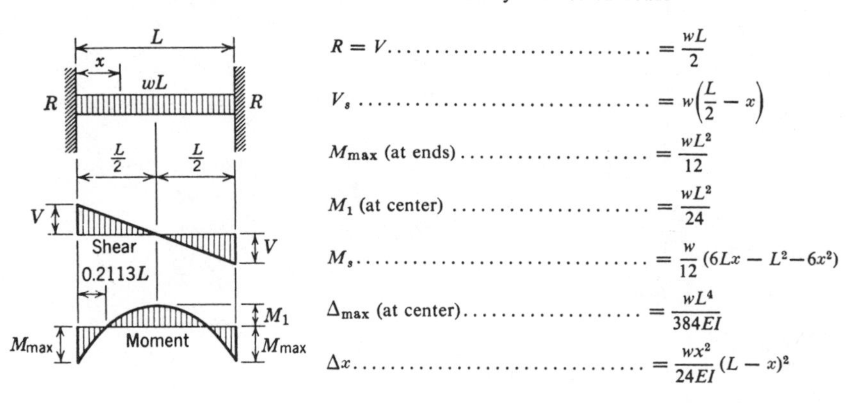

$R = V \ldots = \dfrac{wL}{2}$

$V_x \ldots = w\left(\dfrac{L}{2} - x\right)$

$M_{\max}$ (at ends) $\ldots = \dfrac{wL^2}{12}$

M_1 (at center) $\ldots = \dfrac{wL^2}{24}$

$M_x \ldots = \dfrac{w}{12}(6Lx - L^2 - 6x^2)$

$\Delta_{\max}$ (at center) $\ldots = \dfrac{wL^4}{384EI}$

$\Delta x \ldots = \dfrac{wx^2}{24EI}(L - x)^2$

Beam Fixed at Both Ends—Concentrated Load at Center

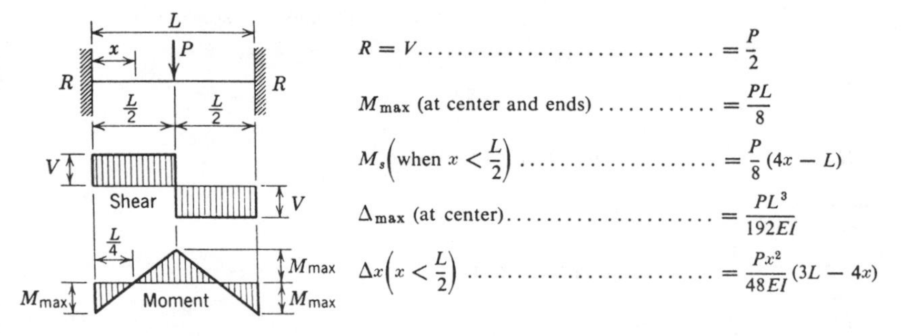

$R = V \ldots = \dfrac{P}{2}$

$M_{\max}$ (at center and ends) $\ldots = \dfrac{PL}{8}$

$M_x\left(\text{when } x < \dfrac{L}{2}\right) \ldots = \dfrac{P}{8}(4x - L)$

$\Delta_{\max}$ (at center) $\ldots = \dfrac{PL^3}{192EI}$

$\Delta x\left(x < \dfrac{L}{2}\right) \ldots = \dfrac{Px^2}{48EI}(3L - 4x)$

Table 2.2 (Continued).

Cantilevered Beam—Concentrated Load at Free End

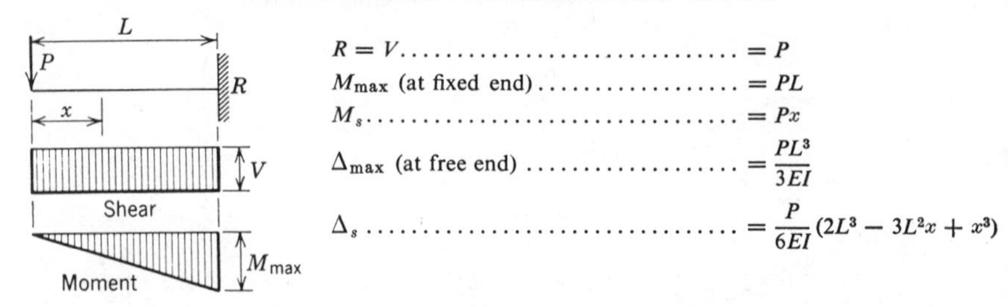

$R = V$ .... $= P$

M_{max} (at fixed end) .... $= PL$

M_x .... $= Px$

Δ_{max} (at free end) .... $= \dfrac{PL^3}{3EI}$

Δ_x .... $= \dfrac{P}{6EI}(2L^3 - 3L^2x + x^3)$

Cantilevered Beam—Concentrated Load at Any Point

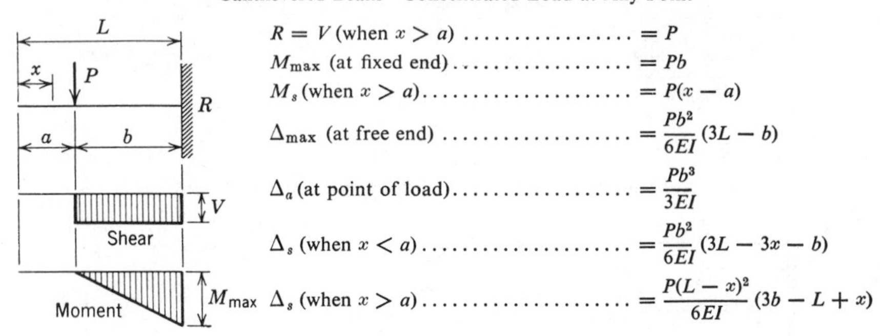

$R = V$ (when $x > a$) .... $= P$

M_{max} (at fixed end) .... $= Pb$

M_x (when $x > a$) .... $= P(x - a)$

Δ_{max} (at free end) .... $= \dfrac{Pb^2}{6EI}(3L - b)$

Δ_a (at point of load) .... $= \dfrac{Pb^3}{3EI}$

Δ_x (when $x < a$) .... $= \dfrac{Pb^2}{6EI}(3L - 3x - b)$

Δ_x (when $x > a$) .... $= \dfrac{P(L - x)^2}{6EI}(3b - L + x)$

Cantilevered Beam—Uniformly Distributed Load

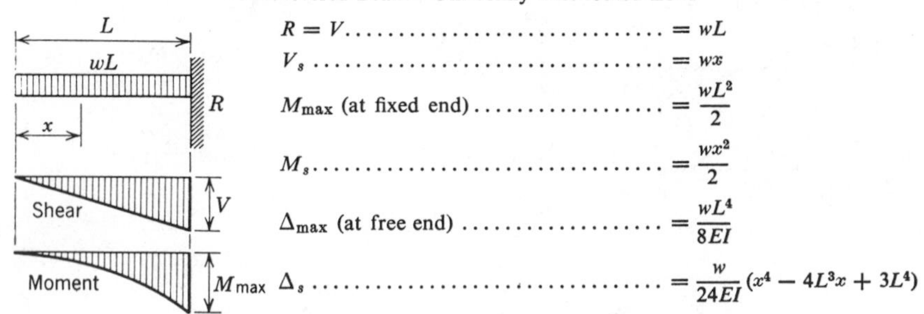

$R = V$ .... $= wL$

V_x .... $= wx$

M_{max} (at fixed end) .... $= \dfrac{wL^2}{2}$

M_x .... $= \dfrac{wx^2}{2}$

Δ_{max} (at free end) .... $= \dfrac{wL^4}{8EI}$

Δ_x .... $= \dfrac{w}{24EI}(x^4 - 4L^3x + 3L^4)$

Cantilevered Beam—Load Increasing Uniformly to Fixed End

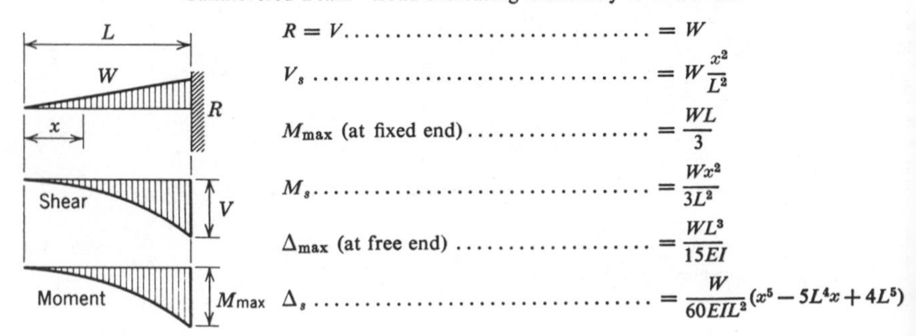

$R = V$ .... $= W$

V_x .... $= W\dfrac{x^2}{L^2}$

M_{max} (at fixed end) .... $= \dfrac{WL}{3}$

M_x .... $= \dfrac{Wx^2}{3L^2}$

Δ_{max} (at free end) .... $= \dfrac{WL^3}{15EI}$

Δ_x .... $= \dfrac{W}{60EIL^2}(x^5 - 5L^4x + 4L^5)$

Table 2.2 (Continued)

Beam With Overhang—Uniformly Loaded

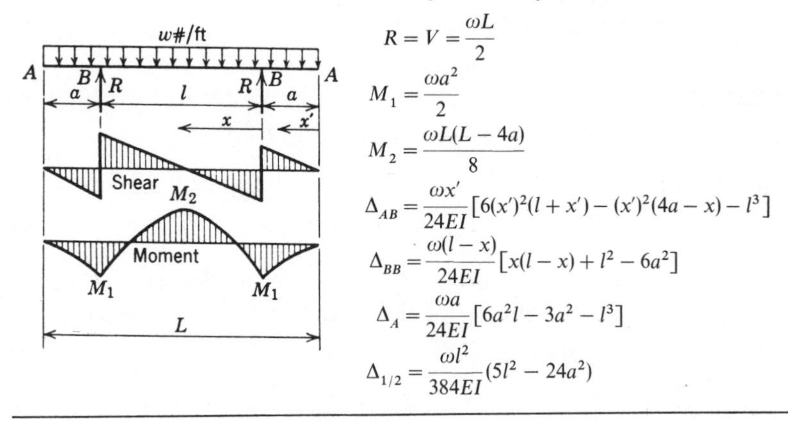

$$R = V = \frac{\omega L}{2}$$

$$M_1 = \frac{\omega a^2}{2}$$

$$M_2 = \frac{\omega L(L - 4a)}{8}$$

$$\Delta_{AB} = \frac{\omega x'}{24EI}[6(x')^2(l + x') - (x')^2(4a - x) - l^3]$$

$$\Delta_{BB} = \frac{\omega(l - x)}{24EI}[x(l - x) + l^2 - 6a^2]$$

$$\Delta_A = \frac{\omega a}{24EI}[6a^2 l - 3a^2 - l^3]$$

$$\Delta_{1/2} = \frac{\omega l^2}{384EI}(5l^2 - 24a^2)$$

Note: An interesting example of a span deflecting under its own weight is the *catenary*, the curve followed by a uniform *flexible* cable under its own weight (Fig. 2.29). In analyzing a simple catenary where the weight of the cable is w lb per unit length it is assumed that the force in the cable is a tension parallel to the length, being horizontal at the lowest point. The horizontal component of the tension in the cable is the same at all locations; thus from Fig. 2.29

$$T^2 = (ws)^2 + H^2 \qquad 2.89$$

where T is the tension force and H its horizontal component. Also $\sin\theta =$

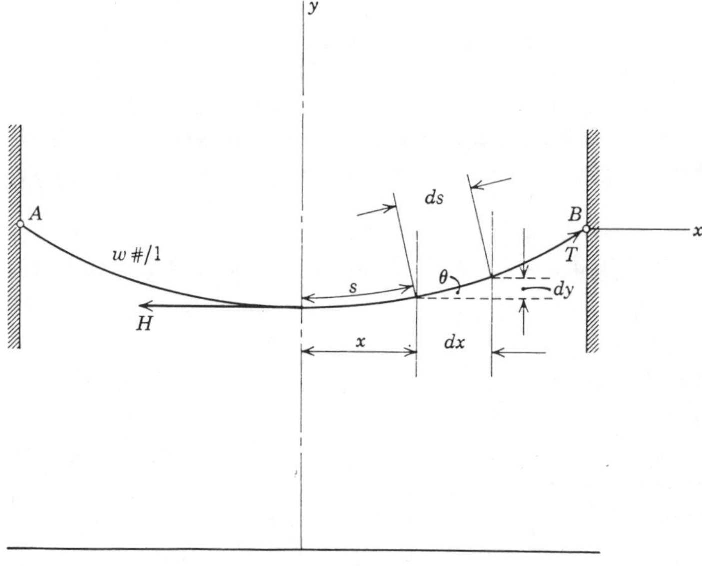

Fig. 2.29 The catenary.

$dy/T = ws/T = dy/ds$ from which

$$T = (ws)(ds)/dy \tag{2.90}$$

Combining equations gives the following:

$$(ws)^2(ds/dy) = (ws)^2 + H^2$$

or

$$dy/ds = ws/(w^2s^2 + H^2)^{1/2}$$

which on integrating and letting $H = wc$ gives

$$y = (s^2 + c^2)^{1/2} + C_1 \tag{2.91}$$

where C_1 is a constant of integration. The value of c is a constant taken such that it is the length of cable whose weight is equal to the horizontal tension. When $y = 0$, $s = 0$; therefore

$$C_1 = -c$$

and

$$y = (s^2 + c^2)^{1/2} - c \tag{2.92}$$

Then from Eqs. 2.89 and 2.92

$$T = wy \tag{2.93}$$

and also

$$y = c \sec \theta \tag{2.94}$$

$$s = y \sin \theta \tag{2.95}$$

or, in terms of the horizontal distance x

$$y = c \cosh x/c \tag{2.96}$$

$$s = c \sinh x/c \tag{2.97}$$

BENDING BY TWO CONCENTRATED LOADS ARRANGED SYMMETRICALLY ABOUT THE CENTER (Fig. 2.30). The end reactions are each equal to P, and the bending moment is

$$M_x = Px \tag{2.98}$$

between $x = 0$ and $x = L - a/2$. Between the loads P the moment remains

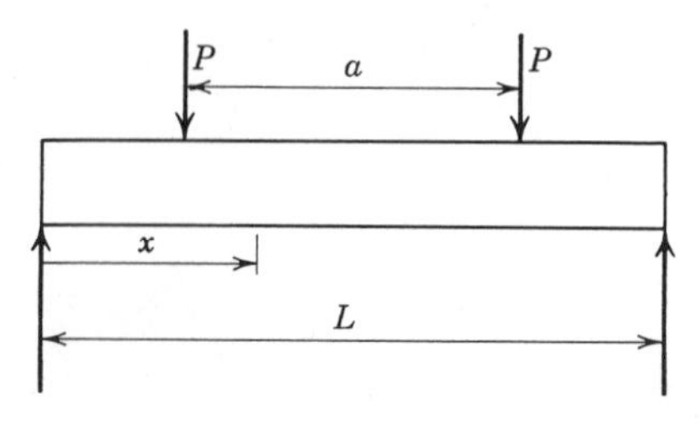

Fig. 2.30 Beam subjected to symmetrical loading.

constant at the value

$$M = P(L - a)/2 \tag{2.99}$$

Therefore, between $x = 0$ and $x = L - a/2$,

$$d^2y/dx^2 = Px/EI \tag{2.100}$$

Between the loads,

$$d^2y/dx^2 = P(L - a)/2 \tag{2.101}$$

Double integration of Eqs. 2.100 and 2.101 gives

$$y = Px^3/6EI + C_1x + C_2 \tag{2.102}$$

and

$$y = P(L - a)x^2/4EI + C_3x + C_4 \tag{2.103}$$

For the condition $y = 0$, $x = 0$ and thus $C_2 = 0$. For the condition dy/dx (slope) $= 0$ for $x = L/2$, $C_3 = -PL(L - a)/4$. The constants C_1 and C_4 are obtained from the condition that at $x = L - a/2$ the deflection and slope must be the same whether calculated from Eq. 2.102 or Eq. 2.103. Consequently,

$$C_1 = -P(L^2 - a^2)/8EI \tag{2.104}$$

and

$$C_4 = P(L - a)^3/48EI \tag{2.105}$$

Therefore, using the values of these constants, Eqs. 2.102 and 2.103 become, respectively,

$$(y)_{0<x<\frac{L-a}{2}} = \frac{Px^3}{6EI} - \frac{Px(L^2 - a^2)}{8EI} \tag{2.106}$$

$$(y)_{x>\frac{L-a}{2}} = \frac{Px^2(L-a)}{4EI} - \frac{PLx(L-a)}{4EI} + \frac{P(L-a)^3}{48EI} \tag{2.107}$$

The maximum deflection occurs between the loads at $L/2$; therefore,

$$y_{\max} = \frac{-P(L-a)(2L^2 + 2La - a^2)}{48EI} \tag{2.108}$$

SIMPLY SUPPORTED BEAM WITH LINEARLY VARYING DISTRIBUTED LOAD. Figure 2.31 shows a simply supported beam carrying a distributed load whose intensity

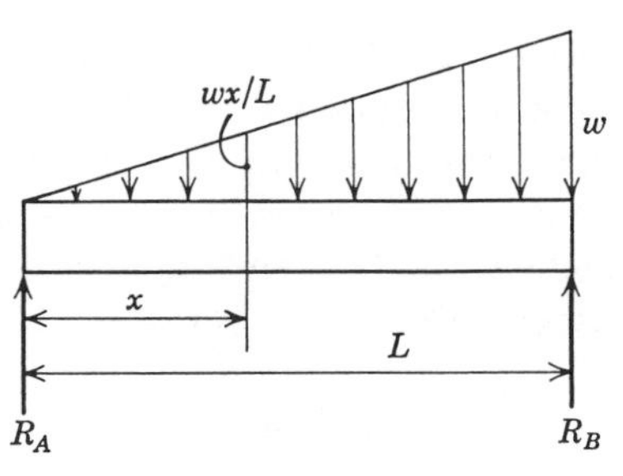

Fig. 2.31 Beam with triangular loading.

varies linearly from zero at one end to w per unit length at the other end. Since the average intensity of loading is $w/2$ per unit length, the total load on the beam is $wL/2$. The center of gravity of the distributed load is at a distance $L/3$ from the right-hand end of the beam; thus

$$R_A = wL/6 \tag{2.109}$$

and

$$R_B = wL/3 \tag{2.110}$$

The intensity of loading at a distance x from the left-hand end of the beam is wx/L, and hence the shearing force equation is

$$S = R_A - (wx/L)(x/2) = wL/6 - wx^2/2L \tag{2.111}$$

or

$$S = (w/6L)(L^2 - 3x^2) \tag{2.112}$$

The shearing force is therefore zero when

$$x = L/\sqrt{3} \tag{2.113}$$

The bending moment at a distance x from the left-hand end of the beam is

$$M_x = R_A x - wx/L(x/2)(x/2)$$

or

$$M_x = wx/6L(L^2 - x^2) \tag{2.114}$$

The bending moment is maximum when the shearing force is zero; thus from Eqs. 2.113 and 2.114

$$M_{max} = wL^2/9\sqrt{3} \tag{2.115}$$

Additional treatment of shear stresses in beams is given in Chapter 3. In the foregoing problem the deflection can be determined as in the previous examples; that is,

$$EI\, d^2y/dx^2 = M_x = wx/6L(L^2 - x^2) \tag{2.116}$$

Double integration of this equation leads to the result

$$y = wx/360EIL(10L^2x^2 - 3x^4 - 7L^4) \tag{2.117}$$

The maximum deflection occurs at a value of $x = 0.519L$ and is

$$y_{max} = -0.00652wL^4/EI \tag{2.118}$$

ELASTICALLY SUPPORTED CANTILEVER BEAM (Fig. 2.32). In the elastically supported cantilever the normally fixed end is provided with either a device allowing the beam to move up and down, to pivot about the centroid, or both. This is shown schematically in Fig. 2.32 where the restraint at pivot C is caused by *equivalent springs* of force constant k_1 and k_2. In the *normal* cantilever beam k_1 and k_2 are infinite (that is, the end is rigidly fixed). The control switch is an example of this type of beam.

In the general case shown in Fig. 2.32 the beam can rotate about C, but C

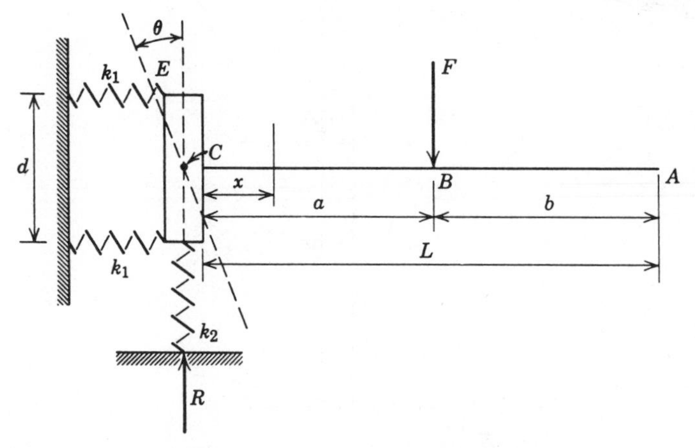

Fig. 2.32 Elastically supported cantilever beam.

does not move horizontally. Shear effects are considered negligible. In order to find the deflection of the beam it is necessary to calculate three deflections, that normally displayed by a cantilever loaded as shown and the deflections associated with the k elements. First consider the *normal* deflection.

Section C-B

$$EI\, d^2y/dx^2 = M = F(a - x) \tag{2.119}$$

$$dy/dx = -(F/EI)(x^2/2 - ax - C_1) \tag{2.120}$$

At $x = 0$, slope $dy/dx = 0$; therefore $C_1 = 0$.

$$y = -\frac{F}{EI}\left(\frac{x^3}{6} - \frac{ax^2}{2} - C_2\right) \tag{2.121}$$

At $x = 0$, $y = 0$; therefore $C_2 = 0$ and

$$y = -(F/EI)(x^3/6 - ax^2/2) \tag{2.122}$$

Section B–A. There is no moment within this part of the beam; therefore deflection is calculated by making use of Eq. 2.120 (Fig. 2.33).

$$\tan\theta = \theta = y'/(x - a)$$

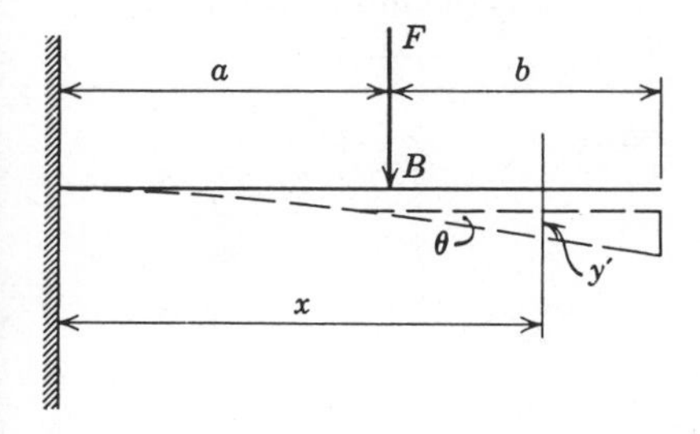

Fig. 2.33 Deflection of beam in Fig. 2.32.

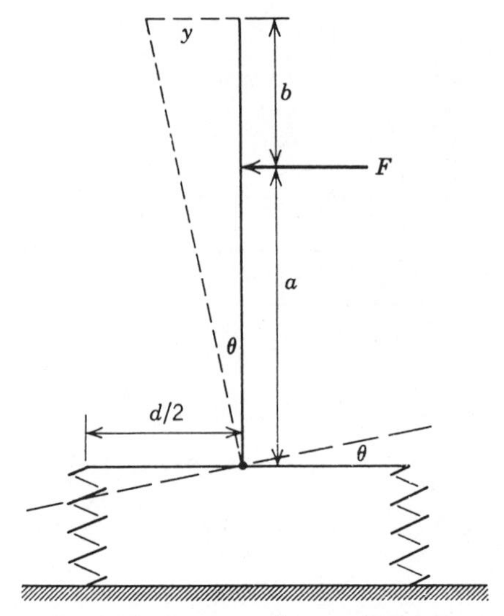

Fig. 2.34 Rotation of beam in Fig. 2.32.

or since $\theta =$ Eq. 2.120

$$y' = \theta(x-a) = (F/EI)(x-a)(a^2/2) \tag{2.123}$$

The total deflection of the curve in the region $B-A$ is thus, from Eqs. 2.122 and 2.123,

$$y = (F/EI)(a^2/2)(x-a) + (F/EI)(a^3/3)$$

or

$$y = -(F/EI)(a^3/2 - a^3/3 - a^2x/2) \tag{2.124}$$

At B the deflection is

$$y_B = Fa^3/3FI \tag{2.125}$$

The maximum deflection is at A, where $x = L$,

$$y_{\max} = y_A = (F/6EI)(3a^2L - a^3)$$

There is a vertical deflection associated with k_2 which is simply

$$y = F/k_2 \tag{2.126}$$

For the k_1 elements the equivalent load is caused by the moment Fa (Fig. 2.34); thus

$$(d\theta/2)k_1(d/2) = Fa$$

or

$$d^2\theta k_1/4 = Fa$$

For the two elements

$$d^2\theta k_1/2 = Fa$$

Therefore at location A,

$$\tan\theta = \theta = y/(a+b) = 2Fa/k_1d^2 \qquad 2.127$$

and

$$y'_A = (2Fa/k_1d^2)(a+b)$$

At location B

$$y'_B = 2Fa^2/k_1d^2$$

$$y'_C = F/k_2$$

and

$$y'_D(\text{horiz}) = Fa/dk_1$$

Therefore the total deflections are as follows:

$$y_A = (F/6EI)(3a^2L - a^3) + (2Fa/k_1d^2)(a+b) + F/k_2 \qquad 2.128$$

$$y_B = Fa^3/3EI + F/k_2 + 2Fa^2/k_1d^2 \qquad 2.129$$

$$y_C = F/k_2 \qquad 2.130$$

Pure bending of a beam. In this special case of bending which is caused by the transmission of a couple in its own plane, no shear forces are involved; hence the name *pure bending*. In the following analysis it is assumed that the cross section of the bar is symmetrical about the y axis and that the plane in which the bending couple acts passes through this axis of symmetry (Fig. 2.35).

For equilibrium the internal stresses at any location of the bar from a couple equal to the external couple, and the bar bends into the arc of a circle. Considering the section shown, the resultant of the normal stresses vanishes, that is,

$$\int_{-h_2}^{h_1} \sigma_x \, dy = 0 \qquad 2.131$$

Also the internal stresses are equivalent to a couple balancing the external

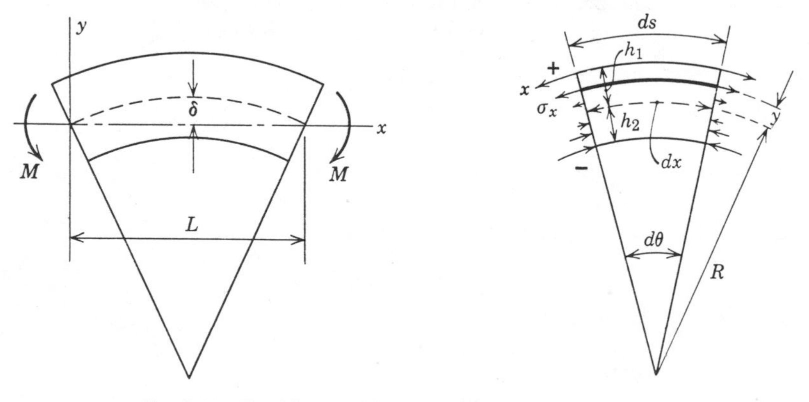

Fig. 2.35 Equilibrium of beam element under pure bending.

couple; that is,

$$\int_{-h_2}^{h_1} \sigma_x y\, dy = M \qquad 2.132$$

The neutral axis remains unchanged; consequently, since the bar is bent to the arc of a circle, the fibers ds are extended or compressed as follows:

$$ds = (R + y)\, d\theta \qquad 2.133$$

or

$$ds - dx = (R + y)\, d\theta - dx \qquad 2.134$$

For small arcs of bending, the bending strain is

$$\varepsilon_b = \frac{ds - dx}{dx} = \frac{(R + y)d\theta}{dx} - 1 \qquad 2.135$$

However,

$$d\theta = dx/R \qquad 2.136$$

Therefore

$$\varepsilon_b = \frac{R + y}{dx}\left(\frac{dx}{R}\right) - 1 = \frac{R + y - R}{R} = \frac{y}{R} \qquad 2.137$$

If Hooke's law is followed, then the bending stress σ_b can be expressed as

$$\sigma_b = \varepsilon_b E = Ey/R \qquad 2.138$$

Now, let the distance from the axis of the outermost fiber in tension be h_1 and the innermost fiber in compression h_2; then the maximum tensile stress is Eh_1/R, and the maximum compressive stress is Eh_2/R. If the cross section of the beam is rectangular of width b, the stresses give a resultant force in the X direction

$$F_x = \frac{(bEh_1/R)h_1}{2} - \frac{(bEh_2/R)h_2}{2} = \frac{bE}{2R}(h_1{}^2 - h_2{}^2) \qquad 2.139$$

Since, however, there was no external force F_x, $F_x = 0$ and $h_1 = h_2$ so that the neutral axis is at the center of the cross section.

From Eqs. 2.131 and 2.132 integration leads to the expression

$$M = \frac{E}{R}\int_{-y}^{y} y^2\, dy \qquad 2.140$$

or

$$M = E/R(2y^3/3) \qquad 2.141$$

However, for a rectangular section of unit width

$$I = 1/R(2d)^3 \qquad 2.142$$

Therefore

$$M = EI/R \qquad 2.143$$

The radius of curvature is thus

$$R = EI/M \qquad 2.144$$

Combining Eqs. 2.138 and 2.139 then leads to

$$\sigma_b = Ey/R = EyM/EI = My/I \qquad 2.145$$

which is the familiar formula for the bending stress in a beam. For pure bending, it is simple to calculate the deflection by geometrical means. From Fig. 2.35

$$(L/2)^2 = \delta\,(2R - \delta) \qquad 2.146$$

For small values of δ

$$L^2/4 = 2R\delta \qquad 2.147$$

and

$$\delta = L^2/8R = L^2 M/8EI \qquad 2.148$$

If the bending couple M is known and L and I are determined and δ is measured, Young's modulus E can be calculated from the relation

$$E = L^2 M/8I\delta \qquad 2.149$$

STATICALLY INDETERMINATE SYSTEMS. It has been shown that in a statically indeterminate system the basic equations of statics (ΣF and $\Sigma M = 0$) are insufficient for obtaining a solution to the problem. The theory of elastic strain energy (Chapter 7) is particularly suited to the solution of such problems as are demonstrated in that chapter. However, some typical problems and their solutions are illustrated here.

STATICALLY INDETERMINATE TENSION FRAME (Fig. 2.36). The first-degree indeterminate system shown in Fig. 2.36 is also used in Chapter 7 to demonstrate the solution of the problem by elastic strain-energy methods. In this system the load F extends bars OB, OC, and OD. Application of Eqs. 2.52 and 2.53 indicates that an additional equation is needed to find the three axial forces. It is assumed here that the bar OC has an area A_c and modulus of elasticity E_c, whereas the inclined members OB and OD have cross-sectional areas A_B and moduli E_B. The length of OC is L; therefore OB and OD have lengths of

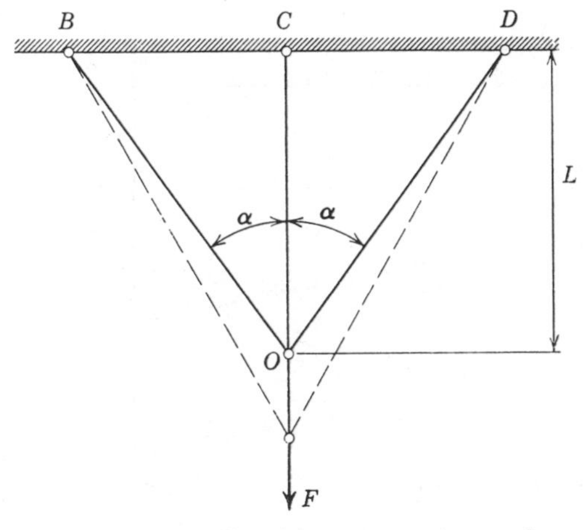

Fig. 2.36 Statically indeterminate tension members.

$L/\cos\alpha$. Now, if X is the force in the bar OC and Y the force in bars OB and OD,

$$F = X + 2Y\cos\alpha \tag{2.150}$$

The deformation of the system must now be investigated. Let δ be the elongation of OC; then the elongation of OB and OD will be

$$\delta_1 = \delta\cos\alpha \tag{2.151}$$

The unit elongation and stresses are for OC

$$\varepsilon_c = \delta/L \qquad \text{and} \qquad \sigma_c = E_c\delta/L \tag{2.152}$$

For OB and OD

$$\varepsilon_B = \delta\cos^2\alpha/L \qquad \text{and} \qquad \sigma_B = E_B(\delta\cos^2\alpha/L) \tag{2.153}$$

Therefore

$$X = \sigma_C A_C = A_C E_C \delta/L \tag{2.154}$$

and

$$Y = \sigma_B A_B = A_B E_B \delta\cos^2\alpha/L \tag{2.155}$$

from which

$$Y = X\cos^2\alpha A_B E_B/A_C E_C \tag{2.156}$$

and substituting this into Eq. 2.150 gives for $A_B E_B = A_C E_C$

$$X = F/(1 + 2\cos^3\alpha) \tag{2.157}$$

END-SUPPORTED CANTILEVER WITH UNIFORM LOAD (Fig. 2.37). For this case Eq. 2.55 shows that, since there are no internal hinges, $n = 3$. There are four unknown reactions, R_A, R_B, R_H, and M_B. Therefore the degree of indeterminacy is $4 - 3 = 1$. The reaction R_H is considered to be zero; therefore one

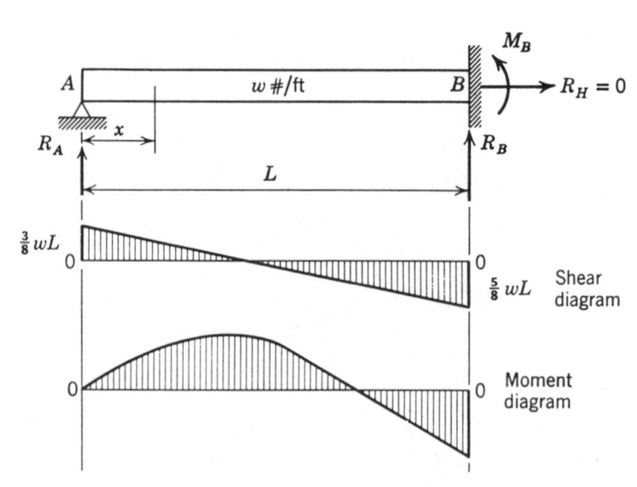

Fig. 2.37 Statically indeterminate beam.

of the other reactions is indeterminate. At any section x,

$$EI\, d^2y/dx^2 = M = R_A x = wx^2/2 \tag{2.158}$$

Integration of Eq. 2.158 gives

$$dy/dx = (1/EI)(R_A x^2/2 - wx^3/6 + C_1) \tag{2.159}$$

where C_1 is a constant of integration. At $x = L$ the slope dy/dx is zero; from this condition

$$C_1 = -R_A L^2/2 + wL^3/6 \tag{2.160}$$

Placing the value of C_1 from Eq. 2.160 in Eq. 2.159 and integrating a second time gives

$$y = (1/EI)(R_A X^3/6 - wx^4/24 - R_A L^2 x/2 + wL^3 x/6 + C_2) \tag{2.161}$$

where C_2 is a constant of integration. At $x = 0, y = 0$; therefore $C_2 = 0$. Since $x = L$ at $y = 0$ and $C_2 = 0$, it is further found that

$$0 = (1/EI)(R_A L^3/6 - wL^4/24 - R_A L^3/2 + wL^4/6) \tag{2.162}$$

from which

$$R_A = \tfrac{3}{8}wL \tag{2.163}$$

Having R_A, M_B by equilibrium is

$$M_B = R_A L - wL(L/2) = \tfrac{3}{8}wL^2 - wL^2/2 = -wL^2/8 \tag{2.164}$$

Also, by statics,

$$R_B = wL - \tfrac{3}{8}wL = \tfrac{5}{8}wL \tag{2.165}$$

From Eq. 2.63 it is seen that when the shear is equal to zero, $dM/dx = 0$; therefore the peak plus moment is at the location of zero shear or at $x = \tfrac{3}{8}L$, and is equal to

$$M_x = \tfrac{3}{8}wL(\tfrac{3}{8}L) - (w/2)(\tfrac{3}{8}L)^2 = (9/128)wL^2 \tag{2.166}$$

The maximum moment is therefore at the fixed end,

$$M_{max} = wL^2/8 \tag{2.167}$$

The general deflection curve is then

$$y = (wx/48EI)(3Lx^2 - 2x^3 - L^3) \tag{2.168}$$

And the maximum deflection is at $dy/dx = 0$ or at $x = 0.4215L$, so that

$$y_{max} = 0.0054wL^4/EI \tag{2.169}$$

BEAM WITH FIXED ENDS CARRYING A CONCENTRATED LOAD (Fig. 2.38). For this case Eq. 2.55 shows that $n = 3$. Since both ends of the beam are fixed, there are six reactions; consequently, the problem is statically indeterminate to the $6 - 3$ or 3rd degree.

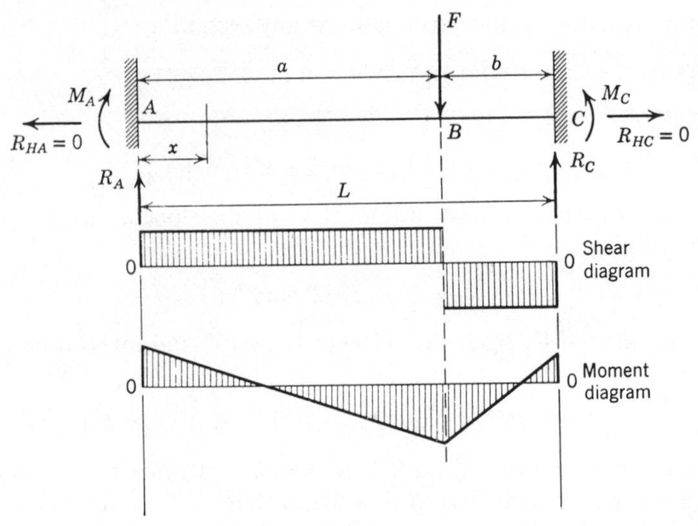

Fig. 2.38 Statically indeterminate structure.

For section AB

$$EI\, d^2y/dx^2 = M = R_A x + M_A \tag{2.170}$$

Integrating twice gives

$$dy/dx = 1/EI(R_A x^2/2 + M_A x + C_1) \tag{2.171}$$

and

$$y = 1/EI(R_A x^3/6 + M_A x^2/2 + C_1 x + C_2) \tag{2.172}$$

where C_1 and C_2 are constants of integration.

For section BC

$$EI\, d^2y/dx^2 = M = R_A x + M_A - F(x - a) \tag{2.173}$$

Integrating twice gives

$$dy/dx = (1/EI)(R_A x^2/2 + M_A x - Fx^2/2 + Fax + C_3) \tag{2.174}$$

and

$$y = (1/EI)(R_A x^3/6 + M_A x^2/2 - Fx^3/6 + Fax^2/2 + C_3 x + C_4) \tag{2.175}$$

where C_3 and C_4 are constants of integration.

The boundary conditions of the problem are as follows:

$$y = 0 \quad \text{at} \quad x = 0, L$$
$$dy/dx = 0 \quad \text{at} \quad x = 0, L$$
$$y_{AB} = y_{BC} \quad \text{at} \quad x = a$$
$$(dy/dx)_{AB} = (dy/dx)_{BC} \quad \text{at} \quad x = a$$

Therefore using these conditions in Eqs. 2.173 to 2.177,

$$C_1 = C_2 = 0 \quad 2.176$$

$$C_3 = FL/2 - FaL - MaL - R_A L^2/2 \quad 2.177$$

$$C_4 = R_A L^3/3 + M_A L^2/2 - FL^3/3 + FaL^2/2 \quad 2.178$$

from which

$$M_A = Fab^2/L^2 \quad 2.179$$

$$M_C = Fa^2b/L^2 \quad 2.180$$

$$R_A = (Fb^2/L^3)(3a + b) \quad 2.181$$

$$R_C = (Fa^2/L^3)(3b + a) \quad 2.182$$

$$M_B = 2Fa^2b^2/L^3 \quad 2.183$$

The maximum value of M_B occurs at $a = L/2$ and equals $FL/8$.

$$y_{AB} = (Fb^2x^2/6EIL^3)(3ax + bx - 3aL) \quad 2.184$$

$$y_{BC} = (Fa^2/6EIL^3)(L - x)^2[(3b + a)(L - x) - 3bL] \quad 2.185$$

$$(y_{max})_{a>b} = (2F/3EI)a^3b^2/(3a + b)^2 \quad \text{at} \quad x = 2aL/(3a + b) \quad 2.186$$

$$(y_{max})_{a<b} = (2F/3EI)a^2/b^3/(3b + a)^2 \quad \text{at} \quad x = L - 2bL/(3b + a) \quad 2.187$$

CONTINUOUS BEAMS. Continuous beams, that is, beams resting on more than *two* supports, are widely used in structural engineering applications. These beams are generally more efficient than other types since, for a given load, the induced moments are less and greater economy of material results.

Beam on three supports (Fig. 2.39) Consider a beam consisting of two spans and a single concentrated load. There are no internal hinges. Application of Eq. 2.55 thus indicates that

$$n = 3 + 0 = 3$$

independent equations available. At the two end supports there are two force reactions and at the intermediate support there is one vertical force reaction

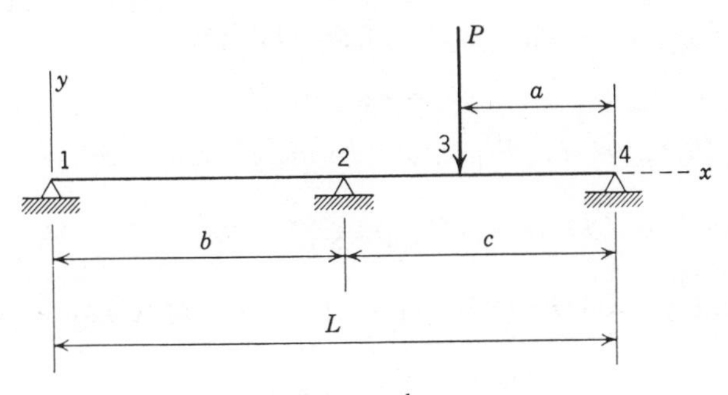

Fig. 2.39 Beam on three supports.

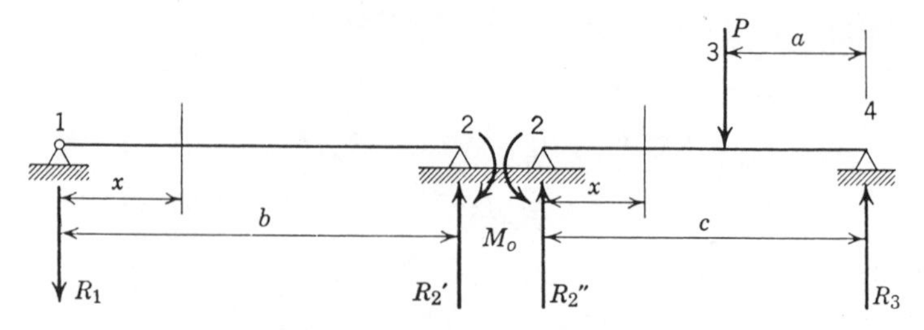

Fig. 2.40 Equilibrium of beam in Fig. 2.39.

and one bending-moment reaction; therefore the degree of indeterminacy is

$$r - n = 6 - 3 = 3$$

The horizontal forces are considered zero; therefore the situation reduces to one degree of indeterminacy. To solve the problem the beam is split into two parts as shown in Fig. 2.40.

For section 1–2

$$EI\, d^2y/dx^2 = M = R_1 x \tag{2.188}$$

By statics, $R_1 b + M_0 = 0$ from which

$$R_1 = -M_0/b \tag{2.189}$$

Substituting Eq. 2.189 into Eq. 2.188 and integrating twice gives the results

$$\text{slope} = \theta = dy/dx = (1/EI)[(-Mx^2/2b) + C_1] \tag{2.190}$$

and

$$\text{deflection} = y = (1/EI)[(-Mx^3/6b) + C_1 x + C_2] \tag{2.191}$$

where C_1 and C_2 are constants of integration. From the boundary condition that the deflection is zero at the supports it follows that

$$C_1 = M_0 b/6 \quad \text{and} \quad C_2 = 0 \tag{2.192}$$

For section 2–3

$$EI\, d^2y/dx^2 = M = (R_2'')x - M_0 \tag{2.193}$$

Again, by statics, $R_2''c - Pa - M_0 = 0$ from which

$$R_2'' = Pa + M_0/c \tag{2.194}$$

Substituting Eq. 2.194 into Eq. 2.193 and integrating twice gives the following results:

$$\text{slope} = \theta = dy/dx = (1/EI)(Pax^2/2c + M_0 x^2/2c - M_0 x + C_3) \tag{2.195}$$

and

$$\text{deflection} = y = (1/EI)(Pax^3/6c + M_0 x^3/6c - M_0 x^2/2 + C_3 x + C_4) \tag{2.196}$$

where C_3 and C_4 are constants of integration. Using the same boundary conditions as mentioned it is seen that $C_4 = 0$. The constant C_3 cannot be evaluated at this point.

For section 3–4

$$EI\, d^2y/dx^2 = M = R_2''x - M_0 - P(x - c + a) \qquad 2.197$$

Double integration of Eq. 2.197 gives the results

$$\text{slope} = \theta = dy/dx = (1/EI) \times (Pax^2/2c + M_0x^2/2c - M_0x - Px^2/2 + Pcx - Pax + C_5) \qquad 2.198$$

$$\text{deflection} = y = (1/EI) \times (Pax^3/6c + M_0x^3/6c - M_0x^2/2 - Px^3/6 + Pcx^2/2 - Pax^2/2 + C_5x + C_6) \qquad 2.199$$

Using the boundary condition $y = 0$ at $x = c$ and noting that at the point of load application the slopes of sections 2-3 and 3-4 are equal,

$$C_3 = -P(c - a)^2/2 + Pc(c - a) - Pa(c - a) + C_5 \qquad 2.200$$

Since the deflection at load P is also equal for both parts of the beam,

$$C_3 = -P(c - a)^2/6 + Pc(c - a)/2 - Pa(c - a)/2 + C_5 + C_6/(c - a) \qquad 2.201$$

and finally,

$$C_3 = P(-c^2/6 + ac/6 + c^2/6 + a^3/6c) + 2M_0c/3 \qquad 2.202$$

Now, noting that the slopes of sections 1-2 and 2-3 are equal at the intermediate support, Eq. 2.190 with $x = c$ is equated to Eq. 2.195 with $x = 0$, with the result

$$M_0 = (Pa/2cL)(c^2 - a^2) \qquad 2.203$$

Having the value of M_0 the other reactions can now be determined by statics, thus

$$R_1 = -M_0/b \qquad 2.204$$

$$R_2 = R_2' + R_2'' = M_0/b + (Pa + M_0)/c \qquad 2.205$$

$$R_3 = [P(c - a) - M_0]/c \qquad 2.206$$

For equilibrium, $R_1 + R_2 + R_3 = P$, which was satisfied previously. Having the moment and reactions, the stress and deflection for any point of the beam length can be determined by the usual methods already presented. The shear and moment diagrams for this beam for specific numerical values are shown in Fig. 2.41.

Beam on four supports (Fig. 2.42). Consider the beam shown in Fig. 2.42, which consists of three equal spans on four supports. There are no internal hinges, and Eq. 2.55 reveals that

$$n = 3 + 0 = 3 \qquad 2.207$$

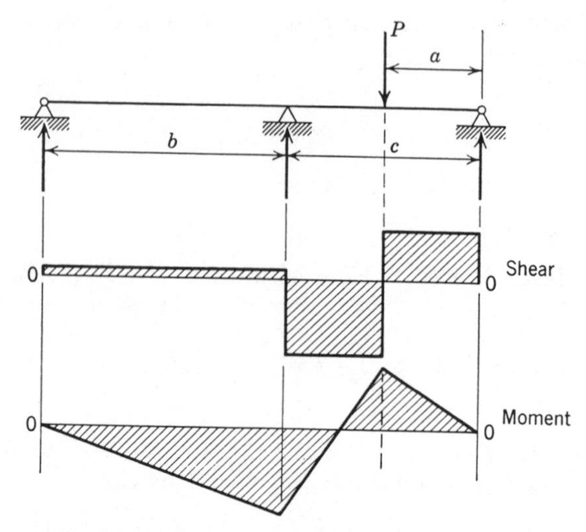

Fig. 2.41 Shear and moment diagrams for a beam on three supports.

independent equations are available. The horizontal reactions are considered to be zero; therefore there are four vertical reactions and one moment so that

$$r - n = 5 - 3 = 2 \tag{2.208}$$

which is seen to be the number of intermediate supports. Actually, for any continuous beam the degree of static indeterminacy is equal to the number of intermediate supports. In solving this problem the same procedure as used before will be applied.

For section 0–1

$$EI\, d^2y/dx^2 = M = R_0x - qx^2/2 \tag{2.209}$$

$$R_0 = qL/2 - M_0/L \tag{2.210}$$

$$dy/dx = (1/EI)(qLx^2/4 - M_0x^2/2L - qx^3/6 + C_1) \tag{2.211}$$

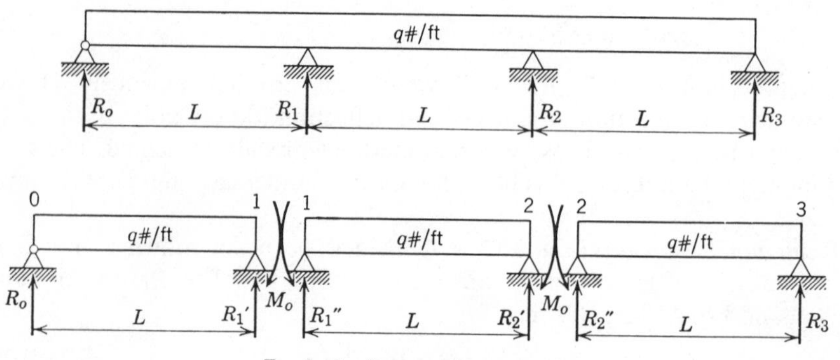

Fig. 2.42 Beam on four supports.

$$y = (1/EI)(qLx^3/12 - M_0x^3/6L - qx^4/24 + C_1x + C_2) \quad 2.212$$

$$y = 0 \quad \text{at} \quad x = 0; \text{therefore } C_2 = 0$$

$$y = 0 \quad \text{at} \quad x = L; \text{therefore } C_1 = -qL^3/24 + M_0L/6$$

For section 1–2

$$EI\, d^2y/dx^2 = M = (R_1''x - M_0 - qx^2/2) \quad 2.213$$

$$R_1'' = qL/2 \quad 2.214$$

$$dy/dx = (1/EI)(qLx^2/4 - M_0x - qx^3/6 + C_3) \quad 2.215$$

$$y = (1/EI)(qLx^3/12 - M_0x^2/2 - qx^4/24 + C_3x + C_4) \quad 2.216$$

$$y = 0 \quad \text{at} \quad x = 0; \quad \text{therefore } C_4 = 0$$

$$y = 0 \quad \text{at} \quad x = L; \quad \text{therefore } C_3 = -qL^3/24 + M_0L/2$$

At the supports the slopes are equal; therefore

$$qL^3/4 - M_0L/2 - qL^3/6 - qL^3/24 + M_0L/6 = qL^3/24 + M_0L/2 \quad 2.217$$

from which

$$M_0 = qL^2/10 \quad 2.218$$

It then follows that

$$R_0 = R_3 = \tfrac{4}{10}(qL) \quad 2.219$$

$$R_1 = R_1' + R_1'' = qL/2 + M_0/L + qL/2 = qL + M_0/L$$

or

$$R_1 = R_2 = \tfrac{11}{10}(qL) \quad 2.220$$

For equilibrium, $2R_1 + 2R_3 = 3qL$, which is satisfied by the preceding values.

These examples illustrate the principle of calculating the indeterminate reactions by making use of common slope and deflection properties for various parts of the beam. This same procedure can be applied to any number of supports and for all types of loading as well as either equal or unequal span lengths. It is thus evident that, although the problem can be solved, the mathematics becomes quite cumbersome for the general case. It can be easily shown, however, by the moment-area method that for any type of loading the indeterminate moments are related by the following expression (23):

$$M_{n-1}L_n + 2M_n(L_n + L_{n+1}) + M_{n+1}L_{n+1} = -6A_na_n/L_n - 6A_{n+1}b_{n+1}/L_{n+1} \quad 2.221$$

where A_n is the area of the bending-moment diagram for the span n and a_n and b_n are the horizontal distances of the centroid of the moment area from the supports $n - 1$ and n. Equation 2.221 is known as the *equation of three moments.* For any particular problem there are as many of these equations as there are intermediate supports. To illustrate the use of this equation consider Fig. 2.43 which shows the shear and bending-moment diagrams for the beam in

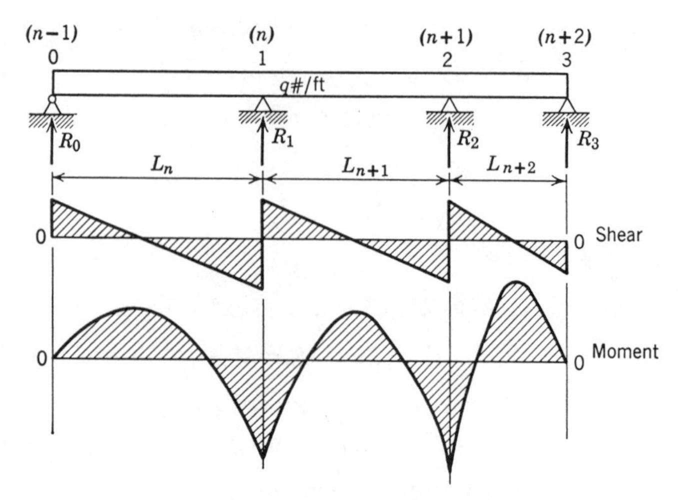

Fig. 2.43 Shear and moment diagrams for a beam on four supports.

Fig. 2.42. From Eq. 2.68 it is seen that the maximum moment for a distributed load is $qL^2/8$; therefore for this case the area of the parabola is

$$A_n = (qL^2/8)(\tfrac{2}{3})(L) = qL^3/12 \tag{2.222}$$

By symmetry, a_n and b_n are both equal to $L/2$; therefore application of Eq. 2.221 gives the result

$$(0)(L) + 2M_1(2L) + M_2(L) = -(6qL^3/12)(L/2)(L) - (6qL^3/12)(L/2)(L)$$

or

$$4M_1L + M_2L = -qL^3/2$$

By symmetry

$$M_1 = M_2$$

therefore

$$M_1 = M_2 = -qL^2/10 \tag{2.223}$$

Shear and moment diagrams for a uniformly loaded beam on six supports are shown in Fig. 2.44. Data and analyses for partially loaded continuous beams are presented in (9 and 15).

Method of moment distribution. The method of three moments just discussed is impractical for many structural engineering applications because of the number of simultaneous equations involved in highly indeterminate systems. A much shorter method is known as the *method of moment distribution.* This method is based on successive approximations following the procedure outlined below.

1. Assume that at each support the beam is rigidly fixed so that it cannot rotate. At each end of each span the fixed-end moment is then calculated.
2. Remove the restraint at each support, one at a time, permitting rotation; permit rotation until the moments balance.

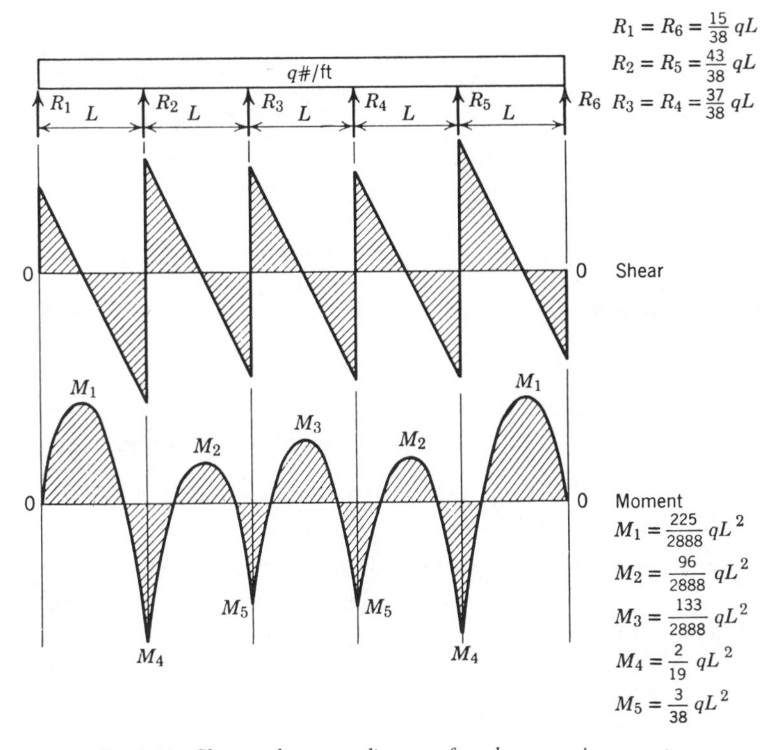

Fig. 2.44 Shear and moment diagrams for a beam on six supports.

3. Repeat step 2 until the slope of the beam at each support has its correct value.

In step 1 the fixed-end moment for a uniformly loaded beam clamped at both ends, for example, is $-wL^2/12$, where w is the unit load and L is the span length. In step 2 the two adjoining beams will seldom have the same moment; therefore, when restraint is removed, the beams act like levers to absorb the unbalance in moment at any support; the unbalance, if any, is distributed in proportion to the relative stiffness (I/L) of the sections. Since each beam section is considered fixed at its far end when the process of restraint removal begins, it is seen that any change in moment at one end will affect the moment at the other end. The magnitude of the change in moment is a function of the shape of the beam. For a prismatic beam, Eq. 2.53 shows that this *carry-over moment* is one-half the moment at the opposite end.

Consider Fig. 2.45. A moment M applied at B produces moments M_1 at A and M_2 at C. By use of either Eq. 2.79 or the moment-area method it can be shown that the relationship between moments M_1 and M_2 is

$$M_1/M_2 = (I_1/L_1)(L_2/I_2) = k_1/k_2 \qquad 2.224$$

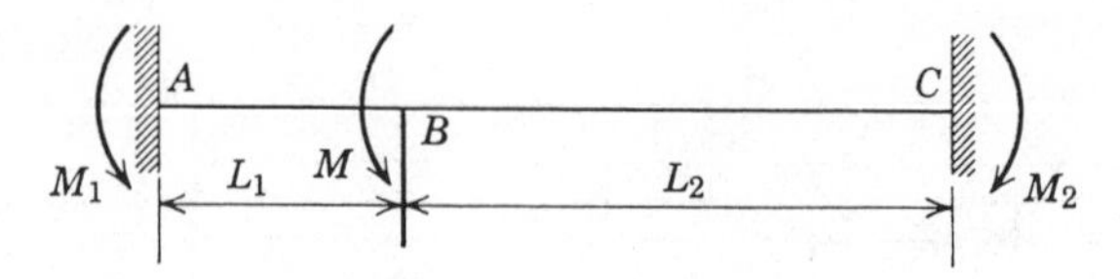

Fig. 2.45 End moments in a beam.

where $k_1 = I_1/L_1$ and $k_2 = I_2/L_2$. These values of k_1 and k_2 are called *stiffness values.* In addition to the moments at A and C_1, moments M'_1 and M'_2 are produced by M from A to B and B to C respectively. Equilibrium shows that

$$M_1 = \alpha_1 M'_1 \qquad \text{and} \qquad M_2 = \alpha_2 M'_2 \tag{2.225}$$

where α_1 and α_2 are the carry-over factors mentioned earlier. Finally, M'_1 and M'_2 can be expressed in terms of M and the stiffness values as follows:

$$M'_1 = Mk_1/(k_1 + k_2) \qquad \text{and} \qquad M'_2 = Mk_2/(k_1 + k_2) \tag{2.226}$$

The quantities $k_1/(k_1 + k_2)$ and $k_2/(k_1 + k_2)$ are called *distribution factors.*

In order to indicate the use of this method of analysis the continuous beam shown in Fig. 2.42 will be considered. This beam was previously calculated by the method of three moments, and the unknown quantities are given by Eqs. 2.218 to 2.220. For the following analysis by the moment-distribution method, refer to Fig. 2.46 which shows the beam of Fig. 2.42 with a distributed load of 2 kip/ft on a beam of uniform moment of inertia, supported at four locations 20 ft apart. The moments at the outer ends of the beam (A and D) are given on line 1 of Fig. 2.46 as zero. At supports B and C the moment is $wL^2/8$ giving values of -100 and $+100$ respectively. At supports B_1 and C_1 the moment is given by $wL^2/12$, giving values of $+66.67$ and -66.67 respectively. For convenience the *stiffness* of the free-end spans is taken as $0.75\ I/L$ [see (15, p. 71)] so that the stiffness factors for the three spans are 0.75, 1.00, and 0.75, respectively, giving distribution factors at supports B and C of 0.43 and 0.57 respectively. The unbalanced moment over support B of 33.33 is distributed as 14.3 and 19 on line 4, Fig. 2.46. Now all joints are in balance and the summation on line 5 is an approximation of the actual results. The "individual beams" are in unbalance, however, because of induced moments on their outer ends; therefore it is necessary to carry over one-half of the moments as shown in line 7. The process is repeated through two more distributions, giving the final result (considered close enough) of 79.14 to 80.7, which compares with the value calculated from Eq. 2.218 of 80.

It is seen that reliable results may be obtained after two or three carry-over cycles; these carry overs show the degree of precision obtained and the process is terminated when the values reach low order. This is no problem when a computer is used so that the cycling can be carried out many times.

Beam with moving load (Fig. 2.47). For certain structural applications such as bridges, it is important to assess the stresses induced in a beam section by moving loads. Consideration is given here only to moving loads with given

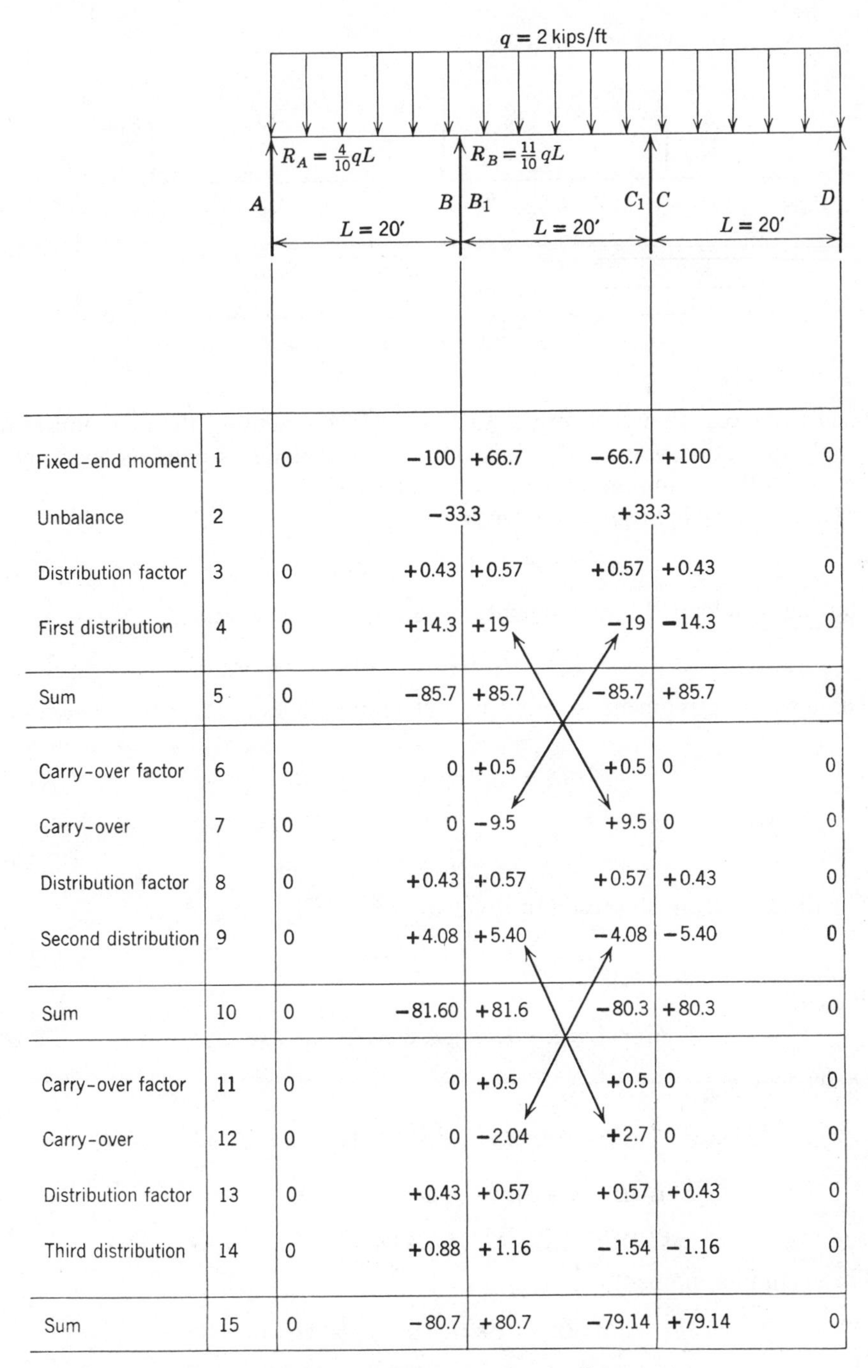

		A	B	B_1	C_1	C	D
Fixed-end moment	1	0	−100	+66.7	−66.7	+100	0
Unbalance	2		−33.3		+33.3		
Distribution factor	3	0	+0.43	+0.57	+0.57	+0.43	0
First distribution	4	0	+14.3	+19	−19	−14.3	0
Sum	5	0	−85.7	+85.7	−85.7	+85.7	0
Carry-over factor	6	0	0	+0.5	+0.5	0	0
Carry-over	7	0	0	−9.5	+9.5	0	0
Distribution factor	8	0	+0.43	+0.57	+0.57	+0.43	0
Second distribution	9	0	+4.08	+5.40	−4.08	−5.40	0
Sum	10	0	−81.60	+81.6	−80.3	+80.3	0
Carry-over factor	11	0	0	+0.5	+0.5	0	0
Carry-over	12	0	0	−2.04	+2.7	0	0
Distribution factor	13	0	+0.43	+0.57	+0.57	+0.43	0
Third distribution	14	0	+0.88	+1.16	−1.54	−1.16	0
Sum	15	0	−80.7	+80.7	−79.14	+79.14	0

Fig. 2.46 Moment distribution for a beam on four supports.

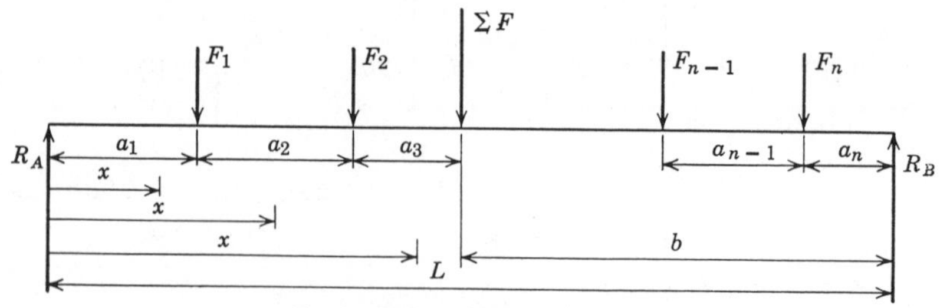

Fig. 2.47 Beam with moving loads.

fixed distances from each other. In Fig. 2.47, representing the situation of a simply supported beam with n loads, the moment diagram consists of straight lines with the maximum moment at one of the loads.

The total load in the beam shown is

$$\Sigma F = F_1 + F_2 + F_3, \text{ and so on} \qquad 2.227$$

Starting with load F_1, the moment is given by

$$M = R_A x = \Sigma F(L - x - a)x/L - F_1 a_1 \qquad 2.228$$

The maximum moment is found by differentiating Eq. 2.228; this operation gives

$$dM/dx = (\sum F/L)(L - 2x - a) = 0 \qquad 2.229$$

from which

$$x = (L - a)/2 \qquad 2.230$$

The distance from the resultant load to B is

$$b = L - x - a \qquad 2.231$$

or

$$b = L - a - L/2 + a/2 = (L - a)/2 = x \qquad 2.232$$

Under load F_n,

$$M = R_A x - F_1(a_2 + a_3 + a) - F_2(a + a_3) - \sum F(a) - (F_{n-1})a_n$$

or

$$M = (\sum F/L)(L - x + a)x - F_1(a + a_2 + a_3) - F_2(a + a_3) - \sum Fa - (F_{n-1})a_n \qquad 2.233$$

The maximum moment is

$$dM/dx = (\sum F/L)(L - 2x + a) = 0$$

or

$$x = (L + a)/2 \qquad 2.234$$

The distance b is

$$b = L - x + a = L - L/2 - a/2 + a = L/2 + a/2 = x \qquad 2.235$$

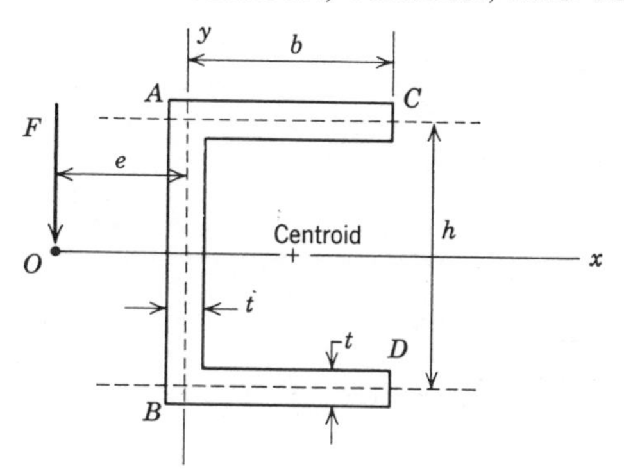

Fig. 2.48 Channel section.

as before. Therefore it is seen that the bending moment is maximum when the load and the resultant load are equidistant from the end supports.

Unsymmetrical bending. Up to this point in the discussion of bending problems it has been assumed that the load acts in a plane of symmetry of the beam cross section. In the more general case loads may act in planes parallel to symmetry axes, causing torsional as well as bending stresses to be set up; also the loads may be applied in a plane that is not a principal plane of bending. The former case is a simple case of combined stresses, which is considered in Chapter 3. However, there are cases such as in channel and elbow sections where loads are applied in planes parallel to a plane of symmetry in such a way that only bending occurs. This critical location of the loading plane relative to a principal plane of bending of the beam cross section is called the *shear center* or *center of twist*. Consider, for example, bending of the channel section shown in Fig. 2.48. For bending without twisting the load F must pass through point O, the shear center. It is assumed that the vertical shear is supported entirely by the web AB. In the flanges AC and BD there is a horizontal shear as shown in Fig. 2.49. By referring to Fig. 2.49, it is seen that

$$\sigma = M \int y\, dA / I_x \qquad 2.236$$

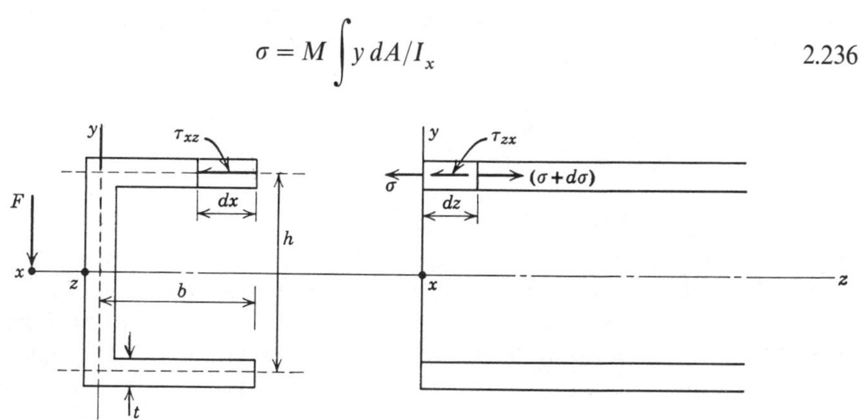

Fig. 2.49 Equilibrium in a channel section.

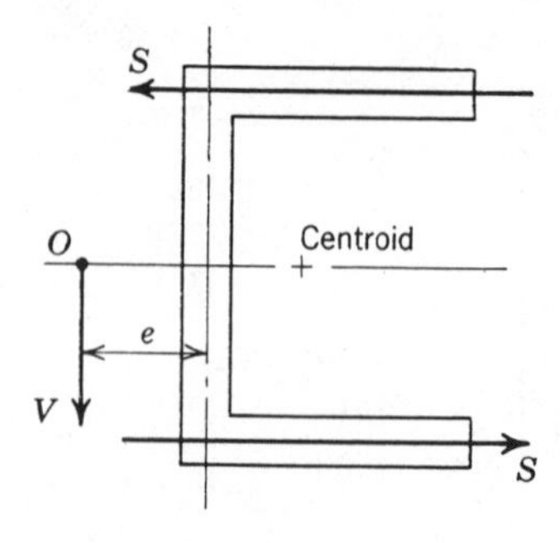

Fig. 2.50 Shear in a channel section.

and

$$\sigma + d\sigma = [M + (dM/dz)\,dz] \int y\,dA/I_x \qquad 2.237$$

The difference between σ and $\sigma + d\sigma$ equals the shear present, which, if assumed uniformly distributed, indicates

$$\tau_{zx}\,dzt = d\sigma = (dM/dz)(dz/I_x) \int y\,dA \qquad 2.238$$

from which

$$\tau_{zx} = [V/(tI_x)] \int y\,dA \qquad 2.239$$

Neglecting complications that arise at the junction of the web and flange and noting that the moment of the section bt with respect to the z axis is $bt(h/2)$, Eq. 2.239 shows that

$$\tau_{zx} = Vbh/2I_x \qquad 2.240$$

The force resulting from the shear τ_{zx} is

$$S = (Vbh/2I_x)(bt/2) = Vb^2ht/4I_x \qquad 2.241$$

The resultant forces S (Fig. 2.50) form a couple Sh which, together with Eq. 2.241, gives the result

$$Sh = Vb^2h^2t/4I_x$$

or

$$Sh/V = b^2h^2t/4I_x = e \qquad 2.242$$

Therefore, for elimination of torsion, the load must pass through O at a distance e from the centroid of the vertical member of the channel. By similar analyses the centers of shear for the sections shown in Fig. 2.51 may be obtained. Tabakman (22) has considered thin closed sections and shows how to shift the center of shear by varying the wall thickness or using two or more materials having different moduli.

Now consider the beam shown in Fig. 2.52, an obliquely loaded cantilever

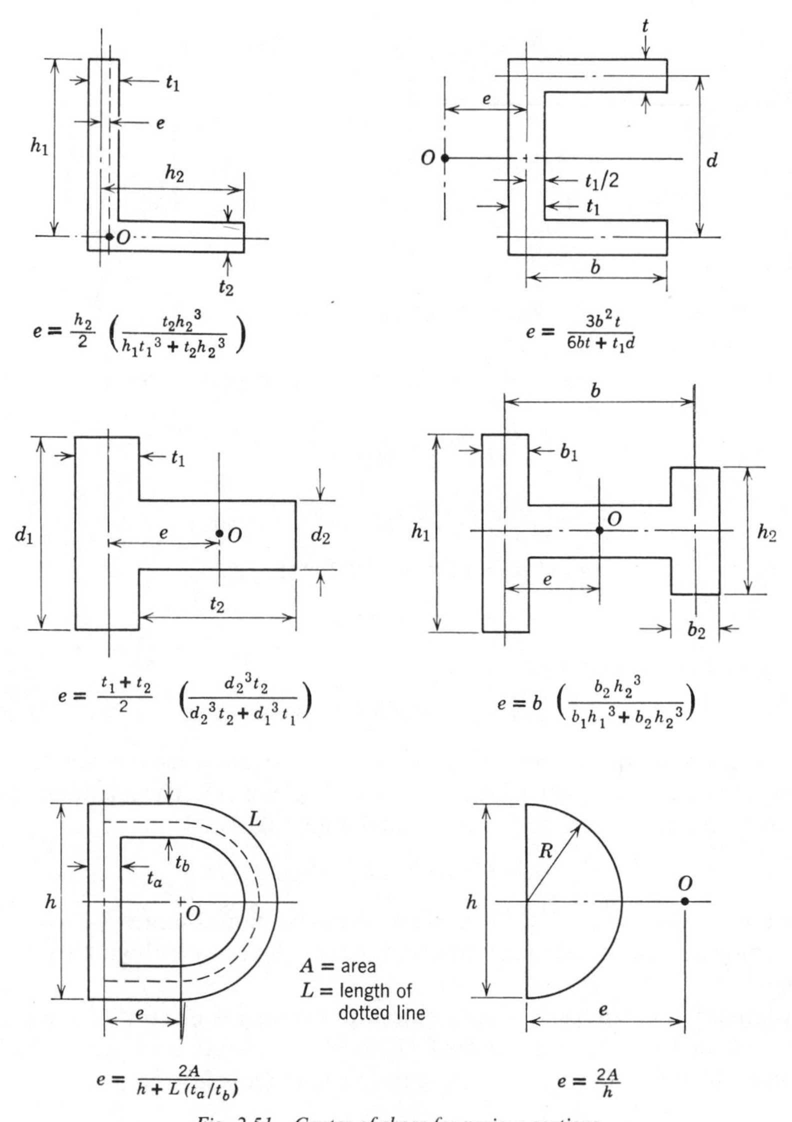

Fig. 2.51 Center of shear for various sections.

with the load plane passing through the center of shear of the cross section (which in this case is also the centroid of the symmetrical cross section). With the loading indicated there is no twist imparted to the beam.

The bending moment at distance x from the load F is Fx, which has two components

$$M_z = Fx(\cos\theta) \tag{2.243}$$

and

$$M_y = Fx(\sin\theta) \tag{2.244}$$

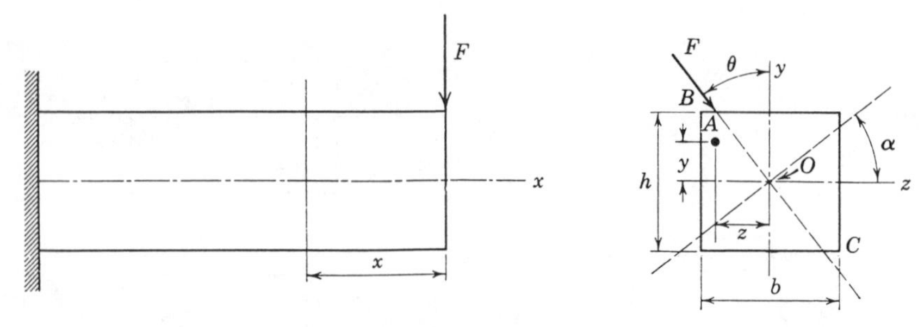

Fig. 2.52 Unsymmetrical bending.

Tensile stresses are produced at A, for example, for positive values of y and z; thus the stress is

$$\sigma_x = M_y z/I_y + M_z y/I_z \tag{2.245}$$

or

$$\sigma_x = Fx[(z \sin\theta)/I_y + (y \cos\theta)/I_z] \tag{2.246}$$

The neutral axis occurs at $\sigma_x = 0$; therefore for this value

$$y/z = -\tan\theta(I_z/I_y) \tag{2.247}$$

and this axis is defined by angle α,

$$\tan\alpha = -y/z = (I_z/I_y)\tan\theta \tag{2.248}$$

Examination of Eq. 2.248 reveals that α and θ are equal when the two moments of inertia are equal or when θ is zero or a multiple of 90°. The maximum bending stress occurs when z and y are a maximum, that is,

$$\sigma_{max} = M[(b \sin\theta)/2I_y + (h \cos\theta)/2I_z] \tag{2.249}$$

with the maximum tensile stress at B and the maximum compressive stress at C. Deflection may be found by superposing the two deflections about the principal planes.

Another important case of unsymmetrical bending is illustrated in Fig. 2.53. Here the load plane passes through the centroid of the cross section but not through the center of shear. To solve this problem it is first necessary to locate

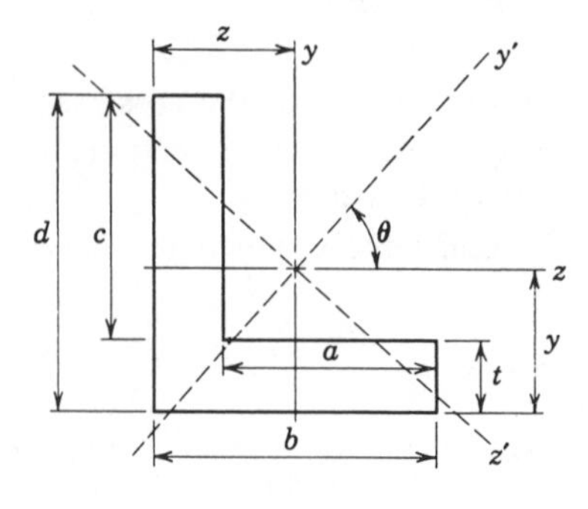

Fig. 2.53 Unsymmetrical section.

the principal axes (see Appendix B), which are indicated in Fig. 2.53 as y' and z'. Having the orientation of these axes it is then necessary to have values for the moments of inertia with respect to these axes so that Eq. 2.249 can be used. It is shown in the Appendix B that

$$I'_y = I_z \sin^2 \theta + I_y \cos^2 \theta + 2I_{yz} \sin \theta \cos \theta \qquad 2.250$$

$$I'_z = I_z \cos^2 \theta + I_y \sin^2 \theta - 2I_{yz} \sin \theta \cos \theta \qquad 2.251$$

where I_y and I_z are moments of inertia about the centroid and are defined by

$$I_y = [t(b - z)^3 + dz^3 - c(z - t)^3]/3 \qquad 2.252$$

$$I_z = [t(d - y)^3 + by^3 - a(y - t)^3]/3 \qquad 2.253$$

and I_{yz} is the product of inertia given by

$$I_{yz} = (I_z \cos^2 \theta - I_y \sin^2 \theta)/\cos 2\theta \qquad 2.254$$

Having I'_y and I'_z Eq. 2.249 can be utilized. It is important to note here that the load does not pass through the center of shear; therefore, in addition to bending stresses, torsional stresses will also be induced. If the load passes through the center of shear, no twisting occurs.

As another example consider the section shown in Fig. 2.54. For this particular geometry the orientation of the principal axis z_1 is at 17° 20′ with respect to the z axis (23). Here the load passes through the center of twist so that no rotation of the section occurs. However, since the load does not occur in a

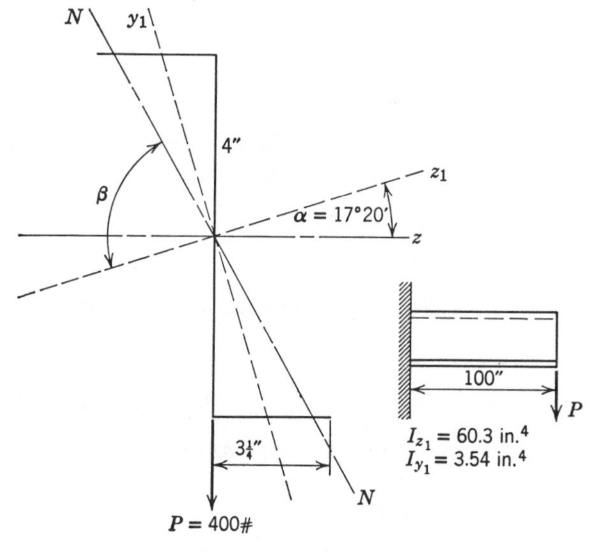

Fig. 2.54 Z-section.

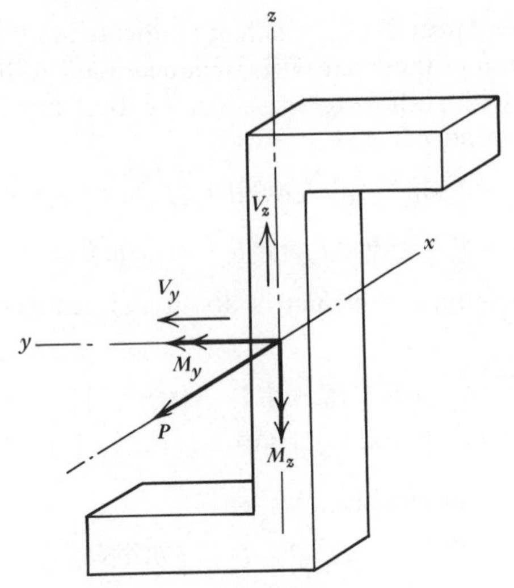

Fig. 2.55 Beam sign convention.

principal plane, there will be both horizontal and vertical deflections. For the case shown by Eq. 2.249.

$$\sigma_{max} = 400(100)[4(0.954)^2/60.3 + 4(0.298)^2/3.54]$$

or

$$\sigma_{max} = 6420 \text{ psi}$$

The deflection in the z_1 direction is

$$\delta_{z_1} = \frac{P_{z_1}L^3}{3EI_{y_1}} = \frac{400(0.298)(100)^3}{(3)(3.54)(30 \times 10^6)}$$
$$= 0.376 \text{ in.}$$

The deflection in the y_1 direction is

$$\delta_{y_1} = \frac{P_{y_1}L^3}{3EI_{z_1}} = \frac{400(0.954)(100)^3}{(3)(60.3)(30 \times 10^6)}$$
$$= 0.070 \text{ in.}$$

Therefore the vertical deflection is

$$\delta_y = 0.07(0.954) + 0.376(0.298) = 0.178 \text{ in.}$$

The horizontal deflection is

$$\delta_z = 0.07(0.298) - 0.376(0.954) = 0.337 \text{ in.}$$

Another unsymmetrical bending equation form (16, 20) can be applied to the example in Fig. 2.54.

$$\sigma_{xx} = \frac{P}{A} - \frac{M_z I_{yy} - M_y I_{yz}}{I_{yy} I_{zz} (I_{yz})^2} y - \frac{M_y I_{zz} - M_z I_{yz}}{I_{yy} I_{zz} (I_{yz})^2} z \qquad 2.255$$

The positive sign conventions are shown in Fig. 2.55. The deflections in the y and z direction are

$$\frac{d^2v}{dx^2} = \frac{M_z I_{yy} - M_y I_{yz}}{E[I_{yy} I_{zz} - (I_{yz})^2]} \tag{2.256}$$

$$\frac{d^2w}{dx^2} = \frac{M_y I_{zz} - M_z I_{yz}}{E[I_{yy} I_{zz} - (I_{yz})^2]} \tag{2.257}$$

These equations eliminate the need for Mohrs circle and allow working from a rough sketch. However, care must be taken with sign conventions.

In the example in Fig. 2.54 the following values are:

$$I_{yy} = 55.28 \text{ in.}^4 \qquad I_{zz} = 8.56 \text{ in.}^4$$
$$M_y = -4 \times 10^4 \text{ in.-lb} \qquad M_z = 0$$
$$I_{yz} = -16.12 \text{ in.}^4 \qquad P = 0$$

which are substituted into Eq. 2.255 yielding

$$\sigma_{xx} = -16.115y - 8.2z = 0$$

When σ_{xx} is zero the equation for the neutral axis is defined

$$z = -1.883y$$

The farthest point from the neutral axis is when z equals four and y is zero, then

$$\sigma_{xx} = -\frac{(-4)(10^4)(8.56)}{[55.28(8.56) - (-16.12)^2]} \times 4$$

$$\sigma_{xx} = 6420 \text{ lb/in.}^2$$

for the deflection Eqs. 2.256 and 2.257

$$My = -400x \qquad M_z = 0 \qquad x = 100 \text{ in.}$$

$$\frac{d^2v}{d^2x} = \frac{-(-400x)(-16.12)}{E[55.28(8.56) - (16.12)^2]}$$

$$v = -\left(\frac{400x^3}{3}\right)\left(\frac{16.12}{213.46E}\right) = -0.336 \text{ in.}$$

$$\frac{d^2w}{dx^2} = \frac{(-400x)(8.56)}{(213.46)E}$$

$$w = \left(-\frac{400x^3}{3}\right)\left(\frac{8.56}{213.46E}\right) = 0.178 \text{ in.}$$

When the cross section is formed of several metals or materials, methods in Chapter 4 must be applied.

Bending-moment deflection and strain curves. A common method used to show bend data is a plot of applied bending moment versus the deflection, usually the maximum deflection. Figure 2.56, for example, shows the moment-

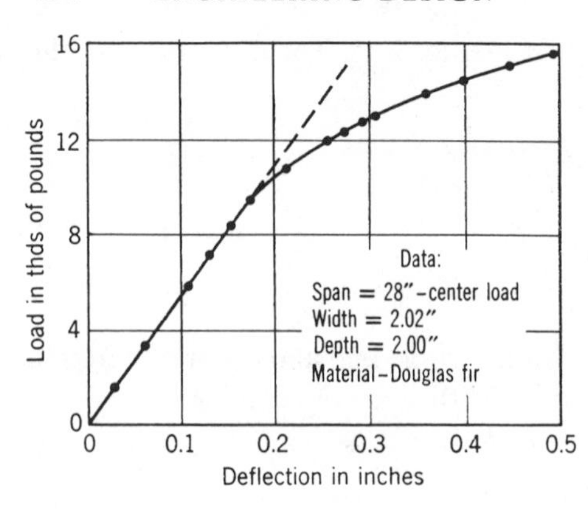

Fig. 2.56 Load-deflection plot for a wood beam. [After Marin (12), courtesy of Prentice-Ha ll.]

deflection curve for a section of Douglas Fir. From a plot of this type an examination of deflection can be made to determine if specifications for a particular material are met.

Bending-moment deflection curves are somewhat difficult to analyze since the conditions of loading have a bearing on the results. For example, in determining mechanical properties by bending tests, two general procedures are used. One method uses a simply supported beam with a centrally applied load (Fig. 2.57). Here the maximum stress is under the load and is given by

$$\sigma = FL/4Z \qquad 2.258$$

in the elastic range. The deflection found by double integration of Eq. 2.258 is

$$y = (Fx/48EI)(3L^2 - 4x^2) \qquad 2.259$$

and the maximum value, which is plotted against σ in Eq. 2.258, is

$$y_{\max} = FL^3/48EI \qquad 2.260$$

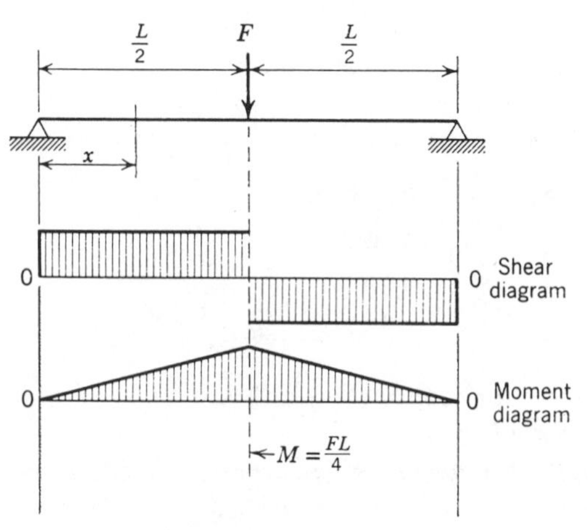

Fig. 2.57 Center-loaded beam.

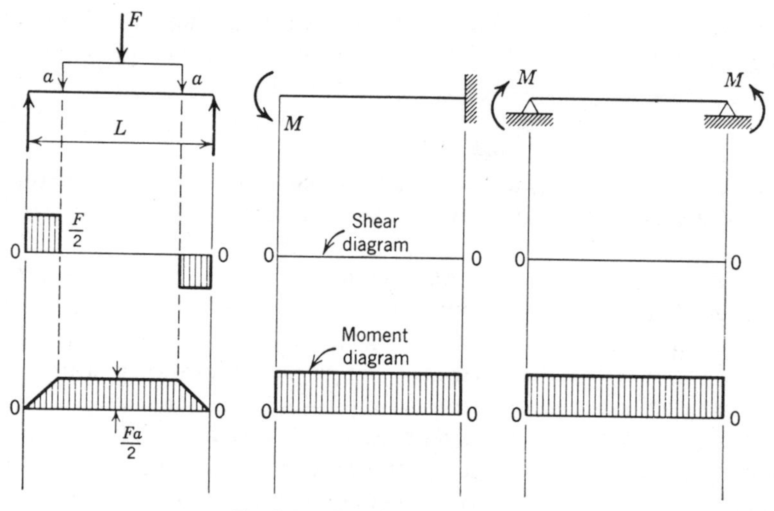

Fig. 2.58 Pure bending of beams.

A more fundamental bend test utilizes the condition of pure bending which can be obtained by the three methods shown in Fig. 2.58. In these procedures the bending moment, and hence the stress, is constant and the shear is zero.

On Fig. 2.56 Eq. 2.260 can be applied only in the range represented by the linear portion of the plot where Hooke's law applies; beyond this range only moment versus deflection can be plotted. On a basis comparable to tension and torsion loading, however, the yield strength in bending is frequently defined by an *offset* deflection as shown in Fig. 2.59, and the yield strength defined as

$$\sigma_y = M_y/Z \qquad 2.261$$

where M_y is the value of the moment at the point of intersection of the offset

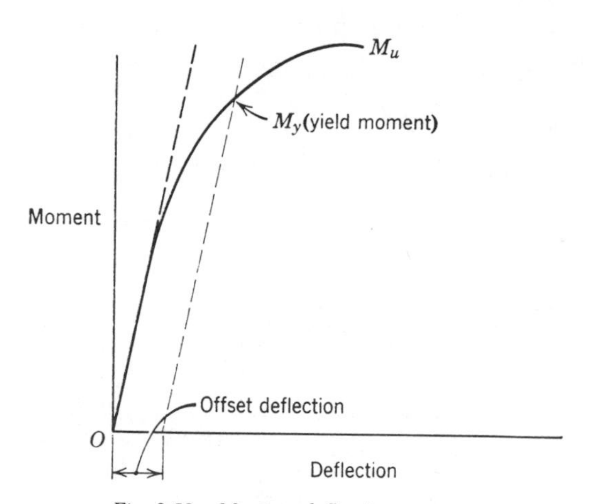

Fig. 2.59 Moment-deflection curve.

parallel with the moment-deflection curve in the elastic range. This is a fictitious value since the formula $\sigma = M/Z$ is valid only in the elastic range. Similarly, the breaking strength or *modulus of rupture* is defined as

$$\sigma = M_u/Z \qquad 2.262$$

where M_u is the bending moment at the time of breaking. Again, this is a fictitious value, but it is frequently used as a measure of the bending strength of materials.

The center load method, although frequently used, seems to be less reliable when applied to materials of low ductility like ceramics. The reason for this is that in the center load test, the maximum stress is located at one cross section, and thus the strength of the material at that particular location is critical. Many brittle materials are not uniformly strong throughout their length, and it is therefore easy to see how quite erratic results could be obtained by this method of evaluation. In addition, brittle materials are more sensitive to stress concentration effects than ductile materials and this condition is magnified in the center load test at the location of maximum stress by the medium transmitting the load to the specimen.

Referring to Fig. 2.60, for example, elastic theory (25) shows that the stress along line AB is approximately

$$\sigma = (F/I)(L/2 - c/\pi)y + F/2\pi c \qquad 2.263$$

Application of Eq. 2.263 to point B shows that the load F required to break the bar is about 2 to 5% greater than calculated from the conventional formula

$$\sigma = M/Z \qquad 2.264$$

The modulus of elasticity is sometimes determined from the bend test; its value is calculated on the basis of the theoretical deflection equation and by assuming Hooke's law. The values obtained do not give the exact modulus, but they are of value in indicating the stiffness in bending for comparative purposes. For a simply supported beam with center load, the modulus of elasticity for round bars is thus

$$E = FL^3/48I_y = \sigma L^2/6\,dy \qquad 2.265$$

where y is the center deflection and d the diameter of the bar. For a rectangular

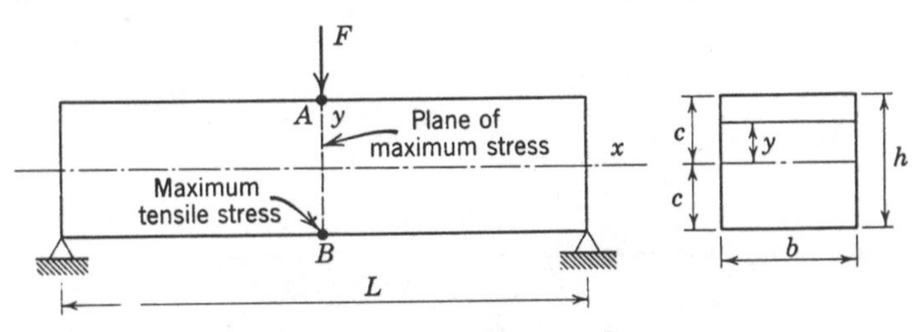

Fig. 2.60 Local stresses in a beam.

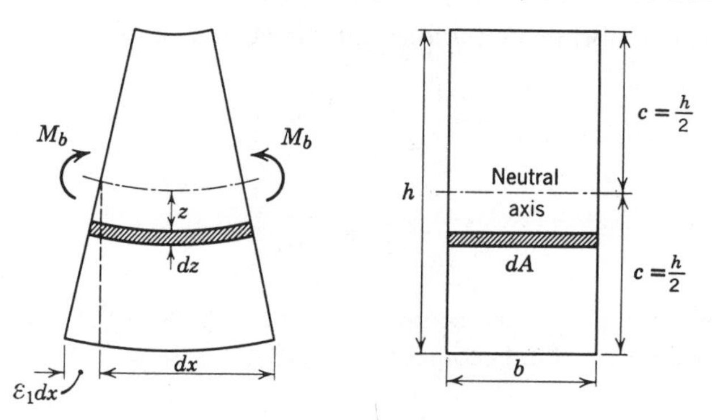

Fig. 2.61 Element of beam in bending.

section

$$E = \sigma b L^2 / 6\, dy \qquad 2.266$$

where d is the depth of the beam and b the width.

Moment-deflection curves are quite useful; however, there are advantages to be gained from having data in the form of tensile stress-strain relationships. Deflection data can be analyzed in terms of tensile strain, however, which will now be considered.

In the following analysis two equations are derived for determining, from a moment-curvature diagram of a straight beam, the stress-strain relation for the most-stressed fibers in tension and compression; it is assumed that the stress-strain relationship for the material is different in tension and compression. In addition the assumption is made that the strains of the longitudinal fibers of the beam are proportional to the distance of the fibers from the neutral surface for both elastic and plastic strains (7). The equation expressing the relation between the stress and the strain for the center fibers in the special case of a beam of rectangular cross section, and made of material for which the tensile and compression stress-strain relations are assumed to be identical, may be obtained as follows (14).

In Fig. 2.61 let the width of the rectangular cross section be b, the depth h, and $c = h/2$. The stress and strain in the outer fibers at any cross section are denoted by σ_1 and ε_1, respectively, and the bending moment at the section is denoted by M_b. At any distance z from the neutral axis the strain is ε and the corresponding stress is σ, where Z is a function of ε, thus

$$\sigma = f(\varepsilon) \qquad 2.267$$

Similarly, for the other fibers,

$$\sigma_1 = f(\varepsilon_1) \qquad 2.268$$

The desired relation between $\sigma_1, \varepsilon_1, M_b$, and the dimensions of the beam may be obtained by using Eqs. 2.267 and 2.268 together with the conditions of equilibrium. In terms of the symbols used in this analysis, Eqs. 2.131, and 2.132, denoting the equilibrium conditions, can be written as follows:

$$\int_{-c}^{c} \sigma \, dA = 0 \qquad 2.269$$

and

$$\int_{-c}^{c} \sigma z \, dA = M_b \qquad 2.270$$

Since the cross section of the beam is symmetrical about the neutral axis, and since $dA = b\,dz$, Eq. 2.270 becomes

$$\int_{0}^{c} \sigma z \, dz = \frac{M_b}{2b} \qquad 2.271$$

The values of z and dz in Eq. 2.271 can be expressed in terms of strain since the strain in any fiber can be obtained in terms of z, c, and ε_1. This relation is

$$z/c = \varepsilon/\varepsilon_1 \qquad 2.272$$

from which

$$dz = c\, d\varepsilon/\varepsilon_1 \qquad 2.273$$

By substituting these two equations in Eq. 2.271, the following equation is obtained:

$$\int_{0}^{\varepsilon_1} \sigma\varepsilon \, d\varepsilon = \frac{\varepsilon_1^{\,2} M_b}{2bc^2} \qquad 2.274$$

By using Eq. 2.267 and letting $F(\varepsilon) = \varepsilon f(\varepsilon)$, Eq. 2.274 may be written in the form

$$\int_{0}^{\varepsilon_1} F(\varepsilon) \, d\varepsilon = \frac{\varepsilon_1^{\,2} M_b}{2bc^2} \qquad 2.275$$

The function $F(\varepsilon)$ is not known and hence the preceding expression cannot be solved directly by integration. A solution can be obtained, however, by making use of the derivative of Eq. 2.275 with respect to ε_1, which may be written as (14)

$$\frac{\partial}{\partial \varepsilon_1} \int_{0}^{\varepsilon_1} F(\varepsilon) \, d\varepsilon = \frac{\partial}{\partial \varepsilon_1} \left(\frac{\varepsilon_1^{\,2} M_b}{2bc^2} \right) \qquad 2.276$$

from which

$$\sigma_1 = \frac{2M_b + \varepsilon_1 \, \partial M_b/\partial \varepsilon_1}{2bc^2} \qquad 2.277$$

Equation 2.277 is the desired equation since all the quantities on the right-hand side can be easily determined from a test of a beam in which a curve giving the relation between the bending moment M_b and the strain ε_1 in the outer fibers is found. The equation is applicable only for a rectangular cross section and for materials having the same tension and compression stress-strain relations. The original equations derived by Herbert for the general case are (7):

In tension

$$\sigma_1 = \frac{2M_b + k\,\partial M_b/\partial k}{bh\,\partial \varepsilon_1/\partial k} \qquad 2.278$$

In compression

$$\sigma_2 = \frac{2M_b + k\,\partial M_b/\partial k}{bh\,\partial \varepsilon_2/\partial k} \qquad 2.279$$

where k is the curvature, σ_1 and ε_1 are the stress and the strain, respectively, in the most-stress tension fiber, and σ_2 and ε_2 are the stress and strain, respectively, for the most-stressed compression fiber.

By an analysis similar to that given, a relation can be found between the shearing stress τ in the most-stressed fibers of a solid cylindrical torsion member, the torque T, angular twist θ per unit length, and the radius of the bar r. The relation is

$$\tau = \frac{3T + \theta\,\partial T/\partial\theta}{2\pi r^3} \qquad 2.280$$

Interpretation of moment-strain diagram. The analysis given here follows that of Morkovin (14). To start with, consider Fig. 2.62 a tensile stress-strain diagram for steel that exhibits a fairly well-defined deviation from a straight line at A, the ordinate which may be designated as the yield stress. To obtain a somewhat more reproducible and significant test value, the yield stress corresponding to a small permanent set is used as the yield stress. Thus in Fig. 2.62 the 0.0001 offset defines a yield stress at D. For a beam of the same material the relation between bending moment at a given section and the strain in the outer fibers at the section is shown in Fig. 2.63. This diagram does not exhibit an abrupt departure from the straight line OA; hence it is difficult to determine a yield stress experimentally. The beginning of yielding in the outer fibers occurs at a bending moment M_y corresponding to some point A, but the curve

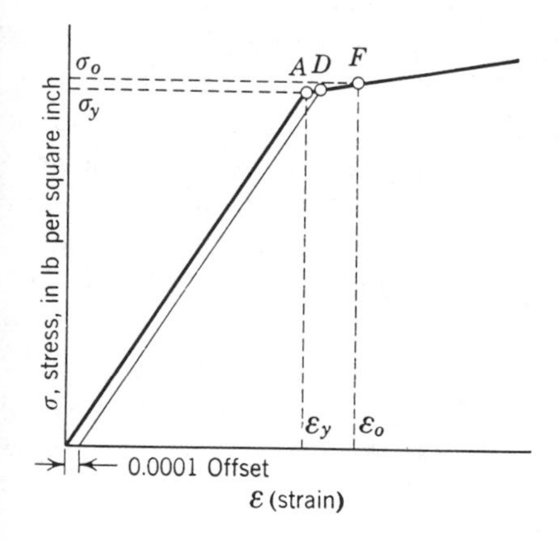

Fig. 2.62 Tensile data.

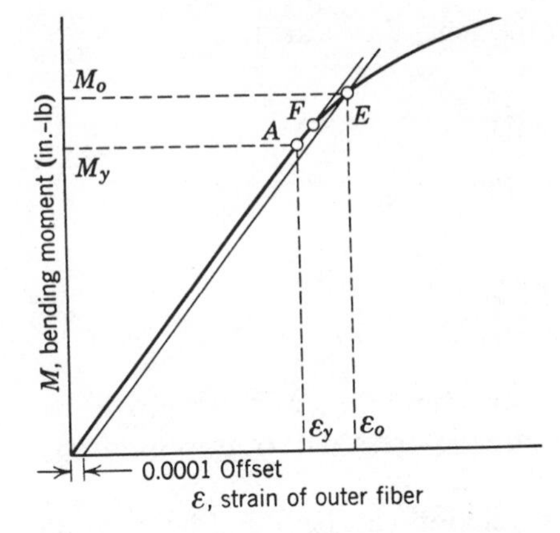

Fig. 2.63 Bend test data.

does not deviate appreciably from a straight line until a bending moment M_0 is reached that is considerably greater than M_y.

By analogy with the tensile test some investigators have considered the material in a beam to act elastically until the moment M_0 is reached since the deviation from the straight line at M_0 is the same as in the tension test (0.0001). Furthermore, the material is considered to act elastically up to this load, making the flexure formula $\sigma = M/Z$ applicable. However, as the bending moment increases from M_y to M_0, the stresses in the fibres change as illustrated in Fig. 2.63. M_0 is greater than M_y not because the elastic stress in the outer fiber increases above σ_y but because the inner understressed fibers are made to resist a greater proportion of the load. The important difference between Figs. 2.62 and 2.63 is that, for identical strains, the ordinates have very different meanings with respect to stress since the ordinates in Fig. 2.63 are not proportional to the stress in the extreme fiber for moments greater than M_y and hence the two diagrams cannot be analyzed in the same manner.

For example, in Fig. 2.63, the stress computed as M_0/Z for the 0.001 offset was found to be 69,000 psi as compared with the actual value of 58,500 psi for the actual yield stress corresponding to point *A*. The error in interpreting the ordinate to the point *E* in Fig. 2.63 as the bending moment that causes the yield stress in the outer fibers is further emphasized by the fact that for an offset of only 0.0001 the abscissa to point *E* in Fig. 2.63 indicates that the extreme fiber has undergone a strain 23% greater than that required to cause initial yielding ($\varepsilon_0 = 1.23\,\varepsilon_y$), whereas in the tension specimen of Fig. 2.62 the yielding that correspond to an offset of 0.0001 is very small. The point *F* on the moment-strain diagram has been added to indicate the actual bending moment corresponding to a plastic strain of 0.0001 in the outer fibers of the beam. Point *F* in Fig. 2.63 corresponds to point *D* in Fig. 2.62.

REFERENCES

1. American Society for Metals, *Properties of Metals in Materials Engineering*, Novelty, Ohio (1949).
2. B. W. Boguslovsky, *J. Eng. Ed.*, **40**, 382–385 (March 1950).

3. W. B. Dobie, *The Structural Engineer*, **50** (2), 34–46 (February 1952).
4. M. F. Gardner, and J. L. Barnes, *Transients in Linear Systems*, John Wiley and Sons, New York, N.Y. (1942).
5. M. Gensamer, *Strength of Metals Under Combined Stresses*, The American Society for Metals, Novelty, Oh. (1941).
6. J. E. Goldberg, *Proc. Am. Soc. Civ. Engrs.*, **78**, Sep. No. 125 (August 1952).
7. H. Herbert, [see page 72 of 14].
8. M. Hetenyi, *J. Franklin Inst.*, **254** (5), 369–380 (November 1952).
9. A. Krivetsky, *Prod. Eng.*, **30** (11), 73–75 (March 16, 1959).
10. J. M. Lessells, *Strength and Resistance of Metals*, John Wiley and Sons, New York, N.Y. (1954).
11. C. W. Macgregor, *J. Franklin Inst.*, **238** (2), 111–135 (August 1944) and **238** (3), 159–176 (September 1944).
12. J. Marin, *Engineering Materials*, Prentice-Hall, Englewood Cliffs, N.J. (1952).
13. J. Marin, *Strength of Materials*, Macmillan Co., New York, N.Y. (1948).
14. D. Morkovin, and O. Sidebottom, *The Effect of Non-Uniform Distribution of Stress on the Yield Strength of Steel*, Univ. of Illinois Bulletin, Vol. 45, No. 26 (December 18, 1947), (Engineering Experimental Station Bulletin Series No. 372), Urbana, Ill.
15. R. C. Reese, Ed., *CRSI Design Handbook*, The Concrete Reinforcing Steel Institute, Chicago, Ill. (1959).
16. R. M. Rivello, *Theory and Analysis of Flight Structures*, McGraw-Hill Book Co., New York, N.Y. (1969).
17. R. J. Roark, and W. C. Young, *Formulas for Stress and Strain*, McGraw-Hill Book Co., New York, N.Y. (1975).
18. F. B. Seely, and N. E. Ensign, *Analytical Mechanics for Engineers*, John Wiley and Sons, New York, N.Y. (1933).
19. F. B. Seely, and J. O. Smith, *Resistance of Materials*, John Wiley and Sons, Inc., New York, N.Y. (1957).
20. F. B. Seely, and J. O. Smith, *Advanced Mechanics of Materials*, 2nd ed., John Wiley and Sons, Inc., New York, N.Y. (1952).
21. F. R. Shanley, *Basic Structures*, John Wiley and Sons, Inc., New York, N.Y. (1947).
22. H. D. Tabakman, *Prod. Eng.*, **25** (6), 137–141 (June 1954).
23. S. Timoshenko, *Strength of Materials*, Vol. 1, 3rd ed., D. Van Nostrand Co., Princeton, N.J. (1955).
24. S. Timoshenko, *Strength of Materials*, Vol. II, 3rd ed., D. Van Nostrand Co., Princeton, N.J. (1956).
25. S. Timoshenko, *Theory of Elasticity*, McGraw-Hill Book Co., New York, N.Y. (1970).
26. E. E. Ungar, *Prod. Eng.*, **30** (11), 50–54 (March 16, 1959).
27. S. B. Williams, *Trans. Soc. for Expl. Stress Anal.*, **12** (1), 135–138 (1954).

PROBLEMS

2.1 The allowable stress in the section is 8 kpsi and it is 36 in. long with a G equal to 6×10^6 lb/in.^{2}. Find the angle of twist of the 36-in. length and the torque applied. (See Fig. P2.1.)

2.2 A lifting tong supports a 2000-lb load by the means of friction. The dimensions on Fig. P2.2 are specifications from a customer. A lifting force F is applied at point O through a cable and the cable travels 3

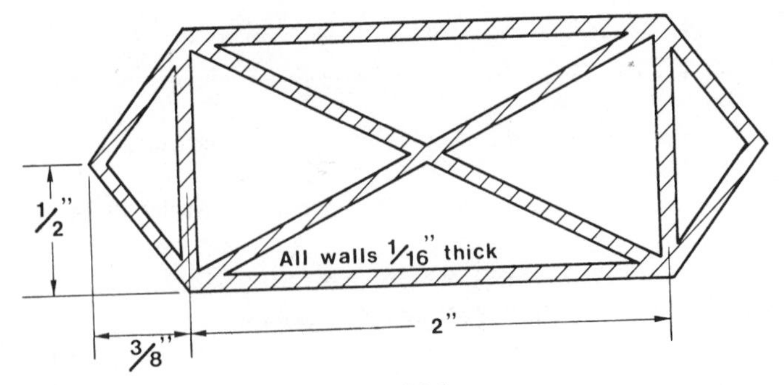

Fig. P2.1

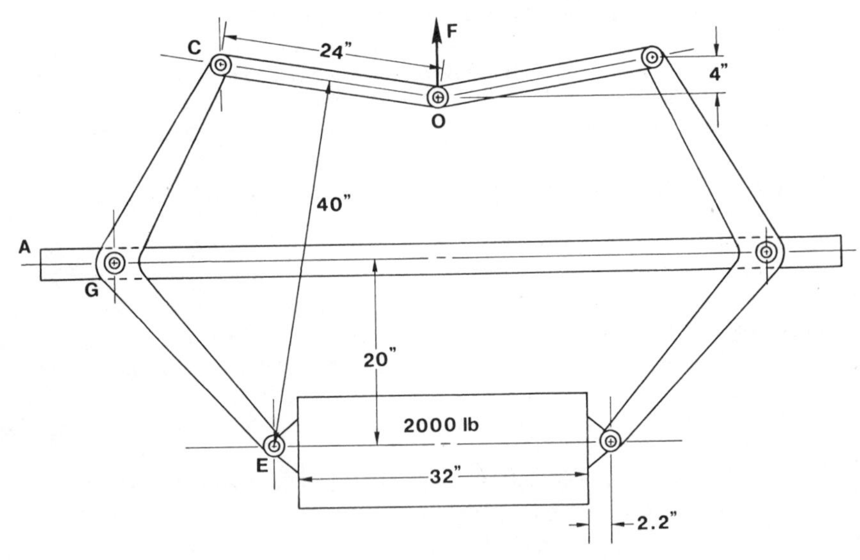

Fig. P2.2

ft/sec as the load is about to be raised. Size the member *CGE* for bending stresses.

2.3 The drive shaft in Fig. P 2.3 has a 963-lb force applied to the chain drive. The torque is equally taken from both sides to drive the wheels. The brake is rated for 3300 in-lb stopping torque. Bearings *A* and *E* are self-aligning, and bearing *C* is 1.25 in. wide and supports the shafting. Find the maximum stresses for the bending and torsional loads and the location of each.

2.4 The stress equation for the torsion of a thick-wall tube and a thin-wall tube can be found in the text. Find the ratio of t/R_0 that defines the boundary between a thin-wall and a thick-wall tube. R_0 is the outside radius.

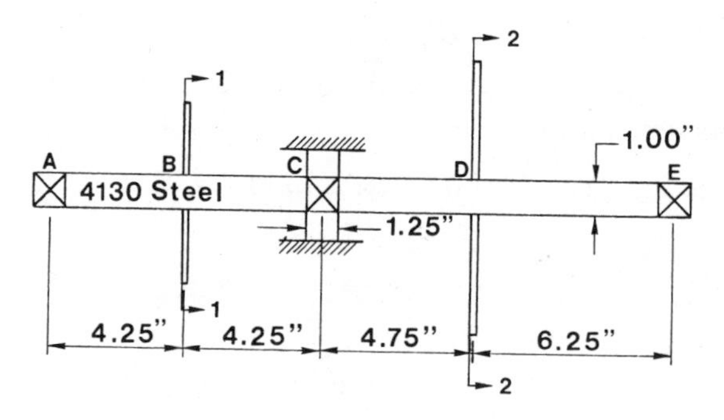

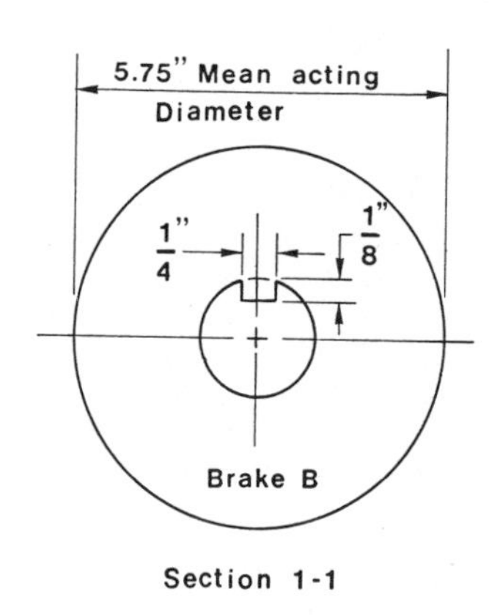

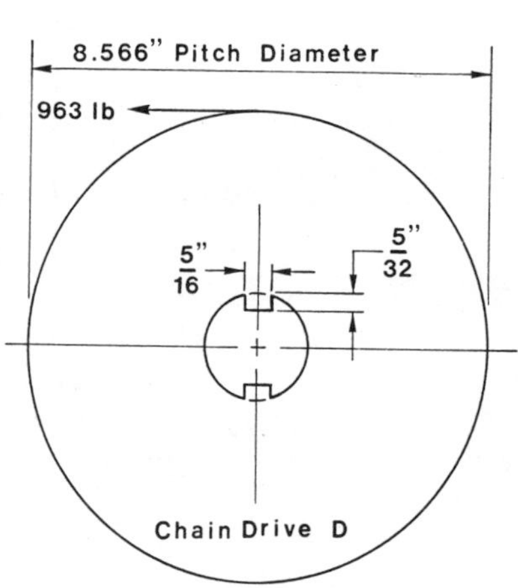

Fig. P2.3

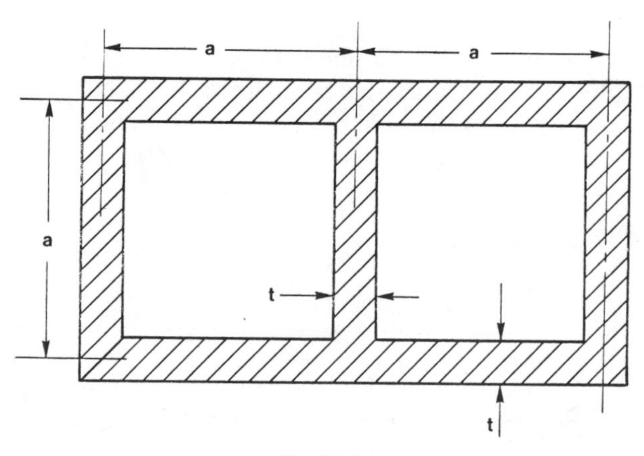

Fig. P2.5

2.5 Find the stress in the central web of the rectangular section of Fig. P2.5 for a torsional load T.

2.6 Find the shear center of the Z section in Fig. P2.6.

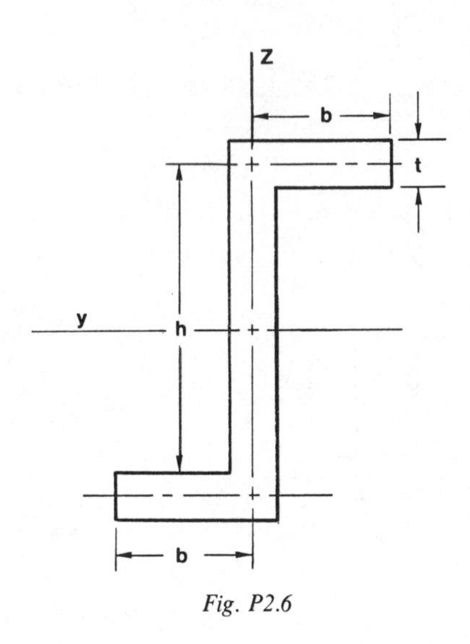

Fig. P2.6

2.7 Find the stress in the upper right-hand corner in Problem 2.6 when b is 0.5 in., t is 0.2 in., h is 1 in., and M_z of 5×10^4 in.-lb acts in a vertical direction.

2.8 The section wall thickness in Fig. P2.8 is 0.25 in. thick and has an applied torque of 1 million in.-lb. Find the stress in the wall and the angle of rotation.

2.9 In the literature the torsional moment of inertia; J; for a square of side A is 0.1406 A^4. How does this compare with $I_{xx} + I_{yy} = I_p$ as given in elementary statics? Also look at an equilateral triangle of side A.

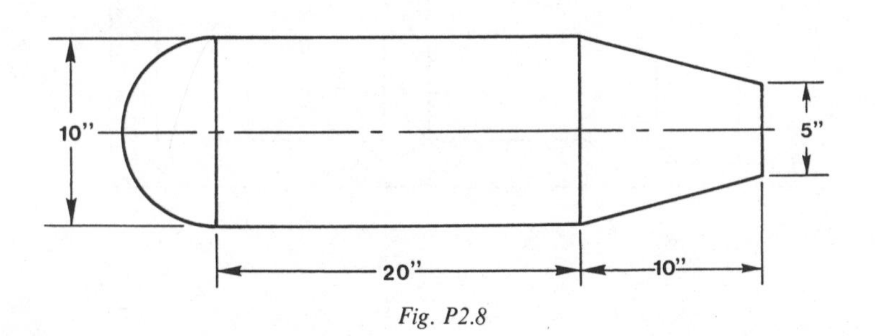

Fig. P2.8

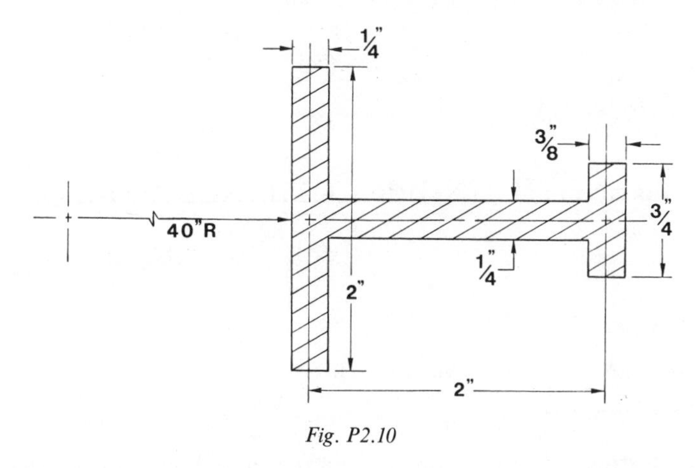

Fig. P2.10

2.10 Shown in Fig. P2.10 is a typical ring section on a tank with a 40 in. inside radius. How must the ring be loaded to minimize torsional loads? Also set up the equations to calculate bending stresses.

chapter 3

STRENGTH UNDER COMBINED STRESS

3-1 GENERAL ANALYSIS OF STRESS AND STRAIN

State of stress

The state of stress within, or on the surface of, a body is described by the magnitude and directions of *principal* and *maximum shear stresses.* In the most general case a body may be subjected to a combination of tension, compression, twisting, and bending loads, all of which contribute to the magnitudes and directions of the stresses. The randomly applied loads (Fig. 3.1) in the general case give rise to *component stresses.* A set of reference axes is chosen and the stresses resulting from direct loads F_1, F_2, and F_3, and the loads producing the bending moment M and the twisting moment T are resolved into six stress components, $\sigma_x, \sigma_y, \sigma_z, \tau_{xy}, \tau_{xz}$, and τ_{yz}, as shown in Fig. 3.2.

In any particular problem there is usually little difficulty in determining the stress components. For example, in a thin-walled tube subjected simultaneously to tension and torsion loads, the axial normal stress component is

$$\sigma_x = F/A \tag{3.1}$$

where F is the tensile load and A is the cross-sectional area of the tube. Also present is a shear stress due to the torque (in the elastic range of strain),

$$\tau = Tr/J \tag{3.2}$$

where T is the torque, r the variable radius in the tube wall, and J the polar moment of inertia. In the plastic range of strain where Hooke's law is not applicable, the axial component stress is still determined by Eq. 3.1; the shear stress, however, cannot be determined from use of Eq. 3.2. To determine the shear stress the condition of static equilibrium is used; from Fig. 3.3,

$$\tau = 2T/[(\pi\, dt)(d+t)] \tag{3.3}$$

Component stresses, however, are not always the critical stresses; consequently, it is necessary to reduce all systems of component stresses to a common critical system of stresses called *principal stresses* and *maximum shear* stress. These latter stresses act in planes that are inclined to the original reference planes and represent critical stresses for the particular point considered.

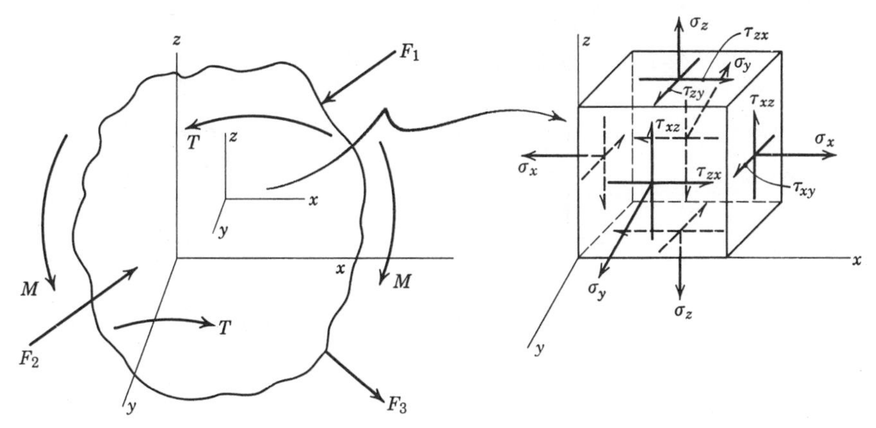

Fig. 3.1 *General three-dimensional loading.*

Fig. 3.2 *Component stresses on element.*

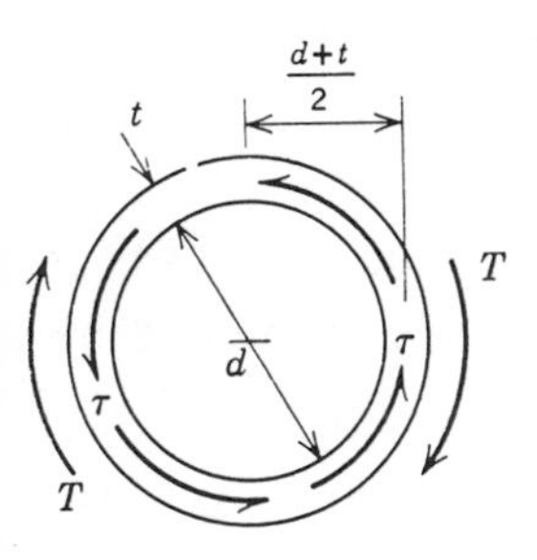

Fig. 3.3 *Equilibrium of torque and shear stress in thin-walled tube.*

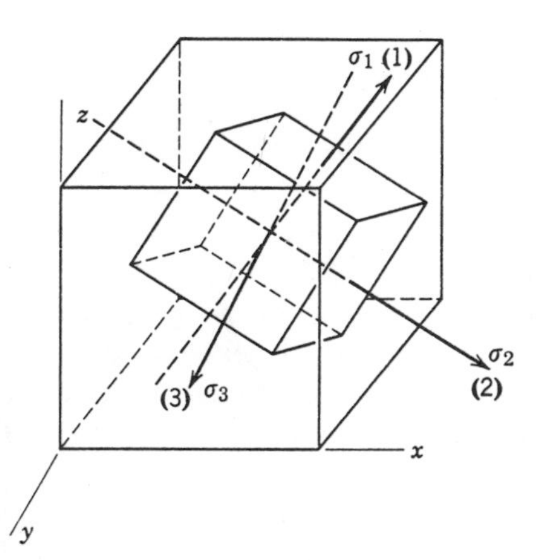

Fig. 3.4 *System of principal stresses.*

In any stressed body, however complicated the loading may be, there are always at least three planes on which the shear stresses are zero; these planes are always mutually perpendicular, and it is in these planes that the principal stresses act, as shown in Fig. 3.4. Thus, by definition, a principal stress is a stress acting normal to a plane of zero shear stress. The planes of maximum shear stress are inclined to those of principal stress; these planes may, depending

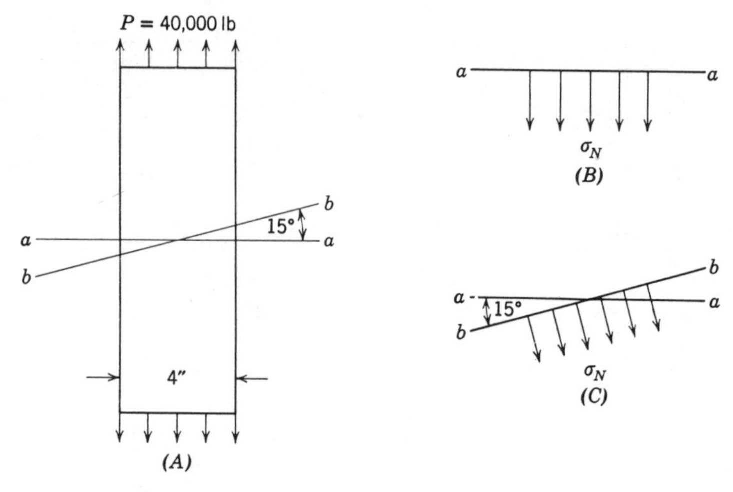

Fig. 3.5 Resolution of axial stress on various planes.

on the situation, have normal stresses acting on them. These normal stresses are, however, not principal stresses. More precise definitions of *normal*, *principal*, and *shear* stresses are now given.

Normal stress. A normal stress is one that acts perpendicularly to a reference plane in a body. In simple tension, Fig. 3.5, for example, the direct axial load is 40,000 lb. If the bar is assumed to have a width of 4 in. and unit thickness, the normal stress on plane *a–a* is (Fig. 3.5*B*)

$$(\sigma_N)_{a-a} = 40{,}000/4 = 10{,}000 \text{ psi}$$

However, on plane *b–b*, inclined 15° to plane *a–a*, the normal component of stress induced by the direct axial load is (Fig. 3.5*C*),

$$(\sigma_N)_{b-b} = \frac{(40{,}000)\cos 15^\circ}{4/\cos 15^\circ} = 9320 \text{ psi}$$

Consequently, when describing a normal stress in a body it is necessary to refer to the normal stress on a particular plane in the body because normal stress is a function of the orientation of the reference plane. For the same reason, it is necessary to specify the direction of the plane.

Shear stress. A shear stress acts parallel to a reference plane in a body; and, as with normal stress, it is necessary to specify orientation of the reference plane. Referring to Fig. 3.5, the shear stress in plane *a–a*, induced by the direct axial load of 40,000 lb, is zero because there is no stress component perpendicular to the direction of the applied load. On plane *b–b*, however, the shear stress is

$$(\tau)_{b-b} = \frac{40{,}000 \sin 15^\circ}{4/\cos 15^\circ} = 2500 \text{ psi}$$

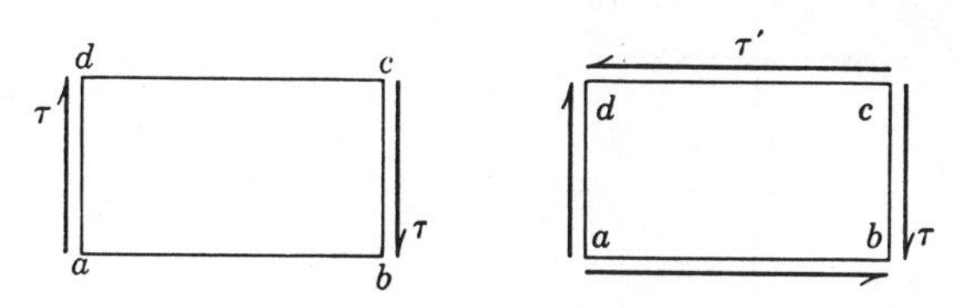

Fig. 3.6 Equilibrium of shear.

From equilibrium considerations, since a shear stress acts parallel to a plane, there must be a shear stress induced in the opposite direction which would tend to produce a rotation. Consequently, a set of shear stresses counterbalancing this effect would have to be induced. Thus, considering Fig. 3.6, if the block *abcd* is assumed to have unit thickness, for equilibrium,

$$(ad)\tau(ab) = (ab)\tau'(ad) \tag{3.4}$$

or

$$\tau = \tau' \tag{3.5}$$

which means that the intensity of shear stress on two sets of mutually perpendicular planes is equal.

Some examples of the use of the preceding relationships are given.

1. Illustrate positive and negative pure shear in a body.

Consider Fig. 3.7 which shows an element of thickness dz stressed in shear. Summing forces in the vertical and horizontal directions in Fig. 3.7*A*,

$$\tau_1 a\,dz - \tau_3 a\,dz = 0; \qquad \text{therefore} \quad \tau_1 = \tau_3$$
$$\tau_2 b\,dz - \tau_4 b\,dz = 0; \qquad \text{therefore} \quad \tau_2 = \tau_4$$

By summing the moments about 0,

$$\tau_1 ab\,dz - \tau_2 ab\,dz = 0; \qquad \text{therefore} \quad \tau_1 = \tau_2 = \tau_3 = \tau_4$$

Two cases may now be illustrated: positive shear, Fig. 3.7*B*, and negative shear, Fig. 3.7*C*. By definition, then, a system of pure shear is positive when the shear diagonal passes through the first quadrant of a coordinate system x–y as shown; otherwise, it is negative.

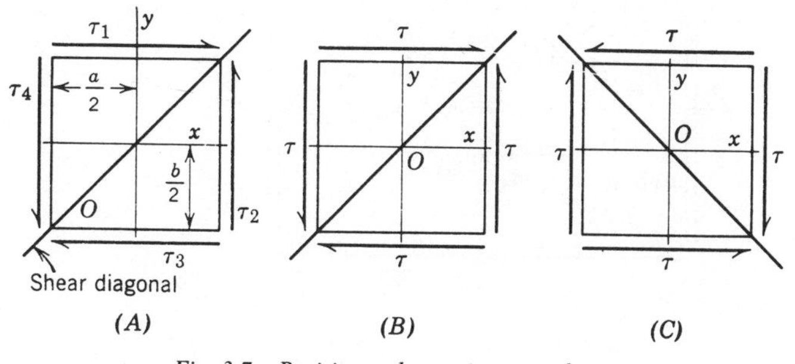

Fig. 3.7 Positive and negative pure shear.

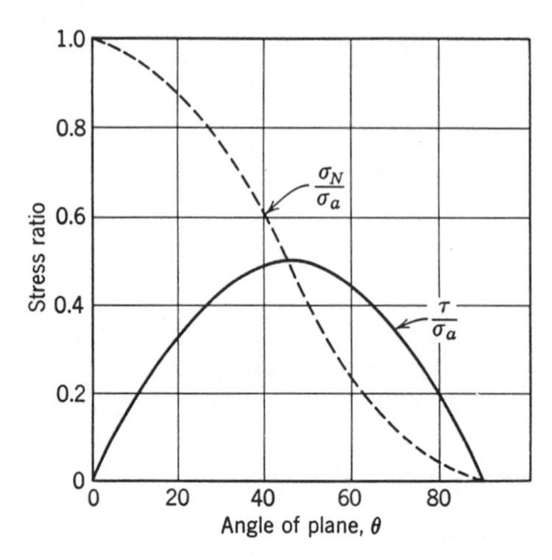

Fig. 3.8 Relation of component, principal, and shear stresses.

2. Plot a graph showing the variation of a normal and shear stress in tension. The expression for normal and shear stress has already been given in terms of a numerical problem. These relationships for the general case are as follows:

$$\text{normal stress} = \sigma_N = P \cos\theta/(A_0/\cos\theta) = (P/A_0)\cos^2\theta \qquad 3.6$$

$$\text{shear stress} = \tau = P \sin\theta/(A_0/\cos\theta) = (P/A_0)\sin\theta\cos\theta \qquad 3.7$$

where A_0 is the cross-sectional area of the bar and θ defines the plane in question (Fig. 3.8). From Eqs. 3.6 and 3.7

$$\sigma_N/(P/A_0) = \sigma_N/\sigma_a = \cos^2\theta \qquad 3.8$$

$$\tau/(P/A_0) = \tau/\sigma_a = \sin\theta\cos\theta \qquad 3.9$$

where σ_a is the normal axial tensile stress; Eqs. 3.8 and 3.9 are plotted in Fig. 3.9. Note that the shear stress is maximum at 45° and that it equals half the tension stress. The normal stress at 45° also equals half the maximum tensile stress.

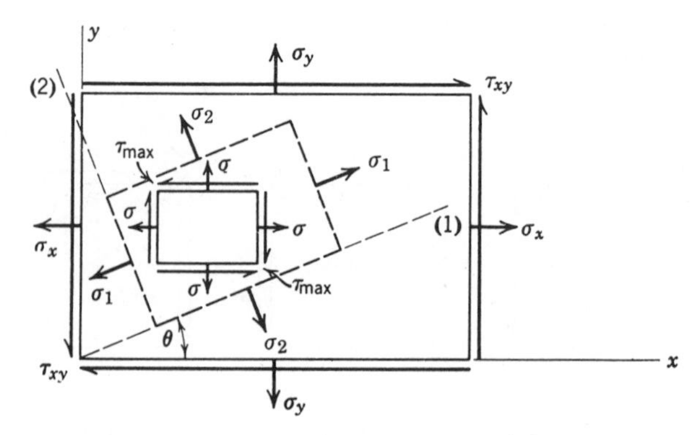

Fig. 3.9 Distribution of normal stress and shear stress on plane in a tension bar.

Principal stress and maximum shear stress for the two-dimensional case

An element subjected to a state of two-dimensional stress is shown in Fig. 3.8. The stress components are σ_x, σ_y, and τ_{xy}. In a thin-walled tube subjected to internal pressure and torsion, for example, the component stress in the longitudinal direction σ_x is due to internal pressure. Also due to pressure is a component stress in the hoop direction σ_y. The component shear stress due to the torsion is given by Eq. 3.2. Component stresses in different planes are not additive, that is, the stress in the tube is not $\sigma_x + \sigma_y + \tau_{xy}$. The tensor of stress, however, which includes all the stress components, can be resolved in the two-dimensional case into two mutually perpendicular normal stresses and one shear stress that act at a particular point—these are the two principal stresses and the maximum shear stress.

In order to determine the principal stresses and the maximum shear stress for this case consider the free-body element in Fig. 3.10. On plane $ACDE$ there are resultant stresses represented by two stress components, σ_N perpendicular to the plane, and τ in the plane, both functions of angle θ; consequently, there is a particular value of θ for which these stresses are maximum. Applying the condition of static equilibrium to forces in the x' direction,

$$\begin{aligned}\sigma_N dw\,dz - \tau_{xy} dy\,dz \cos\theta - \sigma_y dz\,dx \cos\theta \\ - \tau_{xy} dx\,dz \sin\theta - \sigma_x dy\,dz \sin\theta = 0\end{aligned} \tag{3.10}$$

Dividing through by $dw\,dz$ gives the result

$$\begin{aligned}\sigma_N = \tau_{xy}(dy/dw)\cos\theta + \sigma_y(dx/dw)\cos\theta \\ + \tau_{xy}(dx/dw)\sin\theta + \sigma_x(dy/dw)\sin\theta\end{aligned} \tag{3.11}$$

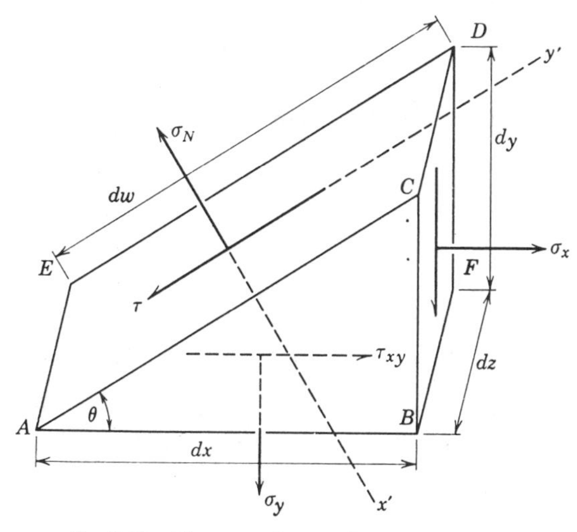

Fig. 3.10 Element under two-dimensional stress.

However, $dy/dw = \sin\theta$ and $dx/dw = \cos\theta$, and therefore

$$\sigma_N - \tau_{xy}\sin\theta\cos\theta - \sigma_y\cos^2\theta - \tau_{xy}\sin\theta\cos\theta - \sigma_x\sin^2\theta = 0$$

or simplifying,

$$\sigma_N = \frac{\sigma_x + \sigma_y}{2} + \frac{\sigma_y - \sigma_x}{2}\cos 2\theta + \tau_{xy}\sin 2\theta \qquad 3.12$$

The maximum value of σ_N is found by differentiating Eq. 3.12 with respect to θ and equating to zero; thus

$$\frac{d\sigma_N}{d\theta} = 0 = \frac{\sigma_x - \sigma_y}{2}(+\sin 2\theta) + \tau_{xy}\cos 2\theta \qquad 3.13$$

from which

$$\tan 2\theta = \frac{2\tau_{xy}}{\sigma_y - \sigma_x} \qquad 3.14$$

Since the second derivative is negative, Eq. 3.14 represents the condition for maximum normal stress. Having the tangent function from Eq. 3.14, trigonometry shows that

$$\sin 2\theta = \frac{\pm\tau_{xy}}{\sqrt{[(\sigma_y - \sigma_x)/2]^2 + \tau_{xy}^2}} \qquad 3.15$$

and

$$\cos 2\theta = \frac{\pm[(\sigma_y - \sigma_x)/2]}{\sqrt{[(\sigma_y - \sigma_x)/2]^2 + \tau_{xy}^2}} \qquad 3.16$$

Substituting Eqs. 3.15 and 3.16 into Eq. 3.12 then gives

$$\sigma_N = \frac{\sigma_x + \sigma_y}{2} \pm \sqrt{[(\sigma_y - \sigma_x)/2]^2 + \tau_{xy}^2} \qquad 3.17$$

or in terms of *principal stress*,

$$\sigma_1 = \frac{\sigma_x + \sigma_y}{2} + \sqrt{[(\sigma_y - \sigma_x)/2]^2 + \tau_{xy}^2} \qquad 3.18$$

$$\sigma_2 = \frac{\sigma_x + \sigma_y}{2} - \sqrt{[(\sigma_y - \sigma_x)/2]^2 + \tau_{xy}^2} \qquad 3.19$$

Similarly, applying the condition of static equilibrium in the y' direction in Fig. 3.10, it can be shown that

$$\tau_{max} = \pm\sqrt{[(\sigma_y - \sigma_x)/2]^2 + \tau_{xy}^2} \qquad 3.20$$

Caution must be exercised in using Eq. 3.20 to calculate the maximum shear stress. There is always a third principal stress σ_3, although it may sometimes equal zero. When the three principal stresses are considered, as is shown later, there are three shear stresses induced, one of which is maximum. For principal stresses of the same sign, the maximum shear stress is one-half the maximum principal stress; for principal stresses of unlike sign, the maximum shear stress is one-half the algebraic sum of the two stresses. Thus, σ_1 and σ_2 were both positive, in Eqs. 3.18 and 3.19, the maximum shear stress would be $\sigma_1/2$ or $\sigma_2/2$, whichever was greater. If σ_1 were positive and σ_2 were negative, the maximum shear stress would be $(\sigma_1 - \sigma_2)/2$.

Equation 3.14 also gives the basis for determining the directions of the principal stresses. For the σ_1 direction.

$$2\theta_1 = \frac{\tan^{-1} 2\tau_{xy}}{\sigma_y - \sigma_x} \tag{3.21}$$

and in the σ_2 direction

$$\theta_2 = \theta_1 + 90^\circ \tag{3.22}$$

The direction of the maximum shear stress is given by the expression

$$2\theta_\tau = \frac{\tan^{-1}(\sigma_x - \sigma_y)}{2\tau_{xy}} \tag{3.23}$$

In Eqs. 3.21 and 3.23 the sign of the shear stress component is determined on the basis of the rotation effected by the stress. If the shear stress forms a couple acting clockwise, the proper sign is plus, if the direction is counterclockwise, the sign is minus. Thus in Fig. 3.9, the planes of principal stress are free of shear stress, whereas the planes of maximum shear stress may have other normal stresses acting on them.

Some examples of the use of these relationships are:

1. Determine the normal stress on any plane in a body subjected to biaxial tension.

An element under biaxial tension is shown in Fig. 3.11. Considering plane A–A, equilibrium shows the following result when forces are summed in the direction of the normal N.

$$dz\sigma_N = \sigma_x \cos\theta \cos\theta + dz\sigma_y ab \sin\theta - dz\sigma_y cd \sin\theta \tag{3.24}$$

from which

$$\sigma_N = \sigma_x \cos^2\theta + \sigma_y \sin\theta(ab - cd) \tag{3.25}$$

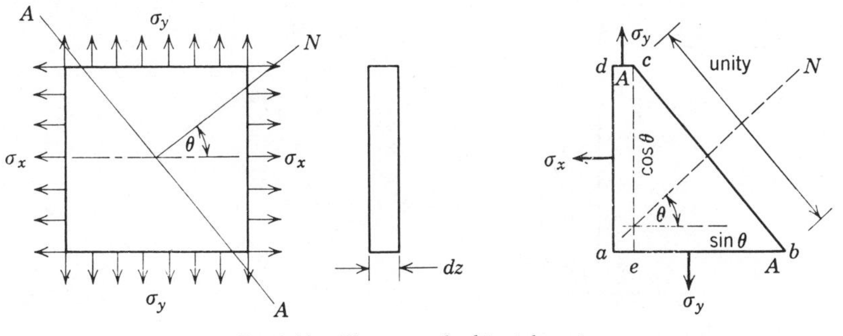

Fig. 3.11 Element under biaxial tension.

However, $ab - cd = eb = \sin\theta$; therefore the normal stress on any plane defined by θ is

$$\sigma_N = \sigma_x \cos^2\theta + \sigma_y \sin^2\theta \qquad 3.26$$

2. Determine the shear stress on any plane in a body subjected to biaxial tension.

From Fig. 3.11 of the previous example, the forces are summed on plane A–A to give the result

$$\tau\, dz = \sigma_x\, dz \cos\theta \sin\theta - \sigma_y ab\, dz \cos\theta + \sigma_y cd\, dz \cos\theta \qquad 3.27$$

Therefore on any plane defined by angle θ the shear stress is given by the expression

$$\tau = \frac{\sigma_x - \sigma_y}{2} \sin 2\theta \qquad 3.28$$

3. Determine the width b of section AB, Fig. 3.12, if the depth d is 2 in. and the allowable stress is 15,000 psi.

This is an example of combined bending and direct axial loading. It is assumed that the material has equal tensile and compressive properties. There are two cases to consider, the tension and compression parts of the structure. At any section AB the total normal stress is

$$\sigma_N = P/A + M/Z \qquad 3.29$$

where P is the axial load, M the bending moment, A the cross-sectional area, and Z the section modulus. On the tension side,

$$(\sigma_N)_T = -P/A + M/Z \qquad 3.30$$

On the compression side,

$$(\sigma_N)_c = -P/A - M/Z \qquad 3.31$$

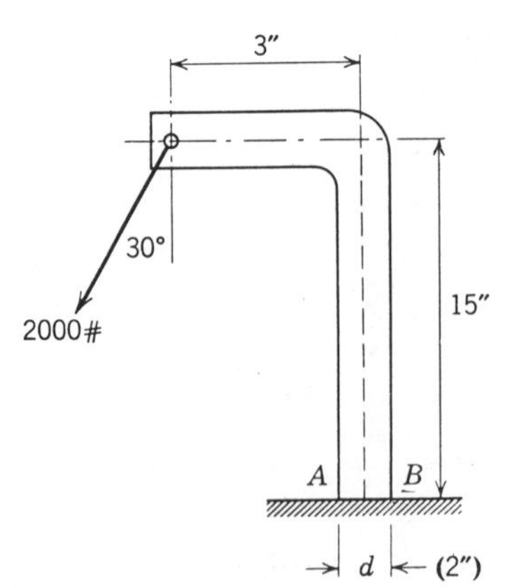

Fig. 3.12 Structural element subjected to biaxial stress.

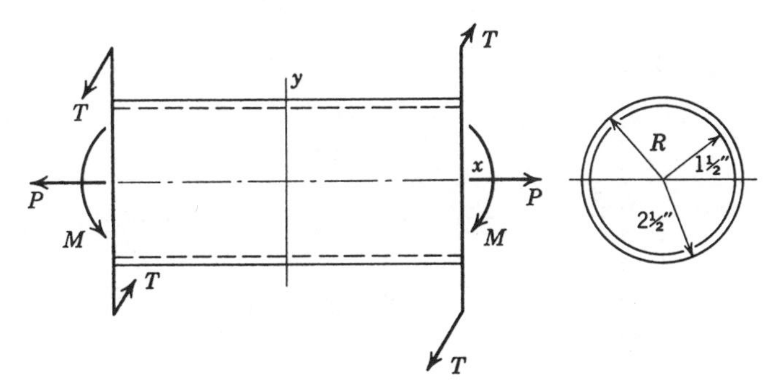

Fig. 3.13 Tubular member under combined loading.

Therefore for the loads and geometry indicated

$$(\sigma_N)_T = -\frac{2000 \cos 30^\circ}{2b} + [2000 \cos 30^\circ(3) + 2000 \sin 30^\circ(15)]\frac{3}{2b}$$

and

$$(\sigma_N)_c = -\frac{2000 \cos 30^\circ}{2b} - [2000 \cos 30^\circ(3) + 2000 \sin 30^\circ(15)]\frac{3}{2b}$$

or

$$(\sigma_N)_T = 29,400/b; \qquad (\sigma_N)_c = -31,170/b$$

The compression side thus governs and

$$(\sigma_N)_c = -15{,}000 = -31,170/b$$

from which the required width b is 2.075 in.

4. A hollow circular shaft (Fig. 3.13) with an outside diameter of 5 in. and an inside diameter of 3 in. is subjected to a bending moment of 20,000 ft-lb, a twisting moment of 12,000 ft-lb, and an axial tensile load of 50,000 lb. Determine the maximum normal and shearing stresses and the orientation of the planes involved.

The component stress in the axial direction of the tube is

$$\sigma_x = \frac{P}{A} + \frac{M}{Z} = \frac{50{,}000}{(\pi/4)(5^2 - 3^2)} + \frac{20{,}000(12)(2.5)}{(\pi/64)(5^4 - 3^4)}$$
$$= 26{,}530 \text{ psi}$$

The component shear stress is

$$\tau_{xy} = \frac{TR}{J} = \frac{12{,}000(12)(2.5)}{(\pi/32)(5^4 - 3^4)} = 6770 \text{ psi}$$

The principal stresses and maximum shear stress are then

$$\sigma_1 = \sigma_x/2 + \sqrt{(\sigma_x/2)^2 + \tau_{xy}{}^2} = 28{,}100 \text{ psi}$$
$$\sigma_2 = \sigma_x/2 - \sqrt{(\sigma_x/2)^2 + \tau_{xy}{}^2} = -1623 \text{ psi}$$
$$\tau_{\max} = \pm\sqrt{(\sigma_x/2)^2 + \tau_{xy}{}^2} = \pm 14{,}900 \text{ psi}$$

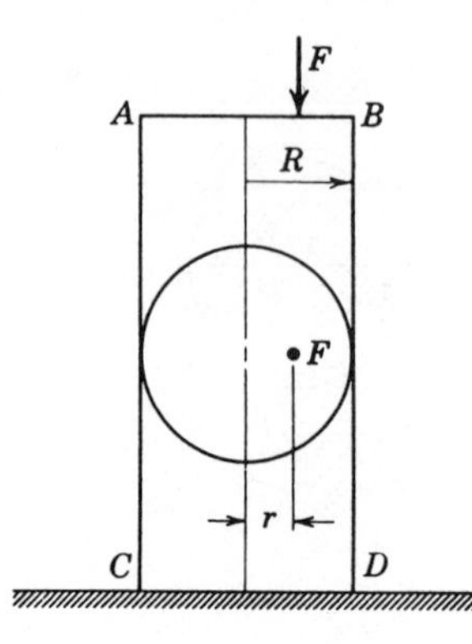

Fig. 3.14 Structural element under combined stresses.

The plane of the maximum principles stress σ_1 is found from Eq. 3.21,

$$2\theta_1 = \tan^{-1} 2\tau/(-\sigma_x) = 153^\circ$$

The plane of σ_2, from Eq. 3.22, is

$$\theta_2 = \theta_1 + 90^\circ = 76^\circ 30' + 90^\circ = 166^\circ 30'$$

The direction of the maximum shear stress by Eq. 3.23 is

$$2\theta\tau_{max} = \tan^{-1} \sigma_x/2\tau_{xy} = 62^\circ$$

5. For the member shown in Fig. 3.14 determine the distance r from the center line for applying the axial load F so that the cross section of the member is subjected to compressive stress at all times.

The direct compression stress is

$$\sigma_c = -F/A = -F/\pi R^2 \qquad 3.32$$

The bending stress is given by

$$\sigma_b = M/Z = \pm Fr/Z \qquad 3.33$$

Since there is no shear involved, σ_c and σ_b may be added; the total stress on the cross section of the member is thus

$$\sigma = -F/\pi R^2 \pm Fr/Z = (F/\pi R^2)(-1 \pm 4r/R) \qquad 3.34$$

The bend stresses are positive on side AC of the member and negative on side BD. Therefore on the positive side of the member the fraction $4r/R$ cannot exceed 1.0 if the total stress is to be negative (compressive). This situation occurs for $r = R/4$. Therefore, if the load F is applied anywhere within the radius $r = R/4$, the entire cross section will always be subjected to compressive loading. The area enclosed by the circumference $2\pi(R/4)$ is called the *kern* or *core* of the section. This type of analysis is particularly important for masonry or brick structures that are weak in tension.

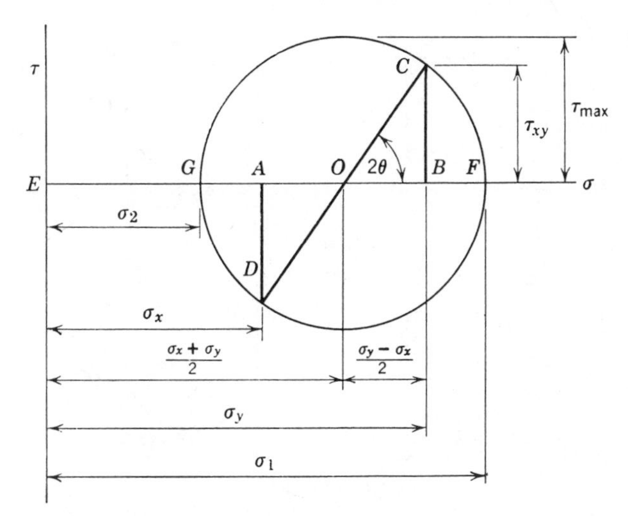

Fig. 3.15 Mohr's circle for two-dimensional tensile stress.

Graphical representation of two-dimensional state of stress

The graphical method for representing combined stresses is popularly known as *Mohr's circle method*. As illustrated in Fig. 3.15, the cartesian coordinate axes represent the normal and shearing stresses. On the horizontal axis, to some convenient scale, the component stresses σ_x and σ_y are scaled off as EA and EB, respectively. In the direction of the ordinate the component shear stress τ_{xy} is scaled off as BC and AD. Then with O as the center of a circle of diameter CD the graph is completed. By geometry it can be seen that $OB = (\sigma_y - \sigma_x)/2$ and that $EO = (\sigma_x + \sigma_y)/2$. Therefore, from the first relationship

$$OC = \sqrt{[(\sigma_y - \sigma_x)/2]^2 + \tau_{xy}{}^2}$$

and $EF = \sigma_1$ and $EG = \sigma_2$. Similarly, if the principal stresses are known, the component stresses can easily by deduced. A number of miscellaneous graphical solutions to combined states of stress are shown in Fig. 3.16.

Analysis of strain for a two-dimensional state of stress

Strains, like stresses, may be of two kinds, component and principal. *Component strains* may be normal strains or shear strains; *principal strains* are the values used in design, for they represent mutually perpendicular strains determined by resolution of the strain tensor that includes all strains imposed on a system.

A normal strain is an extension or contraction in the direction of a perpendicular to the normal stress causing the strain. In tension, for example, if the gauge length along a bar is 2 in. and the gauge length stretches $\frac{1}{2}$ in., the normal

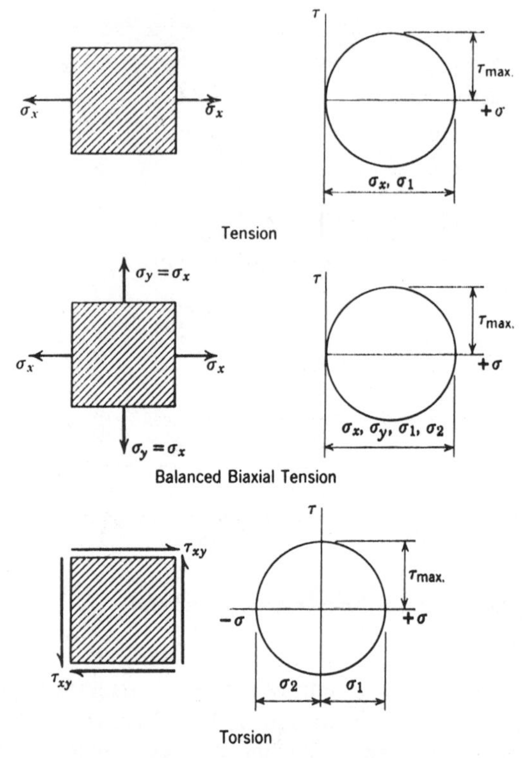

Fig. 3.16 Mohr's circles for three states of stress.

strain is $0.5/2 = 0.25$ in./in. An additional normal strain occurs in tension perpendicular to the direction of loading due to the lengthening of a bar. This phenomenon, called *Poisson effect*, is the ratio of lateral to axial strain (about 0.25 to 0.30 for steel). This was discussed in Chapter 2.

Shear strain is the displacement of one plane relative to another plane that is parallel to it and separated from it by unit distance. For example, in Fig. 3.17 if planes AB and CD are sheared a distance x, the shear strain is x/h or tan α.

In determining the principal strains consider Fig. 3.18 which shows an element subjected to component stresses referred to cartesian coordinates x, y.

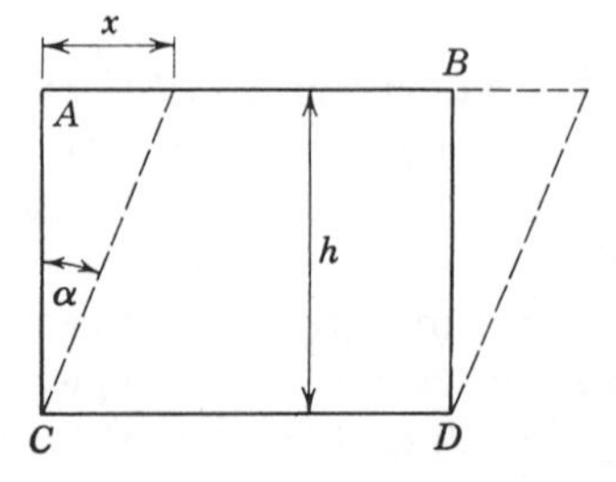

Fig. 3.17 Shear strain.

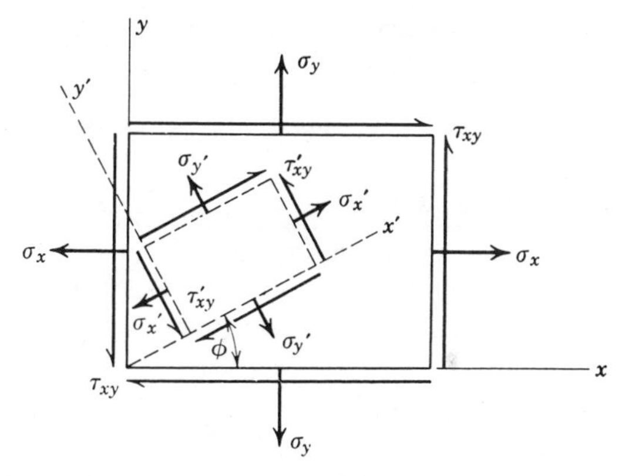

Fig. 3.18 Rotation of axes.

Within this element, and inclined at angle ϕ, another set of coordinate axes x', y' is constructed which orient an element subjected to component stresses $\sigma_{x'}$, $\sigma_{y'}$, and τ'_{xy}. Equilibrium in the x' direction requires that

$$\sigma_{x'} = \sigma_x \cos^2 \phi + \sigma_y \sin^2 \phi + 2\tau_{xy} \sin \phi \cos \phi \tag{3.35}$$

Similarly, in the y' direction,

$$\sigma_{y'} = \sigma_x \sin^2 \phi + \sigma_y \cos^2 \phi - 2\tau_{xy} \cos \phi \sin \phi \tag{3.36}$$

In accordance with Hooke's law, the strains in the x' and y' directions are

$$\varepsilon_{x'} = (1/E)(\sigma_{x'} - v\sigma_{y'}) \tag{3.37}$$

$$\varepsilon_{y'} = (1/E)(\sigma_{y'} - v\sigma_{x'}) \tag{3.38}$$

where v is Poisson's ratio and E is the modulus of elasticity. Combining Eqs. 3.35 to 3.38 gives

$$\varepsilon_{x'} = (1/E)[\sigma_x(\cos^2 \phi - v \sin^2 \phi) + \sigma_y (\sin^2 \phi - v \cos^2 \phi) + (1 + v)\tau_{xy} \sin 2\phi] \tag{3.39}$$

and

$$\varepsilon_{y'} = (1/E)[\sigma_x(\sin^2 \phi - v \cos^2 \phi) + \sigma_y(\cos^2 \phi - v \sin^2 \phi) - (1 + v)\tau_{xy} \sin 2\phi] \tag{3.40}$$

The component stresses σ_x, σ_y, and τ_{xy}, in Eqs. 3.39 and 3.40 are likewise by Hooke's law

$$\sigma_x = \frac{E}{1 - v^2}(\varepsilon_x + v\varepsilon_y) \tag{3.41}$$

$$\sigma_y = \frac{E}{1 - v^2}(\varepsilon_y + v\varepsilon_x) \tag{3.42}$$

$$\tau_{xy} = G\gamma_{xy} = \frac{E\gamma_{xy}}{2(1 + v)} \tag{3.43}$$

where G is the shear modulus of elasticity and γ_{xy} is the shear strain. Substitution of Eqs. 3.41 to 3.43 into Eqs. 3.39 and 3.40 gives

$$\varepsilon_{x'} = \varepsilon_x \cos^2 \phi + \varepsilon_y \sin^2 \phi + \gamma_{xy} \sin \phi \cos \phi \qquad 3.44$$

and

$$\varepsilon_{y'} = \varepsilon_x \sin^2 \phi + \varepsilon_y \cos^2 \phi - \gamma_{xy} \sin \phi \cos \phi \qquad 3.45$$

The *principal strains* are the maximum and minimum values of either $\varepsilon_{x'}$ or $\varepsilon_{y'}$ in preceding equations. Differentiation with respect to ϕ and equating to zero show that the principal strains are defined by

$$\tan 2\phi_1 = \frac{\gamma_{xy}}{\varepsilon_x - \varepsilon_y} \qquad 3.46$$

where ϕ_1 is angle ϕ for maximum or minimum values of strain. The values of principal strain are

$$\varepsilon_1 = \frac{\varepsilon_x + \varepsilon_y}{2} + \tfrac{1}{2}\sqrt{(\varepsilon_y - \varepsilon_x)^2 + \gamma_{xy}^2} \qquad 3.47$$

$$\varepsilon_2 = \frac{\varepsilon_x + \varepsilon_y}{2} - \tfrac{1}{2}\sqrt{(\varepsilon_y - \varepsilon_x)^2 + \gamma_{xy}^2} \qquad 3.48$$

$$\gamma_{\max} = \pm\sqrt{(\varepsilon_y - \varepsilon_x)^2 + \gamma_{xy}^2} \qquad 3.49$$

In the experimental determination of these strains, which is usually the case when the component stress axes are different from the principal strain axes, three strain directions are marked off as shown in Fig. 3.19. Then by substituting ϕ_a, ϕ_b, and ϕ_c for ϕ in Eqs. 3.44 and 3.45,

$$\varepsilon_a = \varepsilon_x \cos^2 \phi_a + \varepsilon_y \sin^2 \phi_a + \gamma_{xy} \sin \phi_a \cos \phi_a \qquad 3.50$$

$$\varepsilon_b = \varepsilon_x \cos^2 \phi_b + \varepsilon_y \sin^2 \phi_b + \gamma_{xy} \sin \phi_b \cos \phi_b \qquad 3.51$$

$$\varepsilon_c = \varepsilon_x \cos^2 \phi_c + \varepsilon_y \sin^2 \phi_c + \gamma_{xy} \sin \phi_c \cos \phi_c \qquad 3.52$$

Since the strains ε_a, ε_b, and ε_c can be measured directly and since their directions are known, it is a simple matter to deduce values of ε_x, ε_y, and γ_{xy} to put into Eqs. 3.47 to 3.49.

The strain analyses given will now be used to show that the axial strain in a

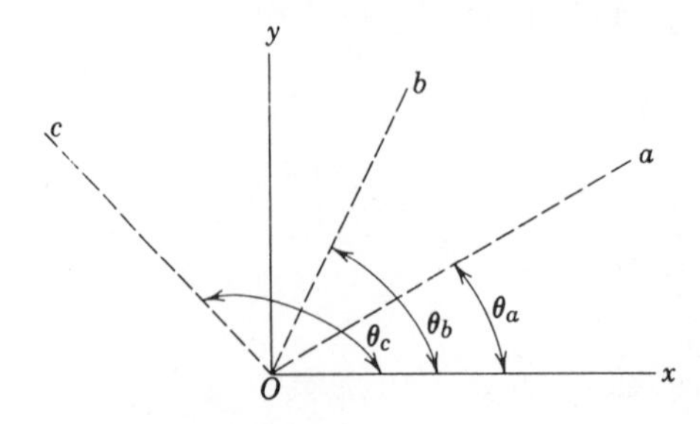

Fig. 3.19 Various strain axes.

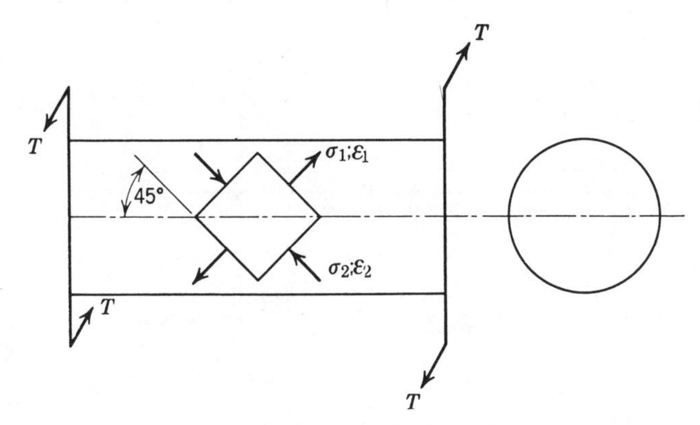

Fig. 3.20 State of stress in torsion.

bar subjected to pure elastic torque is zero. In torsion the only component stress generated is a shear stress

$$\tau_{xy} = Tr/J \tag{3.53}$$

where T is the torque, r the radius of the rod, and J the polar moment of inertia. Thus, from Eqs. 3.18, 3.19, and 3.53

$$\sigma_1 = -\sigma_2 = \tau_{xy} \tag{3.54}$$

and these stresses are oriented at 90° to each other with the σ_1 direction at 45° to the axis of the bar (Fig. 3.20). The principal strains ε_1 and ε_2 are given by Eqs. 3.47 and 3.48. The symmetry shown in Fig. 3.20 then shows that there is no net movement in the axial direction of the bar since $\varepsilon_1 = -\varepsilon_2$. By using the same example, the relationship between Young's modulus of elasticity E and the shear modulus G can be established. Thus from the values just calculated

$$\varepsilon_1 = \frac{1}{E}\left(\frac{Tr}{J} + \nu\frac{Tr}{J}\right) = \frac{\tau}{E}(1+\nu) \tag{3.55}$$

$$\varepsilon_2 = \frac{1}{E}\left(-\frac{Tr}{J} - \nu\frac{Tr}{J}\right) = -\frac{\tau}{E}(1+\nu) \tag{3.56}$$

But shear strain

$$\gamma = (\varepsilon_1 - \varepsilon_2) = \tau/G \tag{3.57}$$

Therefore, by equating Eqs. 3.55 to 3.57, the result is

$$(2\tau/E)(1+\nu) = \tau/G$$

or

$$G = \frac{E}{2(1+\nu)} \tag{3.58}$$

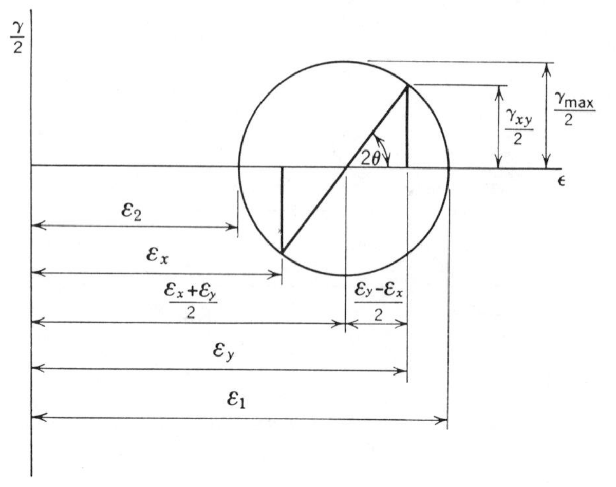

Fig. 3.21 Mohr's circle for strain.

Graphical representation of two-dimensional strain

A graphical representation of two-dimensional strain can be made as described earlier for stress by substituting ε_x for σ_x, ε_y for σ_y, and $\gamma_{xy}/2$ for τ_{xy} as shown in Fig. 3.21.

Stress and strain in two perpendicular directions

The foregoing paragraphs have dealt with the general two-dimensional case where the component and principal stress or strain values were not coincident. Many cases exist, however, where the component and principal axes are the same; in fact, this is always true unless there is an applied shear load. For example, in a thin-walled tube under internal pressure, axial load, and bending, the component stresses simply add up in the two principal directions

$$\sigma_{\text{axial}} = pd/4t + (M/I)[(d/2) + t] + F/A = \sigma_1 \qquad 3.59$$

$$\sigma_{\text{hoop}} = pd/2t = \sigma_2$$

$$\tau = 0 \qquad 3.60$$

where p is the internal pressure, d the inside diameter, t the wall thickness, M the applied bending moment, F the applied axial load, and A the cross-sectional area. The axial strain is

$$\varepsilon_1 = \varepsilon_x = \sigma_1/E \qquad 3.61$$

and

$$\varepsilon_2 = \varepsilon_y = -\nu\varepsilon_x = -\nu\varepsilon_1 \qquad 3.62$$

Equation 3.61, with change in subscripts, also holds for the other directions; therefore by superposition the following expressions for isotropic materials result:

$$\varepsilon_1 = (1/E)(\sigma_1 - \nu\sigma_2) \qquad 3.63$$

$$\varepsilon_2 = (1/E)(\sigma_2 - \nu\sigma_1) \qquad 3.64$$

$$\varepsilon_3 = (-\nu/E)(\sigma_1 + \sigma_2) \qquad 3.65$$

$$\sigma_1 = \frac{E}{1 - \nu^2}(\varepsilon_1 + \nu\varepsilon_2) \qquad 3.66$$

$$\sigma_2 = \frac{E}{1 - \nu^2}(\varepsilon_2 + \nu\varepsilon_1) \qquad 3.67$$

$$\sigma_3 = 0 \qquad 3.68$$

If the material is orthotropic, that is, if it has different properties in different directions, the constants in Eqs. 3.63 to 3.67 change from E and ν to characteristic values. For example, for orthotropic material, Eq. 3.63 would be written as

$$\varepsilon_1 = \sigma_1/E_1 - \nu_{21}\sigma_2/E_2 \qquad 3.69$$

where E_1, E_2, and ν_{21} are the values of modulus and Poisson ratio in the orthotropic directions. The analysis of orthotropic materials is considered in some detail in Chapter 4.

Principal stress and maximum shear stress for the orthogonal three-dimensional case

In this book many of the three-dimensional problems encountered can be resolved in terms of three mutually perpendicular stresses, or the stresses can be deduced, without undue difficulty, from equilibrium considerations (Eq. 3.3, for example). In the absence of applied shear the two-dimensional analysis previously given is converted to a three-dimensional system by adding the Poisson effect in the third direction and applying superposition. Thus, analogous to Eqs. 3.63 to 3.65, for isotropic materials

$$\varepsilon_1 = (1/E)[\sigma_1 - \nu(\sigma_2 + \sigma_3)] \qquad 3.70$$

$$\varepsilon_2 = (1/E)[\sigma_2 - \nu(\sigma_1 + \sigma_3)] \qquad 3.71$$

$$\varepsilon_3 = (1/E)[\sigma_3 - \nu(\sigma_1 + \sigma_2)] \qquad 3.72$$

Associated with the three principal stresses are three shear stresses (21)

$$\tau_1 = \frac{\sigma_1 - \sigma_2}{2} \qquad 3.73$$

$$\tau_2 = \frac{\sigma_2 - \sigma_3}{2} \qquad 3.74$$

$$\tau_3 = \frac{\sigma_3 - \sigma_1}{2} \qquad 3.75$$

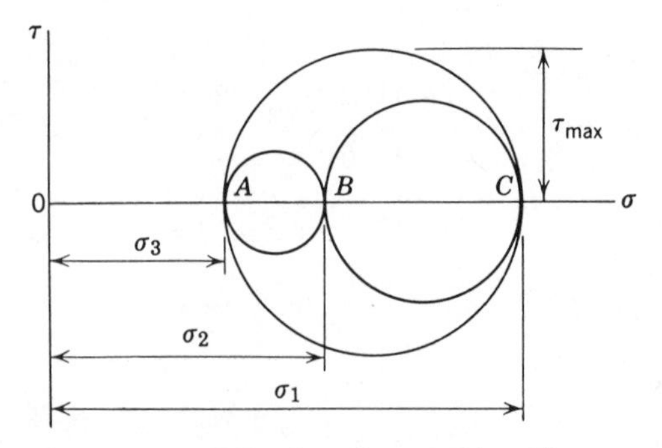

Fig. 3.22 Mohr's circle for three-dimensional tensile stress system.

one of which is maximum depending on the signs of the principal stresses.

The three mutually perpendicular principal stresses can be represented graphically. Figure 3.22 shows the graphical solution for three mutually perpendicular tensile stresses; the solution follows directly from the two-dimensional case. For any section in the σ_1, σ_2 plane there corresponds a circle BC; in the σ_2, σ_3 plane a circle AB, and the σ_3, σ_1 plane a circle AC. Thus $\sigma_1 = OC$, $\sigma_2 = OB$, $\sigma_3 = OA$ and $\tau_{max} = (\sigma_1 - \sigma_2)/2 = OD$. Figure 3.23 shows the graphical

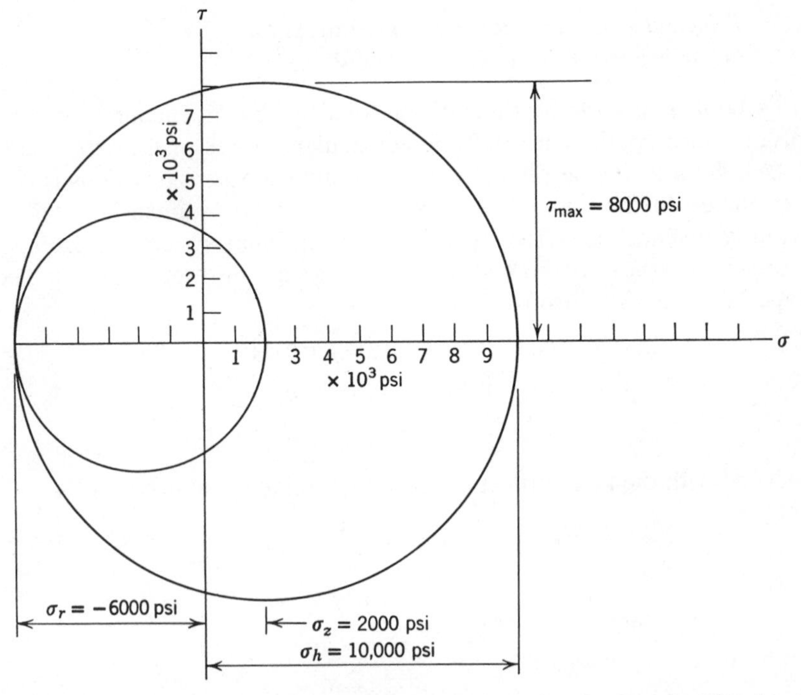

Fig. 3.23 Mohr's circle for stresses at bore of heavy-walled cylinder ($R = 2$) under 6000 psi internal pressure.

solution at the bore for a thick-walled cylinder under an internal pressure of 6000 psi (see Eqs. in Section 3-2). The hoop stress σ_h is positive (10,000 psi—OB); the radial stress is $\sigma_r = -6000$ psi, OA. The longitudinal stress is $\sigma_z = 2000$ psi, OC. The maximum shear stress is ± 8000 psi, CD.

When a three-dimensional cube has shear stresses acting on the faces as well as compressive or tensile stresses, the solution for principal stresses becomes more complex. Graphical solutions (6) as well as an analytical method (7) are available.

Equilibrium and compatability relationships for multidimensional stress and strain

The foregoing analyses have resulted in equations describing the state of stress and strain at a point on or within a body. Consideration will now be given to the equilibrium and compatability that must be maintained in any physical situation. Ordinarily, the considerations to be developed here are treated in the mathematical theory of elasticity. Standard texts may be consulted for a full development of the theory and its applications (27). For the purposes of this book, a brief outline of the essential facts will be presented in order to facilitate the solution of some complex problems that cannot be solved by less elegant methods.

For any problem three conditions must always be fulfilled; force equilibrium, strain compatability, and the particular boundary conditions. Starting with the two-dimensional case, consider Fig. 3.24. Body forces (inertia, magnetic effects, etc.) are considered negligible; therefore for equilibrium in the x direction, for unit thickness,

$$[\sigma_x + (\partial\sigma_x/\partial x)\,dx]\,dy - \sigma_x\,dy + [\tau_{xy} + (\partial\tau_{xy}/\partial y)\,dy]\,dx - \tau_{xy}\,dx = 0$$

or

$$\partial\sigma_x/\partial x + \partial\tau_{xy}/\partial y = 0 \qquad 3.76$$

Similarly, in the direction of the y axis

$$\partial\sigma_y/\partial y + \partial\tau_{xy}/\partial x = 0 \qquad 3.77$$

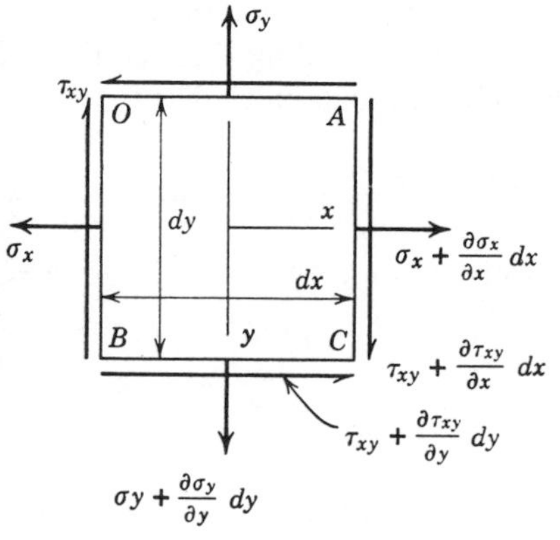

Fig. 3.24 Element under two-dimensional stress.

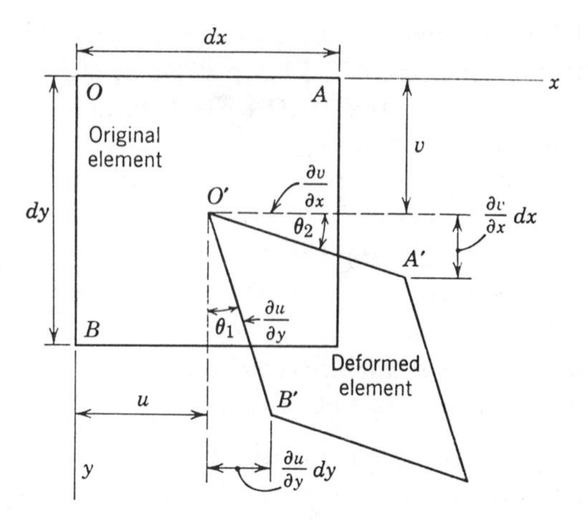

Fig. 3.25 Deformation of element under two-dimensional stress.

Equations 3.76 and 3.77 are the two equations of equilibrium for two-dimensional analysis; there are three unknowns, however, to evaluate the third unknown the deformation involved must be considered (Fig. 3.25). For small angles

$$\tan\theta_1 \doteq (\partial u/\partial y)(dy/dy) = \partial u/\partial y \quad \text{and} \quad \tan\theta_2 = (\partial v/\partial x)(dx/dx) = \partial v/\partial x$$

In the x direction the displacement of point A is $u + (\partial u/\partial x)dx$. The increase in OA is $(\partial u/\partial x)\,dx$, giving a unit strain of

$$\varepsilon_x = (\partial u/\partial x)(dx/dx) = \partial u/\partial x \qquad 3.78$$

Similarly,

$$\varepsilon_y = \partial v/\partial y \qquad 3.79$$

The shearing strain is $\theta_1 + \theta_2$ for small angles. However, since $\theta_1 = \partial u/\partial y$ and $\theta_2 = \partial v/\partial x$, the shearing strain, that is, the change in angle BOA, is

$$\gamma_{xy} = \partial v/\partial x + \partial u/\partial y \qquad 3.80$$

The movement of point A in the y direction is $v + (\partial v/\partial x)\,dx$, and the movement of point B in the x direction is $u + (\partial u/\partial y)\,dy$. The preceding three equations now form the basis for establishing a condition of compatability. The following operations are carried out:

$$(\partial^2/\partial y^2)\varepsilon_x = (\partial^2/\partial y^2)(\partial u/\partial x) = \partial^3 u/\partial y^2 \partial x \qquad 3.81$$

$$(\partial^2/\partial x^2)\varepsilon_y = (\partial^2/\partial x^2)(\partial v/\partial y) = \partial^3 v/\partial x^2 \partial y \qquad 3.82$$

$$(\partial^2/\partial x\,\partial y)(\gamma_{xy}) = (\partial^2/\partial x \partial y)(\partial v/\partial x + \partial u/\partial y) = \partial^3 v/\partial x^2 \partial y + \partial^3 u/\partial x \partial y^2 \qquad 3.83$$

Therefore

$$\partial^2\varepsilon_x/\partial y^2 + \partial^2\varepsilon_y/\partial x^2 = \partial^2\gamma_{xy}/\partial x \partial y \qquad 3.84$$

Equation 3.84 combines normal and shear strains and is known as the *com-*

patability equation. If, instead of ε_x, ε_y, and γ_{xy} in Eq. 3.84, the equivalent stress values are substituted, that is, using Eqs. 3.37, 3.38, and 3.43, it is found that Eq. 3.84 transforms to

$$(\partial^2/\partial x^2 + \partial^2/\partial y^2)(\sigma_x + \sigma_y) = 0 \qquad 3.85$$

which is known as *Laplace's equation.* This equation was derived on the basis that the third principal stress was zero; that is, a condition of *plane stress* was assumed. If, instead of σ_z being zero, ε_z is considered to be zero, giving a condition of *plane strain*, Eq. 3.85 still results.

Equations 3.76, 3.77, and 3.85 may be solved by use of the *Airy stress function.* According to this concept there is some function ϕ of x and y such that if the stresses are expressed in terms of ϕ, equilibrium, compatability, and boundary conditions are satisfied. In terms of the Airy function (27),

$$\sigma_x = \partial^2\phi/\partial y^2 \qquad 3.86$$

$$\sigma_y = \partial^2\phi/\partial x^2 \qquad 3.87$$

$$\tau_{xy} = -\partial^2\phi/\partial x\partial y \qquad 3.88$$

Substitution of Eqs. 3.86 to 3.88 into the equilibrium equations shows that these equations are satisfied. When substituted into the compatability equation, however, the result is

$$\partial^4\phi/\partial x^4 + 2\partial^4\phi/\partial x^2\partial y^2 + \partial^4\phi/\partial y^4 = 0 \qquad 3.89$$

Equation 3.89 represent a multitude of physical situations of which some may be unreal if the boundary conditions cannot be satisfied.

For the three-dimensional case consider Fig. 3.26. In a manner identical to that just discussed for the two-dimensional case, force summations in the three

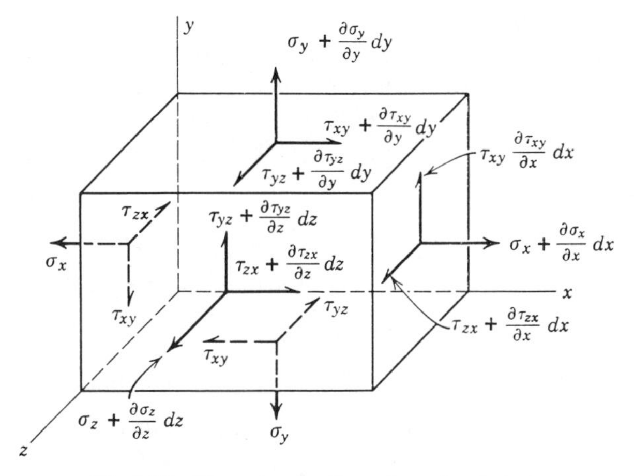

Fig. 3.26 Element under three-dimensional stress.

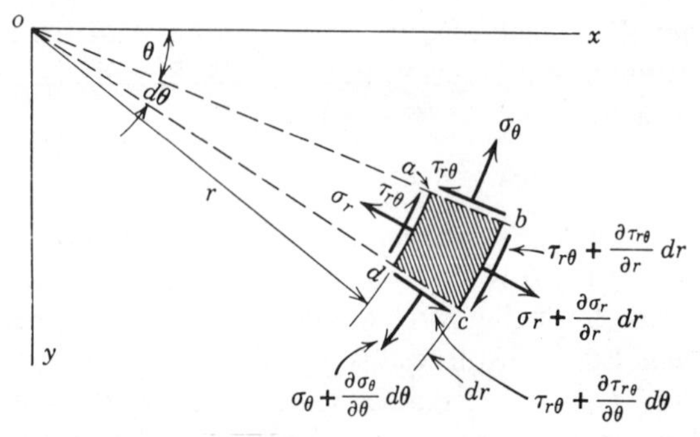

Fig. 3.27 Equilibrium in polar coordinates.

orthogonal directions give the results

$$\partial\sigma_x/\partial x + \partial\tau_{xy}/\partial y + \partial\tau_{zx}/\partial y = 0 \qquad 3.90$$

$$\partial\sigma_y/\partial y + \partial\tau_{yz}/\partial z + \partial\tau_{xy}/\partial x = 0 \qquad 3.91$$

and

$$\partial\sigma_z/\partial z + \partial\tau_{zx}/\partial x + \partial\tau_{yz}/\partial y = 0 \qquad 3.92$$

These three equations are the equilibrium relationships for three-dimensional analysis. Also the various strains are as follows:

$$\varepsilon_x = \partial u/\partial x; \qquad \varepsilon_y = \partial v/\partial y; \qquad \varepsilon_z = \partial w/\partial z$$

and

$$\gamma_{xy} = \partial u/\partial x + \partial u/\partial y; \qquad \gamma_{yz} = \partial w/\partial y + \partial v/\partial z; \qquad \gamma_{zx} = \partial u/\partial z + \partial w/\partial x \qquad 3.93$$

These equations, together with Eqs. 3.63 to 3.65, give the three-dimensional compatability relationship

$$(\partial^2/\partial x^2 + \partial^2/\partial y^2 + \partial^2/\partial x^2)(\sigma_x + \sigma_y + \sigma_z) = 0 \qquad 3.94$$

In many practical problems the situation reduces to one of rotational symmetry; consequently, in order to solve these problems the equilibrium and compatability relationships developed for cartesian coordinates have to be expressed in polar coordinates r and θ. Consider Fig. 3.27. Body forces are neglected so that in the radial direction equilibrium is expressed by the following relationship:

$$\partial\sigma_r/\partial r + (1/r)(\partial\tau_{r\theta}/\partial\theta) + (\sigma_r - \sigma_\theta)/r = 0 \qquad 3.95$$

Similarly, in the hoop (θ) direction, equilibrium requires that

$$(1/r)(\partial\sigma_\theta/\partial\theta) + \partial\tau_{r\theta}/\partial r + 2\tau_{r\theta}/r = 0 \qquad 3.96$$

In terms of Airy's stress function which satisfies these equations

$$\sigma_r = (1/r)(\partial\phi/\partial r) + (1/r^2)(\partial^2\phi/\partial\theta^2) \qquad 3.97$$

$$\sigma_\theta = \partial^2\phi/\partial r^2 \qquad 3.98$$

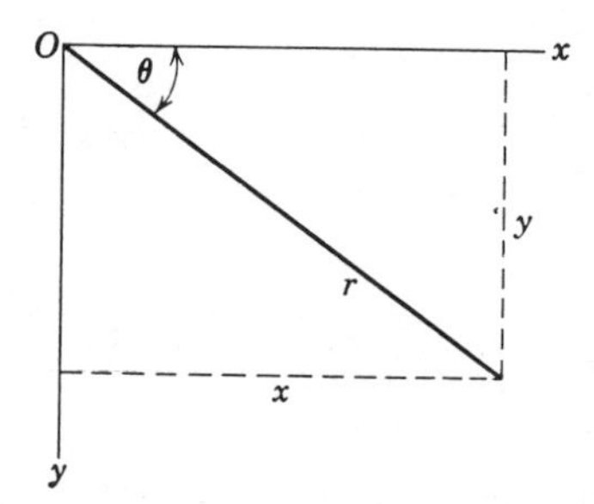

Fig. 3.28 Transformation of coordinates.

$$\tau_{r\theta} = (1/r^2)(\partial\phi/\partial\theta) - (1/r)(\partial^2\phi/\partial r\partial\theta) = (-\partial/\partial r)[(1/r)\,\partial\phi/\partial\theta] \qquad 3.99$$

In order to obtain a possible stress distribution, compatability as well as equilibrium and boundary conditions must be satisfied. The compatability equation in polar coordinates is obtained by transformation of Eq. 3.89. To effect the transformation consider Fig. 3.28 which shows
that

$$x^2 + y^2 = r^2 \quad \text{and} \quad \theta = \arctan y/x \qquad 3.100$$

Differentiation of Eq. 3.100 gives

$$\partial r/\partial x = x/r = \cos\theta \qquad 3.101$$

$$\partial r/\partial y = y/r = \sin\theta \qquad 3.102$$

$$\partial\theta/\partial x = -y/(x^2 + y^2) = -(1/r)\sin\theta \qquad 3.103$$

and

$$\partial\theta/\partial y = x/r^2 = (1/r)\cos\theta \qquad 3.104$$

Equation 3.89 can now be written in the form

$$(\partial^2/\partial x^2 + \partial^2/\partial y^2)(\partial^2\phi/\partial x^2 + \partial^2\phi/\partial y^2) = 0 \qquad 3.105$$

In order to use the terms in the second parentheses of Eq. 3.105 it is necessary to evaluate the derivative of a composite function. Considering ϕ as a function of r and θ

$$\partial\phi/\partial x = (\partial\phi/\partial r)(\partial r/\partial x) + (\partial\phi/\partial\theta)(\partial\theta/\partial x) \qquad 3.106$$

Using Eqs. 3.106, 3.101, and 3.103 leads to

$$\partial\phi/\partial x = \phi[\cos\theta(\partial/\partial r) - (\sin\theta/r)(\partial/\partial\theta)] \qquad 3.107$$

The second derivative is

$$\partial^2\phi/\partial x^2 = [\cos\theta(\partial/\partial r) - (1/r)\sin\theta(\partial/\partial\theta)]\partial\phi/\partial x + \phi(0) \qquad 3.108$$

Use of Eqs. 3.106 and 3.108 gives

$$\partial^2\phi/\partial x^2 = [(\partial/\partial r)\cos\theta - (1/r)\sin\theta(\partial/\partial\theta)][(\partial\phi/\partial r)\cos\theta - (1/r)(\partial\phi/\partial\theta)\sin\theta] \qquad 3.109$$

Similarly,

$$\partial^2\phi/\partial y^2 = (\partial^2\phi/\partial r^2)\sin^2\theta + (2\partial^2\phi/\partial\theta\partial r)(\sin\theta\cos\theta/r) + (\partial\phi/\partial r)(\cos^2\theta/r) - (2\,\partial\phi/\partial\theta)(\sin\theta\cos\theta/r^2) + (\partial^2\phi/\partial\theta^2)(\cos^2\theta/r^2) \qquad 3.110$$

And finally the compatability relationship

$$(\partial^2/\partial r^2 + (1/r)(\partial/\partial r) + 1/r^2\,\partial^2/\partial\theta^2)(\partial^2\phi/\partial r^2 + 1/r\,\partial\phi/\partial r + 1/r^2\,\partial^2\phi/\partial\theta^2) = 0 \qquad 3.111$$

Practical examples of the use of the foregoing theory will be given later in the chapter.

3-2 ANALYSES OF COMPLEX STRESS-STRAIN BEHAVIOR IN THE ELASTIC RANGE

Curved beams and tubes

Beams. Curved bars were mentioned in Chapter 1 in the section on size and shape effects and it was pointed out that in such members the centroid and neutral axes do not coincide. The results of the theory of curved bars are particularly useful in the design of hooks, chain links, clamps, and other curved structural and machine parts. When the curvature is large, the elementary bending stress formula $\sigma = M/Z$ (see Chapter 2) does not apply because of the noncoincidence of the neutral and centroid axes. In curved bars the cross section does not remain plane except with pure bending so that the stress equations developed by the *strength of materials* approach, which neglects this fact, are not as accurate as the corresponding formulas for straight bars. In pure bending the discrepancy between the exact and inexact solutions comes about through neglect of the effect of lateral stress in the inexact theory. For the purposes of this book the "inexact" theory will be developed first and then compared with the solution developed by means available in the mathematical theory of elasticity.

Consider Fig. 3.29, which shows a portion of a curved beam subtended by angle θ. The radius of curvature of the neutral axis is ρ, and of the centroidal axis ρ'. Bending moments M tend to decrease the radius of curvature. In the initial state an element of length L can be expressed as

$$L = R\theta \qquad 3.112$$

After bending, the element is extended by an amount $y\phi$; the strain in the element is thus

$$\varepsilon = \frac{(R\theta + y\phi) - R\theta}{R\theta} = \frac{y\phi}{R\theta} \qquad 3.113$$

By Hooke's law, since $\sigma = E\varepsilon$, Eq. 3.113 becomes

$$\sigma = (y\phi/R\theta)E \qquad 3.114$$

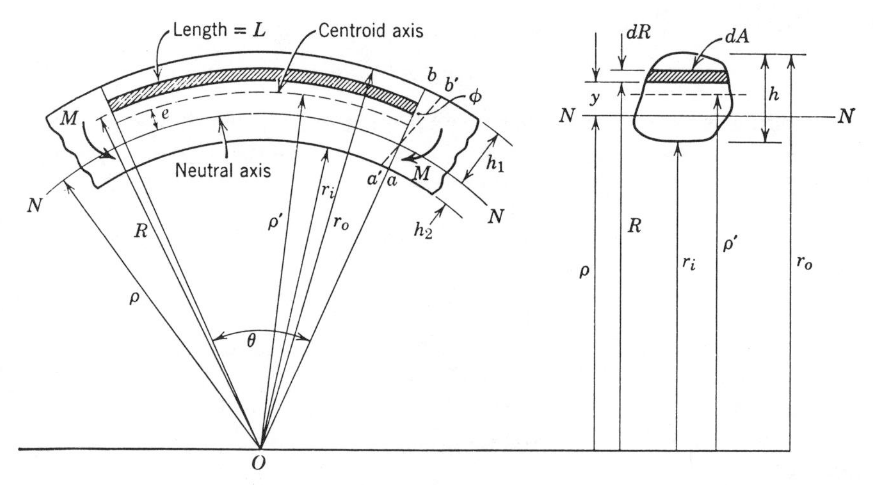

Fig. 3.29 Bending of initially curved bar.

The force on the element dA is $\sigma\, dA$ or

$$dF = \sigma\, dA = (y\phi/R\theta)E\, dA \tag{3.115}$$

For equilibrium the summation of forces over the entire cross section A must vanish; therefore

$$\int_{r_i}^{r_o} \left(\frac{y\phi}{R\theta}\right) E\, dA = 0$$

or

$$0 = \frac{E\phi}{\theta}\int_{r_i}^{r_o} \frac{y}{R}\, dA = \frac{E\phi}{\theta}\int_{r_i}^{r_o} \frac{(R-\rho)}{R}\, dA = \frac{E\phi}{\theta}A - \frac{E\phi\rho}{\theta}\int_{r_i}^{r_o} \frac{dA}{R} \tag{3.116}$$

From Eq. 3.116, ρ which determines the location of the neutral axis is given by

$$\rho = A \bigg/ \int_{r_i}^{r_o} \frac{dA}{R} \tag{3.117}$$

From equilibrium considerations the moment about O of the force acting on dA is

$$dM = (y\phi/R\theta)E\, dAR = (y\phi/\theta)E\, dA \tag{3.118}$$

The total moment ($\Sigma\, M$) is

$$M = \int dM = (\phi E/\theta)\int y\, dA = (E\phi/\theta)eA \tag{3.119}$$

In Eq. 3.119 the quantity $\int y\, dA$ represents the moment of the area about the neutral axis. From Eq. 3.119

$$E\phi/\theta = M/eA \tag{3.120}$$

and from Eq. 3.114, the bending stress is

$$\sigma = (y\phi/R\theta)E = My/eAR \qquad 3.121$$

The bending stresses in the outer fibers are determined by applying Eq. 3.121 with the fiber distance substituted for y; thus, for the geometry indicated in Fig. 3.29

$$\sigma_{\max} = Mh_1/Aer_0 \qquad 3.122$$

and

$$\sigma_{\min} = -Mh_2/Aer_i \qquad 3.123$$

If, in Fig. 3.29, the direction of the moment is reversed, the algebraic signs in Eqs. 3.122 and 3.123 are also reversed.

In addition to the bending stress there is also an axial stress equal to the applied load divided by the cross-sectional area of the member. The total fiber stress is then

$$\sigma = \sigma_b + P/A \qquad 3.124$$

where σ_b is given by Eqs. 3.122 or 3.123, P is applied load, and A is the cross-sectional area.

It is important to note that the various quantities calculated have to be determined very accurately so that in many cases slide-rule computations are inadequate. The calculation of deflections of curved beams is most conveniently handled by elastic energy theory and this is discussed, with examples, in Chapter 7.

Some examples of the use of the preceding relationships are:

1. Determine the stresses in curved bars of various cross-sectional geometry; see also (24 and 30).

Consider a rectangular cross section shown in Fig. 3.30. From Eq. 3.117

$$\rho = A \Big/ \int_{r_i}^{r_0} \frac{dA}{R} = bh \Big/ b \int_{r_i}^{r_0} \frac{dR}{R} = h/(\ln r_0/r_i) \qquad 3.125$$

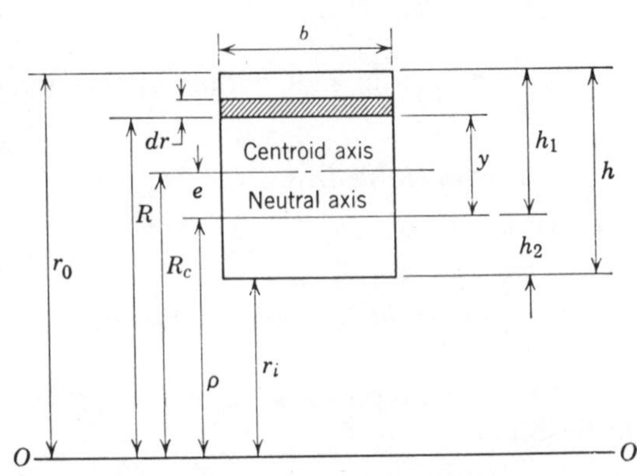

Fig. 3.30 Curved beam with rectangular cross section.

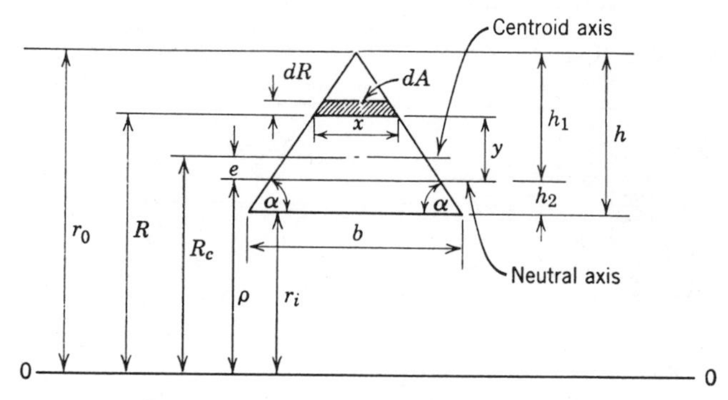

Fig. 3.31 Curved beam of triangular cross section.

However, since

$$e = r_i + h/2 - h/(\ln r_0/r_i) \tag{3.126}$$

the stress is calculated by substituting Eq. 3.126 into Eq. 3.121, which gives, for the bending stress,

$$\sigma = \frac{My}{bhR\left[r_i + \dfrac{h}{2} - \dfrac{h}{\ln r_0/r_i}\right]} \tag{3.127}$$

In order to compare this result with that obtained by application of the elementary theory ($\sigma = M/Z$) Eq. 3.127 is rewritten for $\sigma_{\max}$, using Eq. 3.122. For this particular geometry there is an indicated error of over 30% at $R_c/h = 1.0$ if the elementary theory is used.

For a cross section in the form of an isosceles triangle, Fig. 3.31,

$$\rho = A \Big/ \int_{r_i}^{r_0} \frac{dA}{R} = bh \Big/ 2 \int_{r_i}^{r_0} \frac{x\,dR}{R} = bh \Big/ 2 \int_{r_i}^{r_0} \frac{b(r_0 - R)}{h} \frac{dR}{R} \tag{3.128}$$

Equations defining the quantity $\int dA/R$ for a variety of cross sections are given in Table 3.1. Or

$$\rho = \frac{h^2}{2(r_0 \ln (r_0/r_i) - r_0 + r_i)} \tag{3.129}$$

However, $e = r_i + \frac{1}{3}h - \rho$, or

$$e = r_i + \frac{h}{3} - \frac{h^2}{2(r_0 \ln (r_0/r_i) - h)} \tag{3.130}$$

Therefore from Eq. 3.121

$$\sigma = \frac{2My}{\left[r_i + \dfrac{h}{3} - \dfrac{h^2}{a(r_0 \ln (r_0/r_i) - h)}\right] bhR} \tag{3.131}$$

*Table 3.1 Equations for Curved Bars**

Cross Section	$\int \frac{dA}{R}$
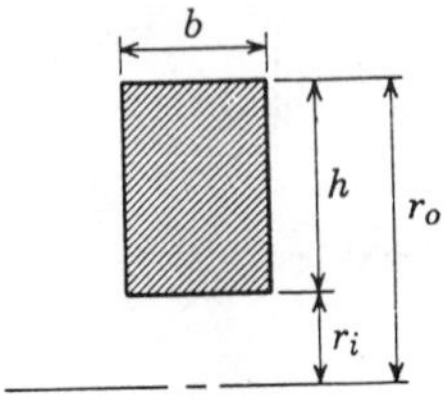 Rectangle	$b \ln \frac{r_0}{r_i}$
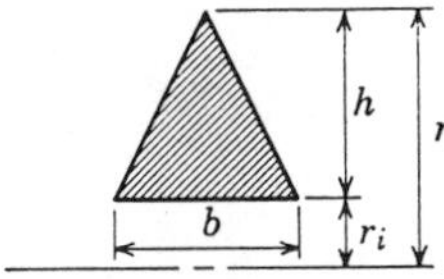 Triangle	$\left(\frac{br_0}{h} \ln \frac{r_0}{r_i}\right) - b$
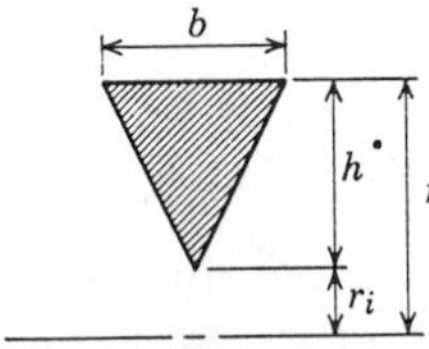 Triangle	$b - \left(\frac{br_i}{h} \ln \frac{r_0}{r_i}\right)$
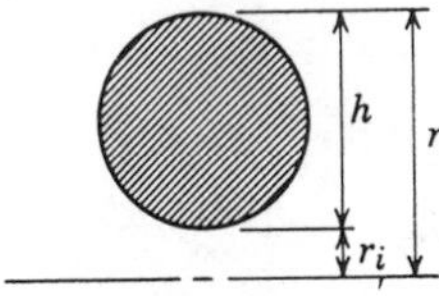 Circle	$2\pi\left\{\left(r_i + \frac{h}{2}\right) - \left[\left(r_i + \frac{h}{2}\right)^2 - \frac{h^2}{4}\right]^{1/2}\right\}$
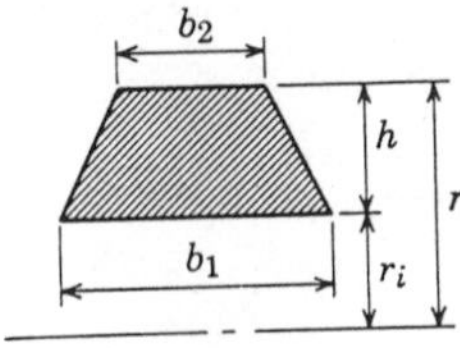 Trapezoid	$\left(\frac{b_1r_0 - b_2r_i}{h} \ln \frac{r_0}{r_i}\right) - b_1 + b_2$

*See Reference 24 for other shapes, including fillets and quarter circles.

Table 3.1 (Continued).

Cross Section	$\int \frac{dA}{R}$
Semicircle	$2h + \pi r_0 - \frac{\pi}{45}(\theta + \beta)(r_0^2 - h^2)^{1/2}$ $\theta = \arctan \frac{r_0 - h}{(r_0^2 - h^2)^{1/2}}$ (in degrees) $\beta = \arctan \frac{h}{(r_0^2 - h^2)^{1/2}}$ (in degrees)
⊥-section	$b_1 \ln \frac{r_i + h_1}{r_i} + b_2 \ln \frac{r_0}{r_i + h_1}$
Modified I-beam	$b_1 \ln \frac{r_i + h_1}{r_i} + b_2 \ln \frac{r_0 - h_3}{r_i + h_1} + b_3 \ln \frac{r_0}{r_0 - h_3}$

By similar analysis, for a circular cross section, Fig. 3.32, it can be shown that

$$\rho = \frac{h^2}{4[2(r_i + h/2) - \sqrt{4(r_i + h/2)^2 - h^2}]} \qquad 3.132$$

$$e = r_i + h/2 - \rho \qquad 3.133$$

and

$$\sigma = \frac{4My}{\pi R h^2 \left[R_c - \frac{h^2}{4(2R_c - \sqrt{4R_c{}^2 - h^2})} \right]} \qquad 3.134$$

2. Determine the stresses in the curved bar of circular cross section (Fig. 3.33) acting as a crane hook.

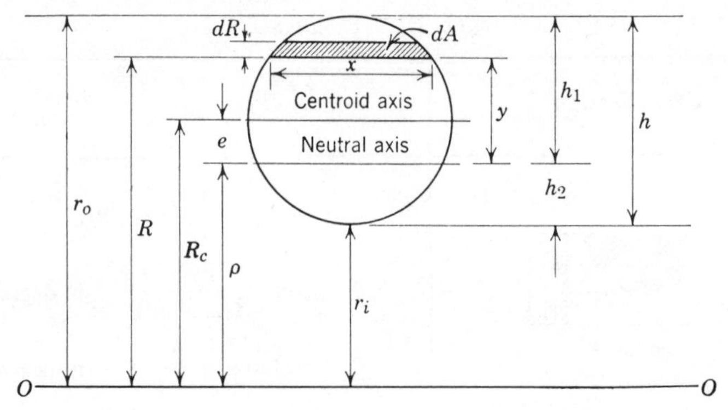

Fig. 3.32 Curved beam of circular cross section.

From Eq. 3.132, the radius of curvature is

$$\rho = \frac{4}{4[2(2+1) - \sqrt{4(2+1)^2 - 4}]} = 2.914 \text{ in.}$$

From Eq. 3.133

$$e = 2 + 1 - 2.914 = 0.086 \text{ in.}$$

and finally from Eqs. 3.122, 3.123, and Table 3.1,

$$\sigma_{\max} = \frac{(4)(5000)(3)(2.94 - 2)}{\pi(2)(4)\left[3 - \dfrac{4}{4[6 - \sqrt{36 - 4}]}\right]}$$

$$\sigma_{\max} = 25{,}435 \text{ psi}$$

$$\sigma_{\min} = \frac{-(4)(5000)(3)(1 + 0.06)}{\pi(4)(4)(0.06)} = -15{,}073 \text{ psi}$$

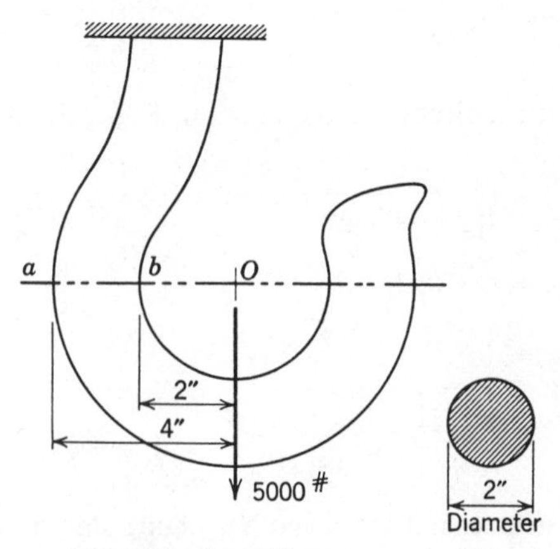

Fig. 3.33 Curved bar as crane hook.

There is also an axial force acting on the cross section equal to 5000/3.14 = 1590 psi. Therefore the total fiber stresses are

$$\sigma_{max} = 25{,}435 + 1590 = 27{,}025 \text{ psi}$$
$$\sigma_{min} = -15{,}073 + 1590 = -13{,}463 \text{ psi}$$

3. There is another approach to the previous example 2 (Figure 3.33) where a correction factor is applied to the basic equations.

$$\sigma_{max} = K_i\left[\frac{P}{A} \pm \frac{Mc}{I}\right] \qquad 3.135$$

$$\sigma_{min} = K_0\left[\frac{P}{A} \pm \frac{Mc}{I}\right] \qquad 3.136$$

The values P and M are taken with respect to the centroid of the cross section ab. Since the moment M creates tension on the inside fiber, the plus sign in Eq. 3.135 and the minus sign in Eq. 3.136 are used in the calculations. The value R for the distance from the center of curvature to the centroid of the cross section ab is determined for this example $R = 3$. C is 1, and R/C equals 3. R/C is used to select values (20) (23) for K_0 and K_i. For the example

$$R/C = 3 \qquad K_i = 1.33 \qquad K_0 = 0.79$$

$$\sigma_{max} = 1.33\left[\frac{5000}{3.14} + \frac{(5000)(3)1}{[\pi(2)^4/64]}\right]$$

$$\sigma_{max} = 27{,}517 \text{ psi}$$

$$\sigma_{min} = 0.790\,[1592 - 19{,}098.59]$$
$$= -13830 \text{ psi}$$

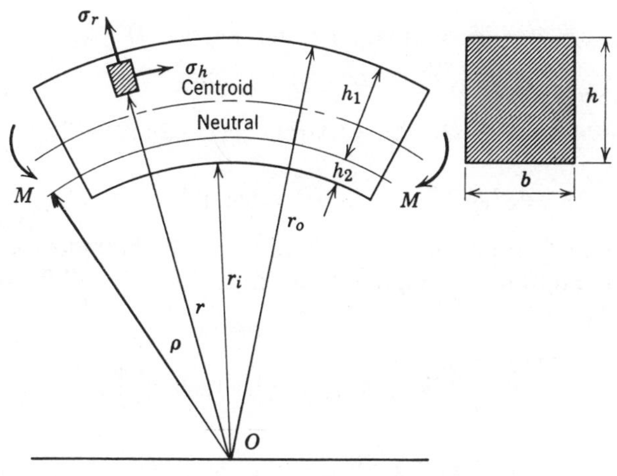

Fig. 3.34 Bending of curved bar—polar coordinates.

This method compares closely with that of example 2 but is in error on the σ_{min} value when compared to an elasticity solution such as presented in example 4.

4. Derive an equation for the stress in a curved beam of rectangular cross section using Airy's stress function.

The beam in question is shown in Fig. 3.34. For the geometry shown the radial stress σ_r is zero at the inner and outer boundaries of the beam; this is one of the boundary conditions. In addition, the summation of forces over the cross section must be zero, and these forces must induce an internal moment.

Symbolically,

$$(\sigma_r)_{r=r_i,r_0} = 0 \tag{3.137}$$

$$\int_{r_i}^{r_0} \sigma_h \, dr = 0 \tag{3.138}$$

$$\int_{r_i}^{r_0} \sigma_h r \, dr = M \tag{3.139}$$

For the symmetrical stress distribution obtained here the stresses are given by Eqs. 3.97 to 3.99 which satisfy the compatability Eq. 3.111. Since there is no boundary shear ($\tau_{r\theta} = 0$), equilibrium is expressed by Eq. 3.95 with zero shear; that is,

$$\partial\sigma_r/\partial r + (1/r)(\sigma_r - \sigma_\theta) = 0 \tag{3.140}$$

Since the stress distribution is symmetrical, the stress function ϕ depends only on r; thus Eq. 3.111 can be written

$$\left(\frac{\partial^2}{\partial r^2} + \frac{1}{r}\frac{\partial}{\partial \theta}\right)\left(\frac{\partial^2 \phi}{\partial r^2} + \frac{1}{r}\frac{\partial \phi}{\partial r}\right) = \frac{\partial^4 \phi}{\partial r^4} + \frac{2}{r}\frac{\partial^3 \phi}{\partial r^3} - \frac{1}{r^2}\frac{\partial^2 \phi}{\partial r^2} + \frac{1}{r^3}\frac{\partial \phi}{\partial r} = 0 \tag{3.141}$$

The stress function satisfying Eq. 3.139 is (21)

$$\phi = A \ln r + Br^2 \ln r + Cr^2 + D \tag{3.142}$$

Substituting Eq. 3.140 into Eqs. 3.97 and 3.98 gives

$$\sigma_r = (1/r)(\partial\phi/\partial r) = (A/r^2) + B(1 + 2 \ln r) + 2C \tag{3.143}$$

and

$$\sigma_h = \partial^2\phi/\partial r^2 = (-A/r^2) + B(3 + 2 \ln r) + 2C \tag{3.144}$$

where A, B, and C are constants determined by the boundary conditions. By using these conditions, as expressed by Eqs. 3.137 to 3.139, the preceding equations reduce to

$$\sigma_r = \frac{4M}{N}\left(\frac{r_i^2 r_0^2}{r^2} \ln \frac{r_0}{r_i} + r_0^2 \ln \frac{r}{r_0} + r_i^2 \ln \frac{r_i}{r}\right) \tag{3.145}$$

$$\sigma_h = \frac{4M}{N}\left(-\frac{r_i^2 r_0^2}{r^2} \ln \frac{r_0}{r_i} + r_0^2 \ln \frac{r}{r_0} + r_i^2 \ln \frac{r_i}{r} + r_0^2 - r_i^2\right) \tag{3.146}$$

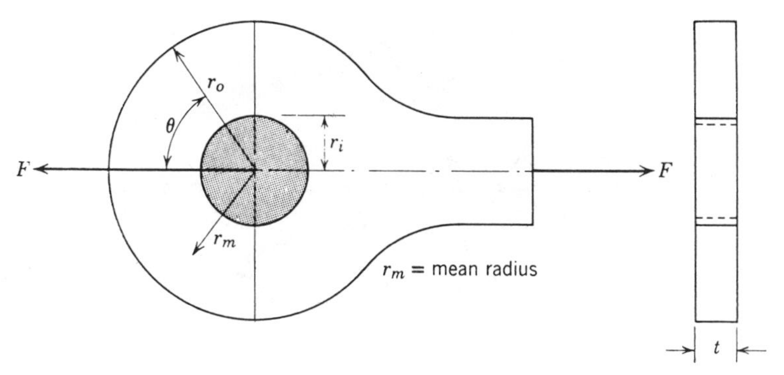

Fig. 3.35 Eye bar with solid insert for pulling.

where

$$N = (r_0{}^2 - r_i{}^2)^2 - 4r_i{}^2 r_0{}^2(\ln r_0/r_i)^2 \qquad 3.147$$

Use of Eq. 3.146 gives results that are quite close to those obtained by using Eq. 3.127.

5. A useful application of curved-bar theory is in the analysis of chain links in which a rod is inserted for pulling as shown in Fig. 3.35. An analysis for the geometry shown has been worked out for the case of $r_0/r_i = 2$ and 4 assuming a perfectly rigid insert of radius r_i(18). From this analysis it has been shown that the tensile stress values in the eye are as follows:

$$\sigma = \frac{\phi 8F}{\pi^2 r_0 t} \qquad 3.148$$

where ϕ is defined by the following tabulation.

r_0/r_i	ϕ at θ equals 0°			45°			90°		
	r_i	r_m	r_0	r_i	r_m	r_0	r_i	r_m	r_0
2	1.781	1.070	1.013	0.935	1.010	1.010	4.296	0.850	0.282
4	1.610	0.519	0.602	1.504	0.583	0.374	4.391	0.589	0.200

The maximum stress is at r_i at $\theta = 90°$.

Tubes. In addition to curved solid bars, curved hollow tubes also find applications in such parts as Bourdon pressure tubes (34, 35) and pipe bends. In general, hollow tubes are much less rigid than solid bars; they approach the rigidity of the solid bar when filled with an incompressible fluid. Consider the curved tube shown in Fig. 3.36; it is assumed that a/R is small enough so that the neutral and centroidal axes coincide. When the tube is bent, there is a

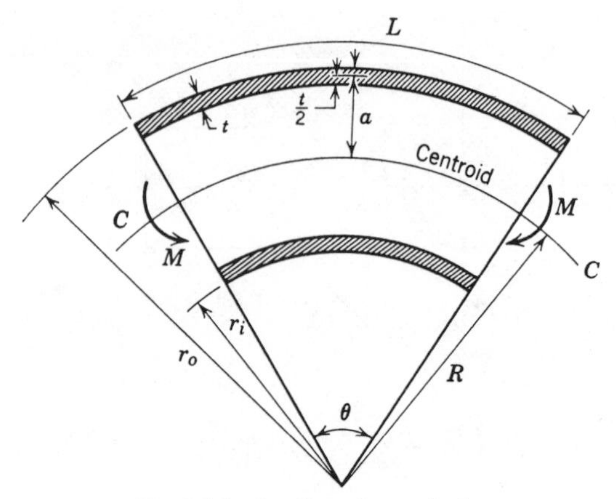

Fig. 3.36 Bending of curved tube.

flattening of the cross section and stresses are developed that are different from that obtained in a solid bar.

FOR CIRCULAR TUBES (31)

$$\sigma = (My/KI)(1 - \beta y^2/a^2) \qquad 3.149$$

where

$$K = 1 - \frac{9}{10 + 12(tR/a^2)^2} \qquad 3.150$$

$$\beta = \frac{6}{5 + 6(tR/a^2)^2} \qquad 3.151$$

y = distance from neutral axis

FOR A HOLLOW SQUARE TUBE (26)

$$\sigma = \frac{My(1 - \beta y^2/a^2)}{KI\left(\dfrac{1 + 0.0270n}{1 - 0.0656n}\right)} \qquad 3.152$$

where $n = b^4/R^2t^2$ and b is the length of a side of the square.

When internal pressure is added to the curved tube, the situation changes considerably. An analysis of this problem gives the following results for hoop stress for curved pipes with internal pressure only and no external bending forces (21).

$$\sigma_h = 0.545\frac{pa}{t}\left[\frac{tR}{a^2}\right]^{1/3}\left\{\frac{1 + (pa/tE)[6(a/t)^{4/3}(R/a)^{1/3}]}{1 + (pa/tE)[3.25(a/t)^{3/2}(R/a)^{2/3}]}\right\} \qquad 3.153$$

where p is the internal pressure.

Twisting of circular rings

Circular rings that are subjected to a continuous twisting moment along the center line (Fig. 3.37) are of interest in many machine design applications, in the

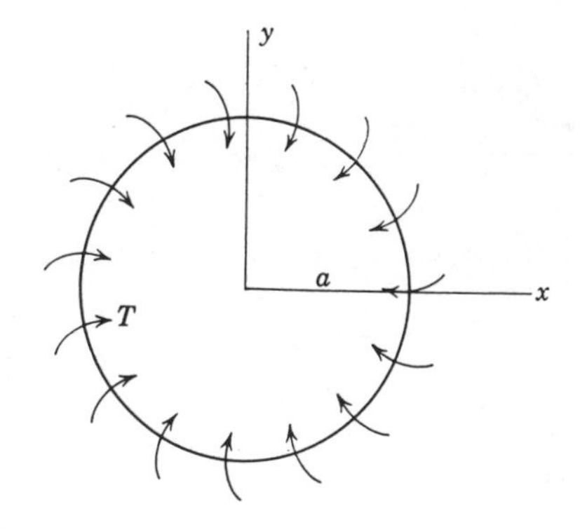

Fig. 3.37 Circular ring under torque.

analysis of pressure seals, and in the design of flanges (28). Some of the results of the theory are given here. Referring to Fig. 3.37, the torque produces a constant bending moment at each section equal to Ta; this moment gives a twist

$$\theta = Ta^2/EI \qquad 3.154$$

and a stress

$$\sigma = Tay/I \qquad 3.155$$

For a rectangular cross section (Fig. 3.38),

$$\theta = \frac{12Ta}{Eh^3 \ln d/c} \qquad 3.156$$

$$\sigma = \frac{12Tay}{h^3 r \ln d/c} \qquad 3.157$$

For a triangular section (Fig. 3.39),

$$\theta = \frac{Ta}{2Eh(a^2 \ln a/2 - r^2/4)} \qquad 3.158$$

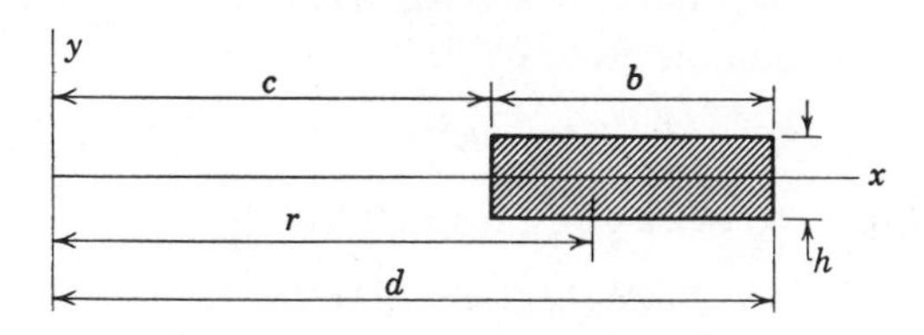

Fig. 3.38 Circular ring with rectangular section.

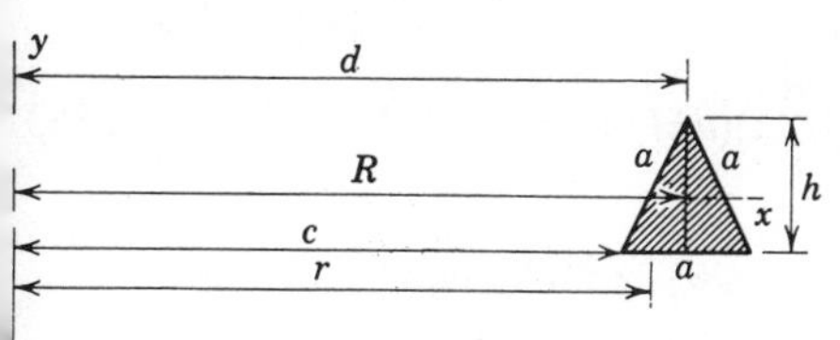

Fig. 3.39 Circular ring with triangular section.

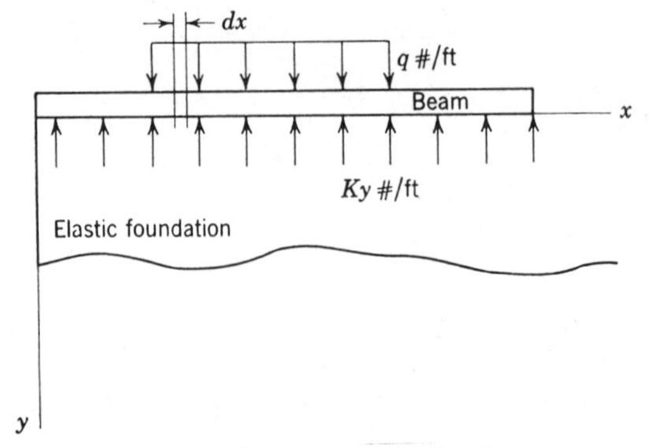

Fig. 3.40 Beam on elastic foundation.

and

$$\sigma = \frac{Tya}{r[2h(a^2 \ln a/2 - r^2/4)]} \tag{3.159}$$

Beams and plates on elastic foundations

Beams and plates on elastic foundations have special application in many structural engineering problems dealing with road beds, railways, piles, and others. The theory has also been used in the analysis of stresses at the juncture of head and shell in pressure vessels and in the design of underwater cameras. A detailed account of the theory has been made available by Hetenyi (9). The subject has also been presented in advanced texts on the strength of materials (1, 20, 28). For the purposes of this book a brief outline of the theory will be given.

Consider Fig. 3.40 which shows a beam resting on an elastic foundation. It is assumed that the reaction offered by the foundation to the deflection of the beam is proportional to the deflection. Thus the reaction per unit length of the beam is Ky, where y is deflection and K is a constant called the *modulus of foundation.** The deflection at any point is given by

$$EI\, d^2y/dx^2 = -M \tag{3.160}$$

Differentiating Eq. 3.160 twice gives, respectively,

$$EI\, d^3y/dx^3 = -dM/dx \tag{3.161}$$

and

$$EI\, d^4y/dx^4 = -d^2M/dx^2 \tag{3.162}$$

From Eq. 3.162

$$EI\, d^4y/dx^4 = -d^2M/dx^2 = -(d/dx)(dM/dx) = -d/dx \quad \text{(shear)} \tag{3.163}$$

*If 1000 lb distributed over 1 in. of length causes a deflection of 1 in., then K = 1000 psi.

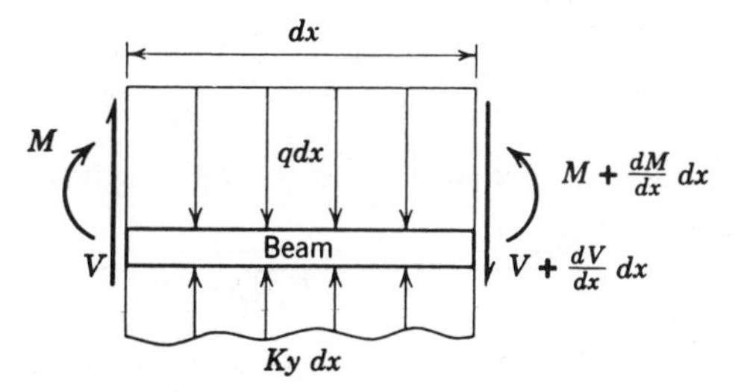

Fig. 3.41 Equilibrium of differential element in Fig. 3.40.

If the vertical shear is called V, Eq. 3.163 becomes

$$EI\, d^4y/dx^4 = -dV/dx \tag{3.164}$$

Now consider a differential element of the beam (Fig. 3.41). By statics,

$$q\,dx - V + [V + (dV/dx)dx] - Ky\,dx = 0 \tag{3.165}$$

from which

$$dV/dx = Ky - q \tag{3.166}$$

Substituting Eq. 3.166 into Eq. 3.164 gives the result

$$EI\, d^4y/dx^4 = -dV/dx = -(Ky - q) = q - Ky \tag{3.167}$$

which is a general equation. For the unloaded part of the beam q is zero; therefore from Eq. 3.167

$$EI\, d^4y/dx^4 = -Ky \tag{3.168}$$

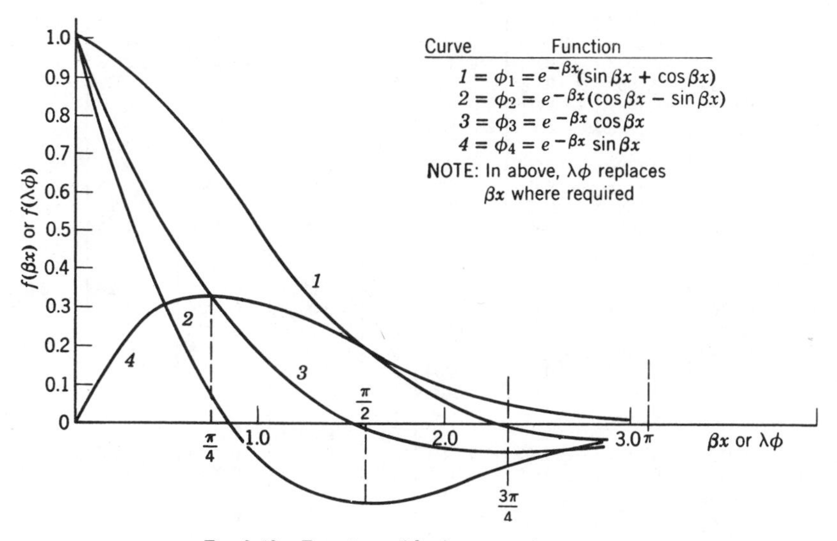

Fig. 3.42 Functions of βx for computing stresses.

The solution for which is

$$y = e^{\beta x}(A \cos \beta x + B \sin \beta x) + e^{-\beta x}(C \cos \beta x + D \sin \beta x) \qquad 3.169$$

where A, B, C, and D are constants of integration determined by the boundary conditions of the problem and

$$\beta = \sqrt[4]{K/4EI} \qquad 3.170$$

The solution of Eq. 3.169 for specific problems is facilitated by Fig. 3.42 and Table 3.2 which give values for the various functions of βx.

Table 3.2 Functions of ϕ. (Adapted From Timoshenko and Woinowsky-Krieger (29), courtesy of McGraw-Hill Book Co.)

βx	ϕ_1	ϕ_2	ϕ_3	ϕ_4
0	1.0000	1.0000	1.0000	0
0.1	0.9907	0.8100	0.9003	0.0903
0.2	0.9651	0.6398	0.8024	0.1627
0.3	0.9267	0.4888	0.7077	0.2189
0.4	0.8784	0.3564	0.6174	0.2610
0.5	0.8231	0.2415	0.5323	0.2908
0.6	0.7628	0.1431	0.4530	0.3099
0.7	0.6997	0.0599	0.3798	0.3199
0.8	0.6354	−0.0093	0.3131	0.3223
0.9	0.5712	−0.0657	0.2527	0.3185
1.0	0.5083	−0.1108	0.1988	0.3096
1.1	0.4476	−0.1457	0.1510	0.2967
1.2	0.3899	−0.1716	0.1091	0.2807
1.3	0.3355	−0.1897	0.0729	0.2626
1.4	0.2849	−0.2011	0.0419	0.2430
1.5	0.2384	−0.2068	0.0158	0.2226
1.6	0.1959	−0.2077	−0.0059	0.2018
1.7	0.1576	−0.2047	−0.0235	0.1812
1.8	0.1234	−0.1985	−0.0376	0.1610
1.9	0.0932	−0.1899	−0.0484	0.1415
2.0	0.0667	−0.1794	−0.0563	0.1230
2.1	0.0439	−0.1675	−0.0618	0.1057
2.2	0.0244	−0.1548	−0.0652	0.0895
2.3	0.0080	−0.1416	−0.0668	0.0748
2.4	−0.0056	−0.1282	−0.0669	0.0613
2.5	−0.0166	−0.1149	−0.0658	0.0492
2.6	−0.0254	−0.1019	−0.0636	0.0383
2.7	−0.0320	−0.0895	−0.0608	0.0287
2.8	−0.0369	−0.0777	−0.0573	0.0204
2.9	−0.0403	−0.0666	−0.0534	0.0132
3.0	−0.0423	−0.0563	−0.0493	0.0071
3.1	−0.0431	−0.0469	−0.0450	0.0019
3.2	−0.0431	−0.0383	−0.0407	−0.0024
3.3	−0.0422	−0.0306	−0.0364	−0.0058
3.4	−0.0408	−0.0237	−0.0323	−0.0085

βx	ϕ_1	ϕ_2	ϕ_3	ϕ_4
3.5	−0.0389	−0.0177	−0.0283	−0.0106
3.6	−0.0366	−0.0124	−0.0245	−0.0121
3.7	−0.0341	−0.0079	−0.0210	−0.0131
3.8	−0.0314	−0.0040	−0.0177	−0.0137
3.9	−0.0286	−0.0008	−0.0147	−0.0140
4.0	−0.0258	0.0019	−0.0120	−0.0139
4.1	−0.0231	0.0040	−0.0095	−0.0136
4.2	−0.0204	0.0057	−0.0074	−0.0131
4.3	−0.0179	0.0070	−0.0054	−0.0125
4.4	−0.0155	0.0079	−0.0038	−0.0117
4.5	−0.0132	0.0085	−0.0023	−0.0108
4.6	−0.0111	0.0089	−0.0011	−0.0100
4.7	−0.0092	0.0090	0.0001	−0.0091
4.8	−0.0075	0.0089	0.0007	−0.0082
4.9	−0.0059	0.0087	0.0014	−0.0073
5.0	−0.0046	0.0084	0.0019	−0.0065
5.1	−0.0033	0.0080	0.0023	−0.0057
5.2	−0.0023	0.0075	0.0026	−0.0049
5.3	−0.0014	0.0069	0.0028	−0.0042
5.4	−0.0006	0.0064	0.0029	−0.0035
5.5	0.0000	0.0058	0.0029	−0.0029
5.6	0.0005	0.0052	0.0029	−0.0023
5.7	0.0010	0.0046	0.0028	−0.0018
5.8	0.0013	0.0041	0.0027	−0.0014
5.9	0.0015	0.0036	0.0026	−0.0010
6.0	0.0017	0.0031	0.0024	−0.0007
6.1	0.0018	0.0026	0.0022	−0.0004
6.2	0.0019	0.0022	0.0020	−0.0002
6.3	0.0019	0.0018	0.0018	+0.0001
6.4	0.0018	0.0015	0.0017	0.0003
6.5	0.0018	0.0012	0.0015	0.0004
6.6	0.0017	0.0009	0.0013	0.0005
6.7	0.0016	0.0006	0.0011	0.0006
6.8	0.0015	0.0004	0.0010	0.0006
6.9	0.0014	0.0002	0.0008	0.0006
7.0	0.0013	0.0001	0.0007	0.0006

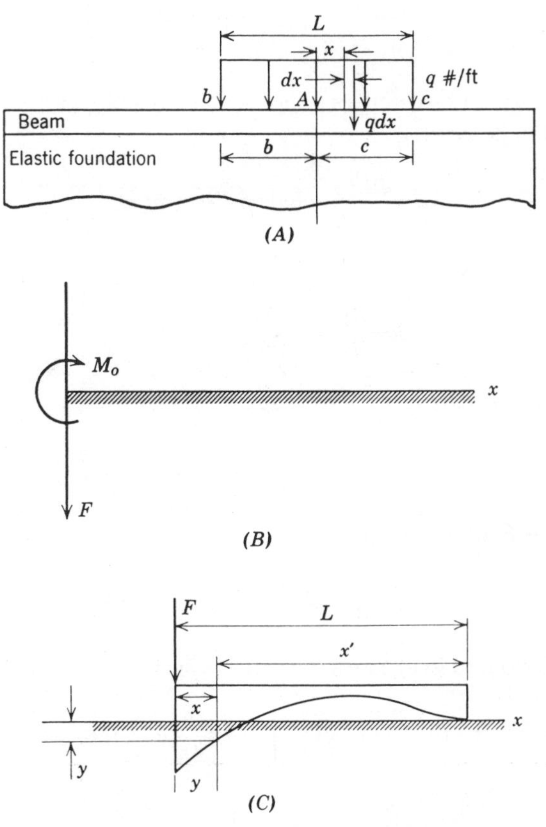

Fig. 3.43 Beam on elastic foundation.

To illustrate the use of the theory consider a uniformly loaded section of a beam resting on an elastic foundation (Fig. 3.43*A*). This problem is solved by considering the distributed load to be made up of a series of concentrated loads. At point A the deflection produced by a load $q\,dx$ will be determined. The basic relationship is given by Eq. 3.169. For points far removed from the load the deflection will be zero; this condition is satisfied if the constants A and B are zero. Therefore

$$y = e^{-\beta x}(C \cos \beta x + D \sin \beta x) \tag{3.171}$$

At the origin $x = 0$ the slope is zero ($dy/dx = 0$), requiring that $C = D$; therefore Eq. 3.171 becomes

$$y = Ce^{-\beta x}(\cos \beta x + \sin \beta x) \tag{3.172}$$

and from the condition that $V = dM/dx = -q\,dx$, when $q\,dx$ is treated as a concentrated load, the constant of integration C using the relation of Eq. 3.161 is

$$C = q\,dx/8\beta^3 EI \tag{3.173}$$

and finally,

$$y = (q\,dx/2K)\beta e^{-\beta x}(\cos\beta x + \sin\beta x) \qquad 3.174$$

The quantity $q\,dx$ is treated as a concentrated load and so Eq. 3.174, for convenience, may be written as

$$y = \frac{\beta F e^{-\beta x}}{2K}(\cos\beta x + \sin\beta x) \qquad 3.175$$

The slope θ is dy/dx so that

$$\theta = \frac{dy}{dx} = \frac{-Fe^{-\beta x}}{4\beta^2 EI}\sin\beta x \qquad 3.176$$

The moment M is $-EI(d^2y/dx^2)$ so that

$$-EI\frac{d^2y}{dx^2} = M = \frac{Fe^{-\beta x}}{4\beta}(\cos\beta x - \sin\beta x) \qquad 3.177$$

The shear Q is $-EI(d^3y/dx^3)$ so that

$$-EI\frac{d^3y}{dx^3} = Q = \frac{-Fe^{-\beta x}}{2}\cos\beta x \qquad 3.178$$

Then for the distributed load, since the deflections for the two parts add up,

$$y = \int_0^b q\,dx\frac{\beta}{2K}e^{-\beta x}(\cos\beta x + \sin\beta x) + \int_0^c q\,dx\frac{\beta}{2K}e^{-\beta x}(\cos\beta x + \sin\beta x) \qquad 3.179$$

or

$$y = (q/2K)(2 - e^{-\beta b}\cos\beta b - e^{-\beta c}\cos\beta c) \qquad 3.180$$

Since the slope in the Ab direction is positive and in the Ac direction is negative,

$$\theta = \int_0^b \frac{-q\,dx\,e^{-\beta x}}{4\beta^2 EI}\sin\beta x - \int_0^c \frac{-q\,dx\,e^{-\beta x}}{4\beta^2 EI}\sin\beta x$$

$$= \frac{q}{8\beta^3 EI}[e^{-\beta b}(\cos\beta b + \sin\beta b) - e^{-\beta c}(\cos\beta c + \sin\beta c)] \qquad 3.181$$

In addition,

$$M = (q/4\beta^2)[e^{-\beta b}\sin\beta b + e^{-\beta c}\sin\beta c] \qquad 3.182$$

$$Q = (q/4\beta)[e^{-\beta b}(\cos\beta b - \sin\beta b) - e^{-\beta c}(\cos\beta c - \sin\beta c)] \qquad 3.183$$

Note in Table 3.2 and Fig. 3.42 that the following notation is used:

$$\phi_1 = e^{-\beta x}(\cos\beta x + \sin\beta x) \qquad 3.184$$

$$\phi_2 = e^{-\beta x}(\cos\beta x - \sin\beta x) \qquad 3.185$$

$$\phi_3 = e^{-\beta x}\cos\beta x \qquad 3.186$$

$$\phi_4 = e^{-\beta x}\sin\beta x \qquad 3.187$$

For a semi-infinite beam such as shown in Fig. 3.43 B the solution given by Eq. 3.169 can be used. At points far removed from the load the deflection is zero, a condition that can only be satisfied if the constants A and B are zero; this gives the result expressed by Eq. 3.171. To determine the value of the constants C and D note that at $x = 0$

$$M_0 = -EI(d^2y/dx^2) \qquad 3.188$$

$$Q = -EI(d^3y/dx^3) \qquad 3.189$$

and

$$F = EI(d^3y/dx^3) \qquad 3.190$$

which, with Eq. 3.171, gives

$$C = (1/2\beta^3 EI)(F - \beta M_0) \qquad 3.191$$

and

$$D = M_0/2\beta^2 EI \qquad 3.192$$

Then substituting these values into Eq. 3.171 gives

$$y = (e^{-\beta x}/2\beta^3 EI)[F \cos \beta x - \beta M_0(\cos \beta x - \sin \beta x)] \qquad 3.193$$

or, using Eqs. 3.183 to 3.186,

$$y = (1/2\beta^3 EI)(F\phi_3 - \beta M_0 \phi_2) \qquad 3.194$$

For a beam of finite length such as shown in Fig. 3.43C, Eq. 3.169 is also used, and the constants of integration are evaluated from the boundary conditions. For this case,

$$\text{at } x = 0, \quad EI(d^2y/dx^2) = 0, \quad \text{and} \quad EI(d^3y/dx^3) = F$$

$$\text{at } x = L, \quad EI(d^2y/dx^2) = 0, \quad \text{and} \quad d^3y/dx^3 = 0$$

Thus differentiating Eq. 3.169 three times and using the preceding values give four linear equations for the constants; substituting these values back into Eq. 3.169 gives the following:

$$y_{x=0} = \frac{F}{2\beta^3 EI} \frac{\sinh \beta L \cosh \beta L - \sin \beta L \cos \beta L}{\sinh^2 \beta L - \sin^2 \beta L} \qquad 3.195$$

$$y_{x=L} = \frac{F}{2\beta^3 EI} \frac{\sinh \beta L \cos \beta L - \cosh \beta L \sin \beta L}{\sinh^2 \beta L - \sin^2 \beta L} \qquad 3.196$$

$$\theta = -\frac{F}{2\beta^2 EI}\left[\frac{\{\sinh \beta L(\sin \beta x \cosh \beta x' + \cos \beta x \sinh \beta x') + \sin \beta L(\sinh \beta x \cos \beta x' + \cosh \beta x \sin \beta x')\}}{\sinh^2 \beta L - \sin^2 \beta L}\right] \qquad 3.197$$

$$M = -\frac{F}{\beta} \frac{\sinh \beta L \sin \beta x \sinh \beta x' - \sin \beta L \sinh \beta x \sin \beta x'}{\sinh^2 \beta L - \sin^2 \beta L} \qquad 3.198$$

$$Q = -\frac{\{F[\sinh \beta L(\cos \beta x \sinh \beta x' - \sin \beta x \cosh \beta x') - \sin \beta L(\cosh \beta x \sin \beta x' - \sinh \beta x \cos \beta x')]\}}{\sinh^2 \beta L - \sin^2 \beta L} \qquad 3.199$$

Table 3.3 Some Solutions for Beams on Elastic Foundations

Geometry	Deflection, y	Moment, M	Shear, Q
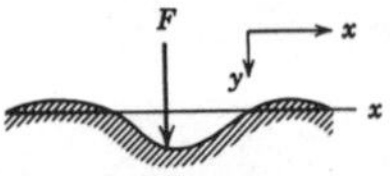 Infinite Beam, Concentrated Load	$\dfrac{F\phi_1}{8\beta^3 EI}$	$\dfrac{F\phi_2}{4\beta}$	$-\dfrac{F\phi_3}{2}$
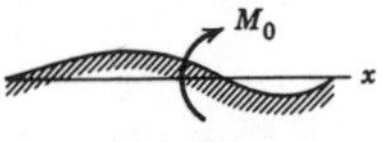 Infinite Beam, Applied Moment	$\dfrac{M_0\phi_4}{4\beta^2 EI}$	$\dfrac{M_0\phi_3}{2}$	$-\dfrac{M_0\beta\phi_1}{2}$
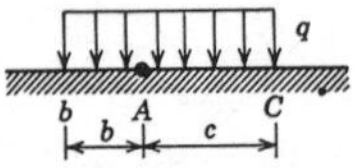 Infinite Beam, Distributed Load	$y_A = \dfrac{q}{8\beta^4 EI}(2 - \phi_{3b} - \phi_{3c})$	$M_A = \dfrac{q}{4\beta^2}[\phi_{4b} + \phi_{4c}]$	$Q_A = \dfrac{q}{4\beta}[\phi_{2b} - \phi_{2c}]$
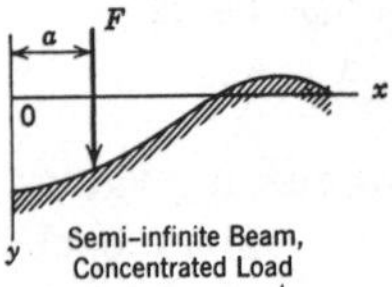 Semi-infinite Beam, Concentrated Load	$\dfrac{F}{8\beta^3 EI}[(\phi_{2a} + 2\phi_{3a})\phi_{1x} - 2(\phi_{2a} + \phi_{3a})\phi_{4x} + \phi_{1(a-x)}]$ *Note:* ϕ_{2a}, for example, in Eq. 3.185, $x = a$.	$\dfrac{F}{4\beta}[\phi_{2x}(\phi_{2a} + 2\phi_{3a}) - 2\phi_{3x}(\phi_{2a} + \phi_{4a}) + \phi_{2(a-x)}]$	$-\dfrac{F}{2}[\phi_{3x}(\phi_{2a} + 2\phi_{3a}) - \phi_{1x}(\phi_{2a} + \phi_{3a}) \pm \phi_{3(a-x)}]$
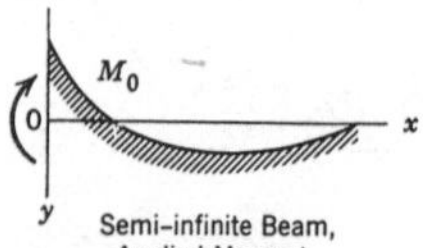 Semi-infinite Beam, Applied Moment	$-\dfrac{M_0\phi_2}{2\beta^2 EI}$	$M_0\phi_1$	$-2M_0\beta\phi_4$

Solutions to some additional problems are given in Table 3.3.

An analysis for thin circular plates on elastic foundations has been given by Galletly (8). In this analysis the plate is assumed to be elastically restrained at the edges against rotation and vertical deflection. Galletly found that it was possible to design plates in which the maximum stress was 25 to 50% smaller than in similar plates with a simply supported rim. For additional information on plates and shells consult the references cited.

Shearing stresses in beams

A fairly common example of combined stresses is that in beams whose lengths are short compared to the other dimensions. When this occurs shearing stresses induced by the loading cannot be ignored in the analysis; the analysis is then concerned with principal stresses since the component stresses resulting from bending and shear are not the critical values.

Consider first a beam of rectangular cross section, Fig. 3.44. Any element of the cross section is subjected to a shearing stress τ and a normal stress σ_x. If the moment on the section varies, that is, if M is at one location and $M + dM$ at another as shown, then the force on an element area of face ab is

$$\sigma_x \, dA = (My/I_z) \, dA \tag{3.200}$$

Similarly, on face cd

$$(\sigma_x + d\sigma_x) \, dA = \left(\frac{M + dM}{I_z}\right) y \, dA \tag{3.201}$$

The resultant of these forces is

$$\left(\frac{M + dM}{I_z}\right) y \, dA - \frac{My}{I_z} dA = \frac{dMy}{I_z} dA \tag{3.202}$$

and the resultant horizontal force above plane bd is

$$\frac{dM}{I_z} \int_{y_1}^{h/2} y \, dA \tag{3.203}$$

For horizontal equilibrium of $abcd$ this force equals the shear force $\tau b \, dx$ on face bd, and

$$\tau b \, dx = \frac{dM}{I_z} \int_{y_1}^{h/2} y \, dA \tag{3.204}$$

Hence

$$\tau = \frac{dM}{dx} \frac{1}{bI_z} \int_{y_1}^{h/2} y \, dA \tag{3.205}$$

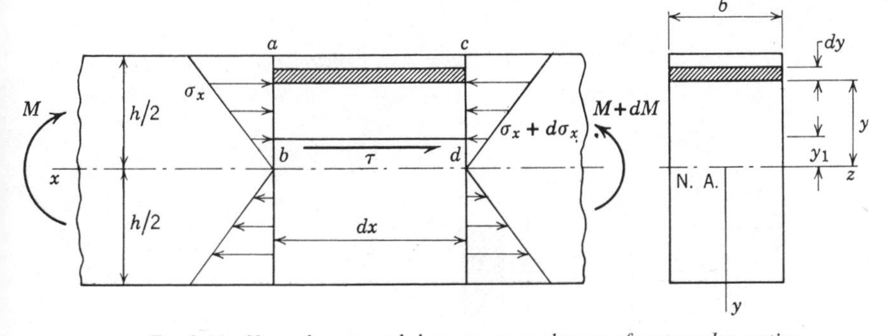

Fig. 3.44 Normal stress and shear stress on element of rectangular section.

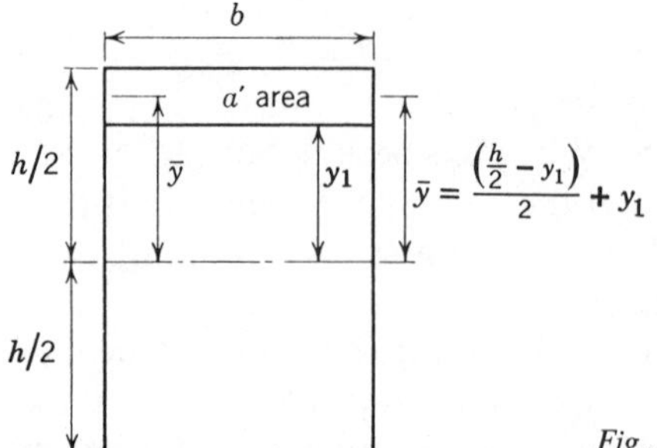

Fig. 3.45 Quantities used in determining vertical shear

However, since the shear force $V = dM/dx$, Eq. 3.205 reduces to

$$\tau = \frac{V}{I_z b} \int_{y_1}^{h/2} y\, dA \tag{3.206}$$

For the rectangular section $dA = b\, dy$; therefore

$$\tau = (V/Ib)a'\bar{y} = (V/2I_z)(h^2/4 - y_1{}^2) \tag{3.207}$$

where a' is the area of the cross section between y_1 and the edge (top or bottom) of the beam, and $\bar{y}$ is the distance from the neutral axis to the centroid of the area a' (Fig. 3.45). Equilibrium requires a horizontal shear stress to balance the vertical shear; this stress accounts for the longitudinal splits that are frequently seen in wooden beams. Equation 3.207 is the equation of a parabola (Fig. 3.46) and the maximum shear occurs at the neutral axis. With *pure bending* there is no differential bending moment so that $\tau = 0$.

For an end-loaded cantilever (Fig. 3.47) the maximum bending stress is at the fixed end

$$\sigma = (6PL/h^2b) = (6P/hb)\beta \tag{3.208}$$

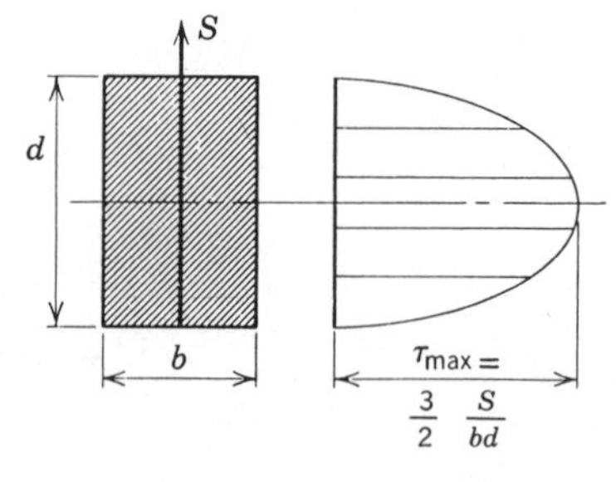

Fig. 3.46 Distribution of shear stress on cross section of rectangular beam.

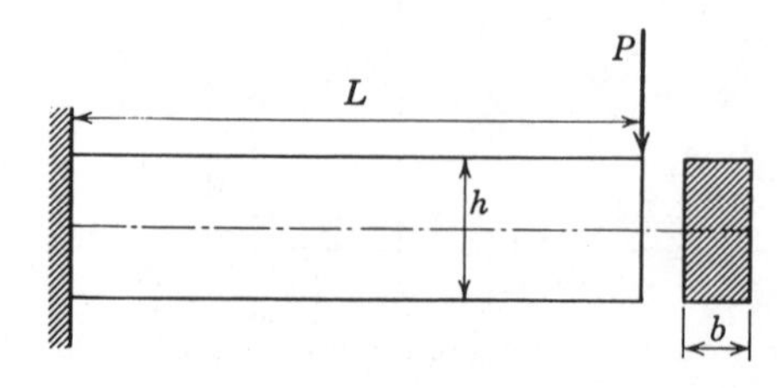

Fig. 3.47 Cantilever with end load.

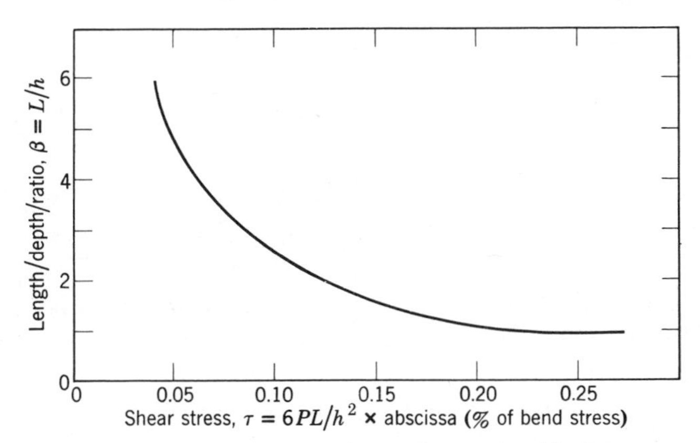

Fig. 3.48 Design curve for cantilever with end load.

where $\beta = L/h$. The maximum vertical shear from Eq. 3.207 is

$$\tau = (3P/2hb) = \sigma(0.25)(1/\beta) \tag{3.209}$$

where σ is given by Eq. 3.208. As can be seen, the maximum normal stress and the maximum shear stress do not necessarily occur at the same location. Equation 3.209 shows the relation between the normal and shear stresses for the particular case of the cantilever (also plotted in Fig. 3.48). Thus with beams of L/h ratios < 4 the effect of shear should be included in the analysis.

For a circular section (Fig. 3.49) it can be shown that

$$\tau = \frac{VR\sqrt{R^2 - {y_1}^2}}{3I_z} \tag{3.210}$$

The importance of shear can be further emphasized by noting that with the rectangular cross section the average shear determined by dividing force by area is

$$\tau_{\text{avg}} = V/A = V/bh \tag{3.211}$$

which compares with Eq. 3.207

$$\tau = 3V/2bh \tag{3.212}$$

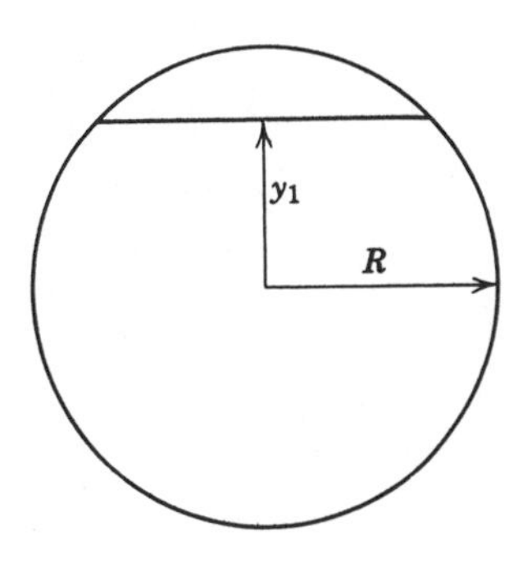

Fig. 3.49 Geometry of circular section subject to shear.

which is 50% greater than for Eq. 3.211.

For a circular cross section,

$$V/A = V/\pi R^2 \tag{3.213}$$

and by Eq. 3.209

$$\tau = \tfrac{4}{3}(V/\pi R^2) \tag{3.214}$$

which is 33% greater than for Eq. 3.213. For a thin circular tube the factor is 100%.

It is clear then that shear is an important factor in "short beams"; consequently, in determining design stresses, use is made of Eqs. 3.18 to 3.20. Thus

$$\sigma_1, \sigma_2 = (\sigma_x/2) \pm \sqrt{(\sigma_x/2)^2 + \tau^2} \tag{3.215}$$

and

$$\tau_{\max} = \pm \sqrt{(\sigma_x/2) + \tau^2} \tag{3.216}$$

where

$$\sigma_x = Mc/I \tag{3.217}$$

and τ is given by Eqs. 3.207 or 3.210 (for rectangular or circular sections). The shear also produces shear strains which must be included in Eqs. 3.47 to 3.49. Deflection resulting from shear is analyzed in Chapter 7.

As another example, consider the development of shearing stresses in an I beam (Fig. 3.50). In order to calculate the shear stress, use Eq. 3.206. For the section at distance y_1 from the neutral axis, the moment is

$$\int_{y_1}^{h/2} y\,dA = \frac{b}{2}\left(\frac{h^2}{4} - \frac{h_1^{\,2}}{4}\right) + \frac{b_1}{2}\left(\frac{h_1^{\,2}}{4} - y_1^{\,2}\right) \tag{3.218}$$

Substitution into Eq. 3.206 gives

$$\tau = \frac{V}{b_1 I_z}\left[\frac{b}{2}\left(\frac{h^2}{4} - \frac{h_1^{\,2}}{4}\right) + \frac{b_1}{2}\left(\frac{h_1^{\,2}}{4} - y_1^{\,2}\right)\right] \tag{3.219}$$

These results are all for the web section. In the flanges, from Eq. 3.207

$$\tau = (V/2I_z)(h^2/4 - y^2) \tag{3.220}$$

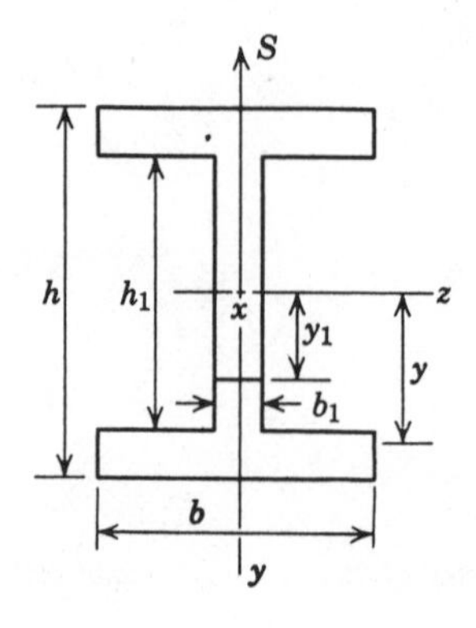

Fig. 3.50 Geometry of I-beam section.

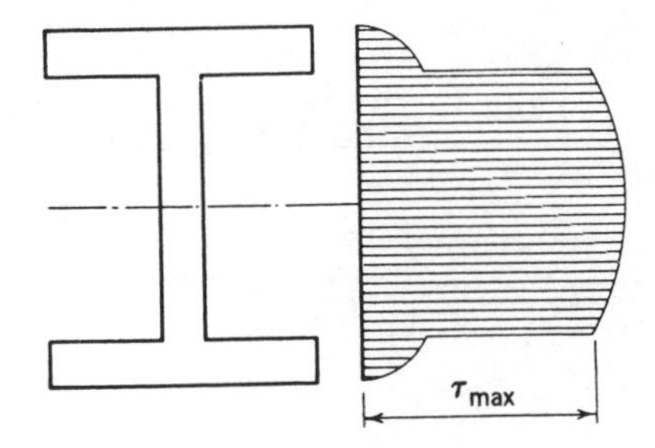

Fig. 3.51 Distribution of shear stress in I-beam.

which is the equation of a parabola. At sections just inside the flange ($y = h_1/2$),

$$\tau = (V/8I_z)(h^2 - h_1^2) \qquad 3.221$$

and at sections just inside the web ($y = h_1/2$ and the width is b_1),

$$\tau = (Vb/8I_z b_1)(h^2 - h_1^2) \qquad 3.222$$

As can be seen in Fig. 3.51, the maximum shear occurs at the neutral axis; the web, in fact, usually can be assumed to support all the shear load, that is,

$$\tau = V/b_1 h_1 \qquad 3.223$$

Some examples illustrating the use of the foregoing results are the following.

1. For the beam shown in Fig. 3.52, the load P is 4000 lb. Determine the principal stresses and maximum shear stress for an element on a cross section 1 ft from the end of the beam and 2 in. below the top fiber. Also determine the planes on which these stresses act. The beam is simply supported at the ends.

The component stresses involved here are a bending stress and a vertical shear stress. The bending stress is

$$\sigma = -\frac{M}{Z} = -\frac{(4000/2)(12)(4)}{(\frac{1}{12})(4)(12^3)} = -167 \text{ psi}$$

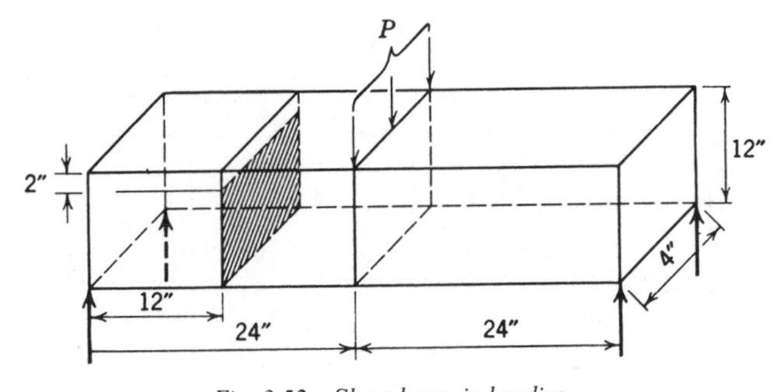

Fig. 3.52 Short beam in bending.

The vertical shear is

$$\tau = \frac{V}{2I}\left(\frac{h^2}{4} - y_1{}^2\right) = \frac{2000}{2(\frac{1}{12})(4)(12^3)}\left(\frac{144}{4} - 16\right) = 34.8 \text{ psi}$$

The principal stresses and maximum shear stress are

$$\sigma_1 = \sigma/2 + \sqrt{(\sigma/2)^2 + \tau^2} = -174 \text{ psi}$$

$$\sigma_2 = \sigma/2 - \sqrt{(\sigma/2)^2 + \tau^2} = 7 \text{ psi}$$

$$\tau_{\max} = \pm\sqrt{(\sigma/2)^2 + \tau^2} = \pm 90.5 \text{ psi}$$

The stress planes are

$$2\theta_1 = \frac{\tan^{-1} 2\tau}{-\sigma} \quad \text{or} \quad \theta_1 = 11° 19'$$

$$\theta_2 = \theta_1 + 90° = 101° 10'$$

$$2\theta_\tau = \frac{\tan^{-1}\sigma}{2\tau} \quad \text{or} \quad \theta_\tau = 56° 18'$$

2. Find the maximum shear stress for the element at A in Fig. 3.53.

The element in question is in the compression zone of the beam; and so the component axial stress is

$$\sigma = -F/A - M/Z$$

or

$$\sigma = -\frac{20{,}000 \sin 60°}{12} - \frac{20{,}000 \sin 30° (24)(2)}{\frac{1}{12}(2)(6^3)} = -24{,}033 \text{ psi}$$

The vertical shear is

$$\tau = (V/2I)(h^2/4 - y_1{}^2) = \frac{20{,}000\sqrt{\frac{3}{2}}(\frac{36}{4} - 4)}{2(\frac{1}{12})(2)(6^3)} = 695 \text{ psi}$$

The maximum shear stress is

$$\tau_{\max} = \pm\sqrt{(\sigma/2)^2 + \tau^2} = \pm 12{,}000 \text{ psi}$$

3. For the short beam shown in Fig. 3.54 determine the section modulus for optimum performance.

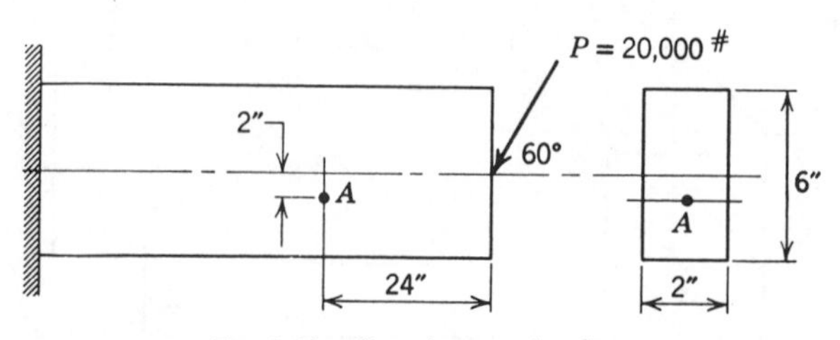

Fig. 3.53 Short beam in bending.

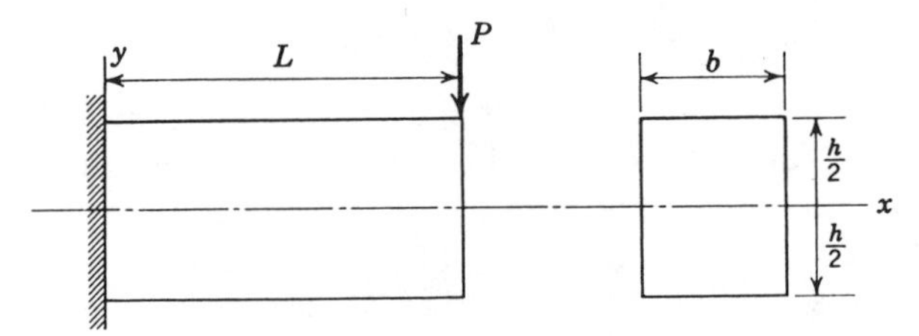

Fig. 3.54 Short beam in bending.

In this example two component stresses are generated. The bending stress is

$$\sigma = M/Z$$

and the shear stress is

$$\tau = (V/2I)(h^2/4 - y_1{}^2)$$

The principal stresses are consequently given by

$$\begin{matrix}\sigma_1 \\ \sigma_2\end{matrix} = \frac{\sigma}{2} \pm \sqrt{(\sigma/2)^2 + \tau^2}$$

For optimum performance it is assumed that the governing factor is stress and that the allowable stress σ_0 is related to the principal stresses by the distortion energy theory [see Section 3-3 and (16)] thus;

$$\sigma_0{}^2 = \sigma_1{}^2 + \sigma_2{}^2 - \sigma_1\sigma_2$$

and replacing the principal stresses by component stresses,

$$\sigma_0{}^2 = \sigma^2 + 3\tau^2$$

The critical section is found by differentiating the preceding equation and setting it equal to zero.

$$d(\sigma_0{}^2)/dy = 0 = 2M^2y - 6V^2h^2y/8 + 3V^2y^3$$

The critical section is thus at $y = 0$ or $y = \sqrt{h^2/4 - 2M^2/3V^2}$ and for the beam shown $V = P$ and $M = PL$; therefore at $y = 0$

$$Z = 0.433\frac{M}{\sigma_0}\left(\frac{h}{L}\right)$$

at

$$y = \sqrt{h^2/4 - 2M^2/3V^2}$$

$$Z = \frac{M}{\sigma_0}\sqrt{1 - \frac{4}{3}\left(\frac{L}{h}\right)^2}$$

The actual critical value thus depends on the specific dimensions of the beam.

Torsion of noncircular and variable section bars

In Chapter 2 it was shown that the bending stress formula $\sigma = M/Z$ was applicable to all forms of cross sections. For torsion, however, the shear formula $\tau = Tr/J$ applies only to round rods. If, for example, in a bar of rectangular cross section, the preceding formula for shear were applicable, there would have to be normal stresses at the outer free surface at the corners, which is impossible. Therefore for torsion of other than circular cross section bars, other means must be employed.

The equilibrium equations, Eqs. 3.90 to 3.92, in the case of torsion where $\sigma_x = \sigma_y = \sigma_z = \tau_{xy} = 0$, reduce to

$$\partial\tau_{zx}/\partial x + \partial\tau_{yz}/\partial y = 0 \qquad 3.224$$

which is satisfied if the stresses are defined in terms of a stress function ϕ such that

$$\tau_{zx} = \partial\phi/\partial y \quad \text{and} \quad \tau_{yz} = -\partial\phi/\partial x \qquad 3.225$$

The displacements (Fig. 3.55) are given by

$$u = -r\theta z \sin\alpha = -y\theta z \qquad 3.226$$

and

$$v = r\theta z \cos\alpha = x\theta z \qquad 3.227$$

With Eqs. 3.225, 3.227, 3.41 to 3.43, 3.93, and 3.94, the following results are obtained:

$$\tau_{yz} = G(x\theta + \partial w/\partial y) \qquad 3.228$$

and

$$\tau_{zx} = G(\partial w/\partial x - y\theta) \qquad 3.229$$

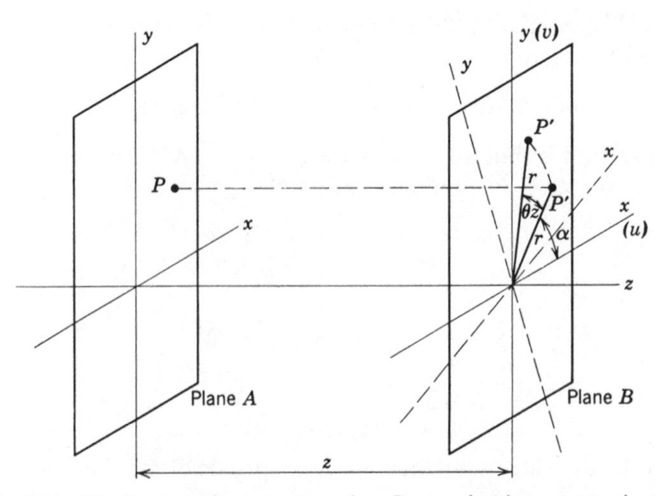

Fig. 3.55 Displacement due to torsion; plane B rotated with respect to plane A.

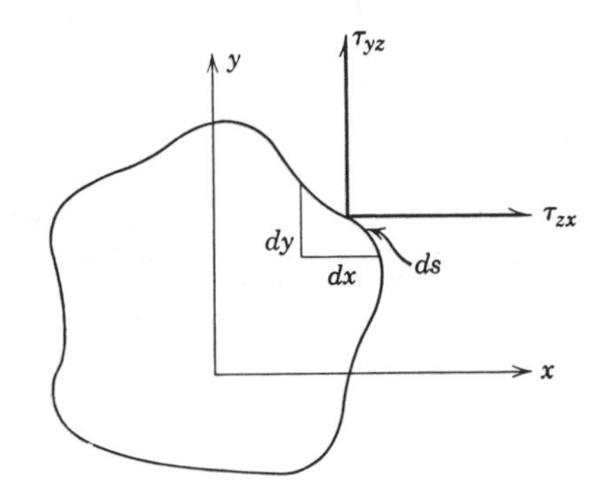

Fig. 3.56 Boundary state of stress.

Equations 3.228 and 3.229 can be rewritten in the form

$$\partial w/\partial x = \tau_{zx}/G + \theta y \qquad 3.230$$

and

$$\partial w/\partial y = \tau_{yz}/G - \theta x \qquad 3.231$$

which on differentiation with respect to x and y respectively and subtracting gives

$$(1/G)(\partial \tau_{xy}/\partial y - \partial \tau_{yz}/\partial x) = -2\theta \qquad 3.232$$

Substituting Eq. 3.225 into Eq. 3.232 then gives

$$\partial^2 \phi/\partial y^2 + \partial^2 \phi/\partial x^2 = -2G\theta = F \qquad 3.233$$

Furthermore, to obtain a result which will be used later, by substituting Eq. 3.225 into Eq. 3.224 and differentiating with respect to s,

$$\partial \phi/\partial s = 0 \qquad 3.234$$

Now consider Fig. 3.56. In torsion, the outer surface of the bar is free of normal stress. Thus, from Fig. 3.56 and the following equations developed from the theory of elasticity (27),

$$X = \sigma_x l + \tau_{xy} m + \tau_{zx} n \qquad 3.235$$

$$Y = \tau_{xy} l + \sigma_y m + \tau_{yz} n \qquad 3.236$$

$$Z = \tau_{zx} l + \tau_{yz} m + \sigma_z n \qquad 3.237$$

where l, m, and n are direction cosines of the normal N with respect to X, Y, and Z. At the outside surface of the bar $X = Y = Z = N = 0$. Equation 3.234 indicates that the stress function ϕ is constant along the outer boundary and may be taken as zero. At the ends of the bar where the twist is applied, $l = m = 0$ and $n = \pm 1.0$, which gives $X = \pm \tau_{zx}$ and $Y = \pm \tau_{yz}$. Thus by summation of forces in the X and Y directions, since

$$\int \partial \phi/\partial y = \phi_1 - \phi_2 = \text{constant} \qquad 3.238$$

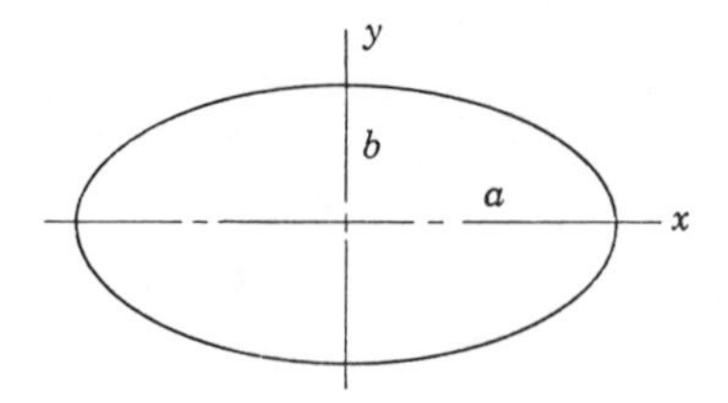

Fig. 3.57 Elliptical cross section.

the following results are obtained.

$$\iint X\,dx\,dy = \iint \tau_{zx}\,dx\,dy = \iint \frac{\partial \phi}{\partial y}\,dx\,dy = \int dx \int \frac{\partial \phi}{\partial y}\,dy = 0 \qquad 3.239$$

and

$$\iint Y\,dx\,dy \quad \iint \tau_{yz}\,dx\,dy = -\iint \frac{\partial \phi}{\partial y}\,dx\,dy = -\int dy \int \frac{\partial \phi}{\partial y}\,dx = 0 \qquad 3.240$$

The twisting moment T is

$$T = \iint (Yx - Xy)\,dx\,dy \qquad 3.241$$

Now substitute Eqs. 3.239 and 3.240 into Eq. 3.241; since, as mentioned, ϕ can be zero on the boundary,

$$T = 2\iint \phi\,dx\,dy \qquad 3.242$$

which is the basic equation for torsion.

Consider the torsion of a solid bar of elliptic cross section, Fig. 3.57. The boundary of the ellipse is defined by the equation

$$x^2/a^2 + y^2/b^2 = 1 \qquad 3.243$$

Equations 3.233 and 3.234 are satisfied by taking the stress function ϕ as follows (27).

$$\phi = \frac{a^2b^2F}{2(a^2+b^2)}\left(\frac{x^2}{a^2} + \frac{y^2}{b^2} - 1\right) \qquad 3.244$$

From Eq. 3.242,

$$T = \frac{a^2b^2F}{a^2+b^2}\left(\frac{1}{a^2}\iint x^2\,dx\,dy + \frac{1}{b^2}\iint y^2\,dx\,dy - \iint dx\,dy\right) \qquad 3.245$$

However,

$$\iint x^2\,dx\,dy = \frac{\pi a^3 b}{4} = I_y \qquad 3.246$$

$$\iint y^2\,dx\,dy = \frac{\pi a b^3}{4} = I_x \qquad 3.247$$

$$\iint dx\,dy = \text{area} = \pi ab \qquad 3.248$$

Therefore Eq. 3.245 becomes

$$T = -\frac{\pi a^3 b^3 F}{2(a^2 + b^2)} \qquad 3.249$$

and Eq. 3.244 becomes

$$\phi = -(T/\pi ab)(x^2/a^2 + y^2/b^2 - 1) \qquad 3.250$$

Substituting Eq. 3.250 into Eq. 3.225 then gives the following:

$$\tau_{zx} = -2Ty/\pi ab^3 \qquad 3.251$$

$$\tau_{yz} = 2Tx/\pi a^3 b \qquad 3.252$$

The maximum value occurs at the ends of the minor axis ($y = b$); therefore

$$\tau_{max} = 2T/\pi ab^2 \qquad 3.253$$

The angle of twist is found by substituting F, from Eq. 3.249 into Eq. 3.233, which gives

$$\theta = \frac{T(a^2 + b^2)}{\pi a^3 b^3 G} \qquad 3.254$$

The torsional properties of other cross-sectional geometries may be found by the method just developed; texts on the mathematical theory of elasticity should be consulted for details. A summary of results is given in Table 2.1.

In addition to solid noncircular bars, other important structural members subjected to torsion and requiring special methods of analysis are bars of variable cross section, thin noncircular members and parts such as channels, split rings, and circular bars with variable diameter. For example, for a bar with the section shown in Fig. 3.58, the shear stress can be calculated from the formula (27),

$$\tau_{r\theta} = \frac{Trz}{2\pi(\frac{2}{3} - \cos\alpha + \frac{1}{3}\cos^3\alpha)(r^2 + z^2)^{5/2}} \qquad 3.255$$

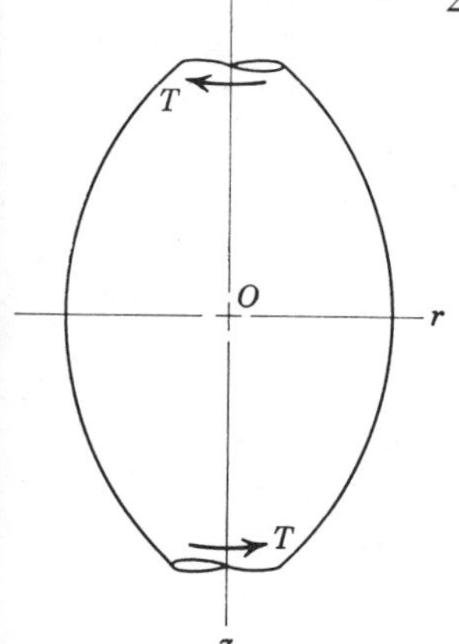

Fig. 3.58 Torsion of bar having a variable diameter.

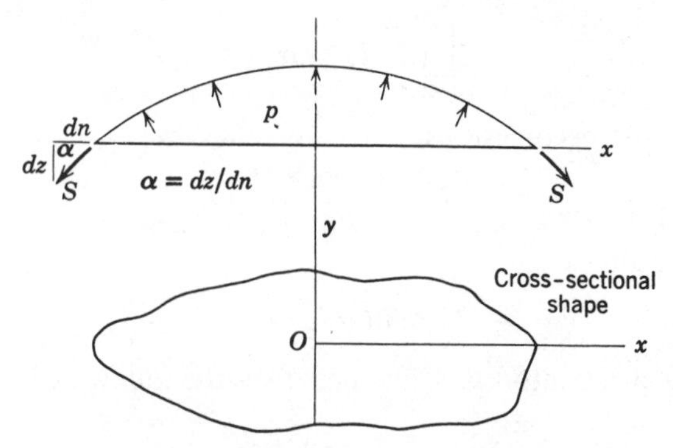

Fig. 3.59 Membrane analogy for torsion.

If the diameter changes gradually, the formula for a uniform bar is sufficiently accurate. If the shaft contains fillets, then the effects of stress concentration must be allowed for; this is treated in Chapter 13.

In determining the torsional properties of irregular shapes and split rings, it is convenient to make use of the ideas developed in the *membrane analogy*. This method of analysis was conceived by Prandtl and has been described by Lee (13). Briefly, the theory makes use of the fact that the maximum slope of a given point on the membrane film obtained by forming a film of soapy water over an opening having the same shape as the cross section to be examined is proportional to the shear stress at that point. The contour lines then represent the direction of stress. Consider Fig. 3.59. Equilibrium in the z direction of the element is

$$S(\partial^2 z/\partial x^2)\,dx\,dy + S(\partial^2 z/\partial y^2)\,dx\,dy + p\,dx\,dy = 0 \qquad 3.256$$

Rewriting Eq. 3.256 gives

$$\partial^2 z/\partial x^2 + \partial^2 z/\partial y^2 = -p/S \qquad 3.257$$

Comparison of Eqs. 3.257 and 3.233 suggests a relationship or correlation between the torsion stress function and the elevation z of the membrane. In the experimental setup, the film boundary lies in a plane, and the area enclosed by the boundary is selected similar to the cross section of the member in question. Thus, since the boundary lies in one plane

$$\partial z/\partial s = 0 \qquad 3.258$$

where s is the length of any contour of the film normal to z. The volume enclosed by the membrane is

$$V = \iint z\,dx\,dy \qquad 3.259$$

which is related to the torsional moment T, Eq. 3.242. Thus

$$T \sim 2\iint z\,dx\,dy \qquad 3.260$$

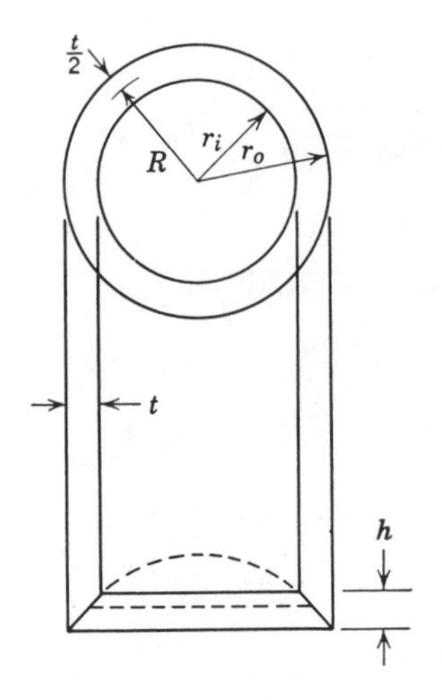

Fig. 3.60 Membrane analogy for thin-walled tube.

that is, the twisting moment is represented by twice the volume enclosed by the membrane over the cross section of the bar in question. Referring again to Fig. 3.59 and letting dz/dn be the tangent of the angle made by a tangent to the membrane and the plane of the boundary, it is seen that

$$\int S \sin \alpha \, ds = \int S(dz/dn)\, ds = \int S(d\phi/dn)\, ds = \int S\tau \, ds \qquad 3.261$$

Equation 3.261 with Eqs. 3.257 and 3.233 then give the following relationship:

$$\int \tau \, ds = 2GA\theta \qquad 3.262$$

where τ, the shear stress, is the slope of the membrane h/t. Therefore in general

$$\tau = T/2At \qquad 3.263$$

where t is variable and A is the average area enclosed by the contour. For a thin-walled hollow tube, Fig. 3.60,

$$\tau = T/[(2t)(\pi R^2)] \qquad 3.264^*$$

and from Eq. 3.262

$$\theta = TsL/4A^2Gt = TL/2t\pi GR^3 \qquad 3.265$$

where R is the midwall radius.

For a narrow rectangular cross section the shear stress τ varies across the section. In this case for the paraboloid formed over the section (Fig. 3.61), the slope of line AB is $h/t = \delta/R$; but, since the membrane is parabolic, which it must be since slope is proportional to stress, the slope of the contour AB is

*For a solid circular bar, $t = 2R$, $R = R_1/2$ giving $\tau = 2T/\pi R_1^3$, the familiar torsion equation. See also Eq. 2.24.

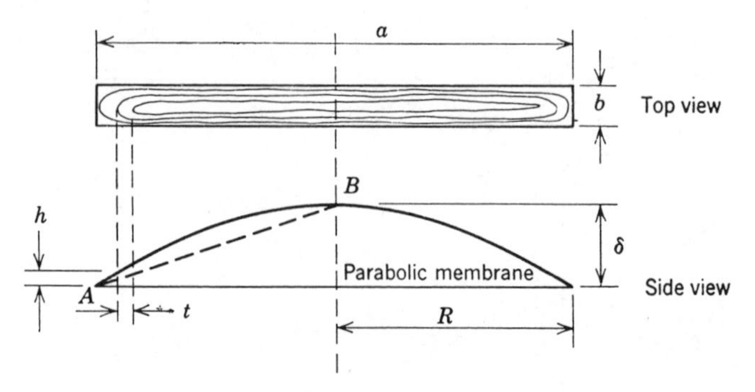

Fig. 3.61 Membrane analogy for thin rectangle.

twice that of line AB so that

$$h/t = 2\delta/R \tag{3.266}$$

Then from Eq. 3.256

$$\tau = \frac{T}{2AhR/2\delta} \tag{3.267}$$

The volume of the paraboloid Ah for the rectangular section is $\frac{2}{3}ab\delta$; therefore

$$\tau = \frac{T}{\frac{2}{3}ab(b/2)} = \frac{3T}{ab^2} \tag{3.268}$$

Also from Eq. 3.263 and the fact that $p/s = 2G\theta$, and $\delta = pb^2/2S$, the angle of twist is expressed by the relation

$$\theta = 3TL/ab^3G \tag{3.269}$$

Equations 3.268 and 3.269 can be used for other thin sections. For example, in the section shown in Fig. 3.62, the maximum shear stress is obtained from Eq. 3.268 and the angle of twist from Eq. 3.269 by substituting $a = r\theta$, the developed length for a. For a split ring, Fig. 3.63, Eq. 3.269 gives

$$\theta = \frac{3TL}{\pi G\left(\frac{d + d_0}{2}\right)\left(\frac{d_0 - d}{2}\right)^3} \tag{3.270}$$

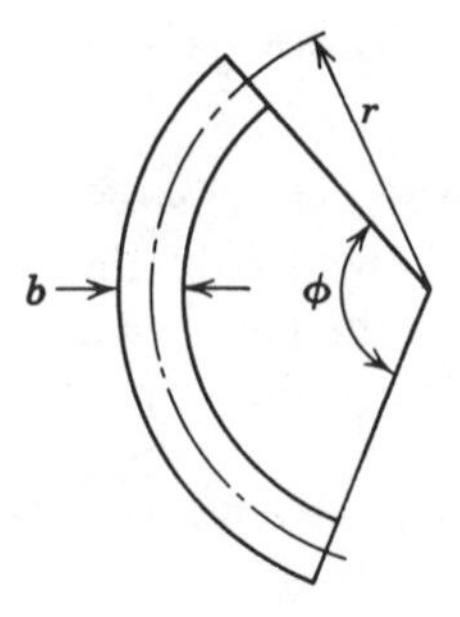

Fig. 3.62 Thin ring sector.

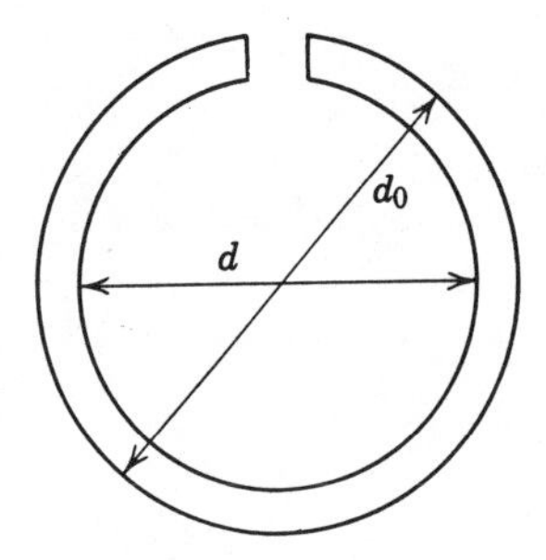

Fig. 3.63 Thin split ring.

*Combined bending and torsion—springs**

The *open-coil* helical spring (Fig. 3.64) is an example of combined bending and torsional stresses. Here it is seen that a tangent to the coil at any point such as A is not perpendicular to the load so that both bending and twisting occur. The torque produced is

$$T = PR\cos\alpha \tag{3.271}$$

and the bending moment is

$$M = PR\sin\alpha \tag{3.272}$$

For design purposes, the principal stresses are (Eqs. 3.18 to 3.20)

$$\sigma_1, \sigma_2 = \frac{16}{\pi d^3}(M \pm \sqrt{M^2 + T^2}) \tag{3.273}$$

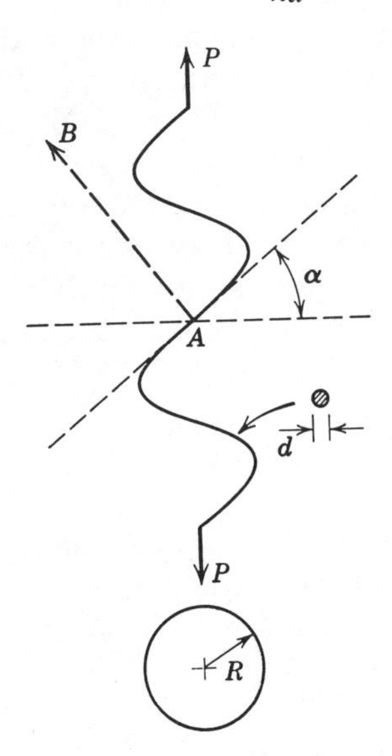

Fig. 3.64 Open-coil helical spring.

*See also Fig. 1.25 for various springs, including leaf springs. The Belleville type is considered in the section of this chapter on plates.

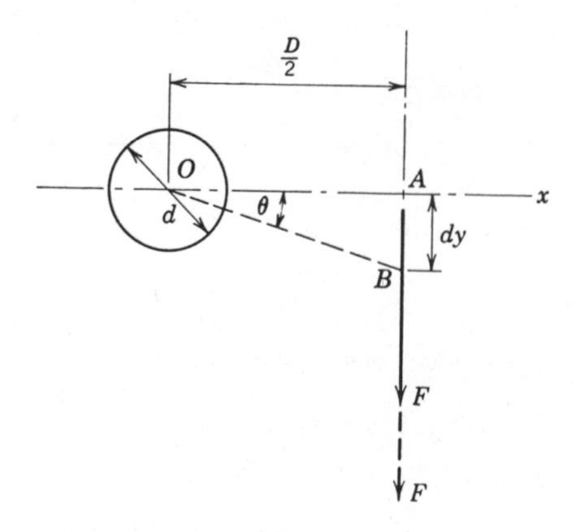

Fig. 3.65 Element of close-coil helical spring.

and the maximum stress is

$$\sigma_{max} = \frac{16PR}{\pi d^3}(1 + \sin\alpha) \qquad 3.274$$

Deflection is calculated from the elastic strain energy (see Chapter 7) and is found to be

$$\delta = PR^2L\left(\frac{\cos^2\alpha}{GJ} + \frac{\sin^2\alpha}{EI}\right) \qquad 3.275$$

where L is the length of the wire in the coil, J the polar moment of inertia, and I the moment of inertia. Additional spring analyses may be found in (32).

In the close-coil helical spring, bending stresses are negligible and the deflection is produced almost entirely by the torsional stresses induced in the coil. It is assumed in the analysis of such springs that the planes of the coils are essentially normal to the spring axis and that the transverse shear is negligible. Consider Fig. 3.65. The spring wire has a diameter d and the coil a diameter D. Load is applied along the spring axis y. The initial position is OA and, after applying the load, point A moves to B covering a distance dy. The angular displacement is θ. Thus the wire is subjected to a torque $FD/2$ and the shear stress (see Chapter 2) is given by

$$\tau = \frac{Tr}{J} = \frac{FD}{2}\left(\frac{d}{2}\right)\left(\frac{1}{\pi d^4/32}\right) = \frac{8FD}{\pi d^3} \qquad 3.276$$

The angle of twist ($\theta = TL/GJ$, Chapter 2) is utilized in determining the deflection of the coil. For a differential length of the coil dL.

$$\theta = \frac{(FD/2)(dL)}{G\pi d^4/32} = \frac{16FD(dL)}{\pi G d^4} \qquad 3.277$$

This angle θ is related to the deflection (Fig. 3.65) by

$$dy = D\theta/2 \qquad 3.278$$

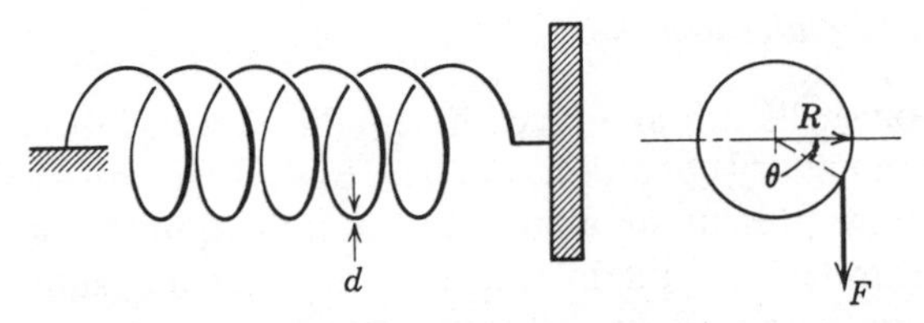

Fig. 3.66 Twisting of a coil spring.

From Eqs. 3.277 and 3.278

$$dy = 8FD^2(dL)/\pi G d^4 \tag{3.279}$$

and integration gives

$$y = 8FD^2L/\pi G d^4 \tag{3.280}$$

The value of L in Eq. 3.280 must be expressed in terms of the diameter D and number of coils n; thus

$$L = \pi D n \tag{3.281}$$

and therefore

$$y = 8FD^3n/Gd^4 \tag{3.282}$$

All of the discussion thus far has been concerned with the stresses developed in, and axial extensions of, springs loaded in the axial direction of the spring. There are many instances, however, like windowshade rollers, where the spring is twisted in a plane normal to the spring axis (Fig. 3.66). In such situations the spring is subjected to bending stresses that are constant throughout the coil (32). Thus the formula $\sigma = M/Z$ is applicable for calculating the stress involved. Usually, however, in twisted springs, the interest is in the total angle of twist, and for a cylindrical helical spring when the total angle of twist (in degrees) is

$$\theta = \pi FRL/180\,EIk \tag{3.283}$$

where k is 1.0 for a loose coil, $(8R - d)/(8R - 4d)$ for a closely-coiled spring of circular wire, and $(6R - h)/(6R - 3h)$ for a coil of rectangular cross-sectional wire of depth h.

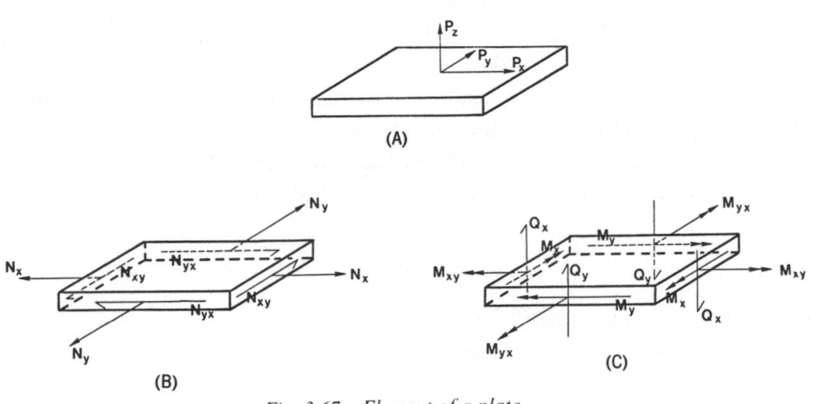

Fig. 3.67 Element of a plate.

Bending of thin plates and slabs

When plates or shells are analyzed the equilibrium equation, the deflection equation, and boundary conditions are needed. In Fig. 3.67(A), (B), and (C) the three elements are present simultaneously and in equilibrium for a plate but are shown separately to aid in understanding further simplifications. The equilibrium equation (5, 19, 25, 29) for a plate with no thermal effect is

$$\frac{\partial^2 M_x}{\partial x^2} + 2\frac{\partial^2 M_{xy}}{\partial x\,\partial y} + \frac{\partial^2 M_y}{\partial y^2} = P_z + N_x\frac{\partial^2 w}{\partial x^2} + N_y\frac{\partial^2 w}{\partial y^2} + 2N_{xy}\frac{\partial^2 w}{\partial x\,\partial y} - P_x\frac{\partial w}{\partial x} - P_y\frac{\partial w}{\partial y} \qquad 3.284$$

Then for D and t constant the deflection equation is

$$\nabla^4 w = \frac{1}{D}\left[P_z + N_x\frac{\partial^2 w}{\partial x^2} + 2N_{xy}\frac{\partial^2 w}{\partial x\,\partial y} + N_y\frac{\partial^2 w}{\partial y^2} - P_x\frac{\partial w}{\partial x} - P_y\frac{\partial w}{\partial y}\right] \qquad 3.285$$

Also

$$\nabla^4 w = \frac{\partial^4 w}{\partial x^4} + 2\frac{\partial^4 w}{\partial x^2\,\partial y^2} + \frac{\partial^4 w}{\partial y^4}$$

The boundary conditions for:

A clamped edge:

$$w = 0 \qquad \frac{\partial w}{\partial \eta} = 0 \qquad 3.286$$

A simply supported edge:

$$w = 0 \qquad \frac{\partial^2 w}{\partial \eta^2} = 0 \qquad 3.287$$

A free edge:

$$\frac{\partial^2 w}{\partial \eta^2} + \nu\frac{\partial^2 w}{\partial s^2} = 0 \qquad 3.288$$

$$\frac{\partial}{\partial \eta}\left[D\left(\frac{\partial^2 w}{\partial \eta^2} + \nu\frac{\partial^2 w}{\partial s^2}\right)\right] + 2(1-\nu)\frac{\partial}{\partial s}\left[D\frac{\partial^2 w}{\partial \eta\,\partial s}\right] = 0 \qquad 3.289$$

The symbol η represents the outward normal to the boundary of a plate, and s represents a direction along the boundary of a plate. For a rectangular plate if η points in an x direction η would become x and s would become y.

In the bending equations for thin plates the following are neglected or are zero:

$$N_x = N_{xy} = N_y = P_x = P_y = 0 \qquad 3.290$$

so that Eqs. 3.284 and 3.285 reduce to

$$\frac{\partial^2 M_x}{\partial x^2} + 2\frac{\partial^2 M_{xy}}{\partial x\,\partial y} + \frac{\partial^2 M_y}{\partial y^2} = P_z \qquad 3.291$$

and

$$\nabla^4 w = \frac{P_z}{D} \tag{3.292}$$

These equations will be derived by other methods as Eqs. 3.300 and 3.319.

In the membrane analysis, bending stiffness is considered to be small so that

$$D = 0 \tag{3.293}$$

Then Eq. 3.285 reduces to

$$P_z + N_x \frac{\partial^2 w}{\partial x^2} + 2N_{xy} \frac{\partial^2 w}{\partial x\, \partial y} + N_y \frac{\partial^2 w}{\partial x^2} - P_x \frac{\partial w}{\partial x} - P_y \frac{\partial w}{\partial y} = 0 \tag{3.294}$$

When further simplifications and substitutions are made such as:

$$P_z = -p; \qquad \frac{1}{R_1} = \frac{\partial^2 w}{\partial x^2}; \qquad \frac{1}{R_2} = \frac{\partial^2 w}{\partial y^2}$$

$$N_x = \sigma_1 t; \qquad N_y = \sigma_2 t$$

and

$$N_{xy} = P_x = P_y = 0$$

This yields

$$\frac{\sigma_1 t}{R_1} + \frac{\sigma_2 t}{R_2} = -p \tag{3.295}$$

which is the same as Eq. 3.367. A more detailed discussion of the various cases of the plate equation follows.

Rectangular coordinates. Consider Fig. 3.68 which shows a plate element under a uniformly distributed normal load. The xy axes are in the neutral plane of the plate. Viewing the element at a larger scale (Fig. 3.68) shows the free-body forces acting in the general case. Bending and twisting moments are produced as well as shear forces S_x and S_y. If w is the applied force per unit of area, the condition of equilibrium in the direction of the z axis is

$$w\,dx\,dy - S_y\,dx + \left(S_y + \frac{dS_y}{dy}dy\right)dx - S_x\,dy + \left(S_x + \frac{dS_x}{dx}dx\right)dy = 0 \tag{3.296}$$

from which

$$dS_x/dx + dS_y/dy = -w \tag{3.297}$$

Taking the summation of moments and disregarding the products of higher differentials give

$$dM_x/dx + dM_{yx}/dy = S_x \tag{3.298}$$

and

$$dM_y/dy + dM_{xy}/dx = S_y \tag{3.299}$$

Because of symmetry $M_{xy} = M_{yx}$; noting this fact and substituting Eqs. 3.298 and 3.299 into Eq. 3.297 give the following result:

$$d^2M_x/dx^2 + 2d^2M_{xy}/dx\,dy + d^2M_y/dy^2 = -w \tag{3.300}$$

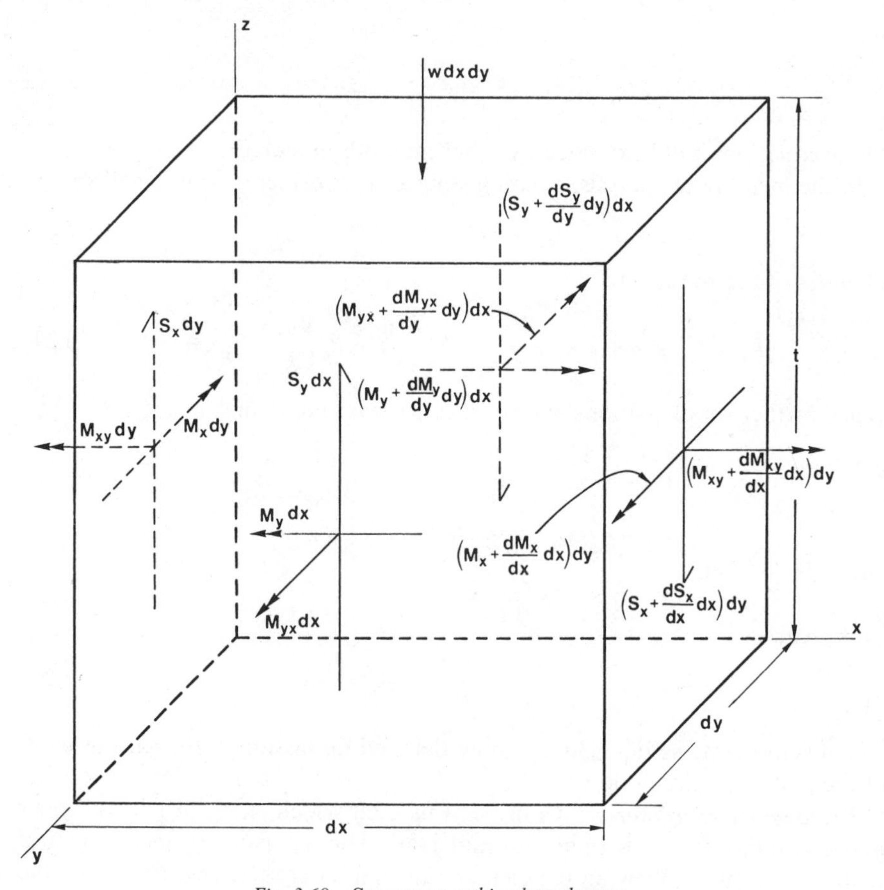

Fig. 3.68 Stresses on a thin plate element.

The condition of equilibrium has now been satisfied. However, in order to complete the solution consideration must be given to the condition of strain compatability which leads to relations between the bending and twisting moments and the plate deflection δ at any point. Figure 3.69 shows the plate with only pure bending moments acting at the boundaries. These moments produce the forces in the differential element *dx dy dz* shown in Fig. 3.70; it is assumed that the lateral faces of the element do not distort and that they rotate

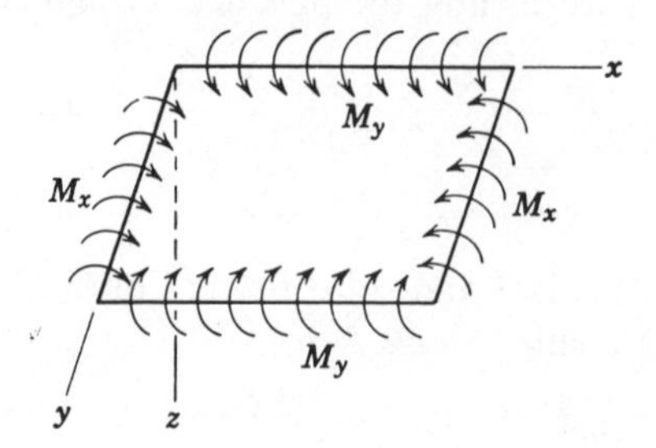

Fig. 3.69 Plate subjected to edge moments.

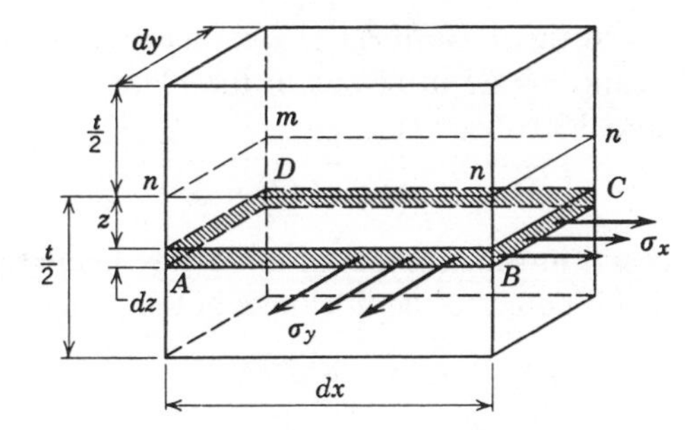

Fig. 3.70 Stress on plate element.

about the neutral axes n–n. The plane at the middle (n, n) thus undergoes no distortion during bending of the plate; therefore this neutral plane has curvatures in sections parallel to the x–z and y–z planes of $1/r_x$ and $1/r_y$. Strains in the x and y directions of the section $ABCD$, distant z from the neutral plane, are

$$\varepsilon_x = z/r_x \tag{3.301}$$

and

$$\varepsilon_y = z/r_y \tag{3.302}$$

The stresses corresponding to these strains are, in accordance with Eqs. 3.66 and 3.67,

$$\sigma_x = \frac{Ez}{1-\nu^2}\left(\frac{1}{r_x} + \frac{\nu}{r_y}\right) \tag{3.303}$$

$$\sigma_y = \frac{Ez}{1-\nu^2}\left(\frac{1}{r_y} + \frac{\nu}{r_x}\right) \tag{3.304}$$

The stresses in Eqs. 3.303 and 3.304 are proportional to the distance z from the neutral axis and form couples that balance the moments M_x and M_y; thus

$$M_x\,dy = \int_{-t/2}^{t/2} \sigma_x z\,dy\,dx \tag{3.305}$$

$$M_y\,dx = \int_{-t/2}^{t/2} \sigma_y z\,dx\,dz \tag{3.306}$$

Substitution of Eqs. 3.303 to 3.304 into Eqs. 3.305 to 3.306 gives

$$M_x = D(1/r_x + \nu/r_y) \tag{3.307}$$

and

$$M_y = D(1/r_y + \nu/r_x) \tag{3.308}$$

where

$$D = \frac{E}{1-\nu^2}\int_{-t/2}^{t/2} z^2\,dz = \frac{Et^3}{12(1-\nu^2)} \tag{3.309}$$

In Eq. 3.309, D is called the *flexural rigidity*, a quantity applied to plates that has the same significance as EI in beams. If the plate is bent to a cylindrical surface, Eqs. 3.307 and 3.308 become

$$M = D/r = -D(d^2y/dx^2) \qquad 3.310$$

Equation 3.310 is useful in the analyses of uniformly loaded plates. When the deflection δ is small compared to the thickness of the plates,

$$1/r_x = -d^2\delta/dx^2 \qquad 3.311$$

and

$$1/r_y = -d^2\delta/dy^2 \qquad 3.312$$

and Eqs. 3.264 to 3.265 become

$$M_x = D(1/r_x + v/r_y) = -D(d^2\delta/dx^2 + d^2\delta/dy^2) \qquad 3.313$$

and

$$M_y = D(1/r_y + v/r_x) = -D(d^2\delta/dy^2 + d^2\delta/dx^2) \qquad 3.314$$

In bending the plate to a cylindrical surface about the x axis, $d^2\delta/dy^2$ is zero and

$$M_x = -D(d^2\delta/dx^2) \qquad 3.315$$

and

$$M_y = -Dv(d^2\delta/dx^2) \qquad 3.316$$

Equations 3.313 to 3.314 are the fundamental moment relations for any plate subjected to pure bending. The twisting-moment relationship M_{xy} is (28)

$$M_{xy} = \frac{-D(1-v)d^2\delta}{dx\,dy} \qquad 3.317$$

Substituting Eqs. 3.313 to 3.314 into Eq. 3.300 then gives the general differential equation for the deflection of a plate; thus

$$d^4\delta/dx^4 + 2d^4\delta/dx^2\,dy^2 + d^4\delta/dy^4 = w/D \qquad 3.318$$

or

$$(\nabla\delta)^2 = w/D \qquad 3.319$$

called *Lagrange's equation*, where

$$\nabla = d^2/dx^2 + d^2/dy^2 \qquad 3.320$$

The analysis for plate deflection is thus governed by Eq. 3.320. The relationship between the shear forces and deflection is found by substituting Eqs. 3.313, 3.314, and 3.317 into Eqs. 3.298 to 3.299; thus

$$S_x = -D(d/dx)\nabla\delta \qquad 3.321$$

and

$$S_v = -D(d/dy)\nabla\delta \qquad 3.322$$

Polar coordinates. The foregoing equations for equilibrium, compatability, and deflection in rectangular coordinates are not suitable for dealing with problems involving rotational symmetry as in circular plates and slabs; it is

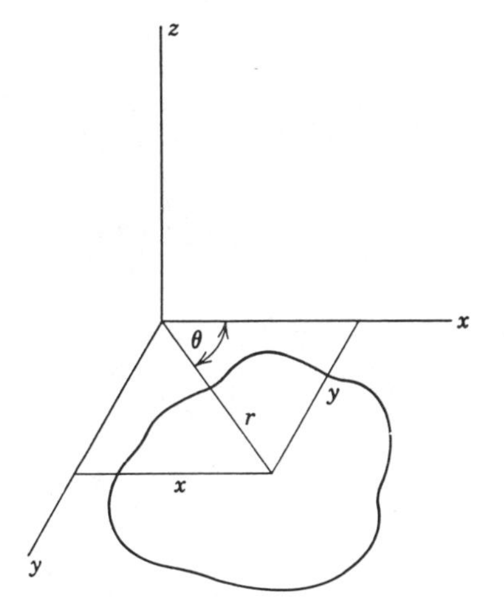

Fig. 3.71 Polar coordinates.

thus convenient to express these relationships in terms of polar coordinates as shown in Fig. 3.71.

$$x = r\cos\theta \tag{3.323}$$

and

$$y = r\sin\theta \tag{3.324}$$

By using the relationship expressed by Eqs. 3.323 to 3.324, Eq. 3.319 becomes

$$\nabla^2\delta = \left(\frac{d^2}{dr^2} + \frac{1}{r}\frac{d}{dr} + \frac{1}{r^2}\frac{d^2}{d\theta^2}\right)^2 \delta = \frac{w}{D}$$

Also, from Eqs. 3.313 to 3.317,

$$M_1 = D\left[\frac{d^2\delta}{dr^2} + v\left(\frac{1}{r}\frac{d\delta}{dr} + \frac{1}{r^2}\frac{d^2\delta}{d\theta^2}\right)\right] \tag{3.325}$$

and

$$M_2 = D\left(\frac{1}{r}\frac{d\delta}{dr} + \frac{1}{r^2}\frac{d^2\delta}{d\theta^2} + v\frac{d^2\delta}{dr^2}\right) \tag{3.326}$$

Equation 3.325 gives the relationship for moments in sections perpendicular to r, and Eq. 3.326 is for moments perpendicular to M_1. Likewise, from Eqs. 3.321 to 3.322, the shear forces are

$$S_1 = -D(d/dr)\nabla\delta \tag{3.327}$$

and

$$S_2 = -(D/r)(d/d\theta)\nabla\delta \tag{3.328}$$

and from Eq. 3.317

$$M_{r\theta} = D(1 - v)\frac{d}{dr}\left(\frac{1}{r}\frac{d\delta}{d\theta}\right) \tag{3.329}$$

If the load on the plate is symmetrical with respect to the center, derivatives

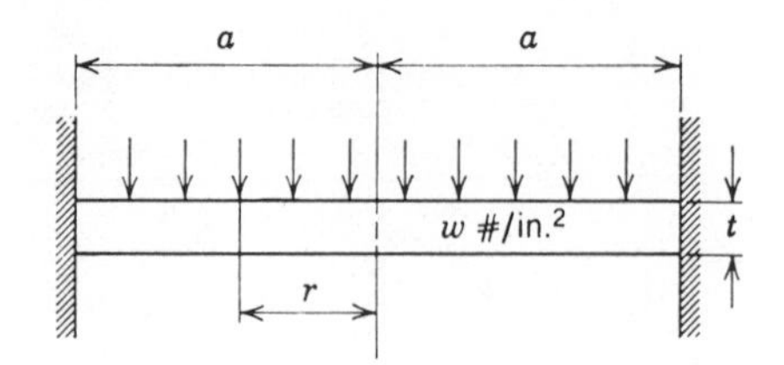

Fig. 3.72 Circular plate with uniform load and clamped at the edge.

involving θ are zero; also $M_{r\theta}$ and S_2 are zero, leaving

$$S_1 = -D\frac{d}{dr}\left[\left(\frac{d^2}{dr^2}+\frac{1}{r}\frac{d}{dr}\right)\delta\right] \qquad 3.330$$

or

$$\frac{d^3\delta}{dr^3}+\frac{1}{r}\frac{d^2\delta}{dr^2}-\frac{d\delta}{dr}\frac{1}{r^2}=-\frac{S_1}{D} \qquad 3.331$$

However, $d\delta/dr$ is equal to the slope ϕ; therefore Eq. 3.299 becomes

$$\frac{d^2\phi}{dr^2}+\frac{1}{r}\frac{d\phi}{dr}-\frac{\phi}{r^2}=-\frac{S_1}{D} \qquad 3.332$$

With these fundamental relationships available a few specific types of plates can now be investigated.

CIRCULAR PLATE WITH UNIFORM LOAD AND CLAMPED AT THE EDGE (Fig. 3.72). At the clamped edge there is a shear force S_1 which, by statics, is determined from the relation

$$2S_1\pi a = w\pi a^2 \qquad 3.333$$

or

$$(S_1)_{r=a} = wa/2 \qquad 3.334$$

At any distance r from the center,

$$(S_1)_r = wr/2 \qquad 3.335$$

Equation 3.335 provides the value of S_1 to be used in Eq. 3.332. Substituting Eq. 3.335 into Eq. 3.332 and performing the indicated integration lead to the expression

$$\phi = -wr^3/16D + C_1r/2 + C_2/r \qquad 3.336$$

where C_1 and C_2 are constants of integration. Substituting for the slope ϕ the value $d\delta/dr$ and integrating a third time lead to the relationship

$$\delta = -wr^4/64D + C_1r^2/4 + C_2\ln r + C_3 \qquad 3.337$$

where C_3 is a third constant of integration. To evaluate the constants of integration the following boundary conditions are utilized:

$$\phi_{r=0,a} = 0$$

and

$$\delta_a = 0$$

From these conditions

$$C_1 = wa^2/8D$$
$$C_2 = 0$$

and

$$C_3 = wa^4/64D$$

Therefore the deflection of the plate at any point is given by

$$\delta_r = (w/64D)(r^2 - a^2)^2 \qquad 3.338$$

and

$$\delta_{max} = \delta_{r=0} = wa^4/64D \qquad 3.339$$

Having the deflection from Eq. 3.338, the moments M_1 and M_2 are easily determined; thus for any point in the plate

$$M_1 = M_r = (w/16)[a^2(1 + \nu) - r^2(3 + \nu)]$$
$$M_2 = M_h = (w/16)[a^2(1 + \nu) - r^2(1 + 3\nu)] \qquad 3.340$$

At the edge of the plate,

$$M_r = -wa^2/8 \qquad 3.341$$

and

$$M_h = -\nu wa^2/8 \qquad 3.342$$

At the center of the plate,

$$M_r = M_h = (wa^2/16)(1 + \nu) \qquad 3.343$$

The maximum moment is M_r at the clamped edge of the plate so that the corresponding maximum stress is

$$\sigma_{max} = (\sigma_r)_{r=a} = (6/t^2)(wa^2/8) = \tfrac{3}{4}(wa^2/t^2) \qquad 3.344$$

$$\sigma_h = 3\nu wa^2/4t^2 \qquad 3.345$$

Design curves are shown in Fig. 3.73.

CIRCULAR PLATE WITH UNIFORM EDGE MOMENT AND NO EDGE SUPPORT (Fig. 3.74). Since the plate is subjected to a uniform moment, $M_x = M_y$ in Eqs. 3.307 to 3.308 so that the radius of curvature of the spherical surface to which the plate deforms is

$$R = D(1 + \nu)/M \qquad 3.346$$

Since the small deflection of a spherical surface at any point measured perpendicular to its chord is given by the relationship

$$\delta = (1/2R)(a^2 - r^2) \qquad 3.347$$

where R is the radius of curvature, a is the half-chord length, and r is the distance from the center of the chord to the point where the deflection is measured, the deflection for the plate at any point is

$$\delta = (1/2R)(a^2 - r^2) = [M/(2D)(1 + \nu)](a^2 - r^2) \qquad 3.348$$

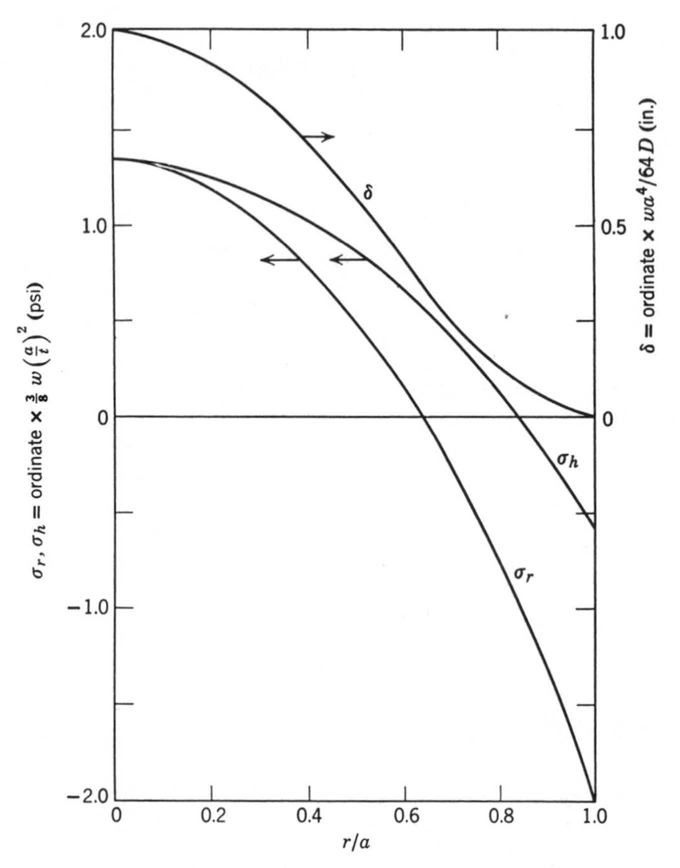

Fig. 3.73 Design curves for uniformly loaded circular plate clamped at the edge.

The maximum deflection is at $r = 0$ or

$$\delta_{\max} = Ma^2/[2D(1+\nu)] \tag{3.349}$$

The maximum stress occurs all over the plate simultaneously and is

$$\sigma_{\max} = (\sigma_r, \sigma_h) = M_1(c/I) = 6M/t^2 \tag{3.350}$$

CIRCULAR PLATE WITH UNIFORM LOAD AND CLAMPED AT THE EDGE (Fig. 3.75). The solution for this problem is obtained by superposing the solutions for the two preceding sections. Thus the maximum deflection is the same as

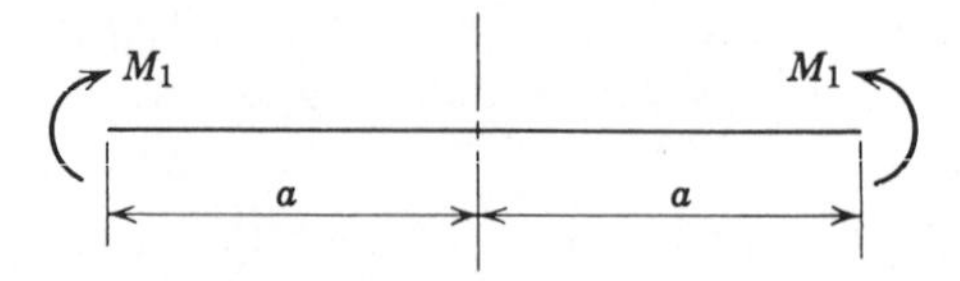

Fig. 3.74 Unsupported circular plate with edge moment applied.

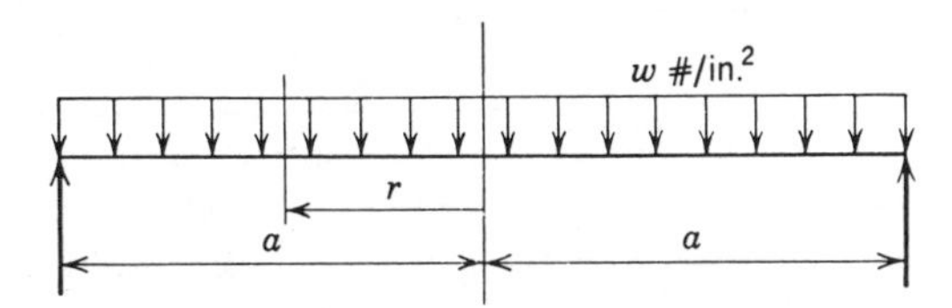

Fig. 3.75 Simply supported, uniformly loaded circular plate.

that for a clamped plate plus the deflection produced by the edge moment (which restricted deflection in the former case); thus

$$\delta = \text{Eq. } 3.338 + \text{Eq. } 3.348 = \frac{w}{64D}\left[r^4 + a^4\left(\frac{5+\nu}{1+\nu}\right) - 2a^2r^2\left(\frac{3+\nu}{1+\nu}\right)\right] \quad 3.351$$

and

$$\delta_{\max} = \frac{wa^4}{64D}\left(\frac{5+\nu}{1+\nu}\right) \quad 3.352$$

Again, by superposition, the moments are

$$M_1 = M_r = (w/16)[(3+\nu)(a^2 - r^2)] \quad 3.353$$

and

$$M_2 = M_h = (w/16)[a^2(3+\nu) - r^2(1+3\nu)] \quad 3.354$$

from which

$$\sigma_r = \tfrac{3}{8}(w/t^2)[(3+\nu)(a^2 - r^2)] \quad 3.355$$

and

$$\sigma_h = \tfrac{3}{8}(w/t^2)[a^2(3+\nu) - r^2(1+3\nu)] \quad 3.356$$

The maximum moment is at $r = 0$, where $M_1 = M_2$, which results in

$$\sigma_{\max} = (\sigma_r, \sigma_h)_{r=0} = 6M_1/t^2 = \tfrac{3}{8}(wa^2/t^2)(3+\nu) \quad 3.357$$

Some results of additional plate analyses are shown in Table 3.4. Analyses of plates of various shapes and edge conditions, plates of variable thickness, and those subjected to large deflections may be found in (20, 28, and 29). Some large deflection analyses are also shown in Appendix C.

An example of the use of the preceding theory is given by the following problem. For the circular plate shown in Fig. 3.76 determine expressions for

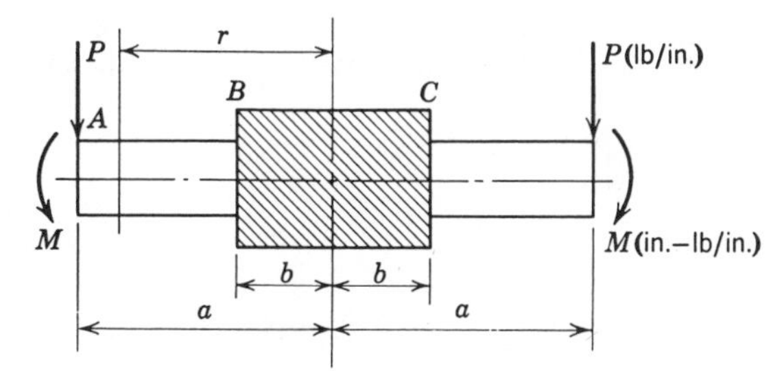

Fig. 3.76 Circular plate with combined loading.

Table 3.4 Some Solutions to Plate Problems

Geometry	Equations
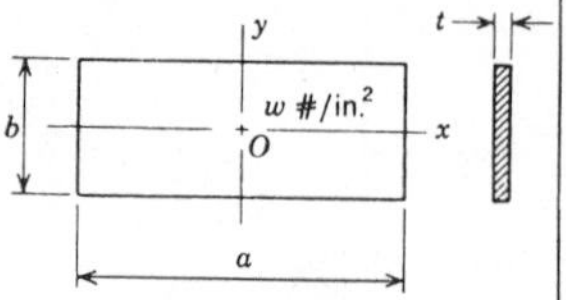 Uniformly Loaded, Simply Supported	$\sigma_{max} = \sigma_{y(center)} = \dfrac{3wb^2}{4t^2}\left[\dfrac{1}{1.61(b/a)^3 + 1}\right]$ $\delta_{max} = \delta_0 = \dfrac{0.142wb^4}{Et^3}\left[\dfrac{1}{2.21(b/a)^3 + 1}\right]$
Rectangular (as above), All Edges Clamped	$\sigma_{max} = \sigma_{y(edge\ at\ a/2)} = \dfrac{wb^2}{2t^2}\left[\dfrac{1}{0.623(b/a)^6 + 1}\right]$ $\delta_{max} = \delta_0 = \dfrac{0.0284wb^4}{Et^3}\left[\dfrac{1}{1.056(b/a)^5 + 1}\right]$
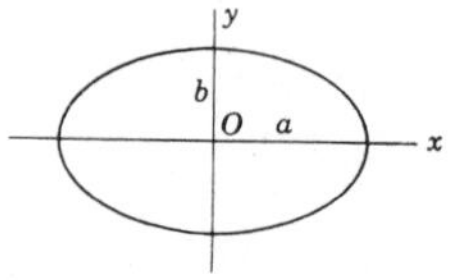 Uniformly Loaded, Simply Supported	$\sigma_{max} = \sigma_{y(center)} = \left(\dfrac{5}{16}\right)\dfrac{wa}{t^2}\left(\dfrac{b}{a}\right)^2(2a - b)$ $\delta_{max} = \delta_0 = \dfrac{wab^2}{10Et^3}\left(\dfrac{b}{a}\right)^2(1.46a - b)$
Elliptic (as above), Perimeter Clamped	$\sigma_{max} = \sigma_{y(edge)} = \dfrac{3wb^2}{2t^2}\left[\dfrac{1}{3(b/a)^4 + 2(b/a)^2 + 3}\right]$ $\delta_{max} = \delta_0 = \dfrac{1.365wb^4}{16Et^2}\left[\dfrac{1}{3(b/a)^4 + 2(b/a)^2 + 3}\right]$
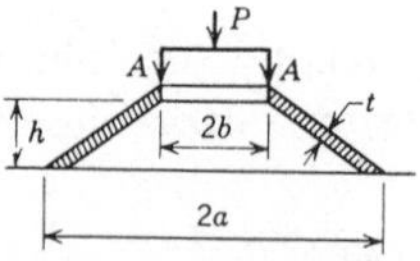 Coned Circular Plate with Hole—Belleville Spring (δ = central deflection)	$P = \dfrac{E\delta}{(1 - \nu)^2 Ka^2}\left[\left(h - \dfrac{\delta}{2}\right)(h - \delta)t + t^3\right]$ $\sigma_A = \dfrac{E\delta}{(1 - \nu^2)Ka^2}\left[C_1\left(h - \dfrac{\delta}{2}\right) + C_2 t\right]$ $K = \dfrac{6}{\pi \ln R}\left(\dfrac{R - 1}{R}\right)^2; \quad C_1 = \dfrac{6}{\pi \ln R}\left(\dfrac{R - 1}{\ln R} - 1\right)$ $R = \dfrac{a}{b}; \quad C_2 = \dfrac{6}{\pi \ln R}\left(\dfrac{R - 1}{2}\right)$

finding the constants of integration required for evaluating the slope and moments. The part BC of the plate is considered to be rigid.

Starting with Eqs. 3.325 and 3.326, the moments are expressed in terms of slope.

$$M_1 = D[(d\phi/dr) + (v/r)\phi] \tag{3.358}$$

$$M_2 = D[(\phi/r) + v(d\phi/dr)] \tag{3.359}$$

Equation 3.332 is rewritten as

$$(d/dr)[(1/r)(d/dr)r\phi] = -S/D = -Pa/Dr \tag{3.360}$$

The first integration gives

$$(d/dr)r\phi = -(Pa/D)r \ln r + C_1 r \tag{3.361}$$

The second integration gives

$$\phi = -(Pa/D)(r/4)(2 \ln r - 1) + C_1 r/2 + C_2/r \tag{3.362}$$

But

$$\phi = ds/dr = (Pa/4D)r(2 \ln r - 1) + C_1 r/2 + C_2/r$$

From the boundary conditions

$$(\phi)_{r=b=0} \quad \text{and} \quad (M_1)_{r=a} = M$$

it is seen that

$$C_1 b/2 + C_2/b - (Pab/4D)(2 \ln b - 1) = 0 \tag{3.363}$$

and

$$M = D\left\{-\frac{Pa}{2D}\ln a - \frac{Pa}{4D} + \frac{C_1}{2} + \frac{v}{a}\left[-\frac{Pa}{4D} - a(2\ln a - 1) + \frac{C_1 v}{2} + \frac{vC_2}{a}\right]\right\} \tag{3.364}$$

which are the required equations.

Thin-walled shells subjected to internal pressure

Membrane stresses only. A general form of shell having a surface of revolution is shown in Fig. 3.77. The two principal radii of curvature are R_1 and R_2. For equilibrium in the radial direction (4, 29),

$$\sigma_1 t R_2\, d\theta_1 + \sigma_2 t R_1\, d\theta_2 = pR_1\, d\theta_2 + pR_2\, d\theta_1 \tag{3.365}$$

Dividing through by $d\theta_1$ gives

$$\sigma_1 t R_2 + \sigma_2 t R_1 (d\theta_2/d\theta_1) = pR_1(d\theta_2/d\theta_1) + pR_2 \tag{3.366}$$

which reduces to

$$\sigma_1/R_1 + \sigma_2/R_2 = p/t \tag{3.367}$$

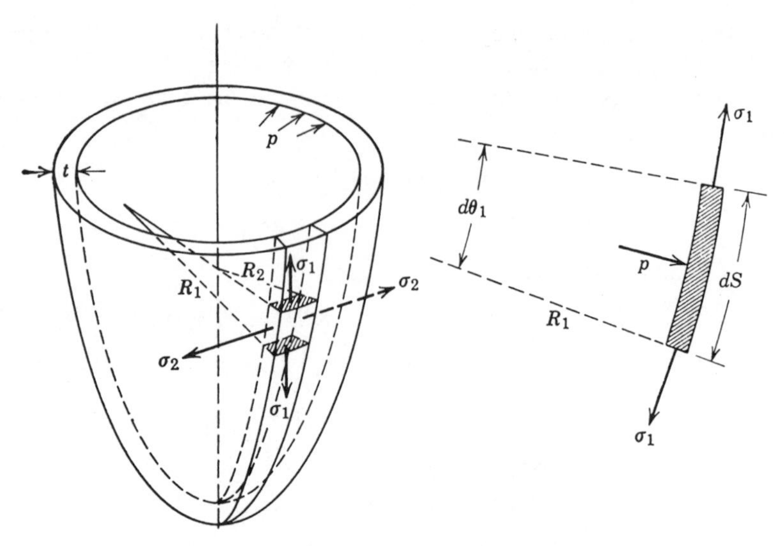

Fig. 3.77 Shell with surface of revolution.

From conditions of equilibrium in the principal directions,

$$\sigma_1 = pR_2/2t \tag{3.368}$$

$$\sigma_2 = (pR_2/2t)(2 - R_2/R_1) \tag{3.369}$$

Some specific cases using the foregoing relationships can now be investigated. Thin-walled shell heads with constant maximum shear stress (Biezeno pressure vessel heads) are discussed in (37).

THIN-WALLED SPHERE. In this instance $R_1 = R_2$ and $\sigma_1 = \sigma_2$, therefore Eqs. 3.368 and 3.369 are identical and

$$\sigma_1 = pR/2t \tag{3.370}$$

where R is the internal radius and t is the wall thickness.

THIN-WALLED CYLINDER (Fig. 1.46). In this instance R_1 is infinite and Eq. 3.369 indicates that the hoop stress developed is

$$\sigma_2 = \sigma_{\text{hoop}} = pR/t \tag{3.371}$$

If the ends of the cylinder are capped, a longitudinal stress is also induced, which by Eq. 3.368 is

$$\sigma_1 = \sigma_{\text{long}} = pR/2t \tag{3.372}$$

In some cases for a more accurate analysis, Eqs. 3.371 and 3.372 are used along with a third stress (radial) which is equal to $-p$ at the bore and zero at the outside surface.

CONICAL VESSEL (Fig. 3.78). From vertical equilibrium considerations.

$$2\pi Rt\sigma/\cos\alpha = p(\pi R^2) \tag{3.373}$$

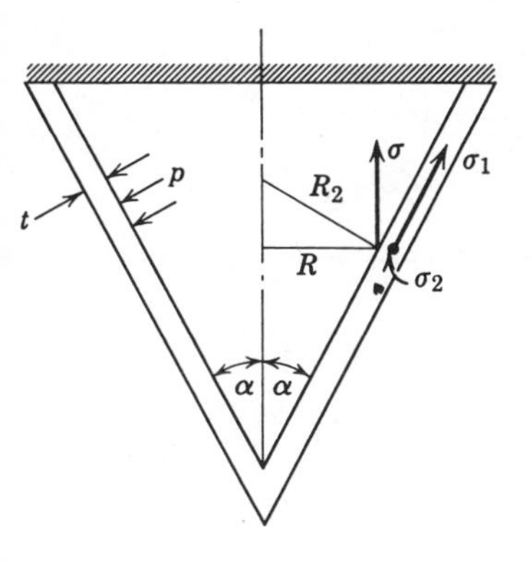

Fig. 3.78 Thin-walled conical shell under internal pressure.

or

$$\sigma = (pR/2t)\cos\alpha \tag{3.374}$$

But $\sigma_1 = \sigma/\cos^2\alpha$ and thus

$$\sigma_1 = \frac{pR}{2t\cos\alpha} \tag{3.375}$$

Since the radius of curvature of the meridian is infinite, Eq. 3.367 shows that,

$$\sigma_2 = \frac{pR}{t\cos\alpha} \tag{3.376}$$

If the conical tank of Fig. 3.78 is loaded by being filled with a liquid of density γ to a height h as shown in Fig. 3.79 at a distance y from the vertex O, the pressure due to the liquid is

$$p = \gamma(h - y) \tag{3.377}$$

One principal radius of curvature R_2 is infinite; from Eq. 3.367

$$\sigma_1 = \frac{pR_1}{t} = \frac{p}{t}\left(y\frac{\tan\alpha}{\cos\alpha}\right) = \frac{\gamma(h-y)}{t\cos\alpha}(y\tan\alpha) \tag{3.378}$$

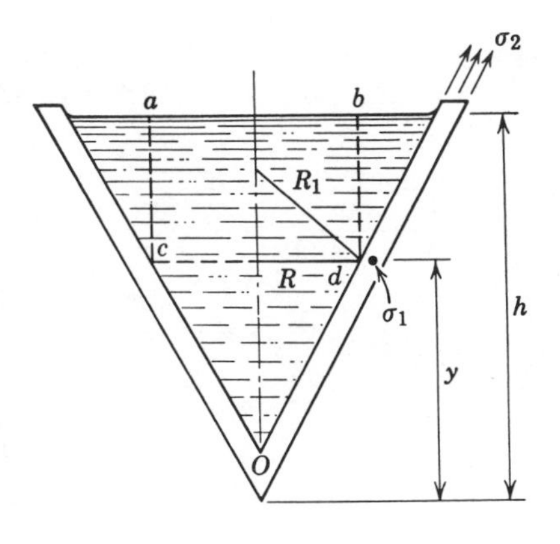

Fig. 3.79 Conical tank with liquid.

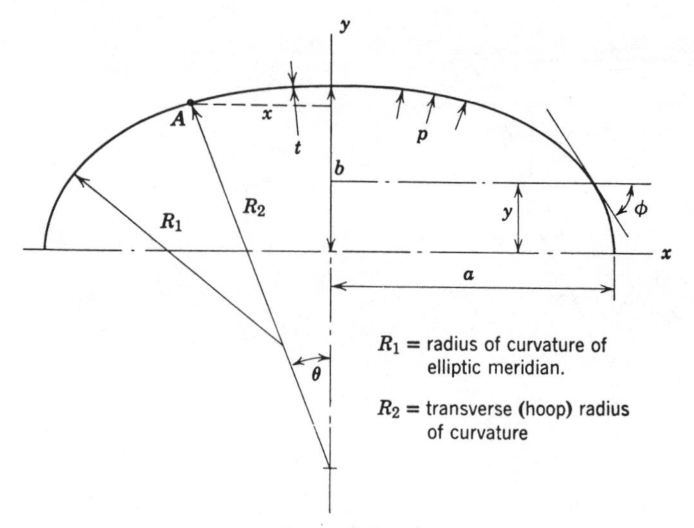

Fig. 3.80 Ellipsoidal shell under pressure.

Differentiation with respect to y and equating to zero reveals that the maximum value of Eq. 3.378 is

$$(\sigma_1)_{\max} = \gamma h^2 \tan \alpha / 4t \cos \alpha \qquad 3.379$$

Again, for equilibrium, to find σ_2, the vertical component of σ_2 supports the volume *acOdb*; thus

$$\sigma_2 t \cos \alpha (2\pi y \tan \alpha) = \pi y^2 (\tan^2 \alpha)(h - y + y/3)\gamma \qquad 3.380$$

or

$$\sigma_2 = y \frac{\tan \alpha (h - \frac{2}{3} y)}{2t \cos \alpha} \gamma \qquad 3.381$$

and the maximum value is

$$(\sigma_2)_{\max} = \frac{3h^2 \gamma \tan \alpha}{16t \cos \alpha} \qquad 3.382$$

ELLIPSOIDAL SHELL (Fig. 3.80). This type of shell is frequently used as a cover for pressure vessels; its principal radii of curvature are defined by the following equations:

$$R_1 = \frac{a^2 b^2}{(a^2 \sin^2 \theta + b^2 \cos^2 \theta)^{3/2}} \qquad 3.383$$

and

$$R_2 = \frac{a^2}{(a^2 \sin^2 \theta + b^2 \cos^2 \theta)^{1/2}} \qquad 3.384$$

At location A, Fig. 3.80, equilibrium demands that

$$2\sigma_1 t \, d\theta = pR_2 \, d\theta \qquad 3.385$$

or the meridian stress is

$$\sigma_1 = pR_2/2t \qquad 3.386$$

Substituting Eq. 3.383 into Eq. 3.367 then gives

$$pR_2/2tR_1 + \sigma_2/R_2 = p/t \qquad 3.387$$

or the hoop stress is

$$\sigma_2 = (p/t - pR_2/2tR_1)R_2 \qquad 3.388$$

Equation 3.388 together with Eqs. 3.383 and 3.384 then give the result

$$\sigma_2 = \frac{pa^2}{2tb^2}\left[\frac{\sin^2\theta(b^2 - a^2) + b^2}{(a^2\sin^2\theta + b^2\cos^2\theta)^{1/2}}\right] \qquad 3.389$$

At the crown where $\theta = 0$, $R_1 = R_2$ so that

$$\sigma_1 = \sigma_2 = pa^2/2bt \qquad 3.390$$

At the extremities where $\theta = 90°$, $R_1 = b^2/a$ and $R_2 = a$, so that

$$\sigma_1 = pa/2t \qquad 3.391$$

and

$$\sigma_2 = (pa/t)(1 - a^2/2b^2) \qquad 3.392$$

ELLIPTICAL CYLINDER. An elliptical cylinder that is under internal pressure and/or stresses caused by liquid within the vessel is a complex problem and solutions are available for both the open-end and closed-end conditions (4). For the open-end type (see Fig. 3.80),

$$R_1 = \frac{a^2b^2}{(a^2\sin^2\phi + b^2\cos^2\phi)^{3/2}} \qquad 3.393$$

$$R_2 = \infty \qquad 3.394$$

The hoop stress is then

$$\sigma_1 = \frac{pR_1}{t} = \frac{p(a^2b^2)}{t(a^2\sin^2\phi + b^2\cos^2\phi)^{3/2}} \qquad 3.395$$

and the longitudinal stress is

$$\sigma_x = -\frac{3p(a^2 - b^2)(L^2 - 4x^2)(a^2\sin^4\phi - b^2\cos^4\phi)}{8(a^2b^2)(a^2\sin^2\phi + b^2\cos^2\phi)^{1/2}} \qquad 3.396$$

CIRCULAR SECTION TORUS (Fig. 3.81). Equilibrium of section $abb'a'$ demands that

$$p\pi(R^2 - a^2) - \sigma_1(2\pi hR\sin\alpha) = 0 \qquad 3.397$$

from which

$$\sigma_1 = \frac{p(R^2 - a^2)}{2Rh\sin\alpha} \qquad 3.398$$

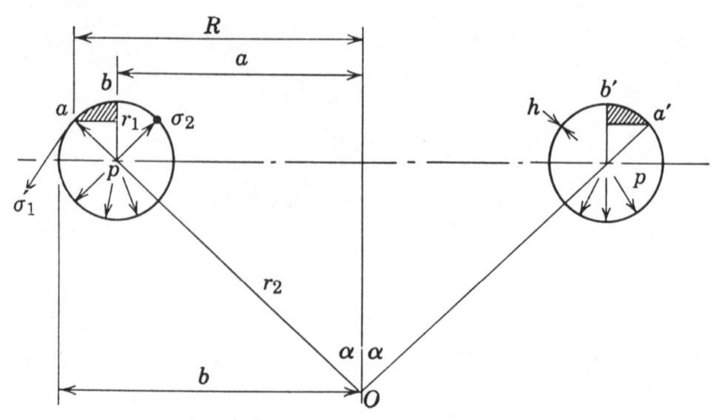

Fig. 3.81 Circular section torus under pressure.

Substituting Eq. 3.397 into Eq. 3.367 with $\sin \alpha = R/r_2$,

$$\sigma_2 = pr_1/2h \tag{3.399}$$

SEMISPHERICAL SHELL (Fig. 3.82). In this case $R_1 = R_2$ and Eq. 3.367 takes the form

$$\frac{\sigma_1 + \sigma_2}{a} = \frac{p}{t} \tag{3.400}$$

But $p = wy$, where w is the weight of the liquid in the tank. Therefore $\sigma_1 + \sigma_2 = wya/t$. The weight of liquid supported by the stress σ_1 is

$$\sigma_1(2\pi x \cos \alpha) = w\left(\pi x^2 \frac{a - y}{2} + \pi x^2 y\right) \tag{3.401}$$

from which

$$\sigma_1 = \frac{(a + y)aw}{4t} \tag{3.402}$$

and also

$$\sigma_2 = (ap/4ty)(4t^2y - a - y) \tag{3.403}$$

COMPOUND TANK (Fig. 3.83). The tank shown in Fig. 3.83 consists of a conical shell with a spherical bottom. In this case, the stresses in the spherical

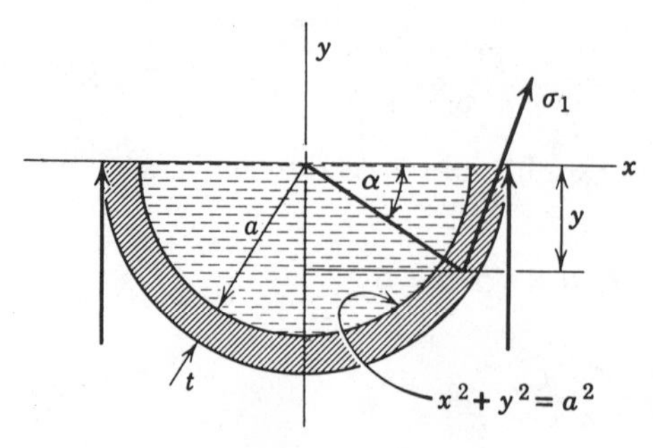

Fig. 3.82 Semispherical shell with fluid.

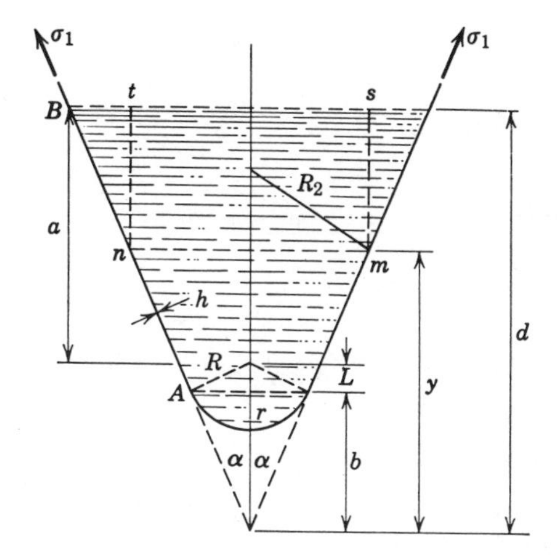

Fig. 3.83 Compound tank with fluid.

part can be calculated from the results in the foregoing examples. In the conical part R_1 is infinite and

$$\sigma_2 = pR_2/h = \gamma(d - y)R_2/h \qquad 3.404$$

From the geometry indicated $R_2 \sin\alpha = y \cot\alpha$. Therefore Eq. 3.404 becomes

$$\sigma_2 = \gamma(d - y)(y/h)(\cot\alpha/\sin\alpha) \qquad 3.405$$

The weight of volume *tnrms* is

$$2\pi\sigma_1 R_2 \sin^2 \alpha h \qquad 3.406$$

or

$$\text{weight} = \gamma(d - y)(\pi R_2{}^2 \sin^2\alpha) + (\pi/3)y\gamma R_2{}^2 \sin^2\alpha - (\pi/3)b\gamma R^2 \sin^2\alpha + \gamma\pi(\tfrac{2}{3}R^3 - R^3\cos\alpha + R^3\cos^3\alpha/3) \qquad 3.407$$

from which

$$\sigma_1 = \frac{\gamma}{6h}\left[(3d - 2y)y\frac{\cos\alpha}{\sin^2\alpha} + \frac{R^2(2R - b)\sin^2\alpha}{y\cos\alpha} + \frac{R^3\sin^2\alpha}{y}\right] \qquad 3.408$$

Discontinuity stresses. In addition to the membrane stresses just considered, additional stresses called *discontinuity stresses* are set up at the juncture of the head and shell in many vessels. The magnitude of these stresses depends on the geometry of the vessel, changes in wall thickness, shape, and curvature, and their calculation is fairly complex. At the location of discontinuity shears and moments give rise to stresses that act in addition to the membrane stresses developed. In most instances the discontinuity stresses damp out rapidly as the distance from the discontinuity increases, and for this reason such stresses are of a local nature, but they can be critical in many instances.

In the calculation of discontinuity stresses use is made of the theory of beams on elastic foundations.

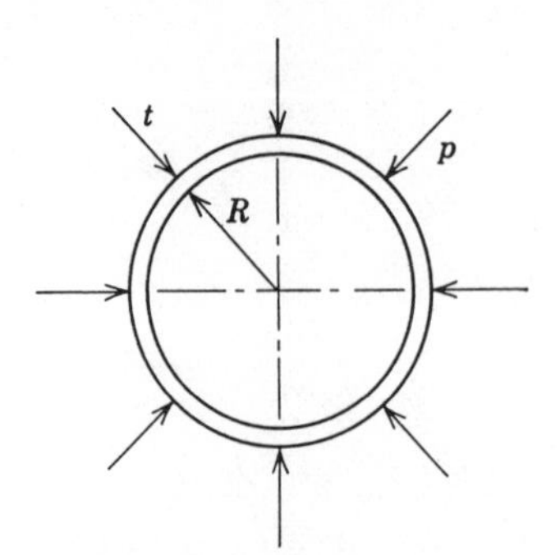

Fig. 3.84 Thin-walled cylinder under external pressure.

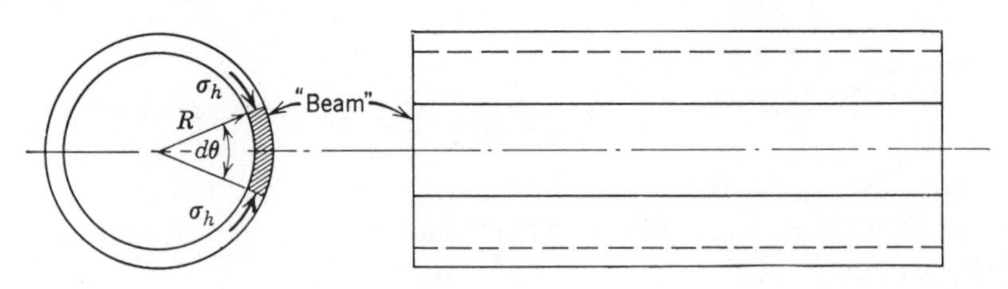

Fig. 3.85 Section in cylinder taken as a beam.

Consider a thin-walled cylinder subjected to external pressure as shown in Fig. 3.84. Here the hoop stress is

$$\sigma_h = pR/t \tag{3.409}$$

and the associated compressive strain is

$$\varepsilon_h = pR/Et \tag{3.410}$$

Since $\varepsilon_h = u/R$, the radial deflection is

$$u = R\varepsilon_h = pR^2/Et \tag{3.411}$$

If now a "beam" is taken from the cylinder as shown in Fig. 3.85 it can be thought of as being supported by the rest of the cylinder, that is, by the compressive hoop stress as shown in Fig. 3.86. Since K, the modulus of foundation,

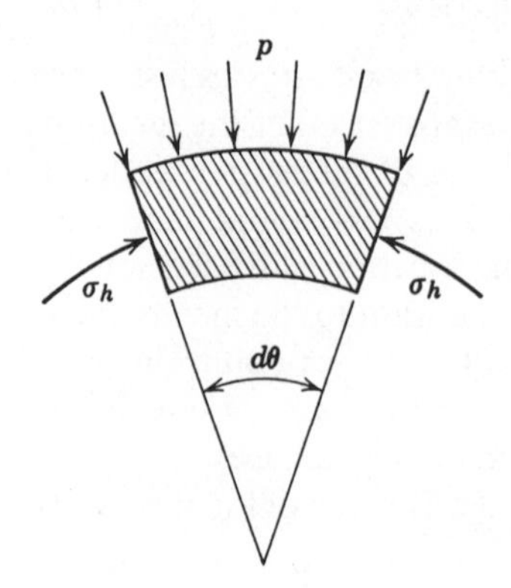

Fig. 3.86 Differential element of equivalent beam.

has been defined as

$$K = q/y \tag{3.412}$$

where q is the reaction per unit length, the value of q is

$$q = pR(d\theta) \tag{3.413}$$

The quantity y is deflection corresponding to u and

$$K = \frac{pR(d\theta)}{pR^2/Et} = \frac{EtR(d\theta)}{R^2} \tag{3.414}$$

and for a "beam" of unit width, $R(d\theta)$ equals unity, so that

$$K = Et/R^2 \tag{3.415}$$

The "beam," however, is restrained by the rest of the shell, and the system acts like a plate having a flexural rigidity D, which replaces the equivalent beam quantity (EI). Consequently, from Eq. 3.168 and 3.415

$$D(d^4y/dx^4) = -(Et/R^2)\delta \tag{3.416}$$

Discontinuity stresses are caused by a differential expansion between two parts. For example, at the location of geometric change a shear and a moment will be induced such as shown in Fig. 3.43*B* and the solution of Eq. 3.416 becomes

$$\delta = (e^{-\beta x}/2\beta^3 D)[Q_0 \cos \beta x - \beta M_0 (\cos \beta x - \sin \beta x)] \tag{3.417}$$

where Q_0 is the shear force, M_0 is the moment at the point of discontinuity, and

$$\beta = \sqrt[4]{Et/4DR^2} \tag{3.418}$$

Several particular cases can now be considered. First, consider a cylinder with a semispherical head as shown in Fig. 3.87. Where the geometry changes at section A–A discontinuity stresses are generated. For the cylinder,

$$\sigma_1 = \sigma_h = pR/t \tag{3.419}$$

$$\sigma_2 = \sigma_x = pR/2t \tag{3.420}$$

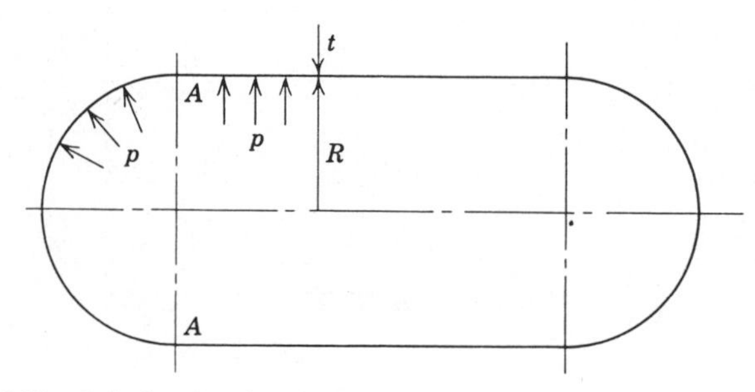

Fig. 3.87 Cylindrical tank with semispherical heads (uniform wall thickness).

and

$$(u_R)_{\text{cyl}} = (R/E)(\sigma_1 - \nu\sigma_2) = (pR^2/2tE)(2 - \nu) \qquad 3.421$$

For the spherical portion,

$$\sigma_1 = \sigma_2 = \sigma_h = pR/2t \qquad 3.422$$

and

$$(u_R)_{\text{sph}} = (pR^2/2tE)(1 - \nu) \qquad 3.423$$

The discontinuity stress arises from the differential radial formation of the two parts, which in this case is

$$\delta = (u_R)_{\text{cyl}} - (u_R)_{\text{sph}} = pR^2/2tE \qquad 3.424$$

In this particular instance the spherical portion at the discontinuity can be considered cylindrical so that the slopes and deformations for both parts are the same. Thus, by symmetry, the induced moment is zero, and Q_0 for each part must produce a deflection $\delta/2$. Thus, from Eq. 3.417, with $x = 0$, M_0 is zero

$$Q_0/2\beta^3 D = \delta/2 \qquad 3.425$$

and

$$Q_0 = \delta\beta^3 D = pR^2 Et/2tE(4\beta R^2) = p/8\beta \qquad 3.426$$

Furthermore,

$$\delta = (e^{-\beta x}/2\beta^3 D)(Q_0 \cos \beta x) \qquad 3.427$$

and

$$M = -D(d^2y/dx^2) \qquad 3.428$$

from which

$$M = -(Q_0/\beta)e^{-\beta x} \sin \beta x \qquad 3.429$$

Finally, the bending stress is

$$\sigma = M/Z \qquad 3.430$$

which must be added to the membrane stress given by Eq. 3.420. There is also a

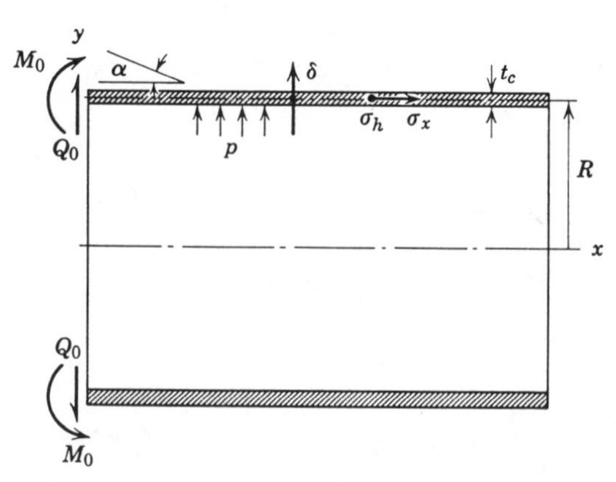

Fig. 3.88 Geometry and loads on circular cylinder.

hoop discontinuity stress that is given here without derivation.

$$(\sigma_h)_{\text{disc}} = Q_0 e^{-\beta x}[(6\nu/\beta t^2)\sin\beta x - (E/2\beta^3 DR)\cos\beta x] \qquad 3.431$$

As an aid to designers, formulas have been derived for the stresses, shears, moments, and displacements associated with various geometries giving rise to discontinuity stresses. In the following paragraphs the results of a few cases are tabulated, as given in (11). For additional analyses (11, 12, 17, 22, and 36) may be consulted. Reference 22 considers stresses in tapered transition joints.

CASE I. CIRCULAR CYLINDER (Fig. 3.88).

$$\text{Axial stress} = \sigma_x = \pm\frac{6}{t_c^2}\left(M_0\phi_1 + \frac{1}{\beta}Q_0\phi_4\right) \qquad 3.432$$

$$\text{Hoop stress} = \sigma_h = \left[2\beta^2\frac{R}{t_c}\phi_2 \pm \frac{6\nu}{t_c^2}\phi_1\right]M_0 + \left(2\beta\frac{R}{t_c}\phi_3 \pm \frac{6\nu}{\beta t_c^2}\phi_4\right)Q_0 \qquad 3.433$$

$$\text{Rotation of meridian} = \alpha = \frac{1}{2\beta^2 D}(2\beta M_0\phi_3 + Q_0\phi_1) \qquad 3.434$$

$$\text{Outward displacement} = \delta = \frac{1}{2\beta^2 D}(\beta M_0\phi_2 + Q_0\phi_3) \qquad 3.435$$

where Q_0 and M_0 are determined by the geometry and the functions ϕ_1, ϕ_2, ϕ_3, and ϕ_4 are given by Eqs. 3.184 to 3.187.

CASE II. HEMISPHERICAL SHELL (Fig. 3.89)

$$\text{Meridian stress} = \sigma_m = -\left[\frac{2\lambda}{R't_s}\cot(\theta-\phi)\phi_4 \mp \frac{6}{t_s^2}\phi_1\right]M_0 - \left[\frac{1}{t_s}\cot(\theta-\phi)\phi_2 \pm \frac{6R'}{\lambda t_s^2}\phi_4\right]Q_0\sin\theta \qquad 3.436$$

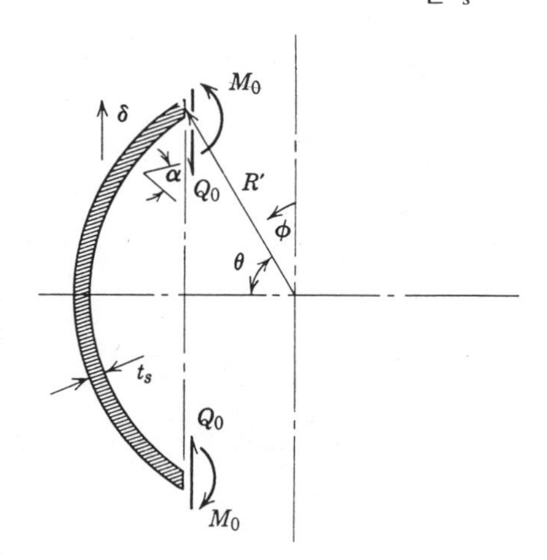

Fig. 3.89 Geometry and loads on a spherical section.

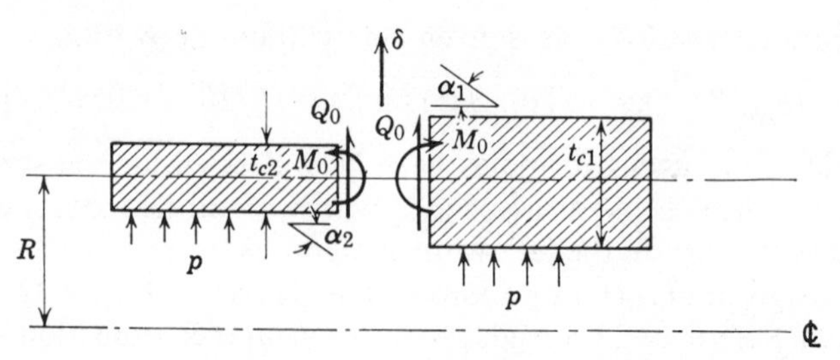

Fig. 3.90 Change of thickness in a cylinder.

$$\text{Hoop stress} = \sigma_h = \left(\frac{2\lambda^2}{R't_s}\phi_2 \pm \frac{6\nu}{t_s^2}\phi_1\right)M_0$$

$$-\left(\frac{2\lambda}{t_s}\phi_3 \pm \frac{6\nu R'}{\lambda t_s^2}\phi_4\right)Q_0 \sin\theta \qquad 3.437$$

$$\text{Rotation of meridian} = \alpha = \frac{2\lambda^2}{Et_s}\left(-\frac{2\lambda}{R'}\phi_3 M_0 + \phi_1 Q_0 \sin\theta\right) \qquad 3.438$$

$$\text{Outward displacement} = \delta = \frac{2\lambda}{Et_s}\sin(\theta - \phi)(\lambda\phi_2 M_0 - R'\phi_3 Q_0 \sin\theta) \qquad 3.439$$

To facilitate the use of the relationships given in the two cases the shears and moments associated with the various vessel and shell designs are given; these quantities have been taken from (11). The analyses for ellipsoidal, conical, and torospherical vessels are not included here because of their length; for these analyses refer to (11, 12, 17, and 36).

Change in thickness of a cylinder (Fig. 3.90)

$$Q_0 = \frac{2-\nu}{2\beta_1}\left[\frac{(c-1)(c^{5/2}+1)}{(c^2+1)^2 + 2c^{3/2}(c+1)}\right]p \qquad 3.440$$

$$M_0 = \frac{2-\nu}{4\beta_1^2}\left[\frac{(c-1)(c^2-1)}{(c^2+1)^2 + 2c^{3/2}(c+1)}\right]p \qquad 3.441$$

$$c = \frac{t_{c1}}{t_{c2}} \qquad 3.442$$

$$\beta_1 = \sqrt[4]{[3(1-\nu^2)]/R^2 t_{c1}^2} \qquad 3.443$$

Junction of a cylinder and a sphere (Fig. 3.91)

$$Q_0 = \left\{\frac{\left(\frac{1-\nu}{\sin\theta} - (2-\nu)c\right)(1 + c^2\sqrt{c\sin\theta})\frac{1}{\lambda_s} - (1 + c^2 + 2c^2\sqrt{c\sin\theta})\cos\theta}{(1+c^2)^2 + (2c^2/\sqrt{c\sin\theta})(1 + c\sin\theta)}\right\}$$

$$\times \frac{pR'}{2} \qquad 3.444$$

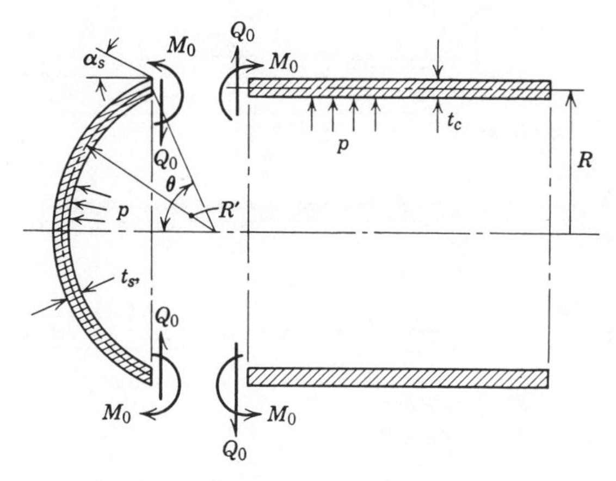

Fig. 3.91 Junction of a cylinder and a sphere.

$$M_0 = \left\{ \frac{\left[\dfrac{1-\nu}{\sin\theta} - (2-\nu)c\right]\left(\dfrac{1-c^2}{2\lambda_s^{\,2}}\right) + \left[\left(1 + \dfrac{1}{\sqrt{c\sin\theta}}\right)\dfrac{c^2}{\lambda_s}\right]\cos\theta}{(1+c^2)^2 + (2c^2/\sqrt{c\sin\theta})(1 + c\sin\theta)} \right\} \frac{pRR'}{2} \quad 3.445$$

$$c = \frac{t_s}{t_c} \quad 3.446$$

$$\lambda_s = \sqrt[4]{[3(1-\nu^2)(R')^2]/t_s^{\,2}} \quad 3.447$$

Junction of a cylinder and a head (Fig. 3.92)

$$Q_0 = -\left\{ \frac{c^3\lambda_c^{\,3} + 2(2-\nu)c^3\lambda_c + 2(2-\nu)(1+\nu)}{2c^3\lambda_c^{\,2} + [(1-\nu)c^4 + (1+\nu)]\lambda_c + (1-\nu^2)c} \right\} \frac{pR}{4} \quad 3.448$$

$$M_0 = \left\{ \frac{2c^3\lambda_c^{\,3} + (1-\nu)c^4\lambda_c^{\,2} + 2(2-\nu)(1+\nu)}{2c^3\lambda_c^{\,2} + [(1-\nu)c^4 + (1+\nu)]\lambda_c + (1-\nu^2)c} \right\} \frac{pR^2}{8\lambda_c} \quad 3.449$$

$$c = \frac{t_c}{t_h} \quad 3.450$$

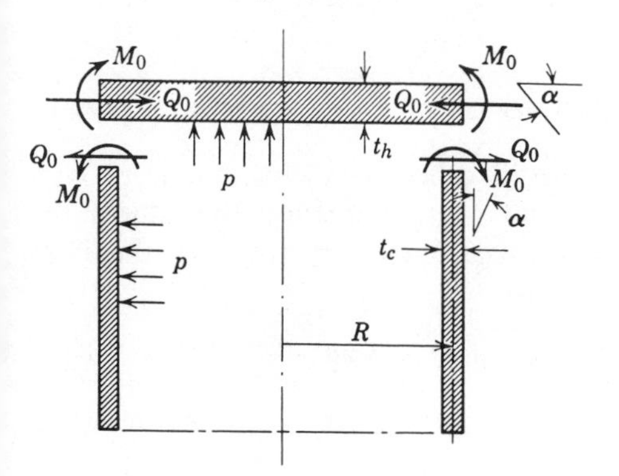

Fig. 3.92 Junction of a cylinder and a flat head.

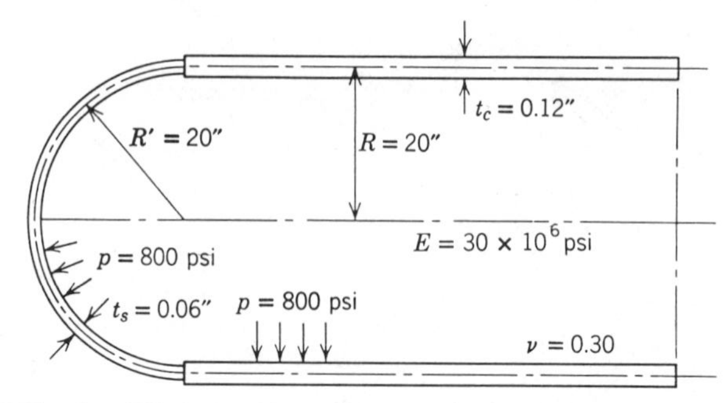

Fig. 3.93 Circular cylinder with semispherical heads of a different thickness.

$$\lambda_c = \sqrt[4]{[3(1-\nu^2)R^2]/t_c^{\,2}} \tag{3.451}$$

As an example of the use of the preceding theory consider the shell structure shown in Fig. 3.93 consisting of a cylinder joined to a spherical head. In this case $\theta = \pi/2$ and from Eq. 3.444,

$$Q_0 = -23 \text{ lb/in.}$$

From Eq. 3.445

$$M_0 = -6.26 \text{ in.-lb/in.}$$

For the cylindrical portion of the vessel, by Eqs. 3.432 and 3.433, the discontinuity stresses are calculated for the shell at various distances from the discontinuity and the results tabulated and plotted.

By using Table 3.2, the results shown in Table 3.5 are obtained. The axial and hoop stress discontinuity values are plotted in Figs. 3.94 and 3.95. To these stresses must be added the membrane stresses which are

$$\sigma_x = 66{,}500 \text{ psi}$$

and

$$\sigma_h = 133{,}000 \text{ psi}$$

The total stresses are also listed in Table 3.5. For the spherical portion of the vessel $\sigma_1 = \sigma_2$ and $\sigma_{90^\circ} = \sigma_m$ (see Fig. 3.89), and the stress values are given by Eqs. 3.436 and 3.437. By using these relationships the results shown in Table 3.6 are obtained.

Heavy-walled cylinder under pressure

Figure 3.96 shows the cross section of a heavy-walled cylinder with uniform wall thickness and subjected to both internal and external pressure. By virtue of the symmetry of the deformation, there are no shearing forces involved;* therefore at location r the condition of equilibrium is

$$\sigma_r r\, d\theta + \sigma_h\, dr\, d\theta - (\sigma_r + d\sigma_r/dr)(r + dr)d\theta = 0 \tag{3.452}$$

*The problem of bending of heavy-walled cylindrical shells by pressure has been treated by MacGregor (15).

Table 3.5 Stresses in Cylindrical Portion of Compound Vessel

Value of	Axial Stress, psi[a]			Hoop Stress, psi[a]		
βx	Discontinuity	Membrane	Total	Discontinuity	Membrane	Total
0	∓2,620	66,500	63,880	−8,590	133,000	124,400
			69,120	−7,022	133,000	125,900
0.2	∓4,410	66,500	62,090	−7,340	133,000	125,660
			70,910	−4,707	133,000	128,293
0.4	∓5,320	66,500	61,180	−6,000	133,000	127,000
			71,820	−2,851	133,000	130,149
0.6	∓5,570	66,500	60,920	−4,762	133,000	128,238
			72,070	+1,428	133,000	134,428
0.8	∓5,400	66,500	61,100	−3,515	133,000	129,485
			71,900	−353	133,000	132,647
1.0	∓4,910	66,500	61,500	−2,578	133,000	130,422
			71,410	+365	133,000	133,365
1.4	∓3,550	66,500	62,950	−1,044	133,000	131,956
			70,050	+1,089	133,000	134,089
1.8	∓2,180	66,500	64,320	−130	133,000	132,870
			68,680	+1,176	133,000	134,176
2.0	∓1,550	66,500	64,950	+135	133,000	133,135
			68,050	+1,095	133,000	134,095
2.4	∓683	66,500	65,816	+379	133,000	133,379
			67,183	+817	133,000	133,817
2.8	∓139	66,500	66,361	+433	133,000	133,433
			66,639	+519	133,000	133,519
4.0	∓227	66,500	66,273	+142	133,000	133,142
			66,727	+5	133,000	133,005

[a] Top figure for outside fiber; bottom figure for inside fiber.

which reduces, on eliminating higher-order differentials, to

$$\sigma_h - \sigma_r - r\frac{d\sigma_r}{dr} = 0 \qquad 3.453$$

Equation 3.453 is a general equilibrium relation valid in both the elastic and plastic ranges of strain. To solve the equation for the stresses, it is necessary to obtain a second relationship, and this is done by considering the deformation involved. Since deformation is symmetrical, it is constant in the σ_h direction and variable in the σ_r direction; thus, if u is the deformation of an element at r, the radial movement at $r + dr$ is

$$u + (du/dr)\,dr \qquad 3.454$$

and the elongation is

$$\varepsilon_r = du/dr \qquad 3.455$$

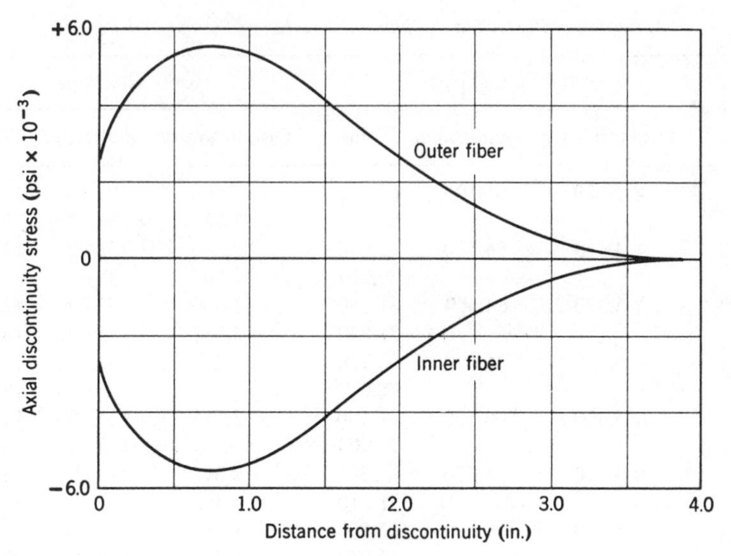

Fig. 3.94 Axial discontinuity stress in cylindrical portion of vessel shown in Fig. 3.93.

Correspondingly, in the σ_h direction,

$$\varepsilon_h = u/r \quad 3.456$$

Combining Eqs. 3.455 and 3.456 with Eqs. 3.70 to 3.72 gives

$$\varepsilon_1 = \varepsilon_h = u/r = (1/E)[\sigma_h - \nu(\sigma_r + \sigma_z)] \quad 3.457$$

$$\varepsilon_2 = \varepsilon_r = du/dr = (1/E)[\sigma_r - \nu(\sigma_h + \sigma_z)] \quad 3.458$$

and

$$\varepsilon_3 = \varepsilon_z = (1/E)[\sigma_z - \nu(\sigma_h + \sigma_r)] \quad 3.459$$

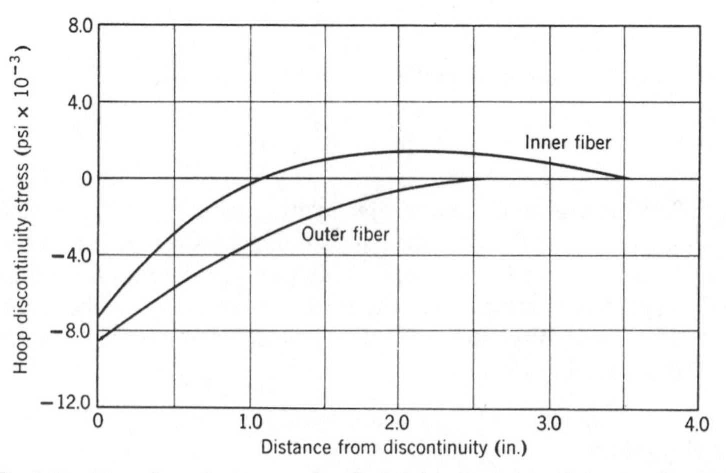

Fig. 3.95 Hoop discontinuity stress in cylindrical portion of vessel shown in Fig. 3.93.

Table 3.6 Stresses in Hemispherical Portion of Compound Vessel

		Meridian Stress, psi[a]			Hoop Stress, psi[a]		
Angle ϕ	Value of $\lambda\phi$	Discontinuity	Membrane	Total	Discontinuity	Membrane	Total
0	0	−10,200	133,000	122,800	+9,070	133,000	142,070
		+10,200	133,000	143,200	+15,330	133,000	148,330
2	0.819	+3,980	133,000	136,980	+6,730	133,000	139,730
		−3,980	133,000	129,020	+4,316	133,000	137,316
4	1.64	+4,435	133,000	137,435	+2,273	133,000	135,273
		−4.435	133,000	128,565	−388	133,000	132,612
6	2.46	+1,884	133,000	134,884	+62	133,000	133,062
		−1,884	133,000	131,116	−1069	133,000	131,931
10	4.10	−221	133,000	132,779	−255	133,000	132,745
		+221	133,000	133,221	−133	133,000	132,867

[a] **Top figure for outside fiber; bottom figure for inside fiber.**

Solving Eqs. 3.457 to 3.458 simultaneously (with $\sigma_z = 0$),

$$\sigma_h = [E/(1 - \nu^2)][u/r + \nu(du/dr)] \tag{3.460}$$

and

$$\sigma_r = [E/(1 - \nu^2)][du/dr + \nu(u/r)] \tag{3.461}$$

Substituting Eqs. 3.460 to 3.461 into Eq. 3.453 gives the following differential equation in u:

$$d^2u/dr^2 + (1/r)(du/dr) - u/r^2 = 0 \tag{3.462}$$

the solution for which is

$$u = C_1 r + C_2/r \tag{3.463}$$

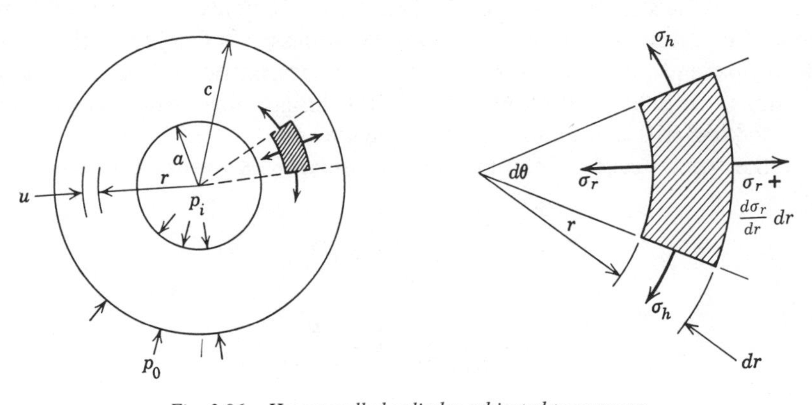

Fig. 3.96 Heavy-walled cylinder subjected to pressure.

The constants C_1 and C_2 are determined from the boundary conditions, which are:

$$(\sigma_r)_{r=c} = -p_o \tag{3.464}$$

and

$$(\sigma_r)_{r=a} = -p_i \tag{3.465}$$

Therefore substituting Eqs. 3.464 to 3.465 into Eqs. 3.460 to 3.461 and 3.463 gives the results

$$\sigma_h = \frac{1}{c^2 - a^2}\left[a^2 p_i - c^2 p_o + \left(\frac{ac}{r}\right)^2 (p_i - p_o)\right] \tag{3.466}$$

and

$$\sigma_r = \frac{1}{c^2 - a^2}\left[a^2 p_i - c^2 p_o - \left(\frac{ac}{r}\right)^2 (p_i - p_o)\right] \tag{3.467}$$

Equations 3.466 and 3.467 may also be obtained from Eqs. 3.143 and 3.144 by letting the constant B in these equations equal zero and using the boundary conditions given by Eqs. 3.464 to 3.465.

A longitudinal stress can also exist under certain conditions. For a cylinder with open ends,

$$(\sigma_z)_0 = 0 \tag{3.468}$$

For a cylinder with closed ends and external pressure acting only on the OD of the cylinder, $p_i \pi a^2 = \sigma_z \pi (c^2 - a^2)$ or

$$(\sigma_z)_c = \frac{p_i a^2}{c^2 - a^2} \tag{3.469}$$

For plane strain, $\varepsilon_z = 0$ in Eq. 3.446 so that

$$(\sigma_z)_r = \frac{2v}{c^2 - a^2}(a^2 p_i - c^2 p_0) \tag{3.470}$$

The hoop and radial stresses are independent of the end conditions of the cylinder, and the longitudinal stress is always intermediate to the other principal stresses. For special situations, which will be discussed in Chapter 6, the choice of the proper end condition has practical advantages, particularly in prestressing operations. Ordinarily, the difference in the longitudinal stress as given by Eqs. 3.468 to 3.470 is not considered significant.

For the special case of internal pressure only, p_0 is zero in Eqs. 3.466 and 3.467 so that

$$\sigma_h = \frac{p_i}{R^2 - 1}\left(1 + \frac{c^2}{r^2}\right) \tag{3.471}$$

$$\sigma_r = \frac{p_i}{R^2 - 1}\left(1 - \frac{c^2}{r^2}\right) \tag{3.472}$$

$$(\sigma_z)_0 = 0 \tag{3.473}$$

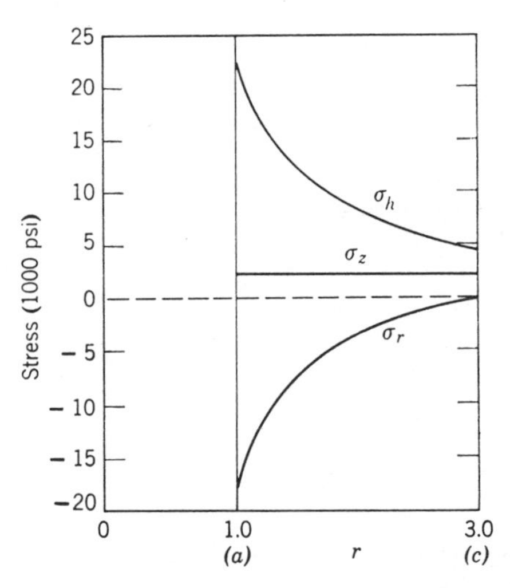

Fig. 3.97 Stress distribution in heavy-walled cylinder.

$$(\sigma_z)_c = p_i/(R^2 - 1) \qquad 3.474$$

and

$$(\sigma_z)_r = 2\nu p_i/(R^2 - 1) \qquad 3.475$$

For a closed-end cylinder of diameter ratio 3.0 and yield strength of 35,000 psi in tension, the stress distributions through the wall at the instant just before yielding occurs under internal pressure are shown in Fig. 3.97. These stress distributions were obtained by Eqs. 3.471 to 3.473

Expressions for the elastic deformation are given in Eqs. 3.457 to 3.459. It is frequently more convenient to express deformation in terms of radial displacement; consequently, from Eqs. 3.457 and 3.466 to 3.470 and limiting consideration to internal pressure only,

$$(u)_0 = \frac{p_i a^2}{Er(c^2 - a^2)}[r^2(1 - \nu) + c^2(1 + \nu)] \qquad 3.476$$

$$(u)_c = \frac{p_i a^2}{Er(c^2 - a^2)}[r^2(1 - 2\nu) + c^2(1 + \nu)] \qquad 3.477$$

and

$$(u)_r = \frac{p_i a^2}{Er(c^2 - a^2)}[r^2(1 - \nu - 2\nu^2) + c^2(1 + \nu)] \qquad 3.478$$

If the bore* of the cylinder is off center (Fig. 3.98), Jeffery (10) has shown that the maximum stress is in the hoop direction at location A with the restriction that $a < r_i/2$. The result is as follows;

$$(\sigma_h)_A = p\left[\frac{2r_0^2(r_0^2 + r_i^2 - 2r_i a - a^2)}{(r_i^2 + r_0^2)(r_0^2 - r_i^2 - 2r_i a - a^2)} - 1\right] \qquad 3.479$$

*A photoelastic analysis of a square plate with a pressurized internal circular bore is given in Reference 2.

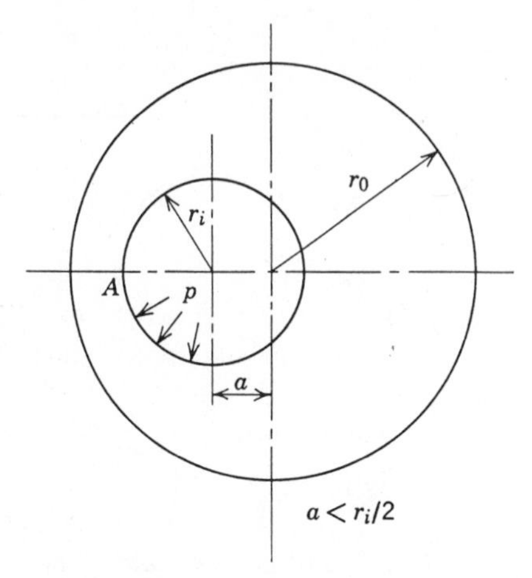

Fig. 3.98 Heavy-walled cylinder with eccentric bore under pressure.

As an illustration of the use of the foregoing theory on heavy-walled cylinders, design charts will be calculated for such vessels under internal pressure. Consideration is given first to elastic action and from Eqs. 3.466, 3.467, and 3.469

$$\sigma_h = \frac{p}{R^2 - 1}\left(1 + \frac{c^2}{r^2}\right) \tag{3.480}$$

$$\sigma_r = \frac{p}{R^2 - 1}\left(1 - \frac{c^2}{r^2}\right) \tag{3.481}$$

$$\sigma_z = \frac{p}{R^2 - 1} \tag{3.482}$$

Dilation is expressed by Eq. 3.457

$$\varepsilon_h = \frac{1}{E}[\sigma_h - \nu(\sigma_r + \sigma_z)] \tag{3.483}$$

The equations given can now be manipulated to produce design charts. In one chart, Fig. 3.99, hoop stress is plotted as a function of location in the wall of the cylinder. This chart shows that as the diameter ratio (O.D. ÷ I.D.) decreases, the distribution of stress becomes more uniform through the wall; for the heaviest wall thickness there is a very steep gradient, the stress being high at the bore and then tapering off to a relatively low value. This condition brings out an important point. If, for example, the cylinder is made of a shallow-hardening steel, it would be possible that the effect of heat treatment would not penetrate far enough into the wall; then the stress could be above the yield strength of the material and local plastic flow could occur in the cylinder wall even though the bore were still elastic. Figure 3.99 also indicates that it is not necessary to have a high-strength (high yield strength) material throughout the wall thickness. In other words, for $R = 10$, a fairly thin sleeve of high-strength material would suffice at the bore and the rest of the cylinder could be cheaper low-strength material.

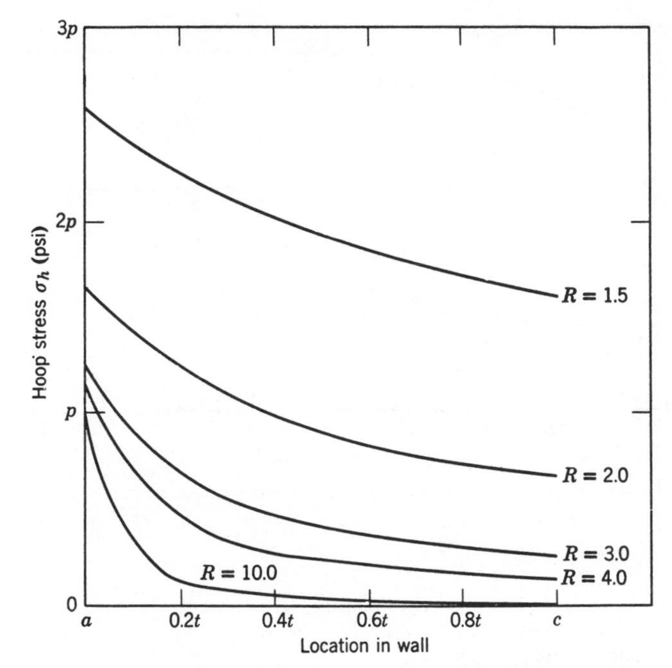

Fig. 3.99 Distribution of hoop stress in cylinder in elastic range.

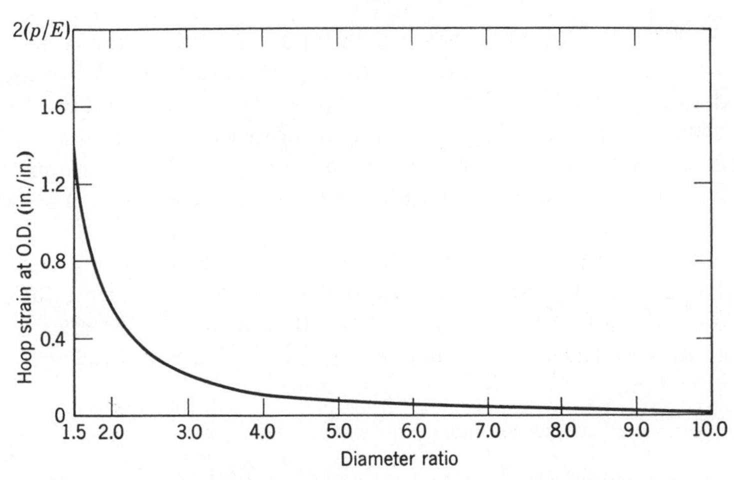

Fig. 3.100 Hoop strain in cylinder in elastic range as a function of diameter ratio and Young's modulus. [Note: Hoop strain at bore = $(\varepsilon_h)_{O.D.} \times (0.765R^2 + 0.2355)$.]

Another chart, Fig. 3.100, gives data on dilation of the cylinder. From the equations already given the hoop strain at the bore (I.D.) and at the outside surface (O.D.) can be calculated. If Poisson's ratio is assumed to be 0.3, then

$$(\varepsilon_h)_{I.D.} = \frac{p}{E(R^2 - 1)}(1.3R^2 + 0.4) \qquad 3.484$$

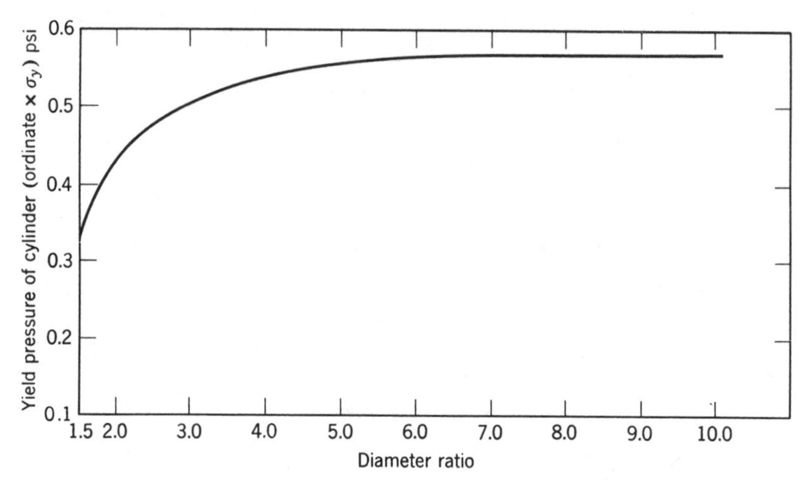

Fig. 3.101 Yield pressure of cylinder as a function of diameter ratio and yield strength of material.

and

$$(\varepsilon_h)_{\text{O.D.}} = \frac{1.7p}{E(R^2 - 1)} \tag{3.485}$$

From these equations it is seen that

$$(\varepsilon_h)_{\text{I.D.}} = (\varepsilon_h)_{\text{O.D.}}(0.765R^2 + 0.2355) \tag{3.486}$$

Figure 3.100 shows the O.D. strain as a function of R (diameter ratio) and Young's modulus E. Again, some important points are revealed. For example, as the diameter ratio increases, the strain decreases very rapidly; it is also evident that to minimize dilation there is little practical effect in using a cylinder of diameter ratio larger than about 5 or 6. In other words, for minimum dilation use a sleeve or linear of high modulus material and the rest of the cylinder can be of a cheaper lower modulus material.

The final chart, Fig. 3.101, shows yield pressure of the cylinder as a function of strength of material and diameter ratio. For any cylinder there is little to be gained by making the diameter ratio more than 4 or 5; it is better to use a higher strength material at lower diameter ratio. The subject of yield pressure is discussed in Section 3-3.

Heavy-walled sphere under pressure

In the heavy-walled sphere, by symmetry, the same state of stress exists at all points of a concentric spherical surface. To establish the equilibrium relationship, consider Fig. 3.102 which shows an element cut from the sphere wall. From the geometry indicated, equilibrium requires that

$$[\sigma_r + (d\sigma_r/dr)\,dr](r + dr)^2(d\theta)^2 - \sigma_r r^2(d\theta)^2 - 2\sigma_h r\,dr\,(d\theta)^2 = 0 \tag{3.487}$$

and neglecting the higher order differentials,

$$(r/2)(d\sigma_r/dr) + (\sigma_r - \sigma_h) = 0 \tag{3.488}$$

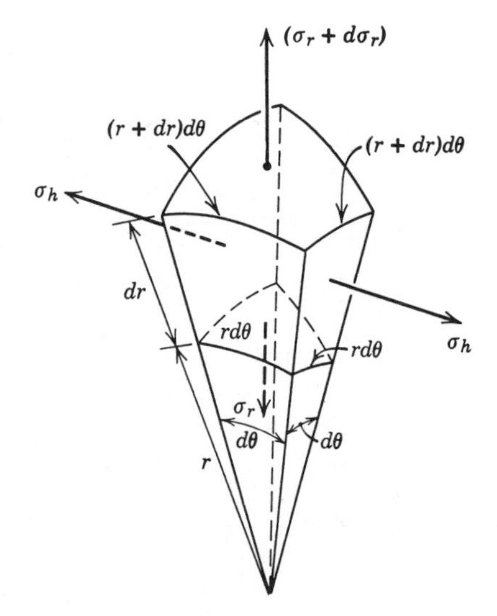

Fig. 3.102 Equilibrium of spherical elements.

Since there are two unknowns and only one equation, it is necessary to obtain another equation by considering the deformation involved, which is the same as in a heavy-walled cylinder; thus

$$\varepsilon_r = du/dr \qquad 3.489$$

and

$$\varepsilon_h = u/r \qquad 3.490$$

Because of symmetry, Eqs. 3.457 to 3.459 reduce to

$$\sigma_1 = \sigma_h = \frac{E}{1 - \nu - 2\nu^2}(\varepsilon_1 + \nu\varepsilon_2) \qquad 3.491$$

and

$$\sigma_2 = \sigma_r = \frac{E}{1 - \nu - 2\nu^2}[2\nu\varepsilon_1 + \varepsilon_2(1 - \nu)] \qquad 3.492$$

Substituting Eqs. 3.489 and 3.490 into Eq. 3.492 leads to the differential equation

$$(d^2u/dr^2) + (2/r)(du/dr) - 2u/r^2 = 0 \qquad 3.493$$

the solution of which is

$$u = C_1 r + C_2/r^2 \qquad 3.494$$

where C_1 and C_2 are constants of integration determined by the boundary conditions, which are

$$(\sigma_r)_{r=a} = -p_i \quad \text{(internal pressure)} \qquad 3.495$$

$$(\sigma_r)_{r=b} = -p_0 \quad \text{(external pressure)} \qquad 3.496$$

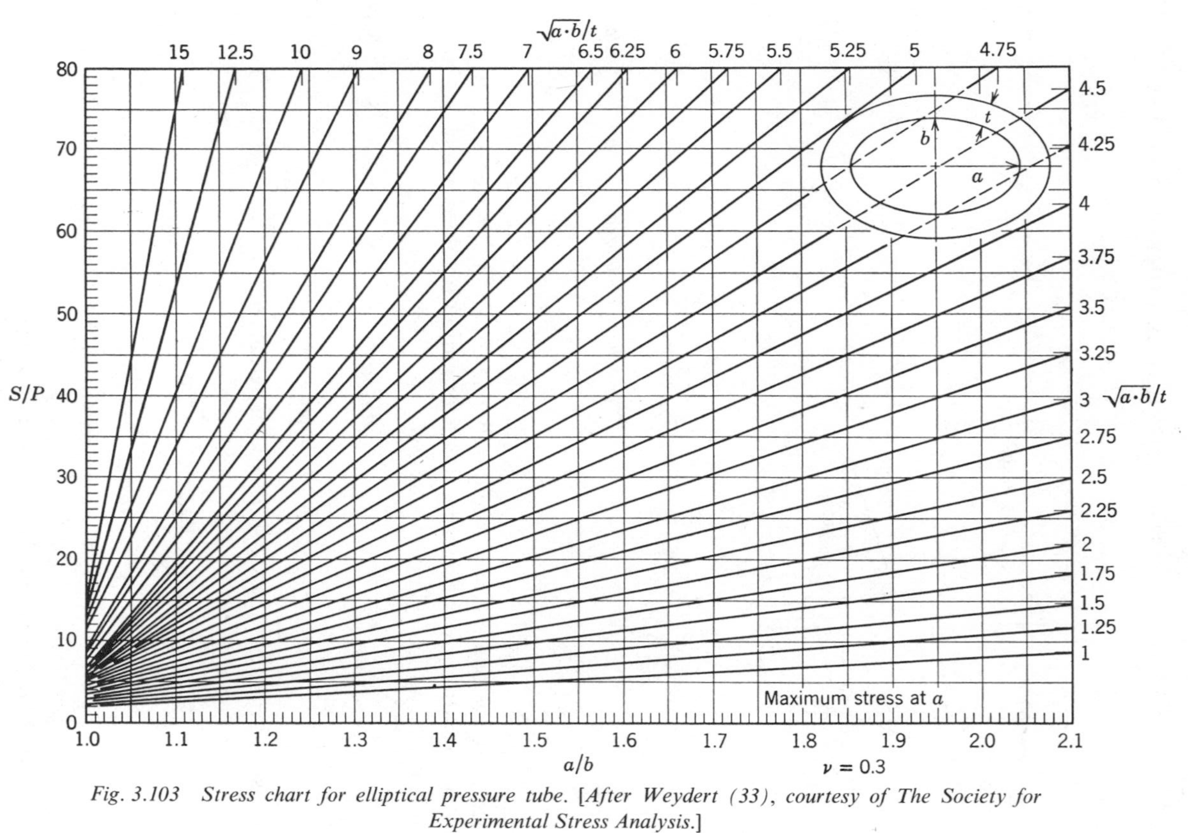

Fig. 3.103 Stress chart for elliptical pressure tube. [After Weydert (33), courtesy of The Society for Experimental Stress Analysis.]

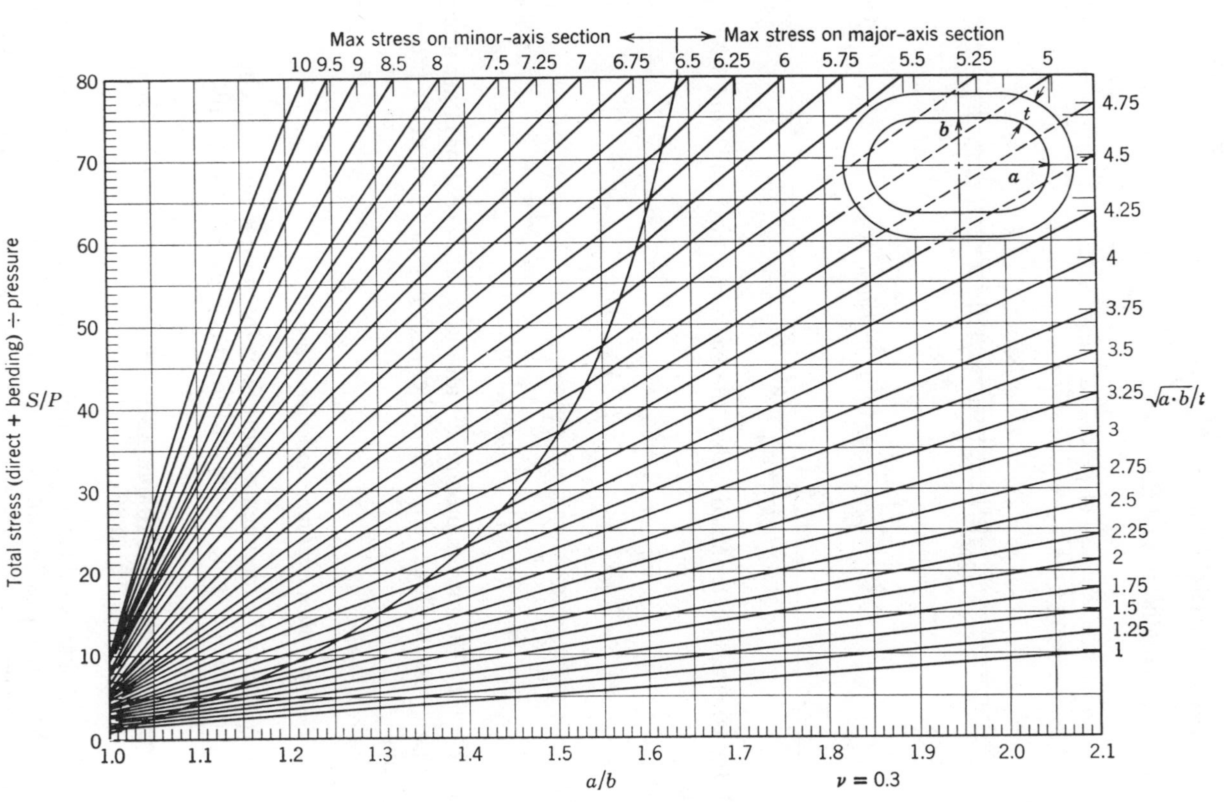

Fig. 3.104 Stress chart for semicircle-tangent pressure tube. [After Wevdert (33), courtesy of The Society for Experimental Stress Analysis.]

By using these conditions and Eq. 3.494

$$\sigma_h = \frac{E}{1 - v - 2v^2}\left[v\left(C_1 - \frac{2C_2}{r^3}\right) + C_1 + \frac{C_2}{r^3}\right] \quad 3.497$$

and

$$\sigma_r = \frac{E}{1 - v - 2v^2}\left[(1 - v)\left(C_1 - \frac{2C_2}{r^3}\right) + 2v\left(C_1 + \frac{C^2}{r^3}\right)\right] \quad 3.498$$

and finally

$$C_1 = \frac{1 - 2v}{E(b^3 - a^3)}(p_1 a^3 - p_0 b^3) \quad 3.499$$

$$C_2 = \frac{1 + v}{2E(b^3 - a^3)}(a^3 b^3)(p_i - p_0) \quad 3.500$$

With the constants determined, substitute back into Eqs. 3.497 and 3.498 and obtain

$$\sigma_r = \frac{p_i}{R^3 - 1}\left(1 - \frac{b^3}{r^3}\right) - \frac{p_0 R^3}{R^3 - 1}\left(1 - \frac{a^3}{r^3}\right) \quad 3.501$$

and

$$\sigma_h = \frac{p_i}{R^3 - 1}\left(1 + \frac{1}{2}\frac{b^3}{r^3}\right) - \frac{p_0 R^3}{R^3 - 1}\left(1 + \frac{1}{2}\frac{a^3}{r^3}\right) \quad 3.502$$

where $R = b/a$. The maximum stress is always σ_h at $r = a$. For either internal or external pressure alone simply let either p_i or p_0 equal zero in the preceding equations.

Heavy-walled oval tube under pressure

Both analytical and experimental work have been done on oval tubes under internal pressure and some results are shown in Figs. 3.103 and 3.104.

3-3 THEORIES OF FAILURE AND APPLICATION TO DESIGN

For structural shapes subjected to simple states of stress like tension or compression the design stress, usually some fraction of the yield or ultimate strength, is readily available from tension test data. On the other hand, for parts subjected to combined stresses the situation is not so simple. In some cases tests can be performed to determine, for example, what the yield pressure is for a cylinder having a particular diameter ratio; this value can then be used as a working criterion. Usually, however, it is not convenient, or even possible, to conduct model tests; it is then necessary to rely on a procedure that makes predictions about performance under one set of conditions from a knowledge of behavior under some other conditions.

For most design procedures it has been established that adequate predictions

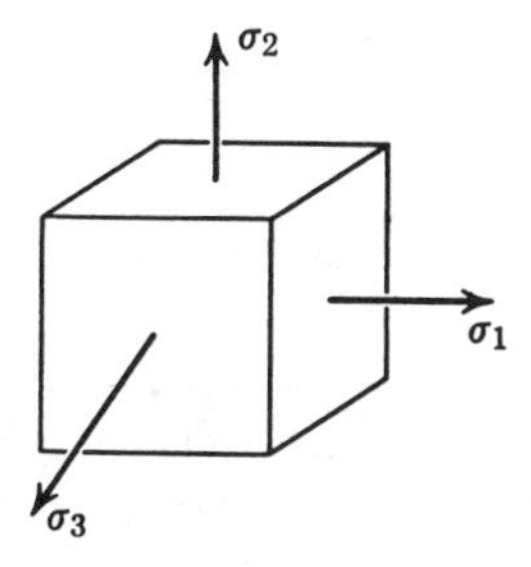

Fig. 3.105 Three-dimensional stress system.

concerning combined stress behavior can be made on the basis of data obtained from tensile or compression tests. In converting the uniaxial to combined stress data three principal theories are used; a brief description follows.

Consider Fig. 3.105 which shows a general state of stress in which it is assumed that $\sigma_1 > \sigma_2 > \sigma_3$.

In the *maximum stress theory*, it is postulated that failure occurs in a member subjected to combined stresses when one of the principal stresses reaches the failure value (yield strength, for example) in simple tension σ_0. Thus, for materials whose tension and compression properties are the same, failure is defined as follows.

If $\sigma_1 > \sigma_2, \sigma_3$ failure occurs when $\sigma_1 = \pm\sigma_0$.
If $\sigma_2 > \sigma_1, \sigma_3$ failure occurs when $\sigma_2 = \pm\sigma_0$.
If $\sigma_3 > \sigma_1, \sigma_2$ failure occurs when $\sigma_3 = \pm\sigma_0$.

A graphical representation in two dimensions is shown in Fig. 3.106. The

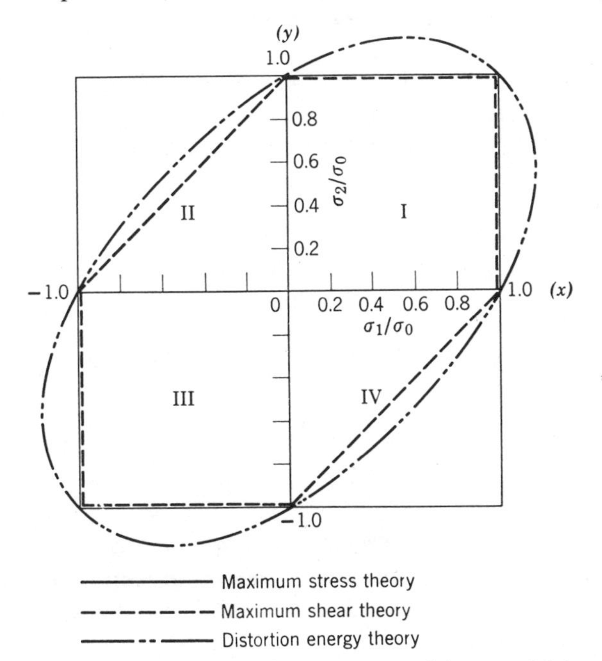

Fig. 3.106 Two-dimensional representation of theories of failure.

plot is made by dividing the preceding expressions by σ_0; thus

$$x = \sigma_1/\sigma_0 \quad \text{and} \quad y = \sigma_2/\sigma_0$$

and the locus of failure points is a square. As will be pointed out later, considerable caution must be exercised in applying this theory to a practical problem.

In the *maximum shear theory*, failure is postulated to occur in a member subjected to combined stresses when the maximum shear stress reaches the value of shear failure stress in tension. Thus for a member under combined stresses there are three shear possibilities.

$$\tau_1 = \pm\frac{\sigma_1 - \sigma_2}{2} \tag{3.503}$$

$$\tau_2 = \pm\frac{\sigma_2 - \sigma_3}{2} \tag{3.504}$$

$$\tau_3 = \pm\frac{\sigma_3 - \sigma_1}{2} \tag{3.505}$$

In tension, $\tau = \sigma_0/2$; therefore, failure under combined stresses is defined by one of the following quantities.

$$\sigma_1 - \sigma_2 = \pm\sigma_0 \tag{3.506}$$

$$\sigma_2 - \sigma_3 = \pm\sigma_0 \tag{3.507}$$

$$\sigma_3 - \sigma_1 = \pm\sigma_0 \tag{3.508}$$

The correct expression depends on the sign of the principal stresses involved. The maximum shear stress for principal stresses of opposite sign is given by one of Eqs. 3.503 to 3.505. For principal stresses of the same sign, failure is defined by

$$\tau_{max} = \pm\sigma_1/2 \quad \text{for} \quad \sigma_1 > \sigma_2, \sigma_3$$
$$\tau_{max} = \pm\sigma_2/2 \quad \text{for} \quad \sigma_2 > \sigma_1, \sigma_3$$
$$\tau_{max} = \pm\sigma_3/2 \quad \text{for} \quad \sigma_3 > \sigma_1, \sigma_2$$

A plot of this theory in two dimensions is also given in Fig. 3.106. The plot is obtained by dividing Eqs. 3.506 to 3.508 by σ_0; thus

$$x - y = \pm 1; \qquad x = \pm 1 \quad \text{and} \quad y = \pm 1$$

where

$$x = \sigma_1/\sigma_0 \quad \text{and} \quad y = \sigma_2/\sigma_0$$

In the *distortion energy theory* failure is postulated to occur in a material subjected to combined stresses when the energy of distortion reaches the same energy for failure in tension. This theory is considerably more involved than the other two theories just developed and will now be considered in some detail. The total strain energy U produced in the element shown in Fig. 3.105 consists of energy producing a volume change U_v and an energy effecting distortion U_d; thus

$$U = U_v + U_d \tag{3.509}$$

For the three-dimensional case

$$U = \sigma_1 \varepsilon_1/2 + \sigma_2 \varepsilon_2/2 + \sigma_3 \varepsilon_3/2 \tag{3.510}$$

and replacing the principal strain values by Eqs. 3.457 to 3.459

$$U = \tfrac{1}{2}E[\sigma_1{}^2 + \sigma_2{}^2 + \sigma_3{}^2 - 2(\sigma_1\sigma_2 + \sigma_2\sigma_3 + \sigma_3\sigma_1)] \tag{3.511}$$

It is now necessary to develop expressions for the two energies U_v and U_d. To obtain an expression for U_v consider that each of the principal stresses is composed of two parts:

$$(\sigma)_{1,2,3} = (\sigma')_{1,2,3} + (\sigma_v)_{1,2,3} \tag{3.512}$$

where σ' stresses are chosen so that the volume change produced by them is zero; their associated strains ε' thus total to zero. Consequently,

$$\sigma_1' + \sigma_2' + \sigma_3' = \varepsilon_1' + \varepsilon_2' + \varepsilon_3' = 0 \tag{3.513}$$

and

$$\sigma_v = \frac{\sigma_1 + \sigma_2 + \sigma_3}{3} \tag{3.514}$$

$$\sigma_1' = \tfrac{1}{3}(2\sigma_1 - \sigma_2 - \sigma_3) \tag{3.515}$$

$$\sigma_2' = \tfrac{1}{3}(2\sigma_2 - \sigma_1 - \sigma_3) \tag{3.516}$$

$$\sigma_3' = \tfrac{1}{3}(2\sigma_3 - \sigma_1 - \sigma_2) \tag{3.517}$$

Equations 3.515 to 3.517 now define the new stress values in terms of the principal stresses. The stress σ_v in Eq. 3.514 represents a condition of hydrostatic loading and hence there is no distortion but only volume change associated with it. The distortion produced is

$$U_v = \tfrac{3}{2}(\sigma_v \varepsilon_v) = \tfrac{1}{2}\varepsilon_v\left(\frac{\sigma_1 + \sigma_2 + \sigma_3}{3}\right) \tag{3.518}$$

But,

$$\varepsilon_v = \frac{1 - 2v}{E}\left(\frac{\sigma_1 + \sigma_2 + \sigma_3}{3}\right) \tag{3.519}$$

Therefore

$$U_v = \frac{1 - 2v}{6E}(\sigma_1 + \sigma_2 + \sigma_3)^2 \tag{3.520}$$

Now, by substituting Eqs. 3.511 to 3.512 into Eq. 3.509

$$U_d = \frac{1 + v}{3E}[\sigma_1{}^2 + \sigma_2{}^2 + \sigma_3{}^2 - \sigma_1\sigma_2 - \sigma_2\sigma_3 - \sigma_3\sigma_1] \tag{3.521}$$

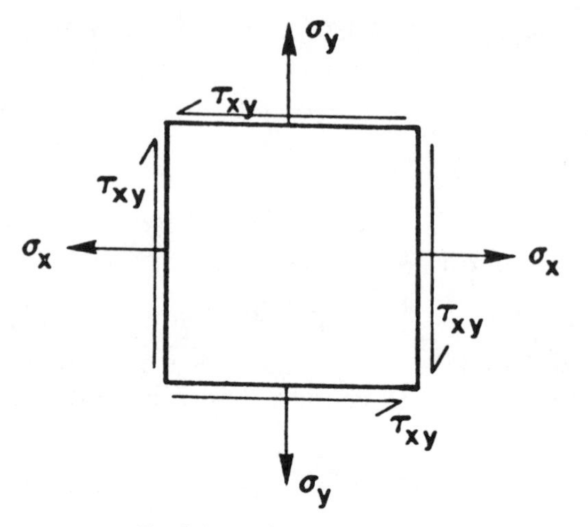

Fig. 3.107 Plane stress element.

For simple tension $\sigma_2 = \sigma_3 = 0$ and

$$(U_d)_{\text{tension}} = \frac{1+\nu}{3E}\sigma_0^2 \tag{3.522}$$

Equating Eqs. 3.521 and 3.522 gives the following expression for the distortion energy theory of failure;

$$\sigma_0^2 = \sigma_1^2 + \sigma_2^2 + \sigma_3^2 - \sigma_1\sigma_2 - \sigma_2\sigma_3 - \sigma_1\sigma_3 \tag{3.523}$$

or

$$2\sigma_0^2 = (\sigma_1 - \sigma_2)^2 + (\sigma_2 - \sigma_3)^2 + (\sigma_3 - \sigma_1)^2 \tag{3.524}$$

In two dimensions, $\sigma_0^2 = \sigma_1^2 + \sigma_2^2 - \sigma_1\sigma_2$, and dividing through by σ_0^2 gives

$$x^2 + y^2 - xy = 1 \tag{3.525}$$

where $x = \sigma_1/\sigma_0$ and $y = \sigma_2/\sigma_0$. This theory is also plotted in Fig. 3.106.

The distortion energy theory, Eq. 3.524, can be simplified for the condition of plane stress, Fig. 3.107, which is often encountered in engineering analysis. In Eq. 3.524 σ_z is zero, and Eq. 3.18 and 3.19 are substituted yielding

$$\sigma_0^2 = \sigma_x^2 - \sigma_x\sigma_y + \sigma_y^2 + 3\tau_{xy}^2 \tag{3.526}$$

Equation 3.526 simplifies failure analysis and does not require the use of Mohr's circle to find the principal stresses.

Validity of the theories. With the exception of some tests on heavy-walled cylinders under internal pressure most of the experimental work has been on biaxial stress systems (3, 14, 16). A comparison of some experimental and theoretical yield stresses for several metals under a variety of stress conditions is shown in Fig. 3.108. Most investigators have used thin-walled tubes under internal pressure combined with axial loading and torsion or with tubes sub-

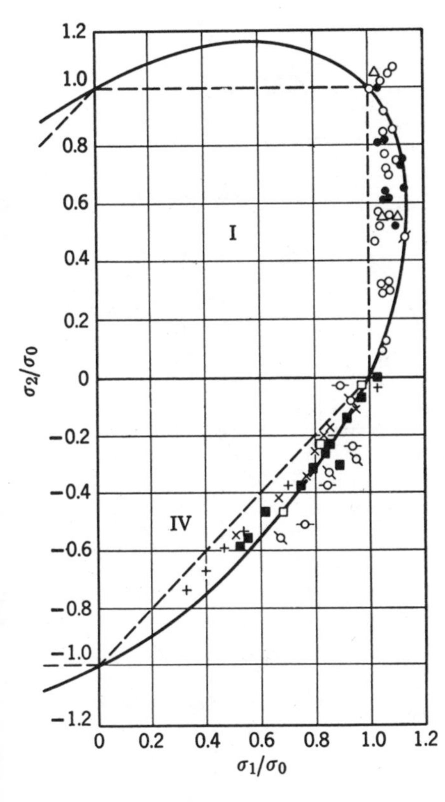

Fig. 3.108 Comparison of various biaxial stress tests on steel, copper, and aluminum with theories of failure. See (14).

jected to a combination of tension plus torsion to obtain the desired state of combined stress. The data shown are all for room temperature and slow loading and indicate slightly better agreement with the distortion energy theory than with the shear theory.

The fact that several theories give about the same result is convenient for design in some instances; however, there has always been an interest in determining which of the theories is the most fundamentally correct. For example, repeated experiments have shown that the yield strength in torsion is about 12 to 15% higher than the shear strength in tension; this discrepancy would not occur if the maximum shear theory were correct. On the other hand, the distortion energy theory fills this gap. Several investigators have calculated the average yield strength in both tension and torsion for face-centered cubic crystals of all orientations on the assumption that a single crystal yields on reaching a critical value of shear stress component in the slip direction in the slip plane. It was found that the shear stress at the yield strength in torsion should be 15.4% greater than the shear stress at the yield strength in tension, which is exactly the difference required by the distortion energy theory. Thus there is considerable theoretical and experimental evidence that the distortion energy theory is the most fundamental of the strength theories.

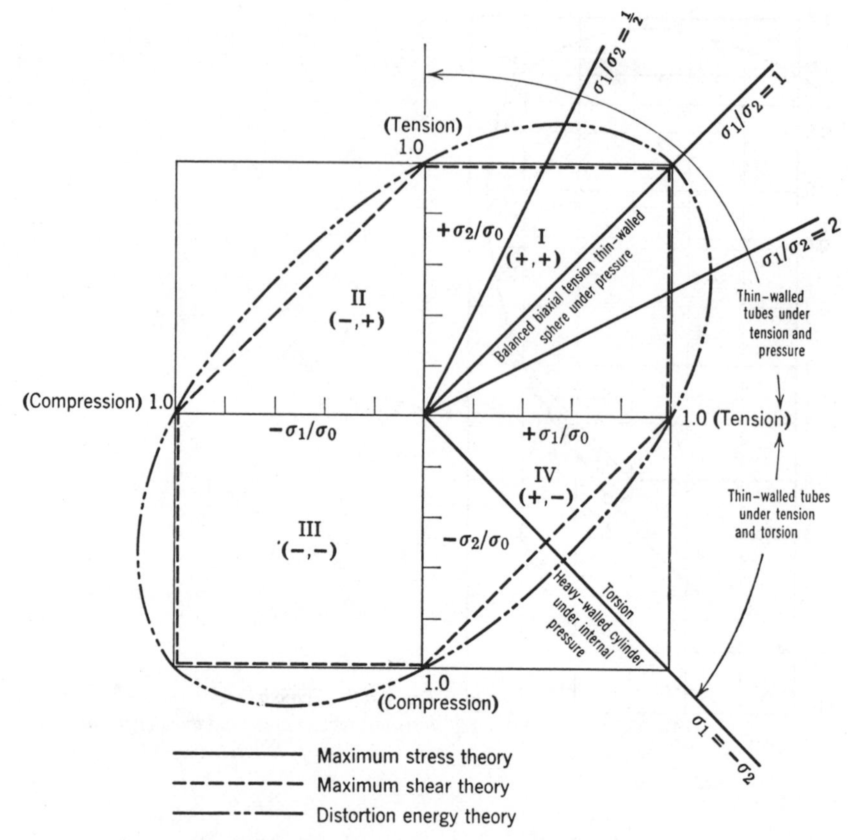

Fig. 3.109 Stress quadrants in two-dimensional cases.

Some tests conducted years ago on cast iron and other brittle materials seemed to indicate that the maximum stress theory was generally more applicable to brittle materials than the other theories. More recent data, however, show clearly that even "brittle" materials can be handled by the distortion energy theory when allowance is made for the stress concentrating effects of inclusions that often give rise to the lack of ductility in a part. This is discussed further in Chapter 13.

From a design point of view it is clear that, for all practical purposes, even the maximum stress theory is permissible as long as the state of stress involved is in the first or third stress quadrants (see Fig. 3.109). In stress quadrant I all stresses are positive, the shear and stress theories predict identical results, and the maximum deviation from the distortion energy theory is about 15%. For balanced biaxial tension (sphere under pressure, for example) all the theories coincide. Moving down into the fourth stress quadrant, it is clear that only the shear and distortion energy theories should be used. Brittle materials like cast iron under quadrant IV states of stress may be analyzed by the stress theory, but its limitations should be kept carefully in mind.

Examples of the use of the failure theories

The conical tank (stress quadrant I). Consider the conical tank shown in Fig. 3.79, which is filled with a liquid of density γ; the wall thickness will be determined by each of the three failure theories just outlined.

Here all stresses are positive (quadrant I, Fig. 3.109) and are given by Eqs. 3.378 and 3.381. The maximum values of these stresses are

$$(\sigma_1)_{\max} = (\gamma h^2/4t)(\tan\alpha/\cos\alpha) \qquad 3.527$$

and

$$(\sigma_2)_{\max} = \tfrac{3}{16}(\gamma h^2/t)(\tan\alpha/\cos\alpha) \qquad 3.528$$

In accordance with the *maximum stress theory* the failure stress is

$$\sigma_0 = (\sigma_1)_{\max} = (\gamma h^2/4t)(\tan\alpha/\cos\alpha)$$

and the required wall thickness of the tank in terms of the geometry and the tensile yield strength σ_0 is

$$t = (h^2\gamma/4\sigma_0)(\tan\alpha/\cos\alpha)$$

In accordance with the maximum shear theory,

$$\tau_{\max} = \sigma_0/2 = \sigma_1/2 = [(\gamma h^2/4t)(\tan\alpha/\cos\alpha)]/2$$

or

$$t = (h^2\gamma/4\sigma_0)(\tan\alpha/\cos\alpha)$$

which is the same result as obtained by use of the maximum stress theory.

Solution of the problem by means of the *distortion energy theory* requires the use of Eq. 3.523 with appropriate values of the principal stresses. From Eqs. 3.526 and 3.527 it is seen that the maximum values of the principal stresses occur at different locations; therefore it is necessary to locate the critical section of the tank where the combined stress has a critical value. This is done by substituting Eqs. 3.378 and 3.381 into Eq. 3.523 which gives the following:

$$\sigma_0{}^2 = \left[\gamma(h-y)y\frac{\tan\alpha}{t\cos\alpha}\right]^2 + \left[\frac{\gamma y\tan\alpha}{2t\cos\alpha}(h-\tfrac{2}{3}y)\right]^2 - \left[\gamma\frac{(h-y)y\tan\alpha}{t\cos\alpha}\right]\left[\frac{\gamma y\tan\alpha}{2t\cos\alpha}(h-\tfrac{2}{3}y)\right]$$

Differentiation of this equation with respect to y and equating to zero then gives the result

$$y = 0.52\,h$$

so that

$$t = 0.225(\gamma h^2/\sigma_0)(\tan\alpha/\cos\alpha)$$

which is 90% of the value calculated by the other two theories.

Elastic-breakdown of a heavy-walled cylinder (stress quadrant IV). When a cylinder containing no residual stresses due to heat treatment, fabrication, or prestress is subjected to a high enough pressure, plastic flow begins at the bore. In order for elastic breakdown to occur, it has been observed that, for all practical purposes, the distortion energy per unit volume of a tensile test bar of the material is equal to the distortion energy per unit volume of the cylinder; consequently, yielding can be predicted by Eq. 3.523. For internal pressure only, the maximum values of the stresses given by Eqs. 3.471 to 3.472 occur at $r = a$; thus, combining Eqs. 3.471, 3.472, 3.468, 3.474, 3.475, and 3.523, expressions for elastic-breakdown become

$$(p_{yi})_0 = \sigma_0\left(\frac{R^2 - 1}{\sqrt{3R^4 + 1}}\right)$$

$$(p_{yi})_c = \frac{\sigma_0}{\sqrt{3}}\left(\frac{R^2 - 1}{R^2}\right)$$

$$(p_{yi})_r = \sigma_0\left(\frac{R^2 - 1}{\sqrt{(3R^4 + 1) + 4\nu(\nu - 1)}}\right)$$

There is little difference in the elastic-breakdown pressures as given by these equations; the differences become very small as R increases, and even for $R = 2$ the differences are less than 2%. The open-end cylinder yields at the lowest pressure. The general relationship between elastic-breakdown pressure and diameter ratio is shown in Fig. 3.101. Note that not much advantage is gained elastically by increasing the size beyond 4 to 5.

For cylinders subjected to radial external pressure only, the critical location is also at $r = a$ and consequently

$$(p_{yo})_0 = \frac{\sigma_0}{2}\left(\frac{R^2 - 1}{R^2}\right)$$

$$(p_{yo})_c = \frac{\sigma_0}{2}\left(\frac{R^2 - 1}{R^2}\right)$$

$$(p_{yo})_r = \frac{\sigma_0}{2R^2}\frac{R^2 - 1}{\sqrt{\nu^2 - \nu + 1}}$$

If residual stresses are present in the cylinder wall before application of service pressure, they will cause either an increase or a decrease in the elastic-breakdown pressure depending on whether they are tension or compression. Residual stresses, if present, must be allowed for in the calculation as additions to the stress terms. An example of such a calculation is given in Chapter 11.

REFERENCES

1. L. E. Brownell, and E. H. Young, *Process Equipment Design (Vessel Design)*, John Wiley and Sons, Inc., New York, N.Y. (1959).
2. A. J. Durelli, and J. Barriage, *J. Appl. Mech.*, **77** (4), 539–554 (December 1955).
3. J. H. Faupel, *Trans. Am. Soc. Mech. Engrs.*, **78** (5), 1031–1064 (July 1956).

4. W. Flügge, *Stresses in Shells*, Springer-Verlag, Berlin (1973).
5. W. Flügge, *Handbook of Engineering Mechanics*, McGraw-Hill Book Co., New York, N.Y. (1962).
6. H. Ford, *Advanced Mechanics of Materials*, John Wiley and Son, Inc. New York, N.Y. (1963).
7. Y. C. Fung, *Foundations of Solid Mechanics*, Prentice-Hall Inc., Engle wood Cliffs, N.J. (1965).
8. G. D. Galletly, *Proc. Inst. Mech. Engrs. (London)*, **173** (27), 687–698 (1959).
9. M. Hetenyi, *Beams on Elastic Foundation*, University of Michigan Press, Ann Arbor, Mich. (1946).
10. G. B. Jeffery, *Phil. Trans. Roy. Soc. London*, **221** (Series A), 265–293 (1921).
11. R. H. Johns, and T. W. Orange, *Theoretical Elastic Stress Distributions Arising from Discontinuities and Edge Loads in Several Shell-Type Structures*, Nat. Aero. and Space Administration Tech. Report R-103, Washington, D.C. (1961).
12. H. Kraus, G. G. Bilodeau, and B. F. Langer, *Trans. Am. Soc. Mech. Engrs. (J. Eng. for Ind.)*, **83** (1), Series B, 29–42 (February 1961).
13. G. H. Lee, *An Introduction to Experimental Stress Analysis*, John Wiley and Sons, Inc., New York, N.Y. (1950).
14. J. R. Low, "Behavior of Metals under Direct or Nonreversed Loading," pp. 17–59 of *Properties of Metals in Materials Engineering*, Am. Soc. for Metals, Novelty, Ohio (1949).
15. C. W. MacGregor, and L. F. Coffin, *J. Appl. Mech.*, **14** (4), A301–A311 (December 1947).
16. J. Marin, *Mechanical Properties of Materials and Design*, McGraw-Hill Book Co., New York, N.Y. (1942).
17. R. R. Meyer, *Discontinuity Stresses in Shells—General Model*, Armed Services Tech. Information Agency (ASTIA), Report AD–264–376, Arlington, Virginia (1961).
18. H. Reissner, and F. Strauch, *Ingenieur Archiv.*, **4**, 481–505 (1933).
19. R. M. Rivello, *Theory and Analysis of Flight Structures*, McGraw-Hill Book Co., New York, N.Y. (1969).
20. R. J. Roark, *Formulas for Stress and Strain*, 3rd ed., McGraw-Hill Book Co., New York, (1954).
21. E. C. Rodabaugh, and H. H. George, *Trans. Am. Soc. Mech. Engrs.*, **79** (4), 939–948 (May 1957).
22. E. C. Rodabaugh, and T. J. Atterbury, *Trans. Am. Soc. Mech. Engrs. (J. Eng. for Ind.)*, **84** (3), Series B, 321–328 (August 1962).
23. F. B. Seely, and J. O. Smith, *Advanced Mechanics of Materials*, 2nd ed., John Wiley and Sons, Inc., New York, N.Y. (1952).
24. W. H. Sparing, *Prod. Eng.*, **33** (26), 62–66 (December 24, 1962).
25. R. Szilard, *Theory and Analysis of Plates*, Prentice-Hall Inc., Englewood Cliffs, N.J. (1974).
26. S. Timoshenko, *Trans. Am. Soc. Mech. Engrs.*, **45**, 135–140 (1923).
27. S. Timoshenko, and J. N. Goodier, *Theory of Elasticity*, McGraw-Hill Book Co., New York, N.Y., 2nd ed. (1951), 3rd ed. (1970).
28. S. Timoshenko, *Strength of Materials, Part II, Advanced Theory and Problems*, D, Van Nostrand Co., Princeton, N.J., 2nd ed. (1941), 3rd ed. (1956).
29. S. Timoshenko, and S. Woinowsky-Krieger, *Theory of Plates and Shells*, 2nd ed., McGraw-Hill Book Co., New York, N.Y. (1959).
30. J. P. Vidosic, *Prod. Eng.*, **22** (11), 180–183 (November 1951).
31. T. Von Karman, *Zeit. Ver. Deut. Ingen.*, **55** (45), 1889–1894 (November 11, 1911).
32. A. M. Wahl, *Mechanical Springs*, Penton Publishing Co., Cleveland, Oh. (1963).
33. J. C. Weydert, *Proc. Soc. Exp. Stress Anal.*, **12** (1), 39–54 (1954).
34. A. Wolf, *J. Appl. Mech.*, **13** (3), A207–A210 (September 1946).

35. W. Wuest, "Theory of High-Pressure Bourdon Tubes," paper No. 58–A–119 (21 pp.). *Am. Soc. Mech. Engrs.*, Translation from *Ingenieur-Archw.*, **24** (2), 92–110 (1956).

36. *Pressure Vessel and Piping Design* (Collected papers 1927–1959), Am. Soc. Mech. Engrs., New York (1960).

37. "Design Data and Methods (Biezeno Pressure Vessel Heads)," *J. Appl. Mech.*, **23** (4), 642–645 (December 1956).

PROBLEMS

3.1 A yellow brass shaft with a yield of 16 kpsi is subjected to combined loads (Fig. P. 3.1). Find the maximum principal stresses and the factor of safety for the yield condition at the wall.

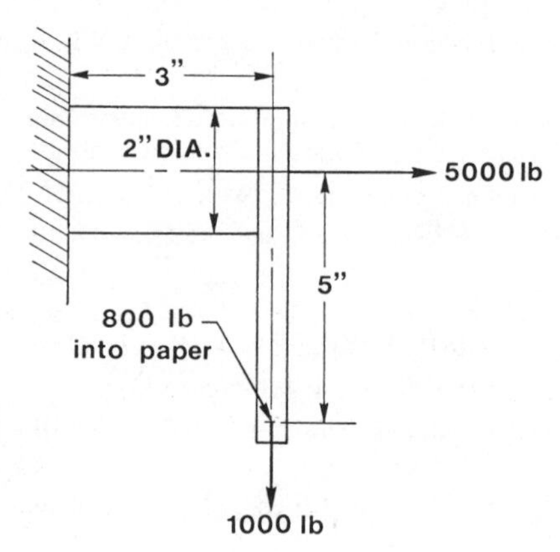

Fig. P3.1

3.2 A sphere is the most economical vessel (weight/unit volume) for containing gas under pressure. However, for space limitations a cylindrical vessel is often required. Find the additional weight required for a cylindrical tank compared to a spherical tank. Assume equal volumes, pressures, and materials (See Fig. P. 3.2).

3.3 Find the largest stress in part *A* and the thickness at *C* and *D* for part *B* of Fig. P. 3.3. Also, find the stress in the pin. The parts are made from 1018 steel.

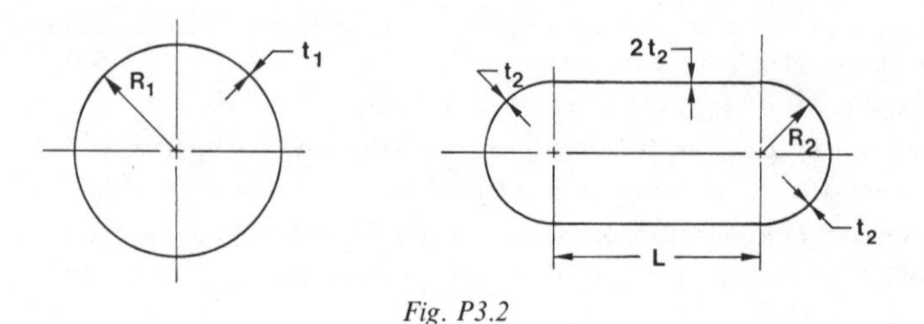

Fig. P3.2

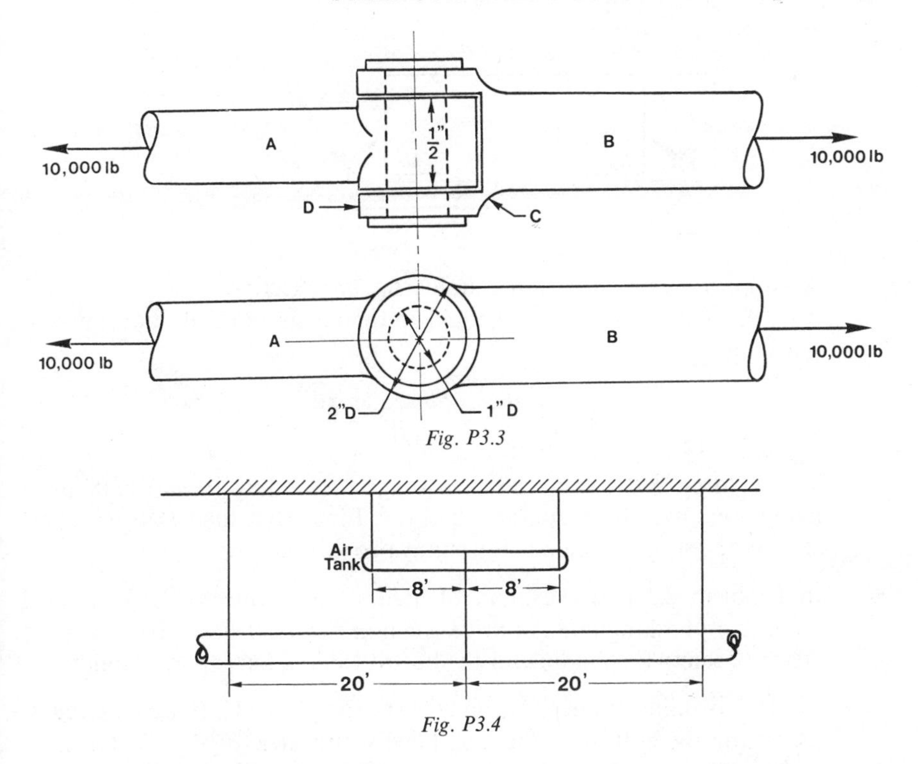

Fig. P3.3

Fig. P3.4

3.4 A 12-in. diameter schedule 80 polyvinyl chloride pipe carries water at 80 psig and is supported every 20 ft. (Fig. P3.4). The pipe is supported at one location from a 302 stainless steel air tank with 200 psig pressure made from an 8-in. diameter schedule 40 pipe capped at both ends. Check the stresses in the air tank to see if the steel yields.

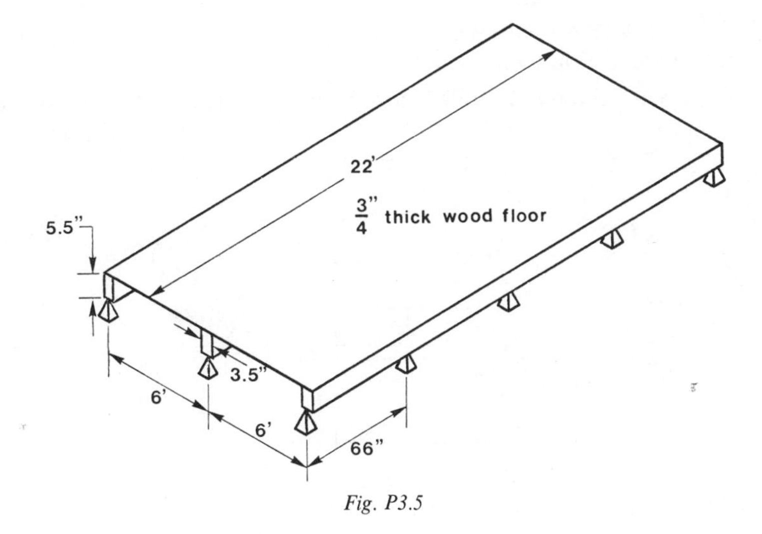

Fig. P3.5

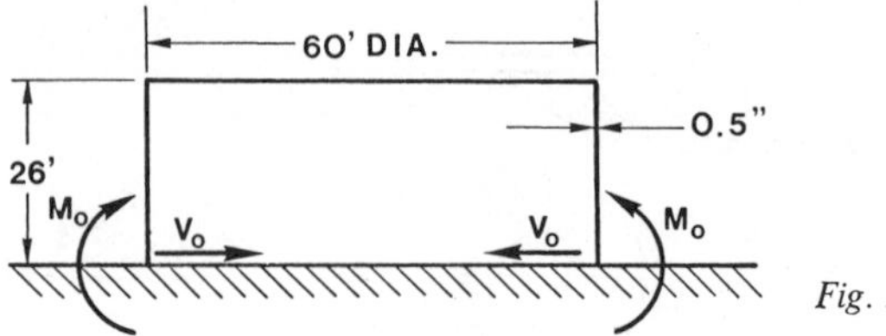

Fig. P3.7

3.5 A support beam in a house holds up an effective plate 12 ft × 22 ft (Fig. P3.5). A tile floor is to be laid so the maximum deflection must be less than

$$Y_{\max} = \frac{\text{unsupported length}}{360} \text{ (inches)}$$

The building codes and references suggest a loading $P = 40 \text{ lb/ft}^2$ for a living area and 100 lb/ft^2 for a hallway. Find what load $P(\text{lb/ft}^2)$ overstresses the support beam? How many people does this represent?

3.6 In Problem 2.3 find the point of highest stress due to the combined loads from braking and the drive chain in the 4130 steel shaft. Find the factor of safety for yielding for the hot-rolled and heat-treated condition.

3.7 Find the bending moment M_0 and the shearing force V_0 along the circumference on the bottom of the tank filled with water (Fig. P3.7). The tank material is steel.

3.8 A steel clamp $E = 30 \times 10^6$ psi is loaded with a 5000-lb load (Fig. P3.8). Calculate the maximum and minimum stress by:

a. Using Eqs. 3.124 and 3.127

b. Exact elasticity methods

c. Using Eqs. 3.135 and 3.136

d. No correction factor to (c)

Then place these in a table using (b) as the base to calculate the percent error.

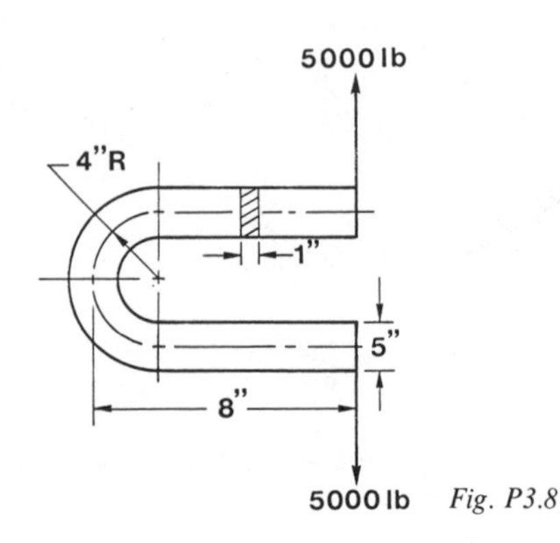

Fig. P3.8

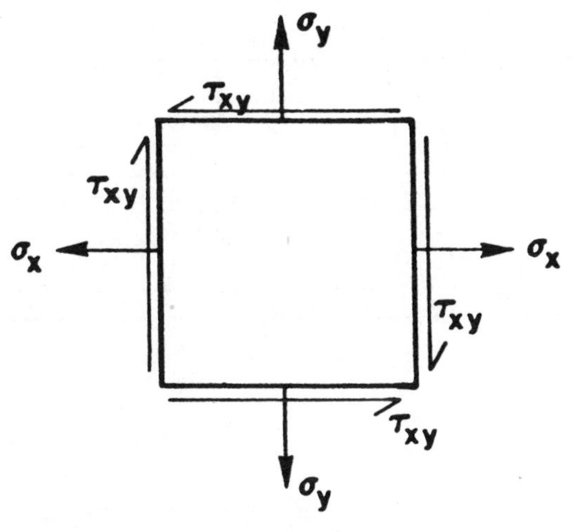

Fig. P3.10

3.9 A thick-walled cylinder with internal pressure P_i has a hoop stress σ_h and as an approximation take

$$\sigma_h = \frac{P_i R_i}{t} \quad \text{and} \quad \sigma_h = \frac{P_i R_{\text{ave}}}{t}$$

for σ_h at the inside of the cylinder and plot percent error using the heavy-wall cylinder equation as a base versus t/R_0 for the two above equations. R_0 is the outside radius.

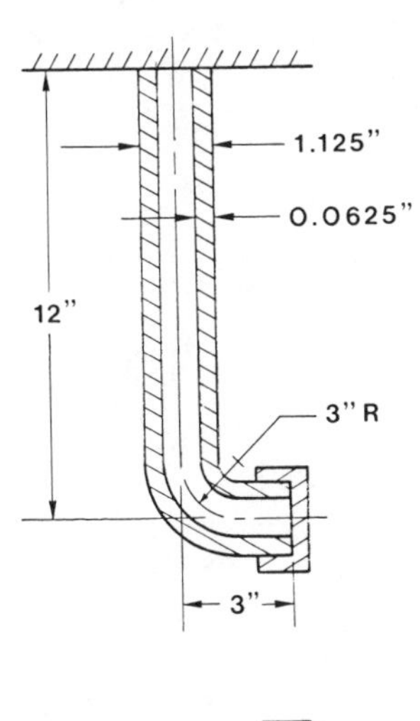

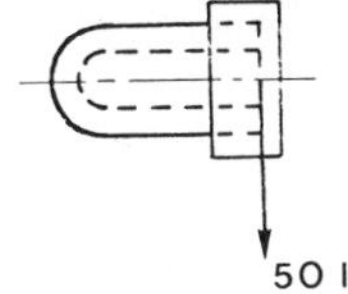

Fig. P3.11

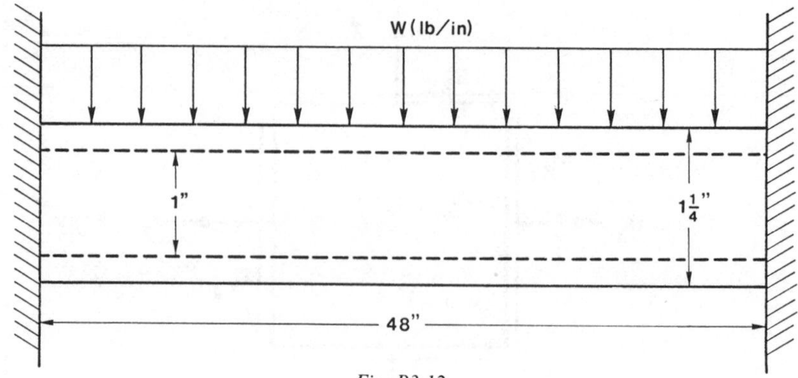

Fig. P3.12

3.10 Derive a distortion energy equation in terms of plane stress. (See Fig. P3.10.)

3.11 A capped pressurized pipe, Fig. P3.11, has an internal pressure of 200 psig. The pipe has an I.D. of 1 in. and O.D. of 1.125 in. Find the principal stresses at the wall. Check to make sure they are the maximum ones.

3.12 A pipe with an internal pressure p is used as a structural member (Fig. P3.12). It is used to hold a uniform load W over its entire length. It is stated if the internal pressure p was zero the uniform load can be doubled to an amount $2W$. What is your answer and why? Assume the pipe is made of steel.

chapter 4

ANALYSIS OF COMPOSITE, HONEYCOMB, AND REINFORCED MATERIALS

The continually increasing need for materials having exceptional strength-weight ratio and other mechanical properties has led to the development of unique combinations of materials (40, 43). Often materials have been so constituted as to possess different properties in different directions; that is, they exhibit orthotropic properties. Furthermore there are numerous cases where different materials are used in combinations in order to gain the maximum advantage of each component part. The material selections often require the use of computer routines for composite materials. In this chapter composites like multiphase alloys and ceramics will not be considered; rather the discussion here will concern isotropic and orthotropic composites like multilayer beams and shells, sandwich and honeycomb structures, reinforced concrete, and filament-reinforced materials. Composites used as thermostatic elements are discussed in Chapter 15; multishell shrink-fit construction and thick-walled vessels prestressed by wire winding are discussed in Chapter 11. The material in this chapter is also primarily concerned with elastic static behavior at ambient temperature. Composite materials in relation to temperature, stability, fatigue, and so on are discussed in other chapters.

4-1 GENERAL THEORY OF STRUCTURAL COMPOSITES

Beams and Compression Members

The general equation (30) for bending and compression is as follows:

$$\sigma_{xx} = \frac{E}{E_1}\left[\frac{P}{A^*} - \frac{M_z I_{yy}^* - M_y I_{yz}^*}{I_{yy}^* I_{zz}^* - (I_{yz}^*)^2}\, y - \frac{M_y I_{zz}^* - M_z I_{yz}^*}{I_{yy}^* I_{zz}^* - (I_{yz}^*)^2}\, z\right] \qquad 4.1$$

The sign convention for the forces, moments, and coordinates is shown in Fig. 4.1. The term E_1 is selected from one of the materials in the cross section. The asterisk values are found with respect to the modulus weighted centroid. When an arbitrary reference axis is selected, the location of the centroid with

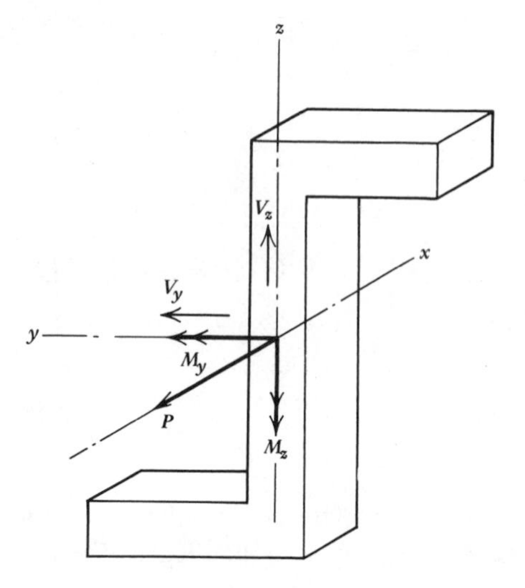

Fig. 4.1 Beam sign convention.

respect to it is located by:

$$\bar{Y}_0^* = \frac{1}{A^*} \sum_{i=1}^{i=n} \bar{Y}_i \left(\frac{E_i}{E_1} A_i \right) \qquad 4.2$$

$$\bar{Z}_0^* = \frac{1}{A^*} \sum_{i=1}^{i=n} \bar{Z}_i \left(\frac{E_i}{E_1} A_i \right) \qquad 4.3$$

$$A^* = \sum_{i=1}^{i=n} \frac{E_i}{E_1} A_i \qquad 4.4$$

The moments of inertia about the modulus weighted centroid are:

$$I_{yy}^* = \sum_{i=1}^{i=n} \frac{E_i}{E_1} (I_{yyi} + \bar{Z}_i^2 A_i) \qquad 4.5$$

$$I_{zz}^* = \sum_{i=1}^{i=n} \frac{E_i}{E_1} (I_{zzi} + \bar{Y}_i{}^2 A_i) \qquad 4.6$$

$$I_{yz}^* = \sum_{i=1}^{i=n} \frac{E_i}{E_1} (\bar{I}_{yzi} + \bar{Y}_i \bar{Z}_i A_i) \qquad 4.7$$

Rivello (30) treats structural members as composites with thermal stresses for more complex situations. Eqs. 4.1 to 4.7 will be illustrated in the examples.

Isotropic duplex structures under simple loading

Compression member. Consider the simple composite system shown in Fig. 4.2*A* consisting of a copper tube over a steel core, the whole assembly being

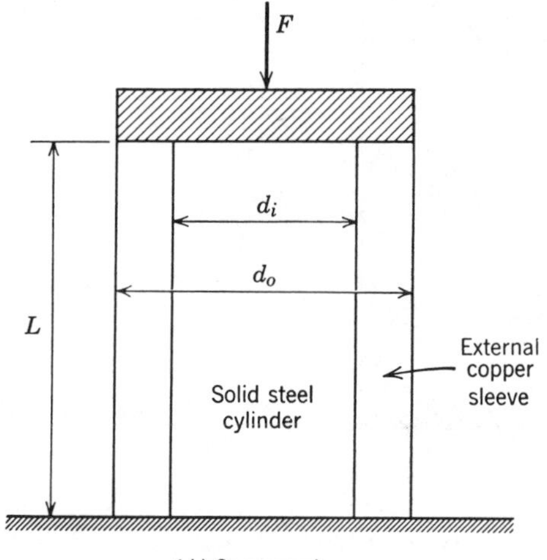

(A) Compression

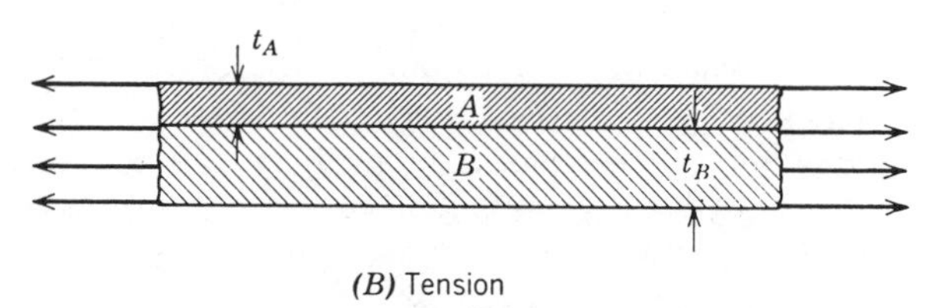

(B) Tension

Fig. 4.2 Simple composite structures.

subjected to compression. Let ε_s and E_s be the strain and modulus of elasticity of the steel and ε_c and E_c the strain and modulus of the copper. For compatability the strains for the two parts must be equal, that is,

$$\varepsilon_s = \varepsilon_c \qquad \text{or} \qquad \sigma_s/E_s = \sigma_c/E_c \tag{4.8}$$

where σ is the stress. In terms of the applied load F,

$$F = \sigma_s A_s + \sigma_c A_c \tag{4.9}$$

where A_s and A_c are the cross-sectional areas of the steel and copper parts respectively. Thus from Eqs. 4.8 and 4.9

$$F = \sigma_c(A_c + E_s A_s/E_c) \tag{4.10}$$

and

$$\sigma_c = \frac{FE_c}{A_c E_c + E_s A_s} \tag{4.11}$$

$$\sigma_s = \frac{FE_s}{A_c E_c + E_s A_s} \tag{4.12}$$

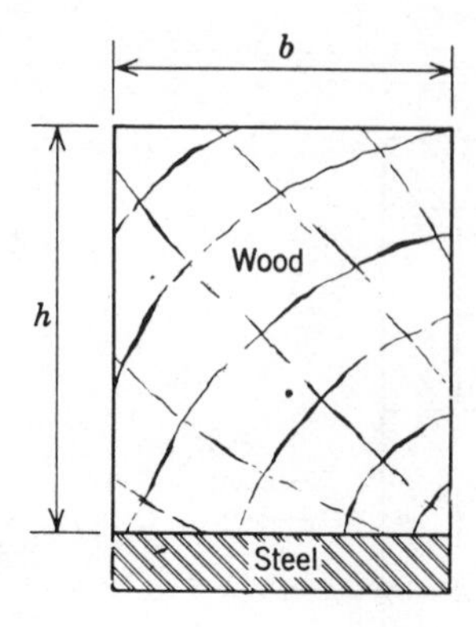

Fig. 4.3 Composite structural section.

The total deformation is εL; therefore

$$\varepsilon = \varepsilon_s L = \varepsilon_c L = \frac{FL}{E_s A_s + E_c A_c} \qquad 4.13$$

Structural element composites. A common type of composite is obtained by stiffening a block of wood with a steel plate (Fig. 4.3). In this situation, assuming no sliding between the parts, the problem can be reduced to that of the bending of the equivalent beam shown in Fig. 4.4, which results because wood has a lower modulus of elasticity than steel. Since the moment of inertia remains the same for a given curvature,

$$b_1 = b(E_w/E_s) \qquad 4.14$$

and the problem is reduced to that of the bending of a T-beam. Aside from relatively simple cases such as illustrated, the analysis of structural stiffening is largely empirical. It is well known, for example, that a flat plate can be effectively stiffened by welding some ribs on the face; this increases the moment of inertia and hence the rigidity. For simple shapes we calculate the moment of inertia

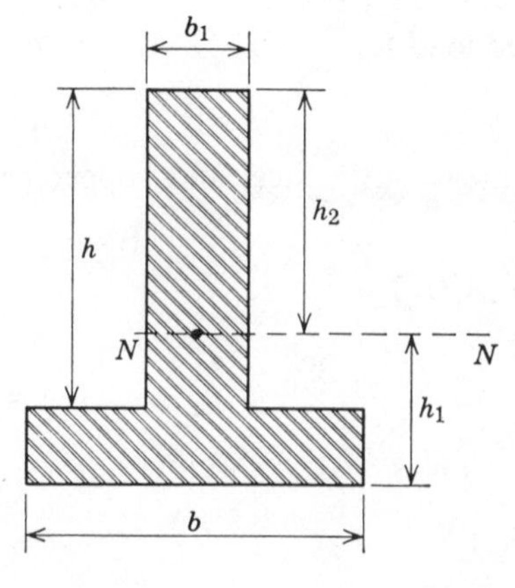

Fig. 4.4 Steel equivalent of section in Fig. 4.3.

for the section and substitute this value in the appropriate deflection formula. For more complicated constructions little information is available. However, Lowenfeld (49) experimented with fifteen different kinds of plate stiffening (Fig. 4.5), and arrived at the results shown in Table 4.1. In the tabulation, the values given represent the stiffness of the various constructions compared to the stiffness of a standard flat plate (stiffness = 1.00). Thus, for example, plate-type 6 in sandwich form is 83 times stiffer in bending than an equivalent flat plate.

General theory. The preceding analysis, resulting in Eq. 4.14, is applicable only where the thickness of the components remains constant, making it relatively simple to locate the neutral axis of the equivalent section. If the cross section is of variable width, such as shown in Fig. 4.6*A*, other means have to be used.

From conventional bending theory (Chapter 2) and Eq. 4.1 with only the axial force P applied, the axial stress is

$$\sigma_{xx} = \frac{E}{E_1}\left[\frac{P}{A^*}\right] \tag{4.15}$$

By applying this relationship to the composite in Fig. 4.6*A*,

$$\sigma_{xx} = \frac{EP}{E_1\left(\dfrac{E_1}{E_1}A_1 + \dfrac{E_2}{E_1}A_2\right)} \tag{4.16}$$

where E_1 and E_2 are the elastic moduli of the component parts and E is the modulus of the point where the stress is calculated. In order to locate the position of the modulus weighted centroid axis, it is convenient to select the base as the reference line so that from Eq. 4.3 with Y_0, Z_0 located so symmetry makes Y_0 zero. Then

$$\bar{Z}_0^*\left[\frac{E_1}{E_1}A_1 + \frac{E_2}{E_1}A_2\right] = \bar{Z}_1\left(\frac{E_1}{E_1}A_1\right) + \bar{Z}_2\left(\frac{E_2}{E_1}A_2\right) \tag{4.17}$$

Further, canceling E_1 in the denominator yields

$$\bar{Z}_0 = \frac{Z_1E_1A_1 + Z_2E_2A_2}{E_1A_1 + E_2A_2} \tag{4.18}$$

The bending stress for the cross section is given by Eq. 4.1 with only M_y applied and I_{yz} zero, because $\bar{y}_i$ is zero resulting in

$$\sigma_{xx} = \frac{E}{E_1}\left[-\frac{M_y}{I_{yy}^*}z\right] \tag{4.19}$$

The term I_{yy}^* expanded gives

$$\sigma_{xx} = -\frac{EM_yz}{E_1\left\{\dfrac{E_1}{E_1}[I_{yy_1} + \bar{Z}_1{}^2A_1{}^2] + \dfrac{E_2}{E_1}[I_{yy_2} + \bar{Z}_2A_2{}^2]\right\}} \tag{4.20}$$

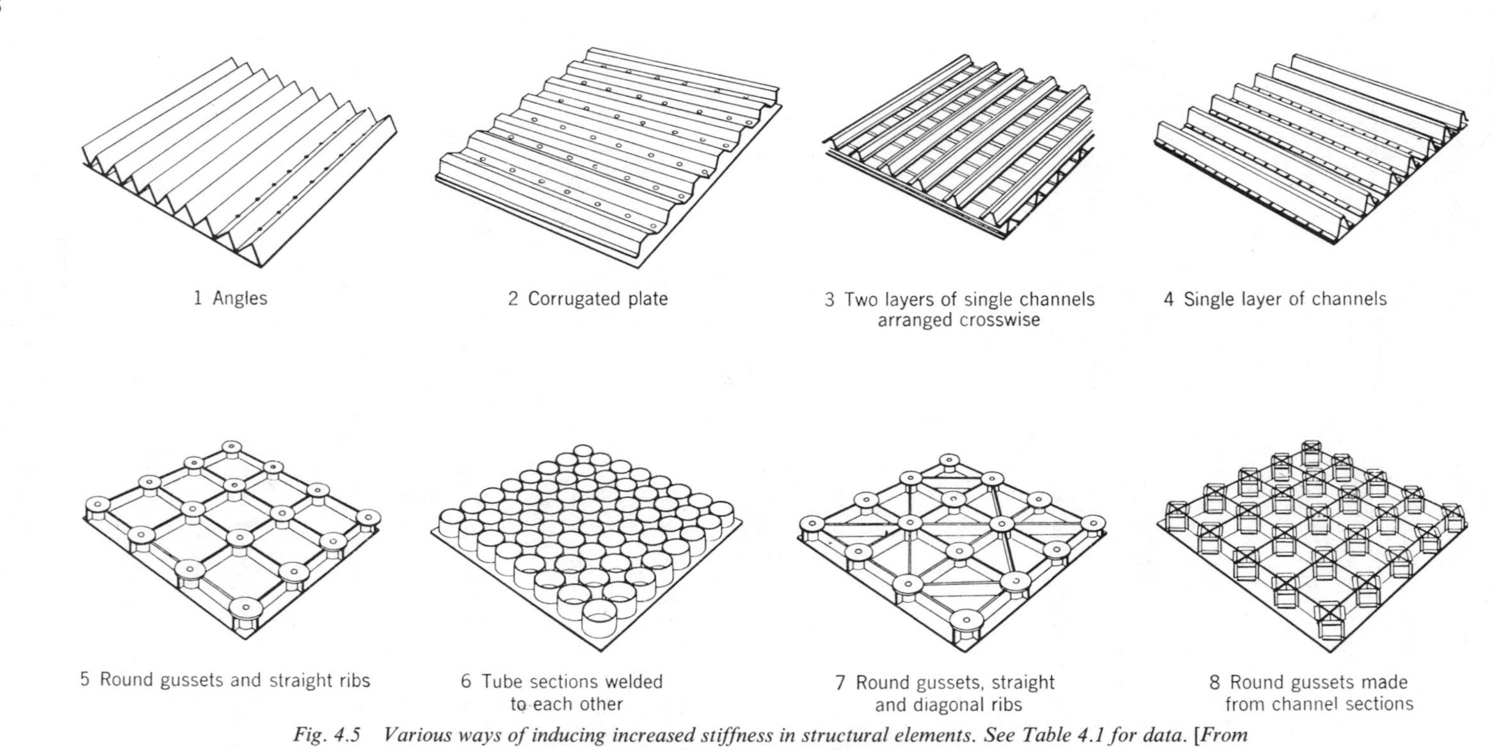

Fig. 4.5 Various ways of inducing increased stiffness in structural elements. See Table 4.1 for data. [From (49), courtesy of McGraw-Hill Publishing Co.]

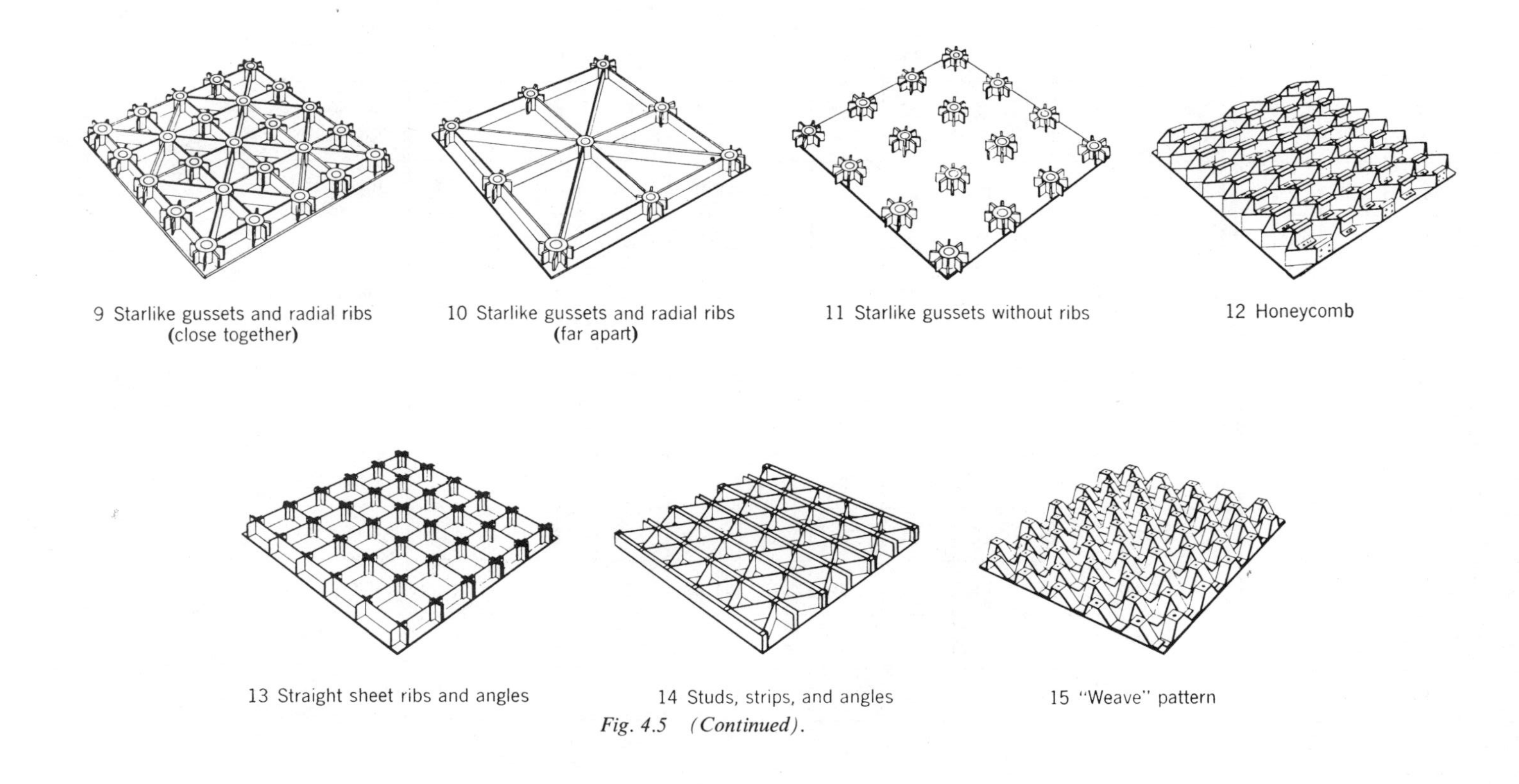

9 Starlike gussets and radial ribs (close together)

10 Starlike gussets and radial ribs (far apart)

11 Starlike gussets without ribs

12 Honeycomb

13 Straight sheet ribs and angles

14 Studs, strips, and angles

15 "Weave" pattern

Fig. 4.5 (Continued).

Table 4.1 Relative Stiffness of Some Structural Members Shown in Fig. 4.5 (From (49), Courtesy of McGraw-Hill Publishing Co.)

Load	Application	Plate Number													
		1	2	3	4	5	6	7	8	9	10	11	12	13	14
Torsion	Single	48	40	28	11.3	3.5	27	33	24	1.3	10.6	1.3	4.2	3.3	7.5
	Sandwich	62	29	20	8.1	29	85	44	42	48	24	4.8	18	56	21
Bending (single)	x Axis	112	58	18	47	10	10.3	12.5	22.5	16.3	8.8	1.0	1.6	27	7.5
	y Axis	1.1	1.3	30	1.2	10	10.3	12.5	7.8	16.3	8.8	1.0	9.3	6.9	65
Bending (sandwich)	x Axis	150	62	19	62	11	83	16	29	32	3.6	3.6	27	52	30
	y Axis	56	13	16	2.0	11	83	16	29	32	3.6	3.6	7.3	31	10
$\frac{\text{Strength}}{\text{Weight}}$	Single, torsion	30.7	26	17	8.1	2.1	10	18	1.5	7.3	6.8	1.0	2.6	2.1	4.1
	Single, bending x	72	28.2	10	34.5	5.8	3.4	6.7	13	9.1	5.7	0.74	1.1	16	3.9
	Single, bending y	0.75	0.8	18	0.85	5.8	3.4	6.7	4.6	9.1	5.7	0.74	5.4	4.2	34
	Sandwich, torsion	48	23	14.5	6.5	23	45	32	32	33	19	40	14	44	15
	Sandwich, bending x	116	50	14	50	8.5	44	11	22	21	2.6	2.6	21	41	21
	Sandwich, bending y	43	10	12	2.3	8.5	44	11	22	21	2.6	2.6	5.6	24	7

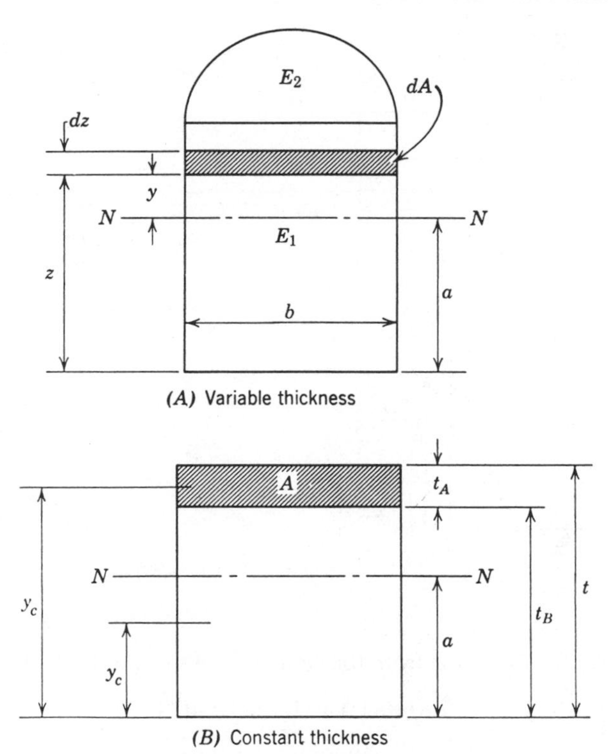

Fig. 4.6 Composite sections.

Again, E is the modulus at the point where the stress is evaluated. Also, from the preceding, the overall rigidity EI is given by the sum of the rigidities of the components of the composite making

$$EI = E_1 I_1 + E_2 I_2 \qquad 4.21$$

where I_1 and I_2 represent properties for the components with respect to the modulus weighted centroid axis.

When calculating unsymmetrical bending deflections, use Eqs. 2.256 and 2.257 with E_1 used in place of E and asterisks on all the area moments of inertia. Many cases will result in both vertical and lateral deflections of the beam.

The use of the preceding theory is illustrated by the beam shown in Fig. 4.7, which is center-loaded and is composed of a length of wood stiffened by a steel plate. Both theories given will be used to calculate stresses in the component parts.

In this symmetrical case, from Eq. 4.18

$$E_1 \int_0^{A_1} z\, dA_1 = E_1 A_1 z_1 \qquad 4.22$$

and

$$E_2 \int_0^{A_2} z\, dA_2 = E_2 A_2 z_2 \qquad 4.23$$

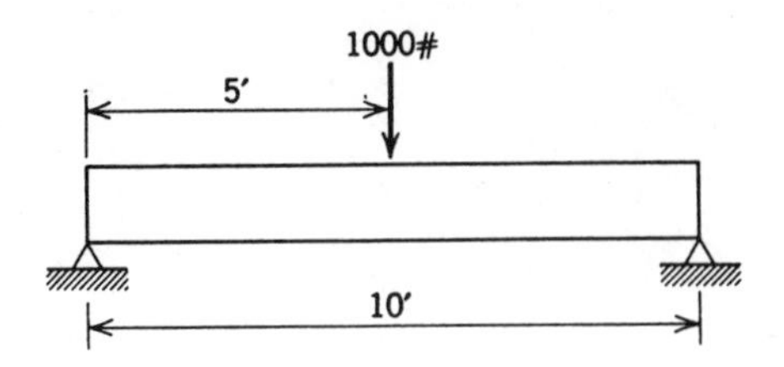

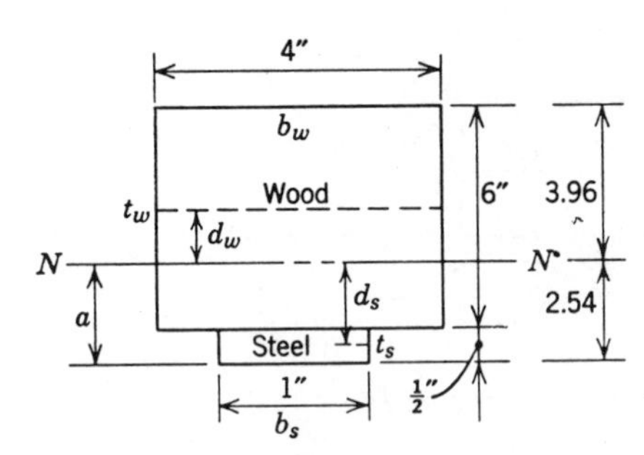

Fig. 4.7 Cross section and loading of a composite beam.

where z_1, z_2 are the distances from the centroids of the part to the base line; thus

$$\bar{z}_0 = \frac{(30 \times 10^6)(0.50)(0.25) + (1.5 \times 10^6)(24)(3.5)}{(30 \times 10^6)(0.50) + (1.5 \times 10^6)(24)} = 2.54 \text{ in.}$$

The maximum stress for the steel from Eq. 4.20 is

$$(\sigma_{\max})_s = \frac{E_s M_y z}{E_s \left\{ \frac{E_s}{E_s}[I_{yys} + \bar{Z}_s{}^2 A_s] + \frac{E_w}{E_s}[I_{yyw} + \bar{Z}_w{}^2 A_w] \right\}} \qquad 4.24$$

or

$(\sigma_{\max})_{\text{steel}}$

$$= \frac{(500)(60)(30 \times 10^6)(2.54)}{(30 \times 10^6)[(\frac{1}{12})(1)(\frac{1}{2})^3 + (1)(\frac{1}{2})(2.29)^2] + (1.5 \times 10^6)[(\frac{1}{12})(4)6^3 + (4)(6)(.96)^2]}$$

$= 10{,}300$ psi

For the wood part

$$(\sigma_{\max})_w = -\frac{E_w M_y z}{E_s I_s + E_w I_w} \qquad 4.25$$

$$(\sigma_{\max})_w = \frac{(-500)(60)(1.5)(3.96)}{221} = -810 \text{ psi}$$

By the first of the methods given, the method of equivalent sections,

$$b_1 = 4(\tfrac{1}{20}) = 0.20 \text{ in.}$$

Therefore the equivalent section is a beam with a web 6×0.20 in. and a flange

$1 \times \frac{1}{2}$ in. For this section, $I = 7.37$ in.^{4} and the distances from the neutral axis to the highest stressed tension and compression fibers is 2.54 and 3.96 in., respectively; thus for the tension side (flange)

$$\sigma_{\max} = \frac{M}{Z} = \frac{(30{,}000)(2.54)}{7.37} = 10{,}300 \text{ psi}$$

For the top of the web;

$$\sigma_{\max} = -\frac{M}{Z} = \frac{(-30{,}000)(3.96)}{7.73} = -810 \text{ psi}$$

the same results are obtained by the alternate method of analysis.

Reinforced concrete. Reinforced concrete continues to be one of the outstanding examples of composite materials. Since the strength of concrete is greater in compression than in tension, it has been customary to reinforce the tension-loaded portion of a concrete structure with steel. For the purposes of this book only a rectangular beam of concrete with tension reinforcement (Fig. 4.8) will be considered; for additional information see (15) of Chapter 2 and (26) of this chapter.

It is assumed that plane sections remain plane. In Fig. 4.8, *abcd* represents the strain diagram. Let

σ_s = maximum tensile stress in the steel

σ_c = compressive stress in the concrete

E_s = modulus of steel

E_c = modulus of concrete

$m = E_s/E_c$

A = total cross-sectional area of steel reinforcement

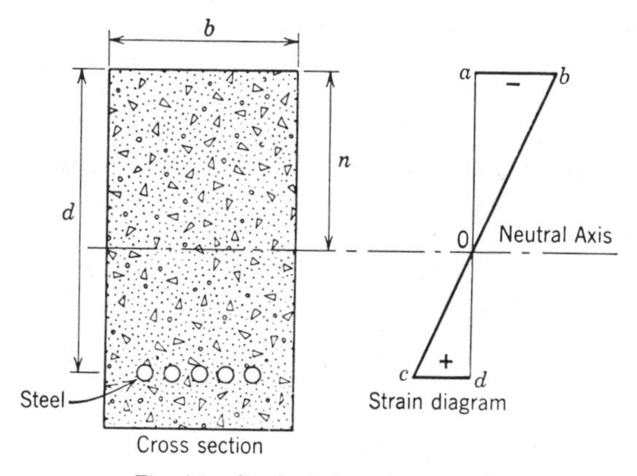

Fig. 4.8 Steel-reinforced concrete beam.

From Fig. 4.8, $oa/od = ab/cd$ or

$$\frac{\sigma_c/E_c}{\sigma_s/E_s} = \frac{n}{d-n} \qquad 4.26$$

from which

$$\sigma_s/\sigma_c = \left(\frac{a-n}{n}\right)m \qquad 4.27$$

Since there is no resultant axial load on the beam, the total compression in the concrete must equal the total tension in the steel. The average compressive stress in the concrete is $\sigma_c/2$ and thus

$$bn\sigma_c/2 = A\sigma_s \qquad 4.28$$

or

$$\sigma_s/\sigma_c = bn/2A \qquad 4.29$$

Since the stress and strain distributions are linear, the center of compression is $n/3$ from the top of the beam and the moment in the cross section is

$$M = (bn\sigma_c/2)(d - n/3) = A\sigma_s(d - n/3) \qquad 4.30$$

which cannot be exceeded for integrity of the beam. The value of n for a given beam depends only on m and A. If these are specified, the stress ratio σ_s/σ_c is determined from Eq. 4.29. Optimum design calls for a value of A which causes the stress in both steel and concrete to reach their allowable limits simultaneously. It should be noted that tension reinforcement as illustrated here is not a case of complete reinforcement since a shear stress also acts in the structure. Any shear in the beam is accompanied by 45° tension and compression stresses and failure could occur. Modern reinforcing methods also include compression reinforcement (see references).

During the last several years concrete shell structures have assumed increased importance since means have been found to obtain strength through geometrical arrangement rather than through bulk of material (47). In particular, hyperbolic paraboloid structures of reinforced prestressed concrete have assumed new importance. In this geometry, with uniform loading, the stress system produced is in uniform shear, tension, and compression. For details see (47) of this chapter and (4) of Chapter 3.

Allowable stress values. In order to emphasize the foregoing theory, the allowable stress (load) values will be calculated for three bimetal systems.

BIMETAL PLATE IN TENSION (Fig. 4.2*B*). In this system material of thickness t_A, cross-sectional area A_A, modulus E_A, and yield strength σ_A is bonded to material of thickness t_B, cross-sectional area A_B, modulus E_B, and yield strength σ_B. From Eqs. 4.8, 4.9, 4.16

$$F = \sigma_A t_A + \sigma_B t_B$$

$$\varepsilon = \varepsilon_A = \varepsilon_B$$

$$\sigma_A = E_A \varepsilon_A = E_A \varepsilon$$

$$\sigma_B = E_B \varepsilon_B = E_B \varepsilon$$

Therefore

$$F = \sigma_B\left(t_B + \frac{\sigma_A t_A}{\sigma_B}\right) = \sigma_B\left(t_B + \frac{E_A t_A}{E_B}\right)$$

and

$$\sigma_A = \frac{FE_A}{E_A t_A + E_B t_B} = \frac{F}{t_A + E_B t_B / E_A}; \qquad \sigma_B = \frac{FE_B}{E_B t_B + E_A t_A} = \frac{F}{t_B + E_B t_A / E_A}$$

The allowable value of F is the smaller of the above. Suppose that the material is steel clad with admiralty metal; the following data are supplied:

$$t_A = 0.20 \text{ in.} \quad \text{(admiralty metal)}$$
$$t_B = 0.80 \text{ in.} \quad \text{(steel)}$$
$$E_A = 16 \times 10^6 \text{ psi}$$
$$E_B = 30 \times 10^6 \text{ psi}$$
$$\sigma_A = 18{,}000 \text{ psi}$$
$$\sigma_B = 30{,}000 \text{ psi}$$

From the preceding equations,

$$\sigma_A = 18{,}000 = \frac{F}{0.20 + \frac{30}{16}(0.80)}$$

and

$$F = 30{,}600 \text{ lb}$$

In addition,

$$\sigma_B = 30{,}000 = \frac{F}{0.80 + \frac{16}{30}(0.20)}$$

and

$$F = 27{,}204 \text{ lb}$$

thus the allowable load is 27,204 lb.

BIMETAL PLATE IN BENDING (Fig. 4.6*B*). From Eq. 4.18

$$\bar{Z}_0 = \frac{Z_B E_B A_B + Z_A E_A A_A}{E_B A_B + E_A A_A}$$

Then,

$$a = \frac{E_B A_B t_B/2 + E_A A_A(t_B + t_A/2)}{E_A A_A + E_B A_B}$$

and

$$(\sigma_B)_{\max} = (M/EI)(E_B)(a)$$
$$(\sigma_A)_{\max} = (M/EI)(E_A)(t - a)$$

where $EI = E_A I_A + E_B I_B$. Consider a steel bar clad with brass, where

$$t_A = 0.10 \text{ in.} \quad \text{(brass)}$$
$$t_B = 0.90 \text{ in.} \quad \text{(steel)}$$
$$E_A = 15 \times 10^6 \text{ psi}$$
$$E_B = 30 \times 10^6 \text{ psi}$$
$$\sigma_A = 18{,}000 \text{ psi}$$
$$\sigma_B = 30{,}000 \text{ psi}$$
$$\text{width of bar} = 0.50 \text{ in.}$$

From the preceding equations

$$a = 0.476 \text{ in.}$$
$$EI = 1{,}064{,}500$$

and

$$\sigma_A = \sigma_B(E_A/E_B)(t/a - 1) = 0.55\sigma_B = 16{,}500 \text{ psi}$$
$$\sigma_B = 1.82\sigma_A = 32{,}760 \text{ psi}$$

The stress in the steel is above the allowable value of 30,000 psi; therefore the stress in the brass has to be limited to 16,500 psi rather than to its allowable value of 18,000 psi. Then the limiting applied bending moment is found using the relationship

$$\sigma_B = 30{,}000 = \frac{M}{1{,}064{,}500}[(30 \times 10^6)(0.476)]$$

from which

$$M_{\text{allowable}} = 2240 \text{ in.-lb}$$

In some cases it is necessary to have a value for the moment of inertia I in order to carry out a computation. In the preceding example if the entire bar were considered to be steel, then

$$EI = 1{,}064{,}500$$

and the "effective" moment of inertia would be

$$I = 1{,}064{,}500/(30 \times 10^6) = 0.0356 \text{ in.}^4$$

BIMETAL TUBE IN BENDING (Fig. 4.9*B*). From Eqs. 4.19 to 4.21

$$M = \frac{EI}{R} = \frac{E_A I_A}{R} + \frac{E_B I_B}{R} = \frac{EI}{Ey/\sigma}$$

from which

$$\sigma_A = \frac{Mc}{I} = \frac{My_A}{I}\left(\frac{E_A}{E}\right) = \frac{M}{EI}(E_A y_A)$$

$$\sigma_B = \frac{Mc}{I} = \frac{My_B}{I}\left(\frac{E_B}{E}\right) = \frac{M}{EI}(E_B y_B)$$

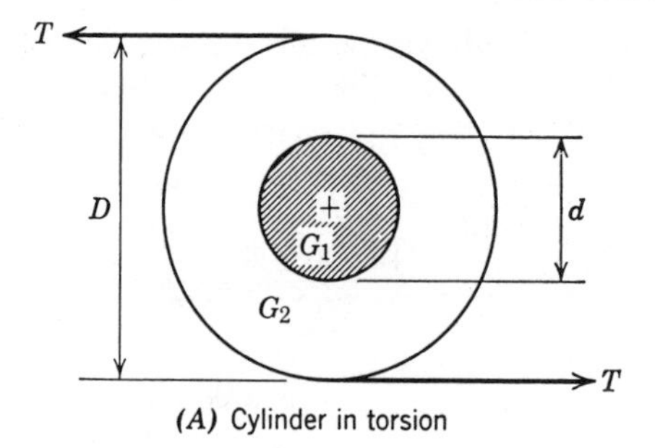

(A) Cylinder in torsion

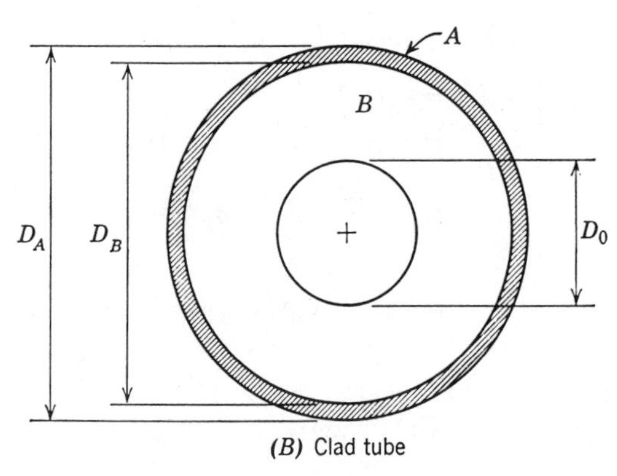

(B) Clad tube

Fig. 4.9 Composite cylinders.

Thus

$$\sigma_B = \sigma_A \left(\frac{E_B}{E_A}\right)\left(\frac{y_B}{y_A}\right)$$

Suppose that the following data are given.

$$D_0 = 0.50 \text{ in.}$$
$$D_A = 1.00 \text{ in.}$$
$$D_B = 0.80 \text{ in.}$$
$$E_A = 15 \times 10^6 \text{ psi}$$
$$E_B = 30 \times 10^6 \text{ psi}$$
$$\sigma_A = 15{,}000 \text{ psi}$$
$$\sigma_B = 30{,}000 \text{ psi}$$

Then

$$EI = E_A I_A + E_B I_B = 14{,}600{,}000$$

$$\sigma_B = \sigma_A \left(\frac{30}{15}\right)\left(\frac{0.80}{1.00}\right) = 1.60\sigma_A = 28{,}800 \text{ psi}$$

$$\sigma_A = 18{,}750 \text{ psi}$$

The stress in the brass is above the allowable value of 18,000 psi; therefore the stress in the steel has to be limited to 28,800 psi. The limiting bending moment is then obtained from the relationship

$$\sigma_A = 18{,}000 = \frac{M}{14{,}600{,}000}[(15 \times 10^6)(0.50)]$$

from which

$$M_{\text{allowable}} = 35{,}000 \text{ in.-lb}$$

Calculation of shear stress. In the calculation of shear stress for points in the cross section of the element shown in Fig. 4.7 use is made of the unsymmetrical bending form for the shear stress (30).

$$\tau = \frac{E}{E_1}\left[\frac{V_y I^*_{yy} - V_z I^*_{yz}}{I^*_{yy} I^*_{zz} - (I^*_{yz})^2}\frac{Q^*_z}{b} + \frac{V_z I^*_{zz} - V_y I^*_{yz}}{I^*_{yy} I^*_{zz} - (I^*_{yz})^2}\frac{Q^*_y}{b}\right] \quad 4.31$$

$$Q^*_y = \sum_{i=1}^{i=n} \bar{Z}_i\left(\frac{E_i}{E_1}A_i\right) \quad 4.32$$

$$Q^*_z = \sum_{i=1}^{i=n} \bar{Y}_i\left(\frac{E_i}{E_1}A_i\right) \quad 4.33$$

In the example in Fig. 4.7, the shear stress for the intersection of the steel and the wood reduces to the following when V_y and I^*_{yz} are zero:

$$\tau_{\text{steel}} = \frac{E_s}{E_s}\frac{V_z}{I^*_{yy}}\frac{Q^*_y}{b_s} = \frac{E_s V_z Q^*_y}{EIb_s} \quad 4.34$$

Q^*_y, the static moment, is that of the steel area between the free edge and the cutting plane about the neutral axis.

$$Q^*_y = (2.54 - 0.25)\frac{E_s}{E_s}(1)(\tfrac{1}{2})$$

$$Q^*_y = 1.15 \text{ in.}^3$$

$$\tau_s = \frac{30 \times 10^6(500)(1.15)}{221 \times 10^6 \times (1)} = 78 \text{ psi}$$

For the wood

$$\tau_w = \frac{E_w V_z Q^*_y}{EIb_w} \quad 4.35$$

$$\tau_w = \frac{1.5 \times 10^6(500)(1.15)}{221 \times 10^6(4)} = 0.98 \text{ psi}$$

Torsion. The torsional behavior of composites is complex for other than a simple cylindrical geometry. Payne (28) has considered composite rectangles and various irregular shapes. For the cylindrical geometry shown in Fig. 4.9*A* shear stresses and strains may be calculated from the equations given in Chapter 2 with GJ as the overall torsional rigidity. For example, for a uniform bar,

$$\tau = Tr/J \quad \text{and} \quad \theta = TL/GJ \tag{4.36}$$

For the composite shown in Fig. 4.9,

$$GJ = G_1 J_1 + G_2 J_2 \tag{4.37}$$

so that Eqs. 4.36 become

$$\tau = TrG/GJ \tag{4.38}$$

and

$$\theta = TL/GJ \tag{4.39}$$

where GJ is defined by Eq. 4.37. For this case,

$$GJ = (\pi/32)[d^4(G_1 - G_2) + D^4 G_2] \tag{4.40}$$

Duplex hydraulic conduit. In this type of application (Fig. 4.10), a thin liner of steel is used in a concrete pipe to prevent penetration of the fluid into the concrete and to absorb the tensile stress that occurs at the bore of a cylinder when it is subjected to internal pressure (see also Chapter 3). Consider the situation depicted in Fig. 4.10. When there is pressure p_i in the cylinder, the steel liner is effectively under an internal pressure p_i and an external pressure p_f, whereas the concrete sleeve is under internal pressure p_f only. Thus for the concrete cylinder the stresses are calculated by Eqs. 3.471 and 3.472, of Chapter 3; thus

$$\text{hoop stress} = \sigma_h = \frac{p_f}{R^2 - 1}\left(1 + \frac{b^2}{r^2}\right) \tag{4.41}$$

$$\text{radial stress} = \sigma_r = \frac{p_f}{R^2 - 1}\left(1 - \frac{b^2}{r^2}\right) \tag{4.42}$$

and, for the steel sleeve (see also Chapter 3),

$$\sigma_h = (p_i - p_f)(a/t - 1) \tag{4.43}$$

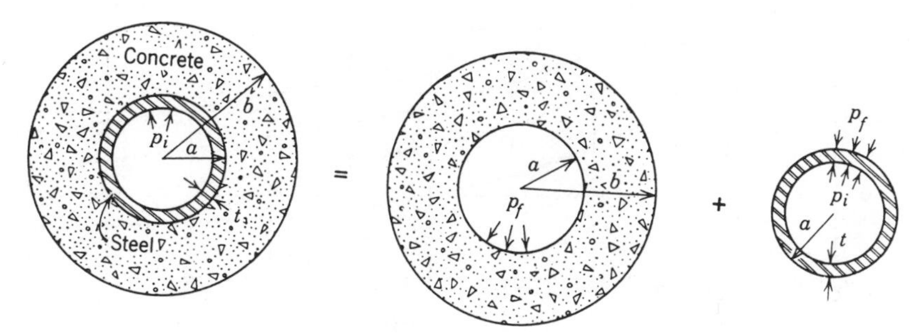

Fig. 4.10 Steel-lined concrete conduit.

The radial displacement at the bore of the concrete cylinder, from Eq. 3.476 of Chapter 3, is

$$(u)_{r=a} = p_f \frac{a}{E_c}\left(\frac{a^2+b^2}{b^2-a^2} + \nu_c\right) \qquad 4.44$$

If the steel also expands by an amount u, assuming the sleeve to be negligibly thin compared with the other dimensions and C is the circumference of the steel ring,

$$\Delta C = 2\pi(a+u) - 2\pi a = 2\pi u \qquad 4.45$$

and, using Eq. 4.43,

$$\sigma_h = E_s \varepsilon = E_s(u/a) = (p_i - p_f)(a/t - 1) \qquad 4.46$$

Substituting in Eq. 4.44 then gives the result

$$p_f = \frac{p_i}{1 + \left(\frac{E_s}{E_c}\right)\left(\frac{t}{a-t}\right)\left(\frac{a^2+b^2}{b^2-a^2} + \nu_c\right)} \qquad 4.47$$

where p_f is the interface pressure transmitted to the concrete shell. For practical purposes $E_s/E_c = 15$ and $\nu_c = 0.20$; also, approximately, $t/(a-t) = t/a$; thus if $R = b/a$, Eq. 4.47 becomes

$$p_f = \frac{p_i}{1 + 15\left(\frac{t}{a}\right)\left(\frac{R^2+1}{R^2-1} + 0.20\right)} \qquad 4.48$$

and a design curve such as shown in Fig. 4.11 can be prepared. This chart shows

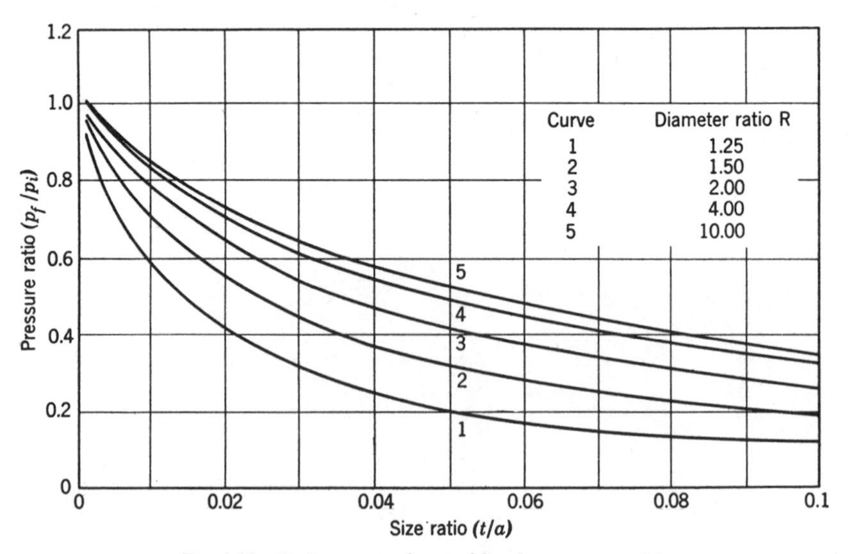

Fig. 4.11 Design curves for steel-lined concrete conduit.

that as the thickness of the liner increases the pressure transmitted to the concrete decreases; however, for any given t/a ratio, the pressure transmitted also increases as the diameter ratio R of the concrete shell increases. To reduce or avoid pressure transmission to the concrete, soft interface liners like pitch or asphalt are frequently used.

Multilayer heavy-walled cylinder. The analysis for a multilayer heavy-walled cylinder is somewhat different from the analysis just given for a thin liner in a vessel. A typical vessel is shown in Fig. 4.12A; it consists of an inner cylinder made of material of elastic modulus E_i and Poisson ratio ν_i and an outer cylinder of material E_0, ν_0. When internal pressure p_i is applied to the bore of the inner

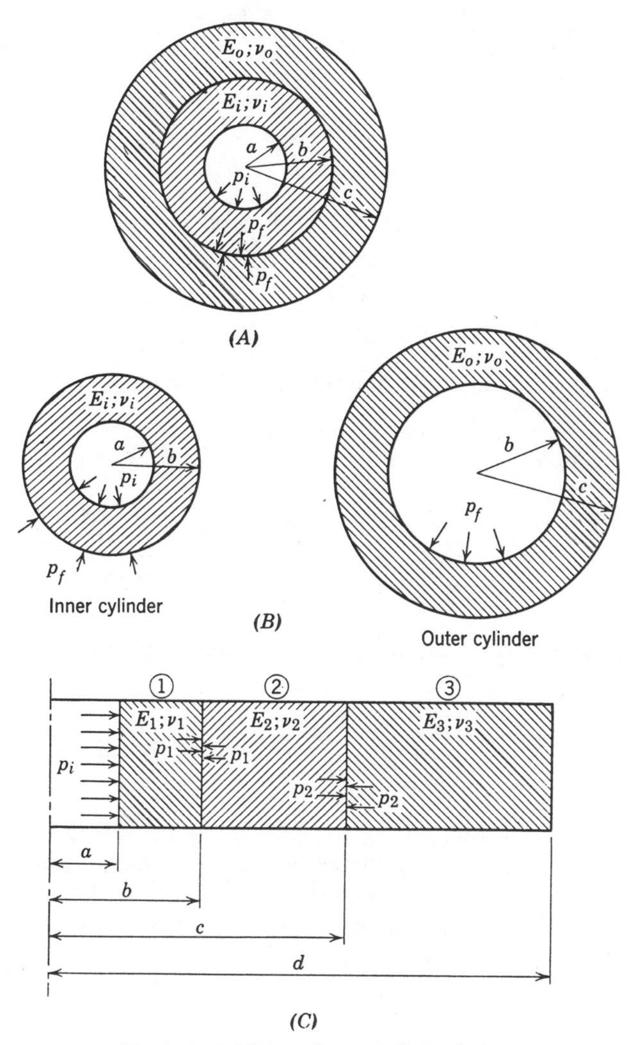

Fig. 4.12 Multilayer heavy-walled cylinders.

cylinder a radial stress distribution is induced in the composite structure, and, at the interface b, the radial stress is equivalent to a pressure p_f; thus, as shown in Fig. 4.12B, the inner cylinder is subjected to an internal pressure p_i and an external pressure p_f, whereas the outer cylinder is subjected to an internal pressure p_f only. The equations for stresses and deformation in heavy-walled cylinders are given in Chapter 3; when applied to the problem here, the following relationships result.

For the inner cylinder:

$$\text{Hoop stress} = (\sigma_h)_i = \frac{p_i}{R^2 - 1}\left(1 + \frac{b^2}{r^2}\right) - \frac{p_f R^2}{R^2 - 1}\left(1 + \frac{a^2}{r^2}\right) \qquad 4.49$$

$$\text{Radial stress} = (\sigma_r)_i = \frac{p_i}{R^2 - 1}\left(1 - \frac{b^2}{r^2}\right) - \frac{p_f R^2}{R^2 - 1}\left(1 - \frac{a^2}{r^2}\right) \qquad 4.50$$

where $R = b/a$.

For the outer cylinder:

$$(\sigma_h)_0 = \frac{p_f}{R_1{}^2 - 1}\left(1 + \frac{c^2}{r^2}\right) \qquad 4.51$$

$$(\sigma_r)_0 = \frac{p_f}{R_1{}^2 - 1}\left(1 - \frac{c^2}{r^2}\right) \qquad 4.52$$

where $R_1 = c/b$. The hoop strain for a cylinder is given by Eq. 3.457. By using this relationship, the radial deformation for the inner cylinder at the interface is

$$u_i = \frac{b}{E_i}\left[\frac{2p_i}{R^2 - 1} - p_f\left(\frac{R^2 + 1}{R^2 - 1}\right) - v_i(-p_f)\right] \qquad 4.53$$

The radial deformation for the outer cylinder at the interface is

$$u_0 = \frac{b}{E_0}\left[p_f\left(\frac{R_1{}^2 + 1}{R_1{}^2 - 1}\right) - v_0(-p_f)\right] \qquad 4.54$$

For compatability, u_0 must equal u_i; therefore, from Eqs. 4.53 and 4.54, $u_0 = u_i$ or

$$p_f = \frac{2p_i}{\left\{E_i(R^2 - 1)\left[\frac{1}{E_i}\left(\frac{R^2 + 1}{R^2 - 1} - v_i\right) + \frac{1}{E_0}\left(\frac{R_1{}^2 + 1}{R_1{}^2 - 1} + v_0\right)\right]\right\}} \qquad 4.55$$

which is a relationship describing the interface pressure p_f in terms of the applied internal pressure p_i, the properties of the materials E and v, and the geometry, R and R_1. Having a value for p_f, Eqs. 4.49 to 4.52 can be used to obtain the stress distribution in the structure. If the two shells are initially shrunk together, an additional interface pressure is developed (see Chapter 11) which must be added to that given by Eq. 4.55. The foregoing theory has proved useful in the design of high-pressure vessels, particularly those containing hard liners like tungsten carbide.

As an example of the use of the preceding theory consider the pressure transmission through a three-layer vessel, each layer having different elastic properties as shown in Fig. 4.12*C*.

For cylinder 1:

$$\sigma_{h1} = \frac{p_i}{R_1^2 - 1}\left(1 + \frac{b^2}{r^2}\right) - \frac{p_1 R_1^2}{R_1^2 - 1}\left(1 + \frac{a^2}{r^2}\right)$$

$$\sigma_{r1} = \frac{p_i}{R_1^2 - 1}\left(1 - \frac{b^2}{r^2}\right) - \frac{p_1 R_1^2}{R_1^2 - 1}\left(1 - \frac{a^2}{r^2}\right).$$

For cylinder 2:

$$\sigma_{h2} = \frac{p_1}{R_2^2 - 1}\left(1 + \frac{c^2}{r^2}\right) - \frac{p_2 R_2^2}{R_2^2 - 1}\left(1 + \frac{b^2}{r^2}\right)$$

$$\sigma_{r2} = \frac{p_1}{R_2^2 - 1}\left(1 - \frac{c^2}{r^2}\right) - \frac{p_2 R_2^2}{R_2^2 - 1}\left(1 - \frac{b^2}{r^2}\right)$$

For cylinder 3:

$$\sigma_{h3} = \frac{p_2}{R_3^2 - 1}\left(1 + \frac{d^2}{r^2}\right)$$

$$\sigma_{r3} = \frac{p_2}{R_3^2 - 1}\left(1 - \frac{d^2}{r^2}\right)$$

$$u_b = \frac{b}{E_1}\left[\frac{2p_i}{R_1^2 - 1} - p_1\left(\frac{R_1^2 + 1}{R_1^2 - 1}\right) - \nu_1(-p_1)\right]$$

$$u_b' = \frac{b}{E_2}\left\{\frac{p_1}{R_2^2 - 1}\left(1 + \frac{c^2}{b^2}\right) - \frac{2p_2 R_2^2}{R_2^2 - 1} - \nu_2\left[\frac{p_1(1 - c^2/b^2)}{R_2^2 - 1}\right]\right\}$$

But $u_b = u_b'$, from which

$$p_1\left[\frac{1}{E_1}\left(\nu_1 - \frac{R_1^2 + 1}{R_1^2 - 1}\right) - \frac{1}{E_2}\left(\frac{R_2^2 + 1}{R_2^2 - 1} + \nu_2\right)\right]$$
$$= 2\left[\frac{p_i}{E_1}\left(\frac{1}{1 - R_1^2}\right) + \frac{p_2 R_2^2}{E_2(1 - R_2^2)}\right]$$

Similarly,

$$u_c = \frac{c}{E_2}\left\{\frac{2p_1}{R_2^2 - 1} - \frac{p_2 R_2^2}{R_2^2 - 1}\left(\frac{R_2^2 + 1}{R_2^2}\right) - \nu_2\left[\frac{-p_2 R_2^2}{R_2^2 - 1}\left(\frac{R_2^2 - 1}{R_2^2}\right)\right]\right\}$$

$$u_c' = \frac{c}{E_3}\left[\frac{p_2}{R_3^2 - 1}(R_3^2 + 1) - \frac{\nu_3 p_2}{R_3^2 - 1}(1 - R_3^2)\right]$$

But $u_c = u_c'$, from which

$$p_1 = \frac{2\{(p_i/E_1)[1/(1 - R_1^2)] + p_2 R_2^2/[E_2(1 - R_2^2)]\}}{\left[\frac{1}{E_1}\left(\nu_1 - \frac{R_1^2 + 1}{R_1^2 - 1}\right) - \frac{1}{E_2}\left(\frac{R_2^2 + 1}{R_2^2 - 1} + \nu_2\right)\right]}$$

and also

$$p_2 = \frac{2p_i}{E_1(1-R_1{}^2)\left[\frac{1}{E_1}\left(\nu_1 - \frac{R_1{}^2+1}{R_1{}^2-1}\right) - \frac{1}{E_2}\left(\frac{R_2{}^2+1}{R_2{}^2-1}+\nu_2\right)\right]}$$
$$\times\left\{\left[\frac{1}{E_3}\left(\frac{R_3{}^2+1}{R_3{}^2-1}+\nu_3\right)+\frac{1}{E_2}\left(\frac{R_2{}^2+1}{R_2{}^2-1}-\nu_2\right)\right]\Big/\frac{2}{E_2(R_2{}^2-1)}\right.$$
$$\left.-\frac{2R_2{}^2}{E_2(1-R_2{}^2)}\Big/\left[\frac{1}{E_1}\left(\nu_1-\frac{R_1{}^2+1}{R_1{}^2-1}\right)-\frac{1}{E_2}\left(\frac{R_2{}^2+1}{R_2{}^2-1}+\nu_2\right)\right]\right\}$$

Having values for the interface pressures the stress distributions can then be calculated.

Multimaterial shells. A relatively new type of structure not a composite in the true sense, but still utilizing a variety of materials, is that typified by the tank shown in Fig. 4.13. In such a tank the hoop stress has the value

$$\sigma_h = (H - y)\gamma R/t \tag{4.56}$$

where γ is the density of the liquid in the tank. In order to design such a tank the maximum stress would be calculated from Eq. 4.56 as

$$(\sigma_h)_{max} = HR\gamma/t \tag{4.57}$$

and the thickness of the material t determined from Eq. 4.57 based on the strength of the material (from Eq. 3.523); thus

$$t = HR\gamma/(\sigma_0)_w \tag{4.58}$$

where $(\sigma_0)_w$ is the design working stress for the material. Equation 4.56 shows that the stress varies from a maximum at the bottom of the tank to a minimum at the top of the tank; therefore, if the tank were designed on the basis of the maximum stress at the bottom, the wall thickness for locations other than at the bottom would be overdesigned. To save both weight and cost of material an alternate procedure would be to select a high-strength material for the

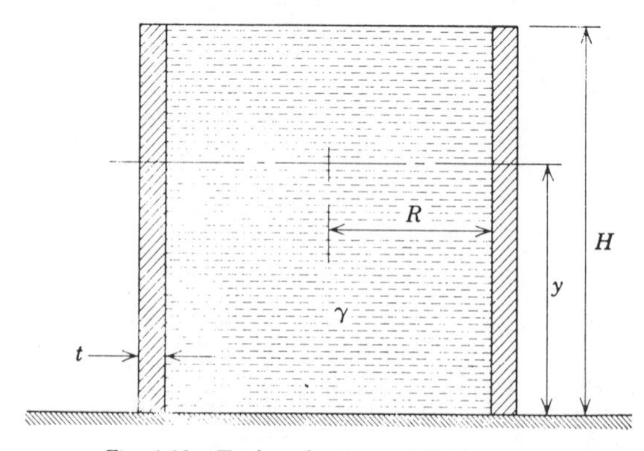

Fig. 4.13 Tank under internal fluid pressure.

bottom part of the tank and then use less costly materials for the upper portions where the stress is less. For example, if the tank shown in Fig. 4.13 were split into three sections, the bottom section could be made of high strength steel having a yield strength of 100,000 psi. For the central section, using the same thickness as for the bottom part, from Eq. 4.56.

$$t = \frac{(H - y)\gamma R}{\sigma_{0w}} = \frac{H\gamma R}{100{,}000} = \left(H - \frac{H}{3}\right)\frac{\gamma R}{\sigma_{0w}} = \text{constant} \qquad 4.59$$

or $(\sigma_0)_w = 67{,}000$ psi for the steel used. For the top part of the tank, by the same procedure, $(\sigma_0)_w = 34{,}000$ psi. In an actual design specific steels would be selected like 100,000 psi steel, 50,000 psi steel, and 30,000 psi steel and the tank proportioned accordingly to accommodate these properties. The same method would also apply to structural shapes like I beams where the web and flanges could be made of different materials. These designs have all been made using the same basic material (steel) with an elastic modulus of approximately 30×10^6 psi. If different basic materials, such as steel and copper or steel and aluminum, are used together, then the difference in deformation brought about by the different moduli of elasticity must be taken into account. (See also the example in Section 3-3.)

Stress-Strain Relations in Orthotropic Materials

In many materials, particularly laminates, the component parts exhibit orthotropic properties, that is, the properties vary with direction in the material. To cope with this situation general stress-strain relationships will be developed that take this orthotropic property of materials into account. It is assumed that straining takes place within the range covered by Hooke's law.

An element of material having longitudinal L and transverse T properties and subjected to a state of biaxial stress is shown in Fig. 4.14*A*. The σ_1 stress is oriented at angle α with respect to the longitudinal direction in the material. Equilibrium of forces in the elements shown in Figs. 4.14*B* and 4.14*C* shows

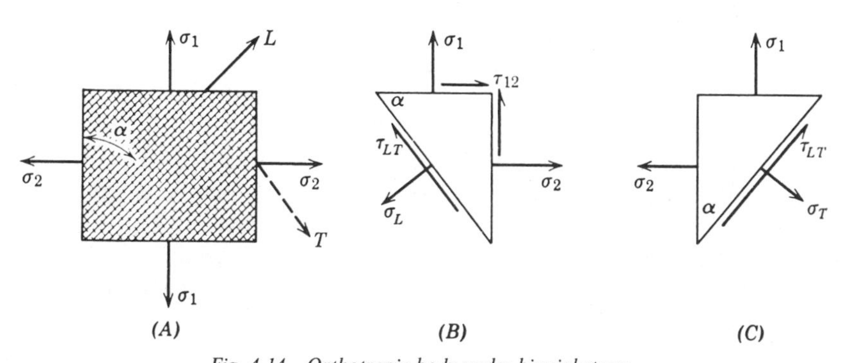

Fig. 4.14 Orthotropic body under biaxial stress.

the following relationship between the applied stresses σ_1 and σ_2 and the L and T directions in the material.

$$\sigma_L = \sigma_1 \cos^2 \alpha + \sigma_2 \sin^2 \alpha + \tau_{12} \sin 2\alpha \tag{4.60}$$

$$\sigma_T = \sigma_1 \sin^2 \alpha + \sigma_2 \cos^2 \alpha - \tau_{12} \sin 2\alpha \tag{4.61}$$

$$\tau_{LT} = \tau_{12} \cos 2\alpha + \frac{\sigma_2 - \sigma_1}{2} \sin 2\alpha \tag{4.62}$$

Similar equations for strain have already been derived in Chapter 3 (Eqs. 3.44 and 3.45). With symbols changed to conform with the nomenclature of this chapter, Eqs. 3.44 and 3.45 may be written as follows:

$$\varepsilon_L = \varepsilon_1 \cos^2 \alpha + \varepsilon_2 \sin^2 \alpha + (\gamma_{12}/2) \sin 2\alpha \tag{4.63}$$

$$\varepsilon_T = \varepsilon_1 \sin^2 \alpha + \varepsilon_2 \cos^2 \alpha - (\gamma_{12}/2) \sin 2\alpha \tag{4.64}$$

The shear strain equation is obtained from Eq. 3.57 and is

$$\gamma_{LT} = \gamma_{12} \cos 2\alpha + (\varepsilon_2 - \varepsilon_1) \sin 2\alpha \tag{4.65}$$

In addition, from Eqs. 3.63 and 3.64, strain equations in the L and T directions can be written in terms of the stresses

$$\varepsilon_L = (\sigma_L/E_L) - \nu_{TL}(\sigma_T/E_T) \tag{4.66}$$

$$\varepsilon_T = (\sigma_T/E_T) - \nu_{LT}(\sigma_L/E_L) \tag{4.67}$$

and

$$\gamma_{LT} = \tau_{LT}/G_{LT} \tag{4.68}$$

Inversion of Eqs. 4.66 to 4.68 gives

$$\sigma_L = E_L\varepsilon_L/k + E_L\nu_{TL}\varepsilon_T/k \tag{4.69}$$

$$\sigma_T = E_T\nu_{LT}\varepsilon_L/k + E_T\varepsilon_T/k \tag{4.70}$$

and

$$\tau_{LT} = G_{LT}\gamma_{LT} \tag{4.71}$$

where $k = 1 - \nu_{LT}\nu_{TL}$. In terms of the 1 and 2 directions

$$\varepsilon_1 = (\sigma_1/E_1) - \nu_{21}(\sigma_2/E_2) \tag{4.72}$$

$$\varepsilon_2 = (\sigma_2/E_2) - \nu_{12}(\sigma_1/E_1) \tag{4.73}$$

and

$$\gamma_{12} = \tau_{12}/G_{12} \tag{4.74}$$

The equations given provide the basis for determining the stress-strain characteristics of orthotropic materials. For example, consider Fig. 4.15 which shows a material with L and T properties subjected to tension in the direction of σ_1 at α degrees to the L direction. The tensile strain ε_1 in the direction of σ_1 is given by Eq. 4.72

$$\varepsilon_1 = \sigma_1/E_1 \tag{4.75}$$

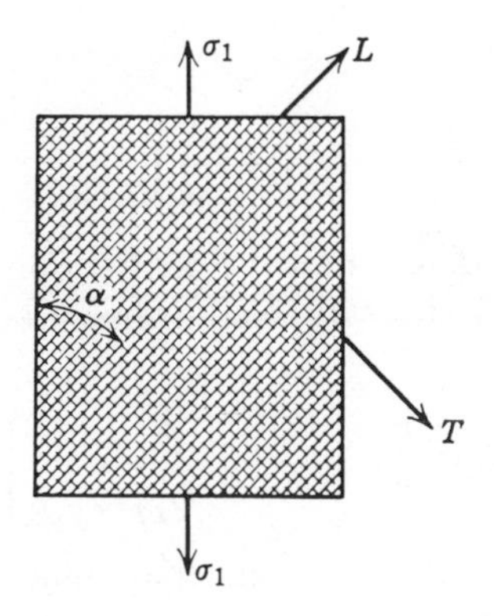

Fig. 4.15 Orthotropic material subjected to tension.

where E_1 now has to be determined as a function of E_L and E_T. With tension, σ_2 and τ_{12} are zero; therefore Eqs. 4.60 to 4.62 become

$$\sigma_L = \sigma_1 \cos^2 \alpha \tag{4.76}$$

$$\sigma_T = \sigma_1 \sin^2 \alpha \tag{4.77}$$

and

$$\tau_{LT} = -\sigma_1(\sin 2\alpha/2) \tag{4.78}$$

Combining the values of σ_L, σ_T, and τ_{LT} given by Eqs. 4.76 to 4.78 with Eqs. 4.72 to 4.74 gives new strain equations in $\varepsilon_1, \varepsilon_2$, and γ_{12}, which, when combined with Eqs. 4.63 to 4.65, gives the result

$$\frac{E_L}{E_1} = \cos^4 \alpha + \left(\frac{E_L}{E_T}\right)\sin^4 \alpha + \frac{\sin^2 2\alpha}{4}\left(\frac{E_L}{G_{LT}} - 2\nu_{LT}\right) \tag{4.79}$$

The value of E_1 in Eq. 4.79 may then be used in Eq. 4.75 to compute the strain ε_1. The strain ε_1 is accompanied by a lateral contraction ε_2 which is given by the relation

$$\varepsilon_2 = -\nu_{12}\varepsilon_1 \tag{4.80}$$

where ν_{12}, Poisson's ratio, is determined as follows. Adding Eqs. 4.63 and 4.64 gives the result

$$\varepsilon_L + \varepsilon_T = \varepsilon_1 + \varepsilon_2 \tag{4.81}$$

or $\varepsilon_L + \varepsilon_T = \varepsilon_1 - \nu_{12}\varepsilon_1$, which, when combined with Eqs. 4.66 and 4.67 gives the result

$$\nu_{12} = 1 - \frac{1}{\varepsilon_1}\left[\frac{\sigma_L}{E_L}(1 - \nu_{LT}) + \frac{\sigma_T}{E_T}(1 - \nu_{TL})\right] \tag{4.82}$$

However, since ν is a function of E and is directly proportional to it,

$$E_L/E_T = \nu_{LT}/\nu_{TL} \tag{4.83}$$

Combining Eqs. 4.82, 4.83, and 4.79 then gives

$$\nu_{12} = \frac{E_1}{E_L}\left[\nu_{LT} - \frac{\sin^2 2\alpha}{4}\left(\frac{E_L}{E_T} - \frac{E_L}{G_{LT}} + 2\nu_{LT} + 1\right)\right] \tag{4.84}$$

Substitution of Eq. 4.84 into Eq. 4.80 then determines the contraction ε_2. A shear strain γ_{12} is also developed by the tensile stress σ_1; by the same procedure as shown the value of this shear strain is

$$\gamma_{12} = -\sigma_1 \frac{\sin 2\alpha}{E_L}\left[\nu_{LT} - \frac{E_L}{2G_{LT}} + \frac{E_L}{E_T} - \cos^2\alpha\left(\frac{E_L}{E_T} + 2\nu_{LT} - \frac{E_L}{G_{LT}} + 1\right)\right] \tag{4.85}$$

Now consider Fig. 4.16 which shows an orthotropic material subjected to *biaxial* stresses. The equations have already been developed for tension loading in the direction of σ_1 (Eqs. 4.79, 4.84, and 4.85). If the tension stress is applied in the direction of σ_2, the preceding equations are corrected for orientation and become

$$\frac{E_T}{E_2} = \cos^4\alpha + \left(\frac{E_T}{E_L}\right)\sin^4\alpha + \frac{\sin^2 2\alpha}{4}\left(\frac{E_T}{G_{TL}} - 2\nu_{TL}\right) \tag{4.86}$$

$$\nu_{21} = \frac{E_2}{E_T}\left[\nu_{TL} - \frac{\sin^2 2\alpha}{4}\left(\frac{E_T}{E_L} - \frac{E_T}{G_{TL}} + 2\nu_{TL} + 1\right)\right] \tag{4.87}$$

and

$$\gamma_{21} = -\frac{\sigma_2}{E_T}\sin 2\alpha\left[\nu_{TL} - \frac{E_T}{2G_{TL}} + \frac{E_T}{E_L} - \cos^2\alpha\left(\frac{E_T}{E_L} + 2\nu_{TL} - \frac{E_T}{G_{TL}} + 1\right)\right] \tag{4.88}$$

where $\gamma_{21} = -\gamma_{12}$. Thus, for an element subjected to the state of stress shown in Fig. 4.16, a more general matrix form can be developed.

Use Eqs. 4.60, 4.61, 4.62; and Eqs. 4.63, 4.64, 4.65 in the following forms:

$$\begin{bmatrix}\sigma_L\\ \sigma_T\\ \tau_{LT}\end{bmatrix} = \begin{bmatrix}Q'_{11} & Q'_{12} & Q'_{13}\\ Q'_{21} & Q'_{22} & Q'_{23}\\ Q'_{31} & Q'_{32} & Q'_{33}\end{bmatrix}\begin{bmatrix}\sigma_1\\ \sigma_2\\ \tau_{12}\end{bmatrix} \tag{4.89}$$

$$\begin{bmatrix}\varepsilon_L\\ \varepsilon_T\\ \gamma_{LT}\end{bmatrix} = \begin{bmatrix}Q''_{11} & Q''_{12} & Q''_{13}\\ Q''_{21} & Q''_{22} & Q''_{23}\\ Q''_{31} & Q''_{32} & Q''_{33}\end{bmatrix}\begin{bmatrix}\varepsilon_1\\ \varepsilon_2\\ \gamma_{12}\end{bmatrix} \tag{4.90}$$

Restate Eqs. 4.69, 4.70, 4.71 in the following form.

$$\begin{bmatrix}\sigma_L\\ \sigma_T\\ \tau_{LT}\end{bmatrix} = \begin{bmatrix}Q_{11} & Q_{12} & 0\\ Q_{12} & Q_{22} & 0\\ 0 & 0 & Q_{33}\end{bmatrix}\begin{bmatrix}\varepsilon_L\\ \varepsilon_T\\ \gamma_{LT}\end{bmatrix} \tag{4.91}$$

Eqs. 4.89 and 4.90 are substituted into Eq. 4.91.

$$[Q']\begin{bmatrix}\sigma_1\\ \sigma_2\\ \tau_2\end{bmatrix} = [Q][Q'']\begin{bmatrix}\varepsilon_1\\ \varepsilon_2\\ \gamma_{12}\end{bmatrix}$$

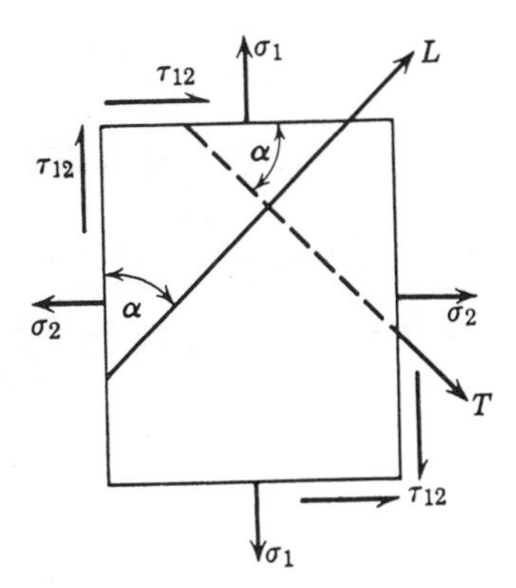

Fig. 4.16 Orthotropic material subjected to combined stress.

Multiply by the inverse $[Q']^{-1}$

$$\begin{bmatrix} \sigma_1 \\ \sigma_2 \\ \tau_{12} \end{bmatrix} = [Q']^{-1}[Q][Q''] \begin{bmatrix} \varepsilon_1 \\ \varepsilon_2 \\ \gamma_{12} \end{bmatrix} \qquad 4.92$$

Also, the values for $\varepsilon_1, \varepsilon_2, \gamma_{12}$ can be obtained as

$$\begin{bmatrix} \varepsilon_1 \\ \varepsilon_2 \\ \gamma_{12} \end{bmatrix} = [[Q']^{-1}[Q][Q'']]^{-1} \begin{bmatrix} \sigma_1 \\ \sigma_2 \\ \tau_{12} \end{bmatrix} \qquad 4.93$$

Eqs. 4.92 and 4.93 can be restated as

$$\sigma_1 = C_{11}\varepsilon_1 + C_{12}\varepsilon_2 + C_{16}\gamma_{12} \qquad 4.94$$

$$\sigma_2 = C_{21}\varepsilon_1 + C_{22}\varepsilon_2 + C_{26}\gamma_{12} \qquad 4.95$$

$$\tau_{12} = C_{61}\varepsilon_1 + C_{62}\varepsilon_2 + C_{66}\gamma_{12} \qquad 4.96$$

$$\varepsilon_1 = S_{11}\sigma_1 + S_{12}\sigma_2 + S_{16}\tau_{12} \qquad 4.97$$

$$\varepsilon_2 = S_{21}\sigma_1 + S_{22}\sigma_2 + S_{26}\tau_{12} \qquad 4.98$$

$$\gamma_{12} = S_{61}\sigma_1 + S_{62}\sigma_2 + S_{66}\tau_{12} \qquad 4.99$$

The numbering of the coefficients comes from a simplification of the equations of elasticity (6, 12).

The individual coefficients are as follows:

$$C_{11} = \frac{1}{C}[E_L \cos^4 \alpha + E_T \sin^4 \alpha + \sin^2 \alpha \cos^2 \alpha(2E_T\nu_{LT} + 4CG_{LT})] \qquad 4.100$$

$$C_{12} = \frac{1}{C}[E_T\nu_{LT}(\cos^4 \alpha + \sin^4 \alpha) + \sin^2 \alpha \cos^2 \alpha(E_T + E_L - 4CG_{LT})] \qquad 4.101$$

$$C_{16} = \frac{1}{C}[\sin\alpha\cos^3\alpha(E_L - \nu_{LT}E_T - 2CG_{LT}) + \sin^3\alpha\cos\alpha(\nu_{LT}E_T - E_T + 2CG_{LT})] \quad 4.102$$

$$C_{21} = C_{12}$$

$$C_{22} = \frac{1}{C}[E_T\cos^4\alpha + E_L\sin^4\alpha + \sin^2\alpha\cos^2\alpha(2\nu_{LT}E_T + 4CG_{LT})] \quad 4.103$$

$$C_{26} = \frac{1}{C}[\sin\alpha\cos^3\alpha(\nu_{LT}E_T - E_T + 2CG_{LT}) + \sin^3\alpha\cos\alpha(E_L - \nu_{LT}E_T - 2CG_{LT})] \quad 4.104$$

$$C_{61} = C_{16}$$

$$C_{62} = C_{26}$$

$$C_{66} = \frac{1}{C}[(E_L + E_T - 2\nu_{LT}E_T - 2CG_{LT})\sin^2\alpha\cos^2\alpha + CG_{LT}(\sin^4\alpha + \cos^4\alpha)] \quad 4.105$$

$$S_{11} = \frac{\cos^4\alpha}{E_L} + \frac{\sin^4\alpha}{E_T} + \left[\frac{1}{G_{LT}} - \frac{2\nu_{LT}}{E_L}\right]\sin^2\alpha\cos^2\alpha \quad 4.106$$

$$S_{12} = \sin^2\alpha\cos^2\alpha\left(\frac{1}{E_L} + \frac{1}{E_T} - \frac{1}{G_{LT}}\right) - \frac{\nu_{LT}}{E_L}(\sin^4\alpha + \cos^4\alpha) \quad 4.107$$

$$S_{16} = \sin\alpha\cos^3\alpha\left(\frac{2}{E_L} + \frac{2\nu_{LT}}{E_L} - \frac{1}{G_{LT}}\right) - \sin^3\alpha\cos\alpha\left(\frac{2}{E_T} + \frac{2\nu_{LT}}{E_L} - \frac{1}{G_{LT}}\right) \quad 4.108$$

$$S_{21} = S_{12}$$

$$S_{22} = \frac{\cos^4\alpha}{E_T} + \frac{\sin^4\alpha}{E_L} + \sin^2\alpha\cos^2\alpha\left[\frac{1}{G_{LT}} - \frac{2\nu_{LT}}{E_L}\right] \quad 4.109$$

$$S_{26} = \sin^3\alpha\cos\alpha\left(\frac{2}{E_L} + \frac{2\nu_{LT}}{E_L} - \frac{1}{G_{LT}}\right) - \sin\alpha\cos^3\alpha\left(\frac{2}{E_T} + \frac{2\nu_{LT}}{E_L} - \frac{1}{G_{LT}}\right) \quad 4.110$$

$$S_{61} = S_{16}$$

$$S_{62} = S_{26}$$

$$S_{66} = \frac{\cos^4\alpha}{G_{LT}} + \frac{\sin^4\alpha}{G_{LT}} + \sin^2\alpha\cos^2\alpha\left[\frac{4}{E_L} + \frac{4}{E_T} + \frac{8\nu_{LT}}{E_L} - \frac{2}{G_{LT}}\right] \quad 4.111$$

$$C = 1 - \nu_{TL}\nu_{LT} \quad 4.112$$

For the inverse, the value of the C_{ij} determinant is

$$D = C_{11}C_{22}C_{33} - C_{16}{}^2C_{22} + 2C_{16}C_{21}C_{26} - C_{12}{}^2C_{66} - C_{11}C_{26}{}^2 \quad 4.113$$

As an example of the use of the preceding theory a design chart will be

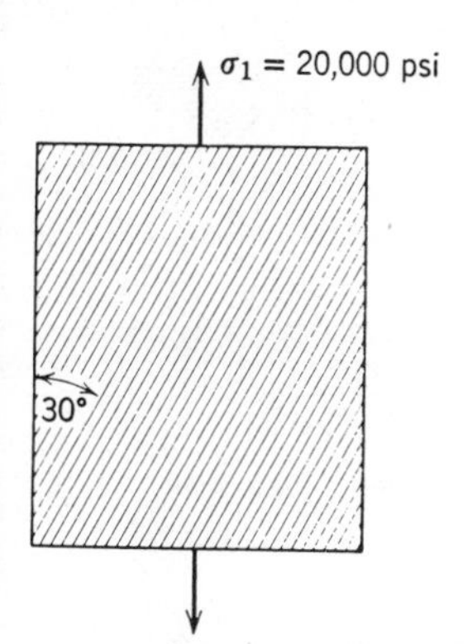

Fig. 4.17 Orthotropic material subjected to tension at 30° to the principal elastic axis of the material.

prepared for the state of stress shown in Fig. 4.17. Different charts would be obtained for various materials depending on their properties. For present purposes the following properties will be used:

$$E_L = 4 \times 10^6 \text{ psi}$$
$$E_T = 4 \times 10^6 \text{ psi}$$
$$G_{LT} = G_{TL} = 5 \times 10^5 \text{ psi}$$
$$\nu_{LT} = \nu_{0^\circ} = 0.40$$
$$\nu_{TL} = \nu_{90^\circ} = 0.10$$

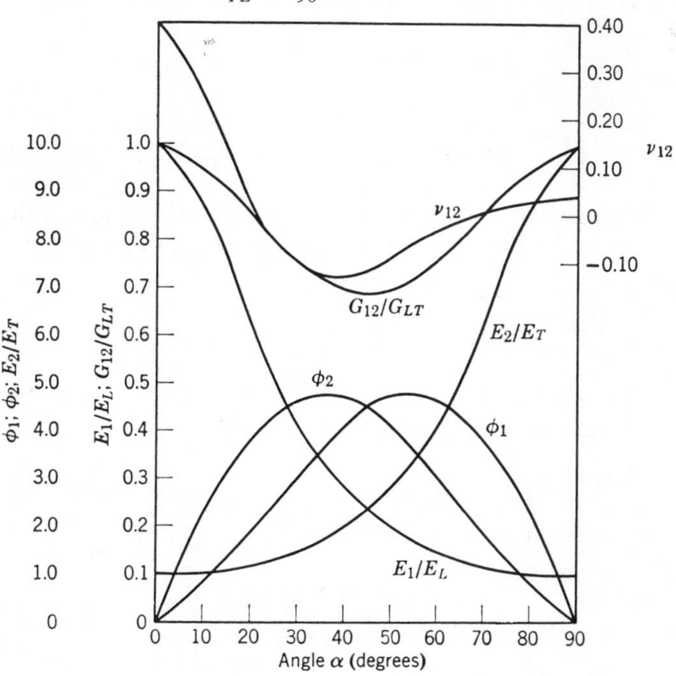

Fig. 4.18 Design chart for orthotropic material under state of stress shown in Fig. 4.17.

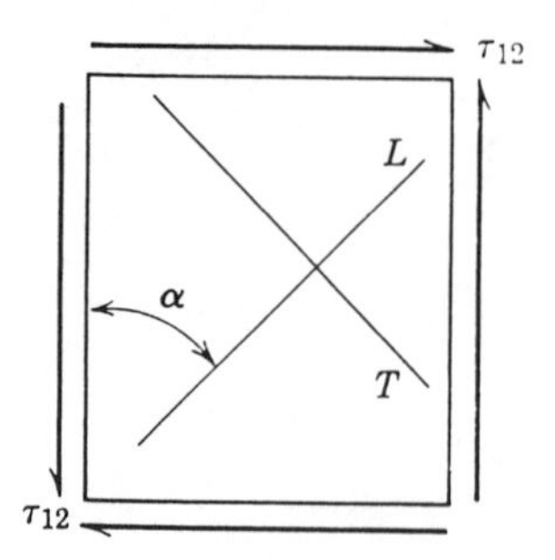

Fig. 4.19 Orthotropic material in pure shear.

Substituting these property values into Eqs. 4.79, 4.84, and 4.85 gives the curves plotted in Fig. 4.18. For the specific case of values shown in Fig. 4.17,

$$E_1/E_L = 0.394 \quad \text{or} \quad E_1 = 1.58 \times 10^6 \text{ psi}$$
$$G_{12}/G_{LT} = 0.757 \quad \text{or} \quad G_{12} = 3.79 \times 10^5 \text{ psi}$$
$$\nu_{12} = -0.122$$
$$\varepsilon_1 = 20{,}000/(1.58 \times 10^6) = 0.0127 \text{ in./in.}$$
$$\varepsilon_2 = -(-0.122)(0.0127) = 0.00155 \text{ in./in.}$$
$$\gamma_{12} = (3.92)(20{,}000)/(4 \times 10^6) = 0.0196 \text{ in./in.}$$

As another example consider pure shear, Fig. 4.19. For this case

$$\gamma_{12} = \tau_{12}/G_{12} \tag{4.114}$$

and

$$\frac{G_{LT}}{G_{12}} = \frac{G_{LT}}{E_L}\left[\left(\frac{E_L}{E_T} + 2\nu_{LT} + 1\right) - \left(1 + 2\nu_{LT} + \frac{E_L}{E_T} - \frac{E_L}{G_{LT}}\right)\cos^2 2\alpha\right] \tag{4.115}$$

which is plotted in Fig. 4.18 as G_{12}/G_{LT}. In isotropic materials there is no axial strain caused by the shear stress τ_{12}; however, in orthotropic materials

$$\varepsilon_1 = S_{16}\tau_{12} \tag{4.116}$$

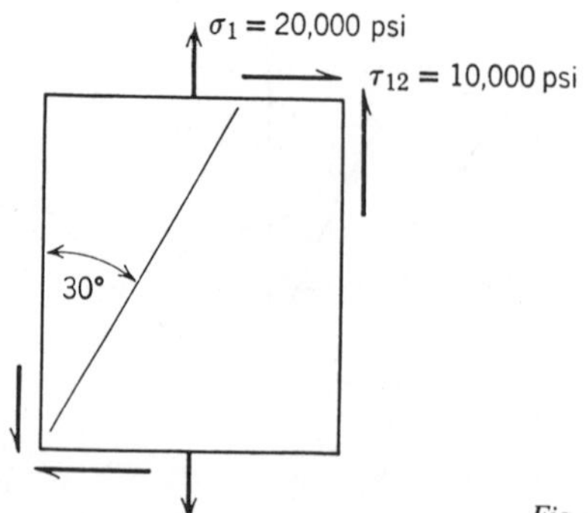

Fig. 4.20 Orthotropic body under combined stresses.

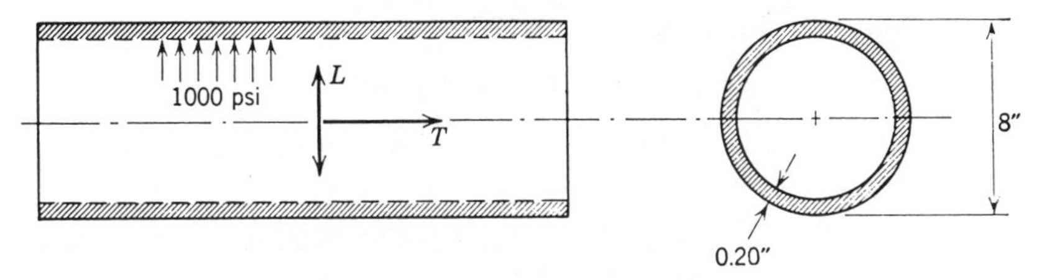

Fig. 4.21 Cylinder of orthotropic material under internal pressure.

and the lateral strain is

$$\varepsilon_2 = S_{26}\tau_{12} \tag{4.117}$$

where S_{16} and S_{26} are defined by Eqs. 4.108 and 4.110. These are also plotted in Fig. 4.18.

Suppose the element shown in Fig. 4.17 is also subjected to a shear stress of 10,000 psi, as shown in Fig. 4.20; then in addition to the values already calculated, τ_{12} causes additional strain as follows. From Eq. 4.110

$$\gamma_{12} = 10{,}000/(5 \times 10^5) = 0.02 \text{ in./in.}$$

$$\varepsilon_1 = -3.92(10{,}000)/(4 \times 10^6) = -0.0098 \text{ in./in.}$$

and

$$\varepsilon_2 = -4.72(10{,}000)/(4 \times 10^6) = -0.0118 \text{ in./in.}$$

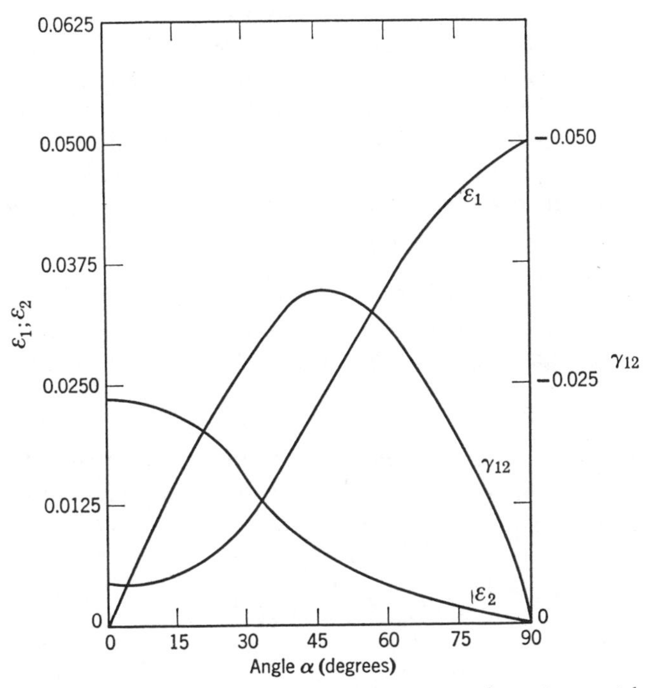

Fig. 4.22 Variation of strain with angle α in an orthotropic material.

The total strains on the element in Fig. 4.20 are thus

$$\gamma_{12} = 0.0196 + 0.02 = 0.0396 \text{ in./in.}$$
$$\varepsilon_1 = 0.0127 - 0.0098 = 0.0029 \text{ in./in.}$$
$$\varepsilon_2 = 0.00155 - 0.0118 = 0.01025 \text{ in./in.}$$

Now consider the cylinder shown in Fig. 4.21 made of material having the properties given in Fig. 4.18. If the longitudinal axis of the material is oriented in the hoop direction (direction of σ_1), then

$$\alpha = 0$$

and

$$\sigma_1 = pd/2t = 1000(8)/[(2)(0.2)] = 20{,}000 \text{ psi}$$
$$\sigma_2 = pd/4t = 1000(8)/[(4)(0.2)] = 10{,}000 \text{ psi}$$
$$\varepsilon_1 = \varepsilon_{\text{hoop}} = 20{,}000/(4 \times 10^6) - 0.1(10{,}000)/(4 \times 10^5) = 0.0025 \text{ in./in.}$$
$$\varepsilon_2 = \varepsilon_{\text{long}} = 10{,}000/(4 \times 10^5) - 0.4(10{,}000)/(4 \times 10^6) = 0.024 \text{ in./in.}$$
$$\gamma_{12} = 0$$

The variation in strains with variation of α in the cylinder is shown in Fig. 4.22.

Criterion of failure for orthotropic materials

In many design applications it is necessary to relate the state of stress and geometry to the condition of failure, that is, to predict under what conditions the structure will no longer behave elastically. It has already been demonstrated in Chapter 3 that the *distortion energy theory* of failure is the best answer to this need; with proper modification of this theory to include directional effects, it is also very useful in the analysis of orthotropic materials.

The distortion energy equation (see Chapter 3) relating the three principal stresses to an equivalent tensile stress $\bar{\sigma}$ is

$$2(\bar{\sigma})^2 = (\sigma_1 - \sigma_2)^2 + (\sigma_2 - \sigma_3)^2 + (\sigma_3 - \sigma_1)^2 \qquad 4.118$$

where $\bar{\sigma}$ is a tensile test value, for example, the yield strength of the material. For orthotropic materials Eq. 4.118 is modified as

$$2(\bar{\sigma})^2 = L(\sigma_1 - \sigma_2)^2 + M(\sigma_2 - \sigma_3)^2 + N(\sigma_3 - \sigma_1)^2 \qquad 4.119$$

where L, M, and N are parameters defining the degree of anisotropy in the material. These constants may be evaluated by successively considering two of the principal stresses to be zero; when this is done it is found that

$$L = 2\left(\frac{\bar{\sigma}}{\sigma_{01}}\right)^2 - \left(\frac{\bar{\sigma}}{\sigma_{03}}\right)^2\left[\beta^2 - \left(\frac{\alpha}{\beta}\right)^2 + 1\right] \qquad 4.120$$

$$M = 2\left(\frac{\bar{\sigma}}{\sigma_{02}}\right)^2 - 2\left(\frac{\bar{\sigma}}{\sigma_{01}}\right)^2 + \left(\frac{\bar{\sigma}}{\sigma_{03}}\right)^2\left[\beta^2 - \left(\frac{\alpha}{\beta}\right)^2 + 1\right] \qquad 4.121$$

$$N = \left(\frac{\bar{\sigma}}{\sigma_{03}}\right)^2 \left[\beta^2 - \left(\frac{\alpha}{\beta}\right)^2 + 1\right] \qquad 4.122$$

where $\alpha = \sigma_{02}/\sigma_{01}$, $\beta = \sigma_{03}/\sigma_{01}$, and σ_{01}, σ_{02}, and σ_{03} are yield strengths of the material in the 1, 2, and 3 directions, respectively. Thus, in using Eq. 4.119 values of σ_{01}, σ_{02}, and σ_{03} must be known and $\bar{\sigma}$ becomes equal to one of these values depending on the direction selected for correlation. The direction selected is unimportant to the solution of the problem; it is just necessary to be consistent. For example, in analyzing a thin-walled cylinder, the following stresses at the bore of the cylinder prevail when internal pressure is applied:

$$\text{hoop stress} = \sigma_h = \sigma_1 = pd/2t \qquad 4.123$$

$$\text{longitudinal stress} = \sigma_z = \sigma_2 = pd/4t \qquad 4.124$$

$$\text{radial stress} = \sigma_r = \sigma_3 = -p \qquad 4.125$$

where p is the internal pressure and d and t are the inside diameter and wall thickness respectively. If the hoop and radial properties are equal, β becomes unity in Eqs. 4.120 to 4.122, and if the longitudinal direction of the cylinder is chosen as the correlation direction (which is the most convenient direction for obtaining test samples), then

$$L = \alpha^4 \qquad 4.126$$

$$M = 2 - \alpha^4 \qquad 4.127$$

$$N = \alpha^2(2 - \alpha^2) \qquad 4.128$$

and Eq. 4.119 becomes

$$(\bar{\sigma})^2 = (\sigma_{02})^2 = p^2 \left\{ \frac{\alpha^4}{8}\left(\frac{d}{2t}\right)^2 + \frac{2-\alpha^4}{2}\left[\left(\frac{d}{4t}\right)^2 + \frac{d}{2t} + 1\right] + \frac{\alpha^2(2-\alpha^2)}{2}\left[\left(\frac{d}{2t}\right)^2 + \frac{d}{t} + 1\right]\right\} \qquad 4.129$$

Additional equations along the same line have been developed by Norris (23) and Fischer (11) for orthotropic materials. For the particular case of a material having longitudinal and transverse properties, Fischer gives the following equation for a material stressed as shown in Fig. 4.16:

$$1 = \left[\frac{\sigma_T}{(\sigma_u)_T}\right]^2 + \left[\frac{\sigma_L}{(\sigma_u)_L}\right]^2 + \left[\frac{\tau_{TL}}{(\tau_u)_{TL}}\right]^2 - K\frac{\sigma_T\sigma_L}{(\sigma_u)_T(\sigma_u)_L} \qquad 4.130$$

where

σ_T = stress in the T direction of the material

σ_L = stress in the L direction of the material

τ_{TL} = shear stress

$$K = \frac{E_T(1 + \nu_{LT}) + E_L(1 + \nu_{TL})}{2[E_T E_L(1 + \nu_{TL})(1 + \nu_{LT})]^{1/2}} \qquad 4.131$$

E = modulus of elasticity

ν = Poissons ratio, and the subscript u refers to the ultimate (elastic) strength of the materials. It is assumed that action is elastic and that failure occurs before the onset of plastic flow. An example of the use of Eq. 4.130 is given later in the discussion of laminated plates in Chapter 8.

4-2 FILAMENT-REINFORCED STRUCTURES

For the purposes of this book filament-reinforced structures will include (*a*) materials where filaments, whiskers, or bars are embedded in a matrix (like copper reinforced with tungsten wire or whiskers and reinforced concrete), (*b*) structures consisting of continuous filaments held together by binders (like glass filament/epoxy rocket motor cases), and (*c*) structures consisting of a base material reinforced by a shroud of continuous filaments (like plastic pipe wound with wire or nylon filaments, or thin-walled metal pipes wound with high-strength wire). In this latter category structures like wire or ribbon-wound heavy-walled pressure vessels, where the winding is applied to achieve primarily elastic strength, will not be considered; this is done in Chapter 11. Also, the mechanics of reinforcement of automobile tires is not considered as this is an "art" in itself and has a voluminous literature.

Filaments and bars in a matrix

If a material having high strength is embedded in a material of lesser strength the over-all strength of the composite is generally increased in proportion to the amount of reinforcing material added. Data for reinforced materials are shown in Tables 4.3 and 4.4. Chadbourne (7) for example, shows in Table 4.2 how several materials are upgraded by incorporating glass filaments in the structure, and Brookfield (4) in Table 4.5 gives typical data for various fibers used as reinforcing agents. Scala (31) as well as Vinson and Chou (37) treat the field of reinforcing a matrix with fibers, whiskers, and other forms. Fiber-reinforced ceramics have been studied by Plant (29). McDanels, Jech, and Weeton (21) in experiments with composites made up of tungsten wires in copper that found that the strengths of the composites were directly proportional to the volume percentage and tensile strength of the tungsten wire. Copper, having a strength of approximately 30,000 psi, was strengthened to more than 120,000 psi by the addition of 35% tungsten wire having a strength of 300,000 psi. These same investigators also found that increased strength could be achieved by short wires that did not run the full length of the test specimen. This finding has added impetus to the search for whisker materials and production techniques to produce composites of unprecedented strength. Hoffman (16), for example, in Table 4.6 lists the strengths of some whisker material that might find applica-

Table 4.2 Properties of Some Materials with Glass Reinforcing (7)

Property	Polyamide (Nylon 66)		Polystyrene		Polycarbonate		Styrene-Acrylonitrile		Zinc (AG-40A)	Aluminum (380)	Magnesium (AZ-91B)
	Unfilled	Filled	Unfilled	Filled	Unfilled	Filled	Unfilled	Filled			
Tensile Strength (1000 psi)	11.8	19.1	6.5	14	9.0	19	11.0	20	41	43	33
Modulus of Elasticity ($\times 10^5$ psi)	4	8.6	4	11	3.2	8.8	5.2	15	—	103	65
Elongation (%)	60	1.5	2	1.1	60–100	1.7	3.2	1.4	10	2	3
Flexural Strength (1000 psi)	11.5	22	11	17.5	26	12	17	28	—	—	—

Table 4.3 Elastic Constants and Design Characteristics of Unidirectional Composites (After Vicario and Toland (36), Courtesy of Academic Press Inc.)

Material	Composite density (lb/in.^{3})	Engineering moduli (10^6 psi)			Poisson's ratio	Maximum strain (%)			10^{-6} in./in./°F	
		E_1	E_2	G_{12}		$\pm\varepsilon_1$	$\pm\varepsilon_2{}^a$	ε_6	α_1	$\alpha_2{}^a$
S-glass–epoxy	0.072	7.5	1.7	0.8	0.25	3.00 1.60	0.33 1.40	1.30	3.5	11
PRD 49-III–epoxy	0.050	11.5	0.6	0.3	0.30	2.00 0.40	0.36 1.20	1.00	−2.0	30
High-strength graphite–epoxy	0.056	18.0	1.5	0.8	0.27	1.20 1.20	0.43 1.60	1.50	−0.2	13
High-modulus graphite–epoxy	0.059	27.6	1.1	0.6	0.30	0.45 0.55	0.70 2.60	0.70	−0.3	14
Ultra high modulus graphite–epoxy	0.058	45.0	0.9	0.6	0.26	0.40 0.21	0.20 1.00	0.35	−0.8	17
Boron–epoxy	0.075	30	3.0	0.7	0.21	0.75 1.13	0.40 1.20	1.90	2.5	8
Boron–aluminum	0.098	33	21.0	7.0	0.23	0.50 0.55	0.10 0.10	0.15	3.2	11

[a] Note that transverse strain and thermal expansion are controlled by the matrix properties. Strains listed in this table are indicative of high-temperature brittle resins. The quoted transverse elongation can generally be increased by a factor of two if ductile low-temperature capability resins are employed (i.e., BP-907). Transverse thermal expansion can be controlled between 10 and 30 with resin selection. Data for 60% fiber volume.

Table 4.4 Properties of Carbon-fiber-reinforced Composites (After Kliguer (19), Courtesy of Industry Media Inc.)

Material	Unidirectional Properties						± 45 Degree Properties				Density
	Elastic moduli (Msi)			Ultimate strength (Ksi)			Elastic moduli		Ultimate strength		(lb/in.^{3})
	Axial E_{11}	Transverse E_{22}	Shear G_{12}	Axial σ_{11}	Transverse σ_{22}	Shear σ_{12}	$E_x = E_y$	G_{xy}	τ_{xy}	$\sigma_x = \sigma_y$	
High-strength carbon-fiber/epoxy	20	1.0	0.65	220	6	14	2.5	4.5	20	50	0.057
High-modulus carbon-fiber/epoxy	32	1.0	0.7	175	5	10	2.5	6.5	18	42	0.058
Ultrahigh-modulus carbon-fiber/epoxy	44	1.0	0.95	110	4	7	3.0	11.5	14	30	0.061
Kevlar 49/epoxy	12.5	0.8	0.3	220	4	6	1.1	3.0	30	32	0.050
Steel	30	30	11.5	60	60	35	30	11.5	55	35	0.284
Aluminium 6061-T6	10.5	10.5	3.8	42	42	28	10.5	3.8	42	28	0.098

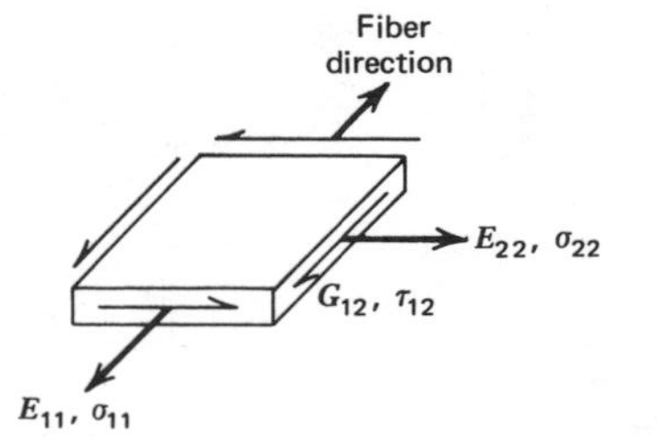

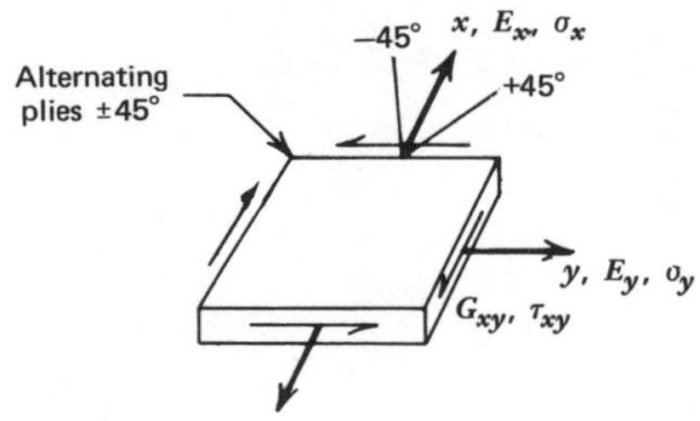

Table 4.5 Tensile Strength of Some Fibers (4)

Fiber	Tensile Strength (1000 psi)
Glass	180–315
Asbestos	200–300
Cotton	42–125
Nylon	65–117
Wool	17–28
Silk	45–83
Viscose rayon	29–88
Dacron	78–124

Table 4.6 Mechanical Properties of Some Whiskers (16)

Material[a]	Modulus of Elasticity (psi × 10^6)	Tensile Strength (psi × 10^6)	Maximum Elastic Strain (in./in.)
Iron	29	1.9	0.05
Copper	18	0.43	0.03
Silver	11	0.24	0.04
Silicon	23	0.55	0.02
Carbon	150	3	0.02
Sapphire	74	1.7	0.025
Quartz	11	0.6	—

[a] Probably 2 to 10 micron size range.

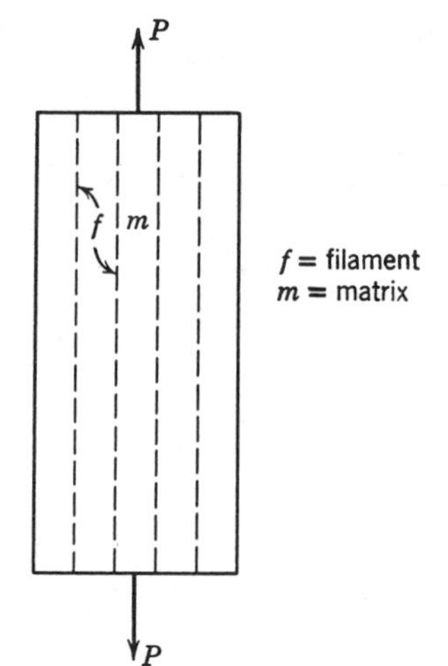

Fig. 4.23 Filament-reinforced matrix system.

tion in structures of unusual strength/weight ratio. The data in Table 4.6 are for materials ranging from 2 to 10 microns (diameter). The atomic cohesive strength is approached in the 1 to 3-micron range. Iron, for example in the form of 15-micron crystals, has a strength of about 100,000 psi, whereas in the $2\frac{1}{2}$-micron form it exhibits strengths as high as 2,000,000 psi.

Theory of reinforcing action. The exact mechanics of reinforcing action is unknown and is probably different for different materials. Some of the variables affecting the reinforcing effect are adhesion, cohesion, strength of components, moduli and Poisson ratio of components, coefficients of expansion, and so on. Despite the difficulties, however, useful engineering theories have been developed that help to give some insight into the problem, which will now be outlined.

Consider the composite in Fig. 4.23, consisting of a bar m (matrix) with f (fiber) reinforcing elements subjected to tensile loading P. For strain compatibility the entire composite elongates the same amount, that is,

$$\varepsilon_m = \varepsilon_f \tag{4.132}$$

or since $\sigma = E\varepsilon$, Eq. 4.132 can be written as

$$\sigma_m/E_m = \sigma_f/E_f \tag{4.133}$$

where σ and E are stress and modulus of elasticity respectively. For equilibrium

$$P = \sigma_m A_m + \sigma_f A_f = \sigma_c A_c \tag{4.134}$$

where A is the cross-sectional area and the subscript c means composite. From these equations it is seen that

$$\sigma_f = PE_f/(A_f E_f + A_m E_m) \tag{4.135}$$

and

$$\sigma_m = PE_m/(A_f E_f + A_m E_m) \tag{4.136}$$

If complete elastic action is assumed, then from Eq. 4.134

$$A_c\sigma_c = A_m\sigma_m + A_f\sigma_f \tag{4.137}$$

or from Eq. 4.1 with only P as an applied load

$$\sigma_c = \frac{E_cP}{E_1A^*} \tag{4.138}$$

$$\sigma_m = \frac{E_mP}{E_1A^*} \tag{4.139}$$

$$\sigma_f = \frac{E_fP}{E_1A^*} \tag{4.140}$$

Now substituting Eqs. 4.138, 4.139, 4.140 into Eq. 4.137 yields

$$A_c\left(\frac{E_cP}{E_1A^*}\right) = A_m\left(\frac{E_mP}{E_1A^*}\right) + A_f\left(\frac{E_fP}{E_1A^*}\right)$$

The term $\dfrac{P}{E_1A^*}$ is divided out leaving

$$E_cA_c = A_mE_m + A_fE_f \tag{4.141}$$

Again, divide by A_c the area of the composite yielding the mixture rule

$$E_c = \frac{A_m}{A_c}E_m + \frac{A_f}{A_c}E_f \tag{4.142}$$

The mixture rule also applies to Poissons ratio (20, 32).

$$\nu_{LT} = -\frac{\varepsilon_T}{\varepsilon_L} \tag{4.143}$$

or

$$\nu_{LT} = \nu_f\frac{A_f}{A_c} + \nu_m\frac{A_m}{A_c} \tag{4.144}$$

If the fibers carry all of the load,

$$E_c = E_fA_f/(A_m + A_f) \tag{4.145}$$

which corresponds to that derived by Outwater (24) for glass filaments in a resin matrix. If the fibers are chopped-up* but still oriented longitudinally with respect to the applied load, Outwater has proposed the following relationship for the glass/resin composite:

$$(E_c)_{\text{chopped fibers}} = E_f[(A_f/A_c) - (1/4\nu)(\sigma/\sigma_y)(D^2/Lt)] \tag{4.146}$$

where ν is Poisson's ratio of the fiber, σ the applied tensile stress, σ_y the strength

*This gives a type of "nonwoven" design. For a discussion of nonwoven fibres in reinforced plastics, see D. V. Rosato, *Ind. and Eng. Chem.*, 54 (8), 31–37 (August 1962).

of the fiber, D the diameter of the fiber, t the uniform distance of one fiber from another at the circumference, and L is the length of fiber.

To find the longitudinal L and transverse T properties of a rod for filaments:

$$E_L = E_C \tag{4.147}$$

$$E_L \nu_{TL} = \nu_{LT} E_T \tag{4.148}$$

$$\frac{1}{E_T} = \frac{A_f}{A_c E_f} + \frac{A_m}{A_c E_m} \tag{4.149}$$

$$\frac{1}{G_{LT}} = \frac{A_f}{A_c G_f} + \frac{A_m}{A_c G_m} \tag{4.150}$$

The properties for whiskers (20, 32) as oriented in Fig. 4.23 are:

$$\frac{E_L}{E_m} = \frac{(1 + \xi \eta \upsilon_f)}{(1 - \eta \upsilon_f)} \tag{4.151}$$

$$\frac{E_T}{E_m} = \frac{(1 + \eta \upsilon_f)}{(1 - \eta \upsilon_f)} \tag{4.152}$$

$$\frac{G_{LT}}{G_m} = \frac{\nu_{LT}}{\nu_m} = \frac{(1 + \lambda \upsilon_f)}{(1 - \lambda \upsilon_f)} \tag{4.153}$$

$$\eta = \frac{(E_f/E_m - 1)}{(E_f/E_m + \xi)} \tag{4.154}$$

$$\lambda = \frac{(G_f/G_m - 1)}{(G_f/G_m + 1)} \tag{4.155}$$

$$\upsilon_f = \frac{A_f}{A_c} \tag{4.156}$$

$$\xi = 2\left(\frac{L}{D}\right) \tag{4.157}$$

where L is the fiber or whisker length and D is the diameter.

In order to calculate the tensile strength of a composite the properties of the materials have to be considered. Equations 4.135 and 4.136 assume that integrity of the composite is maintained. The actual ultimate strength depends on the properties of the components and how much strain each part can withstand. For example, if a few very strong filaments are embedded in a soft matrix, the full potential of the soft material cannot be realized since the filaments will break at a strain corresponding to something less than the ultimate strength of the soft material. Consider, for example, the curves in Fig. 4.24, which represent stress-strain characteristics of a matrix m and a filament f. At any location on the curves, Eq. 4.132 must be satisfied; therefore,

$$\sigma_m/(E_0)_m = \sigma_f/(E_0)_f \tag{4.158}$$

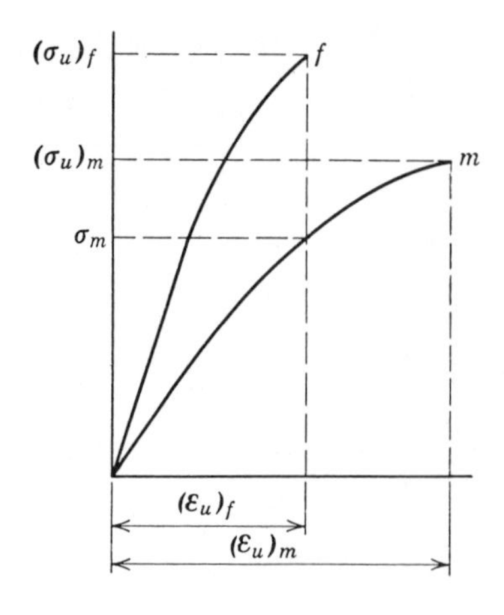

Fig. 4.24 Stress-strain data for system as shown in Fig. 4.23.

where E_0 is a secant modulus. From Eq. 4.134

$$\sigma_c = (\sigma_m A_m + \sigma_f A_f)/A_c = (\sigma_f/A_c)[((E_0)_m A_m/(E_0)_f) + A_f] \qquad 4.159$$

or

$$\sigma_c = \frac{(\sigma_u)_f}{A_c}\left[\frac{(\sigma_m)_{\max}/(\varepsilon_u)_f}{(\sigma_u)_f/(\varepsilon_u)_f}(A_m) + A_f\right] \qquad 4.160$$

and

$$\sigma_c = \frac{(\sigma_u)_f}{A_c}\left[\frac{(\sigma_m)_{\max}}{(\sigma_u)_f}(A_m) + A_f\right] \qquad 4.161$$

If the structure is all filament, $A_m = 0$ and

$$\sigma_c = \frac{(\sigma_u)_f(A_f)}{A_c} = (\sigma_u)_f \qquad 4.162$$

since $A_f = A_c$ for this case and the bar has the same strength as the filament. If there are no filaments in the structure, $A_f = 0$ and

$$\sigma_c = \frac{(\sigma_u)_f}{A_c}\left[\frac{(\sigma_m)_{\max}}{(\sigma_u)_f}(A_m)\right] = (\sigma_m)_{\max} = (\sigma_u)_m \qquad 4.163$$

since $(\varepsilon_u)_f = (\varepsilon_u)_m$, hypothetically, if no filament is present. An example of the use of Eq. 4.161 will now be given. Consider a composite containing 35% tungsten wires ($\sigma_u = 300{,}000$ psi) and 65% copper ($\sigma_u = 30{,}000$ psi). For integrity of this composite the strain must be held to less than 1.5% which is the limit value for tungsten wire; at this strain copper has a strength of approximately 8000 psi; applying Eq. 4.161,

$$\sigma_c = 300{,}000\left[\frac{8 \times 10^3}{30 \times 10^4}(0.65) + 0.35\right] = 110{,}000 \text{ psi}$$

which agrees with the values published by McDanels.

Beam reinforcing. When a beam cross section is reinforced with fibers,

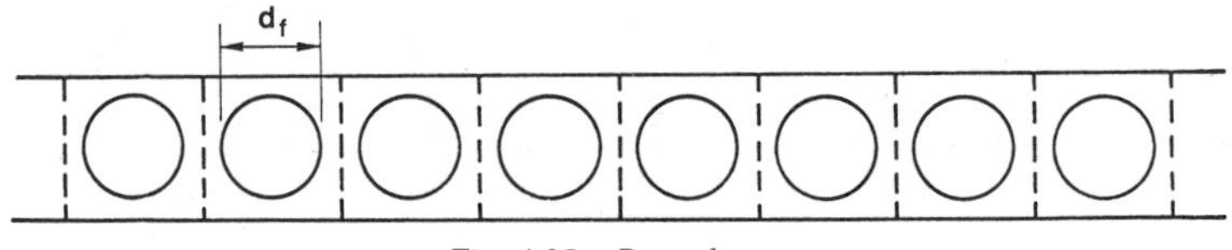

Fig. 4.25 Beam layer.

they are uniformly spaced and oriented in layers as in Fig. 4.25. In order to evaluate the stress in such a beam for bending alone Eq. 4.1 simplifies to

$$\sigma_{xx} = \frac{E}{E_1}\left[-\frac{M_y z}{I^*_{yy}}\right] \qquad 4.164$$

Eq. 4.5 is examined for the layered effect

$$I^*_{yy} = \sum_{i=1}^{i=n} \frac{E_i}{E_1}(I_{yyi} + \bar{Z}_i^2 A_i)$$

If each layer is basically the same in the fiber diameter d_f and the spacing of the fiber such that A_f/A_c and A_m/A_c remain constant from layer to layer then

$$E_i = E = E_1 \qquad 4.165$$

Further, I^*_{yy} reduces to an area moment of inertia dependent on geometry alone. Should E be required for deflections then Eq. 4.142 can be used for fibers and Eq. 4.151 for whiskers.

Beams with fiber reinforcing are produced under a manufacturing process called pultrusion (13, 15). The process spaces the fibers in the matrix, heats it, and cures the beam length. Most common cross sections are available in uniformly spaced fiber or specially packed fiber for special applications. Ultimate strengths of 100,000 psi are obtainable for pultruded products. Pultruded graphite samples for unidirectional flexure have 48–52% fibers, modulus of $16–25 \times 10^6$ psi, and flexural strengths of 147–180 kpsi.

Filament-binder composites

A unique class of composites, developed largely as a result of rocket requirements for exceptional strength/weight ratio, is that formed by winding impregnated filaments over a mandrel to form a shell. When the shell is formed the mandrel is removed and the structure consists of filaments held together by a suitable binder. One of the most useful and popular systems is the glass-filament/epoxy resin combination. Filament structures of this type have been reported to be exceptionally attractive with respect to cost, time of manufacture, weight, strength, reproducibility, and reliability (38, 43).

During winding of the filaments a uniform tension is applied so that a product of uniform density results. One of the primary advantages of such structures is that they can be manufactured with a wide variety of "built-in" mechanical properties through materials selection and geometry of the winding pattern.

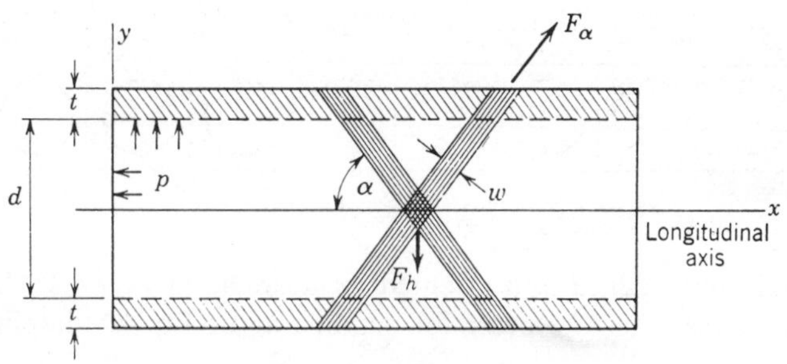

Fig. 4.26 Filament-wound cylindrical pressure vessel.

The first vessels made by filament techniques "*E*" electrical grade glass having the following approximate properties:

Tensile strength	230,000 psi
Flexural strength	280,000 psi
Specific weight	0.078 lbs/in.^{3}
Modulus of elasticity	7.5×10^6 psi
Compressive strength	70,000 psi
Elongation	2–3%

In manufacturing filament vessels three basic systems have been used: longitudinal winding, circumferential winding, and helical winding. In many applications a combination of these methods has been used in order to optimize the vessel behavior. For example, consider a filament vessel in the form of a closed-end cylinder, Fig. 4.26. For such a structure the principal stresses are

$$\text{hoop stress} = \sigma_h = pd/2t \qquad 4.166$$

$$\text{longitudinal stress} = \sigma_L = pd/4t \qquad 4.167$$

If the vessel were longitudinally wound, there would be little or no reinforcement in the hoop direction. Also, if the vessel were circumferentially wound, the longitudinal direction would not be reinforced. Therefore for optimum performance a helical design would be used. In order to determine what the optimum pattern is, consider Fig. 4.26; the vessel is made by winding impregnated filaments over a mandrel. Thus when the vessel is finished the wall thickness t is composed entirely of filament and binder. Now consider a unit of the shell of width w wrapped at angle α. When the vessel is subjected to internal pressure, the force carried by the filament in the α direction is

$$F_\alpha = \sigma_0 wt \qquad 4.168$$

In the hoop direction the force is

$$F_h = F_\alpha \sin \alpha \qquad 4.169$$

where σ_0 is the strength of the filaments. The hoop stress in the vessel wall due to applied internal pressure is

$$\sigma_h = F_h/A = \sigma_0 wt \sin\alpha\left(\frac{1}{wt/\sin\alpha}\right) = \sigma_0 \sin^2\alpha \qquad 4.170$$

where A is the cross-sectional area. In the longitudinal direction by similar analysis

$$\sigma_L = \sigma_0 \cos^2\alpha \qquad 4.171$$

Therefore with Eqs. 4.166, 4.167, 4.170, and 4.171, the optimum winding angle is determined from

$$\sigma_h/\sigma_L = \frac{pd}{2t}\bigg/\left(\frac{pd}{4t}\right) = 2 = \frac{\sigma_0 \sin^2\alpha}{\sigma_0 \cos^2\alpha} = \tan^2\alpha \qquad 4.172$$

or α is approximately 55°. For an open-end cylinder, $\sigma_L = 0$ so that

$$\sigma_L/\sigma_h = 0 = \cos^2\alpha/\sin^2\alpha = \cot^2\alpha \qquad 4.173$$

or $\alpha = 90°$, which is circumferential winding. Young (40) was the first to consider the foregoing analysis.

Another type of filament-wound vessel consists of windings in the longitudinal, hoop, and helical directions. Here if the thicknesses of the preceding layers are t_L, t_H, and t_α respectively, equilibrium requires that

$$\sigma_H = \frac{\sigma'_H t_H + \sigma_{H\alpha} t_\alpha}{t} \qquad 4.174$$

where σ'_H is the stress in the circumferentially wound layer, $\sigma_{H\alpha}$ is the circumferential component of the stress in the helical layer, and t is the total thickness. Similarly,

$$\sigma_L = \frac{\sigma'_L t_L + \sigma_{L\alpha} t_\alpha}{t} \qquad 4.175$$

where $\sigma_{L\alpha}$ is the longitudinal component of the stress in the helical layer. Using the relationships of Eqs. 4.174 and 4.175, Eqs. 4.170 and 4.171 and noting that

$$t = t_H + t_L + t_\alpha \qquad 4.176$$

$$\sigma_H = (\sigma_0/t)(t_H + t_\alpha \sin^2\alpha) \qquad 4.177$$

$$\sigma_L = (\sigma_0/t)(t_L + t_\alpha \cos^2\alpha) \qquad 4.178$$

where σ_0 is the uniform filament stress. However, from Eqs. 4.170 and 4.171

$$\sigma_H + \sigma_L = \sigma_0(\sin^2\alpha + \cos^2\alpha) = \sigma_0 \qquad 4.179$$

and from Eqs. 4.166 and 4.167

$$t = pd/4\sigma_L \qquad 4.180$$

and

$$\sigma_H = pd/2t \qquad 4.181$$

Equations 4.180 and 4.181 together with Eq. 4.179 then give

$$\sigma_H + \sigma_L = 3\sigma_L = \sigma_0 \tag{4.182}$$

or

$$\sigma_L = \sigma_0/3 \tag{4.183}$$

so that for the ideal vessel

$$t = 3pd/4\sigma_0 \tag{4.184}$$

$$t_L = t/3 - t_\alpha \cos^2 \alpha \tag{4.185}$$

$$t_H = 2t/3 - t_\alpha \sin^2 \alpha \tag{4.186}$$

$$t_\alpha = \frac{2t_L - t_H}{1 - 3\cos^2 \alpha} \tag{4.187}$$

The foregoing analysis applies only to the cylindrical portion of the vessel; the ends of the vessel also have to be considered. These may be chosen from any number of ovaloids or surfaces of revolution for which mathematical analyses are available. In helical winding, the filaments have to pass over the ends of the vessel which adds to the weight, because of overlapping and may reduce the operating efficiency. In order to eliminate such objections so-called non-netting structures have been investigated (16) which do not require crossover of filament windings. Unfortunately, the practical range of such vessels is quite small even though the spectrum of possibilities includes elongated "cigarlike" shapes, oblate ovaloids, toroids, and so on. Toroidal shapes are difficult to manufacture and exhibit instability under pressure because of an effective lower modulus of elasticity in the hoop direction of the vessel. The prolate ovaloid has met with the most success [see also (24)].

The structural efficiency of filament wound vessels may be defined as (16)

$$\text{efficiency} = \frac{W}{V_0 p} \tag{4.188}$$

where W is the weight of the vessel, V_0 the enclosed volume, and p the internal pressure. The volumetric expansion of a vessel is related to the first strain invariant ε (equal to $\varepsilon_1 + \varepsilon_2 + \varepsilon_3$) as follows:

$$\Delta V = V_0(1 + \varepsilon)^3 - V_0 \approx 3\varepsilon V_0 \tag{4.189}$$

With the relationship expressed by Eq. 4.189 the efficiencies of several types of vessels can be compared. Consider a spherical vessel where the principal stress is given by Eq. 3.370.

$$\sigma_1 = \sigma_2 = pR/2t \tag{4.190}$$

where R is the inside radius and t the wall thickness of the sphere. The work done by the pressure is $p(\Delta V/2)$, and the strain energy in the shell material by Eq. 4.190 is $\sigma_1(\varepsilon/2) + \sigma_2(\varepsilon/2)$. Equating gives the result

$$p(\Delta V/2) = [\sigma_1(\varepsilon/2) + \sigma_2(\varepsilon/2)]V = 2\sigma\varepsilon V/2 = \sigma\varepsilon V \tag{4.191}$$

where σ is the strength of the material. Substituting Eq. 4.189 into Eq. 4.191 gives

$$\sigma \varepsilon V = (p/2)(3\varepsilon V_0) \qquad 4.192$$

But V, the volume of the shell material, can be expressed by

$$V = W/\gamma \qquad 4.193$$

where γ is the density of the material. Therefore Eq. 4.192 becomes

$$\sigma \varepsilon W/\gamma = (p/2)(3\varepsilon V_0) \qquad 4.194$$

and the structural efficiency is

$$W/V_0 p = \tfrac{3}{2}(\gamma/\sigma) \qquad 4.195$$

Similarly, for a closed-end cylinder where

$$\sigma_1 = pR/t \qquad 4.196$$

and

$$\sigma_2 = pR/2t \qquad 4.197$$

$$W/V_0 p = 2(\gamma/\sigma) \qquad 4.198$$

For a filament-wound isotensoid vessel where the filaments are all uniformly stressed regardless of position, the optimum filament-wound structure is obtained, and

$$W/V_0 p = 3(\gamma/\sigma) \qquad 4.199$$

Since the quantity $W/(V_0 p)$ should be minimized, it is evident that the latter type of vessel has the least structural efficiency and since only γ/σ can be varied it is also clear that the filament-wound vessel (of the type described by Eq. 4.199) has to have γ/σ at least 33 to 50% lower than exhibited by conventional vessels to be competitive strengthwise. Glass-filament/epoxy systems meet this requirement. For further discussion see (42, 46).

In a completely different type of application, Heggernes (14) has found that in glass-reinforced epoxy leaf springs with up to 97% fiber in the bend direction, the spring is twice as efficient as a steel spring of equal thickness; examples of design methods with such structural elements are found in (14).

Filament overlay composites

Filament winding is also accomplished by laying down a pattern over a base material. For example, Fig. 4.27 shows samples of thin-walled polyethylene pipe overlaid with nylon tire cord. The unreinforced pipe had a bursting pressure of about 400 psi. When nylon tire cord was wrapped over the pipe at about $\frac{1}{4}$-in. intervals the burst pressure of the pipe increased 25% to 500 psi. When a complete overlay was applied, the tube could not be burst at 1000 psi.

Winding over a substrait may be, for example, on thick or thin-walled cylinders. In thick-walled cylinders winding is usually applied in order to put a

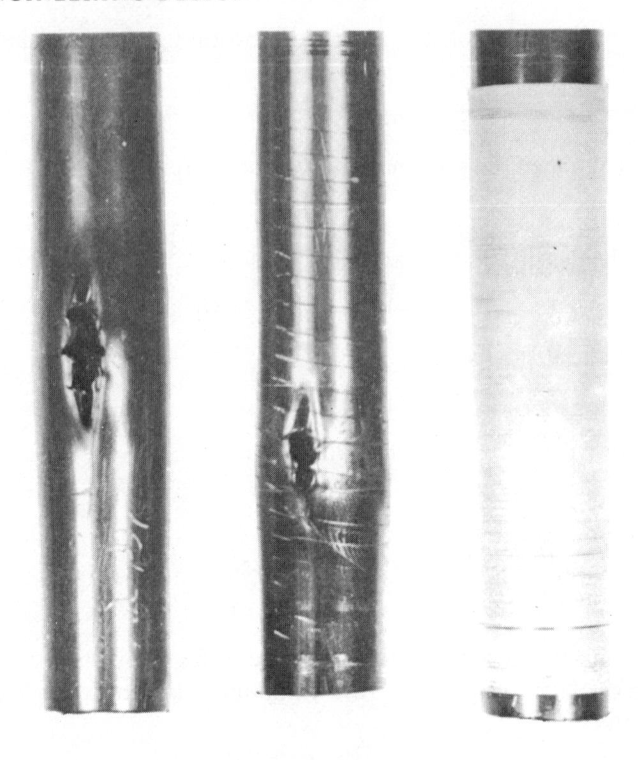

Fig. 4.27 Polyethylene pipe overlaid with nylon tire cord.

compressive stress on the bore of the cylinder as well as to give gross strength; cylinders of this type are considered in Chapter 11. For the purposes of this chapter attention will be directed towards thin-walled structures overlaid with a filament or wire winding. First, consider a thin-walled cylinder wound with wire of the same material as the shell (Fig. 4.28). The stress in the wire is load $(T) \div$ area (uw) or

$$\sigma_{\text{wire}} = T/uw \tag{4.200}$$

For equilibrium over the length of the cylinder,

$$(-\sigma_h)(L)(t) = T(L/w) \tag{4.201}$$

or

$$(\sigma_h)_{\text{cyl}} = -T/wt$$

When internal pressure p is applied, the tension in the wire becomes

$$(\text{tension})_{\text{wire}} = pd/2(t+u) + T/uw \tag{4.202}$$

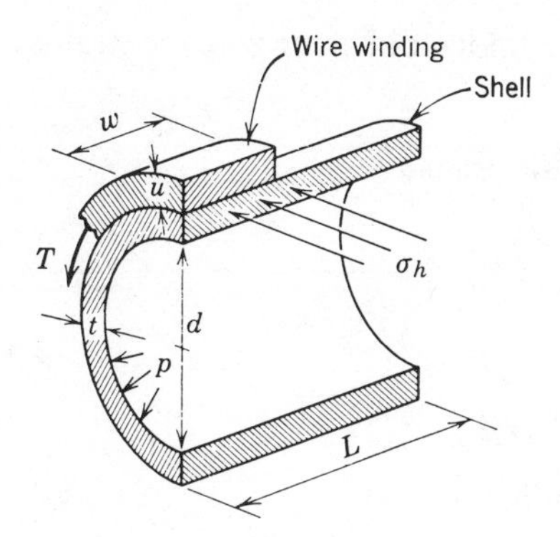

Fig. 4.28 Element of wire-wound cylindrical vessel.

and the tension in the cylinder is

$$(\text{tension})_{\text{cyl}} = pd/2(t+u) - T/wt \qquad 4.203$$

Yielding (plastic flow) of the cylinder occurs at the yield limit of the material; that is, from Eq. 4.201

$$(\sigma_h)_{\text{cyl yield}} = -\,T/wt = -\,\sigma_y \qquad 4.204$$

Thus, for the same winding material, assuming the tension and compression yield limits to be equal,

$$\sigma_{\text{wire}} = T/uw = \sigma_y(wt/uw)$$

or

$$\sigma_{\text{wire}} = \sigma_y(t/u) \qquad 4.205$$

Thus, for the winding to remain elastic, the ratio t/u must be less than unity; that is, the winding has to be heavier than the cylinder it is wound on. However, if the modulus remains the same but the yield strength of the winding material is double that of the cylinder,

$$\sigma_{\text{wire}} = (\sigma_y/2)(t/u) \qquad 4.206$$

and for this case the cylinder can be twice as thick as the winding layer.

Now consider when the winding material is different from that of the cylinder on which it is wound. Equations 4.200 and 4.201 still apply; however, because of the difference in modulus of the materials, corrections have to be applied when internal pressure is added. The mean radius of the cylinder is

$$\bar{r}_{\text{cyl}} = \tfrac{1}{2}(d+t) \qquad 4.207$$

and for the winding

$$\bar{r}_{\text{winding}} = \tfrac{1}{2}(d+u) + t \qquad 4.208$$

The stresses in the cylinder and in the wire layer are

$$\sigma_{\text{cyl}} = \varepsilon_{\text{cyl}} E_{\text{cyl}} \qquad 4.209$$

$$\sigma_{\text{winding}} = \varepsilon_{\text{winding}} E_{\text{winding}} \qquad 4.210$$

For a uniform distribution of stress in the cylinder and in the wire the strains are proportional to the mean radii or

$$\bar{r}_{\text{cyl}}/\bar{r}_{\text{winding}} = \varepsilon_{\text{cyl}}/\varepsilon_{\text{winding}} \qquad 4.211$$

from which

$$\bar{r}_c/\bar{r}_w = \varepsilon_c/\varepsilon_w = \sigma_c E_w/\sigma_w E_c \qquad 4.212$$

and

$$\sigma_c = \sigma_w E_c \bar{r}_c/\bar{r}_w E_w \qquad 4.213$$

$$\sigma_w = \sigma_c E_w \bar{r}_w/\bar{r}_c E_c \qquad 4.214$$

The total load on the cylinder and the winding is

$$\sigma_c(2Lt) + \sigma_w(2Lu) = p\,dL \qquad 4.215$$

Therefore by combining Eq. 4.215 with Eqs. 4.207, 4.208, 4.213, and 4.214,

$$\sigma_c = \frac{pd}{2\left[\dfrac{E_w \bar{r}_w u}{E_c \bar{r}_c} + t\right]} \qquad 4.216$$

and

$$\sigma_w = \frac{pd}{2\left[\dfrac{E_c \bar{r}_c t}{\bar{r}_w E_w} + u\right]} \qquad 4.217$$

The net stresses will be the sums of Eqs. 4.216, 4.201, Eqs. 4.217, 4.200. Further consideration of such vessels is given in Chapter 11.

An example of the use of the preceding theory is the aluminum tube in Fig. 4.29 wound with 0.05-in. diameter steel wire. Assuming that the internal pressure is 300 psi and that the maximum allowable stress in the aluminum is 2000 psi, the wire winding tension and stress in the wire will be calculated.

The mean radius of the aluminum cylinder is

$$\bar{r}_c = \frac{d+t}{2} = \frac{2.8 + 0.1}{2} = 1.450 \text{ in.}$$

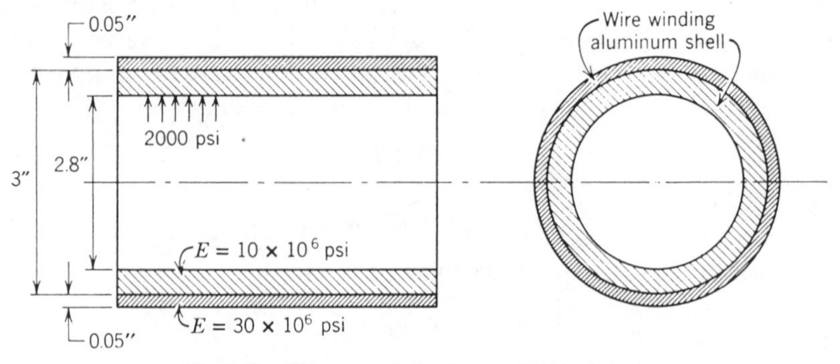

Fig. 4.29 Wire-wound aluminum cylindrical shell.

The mean radius of the winding is

$$\bar{r}_w = \tfrac{1}{2}(d+u) + t = (2.8 + 0.05)/2 + 0.1 = 1.525 \text{ in.}$$

From Eq. 4.212

$$1.45/1.525 = \sigma_c E_w/\sigma_w E_c$$

from which

$$\sigma_c = (1.450\sigma_w/1.525)(E_c/E_w) = 0.317\sigma_w$$

From Eq. 4.215,

$$\sigma_c(2L)(0.10) + \sigma_w(2L)(0.05) = (2000)(2.8)(L)$$

from which

$$\sigma_c = 10{,}830 \text{ psi}$$

The compressive stress in the aluminum cylinder due to the wire winding from Eq. 4.190 is

$$\sigma_c = -T/wt = -T/(0.05)(0.10)$$

The net stress in the cylinder is thus

$$(\sigma_c)_{\text{net}} = 2000 = 10{,}830 - T(200),$$

from which the winding tension is

$$T = 44.15 \text{ lb}$$

The area of the wire is $\pi d^2/4 = 3.14(0.05)^2/4 = 0.001965 \text{ in.}^2$ so that the stress in the wire is

$$\sigma_w = 44.15/0.001965 = 22{,}468 \text{ psi}$$

4-3 LAMINATED AND SANDWICH COMPOSITES

In Section 4-1 only relatively simple composites were considered. In this section consideration will be given to composites made by laminating layers of material together at various orientations with respect to the orthotropic properties and to composites popularly known as *sandwich* (5, 6, 8, 10, 39, 43, 44, 50).

Laminated composites

This type of composite is made by laying down in successive layers sheets of material in a particular pattern. For example, if the material has longitudinal and transverse properties, successive layers of this material might be laid down so that the L or T directions were always oriented at a different angle; the second layer has L at $\alpha°$ with respect to L of the first layer, the third layer at $2\alpha°$, and so on; or the second layer may be at $\alpha°$ and the third layer at $0°$, with the L directions parallel in the two layers. There are an infinite number of com-

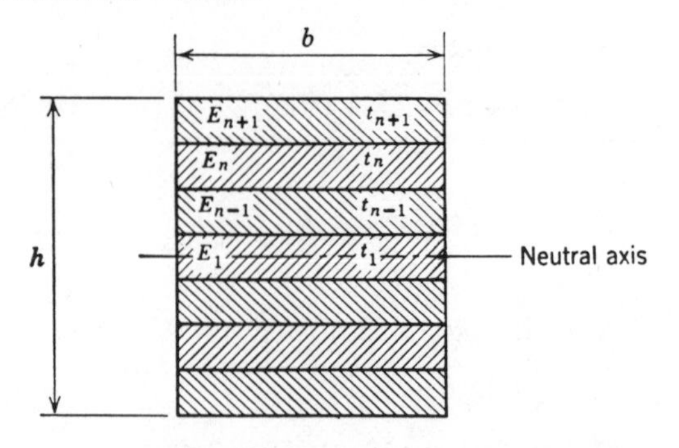

Fig. 4.30 Cross section of multimaterial laminated beam.

binations. Furthermore, the layers may be of the same material, only oriented in different directions, or different materials altogether. First, consider a composite such as shown in Fig. 4.30, consisting of several layers of different materials arranged symmetrically about the centroid axis. For this particular composite, as shown in Section 4-1, the flexural rigidity is

$$EI = \sum E_i I_i \tag{4.218}$$

which when expanded gives the result

$$\begin{aligned} EI = \frac{b}{12}\{ & t_1{}^3 E_1 + 2E_2[t_2{}^3 + 3t_2(t_1 + t_2)^2] \\ & + 2E_{n-1}[t_{n-1}{}^3 + 3t_{n-1}(t_1 + 2t_2 + t_{n-1})^2] \\ & + 2E_n[t_n{}^3 + 3t_n(t_1 + 2t_2 + 2t_{n-1} + t_n)^2] \\ & + 2E_{n+1}[t_{n+1}{}^3 + 3t_{n+1}(t_1 + 2t_2 + 2t_{n-1} + 2t_n + t_{n+1})^2]\} \end{aligned} \tag{4.219}$$

Equation 4.219 may also be written as (17):

$$\begin{aligned} EI = \frac{b}{12}\{ & (2n_1 + 1)t_1{}^3 E_1 + 2n_2 t_2{}^3 E_2 + (t_1 + t_2)^2 \\ & \times [4n_1(n_1 + 1)(2n_1 + 1)t_1 E_1 + (2n_2 - 1)(2n_2)(2n_2 + 1)(t_2 E_2)]\} \end{aligned} \tag{4.220}$$

where n_1 is the number of layers of material E_1 above the centroid axis and n_2 is the number of layers of material E_2 above the axis. The central layer (of material E_1) is not counted; thus the total number of layers is always odd. For example, in a seven-layer laminate, one layer, of material E_1 would be at the center; next would be a layer E_2 followed by a layer E_1 and an outside layer E_2; for this case $n_1 = 1$ and $n_2 = 2$. If the various layers of material have the same thickness, the problem is simplified considerably and the following expressions can be derived from Eq. 4.219 or Eq. 4.220.

System of three layers (Fig. 4.31*A*)

$$EI = (bt^3/12)(E_1 + 26E_2) \tag{4.221}$$

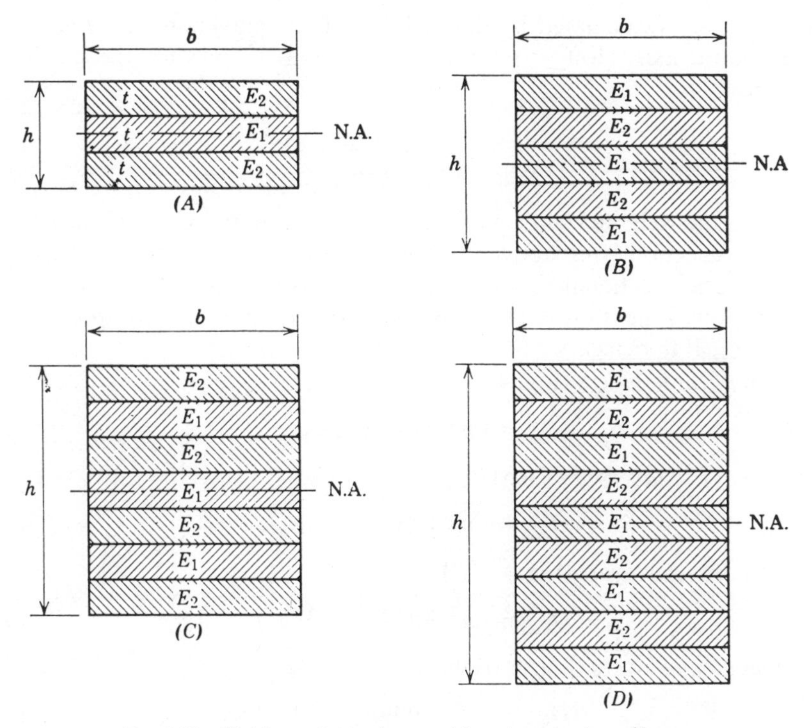

Fig. 4.31 Multimaterial laminates with various number of layers.

System of five layers (Fig. 4.31*B*)

$$EI = (bt^3/12)(99E_1 + 26E_2) \tag{4.222}$$

System of seven layers (Fig. 4.31*C*)

$$EI = (bt^3/12)(99E_1 + 244E_2) \tag{4.223}$$

System of nine layers (Fig. 4.31*D*)

$$EI = (bt^3/12)(485E_1 + 244E_2) \tag{4.224}$$

In order to calculate stresses in such beams it is necessary only to use the standard formulas developed in Chapter 2, with the overall value of EI. That is, the bending stress is given by

$$\sigma = M/Z = McE_c/EI \tag{4.225}$$

where M is the applied bending moment, Z the section modulus (I/c), c the distance from the neutral axis to the location where σ is to be calculated, E_c the modulus of the material at distance c from the neutral axis, and EI the overall flexural rigidity. The maximum stress, according to Eq. 4.225, is not necessarily where c is maximum but where $(E_c)c$ is maximum; thus the maximum stress could be anywhere in the laminate.

Shear stress is calculated by use of Eq. 4.31; its maximum value is always at the neutral axis. Hoff (17) rewrites Eq. 4.31 as follows for the maximum shear stress:

$$\tau_{max} = \frac{SQ'_{max}}{EIb} \tag{4.226}$$

where Q' is the weighted static moment with respect to the x-axis of the portion of the cross section lying above the horizontal line. Q' is calculated by multiplying the cross-sectional area of each layer by its modulus. Thus, for the four cases previously mentioned, Q'_{max} has the following values for composites with layers of equal thickness.

System of three layers (Fig. 4.31*A*)

$$Q'_{max} = (bt^2/8)(E_1 + 8E_2) \tag{4.227}$$

System of five layers (Fig. 4.31*B*)

$$Q'_{max} = (bt^2/8)(17E_1 + 8E_2) \tag{4.228}$$

System of seven layers (Fig. 4.31*C*)

$$Q'_{max} = (bt^2/8)(17E_1 + 32E_2) \tag{4.229}$$

System of nine layers (Fig. 4.31*D*)

$$Q'_{max} = (bt^2/8)(49E_1 + 32E_2) \tag{4.230}$$

As with the two-ply laminate mentioned in Section 4-1, deflection may be calculated by using the standard formulas developed in Chapter 2.

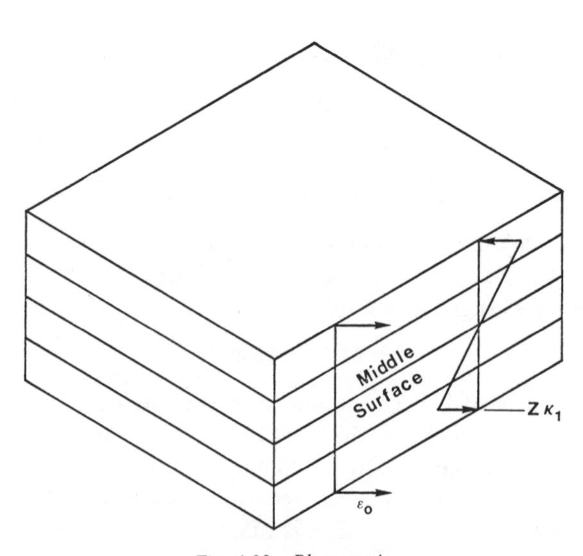

Fig. 4.32 Plate strain.

Laminated plates. In Figure 3.67 a plate element is shown with edge force N_x, N_y, N_{xy} and moments M_x, M_y, M_{xy} applied in a case of plane stress. The strain for a laminated plate element (Figure 4.32) is

$$\varepsilon_x = \varepsilon_1 = \varepsilon_1{}^0 + Z\kappa_1 \quad 4.231$$

$$\varepsilon_y = \varepsilon_2 = \varepsilon_2{}^0 + Z\kappa_2 \quad 4.232$$

$$\gamma_{xy} = \gamma_{12} = \gamma_{12}{}^0 + Z\kappa_{12} \quad 4.233$$

$\varepsilon_1{}^0, \varepsilon_2{}^0, \gamma_{12}{}^0$ are deformations of the middle surface of the plate. $\kappa_1, \kappa_2, \kappa_{12}$ are rotation of a vertical plane about a point in the middle surface and are expressed as

$$\kappa_1 = -\frac{\partial^2 w}{\partial x^2} \quad 4.234$$

$$\kappa_2 = -\frac{\partial^2 w}{\partial y^2} \quad 4.235$$

$$\kappa_{12} = -2\frac{\partial^2 w}{\partial x\,\partial y} \quad 4.236$$

These expressions will be used for the plate deflections later.

Eqs. 4.231, 4.232, 4.233 are substituted into Eqs. 4.94, 4.95, and 4.96 resulting in the following equations:

$$\sigma_x = \sigma_1 = C_{11}\varepsilon_1{}^0 + C_{12}\varepsilon_2{}^0 + C_{16}\gamma_{12}{}^0 + C_{11}Z\kappa_1 + C_{12}Z\kappa_2 + C_{16}Z\kappa_{12} \quad 4.237$$

$$\sigma_y = \sigma_2 = C_{21}\varepsilon_1{}^0 + C_{22}\varepsilon_2{}^0 + C_{26}\gamma_{12}{}^0 + C_{21}Z\kappa_1 + C_{22}Z\kappa_2 + C_{26}Z\kappa_{12} \quad 4.238$$

$$\tau_{xy} = \tau_{12} = C_{61}\varepsilon_1{}^0 + C_{62}\varepsilon_2{}^0 + C_{66}\gamma_{12}{}^0 + C_{61}Z\kappa_1 + C_{62}Z\kappa_2 + C_{66}Z\kappa_{12} \quad 4.239$$

The definitions for $N_x, N_y, N_{xy}, M_x, M_y, M_{xy}$ are the following:

$$N_x = \int_{-h/2}^{h/2} \sigma_x \, dZ \quad 4.240$$

$$N_y = \int_{-h/2}^{h/2} \sigma_y \, dZ \quad 4.241$$

$$N_{xy} = \int_{-h/2}^{h/2} \tau_{xy} \, dZ \quad 4.242$$

$$M_x = \int_{-h/2}^{h/2} \sigma_x Z \, dZ \quad 4.243$$

$$M_y = \int_{-h/2}^{h/2} \sigma_y Z \, dZ \quad 4.244$$

$$M_{xy} = \int_{-h/2}^{h/2} \tau_{xy} Z \, dZ \quad 4.245$$

Equations 4.237, 4.238, and 4.239 for a single laminate oriented at some angle shown in Figures 4.14, 4.15, and 4.16 are substituted into Eqs. 4.240 to 4.245. The equations must be integrated in terms of thickness of the plates (1, 6, 12, 25, 33, and 48) since $\sigma_x, \sigma_y, \sigma_z$, and τ_{xy} represent discontinuous functions as integration passes from laminate to laminate. Therefore, the integral reduces to a summation over the entire area.

$$N_x = A_{11}\varepsilon_1{}^0 + A_{12}\varepsilon_2{}^0 + A_{16}\gamma_{12}{}^0 + B_{11}\kappa_1 + B_{12}\kappa_2 + B_{16}\kappa_{12} \quad 4.246$$

$$N_y = A_{21}\varepsilon_1{}^0 + A_{22}\varepsilon_2{}^0 + A_{26}\gamma_{12}{}^0 + B_{21}\kappa_1 + B_{22}\kappa_2 + B_{26}\kappa_{12} \quad 4.247$$

$$N_{xy} = A_{61}\varepsilon_1{}^0 + A_{62}\varepsilon_2{}^0 + A_{66}\gamma_{12}{}^0 + B_{61}\kappa_1 + B_{62}\kappa_2 + B_{66}\kappa_{12} \quad 4.248$$

$$M_x = B_{11}\varepsilon_1{}^0 + B_{12}\varepsilon_2{}^0 + B_{16}\gamma_{12}{}^0 + D_{11}\kappa_1 + D_{12}\kappa_2 + D_{16}\kappa_{12} \quad 4.249$$

$$M_y = B_{21}\varepsilon_1{}^0 + B_{22}\varepsilon_2{}^0 + B_{26}\gamma_{12}{}^0 + D_{21}\kappa_1 + D_{22}\kappa_2 + D_{26}\kappa_{12} \quad 4.250$$

$$M_{xy} = B_{61}\varepsilon_1{}^0 + B_{62}\varepsilon_2{}^0 + B_{66}\gamma_{12}{}^0 + D_{61}\kappa_1 + D_{62}\kappa_2 + D_{66}\kappa_{12} \quad 4.251$$

$$A_{ij} = \sum_{K=1}^{K=n} (C_{ij})_K (h_K - h_{K-1}) \quad 4.252$$

$$B_{ij} = \frac{1}{2} \sum_{K=1}^{K=n} (C_{ij})_K (h_K{}^2 - h_{K-1}{}^2) \quad 4.253$$

$$D_{ij} = \frac{1}{3} \sum_{K=1}^{K=n} (C_{ij})_K (h_K{}^3 - h_{K-1}{}^3) \quad 4.254$$

The 2 and 3 in Eqs. 4.253 and 4.254 come from integration of Z and Z^2. A sample of the ordering of h_K is shown in Fig. 4.33. The equations also depend on symmetric lamination about the middle surface.

An example will be presented to demonstrate Eqs. 2.246 through 4.254. A composite plate, Figs. 4.34 and 4.35, is built up from a graphite layer 0.010-in. thick with the following properties (from Table 4.4):

$$E_{11} = E_L = 44 \times 10^6 \text{ psi} \qquad E_{22} = E_T = 1 \times 10^6 \text{ psi}$$

$$G_{12} = G_{LT} = 0.95 \times 10^6 \text{ psi} \qquad \nu_{LT} \simeq 0.25$$

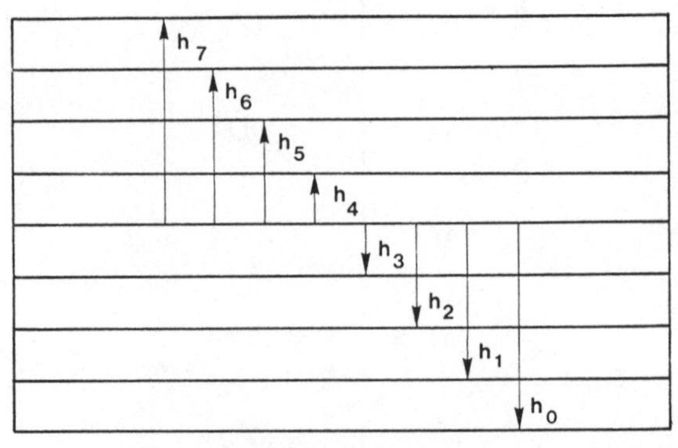

Fig. 4.33 Plate summation convention.

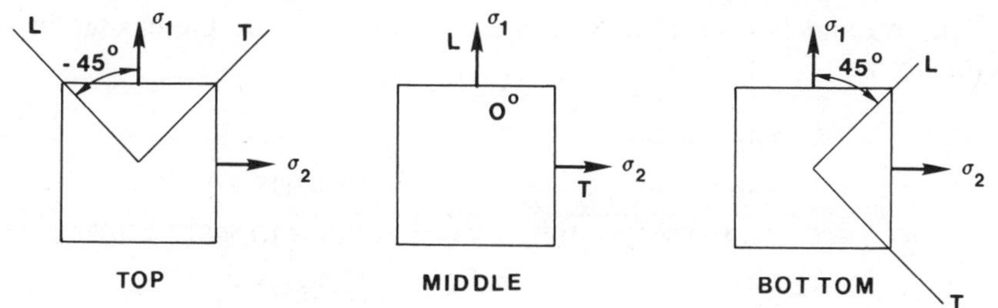

Fig. 4.34 Laminated plate layers.

The first to be calculated are the C_{ij} values Eqs. 4.100 to 4.105. Using Eq. 4.83 to solve for ν_{TL}, the value for C in Eq. 4.112 is 0.9986.

Middle layer:

$\alpha = 0°$, the C_{ij} values reduce to the following by inspection.

$$C_{11} = \frac{E_L}{1 - \nu_{TL}\nu_{LT}} = \frac{44 \times 10^6}{0.9986}$$
$$= 44.06169 \times 10^6 \text{ lb/in.}^2$$

$$C_{21} = C_{12} = \frac{\nu_{LT}E_T}{1 - \nu_{TL}\nu_{LT}} = \frac{0.25(1 \times 10^6)}{0.9986}$$

$$= 0.25035 \times 10^6 \text{ lb/in.}^2$$

$$C_{16} = C_{61} = 0$$

$$C_{22} = \frac{E_T}{1 - \nu_{TL}\nu_{LT}} = \frac{1 \times 10^6}{0.9986}$$
$$= 1.00140 \times 10^6 \text{ lb/in.}^2$$

$$C_{26} = C_{62} = 0$$

$$C_{66} = G_{LT} = 0.95 \times 10^6 \text{ lb/in.}^2$$

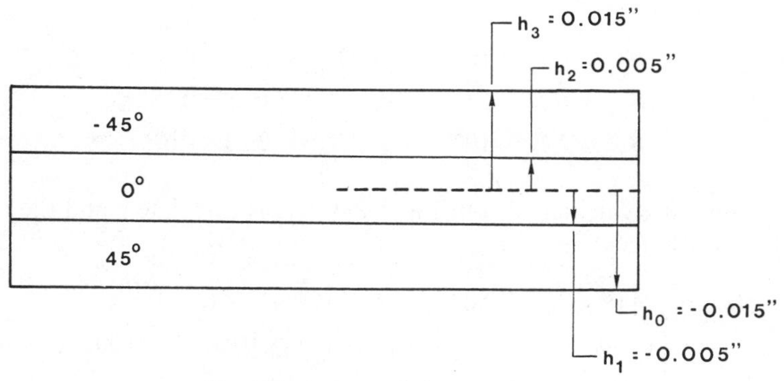

Fig. 4.35 Laminated plate cross section.

The top and bottom layer values are as follows (note the change in sign is due to the $-45°$):

Top Layer	*Bottom Layer*
$C_{11} = 12.34095 \times 10^6$ psi	$C_{11} = 12.34095 \times 10^6$ psi
$C_{12} = C_{21} = 10.56612 \times 10^6$	$C_{12} = C_{21} = 10.56612 \times 10^6$
$C_{22} = 12.34095 \times 10^6$	$C_{22} = 12.34095 \times 10^6$
$C_{16} = C_{61} = -10.76508 \times 10^6$	$C_{16} = C_{61} = 10.76508 \times 10^6$
$C_{26} = C_{62} = -10.71317 \times 10^6$	$C_{66} = C_{62} = 10.71317 \times 10^6$
$C_{66} = 11.1406 \times 10^6$	$C_{66} = 11.1406 \times 10^6$

The A_{ij} values are calculated from Eq. 4.252 and note h_K values have plus and minus signs with a summation over three plates. As an example:

$$A_{11} = (C_{11})_B[(-0.005) - (-0.0015)] + (C_{11})_m[(0.005) - (-0.005)] + (C_{11})_T[0.015 - 0.005]$$

$$A_{11} = 10.87526 \times 10^6 \text{ lb/in.}$$

The rest of the values are stated as follows:

$$A_{11} = 10.87526 \times 10^6 \qquad A_{16} = A_{61} = 0$$
$$A_{12} = A_{21} = 0.21383 \times 10^6 \qquad A_{26} = A_{62} = 0$$
$$A_{22} = 0.25683 \times 10^6 \qquad A_{66} = 0.23231 \times 10^6$$

The B_{ij} values are calculated from Eq. 4.253. B_{11} is calculated, and the rest of the values are stated.

$$B_{11} = \tfrac{1}{2}[(C_{11})_B[(-0.015)^2 - (-0.005)^2] + (C_{11})_m [(+0.005)^2 - (-0.005)^2] + (C_{11})_T(0.015^2 - 0.005^2)]$$
$$= 0.00247 \times 10^6 \text{ in.-lb/in.}$$

$$B_{11} = 2.47 \times 10^3 \qquad B_{16} = B_{61} = 0$$
$$B_{12} = B_{21} = 2.11 \times 10^3 \qquad B_{26} = B_{62} = 0$$
$$B_{22} = 2.47 \times 10^3 \qquad B_{66} = 2.23 \times 10^3$$

The D_{ij} values are calculated from Eq. 4.254. D_{11} is calculated, and the rest of the values are stated.

$$D_{11} = \tfrac{1}{3}\{[(C_{11})_B[(-0.005)^3 - (-0.015)^3] + (C_{11})_m [(+0.005)^3 - (-0.005)^3] + (C_{11})_T[0.015^3 - 0.005^3]\}$$

$$D_{11} = 30.41053 \text{ in.-lb/in.}$$

$$D_{11} = 30.41053 \qquad D_{16} = D_{61} = 0$$
$$D_{12} = D_{21} = 22.91412 \qquad D_{26} = D_{62} = 0$$
$$D_{22} = 26.82206 \qquad D_{66} = 24.21713$$

Eqs. 4.246 to 4.251 appear as follows:

$$\begin{bmatrix} N_x \\ N_y \\ N_{xy} \\ M_x \\ M_y \\ M_{xy} \end{bmatrix} = \left[\begin{array}{ccc|ccc} 10.88 \times 10^6 & 0.22 \times 10^6 & 0 & 2.47 \times 10^3 & 2.11 \times 10^3 & 0 \\ 0.22 \times 10^6 & 0.26 \times 10^6 & 0 & 2.11 \times 10^3 & 2.47 \times 10^3 & 0 \\ 0 & 0 & 0.23 \times 10^6 & 0 & 0 & 2.23 \times 10^3 \\ \hline 2.47 \times 10^3 & 2.11 \times 10^3 & 0 & 30.41 & 22.91 & 0 \\ 2.11 \times 10^3 & 2.47 \times 10^3 & 0 & 22.91 & 26.82 & 0 \\ 0 & 0 & 2.23 \times 10^3 & 0 & 0 & 24.22 \end{array}\right] \begin{bmatrix} \varepsilon_1^0 \\ \varepsilon_2^0 \\ \gamma_{12}^0 \\ \kappa_1 \\ \kappa_2 \\ \kappa_{12} \end{bmatrix} \qquad 4.255$$

It should be noted that the units on N_x, N_y, and N_{xy} are lb/in., and those on the moments M_x, M_y, and M_{xy} are in.-lb/in.

When analyzing a loaded plate, the stresses in the L and T direction for each layer must be checked. The following steps must be taken:

1. Given any or all of the loads in Eq. 4.255, take the inverse and solve for $\varepsilon_1{}^0, \varepsilon_2{}^0, \gamma_{12}{}^0, \kappa_1, \kappa_2, \kappa_{12}$.
2. Substitute the strains into Eqs. 4.231 to 4.233 noting the value Z changes for each layer, and solve for $\varepsilon_1, \varepsilon_2, \gamma_{12}$.
3. Next, for each layer substitute $\varepsilon_1, \varepsilon_2, \gamma_{12}$ into Eqs. 4.63 to 4.65 and solve for $\varepsilon_L, \varepsilon_T, \gamma_{LT}$.
4. Place ε_L, ε_T, γ_{LT} into Eqs. 4.69 to 4.71 and solve for σ_L, σ_T, τ_{LT} in an individual layer.
5. The failure criteria for each layer can be checked using Eq. 4.130. Failure of any layer would mean failure of the laminated plate.

As a second example consider the laminated cylinder shown in Fig. 4.36. The laminates are made of "balanced" orthotropic material; that is, the properties in the L and T directions are the same. Also the lamina are laid at alternate 30° angles in left-hand and right-hand spirals. Geometry and properties are

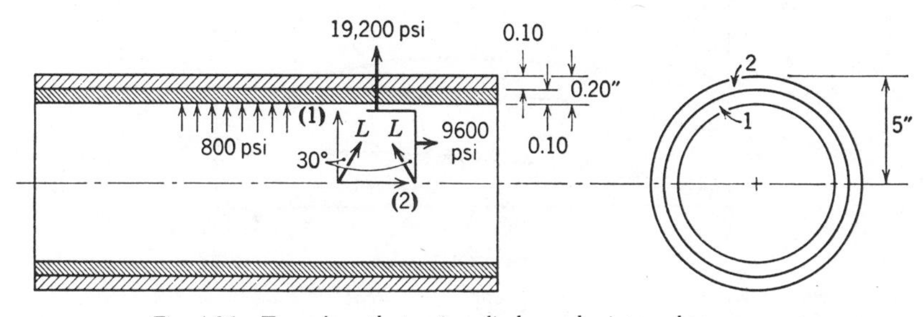

Fig. 4.36 Two-ply orthotropic cylinder under internal pressure.

as follows:

$$\text{thickness of each layer} = 0.10 \text{ in.}$$
$$\text{external radius} = 5.00 \text{ in.}$$
$$\text{total wall thickness} = 0.20 \text{ in.}$$
$$\text{internal pressure} = 800 \text{ psi}$$

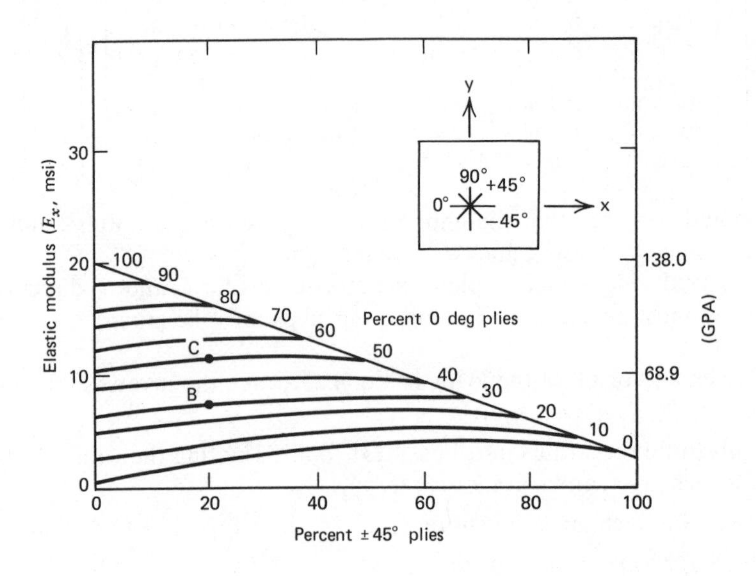

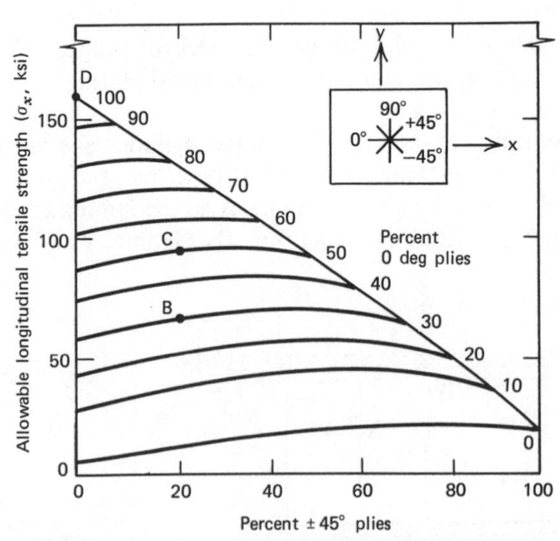

Fig. 4.37 High-strength carbon-fiber/epoxy laminate properties at room temperature. [After Kliger (19), courtesy of Industry Media Inc.]

$$E_L = E_T = 3 \times 10^6 \text{ psi}$$
$$\nu_{LT} = \nu_{TL} = 0.20$$
$$G_{LT} = G_{TL} = 5 \times 10^5 \text{ psi}$$
$$\sigma_1 = pr/t = 800(4.8)/0.20 = 19{,}200 \text{ psi}$$
$$\sigma_2 = pr/2t = \sigma_1/2 = 9600 \text{ psi}$$

With this information and following the procedure outlined previously one can

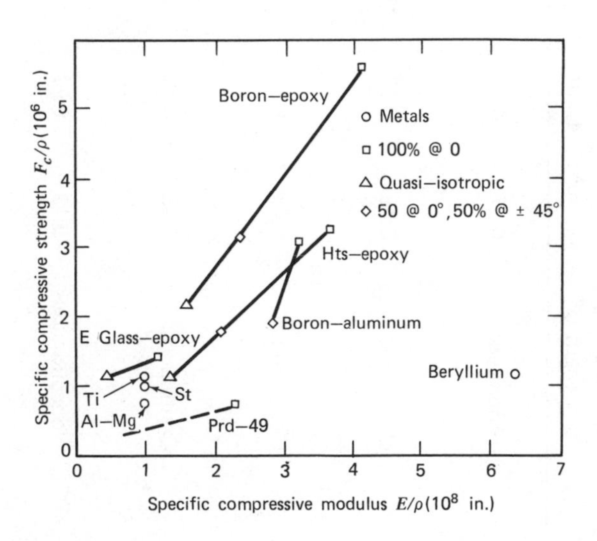

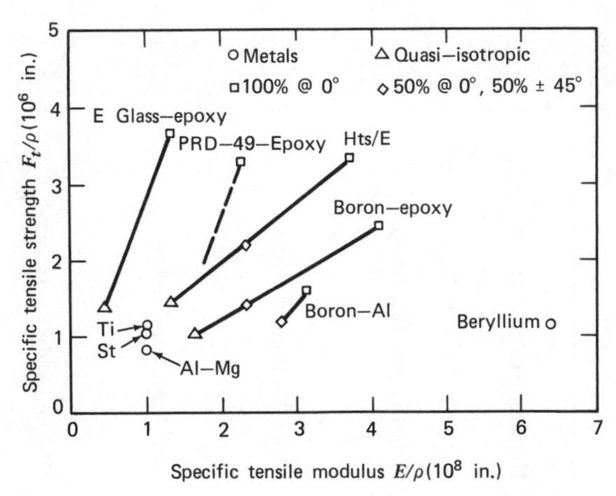

Fig. 4.38 Property comparisons of composites to metals [After Parmley (27), courtesy of Academic Press, Inc.]

Table 4.7 Composite Efficiency of Reinforced Plastics (After Bell (3), Courtesy of the American Rocket Society)

System	Fiber Orientation	Fiber Length	Total Fiber Area, A_{ft}, in.^{2}	Matrix Area, A_m, in.^{2}	Theoretical Strength, lb Eq. 4.134	Actual Strength, lb	Composite Efficiency, Eq. 4.256 (%)
Filament wound fibers		Continuous	0.77	0.23	310,000	180,000	58
Cross-laminated fibers		Continuous	0.48	0.52	197,000	72,500	37
Cloth-laminated fibers		Continuous	0.48	0.52	197,000	43,000	22
Mat-laminated fibers	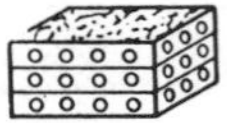	Continuous	0.48	0.52	197,000	57,200	29
Chopped-fiber systems		Noncontinuous	0.13	0.87	60,700	15,000	25
Glass-flake composites	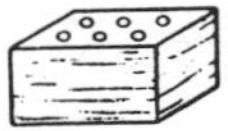	Noncontinuous	0.70	0.30	165,500	20,000	12

Table 4.8 Fiber Efficiency of Reinforced Plastics (After Bell (3), Courtesy of the American Rocket Society)

System	Fiber Orientation	Fiber Length	Area of Fiber, A_f, in.^{2}	(Load on Fiber) ÷ (Load on Composite)	Test Strength, lb	Effective Fiber Stress, numerator of Eq. 4.257	Fiber Efficiency, Eq. 4.257 (%)
Filament wound fibers	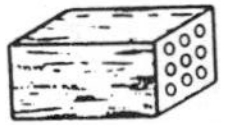	Continuous	0.77	1.00	180,000	234,000	59
Cross-laminated fibers		Continuous	0.24	0.90	72,500	272,000	68
Cloth-laminated fibers	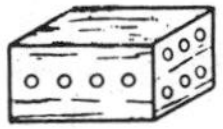	Continuous	0.24	0.90	43,000	161,000	41
Mat-laminated fibers	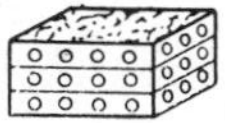	Continuous	0.20	0.88	57,200	252,000	63
Chopped-fiber systems	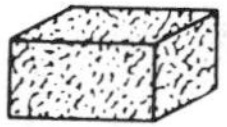	Noncontinuous	0.43	0.38	15,000	132,000	33
Glass flake composites	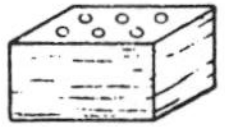	Noncontinuous	0.70	1.00	20,000	28,300	7

show that

$$E_{1(1)} = E_{1(2)} = E_{2(1)} = E_{2(2)} = E_{30} = E_{60} = 1.78 \times 10^6 \text{ psi}$$
$$\nu_{12(1)} = \nu_{12(2)} = \nu_{21(1)} = \nu_{21(2)} = \nu_{30} = \nu_{60} = 0.523$$
$$G_{12(1)} = G_{12(2)} = G_{21(1)} = G_{21(2)} = 9.1 \times 10^5 \text{ psi}$$
$$\sigma_{1(1)} = \sigma_{1(2)} = \sigma_1 = 19{,}200 \text{ psi}$$
$$\sigma_{2(1)} = \sigma_{2(2)} = \sigma_2 = 9600 \text{ psi}$$
$$\tau_{12(1)} = \tau_{12(2)} = 6750 \text{ psi}$$

Note that when the lamina are oriented in this particular way, shear stresses are induced.

Filament geometry. When dealing with laminates (19), design curves, Fig. 4.37*A* and *B*, can be developed. Design point *B* represents a stacking of 20% at $\pm 45°$, 30% at 0°, 50% at 90° giving $E_x = 7$ msi and $\sigma_x = 60$ ksi. From the design curve E_y and σ_y can be found by reversing x–y notation so point *C* is developed with stacking 20% at $\pm 45°$, 50% at 0°, and 30% at 90° giving $E_y = 11$ msi and $\sigma_y = 90$ ksi.

The advantages of filament reinforcing can be further appreciated in Fig. 4.38 where some of the typical values have been compared. The values for the density may be found in Tables 4.3 and 4.4.

The effect of glass-fiber geometry on composite material strength has been investigated by Bell (3). In deriving an equation for load and stress distribution, he found that the highest composite efficiency was 58% for filament wound fibers and the highest fiber efficiency was 68% for cross-laminated fibers (Tables 4.7 and 4.8). His equations for efficiency are

$$\text{composite efficiency} = \frac{\text{test strength of composite}}{\text{theoretical strength}}(100) \qquad 4.256$$

$$\text{fiber efficiency} = \frac{\text{effective fiber stress}}{\text{base fiber strength}}(100)$$

or

$$\text{fiber efficiency} = \frac{\left(\dfrac{\text{test strength of composite}}{\text{area of fiber}}\right)\left(\dfrac{\text{load on fiber}}{\text{load on composite}}\right)}{\text{base fiber strength}}(100) \qquad 4.257$$

4-4 SANDWICH AND HONEYCOMB STRUCTURES

Sandwich construction, of which honeycomb is a special case, has become very popular for applications requiring high strength-weight ratio combined with rigidity. The usual construction consists of high density faces or skins separated by a relatively thick layer of a low density material like plastic foam or honeycomb of various types. Typical constructions are shown in Fig. 4.39. Because

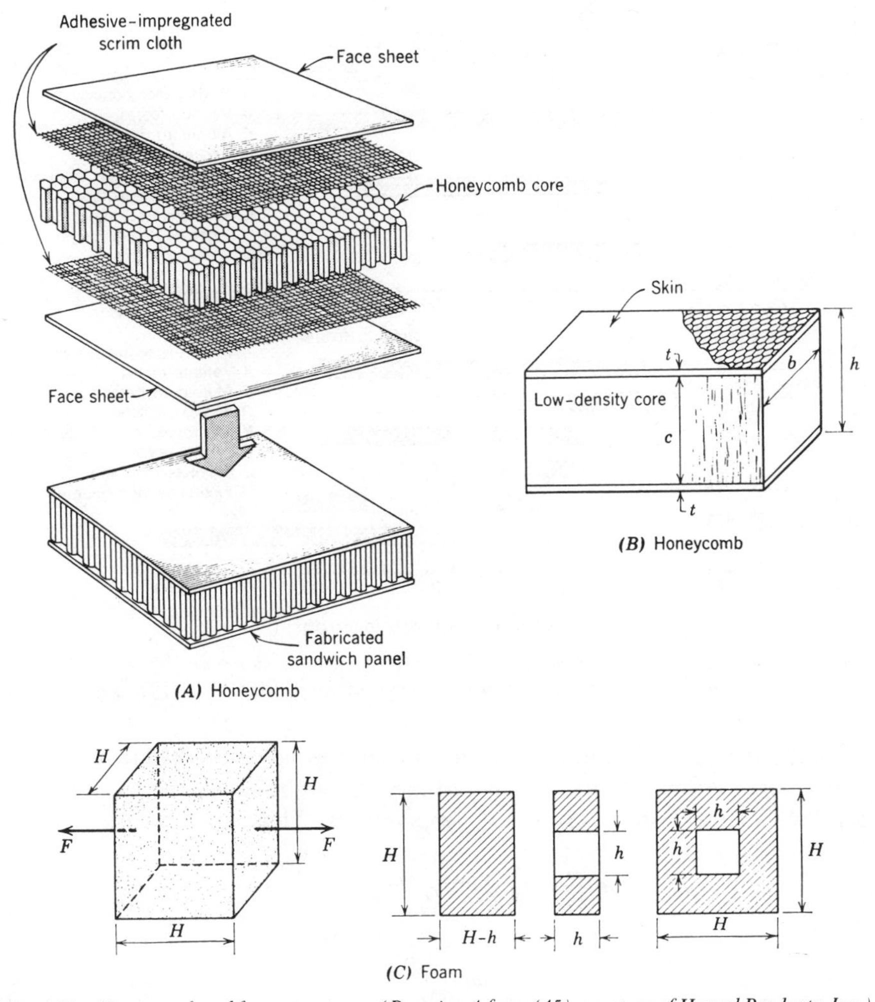

Fig. 4.39 Honeycomb and foam structures. (Drawing A from (45), courtesy of Hexcel Products, Inc.)

of the nature of the construction, sandwich panels act somewhat like I beams; the facing correspond to flanges and carry the tensile and compressive stresses, whereas the core corresponds to the web of an I beam which carries the shear and helps to prevent buckling and wrinkling of the faces. For design applications this type of structure offers almost unlimited possibilities since the properties of the component parts (as well as geometry) can be varied over a wide range. For example, note the range of properties of various core materials as shown in Fig. 4.40. The face sheets and the honeycomb core may be made of cardboard, metal, or even reinforced plastics. In high temperature work, the matrix is a metal with the fibers reinforcing it.

Section properties. In designing with sandwich-type beams and plates the formulas for stress and strain developed in Chapters 2 and 3 can be applied

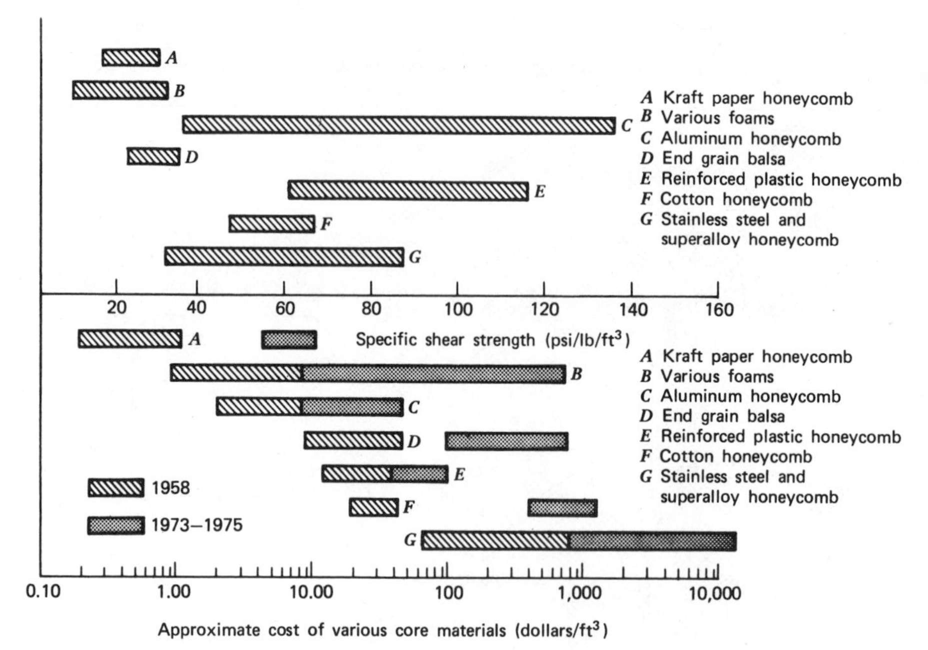

Fig. 4.40 Comparative strengths and costs of honeycomb and other materials. [From (45), courtesy of Hexcel Products, Inc., (1973–1975 price ranges A, C, E (51))].

provided the correct values for section properties are used. Referring to Fig. 4.39, where t is small, I, the moment of inertia, can be calculated by use of the parallel-axis theorem ($I = I_1 + Ad^2$, Appendix B). Thus, for this case

$$I = 2(\tfrac{1}{12}bt^3) + 2bt\left(\frac{c+t}{2}\right)^2 \qquad 4.258$$

or, neglecting powers of t,

$$I = btc(t + c/2) \qquad 4.259$$

The flexural rigidity is thus EI, where E is the modulus of elasticity of the facing material and I is defined by Eq. 4.259. For plates, the flexural rigidity D (Eq. 3.309) of the facings is required. Since the core material is assumed to provide no stiffness to the structure, Eq. 3.296 becomes

$$D = \frac{E}{1 - \nu^2}\left(\int_{-h/2}^{h/2} h^2\,dh - \int_{-c/2}^{c/2} c^2\,dc\right)$$

or

$$D = \frac{E}{12(1 - \nu^2)}(h^3 - c^3) \qquad 4.260$$

where E is the modulus of the facing material, ν its Poisson ratio, and h and c are defined in Fig. 4.39.

Also, from (48)

$$D = \frac{E^1 th^2}{2(1 - \nu_{12}\nu_{21})} \qquad 4.261$$

where E^1 is the effective E of the cross section.

In many cases of sandwich construction $c/t > 5$; therefore Eq. 4.260 can be rewritten as

$$D = \frac{Et(h + c)^2}{8(1 - \nu^2)} \qquad 4.262$$

For an example of stiffness calculations, the graphite face sheets used in the example in Fig. 4.34 will be used. A 1-in. core will be used with the face sheets on the top and bottom as shown in Fig. 4.41. The core stiffness will be neglected. D_{22} will be calculated from Eq. 4.254 and the other values stated.

$$\begin{aligned} D_{22} = \tfrac{1}{3}[(C_{22_1})[(-0.520)^3 - (-0.530)^3] + (C_{22_2}) \\ [(-0.510)^3 - (-0.520)^3] + (C_{22_3})[(-0.5)^3 - (-0.510)^3] \\ + (C_{22_4})[(0.5)^3 - (-0.5)^3] + (C_{22_5})[0.51^3 - 0.50^3] \\ + (C_{22_6})[0.52^3 - 0.51^3] + (C_{22_7})[0.53^3 - 0.52^3] \\ = 1.363 \times 10^5 \end{aligned}$$

$$D_{11} = 3.647 \times 10^5 \qquad D_{61} = D_{16} = -0.045 \times 10^5$$
$$D_{12} = D_{21} = 1.135 \times 10^5 \qquad D_{16} = D_{61} = D_{26} = D_{62}$$
$$D_{22} = 1.363 \times 10^5 \qquad D_{66} = 1.233 \times 10^5$$

The values of D_{26}, D_{62}, D_{61}, and D_{16} are small when compared to the other values.

A comparison of the stiffness to aluminum face sheets can be made first

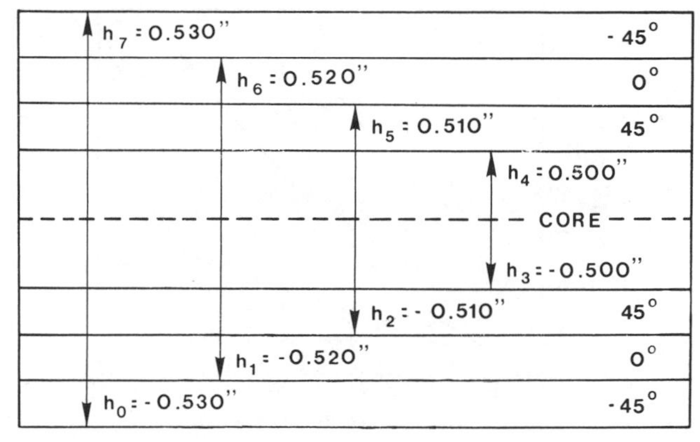

Fig. 4.41 Laminated sandwich plate.

from Eq. 4.260 then from Eq. 4.261. The E is 10×10^6 lb/in.^{2}, t is 0.03 in., C is 1 in., h is 1.6 in., and v is $\frac{1}{3}$. From Eq. 4.260

$$D = 1.791 \times 10^5$$

From Eq. 4.261

$$D = 1.896 \times 10^5$$

Properties of materials. In addition to the section properties just described it is also necessary to have information on certain physical and mechanical properties of the materials. In particular, such properties as elastic and shear moduli are required. The modulus of elasticity of the facing material of a sandwich structure is usually available, but frequently the shear modulus of the core material has to be determined; Jones and Hersch (18) suggest the following formula for determining the shear modulus.

$$G_{\text{core}} = \frac{1.5PLc}{(h+c)^2 b(11\delta_4 - 8\delta_2)} \qquad 4.263$$

where δ_4 and δ_2 are deflections at quarter-span and midspan respectively in a test of a beam of width b in quarter-point loading by a force P over a support span L. If the material is isotropic and the value for modulus of elasticity is available, the shear modulus can be obtained from Eq. 1.13

$$G = \frac{E}{2(1+v)} \qquad 4.264$$

Typical shear modulus data for aluminum honeycomb are shown in Fig. 4.42.

It is frequently desirable to know the modulus of elasticity of the core material; this can be obtained by testing, but it is not always possible. Therefore an approximate procedure is given for the analytical formulation of this quantity. Consider Fig. 4.39*C* which depicts a unit cube of foam subjected to a tension load. In the following discussion E_f is the modulus of elasticity of foam made

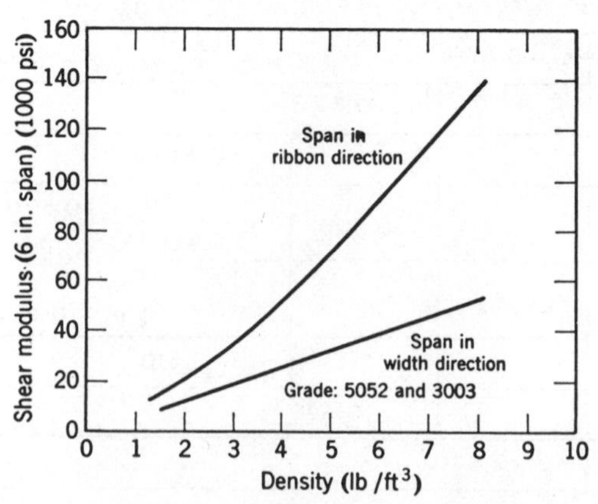

Fig. 4.42 Shear modulus of aluminum honeycomb. [*From (45), courtesy of Hexcel Products, Inc.*]

from a basic solid material having a modulus of elasticity E_m. Tensile stresses and strains are denoted by σ and ε respectively, and the other quantities are as shown in the figure. In accordance with Hooke's law,

$$\sigma_m = E_m \varepsilon_m = \frac{F}{A_m} = E_m\left(\frac{\Delta L}{L_m}\right)$$

$$\Delta L = \left(\frac{F}{A_m}\right)\left(\frac{L_m}{E_m}\right) = \left(\frac{F}{E_m}\right)\left[\left(\frac{L}{A}\right)_A + \left(\frac{L}{A}\right)_B\right]$$

$$= \frac{F}{E_m}\left(\frac{H-h}{H^2} + \frac{h}{H^2 - h^2}\right)$$

But $\sigma_f = E_f \varepsilon_f$ or

$$E_f = \frac{F}{A_f}\left(\frac{L_f}{\Delta L}\right) = \frac{F}{H^2}\left(\frac{H}{\Delta L}\right) = \frac{F}{H(\Delta L)}$$

$$= \frac{F}{H\left[\dfrac{F}{E_m}\left(\dfrac{H-h}{H^2} + \dfrac{h}{H_2 - h^2}\right)\right]}$$

from which

$$\frac{E_f}{E_m} = \frac{1-(h/H)^2}{1-(h/H)^2 + (h/H)^3}$$

If $(h/H)^3 = V$ represents the ratio of volume of holes to volume of foam material, then from the preceding equations, the effective modulus of the foam is

$$E_f = E_m\left(\frac{1 - V^{2/3}}{1 - V^{2/3} + V}\right) \qquad 4.265$$

The value of V may be obtained on a unit basis by knowing the densities of the foam and the solid material of which the foam is made.

In honeycomb structures the compressive strength, shear strength, and shear modulus of the core material are approximately proportional to density; typical data for aluminum honeycomb are shown in Figs. 4.42 and 4.43. These structures also exhibit higher shear strength and shear modulus when the span is parallel to the longitudinal or "ribbon" direction, that is, the lay down direction of the corrugated strips that eventually form the honeycomb structure. In addition, smaller cells lead to increased bond strength and additional resistance to buckling; in aluminum-faced sandwich, for example, a ratio of about 10/1 of cell size to face thickness approaches the optimum (45). A comparison of the shear strength and relative costs for various types of honeycomb is shown in Fig. 4.40.

Stress, strain, and deflection. In the computation of stress, strain, and deflection for sandwich constructions the usual formulas developed in Chapters 2 and 3 are applicable provided the proper values of EI, G, D, and so on, are used. In calculating deflection of sandwich panels the effect of shear in the core

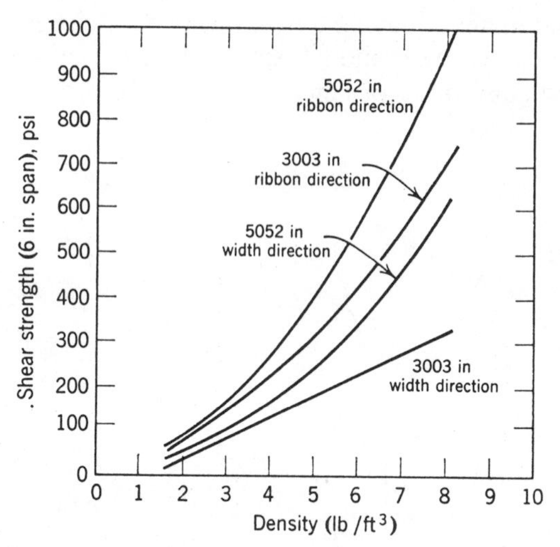

Fig. 4.43 Flexural shear strength of aluminum honey comb. [*From (45), courtesy of Hexcel Products, Inc.*]

material must be allowed for; the method is outlined in Chapter 7. The final result, however, which will be used here is as follows. The deflection for a beam or panel is calculated by Castigliano's theorem which states, mathematically.

$$\delta = \frac{\partial U}{\partial P} = \frac{\partial}{\partial P}\left(\int \frac{M^2\,dx}{2EI} + \int \frac{S^2\,dx}{2GA} \right) \tag{4.266}$$

where P is the load at the point of and in the direction of the calculated deflection, S is the vertical shear load, M the bending moment, and U the elastic strain energy of the system. It frequently happens that the deflection is desired at a location where the direct load P does not occur, or sometimes where there is no direct load P at all, as is the case of a uniformly loaded beam. Here, a *phantom* load Q is applied at the point of calculated deflection and the computation carried out through the integration of Eq. 4.266; in the final step $Q = 0$ and deflection is calculated for the original beam loading. This process is explained in detail in Chapter 7. Consider, for example, the panel shown in Fig. 4.44,

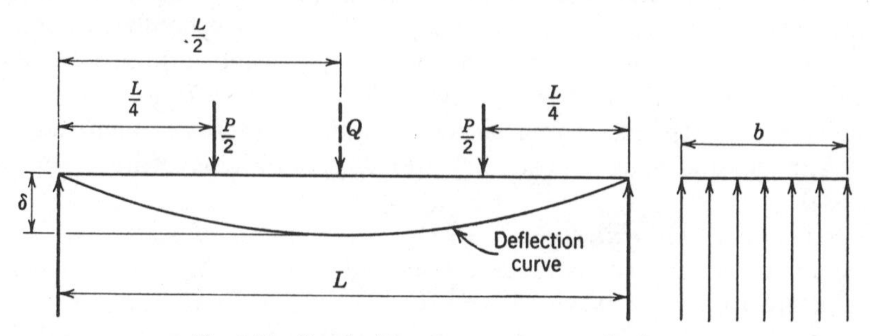

Fig. 4.44 Panel subjected to quarter-point loading.

subjected to quarter-point loading. In order to calculate the deflection at midspan a phantom load Q is applied as shown; then by Castigliano's theorem,

$$\delta_{L/2} = \frac{\partial U}{\partial Q} = \frac{\partial}{\partial Q}\left[\int \frac{M^2\,dx}{2EI} + \int \frac{S^2\,dx}{2GA}\right] \qquad 4.267$$

or, in terms of Fig. 4.44,

$$\delta_{L/2} = 2\int_0^{L/4} \frac{(P/2 + Q/2)x(x/2)\,dx}{EI} + 2\int_0^{L/4}\left[\left(\frac{P}{2}+\frac{Q}{2}\right)\left(\frac{L}{4}+x\right)-\frac{Px}{2}\right]$$
$$\times\left(\frac{L}{8}+\frac{x}{2}\right)\frac{dx}{EI} + 2\int_0^{L/4}\frac{(P/2+Q/2)(\frac{1}{2})\,dx}{GA}$$

which reduces on integration to

$$\delta_{L/2} = 5PL^3/349EI + PL/8GA \qquad 4.268$$

For a panel the deflection per unit width is calculated, EI is replaced by D, and A is the average shear area; therefore from Eq. 4.262 and 4.268

$$\delta_{L/2} = 5PL^3/349Db + PL/8Bb \qquad 4.269$$

where D is defined by Eq. 4.262 and

$$B = G_{\text{core}}[h(h+c)/2c] \qquad 4.270$$

The deflection at the quarter points for the panel shown in Fig. 4.44 is

$$\delta_{L/4} = PL^3/96bD + PL/8Bb \qquad 4.271$$

For a center-loaded panel, Fig. 4.45

$$\delta_{L/2} = PL^3/48bD + PL/4Bb \qquad 4.272$$

The bending stresses induced by the loading are carried primarily by the facings and the shear by the core. For example, for the panel shown in Fig. 4.44 the maximum bending stress is

$$\sigma_{\max} = M/Z = \left(\frac{P}{2}\right)\left(\frac{L}{4}\right)\Big/\left[\frac{btc(t+c/2)}{h/2}\right] \qquad 4.273$$

or

$$\sigma_{\max} = PL/8btc \qquad 4.274$$

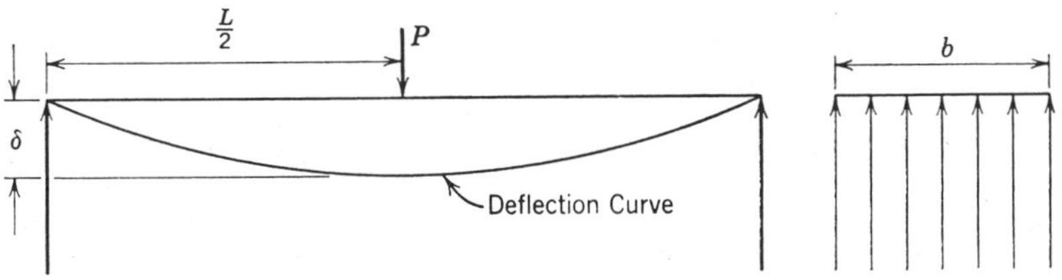

Fig. 4.45 Panel subjected to center loading.

The minimum stress is

$$\sigma_{\min} = PL/8bth \tag{4.275}$$

and the average stress (often used in design) is

$$\sigma_{\text{avg}} = \frac{PL(c+h)}{16btch}$$

or approximately

$$\sigma_{\text{avg}} = \frac{PL}{4bt(h+c)} \tag{4.276}$$

As an approximation the maximum shear stress is calculated on the basis of the average area of the entire cross section and the core material; thus

$$\tau_{\max} = \frac{S}{b(h+c)/2} = \frac{2S}{b(h+c)} \tag{4.277}$$

and the core shear strain is

$$\gamma_{\text{core}} = \tau_{\max}/G_{\text{core}} \tag{4.278}$$

Jones and Hersch (18) suggest that, for sandwich panels, the span-deflection ratio should not exceed 360, that the core shear stress should not exceed 40% of the core shear strength, and that the core shear strain should not exceed 2%.

The foregoing discussion of sandwich-type structures has been concerned entirely with the behavior of such structures in simple structural elements like beams and panels on simple supports. Another important type of "sandwich" structure is the *elastic foundation* discussed in Chapter 3. An extension of this theory has been made by Mehta and Veletsos (22) for multilayer systems frequently found in road building and foundation work; the theory is too involved to include here, but a simple system has been analyzed to illustrate what can be done. A generalized composite system is shown in Fig. 4.46, where E and ν are modulus and Poisson's ratio respectively. The stress and strain distribution is

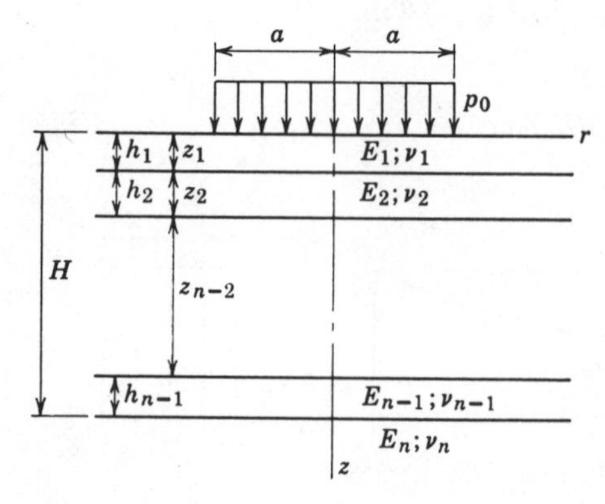

Fig. 4.46 Multilayer composite system.

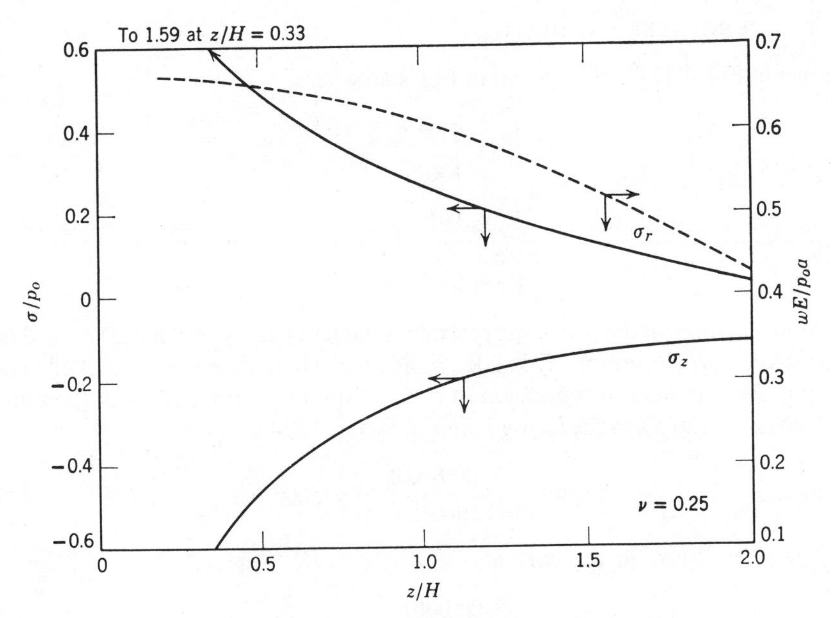

Fig. 4.47 Design chart for multilayer composite structure.

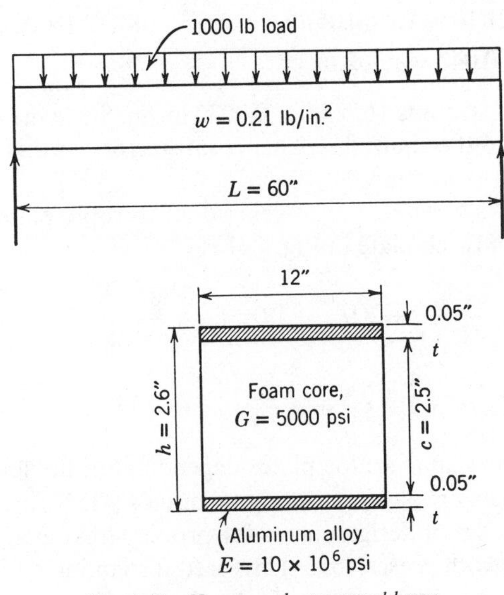

Fig. 4.48 Honeycomb structural beam.

shown in Fig. 4.47 for the system in Fig. 4.46 where

$$h_1 = h_2 = h_3 = H/3$$

$$E_1 = 100E$$

$$E_2 = 10E$$

and

$$E_3 = 5E$$

As an example of the use of the preceding theory stresses and deflection will be calculated for the uniformly loaded sandwich panel shown in Fig. 4.48. The weight of the panel is insignificant in comparison to the applied load; therefore the maximum stress in the facing material by Eq. 4.274 is

$$\sigma_{\max} = \frac{(1000)(60)}{(8)(12)(0.05)(2.5)} = 5000 \text{ psi}$$

The shear stress in the core is

$$\tau_{\text{core}} = \frac{2(1000)}{(2)(12)(2.6 + 2.5)} = 16 \text{ psi}$$

The maximum deflection is

$$\delta_{(L/2)} = 5WL^3/384bD + WL/8bB$$

or

$$\delta_{(L/2)} = \frac{(5)(1000)(60)^3(8)(0.91)}{(384)(12)(10 \times 10^6)(0.05)(2.6 + 2.5)^2} + \frac{(1000)(60)(2)(2.5)}{(8)(12)(5000)(2.6)(2.5 + 2.6)}$$
$$= 1.314 + 0.048 = 1.362 \text{ in.}$$

Finally, core shear strain is $16/5000 = 0.0032$ in./in. Since the deflection should not exceed 1/360 of the span, the panel is satisfactory in this respect; the core shear strain is also less than 2%.

Anisotropic sandwich plate. The general anisotropic plate equation (5, 35) for a laminated sandwich plate in Fig. 4.41 is

$$D_{11}\frac{\partial^4 w}{\partial x^4} + 4D_{16}\frac{\partial^4 w}{\partial x^3\,\partial y} + 2(D_{12} + 2D_{66})\frac{\partial^4 w}{\partial x^2\,\partial y^2} + 4D_{26}\frac{\partial^4 w}{\partial x\,\partial y^3} + D_{22}\frac{\partial^4 w}{\partial y^4}$$
$$= q + N_x\frac{\partial^2 w}{\partial x^2} + N_y\frac{\partial^2 w}{\partial y^2} + 2N_{xy}\frac{\partial^2 w}{\partial x\,\partial y} \qquad 4.279$$

This equation allows analysis for plates depending on the loads. In situations where the face sheets are isotropic, design manuals (41, 50) are available. These manuals contain a set of design curves for proper stress analysis and material selection. In anisotropic cases, (5, 39) are used for laminated face sheets where both the core and face sheets can be fiber reinforced. However, in cases where solutions are not available, the same procedures as outlined for Eq. 4.255 must be formulated for the plate in question. Complex plate problems often require computer solutions. Then the various failure modes (2, 34, 39) must be considered either for a sandwich or honeycomb construction. The face sheets act as plates on an elastic foundation with the honeycomb or the core being the spring elements along with the loading. The face sheet can buckle or the core or honeycomb can crush owing to the lack of stiffness. Whenever a possible failure mode is found, it can be related to the individual layers to find out how each behaves.

REFERENCES

1. J. E. Ashton, J. C. Halpin, and P. H. Petit, *Primer on Composite Material Analysis*, Technomic Publishing Co., Stamford, Conn. (1969).
2. J. E. Ashton, and J. M. Whitney, *Theory of Laminated Plates*, Technomic Publishing Co., Stamford, Conn. (1970).
3. J. E. Bell, *The Effect of Glass Fiber Geometry on Composite Material Strength*, paper No. 1583–60, American Rocket Society (1960).
4. K. J. Brookfield, "Glass Fiber Forms and Properties," Chapter 1, pp. 1–18 in *Glass Reinforced Plastics*, P. Morgan, Ed., Iliffe and Sons, Ltd., London, (1957).
5. L. J. Broutman and R. H. Krock, Ed., *Composite Materials*, Vols. I–VIII, Academic Press, New York, N.Y. (1974).
6. L. R. Calcote, *The Analysis of Laminated Composite Structure*, Van Nostrand Reinhold, New York, N.Y. (1969).
7. W. H. Chadbourne, *Prod. Eng.*, **32** (20), 95–97 (May 15, 1961).
8. A. G. H. Dietz, Ed., *Engineering Laminates*, John Wiley and Sons, Inc., New York, N.Y. (1949).
9. A. G. H. Dietz, Ed., *Composite Engineering Laminates*, Massachusetts Institute of Technology Press, Cambridge, Mass. (1969).
10. H. C. Engell, C. B. Hemming, and H. R. Merriman, *Structural Plastics*, McGraw-Hill Book Co., New York, N.Y. (1950).
11. L. Fischer, *Mod. Plast.*, **37** (10), 120, 122, 127–128, 208–209 (June 1960).
12. S. K. Garg, V. Svalbonas, and G. A. Gurtman, *Analysis of Structural Composite Materials*, Mercel Dekker Inc., New York, N.Y. (1973).
13. R. C. Hazenfield, "Key Design Parameter for Pultruded Products," *Plast. Des. Forum*, **2** (3), 88–93 (May/June 1977).
14. L. A. Heggernes, *Prod. Eng.*, **30** (46), 74–76 (November 9, 1959).
15. J. E. Hill, J. C. Goan, and R. Prescott, "Properties of Pultruded Composites Containing High Modulus Graphite Fiber", *Vol. 4—National SAMPE Technical Conference Series*, Azusa, Cal. (1972).
16. G. A. Hoffman, *Effect of Filamentary Materials on Pressure Vessel Design*, ASTIA Report P1974, U.S. Department of Commerce, Washington, D.C., (April 19, 1960).
17. N. J. Hoff, "The Strength of Laminates and Sandwich Structural Elements," in A.G.H. Dietz, Ed., *Engineering Laminates*, John Wiley and Sons, Inc., New York, N.Y. (1949).
18. R. E. Jones, and P. Hersch, *Prod. Eng.*, **30** (15), 69–71 (April 13, 1960).
19. H. S. Kliger, "Carbon Fibers-Designer's Tool for High-Modulus Parts," *Plast. Des. Forum*, **2** (3), 36–40 (May/June 1977).
20. G. Lubin; Ed., *Handbook of Fiberglass and Advanced Plastics*, Van Nostrand Reinhold, New York, N.Y. (1969).
21. D. L. McDanels, R. W. Jech, and J. W. Weeton, *Met. Prog.*, **78** (6), 118–121 (December 1960).

22. M. R. Mehta, and A. S. Veletsos, *Stresses and Displacements in Layered Systems*, Structural Research Series No. 178, University of Illinois, Urbana, Ill. (June 1959).

23. C. B. Norris, *Strength of Orthotropic Materials Subjected to Combined Stresses*, Report No. 1816, Forest Products Laboratory, U. S. Department of Agriculture, Madison, Wis. (March 1955).

24. J. O. Outwater, *Mod. Plast.*, **33** (7), 156, 158, 160, 162, 245, 248 (March 1956); **40** (7), 135–139 (March 1963).

25. N. J. Pagano, "Exact Modulus of Anisotropic Laminates," in L. J. Broutman and R. H. Krock, Eds., *Composite Materials*, Vol. II, Chap. 2, Academic Press, New York, N.Y. (1974).

26. H. Parker, and H. D. Hauf, *Simplified Design of Reinforced Concrete*, John Wiley and Sons, Inc., New York, N.Y. (1976).

27. P. A. Parmley, "Military Aircraft," in L. J. Broutman and R. H. Krock, Eds., *Composite Materials*, Vol. III, Chap. 4, Academic Press, New York, N.Y. (1974).

28. L. E. Payne, *J. Sci.*, Iowa State College, **23**, 381–395 (1949).

29. H. J. Plant, R. T. Girard, and H. R. Wisely, *Mater. Des. Eng.*, **53** (4), 14–16 (April 1961).

30. R. M. Rivello, *Theory and Analysis of Flight Structures*, McGraw-Hill Book Co., New York, N.Y. (1969).

31. E. Scala, *Composite Materials for Combined Functions*, Hayden Book Co., Rochelle, N.J. (1973).

32. G. P. Sendeckj, "Elastic Behavior of Composites," in L. J. Broutman and R. H. Krock, Eds., *Composite Materials*, Vol. II, Chap. 3, Academic Press, New York, N.Y. (1974).

33. Y. Stavsky and N. J. Hoff, "Mechanics of Composite Structures," in A.G.H. Dietz, *Composite Engineering Laminates*, Chap. 1, Massachusetts Institute of Technology Press, Cambridge, Mass. (1969).

34. J. A. Suarez, J. B. Whiteside, and R. N. Hadcock, "The Influence of Local Failure Modes on the Compressive Strength of Boron/Epoxy Composites," *Composite Materials, Testing and Design*, STP 497, Philadelphia, Pa. (1972).

35. S. W. Tsai, J. C. Halpin, and N. J. Pagano, *Composite Materials Workshop*, Technomic Publishing Co., Stamford, Conn. (1968).

36. A. A., Vicario, Jr. and R. H. Toland, "Failure Criteria and Failure Analysis of Composite Structural Components," in L. J. Broutman and R. H. Krock, Eds., *Composite Materials*, Vol. VII, Chap. 2, Academic Press, New York, N.Y. (1974).

37. J. R. Vinson and T. W. Chou, *Composite Materials and Their Use in Structures*, Halsted Press, New York, N.Y. (1975).

38. G. A. Zimmerman and C. P. Krupp, *Missiles and Rockets*, **7** (22), 28–30 (November 28, 1960).

39. *Advance Composites Design Guide*, Vol. I–V, 3rd ed., Air Force Material Laboratory, Defense Documentation Center, Alexandria, Va. (1973).

40. *Composite Materials and Composite Structures*, OTS Report No. 161443 (549 pp.), Proc. 6th Sagamore Ordance Materials Research Conf., August 1959, U.S. Department of Commerce, Washington, D.C.

41. *Design Handbook for Honeycomb Structures*, TSB 123, Hexcel Products Inc., Dublin, Cal. (1970).

42. *Des. News* **18** (14) 64–65 (July 10, 1963).

43. "Filament Wound Reinforced Plastics: State of the Art," Special Report No. 174, *Mater. Des. Eng.*, **52** (2), 127–146 (August 1960); see also **58** (3), 79–126 (September 1963).

44. *Glass Reinforced Plastics in Naval Applications* (Bureau of Ships Design Manual), PB Report No. 171096, U.S. Department of Commerce, Washington, D.C. (August 1958).

45. *Honeycomb Sandwich Design Data and Test Methods*, Hexcel Products Inc., Dublin, Cal. (1959).

46. *Membrane Analysis of Orthotropic Filament Wound Pressure Vessels*, ASTIA Report AD-273-306, U.S. Department of Commerce, Washington, D.C. (February 1962).

47. "Modern Concrete Design" (Review article by P. Gugliotta), *Int. Sci. Tech.*, No. 24, 54–61 (December 6, 1963).

48. *Plastics for Aerospace Vehicles, Part 1—Reinforced Plastics*, MIL—HDBK-17A, Department of Defense, Washington, D.C. (1971).

49. *Prod. Eng.*, **29** (29), 82–83 (July 21, 1958). Prod. Eng. Digest of article by K. Lowenfeld, *Der Maschinenmarkt* (Wurzburg), Nos. 87 and 98, (October 29 and December 6, 1957).

50. *Structural Sandwich Composites*, MIL-HDBK-23A, Department of Defense, Washington, D.C. (1968).

51. *The Basics on Bonded Sandwich Construction*, TSB 124, Hexcel Products, Inc., Dublin, Cal. (1977).

PROBLEMS

4.1 A simply supported beam 60 in. long and 2 in. wide is loaded in the center with 1000 lb. The cross section is seven layers of $\frac{1}{4}$-in. aluminum and $\frac{1}{8}$-in. steel strips with the steel on the top and bottom. Find the maximum bending stress and shear stress in the aluminum and steel.

4.2 In Problem 4.1 use a design manual (41, 50) and develop a honeycomb design for the stated conditions. Compare the weights of the beams.

4.3 Three bars shown in Fig. P4.3—*A*, *B*, and *C*—are 2 in. thick but vary in width. They are compressed by a rigid plate which remains horizontal. Find the values of *P* and *H* so that the allowable stresses of 20 kpsi of steel bar *A*, 15 kpsi of yellow brass bar *B* and 10 kpsi of the aluminum bar *C* are not exceeded.

4.4 A flywheel is to produce 3.42×10^6 ft-lb of energy to power a vehicle. What would a realistic size of the flywheel be when the flywheel is to be made of a filament winding with a shaft of $0.1\ R_0$. What will a weave pattern look like for a $R_0 = 1$? (Plot for increments of $0.05\ R_0$.) Filament strength is 250 kpsi with $E_L = 20 \times 10^6$ and $\nu = 0.21$.

4.5 In Problem 3.5 the support beam fails in shear around 250 lb/ft.^{2} loading and is 2.77 times over the allowable deflection. Reinforce the wooden

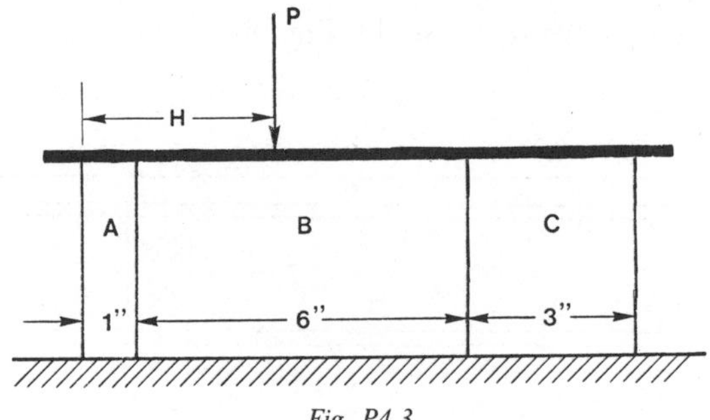

Fig. P4.3

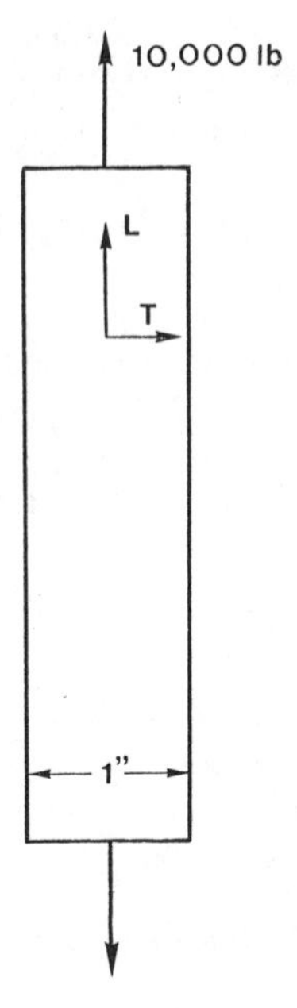

Fig. P4.6

beam with steel plate to reduce the deflections to the allowable limit. Remember it must be reinforced without removing it. Find the plate thickness and screw or nail spacing.

4.6 An orthotropic bar (Fig. P4.6) is loaded by a 10,000-lb tension load in the L direction. The bar is 0.125 in. thick and 1 in. wide with the following properties:

$E_L = 10.3 \times 10^6$ $\quad E_T = 9.5 \times 10^6$ $\quad G_{LT} = 3.9 \times 10^6$

$\nu_{LT} = 0.25$ $\quad \nu_{TL} = 0.3$

Find the total elongation in a 10-in. bar and reduction in cross-sectional area.

4.7 Design a spherical filament pressure vessel, 12 in. I.D. to support 600 psi pressure.

4.8 We have a composite beam in Fig. P4.8 with allowable stresses of

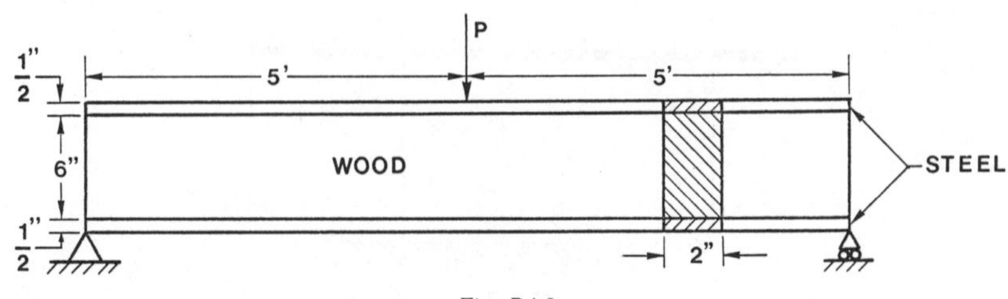

Fig. P4.8

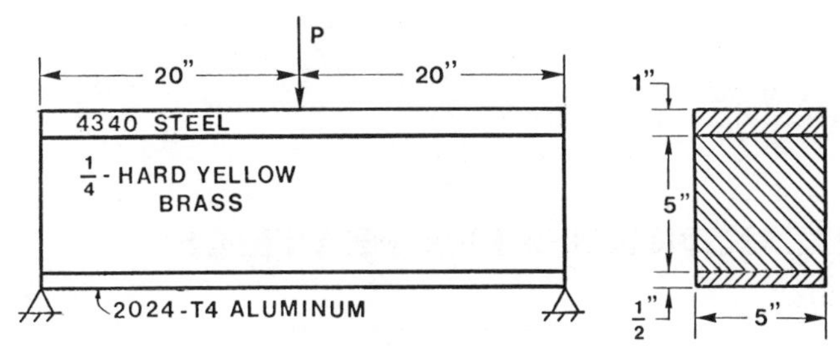

Fig. P4.9

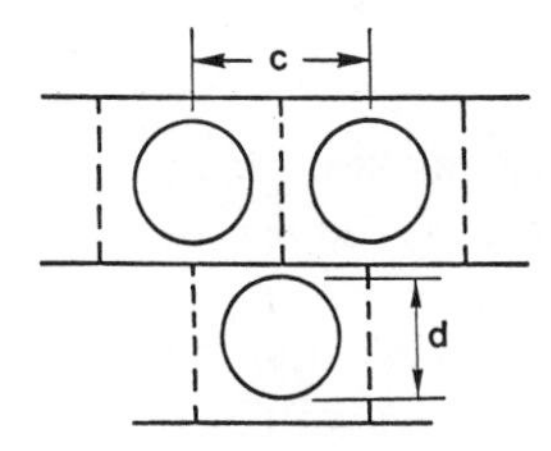

Fig. P4.11

steel 10 kpsi and of wood 1 kpsi with E_s, 30×10^6, and E_w, 1×10^6. Find P_{max}!

4.9 When $y_{max} = \text{span}/360$ (inches) for the loaded beam (Fig. P4.9) find the load P that can be safely supported.

4.10 In Problem 4.9 if the cross section was rotated 90°, what load P could be safely supported?

4.11 Find the A_f/A_c and A_m/A_c (Fig. P4.11) ratios for

a. Fibers touching,

b. For $c = 2d$ spacing

c. For spacing $c = d$.

chapter 5

DESIGNING WITH PLASTICS

5-1 THE NATURE AND PROPERTIES OF VISCOELASTIC MATERIALS

In this chapter the discussion concerning viscoelastic behavior is restricted to a class of engineering materials known as *high polymers*, or, more commonly, as *plastics* (3, 15, 18, 30). The objective here is to present a unified picture of how plastics behave in structural applications, what their properties are, and how these properties are utilized in the design of plastics for structures, machine parts, and other common applications. It is assumed that the engineer is aware of the chemical, electrical, and decorative value of plastics, so that attention here can be directed entirely at the mechanical aspects of viscoelasticity. In this regard only the static behavior of plastics will be discussed; this will include ordinary short-time properties like tensile strength and the time-dependent properties like creep and stress-rupture. Dynamic behavior is discussed in Chapter 10.

Classification and properties of plastics

There are two main classes of plastics, *thermoplasts* and *thermosets*. Thermoplasts or thermoplastic materials are characterized by their ability to flow under application of heat and pressure; thermosets or thermosetting plastics undergo a structural change when subjected to some particular temperature and pressure and are thus rendered incapable of subsequent deformation by temperature and pressure. Thus thermoplastic materials may be formed into a variety of shapes by the simple application of heat and pressure, whereas thermosetting plastics can only be formed by cutting and machining methods or other specialized procedures not generally available to vendors of various plastics compositions. Further classification of plastics is based on chemical structure as shown in Table 5.1, which also lists a few of the common trade names under which plastics are marketed (26).

From a mechanical point of view one of the principal advantages of plastics is in their ability to be readily processed (33); many plastics can be readily molded into complicated shapes. Furthermore, generally, the cost of plastics has been

Table 5.1 Classification of Some Plastics

Chemical Classification	Trade Name
Thermoplastic Materials	
Acetal	Delrin, Celcon
Acrylic	Lucite, Plexiglas
Acrylonitrile-butadiene-styrene	Cycolac, Kralastic, Lustran
Cellulose acetate	Fibestos, Plastacele, Tenite I
Cellulose acetate-butyrate	Tenite II
Cellulose nitrate	Celluloid, Nitron, Pyralin
Ethyl cellulose	Gering, Ethocel
Polyamide	Nylon, Zytel
Polycarbonate	Lexan, Merlon
Polyethylene	Polythene, Alathon
Polypropylene	Avisun, Escon
Polystyrene	Cerex, Lustrex, Styron
Polytetrafluoroethylene	Teflon
Polytrifluorochlorethylene	Kel-F
Polyvinyl acetate	Gelva, Elvacet, Vinylite A
Polyvinyl alcohol	Elvanol, Resistoflex
Polyvinyl butyral	Butacite, Saflex
Polyvinyl chloride	PVC, Boltaron, Tygon, Geon
Polyvinylidene chloride	Saran
Thermosetting Materials	
Epoxy	Araldite, Oxiron
Melamine-formaldehyde	Melmac, Resimene
Phenol-formaldehyde	Bakelite, Catalin, Durez
Phenol-furfural	Durite
Polyester	Beckosol, Glyptal, Teglac
Urea-formaldehyde	Beetle, Plaskon

NOTE: This is not a complete list of chemical types and trade names. For further details consult various plastics catalogs and trade journals.

decreasing while the cost of other materials has increased. In addition to the basic types, thousands of new formulations are being developed and many of these used as insulators have exceptional electrical properties, good thermal insulating properties, low specific gravity, and availability in a wide range of colors and transparencies. Many plastics also show remarkable resistance to wear and corrosive attack by chemicals. On the other hand, most plastics have a relatively low modulus of elasticity and hence are inherently nonrigid. Their properties are also strongly affected by relatively small temperature changes or changes in the rate of load application; most important, most of the mechanical properties of plastics are time-dependent, a subject which will be discussed

Table 5.2 Range of Tensile Strength of Some Plastics at Room Temperature

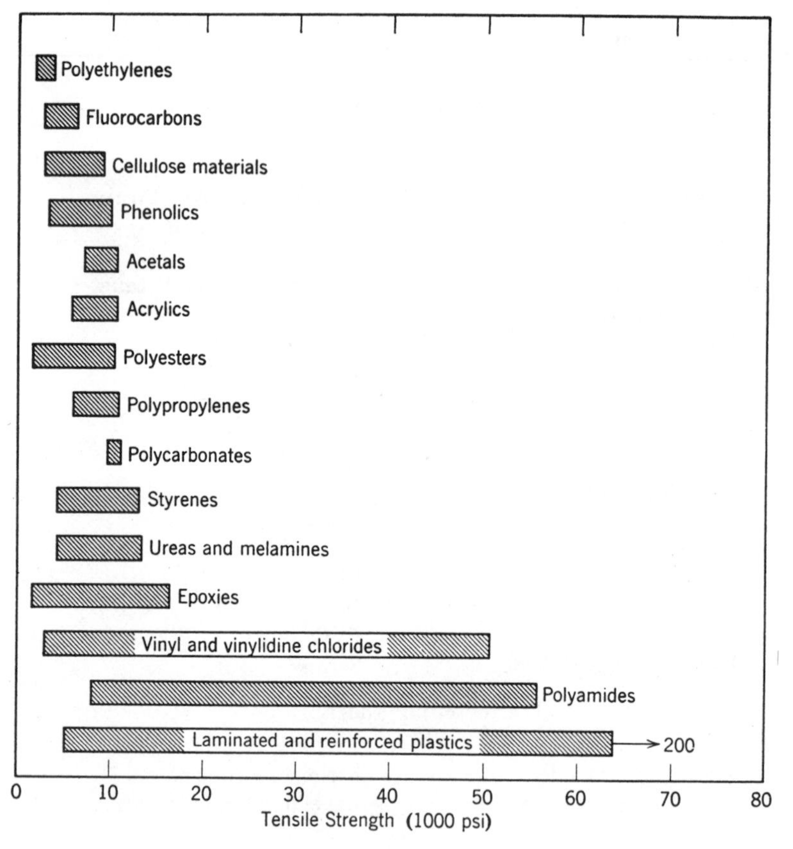

presently. Finally, because of the rate at which new plastics formulations are being introduced, there is often a serious shortage of reliable design data, particularly in the areas of creep, stress-rupture, and complex stressing. From a design viewpoint it is convenient to consider the properties of plastics under two headings; ordinary short-time properties such as tensile strength and the time-dependent properties like creep and stress-rupture. Because of the great importance of time-dependent properties this subject is treated separately in the next section.

As shown in Tables 5.2 to 5.6 plastics exhibit a wide range of properties; this, together with their chemical and electrical properties, makes them important materials of construction. Some specific property values are given in Table 5.7 (7, 9, 18, 24). The general problem of preparing plastics formulations to meet specific properties has been discussed by McGrew (25).

A common method used for presenting properties of plastics is the tensile stress-strain curve. When data are presented in this way, however, considerable caution must be exercised since properties of plastics vary considerably and are

Table 5.3 Range of Modulus of Elasticity for Some Plastics at Room Temperature

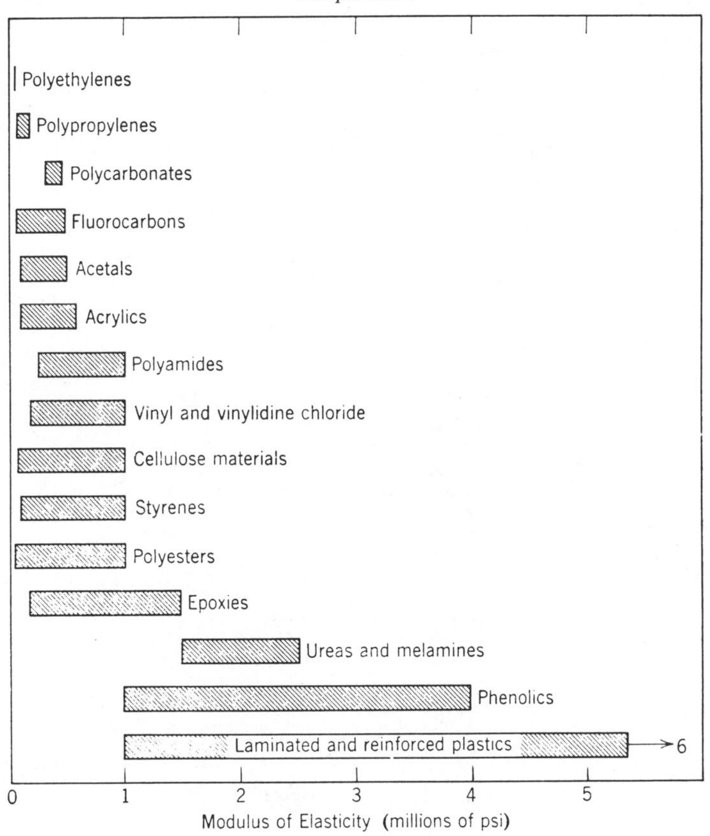

quite sensitive to small changes in temperature, strain rate, and environment and state of stress. In this regard handbook values may often be considered as order of magnitude estimates. For example, consider Fig. 5.1 which gives some tensile stress-strain data for a methacrylate plastic at one rate of strain and several temperatures. Note that even in the normal ambient temperature range of 50 to 100°F. the tensile strength varies nearly 4000 psi for a threefold increase in temperature. In Fig. 5.2 it is seen that the tensile stress of the same material varies nearly 2000 psi at 104°F for an eightfold increase in strain rate. At first examination these figures do not seem to be too far out of line, but it must be remembered that most metals, for example, do not exhibit any appreciable change in properties in the same temperature and speed ranges.

Thus the actual properties of plastics are closely associated with the conditions under which the properties are measured. Note also in Fig. 5.1 that the modulus (stiffness) decreases by a factor of nearly two from 54 to 140°F. These data are for a specific plastic, but the results are typical of the behavior of most plastics materials. As another example consider Fig. 5.3 which shows ordinary tensile

Table 5.4 Maximum Temperature Range of Operation of Some Plastics

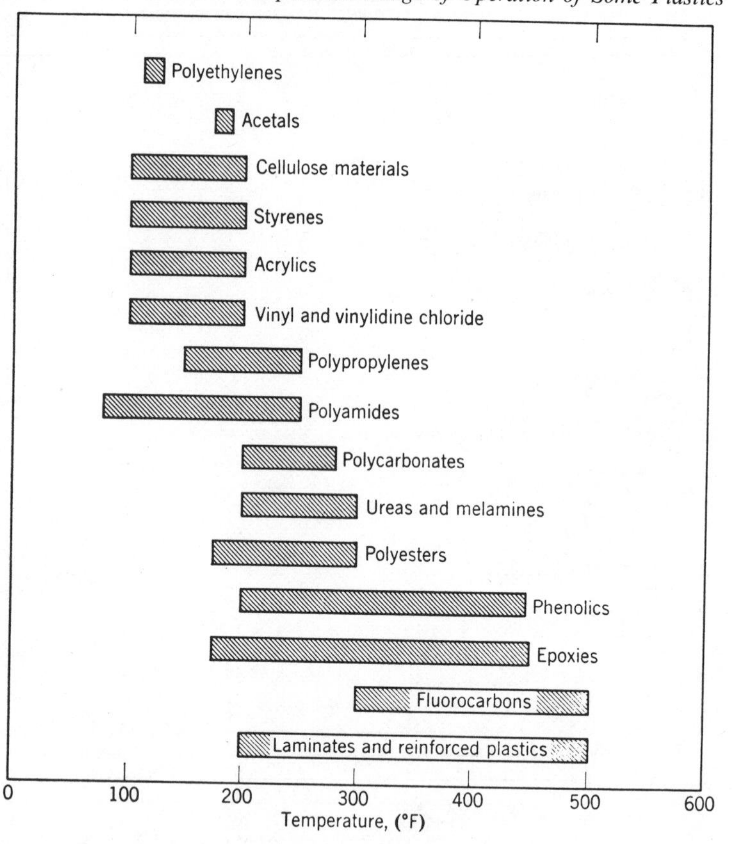

test data for polyvinyl chloride and polyethylene; note the variation in properties for the same material.

As was discussed in Chapter 2 nominal stress-strain data being based on original dimensions of specimens do not give a fundamental picture of how materials deform. True stress-strain data should be used in arriving at basic conclusions about deformation behavior of materials (see, for example, Fig. 2.3 and Eqs. 2.11 and 2.12 defining true values). Since plastics deform to a considerable extent, true stress-strain data should be obtained whenever possible. For example, the data in Fig. 5.3 are *nominal* data; in Fig. 5.4 *nominal* and *true* data are compared for polyvinyl chloride. Note that the true stress recorded at 50% strain is actually nearly twice that exhibited by the nominal plot. Such considerations are not too important for ordinary quality control tests or order of magnitude estimates; their chief utility is in analyzing forming problems and in gaining a deeper insight into the deformation characteristics of these materials.

The effect of temperature on the strength of a methacrylate plastic is shown in Fig. 5.1; some data for several other plastics are shown in Fig. 5.5.

Table 5.5 Range of Coefficient of Thermal Expansion for Some Plastics

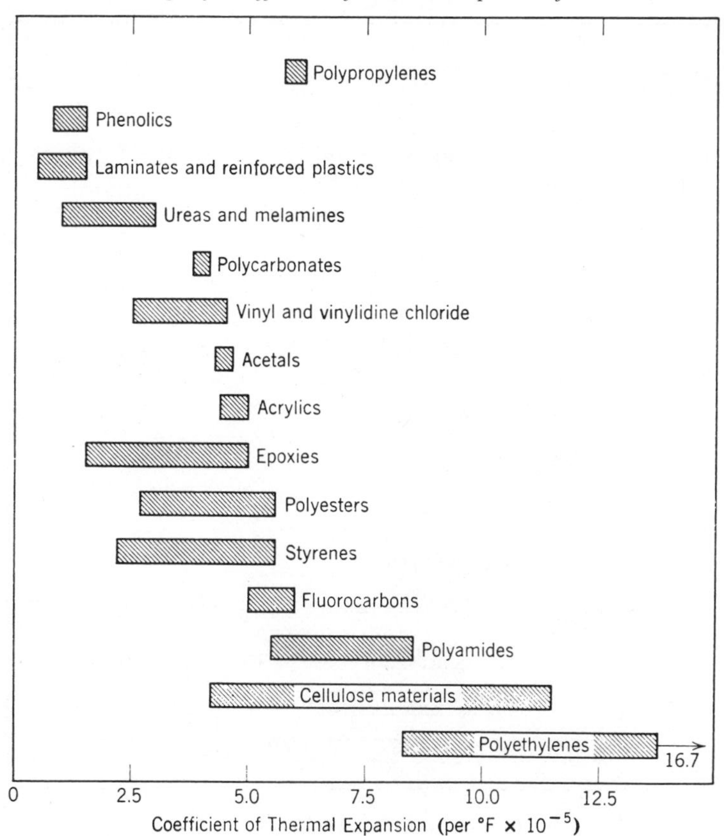

Many plastics are used in chemical plants for piping systems (27, 35) and because this is such an important application it is desirable to ascertain the properties of the plastics used from tests of full size pipes or tubes. For tension tests the apparatus shown in Fig. 5.6 has proved useful (9) for gripping pipe samples for tension tests. The conical expander *A* spreads matched tapered inserts *B*, thus providing a self-tightening system during testing as shown in Fig. 5.7. An extensometer attached to the restraining collars at the ends of the tube section provides a convenient way to obtain deformation data; the data shown in Fig. 5.8 were obtained using this described apparatus. In addition to tension tests it is also desirable to obtain burst data on plastic pipe for use in chemical plants. Typical data for polyvinyl chloride and polyethylene as a function of temperature are shown in Fig. 5.9; additional data on pipe may be found in (27, 28, and 34). A pipe under internal pressure represents a state of combined stresses (Chapter 3). Fundamental data on the effect of combined stresses on plastics are rare; most of the available data are for pipes under internal pressure. In general, a state of combined stress applied to a material

Table 5.6 Range of Thermal Conductivity of Some Plastics

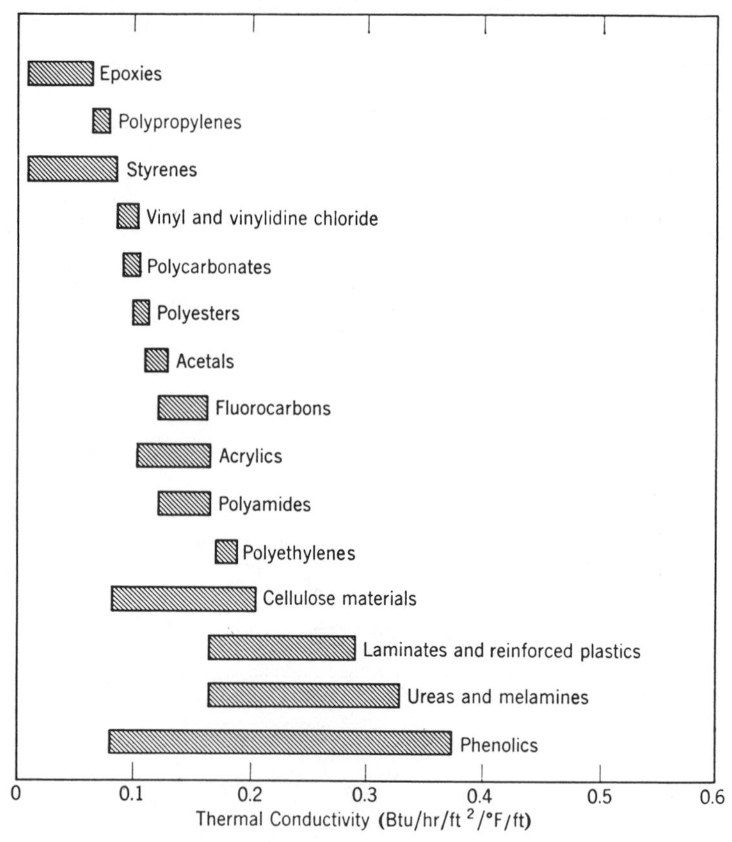

has the effect of increasing its strength and lowering the ductility. This is shown, for example, by the data in Fig. 5.10 for polyethylene (16). Curve *A* is the standard tensile stress-strain curve and curve *B* is for biaxial tension. Note that ductility is drastically reduced by the presence of combined stresses. The ultimate strength is also increased considerably; the dotted part of curve *A* represents a zone of necking down of the material so that the ultimate strength indicated cannot be compared directly with that observed for curve *B*.

Time dependence of mechanical properties

It will be recalled from Chapter 1 that viscoelastic deformation was listed as a *transition* type of behavior that could be characterized by both normal elastic strain and time-dependent flow. It is this time dependence of properties that makes plastics somewhat difficult to assess by mathematical theory. For most plastics even small loads induce a continuous type of *creep behavior* and, depending on the state of stress and environment (16, 17, 29), fracture by stress-rupture can occur at loads considered safe by conventional standards. Thus in

Table 5.7 Properties of Some Plastics at Room Temperature. (In hardness column, D is Shore and R and M are Rockwell numbers. The Table data are representative only; for detailed properties consult the Modern Plastics Encyclopedia, Published annually by the Breskin Publishing Co. This reference also gives properties of various plastics foams; for determination of the elastic modulus of foam see Chapter 4 and Fig. 4.39.)

Trade Name	Tensile Strength, psi	Modulus of Elasticity, psi	Izod Impact, ft-lb/in.	Hardness	Coefficient of Thermal Expansion per °F × 10^{-5}	Specific Gravity
Alathon I	1,750	35,000	—	D48	4.5	1.43
Avisun 1014	5,000	150,000	1.0	R90	6.7	0.91
Bakelite	7,000	800,000	0.6	M110	4.5	1.30
Celcon	8,800	375,000	10.0	M73	4.5	1.40
Cycolac GS	6,500	290,000	6.5	R100	5.2	1.04
Delrin 500	10,000	410,000	1.4	R120	4.5	1.43
Durez 791	7,000	120,000	0.3	M116	3.2	1.36
Escon	4,900	160,000	1.0	R90	4.0	0.91
Kralastic B	5,500	260,000	7.0	R116	3.8	1.06
Lexan	9,000	320,000	14.0	R118	3.9	1.20
Lucite 40	10,500	450,000	0.3	M103	3.0	1.19
Lustran 210	6,800	420,000	1.1	R119	3.4	1.07
Lustrex 55	6,000	450,000	0.3	M75	3.5	1.06
Plexiglas VS	8,000	450,000	0.5	M80	3.0	1.18
PVC, Type I	6,400	450,000	0.8	R118	5.2	1.37
Zytel 101	11,800	410,000	0.9	R118	5.5	1.14

designing with viscoelastic materials a compromise often has to be made between proportions that under a given load will not reach an excessive deformation in a specified time period and proportions that under a given load will not generate stresses that would result in failure by stress-rupture in the given time period. The main difficulty is that these effects can take place at ambient temperature and are intensified at elevated temperature; for metals, for example, creep and stress-rupture are problems only at elevated temperatures (see Chapter 15).

A schematic representation for tensile creep and stress-rupture behavior of most commercial plastics is given in Fig. 5.11. If the material is stressed by an amount σ_A, it creeps until time t_A is reached, at which time failure by stress-rupture occurs at a total deformation ε_A. The same pattern is followed for stresses σ_B and σ_C. Thus, if the design is to stand up for times greater than t_A, it must be proportioned so that the stress induced is less than σ_A. For any particular material both creep and stress-rupture data should be obtained before preparing a design. Sometimes this is not possible if the desired life of the part is to be several months or years; then extrapolations of short-time data have to be made and this presents some difficulties as will be shown later. Another complication is that creep and stress-rupture of plastics may vary considerably

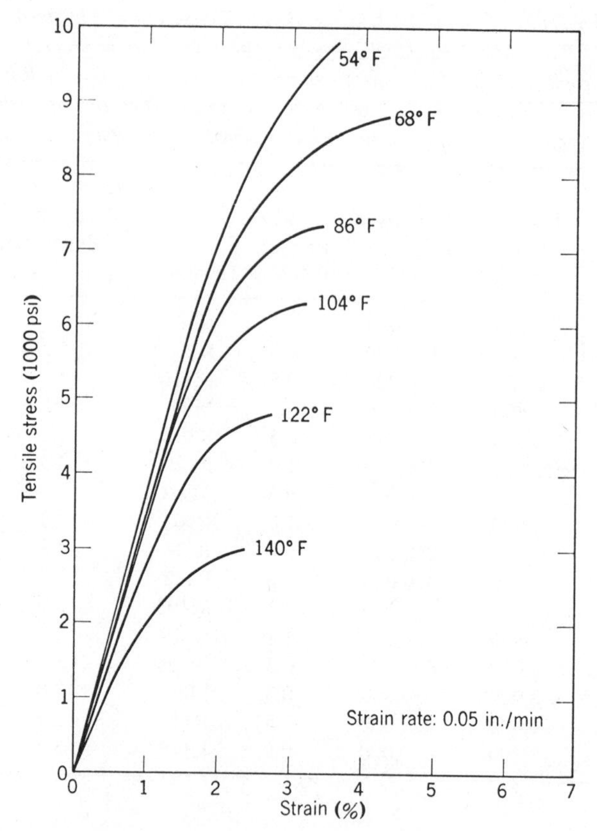

Fig. 5.1 Tensile stress-strain curves as a function of temperature for a methacrylate plastic.

for the same material or different batches obtained from different manufacturers of the material and will also be affected by varying degrees by temperature, rate of loading, and state of stress. However, by careful examination of the material and performing a few control tests satisfactory designs may be obtained and used with confidence. The main point here is that plastics are strongly influenced by time effects that have to be carefully assessed for any particular application of the material.

Restricting attention in this chapter to static loading, the important time effects can be reduced to three principal types: creep, stress relaxation, and stress-rupture. For the present, consideration of stress-rupture will be dismissed by simply stating that its mechanism is largely unknown and that the prediction of this phenomenon is difficult but can be done to a degree by methods discussed later. The other two time effects, creep and stress-relaxation, are closely related and can be predicted with a fair degree of accuracy by mathematical theory. Creep is the deformation that occurs over a period of time in a material subjected to a constant load, and stress-relaxation is the reduction in stress that

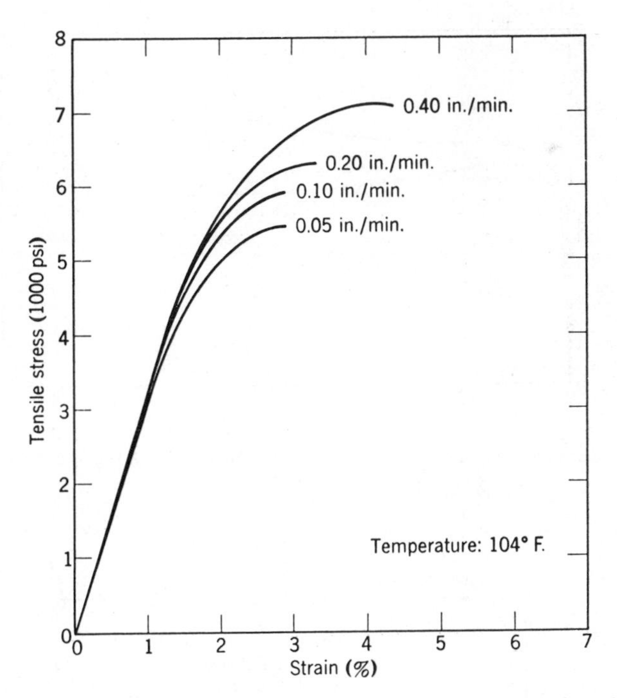

Fig. 5.2 Tensile stress-strain curves as a function of strain rate for a methacrylate plastic.

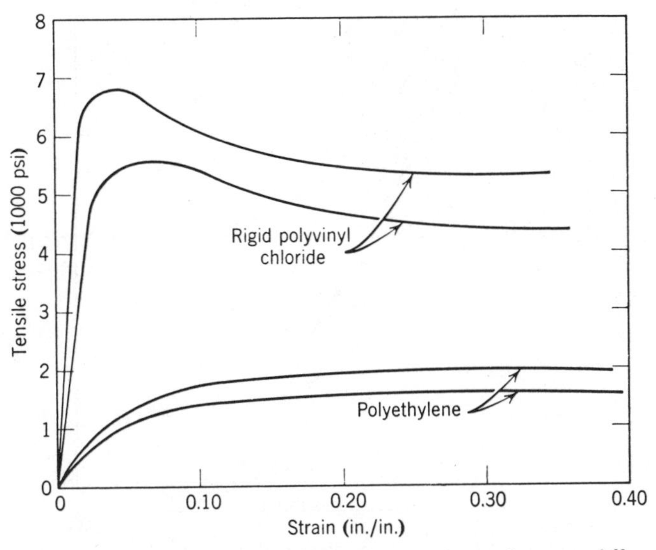

Fig. 5.3 Nominal tensile stress-strain data at 73° F for two plastics from two different batches of resin.

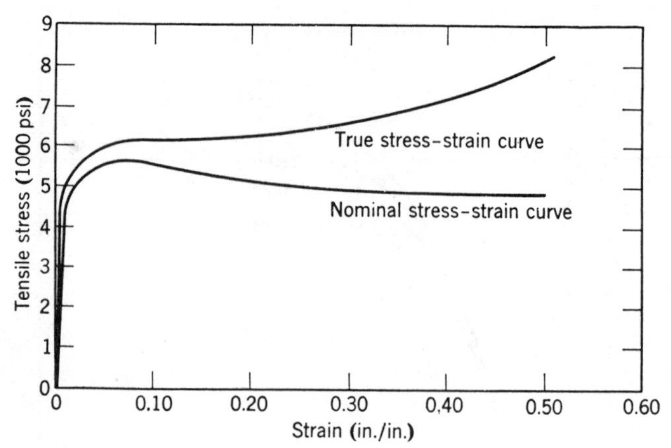

Fig. 5.4 Tensile stress-strain data for test of full size 2-in. schedule 80 rigid polyvinyl chloride pipe at 72° F.

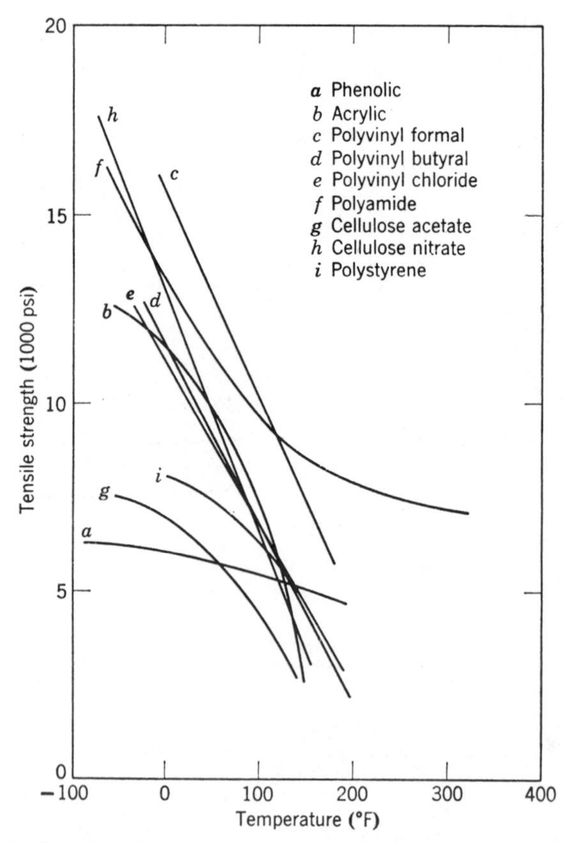

Fig. 5.5 Tensile strength versus temperature for several plastics.

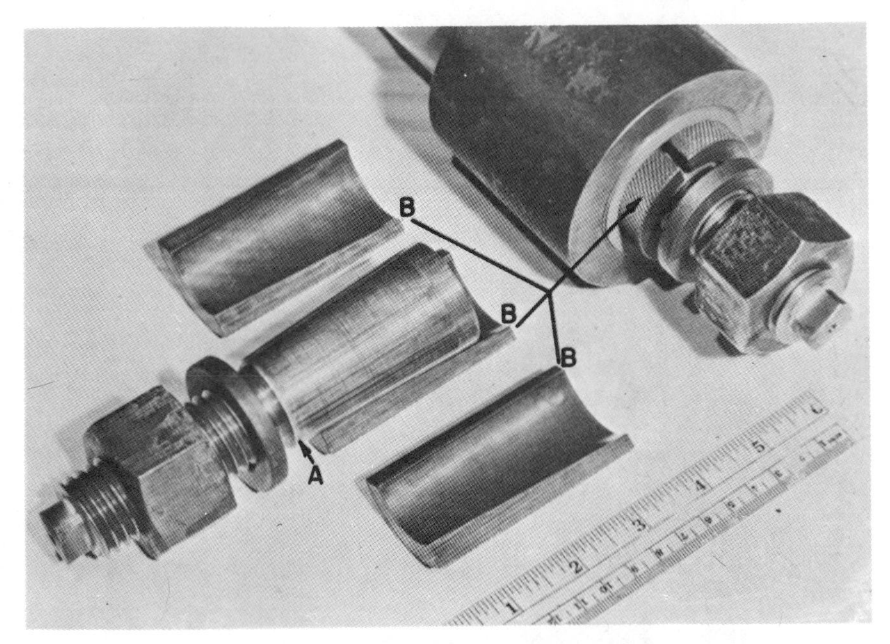

Fig. 5.6 Grips for tubular tension specimens.

occurs in a material when it is deformed to some specified deformation that is held constant. Thus creep is deformation at constant load and stress-relaxation is decay of stress at constant strain. These effects are illustrated in Figs. 5.12*A* and 5.12*B*. In the creep test a sample is instantaneously loaded to some stress σ which is maintained at a constant level. This stress results in a strain (point *A*, Fig. 5.12*A*) which is an elastic response. It is assumed that the tests are carried out at stresses below the short-time "yield strength" of the material. As the stress is maintained, the sample deforms (creeps) viscoelastically to *B*; if the load is removed from the sample at this time, there is an immediate elastic recovery to *C* followed by a final gradual viscoelastic recovery to *D*, leaving a permanent deformation *DE*. Typical apparatus for conducting creep tests is shown in Figs. 5.13 to 5.15. In Figs. 5.13 and 5.14 are shown the details of a constant load creep machine. Loading is transmitted through a circular level arm which maintains a constant moment and hence a constant load on the specimen. For materials undergoing large strains that result in changing the applied stress, the circular lever arm is replaced by one of logarithmic contour. In Fig. 5.13 a specimen under test is shown in the machine against the wall. Figure 5.15 shows apparatus for creep in pure bending.

In the stress-relaxation test, Fig. 5.12*B*, a sample is instantaneously loaded by a stress σ to some fixed deformation which is then held constant. Immediately the material starts to "relax," that is, the stress originally imposed diminishes to *B* over a period of time. If at this time the load is removed, there is an immediate

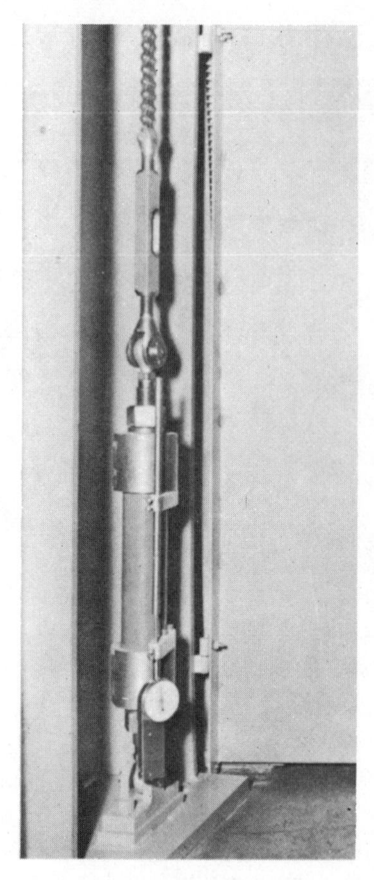

Fig. 5.7 Tension sample with extensometer.

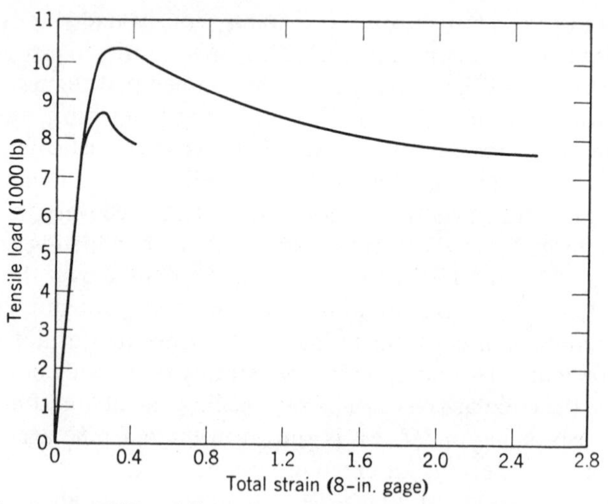

Fig. 5.8 Tension tests at 72° F on full size 2-in. schedule 80 rigid polyvinyl chloride pipe at different strain rates. Curves show range of values for strain rates between 0.005 and 1.0 in./min.

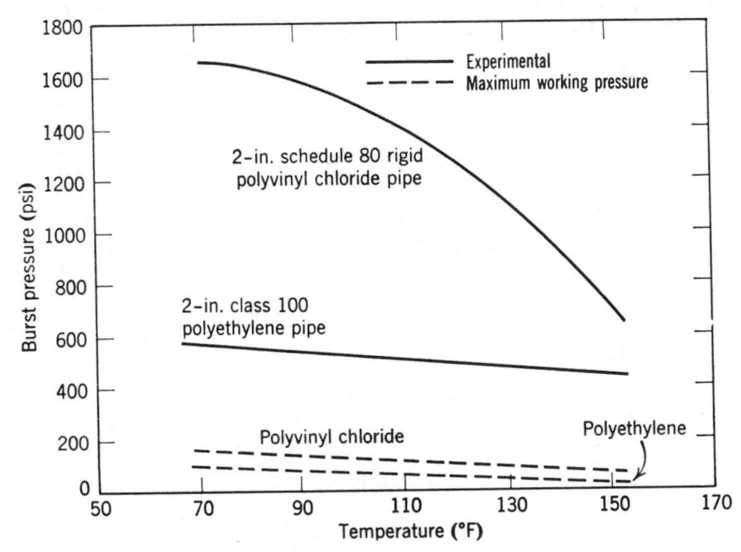

Fig. 5.9 Burst pressure of two types of plastic pipe versus temperature.

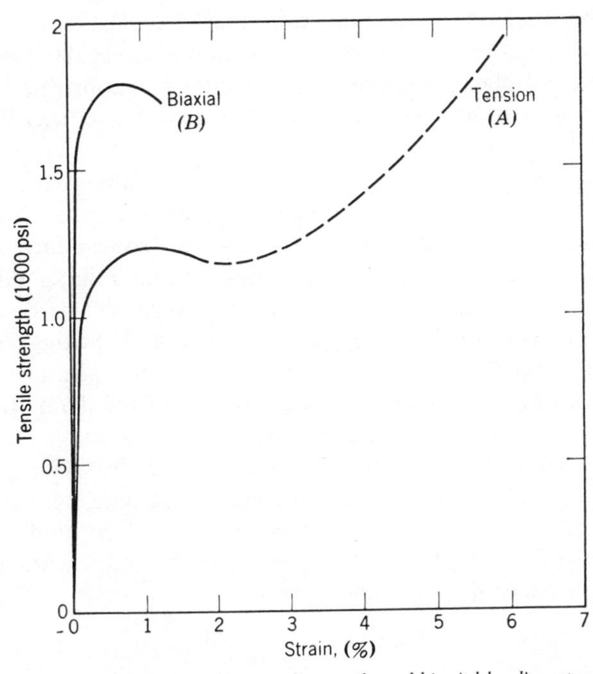

Fig. 5.10 Behavior of a polyethylene plastic under tensile and biaxial loading at room temperature.

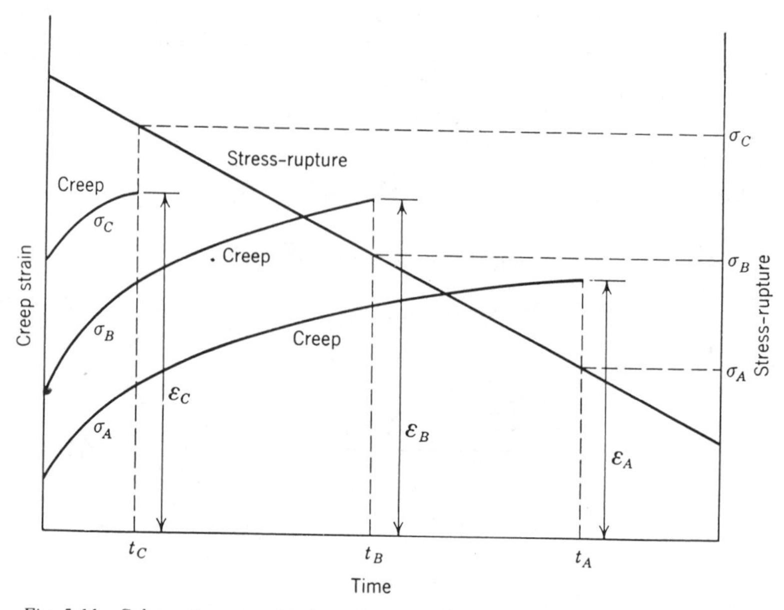

Fig. 5.11 Schematic representation of creep and stresses-rupture behavior of plastics.

elastic recovery to C which may be below zero, depending on the initial deformation; the residual stress then induces further viscoelastic deformation to D. Both creep and stress-relaxation result from internal molecular movements and hence there is a correlation between them. It will be demonstrated later that, in fact, one behavior can be accurately predicted from data concerning the other behavior (2, 4, 9, 13, 14, 20, 29, 31).

Typical tensile creep data for a number of materials are given in Figs. 5.16 to 5.21 and some tensile stress-rupture data are given in Figs. 5.22 to 5.24. Data on creep of plastics under combined stresses are rare; some information is given in (9, 10, 16, 28, 29, and 35). In general, it has been found that good predictions of creep deformation can be made by theory, which will be described later. Typical stress-relaxation data are presented in Fig. 5.25; these data can be used to predict creep behavior as will be explained later. The data for pressure relaxation were obtained by pressurizing tubes to a fixed diametral expansion and then noting the decay of pressure over a period of time as the expansion was held constant. Data concerning the stress-rupture behavior of plastics are presented in Figs. 5.26 and 5.27. In obtaining these data, tubes subjected to tension and internal pressure simultaneously were used, and the ordinate, equivalent stress $\bar{\sigma}$, was calculated by the use of Eq. 3.524 with the following values for the stresses σ_1, σ_2, and σ_3.

$$\text{Hoop stress} = \sigma_h = \sigma_1 = pd/2t \tag{5.1}$$

$$\text{Longitudinal stress} = \sigma_z = \sigma_2 = pd/4t \tag{5.2}$$

$$\text{Radial stress} = \sigma_r = -p \tag{5.3}$$

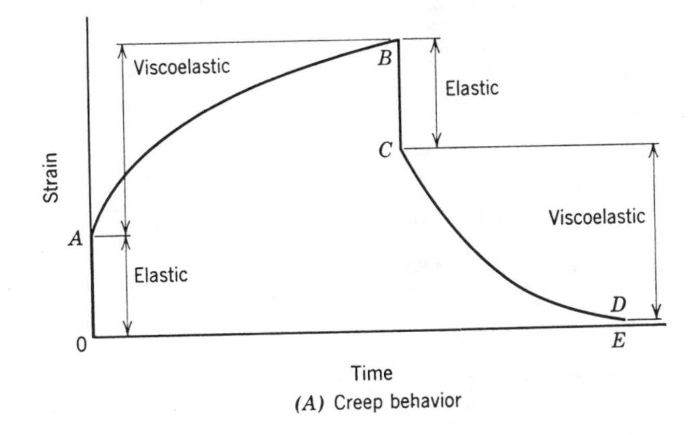

(A) Creep behavior

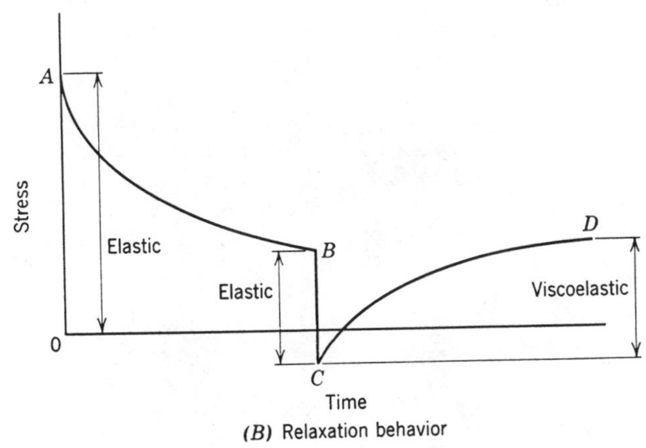

(B) Relaxation behavior

Fig. 5.12 Schematic representation of creep and stress-relaxation characteristics of plastics.

where p is the internal pressure, d the inside diameter of the tube, and t the wall thickness. For a stress ratio $\sigma_z/\sigma_h = 0$, there is no longitudinal stress*, and the equivalent stress becomes

$$p = \frac{\bar{\sigma}}{[\frac{1}{4}(d/t)^2 + \frac{1}{2}(d/t) + 1]^{1/2}} \qquad 5.4$$

The stress ratio $\sigma_z/\sigma_h = \frac{1}{2}$ represents a tube with capped ends under internal pressure with equivalent stress

$$p = \frac{\bar{\sigma}}{[\frac{3}{16}(d/t)^2 + \frac{3}{4}(d/t) + 1]^{1/2}} \qquad 5.5$$

For a stress ratio $\sigma_z/\sigma_h = 1$, the longitudinal stress has to be equal to the hoop stress; this is accomplished by applying an external end load to a capped-end

*Open-end tube with end load supported by floating packings.

Fig. 5.13 Specimen side of tensile creep and stress-rupture machine.

Fig. 5.14 Loading side of tensile creep and stress-rupture machine.

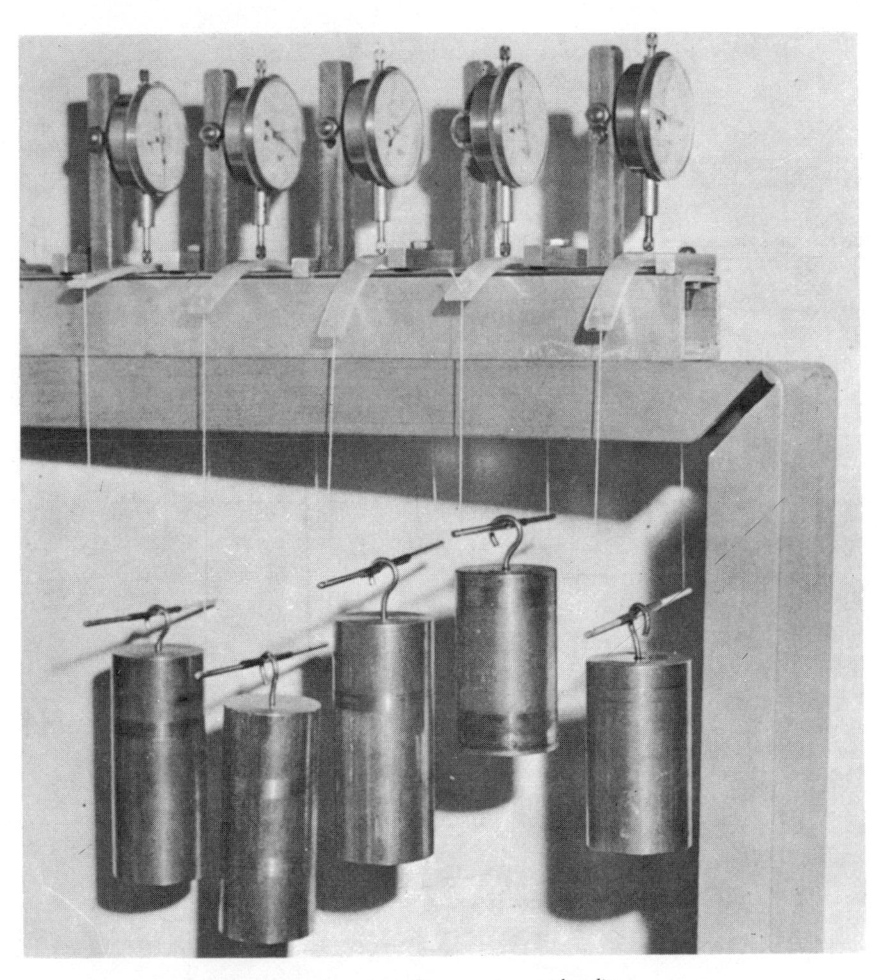

Fig. 5.15 Apparatus for creep in pure bending.

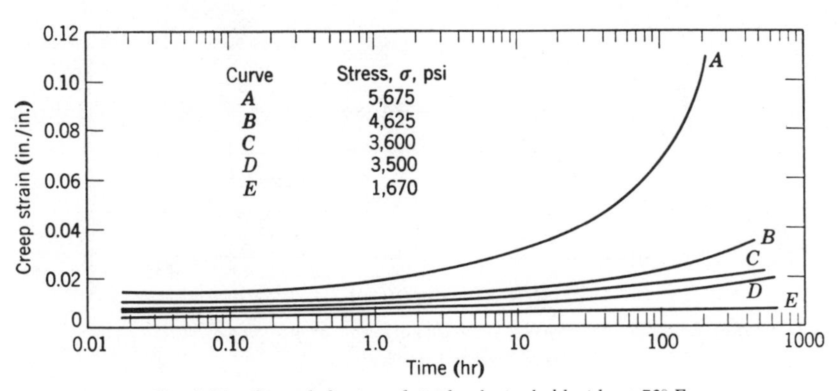

Fig. 5.16 Creep behavior of rigid polyvinyl chloride at 72° F.

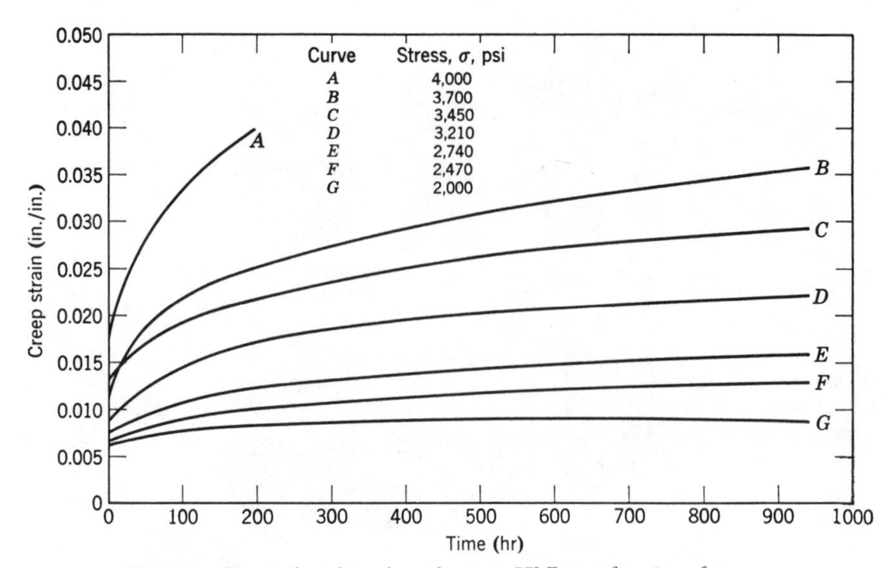

Fig. 5.17 Creep of methacrylate plastic at 77° F as a function of stress.

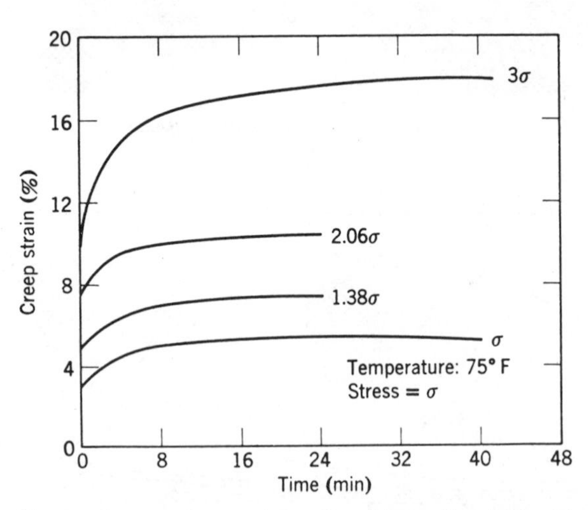

Fig. 5.18 Effect of load on the creep characteristics of a solid propellant. [After Blatz (5), courtesy of The American Chemical Society.]

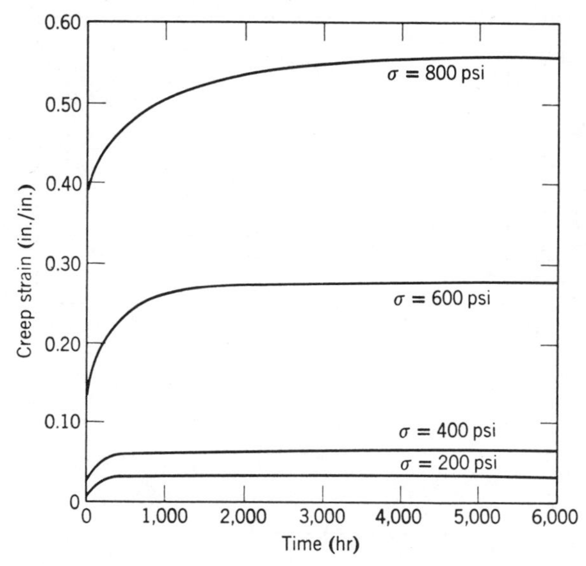

Fig. 5.19 Creep of a polyethylene plastic at 85° F as a function of stress.

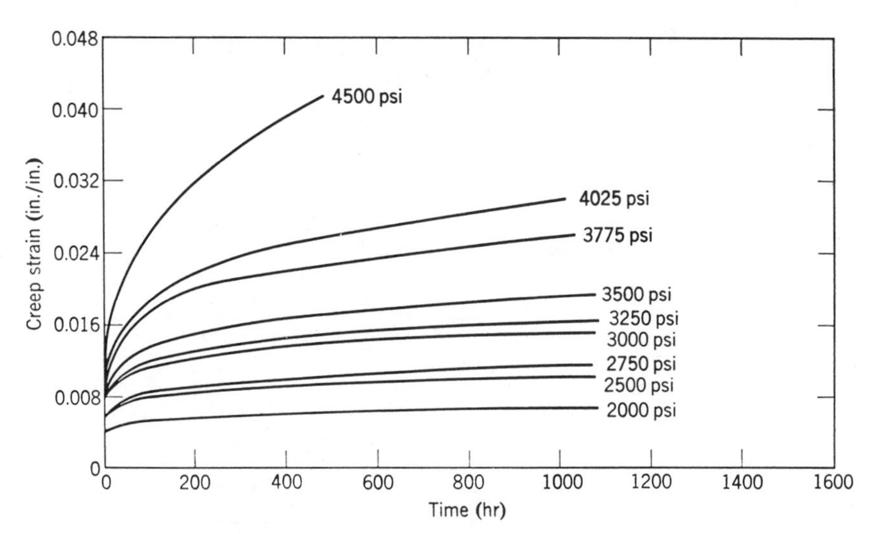

Fig. 5.20 Tension creep of an acrylic plastic at 72° F as a function of stress. [After Marin et. al. (23), courtesy of The American Society of Mechanical Engineers.]

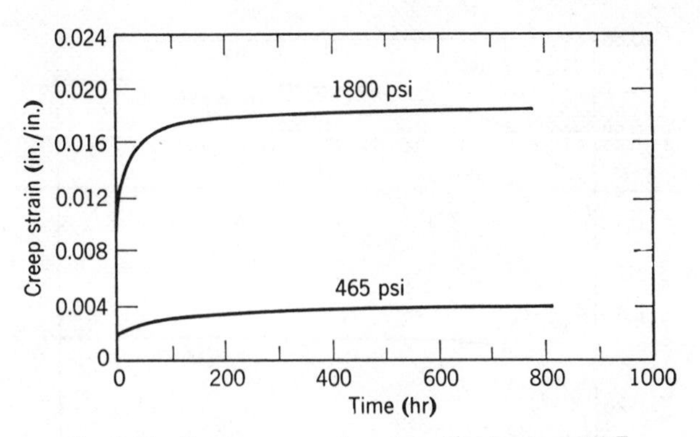

Fig. 5.21 Tension creep of a polyamide plastic at 73° F.

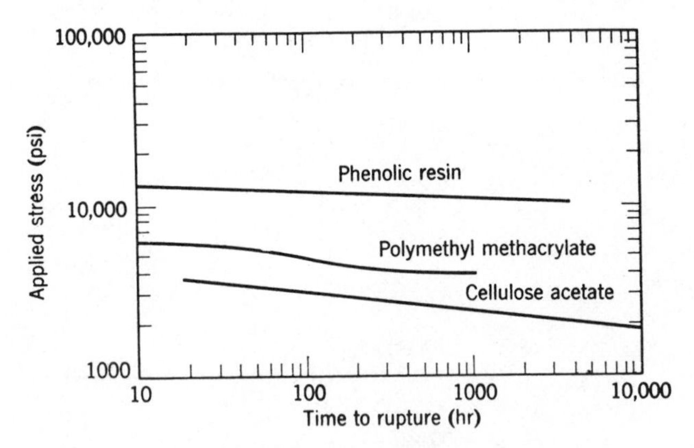

Fig. 5.22 Tensile stress-rupture of some plastics at 73° F.

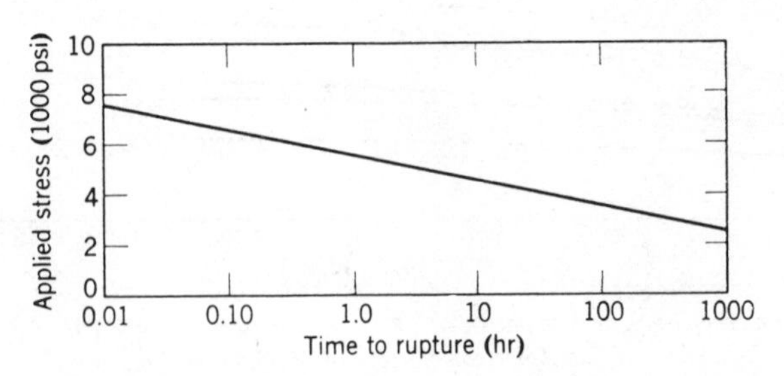

Fig. 5.23 Tensile stress-rupture of rigid polyvinyl chloride at 73° F.

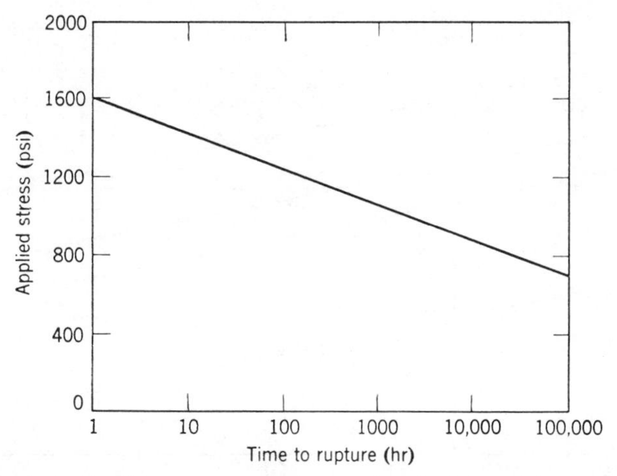

Fig. 5.24 Tensile stress-rupture behavior of a polyethylene plastic at 73°F.

cylinder having the value $pd/4t$; the resulting equivalent stress is then

$$p = \frac{\bar{\sigma}}{[\frac{1}{4}(d/t)^2 + (d/t) + 1]^{1/2}} \tag{5.6}$$

For the stress ratio $\sigma_z/\sigma_h = 2$, the added end load must equal $3pd/4t$ and the equivalent stress is

$$p = \frac{\bar{\sigma}}{[\frac{3}{4}(d/t)^2 + \frac{3}{2}(d/t) + 1]^{1/2}} \tag{5.7}$$

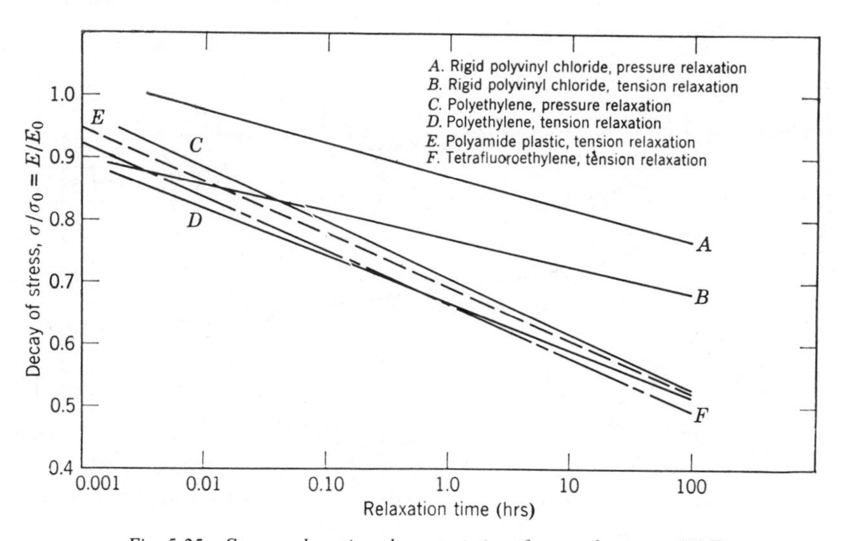

Fig. 5.25 Stress-relaxation characteristics of some plastics at 73° F.

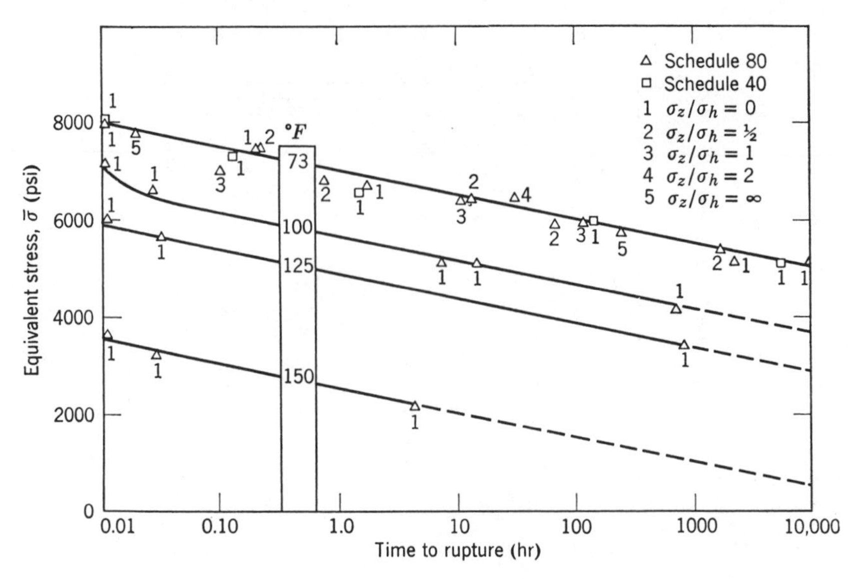

Fig. 5.26 Stress-rupture data for 2-in. rigid polyvinyl chloride pipe as a function of temperature.

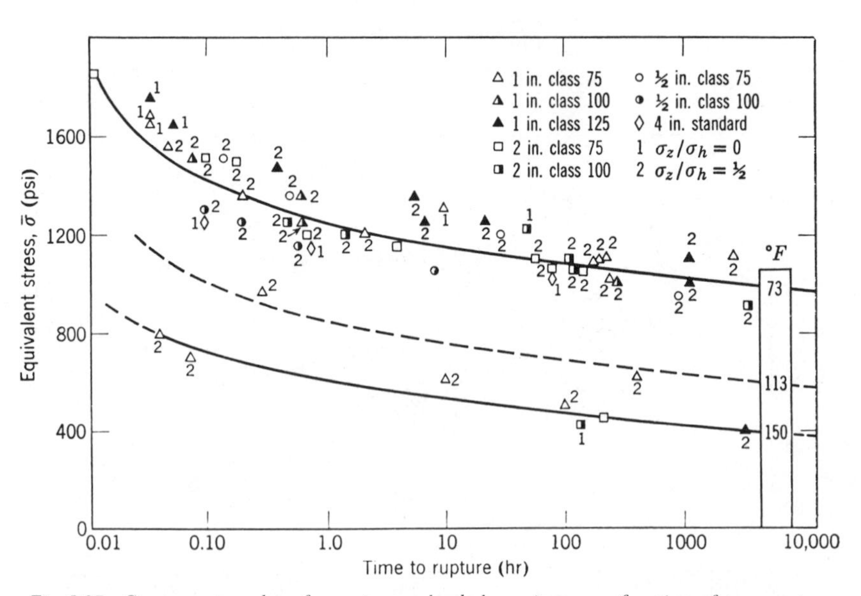

Fig. 5.27 Stress-rupture data for various polyethylene pipes as a function of temperature.

The stress ratio ∞ represents the case of pure tension. For materials known to have orthotropic properties Eq. 4.119 can be used. For example, by taking the cylinder discussed in Chapter 4, Eqs. 4.123–4.125, the following results are obtained.

For $\sigma_z/\sigma_h = 0$,

$$p = \frac{\bar{\sigma}}{[\frac{1}{4}(\alpha d/t)^2 + (d/2t)(2\alpha^2 - \alpha^4) + 1 + \alpha^2 - \alpha^4]^{1/2}} \qquad 5.8$$

For $\sigma_z/\sigma_h = \frac{1}{2}$,

$$p = \frac{\bar{\sigma}}{[\frac{1}{16}(d/t)^2(4\alpha^2 - 2\alpha^4 + 1) + (d/4t)(4\alpha^2 - 3\alpha^4 + 2) + \alpha^2 - \alpha^4 + 1]^{1/2}} \qquad 5.9$$

For $\sigma_z/\sigma_h = 1$,

$$p = \frac{\bar{\sigma}}{[\frac{1}{4}(d/t)^2(\alpha^2 - \alpha^4 + 1) + (d/t)(\alpha^2 - \alpha^4 + 1) + \alpha^2 - \alpha^4 + 1]^{1/2}} \qquad 5.10$$

For $\sigma_z/\sigma_h = 2$,

$$p = \frac{\bar{\sigma}}{[\frac{1}{4}(d/t)^2(\alpha^2 - 2\alpha^4 + 4) + (d/2t)(2\alpha^2 - 3\alpha^4 + 4) + \alpha^2 - \alpha^4 + 1]^{1/2}} \qquad 5.11$$

5-2 MATHEMATICAL DESCRIPTION OF VISCOELASTIC BEHAVIOR

It has already been pointed out that plastics are sensitive to time effects and that “failure” can be considered as being an actual rupture (stress-rupture) or an excessive creep deformation. Since most designs are intended to operate for long periods of time in the prerupture zone, attention will now be given to the problem of characterizing creep or viscoelastic deformation that takes place when the part is subjected to a stress. In the following paragraphs methods used to characterize viscoelastic flow will be considered, which will include a discussion of mathematical models of *linear viscoelastic* behavior, the correlation of creep and stress-relaxation, and a brief treatment of the *equivalent elastic problem.* The method of the equivalent elastic problem is a major assistance to designers of plastics parts since, by knowing the elastic solution to a problem, the viscoelastic solution can be deduced readily by simply replacing elastic physical constants with viscoelastic constants that are experimentally determined.

Mathematical model of prerupture viscoelastic behavior

Although a wide range of properties is exhibited by plastics, they all have in common a structure composed of long, entangled threadlike molecules. Under the action of applied forces these “threads” can uncoil, straighten out, or otherwise accommodate the imposed force, and it is this ability to conform with

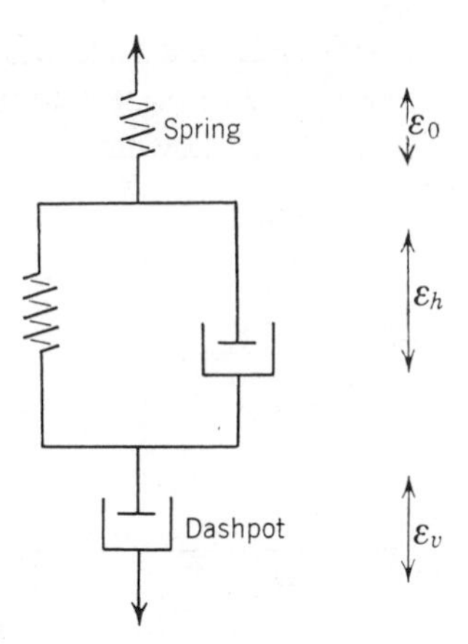

Fig. 5.28 Model for deformation of plastics.

loads that makes plastics interesting as materials of construction. Within the threadlike structure of molecules there are bonds holding the individual molecules together; consequently, from a deformation point of view it is thought that three general types of static deformation can occur. The mathematical model depicting these behaviors is shown in Fig. 5.28.

Ordinary elastic deformation, ε_0. This type of deformation, often called *ideal deformation*, follows Hooke's law of linear proportionality of stress to strain. No time effect is involved, and the deformation is believed to be associated primarily with the molecular bonding forces in the structure. The spring part of the model represents this type of deformation.

Highly elastic deformation. ε_h. This type of deformation depicted by the spring and dashpot part of the model is called *nonideal deformation*. Stress is no longer linearly proportional to strain, and under certain circumstances complete recovery can occur when the imposed load is removed. Physically, this type of deformation is associated with uncoiling of strings of molecules.

Viscous deformation, ε_v. This type of deformation is associated with the slipping of strings of molecules over each other and is depicted by the dashpot in the model. Permanent strains result, and time is a major factor.

The total deformation in a system as discussed is thus

$$\mathscr{E} = \varepsilon_0 + \varepsilon_h + \varepsilon_v \tag{5.12}$$

the individual terms of which will be discussed later. Most plastics exhibit combinations of behavior, and for this reason various models have been devised to interpret observed behavior. The simplest model shown in Fig. 5.29*A* consists of a spring and a dashpot in series; this combination is known as a *Maxwell model* and represents behavior consisting of instantaneous extension (spring) and time-dependent flow (dashpot). In this system only the elastic strain is recoverable; therefore such a model cannot represent materials exhibiting viscoelastic recovery and no permanent set. The model shown in Fig. 5.29*B*, known as a *Voigt-Kelvin system*, consists of a spring and a dashpot in parallel.

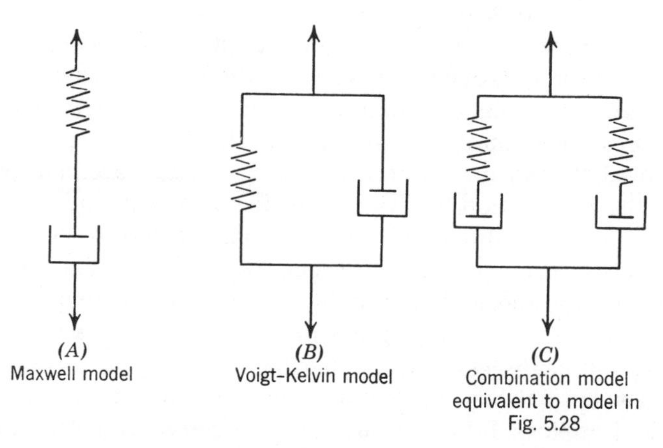

Fig. 5.29 Various models for deformation of plastics.

Again, this model does not give rise to instantaneous elastic deformation and thus cannot represent the deformation behavior of most plastics; it does, however, show recoverable deformation. For most plastics a combination of the Maxwell and Voigt-Kelvin models gives a good approximation of the observed behavior. The model shown in Fig. 5.28, for example, consists of a Voigt-Kelvin unit to which has been added a spring and a dashpot. In response to load this model is equivalent to that shown in Fig. 5.29*C*, two Maxwell elements in parallel (2). In the latter system, when load is imposed, both elements elongate equally according to their spring constants; following this both dashpots react distributing the load, and creep occurs as long as load is applied. When the load is removed, the springs retract, but differently because of their different spring constants; the dashpots move to compensate for this until the load drops to zero, at which time a small permanent set remains. Thus the models in Fig. 5.28 and 5.29*C* represent materials having instantaneous extension, time-dependent creep, strain recovery, stress-relaxation, and permanent set.

The deformations referred to have all been of a prerupture nature. However, since in viscous flow the strings of molecules are envisioned as slipping over one another, some limit would be reached where actual rupture would occur. This is a perplexing problem which has not yet been adequately explained. Alfrey (2) contends that the tensile strength of a plastic increases as molecular weight increases; thus the theory is that low molecular weight plastics rupture by the pulling apart of intertangled groups of molecular string ends, whereas high molecular weight materials break mainly by rupture of the bonds holding the molecules together.

Theory of linear viscoelasticity

Linear viscoelastic theory and its applications to static-stress analysis procedures have received considerable attention over the past few years, and the theory is now highly developed (2, 4, 13, 19, 20, 22, 31, 32, 36). According to this theory the material is linearly viscoelastic if when stressed below some limiting stress (usually around half the short-time yield strength), the strains at any time are approximately linearly proportional to the imposed stresses for small strains. Such behavior is typified by *portions* of the creep data presented in Figs. 5.16–5.21, and it furnishes the basis for fairly accurate predictions concerning the deformation behavior of plastics when subjected to loads over long periods of time. It should be noted that linear behavior, as defined, does not always persist throughout the time span over which the data are acquired; this means that the theory is not valid in the nonlinear regions and other means must be found for carrying out predictions. This is a very important point, and it will be discussed again after the details of the theory have been presented. The basic theory assumes a timewise linear relation between stress and strain and employs the models previously discussed; some of the highlights of the linear theory will now be given.

The basic differential equation for a Maxwell unit is

$$d\varepsilon/dt = (1/E)(d\sigma/dt) + \sigma/\eta \qquad 5.13$$

which integrates to

$$\sigma = \sigma_0 e^{-t/\tau} + Ee^{-t/\tau} \int_0^t e^{t/\tau} \frac{d\varepsilon}{dt} dt \qquad 5.14$$

where σ is the stress at any time t, τ is the relaxation time, σ_0 the initial tensile stress, ε the strain, E the modulus at time t, η the viscosity, and e the base of natural logarithms. For constant strain (relaxation test), $d\varepsilon/dt$ in Eq. 5.14 is zero so that

$$\sigma = \sigma_0 e^{-t/\tau} \qquad 5.15$$

For a single nondegenerate Voigt-Kelvin unit the basic differential equation is

$$\sigma = \eta(d\varepsilon/dt) + E\varepsilon \qquad 5.16$$

which integrates to

$$\varepsilon = \varepsilon_0 e^{-t/\tau} + \frac{1}{\eta} e^{-t/\tau} \int_0^t e^{t/\tau} \sigma t \, dt \qquad 5.17$$

Since the instantaneous deformation ε_0 at constant stress σ is zero, Eq. 5.17 reduces to

$$\varepsilon = (\sigma/E)(1 - e^{-t/\tau}) \qquad 5.18$$

which represents part of the deformation for the model shown in Fig. 5.28. The additional deformation contributed by the spring and dashpot is found by noting that the model actually consists of one nondegenerate Voigt-Kelvin unit and two degenerate elements. When the degenerate element is the spring,

$$\varepsilon = \sigma/E_0 \qquad 5.19$$

where E_0 is the instantaneous modulus. When the dashpot is the degenerate element,

$$\varepsilon = \sigma t/\eta \tag{5.20}$$

Therefore the total deformation experienced is

$$\mathscr{E} = \sigma/E_0 + (\sigma/E)(1 - e^{-t/\tau}) + \sigma t/\eta \tag{5.21}$$

or, excluding permanent set and considering only the creep involved in monotonic loading,

$$\mathscr{E} = \sigma/E_0 + (\sigma/E)(1 - e^{-t/\tau}) \tag{5.22}$$

The term τ in Eqs. 5.15 and 5.22 has different significance since in one equation the term is based on a concept of relaxation and in the other on the basis of creep; thus in the literature these are respectively referred to as a *relaxation time* and a *retardation time*, leading for infinite elements in the deformation model to complex quantities known as *relaxation and retardation functions*. One of the principal accomplishments of viscoelastic theory is the linking of these quantities analytically so that creep deformation can be predicted from relaxation data and vice versa. A concise method for establishing this relationship involves a consideration of Boltzmann's superposition principle (32) combined with a Duhamel-type integral (20) associated with the mathematics of transient phenomena (34).

Let $E_R(t)$ be the relaxation modulus, that is, σ/E in a stress-relaxation test and $J_c(t)$ the creep compliance, that is, ε/σ in a creep test. These are both functions of the time t, therefore, by using the general theory of loading (20),

$$\sigma(t) = \int_0^t \frac{d\varepsilon}{dt}(\tau) E_R(t-\tau)\, d\tau \tag{5.23}$$

and

$$\varepsilon(t) = \int_0^t \frac{d\sigma}{dt}(\tau) J_c(t-\tau)\, d\tau \tag{5.24}$$

Denoting the Laplace transform by a double bar over the variable, Eqs. 5.23 and 5.24 can be transformed by means of the convolution theorem noting that

$$d\bar{\bar{\sigma}}/dt = x\sigma \tag{5.25}$$

where x is a variable in the transform operation. This gives

$$\bar{\bar{\sigma}} = x\bar{\bar{\varepsilon}}\bar{\bar{E}}_R \tag{5.26}$$

and

$$\bar{\bar{\varepsilon}} = x\bar{\bar{\sigma}}\bar{\bar{J}}_c \tag{5.27}$$

Eliminating $\bar{\bar{\sigma}}/\bar{\bar{\varepsilon}}$ gives

$$x^2 \bar{\bar{E}}_R \bar{\bar{J}}_c = 1 \tag{5.28}$$

which establishes a relationship between the transforms of the relaxation modulus and the creep compliance. It is now necessary to select a law that can be used to represent the relaxation behavior, that is, the behavior shown in Fig. 5.25. Two laws that fit published data are the power relations

$$\sigma/\sigma_0 = E_R/E_0 = Kt^{-m} \tag{5.29}$$

and

$$\sigma/\sigma_0 = E_R/E_0 = (t/b)^{-m} \tag{5.30}$$

where K, b, and m are experimental constants obtained from log-log plots of the data. Usually the constants can also be obtained from semilog plots. Taking the Laplace transform of Eq. 5.29 gives the result

$$\bar{\bar{E}}_R = \frac{KE_0\Gamma(1-m)}{x^{(1-m)}} \tag{5.31}$$

where Γ is the gamma function. Thus, combining Eqs. 5.28 and 5.31, the transform of the creep compliance is

$$\bar{\bar{J}}_c = \frac{1}{KE_0\Gamma x^{m+1}(1-m)} \tag{5.32}$$

This can be inverted directly to give

$$J_c = \frac{t^m}{KE_0\Gamma(1-m)(1+m)} \tag{5.33}$$

or

$$J_c = \frac{1}{KE_0} \frac{\sin m\pi}{m\pi} (t^m) \tag{5.34}$$

For small values of m, which is the usual case, Eq. 5.34 is simplified to

$$J_c = E_R^{-1} \tag{5.35}$$

In the preceding analysis which is restricted to simple static cases for essentially nonchanging volume and surface conditions, except the assumed small deformations, the viscoelastic problem is equivalent to the elastic problem with the same initial conditions. Thus, by applying the Laplace transformation, time is eliminated through the transformation parameter x and the analysis of the viscoelastic problem is reduced to that of an elastic problem; inversion then gives the solution to the original viscoelastic problem. Thus, through this theory, the extensive literature in elasticity can be utilized, and any term $1/E$ in the elastic solution is replaced with the viscoelastic term $(1/KE_0)(\sin m\pi/m\pi)(t^m)$ and any term E can be replaced by E_0Kt^{-m} for times greater than about an hour. If the alternate power law is used (Eq. 5.30), the replacement terms are $(1/E_0)(\sin m\pi/m\pi)(t/b)^m$ and $E_0(t/b)^{-m}$ respectively. It is thus seen that the solution to a vast number of viscoelastic problems can be easily solved by having available the elastic solution and relaxation data from which the constants K, b, and m can be obtained. It is emphasized again, however, that this theory applies only for small strains where the principle of linear viscoelasticity is valid. To test the validity for any material it is necessary to conduct stress-relaxation tests at different initial deformations; if the material is linearly viscoelastic, the various resulting plots of stress decay versus time will superpose. An alternate check, of course, is to conduct creep tests at different loads and check for linearity.

By going back to Eq. 5.22 and using the preceding results, it is seen that, in terms of the experimental constants and the instantaneous initial strain ε_0 in a creep problem,

$$\varepsilon = \varepsilon_0(\sin m\pi/m\pi)(t^m) \qquad 5.36$$

or alternately

$$\varepsilon = \varepsilon_0(\sin m\pi/m\pi)(t/b)^m \qquad 5.37$$

For most commercial plastics the trigonometric term in Eqs. 5.36 and 5.37 reduces to unity since m is usually less than about 0.20; therefore for design purposes

$$\varepsilon = \varepsilon_0(t^m/K) = \varepsilon_0(t/b)^m \qquad 5.38$$

which when combined with Eqs. 5.29 and 5.30 gives the result

$$(\varepsilon/\varepsilon_0)_{\text{creep}} = (\sigma_0/\sigma)_{\text{relaxation}} \qquad 5.39$$

Thus, by determining values of σ_0/σ from a relaxation test (Fig. 5.25), creep strains can be calculated by use of Eq. 5.39 in the form

$$\varepsilon = \varepsilon_0(\sigma_0/\sigma) = (\sigma_0/E_0)(\sigma_0/\sigma) \qquad 5.40$$

In Eq. 5.40 the term $(1/E_0)(\sigma_0/\sigma)$ may be thought of as a *time-modified modulus*, that is, $(1/E_0)(\sigma_0/\sigma) = 1/E$, from which E, the modulus at any time t, is

$$E = E_0(\sigma/\sigma_0) \qquad 5.41$$

which is the value used to replace E in the elastic solution to a problem. Where Poisson's ratio appears in the elastic solution it is replaced in the viscoelastic solution by (22)

$$\nu = (3B - E)/6B \qquad 5.42$$

where B is the bulk modulus, a value that remains essentially constant throughout deformation. These relationships will be demonstrated in the next section.

In summary of this part of the discussion, creep and stress-relaxation behavior for plastics are closely related for small strains and one can be predicted from a knowledge of the other. From a practical viewpoint the designer is concerned with the deformation behavior of the material, and deformation can be predicted by the use of standard elastic stress analysis formulas where the elastic constants E and ν can be replaced by their viscoelastic equivalents given in Eqs. 5.41 and 5.42. For many materials data are available on the effect of time, temperature, and strain rate on modulus. If data are not available, creep tests for applications like rockets and missiles data obtained over a time period of 4–5 sec to an hour give the essential information. For structural applications like pipelines, data over a period of years are required. This is usually impossible since the designer cannot wait for several years while data are being collected. This is one of the serious limitations of plastics design; the designer cannot wait for the data and even if he did the chances are that the intended design would be made of some newer formulation. As a compromise, and relying on theory, data obtained in relatively short-time tests are extrapolated to long times. When this is done, however, the limitations of the method should be kept carefully in mind. There is a strong temptation to accept extrapolated data since it usually appears logical and consistent when plotted on log coordinates where a time of hundreds of thousands of hours can be compressed into a plot about an inch long.

Rate theory for creep and stress rupture

The theory of linear viscoelasticity given provides a quick and reliable method for predicting creep deformations when the material in question happens to conform to the restrictions imposed by the theory. An alternate method available to the engineer involves manipulation of *rate theory*. This procedure is somewhat more difficult than the linear theory and requires acquiring considerable test data. However, once the basic data are available the indications are that considerably more latitude is obtained and more materials are included. Also the method can be used to predict both creep and stress-rupture of plastics, which is a strong factor in favor of the method. Most engineers are familiar with the rate concept in connection with the *Larson-Miller parameter* discussed in Chapter 15 for predicting stress-rupture behavior of materials at high temperature. The method appears to have considerable merit and has been popularized for use with plastics particularly by Carey (6), Gloor (11), and Goldfein (12). The elements of the theory have been summarized by Goldfein, (12) and the following description is based on his paper.

Rate theory is based on the Arrhenius equation

$$d \ln K/dT = A/RT^2 \tag{5.43}$$

integration of which leads to the result

$$\ln \frac{K_0}{K} = \frac{A}{R}\left(\frac{1}{T} - \frac{1}{T_0}\right) \tag{5.44}$$

where K is the reaction velocity constant, A the energy of activation of the process, R the gas constant, T the absolute temperature, and the subscript o the upper limit of integration. If T_0 represents the temperature at which the material has no strength, K_o is the rupture rate at that temperature. The activation energy A does not vary appreciably with temperature. By assuming A to be constant, Eq. 5.44 may be rewritten as follows:

$$\frac{A}{R} = K' = \frac{TT_o}{T_o - T}(\log K_0 - \log K) \tag{5.45}$$

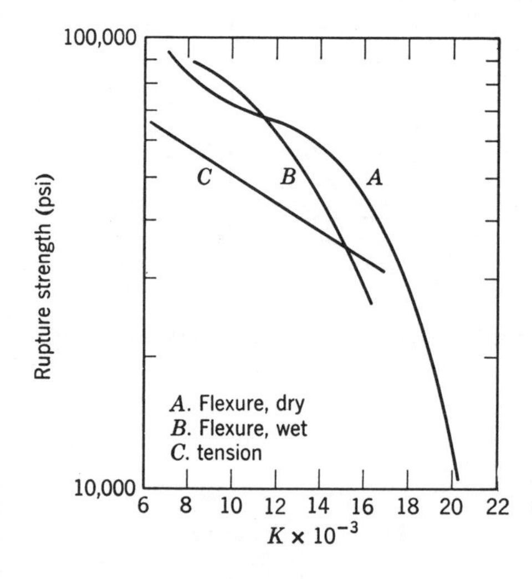

Fig. 5.30 Master rupture data at 73° F for an epoxy plastic. [After Goldfein (12), courtesy of Breskin Publishing Co., Inc.]

Furthermore, if rupture is defined as the separation of molecules in the structure, the maximum rupture rate K_o would be that at which all the molecules in a mole of material separate at the same time, that is, 6.02×10^{23} molecules (Avogadro's number) per unit time, the log of which is 22.78. For practical purposes, discussed by Goldfein, the usable value of K_o is taken as 20 so that Eq. 5.45 becomes

$$K' = \frac{TT_o}{T_o - T}(20 - \log K) \qquad 5.46$$

It is assumed that the properties of the material (physical and chemical) are the same before and after rupture so that the concentration x of material undergoing deformation is related to K as

$$K = x/t \qquad 5.47$$

where t is time. Therefore letting x equal unity and substituting Eq. 5.47 in

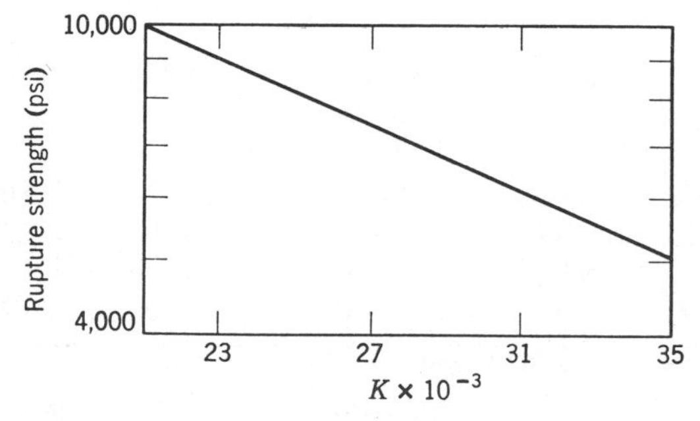

Fig. 5.31 Master rupture data at 73° F for an acetal plastic. [After Goldfein (12), courtesy of Breskin Publishing Co., Inc.]

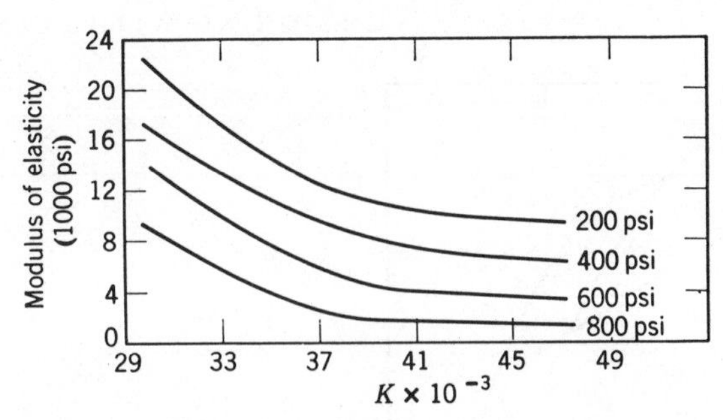

Fig. 5.32 Master modulus curve for a polyethylene plastic. [After Goldfein (12), courtesy of Breskin Publishing Co., Inc.]

Table 5.8 Stress-Rupture Data at 73°F for an Epoxy Plastic

Property	Time, hr	K	Stress from Master Rupture Curve, psi (Fig. 5.30)	Calculated Stress, psi	Observed Stress, psi	Difference, %
Flexural	1	12,660	63,500	58,900	57,000	−3.3
strength,	10	13,300	61,500	57,100	54,000	−5.7
dry	100	13,930	59,100	54,800	51,000	−7.5
	1,000	14,570	56,200	52,200	48,000	−8.8
	10,000	15,200	52,200	48,400	44,500	−9.0
	20,000	15,350	51,200	47,500	43,800	−8.7
	40,000	15,600	49,500	45,900	43,000	−6.7
Flexural	1	12,660	57,500	53,400	54,500	2.0
strength,	10	13,300	52,000	48,300	50,000	3.4
wet	100	13,930	45,500	42,200	45,000	6.2
	1,000	14,570	40,300	37,400	40,500	7.7
	10,000	15,200	35,000	32,500	36,000	9.7
	20,000	15,350	33,700	31,300	34,000	7.9
	40,000	15,600	31,500	29,200	32,800	11.0
Tensile	1	12,660	42,000	43,500	43,000	−1.2
strength,	10	13,300	40,300	41,800	42,000	0.5
dry	100	13,930	38,500	39,900	41,000	2.7
	1,000	14,570	36,800	38,100	39,000	2.3
	10,000	15,200	35,200	36,500	37,500	2.7
	20,000	15,350	34,800	36,100	37,200	3.0
	40,000	15,600	34,100	35,300	36,250	2.6

NOTE: $t = 10^{-3}$ hr, $T_0 = 3360$°F, abs.
After Goldfein, Reference 12, courtesy of Breskin Publishing Co., Inc.

Table 5.9 Stress-Rupture Data at 73° F for an Acetal Plastic

73°F for an Acetal Plastic

Time, hr	K	Tensile Stress-Rupture: Calculated, Fig. 5.31, psi	Observed, psi	Difference, %
1.0	27,410	7,130	7,800	8.6
100	28,800	6,700	7,000	4.3
1,000	30,100	6,320	6,100	−3.6
10,000	31,400	5,950	5,600	−6.3

After Goldfein, Reference 12, courtesy of Breskin Publishing Co., Inc.

Table 5.10 Creep Data at 85° F for a Polyethylene Plastic

Stress level, psi	Time, hr	K	Modulus of Elasticity, E, psi (Fig. 5.32)	Calculated Creep Strain, %	Observed Creep Strain, %	Difference, %
200	100	42,400	9,850	2.03	1.70	−19.4
	300	43,300	9,750	2.05	2.20	6.8
	600	43,800	9,630	2.08	2.03	−2.5
400	100	42,400	7,000	5.71	4.60	−24.1
	300	43,300	6,800	5.88	5.60	−5.0
	500	43,700	6,750	5.93	6.00	1.2
	1,000	44,300	6,700	5.97	6.00	0.5
600	100	42,400	3,800	15.8	13.5	−17.0
	300	43,300	3,650	16.5	14.2	−16.2
	600	43,800	3,600	16.7	15.1	−10.6
	1,000	44,300	3,500	17.2	16.0	−7.5
	2,000	44,850	3,400	17.7	17.0	−4.1
800	100	42,400	1,650	48.5	43.9	−10.5
	500	43,700	1,600	50.0	47.3	−5.9
	750	44,000	1,580	50.5	48.8	−3.5
	1,000	44,300	1,550	51.5	50.0	−3.0
	2,000	44,850	1,500	53.3	53.0	−0.6
	3,000	45,150	1,480	54.0	54.5	0.9
	4,000	45,500	1,430	56.0	55.0	−1.8

NOTE: $t = 10^{-5}$ hr, $T_0 = 760$°F, abs.
After Goldfein, Reference 12, courtesy of Breskin Publishing Co., Inc.

Eq. 5.46 gives the result

$$K' = \frac{TT_0}{T_o - T}(20 + \log t) \qquad 5.48$$

which is the basic equation for plastics. For some materials like glass/epoxy, the term $T_0/(T_0 - T)$ in Eq. 5.48 approaches unity, and the equation reduces to the familiar *Larson-Miller parameter* (Chapter 15).

$$K' = T(20 + \log t) \qquad 5.49$$

The practical utility of Eqs. 5.48 and 5.49 is that, knowing the temperatures involved, the deformation or rupture time can be predicted over a long time period on the basis of short-time tests. For example, for many plastics the time t in Eqs. 5.48 and 5.49, which represents the time the material is under constant load, is approximately 10^{-5} hours for stress-rupture. Thus with this value *master rupture curves* can be computed for all values of T; such curves are shown in Fig. 5.30 and 5.31 for two plastics. Thus, for design purposes, to be illustrated presently, if the various t and T values are known, K' can be computed and rupture stresses read directly from the curves.

Creep deformations are calculated by dividing the stress by the modulus. Briefly, the deformation observed in a tensile test, for example, at an elevated temperature also represents deformation at lower temperature over a longer time span. Thus, from time and temperature, K' can be computed and the associated modulus obtained from a *master modulus curve.* By having the modulus and knowing the stress the deformation can be calculated in the usual way from elastic stress analysis formulas as described earlier. Master modulus curves are shown in Fig. 5.32 for a polyethylene material. The practical utility of this method is demonstrated by the data in Tables 5.8 to 5.10.

5-3 APPLICATION OF VISCOELASTIC PRINCIPLES IN DESIGN

In the preceding sections of this chapter a general introduction to plastics was given; the range of properties exhibited by these materials was noted and general mathematical descriptions were developed to explain and predict mechanical behavior. In this section the objective is to draw from the preceding sections and relate this to practical procedures for performing stress analyses on parts or structures made of viscoelastic materials.

Design criteria

The general design criteria applicable to plastics are the same as those applicable to metals, for example, at elevated temperature (Chapter 15); that is, design is based on a consideration of a limiting deformation on the one hand and a limiting stress for stress-rupture failure on the other. There are, of course, many designs that are concerned primarily with weight (1), but the concern here is about applications like plastic piping systems and tanks where creep and stress-rupture have to be allowed for in the design. The consideration of creep and stress-rupture is the major problem. There are other problems involving the short-time properties of plastics like plastic rocket motor cases, where the short-time strength against bursting is important, and applications like aircraft windows where the distortion characteristics under short-term loading have to be considered (8). Also, for quality control and testing, the short-time properties of plastics are measured. In general then it can be said that design of plastics structures can involve considerations of short-time strength, deformation characteristics, creep, and stress-rupture. Each of these categories will be illustrated by an example.

Methods of analysis

In computing the ordinary short-time characteristics of plastics like deflection of beams or plates, dilation of tubes under pressure, and so on, the standard stress analysis formulas given in the preceding four chapters may be used. For predicting creep and stress-rupture behavior the method will vary depending on the circumstances. If the material is known to be linearly viscoelastic, relaxation data such as shown in Fig. 5.25 together with Eqs. 5.40 to 5.42 can be used to predict creep deformations. For other situations, or even the case just mentioned, the rate theory method can be used. For stress-rupture calculations the rate theory would be used. It may frequently happen that fairly complete data are available for the stress-rupture characteristics of materials in tension; if this is so and the problem involves combined stresses, predictions concerning combined stresses stress-rupture may be obtained through the use of Eq. 3.524.

Solution of problems

Tension (compression) of linearly viscoelastic prismatic bars (Fig. 5.33). This is the simplest type of problem to analyze since most test data on plastics are based on tensile-type tests. Thus the data are directly applicable to the problem without interpretation through theory. Figure 5.33 shows a bar of viscoelastic material subjected to a load F, assumed to give an initial stress below the short-time yield strength of the material. Under this condition the initial modulus E_0 is given by Eq. 1.7,

$$E_0 = \sigma/\varepsilon_0 \qquad 5.50$$

where σ is the applied stress and ε_0 is the initial strain. If the load F is sustained

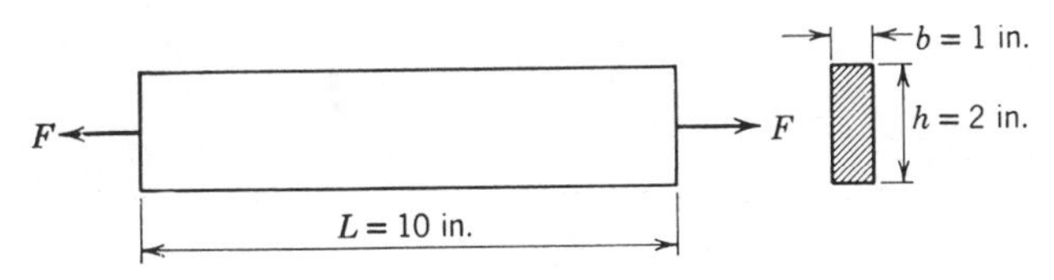

Fig. 5.33 Prismatic plastic bar in tension.

for time t, the strain at the end of this time period will be the sum of the initial strain ε_0 and that due to viscoelastic flow. Therefore, the new or *apparent modulus* at time t is obtained by rewriting Eq. 5.50,

$$E_t = \frac{\sigma}{\varepsilon_0 + \varepsilon_t} \qquad 5.51$$

where ε_t is the creep or viscoelastic strain. Suppose the bar is subjected to a stress of 1000 psi that induces a total strain of 0.10 in.; then from Eq. 5.50

$$E_o = \frac{1000}{0.10/10} = 100{,}000 \text{ psi}$$

If the load is sustained for 1000 hr and the total creep experienced is 0.15 in., the modulus at 1000 hr is

$$E_{t=1000} = \frac{1000}{0.01 + 0.15/10} = 40{,}000 \text{ psi}$$

Thus, for linearly viscoelastic behavior, by measuring creep strains and using Eq. 5.51, *time-modified* modulus curves can be drawn. Having established such curves for a particular material the data can then be used to predict behavior for other conditions. Typical time-modified modulus curves for linearly viscoelastic materials are shown in Fig. 5.34. It should be carefully noted that these data are for a specific temperature and maximum limiting strain. For example, suppose the bar in Fig. 5.33 is made of a polyethylene material conforming to the behavior shown in Fig. 5.34. Equation 5.51 indicates that

$$E_t = \sigma/\varepsilon_t \qquad 5.52$$

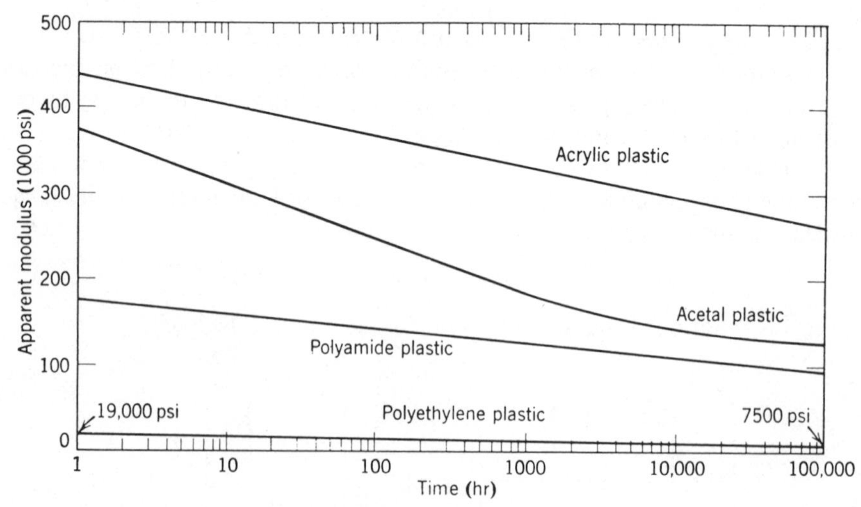

Fig. 5.34 Time-modified modulus curves for four plastics at 73°F (limited to use at maximum of 0.5% strain).

and Fig. 5.34 shows that the limiting modulus values are 19,000 psi and 7500 psi. Thus, by using these values in Eq. 5.52, it is seen that

$$\sigma/\varepsilon_t = 19{,}000 \text{ or } \sigma/\varepsilon_t = 7500$$

Suppose the bar is stressed at 300 psi. The initial elastic modulus of this material is 20,000 psi at 73°F., therefore the initial strain ε_0 is, from Eq. 5.50,

$$\varepsilon_0 = \sigma/E_0 = 300/20{,}000 = 0.015 \text{ in./in.}$$

Then, from Eqs. 5.40 and 5.41, the creep strain ε_t is

$$\varepsilon_t = \varepsilon_0(E_0/E) \tag{5.53}$$

Thus, for the conditions at hand, the following predictions can be made:

Time (hr)	Apparent Modulus (psi) (from Fig. 5.34)	Calculated Creep Strain (Eq. 5.53) (in./in.)
1	19,000	0.0158
10	16,500	0.0182
100	14,000	0.0212
1,000	12,000	0.0250
10,000	10,000	0.0300
100,000	7,500	0.0400

Since these strains are all in excess of 0.5% they are inaccurate for the material used. In order to stay below the maximum strain of 0.5% the imposed load could not exceed 100 psi. If the stress were 50 psi, the value of ε_0 would be 0.0025 in./in. For a strain of 0.5%, ε_0 would be multiplied by 2 which corresponds (see preceding table) to 10,000 hr operation. Thus, with a load of 50 psi accurate creep predictions could be made up to 10,000 hr service for the material and conditions imposed. For loads in excess of 100 psi creep predictions would have to be made by the rate theory method (see Table 5.10, for example, for data on another polyethylene material).

Bending (Fig. 5.35). The bars shown are made of an acrylic material conforming to the behavior shown in Fig. 5.34. For both cases a limiting deflection of 0.15 in. for one year's service is specified; find the limiting value of the applied moment M or load W.

For the bar shown in Fig. 5.35A (pure bending), the maximum deflection occurs at midspan and is given by Eq. 1.45.

$$\delta_{\max} = ML^2/8EI \tag{5.54}$$

From Fig. 5.34 the modulus of the material at one year is 300,000 psi and for the conditions imposed, Eq. 5.54 becomes

$$0.15 = \frac{M(30)^2}{(8)(300{,}000)(\frac{1}{12})(8)}$$

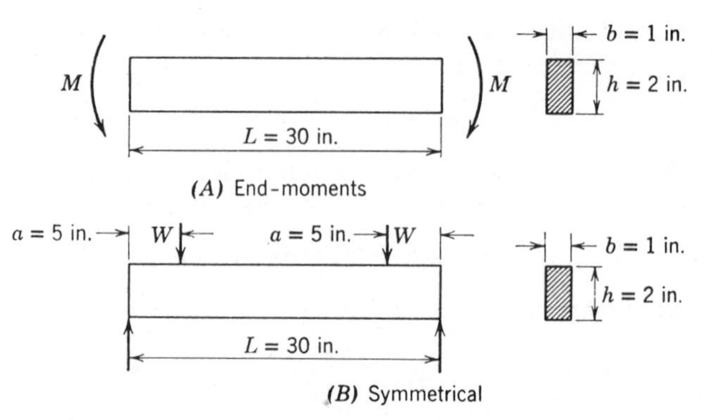

Fig. 5.35 Plastic beams in pure bending.

or

$$M = 267 \text{ in.-lb}$$

It is now necessary to check the tensile strain involved for the stress caused by the moment of 267 in.-lb. From Eq. 15.189 the bending stress for a bar of rectangular cross section in pure bending is

$$\sigma_{\max} = \frac{Mh}{2I}\left(\frac{2n+1}{3n}\right) \tag{5.55}$$

where n is a material constant discussed in the next example, assumed to be 3 for the problem at hand. Thus, from Eq. 5.55

$$\sigma_{\max} = \frac{(267)(1)(7)}{9(\frac{1}{12})(8)(2)} = 156 \text{ psi}$$

The tensile strain is $\varepsilon = \sigma/E$ or

$$\varepsilon = 156/300{,}000 = 0.00052 \text{ in./in.}$$

The data in Fig. 5.34 are valid for strains up to 0.5%; therefore the above calculations are valid.

For the bar in Fig. 5.35*B*, the maximum deflection occurs at midspan and is given by Eq. 1.46

$$\delta_{\max} = \frac{Wa(3L^2 - 4a^2)}{24EI} \tag{5.56}$$

or in terms of the problem here

$$0.15 = \frac{W(5)[3(900) - 4(25)]}{(24)(300{,}000)(\frac{1}{12})(8)}$$

or $W = 56$ lb. Applying Eq. 5.55 shows that

$$\sigma_{\max} = \frac{(56)(5)(7)}{(9)(2)(\frac{1}{12})(8)} = 163 \text{ psi}$$

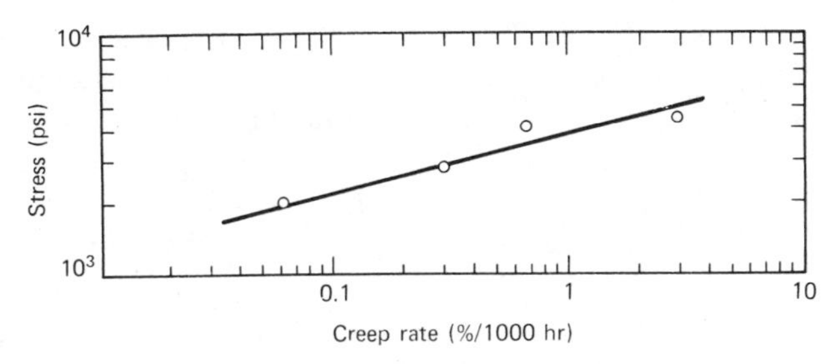

Fig. 5.36 Stress versus creep rate for Fig. 5.20.

which corresponds to a tensile strain of

$$163/300{,}000 = 0.000543 \text{ in./in.}$$

which is less than 0.5% and hence valid.

The value for n can be found by using creep behavior of metals in Chapter 15. In Eq. 15.175, where $C = B\sigma^n$, C is the slope of the strain time curve for constant values of stress. As an example take the straight-line slopes of the constant stress lines in Figure 5.20 for 2000 psi, 2750 psi, 4025 psi, and 4500. For 2000 psi:

$$\frac{d\varepsilon}{dt} = \frac{(0.0075 - 0.0065)}{1600 \text{ Hrs.}} \text{ in./in.} \times 100$$

$$= 0.065\% \frac{\text{strain}}{1000 \text{ hr}}$$

The other values are:

$$C(2000) = 0.065 \qquad C(4025) = 0.666$$

$$C(2750) = 0.3 \qquad C(4500) = 2.93$$

These values are plotted on log paper, Figure 5.36, where n is the slope and B is the intercept at C equal to 1. The equation is

$$C = \left(\frac{\sigma}{3750}\right)^{4.1}$$

When n is substituted into Eq. 5.55

$$\sigma = 150 \text{ psi}$$

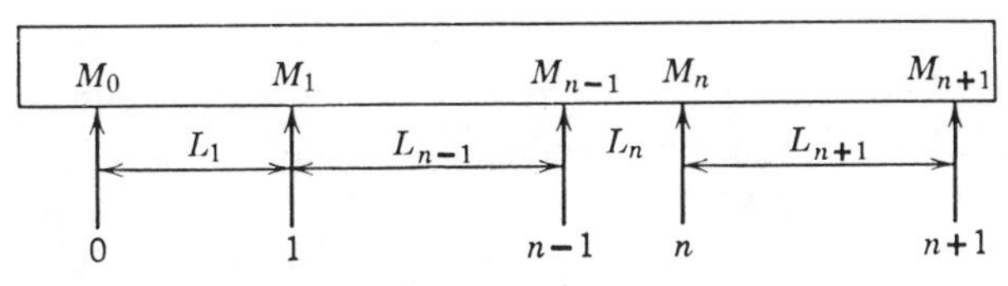

Fig. 5.37 Plastic pipeline on supports.

Pipeline on supports (Fig. 5.37). A pipeline on supports represents a problem of a continuous beam. The bending moments at the supports can be calculated using the theorem of three moments, Eq. 2.259

$$M_{n-1}L_n + 2M_n(L_n + L_{n+1}) + M_{n+1}L_{n+1} = -\frac{6A_n a_n}{L_n} - \frac{6A_{n+1}b_{n+1}}{L_{n+1}} \quad 5.57$$

where A_n is the area of the bending-moment diagram for the span n and a_n is the horizontal distance of the centroid of the moment area from the support $n - 1$, and so on. For a system having a large number of equal spans, it is easily shown by Eq. 5.57 that the maximum moment is equal to $WL/12$. Therefore

$$L = (12\sigma_{max}/W)(I/c) \quad 5.58$$

Suppose the problem involves a length of 2 in. schedule 80 polyvinyl chloride pipe holding water at 50 psi at room temperature; it is desired to examine the stress and deflection at 10,000-hr service and to determine a suitable span length for the pipe supports. It is assumed that there are no wind loads and that the material has an initial elastic modulus of 450,000 psi.

The 2-in. pipe with water has a unit weight of 0.194 lb and is stressed as follows:

$$\text{hoop stress} = \sigma_h = pd/2t = \sigma_1 \quad 5.59$$

$$\text{longitudinal stress} = \sigma_z = pd/4t + M/Z = \sigma_2 \quad 5.60$$

$$\text{radial stress} = \sigma_r = -p_{\text{I.D.}} \text{ and } 0_{\text{O.D.}} = \sigma_3 \quad 5.61$$

From Eqs. 5.59 to 5.61 and the failure criterion equation

$$2(\bar{\sigma})^2 = (\sigma_1 - \sigma_2)^2 + (\sigma_2 - \sigma_3)^2 + (\sigma_3 - \sigma_1)^2 \quad 5.62$$

it is concluded that the approximate *equivalent* wall stresses are as follows for various span lengths:

Span Length (ft)	Equivalent Stress, $\bar{\sigma}$ (psi)	
	O.D.	I.D.
2	193	267
4	199	284
8	279	366
10	372	442
15	740	738

From Fig. 5.26 it is concluded that stress-rupture failure is not likely even for 15-ft spans. However, using a conventional factor of safety of 4.0 the maximum design stress would be 2960 psi, which is still below the expected rupture-stress. It is concluded, therefore, that even for 15-ft spans a safe design could be effected based on stress alone.

Now consider creep. The elastic deflection is given by the relation

$$\delta_{max} = WL^3/384EI \qquad 5.63$$

which indicates the following deflection values:

Span Length (ft)	Initial Elastic Deflection, δ_0 (in.) (Based on O.D. Measurement)
2	0.000428
4	0.00683
8	0.1090
10	0.268
15	0.418

Now, to determine the creep deflections, insert in Eq. 5.41 the appropriate values of σ_0/σ from Fig. 5.25 which gives the following results:

Span Length (ft)	Maximum Deflections (in.)*	
	Initial	After 10,000 hr
2	0.000428	0.00072
4	0.00683	0.01147
8	0.1090	0.1835
10	0.2680	0.4500
15	0.4180	0.7050

*Any deflection due to internal pressure has been omitted.

Thus, from this tabulation, if the maximum allowable deflection is 0.25 in., which it frequently is to avoid excessive trouble in line drainage, a span length for the conditions imposed of not more than about 6 to 8 ft would be used.

Increasing stiffness in a plastics structure. Most plastics materials have a relatively low modulus of elasticity as compared with metals. This is a decided disadvantage when a structure requires high rigidity while at the same time requiring high resistance to corrosion, minimum weight, and ease of formability. To take advantage of the chemical and formability properties of plastics it is desirable to find a way to increase rigidity without sacrificing the desirable properties of plastics. This problem has been treated by Levy (21) and the following example is taken from his paper.

From Eq. 3.309 the flexural rigidity of a plate is given by

$$D = \frac{Et^3}{12(1 - \nu^2)} \qquad 5.64$$

Thus, if a plastic having a modulus of 300,000 psi and a metal with a modulus of 10,000,000 psi are compared, Eq. 5.64 shows that for the same rigidity the plastic

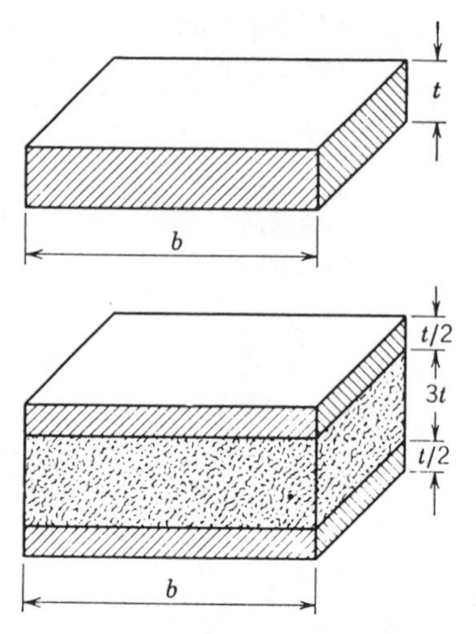

Fig. 5.38 Plates of equivalent stiffness.

plate would have to be about three times as thick as the metal plate. If the plastic is relatively inexpensive and easy to mold, the plastic plate having three times the thickness of the metal plate could be used. However, very often, particularly if the loading conditions are such that a fairly thick metal plate would be required, the triple thickness plastic plate would be impractical. Consider Fig. 5.38 which shows two plates, one of thickness t and the other of the same total thickness but made of two plates of thickness $t/2$ separated by a distance of $3t$. The moment of inertia of the single plate is

$$I = \tfrac{1}{12}bt^3 \qquad 5.65$$

and the moment of inertia of the split plate from Eq. 4.258 is

$$I = \tfrac{37}{12}bt^3 \qquad 5.66$$

which shows that this arrangement has a moment of inertia which is 37 times that of the original plate of thickness t. Thus, if the two plates are held apart, the EI values are 833,000 bt^3 for the single metal plate and 925,000 bt^3 for the double plastic plate, the plastic structure being more rigid than the metal structure. Practical implementation of such arrangements is discussed in (21).

Expansion of pipe under internal pressure (Fig. 5. 39). The stresses in a thin-walled tube under internal pressure are given by Eqs. 5.1 to 5.3, and the strains are calculated from Eqs. 3.70 to 3.72 (assuming isotropic materials). From the latter equations the hoop strain for a cylinder under pressure is given by

$$\varepsilon_1 = \varepsilon_h = \frac{1}{E}\left[\frac{pd}{2t} - \nu\left(\frac{pd}{4t} - p\right)\right] \qquad 5.67$$

from which

$$\varepsilon_h = \frac{p}{E}\left[\frac{d}{2t}\left(1 - \frac{\nu}{2}\right) + \nu\right] \qquad 5.68$$

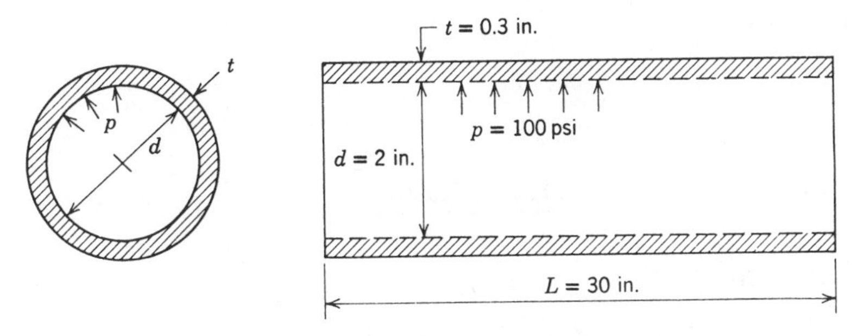

Fig. 5.39 Thin-walled plastic tube under internal pressure.

Since the radial expansion $u = (\varepsilon_h)(d/2)$, Eq. 5.68 can be rewritten as

$$u = \frac{pd}{2E}\left[\frac{d}{2t}\left(1 - \frac{v}{2}\right) + v\right] \tag{5.69}$$

Assuming linear viscoelastic behavior E and v in Eq. 5.69 are replaced by Eqs. 5.41 and 5.42; therefore

$$u = \frac{pd}{2E_o}\left(\frac{\sigma_o}{\sigma}\right)\left\{\frac{d}{4t}\left[2 - \left(\frac{3B - E_o(\sigma/\sigma_o)}{6B}\right)\right] + \frac{3B - E_0(\sigma/\sigma_0)}{6B}\right\} \tag{5.70}$$

where $\sigma/\sigma_0 = E/E_0$ and can be obtained from plots similar to those in Fig. 5.34, and B is bulk modulus of the material. In Eq. 5.70, u is the radial expansion at the bore of the cylinder. The longitudinal extension is given by

$$\varepsilon_2 = \varepsilon_z = \frac{1}{E}\left[\frac{pd}{4t} - v\left(\frac{pd}{2t} - p\right)\right] \tag{5.71}$$

or a total length change of

$$\mathscr{E} = \frac{Lp}{E}\left[\frac{d}{2t}\left(\frac{1}{2} - v\right) + v\right] \tag{5.72}$$

and for viscoelastic behavior

$$\mathscr{E} = \frac{Lp}{E_o}\left(\frac{\sigma_o}{\sigma}\right)\left\{\frac{d}{4t}\left[1 - \left(\frac{3B - E_o\dfrac{\sigma}{\sigma_o}}{3B}\right)\right] + v\right\} \tag{5.73}$$

Assume that the pipe is made of polyamide material conforming to the behavior in Fig. 5.34. The pipe has an inside diameter d of 2 in.; a wall thickness t of 0.30 in. is 30 in. long and is subjected to an internal pressure p of 100 psi. The material has a yield strength at room temperature of approximately 6000 psi and a modulus of about 210,000 psi. With this information and assuming five-year service the deformations can be calculated. From Fig. 5.34 the material has an effective modulus after five years of about 80,000 psi; therefore

$$\frac{\sigma_o}{\sigma} = \frac{E_o}{E} = \frac{210{,}000}{80{,}000} = 2.62$$

From Eq. 1.16

$$B = \frac{E}{3(1-2\nu)} = \frac{\sigma}{\sigma_0} \frac{E_o}{3(1-0.6)}$$

or

$$B = \frac{(0.38)(210{,}000)}{1.2} = 66{,}500 \text{ psi}$$

From Eq. 5.70

$$u = \frac{(100)(2)(2.62)}{(2)(210{,}000)}\left[\frac{2}{1.2}(2-0.298)+0.3\right] = 0.00695 \text{ in.}$$

and from Eq. 5.73

$$\mathscr{E} = \frac{(30)(100)(2.62)}{210{,}000}\left[\frac{2}{1.2}(1-0.6)+0.3\right] = 0.0237 \text{ in.}$$

In order to compute the external expansion the condition of constancy of volume is assumed, that is,

$$\text{volume} = \text{constant} = (\pi/4)[2.6^2 - 2^2](30) = (\pi/4)[d^2 - (2+0.0139)^2](30.0237)$$

from which the expanded outside diameter d is 2.61 in. The original outside diameter was 2.60 in.; therefore the outside radial expansion is 0.5%.

In addition to the deformation it is necessary to check the failure of the material by yielding at the end of five years, which is done by application of Eq. 5.5, thus,

$$p = \frac{6000/2.62}{[\frac{3}{16}(2/0.3)^2 + \frac{3}{4}(2/0.3) + 1]^{1/2}}$$

or

$$p = 610 \text{ psi}$$

Therefore there is no danger for failure.

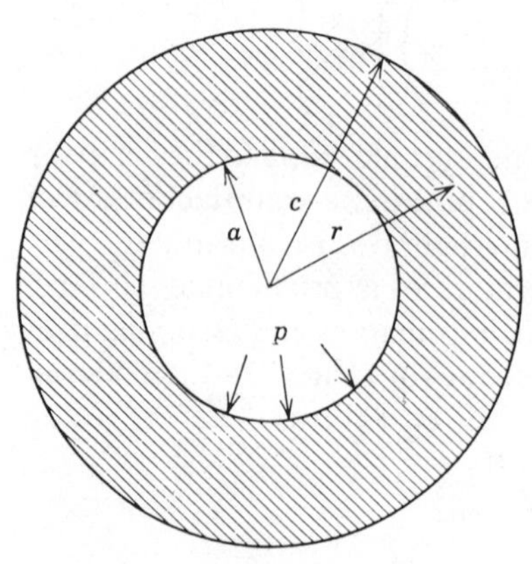

Fig. 5.40 Heavy-walled plastic cylinder under internal pressure.

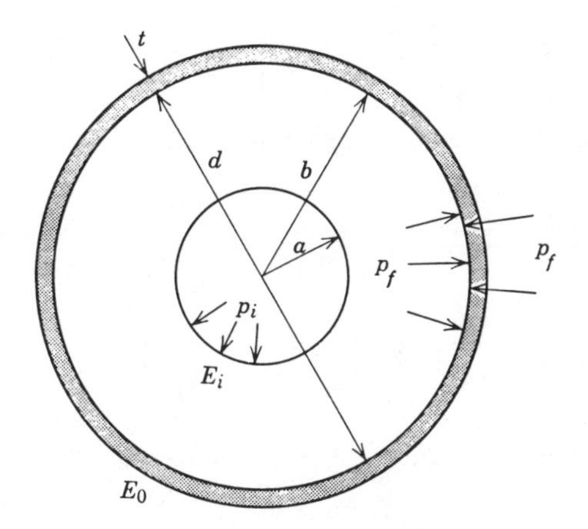

Fig. 5.41 Heavy-walled viscoelastic cylinder restrained by a rigid outside shell.

Heavy-walled cylinder under internal pressure (Fig. 5.40). The elastic stress and strain analysis for a heavy-walled cylinder is given by Eqs. 3.471, 3.472, 3.474, and 3.477. Creep strains are calculated using Eq. 3.477 with E and ν replaced by their viscoelastic equivalents. Thus the creep radial expansion is given by

$$u = \frac{pa^2}{r(c^2 - a^2)E_o}\left(\frac{\sigma_o}{\sigma}\right) \times \left\{r^2\left[1 - 2\left(\frac{3B - E_o(\sigma/\sigma_o)}{6B}\right)\right] + c^2\left[1 + \left(\frac{3B - E_o(\sigma/\sigma_o)}{6B}\right)\right]\right\} \quad 5.74$$

In order to predict stress-rupture, the stress values for creep of a heavy-walled cylinder given by Eqs. 15.275 to 15.277 would be used in Eq. 5.62 with $\bar{\sigma}$ being the tensile stress-rupture value for the particular time in question.

Heavy-walled viscoelastic cylinder with rigid support band (Fig. 5.41). An exact solution to this problem has been given by Lee (20). Since even an approximate solution is very complicated, only an outline of a method is given here. Consider Fig. 5.41 which shows a heavy-walled viscoelastic cylinder supported by an outside rigid band. Such a system might be a propellant charge in a rocket case and the problem might be to arrange the geometry and properties of the materials in such a way as to maintain a particular set of conditions during firing. For the geometry indicated the outside rigid band is under an internal pressure p_f which gives rise to a hoop stress

$$\sigma_h = p_f(d/2t) \quad 5.75$$

The viscoelastic cylinder is under internal pressure p_i and external pressure p_f giving rise to the following stresses for an open-end cylinder:

$$\sigma_h = \frac{1}{b^2 - a^2}\left[a^2 p_i - b^2 p_f + \left(\frac{ab}{r}\right)^2 (p_i - p_f)\right] \quad 5.76$$

$$\sigma_r = \frac{1}{b^2 - a^2}\left[a^2 p_i - b^2 p_f - \left(\frac{ab}{r}\right)^2 (p_i - p_f)\right] \qquad 5.77$$

The radial displacement at the interface for the outside shell is

$$u_b = \frac{b}{E}(\sigma_h - \nu\sigma_r) = \frac{p_f b}{E}\left(\frac{d}{2t} + \nu\right) \qquad 5.78$$

For the viscoelastic cylinder

$$u_b = \frac{b}{E(b^2 - a^2)}\left[(1 - \nu)(a^2 p_i - b^2 p_f) + \left(\frac{ab}{r}\right)^2 (1 + \nu)(p_i - p_f)\right] \qquad 5.79$$

For compatability the interface deformations must be the same and by equating Eqs. 5.78 and 5.79 it is found that

$$p_f = \frac{p_i[a^2(1 - \nu) + (ab/r)^2(1 + \nu)]}{d/2t + \nu + b^2(1 - \nu) + (ab/r)^2(1 + \nu)} \qquad 5.80$$

The complete elastic stress distribution can now be calculated by substituting the value of p_f from Eq. 5.80 into Eqs. 5.75 to 5.77.

Since the viscoelastic cylinder creeps, the bore enlarges, giving rise to a new geometry which affects the stress distribution. From 5.41 the radial deformation at any time t is given by

$$(u_r)_t = \left(\frac{\sigma_o}{\sigma}\right)\left(\frac{r}{E_o}\right)\left(\frac{1}{b^2 - a^2}\right)\left[(1 - \nu)(a^2 p_i - b^2 p_f) + \left(\frac{ab}{r}\right)^2 (1 + \nu)(p_i - p_f)\right] \qquad 5.81$$

at the bore, where $r = a$

$$(u_a)_t = \left(\frac{\sigma_o}{\sigma}\right)\left(\frac{a}{E_o}\right)\left(\frac{1}{b^2 - a^2}\right)[a^2 p_i(1 - \nu) + b^2 p_i(1 + \nu) - 2b^2 p_f] \qquad 5.82$$

Therefore, Eq. 5.80 can be rewritten to include the effect of an increasing bore size

$$p_f = \frac{p_i\{(a + u_a)^2(1 - \nu) + [(a + u_a)b/r]^2(1 + \nu)\}}{d/2t + \nu + b^2(1 - \nu) + [(a + u_a)b/r]^2(1 + \nu)} \qquad 5.83$$

Thus, in order to obtain the new stress distribution at each increment of time, values of $(u_a)_t$ from Eq. 5.82 are substituted into Eq. 5.83 and, in turn, Eq. 5.83 is substituted into Eqs. 5.75 to 5.77. With the stress distribution known, Eq. 5.62 can be applied as previously described.

REFERENCES

1. C. H. Adams, W. N. Findley, and F. D. Stockton, *Mod. Plast.*, **32** (12), 139–148, 216, (August 1955).
2. T. Alfrey, *Mechanical Behavior of High Polymers*, Interscience Publishers, New York, N.Y. (1948).

3. F. W. Billmeyer, *Textbook of Polymer Chemistry*, Interscience Publishers, New York, N.Y. (1957).

4. D. R. Bland, *The Theory of Linear Viscoelasticity*, Pergamon Press, New York, N.Y. (1960).

5. P. T. Blatz, *Ind. and Eng. Chem.*, **48** (4), 727–729 (April 1956).

6. R. H. Carey, and E. T. Oskin, *SPE J.*, **12** (3), 21–23, 55 (March 1956).

7. A. G. H. Dietz, *Archit. Rec.*, **117** (4), 225–231 (April 1955).

8. A. G. H. Dietz, and F. J. McGarry, *Mod. Plast.* **36** (1), 135–139 (September 1958).

9. J. H. Faupel, *Mod. Plast.*, **35** (11), 120–127, 188 (July 1958), and **35** (12), 132–139, 202 (August 1958).

10. W. N. Findley, *SPE J.*, **16**, (1), 57–65 (January 1960), and **16** (2), 192–198 (February 1960).

11. W. E. Gloor, *Mod. Plast.*, **36** (2), 144–148, 214 (October 1958).

12. S. Goldfein, *Mod. Plast.*, **37** (8), 127–132, 194, 198, 200 (April 1960).

13. B. Gross, *Mathematical Structure of the Theories of Viscoelasticity*, Hermann and Co., Paris (1953).

14. W. D. Harris, W. W. Burlew, and F. J. McGarry, *SPE. J.*, **16** (11), 1231–1234 (November 1960).

15. R. N. Haward, *The Strength of Plastics and Glass*, Cleaver-Hume Press, Ltd., London (1949).

16. I. L. Hopkins, W. O. Baker, and J. B. Howard, *J. Appl. Phys.*, **21** (3), 206–213 (March 1950).

17. J. B. Howard, *SPE J.*, **15** (5), 397–408 (May 1959).

18. G. F. Kinney, *Engineering Properties and Applications of Plastics*, John Wiley and Sons, Inc., New York, N.Y. (1957).

19. E. H. Lee, *Quart. Appl. Math.*, **13** (2), 183–190 (July 1955).

20. E. H. Lee, "Viscoelastic Stress Analysis," pp. 456–482 of *Structural Mechanics*, Pergamon Press, New York, N.Y. (1960).

21. S. Levy, *Mod. Plast.*, **36** (1), 123–133, 212 (September 1958).

22. A. A. MacLeod, *Ind. and Eng. Chem.*, **47** (7), 1319–1323 (July 1955).

23. J. Marin, Y. H. Pao, and G. Guff, *Trans. Am. Soc. Mech. Engrs.*, **73** (5), 705–719 (July 1951).

24. B. Maxwell, *Elect. Manuf.*, **58** (3), 146–153 (September 1956).

25. F. C. McGrew, *Mod. Plast.*, **35** (3), 155–162, 273–275 (November 1957).

26. D. F. Miner, and J. B. Seastone, Eds., *Handbook of Engineering Materials*, John Wiley and Sons, Inc., New York, N.Y. (1955).

27. J. D. D. Morgan, "Thermoplastics in Service with Particular Reference to the Application of Unplasticized Polyvinyl Chlorides," *Plas. Prog.* (Britain) (39–55) (1951).

28. R. K. Multer, and R. H. Rayfield, *Mater. Methods*, **43** (4), 132–135 (April 1956).

29. R. M. Ogorkiewicz, Ed., *Engineering Properties of Thermoplastics*, John Wiley and Son, Inc., New York, N.Y. (1970).

30. Y. H. Pao, and W. Brandt, *Appl. Mech. Rev.*, **9** (6), 233–236 (June 1956).

31. R. Slips, *J. Polym. Sci.*, **7** (2/3), 191–205, August/September (1951).

32. H. A. Stuart, *Die Physik Der Hochpolymeren*, Springer-Verlag, Berlin (1956).

33. R. L. Thorkildsen, and J. V. Schmitz, *Des. News*, **16** (20), 120–122 (September 29, 1961).

34. T. Von Karman and M. A. Biot, *Mathematical Methods in Engineering*, McGraw-Hill Book Co., New York, N.Y. (1940).

35. "Plastics as Materials of Construction," Symposium of Am. Chem. Soc., *Ind. Eng. Chem.*, **47** (7), 1292–1367 July 1955).

36. "Special Issue on High Polymers," J. *Appl. Phys.*, **27** (7), 665–696 (July 1956).

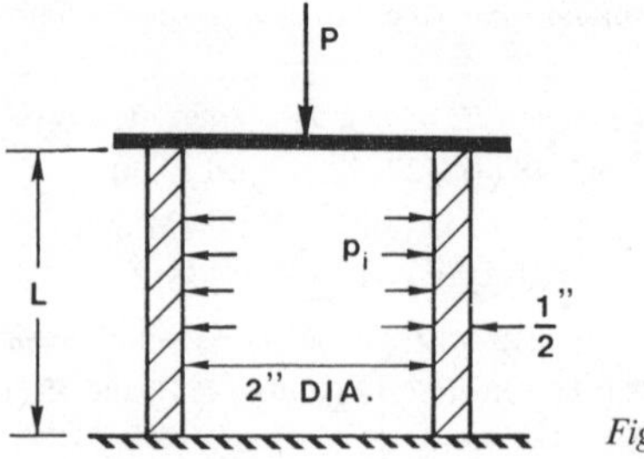

Fig. P5.1

PROBLEMS

5.1 A polyvinyl chloride cylinder (Fig. P5.1) is subjected to an internal pressure of 300 psi and an axial load of 2000 lb. Find:

a. The effective stress in the cylinder.

b. Find how long the cylinder can withstand this load without rupture.

c. If the temperature is raised to 100°F for one hour, what is the allowable internal pressure?

d. For a safety factor of 3 and a one-year life at 100°F, what is the allowable internal pressure?

5.2 A 0.875-in. I.D. cylindrical tank stores gas at a design pressure of 15 psig. A section the length of the tank shows a hole in one of the capped ends (Fig. P5.2). If the tank is made of plaskon 22 (PVC) ASTM D 1784 Type II GR I find:

a. t for a one-week life at 70°F with a safety factor of 1.

b. t for a five-year life at 100°F and 0.5% maximum strain.

c. t for rupture in five years at 100°F.

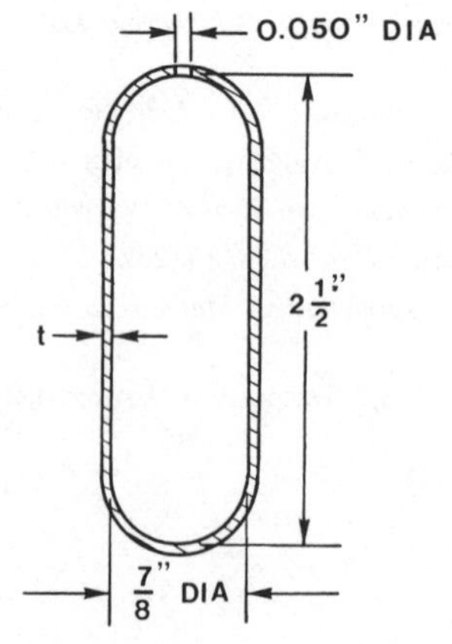

Fig. P5.2

5.3 A plastic flywheel rotates at 3600 rpm and weighs 25 lb. to give the proper angular momentum to the shaft it is attached to. Select a plastic material and design the flywheel for demensions a inside radius, b outside radius, and the thickness h. Assume a life of two years at the rated loading.

5.4 A plastic container is to be designed to hold 5 gal water and is to be stored in disaster centers. The centers are for earthquakes, severe storms, and other events that disturb fresh water supply and other services. The containers are to be stored, flushed, treated, and refilled every five years and discarded at the end of 15 years. The containers when empty should be easy to ship and store. Select the shape, type of plastic, and the thickness for the container.

5.5 Design a polyethylene sphere for 50 psig pressure with a 5-in. inside diameter for one-year life at 93°F. Size the design from the following considerations:

a. From stress rupture data with $\bar{\sigma} = 800$ psi at 93°F for one-year life.

b. From the linear viscoelastic approach.

5.6 A polyethylene bar 1 in^2 in cross-sectional area is to support 1000 lb at 80°F. How long can it support the load in tension before it ruptures?

5.7 A plastic acetal disk 2 in. in diameter, simple supported, must support 200 psi for one year. How thick must the disk be?

5.8 In Figure 5.17, plot and find all parameters for $C = B\sigma^n$ and work the bending example in Figure 5.35.

5.9 A polyethylene pipe is used in a sprinkler system. The pipe is supposed to be straight, but because of errors, it is sometimes bent. Find and plot the range of radii a hose can be bent to versus its life at each radius. Use $\bar{\sigma} =$ 1000 psi.

chapter **6**

BEYOND THE ELASTIC RANGE

For convenience the theory of plasticity will be discussed under three headings. First a discussion of some fundamental principles will be given; this will include a discussion of stress-strain relationships in the plastic range, an extension of the theory given for uniaxial loading in Section 2-1, and a brief introduction to the theory of *plane plastic strain* frequently used in metal-working applications and familiar to most engineers as *slip line theory*. Next the principles developed will be applied to problems of tension, torsion, bending, and combined stresses in the plastic range with particular emphasis on ultimate load theory commonly referred to as *plastic design* or *limit design*. Finally, problems of plasticity associated with process equipment like cylinders and spheres and metal working analyses will be discussed. The reader is referred to other more detailed sources for further study: Baker (1), Drucker (8), Freudenthal (13), Heyman (15), Hill (17), Hoffman and Sachs (19), Houwink (22), Johnson and Mellor (24), Martin (31), Nadai (32), Neal (33), Phillips (38), Prager (39), Prager and Hodge (40), Reiner (41), Van Den Broek (46), Van Iterson (47).

6-1 BASIC CONCEPTS AND STRESS-STRAIN RELATIONS

Uniaxial loading

It is convenient to begin the discussion by referring to the analysis of true stress and true strain outlined in Chapter 2. First, however, consider the hypothetical stress-strain diagram shown in Fig. 6.1. The elastic range covered by Hooke's law terminates at the stress σ_y. For present purposes it is assumed that the elastic limit, proportional limit, and yield stress are all identical, although this is not actually the case as was discussed in Chapter 1. Beyond the yield stress σ_y and up to the ultimate stress σ_u there is a range identified as the *range of uniform elongation*, that is, the range in which the cross section of the test specimen is under pure tension. At the stress σ_u, the load-carrying capacity of the material reaches the ultimate; that is, the increase in strength brought about by work-hardening just balances the decrease in strength brought about by reduction in area; at this stress the specimen *necks* locally, instability is introduced, and finally fracture results at a stress σ_f. In the range from σ_u to σ_f the

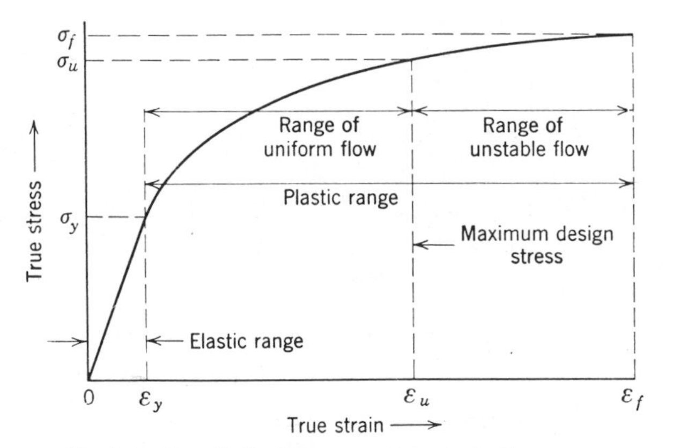

Fig. 6.1 Hypothetical true stress-true strain diagram.

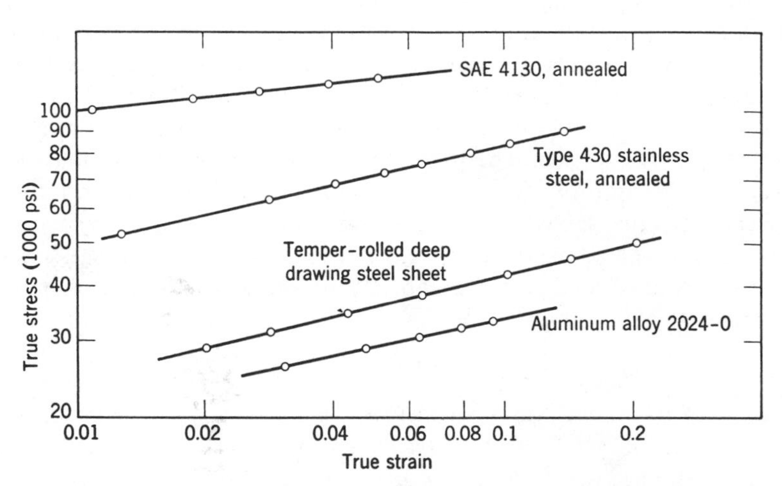

Fig. 6.2 True stress-true strain data for some materials.

specimen is no longer under pure tension, and therefore for obvious reasons the fundamental design utility of the plot ends at σ_u. A comparison of a true and a nominal stress-strain plot is given in Fig. 2.3.

The form of the true stress-strain curve is of particular interest. For many materials the true stress-strain relationship is given by the equation

$$\sigma = K\varepsilon^n \tag{6.1}$$

where σ is the true stress at the beginning of plastic flow (σ_y in Fig. 6.1) and ε is the true strain. The constants K and n in Eq. 6.1 are easily evaluated since the equations is parabolic and plots a straight line on log coordinates, as shown for several materials in Fig. 6.2. Thus Eq. 6.1 can be rewritten as

$$\log \sigma = \log K + n \log \varepsilon \tag{6.2}$$

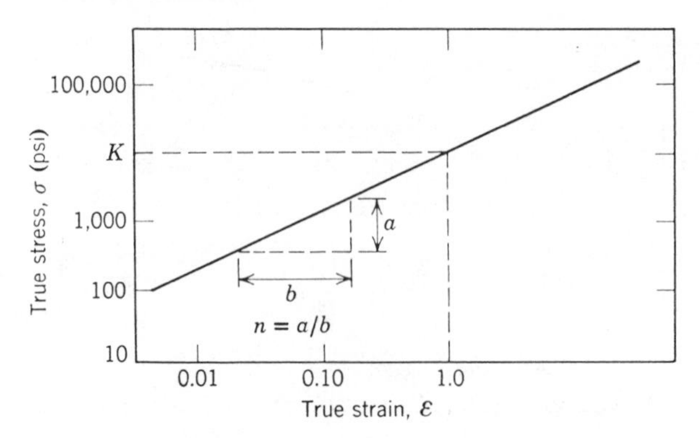

Fig. 6.3 Plot of true stress-true strain on log-log coordinates.

and n, the *strain-hardening coefficient*, is the slope of the plotted line and K, the *strength coefficient*, is the true stress corresponding to true strain at 1.0 on the log-log plot. This is shown schematically on Fig. 6.3. Referring again to Fig. 6.1, the initiation of unstable flow occurs at the ultimate stress σ_u. This means that Eq. 6.1 is valid only in the region of uniform elongation from σ_y to σ_u. At the stress σ_u there is a balance between the effect of strain-hardening that tends to increase the strength of the material and the decreasing cross-sectional area of the specimen that tends to weaken the material. When instability occurs, the rate of increase of load-carrying capacity due to strain-hardening is less than the rate of decrease of load-carrying capacity produced by the decreased cross-sectional area. This point is defined as

$$dF = 0 \qquad 6.3$$

This point of maximum load is of considerable interest in forming problems and is calculated as follows. Consider a tensile specimen of cross-sectional area A at any instant stressed to σ by load F. For equilibrium

$$F = A\sigma \qquad 6.4$$

From Eqs. 2.12 and 2.14 the true strain is given by

$$\varepsilon = \ln (A_0/A) \qquad 6.5$$

from which the instantaneous area A is

$$A = A_0 e^{-\varepsilon} \qquad 6.6$$

where e is the base of natural logarithms.
From Eqs. 6.4 and 6.6

$$F = \sigma A_0 e^{-\varepsilon} \qquad 6.7$$

The load F is a function of both the true stress and the true strain; therefore

$$dF = (\partial F/\partial \sigma)\, d\sigma + (\partial F/\partial \varepsilon) d\varepsilon \qquad 6.8$$

From Eq. 6.7

$$\partial F/\partial \sigma = A_0 e^{-\varepsilon} \qquad 6.9$$

and

$$\partial F/\partial \varepsilon = -A_0 \sigma e^{-\varepsilon} \qquad 6.10$$

Placing the values given by Eqs. 6.9 and 6.10 in Eq. 6.8 then gives the result

$$dF = (A_0 e^{-\varepsilon})(d\sigma - \sigma\, d\varepsilon) \qquad 6.11$$

Since the instant of instability is defined by Eq. 6.3, Eq. 6.11 becomes

$$0 = (A_0 e^{-\varepsilon})(d\sigma - \sigma\, d\varepsilon) \qquad 6.12$$

In Eq. 6.12 the term $A_0 e^{-\varepsilon}$ cannot equal zero; therefore

$$d\sigma - \sigma\, d\varepsilon = 0 \qquad 6.13$$

or instability is defined by the relation

$$\sigma = d\sigma/d\varepsilon \qquad 6.14$$

The true stress σ is also defined by Eq. 6.1. Equation 6.14 now becomes

$$\sigma = K\varepsilon^n = d\sigma/d\varepsilon = (d/d\varepsilon)(K\varepsilon^n) = nK\varepsilon^{n-1} \qquad 6.15$$

from which

$$\varepsilon = n \qquad 6.16$$

Equation 6.16 states that at the instant of instability of flow in tension, the true strain ε is numerically equal to the strain-hardening coefficient.

Multiaxial loading

In Chapter 3 a consideration of the distortion energy theory of failure led to the results expressed by Eqs. 3.523 and 3.524,

$$\sigma_0 = \sqrt{\sigma_1^2 + \sigma_2^2 + \sigma_3^2 - \sigma_1\sigma_2 - \sigma_2\sigma_3 - \sigma_3\sigma_1} \qquad 6.17$$

and

$$\sigma_0 = (\sqrt{2}/2)\sqrt{(\sigma_1 - \sigma_2)^2 + (\sigma_2 - \sigma_3)^2 + (\sigma_3 - \sigma_1)^2} \qquad 6.18$$

where σ_0 is the flow stress in tension and σ_1, σ_2, and σ_3 are the true principal stresses in combined loading. If it is assumed that the material continues to flow at constant strain energy, then Eqs. 6.17 and 6.18 describe the stresses from initial yielding to the time local instability occurs. The principal utility of the theory described by these equations is that σ_0, representing tensile behavior, is equated to a system of combined stresses. Thus the ordinary true stress-strain diagram of a material is also the *equivalent* combined stress-strain diagram and predictions concerning combined stress behavior can be made on the basis of tensile test data alone. Proof concerning the validity of this theory for initial yielding was given in Chapter 3. Post-yielding behavior will now be examined.

First it is necessary to obtain expressions for the strains involved. A convenient way to do this is to start with Eqs. 3.70 to 3.72 and replace the elastic constants $1/E$ and v by θ and $\frac{1}{2}$ respectively. The quantity θ now represents a variable compliance and $\frac{1}{2}$ is Poisson's ratio in the plastic range by virtue of the condition of constancy of volume; thus

$$\varepsilon_1 = \theta[\sigma_1 - \tfrac{1}{2}(\sigma_2 + \sigma_3)] \qquad 6.19$$

$$\varepsilon_2 = \theta[\sigma_2 - \tfrac{1}{2}(\sigma_1 + \sigma_3)] \qquad 6.20$$

$$\varepsilon_3 = \theta[\sigma_3 - \tfrac{1}{2}(\sigma_1 + \sigma_2)] \qquad 6.21$$

Collecting terms in Eqs. 6.19 to 6.21 leads to the result

$$\sqrt{\tfrac{2}{3}(\varepsilon_1{}^2 + \varepsilon_2{}^2 + \varepsilon_3{}^2)} = (\sqrt{2}/2)\theta\sqrt{(\sigma_1 - \sigma_2)^2 + (\sigma_2 - \sigma_3)^2 + (\sigma_3 - \sigma_1)^2} \qquad 6.22$$

In Eq. 6.22 the right-hand term is 6.18 multiplied by θ. For tension loading, $\sigma_2 = \sigma_3 = 0$, and

$$\varepsilon_2 = \varepsilon_3 = -\varepsilon_1/2 \qquad 6.23$$

Therefore, Eq. 6.22 becomes

$$\varepsilon_1 = \theta\sigma_1 \qquad 6.24$$

But since $\sigma_0 = (\sqrt{2}/2)\sqrt{(\sigma_1 - \sigma_2)^2 + (\sigma_2 - \sigma_3)^2 + (\sigma_3 - \sigma_1)^2}$,

$$\theta\sigma_1 = \theta\sigma_0 \qquad 6.25$$

or finally

$$\sigma_1 = \sigma_0 \qquad 6.26$$

Thus by definition and by the derivation of Eq. 3.524 the following useful combined stress and strain relationships can be set down.

$$\sigma_0 = \sqrt{\sigma_1{}^2 + \sigma_2{}^2 + \sigma_3{}^2 - \sigma_1\sigma_2 - \sigma_2\sigma_3 - \sigma_1\sigma_3} \qquad 6.27$$

or

$$\sigma_0 = (\sqrt{2}/2)\sqrt{(\sigma_1 - \sigma_2)^2 + (\sigma_2 - \sigma_3)^2 + (\sigma_3 - \sigma_1)^2} \qquad 6.28$$

$$\varepsilon_0 = \sqrt{\tfrac{2}{3}(\varepsilon_1{}^2 + \varepsilon_2{}^2 + \varepsilon_3{}^2)} \qquad 6.29$$

or

$$\varepsilon_0 = (\sqrt{2}/3)\sqrt{(\varepsilon_1 - \varepsilon_2)^2 + (\varepsilon_2 - \varepsilon_3)^2 + (\varepsilon_3 - \varepsilon_1)^2} \qquad 6.30$$

Note that the symbol ε_0 is used to express equivalent combined strain; care must be exercised to insure that this strain is not confused with the nominal tensile strain also written ε_0.

The quantities expressed by Eqs. 6.27 to 6.30 are known as *equivalent stress* (σ_0) and *equivalent strain* (ε_0). In the literature these same quantities have also been referred to as *effective* stress and strain and *significant* stress and strain. The original theory developed by Nadai (32) resulted in what Nadai called *octahedral shear stress* and *octahedral shear strain* and his equations for these quantities are as follows:

$$\tau_0 = (\sqrt{2}/3)\sqrt{\sigma_1{}^2 + \sigma_2{}^2 + \sigma_3{}^2 - \sigma_1\sigma_2 - \sigma_2\sigma_3 - \sigma_1\sigma_3} \qquad 6.31$$

or

$$\tau_0 = \tfrac{1}{3}\sqrt{(\sigma_1 - \sigma_2)^2 + (\sigma_2 - \sigma_3)^2 + (\sigma_3 - \sigma_1)^2} \qquad 6.32$$

$$\gamma_0 = \sqrt{\tfrac{4}{3}(\varepsilon_1{}^2 + \varepsilon_2{}^2 + \varepsilon_3{}^2)} \qquad 6.33$$

or

$$\gamma_0 = \tfrac{2}{3}\sqrt{(\varepsilon_1 - \varepsilon_2)^2 + (\varepsilon_2 - \varepsilon_3)^2 + (\varepsilon_3 - \varepsilon_1)^2} \qquad 6.34$$

where τ_0 is octahedral shear stress, γ_0 is octahedral shear strain, and the other quantities are as previously defined. Combining the preceding equations it is seen that

$$\tau_0 = (\sqrt{2}/3)\sigma_0 \qquad 6.35$$

and

$$\gamma_0 = \sqrt{2}\,\varepsilon_0 \qquad 6.36$$

Proof of the validity of this theory was demonstrated by Sachs (42, p. 456) in tests of aluminum, copper, and nickel. Sachs observed that the tensile yield point σ_y for polycrystalline specimens was related to the critical shear stress for yield τ_{cr}, for crystals of all orientations by the relation

$$\sigma_y = 2.238\,\tau_{cr}, \qquad 6.37$$

In torsion tests Sachs found that

$$\tau_y = 1.293\,\tau_{cr}, \qquad 6.38$$

Thus equating Eqs. 6.37 and 6.38,

$$\tau_y = \frac{1.293}{2.238}\sigma_y = \frac{\sigma_y}{\sqrt{3}} \qquad 6.39$$

which is in agreement with the distortion energy theory, Eq. 6.17, since for tension

$$\sigma_0 = \sigma_y = \sigma_1 \qquad 6.40$$

and for torsion

$$\sigma_0 = \sigma_y = \sqrt{\sigma_1{}^2 + \sigma_1{}^2 + \sigma_1{}^2} = \sqrt{3}\,\tau_y \qquad 6.41$$

which agrees with Eq. 6.39.

Equation 6.26 shows that for tension the equivalent stress is simply equal to the tensile stress σ_1 which at failure in yielding becomes σ_y. Similarly, for tension the equivalent strain, Eq. 6.29, reduces to

$$\varepsilon_0 = \varepsilon_1 \qquad 6.42$$

by virtue of Eq. 6.23. Since σ_0 and ε_0 are the same for both tension and combined stresses, simple tension data can be used to interpret or predict the behavior of materials under conditions of combined stress. In other words, a true stress-strain diagram for tension is also the $\sigma_0 - \varepsilon_0$ plot for combined stresses. The practical utility of this theory is that tension test data are relatively easy to obtain. Consequently, from a practical point of view it is possible to design in the plastic range for states of combined stress on the basis of the tension stress-strain diagram for the material. In order to facilitate use of the theory *flow equations* have been developed which relate tension to combined stresses for various ratios of the principal stresses as follows. In terms of simple tension, Eqs. 6.19 to 6.21, with the aid of Eq. 6.24, can be rewritten as

$$\varepsilon_1 = \left(\frac{\varepsilon}{\sigma}\right)[\sigma_1 - \tfrac{1}{2}(\sigma_2 + \sigma_3)] \tag{6.43}$$

$$\varepsilon_2 = \left(\frac{\varepsilon}{\sigma}\right)[\sigma_2 - \tfrac{1}{2}(\sigma_1 + \sigma_3)] \tag{6.44}$$

$$\varepsilon_3 = \left(\frac{\varepsilon}{\sigma}\right)[\sigma_3 - \tfrac{1}{2}(\sigma_1 + \sigma_2)] \tag{6.45}$$

where σ and ε are true stress and true strain for simple tension. In this form, however, the equations have little utility and must be converted to a more useful form. From Eqs. 6.1, 6.26 and 6.42,

$$\sigma_0 = \sigma = K\varepsilon_0{}^n \tag{6.46}$$

or

$$\frac{\varepsilon_0}{\sigma} = \frac{\sigma_0^{(1-n)/n}}{K^{1/n}} \tag{6.47}$$

and Eqs. 6.43 to 6.45 become

$$\varepsilon_1 = \left(\frac{\sigma_1}{K}\right)^{1/n}(\alpha^2 + \beta^2 - \alpha\beta - \alpha - \beta + 1)^{(1-n)/2n}\left(1 - \frac{\alpha}{2} - \frac{\beta}{2}\right) \tag{6.48}$$

$$\varepsilon_2 = \left(\frac{\sigma_1}{K}\right)^{1/n}(\alpha^2 + \beta^2 - \alpha\beta - \alpha - \beta + 1)^{(1-n)/2n}\left(\alpha - \frac{\beta}{2} - \frac{1}{2}\right) \tag{6.49}$$

$$\varepsilon_3 = \left(\frac{\sigma_1}{K}\right)^{1/n}(\alpha^2 + \beta^2 - \alpha\beta - \alpha - \beta + 1)^{(1-n)/2n}\left(\beta - \frac{\alpha}{2} - \frac{1}{2}\right) \tag{6.50}$$

where $\alpha = \sigma_2/\sigma_1$, $\beta = \sigma_3/\sigma_1$, σ_1 is the true maximum principal stress, and K and n are constants as previously defined. The flow equations completely define the principal plastic strains for combined stresses and the tension material constants K and n.

Examples of application of theory

In order to emphasize the utility of the foregoing developments, a few examples illustrating the use of the theory will now be given.

True stress for a thin-walled tube under internal pressure (Fig. 6.4). A tube of initial dimensions d_0 and t_0 is expanded by application of internal pressure p to dimensions d_p and t_p, where d is diameter, t, wall thickness, and the subscript

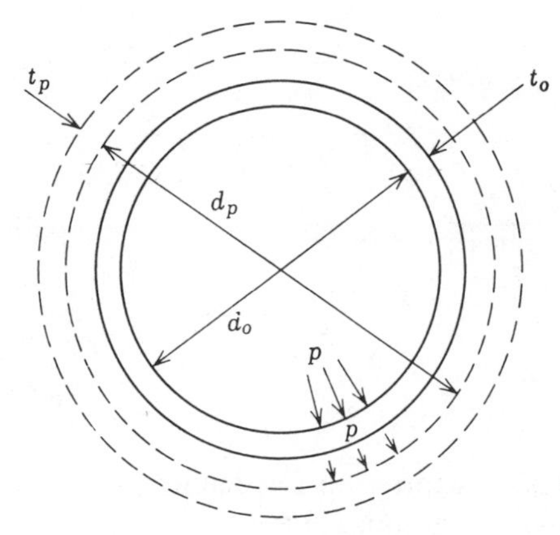

Fig. 6.4 Expansion of a cylindrical tube in the plastic range.

p means plastic. If the tube is open at the ends, two principal stresses are developed, a hoop stress σ_h and a radial stress σ_r; thus

$$\sigma_h = pd/2t \tag{6.51}$$

and

$$\sigma_r \approx -p \tag{6.52}$$

where d and t are instantaneous values. Since the volume is assumed to remain constant,

$$\varepsilon_h + \varepsilon_r + \varepsilon_z = 0 \tag{6.53}$$

from which

$$\varepsilon_h = -(\varepsilon_r + \varepsilon_z) \tag{6.54}$$

where ε represents strain and the subscript z denotes axial direction. Equation 6.54 expresses the radial strain in terms of the hoop and axial strain. However, the true radial strain may also be expressed by Eq. 2.6,

$$\varepsilon_r = \ln(t/t_0) \tag{6.55}$$

Combining Eqs. 6.54 and 6.55 gives the result

$$t = t_0 e^{-(\varepsilon_z + \varepsilon_h)} \tag{6.56}$$

The change in diameter on the basis of true strain is likewise

$$\varepsilon_h = \ln(d/d_0) \tag{6.57}$$

from which

$$d = d_0 e^{\varepsilon_h} \tag{6.58}$$

Substituting Eq. 6.56 and 6.58 into 6.51 gives an expression for the true hoop stress in terms of the original dimensions

$$\sigma_h = \frac{p}{2} d_0 e^{\varepsilon_h} \left[\frac{1}{t_0 e^{-(\varepsilon_h + \varepsilon_z)}} \right] \tag{6.59}$$

from which

$$p = 2\sigma_h(t_0/d_0)e^{-\varepsilon_h(2+\varepsilon_z/\varepsilon_h)} \qquad 6.60$$

However,

$$\frac{\varepsilon_z}{\varepsilon_h} = \frac{2 - \sigma_h/\sigma_z}{(2\sigma_h/\sigma_z) - 1} \qquad 6.61$$

For this case $\sigma_z = 0$; therefore

$$\frac{\varepsilon_z}{\varepsilon_h} = \frac{2 - \infty}{2\infty - 1} = -\tfrac{1}{2} \qquad 6.62$$

Substituting Eq. 6.62 into Eq. 6.60 gives an expression for defining how the true hoop stress is related to the original dimensions; thus

$$\sigma_h = \frac{p}{2(t_0/d_0)}e^{(3/2)\varepsilon_h} \qquad 6.63$$

In the elastic range ε_h is nearly zero; therefore Eq. 6.63 reduces to the familiar form

$$\sigma_h = pd_0/2t_0 \qquad 6.64$$

For a closed-end tube, the longitudinal stress

$$\sigma_z = \sigma_h/2 \qquad 6.65$$

and Eq. 6.62 becomes

$$\varepsilon_z/\varepsilon_h = (2-2)/3 = 0 \qquad 6.66$$

Similarly, Eq. 6.60 is

$$\sigma_h = \frac{pe^{2\varepsilon_h}}{2(t_0/d_0)} \qquad 6.67$$

Equations 6.63 and 6.67 show that the true hoop stress increases directly as a function of ε_h; thus, for 100% ductility, for example,

$$\varepsilon_h = \ln(2d_0/d_0) = 0.693$$

and by Eq. 6.67

$$\sigma_h = (\sigma_h)_{\text{nominal}} \times 3.998$$

For a 500% change in diameter

$$\sigma_h = (\sigma_h)_{\text{nominal}} \times 25$$

Thus for a closed-end tube the true hoop stress increases by a factor of about 4 for 100% ductility and by a factor of 25 for 500% ductility. For an open-end tube the factors are approximately 3 and 11 respectively. Therefore it is indicated that some advantage may be gained by using the principle of an open-end tube if conditions allow.

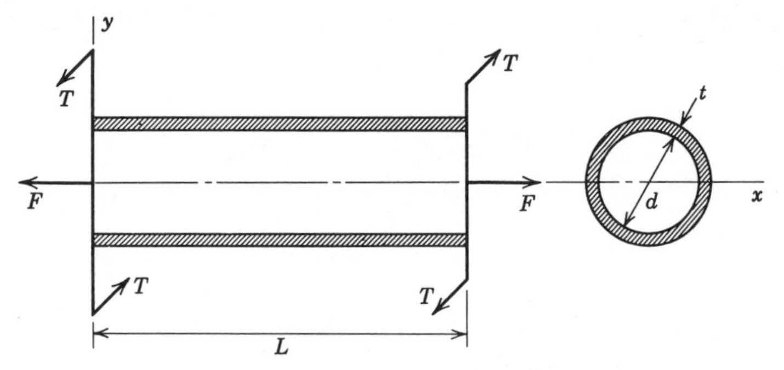

Fig. 6.5 Thin-walled tube subjected to combined tension and torsion.

Combined tension and torsion of the thin-walled tube (Fig. 6.5). The thin-walled tube shown in the figure is simultaneously subjected to a tensile force F and a twisting moment T. The tension load induces axial stresses and strains and a radial strain while the twisting moment induces shear stresses and strains in the member. From conditions of equilibrium, it is seen that the nominal axial and shearing stresses are defined by

$$(\sigma_x)_{\text{nominal}} = \frac{F}{\pi t(d + t)} \qquad 6.68$$

$$(\tau_{xy})_{\text{nominal}} = \frac{2T}{\pi \, dt(d + 2t)} \qquad 6.69$$

The nominal axial and shear strains are

$$(\varepsilon_x)_{\text{nominal}} = (L - L_0)/L_0 \qquad 6.70$$

$$(\gamma_{xy})_{\text{nomnal}} = (\pi/360)(d + 2t)\theta \qquad 6.71$$

where L and L_0 are final and initial gauge lengths and θ is the angle of twist in degrees. In the plastic range the true strain in the direction of the wall thickness is

$$\varepsilon_r = \ln{(t_p/t_0)} \qquad 6.72$$

where t_0 is the initial wall thickness and t_p is the instantaneous value in the plastic range. Since the volume remains constant and the twisting produces no axial strain,

$$\varepsilon_r = -\varepsilon_x/2 \qquad 6.73$$

and

$$t_p = \frac{t}{(1 + \varepsilon_x)^{1/2}} \qquad 6.74$$

The true hoop strain is given by

$$\varepsilon_h = \ln{(d_p/d)} \qquad 6.75$$

so that in terms of the original dimensions of the tube the true stress components are

$$\sigma_x = \frac{F}{\pi t_0(d_0 + t_0)}(1 + \varepsilon_x) \tag{6.76}$$

and

$$\tau_{xy} = \frac{(2T/\pi d_0 t_0)(1 + \varepsilon_x)^{3/2}}{d_0 + 2t_0} \tag{6.77}$$

Having the component stresses from Eqs. 6.76 and 6.77, the principal stresses are determined by Eqs. 3.18 and 3.19.

In the plastic range the values of the *nominal principal strains* are as follows (11):

$$(\varepsilon_1)_n = \frac{1}{4}\left\{\varepsilon_x + \left[\left(\frac{d_0 + 2t_0}{L_0}\right)^2\left(\frac{\pi\theta}{180}\right)^2 + 9\varepsilon_x^{\,2}\right]^{1/2}\right\} \tag{6.78}$$

$$(\varepsilon_2)_n = \frac{1}{4}\left\{\varepsilon_x - \left[\left(\frac{d_0 + 2t_0}{L_0}\right)^2\left(\frac{\pi\theta}{180}\right)^2 + 9\varepsilon_x^{\,2}\right]^{1/2}\right\} \tag{6.79}$$

$$(\varepsilon_3)_n = -(\varepsilon_1 + \varepsilon_2) \tag{6.80}$$

The *true principal strains* are as follows:

$$\varepsilon_1 = \ln\left[1 + (\varepsilon_1)_n\right] \tag{6.81}$$

$$\varepsilon_2 = \ln\left[1 + (\varepsilon_2)_n\right] \tag{6.82}$$

and

$$\varepsilon_2 = -\left[(\varepsilon_1)_n + (\varepsilon_2)_n\right] \tag{6.83}$$

Having defined the true principal stresses and strains, the effective stress-strain plot can then be made using Eqs. 6.17 and 6.29. Faupel and Marin (11) conducted combined tension-torsion tests on thin-walled tubes of commercially

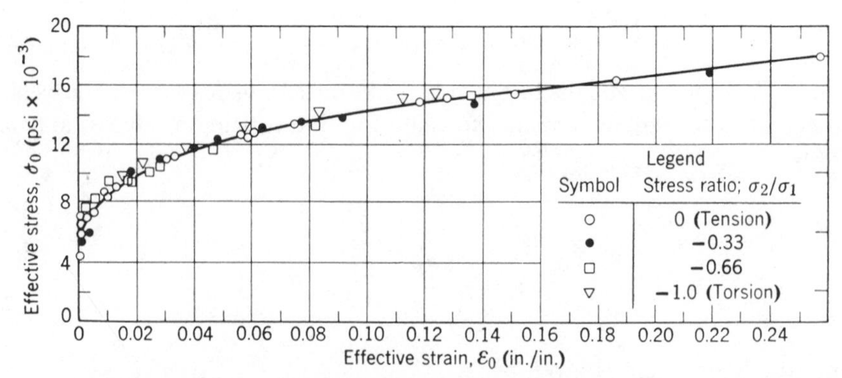

Fig. 6.6 Stress-strain relations for thin-walled tubes of aluminum alloy 1100–0 subjected to various states of constant biaxial stress.

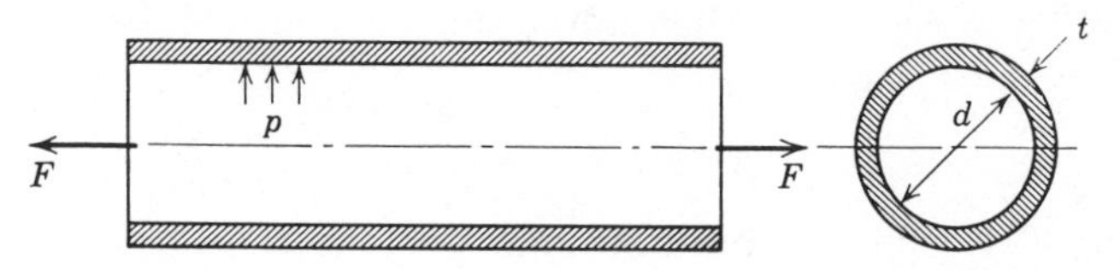

Fig. 6.7 Thin-walled tube subjected to combined tension and internal pressure.

pure aluminum and obtained the results shown in Fig. 6.6; note that the $\sigma_0 = \varepsilon_0$ curve is identical for tension, torison, and other stress-ratios employed in the tests.

Determination of wall thickness at instability of a thin-walled cylinder subjected to internal pressure and tension (Fig. 6.7). Two cases are considered: (*a*) when the hoop stress is greater than the axial stress and (*b*) when the reverse is true.

CASE 1. HOOP STRESS > AXIAL STRESS ($\sigma_h > \sigma_z$). In this case there is a limiting internal pressure that determines the instability pressure. In the plastic range the principal true stresses are as follows:

$$\text{hoop stress} = \sigma_h = \sigma_1 = pd_p/2t_p \qquad 6.84$$

$$\text{axial stress} = \sigma_z = \sigma_2 = pd_p/4t_p + F/\pi d_p t_p \qquad 6.85$$

Since the hoop stress is critical for this case, it is necessary to express Eq. 6.84 in terms of true strain. First consider the strain in the direction of the wall thickness (Fig. 6.4). The hoop strain is $\varepsilon_h = \varepsilon_1$ the axial strain is $\varepsilon_z = \varepsilon_2$, and the radial strain $\varepsilon_r = \varepsilon_3$. With this notation and assuming constancy of volume,

$$\varepsilon_3 = -(\varepsilon_1 + \varepsilon_2) = \ln(t_p/t_0) \qquad 6.86$$

from which

$$t_p = t_0 e^{-\varepsilon_1 - \varepsilon_2} \qquad 6.87$$

In addition, in the hoop direction

$$\varepsilon_1 = \ln(d_p/d_0) \qquad 6.88$$

or

$$d_p = d_0 e^{\varepsilon_1} \qquad 6.89$$

Since expressions for the dimensions d_p and t_p are available, Eq. 6.84 can be written as

$$\sigma_1 = \sigma_h = \frac{p}{2}(d_0 e^{\varepsilon_1})\left(\frac{1}{te^{-\varepsilon_1 - \varepsilon_2}}\right) \qquad 6.90$$

from which

$$p = 2\sigma_1 (t_0/d_0) e^{-\varepsilon_1(2 + \varepsilon_2/\varepsilon_1)} \qquad 6.91$$

The strain ratio $\varepsilon_2/\varepsilon_1$ may be evaluated by reference to the flow equations,

6.48 to 6.50; these equations when solved simultaneously show that

$$\varepsilon_1/\varepsilon_2 = (2-\alpha)/(2\alpha-1) \tag{6.92}$$

where $\alpha = \sigma_2/\sigma_1$.

Equation 6.91 defining p can now be written in the form

$$p = 2\sigma_1(t_0/d_0)e^{-\varepsilon_1[3/(2-\alpha)]} \tag{6.93}$$

The criterion for instability is taken as that pressure where strain increases with no increase in pressure; that is,

$$dp = 0 \tag{6.94}$$

where d is the differential operator. Since the pressure is a function of both σ_1 and ε_1, the following partial differential equation expresses the instability condition:

$$dp = (\partial p/\partial \sigma_1)d\sigma_1 + (\partial p/\partial \varepsilon_1)\,d\varepsilon_1 = 0 \tag{6.95}$$

From Eq. 6.93

$$\partial p/\partial \sigma_1 = 2(t_0/d_0)e^{-\varepsilon_1[3/(2-\alpha)]} \tag{6.96}$$

and

$$\frac{\partial p}{\partial \varepsilon_1} = 2\sigma_1\left(\frac{t_0}{d_0}\right)e^{-\varepsilon_1[3/(2-\alpha)]}\left(-\frac{3}{2-\alpha}\right) \tag{6.97}$$

Substituting Eqs. 6.96 and 6.97 into Eq. 6.95 gives the result

$$dp = 2\left(\frac{t_0}{d_0}\right)e^{-\varepsilon_1[3/(2-\alpha)]}d\sigma_1 + 2\sigma_1(t_0/d_0)e^{-\varepsilon_1[3/(2-\alpha)]}\left(-\frac{3}{2-\alpha}\right)d\varepsilon_1 = 0 \tag{6.98}$$

and from this relationship

$$\frac{d\sigma_1}{d\varepsilon_1} = \sigma_1\left(\frac{3}{2-\alpha}\right) \tag{6.99}$$

An examination of the flow equations (6.48 to 6.50) shows that

$$\sigma_1 = f(\alpha)\varepsilon_1{}^n \tag{6.100}$$

Consequently

$$d\sigma_1/d\varepsilon_1 = nf(\alpha)\varepsilon_n{}^{n-1}$$

from which

$$\varepsilon_1 = \left(\frac{2-\alpha}{3}\right)n \tag{6.101}$$

Equation 6.101 defines the instability condition when σ_h dominates.

The bursting pressure for a thin-walled cylindrical tube is found by using Eqs. 5.4 to 5.7 with true values for d and t. See, for example, for a closed-end tube, Eqs. 6.65 and 6.67.

For simplicity, the radial stress can be eliminated and Eqs. 6.65 and 6.67 together with Eqs. 6.17 and 6.101 show that

$$p_b = \frac{4\sigma_u t_0/d_0}{\sqrt{3}e^n} \qquad 6.102$$

where σ_u, the ultimate tensile strength, replaces σ_0 in Eq. 6.17. For less exact calculations, Eqs. 5.4 to 5.7 can be used directly with σ_u replacing $\bar{\sigma}$.

CASE 2. AXIAL STRESS > HOOP STRESS ($\sigma_z > \sigma_h$). In this case, there is a limiting axial force that determines instability. The total axial force is given by

$$F = \pi d_p t_p \sigma_1 \qquad 6.103$$

where σ_1 is now the maximum principal stress in the axial direction. Substituting the values of d_p and t_p from Eqs. 6.89 and 6.87 into Eq. 6.103 gives

$$F = \pi d_0 t_0 e^{-\varepsilon_1}\sigma_1 \qquad 6.104$$

For instability

$$dF = (\partial F/\partial\sigma_1)\,d\sigma_1 + (\partial F/\partial\varepsilon_1)\,d\varepsilon_1 = 0 \qquad 6.105$$

from which

$$d\sigma_1/d\varepsilon_1 = \sigma_1 \qquad 6.106$$

From the flow equations, as before,

$$\sigma_1 = f(\alpha)\varepsilon_1{}^n \qquad 6.107$$

from which

$$\varepsilon_1 = n \qquad 6.108$$

Having established the instability conditions for the two cases noted, it now remains to actually determine the wall thickness at instability. For Case 1 the instability condition is given by Eq. 6.101. By equating this value of ε_1 to the value of ε_1 defined by the flow equations

$$\left(\frac{2-\alpha}{3}\right)n = \left(\frac{\sigma_1}{K}\right)^{1/n}(\alpha^2-\alpha+1)^{(1-n)/2n}\left(\frac{2-\alpha}{2}\right) \qquad 6.109$$

from which

$$\sigma_1 = K\left(\frac{2n}{3}\right)^n\left(\frac{1}{\alpha^2-\alpha+1}\right)^{(1-n)/2} \qquad 6.110$$

Combining Eqs. 6.84 and 6.110 gives

$$\sigma_1 = \frac{pd_p}{2t_p} = K\left(\frac{2n}{3}\right)^n\left(\frac{1}{\alpha^2-\alpha+1}\right)^{(1-n)/2} \qquad 6.111$$

For Case 2 the condition of instability is given by Eq. 6.108. The principal strain ε_1 is now in the axial direction and given by Eq. 6.48; therefore, as in the preceding,

$$n = \left(\frac{\sigma_1}{K}\right)^{1/n} (\alpha^2 - \alpha + 1)^{(1-n)/2n} \left(\frac{2-\alpha}{2}\right) \qquad 6.112$$

from which

$$\sigma_1 = K(2n)^n \left(\frac{1}{\alpha^2 - \alpha + 1}\right)^{(1-n)/2} \left(\frac{1}{2-\alpha}\right)^n \qquad 6.113$$

Combining Eq. 6.113 with Eq. 6.103 then gives

$$\sigma_1 = \frac{F}{\pi d_p t_p} = K(2n)^n \left(\frac{1}{\alpha^2 - \alpha + 1}\right)^{(1-n)/2} \left(\frac{1}{2-\alpha}\right)^n \qquad 6.114$$

Equations 6.111 and 6.114 for σ_1 contain values of diameter and wall thickness in the plastic range. In order to express σ_1 in terms of the original values note that

$$\text{hoop strain} = \ln(1 + \varepsilon_h) = \ln(d_p/d_0) \qquad 6.115$$

$$\text{radial strain} = \ln(1 + \varepsilon_r) = \ln(t_p/t_0) \qquad 6.116$$

$$\text{axial strain} = \ln(1 + \varepsilon_z) = \ln(L/L_0) \qquad 6.117$$

When internal pressure governs the maximum principal strain is expressed by Eq. 6.115; thus

$$\varepsilon_1 = \ln(d_p/d_0) \qquad \text{or} \qquad d_p = d_0 \ln^{-1} \varepsilon_1 \qquad 6.118$$

where ε_1 is defined by Eq. 6.48. An evaluation of t_p depends on whether ε_z or ε_r is the minimum principal strain. If ε_r is the minimum strain, then

$$\varepsilon_3 = \ln(t_p/t_0) \qquad \text{or} \qquad t_p = t_0 \ln^{-1} \varepsilon_3 \qquad 6.119$$

where ε_3 is defined by Eq. 6.50. If ε_z is the minimum strain, then

$$\varepsilon_2 = \ln(t_p/t_0) \qquad \text{or} \qquad t_p = t_0 \ln^{-1} \varepsilon_2 \qquad 6.120$$

where ε_2 is defined by Eq. 6.49. If it is assumed, for example, that ε_r is the minimum strain, then for Case 1,

$$\sigma_1 = \frac{pd_0}{2t_0}\left[\frac{\ln^{-1}(\sigma_1/K)^{1/n}(\alpha^2 - \alpha + 1)^{(1-n)/2n}[(2-\alpha)/2]}{\ln^{-1}(\sigma_1/K)^{1/n}(\alpha^2 - \alpha + 1)^{(1-n)/2n}(-1-\alpha)}\right] \qquad 6.121$$

from which

$$t_0 = \frac{pd_0}{2\sigma_1} \ln^{-1}\left[\frac{\alpha - 2}{2(1+\alpha)}\right] \qquad 6.122$$

where σ_1 is defined by Eq. 6.110. When the axial load governs, the maximum principal strain is expressed by Eq. 6.117, that is,

$$\varepsilon_1 = \ln(L/L_0) = -[\ln(t_p/t_0) + \ln(d_p/d_0)] \qquad 6.123$$

which results from Eq. 6.53. If it is assumed that the radial strain is minimum, then

$$\varepsilon_3 = \ln(t_p/t_0) \qquad \text{and} \quad t_p = t_0 \ln^{-1}\varepsilon_3 \tag{6.124}$$

where ε_3 is defined by Eq. 6.50. Then, under these conditions, for Case 2,

$$\sigma_1 = \frac{F}{\pi d_p t_p} = \frac{F}{\pi d_0 \ln^{-1}\varepsilon_2 t_0 \ln^{-1}\varepsilon_3} \tag{6.125}$$

where ε_2 and ε_3 are defined by Eqs. 6.49 and 6.50 respectively. Therefore by substituting in Eq. 6.125 the proper values of ε_2 and ε_3,

$$\sigma_1 = \frac{F}{\pi d_0 t_0 \ln^{-1}\{[\sigma_1/K)^{1/n}(\alpha^2 - \alpha + 1)^{(1-n)/2n}](\alpha/2 - 1)\}} \tag{6.126}$$

from which

$$t_0 = \frac{F}{\pi d_0 \sigma_1 \ln^{-1}\{[(\sigma_1/K)^{1/n}(\alpha^2 - \alpha + 1)^{(1-n)/2n}(\alpha/2 - 1)]\}} \tag{6.127}$$

where σ_1 is defined by Eq. 6.113.

Consider a cylinder for which the following data are given:

$$\sigma_y = \text{yield strength} = 50{,}000 \text{ psi}$$
$$K = 114{,}000 \text{ psi}$$
$$d_0 = 20 \text{ in}$$
$$n = 0.16$$
$$p = 2000 \text{ psi}$$

The problem is to determine the required wall thickness for elastic behavior and also to determine the wall thickness at instability for Case 1. Consider a closed-end cylinder; in this case,

$$\sigma_1 = \sigma_h = \frac{pd}{2t}$$

$$\sigma_2 = \sigma_z = \frac{pd}{4t}$$

For elastic behavior, in accordance with Eq. 6.27

$$\sigma_0{}^2 = \sigma_y{}^2 = \sigma_h{}^2 + \sigma_z{}^2 - \sigma_h\sigma_z = \tfrac{3}{4}\sigma_h{}^2$$

from which

$$t = \frac{\sqrt{3}\,pd}{4\ \sigma_y} = \frac{(\sqrt{3})(2000)(20)}{(4)(50{,}000)} = 0.346 \text{ in.}$$

For instability under the conditions given, σ_h is calculated using Eq. 6.110 which gives a value of 45,000 psi; then, from Eq. 6.122, with $\alpha = \frac{1}{2}$, the wall thickness is 0.271 in. That is, under the conditions given, a cylinder with an initial wall thickness of 0.271 in. would become unstable at an internal pressure of 2000 psi. For elastic behavior the cylinder diameter ratio was $20.692/20 = 1.035$, whereas for instability the ratio is $20.542/20 = 1.028$ (about 1% less).

It should be noted in the preceding analyses that the condition of instability depended on a single load factor like pressure or force. As long as one of these factors dominates, the analysis, as presented, is valid. However, for the particular case of $\alpha = 1.0$ the condition of instability depends on both a load and a pressure in the problem discussed; therefore, for $\alpha = 1.0$, Eqs. 6.110 and 6.113 give different results and hence are not valid. For $\alpha = 1.0$ recourse may be had in analyzing a thin, walled sphere under internal pressure. In this case, $\sigma_1 = \sigma_2$ and instability can be defined by $dp = 0$. This case will be analyzed next.

Thin-walled spherical shell under internal pressure. For this type of vessel the two principal stresses are equal, that is,

$$\sigma_1 = \sigma_2 = pd_0/4t_0 \qquad 6.128$$

where p is internal pressure, d_0 is the original inside diameter, and t_0 is the original wall thickness. From Eq. 6.128, in the plastic range

$$p = 4\sigma_1 t_p/d_p \qquad 6.129$$

and instability is defined as $dp = 0$, where d is the differential operator. Since p is a function of both σ_1 and ε_1,

$$(\partial p/\partial \sigma_1)\, d\sigma_1 + (\partial p/\partial \varepsilon_1)\, d\varepsilon_1 = 0 \qquad 6.130$$

Expressions for d_p and t_p have already been given; therefore from Eq. 6.129

$$p = \frac{4\sigma_1 t_0 e^{-2\varepsilon_1}}{4e^{\varepsilon_1}} = 4\sigma_1 \frac{t_0}{d_0} e^{-3\varepsilon_1} \qquad 6.131$$

Combining Eqs. 6.130 and 6.131 then gives the result

$$\frac{\partial p}{\partial \sigma_1} = 4\frac{t_0}{d_0} e^{-3\varepsilon_1}, \qquad \frac{\partial p}{\partial \varepsilon_1} = 4\sigma_1 \frac{t_0}{d_0} e^{-3\varepsilon_1}(-3)$$

and

$$4\frac{t_0}{d_0} e^{-3\varepsilon_1}\, d\sigma_1 + 4\sigma_1 \frac{t_0}{d_0} e^{-3\varepsilon_1}(-3)\, d\varepsilon_1 = 0$$

from which

$$\frac{d\sigma_1}{d\varepsilon_1} = 3\sigma_1 \qquad 6.132$$

From the flow equations $\sigma_1 = f(\alpha)\varepsilon_1{}^n$; therefore

$$\frac{d\sigma_1}{d\varepsilon_1} = nf(\alpha)\varepsilon_1{}^{n-1}$$

and finally,

$$\varepsilon_1 = n/3 \qquad 6.133$$

which defines the instability condition for $\alpha = 1.0$. Tests of spherical shells were carried out by Marin, Dutton, and Faupel (29). The shells, approximately 20 in. diameter and 13/16 in. thick, were made of steel and were tested by internal pressure at temperatures from -35 to $+80°F$. The results were analyzed in the same manner as outlined for the tension-compression tests of tubes, and curves similar to those shown in Fig. 6.6 were obtained. The bursting pressure is found by the same procedure used for the thin-walled cylinder and is expressed as

$$p_u = \frac{4\sigma_u(t_0/d_0)}{e^n} \tag{6.134}$$

Relation defining instability for two-dimensional tension (Fig. 6.8). Consider the plate shown in Fig. 6.8 subjected to two-dimensional tension. It is assumed that instability will occur in the direction of the greatest load; thus, if F_1 is the load in the σ_1 direction,

$$F_1 = \sigma_1 H t \tag{6.135}$$

and the corresponding strain is

$$\varepsilon_1 = \ln (L/L_0) \tag{6.136}$$

From Eq. 6.136

$$L = L_o e^{\varepsilon_1} \tag{6.137}$$

Similarly,

$$H = H_o e^{\varepsilon_2} \tag{6.138}$$

and

$$t = t_o e^{\varepsilon_3} \tag{6.139}$$

where the subscript o denotes the original value. Substituting Eqs. 6.138 and 6.139 into Eq. 6.135 and using the condition of constancy of volume

$$F_1 = \sigma_1 H_0 e^{\varepsilon_2} t_0 e^{\varepsilon_3} \tag{6.140}$$

or

$$F_1 = \sigma_1 H_0 t_0 e^{-\varepsilon_1} \tag{6.141}$$

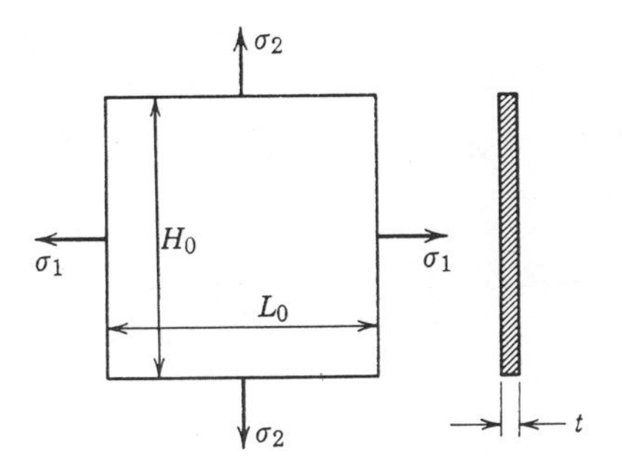

Fig. 6.8 Plate element subjected to two-dimensional tension.

For instability $dF_1 = 0$; therefore, since F_1 is a function of both σ_1 and ε_1,

$$dF_1 = \frac{\partial F_1}{\partial \sigma_1} d\sigma_1 + \frac{\partial F_1}{\partial \varepsilon_1} d\varepsilon_1 = 0 \qquad 6.142$$

from which

$$\frac{d\sigma_1}{d\varepsilon_1} = \sigma_1 \qquad 6.143$$

From the flow equations, with $\beta = 0$,

$$\sigma_1 = f(\alpha)\varepsilon_1{}^n \qquad 6.144$$

and

$$\frac{d\sigma_1}{d\varepsilon_1} = nf(\alpha)\varepsilon^{n-1} \qquad 6.145$$

from which

$$\varepsilon_1 = n \qquad 6.146$$

which defines the condition of instability.

Theory of plane plastic strain

In many problems involving plastic flow the elastic-plastic boundary can be specified because of conditions of rotational symmetry. The plastic expansion of cylinders, spheres, and so on are examples. In many other cases this condition does not prevail, and other methods have to be used to determine stress and strain distributions. Except for even somewhat simplified cases like plane-strain, solutions are not available at all. The theory of plane plastic strain has had particular application in problems like rolling and drawing of metals and other fabrication processes. Some examples are given following a brief introduction to the theory. In the simplest cases the theory of plane plastic strain is rather difficult to apply; therefore it is essential that the references cited are carefully studied. The following discussion is brief and is based largely on the account given by Prager and Hodge (40).

In plane strain ε_3 is zero and the strain equations for the case of combined stresses are written as follows:

$$\varepsilon_1 = (1/D)[\sigma_1 - \tfrac{1}{2}(\sigma_2 + \sigma_3)] \qquad 6.147$$

$$\varepsilon_2 = (1/D)[\sigma_2 - \tfrac{1}{2}(\sigma_1 + \sigma_3)] \qquad 6.148$$

$$\varepsilon_3 = (1/D)[\sigma_3 - \tfrac{1}{2}(\sigma_1 + \sigma_2)] = 0 \qquad 6.149$$

where ε is the principal plastic strain, σ the principal stress, $1/D$ is the variable plastic compliance, and $\frac{1}{2}$ is Poisson's ratio based on constancy of volume. From Eq. 6.149 it is seen that

$$\sigma_3 = \tfrac{1}{2}(\sigma_1 + \sigma_2) \qquad 6.150$$

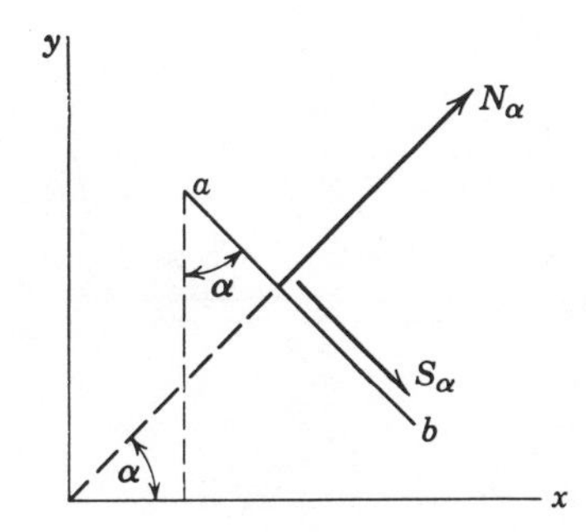

Fig. 6.9 Plane in a two-dimensional stress field.

The term σ_3 in Eq. 6.150 is called the *mean normal stress.* The general flow rule is given by Eq. 6.27, which for plane strain, by virtue of Eq. 6.150 becomes

$$\sigma_0{}^2 = \tfrac{3}{4}(\sigma_1{}^2 + \sigma_2{}^2) - \tfrac{3}{2}\sigma_1\sigma_2 \qquad 6.151$$

But σ_1 and σ_2 may also be expressed by Eqs. 3.18 and 3.19; therefore Eq. 6.151 in terms of the component stresses may be written as

$$(\sigma_0/\sqrt{3})^2 = \tfrac{1}{4}(\sigma_x + \sigma_y)^2 + \tau_{xy}{}^2 \qquad 6.152$$

Consider now Fig. 6.9 which shows a plane a–b in a plastic material perpendicular to the plane of flow. This plane has acting on it a normal stress N_α and a shear stress S_α, which may be expressed as follows in terms of the notation in the figure; (see also 40).

$$N_\alpha = \sigma_x \cos^2\alpha + \sigma_y \sin^2\alpha + 2\tau_{xy} \sin\alpha\cos\alpha \qquad 6.153$$

$$S_\alpha = (\sigma_x - \sigma_y)\sin\alpha\cos\alpha + \tau_{xy}(\sin^2\alpha - \cos^2\alpha) \qquad 6.154$$

Equations 6.153 and 6.154 are shown in Mohr's circle notation in Fig. 6.10,

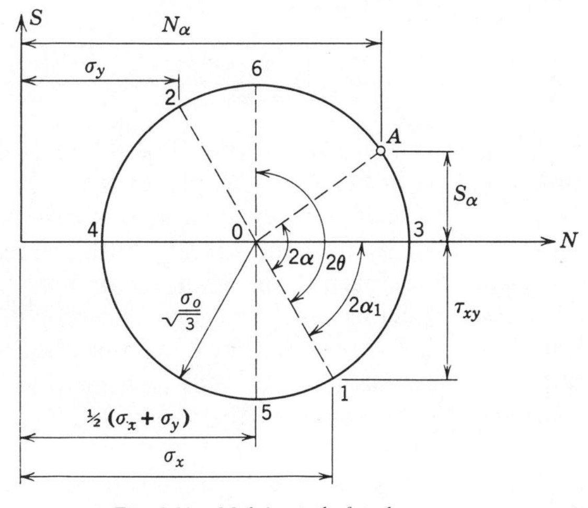

Fig. 6.10 Mohr's circle for plane stress.

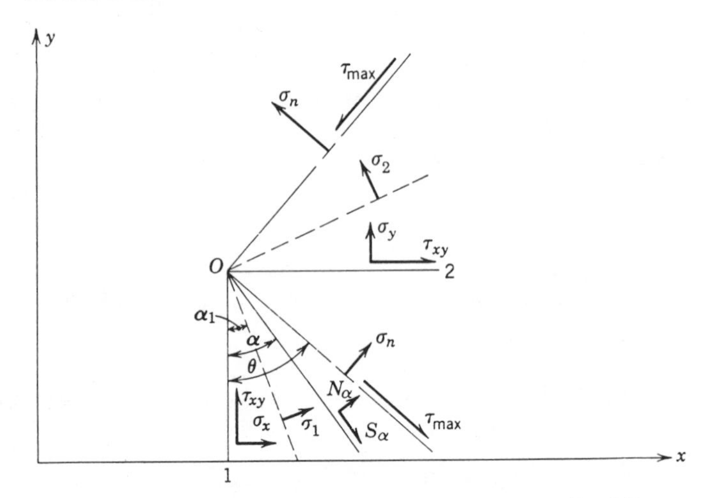

Fig. 6.11 Physical plane corresponding to stresses in Fig. 6.10

where the coordinates of the circle represent normal and shear stress transmitted at a fixed point across the surface elements that are perpendicular to the plane of flow. A physical representation proposed by Prager and Hodge (33) is shown in Fig. 6.11. Planes $O1$ and $O2$ are for $\alpha = 0$ and $90°$ respectively; in Fig. 6.10 it is indicated that these planes have coordinates σ_x and $-\tau_{xy}$ and σ_y and $+\tau_{xy}$ respectively. When these stress components are known, a Mohr's circle can be constructed and from the discussion of this method in Chapter 3 it will be recalled that

$$\sigma_{max} = \sigma_1 = \frac{\sigma_x + \sigma_y}{2} + \left[\left(\frac{\sigma_x - \sigma_y}{2}\right)^2 + \tau_{xy}^{\ 2}\right]^{1/2} \qquad 6.155$$

which is at point 3, Fig. 6.10. Furthermore,

$$\sigma_{min} = \sigma_2 = \frac{\sigma_x + \sigma_y}{2} - \left[\left(\frac{\sigma_x - \sigma_y}{2}\right)^2 + \tau_{xy}^{\ 2}\right]^{1/2} \qquad 6.156$$

which is at point 4, Fig. 6.10. Plane OA at $\alpha°$ in Fig. 6.11 corresponds to the angle 2α between OA and $O1$ in Fig. 6.10. The planes of maximum stress and minimum stress (σ_1 and σ_2) which are at an angle α_1 as shown in Fig. 6.11 are orthogonal and correspond to points 3 and 4 in Fig. 6.10. The planes of maximum shear stress, 3–4 and 5–6, Fig. 6.10, are at angle θ to plane $O1$ in Fig. 6.11 and have normal stresses σ_n acting on them. Note that σ_n is not a principal stress. Finally, from Fig. 6.10 it is seen that the maximum shear stress has the value $\sigma_0/\sqrt{3}$ (or $\sigma_y/\sqrt{3}$). The foregoing analysis now provides the basis for calculations of *slip lines*. In the literature, the theory is often referred to as *slip line theory*.

The maximum shear stresses are always 45 degrees to the principal stresses, and since the *first shear direction* is obtained by a 45° rotation from the σ_1

direction and the *second shear direction* by a similar rotation from the σ_2 direction, an orthogonal net is formed. In the literature these shear lines are called first and second shear lines or sometimes α and β lines. The flow law, Eq. 6.152, shows that the stress components are not independent, and thus it is necessary to be able to express these stresses in terms of the known quantities. This is done by defining

$$\phi = \frac{\text{mean normal stress}}{\text{yield function}} = \frac{\frac{1}{2}(\sigma_x + \sigma_y)}{2\sigma_0/\sqrt{3}} \tag{6.157}$$

and

$$\theta = \tfrac{1}{2} \arctan \frac{2\tau_{xy}}{\sigma_x - \sigma_y} + \frac{\pi}{4} \tag{6.158}$$

where θ is the angle through which the y axis must be rotated to assume the first shear direction (Fig. 6.11). Then

$$\sigma_x = (2\sigma_0/\sqrt{3})\phi + (\sigma_0/\sqrt{3}) \sin 2\theta \tag{6.159}$$

$$\sigma_y = (2\sigma_0/\sqrt{3})\phi - (\sigma_0/\sqrt{3}) \sin 2\theta \tag{6.160}$$

$$\tau_{xy} = -(\sigma_0/\sqrt{3}) \cos 2\theta \tag{6.161}$$

$$N_\alpha = (2\sigma_0/\sqrt{3})\phi + (\sigma_0/\sqrt{3}) \sin 2(\theta - \alpha) \tag{6.162}$$

$$S_\alpha = -(\sigma_0/\sqrt{3}) \cos 2(\theta - \alpha) \tag{6.163}$$

These equations, together with the equilibrium equations (Eqs. 3.76 and 3.77),

$$\partial\sigma_y/\partial y + \partial\tau_{xy}/\partial x = 0 \tag{6.164}$$

$$\partial\sigma_x/\partial x + \partial\tau_{xy}/\partial y = 0 \tag{6.165}$$

and the yield condition, Eq. 6.152, with appropriate boundary conditions, allow for a partial solution to many practical problems. In the last section of this chapter the method will be applied to an extrusion problem and the result compared with another method of analysis.

6-2 PLASTICITY IN MACHINE AND STRUCTURAL DESIGN

In this section the plastic behavior of various machine and structural parts will be considered. In particular, reference will be made to the subject of *limit analysis* or *limit design* which is a system of analysis based on a material's total ability to carry service loads.

Tension member

Tension (compression) of prismatic bars is the simplest type of plasticity problem and quite accurate solutions may be obtained for both work-hardening and non-work-hardening (ideally plastic) materials. Consider a bar of constant

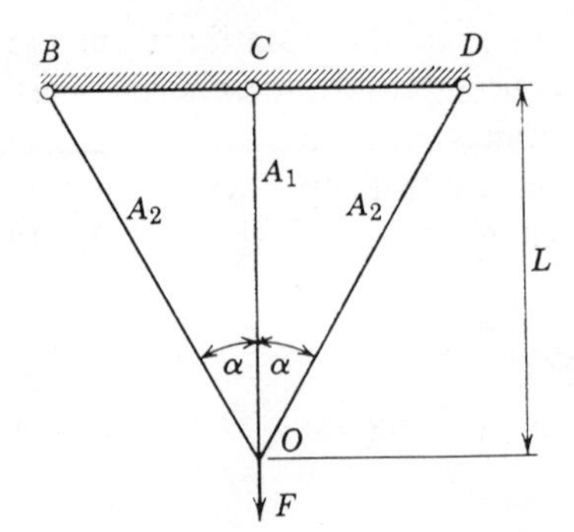

Fig. 6.12 Three-bar tension member.

cross-sectional area A subjected to a tensile force F. In the plastic range

$$\sigma = \frac{F}{A} = \sigma_y = \frac{F}{A_0}(1 + \varepsilon_0) = K\varepsilon^n \tag{6.166}$$

where σ_y is yield stress, ε_0 is nominal tensile strain, A_0 the initial cross-sectional area, and K and n the constants previously defined. The corresponding strain is given by

$$\varepsilon = \ln(L/L_0) = \ln(1 + \varepsilon_0) \tag{6.167}$$

Another type of tension problem is illustrated by the framework in Fig. 6.12 (see also Fig. 2.39). This frame is elastically statically indeterminate and the solution is given in Chapter 2. In the plastic range it is assumed that the bars yield at a stress σ_y, the yield strength of the material, so that for equilibrium

$$F = \sigma_y A_1 + 2\sigma_y A_2 \cos\alpha \tag{6.168}$$

If stress-strain data are available for the material, the maximum allowable stress and strain before instability sets in may be computed from Eqs. 6.1 and 6.16, thus

$$(\sigma/K)^{1/n} = \varepsilon = n \tag{6.169}$$

Suppose the frame is made of an aluminum alloy and the following data are supplied:

$$\sigma_y = 50{,}000 \text{ psi}$$
$$K = 115{,}000 \text{ psi}$$
$$n = 0.15$$
$$L_0 = 10 \text{ in.}$$
$$A_1 = 0.25 \text{ in.}^2$$
$$A_2 = 0.15 \text{ in.}^2$$
$$\alpha = 60^\circ$$

from Eq. 6.168

$$F = (50{,}000)(0.25) + 2(50{,}000)(0.15)(0.50) = 20{,}000 \text{ lb}$$

From Eq. 6.169

$$\sigma/K = n^n \qquad \text{or} \qquad \sigma = 115{,}000(0.15)^{0.15} = 86{,}480 \text{ psi}$$

That is, the stress in the material at the instant of instability would be 86,000 psi at a strain of 0.15 in./in. Since the length of the central bar in Fig. 6.12 is 10 in. the total elongation is 10(0.15) = 1.5 in. for instability.

Pure bending of bars

Symmetrical cross section. Elastic bending was analyzed in Chapter 2 on the basis of Hooke's law $\sigma = E\varepsilon$. Consider now pure bending where the relation between stress and strain is given by the expression

$$\sigma = (E\varepsilon)^{1/m} \tag{6.170}$$

An expression will be developed for bending stress based on Eq. 6.170 with the following assumptions:

1. Plane sections remain plane.
2. Material is homogeneous and isotropic.
3. No lateral stresses are present.
4. The bar will not buckle.
5. The modulus is the same in tension and compression.

Considering Fig. 6.13, where it can be seen that

$$\varepsilon/\varepsilon_y = c/y \tag{6.171}$$

From Eq. 6.170

$$E = \sigma^m/\varepsilon = \sigma_y{}^m/\varepsilon_y \tag{6.172}$$

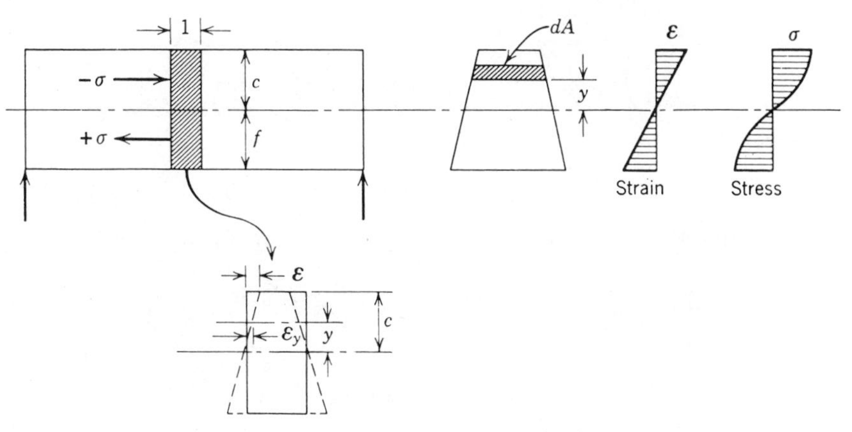

Fig. 6.13 Beam element in plastic bending.

from which

$$\frac{\sigma^m}{\varepsilon} = \sigma_y{}^m/\varepsilon_y \qquad \text{or} \qquad \frac{\varepsilon}{\varepsilon_y} = (\sigma/\sigma_y)^m \qquad 6.173$$

However, from Eq. 6.171

$$(\sigma/\sigma_y)^m = c/y \qquad 6.174$$

The load on a fiber is $\sigma_y\,dA$ and the moment of the total stress is

$$M = \int_{-f}^{c} (\sigma_y dA)y \qquad 6.175$$

But from Eq. 6.174

$$\sigma_y = \left(\frac{y}{c}\right)^{1/m} \sigma \qquad 6.176$$

Therefore

$$M = \int_{-f}^{c} \sigma\left(\frac{y}{c}\right)^{1/m} y\,dA \qquad 6.177$$

For a bar with a rectangular cross section $dA = b\,dy$, where b is the width and

$$M = \sigma \int_{-f}^{c} \left(\frac{y}{c}\right)^{1/m} yb\,dy \qquad 6.178$$

or

$$M = \frac{\sigma b}{c^{1/m}} \int_{-f}^{c} y^{1/m} y\,dy = \frac{\sigma b}{c^{1/m}} \int_{-f}^{c} y^{(m+1)/m}\,dy \qquad 6.179$$

For equilibrium the sum of the normal forces has to be zero; therefore

$$\int_{-f}^{c} \sigma_y b\,dy = 0 \qquad 6.180$$

Substituting Eq. 6.176 into Eq. 6.180 gives

$$\int_{-f}^{c} \left(\frac{y}{c}\right)^{1/m} \sigma b\,dy = 0 = \frac{\sigma b}{c^{1/m}} \int_{-f}^{c} y^{1/m}\,dy = \int_{-f}^{c} y^{1/m}\,dy \qquad 6.181$$

from which

$$c^{(m+1)/m} - f^{(m+1)/m} = 0 \qquad \text{or} \quad c = f \qquad 6.182$$

which says that the neutral and centroidal axes coincide. From Eq. 6.179

$$M = \frac{2\sigma b}{(d/2)^{1/m}} \int_{0}^{d/2} y^{1/m} y\,dy = \frac{2\sigma b}{(d/2)^{1/m}} \int_{0}^{d/2} y^{(m+1)/m}\,dy \qquad 6.183$$

from which

$$\sigma = \frac{1 + 2m}{m}\left(\frac{2M}{bd^2}\right) \qquad 6.184$$

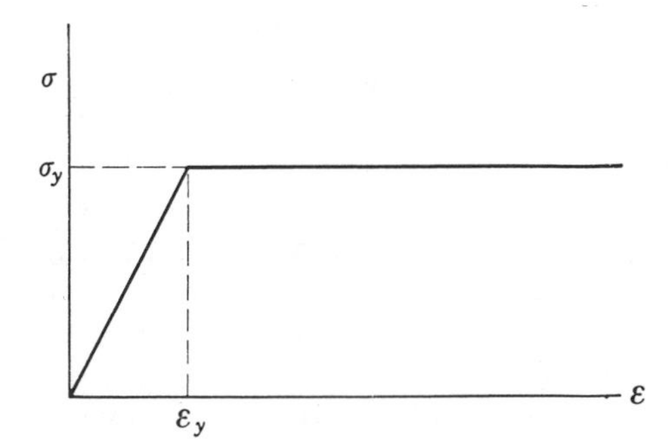

Fig. 6.14 Stress-strain diagram for perfect plasticity (yielding at constant stress).

where d is the depth of the beam. When $m = 1$, Eq. 6.184 becomes for a rectangular cross section

$$\sigma = 6M/bd^2 \qquad 6.185$$

which is the familiar elastic solution.

By the same procedure it is easily shown that if the stress-strain law is given by Eq. 6.1, the bending stress takes the form (for a rectangular cross section)

$$\sigma = (2M/bd^2)(n + 2) \qquad 6.186$$

Again, when $n = 1$, the elastic solution is obtained,

$$\sigma = 6M/bd^2 \qquad 6.187$$

Suppose the material has the stress-strain characteristics shown in Fig. 6.14; this is commonly referred to as a *perfectly plastic* material; that is, it is non-work-hardening. For many practical applications steel exhibits this type of behavior in its useful working range. If a material having such characteristics is used as a beam, for example, Eq. 6.187 becomes

$$\sigma_y = 6M_y/bd^2 \qquad 6.188$$

at the time the bending stress σ reaches the yield strength σ_y of the material.

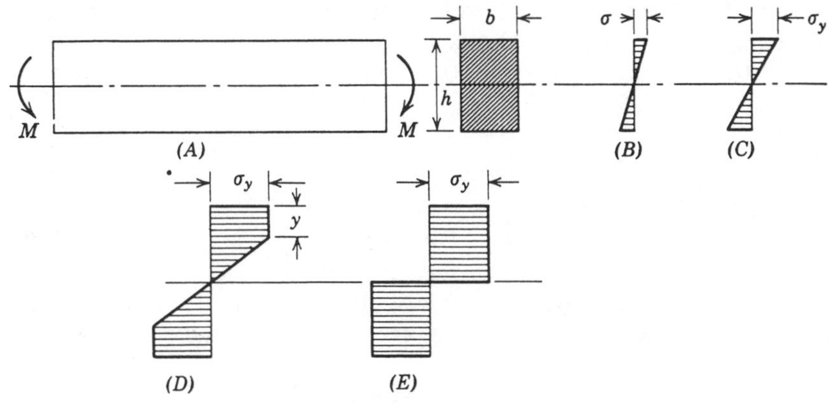

Fig. 6.15 Pure bending of a beam in the plastic range.

Equation 6.188 is for a beam of rectangular cross section. In general, for any cross section the yield moment for a beam is

$$M_y = \sigma_y Z \qquad 6.189$$

where Z is the section modulus. The stress distribution across the beam section is shown in Fig. 6.15 for a rectangular cross section. The elastic distribution is shown in Fig. 6.15B, where σ is the stress in the elastic range. When σ equals σ_y, the distribution shown in Fig. 6.15C is reached (the outside fibers are under a stress σ_y). As flow continues and the material yields at constant stress the distribution shown in Fig. 6.15D is reached. Finally, for full plasticity the distribution shown in Fig. 6.15E is obtained. The bending moment at the beginning of yielding is given by Eq. 6.188. When plastic flow has progressed as shown in Fig. 6.15D, equilibrium shows that

$$M = 2\left[\sigma_y by\left(\frac{h}{2} - \frac{y}{2}\right)\right] + 2\left[\sigma_y \frac{b}{2}\left(\frac{h}{2} - y\right)\frac{2}{3}\left(\frac{h}{2} - y\right)\right]$$

from which

$$M = \sigma_y \frac{bh^2}{6}\left[1 + \frac{2y}{h}\left(1 - \frac{y}{h}\right)\right] \qquad 6.190$$

when the beam is fully plastic, as in Fig. 6.15E, $y = h/2$, so that Eq. 6.190 becomes

$$M = \sigma_y (bh^2/4) \qquad 6.191$$

Thus it is seen by comparing Eqs. 6.188 and 6.191 that for full plasticity the bending moment is 50% greater than that required to initiate plastic flow. Similar relationships can be developed for other cross-section shapes by using the foregoing analysis and taking into account the effect of Z. A summary of results for several cross sections is given in Table 6.1.

Nonsymmetrical cross section. In a nonsymmetrical section the neutral and centroidal axes do not occur at the center of the section, which complicates the situation. The general method of solving the problem, however, is the same as outlined for the symmetrical section. Consider, for example, the section shown in Fig. 6.16. For this particular cross section,

$$I = \tfrac{1}{3}(bc_1{}^3 - b_1 d^3 + tc_2{}^3) \qquad 6.192$$

where

$$c_1 = \frac{1}{2}\left(\frac{d^2 + b_1 t}{d + b_1}\right) \qquad 6.193$$

At the instant of yielding of the outermost fibers the neutral and centroidal axes coincide and the distribution of stress shown in Fig. 6.16B is obtained. Then from Eqs. 6.189 and 6.192,

$$M_y = (\sigma_y/3c_2)(bc_1{}^3 - b_1 d^3 + tc_2{}^3) \qquad 6.194$$

Table 6.1 Bending-Moment Relationships for Beams of Various Symmetrical Cross Sections and Yielding at Constant Stress

Beam Section	Moment of Inertia, I	Moment at Instant of Yielding, M_y	Moment for Full Plasticity, M_f	Moment Ratio, M_f/M_y
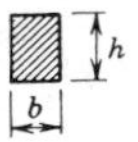 b, h	$\frac{bh^3}{12}$	$\sigma_y \frac{bd^2}{6}$	$\sigma_y \frac{bd^2}{4}$	1.5
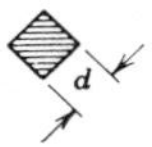 d	$\frac{d^4}{12}$	$\sigma_y \frac{\sqrt{2}\,d^3}{12}$	$\sigma_y \frac{\sqrt{2}\,d^3}{6}$	2.0
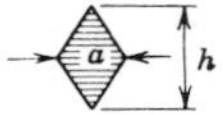 a, h	$\frac{ah^3}{48}$	$\sigma_y \frac{ah^2}{24}$	$\sigma_y \frac{ah^2}{12}$	2.0
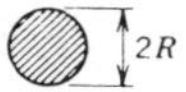 $2R$	$\frac{\pi R^4}{4}$	$\sigma_y \frac{\pi R^3}{4}$	$\sigma_y(0.4244)\pi R^3$	1.70
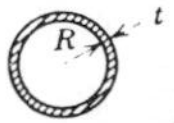 R, t	$\pi R^3 t$	$\sigma_y \frac{2\pi R^3 t}{R+t}$	$\sigma_y(1.274)\pi R^2 t$	$1.274\left(1+\frac{t}{R}\right)$
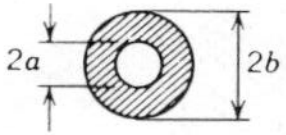 $2a$, $2b$	$\frac{\pi(b^4-a^4)}{4}$	$\sigma_y \frac{\pi(b^4-a^4)}{4b}$	$\sigma_y \frac{4(b^3-a^3)}{3}$	$\frac{16b(b^3-a^3)}{3\pi(b^4-a^4)}$
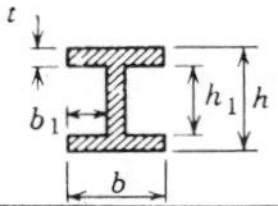 t, b_1, b, h_1, h	$\frac{bh^3-2b_1h_1^3}{12}$	$\sigma_y \frac{(bh^3-2b_1h_1^3)}{6h}$	$\sigma_y \frac{(bh-b_1h_1)(bh^2-2b_1h_1^2)}{4(bh-b_1h_1)}$	$\frac{3h}{2}\left(\frac{bh^2-2b_1h_1^2}{bh^3-2b_1h_1^3}\right)$

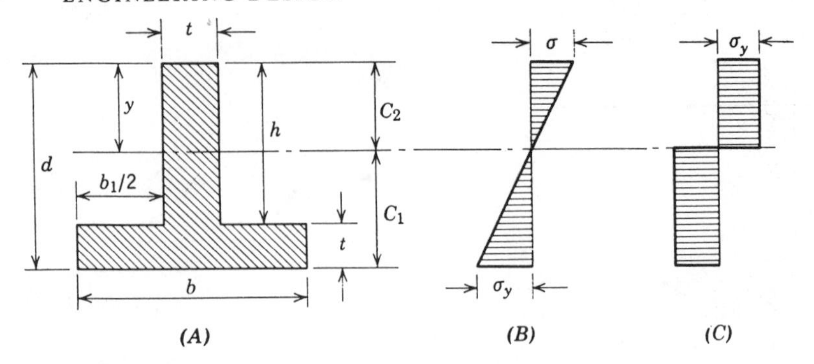

Fig. 6.16 Nonsymmetrical section in pure bending in the plastic range.

For simplicity, if h equals b, the fully plastic moment is obtained by noting that the neutral axis must divide the area of the cross section, that is, for b equals h the neutral axis is in the plane of the upper part of the horizontal part of the section, that is, bt equals ht. From this condition;

$$\int \sigma \, dA = 0 \qquad 6.195$$

and

$$M = \sigma_y \left[bt\left(\frac{t}{2}\right) + bt\left(\frac{b}{2}\right) \right] = \sigma_y bt\left(\frac{b+t}{2}\right) \qquad 6.196$$

Case of work-hardening material. If the material is not perfectly plastic but work-hardens, this factor has to be allowed for in the analysis. In the following paragraphs two methods will be discussed, the first due to Timoshenko (45) which utilizes the actual stress-strain diagram of the material and the second, due to Graziano (14), which is an intermediate between perfect plasticity and the method devised by Timoshenko. Consider bending of a beam of rectangular cross section as shown in Fig. 6.17. From Eq. 2.137 the bending strain is given by the relationship

$$\varepsilon_b = y/R \qquad 6.197$$

where R is the radius of curvature. For the outermost fibers of the bar, in accordance with Eq. 6.197,

$$\varepsilon_1 = \frac{h_1}{R} \quad \text{and} \quad \varepsilon_2 = \frac{-h_2}{R} \qquad 6.198$$

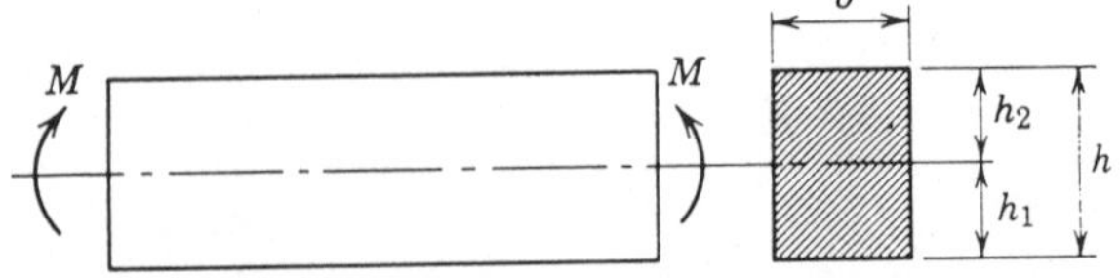

Fig. 6.17 Beam subjected to pure bending.

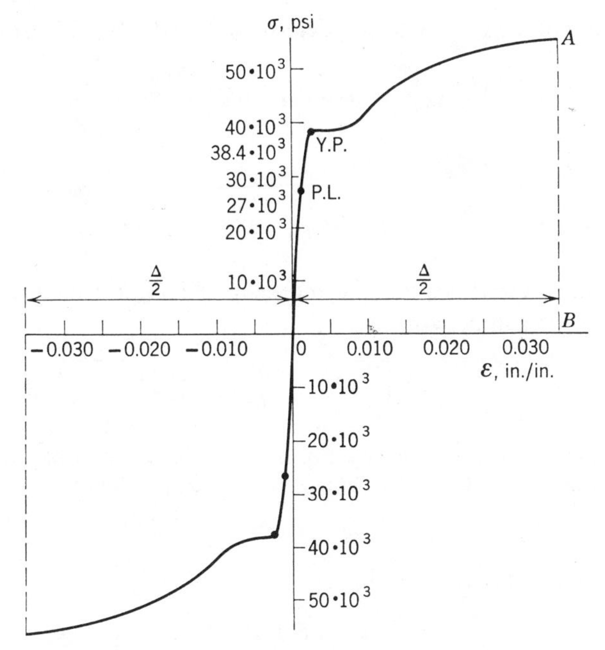

Fig. 6.18 Stress-strain diagram for structural steel. [*After Timoshenko and Gere (45), courtesy of McGraw-Hill Book Co.*]

Equation 6.197 also indicates that

$$y = R\varepsilon_b \qquad \text{and} \quad dy = R\,d\varepsilon_b \tag{6.199}$$

Substituting Eq. 6.199 into Eq. 2.131 gives

$$\int_{-h_2}^{h_1} \sigma\,dy = R \int_{\varepsilon_2}^{\varepsilon_1} \sigma\,d\varepsilon = 0 \tag{6.200}$$

The position of the neutral axis is such that

$$\int_{\varepsilon_2}^{\varepsilon_1} \sigma\,d\varepsilon = 0 \tag{6.201}$$

and this can be determined graphically by reference to Fig. 6.18, the stress-strain curve of the material. From Eqs. 6.198 and Fig. 6.18 it is seen that

$$\Delta = |\varepsilon_1 + \varepsilon_2| = \frac{h_1}{R} + \frac{h_2}{R} = \frac{h}{R} \tag{6.202}$$

from which

$$\frac{h_1}{h_2} = \left|\frac{\varepsilon_1}{\varepsilon_2}\right| \tag{6.203}$$

and the position of the neutral axis is determined. From Eq. 2.132 and Eq. 6.199

$$M = b\int_{h_2}^{h_1} \sigma R\varepsilon(R\,d\varepsilon) = bR^2\int_{\varepsilon_2}^{\varepsilon_1} \sigma\varepsilon\,d\varepsilon \qquad 6.204$$

which with Eq. 6.202 gives the result

$$M = \frac{bh^2}{\Delta^2}\int_{\varepsilon_2}^{\varepsilon_1} \sigma\varepsilon\,d\varepsilon \qquad 6.205$$

For a rectangular cross section the moment of inertia I equals $bh^3/12$ and therefore Eq. 6.205 can be written as

$$M = \frac{bh^3}{12}\left(\frac{12}{\Delta^2 h}\right)\int_{\varepsilon_2}^{\varepsilon_1} \sigma\varepsilon\,d\varepsilon = \frac{bh^3}{12}\left(\frac{1}{R}\right)\left(\frac{12}{\Delta^3}\right)\int_{\varepsilon_2}^{\varepsilon_1} \sigma\varepsilon\,d\varepsilon \qquad 6.206$$

The radius of curvature is given by Eq. 2.77, which when substituted into the moment equation gives Eq. 2.76, that is, the conventional elastic solution

$$M = EI/R \qquad 6.207$$

For plastic bending, Eq. 6.207 becomes

$$M = E_r I/R \qquad 6.208$$

where E_r is called the *reduced modulus*. Rewriting Eq. 6.206 and using Eq. 6.208 then gives the value of the reduced modulus as

$$E_r = \frac{12}{\Delta^3}\int_{\varepsilon_2}^{\varepsilon_1} \sigma\varepsilon\,d\varepsilon \qquad 6.209$$

The integral in Eq. 6.209 represents the moment of the area of the curve in Fig. 6.18 with respect to a vertical through the origin. For example, in Fig. 6.18, if $\Delta/2 = 0.035$ the area under the curve (by planimeter measurement) is 1620

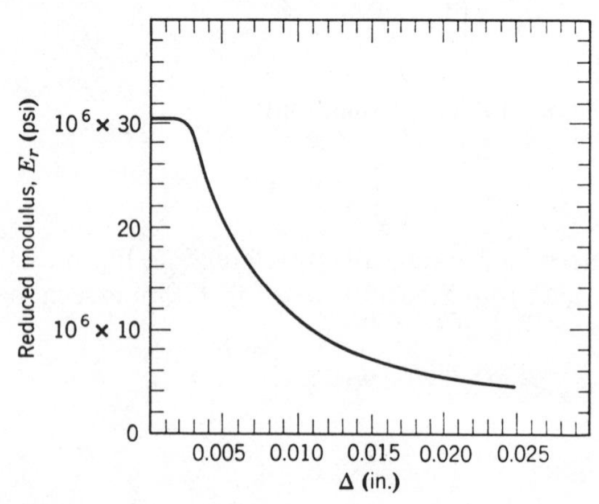

Fig. 6.19 Plot of deflection versus reduced modulus. [*After Timoshenko and Gere (45), courtesy of McGraw-Hill Book Co.*]

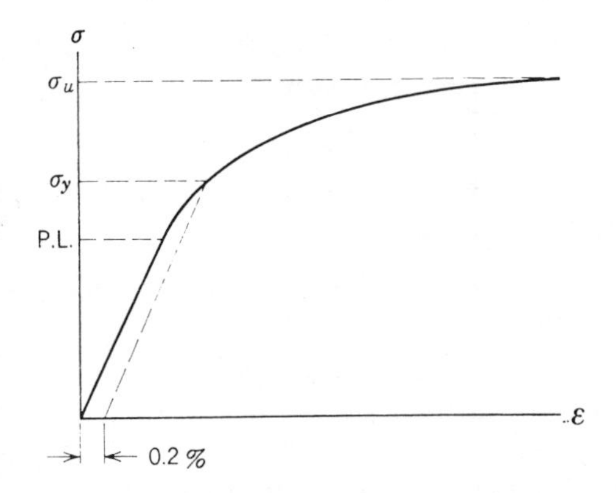

Fig. 6.20 General tensile stress-strain curve.

tension units and 1639 compression units; therefore

$$E_r = \frac{12}{0.000343}(1620)(0.02) + (1639)(0.0191)$$

or

$$E_r = 2{,}230{,}000 \text{ psi}$$

Additional values of E_r are plotted in Fig. 6.19, which, together with Eq. 6.208, can be used to calculate the bending moment for various degrees of plasticity.

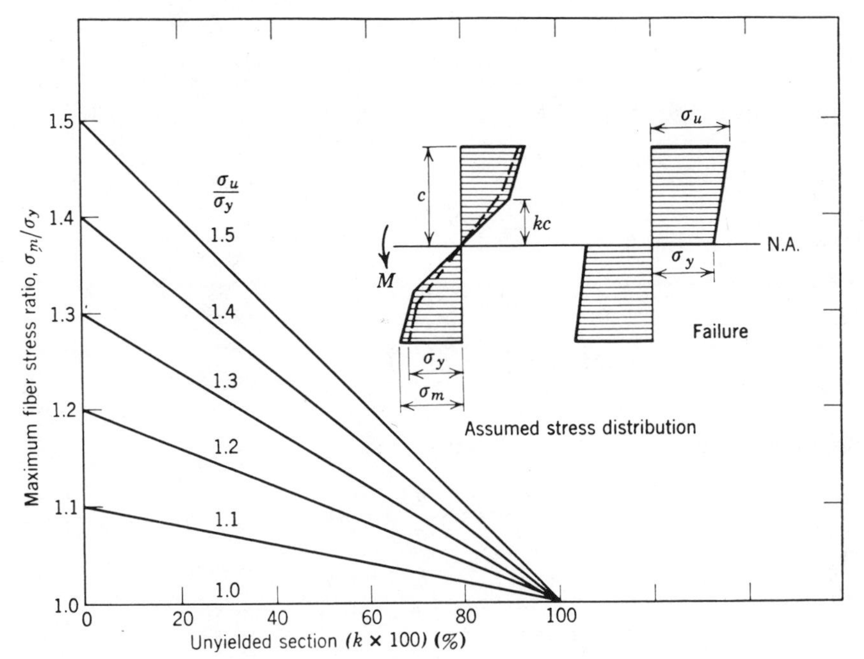

Fig. 6.21 Stress relationships in plastic bending. [After Graziano (14), courtesy of McGraw-Hill Publishing Co.]

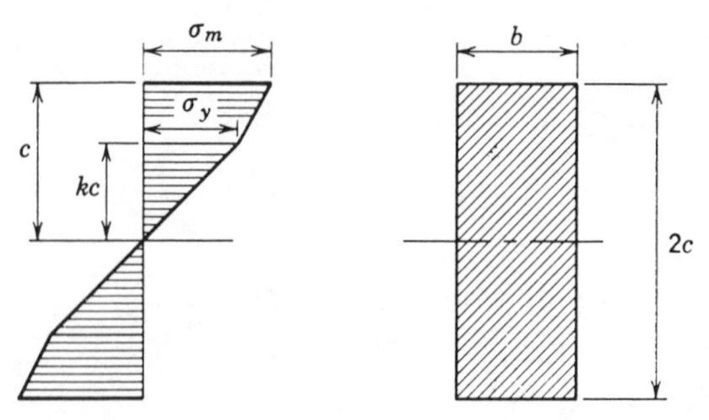

Fig. 6.22 Stress distribution in bent beam.

The same general procedure as outlined can also be applied to beams of other than rectangular cross section.

Now consider an analysis due to Graziano (14), who assumes a general type of stress-strain curve (Fig. 6.20) and the stress distribution shown in Fig. 6.21 for various ratios of ultimate to yield stress and ratios of the maximum bending stress σ_m to the yield strength of the material. For the general case, as previously outlined, the bending moment is calculated on the basis of force equilibrium. Consider, for example, Fig. 6.22 for a beam of rectangular cross section. For equilibrium,

$$M = 2\left\{\sigma_m b(c - kc)\frac{2kc + c - kc}{2} - b(c - kc)\tfrac{1}{2}(\sigma_m - \sigma_y) \times [kc + \tfrac{1}{3}(c - kc)] + bkc(\tfrac{1}{2}\sigma_y)(\tfrac{2}{3}kc)\right\} \qquad 6.210$$

which simplifies to

$$M = 2bc^2\left[\frac{\sigma_m(1 - k^2)}{2} - \tfrac{1}{6}(1 - k)(\sigma_m - \sigma_y)(1 + 2k) + \frac{k^2\sigma_y}{3}\right] \qquad 6.211$$

However, from Eq. 6.188

$$M_y = 2bc^2\sigma_y/3 \qquad 6.212$$

Therefore

$$\frac{M}{M_y} = \frac{\sigma_m}{\sigma_y}\left(1 - \frac{k^2}{2} - \frac{k}{2}\right) + \tfrac{1}{2}(1 + k) \qquad 6.213$$

But from Fig. 6.21

$$\frac{\sigma_m}{\sigma_y} = k + \frac{\sigma_u}{\sigma_y}(1 - k) \qquad 6.214$$

Therefore

$$\frac{M}{M_y} = \frac{3}{2} - \frac{k^2}{2} + \left(1 + \frac{k^3}{2} - \frac{3}{2}k\right)\left(\frac{\sigma_u}{\sigma_y} - 1\right) \qquad 6.215$$

which is plotted in Fig. 6.23 as a function of k. Several other cross-sectional

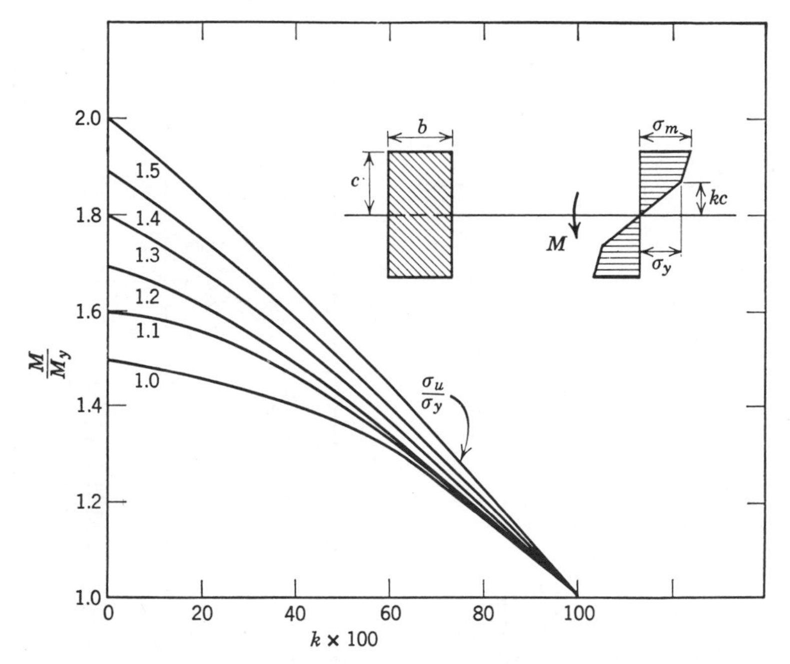

Fig. 6.23 Moment curve for beam of rectangular cross section. [After Graziano (14), courtesy of McGraw-Hill Publishing Co.]

shapes have been analyzed by Graziano, and reference is made to his paper for moment curves similar to those shown in Fig. 6.23. The analytical results are summarized in Table 6.2.

Deflection analysis for pure plastic bending. The computation of deflection for pure bending is simplified since the beam deforms to a circular arc. For example, suppose the maximum deflection is to be computed for the beam shown in Fig. 6.24. From Eq. 2.137

$$\varepsilon = y/R \tag{6.216}$$

where R is the radius of curvature. When the cross section of the beam is plastic to a depth a, as shown by the stress distribution in Fig. 6.24,

$$\sigma_y = E\varepsilon = E\left(\frac{y}{R}\right) = E\left(\frac{h/2 - a}{R}\right) \tag{6.217}$$

from which

$$R = (E/\sigma_y)(h/2 - a) \tag{6.218}$$

and by Eq. 2.147 for a circular arc the maximum deflection is

$$\delta = \frac{L^2}{8R} = \frac{L^2\sigma_y}{8E(h/2 - a)} \tag{6.219}$$

At full plasticity $a = h/2$ and the deflection increases to infinity. For comparison with deflection of transversely loaded beams see Table 6.3.

Table 6.2 Bending-Moment Relationships for Beams of Various Symmetrical Cross Sections and Yielding under Conditions of Strain Hardening [After Graziano (14), courtesy of McGraw-Hill Publishing Co.]

Beam Section	Moment at Yield, M_y	Bending Moment Ratio, M/M_y
$2c$, b	$\sigma_y(\frac{2}{3}bc^2)$	$\frac{3}{2} - \frac{1}{2}k^2 + \left(1 + \frac{k^3}{2} - \frac{3k}{2}\right)\left(\frac{\sigma_u}{\sigma_y} - 1\right)$
r, α	$\sigma_y\left(\frac{\pi r^3}{4}\right)$	$\frac{2\alpha}{\pi \sin\alpha} + \frac{2\cos\alpha}{3\pi}(2\cos^2\alpha + 3) + \left[1 - \frac{2\alpha}{\pi} - \frac{2\sin\alpha\cos\alpha}{3\pi}(2\cos^2\alpha + 3)\right]\left(\frac{\sigma_u}{\sigma_y} - 1\right)$ $k = \sin\alpha$ $c = r$
t, r, α	$\sigma_y(\pi r^2 t)$	$\frac{2}{\pi}\left[\frac{\alpha}{\sin\alpha} + \cos\alpha + \left(\frac{\pi}{2} - \alpha - \sin\alpha\cos\alpha\right)\left(\frac{\sigma_u}{\sigma_y} - 1\right)\right]$ $k = \sin\alpha$ $c = r$
b, $2c$	$\sigma_y\left(\frac{bc^2}{6}\right)$	$2 - k^2(2 - k) + (1 + k)(1 - k)^3\left(\frac{\sigma_u}{\sigma_y} - 1\right)$

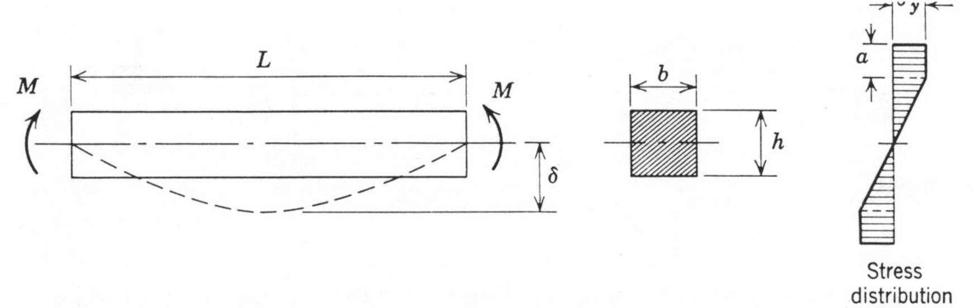

Fig. 6.24 Deflection of beam in plastic bending.

Bending by transverse loads

In pure bending the stress was uniform in the region between the supports. In the general case the bending moment varies continuously across the length of the beam giving rise to a stress gradient, which means that the analyses made for pure bending still apply here but to only one particular plane of the section. In the following simplified analysis the effect of transverse shear is neglected and the cross section of the beam is assumed to be rectangular.

Consider a simply supported beam of ideally plastic material with central concentrated load as shown in Fig. 6.25. For this case the maximum bending is under the load and has the value

$$M_{\max} = M_{L/2} = (P/2)(L/2) = PL/4 \qquad 6.220$$

Table 6.3 Maximum Deflection of Beams Bending at Constant Stress

Beam	Elastic Deflection, δ		Deflection at Instant of Plastic Collapse
	General Form	At Instant of Yielding	
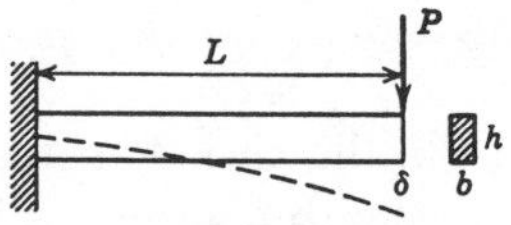	$\dfrac{PL^3}{3EI}$	$\frac{2}{3}\sigma_y\left(\dfrac{L^2}{Eh}\right)$	$\frac{40}{27}\sigma_y\left(\dfrac{L^2}{Eh}\right)$
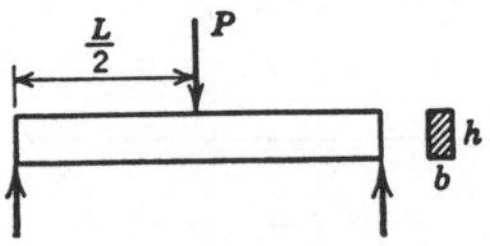	$\dfrac{PL^3}{48EI}$	$\frac{1}{6}\sigma_y\left(\dfrac{L^2}{Eh}\right)$	$\frac{1}{4}\sigma_y\left(\dfrac{L^2}{Eh}\right)$
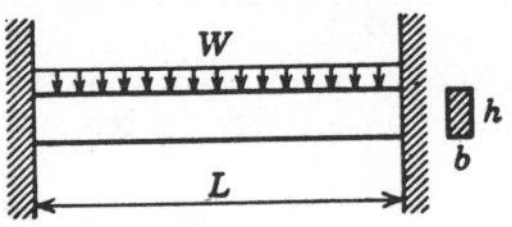	$\dfrac{WL^3}{384EI}$	$\frac{1}{16}\sigma_y\left(\dfrac{L^2}{Eh}\right)$	$\frac{1}{6}\sigma_y\left(\dfrac{L^2}{Eh}\right)$

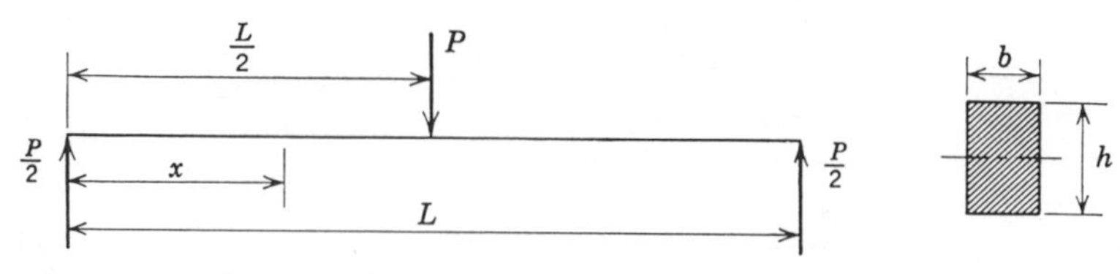

Fig. 6.25 Beam on simple supports and center loaded.

The bending-moment curve for a simply supported beam is a straight line; therefore, since the fully plastic moment given by Eq. 6.191 is

$$M_f = \sigma_y bh^2/4 \tag{6.221}$$

the moment at intermediate locations along the beam is

$$M_x = 2M_f(x/L) \tag{6.222}$$

When the center of the beam ($x = L/2$) is fully plastic, Eq. 6.222 shows that the top fibers are just starting to yield at $x = L/3$. If the center location is only partially plastic, then, using Eq. 6.190, Eq. 6.222 becomes

$$M_x = (\sigma_y bh^2/3)[1 + (2y/h)(1 - y/h)](x/L) \tag{6.223}$$

For example, if at $x = L/2$ the beam is 50% plastic, yielding is just beginning at $M_x = \sigma_y bh^2/6$ and Eq. 6.223 shows that $x = \frac{4}{11}L$. The plastic zones for the two cases are shown in Fig. 6.26; the plastic zones are parabolic in shape. In the foregoing discussion it was assumed that the material was perfectly plastic. However, by the same analysis the equations given in Table 6.2 could be used to describe the behavior of beams made of strain-hardening material. Additional problems are to be solved such as fixed-end beams and deflections of laterally loaded beams. Before doing this, however, it is necessary to discuss the theory of *limit design* since some of the principles developed in limit design are required in the solution of the previously mentioned problems.

Principles and application of limit design

The basic concepts of limit design were developed in the period around 1900 (1, 2, 7, 8, 15, 33, 39, 46). The earliest applications were to continuous beams;

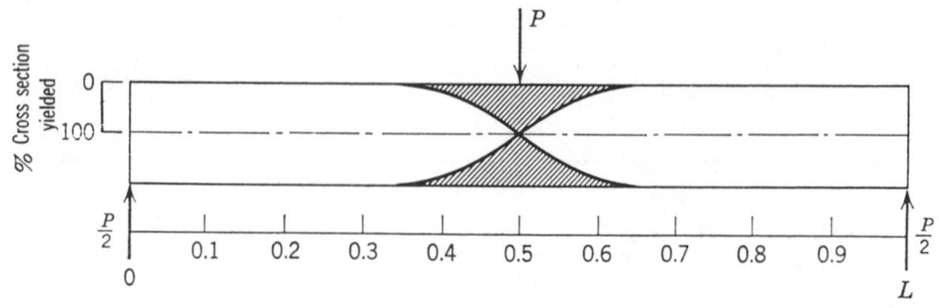

Fig. 6.26 Zone of plasticity in bent beam.

later the methods were extended to include frames, structural elements, soil (8), plates, shells, and arches. By far the widest application of limit design has been in the analysis of statically indeterminate structures, which, in the usual procedure, are analyzed by means of elasticity theory. The conventional stress analysis of various designs is performed by elasticity methods and most machine or structural elements are designed so that nominal working stresses are well within the elastic range of strain. For this reason, measured deflections and strains often check theory quite well. It is known, however, that many structural parts may remain entirely serviceable long after they have ceased to conform to Hooke's law. Thus the limits of their usefulness cannot be determined by the methods employing strictly elastic analyses. A plastic analysis is therefore essential for an understanding of what happens and for achievement of the best design. Specifically, limit design is concerned with estimating the load intensity at which a given part ceases to be serviceable, or conversely, with allocating local yield strength to members of a structure in such a manner that the structure remains serviceable under given conditions of loading.

The basic idea behind limit design is best illustrated in comparison with elastic theory. Suppose a beam made of material having a stress-strain curve such as shown in Fig. 6.14 is loaded as shown in Fig. 6.25. The maximum bending stresses are in the outer fibers of the beam. On the basis of an elastic analysis, load P would be safe if

$$\sigma_{\max} < \sigma_y \tag{6.224}$$

If load P is increased, yielding starts in the outer fibers of the beam, as previously discussed. This process can continue until the entire cross section of the beam is stressed to σ_y (corresponding to Eq. 6.188). Since the entire critical section is stressed to the yield strength of the material, the effect of adding more load would be like placing a *hinge* beneath the load. With this *plastic hinge* the beam becomes a mechanism and collapses as shown in Fig. 6.27. The load P which

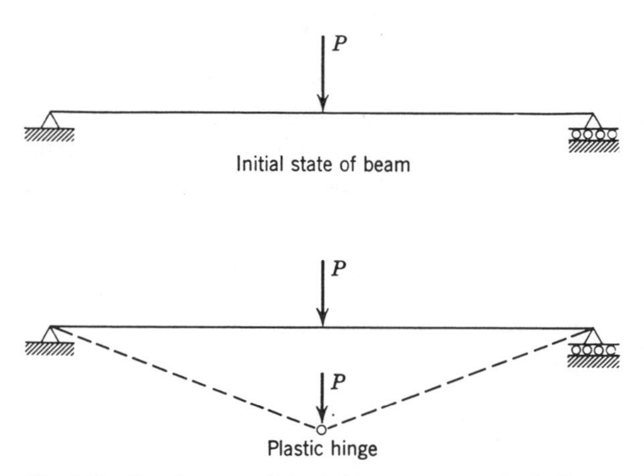

Fig. 6.27 Development of plastic hinge in a center-loaded beam.

is just large enough to cause collapse is called the *limit load* (corresponding to Eq. 6.191). Allowable loading is always some fraction of the collapse or limit load. Based on this concept there are two fundamental theorems in limit design theory as follows.

1. *The statical principle.* Collapse will not occur if, at any stage of loading, a statically admissible state can be found which is below the yield stress of the material. In other words, the collapse load is the largest load at which it is possible to find a state of yield stress satisfying equilibrium and boundary conditions. This theorem gives lower bounds on the limit load.

2. *The kinematic principle.* Collapse must occur if a kinematically admissible collapse state can be found. The collapse state is reached when a sufficient number of plastic hinges form in the design to reduce it to a mechanism. Kinematically admissible means that the rate at which the external forces do work at least equals the rate of internal dissipation. This theorem gives upper bounds on the limit load.

In many applications of limit design only the first theorem is used as stated. For example, in a heavy-walled cylinder under internal pressure the limit load is the pressure causing full overstrain of the cylinder wall. On the other hand, for flat plates, for example, theorem 1 is used directly and upper bounds are imposed by considering the rate of deflection of the plate (modification of theorem 2). In column design the limit load is that load which initiates plastic buckling for columns that do not fail first by elastic buckling. For all types of beams, frames, and trusses, theorems 1 and 2 are both used directly as stated.

The elementary cases of pure bending and beams on simple supports have already been discussed. In these cases the limit moment is at a plastic hinge where

$$M_f = \sigma_y bh^2/4 \qquad 6.225$$

for a beam of rectangular cross section. The same analysis also holds true for a cantilever beam (Fig. 6.28); that is, the moment required to just start yielding

$$M_y = \sigma_y bh^2/6 \qquad 6.226$$

and the moment required for full plasticity is given by Eq. 6.225. Consider now the fixed-end beam shown in Fig. 6.29. If the load applied to this beam exceeds that which the structure can carry elastically, *plastic hinges* form at the ends of the beam and directly beneath the load P. Eventually, the entire cross section

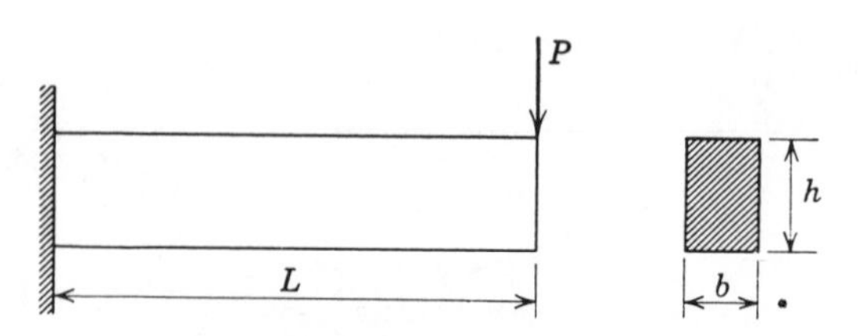

Fig. 6.28 Cantilever beam with end load.

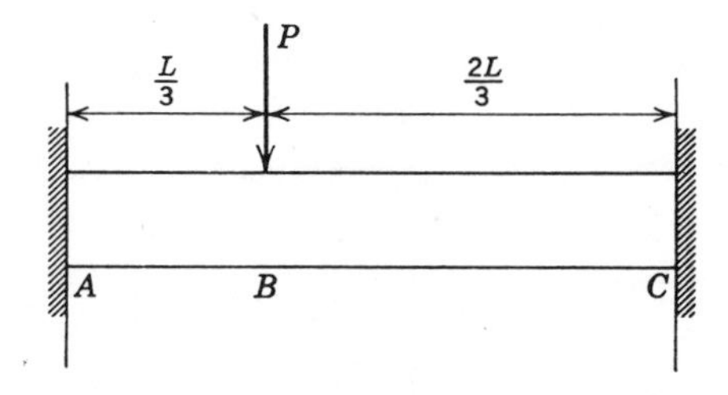

Fig. 6.29 Beam with built-in ends.

becomes plastic and collapse occurs. In this statically indeterminate case the conventional elastic solution shows that the following values persist:

vertical reaction at A	$0.74P$	6.227
vertical reaction at C	$0.26P$	6.228
bending moment at A	$0.148PL$	6.229
bending moment at C	$0.074PL$	6.230

At the instant of initiation of plastic flow the stress distribution in the beam is very complicated and a complete analysis for each additional increment of load would require an enormous amount of calculation. However, by limit design only the fully plastic condition is considered and the problem can be solved by statics. Thus for full plasticity

$$R_A L - M_L - P(2L/3) + M_L = 0 \tag{6.231}$$

from which

$$R_A = 2P/3 \quad \text{and} \quad R_C = P/3 \tag{6.232}$$

Also

$$R_A(L/3) - M_L - M_L = 0 \tag{6.233}$$

and

$$M_L = PL/9 \tag{6.234}$$

Thus, by limit design, a statically indeterminable problem has been reduced to determinacy and limiting loads calculated with little effort. It is interesting to compare the fixed-end beam of Fig. 6.29 with a beam similarly loaded but resting on simple supports. For this latter case the allowable load at the elastic limit is

$$P_y = \tfrac{9}{2}\sigma_y(Z/L) \tag{6.235}$$

where Z is the section modulus. For the fixed-end beam

$$P_y = \tfrac{27}{4}(\sigma_y Z/L) \tag{6.236}$$

Thus the factor of safety for the fixed-end beam is

$$\left(\frac{P_L}{P_y}\right)_{\text{fixed-end}} = \frac{4}{3}\frac{M_L}{Z\sigma_y} \tag{6.237}$$

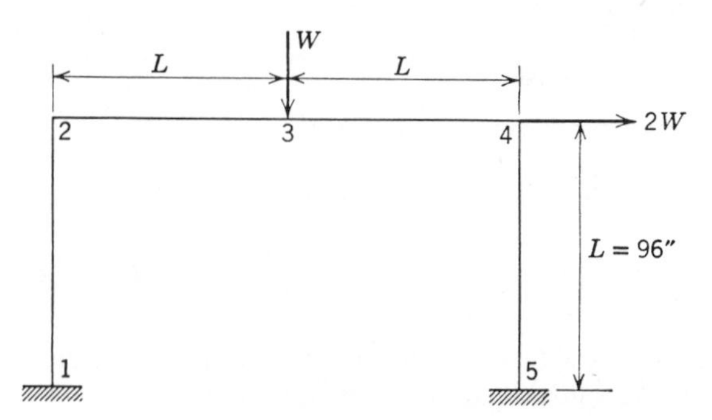

Fig. 6.30 An example of structural framework with multiple loading.

and for the simply supported beam

$$\left(\frac{P_L}{P_y}\right)_{\text{simple}} = \frac{M_L}{Z\sigma_y} \tag{6.238}$$

Since the bending moment required to induce full section plasticity is about the same for both beams (since it is assumed that they have the same shape and are made of the same material), the quantity $M_L/Z\sigma_y$ is the same for both beams. Equations 6.237 and 6.238 then show that the fixed-end beam has been designed with a factor of safety of four-thirds that of the simply supported beam.

Application to structural framework. Limit design methods have had particular application to design of structural framework. As one example, consider the frame shown in Fig. 6.30, which was analyzed by Baker and Neal (3). All members of this frame are 4 in. × 8 in.-18 lb steel joists. Tests of control bend

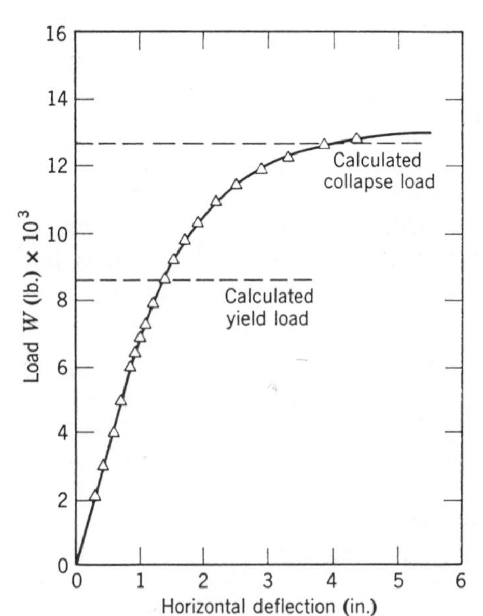

Fig. 6.31 Test data for full-scale frame test of Fig. 6.30. [After Baker and Neal, (3), courtesy of Butterworths, Inc.]

specimens of the material gave a value for the fully plastic bending moment of 592,000 in.-lb. In Fig. 6.31 the value of the load W is plotted against the observed horizontal deflection on the top member of the frame. As W increases from zero the frame behaves elastically up to a load W of about 8000 lb; at this load yielding begins in the most highly stressed part of the frame. The yield strength value corresponding to the experimentally determined M_y of 592,000 in.-lb can be calculated from Eq. 6.189. For the material in question the value of σ_y is 37,000 psi. This value used in an elastic calculation for the load W to just cause yielding (at the upper right-hand corner of the frame which is most highly stressed) is 8550 lb (Fig. 6.31). In the elastic analysis it is assumed that the structure becomes unsafe when the bending moment to cause yielding is reached. Figure 6.31 shows this to be untrue since the collapse load is shown to be over 12,000 lb.

Final collapse of the structure in Fig. 6.30 takes place by the formation of plastic hinges which reduces the structure to a mechanism. The actual mode of collapse corresponds to the mode giving the lowest value of internal work. Various possible collapse mechanisms are shown in Fig. 6.32; the actual collapse mechanism, which is shown in Fig. 6.32*A*, can also be calculated

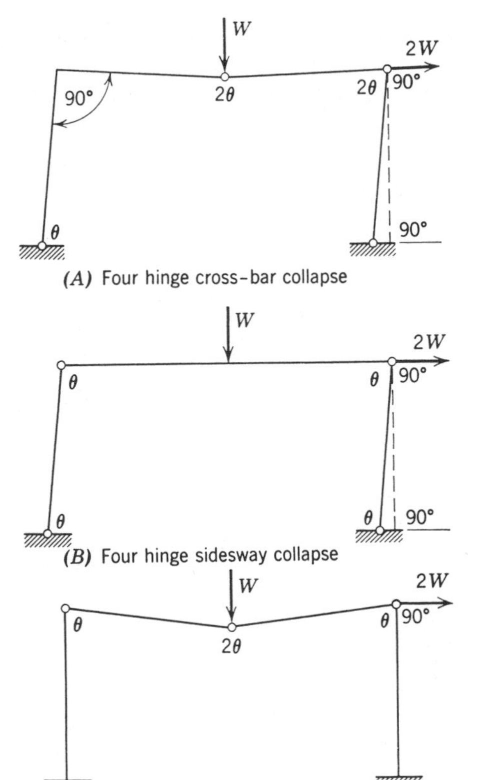

Fig. 6.32 Collapse mechanisms for the frame in Fig. 6.30. [After Baker and Neal, (3), courtesy of Butterworths, Inc.]

(predicted) by using the principle of virtual work

$$\sum P\delta = \sum M\theta \tag{6.239}$$

For the frame in question,

$$W(L\theta) + 2W(L\theta) = M(6\theta) \tag{6.240}$$

or

$$3LW = 6M \tag{6.241}$$

from which

$$W = \tfrac{6}{288}(592{,}000) = 12{,}300 \text{ lb} \tag{6.242}$$

The value of 12,300 lb is the lowest load that can be calculated from any of the mechanisms shown in Fig. 6.32; consequently, 12,300 lb is the collapse load. According to elastic theory the structure would be unsafe for a load $W = 8550$ lb, whereas, in fact, the frame becomes unsafe only when $W = 12{,}300$ lb, an increase in load-carrying capacity of over 40%. The elastic design method underestimates the actual strength of the structure. Several structures built using limit design have proven entirely satisfactory in service and have resulted in weight savings in steel framework from 30 to 50%. Additional examples of analyses of frame structures by limit design methods are given in (1, 4, 33, 39, 46 and 49).

In limit analysis the collapse loads can be located by examining an existing elastic solution (26, 28) and noting that the hinges occur at the same position as the range of the maximum moments. Then using enough hinges to obtain collapse the corresponding load can be solved for.

Right-angle bent and grillage

When looking at a fixed-end right angle frame (Fig. 6.33) another load consideration (15) must be taken into account; namely, a plastic torque. The virtual

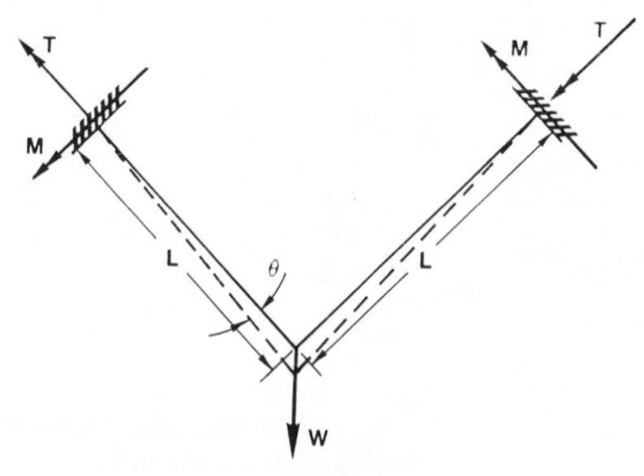

Fig. 6.33 Right angle bent.

work, Eq. 6.239 for a simple right angle in Fig. 6.33 is

$$WL\theta = 2M\theta + 2T\theta \tag{6.243}$$

or

$$\frac{WL}{2} = M + T \tag{6.244}$$

Also, for collapse the yield condition is

$$\left(\frac{M}{M_0}\right)^2 + \left(\frac{T}{T_0}\right)^2 = 1 \tag{6.245}$$

Further, with the same amount of twisting

$$\frac{T}{T_0^{\,2}} = \frac{M}{M_0^{\,2}} \tag{6.246}$$

Substituting Eq. 6.246 into Eq. 6.245 gives

$$M = \frac{M_0^{\,2}}{(M_0^{\,2} + T_0^{\,2})^{1/2}} \tag{6.247}$$

$$T = \frac{T_0^{\,2}}{(M_0^{\,2} + T_0^{\,2})^{1/2}} \tag{6.248}$$

so that

$$\frac{WL}{2} = (M_0^{\,2} + T_0^{\,2})^{1/2}$$

Then for collapse

$$\frac{W_c L}{2} = (M_0^{\,2} + T_0^{\,2})^{1/2} \tag{6.249}$$

Values for the collapse moment M_0 and torque T_0 can be substituted for the corresponding section, and a value for W_c can be obtained.

The right angle frame can be formed as in Fig. 6.34 to develop grillage to

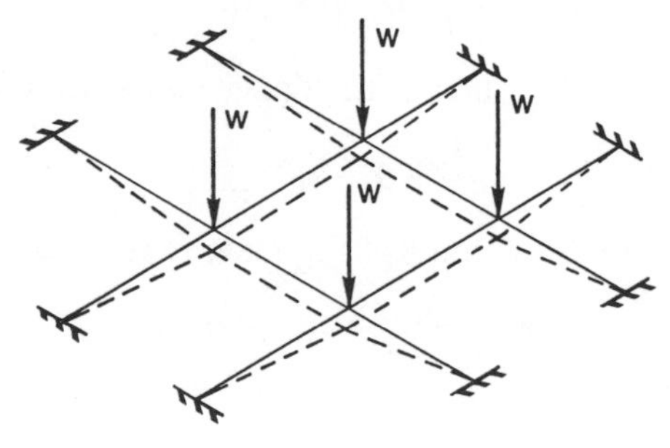

Fig. 6.34 Grillage.

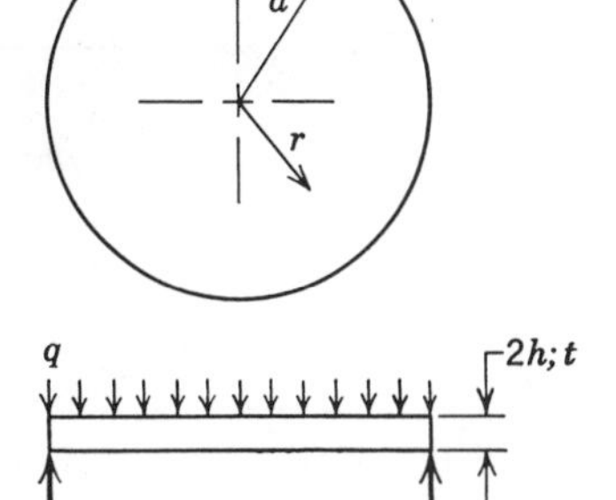

Fig. 6.35 Uniformly loaded circular plate simply supported along the rim.

cover openings. In solutions of this case, there are no torques at the fixed ends for symmetric cases. Therefore, the W_c can be solved for in terms of the dimensions and the plastic bending moment.

Flat plates. Consider Fig. 6.35 which shows a uniform circular plate carrying a distributed load. For this combined stress problem the two basic theorems of limit design will be applied as follows. A statically admissible stress distribution will be described in terms of bending moments and a kinematically admissible state of strain will be described in terms of the rate of deflection of the middle surface of the plate.

For the case shown in Fig. 6.35 the principal bending moments are

$$M_r = (q/16)(3 + \nu)(a^2 - r^2) \qquad 6.250$$

$$M_h = (q/16)[a^2(3 + \nu) - r^2(1 + 3\nu)] \qquad 6.251$$

and for full plasticity of the cross section of the plate, Eq. 6.17, which relates the principal stresses and the yield strength of the material, must be satisfied for all values of stress; that is,

$$\sigma_y{}^2 = \sigma_h{}^2 + \sigma_r{}^2 - \sigma_h\sigma_r \qquad 6.252$$

By using Eqs. 6.250 to 6.252 and conforming to statical conditions of equilibrium, it can be shown that the lowest statically admissible load that the plate can carry is given by the relation (20).

$$q = 6\sigma_y(h^2/a^2) \qquad 6.253$$

By applying the kinematic principle, the upper bound for the flow limit of the plate is

$$q = 6.67\sigma_y(h^2/a^2) \qquad 6.254$$

Thus, within $\pm 5.3\%$, the flow limit of the plate is given by

$$q = 6.33\sigma_y(h^2/a^2) \qquad 6.255$$

Letting the thickness of the plate equal t, from Eq. 6.255

$$\sigma_y = 0.633q(a^2/t^2) \qquad 6.256$$

The conventional elastic solution, which defines failure of the plate at first yielding, indicates that

$$\sigma_y = 1.24q(a^2/t^2) \qquad 6.257$$

Thus, for the design thickness, from Eqs. 6.256 and 6.257

$$t_{\text{limit design}} = 0.795a\sqrt{q/\sigma_y} \qquad 6.258$$

$$t_{\text{conventional}} = 1.113a\sqrt{q/\sigma_y} \qquad 6.259$$

or

$$t_{\text{conventional}} = t_{\text{limit design}} \times 1.4 \qquad 6.260$$

or the conventional design is 53% thicker than the limit design for the same load. This does not mean that all plates of this type should be proportioned by using limit design. For example, limit design would not be advisable for a rupture disk under steady load at a high temperature since creep and stress rupture might occur. Additional information on the limit design of plates is given in (16, 20, 21, 23, and 37); a few of the results presented by Hu (23) are given in Table 6.4.

Columns. Paris (36), following the analysis given later, determined an approximate load-axial deformation relationship for pin-jointed columns by limit design procedure; see also (4).

If a column is shorter than its critical length (length at which the critical buckling load equals $\pi^2 EI/L^2$ equals $\sigma_y A$), it will behave linearly elastic until the stress σ_y is reached, whereupon the column will fail by yielding and buckle plastically. A column longer than its critical length will behave linearly elastic with no buckling until the load approaches the critical buckling load, at which time an *elastic* buckle forms.

Referring to Fig. 6.36, under load, as the first fiber yields, buckling is said

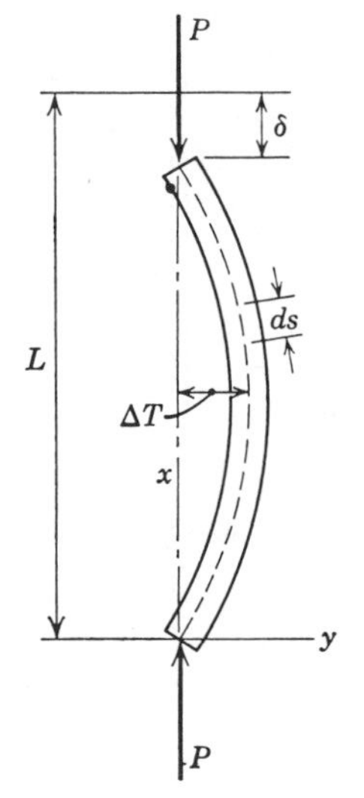

Fig. 6.36 Deflection of a column with pinned ends.

Table 6.4 *Limit Loads for Flat Circular Plates (After Hu (23), courtesy of the American Society of Civil Engineers)*

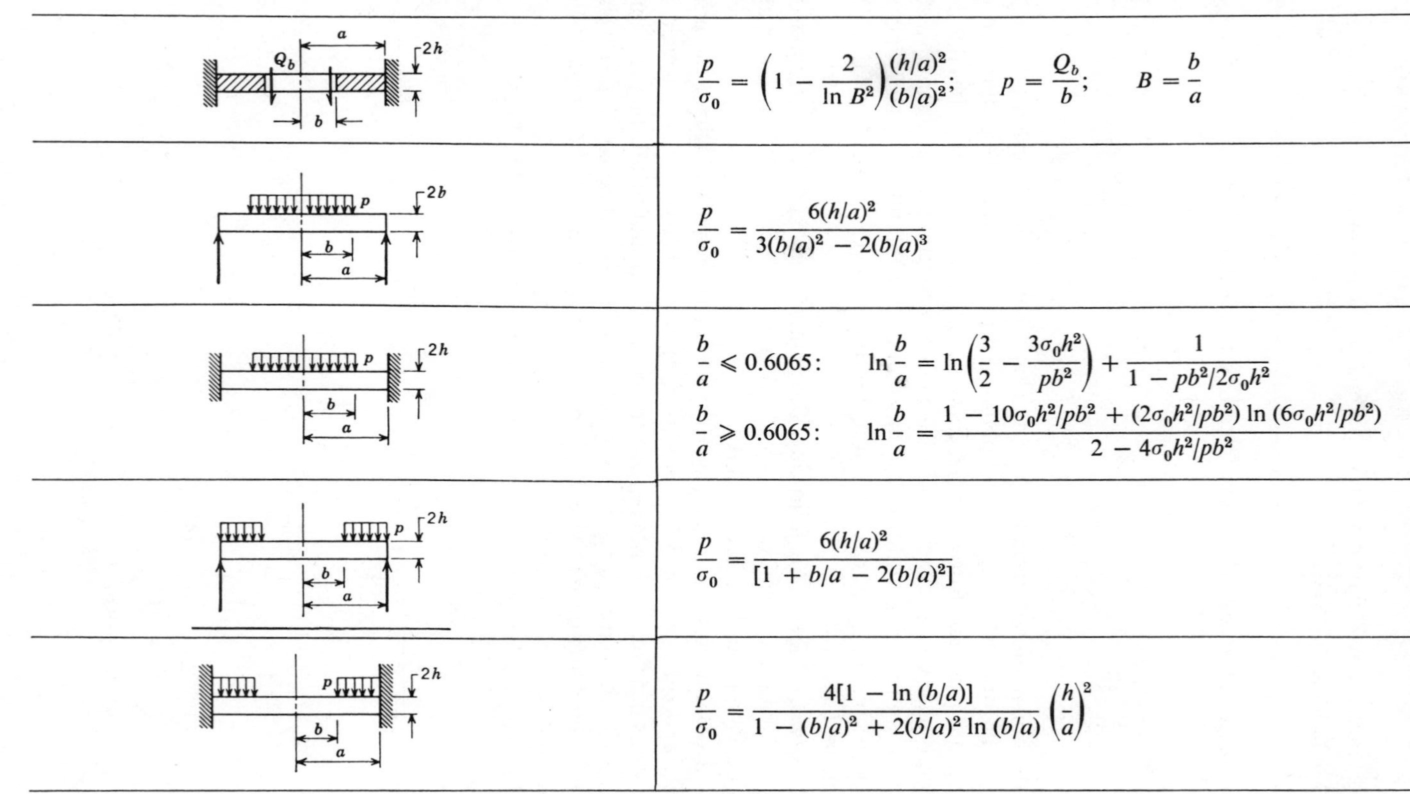

Plate	Limit load
	$\dfrac{p}{\sigma_0} = \left(1 - \dfrac{2}{\ln B^2}\right)\dfrac{(h/a)^2}{(b/a)^2}; \qquad p = \dfrac{Q_b}{b}; \qquad B = \dfrac{b}{a}$
	$\dfrac{p}{\sigma_0} = \dfrac{6(h/a)^2}{3(b/a)^2 - 2(b/a)^3}$
	$\dfrac{b}{a} \leqslant 0.6065: \quad \ln\dfrac{b}{a} = \ln\left(\dfrac{3}{2} - \dfrac{3\sigma_0 h^2}{pb^2}\right) + \dfrac{1}{1 - pb^2/2\sigma_0 h^2}$ $\dfrac{b}{a} \geqslant 0.6065: \quad \ln\dfrac{b}{a} = \dfrac{1 - 10\sigma_0 h^2/pb^2 + (2\sigma_0 h^2/pb^2)\ln(6\sigma_0 h^2/pb^2)}{2 - 4\sigma_0 h^2/pb^2}$
	$\dfrac{p}{\sigma_0} = \dfrac{6(h/a)^2}{[1 + b/a - 2(b/a)^2]}$
	$\dfrac{p}{\sigma_0} = \dfrac{4[1 - \ln(b/a)]}{1 - (b/a)^2 + 2(b/a)^2 \ln(b/a)}\left(\dfrac{h}{a}\right)^2$

to be plastic; as axial deformation increases, the axial load decreases. Within the elastic range,

$$\delta = PL/AE \tag{6.261}$$

and the critical buckling load (Euler's load) is

$$P_{cr} = \pi^2 EI/L^2 \tag{6.262}$$

The axial deformation of a column in the plastic range is represented as the sum of two causes, the shortening due to direct stress and that due to flexure. Mathematically (36)

$$\delta = \frac{PL}{AE} + \frac{L}{2}\int_0^L \left(\frac{dy}{dx}\right)^2 dx \tag{6.263}$$

which can be integrated by assuming a sinusoidal deflection curve for the columns so that

$$\delta = PL/AE + \pi^2(\Delta T)^2/4L \tag{6.264}$$

However, from statics

$$P = \int \sigma\, dz \qquad \text{and} \qquad M = P(\Delta T) = \int \sigma z\, dz \tag{6.265}$$

from which

$$\Delta T = \frac{\int \sigma z\, dz}{\int \sigma\, dz} \tag{6.266}$$

Substituting Eq. 6.266 into Eq. 6.264 and solving for a rectangular cross section gives

$$\delta = \frac{PL}{AE} + \frac{\pi^2}{4L}\left[\frac{h}{4}\left(\frac{\sigma_y A}{P} - \frac{P}{\sigma_y A}\right)\right] \tag{6.267}$$

Equation 6.267, which relates limit load P and deflection, is reported to give results that check experiment to less than 7% deviation for pin-end columns of $L/k > 35$. For values less than 35 no correlation was obtained, but designs based on Eq. 6.267 were on the safe side. In actual design, using Eq. 6.267, the critical load for crushing or elastic buckling is obtained from $P = \sigma_y A$ or Eq. 6.262. Then for larger deflections than calculated from Eq. 6.261, the desired value is substituted into Eq. 6.267 and the critical (maximum) load P is determined [see also (48)].

Arches. The limit analysis of arches has been discussed by Kooharian (27) and by Onat and Prager (35). The analysis considered here is for the single-span two-hinge arch. The horizontal thrust is chosen as the redundant quantity and the determinate base structure is an arch with a fixed hinge at one end and a

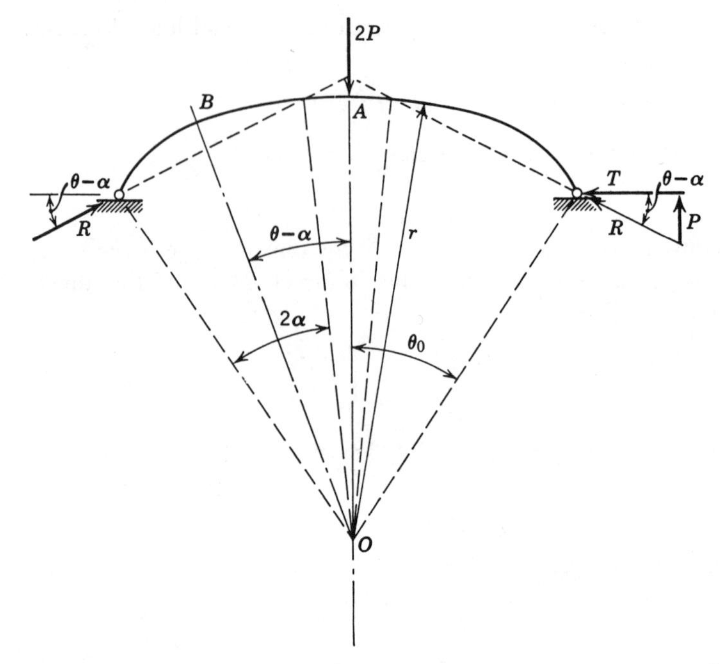

Fig. 6.37 Central loading of a two-hinge arch. [*After Onat and Prager (35), courtesy of Pergamon Press, Inc.*]

movable hinge at the other. All distributions of axial force and bending moment that are produced in the base structure by the given loading together with assumed intensity of the redundant quantity will be referred to as *statically admissible states of stress.* Thus by the first limit design theorem the given loads cannot be critical if a safe statically admissible state of stress can be found for these loads.

Consider the arch shown in Fig. 6.37 which is rectangular in cross section and carries a single vertical load at the center of the span. The vertical component of the reaction R at each support equals P and the horizontal component is the redundant thrust T. According to the limit design theorem, the load $2P$ cannot be critical if a value of T can be found such that the corresponding value of axial force N and bending moment M at the dangerous sections result in stresses that do not exceed the yield strength of the material σ_y. From Fig. 6.37 at the dangerous sections A and B

$$N_1 = -R\cos(\theta_0 - \alpha) \qquad 6.268$$

$$N_2 = -R \qquad 6.269$$

$$M_1 = rR[\cos\alpha - \cos(\theta - \alpha)] \qquad 6.270$$

$$M_2 = -rR(1 - \cos\alpha) \qquad 6.271$$

$$R = P\csc(\theta_0 - \alpha) \qquad 6.272$$

Since $T = P \cot(\theta - \alpha)$, the angle α is determined by T and by approximation it can be shown that

$$1 - \cos\alpha = \cos\alpha - \cos(\theta_0 - \alpha) \qquad 6.273$$

After α is determined from Eq. 6.273 the limit moment M_L is substituted (absolute value) into, say Eq. 6.271, so that

$$P = \frac{M_L \sin(\theta_0 - \alpha)}{r(1 - \cos\alpha)} \qquad 6.274$$

For $\theta_0 = 45°$ and $r = 30$ ft, for example, $\alpha = 18°35.5'$ so that

$$P = 850 M_L/r \qquad 6.275$$

If the section 1 and 2 are considered equally dangerous, it can be shown that

$$P = \frac{M_L}{r}\left[\frac{\sin(\theta_0 - \alpha)}{1 - \cos\alpha + H/8r}\right] \qquad 6.276$$

For $\theta_0 = 45°$, $r = 30$ ft and $H = 2$ ft, $\alpha = 18°32.6'$, and

$$P = 7.40 M_L/r \qquad 6.277$$

Equation 6.275 gives a statically admissible stress distribution which is greater than that given by Eq. 6.277. However, by applying the kinematic principle of limit design, that is, the given loads exceed the critical load if an unsafe kinematically admissible state of strain can be found, it can be shown that the loading expressed by Eq. 6.277 is the largest admissible loading. The evaluation of M_L for the problem is found by application of Eq. 6.189.

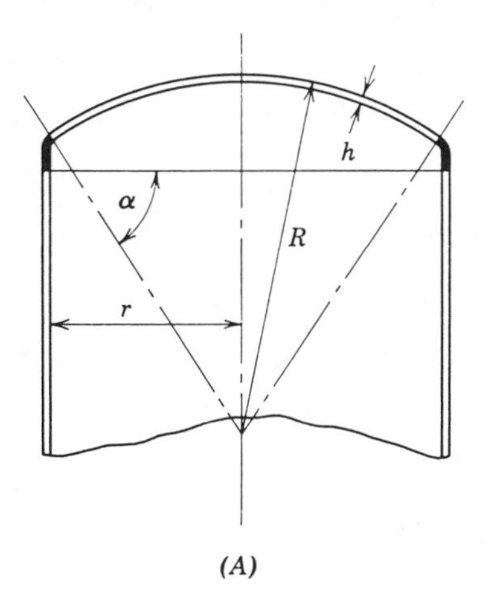

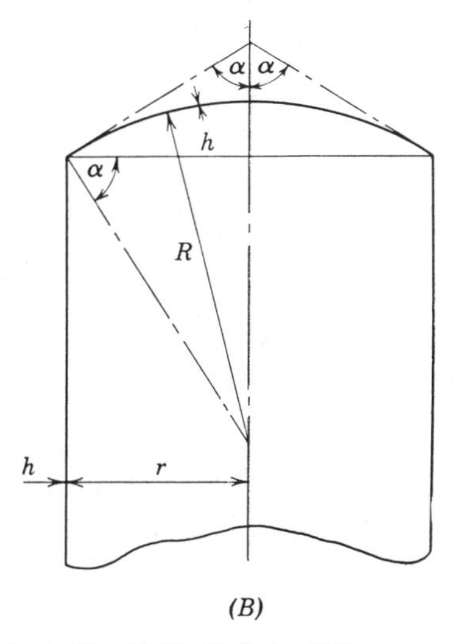

Fig. 6.38 Thin-walled cylindrical pressure vessel with spherical head. [*After Odqvist (34), courtesy of Edwards Brothers, Inc.*]

Thin shells and pressure vessels. Consider a thin-walled pressure vessel of the conventional type such as shown in Fig. 6.38*A*, consisting of a cylindrical part and two dished heads. In the geometry of the vessel $\sqrt{h/R}$ and $\sqrt{h/r}$ are small compared with unity and the shaded portions at the head-shell junction are entirely neglected. The idealized form of the vessel is shown in Fig. 6.38*B*. If such a vessel is subjected to an internal pressure p and Hooke's law is assumed, the stresses and strains in the vessel can be calculated by the conventional means discussed in Chapter 3. If the vessel is made of material having the characteristics shown in Fig. 6.14, the material will yield in bending at the joints when p exceeds a yield value σ_y and a plastic hinge (or hinges) form. The yield loading moment is given by Eq. 6.189 or

$$M_L = \sigma_y(h^2/6) \qquad 6.278$$

from which it can be deduced by the analysis given in (34) that the limit internal pressure that the vessel can withstand is

$$p_L = 2.43\,p_y \qquad 6.279$$

or nearly five times the ordinary working pressure for a vessel operating at half the yield pressure [see also (8 and 40)].

Deflection of transversely loaded beams

It has already been shown that the calculation of deflection for beams subjected to pure bending is relatively simple. Consider now the deflection calculation for a transversely loaded beam.

Simply supported beam with center load (Fig. 6.25). For this case the maximum moment is under the load P and has the value

$$M_{\max} = PL/4 \qquad 6.280$$

and the maximum deflection $\delta_{\max}$ at $L/2$ in the elastic range is given by

$$\delta_{\max} = PL^3/48EI \qquad 6.281$$

At the instant of yielding P equals P_y in Eq. 6.280 and Eq. 6.281 becomes

$$\delta = P_yL^3/48EI = M_yL^2/12EI \qquad 6.282$$

and for a perfectly plastic material having a rectangular cross section, from Eq. 6.189,

$$\delta_{\max} = \sigma_yL^2/6Eh \qquad 6.283$$

Similarly, at the instant of plastic collapse when there is a plastic hinge directly beneath the load

$$\delta = M_fL^2/12EI = \sigma_yL^2/4Eh \qquad 6.284$$

Fixed-end beam with uniform load (Fig. 6.39). In this case there is a bending moment across the entire beam; the critical values are at the center and ends

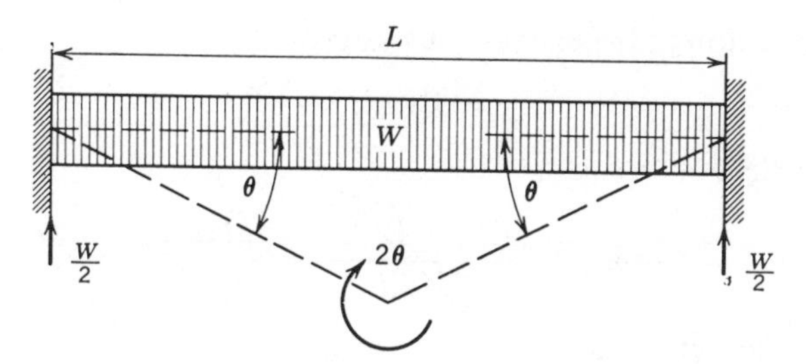

Fig. 6.39 Plastic bending of beam with built-in ends.

of the beam and are as follows:

$$\text{at beam ends} \qquad M = WL/12 \tag{6.285}$$

$$\text{at beam center} \qquad M = WL/24 \tag{6.286}$$

The maximum elastic deflection occurs at the center of the beam and has the value

$$\delta_{\max} = WL^3/384EI \tag{6.287}$$

At the instant of yielding the maximum moment is given by Eq. 6.285; therefore substituting this value into Eq. 6.287 gives the result

$$\delta_{\max} = M_y L^2/32EI \tag{6.288}$$

In order to calculate the load causing full plasticity consider Fig. 6.40 which shows the three plastic hinges M_p. From this geometry by virtual work

$$\delta = L\theta/2 \tag{6.289}$$

The external work is equal to

$$(W_f/2)(L\theta/2) = W_f L\theta/4 \tag{6.290}$$

where W_f is the load for full plasticity. At the hinges the work is equal to $M_p\theta$ for the ends of the beam and $2M_p\theta$ for the center; therefore the internal work is $4M_p\theta$, which when equated to Eq. 6.290 gives the value

$$M_f = W_f L/16 \tag{6.291}$$

The difference in moment between yield and collapse is thus $4M/L = W_f - W_y$. This load increment may be thought to act on a simply supported beam whose

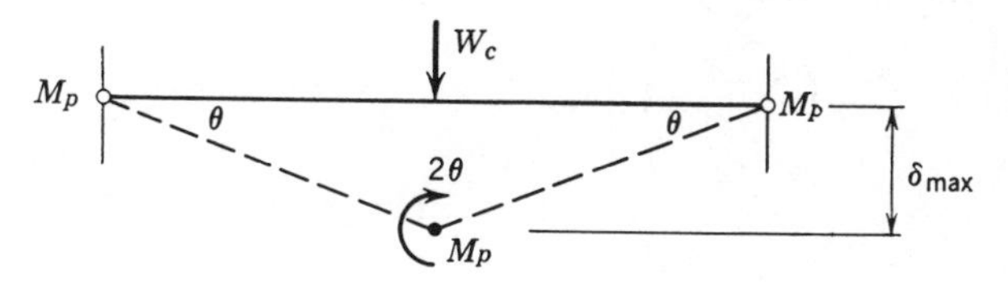

Fig. 6.40 Development of plastic hinges in a beam.

maximum deflection in the elastic range is given by

$$\delta_{\max} = 5WL^3/384EI \qquad 6.292$$

The plastic deflection is thus

$$\delta_{\max} = \frac{5(W_f - W_y)L^3}{384EI} = \frac{5M_yL^2}{96EI} \qquad 6.293$$

and the total deflection is

$$\delta_{\max} = \frac{M_yL^2}{32EI} + \frac{5M_yL^2}{96EI} = \frac{M_yL^2}{12EI} \qquad 6.294$$

A summary of some beam deflection values is given in Table 6.3.

If the load on a beam in the plastic range is removed, a residual deformation is left as a result of the plastic action. For example, in the previous problem (Fig. 6.39) the maximum deflection is given by Eq. 6.294. If the load is removed at this instant, there is an elastic recovery induced by the maximum moment. Actually, the residual deformation equals the strain caused by this moment; thus, from Eqs. 6.287 and 6.291, the residual deflection is

$$\delta = \frac{WL^3}{384EI} = \frac{16M_fL^3}{L(384EI)} = \frac{M_yL^2}{24EI} = \frac{\sigma_yL^2}{12Eh} \qquad 6.295$$

Plastic Torsion

Solid circular shaft. The cross section of a solid circular shaft subjected to torsion is shown in Fig. 6.41 and the torque-angle of twist curve in Fig. 6.42. For equilibrium

$$T = \int_0^a (2\pi r\, dr)\tau r = 2\pi \int_0^a \tau r^2\, dr \qquad 6.296$$

It is assumed that plane sections remain plane; therefore for shearing strain γ

$$r/a = \gamma_r/\gamma_a \qquad 6.297$$

and the unit shearing strain is given by

$$\gamma = r\theta \qquad 6.298$$

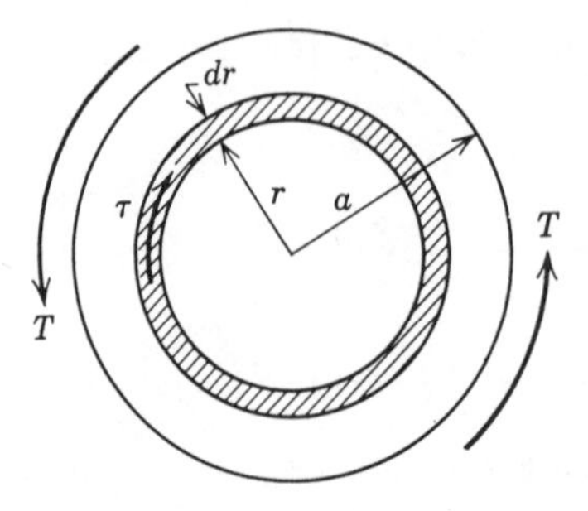

Fig. 6.41 Plastic torsion of circular bar.

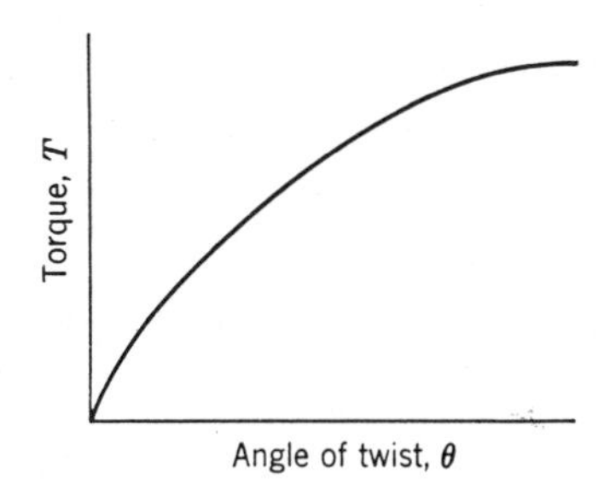

Fig. 6.42 Torque-angle of twist curve.

from which

$$\theta_a = \gamma_a/a \quad 6.299$$

and

$$\theta_r = \gamma_r/r \quad 6.300$$

Thus, expressing Eq. 6.296 in terms of shearing strain γ,

$$T = 2\pi \int_0^{\gamma_a} \frac{f(\gamma)\gamma^2\, d\gamma}{\theta^3} \quad 6.301$$

from which

$$T\theta^3 = 2\pi \int_0^{\gamma_a} f(\gamma)\gamma^2\, d\gamma \quad 6.302$$

From Eq. 6.302 by differentiation

$$(d/d\theta)T\theta^3 = 2\pi f(\gamma)(\gamma^3/3) = 2\pi f(a\theta)(a^3\theta^3/3) \quad 6.303$$

and

$$T\theta^3 = 2\pi f(\gamma_a)(\gamma_a{}^3/3) \quad 6.304$$

Furthermore,

$$(d/d\theta)(T\theta^3) = 2\pi f(\gamma_a)(\gamma_a)^2 = 2\pi a^3\theta^2\tau_a \quad 6.305$$

or

$$\tau_a = (d/d\theta)(T\theta^3)(1/2\pi a^3\theta^2) \quad 6.306$$

and finally

$$\tau_a = (1/2\pi a^3)[3T + \theta(dT/d\theta)] \quad 6.307$$

Thus from Eq. 6.307 and Fig. 6.42 stress-strain relationships can be obtained. The shear stress is given by Eq. 6.307 where T is the applied torque, θ the angle of twist, and $dT/d\theta$ the slope of the curve at any particular angle of twist. The shearing strain γ is simply

$$\gamma_a = a\theta \quad 6.308$$

If the material responds in a perfectly plastic manner, yielding takes place at constant shear stress τ_y and from Eq. 6.296 the torque required to put the bar in

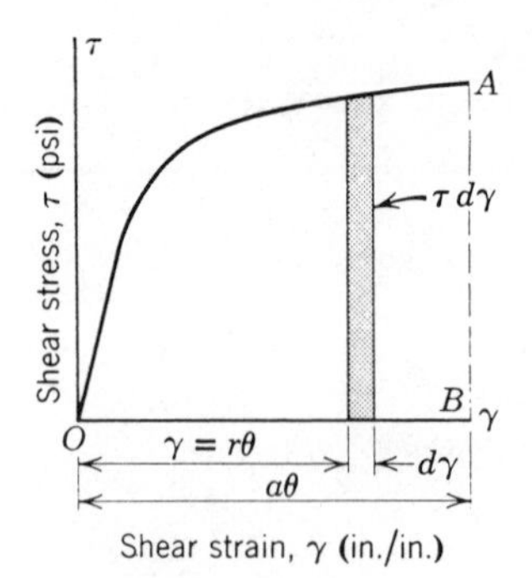

Fig. 6.43 Shear stress-shear strain curve.

a full yield condition is

$$T = \int_0^a 2\pi r^2 (\tau_y)\, dr = \frac{2\pi a^3}{3}\tau_y \qquad 6.309$$

The shear stress for the beginning of yielding is found from the elastic equation for shear stress

$$\tau_y = Tr/J \qquad 6.310$$

where T is the torque at yield and J is the polar moment of inertia $\pi a^4/2$ so that

$$T_y = (\pi a^3/2)\tau_y \qquad 6.311$$

In some instances the shear stress-shear strain diagram of a material may be obtained by a torsion test of a thin-walled tube. In this case the stress is uniform across the wall thickness and Eq. 6.301 becomes

$$T = \frac{2\pi}{\theta^3}\int_0^{a\theta} \gamma^2 \tau\, d\gamma \qquad 6.312$$

where $\int_0^{a\theta} \gamma^2 \tau\, d\gamma$ is the moment of inertia with respect to the τ axis of area $OABO$, Fig. 6.43. This area may be measured by planimeter. With the value of the moment of inertia calculated the $T - \theta$ curve can be plotted.

Bars of noncircular section. The torsional analysis of bars of other than circular cross section is complicated, but some solutions are available (32, 40). Experimentally there is a plastic analog of the *membrane analogy* described in Chapter 3 for elastic torsion of noncircular sections. This plastic analog, which was also devised by Prandtl and discussed by Nadai (32), is known as the *sand heap analogy*. In this method the cross section of the bar to be analyzed is reproduced in a piece of sheet metal or other stiff material and then sand is poured over this template. A sand heap results which has a shape particular to that of the cross section of the bar being analyzed and its volume is equal to half the twisting moment required to bring the bar into a condition of full plasticity.*

*For a circular bar of diameter $2a$, the sand heap is a right circular cone whose slope is τ (shear stress $= \sigma_y/\sqrt{3}$) and whose height is $\tau = h/a$, where h is the altitude. The plastic moment is $2V = \frac{2}{3}(\pi a^2 h)$. For sections with holes, the sand heap that would be formed by the hole is subtracted.

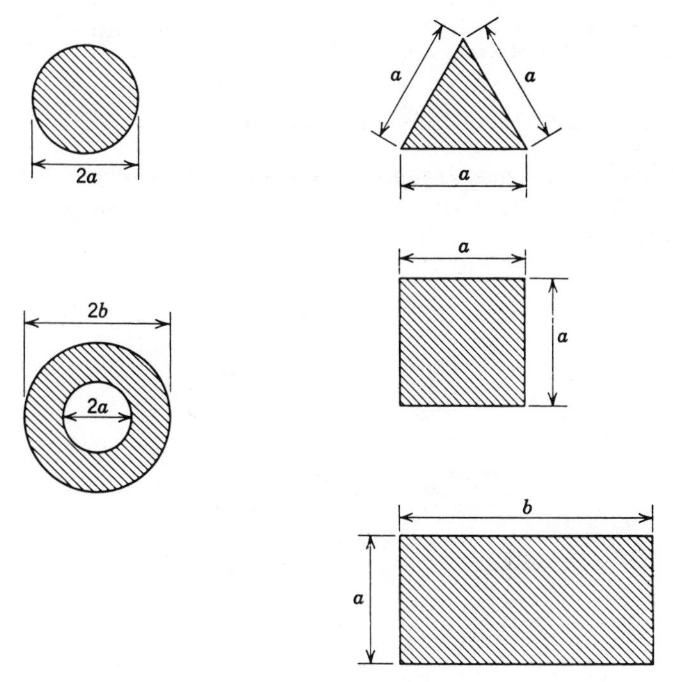

Fig. 6.44 Geometry of various sections.

Thus (see Fig. 6.44) the fully plastic torques are as follows for the cross sections shown.

Cross Section Shape	Torque for Full Plasticity, (M_f)
Circular	$\left(\dfrac{\sigma_y}{\sqrt{3}}\right)\left(\dfrac{2\pi a^3}{3}\right)$
Equilateral triangle	$\left(\dfrac{\sigma_y}{\sqrt{3}}\right)\left(\dfrac{a^3}{12}\right)$
Rectangle	$\left(\dfrac{\sigma_y}{\sqrt{3}}\right)\left(\dfrac{a^2}{6}\right)(3b - a)$
Square	$\left(\dfrac{\sigma_y}{\sqrt{3}}\right)\left(\dfrac{a^3}{3}\right)$
Thick-walled tube	$\left(\dfrac{\sigma_y}{\sqrt{3}}\right)\left[\dfrac{2\pi}{3}(b^3 - a^3)\right]$

6-3 PLASTICITY IN HEAVY-WALLED SPHERES, CYLINDERS, AND METAL-WORKING APPLICATIONS

Heavy-walled sphere under internal pressure

The cross section of a heavy-walled sphere subjected to internal pressure is shown in Fig. 6.45. In the elastic range in accordance with Eqs. 3.501 and 3.502 the principal stresses are as follows:

$$\sigma_1 = \sigma_h = \frac{p}{R^3 - 1}\left(1 + \frac{1}{2}\frac{c^3}{r^3}\right) \qquad 6.313$$

$$\sigma_2 = \sigma_h = \frac{p}{R^3 - 1}\left(1 + \frac{1}{2}\frac{c^3}{r^3}\right) \qquad 6.314$$

$$\sigma_3 = \frac{p}{R^3 - 1}\left(1 - \frac{c^3}{r^3}\right) \qquad 6.315$$

$$(\sigma_h)_{\max} = \frac{p}{2}\left(\frac{R^3 + 2}{R^3 - 1}\right) = (\sigma_1)_{\max} = (\sigma_2) \qquad 6.316$$

$$(\sigma_r)_{\max} = -p \qquad 6.317$$

where $R = \dfrac{c}{a}$

At the instant of yielding, Eqs. 6.316 and 6.317 are related by Eq. 6.17, that is,

$$\sigma_0{}^2 = \sigma_h{}^2 - 2\sigma_h\sigma_r + \sigma_r{}^2 \qquad 6.318$$

from which

$$\sigma_0 = \frac{3p}{2}\left(\frac{R^3}{R^3 - 1}\right) \qquad 6.319$$

and the yield pressure is*

$$p_y = \frac{2\sigma_0}{3}\left(\frac{R^3 - 1}{R^3}\right) \qquad 6.320$$

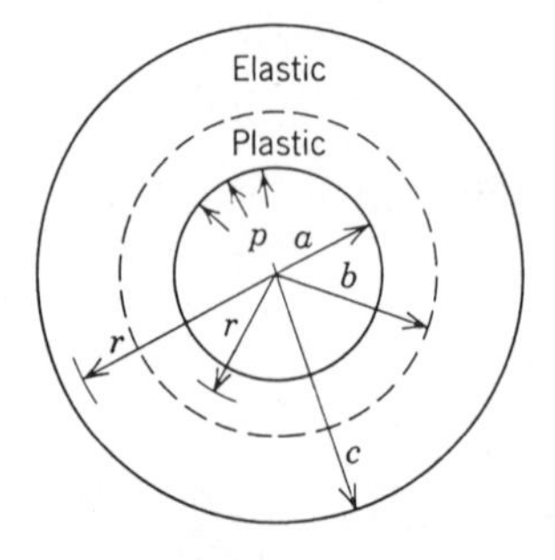

Fig. 6.45 Plastic expansion of a spherical shell.

*σ_0 and σ_y have the same meaning and are used interchangeably.

At the elastic limit of the shell the radial displacements are calculated using Eq. 3.70; thus for the sphere in Fig. 6.45

$$u_a = \varepsilon_h(a) = \frac{\sigma_0 a}{E}\left[\frac{2(1-2\nu)a^3}{3c^3} + \frac{1+\nu}{3}\right] \tag{6.321}$$

and

$$u_c = \varepsilon_h(c) = (\sigma_0 a^3/Ec^2)(1-\nu) \tag{6.322}$$

When the yield pressure is reached, plastic flow of the wall of the sphere starts and at some pressure the plastic boundary reaches some radius b(Fig. 6.45); at this point the sphere is plastic in the region a to b and elastic from b to c. In the elastic portion of the shell the stresses are still described by equations of the type given by Eqs. 6.314 and 6.315; thus

$$(\sigma_h)_{b<r<c} = \phi[1 + \tfrac{1}{2}(c^3/r^3)] \tag{6.323}$$

$$(\sigma_r)_{b<r<c} = \phi(1 - c^3/r^3) \tag{6.324}$$

where ϕ is an interface pressure function to be determined. At the plastic-elastic interface Eq. 6.318 is valid so that

$$\sigma_0 = \phi[1 + \tfrac{1}{2}(c^3/b^3)] - \phi(1 - c^3/b^3) \tag{6.325}$$

from which

$$\phi = 2b^3\sigma_0/3c^3 \tag{6.326}$$

Then substituting Eq. 6.326 into Eqs. 6.323 and 6. 324 gives

$$(\sigma_h)_{b<r<c} = (2\sigma_0 b^3/3c^3)[1 + \tfrac{1}{2}(c^3/r^3)] \tag{6.327}$$

and

$$(\sigma_r)_{b<r<c} = (2\sigma_0 b^3/3c^3)(1 - c^3/r^3) \tag{6.328}$$

and the displacement is

$$(u)_{b<r<c} = \frac{2\sigma_0 b^3}{3Ec^3}\left[(1-2\nu)r + \frac{(1+\nu)c^3}{2r^2}\right] \tag{6.329}$$

In the plastic zone of the sphere ($a<r<b$), by equilibrium (Eq. 3.488)

$$(r/2)(d\sigma_r/dr) + (\sigma_r - \sigma_h) = 0 \tag{6.330}$$

However, Eq. 6.318 can be written in the form

$$\sigma_0 = \sigma_h - \sigma_r \tag{6.331}$$

Therefore, using Eq. 6.331, Eq. 6.330 becomes

$$(r/2)(d\sigma_r/dr) = \sigma_0 \tag{6.332}$$

which when integrated becomes

$$\sigma_r = 2\sigma_0 \ln r + C_1 \tag{6.333}$$

where C_1 is a constant of integration. At $r = b$, σ_r is given by Eq. 6.328, that is,

$$(\sigma_r)_{r=b} = (2\sigma_0 b^3/3c^3)(1 - c^3/b^3) \qquad 6.334$$

which when substituted into Eq. 6.333 gives the value of C_1 which is

$$C_1 = \frac{2\sigma_0 b^3}{3c^3}\left(1 - \frac{c^3}{b^3}\right) - 2\sigma_0 \ln b \qquad 6.335$$

and then

$$(\sigma_h)_{a<r<b} = \sigma_0\left[1 + 2\ln\frac{r}{b} + \frac{2b^3}{3c^3}\left(1 - \frac{c^3}{b^3}\right)\right] \qquad 6.336$$

and

$$(\sigma_r)_{a<r<b} = \sigma_0\left[2\ln\frac{r}{b} + \frac{2b^3}{3c^3}\left(1 - \frac{c^3}{b^3}\right)\right] \qquad 6.337$$

The pressure required to cause plastic flow to radius b is

$$p_b = -(\sigma_r)_{r=a} = 2\sigma_0\left[\ln\frac{b}{a} + \frac{b^3}{3c^3}\left(\frac{c^3}{b^3} - 1\right)\right] \qquad 6.338$$

The pressure to cause full overstrain, that is, for the entire wall to become plastic, is obtained by letting $b = c$; therefore, noting that $R = c/a$,

$$p_0 = 2\sigma_0 \ln R \qquad 6.339$$

The bursting pressure of a heavy-walled sphere is calculated by the procedure developed by Faupel (10) for heavy-walled cylinders.

$$p = 2\sigma_0 \ln R(2 - \sigma_0/\sigma_u) \qquad 6.340$$

where σ_0 is the tensile yield strength of the material at 0.2% offset and σ_u is the ultimate tensile strength. An alternate method for determining the bursting pressure is given by Svensson (43).

Heavy-walled cylinder under internal pressure

The analysis for a heavy-walled cylinder (Fig. 6.46) subjected to either internal or external pressure in the elastic range was outlined in Chapter 3; also equa-

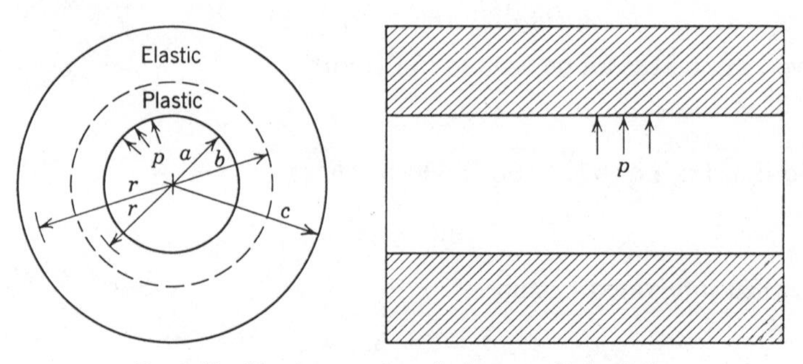

Fig. 6.46 Plastic expansion of a heavy-walled cylinder.

tions were given for the elastic-breakdown pressure. This chapter is concerned with the behavior of the cylinder in the plastic range. The condition for radial equilibrium is given by Eq. 3.453. If the assumption is made that flow is governed by the distortion-energy criterion, the shear stress $(\sigma_h - \sigma_r)/2$ can be replaced by the tensile equivalent $\sigma_y/\sqrt{3}$. Thus Eq. 3.453 becomes

$$\frac{d\sigma_r}{dr} = \frac{1}{r}\left(\frac{2\sigma_y}{\sqrt{3}}\right) = \frac{1}{r}\left(\frac{2\sigma_0}{\sqrt{3}}\right) \qquad 6.341$$

which on integration leads to the result

$$\sigma_r = \frac{2\sigma_y}{\sqrt{3}} \ln r + C_1 \qquad 6.342$$

where C_1 is a constant of integration determinable from the boundary conditions. At $a < r < c$, that is, at $r = b$, the interface,

$$\frac{-\sigma_y}{\sqrt{3}}\left(\frac{c^2 - b^2}{c^2}\right) = \frac{2\sigma_y}{\sqrt{3}} \ln b + C_1 \qquad 6.343$$

so that, on solving for C_1, and substituting back into Eq. 6.342 with $p_r = -\sigma_r$ at $r = a$,

$$p_r = \frac{\sigma_y}{\sqrt{3}}\left[1 - \left(\frac{b}{c}\right)^2 + 2 \ln \frac{b}{a}\right] \qquad 6.344$$

Equation 6.344 indicates the internal pressure required to cause plastic flow in the cylinder wall to some radius b. At the interface b the pressure (Eq. 6.343) is

$$p = \frac{\sigma_y}{\sqrt{3}}\left(\frac{c^2 - b^2}{c^2}\right) \qquad 6.345$$

The elastic portion of the wall is now effectively under an internal pressure given by Eq. 6.345 and therefore the stresses from Eqs. 3.468, 3.471 to 3.472, and 6.345 are as follows:

$$(\sigma_h)_\varepsilon = \frac{\sigma_y}{\sqrt{3}}\left(\frac{b}{c}\right)^2\left[1 + \left(\frac{c}{r}\right)^2\right] \qquad 6.346$$

$$(\sigma_r)_\varepsilon = \frac{\sigma_y}{\sqrt{3}}\left(\frac{b}{c}\right)^2\left[1 - \left(\frac{c}{r}\right)^2\right] \qquad 6.347$$

$$(\sigma_z)_{\varepsilon o} = 0 \qquad 6.348^*$$

$$(\sigma_z)_{\varepsilon c} = \frac{\sigma_y}{\sqrt{3}}\left(\frac{b}{c}\right)^2 \qquad 6.349^*$$

*Both open- and closed-end cylinders are considered from here on since both types of cylinders are used for the autofrettage type of construction discussed in Chapter 11.

where the subscript ε means elastic and o or c means open- or closed-end cylinder, respectively. The strains within the elastic portion of the wall can be calculated from Eqs. 3.457 to 3.459 and 6.346 to 6.349, thus

$$(\varepsilon_h)_{\varepsilon o} = \frac{\sigma_y}{E\sqrt{3}}\left(\frac{b}{c}\right)^2\left[(1-\nu)+\left(\frac{c}{r}\right)^2(1+\nu)\right] \quad 6.350$$

$$(\varepsilon_h)_{\varepsilon c} = \frac{\sigma_y}{E\sqrt{3}}\left(\frac{b}{c}\right)^2\left[(1-2\nu)+\left(\frac{c}{r}\right)^2(1+\nu)\right] \quad 6.351$$

$$(\varepsilon_r)_{\varepsilon o} = \frac{\sigma_y}{E\sqrt{3}}\left(\frac{b}{c}\right)^2\left[(1-\nu)-\left(\frac{c}{r}\right)^2(1+\nu)\right] \quad 6.352$$

$$(\varepsilon_r)_{\varepsilon c} = \frac{\sigma_y}{E\sqrt{3}}\left(\frac{b}{c}\right)^2\left[(1-2\nu)-\left(\frac{c}{r}\right)^2(1+\nu)\right] \quad 6.353$$

$$(\varepsilon_z)_{\varepsilon o} = -\frac{2\nu\sigma_y}{E\sqrt{3}}\left(\frac{b}{c}\right)^2 \quad 6.354$$

$$(\varepsilon_z)_{\varepsilon c} = \frac{\sigma_y}{E\sqrt{3}}\left(\frac{b}{c}\right)^2(1-2\nu) \quad 6.355$$

The corresponding radial displacements from Eqs. 3.456, 6.350, and 6.351 are

$$(u)_{\varepsilon o} = \frac{\sigma_y r}{E\sqrt{3}}\left(\frac{b}{c}\right)^2\left[(1-\nu)+\left(\frac{c}{r}\right)^2(1+\nu)\right] \quad 6.356$$

and

$$(u)_{\varepsilon c} = \frac{\sigma_y r}{E\sqrt{3}}\left(\frac{b}{c}\right)^2\left[(1-2\nu)+\left(\frac{c}{r}\right)^2(1+\nu)\right] \quad 6.357$$

In the plastic portion of the cylinder wall, the ring of material is subjected (effectively) to both an internal and an external pressure. Since, as previously noted,

$$(\sigma_h - \sigma_r) = 2\sigma_y/\sqrt{3} \quad 6.358$$

the stresses in the plastic section can be calculated through the use of Eqs. 6.342, 6.343, and 6.358; thus,

$$(\sigma_h)_p = \frac{\sigma_y}{\sqrt{3}}\left[1+\left(\frac{b}{c}\right)^2+2\ln\frac{r}{b}\right] \quad 6.359$$

$$(\sigma_r)_p = \frac{\sigma_y}{\sqrt{3}}\left[\left(\frac{b}{c}\right)^2-1+2\ln\frac{r}{b}\right] \quad 6.360$$

$$(\sigma_z)_{po} = 0 \quad 6.361$$

$$(\sigma_z)_{pc} = \frac{\sigma_y}{\sqrt{3}}\left[\left(\frac{b}{c}\right)^2+2\ln\frac{r}{b}\right] \quad 6.362$$

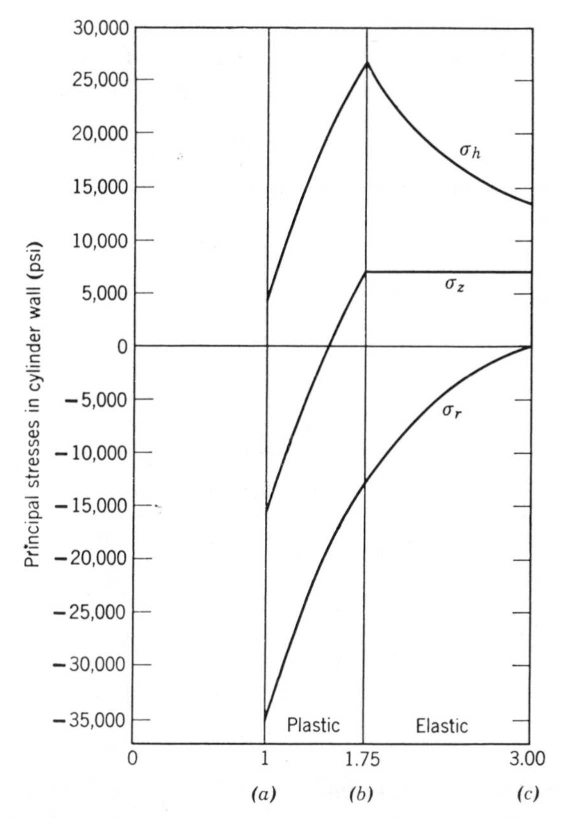

Fig. 6.47 Stress distribution in wall of closed-end, heavy-walled cylinder at pressure just sufficient to induce reverse yielding (35,000 psi).

where the subscript p indicates plastic range. A typical stress-distribution plot for a closed-end cylinder having a diameter ratio of 3.00 and a yield strength (tension) of 35,000 psi is shown in Fig. 6.47. The plastic-elastic boundary was calculated by Eq. 6.344, stresses in the elastic zone were deduced from Eqs. 6.346, 6.347, and 6.348, plastic-zone stresses were calculated from Eqs. 6.359, 6.360, and 6.362. For this particular example the stress distribution is that obtained at a pressure of 35,000 psi, the pressure just sufficient to cause reverse yielding of the material when the pressure is released (see Chapter 11).

Strain values within the plastic zone are difficult to obtain by calculation but, as a first-order approximation, the bore strain, which is important in autofrettage calculations (Chapter 11), can be computed from considering the constancy of volume. Thus, if the strain at the plastic-elastic boundary is obtained by Eq. 6.357, then, at the bore

$$(u_a)_{pc} = (u_b)_{\varepsilon c}(b/a) \tag{6.363}$$

are also

$$(\varepsilon_h)_{r=a} = (\varepsilon_h)_{r=b}(b/a)^2 \tag{6.364}$$

When the wall of the vessel is completely plastic (full overstrain), the values of b and c become equal in the foregoing equations. Thus it is found from Eq. 6.344 that the internal pressure required to cause full overstrain is

$$p_r = (2\sigma_y/\sqrt{3}) \ln R \qquad 6.365$$

Similarly, the equations for stress and strain become

$$(\sigma_h)_p = (2\sigma_y/\sqrt{3})(1 + \ln r/c) \qquad 6.366$$

$$(\sigma_r)_p = (2\sigma_y/\sqrt{3}) \ln r/c \qquad 6.367$$

$$(\sigma_z)_{po} = 0 \qquad 6.368$$

$$(\sigma_z)_{pc} = (2\sigma_y/\sqrt{3})(\tfrac{1}{2} + \ln r/c) \qquad 6.369$$

$$(\varepsilon_h)_{po,r=c} = 2\sigma_y/E\sqrt{3} \qquad 6.370$$

$$(\varepsilon_h)_{pc,r=c} = (\sigma_y/E\sqrt{3})(2 - v) \qquad 6.371$$

$$(\varepsilon_h)_{po,r=a} = (\sigma_y/E\sqrt{3})(2R^2) \qquad 6.372$$

$$(\varepsilon_h)_{pc,r=a} = (\sigma_y/E\sqrt{3})(2 - v)R^2 \qquad 6.373$$

$$(\varepsilon_z)_{po,r=a,c} = -2v\sigma_y/E\sqrt{3} \qquad 6.374$$

$$(\varepsilon_z)_{pc,r=a,c} = (\sigma_y/E\sqrt{3})(1 - 2v) \qquad 6.375$$

$$(\varepsilon_r)_p = -[(\varepsilon_h)_p + (\varepsilon_z)_p] \qquad 6.376$$

The pressure-expansion curve and unrestricted plastic radial expansion. In the design of a vessel, deformation as well as stress is an important item and should be allowed for in the calculations. In particular, for an autofrettaged vessel, it is important to be able to say, with some degree of confidence, that the vessel will not rupture during the course of autofrettage. In addition, post-autofrettage machining has to be allowed for; thus it is necessary to have the proper strain incorporated in the vessel before such machining has to take place. In general, analyses involving strain are much more difficult and unreliable than those involving only stress. There are at least two reasons for this; first, a glance at most stress-strain or pressure-strain plots shows that a very small increase in load in the plastic range can result in a deformation which might change by an order of magnitude; thus, depending on the reliability of the experiment, the accuracy of plotting, etc., serious errors could result if the criterion were strain alone. Secondly, in the plastic range, more assumptions are involved—constancy of volume, etc. Therefore, in using pressure-strain plots, great care and discretion must be employed. Depending on the situation, different methods will be indicated that are consistent with the seriousness of the problem. A laboratory reactor costing a few hundred dollars will not justify the detail and extent of analysis required for a major reactor costing hundreds of thousands of dollars. No general rules can be set down since different approaches will necessarily be taken by different individuals and groups as the situation dictates. Some procedures which may assist in arriving at a conclusion will be recorded here, but none will be recommended as being entirely adequate.

In developing the pressure-strain curve, it will be noted that the curve may be thought of as consisting of three parts.

1. Elastic-range.
2. Elastic-breakdown to full overstrain.
3. Full overstrain to instability or fracture.

In the elastic-range no particular difficulty is encountered; pressure is plotted against the elastic hoop strain given by Eq. 3.457. Since Hooke's law is generally considered reliable, an accurate plot for either the bore or surface results. Difficulty starts when the region of elastic-breakdown is approached, and from here on various choices of analysis can be made. It is convenient to introduce Fig. 6.48 at this point, which shows various degrees of agreement of theory and experiment for some of the methods proposed. For details the original papers should be consulted.

In Fig. 6.48*A* are shown the results of an analysis proposed by the author a few years ago (10). In this method the pressure-strain curve is approximated by calculating three points and then roughing in a curve. This method may be used when limited data concerning the material are available and only a gross

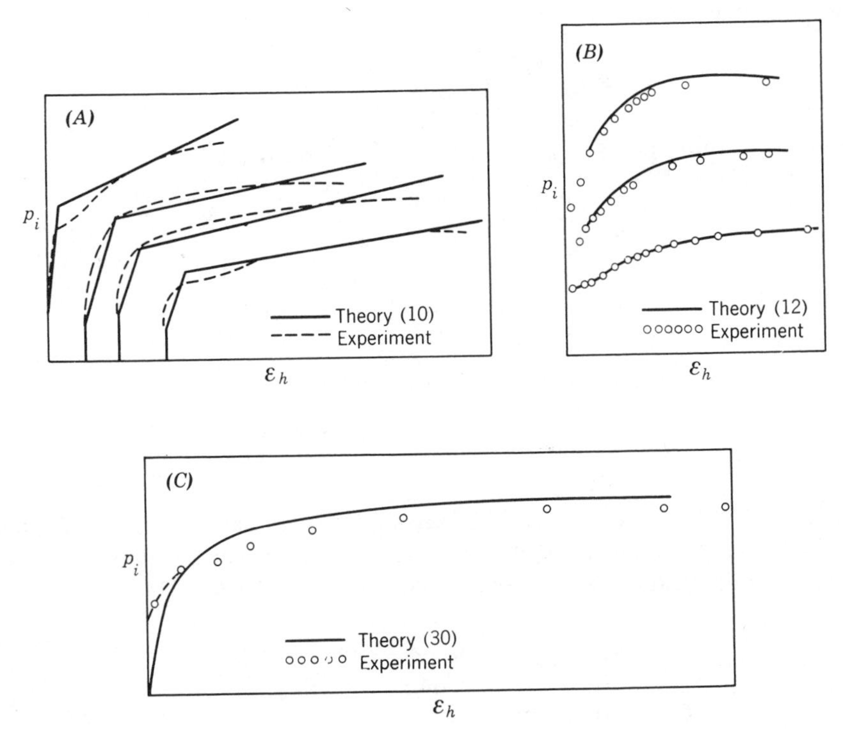

Fig. 6.48 Pressure-strain curves for heavy-walled cylinders according to various types of analyses.

estimate of a pressure-strain curve is needed for preliminary design consideration. However, it is obvious that this method may not be suitable as a final criterion of design. The first point on the plot involves a calculation of the elastic-breakdown pressure for a closed-end cylinder (on page) and the hoop strain at elastic-breakdown using Eq. 3.477, with p_i replaced by $(p_{yi})_c$. The second point involves a calculation of the overstrain pressure using Eq. 6.365, and the hoop strain at full overstrain using Eq. 6.370. The final point is calculated using an expression for the bursting pressure [see (10)].

$$p_b = (2\sigma_y/\sqrt{3}) \ln R(2 - \sigma_y/\sigma_u) \tag{6.377}$$

and the hoop strain at the burst pressure. This strain is approximated from the strain at the ultimate stress in tension ε_u. It is postulated that failure starts at the bore when the hoop strain equals ε_u; this seems reasonable since the maximum strain theory (44) is in close agreement with the Mises theory in the tension-compression stress quadrant, the stress location for a cylinder under internal pressure. To get the surface strain use ε_u together with Eq. 6.364.

Another somewhat similar method may be developed using the analysis due to Prager and Hodge for stresses in the region of unrestricted plastic flow following full overstrain (40). In this analysis the initial bore and O.D. radii a_0 and c_0 increase to a and c with time. A particle that originally had a distance r_0 from the axis of the cylinder is now under the following stresses when the bore radius a_0 reaches the value a.

$$(\sigma_h)_{uf} = \frac{\sigma_y}{\sqrt{3}}\left(\ln\frac{r_0^2 + a^2 - a_0^2}{c_0^2 + a^2 - a_0^2} + 2\right) \tag{6.378}$$

where the subscript uf means unrestricted plastic flow.

$$(\sigma_r)_{uf} = \frac{\sigma_y}{\sqrt{3}}\left(\ln\frac{r_0^2 + a^2 - a_0^2}{c_0^2 + a^2 - a_0^2}\right) \tag{6.379}$$

$$(\sigma_u)_{uf} = \frac{\sigma_y}{\sqrt{3}}\left(\ln\frac{r_0^2 + a^2 - a_0^2}{c_0^2 + a^2 - a_0^2} + 1\right) \tag{6.380}$$

The pressure necessary to maintain flow becomes

$$(p_i)_{uf} = \frac{\sigma_y}{\sqrt{3}}\ln\left(1 + \frac{c_0^2 - a_0^2}{a^2}\right) \tag{6.381}$$

Equation 6.378 refers only to stresses beyond full overstrain, so that some other method has to be employed to obtain the pressure-strain curve up to the point of full overstrain. Beyond this, however, note that by Eq. 3.456.

$$(\varepsilon_h)_{uf,r=a} = \frac{u}{r} = \frac{a - a_0}{a_0} = \left(\frac{a}{a_0} - 1\right) \tag{6.382}$$

Thus, from Eq. 6.382, bore hoop strains can be calculated (and corrected to the O.D. by Eq. 6.364 if desired) and plotted against the pressure given by Eq.

6.381. One difficulty arises here with the value for σ_y. In Eq. 6.381 σ_y equals σ_0 of Eq. 6.27, and it is implied that only the yield strength of the material controls flow and eventual fracture; this may be true if the pressure is held constant as in a creep test. But in monotonic loading it is believed that the ultimate strength of the material σ_u also exerts its influence. One possible method of making a correction to σ_y in Eq. 6.381 is to recall the expression for "equivalent strain," the abscissa for the ordinate, "equivalent stress," given by Eq. 6.17. This strain value is given by Eq. 6.29.

$$\varepsilon_0 = \sqrt{\tfrac{2}{3}(\varepsilon_1{}^2 + \varepsilon_2{}^2 + \varepsilon_3{}^2)} \qquad 6.383$$

Assuming the longitudinal strain to be inconsequential and calling it zero, Eq. 6.383, because of constancy of volume, reduces to

$$\varepsilon_0 = (2/\sqrt{3})\varepsilon_h \qquad 6.384$$

Having Eq. 6.384, where ε_0 represents the tension strain, the proper value of σ_y can be deduced from an ordinary tensile stress-strain curve; for any particular value of ε_h, ε_0 is calculated and then σ_0 is read directly from the stress-strain curve and this σ_0 value is put in Eq. 6.381 as a replacement for σ_y. Fracture would presumably occur when σ_0 equaled σ_u.

In another method due to Nadai and described by Franklin and Morrison(12), torsion test data are analyzed in terms of cylinder behavior. In this theory, it is assumed that the state of stress is one of simple shear with superposed uniform volumetric stress and that the material is incompressible. Thus

$$2ur + u^2 = 2u_a(a) + u_a{}^2 \qquad 6.385$$

and the shear strain equivalent to a displacement of u at radius r is

$$\gamma = 2 \ln \frac{r+u}{u} \qquad 6.386$$

The internal pressure is obtained by integrating the equation for radial equilibrium, allowing for dimensional changes; thus

$$p_i = 2 \int_{a+u_a}^{c+u_c} \frac{\tau}{r+u}\, d(r+u) \qquad 6.387$$

where τ is the shear stress obtained from torsion tests. Equations 6.386 and 6.387 then provide values for a pressure-strain plot and also for the ultimate pressure or burst pressure. The results of a typical analysis using this method are shown in Fig. 6.48*B*.

Jorgensen (25), utilizing the basic analysis developed by Nadai et al., suggested the use of tension rather than torsional data in the computations. In this method the strain coordinate is supplied by Eq. 6.384 and the pressure coordinate by a relation similar to that in Eq. 6.387.

$$p_i = \frac{2\sigma_0}{\sqrt{3}} \int_{a+u_a}^{c+u_c} \frac{d(r+u)}{r+u} \qquad 6.388$$

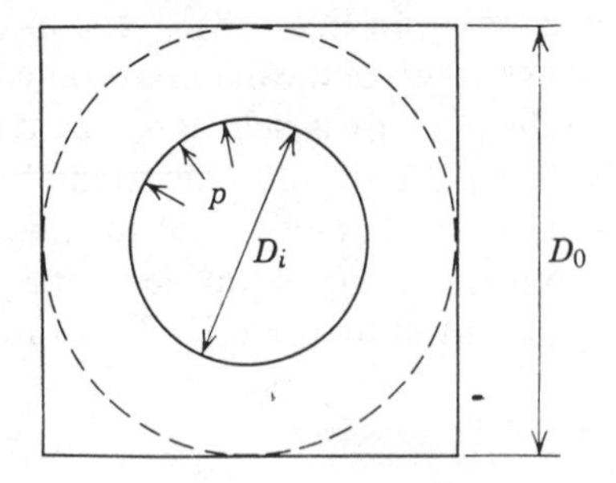

Fig. 6.49 Square tube with circular bore.

Another method is that due to Marin and Rimrott (30), who developed equations for a pressure-strain curve based on Eqs. 6.17, 6.383, and the logarithmic expression for a tensile stress-strain diagram (Eq. 6.1),

$$\sigma = K\varepsilon_t^{\,n} \qquad 6.389$$

Figure 6.48*C* shows a result based on this method of analysis.

The bursting pressure of a heavy-walled cylinder may be calculated by use of Eq. 6.377 if data are available for the yield and ultimate tensile strengths of the material. The integration methods of Nadai and Jorgensen may also be used if time permits and more basic information is desired concerning the deformation behavior of the material. For a square tube with a circular bore (Fig. 6.49) the burst pressure would be the same as for a circular cylinder of dimensions D_i and D_0. For a circular cylinder with an elliptic bore (Fig. 6.50) the burst pressure may be approximated by the formula

$$p = 1.2\,\sigma_u L/C \qquad 6.390$$

where it is assumed that failure takes place by a sliding motion along the lines marked L and the ultimate shear strength of the material is 0.6 times the ultimate tensile strength σ_u of the material. The analysis of rotating cylinders and disks is outlined in Chapter 10.

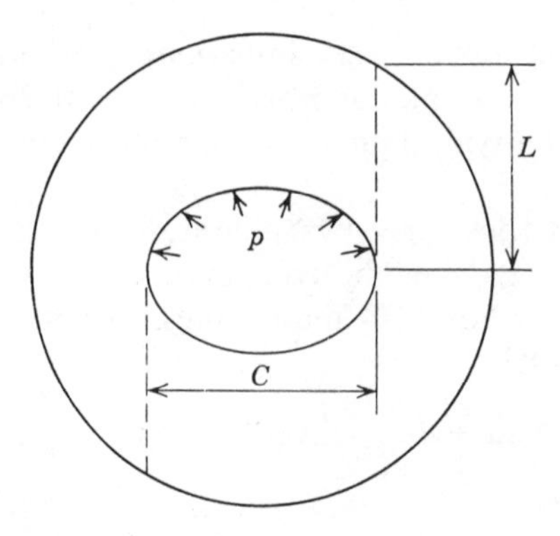

Fig. 6.50 Circular tube with elliptical bore.

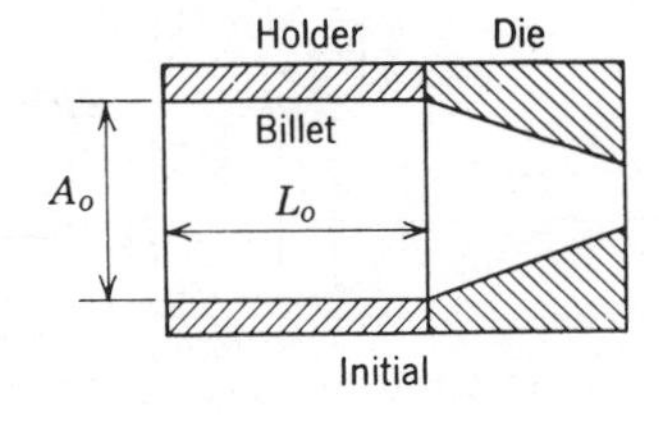

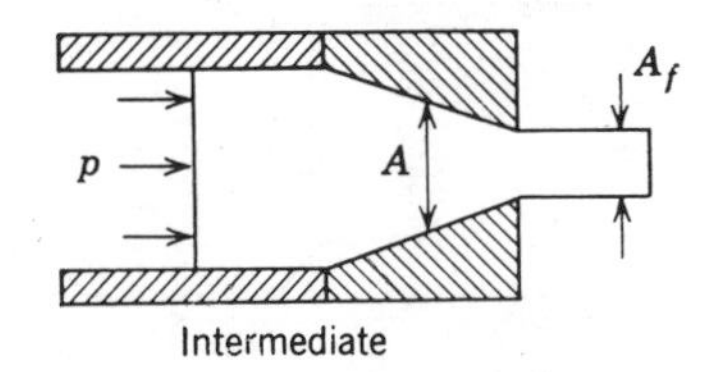

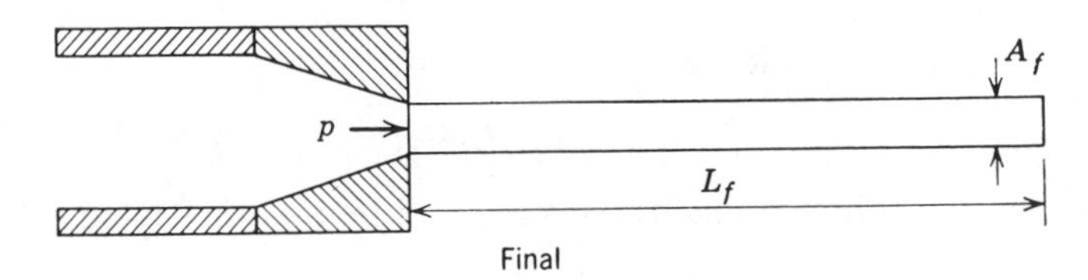

Fig. 6.51 Extrusion of billet through a conical die.

Metal-working applications

Since the theory of plasticity is concerned with flow of material in the post-yield region, it is natural that its applications (24) to various metal-working processes would be extensive. The analysis of material flow in practically all metal-working processes has been made by some application of plasticity theory.

Mechanics of extrusion. Consider first the simple case of a circular billet where all of the energy expended goes into deforming the metal (Fig. 6.51). The billet has an initial length L_0 and cross-sectional area A_0. Its final dimensions are L_f and A_f. At any intermediate time the dimensions are L and A. If the billet is pushed by the pressure p, the amount of work done in extruding the rod an amount ΔL is

$$\Delta W = \sigma_y A(\Delta L) \qquad 6.391$$

where σ_y is the yield strength of the material. The total work done is obtained by integrating Eq. 6.391,

$$W = \int_{L_0}^{L_f} \sigma_y(AL)\frac{dL}{L} \qquad 6.392$$

where AL is the volume of the material, which remains constant, that is,

$$AL = A_0L_0 = A_fL_f = \text{constant} \qquad 6.393$$

Solving Eq. 6.392 gives

$$W = \sigma_y A_0 L_0 \ln (L_f/L_0) \qquad 6.394$$

Since work equals force times distance,

$$W = pA_0L_0 = \sigma_y AL \ln (L_f/L_0) \qquad 6.395$$

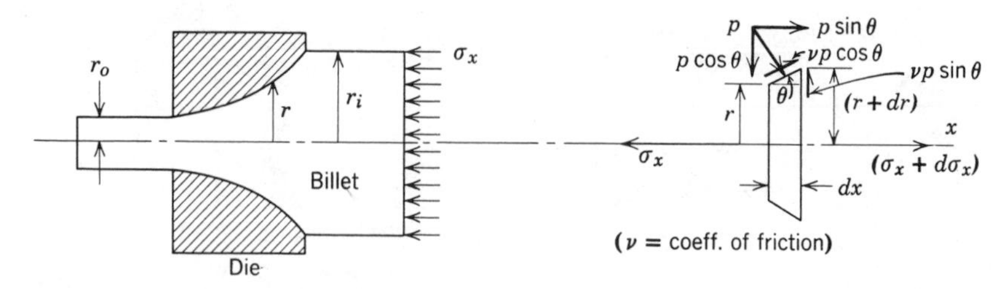

Fig. 6.52 Generalized model for process of extrusion.

from which the extrusion pressure p is

$$p = \sigma_y (A/A_0)(L/L_0) \ln (L_f/L_0) \qquad 6.396$$

However, the volume remains constant and by Eqs. 6.393 and 6.396,

$$p = \sigma_y \ln (L_f/L_0) = \sigma_y \ln (A_0/A_f) \qquad 6.397$$

This formula has proven quite useful for order of magnitude estimates of the extrusion pressure. Actually it predicts about 30 to 40% on the low side. Increasingly accurate results are obtained by taking into account friction and the state of stress existing in the die zone. Consider, for example, extrusion of the billet shown in Fig. 6.52. For equilibrium, with v for coefficient of friction,

$$-\sigma_x \pi r^2 + \left(\sigma_x + \frac{d\sigma_x}{dx} dx\right)(\pi)(r + dr)^2 + p \sin\theta \frac{dx}{\cos\theta}(2\pi r)$$
$$+ vp \cos\theta \frac{dx}{\cos\theta}(2\pi r) = 0 \qquad 6.398$$

from which

$$2\sigma_x dr + r\, d\sigma_x + 2p(\tan\theta + v)\, dx = 0 \qquad 6.399$$

However, $\tan\theta = dr/dx$; therefore

$$2\sigma_x dr + r\, d\sigma_x + 2p(dr + v\, dx) = 0 \qquad 6.400$$

For equilibrium in the radial direction,

$$-\sigma_r(2\pi r)dx - p \cos\theta(dx/\cos\theta)(2\pi r) + vp \sin\theta(dx/\cos\theta)(2\pi r) = 0 \qquad 6.401$$

from which

$$\sigma_r = -p(1 - v \tan\theta) \qquad 6.402$$

or since $\tan\theta = dr/dx$,

$$\sigma_r = -p[1 - v(dr/dx)] \qquad 6.403$$

By symmetry the hoop stress, $\sigma_h = \sigma_r$; therefore by Eq. 6.403

$$\sigma_h = \sigma_r = -p[1 - v(dr/dx)] \qquad 6.404$$

Yielding is defined by Eq. 6.17, which, together with Eq. 6.404, gives

$$\sigma_0 = \sigma_x - \sigma_h \tag{6.405}$$

However, $\sigma_r = \sigma_h$ is maximum at the die interface where $\sigma_r = -p$; therefore Eq. 6.405 can be written as

$$\sigma_0 = \sigma_x + p \tag{6.406}$$

Note that the notation here shows p as a radial die pressure and not the extrusion pressure. Substituting Eq. 6.406 into Eq. 6.400 then results in

$$2\sigma_x\,dr + r\,d\sigma_x + 2(dr + v\,dx)(\sigma_0 - \sigma_x) = 0 \tag{6.407}$$

or

$$2\sigma_x\,dr + r\,d\sigma_x + 2B(\sigma_0 - \sigma_x)\,dr = 0 \tag{6.408}$$

where

$$B = (1 + v/\tan\theta) = [1 + v(dx/dr)] \tag{6.409}$$

Rearranging then gives the result

$$\frac{2\,dr}{r} = \frac{d\sigma_x}{\sigma_x(B-1) - \sigma_0 B} \tag{6.410}$$

or

$$\frac{2\,dr}{r} = \frac{d\sigma_x}{\sigma_x(v/\tan\theta) - \sigma_0[1 + (v/\tan\theta)]} \tag{6.411}$$

Letting $\alpha = 1/\tan\theta$, Eq. 6.411 becomes

$$\frac{2\,dr}{r} = \frac{d\sigma_x}{\sigma_x(v\alpha) - \sigma_0(1 + v\alpha)} = \frac{d\sigma_x}{\sigma_x M - \sigma_0(1 + M)} \tag{6.412}$$

where $M = B - 1$. Integrating Eq. 6.412 gives the result

$$(1/M)\ln[\sigma_x M - \sigma_0(1 + M)] = 2\ln r + C_1 \tag{6.413}$$

where C_1 is a constant of integration. Solving for σ_x,

$$\sigma_x = (e^{C_1 M})\frac{r^{2M}}{M} + \sigma_0\left(\frac{1 + M}{M}\right) \tag{6.414}$$

In order to solve for the constant C_1 note that at the die exit $r = r_0$ and there is no extrusion pressure, that is, $\sigma_x = (\sigma_x)_{r0} = 0$; therefore

$$0 = (e^{C_1 M})\frac{r_0^{\,2M}}{M} + \sigma_0\left(\frac{1 + M}{M}\right) \tag{6.415}$$

from which

$$C_1 = \ln\left[\frac{-\sigma_0(1 + M)}{r_0^{\,2} M}\right]\frac{1}{M} \tag{6.416}$$

Substituting Eq. 6.416 into Eq. 6.414 then gives the result

$$\frac{\sigma_x}{\sigma_0} = \left(\frac{1+M}{M}\right)\left[1 - \left(\frac{r^2}{r_0^{\,2}}\right)^M\right] + \frac{(\sigma_x)_{r0}}{\sigma_0}\left(\frac{r^2}{r_0^{\,2}}\right)^M \qquad 6.417$$

However, from Eq. 6.400, $p = \sigma_0 - \sigma_x$; therefore using this value of p in Eq. 6.417 gives the result

$$\frac{p}{\sigma_0} = 1 - \frac{\sigma_x}{\sigma_0} = \frac{1}{M}\left[1 + (1+M)\left(\frac{r^2}{r_0^{\,2}}\right)^M\right] - \frac{(\sigma_x)_{r0}}{\sigma_0}\left(\frac{r^2}{r_0^{\,2}}\right)^M \qquad 6.418$$

Equation 6.418 gives the radial die pressure distribution. In order to find the extrusion pressure (σ_x at $r = r_i$), let $(\sigma_x)_{r_0} = 0$ and $r = r_i$ in Eq. 6.417; thus

$$\frac{(\sigma_x)_{ri}}{\sigma_0} = \frac{1+M}{M}\left[1 - \left(\frac{r_i^{\,2}}{r_0^{\,2}}\right)^M\right] \qquad 6.419$$

Extrusion reduction for round rods is expressed as

$$\text{reduction} = R = \frac{\text{finish area}}{\text{initial area}} = \frac{A_0}{A_i} = \frac{r_0^{\,2}}{r_i^{\,2}} \qquad 6.420$$

The percentage reduction is $(1 - R)(100)$; therefore, Eq. 6.419 can be rewritten as

$$(\sigma_x)_{r_i} = \frac{1+M}{M}(1 - R^{(1-B)})\sigma_0 \qquad 6.421$$

For the special case of no friction, that is, when $B = 1.0$, the analysis begins with Eq. 6.410—this is necessary since B is a function of both x and r; thus,

$$2\,dr/r = d\sigma_x/(-\sigma_0) \qquad 6.422$$

which on integration gives

$$\sigma_x = -\sigma_0(2\ln r + C_2) \qquad 6.423$$

where C_2 is a constant of integration. At $r = r_0$ the pressure is zero; therefore

$$0 = -\sigma_0(2\ln r_0 + C_2) \qquad 6.424$$

from which

$$C_2 = -2\ln \mathrm{r}_0 \qquad 6.425$$

Substituting Eq. 6.425 into Eq. 6.423 gives the result

$$\sigma_x = -\sigma_0(2\ln r - 2\ln r_0) \qquad 6.426$$

or

$$\frac{\sigma_x}{\sigma_0} = 2\ln\frac{r_0}{r} = \ln\frac{r_0^{\,2}}{r^2} \qquad 6.427$$

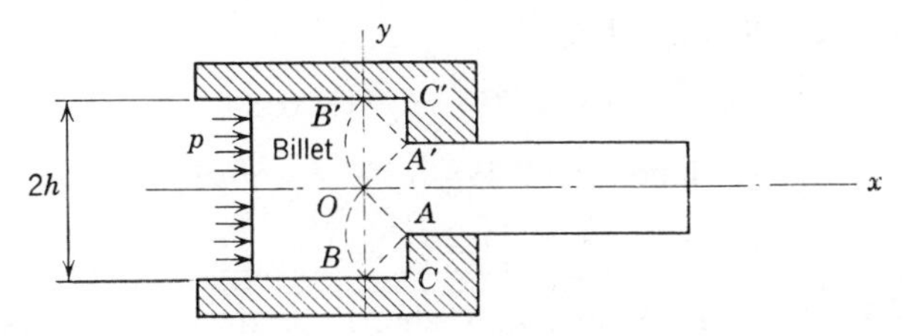

Fig. 6.53 Assumed stress field and process for extrusion in plane strain.

In addition,

$$\frac{p}{\sigma_0} = 1 - \frac{\sigma_x}{\sigma_0} = 1 - \ln \frac{r_0^2}{r^2} \qquad 6.428$$

and the extrusion pressure at $r = r_i$ is

$$(\sigma_x)_{r_i} = \sigma_0 \ln \frac{r_0^2}{r_i^2} = \sigma_0 \ln R \qquad 6.429$$

which is the same result as obtained by the simplified method resulting in Eq. 6.397. It is also of interest to compare these results with those obtained by the theory of plane strain (slip line theory) presented in the first section of this chapter. The solution is already available for sheet extrusion at 50% reduction (17, 40) and is outlined briefly later. The physical setup is shown in Fig. 6.53. For this problem in plane strain, friction is assumed to be zero and the assumed stress field (Fig. 6.53) consists of centered fans OAB and $OA'B'$ which intersect at the centroid axis and at the boundaries at 45° since no shearing can be transmitted in frictionless extrusion. For equilibrium the force across AOA' must be zero, that is, $\sigma_x = 0$. The shear lines OA and OA' are at 45°; therefore there is no shear and from Eq. 6.152

$$(\sigma_0/\sqrt{3})^2 = \tfrac{1}{4}(0 - \sigma_y)^2 + 0 \qquad 6.430$$

or

$$\sigma_y = 2\sigma_0/\sqrt{3} \quad \text{(compression)} \qquad 6.431$$

From Eqs. 6.159 to 6.161

$$\phi = -\tfrac{1}{2} \quad \text{and} \quad \theta = \pi/4 \quad \text{along } OA \qquad 6.432$$

Thus OA is a first shear line and the fan OAB is formed by straight first (α) shear lines. Since $\phi + \theta$ is constant throughout the fan and since θ increases by $\pi/2$ from AO to AB

$$\phi = -\tfrac{1}{2}(1 + \pi) \quad \text{and} \quad \phi = 3\pi/4 \text{ along } AB \qquad 6.433$$

Equations 6.159 to 6.161 show that

$$\sigma_x = -(2 + \pi)\sigma_0/\sqrt{3}, \qquad \sigma_y = -\pi\sigma_0/\sqrt{3}, \qquad \tau_{xy} = 0 \quad \text{along } AB \qquad 6.434$$

The force for extrusion equals one-half the pressure transmitted in the

x-direction across AB and $A'B'$ or

$$p = -(1 + \pi/2)(\sigma_0/\sqrt{3}) = 0.908\ \sigma_0 \qquad 6.435$$

For 50% reduction in plane strain, Eq. 6.397 becomes

$$p = \sigma_0 \ln (A_0/A_f) = \sigma_0 \ln 2 = \sigma_0(0.693) \qquad 6.436$$

The result of Eq. 6.435 is about 30% higher than given by Eq. 6.436 and thus gives a prediction more closely approaching experimental results.

In this analysis only the total reduction was considered, that is, the ratio of die opening and die exit. It is known, however, that (mechanically) there are at least two additional considerations, the shape of the die or its contour profile and its length. The effect of die profile on die pressure is quite marked as will be shown, although the die pressure at both entrance and exit is the same for a given reduction, regardless of the die profile. The effect of profile on extrusion pressure, for a given reduction, is small but important in very high pressure extrusions. It is known from experience (4) that the die shape does have an effect on pressure as well as the die length, particularly for reductions in the range 20 to 50%. Practical considerations may dictate the length of the die so that a certain amount of inefficiency is often unavoidable. On the other hand, some control can be exercised on the die profile to facilitate better working of the material or to improve die life. Very little information is available on the latter effects, but from a design point of view the problem of establishing profiles in extrusion dies is straightforward and some typical analyses are given.

Although many possible die profiles are possible, only three will be considered here to illustrate the mechanics involved. A typical *conical* die is shown in Fig. 6.54, where the profile is expressed as

$$x = f(r) \qquad 6.437$$

Equation 6.437 can be combined with Eq. 6.400, since

$$d[f(r)] = \frac{dx}{dr} = \frac{1}{\tan\theta} = \frac{1}{r_i - r_0} = \alpha \qquad 6.438$$

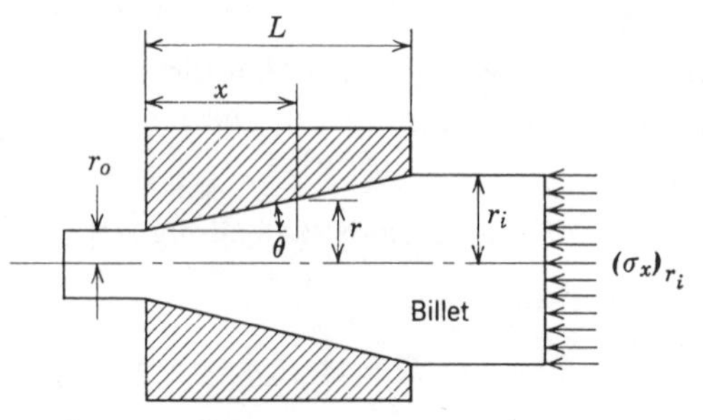

Fig. 6.54 Extrusion die with conical profile.

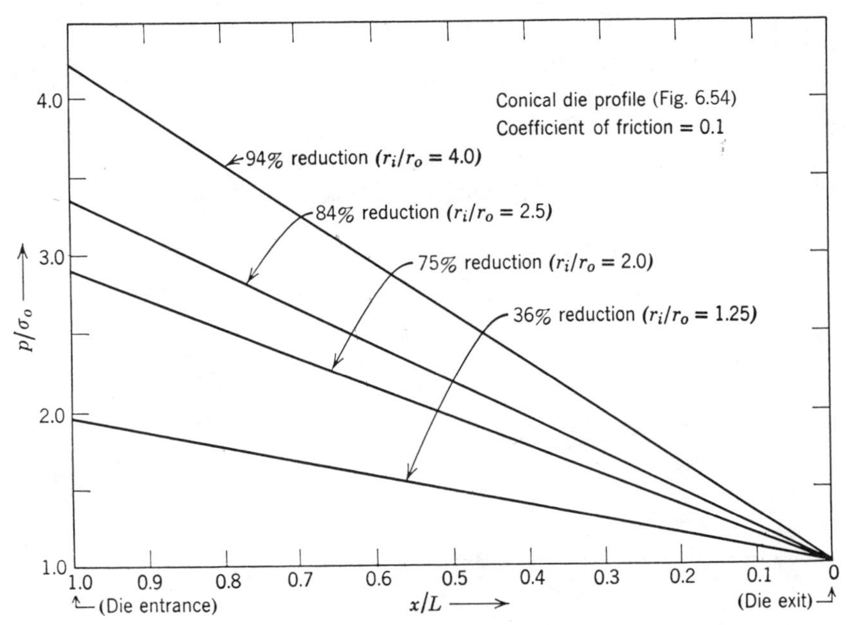

Fig. 6.55 Radial die pressure versus location along die profile.

By using Eq. 6.438, Eq. 6.410 becomes

$$\frac{2\,dr}{r} = \frac{d\sigma_x}{(v\alpha) - \sigma_0(1 + va)} \tag{6.439}$$

which corresponds to Eq. 6.412. The variation of die pressure along the die is given by Eq. 6.418 with $(\sigma_x)_{r0} = 0$; thus

$$\frac{p}{\sigma_0} = \frac{1}{M}\left[1 + (1 + M)\left(\frac{r^2}{r_0{}^2}\right)^M\right] \tag{6.440}$$

Equation 6.440 is plotted in Fig. 6.55 for $v = 0.10$. On the same plot is the case where $v = 0$, expressed by Eq. 6.428. A *concave* die is shown in Fig. 6.56

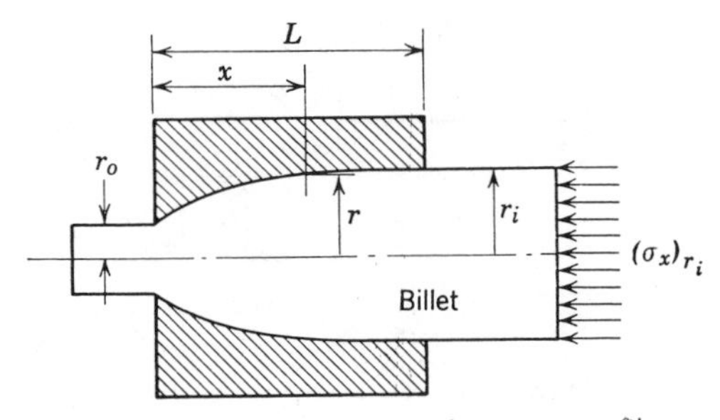

Fig. 6.56 Extrusion die with concave profile.

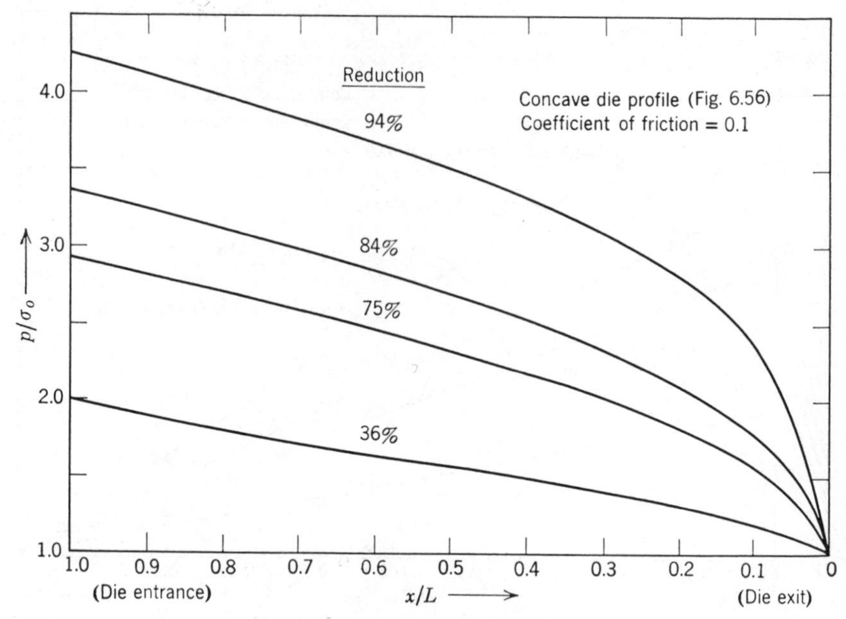

Fig. 6.57 Radial die pressure versus location along die profile.

where the profile is expressed as

$$x = L\left(\frac{r - r_0}{r_i - r_0}\right)^2 \tag{6.441}$$

Following Eq. 6.437 and differentiating,

$$d[f(r)] = \frac{dx}{dr} = \frac{2L(r - r_0)}{(r_i - r_0)^2} = K(r - r_0) \tag{6.442}$$

and substituting into Eq. 6.410 gives the result

$$\frac{2\,dr}{r} = \frac{d\sigma_x}{\sigma_x[\nu K(r - r_0)] - \sigma_0[1 + \nu K(r - r_0)]} \tag{6.443}$$

whose solution is plotted in Fig. 6.57. A typical *convex* die is shown in Fig. 6.58

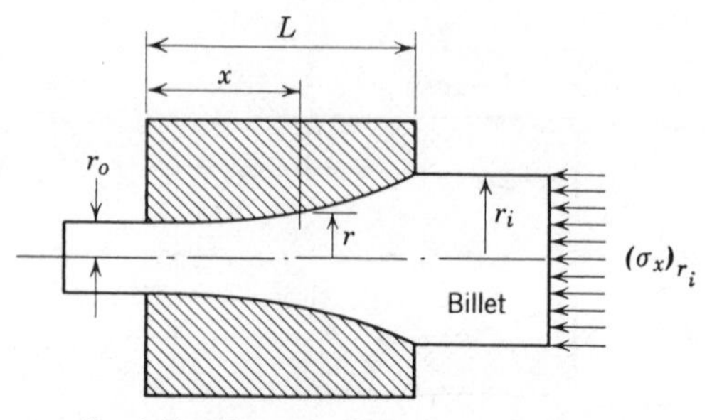

Fig. 6.58 Extrusion die with convex profile.

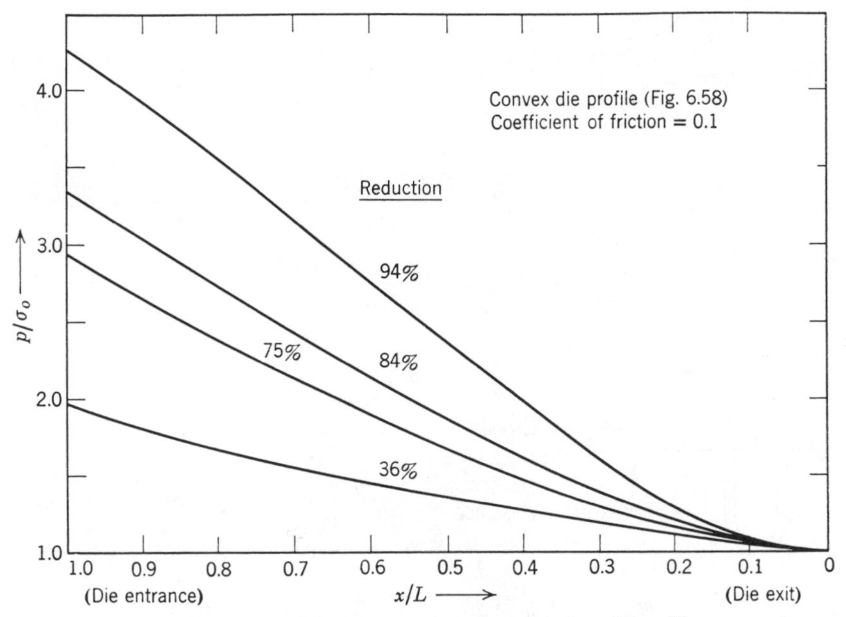

Fig. 6.59 Radial die pressure versus location along die profile.

whose contour is given by

$$x = L\left(\frac{r - r_0}{r_i - r_0}\right)^{1/2} \qquad 6.444$$

which when substituted into Eq. 6.410 gives the results plotted in Fig. 6.59. All these analyses can be applied to wire-drawing by re-evaluating the constants of integration in Eqs. 6.414 and 6.423 to conform to the boundary conditions. One boundary condition is, for example, at the die exit the stress σ_x equals the pull divided by the cross-sectional area of the wire.

A few additional results of plasticity theory as applied to metal-forming or working are presented in Fig. 6.60.

Bulge testing for formability criteria. The successful correlation of uniaxial test data with states of combined stress led many investigators to study the application of such data to metal-working operations. In particular, the deep-drawing of sheet materials presented unique problems in that very large strains were encountered in practice that put rather severe limitations on test methods. Thus, in order to correlate uniaxial test data with combined stress test data two schemes were developed (6); in one test a circular diaphragm of material was clamped over a reservoir of oil, which when pressurized caused the sheet to assume a circular bulge that could be analyzed in terms of stress and strain. Strains were measured with various types of calipers, strain gauges, and by observing deformation of grids marked on the sheet surface before testing. The second type of test employed an elliptical bulge that gave various stress ratios

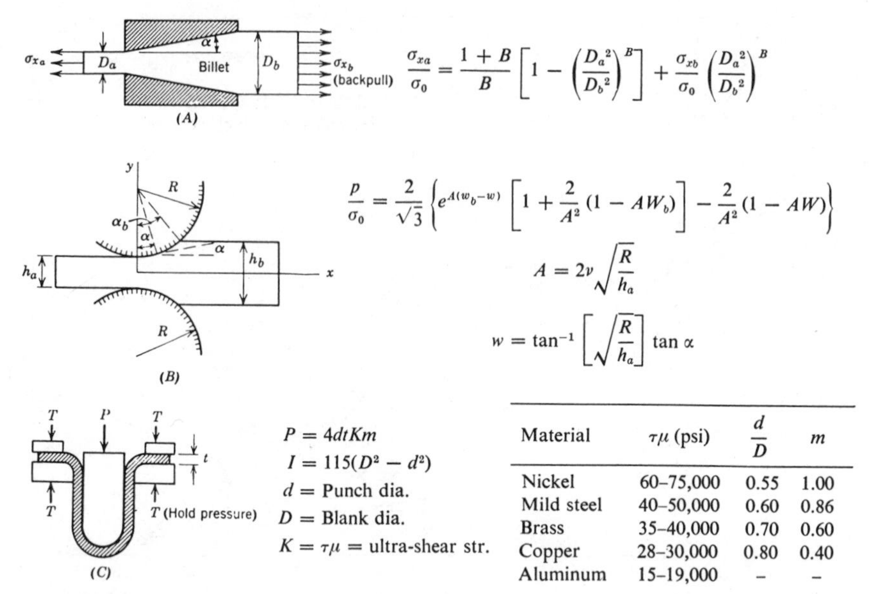

Material	$\tau\mu$ (psi)	$\frac{d}{D}$	m
Nickel	60–75,000	0.55	1.00
Mild steel	40–50,000	0.60	0.86
Brass	35–40,000	0.70	0.60
Copper	28–30,000	0.80	0.40
Aluminum	15–19,000	–	–

Fig. 6.60 Plasticity applied to metal-working. (A) Wire-drawing with backpull [according to Hoffman and Sachs (19)]. (B) Strip-rolling with no tension [according to Hoffman and Sachs (19)]. (C) Deep-drawing of thin sheet [according to Strasser (42), data courtesy of McGraw-Hill Publishing Co.].

depending on the ellipse axis ratio. Tests of this type, in general, were more satisfactory than tube tests since the strains generated in the bulges were considerably larger. In the general case, Fig. 6.61, the equilibrium relationship is given by Eq. 3.367,

$$p/t = \sigma_1/R_1 + \sigma_2/R_2 \qquad 6.445$$

where σ_1 and σ_2 are principal stresses and R_1 and R_2 are the corresponding

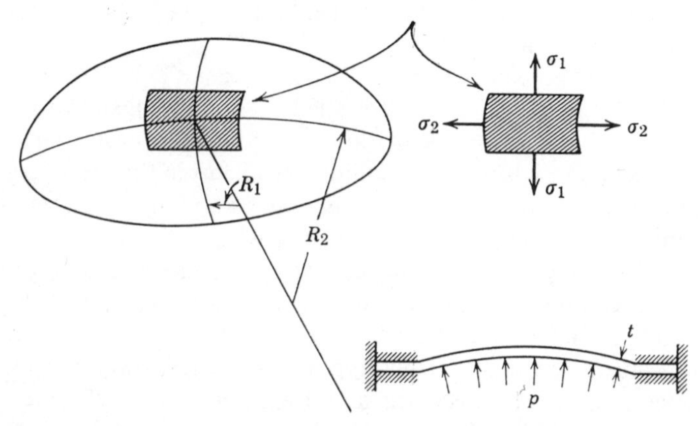

Fig. 6.61 Elliptical bulge in sheet metal.

Table 6.5 Miscellaneous Solutions to Plasticity Problems Obtained by Application of the Theory of Plane Plastic Strain (Slip Line Theory)

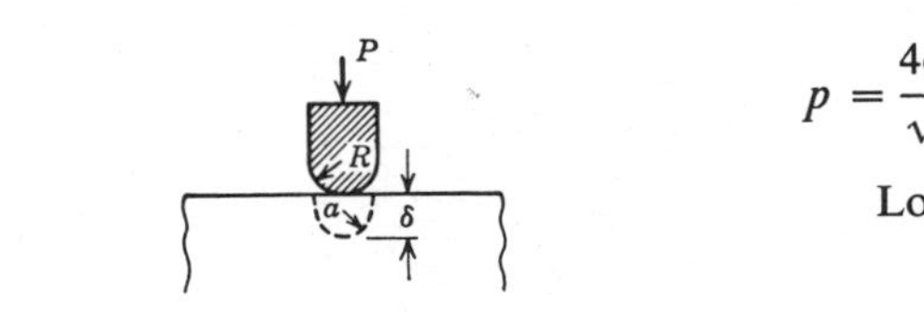	$p = \frac{4\sigma_y}{\sqrt{3}}(a)\left[1 + \frac{\pi}{2} - \sqrt{\delta/2R}\right]$ Load to indent distance δ by punch	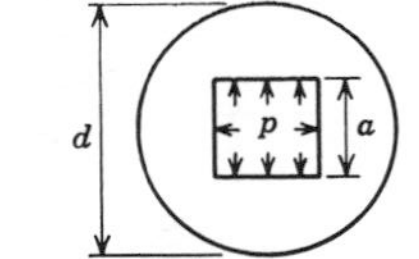	$p_y = \frac{2\sigma_y}{\sqrt{3}} \ln\left(\frac{d\sqrt{2}}{2a}\right)$ Yield pressure of cylinder with square bore
	$p = \frac{2\sigma_y}{\sqrt{3}}\left(\frac{1}{2} + \frac{\pi}{4} + \alpha + \frac{\cot\alpha}{2}\right)$ Pressure to indent		$p_y = \frac{2\sigma_y}{\sqrt{3}}(1 + \sin\alpha)$ Yield pressure of wedge
	$p_y = \frac{2\sigma_y}{\sqrt{3}}\left(\frac{b - a}{a}\right)$ Yield pressure of square tube	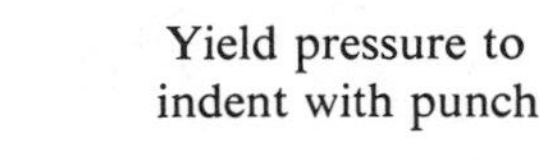	$p_y = \frac{\sigma_y}{\sqrt{3}}(2 + \pi)$ Yield pressure to indent with punch
(Plastic mass compressed between two plates)	Stresses: $\sigma_x'; \sigma_y'; \tau_{xy}$ Stresses in plastic mass $\sigma_x' = \frac{\sigma_y}{2}\left[\frac{x}{a} - \frac{\pi}{2} + 2\sqrt{1 - \left(\frac{y}{a}\right)^2}\right]$ $\sigma_y' = \frac{\sigma_y}{2}\left(\frac{x}{a} - \frac{\pi}{2}\right); \ \tau_{xy} = -\frac{\sigma_y}{2}\left(\frac{y}{a}\right)$	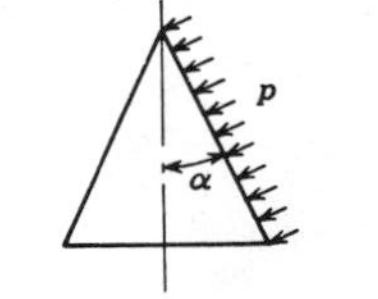	$p_y = \frac{2\sigma_y}{\sqrt{3}}(1 - \cos 2\alpha)$ Yield pressure of wedge

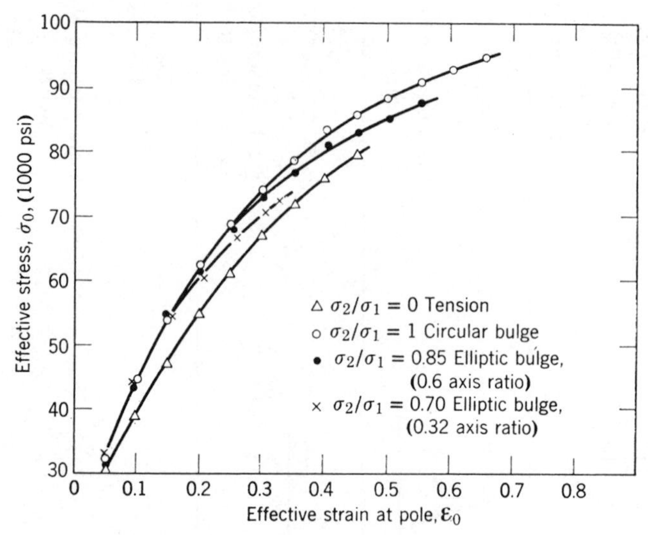

Fig. 6.62 Effective stress versus effective strain in tension and bulge tests on annealed 70/30 brass. [After Chow, Dana, and Sachs (6), courtesy of the American Institute of Mining and Metallurgical Engineers.]

principal radii of curvature. By constancy of volume

$$\varepsilon_1 + \varepsilon_2 + \varepsilon_3 = 0 \tag{6.446}$$

Biaxiality is assumed; therefore $\sigma_3 = 0$ so that Eq. 6.17 becomes

$$\sigma = \sqrt{\sigma_1{}^2 - \sigma_1\sigma_2 + \sigma_2{}^2} \tag{6.447}$$

The strain is computed from Eq. 6.30,

$$\varepsilon_0 = (\sqrt{2}/3)\sqrt{(\varepsilon_1 - \varepsilon_2)^2 + (\varepsilon_2 - \varepsilon_3)^2 + (\varepsilon_3 - \varepsilon_1)^2} \tag{6.448}$$

In the experiments ε_1 and ε_2 are measured so that R_1 and R_2 can be computed. With this accomplished, Eq. 6.445 can be solved and the values put into Eq. 6.447 in order to compute σ_0 the equivalent stress. Typical results obtained by Chow, Dana, and Sachs (6) are shown in Fig. 6.62.

Miscellaneous applications solved by plane strain (slip line) theory. It has been mentioned before that the slip line theory has proven very useful in the solution of metal-working problems and many examples are given by Hill (17), Nadai (32), Prager and Hodge (40), and Van Iterson (47). An example of extrusion was given and the results of a few analyses taken from the references cited are given in Table 6.5.

REFERENCES

1. J. F. Baker, *The Steel Skeleton*, Vol. 1, "Elastic Behavior and Design," Cambridge University Press, London (1954), and J. F. Baker, M. R. Horne, and J. Heyman, *The Steel Skelton*, Vol. 2, "Plastic Behavior and Design," Cambridge University Press, London (1956).

2. J. F. Baker and J. Heyman, *Plastic Design of Frames*, Vol. 1, "Fundamentals," Cambridge University Press, London (1969).

3. J. F. Baker and B. G. Neal, *Research*, **5**, 412–18 (September 1952).

4. L. S. Beedle, B. Thurlmann, and R. L. Ketter, *Plastic Design in Structural Steel*, American Institute of Steel Construction, Inc., New York, N.Y. (1955).

5. J. F. W. Bishop, *Metall. Rev.*, **2** (8), 361–90 (1957).

6. C. C. Chow, A. W. Dana, and G. Sachs, *Trans. Am. Inst. Min. Met. Eng. (Met. Branch)*, **185**, 49–58 (January 1949).

7. D. C. Drucker, *Appl. Mech. Rev.*, **7** (10), 421–423 (October 1954).

8. D. C. Drucker, "Plasticity," in *Structural Mechanics*, 407–455, Pergamon Press, New York, N.Y. (1960).

9. D. C. Drucker and W. Prager, *Quart. Appl. Math.*, **10** (2), 157–65 (July 1952).

10. J. H. Faupel, *Trans. Am. Soc. Mech. Eng.*, **78** (5), 1031–64 (July 1956). See also J. H. Faupel and A. R. Furbeck, *Trans. Am. Soc. Mech. Eng.*, **75** (3), 345–54 (April 1953).

11. J. H. Faupel, and J. Marin, *Trans. Am. Soc. Met.*, **43**, 993–1012 (1951).

12. G. J. Franklin, and J. L. M. Morrison, *Proc. Inst. Mech. Eng. (London)*, **174** (35), 947–974 (1960).

13. A. M. Freudenthal, *The Inelastic Behavior of Engineering Materials and Structures*, John Wiley and Sons, Inc., New York, N.Y. (1950).

14. D. J. Graziano, *Prod. Eng.*, **29** (40), 44–47 (September 29, 1958); **30** (52), 45–48 (December 21, 1959).

15. J. Heyman, *Plastic Design of Frames*, Vol. 2, "Applications," Cambridge University Press, London (1971).

16. R. Hicks, *Engineering*, **177** (4593), 335–336 (March 1954).

17. R. Hill, *The Mathematical Theory of Plasticity*, Clarendon Press, Oxford (1950).

18. P. G. Hodge, *Plastic Analysis of Structures*, McGraw-Hill Book Co., New York, N.Y. (1959).

19. O. Hoffman and G. Sachs, *Introduction to the Theory of Plasticity for Engineers*, McGraw-Hill Book Co., New York, N.Y. (1953).

20. H. G. Hopkins and W. Prager, *J. Mech. and Phys. Solids*, **2** (1), 1–13 (October 1953).

21. H. G. Hopkins and A. J. Wang, *J. Mech. and Phys. Solids*, **3** (2), 117–29, (January 1955).

22. R. Houwink, *Elasticity, Plasticity, and the Structure of Matter*, Cambridge University Press, London (1971).

23. L. W. Hu, *J. Eng. Mech. Div., Am. Soc. Civ. Eng.*, Paper 2338, 91–115 (January 1960).

24. W. Johnson and P. B. Mellor, *Engineering Plasticity*, Van Nostrand Reinhold Co., New York (1973).

25. S. M. Jorgensen, *Trans. Am. Soc. Mech. Eng.*, **80** (3), 561–70 (April 1958).

26. A. Kleinlogel, *Rigid Frame Formulas*, Frederick Ungar Publishing Co., New York, N.Y. (1964).

27. A. Kooharian, *J. Am. Concr. Inst.*, **24** (4), 317–28 (December 1952).

28. V. Leontovich, *Frames and Arches*, McGraw-Hill Book Co., New York, N.Y. (1959).

29. J. Marin, V. L. Dutton, and J. H. Faupel, *Weld. J. Res. Suppl.*, **27** (12), 593S–607S (December 1948).

30. J. Marin and F. P. J. Rimrott, *Welding Research Council Report*, No. 41 (1958).

31. J. B. Martin, *Plasticity*, Massachusetts Institute of Technology Press, Cambridge, Mass. (1975).

32. A. Nadai, *Theory of Flow and Fracture of Solids*, McGraw-Hill Book Co., New York, N.Y. (1950).

33. B. G. Neal, *The Plastic Methods of Structural Analysis*, John Wiley and Sons, Inc., New York, N.Y. (1965).

34. F. K. G. Odqvist, "Plasticity Applied to the Theory of Thin Shells and Pressure Vessels," in *Reissner Anniversary Volume*, 449–460, Edwards Brothers, Inc., Ann Arbor, Mich. (1949).

35. E. T. Onat and W. Prager, *J. Mech. and Phys. Solids*, **1** (2), 77–89 (January 1953).

36. P. C. Paris, *J. Aeronaut. Sci.*, **21** (1), 43–49 (January 1954).

37. W. H. Pell and W. Prager, "Limit Design of Plates," in *Proc. 1st Nat. Congr. Appl. Mech.*, 547–50 Edwards Brothers, Inc., Ann Arbor, Mich. (1952).

38. A. Phillips, *Introduction to Plasticity*, Ronald Press Co., New York, N.Y. (1956).

39. W. Prager, *Introduction to Plasticity*, Addison-Wesley Publishing Co., Reading, Mass. (1959).

40. W. Prager and P. G. Hodge, *Theory of Perfectly Plastic Solids*, John Wiley and Sons, Inc., New York, N.Y. (1951). See also P. G. Hodge, *Limit Analysis of Rotationally Symmetric Plates and Shells*, Prentice-Hall, Englewood Cliffs, N.J. (1963).

41. M. Reiner, *Deformation and Flow*, H. K. Lewis and Co., Ltd., London (1969).

42. G. Sachs and K. R. Van Horn, *Practical Metallurgy*, American Society for Metals, Novelty, Oh. (1940). See also F. Strasser, *American Machinist*, 94, (October 30, 1950).

43. N. L. Svensson, *J. Appl. Mech.*, **25** (1), 89–96 (March 1958).

44. S. Timoshenko, *Strength of Materials*, Part II, 3rd ed., D. van Nostrand Co., Princeton, N.J. (1956).

45. S. Timoshenko and J. M. Gere, *Theory of Elastic Stability*, 2nd ed., McGraw-Hill Book Co., New York, N.Y. (1961).

46. J. A. Van Den Broek, *Theory of Limit Design*, John Wiley and Sons, Inc., New York, N.Y. (1948).

47. F. K. Van Iterson, *Plasticity in Engineering*, Blackie and Son, Ltd., London (1947).

48. W. H. Weiskopf, *Weld. J. Res. Suppl.*, **31**, (7), 353s–360s (July 1952).

49. *Commentary on Plastic Design in Steel*, Welding Research Council and American Society of Civil Engeers, New York, N.Y. (1961).

PROBLEMS

6.1 A sphere is made of a perfectly plastic material with a yield stress of 100 kpsi. If the inside radius is 2 in. and the outside radius is 6 in. find the internal pressure for (a) yielding at the inside radius, (b) yielding midway through the wall, and (c) yielding throughout the wall.

6.2 A ring with a mean radius of 5 in. has loads P applied at the top and bottom. The ring is made of one inch diameter 2330 steel drawn at 400°F with a yield of 195 kpsi and an ultimate of 225 kpsi. Find the safety factor by comparing the collapse load to that obtained for an elastic solution.

6.3 In Problem 3.5 find the collapse loading in pounds/square foot from limit design for the unreinforced wooden beam.

6.4 Find q_c in Fig. P6.4.

6.5 Find q_c in Fig. P6.5.

6.6 Find P_c in Fig. P6.6.

6.7 Given a thin-wall cylinder with closed ends having the following dimensions: $D = 10$ in., $t = 0.1$ in., $L = 20$ in. If the cylinder is subjected

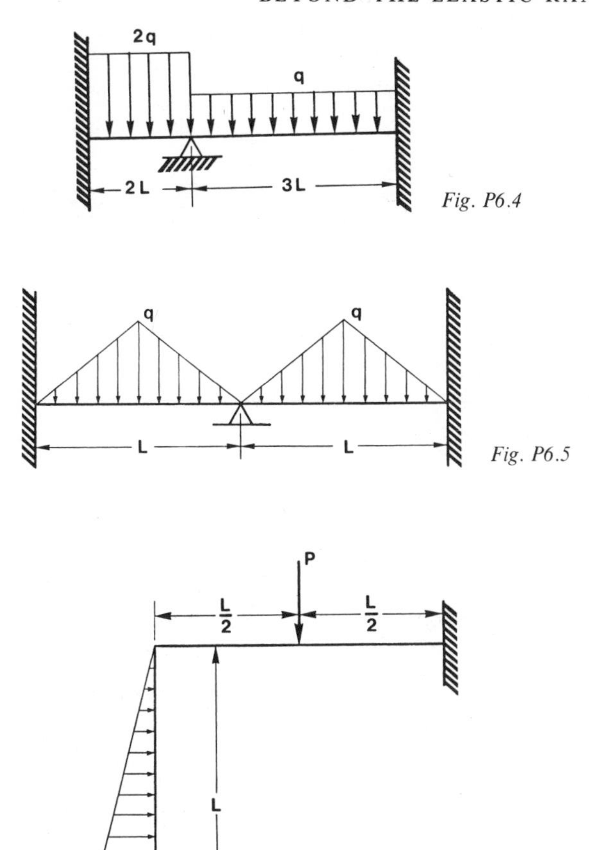

Fig. P6.4

Fig. P6.5

Fig. P6.6

to an internal pressure of 1000 psi, determine whether it has yielded according to either the Tresca or Von Mises criterion with $\sigma_y = 50$ kpsi.

6.8 In the above cylinder, Problem 6.7, determine the pressure that the cylinder can withstand if the diameter is allowed to increase by 1%. Consider two types of materials—(a) perfectly plastic material and (b) a material whose stress-strain obeys $\sigma = 143{,}000\ \varepsilon^{0.229}$.

6.9 Determine the maximum pressure in the above cylinder at the point of instability in the material whose stress-strain curve is that of Problem 6.8 (b).

6.10 A load L is supported by three rods AO, BO, CO (Fig. P610). The dia-

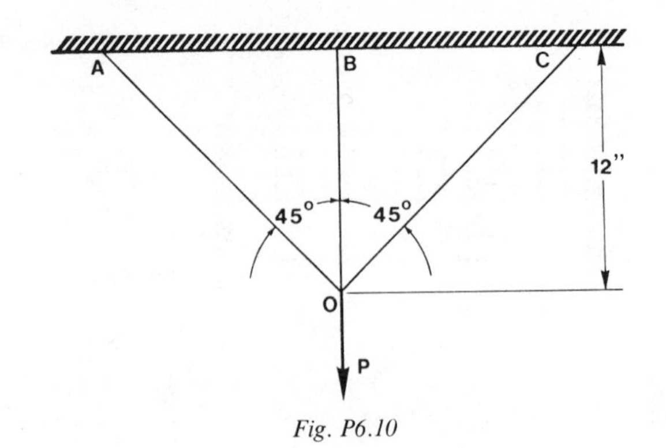

Fig. P6.10

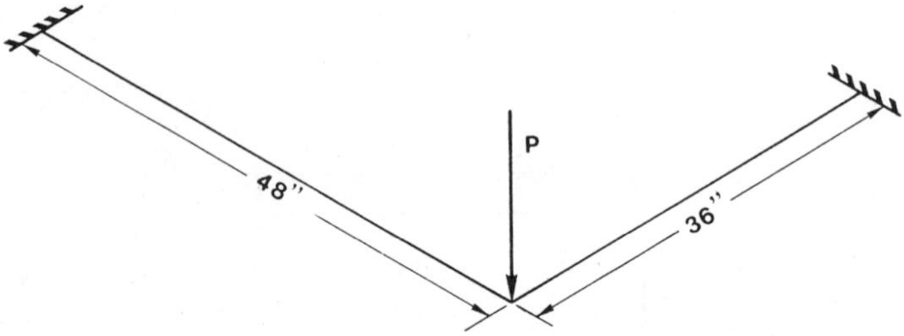

Fig. P6.11

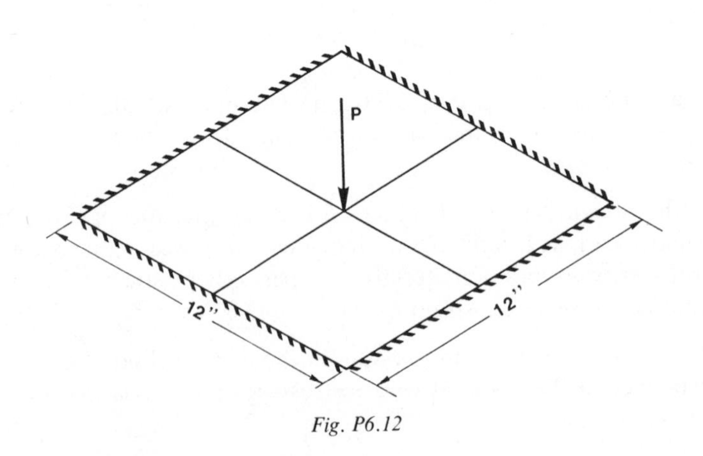

Fig. P6.12

meter of AO, CO is $\frac{1}{2}$ in., with the diameter of BO being $\frac{3}{4}$ in., and the E of all rods being 30×10^6 psi.

a. If the material in the three rods is a perfectly plastic one, $\sigma_y = 40$ kpsi, find the load L at the start of yielding and collapse.

b. If $\sigma_y = 40$ kpsi for rods AO and CO, and BO has a true stress property of $\sigma = 150{,}000\ \varepsilon^{0.25}$, find the load at start of yielding and at the point of instability.

6.11 A 0.5-in. square steel bar right angle frame is loaded so that plastic moments and torques are developed at the wall (Fig. P6.11). Find P_c.

6.12 A grill of 0.25-in. round steel rods covers a 12-in. square opening. Find the collapse load that is perpendicular to the opening and applied at the center. (Fig. P6.12).

chapter 7

ENERGY METHODS IN DESIGN

This chapter deals with applications of elastic strain energy in structural problems. *Toughness* is frequently expressed in terms of energy; this was discussed in Chapter 1 where it was shown that toughness could be interpreted in terms of the area enclosed by a stress-strain curve. In Chapter 3 failure criteria for materials were discussed in terms of *distortion energy*. In Chapter 6 energy as expressed by *virtual work* was associated with the development of plastic hinges in bent structures. In Chapter 10 energy is discussed in connection with impact, rapid loading, and vibration problems and, in Chapter 15, *complementary energy* is used in the solution of nonlinear creep and high-temperature problems.

7-1 FUNDAMENTAL STRAIN-ENERGY RELATIONSHIPS

Energy is defined as capacity to do work and, in Fig. 7.1, which shows a bar strained an amount δ by a gradually applied load F, the work done and potential energy are $F\delta/2$; it is assumed that the deformations are small, that Hooke's law applies, and that the strains or displacements are proportional to the loads.

Consider Fig. 7.2 which shows a body subjected to several external loads; for the configuration shown, and following the definition given, the potential or elastic strain energy in the body as a consequence of the loading is

$$U = \frac{F_1\delta_1}{2} + \frac{F_2\delta_2}{2} + \frac{F_3\delta_3}{2} + \frac{F_4\delta_4}{2} + \frac{F_n\delta_n}{2} \tag{7.1}$$

From the simple relationship of Eq. 7.1 the elastic strain energy of a number of systems can be computed. For example, suppose an elemental cube of material is subjected to tensile stresses as shown in Fig. 7.3. For tensile stresses σ_x, σ_y,

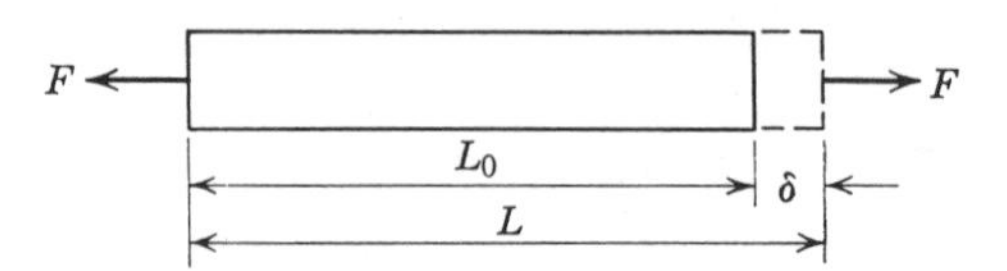

Fig. 7.1 Prismatic bar subjected to a tension load.

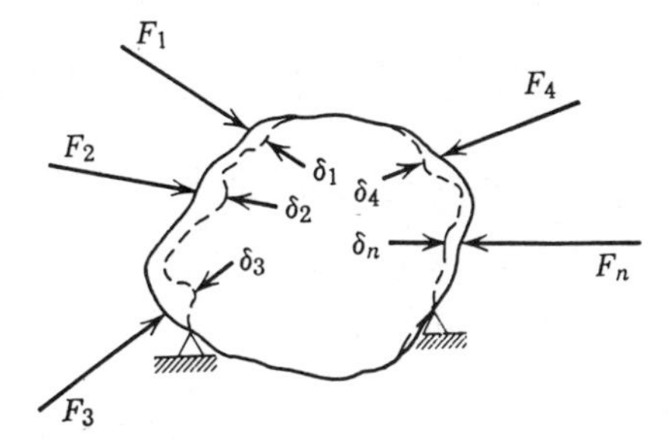

Fig. 7.2 Body subjected to several external loads.

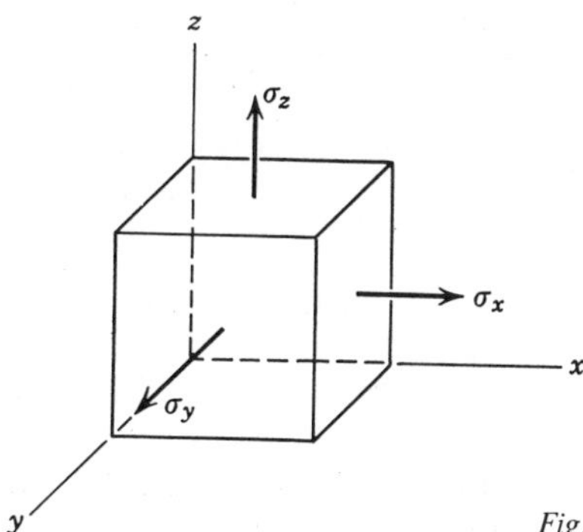

Fig. 7.3 Elemental cube of material under triaxial loading.

and σ_z, the corresponding elastic strains are given by Eqs. 3.70 to 3.72; therefore, for the cube in Fig. 7.3, the strain energy is

$$U = \frac{\sigma_x \varepsilon_x}{2} + \frac{\sigma_y \varepsilon_y}{2} + \frac{\sigma_z \varepsilon_z}{2} \tag{7.2}$$

or in terms of strain energy per unit volume w,

$$w = \frac{1}{2E}[\sigma_x^{\,2} + \sigma_y^{\,2} + \sigma_z^{\,2}] - \frac{\nu}{E}[\sigma_x \sigma_y + \sigma_y \sigma_z + \sigma_z \sigma_x] \tag{7.3}$$

If shearing stresses are involved (Fig. 7.4), the work done by a gradually applied load is $F\delta/2$. By definition, shear strain is

$$\gamma = \delta / L \tag{7.4}$$

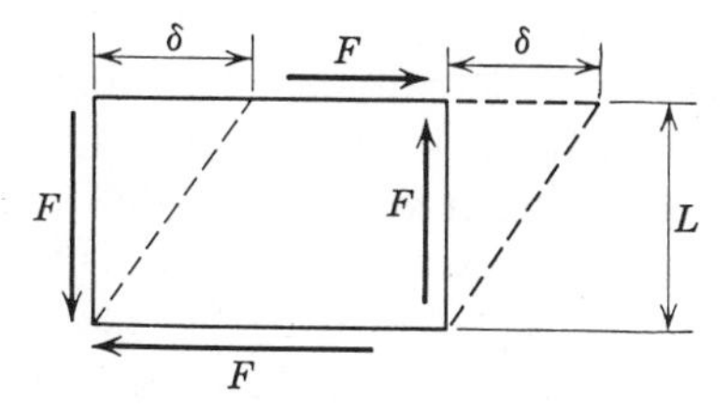

Fig. 7.4 Block of material subjected to shearing forces.

and by Hooke's law

$$\gamma = \tau/G \qquad 7.5$$

where τ is the shear stress and G is the shear modulus. Therefore, from Eq. 7.5, if A is the area subjected to the stress,

$$\gamma = \tau/G = F/AG \qquad 7.6$$

and, using Eq. 7.4,

$$\delta = L\gamma = FL/AG \qquad 7.7$$

Since the strain energy is $F\delta/2$ the combination of Eqs. 7.1 and 7.7 shows that

$$U = F^2L/2AG \qquad 7.8$$

or in terms of strain energy per unit volume w

$$w = \frac{(F^2L/2AG)}{AL} \qquad 7.9$$

or

$$w = \tau^2/2G \qquad 7.10$$

The result expressed by Eq. 7.10 can be applied to the case where shear forces are applied on all faces of the three-dimensional body; here the unit energy is simply

$$w = \frac{1}{2G}(\tau_{xy}{}^2 + \tau_{yz}{}^2 + \tau_{zx}{}^2) \qquad 7.11$$

The combination of normal and shear stresses is solved by adding Eqs. 7.3 and 7.11; this is allowable because the shear stress does not change the lengths of the sides of the element, and the normal forces do no work during the process of shearing. Thus, for combined normal and shearing stresses,

$$w = \frac{1}{2E}(\sigma x^2 + \sigma_y{}^2 + \sigma_z{}^2) - \frac{v}{E}(\sigma_x\sigma_y + \sigma_y\sigma_z + \sigma_z\sigma_x) + \frac{1}{2G}(\tau_{xy}{}^2 + \tau_{yz}{}^2 + \tau_{zx}{}^2) \qquad 7.12$$

Strain energy in tension and compression

Consider a bar of material of length L_0 and cross-sectional area A gradually loaded by a force F, as shown in Fig. 7.1. As the bar is loaded, work is done which is transformed into potential energy of strain. It is assumed that the strains are within the elastic limit and that the deformations are small. As the bar is loaded, the load-strain relationship can be represented by the plot in Fig. 7.5. The final load F is associated with a final extension δ. At some inter-

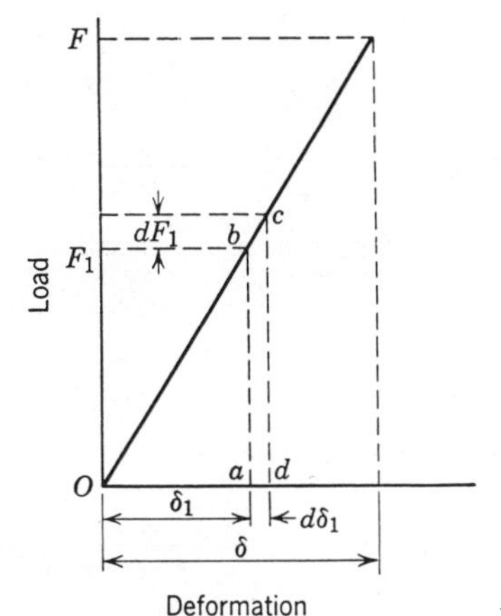

Fig. 7.5 Load-deformation plot for tension loading.

mediate extension δ_1 the load is F_1; now, if the load F_1 is increased by a differential amount, the extension also increases by a differential amount $d\delta_1$. Since work is average force times distance, the differential work done by the force dF_1 is

$$dw = dF_1 d\delta_1/2 \tag{7.13}$$

and the work stored up is represented by the area *abcd* in Fig. 7.5. The total work involved is the integral of Eq. 7.13 or

$$W = \int dw = \int_0^F \int_0^\delta \frac{dF_1\, d\delta_1}{2} = \frac{F\delta}{2} \tag{7.14}$$

In accordance with Hooke's law, Eq. 2.4,

$$\delta = FL/AE \tag{7.15}$$

Therefore substitution of Eq. 7.15 into Eq. 7.14 gives the result

$$U = F^2 L/2AE \tag{7.16}$$

where E is the modulus of elasticity. Equation 7.16 was also discussed in Chapter 1 where it was noted that the resilience was given by

$$U = \sigma^2/2E \tag{7.17}$$

Strain energy in uniaxial bending

Consider Fig. 7.6 which shows a differential length of a beam subjected to bending moments. Here any effect of shear is ignored. Furthermore, assuming that plane sections remain plane, the edges of the differential element rotate by an amount $d\theta/2$ to positions $a'b'$ and $c'd'$. The work done is the average applied moment multiplied by the angular displacement, thus

$$dw = \tfrac{1}{2}(M\, d\theta/2) + \tfrac{1}{2}[(M + dM)d\theta]/2 = M\, d\theta/2 + dM\, d\theta/4 \tag{7.18}$$

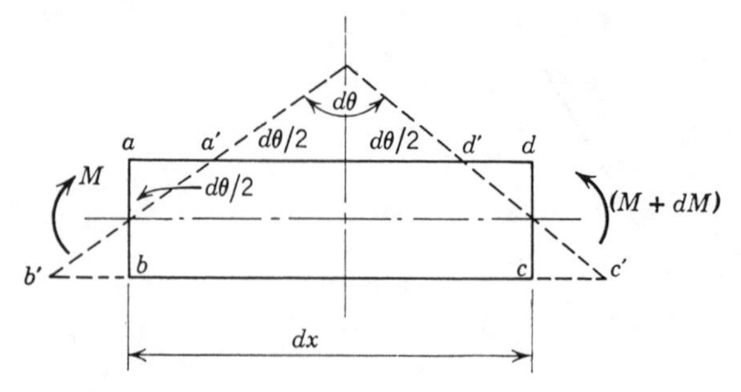

Fig. 7.6 Differential length of a beam subjected to bending.

and, neglecting products of differentials,

$$dw = M\,d\theta/2 \tag{7.19}$$

The total work and strain energy is the integral of Eq. 7.19 or

$$W = \int dw = U = M\theta/2 \tag{7.20}$$

The general differential equation for beams in bending is given by Eq. 2.82,

$$EI\,d^2y/dx^2 = M \tag{7.21}$$

or

$$M = EI(d/dx)(dy/dx) = EI(d/dx)\theta \tag{7.22}$$

where θ is the slope. Substituting Eq. 7.20 into Eq. 7.22 gives the result

$$M = EI(d/dx)(2U/M) \tag{7.23}$$

which on integrating gives the strain energy as

$$U = \int \frac{M^2\,dx}{2EI} \tag{7.24}$$

By substituting Eq. 7.21 into Eq. 7.24 an expression for strain energy in bending in terms of deflection can be obtained,

$$U = \int \frac{(EI)^2(d^2y/dx^2)^2\,dx}{2EI} = \int \frac{EI(d^2y/dx^2)^2\,dx}{2} \tag{7.25}$$

Strain energy in direct shear

Consider the element shown in Fig. 7.4 where a gradually applied shear force F causes a deformation δ. The work and strain energy may be expressed as

$$U = F\delta/2 \tag{7.26}$$

The deformation δ may be expressed in terms of shear stress and shear modulus

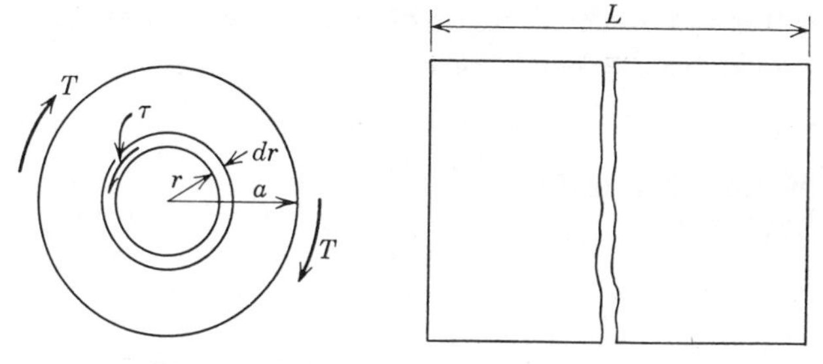

Fig. 7.7 Circular bar subjected to torsion load.

which converts the expression given in Eq. 7.26 to that given in Eq. 7.8, namely,

$$U = F^2L/2AG \tag{7.27}$$

Strain energy in torsion

Consider the bar of circular cross section shown in Fig. 7.7. In the elastic range the shear stress is maximum at the outside surface of the bar; therefore at any intermediate location, since the shearing stress varies directly as the distance from the axis,

$$\tau = \tau_{\max}(r/a) \tag{7.28}$$

However, the strain energy per unit volume is given by Eq. 7.10 and is

$$w = \tau^2/2G \tag{7.29}$$

so that the strain energy per unit volume at radius r is

$$(w)_r = (\tau_{\max}^2/2G)(r/a)^2 \tag{7.30}$$

In the annulus of thickness dr the total energy is

$$U = \int_0^a \frac{\tau_{\max}}{2G}(r/a)^2(2\pi rL)\,dr \tag{7.31}$$

However, in elastic torsion of circular bars

$$\tau = Tr/J \tag{7.32}$$

where

$$J = \pi r^4/2 \tag{7.33}$$

Therefore, using Eqs. 7.31 to 7.33,

$$U = T^2L/2GJ \tag{7.34}$$

In terms of the angle of twist θ, Eq. 7.34 becomes

$$U = \theta^2GJ/2L \tag{7.35}$$

In Chapter 3 several cases of torsion of noncircular bars were considered. In general, the angle of twist is given by

$$\theta = T/C \tag{7.36}$$

where C is the torsional rigidity of the member. The total strain energy is (13)

$$U = \frac{C}{2}\int_0^L \left(\frac{d\theta}{dx}\right)^2 dx \tag{7.37}$$

7-2 ENERGY METHODS IN STRUCTURAL ANALYSIS

In general, energy methods are concise and simple to use (1, 15, 26). On the other hand, energy methods do not convey a physical picture of the process involved and hence do not give as clear an over-all understanding as some other methods.

Conservation of energy

This law states that the energy of a system remains constant if the system is isolated so that it neither receives nor gives out energy. Consider, for example, the tension bar shown in Fig. 7.1. If the load is gradually applied, the work involved is

$$W_{\text{external}} = F\delta/2 \tag{7.38}$$

The internal energy is given by Eq. 7.17 and, in accordance with the law of conservation of energy

$$F\delta/2 = \sigma^2/2E = F^2/2EA^2 = F^2L_0/2EA^2L_0 \tag{7.39}$$

The volume equals AL_0; therefore, per unit volume,

$$\delta = FL_0/AE \tag{7.40}$$

which is the equation for elastic tension loading. Similarly, for an end-loaded cantilever beam (Fig. 7.8*A*), the external work is $F\delta/2$ and the internal work is

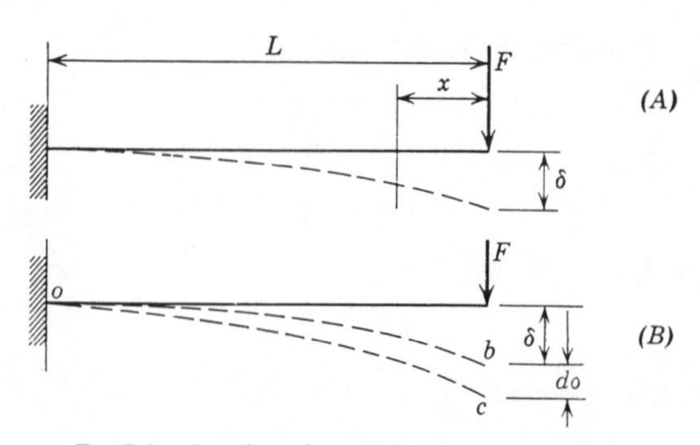

Fig. 7.8 Cantilever beam subjected to an end load.

given by Eq. 7.24, thus

$$\frac{F\delta}{2} = \int_0^L \frac{M^2\,dx}{2EI} = \frac{F^2L^3}{6EI} \qquad 7.41$$

or

$$\delta = FL^3/3EI \qquad 7.42$$

which is the deflection equation for an end-loaded cantilever beam.

Virtual work

The principle of virtual work is closely related to conservation of energy and may be stated as follows: if a structural system is given a small (virtual) deformation, the work done by the external loads during the virtual deformation equals the increase in strain energy of the system. For example, in the cantilever of Fig. 7.8*B*, the line *ob* represents the equilibrium position of the structure when loaded by the force F. If now the beam is given the additional small deformation $d\delta$, a new curve *oc* is obtained and the change in potential energy is

$$dU = F\,d\delta \qquad 7.43$$

from which

$$F = dU/d\delta \qquad 7.44$$

Equation 7.44 is frequently used, as will be shown later, in the calculation of statically indeterminate systems. Rather than a differential deformation, a differential increase in load could have been applied; this results in the inversion of Eq. 7.44 so that

$$\delta = dU/dF \qquad 7.45$$

which, when applied to Eq. 7.41, gives the result

$$\delta = FL^3/3EI \qquad 7.46$$

Other applications of this method are given in the references cited.

Minimum potential energy

The law of minimum potential energy states that of all displacements satisfying the boundary conditions, those that satisfy equilibrium conditions give minimum potential energy. A rigorous proof of this is given by Sokolnikoff (22), and the following simplified analysis illustrates the physical meaning involved (4). Consider a beam loaded by a single force F, like the cantilever in Fig. 7.8, which results in a plot such as shown in Fig. 7.9 in the general case; an elastic beam gives a straight-line plot. At load F_1 the deflection is δ_1. If now the deflection is increased by an amount $d\delta$, the resulting decrease in the potential energy of the load is $-F_1\,d\delta$ (area *AECD*, Fig. 7.9). The resulting increase in strain

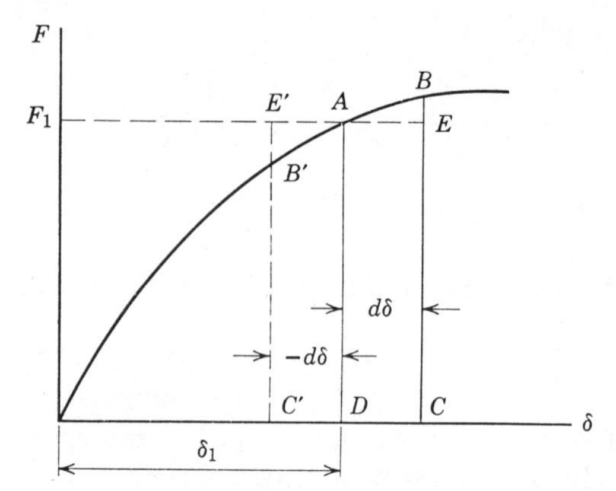

Fig. 7.9 General load-deformation diagram in bending.

energy of the beam is dU (area $ABCD$, Fig. 7.9), and the change in total energy (area ABE) is

$$dU = \text{change in energy of load} + \text{energy in beam}$$

or

$$dU = U_{\text{change}} + U_{\text{beam}} \tag{7.47}$$

If $d\delta$ is small, area ABE tends toward zero and the change in strain energy equals the change of potential energy of the load (principle of virtual work). In addition, if the beam is additionally deformed so that δ decreases by an amount $d\delta$, the area $AB'E'$ is involved. Therefore, since a small variation in δ, either positive or negative, produces a variation in the total energy consisting of positive higher-order differentials, the total energy U at equilibrium must be a minimum. The principles of minimum potential energy will now be used to develop some practical applications.

Going back to Eq. 7.1, suppose that the force F_n is increased a differential amount (Fig. 7.10). Figures 7.2 plus 7.10 then give the result shown in Fig. 7.11, for which the strain energy is

$$U' = U + (\partial U/\partial F_n)\, dF_n \tag{7.48}$$

If the load sequence is altered so that dF_n is applied first and then the other loads, the load dF_n is associated with an energy

$$w_n = dF_n\, d\delta_n/2 \tag{7.49}$$

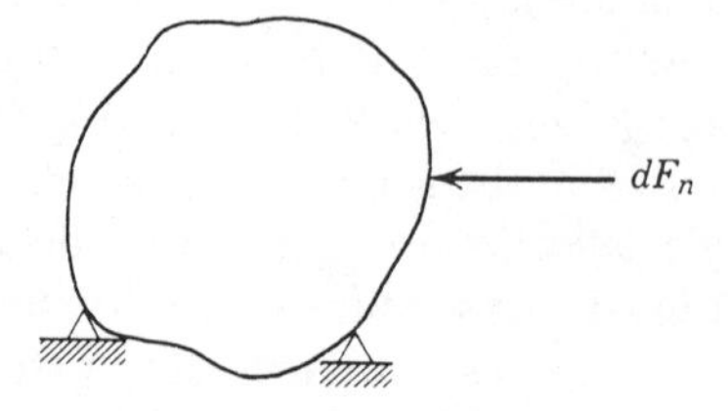

Fig. 7.10 Body subjected to differential load.

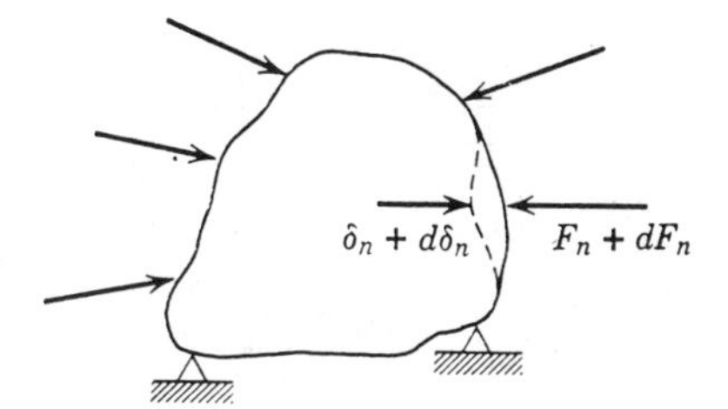

Fig. 7.11 Superposition of Figs. 7.2 and 7.10.

Equation 7.49 is allowable since strain energy does not depend on the order in which the loads are applied. When the rest of the loads are applied, the force of dF_n moves through a distance δ_n; therefore

$$w = (dF_n + dF_n)\delta_n/2 = dF_n\delta_n \qquad 7.50$$

and the total energy is

$$U' = U + dF_n\delta_n + dF_n\,d\delta_n/2 \qquad 7.51$$

In Eq. 7.51 products of differentials can be neglected so that

$$U + (\partial U/\partial F_n)\,dF_n = U + dF_n\delta_n \qquad 7.52$$

from which

$$\delta_n = \partial U/\partial F_n \qquad 7.53$$

Equation 7.53 is known *Castigliano's first theorem* (24) and states that the deflection in the direction of a load is a direct function of strain energy. The inverse of this result was obtained from the principle of virtual work (Eq. 7.44).

Castigliano's theorem. Equation 7.53 states mathematically that the partial derivatives of the total strain energy of any structure with respect to any one of the applied loads is equal to the displacement of the point of application of the force in the direction of the force. Castigliano's works were published around 1870 and have been widely applied in structural engineering work (27). It has already been shown by Eq. 7.53 that deformations may be easily calculated from a knowledge of the elastic strain energy in a structural system; several examples illustrating this are given in the next section. It should be also noted from the integration of Eq. 2.82 that the slope is

$$\theta = ML/EI \qquad 7.54$$

which, when combined with Eq. 7.24, gives the result

$$\theta = \partial U/\partial M \qquad 7.55$$

Equation 7.55 indicates that the slope of a beam can be computed from a knowledge of the strain energy in the system. Similarly, the angle of twist in torsion can be computed from the relationship

$$\theta = \partial U/\partial T \qquad 7.56$$

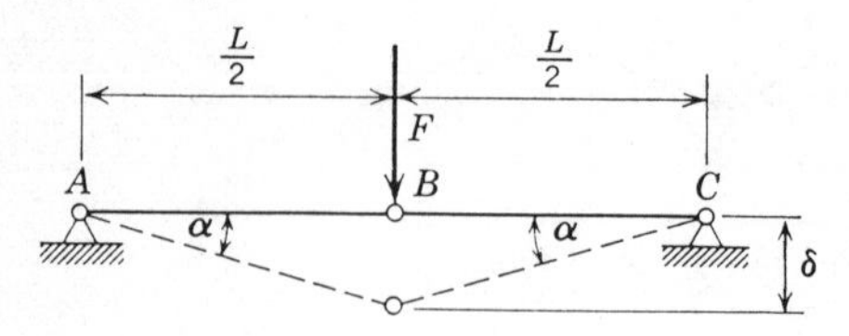

Fig. 7.12 Two-bar pin-jointed structure subjected to central loading.

Therefore, having Eqs. 7.53 to 7.56, it is possible to compute deflections, slopes, and twists of structural members. The examples given later illustrate the application of these relationships to both statically determinate and statically indeterminate structures.

It has already been mentioned, but needs emphasis, that Castigliano's theorem applies only when displacements are proportional to the loads; the central deflection for the hinged structure shown in Fig. 7.12, for example, cannot be computed by the Castigliano theorem. In the system shown, since the members are hinged, there are no bending moments present and the two bars are subjected to tension forces only. In each bar, for small deflections, the strain induced is

$$\varepsilon = \frac{\left(\dfrac{L}{2\cos\alpha} - \dfrac{L}{2}\right)}{L/2} = \frac{1-\cos\alpha}{\cos\alpha} \qquad 7.57$$

from which

$$\varepsilon \approx \alpha^2/2 \qquad 7.58$$

The tensile force in the bars is given by

$$F_t = AE\varepsilon = AE\alpha^2/2 \qquad 7.59$$

From conditions of static equilibrium

$$F = 2\alpha F_t \qquad 7.60$$

Therefore

$$F = AE\alpha^3 \qquad 7.61$$

and the deflection δ is

$$\delta = L(F/AE)^{1/3} \qquad 7.62$$

which is not proportional to F. It is shown presently that the method of complementary energy is applicable to such cases.

Complementary energy

When a material is strained, the load-deformation relationship may assume various forms depending on the nature of the loading and properties of the materials; several examples are shown in Fig. 7.13, which may be for tension,

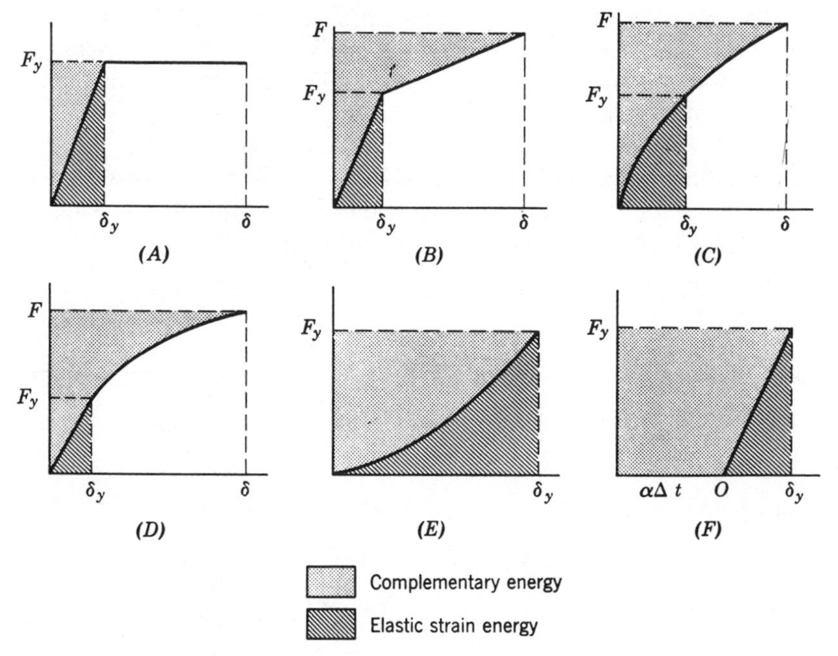

Fig. 7.13 Various types of load-deformation diagrams.

bending, or torsional loading. The elastic strain energy is represented by the area under the curve up to the elastic limit load (hatched area in the figure) and is mathematically defined

$$\text{elastic energy} = \int F\, d\delta \tag{7.63}$$

The area of the plot enclosed by the curve and the load axis (dotted area in the figure) is called *complementary energy* and is defined as

$$\text{complementary energy} = \int \delta\, dF \tag{7.64}$$

As can be seen, the sum of the two energies defined by Eqs. 7.63 and 7.64 equals $F\delta$.

The basic laws concerning complementary energy were given in 1889 (7); Engesser extended Castigliano's theorem to nonlinear behavior. A rigorous proof of the method is given by Sokolnikoff (22) and discussions concerning applications to structural problems are given in (13 and 27). The method has not been used much in structural analysis principally because most computations have been concerned only with elastic linear behavior. Furthermore, the development of the theory of plasticity and limit design (see Chapter 6) has provided means for calculating structural behavior that are frequently easier to apply than complementary energy. Perhaps the greatest application of complementary energy in recent times has been to nonlinear mechanics associated with problems of thermal stress analysis (9) which is discussed in Chapter 15. For present purposes, however, to illustrate what the method can do, the following examples are given.

***Deflection of structure in Fig.* 7.12.** The structure shown in Fig. 7.12 has a load-deformation diagram of the type shown in Fig. 7.13E; for this particular case the curve is elliptic and the area enclosed by it and the load axis is approximately $3F\delta/4$, which means that the complementary energy is

$$U_c = \tfrac{3}{4}F\delta \qquad 7.65$$

From Eq. 7.62

$$\delta = L(F/AE)^{1/3} \qquad 7.66$$

Therefore Eq. 7.65 is rewritten as

$$U_c = \tfrac{3}{4}FL(F/AE)^{1/3} \qquad 7.67$$

from which

$$U_c = \tfrac{3}{4}LF^{4/3}/(AE)^{1/3} \qquad 7.68$$

However,

$$\delta = \partial U_c/\partial F \qquad 7.69$$

Substitution of Eq. 7.68 into Eq. 7.69 and differentiating gives

$$\delta = L(F/AE)^{1/3} \qquad 7.70$$

which is the same result obtained by the geometrical method (Eq. 7.62).

***Deflection of structure in Fig.* 6.25.** For purposes of comparing results of complementary energy theory and limit design consider the simply supported beam in Fig. 6.25. Limit analysis shows (see Table 6.3) that the center deflection for this beam at the instant of plastic collapse is

$$\delta = (\sigma_y/4)(L^2/Eh) = F_yL^3/32EI \qquad 7.71$$

since the material follows the behavior expressed by the stress-strain diagram of Fig. 6.14. In terms of complementary energy

$$U_c = F_c\delta_c/2 \qquad 7.72$$

However, δ_y, the deflection at the elastic limit for the beam, is given by

$$\delta_y = F_yL^3/48EI \qquad 7.73$$

Equation 7.72 becomes

$$U_c = \frac{F_c}{2}\left(\frac{F_yL^3}{32EI}\right) \qquad 7.74$$

In Chapter 6 it was also shown that in a beam yielding at constant stress

$$F_c = \tfrac{3}{2}F_y \qquad 7.75$$

where F_c is the *collapse* load and F_y is the *elastic limit* load; using this relationship in Eq. 7.74 gives the result

$$U_c = F_c^2 L^3/96\,EI \qquad 7.76$$

Finally, application of Eq. 7.69 shows that

$$\delta_c = \partial U_c/\partial F_c = F_c L^3/48EI \qquad 7.77$$

But again, using Eq. 7.75, Eq. 7.77 becomes

$$\delta_c = (F_y L^3/48EI)(\tfrac{3}{2}) = F_y L^3/32EI \qquad 7.78$$

which checks Eq. 7.71. If the load-deflection curve was parabolic as shown in Fig. 7.13*C*,

$$U_c = \tfrac{1}{3} F_c \delta_c \qquad 7.79$$

The elastic deflection is given by Eq. 7.73 and the relation between F_c and F_y is given by Eq. 7.75. The deflection δ_c, however, for a parabola is related to the elastic deflection by the formula

$$\delta_c = \delta_y (F_c/F_y)^2 \qquad 7.80$$

By using the preceding relationships

$$U_c = \frac{F_c}{3}\delta_y\left(\frac{F_c}{F_y}\right)^2 = \frac{F_c}{3}\left(\frac{9}{4}\right)\left(\frac{F_y L^3}{48EI}\right) = \frac{F_c^2 L^3}{96EI} \qquad 7.81$$

and finally

$$\delta_c = \frac{\partial U}{\partial F_c} = \frac{F_c L^3}{48EI} = \frac{F_y L^3}{32EI} \qquad 7.82$$

The stress-strain relationship shown in Fig. 7.13*F* is characteristic of problems in which thermal expansion have to be allowed for; this is discussed in Chapter 15.

7-3 APPLICATION OF STRAIN-ENERGY THEORY

In this section attention is directed toward the solution of some machine design and structural problems, of which all make use of the theory of elastic-strain energy and Castigliano's theorem. Solution to many other problems are given in (17), Chapter 2, and in the references cited.

Analysis of tension loading

Tensile bar. The calculation of simple tensile elongation has already been described. Suppose the strain δ for the tension bar in Fig. 7.1 is to be calculated by the strain-energy method. In order to do this it is noted from Eq. 7.53 that

$$\delta = \partial U/\partial F \qquad 7.83$$

The strain energy in tension is given by Eq. 7.16; therefore from these two relationships

$$\delta = \frac{\partial}{\partial F}\left(\frac{F^2L}{2AE}\right) = \frac{FL}{AE} \qquad 7.84$$

The result expressed by Eq. 7.84 simply illustrates the mechanics involved; the same result was obtained in Chapter 2 (Eq. 2.4) by somewhat simpler means.

Simple frames. Consider the frame shown in Fig. 2.36 which is statically indeterminate. If the force in bar OC is X and the force in the bars OB and OD is Y, then

$$F = X + 2Y\cos\alpha \qquad 7.85$$

Since the bars are all hinged, there are no bending stresses and the strain energy can be computed from Eq. 7.16,

$$U = \frac{F^2L}{2AE} = \frac{X^2L}{2A_cE_c} + \frac{2Y^2L}{2A_BE_B\cos\alpha} \qquad 7.86$$

or

$$U = \frac{X^2L}{2A_cE_c} + \left(\frac{F-X}{2\cos\alpha}\right)^2\left(\frac{L}{A_BE_B\cos\alpha}\right) \qquad 7.87$$

Since the principle of least work states that $\partial U/\partial X = 0$, Eq. 7.87 shows that

$$0 = \frac{XL}{A_cE_c} + \frac{L}{A_BE_B\cos\alpha}\left(\frac{1}{4\cos^2\alpha}\right)(2X - 2F) \qquad 7.88$$

from which

$$X = \frac{F}{1 + \dfrac{2A_BE_B\cos^3\alpha}{A_cE_c}} \qquad 7.89$$

The deflection at the load F is found from the expression

$$\delta_F = \frac{\partial U}{\partial F} = \frac{FL}{2A_BE_B\cos^3\alpha} - \frac{X}{2A_BE_B\cos^3\alpha} \qquad 7.90$$

or using Eq. 7.89

$$\delta_F = \frac{FL}{A_cE_c + 2A_BE_B\cos^3\alpha} \qquad 7.91$$

It is evident that the results of Eqs. 7.89 and 7.91 agree with the values calculated by geometrical means in Chapter 2.

As another example consider the system shown in Fig. 7.14 consisting of two bars of equal length L and cross-sectional area A, loaded by a single force F at location A. By simple statics the force in bars AB and AC is F. The vertical

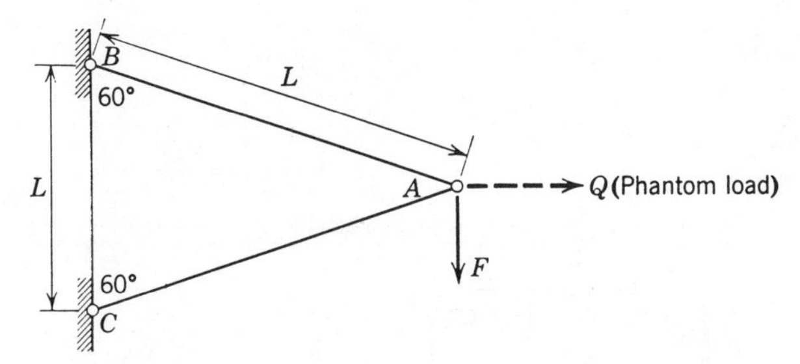

Fig. 7.14 Frame structure subjected to a single vertical load F.

displacement of load F is then

$$\delta_v = \frac{\partial U}{\partial F} = \frac{\partial}{\partial F}\left(\frac{2F^2L}{2AE}\right) = \frac{2FL}{AE} \qquad 7.92$$

In order to find the horizontal movement of point A it is necessary to add a *phantom* load Q in the horizontal plane at A since Castigliano's theorem can be used for a deflection calculation only when the load is applied in the direction of the deflection; the horizontal deflection is thus

$$\delta_H = \frac{\partial U}{\partial Q} = \frac{\partial}{\partial Q}\left[\frac{(F + Q/\sqrt{3})^2 L}{2AE} + \frac{(-F + Q/\sqrt{3})^2 L}{2AE}\right] = 0 \qquad 7.93$$

In arriving at the result given by Eq. 7.93 it was necessary to perform the indicated differentiations; after this operation was completed the phantom load, which is actually zero, was set equal to zero, giving the value of horizontal deflection.

Another type of simple tension frame is shown in Fig. 7.15, consisting of three bars of equal length and cross-sectional area loaded by a single force F. Since this system is statically indeterminate, it is assumed that the force in bar BD equals X; by statics the forces in bars AB and BC are each $-(F - X)/\sqrt{2}$. In order to find the value of the unknown force X it is noted that the principle of minimum work is applicable; that is,

$$\frac{\partial U}{\partial X} = 0 = \frac{\partial}{\partial X}\left[\frac{X^2 L}{2AE} + \frac{(F - X)^2 L}{2AE}\right] \qquad 7.94$$

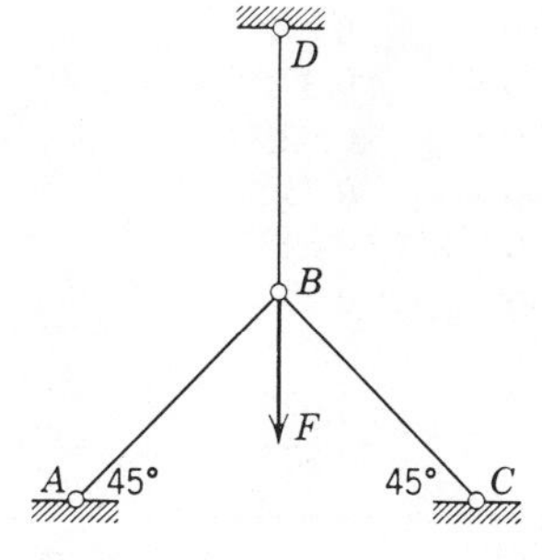

7.15 Three-bar pin-jointed structure subjected to a single vertical load F.

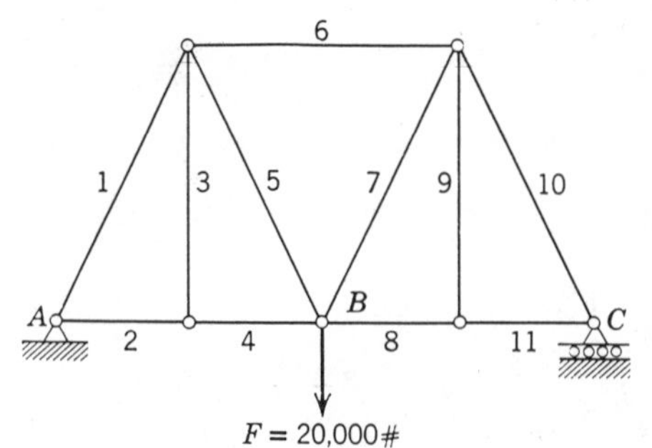

Fig. 7.16 Steel truss subjected to center loading.

from which

$$X = F/2 \tag{7.95}$$

The vertical deflection of point B is

$$\delta_v = \frac{\partial U}{\partial F} = \frac{\partial}{\partial F}\left[\frac{X^2L}{2AE} + \frac{(F-X)^2L}{2AE}\right] = -\frac{XL}{AE} \tag{7.96}$$

or using Eq. 7.95

$$\delta_v = -FL/2AE \tag{7.97}$$

Trusses. Elastic strain-energy methods are particularly useful in determining deflections of trusses. In the following examples all joints are pin-connected so that there is no bending; it is assumed further that no twisting occurs. Consider the truss shown in Fig. 7.16; each member is numbered and application of Eq. 2.55 shows that the problem is statically determinate. Pertinent data concerning the truss are given in Table 7.1. Since the various members are

Table 7.1 Data for Truss Shown in Fig. 7.16

Member	Length, in.	Area, sq in.	Load, lb	K (unit load)
1	200	3.0	−12,500	−0.625
2	120	1.5	7,500	0.375
3	160	1.0	0	0
4	120	1.5	7,500	0.375
5	200	1.0	12,500	0.625
6	240	2.0	−15,000	−0.750
7	200	1.0	12,500	0.625
8	120	1.5	7,500	0.375
9	160	1.0	0	0
10	200	3.0	−12,500	−0.625
11	120	1.5	7,500	0.375

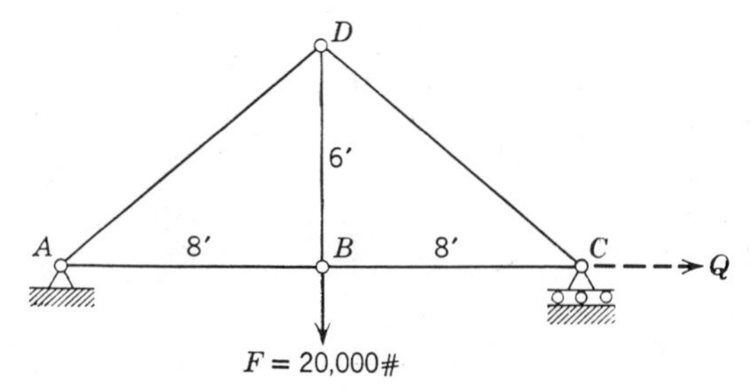

Fig. 7.17 Steel truss subjected to center loading. All members are made of steel bars having cross-sectional areas of 2 in².

axially loaded, the strain energy in the system is the summation of the strain energies of the individual bars; thus

$$U = \sum\left(\frac{F^2 L}{2AE}\right) \tag{7.98}$$

where $\sum$ denotes a summation and E equals 30×10^6 psi for a steel truss. The deflection at the point of application of the load is

$$\delta_v = \frac{\partial U}{\partial F_B} = \sum\frac{FL}{AE}\frac{\partial F}{\partial F_B} \tag{7.99}$$

where $\partial F/\partial F_B$ is the rate of increase of force F with increase in load F_B. Numerically it is equal to the force in any bar by unit load at the F_B position and is designated by K in Table 7.1. Thus at B

$$\delta_B = \frac{\sum(FLK)}{AE} \tag{7.100}$$

or

$$\delta_B = 0.214 \text{ in.} \tag{7.101}$$

If the deflection at other locations is desired, the same procedure as just outlined is followed. However, since by these methods the deflection can be determined only in the direction of the load and in the loading plane, a *phantom* load is added as explained in a previous example. For example, in the truss shown in Fig. 7.17 the same method of analysis is used as for the previous example, and here the deflection at B, vertically, is 0.108 in. and the horizontal deflection at C is 0.0418 in. See also (6).

Analysis of direct shear loading

Consider Fig. 7.18 which shows a beam of rectangular cross section in which transverse shear is a significant factor. For the geometry shown the shear stress

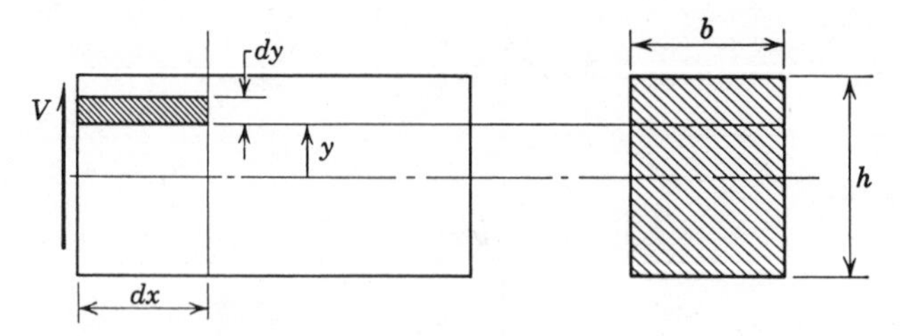

Fig. 7.18 Beam of rectangular cross section with transverse shear.

generated by the transverse shear V is given by Eq. 3.207

$$\tau = \frac{V}{Ib} a' \bar{y} = \frac{V}{2I}\left(\frac{h^2}{4} - y^2\right) \tag{7.102}$$

The strain energy per unit volume is given by Eq. 7.10, and the strain energy for the element $b\,dx\,dy$ in Fig. 7.18 is

$$dU = wb\,dx\,dy \tag{7.103}$$

or

$$dU = (\tau^2/2G)b\,dx\,dy \tag{7.104}$$

From Eqs. 7.102 and 7.104

$$dU = \left[\frac{V}{2I}\left(\frac{h^2}{4} - y^2\right)\right]^2 \frac{b}{2G} dx\,dy \tag{7.105}$$

and integrating

$$U = \int dU = \int\int \frac{V^2}{8GI^2}\left(\frac{h^2}{4} - y^2\right)^2 b\,dx\,dy \tag{7.106}$$

In order to evaluate the integral of Eq. 7.106 it is necessary to specify a particular problem; let that problem be an end-loaded cantilever beam such as shown in Fig. 7.19. For a rectangular cross section, from Eq. 7.106.

$$U = \int_0^L \int_{-h/2}^{+h/2} \frac{F^2}{8GI^2}\left(\frac{h^2}{4} - y^2\right)^2 b\,dx\,dy = \frac{F^2Lh^2}{20GI} \tag{7.107}$$

Consider now the maximum deflection of this beam. When the effect of shear is not included in the analysis, the end deflection δ_{max} is calculated from Eq. 7.53 where U is defined by Eq. 7.24, thus,

$$\delta_{max} = \frac{\partial U}{\partial F} = \frac{\partial}{\partial F}\int_0^L \frac{M^2\,dx}{2EI} = \int_0^L M\frac{\partial M}{\partial F}\frac{dx}{EI} = \int_0^L \frac{(Fx)(x)\,dx}{EI} \tag{7.108}$$

or

$$\delta_{max} = \frac{FL^3}{3EI} \tag{7.109}$$

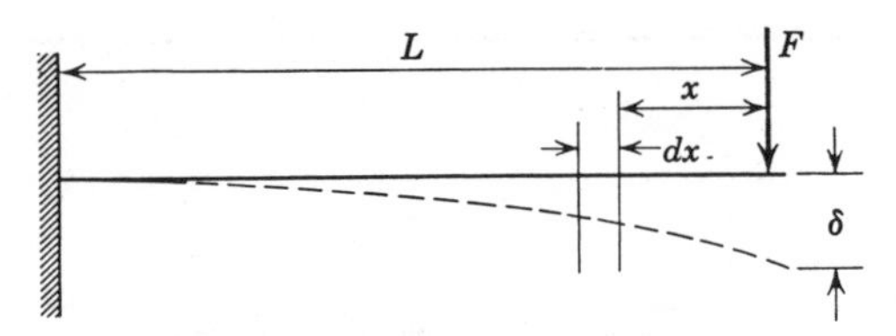

Fig. 7.19 Cantilever beam with end load.

which is the elastic solution. If the effect of shear is included in the analysis

$$U_{\text{total}} = \text{Eq. 7.24} + \text{Eq. 7.107}$$

from which

$$\delta_{\max} = \frac{\partial U}{\partial F}\left(\int_0^L \frac{M^2\,dx}{2EI} + \frac{F^2Lh^2}{20GI}\right) = \frac{FL^3}{3EI}\left[1 + \frac{3}{10}\left(\frac{E}{G}\right)\left(\frac{h}{L}\right)^2\right] \quad 7.110$$

In Eq. 7.110 the second term in the bracket represents a correction to the conventional solution. A design chart showing this effect is given in Fig. 7.20. As would be expected the largest corrections are involved for short beams where h/L is high; see also, for example, Fig. 3.48.

Analysis of torsion loading

Deflection of spring. Consider the close-coil helical spring shown in Fig. 3.65. In this type of spring bending stresses are negligible and the deflection is due almost entirely to the torsional stresses induced in the coil. The strain energy for the coil is determined from Eq. 7.34. The deflection in the direction of the applied load F is thus

$$\delta = \frac{\partial U}{\partial F} = \frac{\partial}{\partial F}\int_0^L \frac{T^2\,dx}{2GJ} = \int_0^L T\frac{\partial T}{\partial F}\frac{dx}{GJ} \quad 7.111$$

or in terms of angular distortion

$$\delta = \int_0^{2\pi} \frac{FD}{2}\left(\frac{D}{2}\right)\left(\frac{D\,d\theta}{2}\right)\frac{n}{GJ} = \frac{FD^3n\pi}{4GJ} \quad 7.112$$

Since $J = \pi d^4/32$ for the wire in the coil, Eq. 7.112 becomes

$$\delta = 8nFD^3/Gd^4 \quad 7.113$$

Torsion of rod. Results for a torsional analysis of the system shown in Fig. 1.44 are given by Eqs. 1.118 to 1.120. The same results may be obtained by the strain-energy method by using Eq. 7.56. The total strain energy of the system is

$$U = \frac{(T_a)^2a}{2GJ} + \frac{(T_a - T_1)^2b}{2GJ} + \frac{(T_a - T_1 - T_2)^2c}{2GJ} \quad 7.114$$

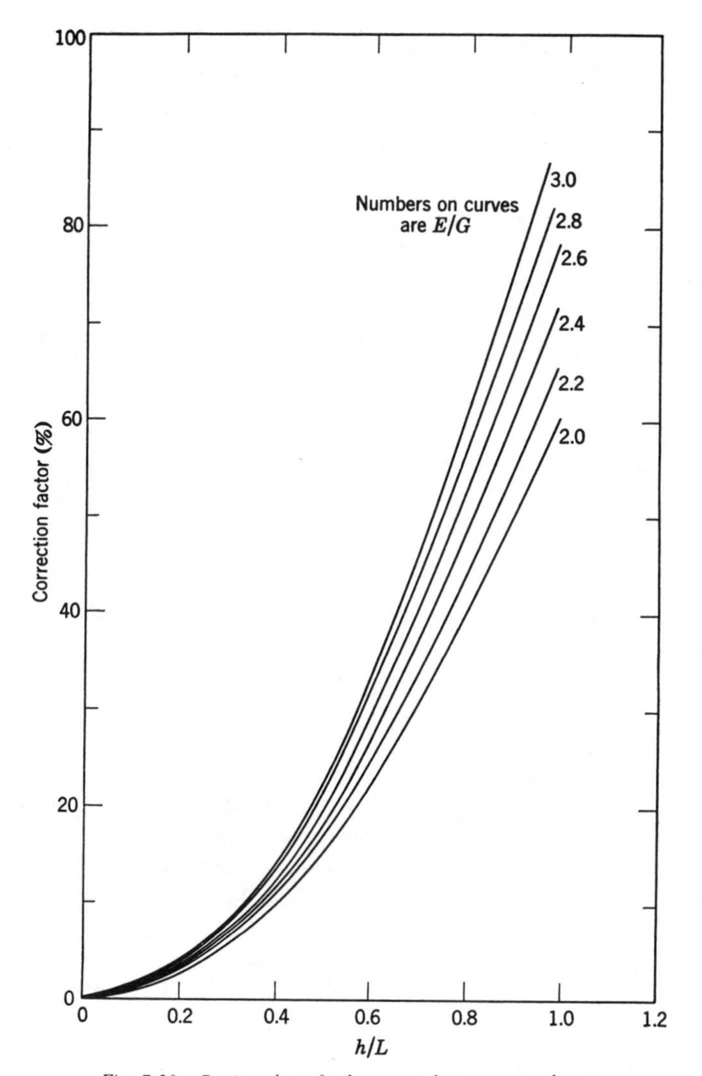

Fig. 7.20 Design chart for beams with transverse shear.

At the left-hand support the angle of twist is zero so that

$$\theta = \partial U / \partial T = 0 \tag{7.115}$$

Thus by substituting Eq. 7.114 into Eq. 7.115

$$0 = T_a a + T_a b - T_1 b + T_a c - T_1 c - T_2 c \tag{7.116}$$

or

$$T_a = \frac{T_1(b + c) + T_2 c}{L} \tag{7.117}$$

The other values are obtained in a similar manner.

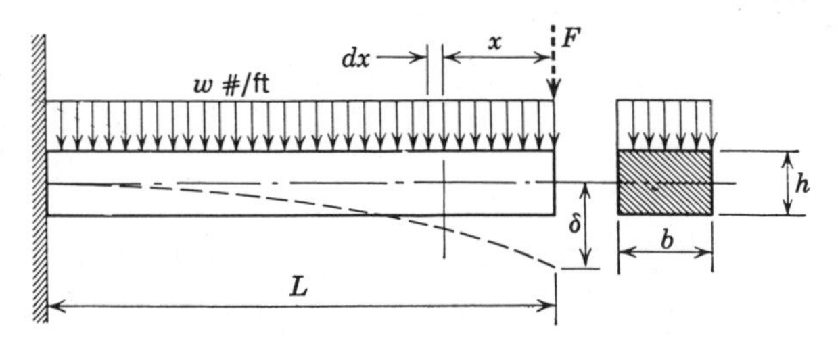

Fig. 7.21 Uniformly loaded cantilever beam.

Analysis of bending

Deflection of beams. It was shown in the preceding example of an end-loaded cantilever beam (Fig. 7.19) that the deflection could be calculated very easily by the strain-energy method; the result was given by Eq. 7.109. Suppose now that the cantilever beam is uniformly loaded as shown in Fig. 7.21. In order to calculate the end deflection it is necessary to add a *phantom* load F at the end of the beam; when this is done, the end deflection is

$$\delta = \frac{\partial U}{\partial F} = \frac{\partial}{\partial F}\int_0^L \frac{M^2\,dx}{2EI} = \int_0^L M\frac{\partial M}{\partial F}\frac{dx}{EI} \qquad 7.118$$

or

$$\delta = \int_0^L \left[wx\left(\frac{x}{2}\right) + Fx\right](x)\frac{dx}{EI} = \frac{1}{EI}\left(\frac{wL^4}{8} + \frac{FL^3}{3}\right) \qquad 7.119$$

The phantom load F, however, is zero; therefore Eq. 7.119 becomes

$$\delta = wL^4/8EI \qquad 7.120$$

which is recognized as the standard elastic solution. Consider now in Fig. 7.22 the cantilever beam with an end moment, an end load, and also uniformly loaded. The end deflection will be

$$\delta = \frac{\partial U}{\partial F} = \frac{\partial}{\partial F}\int_0^L \frac{M^2\,dx}{2EI} = \int_0^L M\frac{\partial M}{\partial F}\frac{dx}{EI} \qquad 7.121$$

or

$$\delta = \int_0^L \left(-Fx - \frac{wx^2}{2} - M_1\right)(-x)\frac{dx}{EI} = \frac{1}{EI}\left(\frac{FL^3}{3} + \frac{wL^4}{8} + \frac{M_1L^2}{2}\right) \qquad 7.122$$

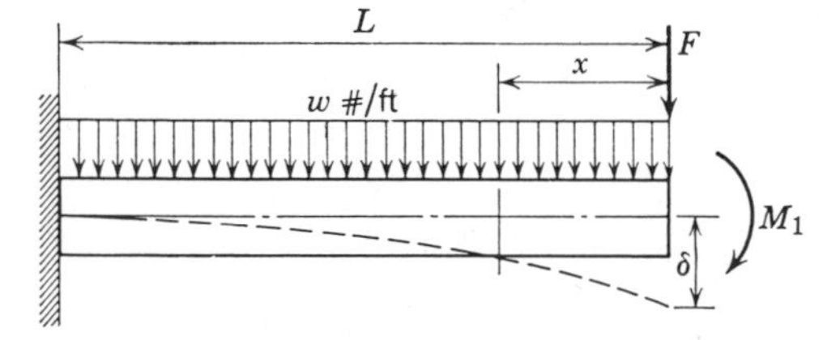

Fig. 7.22 Cantilever beam subjected to uniform loading, end load, and end moment.

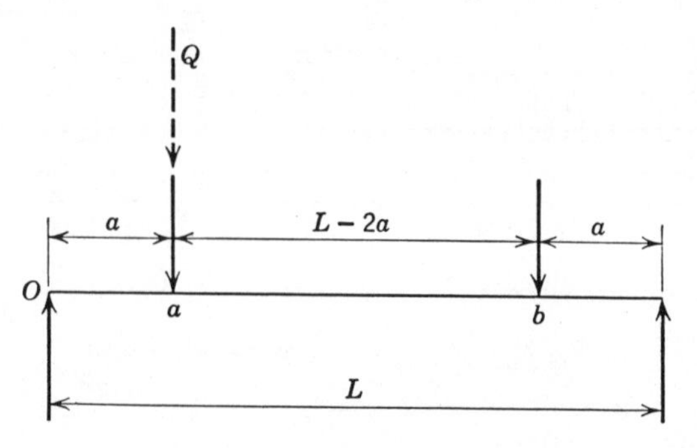

Fig. 7.23 Beam subjected to pure bending.

The slope at the end of the beam is given by Eq. 7.55, that is,

$$\theta = \frac{\partial U}{\partial M_1} = \frac{\partial}{\partial M_1}\int_0^L \frac{M^2\,dx}{2EI} = \int_0^L M\frac{\partial M}{\partial M_1}\frac{dx}{EI} \qquad 7.123$$

or

$$\theta = \int_0^L \left(-Fx - \frac{wx^2}{2} - M_1\right)(-1)\frac{dx}{EI} = \frac{1}{EI}\left(\frac{FL^2}{2} + \frac{wL^3}{6} + M_1 L\right) \qquad 7.124$$

As another example of beam deflection, consider the system shown in Fig. 7.23; this is a representation of pure bending, and the deflections at the points of loading and at midspan will be computed. Since the deflection can be calculated only for one load, it is necessary to add a phantom load Q as shown in Fig. 7.23; the deflection under the load P is then

$$\delta_p = \frac{\partial U}{\partial Q} = \frac{\partial}{\partial Q}\int_0^L \frac{M^2\,dx}{2EI} = \frac{1}{EI}\int_0^L M\frac{\partial M}{\partial Q}\,dx \qquad 7.125$$

or

$$\delta_p = \frac{1}{EI}\left\{\int_0^a \left(P + \frac{Q}{L}(L-a)x\right)\frac{\partial M_{ac}}{\partial Q}\,dx\right.$$
$$+ \int_a^b \left\{\left[P + \frac{Q}{L}(L-a)\right](x) - (P+Q)(x-a)\right\}\frac{\partial M_{ab}}{\partial Q}\,dx$$
$$+ \int_b^L \left\{\left[P + \frac{Q}{L}(L-a)\right](x) - (P+Q)(x-a) - P(x-L+a)\right\}$$
$$\left.\times \frac{\partial M_{bL}}{\partial Q}\,dx\right\} \qquad 7.126$$

from which (after setting Q equal to zero),

$$\delta_p = \frac{Pa^2}{6EI}(3L - 4a) \qquad 7.127$$

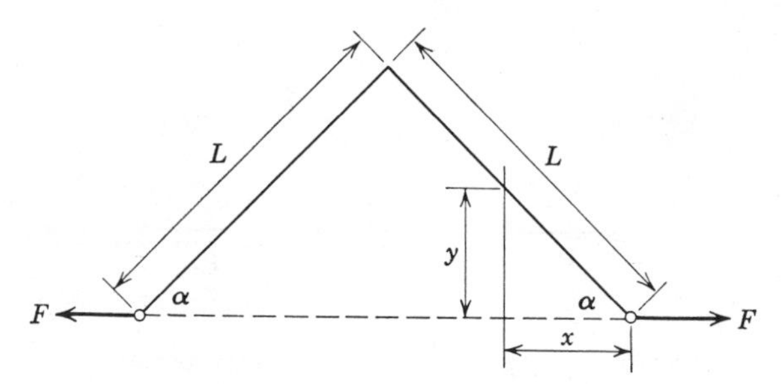

Fig. 7.24 Structure subjected to bending loads.

The midspan deflection for the beam is found by adding a phantom load Q at midspan; then

$$\delta_{L/2} = \frac{\partial U}{\partial Q} = \frac{\partial}{\partial Q} \int_0^L \frac{M^2\, dx}{2EI} = \frac{1}{EI} \int_0^L M \frac{\partial M}{\partial Q} dx \tag{7.128}$$

or

$$\delta_{L/2} = \frac{2}{EI} \left\{ \int_0^a \left(P + \frac{Q}{2}\right) x \left(\frac{x}{2}\right) dx + \int_a^{L/2} \left[\left(P + \frac{Q}{2}\right)(x) - Px + Pa\right] \left(\frac{x}{2}\right) dx \right\} \tag{7.129}$$

from which

$$\delta_{L/2} = (Pa/24EI)(3L^2 - 4a^2) \tag{7.130}$$

Suppose a structural member has the shape shown in Fig. 7.24; in this system the two bars are made of the same material, have the same length, and same cross-sectional area. The horizontal deflection is given by

$$\delta_F = \frac{\partial U}{\partial F} = \frac{\partial}{\partial F} \int_0^{2L} \frac{M^2\, dx}{2EI} = \frac{2}{EI} \int_0^L M \frac{\partial M}{\partial F} dx \tag{7.131}$$

or

$$\delta_F = \frac{2}{EI} \int_0^L Fy(y)\, dx = \frac{2}{EI} \int_0^L Fx^2 \sin^2 \alpha\, dx = \frac{2}{3} \frac{FL^3 \sin^2 \alpha}{EI} \tag{7.132}$$

Rings and loops. A number of geometrically simple structural members are made in the form of rings and loops which will now be considered (2, 6, 8, 10, 15). For the semicircular ring shown in Fig. 7.25A, for example, the vertical deflection at B is given by

$$\delta_B = \frac{\partial U}{\partial F} = \frac{\partial}{\partial F} \int_0^\pi \frac{M^2\, dx}{2EI} = \frac{1}{EI} \int_0^\pi M \frac{\partial M}{\partial F} dx \tag{7.133}$$

The bending moment M for the ring is

$$M = FR(1 - \cos\theta) \tag{7.134}$$

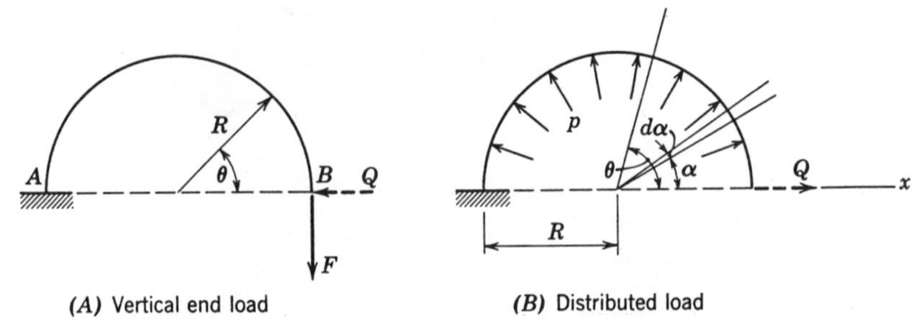

(A) Vertical end load (B) Distributed load

Fig. 7.25 Semicircular ring subjected to applied loads.

Therefore, using Eq. 7.134 in Eq. 7.133,

$$\delta_B = \frac{1}{EI}\int_0^{\pi} [FR(1-\cos\theta)]R(1-\cos\theta)(R\,d\theta)$$

or

$$\delta_B = 3FR^3\pi/2EI \tag{7.135}$$

In order to find the horizontal deflection at B a phantom load Q must be added.

The moment for the ring is then

$$M = FR(1-\cos\theta) + QR\sin\theta \tag{7.136}$$

and the deflection is

$$\delta_H = \frac{\partial U}{\partial Q} = \frac{\partial}{\partial Q}\int_0^{\pi} \frac{M^2 R\,d\theta}{2EI} = \frac{1}{EI}\int_0^{\pi} MR\frac{\partial M}{\partial Q}\,d\theta$$

from which (after letting Q equal zero)

$$\delta_H = 2FR^3/EI \tag{7.137}$$

If, as shown in Fig. 7.25B, the loop is subjected to a distributed load p, the

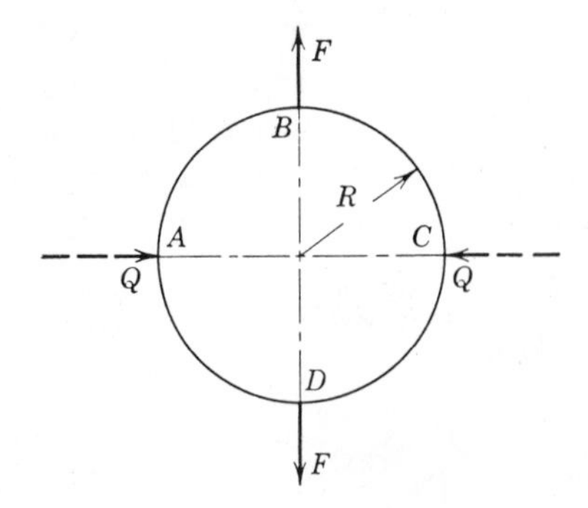

Fig. 7.26 Circular ring subjected to diametral loading.

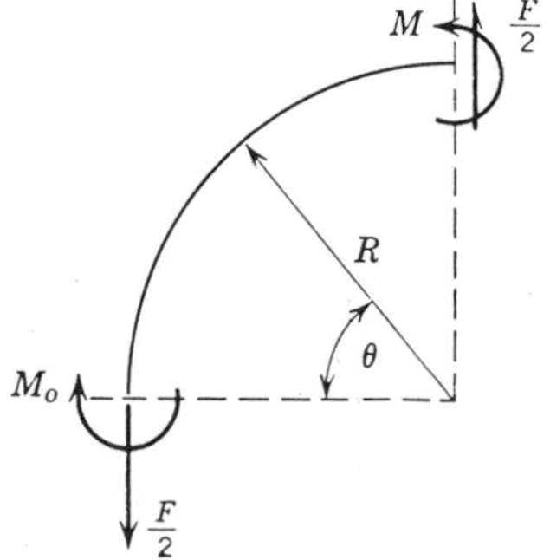

Fig. 7.27 Free body for structure shown in Fig. 7.26.

deflection in the x direction is

$$\delta_x = \frac{\partial U}{\partial Q} = \frac{\partial}{\partial Q}\int_0^{\pi} \frac{M^2\,ds}{2EI} = \frac{1}{EI}\int_0^{\pi} M\frac{\partial M}{\partial Q}\,ds = \frac{R}{EI}\int_0^{\pi} M\frac{\partial M}{\partial Q}\,d\theta$$

However,

$$M = QR\sin\theta + \int_0^{\theta} (pR\,d\alpha)R\sin(\theta-\alpha) \qquad 7.138$$

which on integrating and setting $Q = 0$ gives

$$\delta_x = 2pR^4/EI \qquad 7.139$$

Consider the circular ring shown in Fig. 7.26. This ring is symmetrically loaded by forces F along a diameter causing the diameter to increase in the vertical direction and to decrease in the horizontal direction. Since the structure is symmetrical, only one quadrant need be considered, the free body of which is shown in Fig. 7.27. The moment M_0 is statically indeterminate, but it can be computed using Eq. 7.55 since, by symmetry, there is no rotation at A. At any location defined by θ the bending moment is given by

$$M = M_0 - (F/2)(R - R\cos\theta) \qquad 7.140$$

From Eq. 7.55,

$$\theta = 0 = \frac{\partial U}{\partial M_0} = \frac{\partial}{\partial M_0}\int_0^{\pi/2} \frac{M^2\,ds}{2EI} = \frac{1}{EI}\int_0^{\pi/2} M\frac{\partial M}{\partial M_0}\,d\theta \qquad 7.141$$

Substituting Eq. 7.140 into Eq. 7.141 and performing the indicated differentiation and integration gives

$$0 = \int_0^{\pi/2}\left[M_0 - \frac{FR}{2}(1-\cos\theta)\right]d\theta \qquad 7.142$$

from which

$$M_0 = (FR/2)(1 - 2/\pi) \qquad 7.143$$

Combining Eqs. 7.140 and 7.143 gives

$$M = (FR/2)(1 - 2/\pi) - (FR/2)(1 - \cos\theta) \qquad 7.144$$

the maximum value of which is for $\theta = \pi/2$, the point of application of the load, or

$$M_{max} = (FR/2)(0 - 2/\pi) \qquad 7.145$$

The radial deflection at B in Fig. 7.26 is then

$$\delta_B = \frac{\partial U}{\partial F} = \frac{\partial}{\partial F}\left[2\int_0^{\pi/2} \frac{M^2 R\, d\theta}{2EI}\right] = \frac{2}{EI}\int_0^{\pi/2} M\frac{\partial M}{\partial F} R\, d\theta \qquad 7.146$$

Substituting Eq. 7.144 into Eq. 7.146 gives

$$\delta_B = \frac{2}{EI}\int_0^{\pi/2} \frac{FB}{2}\left(\cos\theta - \frac{2}{\pi}\right)\left(\frac{R}{2}\right)\left(\cos\theta - \frac{2}{\pi}\right) R\, d\theta \qquad 7.147$$

or

$$\delta_B = (FR^3/EI)(\pi/8 - 1/\pi) \qquad 7.148$$

The total diametral deformation is $2\delta_B$ or

$$2\delta_B = (FR^3/EI)(\pi/4 - 2/\pi) \qquad 7.149$$

The decrease in diameter AC is found by adding a phantom load Q as shown in Fig. 7.26. The moment in the ring is then

$$M = M_0 + (QR/2)\sin\theta - (FR/2)(1 - \cos\theta) \qquad 7.150$$

and by the same procedure as outlined

$$\delta_{AC} = (FR^3/EI)(2/\pi - \tfrac{1}{2}) \qquad 7.151$$

Consider now the loop structure shown in Fig. 7.28. As in the previous problem only one-quarter of the structure need be considered in the analysis because of symmetry. The free body is shown in Fig. 7.29. The moment M_0 is statically indeterminate and may be computed from Eq. 7.55. Thus, as in the previous

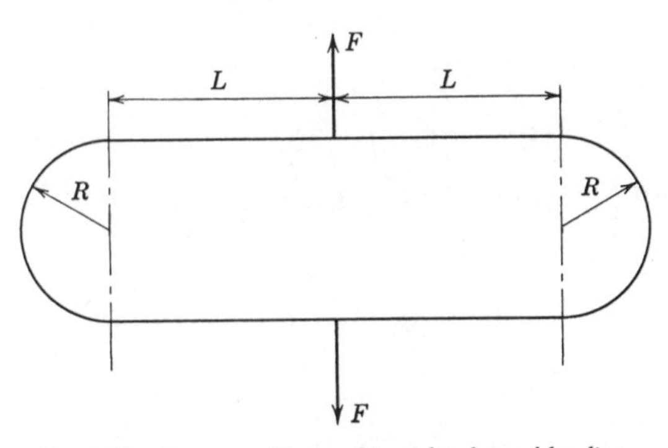

Fig. 7.28 Compound loop subjected to lateral loading.

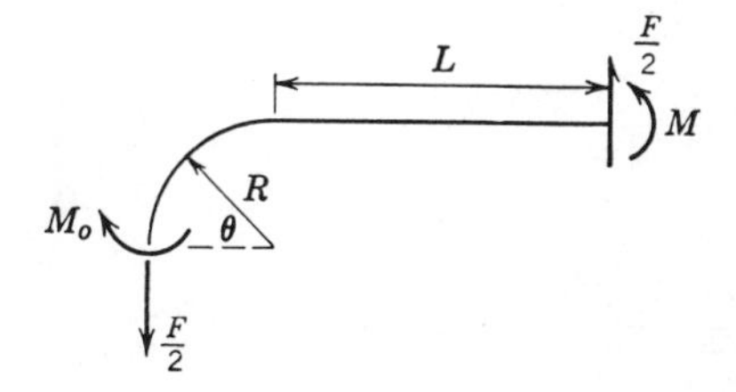

Fig. 7.29 Free body for structure shown in Fig. 7.28.

problem, application of Eqs. 7.24 and 7.55 gives

$$0 = M_0 R\frac{\pi}{2} - \frac{FR^2}{2}\left(\frac{\pi}{2}\right) + \frac{FR^2}{2} + M_0 L - \frac{FRL}{2} - \frac{FL^2}{4} \qquad 7.152$$

from which

$$M_0 = \frac{F}{2}\left[\frac{R^2(\pi - 2) + 2RL + L^2}{\pi R + 2L}\right] \qquad 7.153$$

Having the statically indeterminate quantity M_0, the total strain energy for the system can be computed and deflections obtained.

Another example of a ring structure is shown in Fig. 7.30; the circular ring of radius R is loaded by three forces F, and by symmetry only half of the ring has to be considered as shown in Fig. 7.31. The bending moment in arc AB is given by

$$M_{AB} = M_0 + HR(1 - \cos\phi) \qquad 7.154$$

and the moment in arc BC is given by

$$M_{BC} = M_0 + HR(1 - \cos\phi) - (FR/2\cos\alpha)(\cos\alpha\sin\phi - \sin\alpha\cos\phi) \qquad 7.155$$

There is no rotation at A; therefore

$$\theta = 0 = \partial U/\partial M_0 \qquad 7.156$$

and the horizontal deflection at A is

$$\delta_A = \partial U/\partial H \qquad 7.157$$

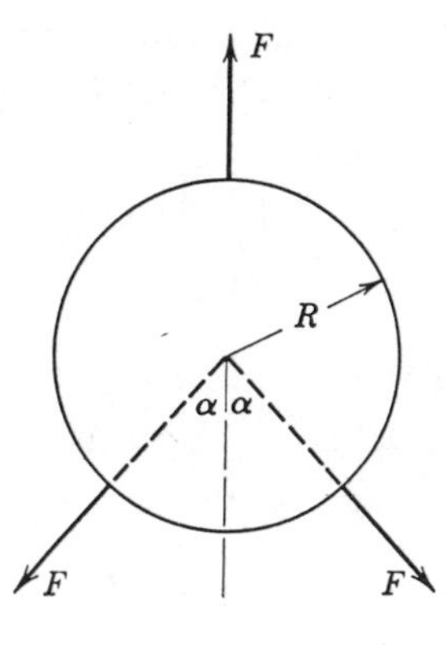

Fig. 7.30 Circular ring subjected to symmetrical radial loads.

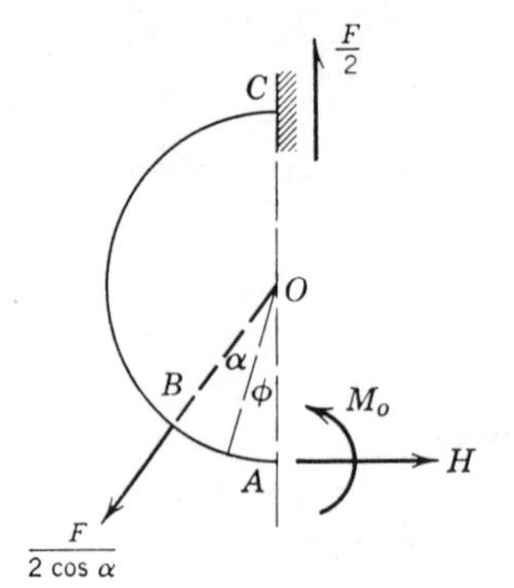

Fig. 7.31 Free body for structure shown in Fig. 7.30.

From Eqs. 7.156 and 7.157

$$0 = \int_A^B M_{AB} \frac{\partial M_{AB}}{\partial M_0} R\, d\phi + \int_B^C M_{BC} \frac{\partial M_{BC}}{\partial M_0} R\, d\phi \tag{7.158}$$

and

$$0 = \int_A^B M_{AB} \frac{\partial M_{AB}}{\partial H} R\, d\phi + \int_B^C M_{BC} \frac{\partial M_{BC}}{\partial H} R\, d\phi \tag{7.159}$$

from which

$$M_0 = (FR/2\pi \cos\alpha)(1 + \cos\alpha) - HR \tag{7.160}$$

and

$$H = (F/2\pi)\,[\tan\alpha(\pi - \alpha)] \tag{7.161}$$

Substituting Eq. 7.161 into Eq. 7.160 gives

$$M_0 = (FR/2\pi)[\sec\alpha + 1 - \tan\alpha(\pi - \alpha)] \tag{7.162}$$

Bends and expansion loops in piping systems are examples of structures that lend themselves to analysis by strain-energy methods. Consider, for example, the loop shown in Fig. 7.32*A* which is made of constant pipe size and is subjected to a temperature differential. At each end of the loop (*A* and *K*) a restraining bending moment M_1 and a force Q are induced whose values will now be determined by strain-energy theory. The movement at *A* and *K* is given by

$$\delta_Q = \frac{\partial U}{\partial Q} = \frac{\partial}{\partial Q}\left[\int_{\substack{BC,\\HG}} \frac{M^2\,dx}{2EI} + \int_{\substack{CD,\\FG}} \frac{M^2\,dx}{2EI} + \int_{\substack{DE,\\EF}} \frac{M^2\,dx}{2EI}\right] \tag{7.163}$$

or

$$\delta_Q = 2\int_0^{\phi} M_{BC} \frac{\partial M}{\partial Q}\frac{dx}{EI} + 2\int_0^{\phi} M_{CD} \frac{\partial M}{\partial Q}\frac{dx}{EI} + M_{DE}\frac{\partial M}{\partial Q}\frac{S}{EI} \tag{7.164}$$

For sections *BC* and *HG*,

$$M = Q(R_1 - R_1\cos\theta) - M_0 = QR_1(1 - \cos\theta) - M_0 \tag{7.165}$$

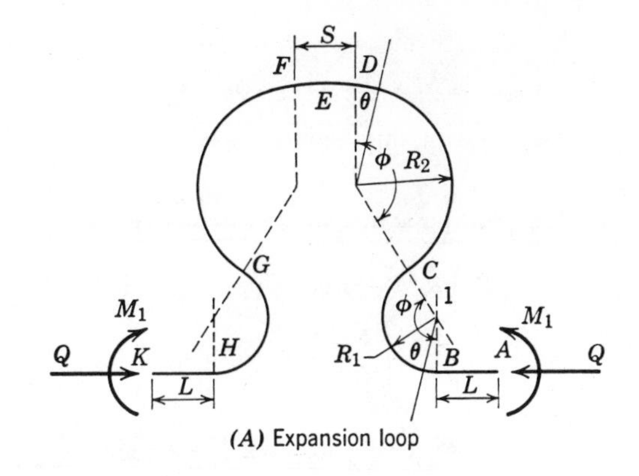

(A) Expansion loop

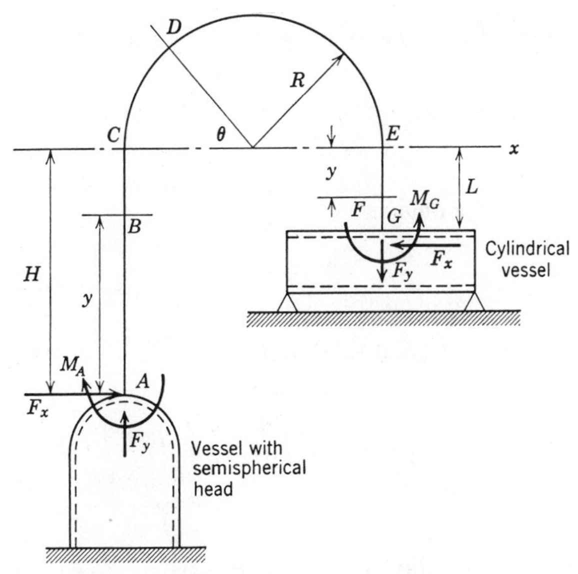

(B) Pipe and pipe bend connecting equipment (See Chapter 15)

Fig. 7.32 Examples of piping structures.

and

$$\partial M/\partial Q = R_1(1 - \cos\theta) \tag{7.166}$$

For sections CD and FG,

$$M = Q[R_1 + (R_1 + R_2)(-\cos\phi) + R_2 \cos\theta] - M_0 \tag{7.167}$$

and

$$\partial M/\partial Q = R_1(1 - \cos\phi) - R_2(\cos\phi - \cos\theta) \tag{7.168}$$

For section DE and EF,

$$M = Q[R_1 + R_2 + (R_1 + R_2)(-\cos\phi)] - M_1 \tag{7.169}$$

and

$$\partial M/\partial Q = (1 - \cos\phi)(R_1 + R_2) \qquad 7.170$$

Combining these relationships gives

$$\begin{aligned}\partial_Q = \frac{2Q}{EI}\Bigg\{ & R_1{}^3\left(\frac{3}{2}\phi - 2\sin\phi + \frac{\sin 2\phi}{4}\right) \\ & + R_1R_2[R_1\phi - 2\cos\phi(\phi)(R_1 + R_2) + 2R_2\sin\phi] \\ & + (R_1 + R_2)^2\left[\frac{S}{2}(1 - \cos\phi)^2 + R_2(\cos^2\phi)\phi\right] \\ & + R_2{}^2\left(R_2\frac{\phi}{2} - \frac{3}{4}R_2\sin 2\phi - R_1\sin 2\phi\right)\Bigg\} \\ - \frac{2M_0}{EI}\Bigg\{ & R_1{}^2(\phi - \sin\phi) + R_2{}^2[\sin\phi - \cos\phi(\phi)] \\ & + R_1R_2(\phi - \phi\cos\phi) + \frac{S}{2}(R_1 + R_2)(1 - \cos\phi)\Bigg\}\end{aligned} \qquad 7.171$$

If the over-all length of the loop is X_1, the coefficient of thermal expansion is α, and t is the temperature differential,

$$\delta_x = \alpha t X_1 = \delta_Q \qquad 7.172$$

At location A and K there is no rotation; therefore

$$0 = \partial U/\partial M_0 \qquad 7.173$$

from which

$$M_1 = \frac{Q[\phi(R_1{}^2 + R_1R_2) + \sin\phi(R_2{}^2 - R_1{}^2) + (S/2)(R_1 + R_2) - \cos\phi(R_1 + R_2)(S/2 + R_2\phi)]}{L + S/2 + \phi(R_1 + R_2)} \qquad 7.174$$

Then, from Eqs. 7.171, 7.172, and 7.174 the individual quantities can be obtained.
Consider the hanger shown in Fig. 7.33 for which the deflection at C will be

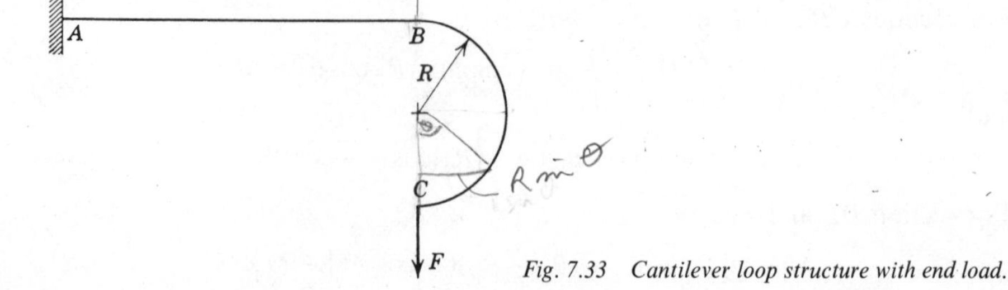

Fig. 7.33 Cantilever loop structure with end load.

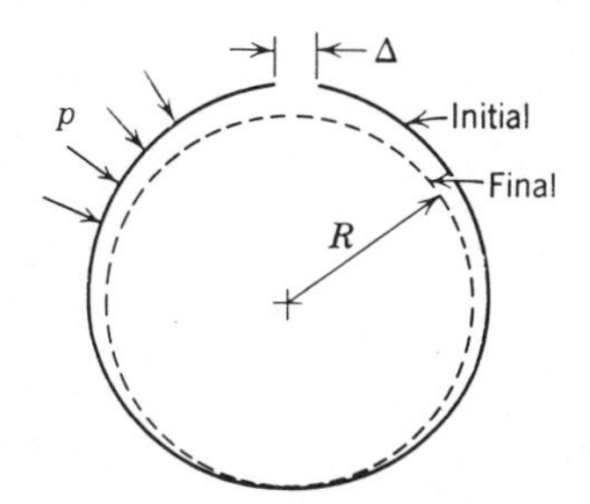

Fig. 7.34 Example of a type of piston ring.

computed. For the geometry indicated

$$\delta_F = \frac{\partial U}{\partial F} = \frac{\partial}{\partial F}\left[\int_0^\pi \frac{M^2\,ds}{2EI} + \int_0^L \frac{M^2\,dx}{2EI}\right] \qquad 7.175$$

$$M_{CB} = FR \sin\theta \qquad \text{and} \quad \partial M/\partial F = R\sin\theta \qquad 7.176$$

$$M_{AB} = Fx \qquad \text{and} \quad \partial M/\partial F = x \qquad 7.177$$

Therefore from Eqs. 7.175 to 7.177

$$\delta_F = \frac{F}{EI}\left(\int_0^\pi R^3 \sin^2\theta\,d\theta + \int_0^L x^2\,dx\right) \qquad 7.178$$

from which

$$\delta_F = (F/6EI)(3\pi R^3 + 2L^3) \qquad 7.179$$

Another interesting application of strain energy is in the design of piston rings. In the following discussion only circular rings subjected to uniform radial pressure will be considered. One approach is to start with a split ring such as shown in Fig. 7.34 which has a constant width b and thickness t. When the ring is assembled and under pressure, it assumes a perfectly circular shape; the problem is to determine what the initial gap Δ should be. Consider that the circular ring is loaded by forces F as shown in Fig. 7.35. With this loading the bending moment at any location of the ring is

$$M = FR(1 - \cos\theta) \qquad 7.180$$

The force exerted on the ring by the pressure p is $2\pi Rtp$. The horizontal compo-

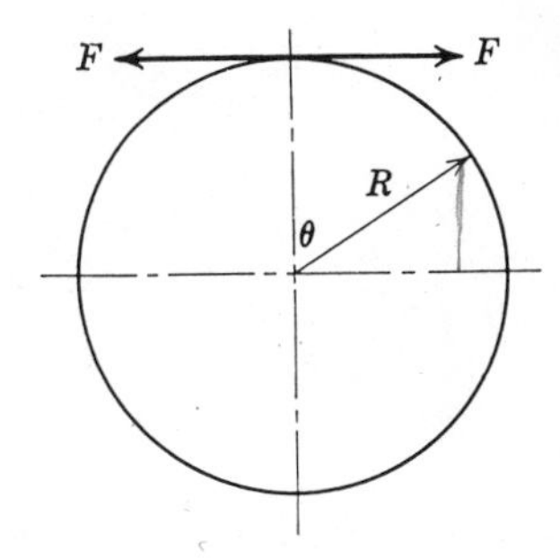

Fig. 7.35 Analysis for piston ring of Fig. 7.34.

nent of this force, however, is

$$F_h = 4\int_0^{\pi/2} ptR\cos\theta\, d\theta \qquad 7.181$$

from which

$$(F_h)_{\text{total}} = 2ptR \qquad 7.182$$

Thus from Eq. 7.182 each force F is ptR and substituting this value into Eq. 7.180 gives the result

$$M = ptR^2(1-\cos\theta) \qquad 7.183$$

The gap Δ is given by

$$\Delta = \frac{\partial U}{\partial F} = \frac{\partial}{\partial F}\int_0^{2\pi}\frac{M^2\, ds}{2EI} = \int_0^{2\pi} M\frac{\partial M}{\partial F}\frac{ds}{EI} \qquad 7.184$$

or using Eq. 7.182

$$\Delta = \frac{1}{EI}\int_0^{2\pi} FR(1-\cos\theta)R(1-\cos\theta)R\, d\theta \qquad 7.185$$

or

$$\Delta = 3\pi ptR^4/EI \qquad 7.186$$

The stress in bending is given by the equation

$$\sigma = M/Z = 12ptR^2/tb^2 = 12pR^2/b^2 \qquad 7.187$$

Therefore the gap Δ can also be expressed as

$$\Delta = \frac{3\pi R^2\sigma_w}{bE} \qquad 7.188$$

where σ_w is the working stress of the material. The foregoing method is essentially that suggested by Van Den Broek (26). Another analysis along different lines by Timoshenko (25) starts with a variable width split ring such as shown in Fig. 7.36; this ring is also rectangular in cross section with constant

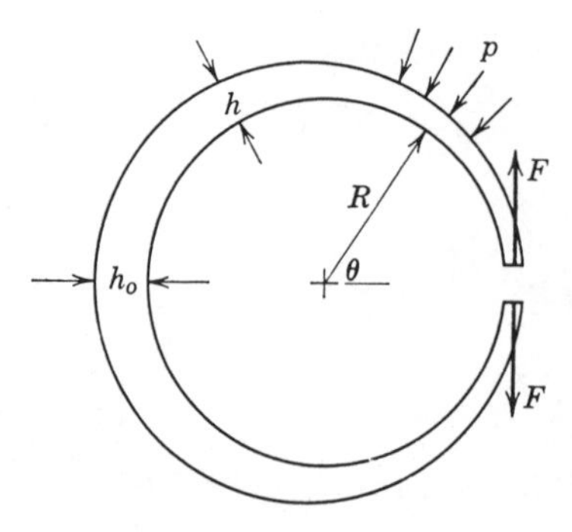

Fig. 7.36 Another example of a type of piston ring.

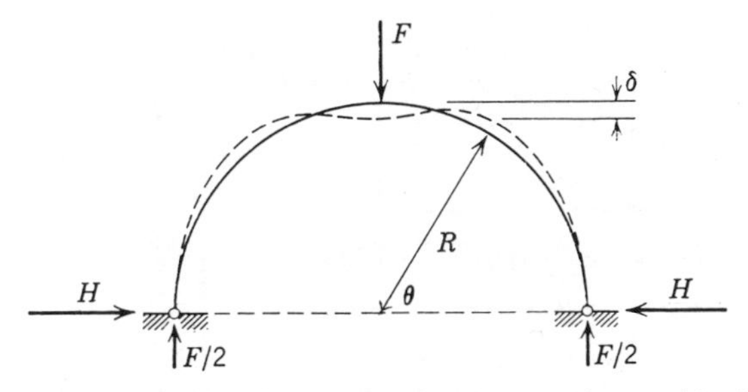

Fig. 7.37 Semicircular hinged arch structure with central load.

thickness t and variable width h. Again, when assembled, it is desired to have a ring of circular boundary producing a uniform radial pressure. Again, as before, forces F applied as shown produce a bending moment in the ring which is

$$M = FR(1 - \cos\theta) = ptR^2(1 - \cos\theta) \qquad 7.189$$

Then, taking the variable width of the ring into account Timoshenko shows that the gap

$$\Delta = 24pR^4/Eh_0^3 \qquad 7.190$$

and that maximum stress is

$$\sigma_{max} = 12pR^2/h_0^2 \qquad 7.191$$

which is the same as given by Eq. 7.187.

Arch structures. Consider the simple semicircular arch shown in Fig. 7.37; the ends of the arch are pin-connected and the load F is at midspan. At each end there are two reactions, a vertical component equal to $F/2$ and a horizontal component H, which may be ascertained by application of strain-energy theory. Since the horizontal deflection at the support is zero,

$$0 = \frac{\partial U}{\partial H} = \frac{\partial}{\partial H}\left[2\int_0^{\pi/2} \frac{M^2\, ds}{2EI}\right] = \frac{2}{EI}\int_0^{\pi/2} M\frac{\partial M}{\partial H} R\, d\theta \qquad 7.192$$

at any location on the arch the moment is given by

$$M = FR\cos\theta - (FR/2)(1 - \cos\theta) + HR\sin\theta \qquad 7.193$$

which when substituted into Eq. 7.192 gives the value of H,

$$H = -F/\pi \qquad 7.194$$

The central deflection δ is then found by application of Eq. 7.192 with $\partial/\partial F$ substituted for $\partial/\partial H$, that is,

$$\delta = \frac{\partial U}{\partial F} = \frac{\partial}{\partial F}\left[2\int_0^{\pi/2} \frac{M^2\, ds}{2EI}\right] = \frac{2}{EI}\int_0^{\pi/2} M\frac{\partial M}{\partial F} R\, d\theta \qquad 7.195$$

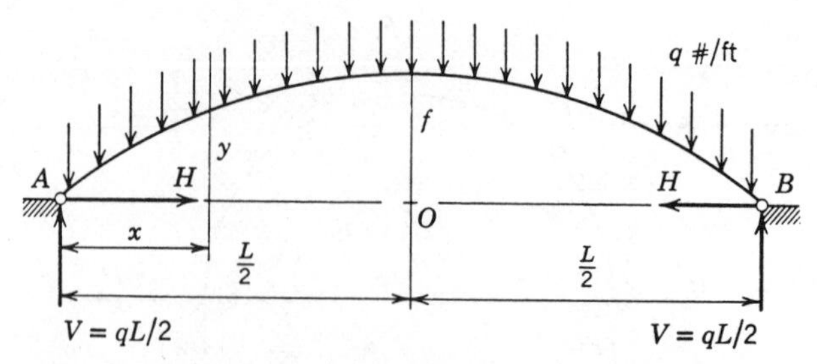

Fig. 7.38 Parabolic hinged arch structure with uniform load.

from which

$$\delta = \frac{ER^3}{8EI}\left(\frac{3\pi^2 - 8\pi - 4}{\pi}\right) \tag{7.196}$$

If the supports of the arch are fixed rather than being pinned, then (26)

$$H = 0.46F \tag{7.197}$$

$$\Delta = 0.0117(FR^3/EI) \tag{7.198}$$

and the end moment at the support is

$$M = 0.111FR \tag{7.199}$$

A parabolic arch, hinged at the supports, is shown in Fig. 7.38. The general equation for a parabola is

$$x^2 = 4ay \tag{7.200}$$

For the parabola shown in Fig. 7.38, when $x = L/2$ and $y = f$, the constant a assumes the value $L^2/16f$ so that Eq. 7.200 becomes

$$x^2 = 4L^2y/16f \tag{7.201}$$

which is the equation for a parabola with origin at O in Fig. 7.38. In order to compute the thrust at the supports, Eq. 7.201 is transformed so that the x axis is AB and the y axis is at A. This operation requires the new axes to be as follows:

$$x' = x + L/2 \tag{7.202}$$

and

$$y' = -y + f \tag{7.203}$$

Substituting Eqs. 7.202 and 7.203 into Eq. 7.201 then gives the result

$$y' = \frac{4fx'(L - x')}{L^2} \tag{7.204}$$

and for convenience the primes are eliminated. At any location of the arch the

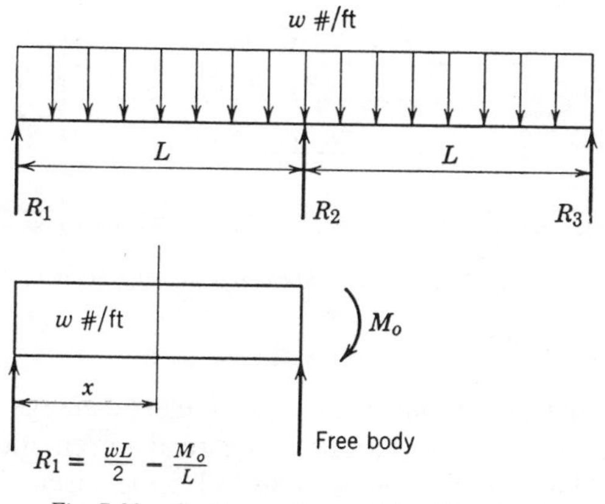

Fig. 7.39 Continuous beam with uniform load.

moment is

$$M = q\frac{Lx}{2} - q\frac{x^2}{4} - H\left[\frac{4fx(L-x)}{L^2}\right] \qquad 7.205$$

The deflection, horizontally, at the supports is zero; therefore

$$0 = \frac{\partial U}{\partial H} = \frac{\partial}{\partial H}\left[2\int_0^{L/2}\frac{M^2\,dx}{2EI}\right] \qquad 7.206$$

from which

$$H = qL^2/8f \qquad 7.207$$

If, instead of a uniformly distributed load being applied, the arch were loaded centrally by a single force, F, then

$$H = 25FL/128f \qquad 7.208$$

Continuous beam. Consider the continuous beam shown in Fig. 7.39. In order to calculate the internal moment M_0 at the center support it is noted that the slope is zero at this location so that

$$\theta = \partial U/\partial M_0 = 0 \qquad 7.209$$

Therefore from Eq. 7.209

$$0 = \frac{\partial}{\partial M_0}\left[2\int_0^L \frac{M^2\,dx}{2EI}\right] = \frac{2}{EI}\int_0^L M(\partial M/\partial M_0)\,dx \qquad 7.210$$

at any location x in the beam

$$M = (wL/2 - M_0/L)x - wx^2/2 \qquad 7.211$$

Substituting Eq. 7.211 into Eq. 7.210 then gives

$$M_0 = -wL^2/8 \qquad 7.212$$

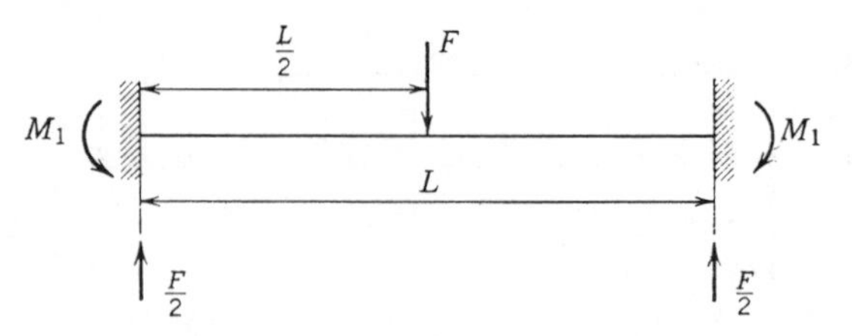

Fig. 7.40 Fixed-end beam with center load.

Statically indeterminate beams. The use of elastic-energy theory in solving for statically indeterminate reactions has already been illustrated for some structures other than simple beams. The following discussion illustrates the application of the energy method to the determination of indeterminate reactions for two simple beams. First consider the center-loaded, fixed-end beam shown in Fig. 7.40. At each support there is a vertical reaction $F/2$ and an indeterminate bending moment M_1. This moment is easily calculated from strain-energy theory since the slope is zero at the supports, that is,

$$\theta = 0 = \frac{\partial U}{\partial M_1} = \frac{\partial}{\partial M_1}\left[2\int_0^{L/2} \frac{M^2\,dx}{2EI}\right] = \frac{2}{EI}\int_0^{L/2} M\frac{\partial M}{\partial M_1}dx \qquad 7.213$$

The moment at any location x in the beam is

$$M = Fx/2 - M_1 \qquad 7.214$$

Therefore, by substituting Eq. 7.214 into Eq. 7.213, the result is

$$M_1 = FL/8 \qquad 7.215$$

As a second example consider the beam shown in Fig. 7.41. This beam has one indeterminate reaction. If the vertical reaction at A is taken as the indeterminate quantity,

$$0 = \frac{\partial U}{\partial R} = \frac{\partial}{\partial R}\int_0^L \frac{M^2\,dx}{2EI} = \frac{1}{EI}\int_0^L M\frac{\partial M}{\partial R}dx \qquad 7.216$$

At any location x in the beam the moment is given by

$$M = Rx - wx^2/2 \qquad 7.217$$

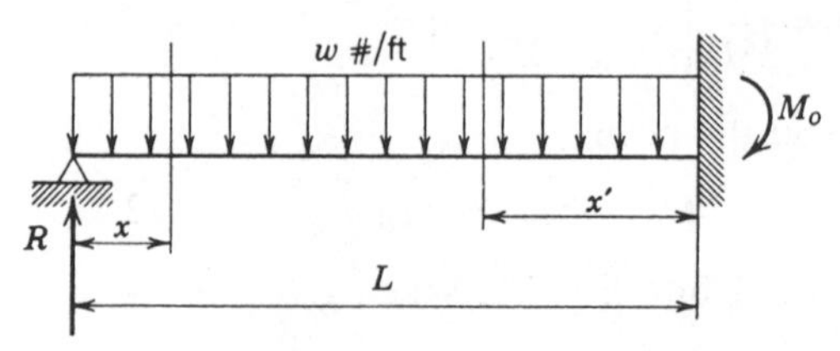

Fig. 7.41 End-supported uniformly loaded cantilever.

which, when substituted into Eq. 7.216, gives the result

$$R = \tfrac{3}{8}wL \qquad 7.218$$

As an alternative the moment M_0 at B could have been selected as the indeterminate quantity; then

$$0 = \frac{\partial U}{\partial M_0} = \frac{\partial}{\partial M_0}\int_0^L \frac{M^2\,dx}{2EI} = \frac{1}{EI}\int_0^L M\frac{\partial M}{\partial M_0}\,dx \qquad 7.219$$

At any section x' in the beam

$$M = (wL/2 - M_0/L)x - w(x')^2/2 \qquad 7.220$$

which, when substituted into Eq. 7.219, gives the result

$$M_0 = wL^2/8 \qquad 7.221$$

Analysis of combined loading

In many machine and structural parts the loads are so applied that various combinations of axial load, bending, and twisting are obtained. In the following examples the use of elastic-energy theory in solving for reactions and deformations is illustrated.

Split circular rings The ring shown in Fig. 7.42 has a radius R and is loaded at a distance 2π from the support by a force F applied perpendicular to the plane of the ring. Such a loading induces both bending and torsional stresses in the ring. The deflection at the point of load application is

$$\delta = \frac{\partial U}{\partial F} = \frac{\partial}{\partial F}\left[\int_0^{2\pi}\frac{M^2\,ds}{2EI} + \int_0^{2\pi}\frac{T^2\,ds}{2GJ}\right] \qquad 7.222$$

from which

$$\delta = \frac{FR^3}{EI}\int_0^{2\pi}\sin^2\theta\,d\theta + \frac{FR^3}{GJ}\int_0^{2\pi}(1 - 2\cos\theta + \cos^2\theta)\,d\theta \qquad 7.223$$

or

$$\delta = FR^3\pi(1/EI + 3/GJ) \qquad 7.224$$

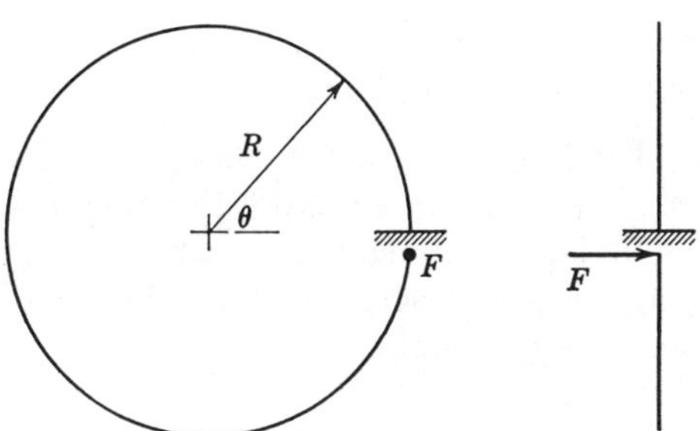

Fig. 7.42 Circular ring loaded normal to its plane.

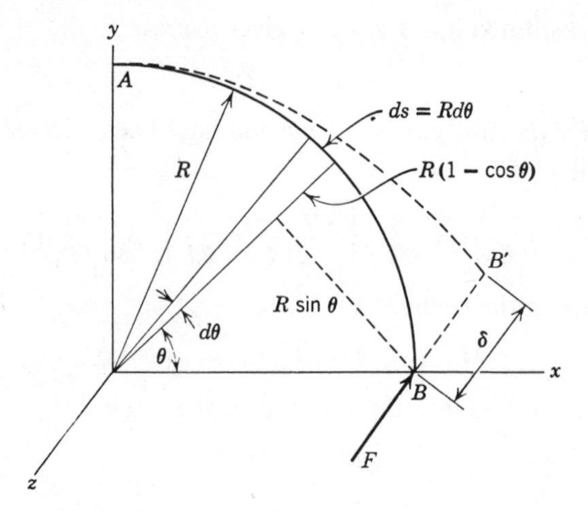

Fig. 7.43 Quarter circle loop loaded normal to its plane.

Quarter ring. The structure shown in Fig. 7.43, a quarter circle, is loaded by a force F perpendicular to its plane inducing both bending and torsional stresses in the ring. At the point of application of the load

$$\delta = \frac{\partial U}{\partial F} = \frac{\partial}{\partial F}\left[\int_0^{\pi/2} \frac{M^2\,ds}{2EI} + \int_0^{\pi/2} \frac{T^2\,ds}{2GJ}\right] \tag{7.225}$$

or

$$\delta = \frac{R}{EI}\int_0^{\pi/2} M\frac{\partial M}{\partial F}\,d\theta + \frac{R}{GJ}\int_0^{\pi/2} T\frac{\partial T}{\partial F}\,d\theta \tag{7.226}$$

At any location in the ring

$$M = FR\sin\theta \tag{7.227}$$

and

$$T = FR(1 - \cos\theta) \tag{7.228}$$

Substituting Eqs. 7.227 and 7.228 into Eq. 7.226 then gives

$$\delta = \frac{FR^3}{4}\left(\frac{\pi}{EI} + \frac{3\pi - 8}{GJ}\right) \tag{7.229}$$

Compound loop. The loop structure shown in Fig. 7.28 was subjected only to bending stresses. If now in that structure the force is located horizontally as shown in Fig. 7.44 the structure is subjected to both axial tension and bending stresses. The free body for this structure is shown in Fig. 7.45. At the midpoint of the loop the slope is zero; therefore

$$\theta = 0 = \frac{\partial U}{\partial M_0} = \frac{\partial}{\partial M_0}\left[4\left(\int_0^{\pi/2} \frac{M^2\,ds}{2EI} + \int_0^{L} \frac{{M_0}^2\,dx}{2EI} + \int_0^{L} \frac{F^2\,dx}{8EA}\right)\right] \tag{7.230}$$

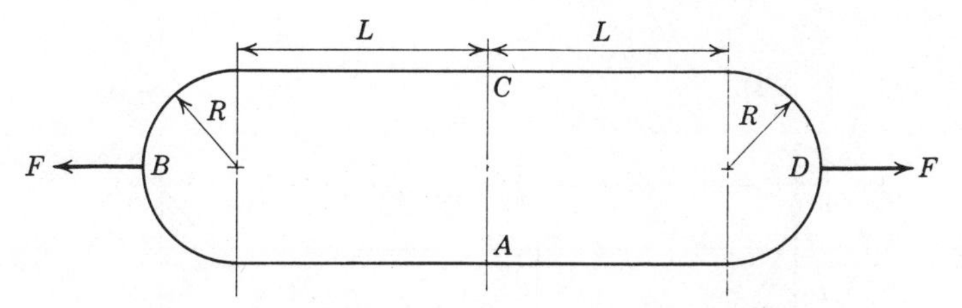

Fig. 7.44 Compound loop subjected to axial loading.

from which

$$0 = 4\left[\int_0^{\pi/2} M \frac{\partial M}{\partial M_0} \frac{R\,d\theta}{EI} + \int_0^L \frac{M_0\,dx}{EI} + \int_0^L F \frac{\partial F}{\partial M_0} \frac{dx}{4EA}\right] \qquad 7.231$$

At the end sections of the loop

$$M = M_0 - (FR/2)(1 - \cos\theta) \qquad 7.232$$

Therefore, using this value of M in Eq. 7.231 leads to the result

$$M_0 = \frac{FR^2(\pi - 2)}{2(\pi R + 2L)} \qquad 7.233$$

Having the value of M_0, the increase in BD, or δ, can now be computed as follows:

$$\delta_{BD} = \frac{\partial U}{\partial F} = \frac{\partial}{\partial F}\left[4\left(\int_0^{\pi/2} \frac{M^2\,ds}{2EI} + \int_0^L \frac{M_0^{\,2}\,dx}{2EI} + \int_0^L \frac{F^2\,dx}{8EA}\right)\right] \qquad 7.234$$

Substituting the values of M and M_0 from Eqs. 7.232 and 7.233 into Eq. 7.234 gives

$$\delta_{BD} = \frac{FR^3(\pi^2 R + 6\pi L - 16L - 8R)}{4EI(\pi R + 2L)} + \frac{FL}{AE} \qquad 7.235$$

Balcony structure. Two types of balcony structures will be considered, both types being semicircular arches with fixed ends loaded perpendicular to the plane of the arch. First consider the type shown in Fig. 7.46 which is loaded

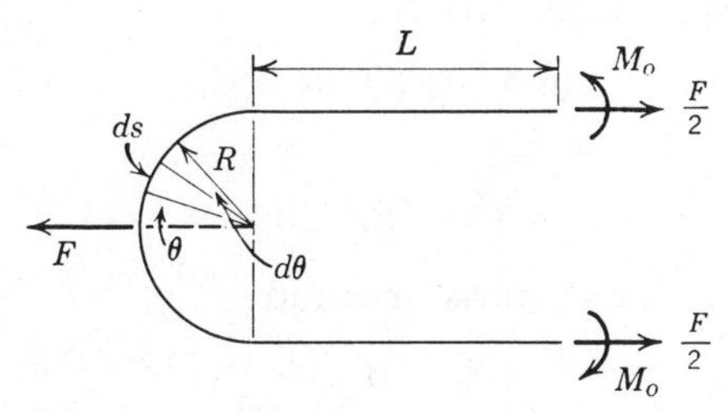

Fig. 7.45 Free body for structure in Fig. 7.44.

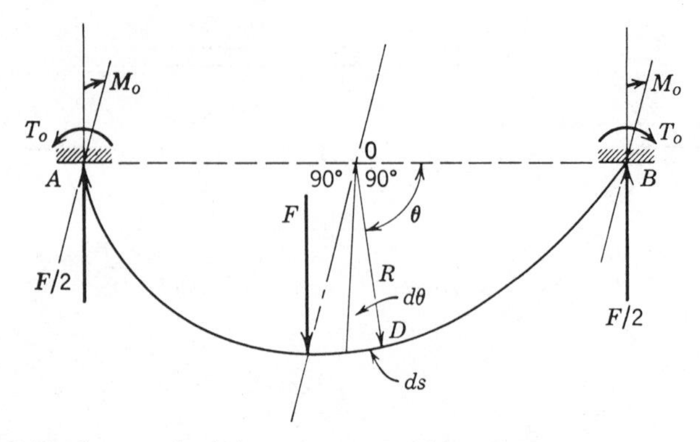

Fig. 7.46 Semicircular balcony structure with center load normal to its plane.

at the center with a single force F. At the fixed ends there is a vertical reaction $F/2$ and a bending moment $FR/2$. A third reaction, a twisting moment T_0, is statically indeterminate. For the section BD the torque T_0 produces (at D) a torque and a bending moment

$$T_{D_1} = T_0 \cos\theta \tag{7.236}$$

and

$$M_{D_1} = T_0 \sin\theta \tag{7.237}$$

Similarly, the moment M_0 produces

$$T_{D_2} = -M_0 \sin\theta \tag{7.238}$$

and

$$M_{D_2} = -M_0 \cos\theta \tag{7.239}$$

The vertical reaction $F/2$ produces

$$T_{D_3} = (F/2)R(1 - \cos\theta) \tag{7.240}$$

and

$$M_{D_3} = (F/2)R \sin\theta \tag{7.241}$$

Therefore at D the total effective torque and bending moments are obtained by summation of Eqs. 7.236 to 7.241, thus,

$$T_D = T_0 \cos\theta + (FR/2)(1 - \sin\theta - \cos\theta) \tag{7.242}$$

and

$$M_D = -T_0 \sin\theta - (FR/2)(\sin\theta - \cos\theta) \tag{7.243}$$

There is no rotation at the supports; therefore

$$0 = \frac{\partial U}{\partial T_0} = \frac{\partial}{\partial T_0}\left[2\left(\int_0^{\pi/2} \frac{M_D{}^2\,ds}{2EI} + \int_0^{\pi/2} \frac{T_D{}^2\,ds}{2GJ}\right)\right]$$

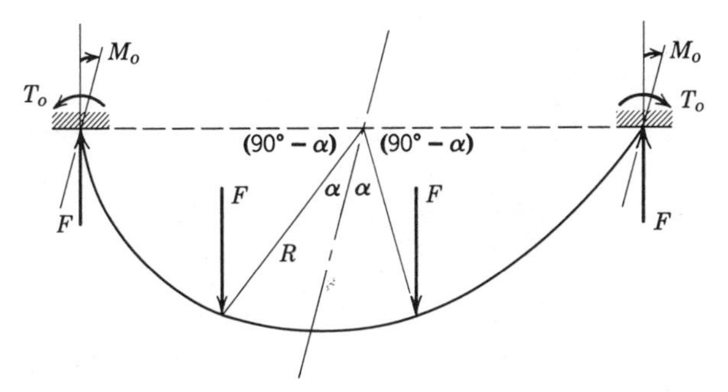

Fig. 7.47 Semicircular balcony structure with symmetrical loads normal to its plane.

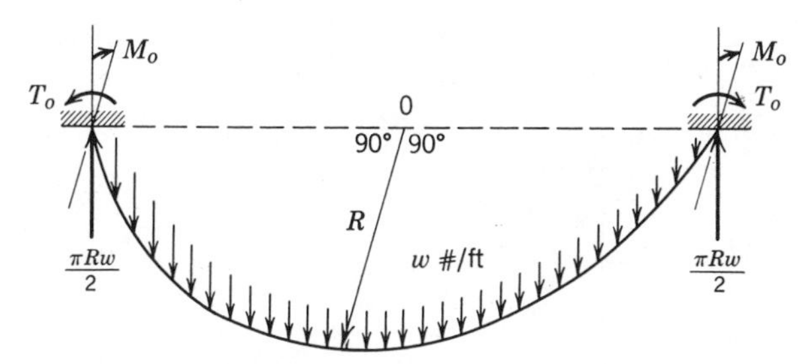

Fig. 7.48 Semicircular balcony structure with uniformly distributed load normal to its plane.

or

$$0 = \frac{2R}{EI}\int_0^{\pi/2} M_D \frac{\partial M_D}{\partial T_0} d\theta + \frac{2R}{GJ}\int_0^{\pi/2} T_D \frac{\partial T_D}{\partial T_0} d\theta \qquad 7.244$$

Substituting the values given by Eqs. 7.242 and 7.243 into Eq. 7.244 then gives the result

$$T_0 = (FR/2\pi)(\pi - 2) \qquad 7.245$$

By using the same procedure as outlined the indeterminate twisting moment T_0 for double loading (Fig. 7.47) is (25)

$$T_0 = (2FR/\pi)(\pi/2 - \cos\alpha - \alpha\sin\alpha) \qquad 7.246$$

Similarly (Fig. 7.48), if the load is uniformly distributed along the balcony,

$$T_0 = (wR^2/2\pi)(\pi^2 - 8) \qquad 7.247$$

Structural frames. The frame shown in Fig. 7.49 is loaded by a single horizontal force F at B; the free body is shown in Fig. 7.50. Since the load is in the plane of the frame there are no twisting moments induced, and, neglecting any shear effect, the deflection at B is

$$\delta_B = \frac{\partial U}{\partial F} = \frac{\partial}{\partial F}\left[\int_0^L \frac{M^2\,dx}{2EI} + \int_0^L \frac{F^2\,dx}{2EA} + 2\int_0^H \frac{F^2H^2\,dx}{2EI_1}\right] \qquad 7.248$$

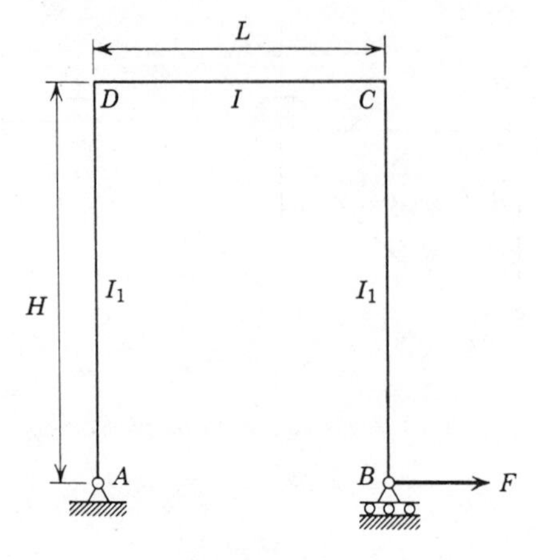

Fig. 7.49 Structural frame subjected to horizontal loading.

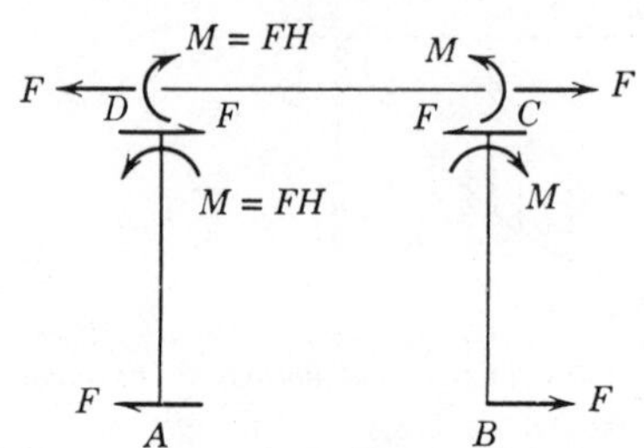

Fig. 7.50 Free body for the structure shown in Fig. 7.49.

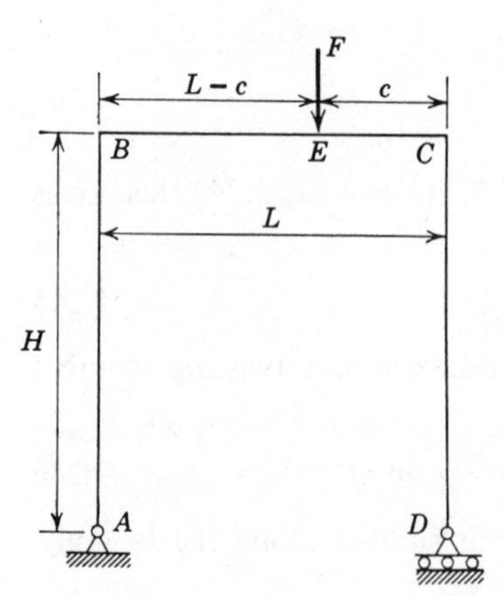

Fig. 7.51 Structural frame subjected to lateral loading.

The moment involved is equal to FH; therefore

$$\delta_B = \frac{F}{E}\left[\frac{H^2L}{I} + \frac{L}{A} + \frac{2}{3}\frac{H^3}{I_1}\right] \qquad 7.249$$

Now consider the frame shown in Fig. 7.51 and its free body, Fig. 7.52. As in the previous problem shear effects are neglected and the deflection at E is

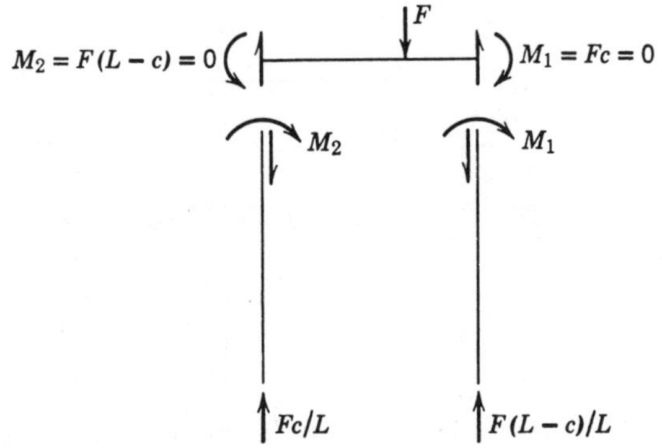

Fig. 7.52 *Free body for the structure shown in Fig. 7.51.*

$$\delta_E = \frac{\partial U}{\partial F} = \frac{\partial}{\partial F}\left[\int_0^H \frac{[F(c/L)]^2\,dy}{2EA} + \int_0^{L-c} \frac{[F(c/L)x]^2\,dx}{2EI} + \int_0^c \frac{[(F/L)(L-c)x]^2\,dx}{2EI} + \int_0^H \frac{[(F/L)(L-c)]^2\,dy}{2EA}\right] \qquad 7.250$$

from which

$$\delta_E = \frac{FH}{EAL^2}(2c^2 + L^2 - 2cL) + \frac{Fc^2(L-c)^2}{3EIL} \qquad 7.251$$

In the frame shown in Fig. 7.53 all members carry bending and axial stress; at the supports A and B there is no movement vertically or horizontally; there-

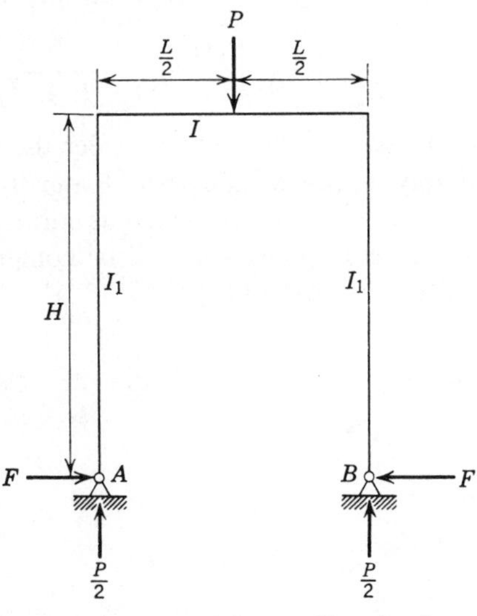

Fig. 7.53 *Structural frame subjected to lateral load.*

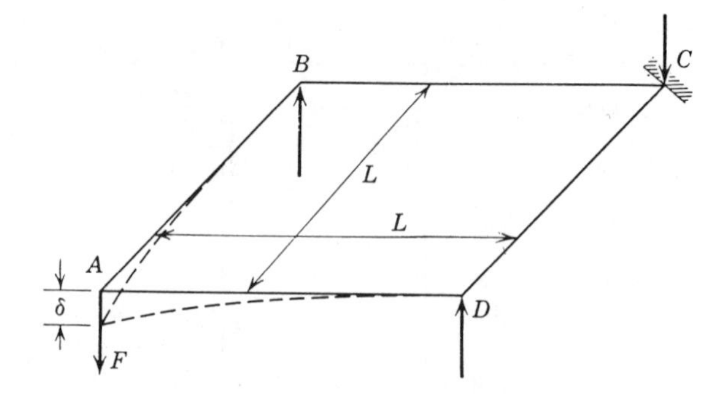

Fig. 7.54 Square frame subjected to loading normal to its plane.

fore

$$0 = \partial U/\partial F \tag{7.252}$$

or

$$0 = \frac{\partial}{\partial F}\left[2\int_0^{L/2}\frac{M_{DC}{}^2\,dx}{2EI} + 2\int_0^H \frac{M_{BC}{}^2\,dy}{2EI_1} + 2\int_0^H \frac{(P/2)^2\,dy}{2EA} + \int_0^L \frac{F^2\,dx}{2EA}\right] \tag{7.253}$$

However,

$$M_{DC} = Px/2 - FH \tag{7.524}$$

and

$$M_{BC} = Fy \tag{7.255}$$

Therefore placing the values of these moments into Eq. 7.253 gives the result

$$F = \frac{PHL^2}{8H^2L[1 + \frac{2}{3}(H/L)(I/I_1) + (I/AH^2)]} \tag{7.256}$$

As a final example of this type of structure consider the square frame shown in Fig. 7.54 which is simply supported at corners B and D, is restrained against vertical displacement at corner C, and is loaded at corner A by a load F. The deflection at the point of load application will be computed. Free-bodies of the two symmetrical sides are shown in Fig. 7.55.

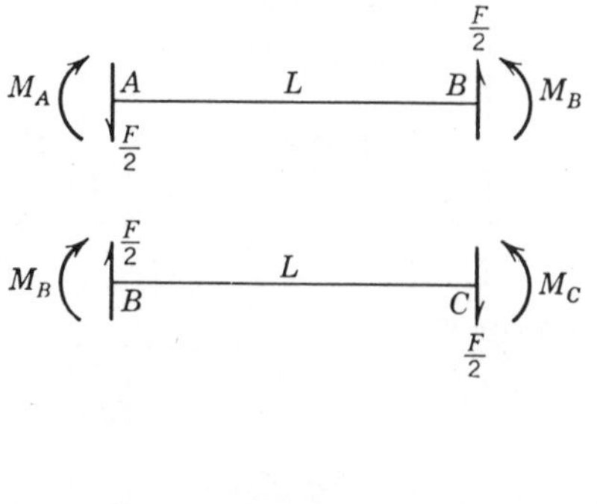

Fig. 7.55 Free body for the structure shown in Fig. 7.54.

By statics

$$M_A - FL/2 - M_B = 0 \qquad 7.257$$

from which

$$M_A = M_B + FL/2 \qquad 7.258$$

In addition

$$M_x = M_A - Fx/2 \qquad 7.259$$

and

$$M_{x1} = M_B + Fx_1/2 \qquad 7.260$$

The corresponding torques are

$$T_x = M_B L/x \qquad 7.261$$

and

$$T_{x1} = M_A L/x_1 \qquad 7.262$$

Then at A

$$0 = \frac{\partial U}{\partial M_A} = 2\int_0^L \left(M_A - \frac{Fx}{2}\right)\frac{dx}{EI} + 2\int_0^L \left(M_A - \frac{FL}{2} + \frac{Fx}{2}\right)\frac{dx}{EI}$$
$$+ 2\int_0^L \left(M_A - \frac{FL}{2}\right)\frac{L}{x}\frac{dx}{GJ} + 2\int_0^L M_A \frac{L}{x}\frac{dx}{GJ} \qquad 7.263$$

from which

$$M_A = FL/4 \qquad 7.264$$

Substituting Eq. 7.264 into Eq. 7.258 shows that

$$M_B = -FL/4 \qquad 7.265$$

indicating constant torque through the system. At the point of load application

$$\delta_A = \frac{\partial U}{\partial F} = \frac{2}{EI}\int_0^L \left[\left(\frac{FL}{4} - \frac{Fx}{2}\right)\left(\frac{L}{4} - \frac{x}{2}\right) + \left(-\frac{FL}{4} + \frac{Fx}{2}\right)\left(-\frac{L}{4} + \frac{x}{2}\right)\right] dx$$
$$+ \frac{4}{GJ}\int_0^L \frac{FL}{4}\left(\frac{L}{4}\right) dx \qquad 7.266$$

from which

$$\delta_A = FL^3(1/12EI + 1/4GJ) \qquad 7.267$$

Uniformly loaded loops. In the following discussion two cases of uniformly loaded structures are considered; first, the rectangle shown in Fig. 7.56 will be analyzed. The free body for one of the symmetrical sides is shown in Fig. 7.57, where it is indicated that the moment at the corner, M_B is

$$M_B = M_0 - wa^2/8 \qquad 7.268$$

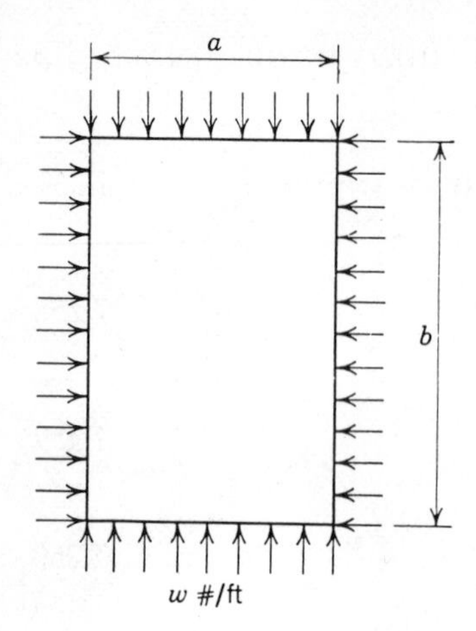

Fig. 7.56 Rectangular frame subjected to uniform external loading.

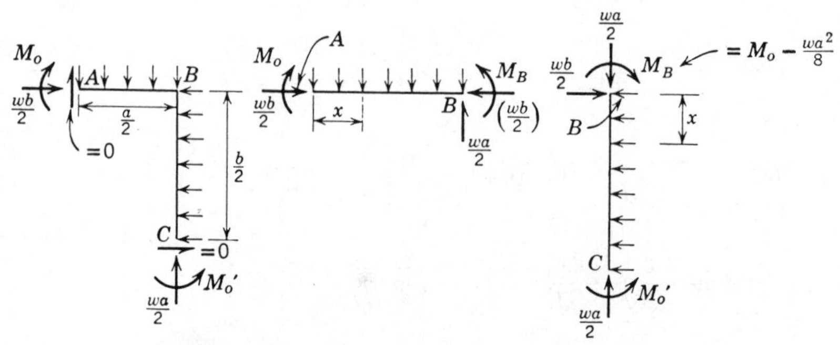

Fig. 7.57 Free bodies for the structure shown in Fig. 7.56.

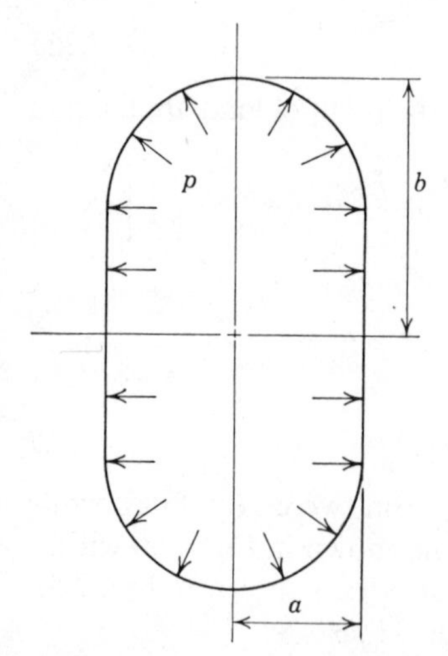

Fig. 7.58 Compound tank subjected to internal pressure.

There is no rotation at the center of the sides; therefore

$$0 = \frac{\partial U}{\partial M_0} = \frac{\partial}{\partial M_0}\left[\int_0^{a/2,b/2} \frac{M^2\,dx}{2EI} + \int_0^{a/2,b/2} \frac{F^2\,dx}{2EA}\right] \qquad 7.269$$

The moment M_{AB} is given by

$$M_{AB} = M_0 - wx^2/2 \qquad 7.270$$

and the axial load is

$$F_{AB} = wb/2 \qquad 7.271$$

The moment M_B is given by Eq. 7.268 and the moment M_{BC} is given by

$$M_{BC} = wbx/2 - wx^2/2 + M_0 - wa^2/8 \qquad 7.272$$

Therefore, using these values

$$0 = \frac{1}{EI}\int_0^{a/2}\left(M_0 - \frac{wx^2}{2}\right)dx + \frac{1}{EI}\int_0^{b/2}\left(\frac{wbx}{2} - \frac{wx^2}{2} + M_0 - \frac{wa^2}{8}\right)dx \qquad 7.273$$

from which

$$M_0 = \frac{w}{24}\left(\frac{a^3 - 2b^3 + 3a^2b}{a+b}\right) \qquad 7.274$$

Substituting M_0, into Eq. 7.268 gives

$$M_B = \frac{w}{12}\left(\frac{a^3 + b^3}{a+b}\right) \qquad 7.275$$

The second case concerns the duplex structure shown in Fig. 7.58, a loop with two axes of symmetry subjected to internal pressure p. A free body of one quadrant of the structure is shown in Fig. 7.59. At any section defined by r the bending moment is

$$M_r = M_0 - pa(a - x) + p\frac{(a-x)^2}{2} + \frac{py^2}{2} \qquad 7.276$$

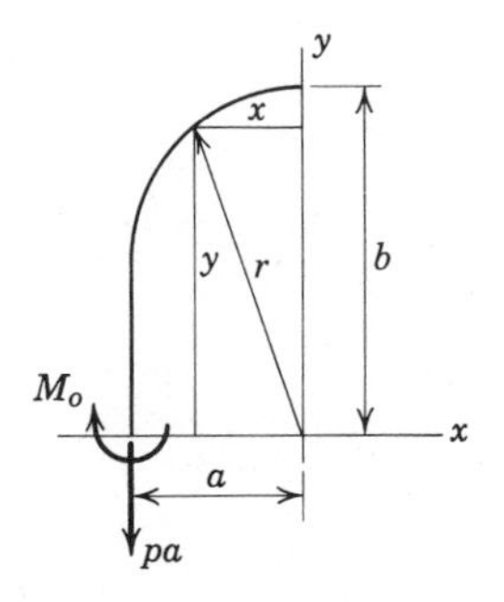

Fig. 7.59 Free body for the tank shown in Fig. 7.58.

Since there is no rotation at M_0,

$$0 = \frac{\partial U}{\partial M_0} = \frac{\partial}{\partial M_0}\left(4\int_0^{S/4} \frac{M^2\,ds}{2EI} + fp\right) \qquad 7.277$$

The term fp in Eq. 7.277 represents a strain energy term due to axial loading by the force pa; however, since the expression is differentiated with respect to M_0, the term drops out and is of no further interest. Therefore Eq. 7.277 becomes

$$0 = \frac{2}{EI}\left(\frac{M_0 S}{4} - \frac{pa^2 S}{8} + \frac{1}{2}\int_0^{S/4} px^2\,dS + \frac{1}{2}\int_0^{S/4} py^2\,dS\right) \qquad 7.278$$

or

$$0 = \left(M_0 - \frac{pa^2}{2}\right)\frac{S}{4} + \frac{p}{2}(I_x + I_y) \qquad 7.279$$

from which

$$M_0 = \frac{pa^2}{2} - \frac{2p}{S}(I_x + I_y) \qquad 7.280$$

where S is the length of the loop and

$$I_x = \int_0^{S/4} y^2\,dS \qquad 7.281$$

$$I_y = \int_0^{S/4} x^2\,dS \qquad 7.282$$

With the results given by Eqs. 7.280 to 7.282 various reactions for different shapes of loops can be obtained. For example, suppose that the loop in Fig. 7.60 consists of straight sides and semicircular ends commonly known as an obround. Then, using the above relationships, the stress can be computed at the various sections as follows.

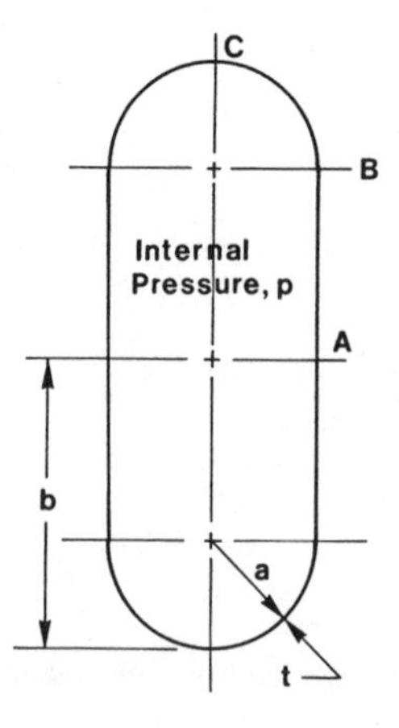

Fig. 7.60 Obround section.

Semicircular sections: In Fig. 7.60 at location C, the top and bottom of the loop, there is a membrane stress

$$\sigma_m = p\left(\frac{b}{t}\right) \tag{7.283}$$

Also, at C there is a bending stress

$$\sigma = \frac{p(b-a)}{t^2}\left[3(b+a) - \frac{14a^2 + 2b^2 - 4ab + 3\pi a(b-a)}{2(b-a) + \pi a}\right] \tag{7.284}$$

The total stress at C is Eq. 7.283 added to Eq. 7.284.

$$\sigma_T = \sigma_m + \sigma \tag{7.285}$$

The membrane stress at location B, the junction of the sides with the semicircular ends is

$$\sigma_m = p\left(\frac{a}{t}\right) \tag{7.286}$$

The bending stress at B is

$$\sigma = \frac{p(b-a)}{t^2}\left[3(b-a) - \frac{14a^2 + 2b^2 - 4ab + 3\pi a(b-a)}{2(b-a) + \pi a}\right] \tag{7.287}$$

The total stress at B is the sum of Eqs. 7.286 and 7.287.

$$\sigma_T = \sigma_m + \sigma \tag{7.288}$$

The straight sides: In Fig. 7.60 when b is zero or the middle of the straight sides at point A, the membrane stress is

$$\sigma_m = p\left(\frac{a}{t}\right) \tag{7.289}$$

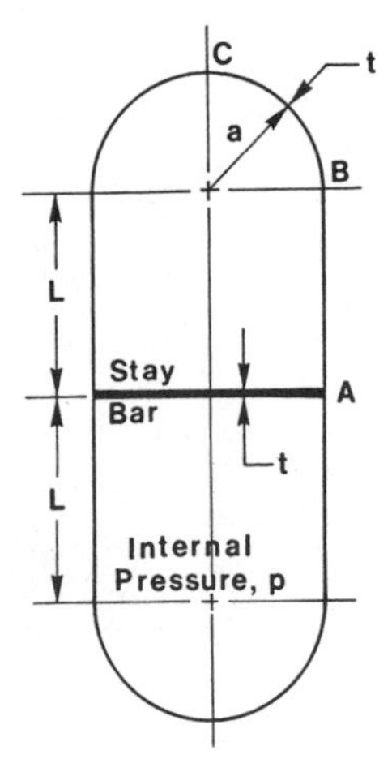

Fig. 7.61 Obround section with central stay bar.

At location A the bending stress is

$$\sigma = \frac{p(b-a)}{t^2}\left[\frac{14a^2 + 2b^2 - 4ab + 3\pi a(b-a)}{2(b-a) + \pi a}\right] \qquad 7.290$$

As before, the total stress is the sum of Eqs. 7.289 and 7.290.

$$\sigma_T = \sigma_m + \sigma \qquad 7.291$$

An obround vessel in Fig. 7.61 with a single central stay bar at location A has membrane and bending stresses. The total stress will be stated as sum of the membrane and bending stress.

$$\sigma_T = \sigma_m + \sigma \qquad 7.292$$

The total stress at location C, the top and bottom of the loop, is

$$\sigma_T = \frac{p}{4t}\left[4(a+L) - L\left(\frac{3GJ - 2HI}{GK - 6H^2}\right)\right] + \frac{3pL}{Gt^2}\left[M(H - GL - Ga) + G(L + 2a) - \frac{I}{3}\right] \qquad 7.293$$

At location B, the total stress is the same for the circular portion and straight portion.

$$\sigma_T = p\left(\frac{a}{t}\right) + \frac{3pL}{Gt^2}\left[M(H - GL) - \frac{I}{3} + GL\right] \qquad 7.294$$

The total stress at location A, the middle of the straight side, is

$$\sigma_T = p\left(\frac{a}{t}\right) + \frac{3pL}{Gt^2}\left[HM - \frac{I}{3}\right] \qquad 7.295$$

In the staying member there is only a membrane stress

$$\sigma_T = pM\left(\frac{L}{t_1}\right) \qquad 7.296$$

In Eqs. 7.293 to 7.296 the following terms are defined as:

$$G = a(2\gamma + \pi) \qquad 7.297$$

$$H = a^2(\gamma^2 + \gamma\pi + 2) \qquad 7.298$$

$$I = a^2(2\gamma^2 + 3\gamma\pi + 12) \qquad 7.299$$

$$J = a^3(\gamma^3 + 2\gamma^2\pi + 12\gamma + 2\pi) \qquad 7.300$$

$$K = a^3(4\gamma^3 + 6\gamma^2\pi + 24\gamma + 3\pi) \qquad 7.301$$

$$M = \frac{(3GJ - 2HI)}{(GK - 6H^2)} \qquad 7.302$$

$$\gamma = \frac{L}{a} \qquad 7.303$$

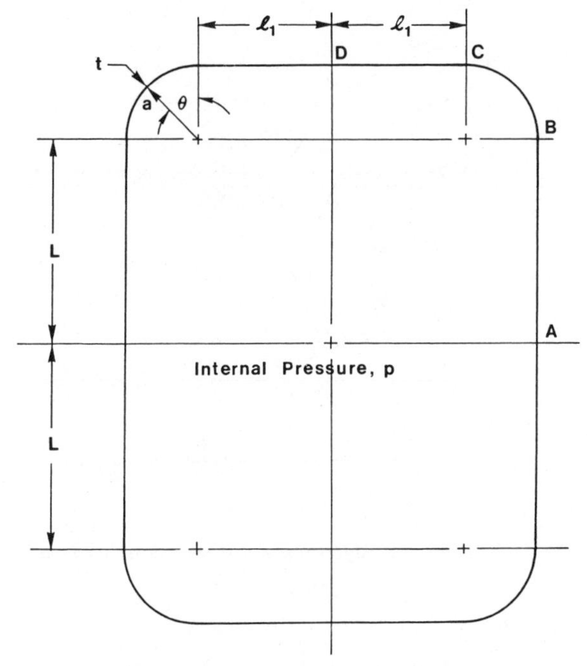

Fig. 7.62 Rectangular tank with circular corners.

For rectangular tanks as in Fig. 7.62, the total stress at locations A, B, C, and D is the sum of the membrane and bending stress. At location D in the short side

$$\sigma_T = p\frac{(a+L)}{t} + \frac{3}{t^2}[2M_a + p(L^2 + 2aL - 2al_1 - l_1{}^2)] \tag{7.304}$$

Location C:

$$\sigma_T = p\frac{(a+L)}{t} + \frac{3}{t^2}[2M_a + p(2aL - 2al_1 + L^2)] \tag{7.305}$$

In the corner section from C to B the maximum stress is developed when $\tan^{-1}\theta$ is l_1/L. The total stress in the corner section is

$$\sigma_T = \frac{p}{t}[(L^2 + l_1{}^2)^{1/2} + a] + \frac{6M_r}{t^2} \tag{7.306}$$

The stress in the long side at location B is

$$\sigma_T = p\frac{(l_1 + a)}{t} + \frac{3}{t^2}[2M_A + pL^2] \tag{7.307}$$

In the middle of the long side at location A

$$\sigma_T = p\frac{(l_1 + a)}{t} + \frac{6M_A}{t^2} \tag{7.308}$$

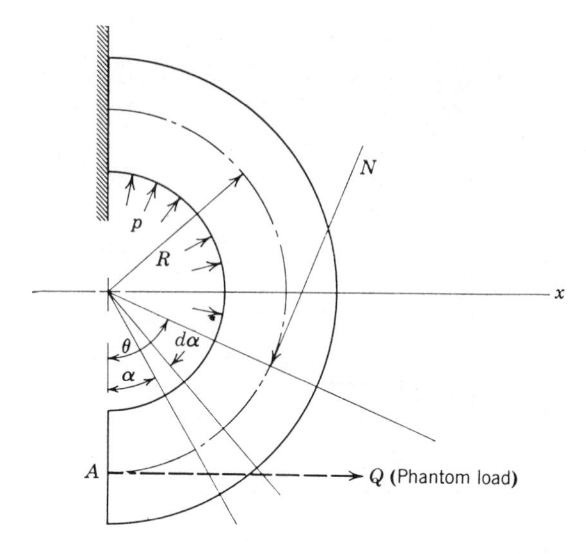

Fig. 7.63 Thick ring subjected to a uniformly distributed load.

The terms used in Eqs. 7.303 to 7.308 are

$$M_A = -pK_3 \tag{7.309}$$

$$M_r = M_A + p\left\{a[L\cos\theta - l_1(1-\sin\theta)] + \frac{L^2}{2}\right\} \tag{7.310}$$

$$K_3 = \frac{l_1{}^2[6\phi^2\alpha_3 - 3\pi\phi^2 + 6\phi^2 + \alpha_3{}^3 + \alpha_3{}^2 - 6\phi - 3 + 1.5\pi\phi\alpha_3{}^2 + 6\phi\alpha_3]}{3(2\alpha_3 + \pi\phi + 2)} \tag{7.311}$$

$$\alpha_3 = \frac{L}{l_1} \tag{7.312}$$

$$\phi = \frac{a}{l_1} \tag{7.313}$$

Results for the compound and elliptic loops are also found in (25).

Deflection of thick rings. If the cross-sectional dimensions of a ring are large, normal and shear stresses as well as bending stresses have to be allowed for, Consider the ring shown in Fig. 7.63, which is subjected to a distributed load on the inner radius. The deflection in the x direction at A will be calculated. At any section θ the strain energy is (21) (24)

$$U = \int_0^S \left(\frac{M^2}{2AEeR} + \frac{N^2}{2AE} - \frac{MN}{AER} + \frac{\phi V^2}{2AG}\right) dS \tag{7.314}$$

where

$$M = QR(1 - \cos\theta) + \int_0^\theta (pR\,d\alpha)R\sin(\theta - \alpha)$$

$$= QR(1 - \cos\theta) + pR^2(1 - \cos\theta) \qquad 7.315$$

$$N = -Q\sin(90^\circ - \theta) - \int_0^\theta (pR\,d\alpha)\sin(\theta - \alpha)$$

$$= Q\cos\theta - pR(1 - \cos\theta) \qquad 7.316$$

$$V = Q\cos(90^\circ - \theta) + \int_0^\theta (pR\,d\alpha)\cos(\theta - \alpha)$$

$$= Q\sin\theta + pR\sin\theta \qquad 7.317$$

and ϕ is a factor dependent on the cross section. As shown in Chapter 3 this factor is 1.5 for a rectangular cross section (Eq. 3.212), 1.33 for a circular cross section (Eq. 3.214), and 2 for a thin, circular tube. Using these relationships, the deflection is

$$\delta = \frac{\partial U}{\partial Q} = \int_0^\pi \left[M\frac{\partial M}{\partial Q}\left(\frac{1}{AeER}\right) + N\frac{\partial N}{\partial Q}\left(\frac{1}{AE}\right) - \frac{1}{AER}\left(M\frac{\partial N}{\partial Q} + N\frac{\partial M}{\partial Q}\right) + \phi V\frac{\partial V}{\partial Q}\left(\frac{1}{AG}\right)\right] dS \qquad 7.318$$

or

$$\delta = \frac{pR^2}{AE}\left[\left(1 + \frac{R}{e}\right)\left(\frac{3\pi}{2}\right) + \pi\left(1 + \frac{\phi E}{2G}\right)\right] \qquad 7.319$$

However, $A = bh$ and $e = h^2/12R$ (see Chapter 3); therefore

$$\delta = \frac{pR^2}{bhE}\left(\frac{5\pi}{2} + \frac{\pi\phi E}{2G} + \frac{18\pi R^2}{h^2}\right) \qquad 7.320$$

When h is small compared to R, $5\pi/2 + \pi\phi E/2G$ vanishes and

$$\delta = 3\pi pR^4/2EI \qquad 7.321$$

which is the deflection equation for a thin ring. See also Fig. 7.25.

7-4 UNIT LOAD OR DUMMY LOAD METHOD

The unit load method is an extension of Castigaliano's theorem, which is shown by an example. A slender cantilever beam, Fig. 7.64, will first be analyzed by Castigliano's theorem. The phantom load Q is applied, and the bending

Fig. 7.64 Cantilever beam deflection.

moment is

$$M = (F + Q)x$$

$$\frac{\partial M}{\partial Q} = x$$

and

$$\delta = \int_0^L \frac{M(\partial M/\partial Q)}{EI} dx$$

$$= \left[\frac{Fx^3}{3EI} + \frac{Qx^3}{3EI}\right]\Bigg|_0^L$$

setting Q equal to zero

$$\delta = \frac{FL^3}{3EI}$$

When applying the unit load method, the two models in Fig. 7.65 are used—one for the unit load and the other for real loads. The bending moment from the real loads is

$$M = Fx$$

and from the unit load

$$\frac{\partial M}{\partial Q} = m = x$$

substituting into the bending equation

$$\delta = \int_0^L \frac{M(\partial M/\partial Q)}{EI} dx$$

$$\delta = \frac{Fx^3}{3EI}\Bigg|_0^L$$

$$\delta = \frac{FL^3}{3EI}$$

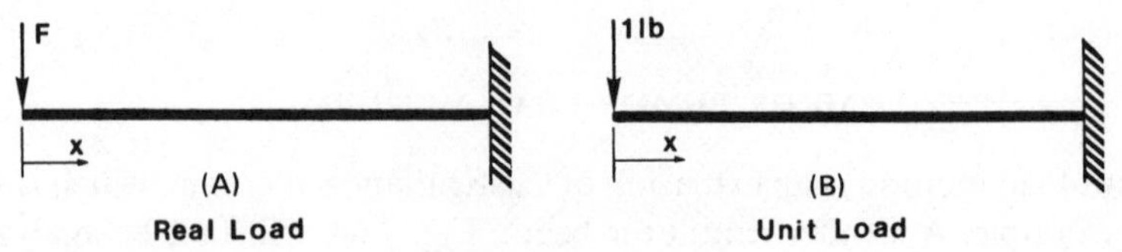

Fig. 7.65 Deflection models.

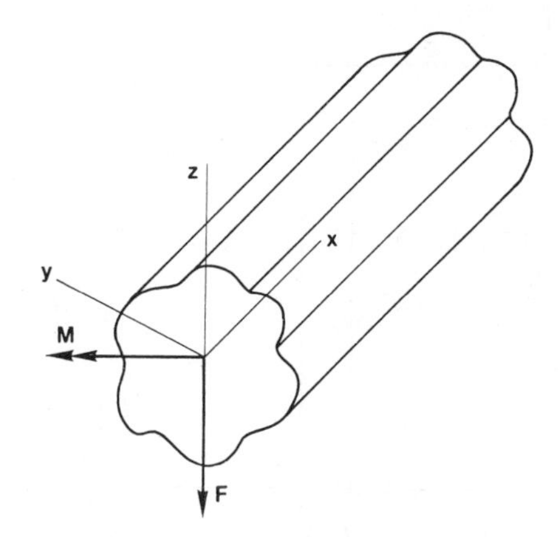

Fig. 7.66 Beam element.

The unit load method is used to obtain solutions to many engineering problems. The method is widely used in the analysis of structures.

In the previous examples only the bending energy is considered in finding the deflections. When a beam element is cut as in Fig. 7.66, a moment vector **M** and a force vector **F** are the resultants.

$$\mathbf{M} = T\mathbf{i} + M_y\mathbf{j} + M_z\mathbf{k} \tag{7.322}$$

$$\mathbf{F} = P\mathbf{i} + V_y\mathbf{j} + V_z\mathbf{k} \tag{7.323}$$

The term T is the torque with M_y and M_z the bending moments for the cross section. Then, in Eq. 7.323, P is the axial force with V_y and V_z as the shear forces on the beam cross section. These terms hold the beam in equilibrium and contribute to the total deflection and rotation at the cut in Fig. 7.66 such that

$$\boldsymbol{\delta}_t = \delta_x\mathbf{i} + \delta_y\mathbf{j} + \delta_k\mathbf{k}$$
$$\boldsymbol{\theta}_t = \theta_x\mathbf{i} + \theta_y\mathbf{j} + \theta_k\mathbf{k}$$

The beam deflections δ_x, δ_y, and δ_z at a point are developed from a single $1 - \text{lb}$ load applied at the point on a structural model and in a desired direction. Also, the beam rotations θ_x, θ_y, and θ_z at a point are developed from a single 1-in.-lb moment or torque applied at the point on a structural model and in a desired sense of rotation. The deflection or rotation for a single beam element is

$$\delta \text{ or } \theta = \int_0^L \frac{M_y m_y}{EI_{yy}} dx + \int_0^L \frac{M_z m_z}{EI_{zz}} dx + \int_0^L \frac{Tt}{GJ} dx + \int_0^L \frac{Pp}{EA} dx + \int_0^L \frac{F V_y v_y}{GA_{zz}} dx + \int_0^L \frac{F V_z v_z}{GA} dx \tag{7.324}$$

The capital letters denote the terms bending moment, shears, torque, and axial load developed from the real load model; the small letters denote the same terms developed from the application of a single unit load. When one or both of the terms in the numerator of the integral is zero, the integral value is also zero.

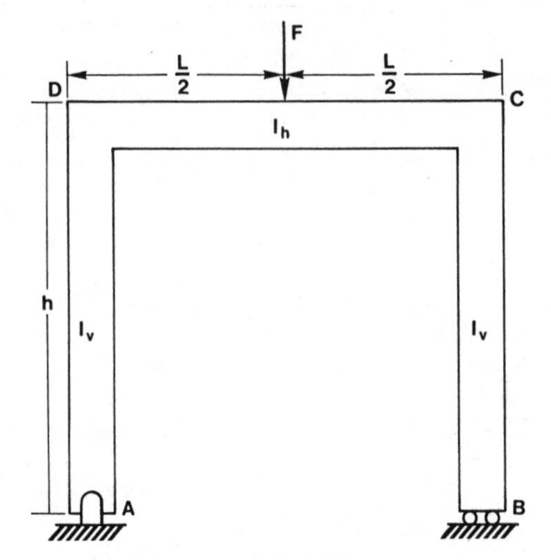

Fig. 7.67 Loaded frame.

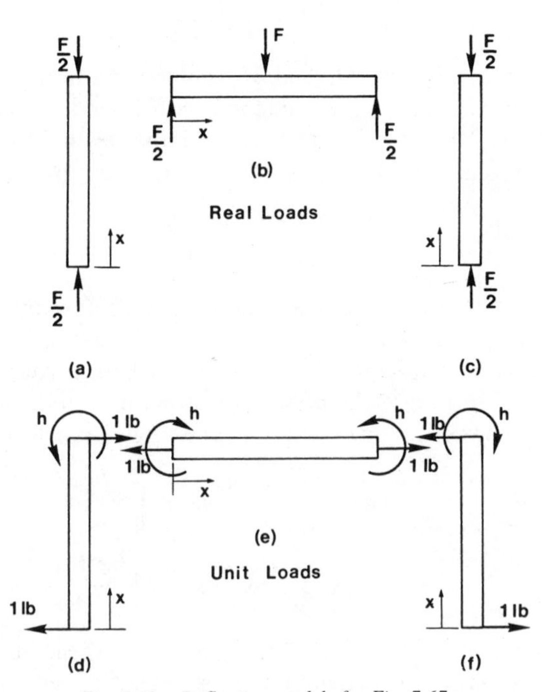

Fig. 7.68 Deflection models for Fig. 7.67.

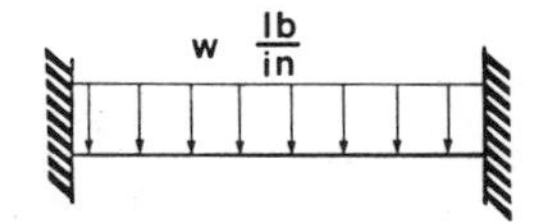

Fig. 7.69 Fixed-end beam.

The shear integral terms are often dropped for long slender beams. However, when dealing with composites or reinforced elements, shear deflections become important as the shear modulus G decreases. The term F is dependent on the shape of the beam and is often taken as one, unless shear deflections become important. The term J is the polar moment of inertia for torsion. The F and J terms are available (18) in handbooks.

The unit load method will be used to find the deflection of point B in Fig. 7.67. The single-beam elements are shown in Fig. 7.68 for Eq. 7.324, which shows the contribution of each beam to the total deflection at B. When the terms in Eq. 7.324 are developed from Fig. 7.68, either the real loads or unit loads are zero except in beams b and e where

$$M_y = -\frac{F}{2}x \qquad 0 \le x \le L/2$$

$$m_y = -h$$

then

$$\delta_B = 2\int_0^{L/2} \frac{M_y m_y}{EI_H} dx$$

$$\delta_B = \frac{FhL^2}{EI_H}$$

This solution is valid for short, thick beams as shear deflections are not neglected.

In statically indeterminate problems, Eq. 7.324 is used to solve for unknown reactions. As an example, find the deflection at the middle of the fixed-end slender beam in Fig. 7.69. The beam is twice statically indeterminate in Fig. 7.70(A), and this problem must be solved first before finding the center deflection. In Fig. 7.70(A) an unknown force R and an unknown moment S are applied to make δ_R and θ_S both zero for the fixed-end condition. The real load

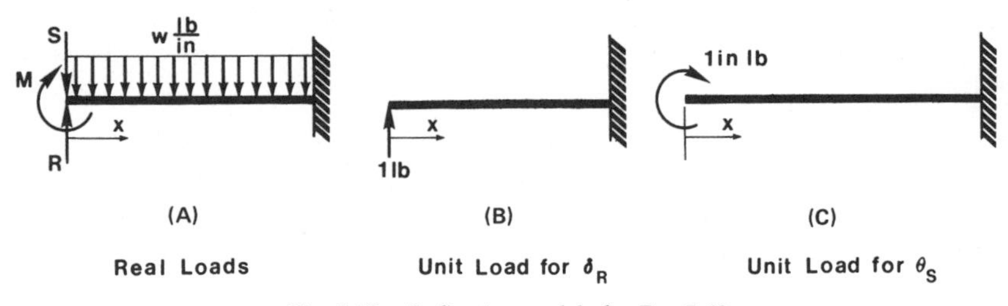

Fig. 7.70 Deflection models for Fig. 7.69.

bending moment is

$$M_y = S + Rx - \frac{\omega x^2}{2} \qquad 0 \le x \le L$$

In Fig. 7.70(*B*) the unit bending moment for δ_R is

$$m_y = x$$

In Fig. 7.70(*C*) the unit bending moment for θ_S is

$$m_y = 1$$

The only terms in Eq. 7.324 used for the slender beam are

$$\delta_R = 0 = \int_0^L \frac{M_y m_y}{EI} dx$$

$$\theta_R = 0 = \int_0^L \frac{M_y m_y}{EI} dx$$

The integral when evaluated yields two equations with two unknowns.

$$\frac{SL^2}{2} + \frac{RL^3}{3} - \frac{\omega L^4}{8} = 0$$

$$SL + \frac{RL^2}{2} - \frac{\omega L^3}{6} = 0$$

Solving these equations

$$S = -\frac{\omega L^2}{2} \qquad R = \frac{\omega L}{2}$$

The sign of *S* means the direction of the moment is reversed from that shown in Fig. 7.70(*A*). Once the values of *S* and *R* are known, the bending moment due to the real loading is

$$M_y = \frac{\omega L^2}{12} + \frac{\omega L x}{2} - \frac{\omega x^2}{2}$$

The unit load, Fig. 7.71, for the midpoint deflection of the beam gives a bending moment of

$$m_y = 0 \qquad 0 \le x \le \frac{L}{2}$$

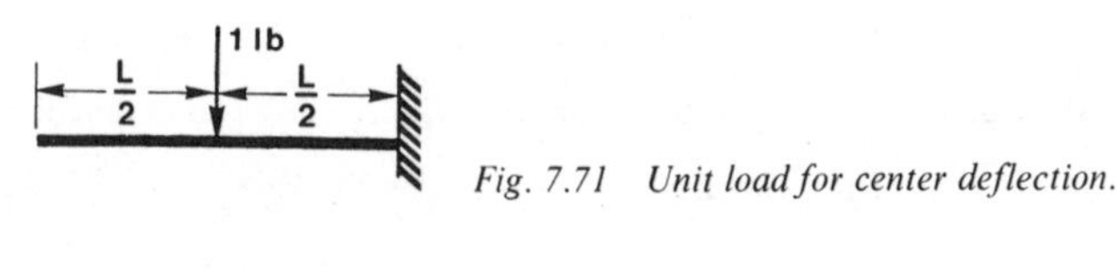

Fig. 7.71 Unit load for center deflection.

$$m_y = 1\left(x - \frac{L}{2}\right) \quad \frac{L}{2} \leq x \leq L$$

These equations are substituted into Eq. 7.324 where the midpoint deflection is

$$\delta = \int_{L/2}^{L} \frac{M_y m_y}{EI}\, dx$$

yielding a solution of

$$\delta = \frac{\omega L^4}{384EI}$$

It should be noted that three rotations and three deflections can be used to solve for six unknown reactions at a point on a beam.

7-5 FLEXIBILITY MATRIX AND STIFFNESS MATRIX

In Eq. 7.324 the unit load method (1,17) gives the linear deflection at a point i due to a load at another point j. Also, the rotation can be found from an applied moment. When the loads are applied at several points

$$r_i = C_{ij} R_j \qquad 7.325$$

The loads R_j can be applied one at a time giving C_{ij} the flexibility matrix.

A cantilever beam, Fig. 7.72, is used to obtain the C_{ij} values for the tip. The unit load models are also used for the real loads. The tip has three deflections in the x, y, z directions and three rotations, θ_x, θ_y, θ_z. The displacements and rotations in order will be $r_1, r_2, r_3, r_4, r_5, r_6$ for the tip with the positive coordinate system shown in Fig. 7.72. The values for the positive loads, Fig. 7.73,

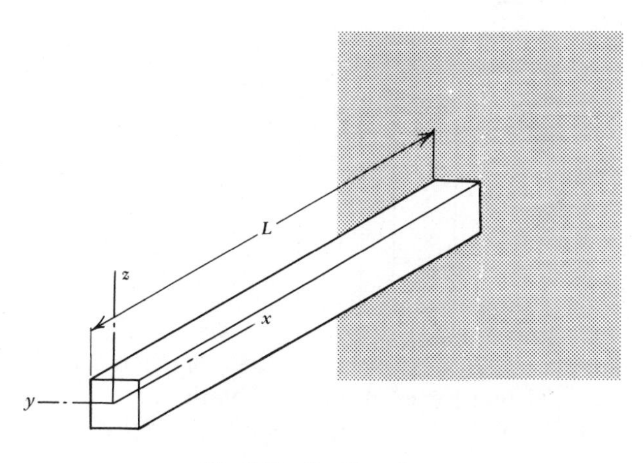

Fig. 7.72 Cantilever beam.

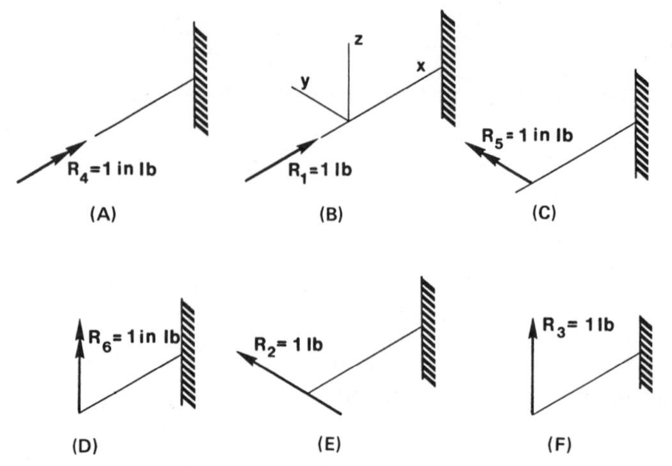

Fig. 7.73 Real and unit load models for Fig. 7.72.

are substituted into Eq. 7.325 with the following results.

$$\begin{bmatrix} x \\ y \\ z \\ \theta_x \\ \theta_y \\ \theta_z \end{bmatrix} = \begin{bmatrix} r_1 \\ r_2 \\ r_3 \\ r_4 \\ r_5 \\ r_6 \end{bmatrix} = \begin{bmatrix} \frac{L}{EA} & 0 & 0 & 0 & 0 & 0 \\ 0 & \left(\frac{L^3}{3EI_{zz}} + \frac{FL}{GA}\right) & 0 & 0 & 0 & \frac{-L^2}{2EI_{zz}} \\ 0 & 0 & \left(\frac{L^3}{3EI_{yy}} + \frac{FL}{GA}\right) & 0 & \frac{L}{2EI_{yy}} & 0 \\ 0 & 0 & 0 & \frac{L}{GJ} & 0 & 0 \\ 0 & 0 & \frac{L^2}{2EI_{yy}} & 0 & \frac{L}{EI_{yy}} & 0 \\ 0 & \frac{-L^2}{2EI_{zz}} & 0 & 0 & 0 & \frac{L}{EI_{zz}} \end{bmatrix} \begin{bmatrix} R_1 \\ R_2 \\ R_3 \\ R_4 \\ R_5 \\ R_6 \end{bmatrix} \qquad 7.326$$

where

$$\begin{bmatrix} R_1 \\ R_2 \\ R_3 \\ R_4 \\ R_5 \\ R_6 \end{bmatrix} = \begin{bmatrix} P \\ V_y \\ V_z \\ T \\ M_y \\ M_z \end{bmatrix} \qquad 7.327$$

When examining Eqs. 7.326 and 7.327, it can be seen that torques and moments cause deflections and that forces also cause rotations.

Eq. 7.325 can be multiplied by the inverse of $[C_{ij}]$ so that

$$[C_{ij}]^{-1} r_i = [C_{ij}]^{-1} [C_{ij}] R_j \qquad 7.328$$

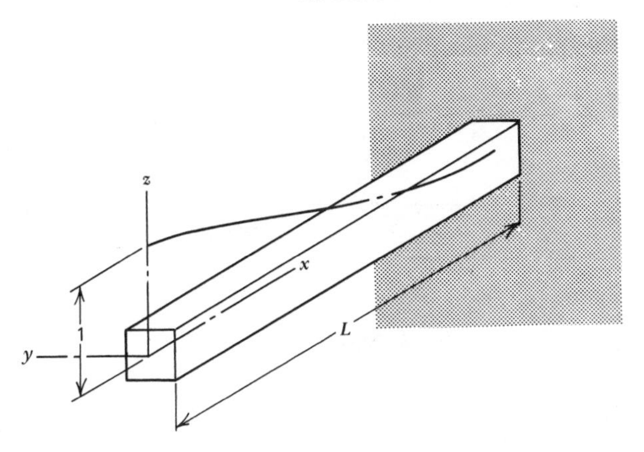

Fig. 7.74 Unit deflection of cantilever beam.

Also, it is noted that

$$[C_{ij}]^{-1}[C_{ij}]=1$$

And, defining the stiffness matrix as

$$K_{ij}=[C_{ij}]^{-1} \tag{7.329}$$

then, Eq. 7.297 can be rewritten as

$$R_i=K_{ij}r_j \tag{7.330}$$

The calculations of K_{ij} values are made more complicated than the C_{ij} values. In Eq. 7.330 when a single rotation or deflection is unity then all the other r_j's are zero or fixed so that the resulting R_i forces can be calculated.

In Fig. 7.72 deflect the tip 1 unit in the z direction resulting in Fig. 7.74. However, note that θ_y or r_5 must also be zero, as are the other values. The rotation at r_5 is made zero by applying a moment S. In Fig. 7.75 the real loads and unit loads are shown with positive signs taken from Fig. 7.72 and 7.73. The use of Eq. 7.324 results in

$$\theta_S=\int_0^L \frac{M_y m_y}{EI_{yy}}dx=0$$

$$\delta_R=1=r_3=\int_0^L \frac{M_y m_y}{EI_{yy}}\mathrm{d}x+\int_0^L \frac{FV_z v_z}{GA}dx$$

The real loads from Fig. 7.75(*A*) are

$$M_y=Rx+S \qquad V_z=R$$

The unit load for θ_S from Fig. 7.75(*C*) is

$$m_y=1$$

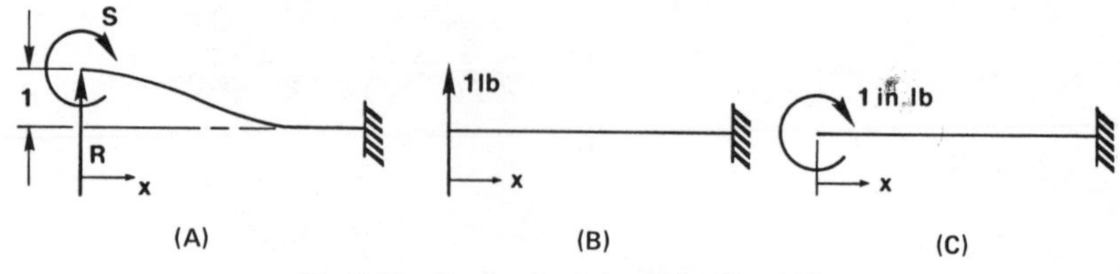

Fig. 7.75 Real and unit loads for Fig. 7.74.

The unit loads for r_3 from Fig. 7.75(B) are

$$m_y = x \qquad v_z = 1$$

These values when substituted in r_3 and θ_S yield

$$\frac{RL^2}{2EI_{yy}} + \frac{SL}{EI_{yy}} = 0$$

$$\frac{RL^3}{3EI_{yy}} + \frac{SL^2}{2EI_{yy}} + \frac{FRL}{GA} = 1$$

The solution for R and S gives

$$R = \frac{12EI_{yy}}{L^3}\left\{\frac{1}{\left[1 + 12F\left(\frac{EI_{yy}}{GAL^2}\right)\right]}\right\} \tag{7.331}$$

$$S = -\frac{6EI_{yy}}{L^2}\left\{\frac{1}{\left[1 + 12F\left(\frac{EI_{yy}}{GAL^2}\right)\right]}\right\} \tag{7.332}$$

The last two terms in the denominator should be less than 10% when neglecting the deflection due to shear.

When unit deflections are given to each coordinate separately for a slender cantilever beam in Fig. 7.72 the results are

$$\begin{bmatrix} R_1 \\ R_2 \\ R_3 \\ R_4 \\ R_5 \\ R_6 \end{bmatrix} = \begin{bmatrix} P \\ V_y \\ V_z \\ T \\ M_y \\ M_z \end{bmatrix} = \begin{bmatrix} \frac{EA}{L} & 0 & 0 & 0 & 0 & 0 \\ 0 & \frac{12EI_{zz}}{L^3} & 0 & 0 & 0 & \frac{6EI_{zz}}{L^2} \\ 0 & 0 & \frac{12EI_{yy}}{L^3} & 0 & -\frac{6EI_{yy}}{L^2} & 0 \\ 0 & 0 & 0 & \frac{GJ}{L} & 0 & 0 \\ 0 & 0 & -\frac{6EI_{yy}}{L^2} & 0 & 4\frac{EI_{yy}}{L} & 0 \\ 0 & \frac{6EI_{zz}}{L^2} & 0 & 0 & 0 & 4\frac{EI_{zz}}{L} \end{bmatrix} \begin{bmatrix} r_1 \\ r_2 \\ r_3 \\ r_4 \\ r_5 \\ r_6 \end{bmatrix} \tag{7.333}$$

$$\begin{bmatrix} r_1 \\ r_2 \\ r_3 \\ r_4 \\ r_5 \\ r_6 \end{bmatrix} = \begin{bmatrix} x \\ y \\ z \\ \theta_x \\ \theta_y \\ \theta_z \end{bmatrix} \tag{7.334}$$

The K_{ij} stiffness matrix in Eq. 7.333 is one-fourth the expression for the stiffness matrix of a slender beam finite element to be discussed in Chapter 8.

The flexibility matrix C_{ij} is the easier to compute by hand while the stiffness matrix K_{ij} is harder for hand calculations. However, in large computer programs the reverse is true in that the stiffness matrix K_{ij} and the concept of finite elements in Chapter 8 work out better for the design analyst or the designer. It should be noted that for crude design checks the flexibility matrix C_{ij} and a hand calculator that can operate on three or four variables will yield quick and economical answers. These answers can be in the structural analysis area or in obtaining vibration frequencies (Chapter 10), which must always be checked.

A detailed discussion of computer matrix structural analysis is not presented. In both the determinate and indeterminate structures problems the user, for most cases, is not concerned with the functions performed inside the computer but what information must be supplied, how to idealize the structure, interpret the answers, and modify the structures to meet design requirements. However, in order to obtain some basic understanding of matrix analysis, the work by Pestel and Leckie (12) gives an overall fundamental approach. The references (2, 5, 10, and 28) develop structures in terms of stiffness and flexibility for computer usage. The references (16, 19, and 20) use the stiffness and flexibility matrices for solutions of structural problems in vibration, dynamics, elastic stability, and stress.

REFERENCES

1. J. H. Argyris and S. Kelsey, *Energy Theorems and Structural Analysis*, Butterworths and Co., Ltd., London (1960).
2. F. W. Beaufait, W. H. Rowan, Jr., P. G. Hoadley, and R. M. Hackett, *Computer Methods of Structural Analysis*, Prentice-Hall, Englewood Cliffs, N.J. (1970).
3. A. Blake, *Prod. Eng.*, **34** (1), 70–81 (January 7, 1963).
4. S. F. Borg and J. J. Gennaro, *Modern Structural Analysis*, D. Van Nostrand Co., Princeton, N.J. (1969).
5. J. M. Gere, and W. Weaver, Jr., *Analysis of Frame Structures*, D. Van Nostrand, Princeton, N.J. (1965).
6. A. Kleinlogel, *Rigid Frame Formulas*, Frederick Ungar Publishing Co., New York, N.Y. (1952).
7. H. L. Langhaar, *Energy Methods in Applied Mechanics*, John Wiley and Sons, Inc., New York, N.Y. (1962). Also *J. Franklin Inst.*, **256** (9), 255–264 (September 1953).
8. C. E. Larard, *Phil Mag.*, **13**, 705–710 (1932).

9. S. S. Manson, *Mach. Des.*, **31** (4), 156–160 (February 19, 1959).

10. H. C. Martin, *Matrix Methods of Structural Analysis*, McGraw-Hill Book Co., New York, N.Y. (1966).

11. A. F. Menton, *Space/Aeronaut.*, **37** (6), 91–96 (June 1962).

12. E. C. Pestel and F. A. Leckie, *Matrix Methods in Elastomechanics*, McGraw-Hill Book Co., New York, N.Y. (1963).

13. L. A. Pipes, *J. Franklin Inst.*, **274** (3), 198–226 (September 1962).

14. A. J. S. Pippard, *Studies in Elastic Structures*, Edward Arnold Co., London (1952).

15. A. J. S. Pippard and J. Baker, *The Analysis of Engineering Structures*, American Elsevier, New York, N.Y. (1968).

16. J. S. Przemieniecki *Theory of Matrix Structural Analysis*, McGraw-Hill Book Co., New York, N.Y. (1968).

17. R. M. Rivello, *Theory and Analysis of Flight Structures*, McGraw-Hill Book Co., New York, N.Y. (1969).

18. R. J. Roark and W. C. Young, *Formulas for Stress and Strain*, 5th ed., McGraw-Hill Book Co., New York, N.Y. (1975).

19. M. F. Rubenstein, *Structural Systems—Statics, Dynamics, and Stability*, Prentice-Hall, Englewood Cliffs, N.J. (1970).

20. M. F. Rubenstein, *Matrix Computer Analysis of Structures*, Prentice-Hall, Englewood Cliffs, N.J. (1966).

21. F. B. Seely and J. O. Smith, *Advanced Mechanics of Materials*, John Wiley and Sons, Inc., New York, N.Y. (1952).

22. I. S. Sokolnikoff, *Mathematical Theory of Elasticity*, McGraw-Hill Book Co., New York, N.Y. (1956).

23. S. Timoshenko and J. N. Goodier, *Theory of Elasticity*, 3rd ed., McGraw-Hill Book Co., New York, N.Y. (1970).

24. S. Timoshenko, *Strength of Materials*, Vol. 1, 2nd and 3rd eds., D. Van Nostrand Co., Princeton, N.J. (1940, 1955).

25. S. Timoshenko, *Strength of Materials*, Vol. 2, 2nd and 3rd eds, D. Van Nostrand Co., Princeton, N.J. (1941, 1956).

26. J. A. Van Den Broek, *Elastic Energy Theory*, John Wiley and Sons, Inc., New York, N.Y. (1942).

27. H. M. Westergaard, *Trans. Am. Soc. Civ. Eng.*, **107**, Paper No. 2145, 765–803 (1942).

28. N., Willems and W. M. Lucas, Jr., *Matrix Analysis for Structural Engineers*, Prentice-Hall, Englewood Cliffs, N.J. (1968).

PROBLEMS

7.1 An SAE 1020 steel helical lock washer 2.43 in. O.D., 1.56 in. I.D., and 0.375 in. thick is loaded as shown in Fig. P7.1. Find the maximum deflection of one force relative to the other before yielding of 30 kpsi occurs. Assume the initial gap of the washer is zero.

7.2 The same lock washer in Problem 7.1 is loaded as shown in Fig. P7.2. Find the maximum relative deflection before yielding occurs.

Fig. P7.1

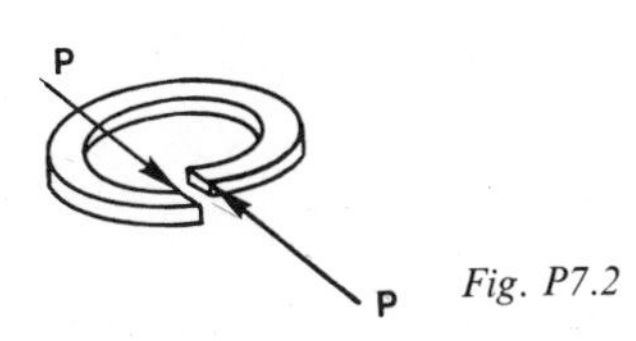

Fig. P7.2

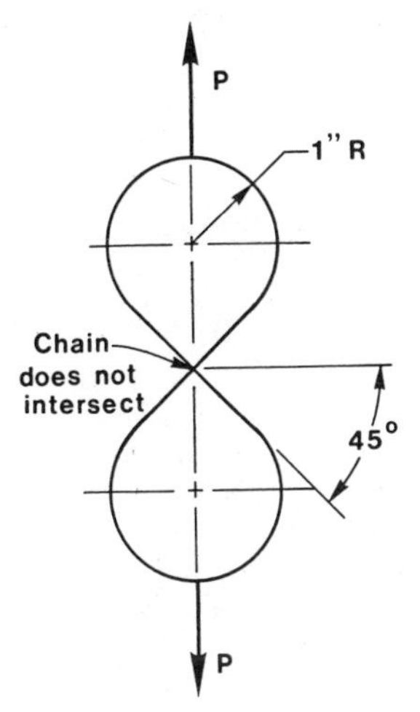

Fig. P7.3

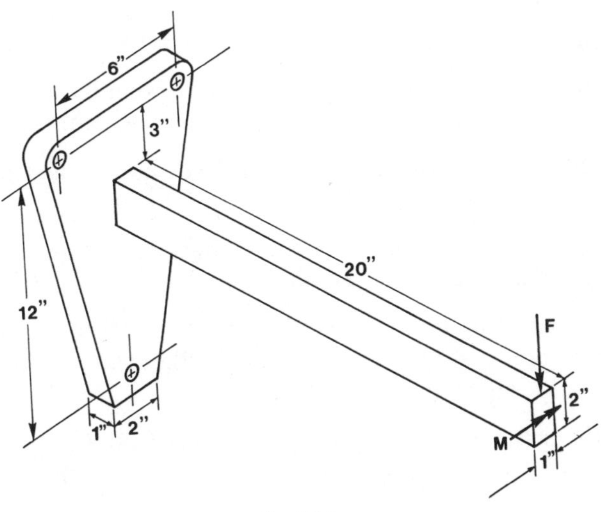

Fig. P7.7

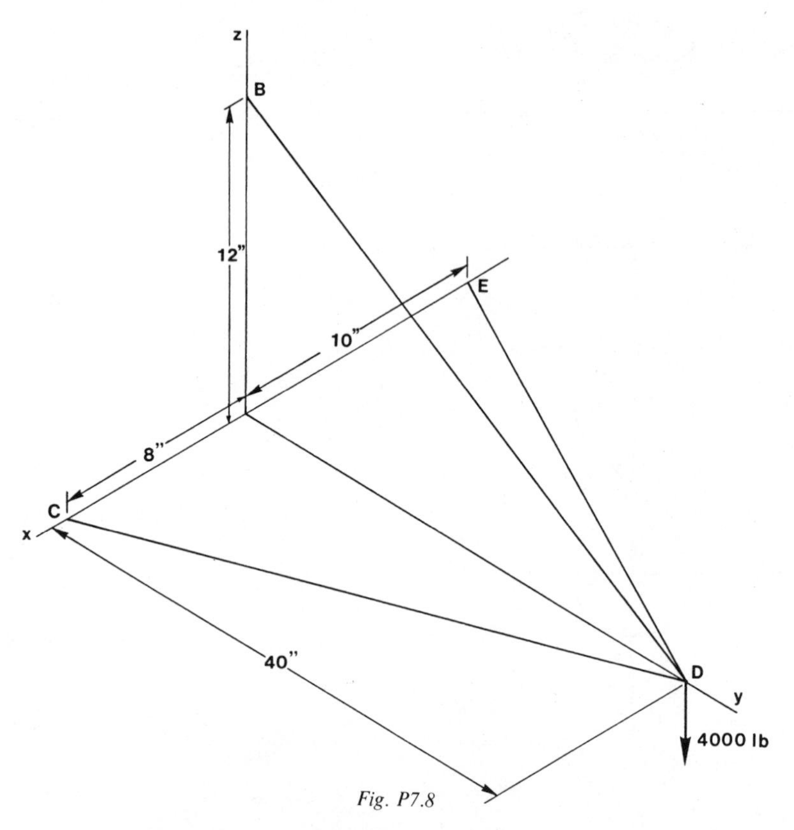

Fig. P7.8

7.3 A chain is made of SAE 1020 0.250-in. diameter steel links (see Fig. P7.3). What is the maximum load for one link and a chain 100 ft in length? What is the deflection of one link and of the chain of 100 ft?

7.4 In Problem 2.2 find the deflection relative to point 0 using the unit load method. Does the deflection exceed 4 in. and release the load?

7.5 In Problem 2.3 if bearing C acts as a fixed support, find the reactions at A, C, and E for the chain drive only.

7.6 Refer to Problem 3.8. Find the deflection of one 5000-lb load relative to the other. Also, find the slope of the bar at that point if possible.

7.7 Find the tip deflections if $F = 1000$ lb, $M = 1000$ in.-lb, and the steel bracket is bolted at the top and bottom in Fig. P7.7. Assume the bottom hole carries all of the vertical reaction.

7.8 Find the deflections of the steel frame (Fig. P7.8) with a σ_y of 40 kpsi at point D. All joints are pinned. Find A, the area, so the frame does not buckle.

7.9 In problem 6.10 find the load in each member by the unit load method.

7.10 Solve Problem 6.6 for the shear and bending moments in the frame.

chapter 8

FINITE ELEMENTS AND FINITE DIFFERENCE METHODS

The advances in hand calculators and large structural computer programs have reduced the need for many graphical and analytical methods commonly used. Instead, many deflection and stress calculations are available as standard interaction computer routines. Complex problems are modeled by finite elements using large computer programs. Finite element method concepts have been used since the early 1940s and were finally named as such by R. W. Clough in 1960 (4). The finite elements are the basic modeling tools in matrix structural analysis (15).

The finite difference method, as opposed to the finite elements, is used for equilibrium equations with partial derivatives of deflections. The solutions give the desired deflections and other information. These two techniques are used for stress, deflection, vibration, creep, viscoelastic, buckling, and thermal effects in both composite and other complex systems.

8-1 FINITE ELEMENTS

Complex structural systems are often made of beams, plates, and shells tangent to and intersecting each other. The intersections present boundary conditions, many of which have never been analyzed. The plates and shells have irregular boundaries, varying thicknesses, and often cutouts. The hand calculation of stiffness or influence coefficients represents a large task since the systems are often statically indeterminate. Further, the requirements of meeting shock, vibration, and thermal environments complicate the analysis. Accurate calculations are required so that a component is not highly stressed and develops cracks, thus failing the system. Deflection requirements to four decimals can require different considerations from the stress calculations, such as individual component stiffness.

All of the above requirements could be met if the input to a computer routine is simple enough. Large computer routines (Chapter 7) have been developed to save time and simplify an analysis. The input is developed from drawings partitioned into the finite elements available in a computer routine. The finite elements are beams, plates, cubes, and wedges. These elements form basic

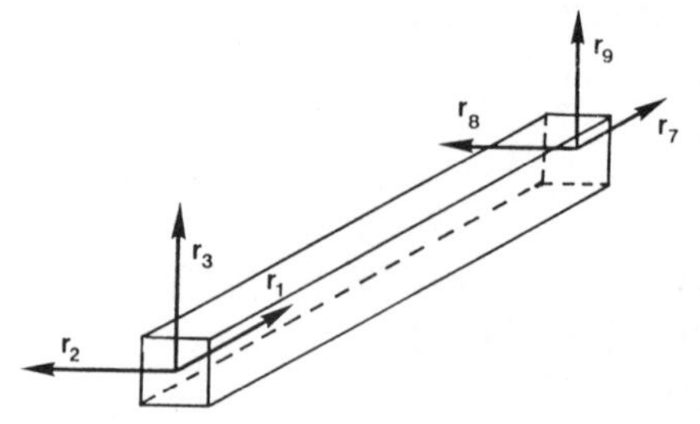

Fig. 8.1 Beam element.

building blocks to model the structural system. For the applications of these principles to various problems see (3, 6, 7, 11, 13, 17, 22, 27).

Beams

Beams have been used in analysis of frame structures (8) in civil engineering, and many references (see Chapter 7) used beams as finite elements prior to the introduction of the name in 1960 (4). Therefore, beams are one of the easier finite elements to understand. A beam element in Fig. 8.1 will have, in general, two moments, two shears, a torque, and an axial force acting on each end. These forces are found with the use of Eq. 7.330 by unit displacements as discussed in Chapter 7. A stiffness matrix is developed with terms for EA, EI_{zz}, EI_{yy}, GJ, GA, and L. This information is to be supplied for each beam element. Eq. 7.333 is one fourth of the stiffness matrix for a slender beam. When beam elements become short, thick beams as can be seen in Fig. 7.20 and Eq. 7.110, it requires the use of plate elements to develop the accuracy desired.

The stiffness of bearings can be modeled using equal length slender beams equally spaced in a circle as in Fig. 8.2. The center represents the center line of a shaft or the bearing. The fixed ends serve as nodes for plate elements. From Eq. 7.330

$$R_i = K_{ij} r_j \qquad 8.1$$

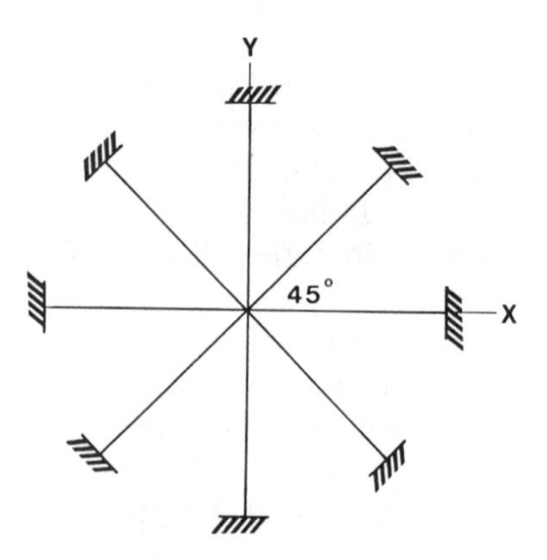

Fig. 8.2 Bearing model.

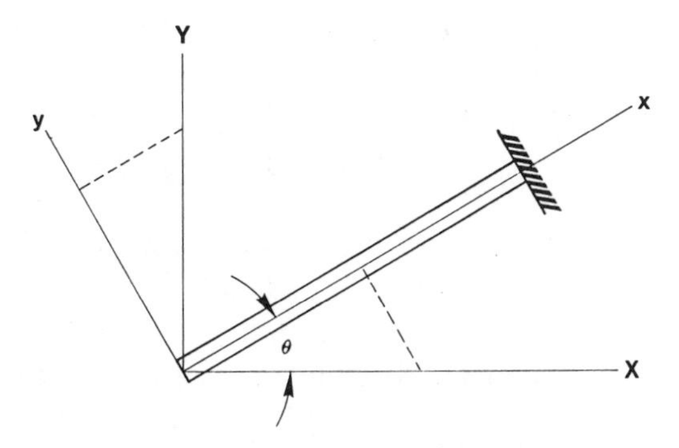

Fig. 8.3 Coordinate transformation.

where K_{ij} is Eq. 7.333, the stiffness of each beam element in its local coordinate system. A transformation (Fig. 8.3) from the global coordinate system X, Y, Z must be made [See (10) in Chapter 7] (8, 15) in terms of the local coordinate system x, y, z, or r_1, r_2, r_3. Let

$$\lambda = \cos\theta \qquad \mu = \sin\theta \tag{8.2}$$

The results are

$$\begin{bmatrix} r_1 \\ r_2 \\ r_3 \end{bmatrix} = \begin{bmatrix} \lambda & \mu & 0 \\ -\mu & \lambda & 0 \\ 0 & 0 & 1 \end{bmatrix} \begin{bmatrix} X \\ Y \\ Z \end{bmatrix} \tag{8.3}$$

The same transformation applies to the global forces, moments, and rotations.

$$[R_i] = \left[\begin{array}{ccc|ccc} \lambda & \mu & 0 & 0 & 0 & 0 \\ -\mu & \lambda & 0 & 0 & 0 & 0 \\ 0 & 0 & 1 & 0 & 0 & 0 \\ \hline 0 & 0 & 0 & \lambda & \mu & 0 \\ 0 & 0 & 0 & -\mu & \lambda & 0 \\ 0 & 0 & 0 & 0 & 0 & 1 \end{array}\right] \left[\begin{array}{c} F_x \\ F_y \\ F_z \\ \hline M_x \\ M_y \\ M_z \end{array}\right] \tag{8.4}$$

or

$$[R_i] = [T]\left[\frac{F}{M}\right] \tag{8.5}$$

Also

$$[r_j] = [T]\left[\frac{\delta}{\theta}\right] \tag{8.6}$$

Substituting Eq. 8.5 and 8.6 into Eq. 8.1

$$[T]\left[\frac{F}{M}\right] = [K_{ij}][T]\left[\frac{\delta}{\theta}\right] \tag{8.7}$$

Multiplying Eq. 8.7 by $[T]^{-1}$, which reduces to $[T]^T$, the transpose of the matrix, yields

$$\left[\frac{F}{M}\right] = [T]^T [K_{ij}][T]\left[\frac{\delta}{\theta}\right] \tag{8.8}$$

where

$$[K_l] = [T]^T [K_{ij}][T] \tag{8.9}$$

is the stiffness of the single beam element in Fig. 8.3 in the global system. The stiffness for the bearing is the matrix addition of each of the beam stiffnesses

$$[K] = [K_1] + [K_2] + \dots + [K_N] \tag{8.10}$$

For a single beam element using $[T]$ in Eq. 8.4 and K_{ij} in Eq. 7.333 for a slender beam Eq. 8.9 becomes

$$[K_l] = \frac{E}{L^3}\begin{bmatrix} (AL^2\lambda^2 + 12I_{zz}\mu^2) & \mu\lambda(AL^2 - 12I_{zz}) & 0 & 0 & 0 & 0 \\ \lambda\mu(AL^2 - 12I_{zz}) & (AL^2\mu^2 + 12\lambda^2 I_{zz}) & 0 & 0 & 0 & 0 \\ 0 & 0 & 12I_{yy} & & & \\ 0 & 0 & 0 & \left(\lambda^2\frac{GJL^2}{E} + 4L^2\mu^2 I_{yy}\right) & \lambda\mu\left(\frac{GJL^2}{E} - 4L^2 I_{yy}\right) & 0 \\ 0 & 0 & 0 & \lambda\mu\left(\frac{GJL^2}{E} - 4L^2 I_{yy}\right) & \left(\mu^2\frac{GJL^2}{E} + 4L^2\lambda^2 I_{yy}\right) & 0 \\ 0 & 0 & 0 & 0 & 0 & 4L^2 I_{zz} \end{bmatrix} \tag{8.11}$$

Let $X = 1$ and the other deflections and rotations in Eq. 8.8 be zero. The results for each beam element are

$$F_x = (AL^2\lambda^2 + 12I_{zz}\mu^2)\frac{E}{L^3} = K_{lx} \tag{8.12}$$

$$F_y = \lambda\mu(AL^2 - 12I_{zz})\frac{E}{L^3} = K_{ly} \tag{8.13}$$

For $Z = 1$ with the other coordinates in Eq. 8.8 zero

$$F_z = \frac{12EI_{yy}}{L^3} = K_{lz} \tag{8.14}$$

For $\theta_x = 1$ and the rest zero

$$M_x = \left(\lambda^2\frac{GJ}{L} + 4\mu^2\frac{EI_{yy}}{L}\right) = K_{l\theta x} \tag{8.15}$$

$$M_y = \lambda\mu\left(\frac{GJ}{L} - \frac{4EI_{yy}}{L}\right) = K_{l\theta y} \tag{8.16}$$

Using the model in Fig. 8.2 and summing Eqs. 8.12–8.16 for the eight beams, the results are

$$\sum F_x = 4\left[\frac{EA}{L} + \frac{12EI_{zz}}{L^3}\right] = K_R \qquad 8.17$$

$$\sum F_y = 0 \qquad 8.18$$

$$\sum F_z = 8\left(\frac{12EI_{yy}}{L^3}\right) = K_A \qquad 8.19$$

$$\sum M_x = 4\left[\frac{GJ}{L} + \frac{4EI_{yy}}{L}\right] = K_\theta \qquad 8.20$$

$$\sum M_y = 0 \qquad 8.21$$

$$\sum M_z = 0 \qquad 8.22$$

In Eqs. 8.18 and 8.21 the terms cancel while in Eq. 8.22 a bearing does not offer torque constraint. K_R, K_A, K_θ are the radial, axial, and rotational stiffnesses of the bearing.

For an example, take the following bearing:

$K_A = 3 \times 10^6$ lb/in. $\qquad K_R = 2.5 \times 10^6$ lb/in.
$K_\theta = 4 \times 10^6$ in.-lb/rad $\qquad L = 2$ in.
$E = 30 \times 10^6$ psi

Let the two terms in Eq. 8.17 be equal yielding

$$A = 0.02083 \text{ in.}^2 \qquad I_{zz} = 0.00694 \text{ in.}^4$$

From Eq. 8.19 solve for

$$I_{yy} = 0.00833 \text{ in.}^4$$

Substitute into Eq. 8.20 and find

$$GJ = 1 \times 10^6 \text{ lb/in.}^2$$

Before finding J, Eq. 7.331 must be checked for shear through the following term with F unity.

$$F\frac{12EI_{yy}}{GAL^2} = \frac{12(30 \times 10^6)(0.00833)}{G\,(0.02083)\,(4)} \qquad 8.23$$

$$F\frac{12EI_{yy}}{GAL^2} = \frac{35.99 \times 10^6}{G}$$

Eq. 8.23 must be much less than 1—say 0.1—for shear to be neglected, or

$$G = 360 \times 10^6 \text{ psi}$$

While this material does not exist, the computer will accept it for modeling. Then, J becomes

$$J = 0.00278 \text{ in.}^4$$

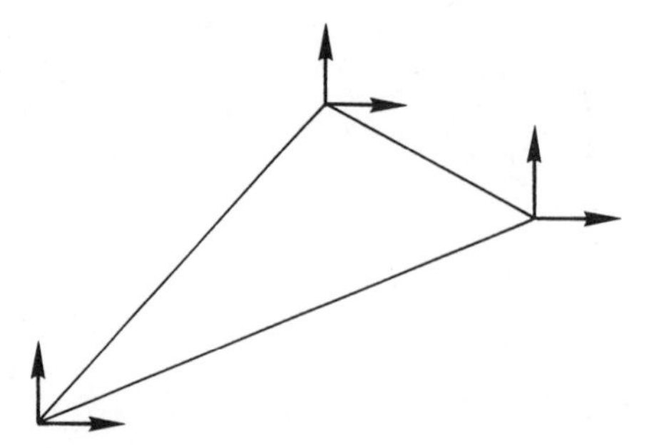

Fig. 8.4 Membrane element.

When a frame structure exists where shear deflections are important, (15) gives the K_{ij} for this condition.

Plates

Triangular and rectangular plates are popular elements for plane stresses (7, 17, 22) and plate bending. These elements are loaded with forces only at the corners (Fig. 8.4) with the sides free and unable to support any load. Plate elements of this type are used for plane stress and membrane problems. In Fig. 8.5 if bending moments are added in the plane, this type of element can be used for plate problems. Both of the elements can be used for linear and nonlinear problems (6, 11, 20, 27). Nonlinear problems occur in instability, large membrane deflections, and when the limit of thin-plate analysis is exceeded.

The interesting question has always been: when is a plate thin and when should large deflections be included? Comparing small deflection and large deflection theory (6, 24) at a maximum deflection/plate thickness δ/t of 0.8, the percent error varies from 60% for a simply supported plate with uniform loading to 14.3% for a fixed-edge plate with the same loading. When δ/t is 1, the error varies from 100% to 33.3% for the corresponding conditions. When δ/t is 0.4 and less in both cases, the agreement is very close. In many cases in a

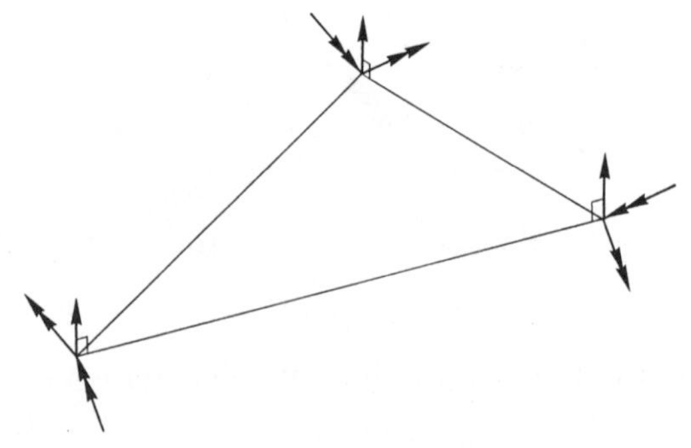

Fig. 8.5 Plate bending element.

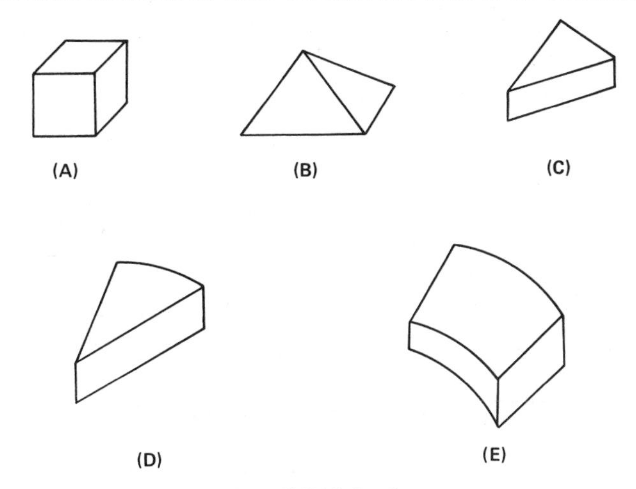

Fig. 8.6 Solid finite elements.

design, true simply supported conditions are only obtained with considerable effort. Therefore, δ/t of 1 from a finite solution would be a fair rule for the upper bound on a thin plate since the finite element solution makes the plate more flexible than it really is. Further, most engineering cases tend to be more fixed than simply supported.

The number of finite elements used affects the accuracy of the results for most problems. In some cases, for example (12, 17, 24), the use of as few as 16 elements yielded deflection and stress results with less than 10% error. On the other hand, use of only four elements for a uniformly loaded clamped plate (17) yielded 18% error in deflection and 30% in stress. It should be noted that computer routines for the elements vary so that accuracy will also depend on the finite element routine available. Further, the manner in which a problem is partitioned will affect the accuracy (7, 30) such as a crude grid where the stress is low and a finer grid where the stress is higher.

Solids and Shells

When stress calculations are performed on complicated shapes, the elements become much smaller with shapes in Fig. 8.6 such as rectangular solids, tetrahedrons, and wedges for axisymmetric work (1, 13, 17, 27). The special shapes in Fig. 8.6B–E are used for modeling around holes and spheres. The elements generally have forces applied only at the corners. However, there are finite elements with loading on the surfaces so that a loading condition can be modeled as in Fig. 8.7. This particular situation is one where the loading is nearly over the entire surface, and it definitely is not loaded as a plate. The thickness of the window is also of the magnitude as the other dimensions, which also makes analysis as a plate a poor selection. The stresses are important to prevent cracking of the material and, hence, a finer mesh is expected.

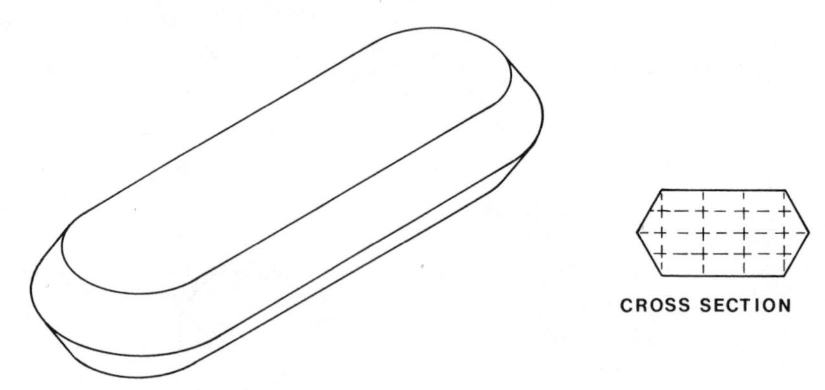

Fig. 8.7 Diving sphere window.

8-2 FINITE ELEMENT METHOD OF STRESS ANALYSIS

Structural and process equipment design problems are being solved increasingly by computer programs utilizing the finite element method of stress analysis. This method is a very powerful one and reserved for problems that can be programmed and solved by computer techniques. The method is never used in a "hand" type of calculation because of the enormous amount of computation required.

The literature, particularly during the past 5–10 years abounds in references to both development and application of finite element programs (5, 18, 28, 31–35) to solve a variety of problems, both static and dynamic. Many programs have both linear and nonlinear options so that, in addition to ordinary elasticity problems, special problems involving creep, viscoelastic, plasticity, and stress rupture can be solved. The field is progressing so rapidly that the analyst must keep pace with developments and availability of new finite element computer programs.

In this chapter only the basic elements of the finite element method will be touched on and illustrated by a trivial four-element problem. A couple of equipment design problems will also be presented to illustrate typical application.

As the name of the method implies, the basic building block is an element of finite dimensions—as contrasted to classical elasticity theory which has its basis in infinitesimal elements. In the finite element method the object analyzed is divided into small segments like triangles, for example. The triangles are joined at the corners, and it is usually assumed that the stress is uniform throughout the element. The element distortions are computed by conventional theory—a triangular element, for example, becomes a wedge-shaped cantilever beam. The total behavior of the structure depends on the integrated effects of each of the parts; thus, since a part is usually divided into a multitude of elements, a solution is possible only on a computer. Hand solution of a trivial

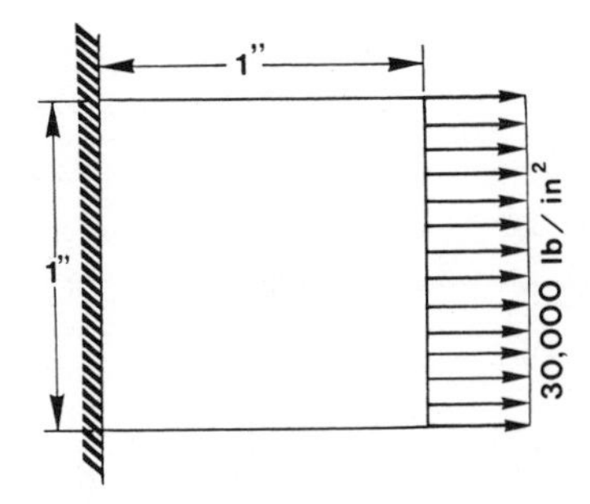

Fig. 8.8 Tension-loaded bar.

problem with only four elements results in the analyst having to solve 10 simultaneous equations.

In general, the sequence of steps (3, 14, 17, 19, 27) in a finite element program is:

1. Divide part into segments or "elements" (triangles, for example).
2. Superpose elements on grid coordinate system in order to establish geometrical orientation of corners (nodes) of elements; this eventually leads to a B matrix.
3. From elastic properties of material form a D matrix.
4. Transform B and D matrices to a K (stiffness) matrix.
5. Solve for stresses and strains.

The above steps are illustrated in the following trivial example.

Step 1: Sketch the structure, Fig. 8.8.

Step 2: Divide structure into elements, Fig. 8.9.

Step 3: Indicate loads and distortions at the element nodes, Fig. 8.10. At node 1, for example, the motion in the x direction is u_1 with force F_1x.

Step 4: Sketch individual elements and orientation, Fig. 8.11. Each element is identified on the x–y coordinate system as ijm, counterclockwise. From this

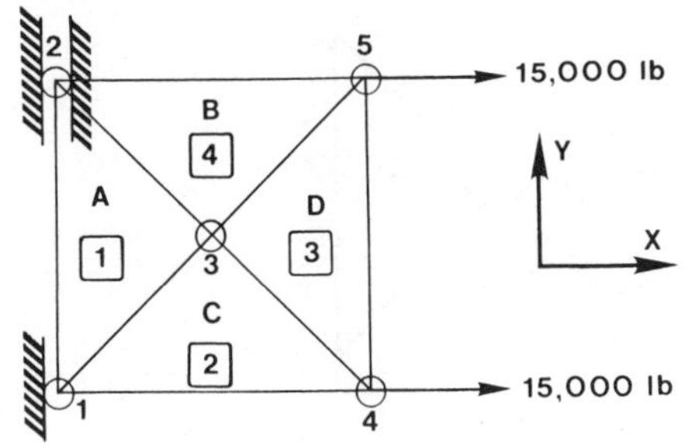

Fig. 8.9 Finite element representation of Fig. 8.8.

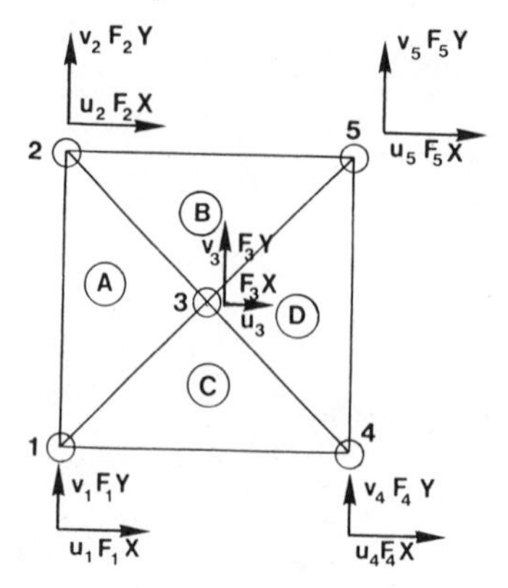

Fig. 8.10 Force and displacement coordinates for Fig. 8.9.

system comes the B matrix.

$$[B] = \frac{1}{2\Delta}\begin{bmatrix} b_i & 0 & b_j & 0 & b_m & 0 \\ 0 & c_i & 0 & c_j & 0 & c_m \\ c_i & b_i & c_j & b_j & c_m & b_m \end{bmatrix} \qquad 8.24$$

where:

$$b_i = y_j - y_m \quad c_i = x_m - x_j$$
$$b_j = y_m - y_i \quad c_j = x_i - x_m$$
$$b_m = y_i - y_j \quad c_m = x_j - x_i$$

Δ = area of an element.

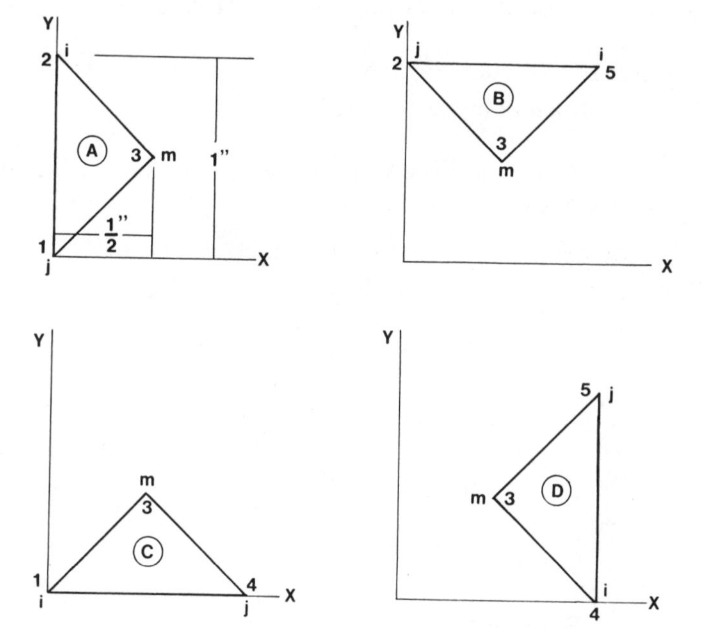

Fig. 8.11 Element orientation for Fig. 8.10.

Step 5: Form the D matrix, which describes the material physical properties.

$$[D] = \frac{E}{1-v^2}\begin{bmatrix} 1 & v & 0 \\ v & 1 & 0 \\ 0 & 0 & \dfrac{1-v}{2} \end{bmatrix} \qquad 8.25$$

where

E = modulus of elasticity
v = Poisson's ratio

Step 6: Form the transpose of Eq. 8.24, the B matrix. For example, from Eq. 8.24, for element A,

$$[B]_A = \frac{1}{2\Delta}\begin{bmatrix} -0.5 & 0 & -0.5 & 0 & 1 & 0 \\ 0 & 0.5 & 0 & -0.5 & 0 & 0 \\ 0.5 & -0.5 & -0.5 & -0.5 & 0 & 1 \end{bmatrix}$$

and its transpose is

$$[B]_A^T = \frac{1}{2\Delta}\begin{bmatrix} -0.5 & 0 & 0.5 \\ 0 & 0.5 & -0.5 \\ -0.5 & 0 & -0.5 \\ 0 & -0.5 & -0.5 \\ 1 & 0 & 0 \\ 0 & 0 & 1 \end{bmatrix}$$

In the transpose matrix $[B]^T$ the vertical rows are the same as the horizontal rows in the $[B]$ matrix. The transpose matrix is formed for each element.

Step 7: Form the K (stiffness) matrix for each element from Eqs. 8.24 and 8.26.

$$[K] = [B^T][D][B]t\Delta \qquad 8.26$$

where t is thickness of the element. For element A, for example,

$$[K]_A = \frac{1}{2\Delta}\begin{bmatrix} -0.5 & 0 & 0.5 \\ 0 & 0.5 & -0.5 \\ -0.5 & 0 & -0.5 \\ 0 & -0.5 & -0.5 \\ 1 & 0 & 0 \\ 0 & 0 & 1 \end{bmatrix} \frac{E}{1-v^2}\begin{bmatrix} 1 & 0.3 & 0 \\ 0.3 & 1 & 0 \\ 0 & 0 & 0.35 \end{bmatrix}[B]_A t\Delta$$

$$[K]_A = \frac{Et}{0.91}\begin{bmatrix} 0.3375 & -0.1625 & 0.1625 & -0.0125 & -0.5 & 0.175 \\ -0.1625 & 0.3375 & 0.0125 & -0.1625 & 0.15 & -0.175 \\ 0.1625 & 0.0125 & 0.3375 & 0.1625 & -0.5 & -0.175 \\ -0.0125 & -0.1625 & 0.1625 & 0.3375 & -0.15 & -0.175 \\ -0.5 & 0.15 & -0.5 & -0.15 & 1 & 0 \\ 0.175 & -0.175 & -0.175 & -0.175 & 0 & 0.35 \end{bmatrix}$$

In the matrix multiplication, the number of entries in the horizontal rows must equal the number of entries in the vertical rows of the matrix multiplied. For example, the first multiplication is a 3×6 matrix multiplied by a 3×3 matrix.

Step 8: Combine the element stiffness matrices and compute stresses and strains. Strains are computed from

$$[\varepsilon] = [B][\delta] \tag{8.27}$$

$$[\delta] = \begin{bmatrix} u_i \\ v_i \\ u_j \\ v_j \\ u_m \\ v_m \end{bmatrix} \tag{8.28}$$

For element A

$$[\varepsilon]_A = \begin{bmatrix} -0.5 & 0 & -0.5 & 0 & 1 & 0 \\ 0 & 0.5 & 0 & -0.5 & 0 & 0 \\ 0.5 & -0.5 & -0.5 & -0.5 & 0 & 1 \end{bmatrix} \frac{1}{2\Delta} \begin{bmatrix} u_2 \\ v_2 \\ u_1 \\ v_1 \\ u_3 \\ v_3 \end{bmatrix}_A$$

$$\begin{bmatrix} \varepsilon_x \\ \varepsilon_y \\ \gamma_{xy} \end{bmatrix}_A = \begin{bmatrix} -0.5 & 0 & -0.5 & 0 & 1 & 0 \\ 0 & 0.5 & 0 & -0.5 & 0 & 0 \\ 0.5 & -0.5 & -0.5 & -0.5 & 0 & 1 \end{bmatrix} 2 \begin{bmatrix} 0 \\ -0.0003 \\ 0 \\ 0 \\ 0.0005 \\ -0.00015 \end{bmatrix}$$

from which

$$\varepsilon_{xA} = 0.001 \text{ in./in.}$$
$$\varepsilon_{yA} = -0.0003 \text{ in./in.}$$
$$\gamma_{xyA} = 1.6 \times 10^{-9} \text{ in./in.}$$

Stresses are obtained from

$$[\sigma] = [D][B][\delta] \tag{8.29}$$

$$[\sigma] = \begin{bmatrix} \sigma_x \\ \sigma_y \\ \tau_{xy} \end{bmatrix} \tag{8.30}$$

For element A substituting Eqs. 8.24, 8.25, and 8.28 into Eq. 8.29

$$\begin{bmatrix} \sigma_x \\ \sigma_y \\ \tau_{xy} \end{bmatrix} = \begin{bmatrix} 1 & 0.3 & 0 \\ 0.3 & 1 & 0 \\ 0 & 0 & 0.35 \end{bmatrix} \frac{30 \times 10^6}{0.91} \begin{bmatrix} -0.5 & 0 & -0.5 & 0 & 1 & 0 \\ 0 & 0.5 & 0 & -0.5 & 0 & 0 \\ 0.5 & -0.5 & -0.5 & -0.5 & 0 & 1 \end{bmatrix} 2 \begin{bmatrix} u_2 \\ v_2 \\ u_1 \\ v_1 \\ u_3 \\ v_3 \end{bmatrix}$$

from which

$$\sigma_{xA} = 30{,}000 \text{ psi}$$
$$\sigma_{yA} = 0$$
$$\tau_{xyA} = 0$$

By conventional theory in Fig. 8.8

$$\sigma_x = \text{force/area} = 30{,}000/1 = 30{,}000 \text{ psi}$$
$$\sigma_y = \tau_{xy} = 0$$
$$\varepsilon_x = \frac{\sigma_x}{E} = \frac{30{,}000}{30 \times 10^6} = 0.001 \text{ in./in.}$$
$$\varepsilon_y = \frac{-\nu\sigma_x}{E} = \frac{-0.3(30{,}000)}{30 \times 10^6} = -0.003 \text{ in./in.}$$
$$\gamma_{xy} = \frac{\tau_{xy}}{G} = 0$$

The accuracy of the finite element method depends both on the type of problem and upon the number and type of elements selected. For example, consider Fig. 8.12, which shows a simple cantilever beam with an end load. The theoretical maximum deflection for this beam is 0.00853 in. If the beam is split into four elements, Fig. 8.13A, the maximum deflection by the finite element method is 0.000969 in., a very inaccurate result. For the case of 16 elements (Fig. 8.13*B*), the maximum deflection is 0.005 in., a considerable improvement. In order to obtain the theoretical value about 30–90 elements would be required. This particular example was solved by the finite element option of the computer program ICES-STRUDL (31).

Fig. 8.14 shows the deformed geometry of a standard 12-in. 300-lb flange

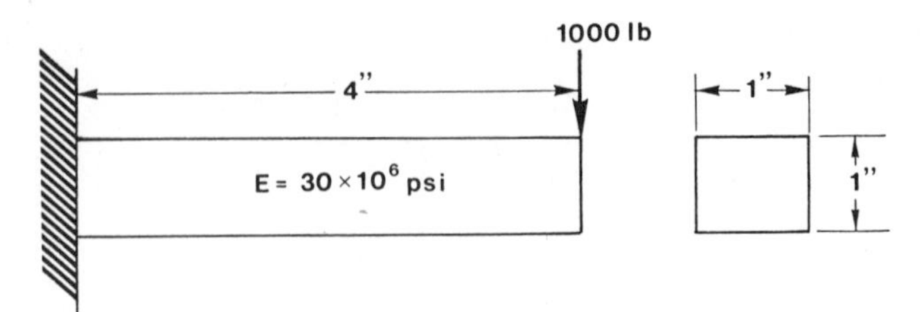

Fig. 8.12 Loaded cantilever beam.

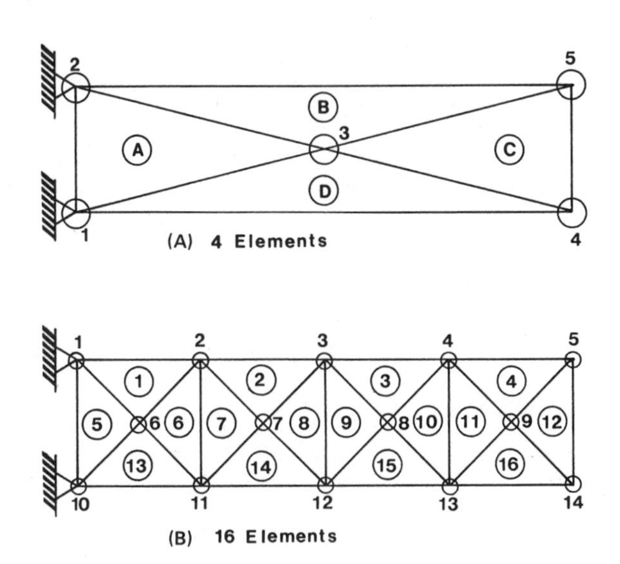

Fig. 8.13 Cantilever beam with triangular elements.

subject to gasket, bolt, and internal pressure load. The common classical theoretical treatment provides no information on deformations—only stresses are calculated—and only at two or three locations like at the junction of the flange hub with the shell (location *A*) or at the junction of the hub with the flange ring (location *B*). Recent results have corrected some of these deficiencies (18), but only the finite element method provides a complete stress and deformation account of what takes place.

Another example is shown in Fig. 8.15 which represents a three-part pressure vessel subject to internal pressure. Classical treatment requires a discontinuity analysis by computing conditions at the junction of the parts (see Chapter 3). In the finite element method, the exact geometry can be considered and a complete stress-deformation account provides for all locations in the structure. The analysis for this example is detailed in (29); the finite element solution confirmed the analytical solution as far as it went but also provided much more detail about overall behavior.

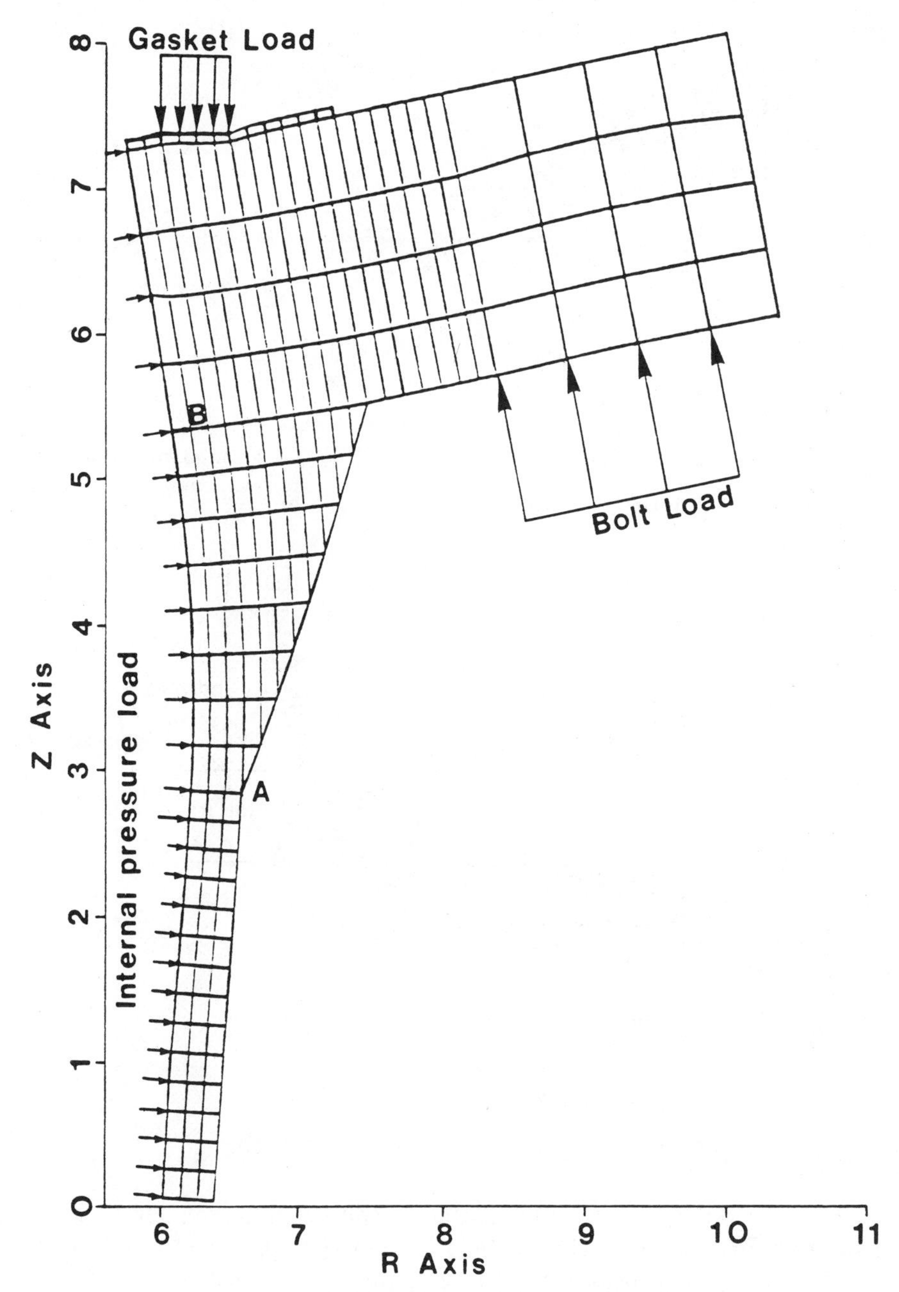

Fig. 8.14 12-in., 300-lb steel flange deformed geometry (× 100).

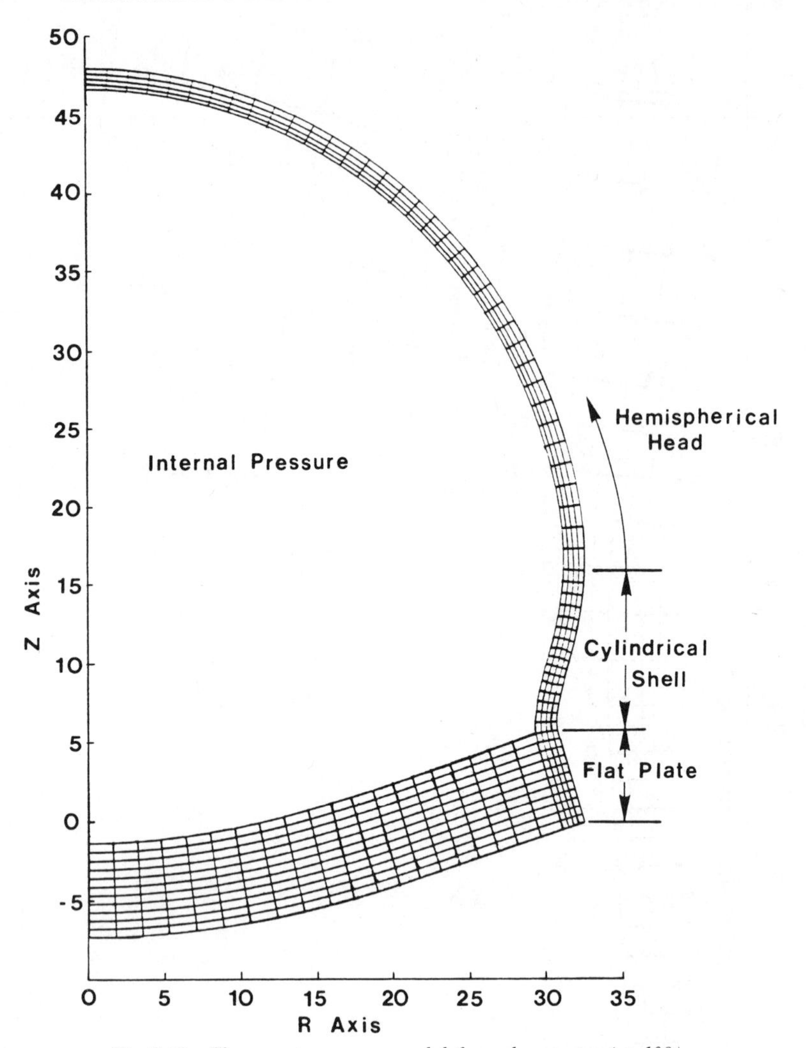

Fig. 8.15 Three-part pressure vessel deformed geometry (× 100).

8-3 FINITE DIFFERENCE

In this method the derivatives and partial derivatives in the equilibrium equations for beams (16), plates (16, 24), and shells (23) are represented as deflections in a Taylor series expansion (25). The derivatives are obtained from

$$\frac{df(x_i)}{dx} = \frac{f(x_i + h) - f(x_i)}{h} + E(h) \tag{8.31}$$

In Eq. 8.31 the term h is the spacing between two points on the x axis. When h is small enough an error $E(h)$ is left and represents the remainder of the series in Eq. 8.31.

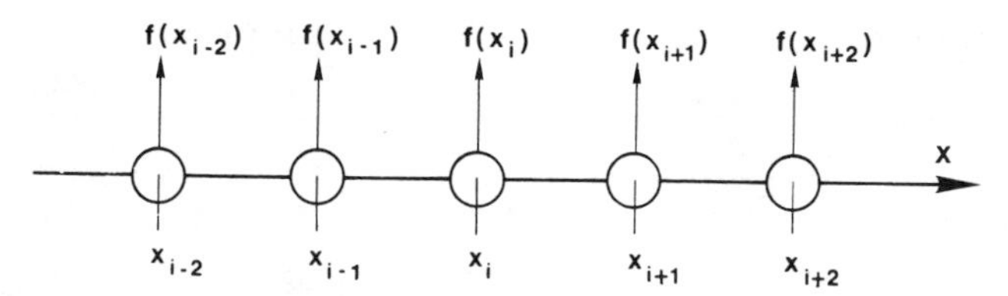

Fig. 8.16 Finite difference grid for x axis.

Most of the finite differences used are of error magnitude h^2. However, with the use of more coordinate points, errors of magnitude h^4 are possible (2). The h^2 error equation for Eq. 8.31 is

$$\frac{df(x_i)}{dx} = \frac{f(x_i + h) - f(x_i - h)}{2h} + E(h^2) \qquad 8.32$$

A form that is useful as a template from Fig. 8.16 for calculations is

$$\frac{df(x_i)}{dx} = \frac{1}{2h}\left[(-1) — (0) — (1) \right] + E(h^2) \qquad 8.33$$

The center circle is the coefficient for $f(x_i)$. Later various forms of templates for the different structure forms of beams, plates, and shells will be presented.

Beams

When the beam deflection equations are properly formulated, the following central finite difference forms, Fig. 8.16, are used:

$$\frac{d^2f(x_i)}{dx^2} = \frac{1}{h^2}\left[(1) — (-2) — (1) \right] + E(h^2) \qquad 8.34$$

$$\frac{d^3f(x_i)}{dx^2} = \frac{1}{2h^3}\left[(-1) — (2) — (0) — (-2) — (1) \right] + E(h^2) \qquad 8.35$$

$$\frac{d^4f(x_i)}{dx^4} = \frac{1}{h^4}\left[(1) — (-4) — (6) — (-4) — (1) \right] + E(h^2) \qquad 8.36$$

The center circle is the coefficient for $f(x_i)$. There are $E(h^4)$ forms (2) for Eqs. 8.34–8.36 for faster converging solutions. The Eqs. 8.32 and 8.34 also have forward and backward finite difference forms (10) that allow representation of the boundary conditions on the end of a beam. The following examples will demonstrate this method.

A 2-in. diameter simply supported steel shaft in Fig. 8.17*A* is loaded with 4-kip and 2-kip loads. The maximum deflections and stress are desired. The shear and bending moment curves have been calculated and are shown in Fig. 8.17*B* and *C*. The beam length is divided so that h, the beam point spacing, is 5 in. Points 1 and 8 require forward and backward difference; however, since ω_3 and ω_8 are zero, the equations for points 1 and 8 can be dropped, which

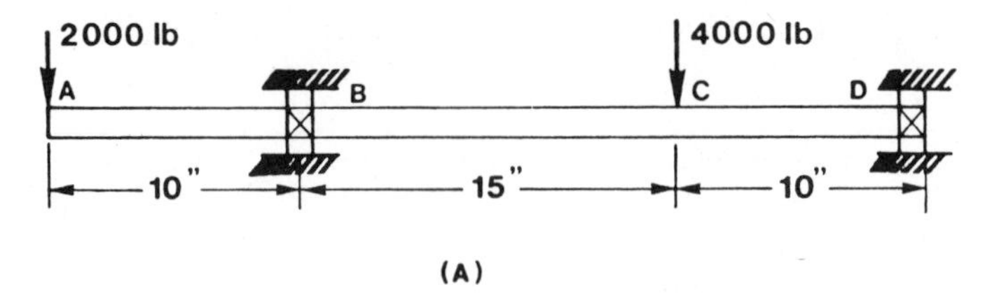

(A)

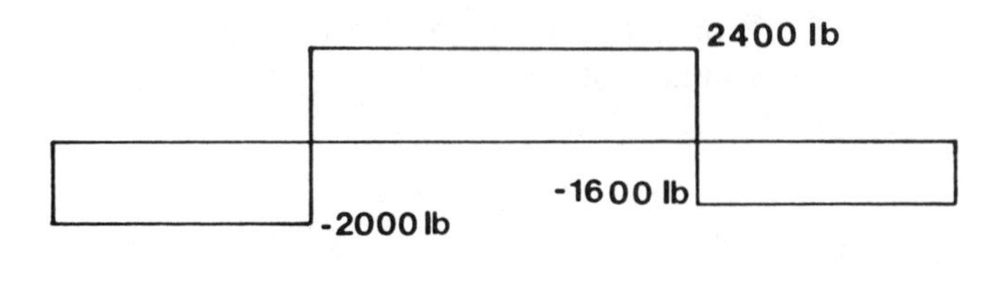

(B)

16000 in lb

0

-20000 in lb

(C)

0

-0.02562

-0.04444

(D)

Fig. 8.17 Overhanging beam

leaves a 6×6 determinant for computer solution. The central difference, Eq. 8.34, is used so that

$$\left(\frac{d^2\omega}{dx^2}\right)_i = \left(\frac{M}{EI}\right)_i$$

For point 2

$$\frac{1}{(5)^2}[\omega_1 - 2\omega_2 + \omega_3] = \left(\frac{-10{,}000}{2.355 \times 10^7}\right)$$

The rest of the equations are

$$\begin{bmatrix} 1 & -2 & 0 & 0 & 0 & 0 \\ 0 & 1 & 1 & 0 & 0 & 0 \\ 0 & 0 & -2 & 1 & 0 & 0 \\ 0 & 0 & 1 & -2 & 1 & 0 \\ 0 & 0 & 0 & 1 & -2 & 1 \\ 0 & 0 & 0 & 0 & 1 & -2 \end{bmatrix} \begin{bmatrix} \omega_1 \\ \omega_2 \\ \omega_4 \\ \omega_5 \\ \omega_6 \\ \omega_7 \end{bmatrix} = \begin{bmatrix} -0.01062 \\ -0.02125 \\ -0.00850 \\ +0.00440 \\ +0.01700 \\ +0.00850 \end{bmatrix}$$

The ω_i for the deflection curve are shown in Fig. 8.17*D* with the maximum deflection under the 2-kip load.

Torsion

In chapter 3 equations are developed for torsional stress and rotation of bars. These equations are for solid cross sections that have no holes inside the boundary. Eq. 3.233 is restated

$$\nabla^2 \phi = \frac{\partial^2 \phi}{\partial x^2} + \frac{\partial^2 \phi}{\partial y^2} = -2G\theta \qquad 8.37$$

where θ is the angle of twist (rad/in.). Also, from Eq. 2.28

$$T = \theta G J \qquad 8.38$$

If these equations are solved for $\theta = \dfrac{1}{G}$ then

$$T = J$$

Then, from Eq. 3.242, the solution for J is

$$J = 2\int\int \phi \, dx \, dy \qquad 8.39$$

The function ϕ is zero on the boundary and not known outside the shape. Therefore, forward and backward finite difference forms for stress must be used for Eq. 3.225 since the maximum stress is generally on the boundary.

$$\tau_{xz} = \frac{\partial \phi}{\partial y} \qquad \tau_{yz} = -\frac{\partial \phi}{\partial x} \qquad 8.40$$

When values for torques, other than T equals J, are required (16), the results can be scaled by T/J. The term J is found from Eq. 8.39; and the torque T is a variable.

When locating the largest stress, think of the cross section as a rubber membrane and pressure applied to one side. The ϕ value is a height, and $\partial\phi/\partial y$ and $\partial\phi/\partial x$ are slopes where the steepest of these defines the maximum stresses.

The Eqs. 8.37, 8.39, and 8.40 are solved with equally spaced finite difference

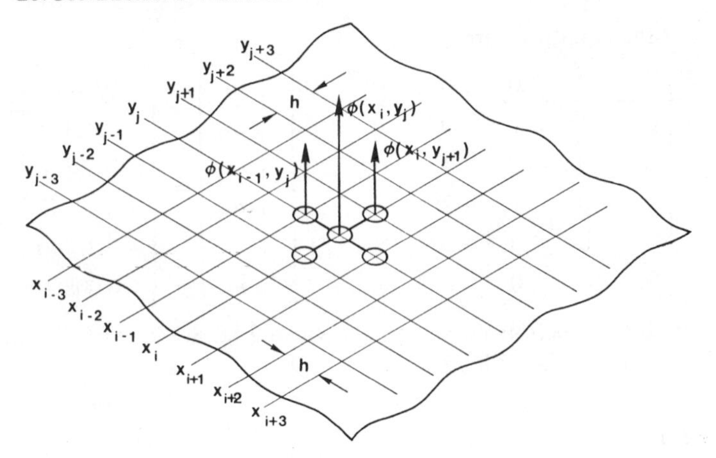

Fig. 8.18 Finite difference grid for xy plane.

forms represented in Fig. 8.18.

$$\nabla^2 \phi(x_i, y_j) = \frac{1}{h^2}\left[\begin{array}{ccc} & 1 & \\ 1 & -4 & 1 \\ & 1 & \end{array}\right] + E(h^2) \tag{8.41}$$

The volume integral is

$$\iint \phi \, dx \, dy = \frac{h^2}{9}\left[\begin{array}{ccc} 1 & 4 & 1 \\ 4 & 16 & 4 \\ 1 & 4 & 1 \end{array}\right] + E(h^4) \tag{8.42}$$

The center circle is the coefficient for $\phi(x_i, y_j)$. The forward difference is

$$\left(\frac{d\phi}{dx}\right)_j = -\frac{1}{2h}\left[\begin{array}{ccc} 3 & -4 & 1 \end{array}\right] + E(h^2) \tag{8.43}$$

The left circle is the coefficient for $\phi(x_i, y_j)$. The backward difference is

$$\left(\frac{d\phi}{dx}\right)_j = \frac{1}{2h}\left[\begin{array}{ccc} 1 & -4 & 3 \end{array}\right] + E(h^2) \tag{8.44}$$

where the right circle is the coefficient for $\phi(x_i, y_j)$.

There are other forms (10) for Eq. 8.41 such as polar, unequal spacing, skewed,

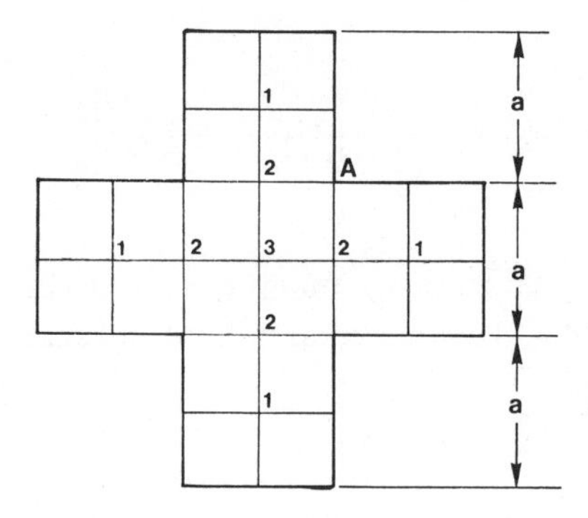

Fig. 8.19 Torsion cross section.

and hexagonal coordinates. The volume integral, Eq. 8.42, has other forms (2) for the square, triangle, and hexagon with $E(h^4)$ or better.

In the following example the torsional constant J and the stress equation will be found for torque applied to the cross section in Fig. 8.19.

The cross section in Fig. 8.19 is divided into $a/2$ square divisions for this example. However, the divisions can be smaller for a computer solution. The ϕ values are solved using Eq. 8.37 with θ equal to $1/G$ and Eq. 8.41.

$$\nabla^2\phi = -2$$

The $\nabla^2\phi$ for point 3 will be written out and the others stated.

$$(\nabla^2\phi)_3 = \frac{1}{(a/2)^2}[\phi_2 - 4\phi_3 + \phi_2 + \phi_2 + \phi_2] = -2$$

or

$$\begin{bmatrix} -4 & 1 & 0 \\ 1 & -4 & 1 \\ 0 & 4 & -4 \end{bmatrix} \begin{bmatrix} \phi_1 \\ \phi_2 \\ \phi_3 \end{bmatrix} = -\frac{1}{2}\begin{bmatrix} a^2 \\ a^2 \\ a^2 \end{bmatrix}$$

solving

$$\phi_1 = \frac{17}{88}a^2 \qquad \phi_2 = \frac{24}{88}a^2 \qquad \phi_3 = \frac{35}{88}a^2$$

The value for J is found using Eq. 8.42 once in the center portion and four

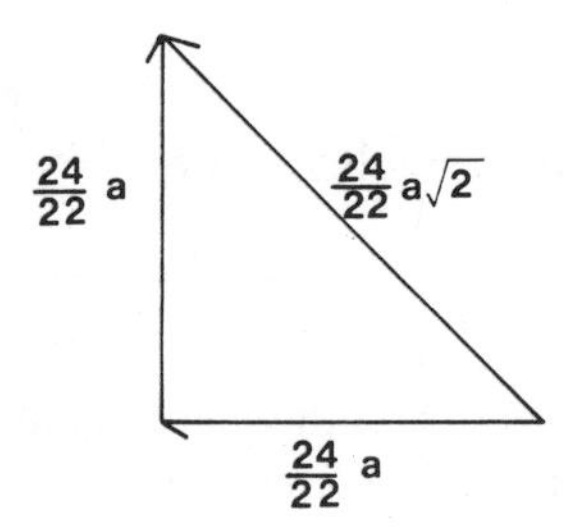

Fig. 8.20 Shear resultant.

times for the side tabs.

$$J = 2\left\{\left(\frac{a}{2}\right)^2 \frac{1}{9}(16\phi_3 + 16\phi_2) + 4\left(\frac{a}{2}\right)^2 \frac{1}{9}(16\phi_1 + 4\phi_2)\right\}$$

$$J = 1.54a^4$$

The values for the shear stresses (Eq. 8.40) are found for $T = J$. The values for the slope are found from Eq. 8.44 and evaluated at point A in Fig. 8.19.

$$\frac{\partial \phi}{\partial y} = \frac{1}{2(a/2)}[-4\phi_2] = \tau_{xz}$$

$$\frac{\partial \phi}{\partial x} = \frac{1}{2(a/2)}[-4\phi_2] = \tau_{yz}$$

The resultant shear stress in Fig. 8.20 is

$$\tau_T = \frac{33.94}{22}a$$

The stress can be evaluated by a backward difference through points A and 3. The stress value with

$$h = \frac{a}{2}\sqrt{2}$$

yields

$$\tau_T = \frac{24.75}{22}a$$

The error in this calculation, 23%, is due to the h value. And even then, the value in Fig. 8.20 is crude because values in the calculation are on the boundary.

The present value for the stress is found by multiplying τ_T by T/J.

$$\tau_T = \frac{33.94}{22}a\frac{T}{J}$$

$$J = 1.54a^4$$

$$\tau_T = \frac{33.94}{22}a\frac{T}{(1.54a^4)}$$

$$\tau_T = \frac{T}{a^3}$$

An accurate equation for the stress requires smaller values of h—say $a/4$ or $a/8$.

Plates

The types of plates analyzed for deflections are thin where the maximum deflection is less than the thickness of the plate. The homogeneous isotropic plates such as steel, and aluminum are represented by Eq. 3.292.

$$\nabla^4 \omega = \frac{\partial^4 \omega}{\partial x^4} + 2\frac{\partial^4 \omega}{\partial x^2 \partial y^2} + \frac{\partial^4 \omega}{\partial y^4} = \frac{p_z}{D} \qquad 8.45$$

where p_z is the uniform loading perpendicular to the plate, and D is from Eq. 3.309. The boundary conditions are represented by Eqs. 3.286–3.289. The moment equations are represented by Eqs. 3.313, 3.314, and 3.317.

The stress equations σ_x, σ_y, and τ_{xy} can be developed from Eq. 3.307, 3.308, 3.303, and 3.304 for stress on the top surface of the plate.
The signs conform to the conventions in Figs. 3.67 and 3.68. The finite difference forms required, Fig. 8.18, are

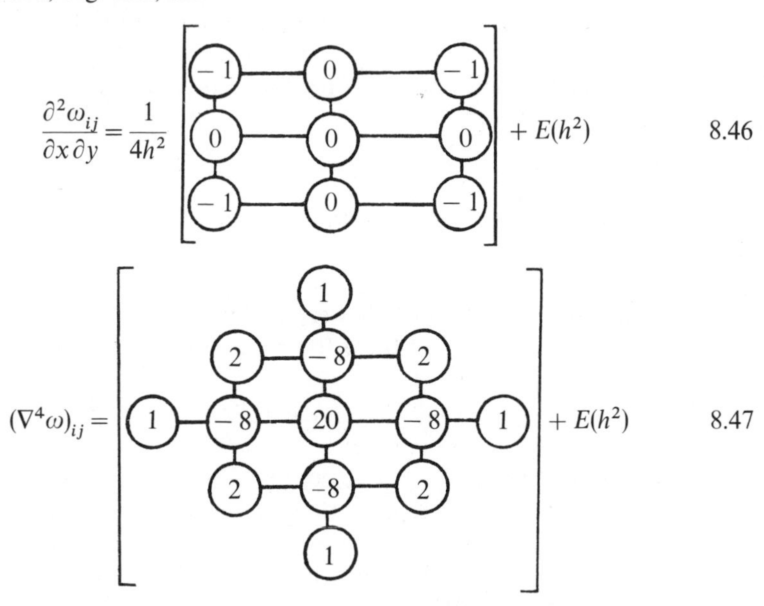

The center circle is the coefficient for $\phi(x_i, y_j)$ in Eqs. 8.46 and 8.47.

Most of the forms needed for the moments and the boundary conditions can be developed from Eqs. 8.33, 8.34, and 8.36. The forms for the partial with respect to y are basically the same.

There are finite difference forms (24) for boundary conditions in Eqs. 3.286–3.289. These are developed for ease of use and higher accuracy. The free edge and free-edge corner, Eqs. 3.288 and 3.289, are difficult to work with in obtaining good results. Heins (9) develops the finite difference equations in polar form for circular type plates and demonstrates several examples. Skewed and hexagonal forms are presented in (10, 24).

There are other orthotropic plates that can be analyzed and fall into three classes:

1. A plate that has stiffeners such as angles or beams forming a grid pattern under it and used for decks of bridges, aircraft fuselage and flooring,

and to some extent the flooring of a house or building. These are presented in (9, 24).

2. Rolled plates of aluminum, steel, and so on, can exhibit orthotropic behavior such as Poissons ratio, v, and the modulus of elasticity E, varying with direction.
3. The other orthotropic form is treated in Chapter 4 where graphite fibers or wood sheets are used as laminations to build up a plate.

In viewing the first two forms, they appear homogeneous and isotropic. However, in the first case the stiffness D of the plate is not a constant with direction because of the stiffeners. The second case is developed because of the fabrication process. The third case is truly a natural orthotropic form laminated into an orthotropic plate. The third case is the only one that will be discussed.

The general orthotropic or anisotropic plate equation, 4.279, can be simplified by dropping the membrane terms N_x, N_y, and N_{xy} to form the bending equation for the plate.

$$D_{11}\frac{\partial^4\omega}{\partial x^4}+4D_{16}\frac{\partial^4\omega}{\partial x^3\,\partial y}+2(D_{12}+2D_{66})\frac{\partial^4\omega}{\partial x^2\,\partial y^2}+4D_{26}\frac{\partial^4\omega}{\partial x\,\partial y^3}+D_{22}\frac{\partial^4\omega}{\partial y^4}=q \qquad 8.48$$

In the example in Fig. 4.40, the D's were calculated. When Eq. 8.48 is divided by D_{11} the following results for the orthotropic laminated plate are

$$\frac{\partial^4\omega}{\partial x^4}+1.97477\frac{\partial^4\omega}{\partial x^2\,\partial y^2}+0.37373\frac{\partial^4\omega}{\partial y^4}=\frac{q}{D_{11}} \qquad 8.49$$

The equation on the left side can be compared with that of Eq. 8.45 term for term, and they are similar except for the third coefficient. It is possible that an orthotropic laminated plate could be treated as homogeneous and isotropic to obtain a solution. Plates can be laminated that are quasi-isotropic, or Eqs. 8.49 and 8.45 would appear nearly the same.

When obtaining a solution, the forms for the partial derivatives, Eqs. 8.34, 8.36, 8.47 and the following can be used:

$$\frac{\partial^4\omega_{ij}}{\partial x^2\,\partial y^2}=\begin{bmatrix} 1 & -2 & 1 \\ -2 & 4 & -2 \\ 1 & -2 & 1 \end{bmatrix}+E(h^2) \qquad 8.50$$

The center circle is the coefficient for $\omega(x_i, y_j)$. Then the ω's can be found, and Eqs. 4.234–4.236 can be evaluated. Note in Eq. 4.255, even when N_x, N_y, and N_{xy} are neglected in the plate solution, that unless the B_{ij} values in Eqs. 4.236–4.254 are zero for the entire plate there will be values for $\varepsilon_1{}^0$, $\varepsilon_2{}^0$, and $\gamma_2{}^0$. The B_{ij}

values are zero for laminates symmetric to the z axis, as discussed in (1) in Chapter 4. The stress values for an individual laminate plate, Eqs. 4.257–4.239, can then be evaluated.

REFERENCES

1. D. G. Ashwell and R. H. Gallagher, *Thin Elements for Thin Shells and Curved Members*, John Wiley and Sons, Inc., New York, N.Y. (1976).
2. M. Abramowitz and I. A. Stegun, *Handbook of Mathematical Functions*, Dover, New York, N.Y. (1965).
3. K. J. Bathe and E. L. Wilson, *Numerical Methods in Finite Element Analysis*, Prentice-Hall Inc., Englewood Cliffs, N.J. (1976).
4. R. W. Clough, "The Finite Element Method in Plane Stress," *Proceedings 2nd ASCE Conference on Electronic Computation*, ASCE, New York, N.Y. (1960).
5. G. Crose et. al., *SAAS III, Finite Element Stress Analysis of Axisymmetric and Plane Solids with Different Orthotropic, Temperature-Dependent Material Properties in Tension and Compression*, Report AD–729–188, U.S. Department of Commerce (June 1971).
6. C. S. Desai and J. F. Abel, *Introduction to the Finite Element Method*, Van Nostrand Reinhold, New York, N.Y. (1972).
7. R. H. Gallagher, *Finite Element Analysis*, Prentice-Hall, Englewood Cliffs, N.J. (1975).
8. J. M. Gere and W. Weaver, Jr., *Analysis of Framed Structures*, D. Van Nostrand, Princeton, N.J. (1965).
9. C. P. Heins, *Applied Plate Theory for the Engineer*, D. C. Heath and Co., Lexington, Mass. (1976).
10. G. A. Korn, and T. M. Korn, *Mathematical Handbook for Scientists and Engineers*, McGraw-Hill Book Co., New York (1961).
11. H. C. Martin, and G. F. Carey, *Introduction to Finite Element Analysis*, McGraw-Hill Book Co., New York, N.Y. (1973).
12. D. McFarland, B. L. Smith, and W. D. Bernhart, *Analysis of Plates*, Spartan Books, New York, N.Y. (1972).
13. J. T. Oden, R. W. Clough, and Y. Yamamoto, *Advances in Computational Methods in Structural Mechanics and Design*, UAH Press, Huntsville, Ala. (1972).
14. W. C. Paulsen, "Finite Element Stress Analysis," *Mach. Des.*, **43** (24), 46–52 (September 30, 1971); **43** (25), 146–150 (October 14, 1971); **43** (26), 90–94 (October 28, 1971).
15. J. S. Przemieniecki, *Theory of Matrix Structural Analysis*, McGraw-Hill Book Co., New York, N.Y. (1968).
16. R. M. Rivello, *Theory and Analysis of Flight Structures*, McGraw-Hill Book Co., New York, N.Y. (1969).
17. K. C. Rockey, H. R. Evans, D. W. Griffith, and D. A. Nethercot, *The Finite Element Method*, John Wiley and Sons, Inc., New York, N.Y. (1975).
18. E. C. Rodabaugh et al., *Flange, A Computer Program for the Analysis of Flanged Joints with Ring-Type Gaskets*, Report No. ORNL-5035, Oak Ridge National Laboratory (January 1976).
19. R. Rosen, "*Finite Element Analysis—an Introduction*," *Mech. Eng.*, **96** (1), 21–26 (January 1974).
20. W. H. Rowan and R. M. Hackett, *Applications of Finite Element Methods in Civil Engineering*, ASCE Symposium, Nashville, Tenn. (November 1969).
21. M. F. Rubinstein, *Matrix Computer Analysis of Structures*, Prentice-Hall, Englewood Cliffs, N.J. (1966).
22. L. J. Segerlind, *Applied Finite Element Analysis*, John Wiley and Sons, Inc. (1976).

23. M. Soare, *Application of Finite Difference Equations to Shell Analysis*, Pergamon Press, Oxford (1967).

24. R. Szilard, *Theory and Analysis of Plates*, Prentice-Hall, Englewood Cliffs, N.J. (1974).

25. A. E. Taylor, *Advanced Calculus*, Ginn and Co., Boston, Mass. (1955).

26. P. Tong and J. N. Rossettos, *Finite Element Method*, Massachusetts Institute of Technology Press, Cambridge, Mass. (1977).

27. O. C. Zienkiewicz, *The Finite Element in Engineering Science*, McGraw-Hill Book Co., New York, N.Y. (1971).

28. "Ansys," *Finite Element Program*, Swanson Analysis Systems Inc., Elizabeth, Pa.

29. *ASME Boiler and Pressure Vessel Code*, Section VIII, Div. 2, American Society of Mechanical Engineers, New York, N.Y. (1977).

30. "Finite Element Stress Analysis," *Automot. Eng.* **85** (4), 44–48 (April 1977).

31. "ICES—Strudl," *The Structural Design Language*, Bulletin R67-56, Massachusetts Institute of Technology School of Engineering, Cambridge, Mass. (September 1967).

32. "MARC," *Finite Element Program*, MARC Analysis Research Corp., Providence, R.I.

33. "NASTRAN," *Computer Program*, MacNeal Schwendler Co., Los Angeles, Cal.

34. "SAP V," *Computer Program*, Report No. EERC-73-11, College of Engineering, University of California, Berkeley, Cal. (April 1974).

35. "STARDYNE," *Finite Element Computer Program*, Control Data Corporation, Minneapolis, Minn.

PROBLEMS

8.1 The entire plate area supports a 325 kips pressure vessel (Fig. P8.1). Using a finite difference or finite element solution, find a plate thickness that will not exceed 12 kpsi (allowable stress). The dotted lines represent simple supports for the plate. Also, place the solution for σ in terms of

$$\sigma_{\max} = K \frac{pa^2}{t^2}$$

where

$$p = \text{uniform load}$$
$$t = \text{thickness of plate}$$
$$a = 21''$$

Compare K with other plate solutions in handbooks.

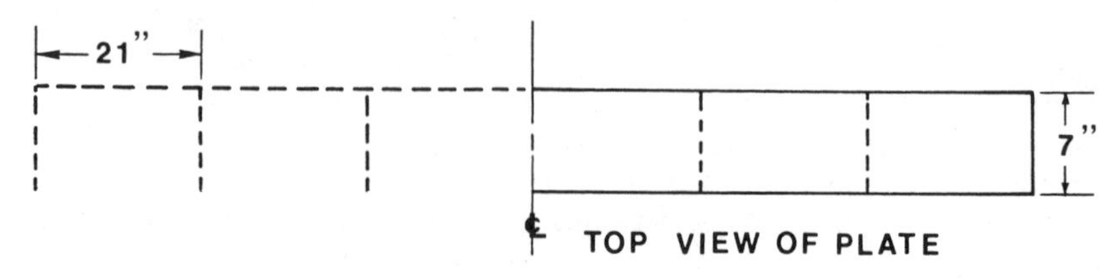

Fig. P8.1

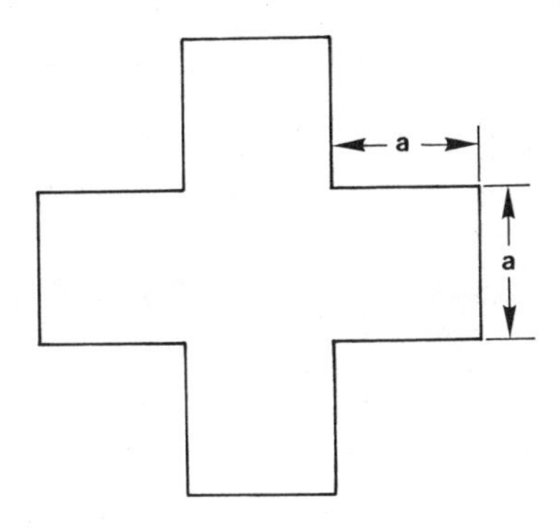

Fig. P8.3

8.2 In Problem 3.8 find the deflection curve for the 22-ft support beam. For the maximum deflection allowed, what is p in pounds/square foot?

8.3 A simply supported thin plate is loaded with a constant pressure p in pounds/square inch (Fig. P8.3). Find:

a The deflection pattern and show this on a large sketch.

b Using the deflection pattern, estimate the maximum bending moments and stresses.

8.4 A fixed-edge square plate, sides a, is loaded with a uniform load p (psi). Find the center deflection by finite difference and compare it with the exact solution.

8.5 Find the deflection curve for the loading shown in Fig. P8.5. Find the maximum moment from the finite difference form of $M = EI\, d^2\omega/dx^2$ when EI is 45×10^6 lb-in.^{2} for a round steel shaft.

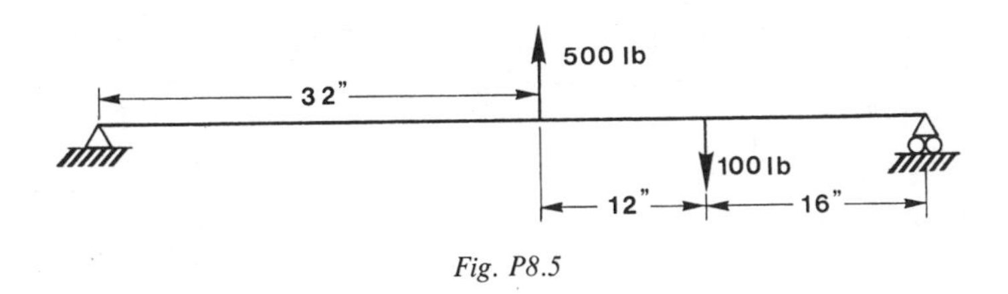

Fig. P8.5

8.6 Find the deflection curve and $M = EI(d^2\omega/dx^2)$ using the EI from Problem 8.5 by using finite difference or finite elements. (See Fig. P8.6.)

8.7 Find the deflection curve and the maximum deflection for the simply

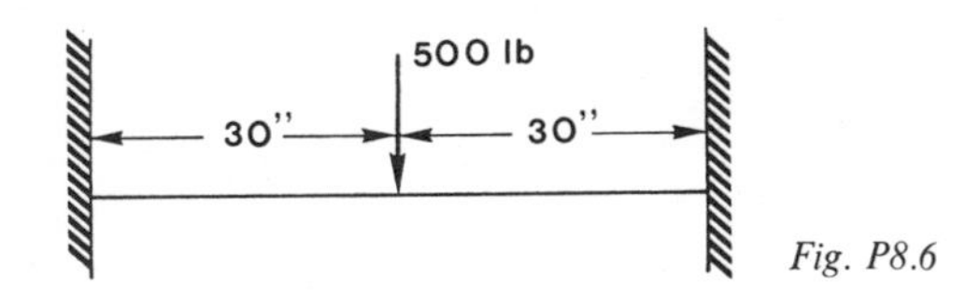

Fig. P8.6

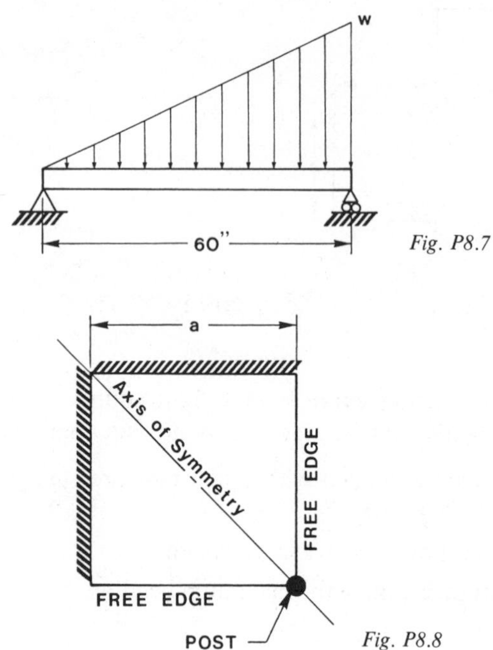

Fig. P8.7

Fig. P8.8

supported beam shown in Fig. P8.7 when the loading increases from zero to 600 lb/in. Use the same EI as in Problem 8.5.

8.8 A square plate, sides a, has a pressure loading p. (Fig. P8.8.) Find the deflection pattern for the conditions shown. First take a crude grid and then cut h in half to see if the deflections differ. Using the finer grid, try to find the bending stresses for a thickness t and a modulus E. Compare the stress with other end conditions of the form

$$\sigma = \beta \frac{pa^2}{t^2}$$

where β is a constant.

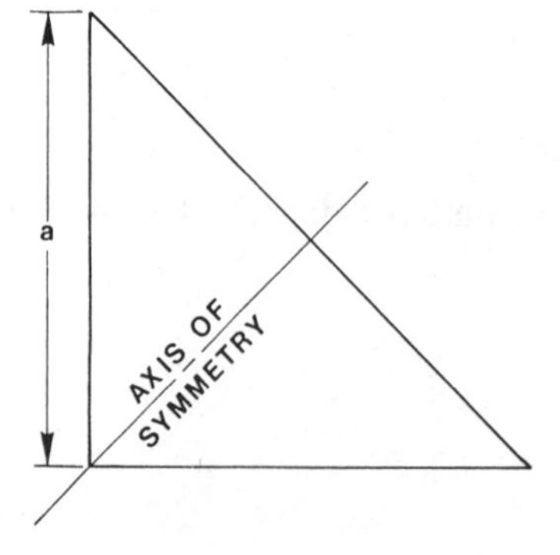

Fig. P8.9

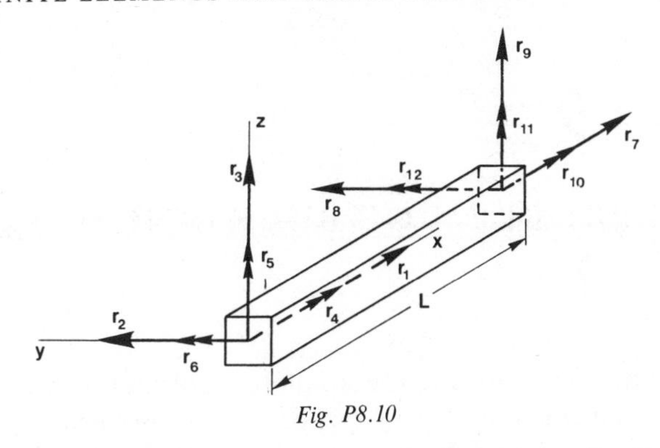

Fig. P8.10

8.9 Find the value of J for the triangular section (See Fig. P8.9.) Also, find the equation for the maximum stress and its location.

8.10 Develop the stiffness matrix for a beam finite element (Fig. P8.10). Let a single r_i equal 1 with the rest zero as shown in Figs. 7.74 and 7.75. The result will be a 12×12 matrix.

8.11 A clamped plate is uniformly loaded with a pressure p. The shape is that of Problem P8.9. Find the maximum deflection and stress and the location of each.

chapter **9**

THE PROBLEM OF BUCKLING

In this chapter attention is directed toward those problems in structural stability arising mainly from situations where applied loads cause the body to wrinkle, buckle, or otherwise collapse. In this sense instability, for the purposes of this chapter, implies a change in the geometrical form of the body in question; plastic instability caused by overloading, where the structure simply changes dimensions while essentially retaining its original form, was discussed in Chapter 6 (plastic expansion of tubes under internal pressure, for example). In that chapter, the situation of plastic collapse by formation of *plastic hinges* in the structure was also discussed (limit design) and will not be repeated here.

9-1 BUCKLING OF SIMPLE PRISMATIC BARS AND BEAM COLUMNS

Elementary column theory

Euler's critical buckling load. In considering the elementary theory of columns the applied load (compression) acts along the axis of the bar; the bar has a length that is large compared with its other dimensions. Ordinarily, in compression loading of short bars the stresses are considered as being uniformly distributed over the cross section of the bar. In long columns, however, there are always slight irregularities or a nonstraight condition which gives rise, under load, to lateral deflections, which at a critical amount cause a *buckle* to form in the column. In this discussion only direct compression loads on prismatic bars are considered; beam columns, that is, columns that have lateral as well as axial loading, are discussed later in the chapter. It should also be noted that the theory developed, although in itself is mainly of academic interest, has been of great practical importance in assisting in the development of empirical formulas applicable to particular practical cases. Elementary column theory has been credited mainly to Euler (32).

Consider the column shown in Fig. 9.1; one end is fixed and the other end is loaded by a force F at a point away from the center of gravity of the column (eccentricity e). After application of the load, a deflection δ is produced. At any cross section of the column A–B, a distant x from the base, the bending moment is

$$M = F(\delta + e - y) \tag{9.1}$$

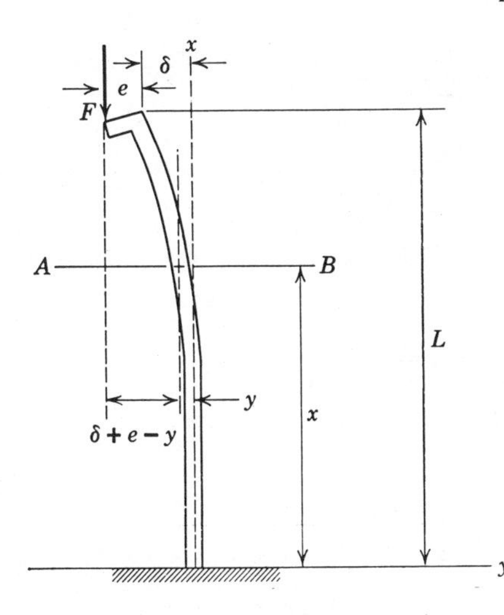

Fig. 9.1 Column with one end fixed.

where y is the deflection of the column at A–B. From Eq. 9.1 and the deflection equation 2.82,

$$EI(d^2y/dx^2) = M = F(\delta + e - y) \tag{9.2}$$

or

$$d^2y/dx^2 + Fy/EI = (F/EI)(\delta + e) \tag{9.3}$$

In order to integrate Eq. 9.3 it is convenient to transform it to equation for which a solution exists; this can be done by letting

$$F/EI = k^2 \tag{9.4}$$

so that Eqs. 9.3 becomes

$$d^2y/dx^2 + k^2y = k^2(\delta + e) \tag{9.5}$$

whose solution is

$$y = A \sin kx + B \cos kx + e + \delta \tag{9.6}$$

where A and B are constants of integration. At $x = 0$, $y = 0$, so that in Eq. 9.6

$$B = -(e + \delta) \tag{9.7}$$

In addition, at $y = 0$, $dy/dx = 0$, so that

$$A = 0 \tag{9.8}$$

Substituting Eqs. 9.7 and 9.8 into Eq. 9.6 gives the result

$$y = -(e + \delta)(\cos kx) + e + \delta \tag{9.9}$$

or

$$y = (e + \delta)(1 - \cos kx) \tag{9.10}$$

At $x = L$, $y = \delta$; therefore from these conditions and Eq. 9.10

$$\delta = e(\sec kL - 1) \tag{9.11}$$

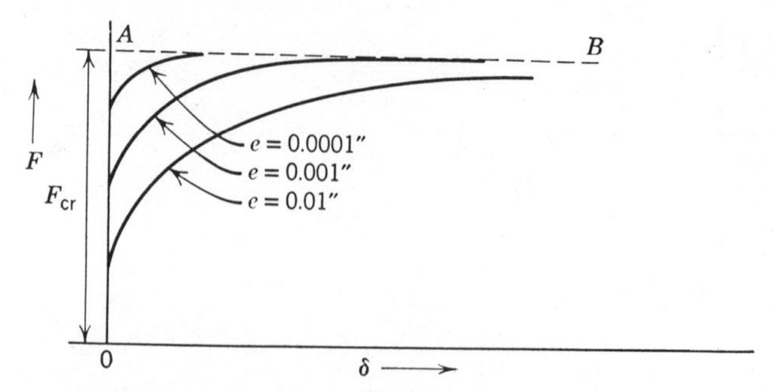

Fig. 9.2 Effect of eccentricity on the load-deflection characteristics of a column.

or

$$\delta = e(\sec L\sqrt{F/EI} - 1) \qquad 9.12$$

From Eq. 9.12 and assuming small values of e, a family of curves of F versus δ can be plotted as shown in Fig. 9.2; it is clearly evident in Fig. 9.2 that when δ is large, the curves superpose—this is called the *Euler critical load*, or the *buckling load* F_{cr}. It is also evident that if e, the eccentricity, is very small, a plot OAB would result, signifying that no instability in the loading process occurs until point A is reached, at which instant sudden buckling collapse occurs. To determine the Euler critical load, note in Eq. 9.11 that as δ increases the quantity, sec kL also increases, that is,

$$kL = \frac{\pi}{2}, \frac{3\pi}{2} \ldots (2n+1)\frac{\pi}{2} \qquad 9.13$$

where n is a positive integer. The smallest value of k, corresponding to the critical load F_{cr} to cause buckling, is found by letting $n = 0$ in Eq. 9.13, giving $\pi/2$; therefore

$$k^2L^2 = F_{cr}L^2/EI = \pi^2/4 \qquad 9.14$$

from which

$$F_{cr} = \pi^2 EI/4L^2 \qquad 9.15$$

Equation 9.15 is known as *Euler's column formula* and indicates that the critical buckling load is not a function of the strength of the material (yield and ultimate strengths are not involved) but only of the elastic modulus and geometry. For other values of n in Eq. 9.13 buckling modes shown in Fig. 9.3 are obtained but these are mainly of academic interest. If both ends of the column are pinned, as in Fig. 9.4, dy/dx is zero at $x = L/2$ and Eq. 9.11 becomes

$$\delta = e[\sec(kL/2) - 1] \qquad 9.16$$

and the Euler buckling load is

$$F_{cr} = \pi^2 EI/L^2 \qquad 9.17$$

If both ends of the column are fixed, as in Fig. 9.5, a bending moment M is

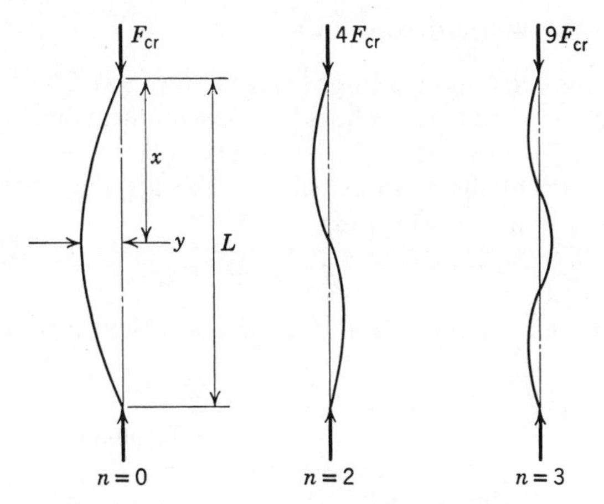

Fig. 9.3 Various modes of buckling of a column.

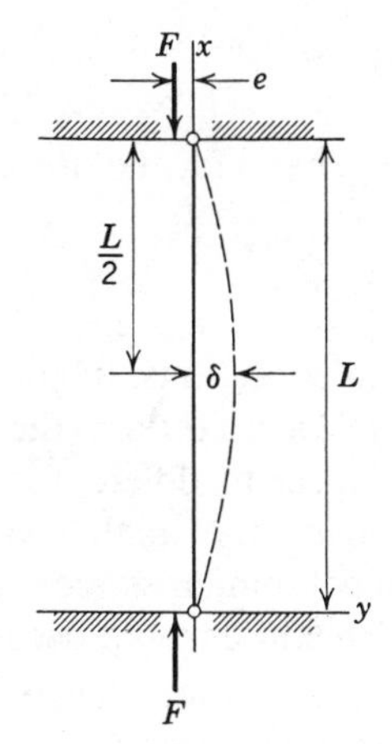

Fig. 9.4 Pin-connected column.

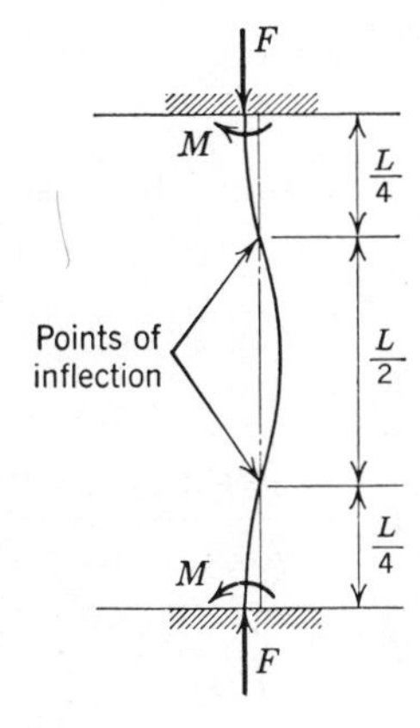

Fig. 9.5 Column with both ends fixed.

induced at the ends, and the slope is zero at the ends. The end-moment opposes the moment required to cause buckling. The moment diagram for this condition (see Chapter 2) shows that the moments balance out at $L/4$; therefore the central portion of the column behaves like a pinned-end column, and L is replaced by $L/2$ in Eq. 9.17; thus

$$F_{cr} = 4\pi^2 EI/L^2 \qquad 9.18$$

In summary, for the three fundamental cases, the Euler critical buckling load is as follows:

End Condition	Figure	Equation for F_{cr}	Equation No.
Both ends pinned	9.4	$F_{cr} = \pi^2 EI/L^2$	9.19
One end fixed	9.1	$F_{cr} = \pi^2 EI/4L^2$	9.20
Both ends fixed	9.5	$F_{cr} = 4\pi^2 EI/L^2$	9.21

From Eqs. 9.19 to 9.21 it is seen that the general form of the critical buckling load equation is

$$F_{cr} = K(\pi^2 EI/L^2) \qquad 9.22$$

where K is a constant depending on the end conditions. In addition to the end conditions, it is important to note that Eq. 9.22 also contains the term I, moment of inertia, which may be different with respect to different axes of the cross section; the critical buckling load is based on the smaller value of I if it is different for the two symmetry axes of the column cross section. For example, for a column of rectangular cross section, buckling would occur about the axis of the longest dimension. This particular condition suggests that circular tubular columns should give optimum performance since I is constant and all of the material is concentrated away from the column axis. In column work it is customary to express the moment of inertia as

$$I = Ar^2 \qquad 9.23$$

where r is the least radius of gyration of the cross section. Thus, using Eq. 9.23 in Eqs. 9.19 to 9.21,

$$\text{Pinned ends (Fig. 9.4)} \qquad F_{cr} = \frac{\pi^2 EA}{(L/r)^2} \qquad 9.24$$

$$\text{One end fixed (Fig. 9.1)} \qquad F_{cr} = \frac{\pi^2 EA}{4(L/r)^2} \qquad 9.25$$

$$\text{Both ends fixed (Fig. 9.5)} \qquad F_{cr} = \frac{4\pi^2 EA}{(L/r)^2} \qquad 9.26$$

where L/r is called the *slenderness ratio*. Another quantity frequently used in column work is the *equivalent column length* L_e. For example, using this quantity, Eq. 9.22 takes the form

$$F_{cr} = \pi^2 EI/(L_e)^2 \tag{9.27}$$

The equivalent length of a column is that used in the Euler formula for a pinned-end column to give the buckling load for the actual column. For example, the equivalent length of a column with both ends fixed is $L/2$.

Limitation of Euler's theory. It was assumed in the foregoing development that Hooke's law applied and that the failure by buckling was independent of the strength of the material. As the length decreases for the same cross section a point is reached where failure would occur, not by buckling, but by compression yielding of the material. According to Euler's formulas the critical load for short columns is infinite, which is impossible. For example, for a pinned-end column, failure by yielding occurs when

$$\frac{F_{cr}}{A} = \frac{\pi^2 E}{(L/r)^2} \tag{9.28}$$

and Euler's theory does not hold beyond this stress value. Suppose a column is made of steel having an elastic modulus of 30×10^6 psi and a proportional limit of 30,000 psi. Here Eq. 9.28 becomes

$$30{,}000 = \frac{\pi^2(30 \times 10^6)}{(L/r)^2} \tag{9.29}$$

from which L/r has a value of 99.5. Equation 9.28 can therefore be used to obtain buckling loads for values of slenderness ratio greater than 99.5 since buckling for this set of conditions occurs before yielding. A plot of failure stress versus slenderness ratio is given in Fig. 9.6; in this plot AB represents the condition for failure by direct compression. For a perfect column a plot ABD would be obtained. Curve DE represents Euler's theory without regard

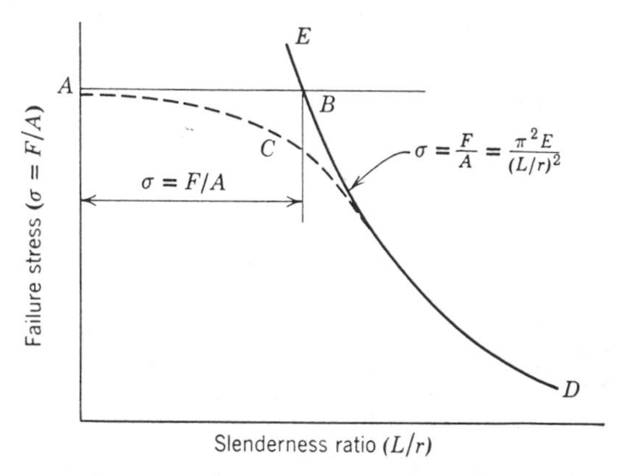

Fig. 9.6 Failure stress versus slenderness ratio.

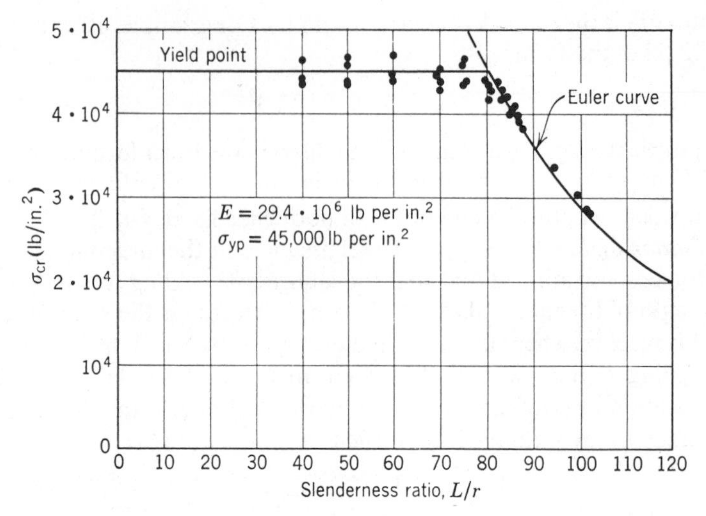

Fig. 9.7 Column behavior of mild steel. [*After Timoshenko (30), courtesy of D. Van Nostrand Co.*]

to strength of material, and curve *ACD* represents the general trend established by experimentation. The actual form of the curve depends on the material. For mild steel, for example, that yields at constant stress data as shown in Fig. 9.7 are obtained, corresponding to a *perfect column* (30). Draffin and Collins (11) classify columns as *short and stocky, intermediate, and long and slender* and show data for several materials on a plot similar to Fig. 9.6 (their data are shown in Fig. 9.8 for pinned-end columns). In the short and stocky classification the column length does not exceed 10 to 50 times the least radius of gyration. For intermediate columns, the length does not exceed 80 to 160 times the least radius of gyration, and for long slender columns the length exceeds 80 to 160 times the least radius of gyration. The data in Fig. 9.8 for short columns are for piers, short concrete columns, and various types of

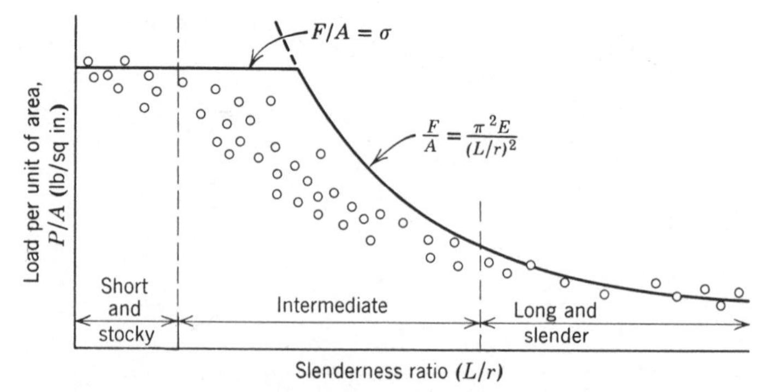

Fig. 9.8 Actual versus ideal column strength. [*After Draffin and Collins (11), courtesy of Ronald Press Co.*]

supports where failure is usually by crushing (wood), yielding (ductile metal), or shearing (brittle material), and ultimate separation of parts occurs. The intermediate column data are for all kinds of metal and timber columns used in bridges and building structures, and failure is intermediate between buckling and rupture. The third group of data consists mainly of sheathing and long aircraft parts where failure is by elastic buckling. There are numerous data on column behavior for a wide range of materials which are given in the *Military Handbook* of aircraft elements (37), the *Guide for Compression Members* by the Column Research Council (16), and numerous bulletin reports of the Welding Research Council of the Engineering Foundation (38, 39) as well as the standard texts and papers cited as references, particularly (3, 4, 11, 18, 24, 25, 31 to 33, and 36).

Column efficiency. Various combinations of cross-sectional shapes, materials, and slenderness ratios may be used for columns depending on the situation. As an example here the structural efficiency of a column of square cross section is considered; also the columns are simple pinned-end structures acting in the elastic range and conforming to Eulers critical buckling load, Eq. 9.17. For an element of square cross section the moment of inertia is given by

$$I = a^4/12 \tag{9.30}$$

where a is the length of a side. By letting the cross-sectional area equal A, this equation can be written as

$$I = A^2/12 \tag{9.31}$$

which, when substituted in to Eq. 9.17, gives an expression for A

$$A = (12F_{cr}L^2/\pi^2 E)^{1/2} \tag{9.32}$$

The critical *stress* at the instant of buckling is given by

$$\sigma_{cr} = F_{cr}/A \tag{9.33}$$

which, utilizing Eq. 9.32, becomes

$$\sigma_{cr} = (\pi^2 E F_{cr}/12L^2)^{1/2} \tag{9.34}$$

For present purposes column efficiency will be defined as the weight W, per unit length L, and load F; weight equals density γ times volume, which with Eq. 9.34 gives the result

$$\frac{W}{FL} = \frac{\gamma}{\sigma_{cr}} = \frac{\gamma}{E^{1/2}}\left(\frac{12L^2}{\pi^2 F_{cr}}\right)^{1/2} \tag{9.35}$$

which is plotted in Fig. 9.9 for various materials. Figure 9.9 shows that for low values of loading parameter the materials are most efficient. Aluminum and magnesium have relatively low yield strengths and hence are efficient only for low values of loading parameter. Figure 9.9 also illustrates the importance of both weight and modulus. Fiberglass, for example, on the basis of Eq. 9.17 would be only half as good as magnesium, for example, but when its density

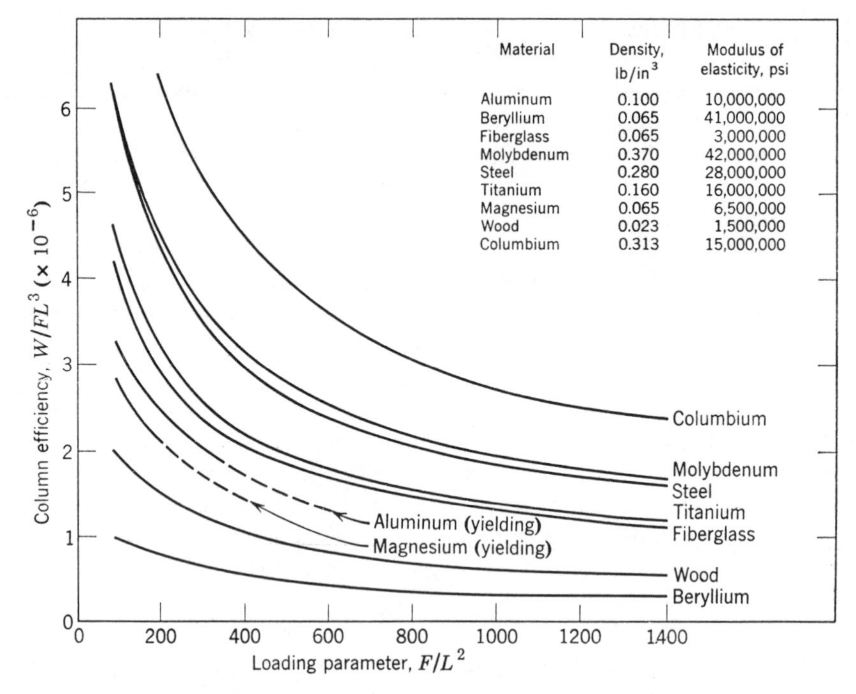

Fig. 9.9 Column efficiency of various materials.

is figured into the problem, fiberglass becomes more *efficient* than magnesium.

Extension of Euler's theory. It has already been shown that Euler's theory is generally inadequate for short columns. Therefore for design convenience various means have been attempted to provide a formula that would reduce to Euler's formula for large values of slenderness ratio and still agree with the failure-by-compression concept for short columns. Of the variations developed over the last several years the following are reviewed here for illustration.

STRAIGHT-LINE FORMULA. In place of the Euler curve *ABC* in Fig. 9.10 it

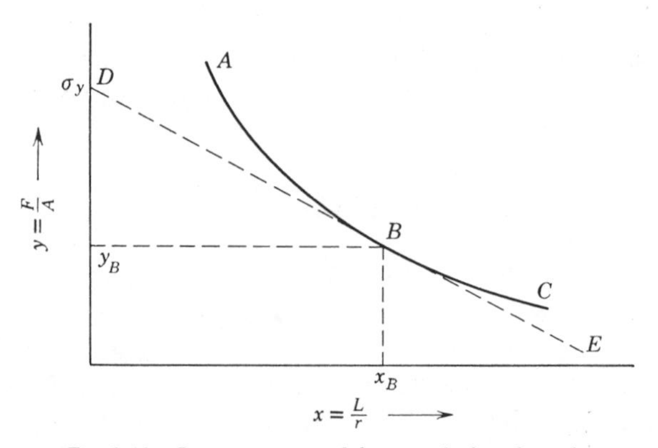

Fig. 9.10 Representation of the straight-line formula.

is assumed that failure is defined by the straight line *DE* drawn tangent to the Euler curve from point *D*, the yield stress in compression for the material. For line *DE*

$$y = \frac{F}{A} = \sigma_y + m\left(\frac{L}{r}\right) = \sigma_y + \frac{dy}{dx}\left(\frac{L}{r}\right) \qquad 9.36$$

where σ_y is the compression yield strength of the material and m is the slope of the line *DE*. Equation 9.36 gives the critical stress y for columns of various slenderness ratio. The slope m is found by noting from Eq. 9.22 and Fig. 9.10 that

$$y = \frac{F_{cr}}{A} = K\pi^2 E/(L/r)^2 = K\pi^2 E/x^2 \qquad 9.37$$

from which

$$dy/dx = -(2K\pi^2 E/x^3) \qquad 9.38$$

When $x = x_B$ (Fig. 9.10),

$$dy/dx = m = -[2K\pi^2 E/(x_B)^3] \qquad 9.39$$

so that

$$y_B = \sigma_y - [2K\pi^2 E/(x_B)^2] \qquad 9.40$$

From Eqs. 9.37 and 9.40

$$x_B = (3K\pi^2 E/\sigma_y)^{1/2} \qquad 9.41$$

and

$$y_B = \sigma_y/3 \qquad 9.42$$

Placing Eq. 9.41 into Eq. 9.39 then gives the slope as

$$m = -\frac{2\sigma_y^{3/2}}{3\pi\sqrt{3KE}} \qquad 9.43$$

and using this value in Eq. 9.36 gives the straight-line formula

$$\left(\frac{F_{cr}}{A}\right)_{\text{straight line}} = \sigma_y - \left(\frac{L}{r}\right)\left(\frac{2\sigma_y^{3/2}}{3\pi\sqrt{3KE}}\right) \qquad 9.44$$

In design, Eq. 9.44 is multiplied by a factor of safety, dictated by the particular situation or by the building code in force at the particular location.

PARABOLIC FORMULA. If curve *ABC* in Fig. 9.11 represents Euler's theory, the parabola *DBE* is taken to represent actual behavior and it is constructed so that it is tangent to the Euler curve at point *B* from the compressive yield strength of the material σ_y at *D*. The equation for the parabola *DBE* is

$$y = \sigma_y - ax^2 \qquad 9.45$$

where a is a constant. At *B*, Fig. 9.11, the tangents of the two curves are

For *ABC*,
$$\frac{dy}{dx} = -\frac{2K\pi^2 E}{(x_B)^3} \qquad 9.46$$

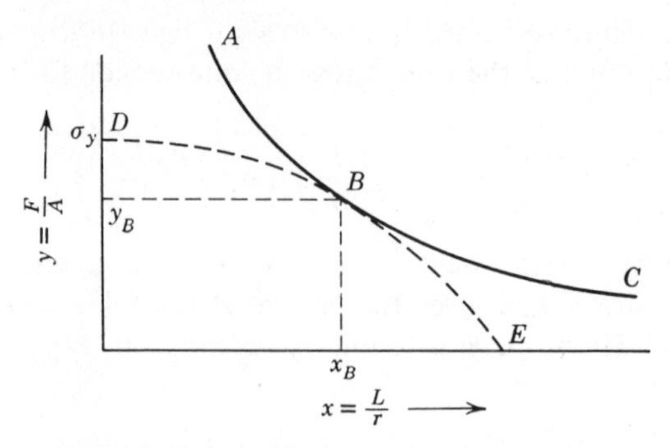

Fig. 9.11 Representation of the parabolic formula.

For DBE,
$$\frac{dy}{dx} = -2ax_B \tag{9.47}$$

Since the slopes are equal,
$$\frac{-2K\pi^2 E}{(x_B)^3} = -2ax_B \tag{9.48}$$

or
$$a = \frac{K\pi^2 E}{(x_B)^4} \tag{9.49}$$

However, from Eqs 9.37 and 9.45,
$$y_B = \sigma_y - ax_B{}^2 = K\pi^2 E/(x_B)^2 \tag{9.50}$$

and then from Eq. 9.48
$$a = \sigma_y{}^2/4K\pi^2 E \tag{9.51}$$

which when substituted into Eq. 9.45 gives the parabolic formula
$$(F_{\text{cr}}/A)_{\text{parabolic}} = \sigma_y - [(\sigma_y{}^2/4K\pi^2 E)(L/r)^2] \tag{9.52}$$

Again, as with Eq. 9.44, a factor of safety is applied.

RANKINE FORMULA. This is an empirical formula given in various texts on strength of materials as
$$\left(\frac{F_{\text{cr}}}{A}\right)_{\text{Rankine}} = \frac{\sigma_y}{1 + a(L/r)^2} \tag{9.53}$$

where a is a constant depending on the material. This formula is frequently used to compute a safe working stress rather than a buckling load. For example, one form of Eq. 9.53 is
$$\sigma_w = \frac{F}{A} = \frac{18{,}000}{1 + (1/18{,}000)(L/r)^2} \tag{9.54}$$

for L/r ratios between about 120 and 200 of secondary compression members

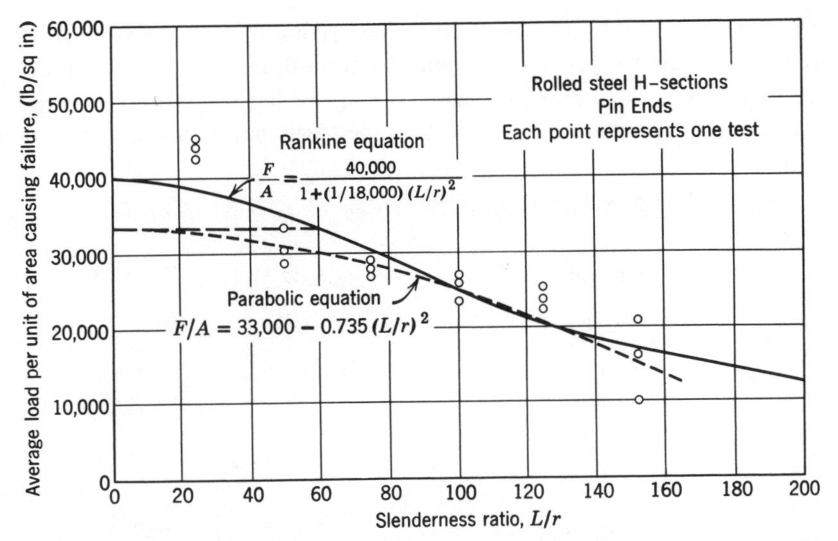

Fig. 9.12 Comparison of Rankine and parabolic formulas with tests of steel H-sections. [*After Draffin and Collins (11), courtesy of Ronald Press Co.*]

of structural steel; in this case F/A is generally limited to about 15,000 psi. For structural steel columns of L/r ratios in the range of about 60 to 160,

$$\sigma_w = \frac{F}{A} = \frac{40{,}000}{1 + (1/18{,}000)(L/r)^2} \tag{9.55}$$

with a maximum F/A of about 33,000 psi. Typical data are shown in Fig. 9.12(11).

WOODEN COLUMN FORMULA. Various empirical column formulas have been derived by the Forest Products Laboratory (U. S. Department of Agriculture) at Madison, Wisconsin. For a column of rectangular cross section a *short column* is defined as one whose unsupported length L is not greater than 10 times the shortest side b, and the working stress σ_w is taken as the safe compression strength parallel to the wood grain. In a *medium length* column, L/b varies between about 10 and 50

$$K = (\pi/2)\sqrt{E/6\sigma_y} \tag{9.56}$$

so that

$$\sigma_w = \sigma_y[1 - \tfrac{1}{3}(L/Kb)^4] \tag{9.57}$$

where K is the value of L/b when Eq. 9.57 is tangent to the Euler curve; σ_w equals $2\sigma_y/3$ at this point. In a *long column*, L/b lies between 50 and 120

$$K = (\pi/2)\sqrt{E/6\sigma_y} \tag{9.58}$$

so that

$$\sigma_\omega = \frac{\pi^2 E}{36(L/b)^2} \tag{9.59}$$

MISCELLANEOUS COLUMN FORMULAE. Numerous formulas have been proposed for specific materials under specific conditions, and a few of these are listed. For design, one should consult the latest design literature distributed by the fabricator or building code body in the particular area where the column is used.

Cast iron column $$\sigma_\omega = 10{,}000 - 60(L/r) \qquad 9.60$$

for $L/r < 70$.

Aluminium alloys $$F_{cr}/A = \sigma_y(1 + \sigma_y/200{,}000) \qquad 9.61$$

Magnesium alloys $$F_{cr}/A = K\sigma_y{}^n/(L_e/\rho)^m \qquad 9.62$$

where L_e is the effective column length, ρ is the radius of gyration, and the constants are defined by

Alloy	K	n	m	$(F_{cr}/A)_{max}$
M1A	180	0.50	1.0	$0.90\sigma_y$
AZ31B	2,900	0.25	1.5	σ_y
AZ61A	2,900	0.25	1.5	σ_y
AZ80A	2,900	0.25	1.5	σ_y
AZ80A-T6	3,300	0.25	1.5	$0.96\sigma_y$
ZK60A-T5	3,300	0.25	1.5	$0.96\sigma_y$

Steel (55,000 psi yield) $$\frac{F_{cr}}{A} = 25{,}000 - 0.902(L/r)^2 \qquad 9.63$$

for $L/r < 105$ and

$$\frac{F_{cr}}{A} = \frac{25{,}000}{0.5 + (1/9510)(L/r)^2} \qquad 9.64$$

for $105 < L/r < 200$.

Application of theory. The following examples are given to illustrate application of the foregoing theory.

COMPARISON OF COLUMN STRENGTH. In this example the relative strength of two circular columns will be determined. One column is 2 in. in diameter and is solid, whereas the other column has the same cross-sectional area but different values for wall thickness. All columns are pin-connected. From Eq. 9.28, for the solid column

$$(F_{cr})_{solid} = \frac{\pi^2 EA}{(L/r_{solid})^2} \qquad 9.65$$

and for the hollow tube of equal area

$$(F_{cr})_{tube} = \frac{\pi^2 EA}{(L/r_{tube})^2} \qquad 9.66$$

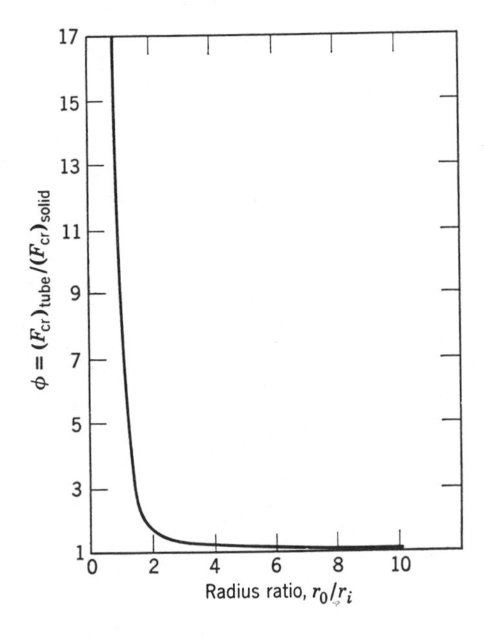

Fig. 9.13 Relative column efficiency of solid versus hollow tubes.

From the two equations

$$\phi = \frac{(F_{cr})_{tube}}{(F_{cr})_{solid}} = \frac{(r_{tube})^2}{(r_{solid})^2} \tag{9.67}$$

or

$$\phi = \left(\frac{I_0 - I_i}{A_0 - A_i}\right) \Big/ \left(\frac{r_{solid}}{2}\right)^2 \tag{9.68}$$

where I_0 and I_i are moments of inertia referred to the outside and inside of the tube respectively. Equation 9.68 can be rewritten in terms of radii alone; thus

$$\phi = \left(\frac{r_0^2 + r_i^2}{4}\right) \Big/ \left(\frac{r_{solid}}{2}\right)^2 = \frac{r_0^2 + r_i^2}{4} \Big/ \frac{r^2_{solid}}{4} \tag{9.69}$$

or

$$\phi = (r_0^2 + r_i^2)/r^2_{solid} \tag{9.70}$$

which is plotted in Fig. 9.13 for various ratios of r_0/r_i of the hollow tube. It can be seen that for the same cross-sectional area the hollow tube is always more efficient and for progressively thinner walls the difference is exceptional. For example, at $r_0/r_i = 1.12$ the hollow tube is nine times as efficient as the solid tube.

COMPOSITE COLUMN. Consider the pin-connected column of length L shown in Fig. 9.14*A*, made up of steel-reinforced wood. As described in Chapter 4, the equivalent of the composite column is a column as shown in Fig. 9.14*B*, where

$$b_1 = b(E_\omega/E_s) \tag{9.71}$$

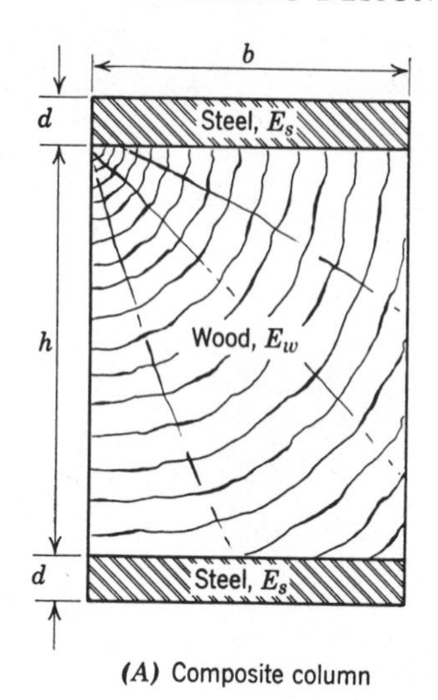

(A) Composite column

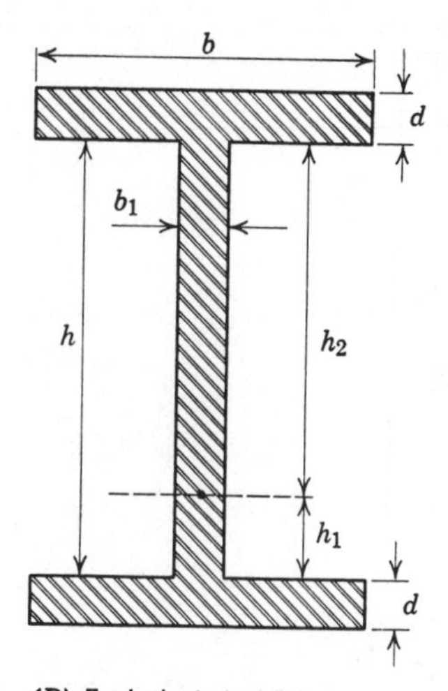

(B) Equivalent steel I-beam

Fig. 9.14 Example of equivalent columns.

where E_ω is the modulus of elasticity of the wood and E_s is the modulus for steel. For a pin-connected column,

$$F_{cr} = \frac{\pi^2 EI}{L^2} = \frac{\pi^2 E}{L^2}\left[2bd\left(h_1 + \frac{d}{2}\right) + 2b_1\frac{h}{2}\left(\frac{h}{4}\right)\right] \qquad 9.72$$

or

$$F_{cr} = \frac{\pi^2 E}{L^2}\left(2b\,dh_1 + bd^2 + \frac{b_1 h^2}{4}\right) \qquad 9.73$$

where $E = E_s$. The loading in the wood is found by multiplying F_{cr} by the ratio E_ω/E_s.

DESIGN OF A BUILDING COLUMN. The following example from (11) illustrates a typical procedure for designing an interior column. For the building in question the columns are spaced 15 ft apart in rows at right angles to each other. The dead load is taken as 100 lb/ft^2 for each of three floors and the roof, the live load is 150 lb/ft^2, and the snow load on the roof is 50 lb/ft^2. The column length is 15 ft, and the following formula is specified by the building code for the particular area

$$\frac{F}{A} = \frac{18{,}000}{1 + (1/18{,}000\,(L/r)^2} \qquad 9.74$$

with a maximum stress of 15,000 psi. From these conditions the total load the column must support is

$$\begin{aligned} \text{dead load} &= 225 \times 100 \times 4 = 90{,}000 \text{ lb} \\ \text{live load} &= 225 \times 150 \times 3 = 101{,}250 \text{ lb} \\ \text{snow load} &= 225 \times 50 \quad\;\; = 11{,}250 \text{ lb} \end{aligned}$$

giving a total load of 202,500 lb. The most economical column is the one with the least weight; therefore as a trial since wide flange beams have radii of gyration more nearly equal to each other than other sections, an 8 × 8 in., 48 lb section is selected. This choice gives the result

$$\frac{F}{A} = \frac{18{,}000}{1 + (1/18{,}000)(86.5)^2} = 12{,}700 \text{ psi}$$

from which

$$F = (14.11)(12{,}700) = 179{,}200 \text{ lb}$$

This choice gives a column inadequate to perform the assigned job. Next, an 8 × 8 in., 67-lb section will be tried; this gives

$$\frac{F}{A} = \frac{18{,}000}{1 + (1/18{,}000)(85.0)^2} = 12{,}850 \text{ psi}$$

from which

$$F = (19.70)(12{,}850) = 253{,}100 \text{ lb}$$

This choice gives a column larger than required. As a third choice a 10 × 10 in., 49-lb section is tried; this gives

$$\frac{F}{A} = \frac{18{,}000}{1 + (1/18{,}000)(70.9)^2} = 14{,}080 \text{ psi}$$

from which

$$F = (14.40)(14{,}080) = 202{,}750 \text{ lb}$$

which is a satisfactory solution to the problem.

Eccentrically loaded columns

In an eccentrically loaded column the stresses acting on the column cross section are derived from loads acting parallel to and distant e from the centroid axis. Consider the column shown in Fig. 9.15, which is eccentrically loaded and pin-

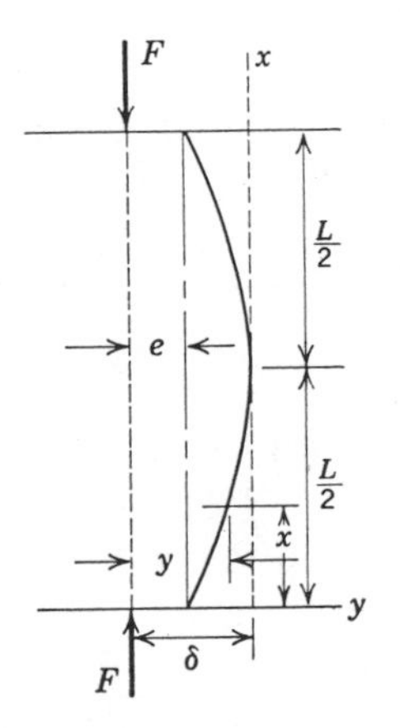

Fig. 9.15 Eccentrically loaded column.

connected. In all practical cases columns have some eccentricity either due to imperfections in the column itself or by virtue of the loading placed upon it. In the analysis of such a column the maximum resultant compressive stress on the cross section is given by

$$\sigma_{max} = \sigma + \frac{M_{max}}{Z} \tag{9.75}$$

where σ is the direct stress and M_{max} occurs at the middle of the column and is equal to $F\delta$. For a section distant x from the base

$$M = Fy \tag{9.76}$$

and

$$EI(d^2y/dx^2) = M = Fy \tag{9.77}$$

In order to solve Eq. 9.77 it must be converted to an equation that is integrable; this is done by multiplying through by the factor $2(dy/dx)$, so that

$$2\frac{dy}{dx}\frac{d^2y}{dx^2} - 2\frac{Fy}{EI}\frac{dy}{dx} = 0 \tag{9.78}$$

Integrating Eq. 9.78 gives the result

$$(dy/dx)^2 - Fy^2/EI = C_1 \tag{9.79}$$

where C_1 is a constant of integration. At $x = L/2$, $y = \delta$, and the slope dy/dx is zero; therefore

$$C_1 = F\delta^2/EI \tag{9.80}$$

and Eq. 9.79 becomes

$$\frac{dy}{(\delta^2 - y^2)^{1/2}} = \left(\frac{F}{EI}\right)^{1/2} dx \tag{9.81}$$

Integrating gives the result

$$\sin^{-1}\frac{y}{\delta} = \left(\frac{F}{EI}\right)^{1/2} x + C_2 \tag{9.82}$$

where C_2 is another constant of integration. At $x = L/2$, $y = \delta$; therefore

$$C_2 = \frac{\pi}{2} - \left[\left(\frac{F}{EI}\right)^{1/2}\left(\frac{L}{2}\right)\right] \tag{9.83}$$

and finally

$$y = \delta \sin\left[\frac{\pi}{2} - \left(\frac{F}{EI}\right)^{1/2}\left(\frac{L}{2} - x\right)\right] \tag{9.84}$$

When $x = 0$, $y = e$; therefore from Eq. 9.84

$$\delta = e \sec\left[\left(\frac{F}{EI}\right)^{1/2}\left(\frac{L}{2}\right)\right] \tag{9.85}$$

Substituting Eq. 9.85 into Eq. 9.75 then gives

$$\sigma_{max} = \sigma + Fe \sec\left[\sqrt{F/EI}(L/2)\right]c/Ar^2 \qquad 9.86$$

or

$$\sigma_{max} = \sigma\left[1 + (ec/r^2)\sec(L/2r)\sqrt{\sigma/E}\right] \qquad 9.87$$

which is called the *secant formula* for eccentrically loaded columns. In using the formula for a particular case, if F and δ are known σ_{max} is found and compared with the allowable value. The value of σ_{max} does not vary linearly with σ in Eq. 9.87; therefore superposition cannot be applied. The working stress σ_w cannot be determined simply by multiplying σ_y by the factor of safety. To include a factor of safety it is customary to assume that if the value of σ is multiplied by a factor of safety n, the maximum stress developed is σ_y; therefore

$$\sigma_y = n\sigma\left[1 + (ec/r^2)\sec(L/2r)\sqrt{n\sigma/E}\right] \qquad 9.88$$

The secant formula is also difficult to use in practice because the value of σ has to be determined by trial and error. Thus, to design an eccentrically loaded column by Eq. 9.88 when F, L, e, σ_y and n are known, σ (or F/A) has to be determined by trial and error since $\sqrt{\sigma/E}$ depends on the dimensions of the cross section. The secant formula is shown graphically in Figs. 9.16 to 9.18 for various columns; it is indicated that the influence of eccentricity is small for

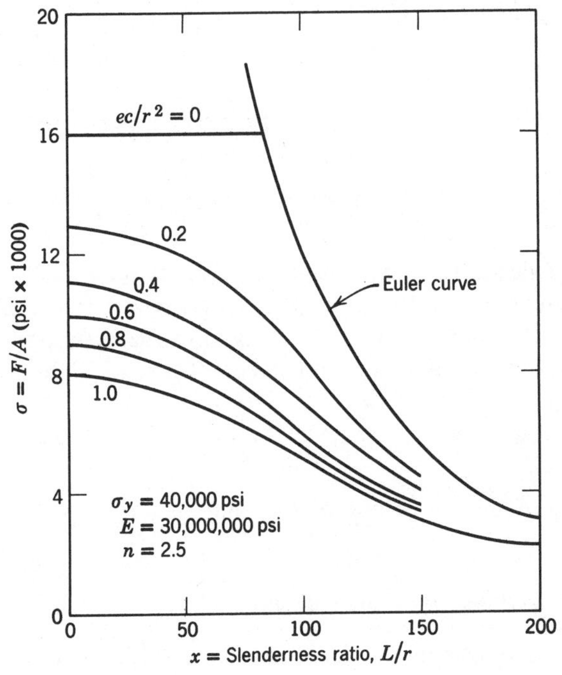

Fig. 9.16 Graphical representation of the secant formula.

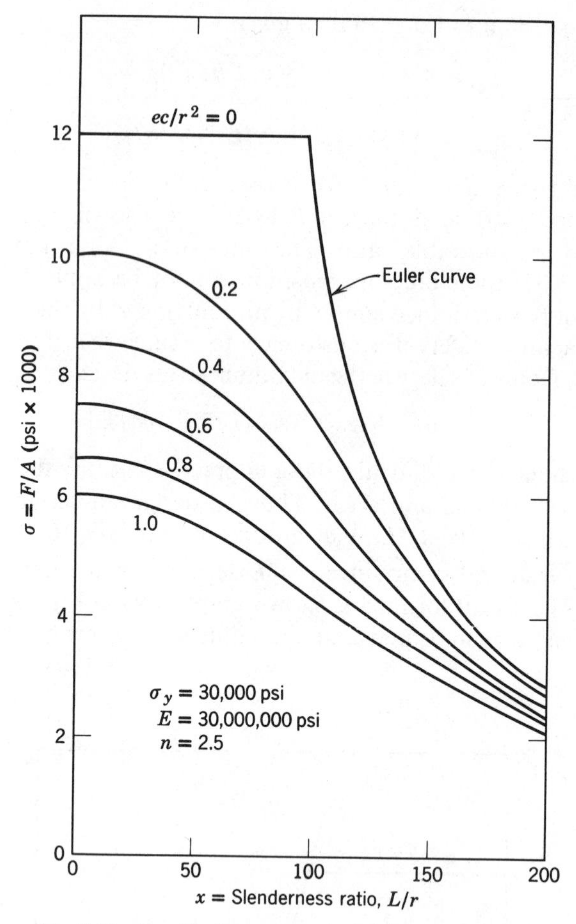

Fig. 9.17 Graphical representation of the secant formula.

values of large slenderness ratio. With computers now available there is no difficulty in using the secant formula in a computerized design office. However, an alternative is available. This method ignores the deflections so that for a column with both an axial load F and an eccentric load F_e the stress developed is

$$\sigma = \frac{F + F_e}{A} + \frac{F_e}{Z} \tag{9.89}$$

where σ is determined from one of the column formulas for axially loaded columns.

Plastic buckling of simple columns

The plastic buckling of columns has been discussed by many authors; for the purposes of this book the methods proposed by Timoshenko (31), Timoshenko

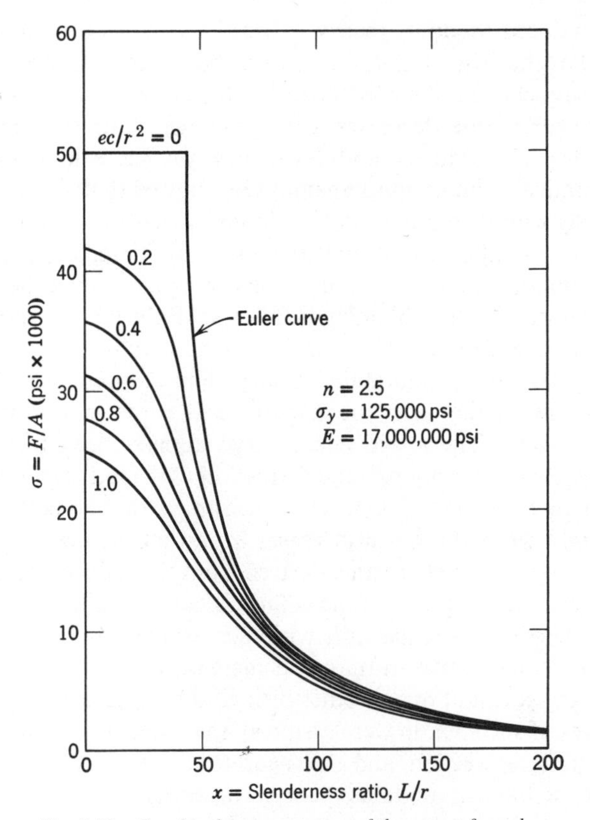

Fig. 9.18 Graphical representation of the secant formula.

and Gere (33), Shanley (24, 25), and Lin (18) will be discussed. In the elastic range Euler's theory and various modifications of that theory have proven quite successful. The "pure" form of Euler's theory is adequate for those cases involving purely Hookean elastic behavior; inelastic effects have been taken care of largely by various empirical methods like the *straight-line formula* and others already discussed. However, through use of the theory of plasticity more fundamental column formulas have been developed which are founded on physical principles and are less empirical in nature than the popular *working* or *design formulas*. One of the first formulas proposed for inelastic columns was proposed by Engesser (18, 32) and had the form

$$F_{cr} = \pi^2 E_t I / L^2 \qquad 9.90$$

where E_t is the *tangent modulus* at that load. Another formulation from the *Considére-Engesser* or *Kármán theory* (18) has the form

$$F_{cr} = \pi^2 E_r I / L^2 \qquad 9.91$$

where E_r is the *reduced modulus* which was discussed in Chapter 6. As explained by Lin, the reduced modulus theory is based on the assumption that after a column reaches the critical uniform stress, it bends and the strain on one side increases, giving change of stress according to the tangent modulus while the strain on the other side decreases, giving change of stress according to the elastic modulus. The reduced modulus theory was generally considered to be the true theoretical solution until Shanley (24) showed that a column may bend simultaneously with increasing axial loading and that the maximum load for a perfect column probably lies somewhere between that prediction by the tangent and reduced modulus theories. This conclusion has been called the *column paradox*, and is discussed by Shanley as follows: "If the tangent modulus is used directly in the Euler formula the resulting critical load is somewhat lower than that given by the reduced modulus theory. The reduced modulus theory is derived by assuming that *after* the column reaches the critical uniform stress, the column bends. [The strain effects have already been mentioned.] But there is an implied assumption in the derivation of the reduced modulus theory that is open to question. It is, in effect, assumed that something keeps the column straight while the strain increases from that predicted by the tangent modulus theory to the higher value derived from the reduced modulus theory. Actually, there is nothing (except the column's bending stiffness) to prevent the column from bending *simultaneously* with increasing axial loading. Under such conditions the compressive strain could increase on one edge of the column while remaining constant on the other or it could increase at different rates on opposite edges. If such action were assumed, the tangent modulus would apply over the entire cross section, and the theoretical buckling load would be that predicted by the tangent modulus theory. This creates a paradox because if all the strains equal or exceed the tangent modulus value, the average strain will be greater than that predicted by the tangent modulus theory. The assumption involved in the reduced modulus theory also represents a paradox. The theory predicts that the column will remain straight up to the calculated maximum load, but it also shows that some strain reversal is needed to provide the additional column stiffness required beyond the tangent modulus load. It is impossible to have strain reversal in a straight column" (24)*. In order to find the maximum load for a perfect column it is necessary to analyze each case separately as will be indicated. For example, starting with the Euler-Engesser theory (Eq. 9.90) it is necessary to obtain a compression (or tension) stress-strain curve for the material and from the plot of that data determine the tangent modulus at various loads by use of Eq. 1.9. Next, Eq. 9.90 is written in terms of the equivalent length, slenderness ratio and stress, thus

$$(L_e/r)_{cr} = \pi\sqrt{E_t/\sigma} \qquad 9.92$$

Equation 9.92 is then solved for the slenderness ratio at which σ is critical. The following example is given for illustration (25)†.

*Quotation by permission of The Institute of Aeronautical Sciences.

†Courtesy of McGraw-Hill Book Co.

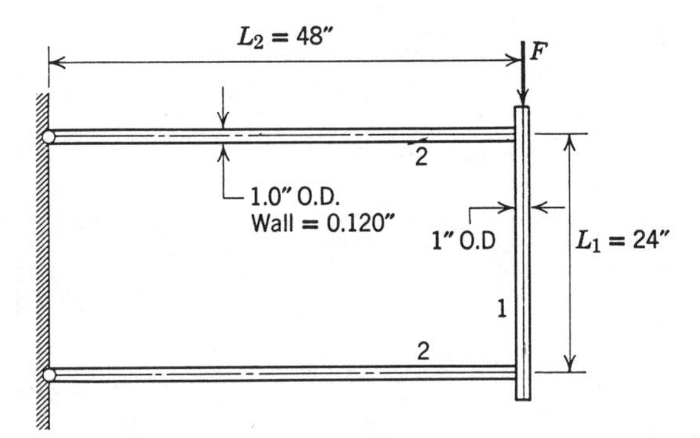

Fig. 9.19 Column attached to a framework.

Consider the structure in Fig. 9.19 which is composed of an aluminum alloy for which the column-stress curve is shown in Fig. 9.20. The horizontal tubes are welded to the column at each end and thereby supply an elastic constraint. The ordinary modulus of elasticity for the material is 10×10^6 psi. For the geometry indicated the following quantities are obtained:

cross-sectional area A	0.332 in.^{2}
moment of inertia I	0.0327 in.^{4}
radius of gyration r	0.314 in.

The effective length factor corresponding to the elastic end constraint is found by successive approximations. The spring constant is

$$K = M/\phi \tag{9.93}$$

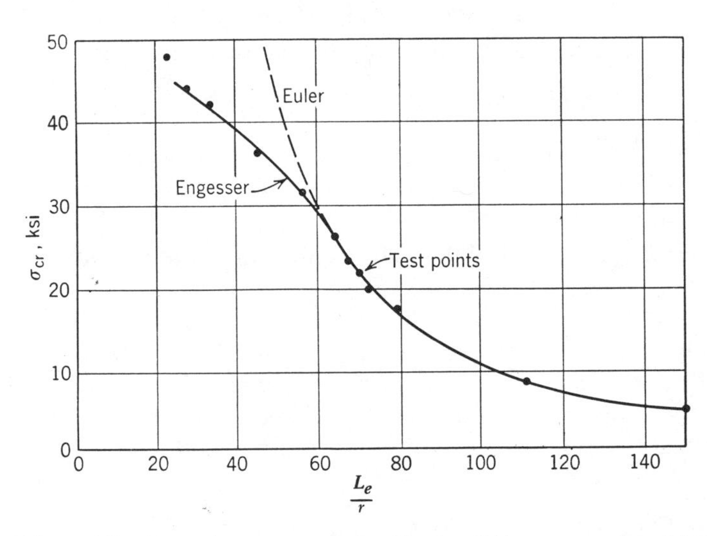

Fig. 9.20 Euler and Engesser column curves. [After Shanley (25), courtesy of the McGraw-Hill Book Co.]

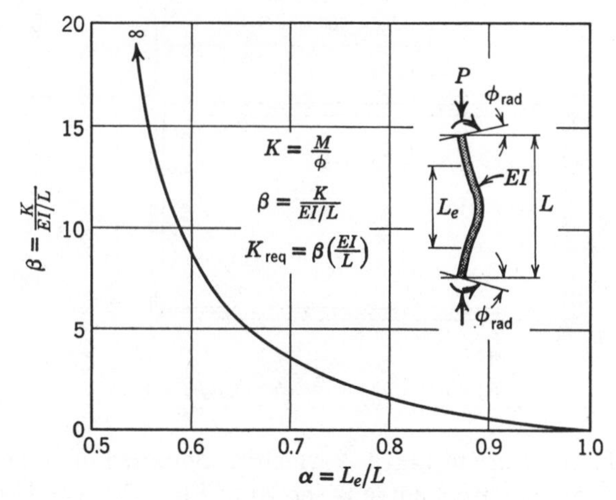

Fig. 9.21 Spring constant required to reduce effective column length. [After Shanley (25), courtesy of the McGraw-Hill Book Co.]

where M is the bending moment and ϕ is slope equal to

$$\phi = 3E_2I_2/L_2 \qquad 9.94$$

or for the geometry indicated $K = 2.04 \times 10^4$ in. lb/radian. If the end constraint is entirely neglected,

$$L_e = L \qquad 9.95$$

and $\alpha = L_e/L = 1.0$. This gives

$$L/r = 24/0.314 = 76.4$$

and from Fig. 9.20 $\sigma_{cr} = 17{,}500$ psi, from which $F_{cr} = \sigma_{cr}A = (17{,}500)(0.332) = 5810$ lb. The column stiffness EI/L for this case then becomes

$$EI/L = E_tI/L = F_{cr}La^2/\pi^2 = 1.41 \times 10^4 \text{ in. lb} \qquad 9.96$$

and β (Fig. 9.21) is

$$\beta = \frac{K}{E_tI/L} = 1.444 \qquad 9.97$$

From Fig. 9.21 $\alpha = 0.81$, and this value is used for a second trial; thus

$$L_e/r = \alpha(L/r) = (0.81)(76.4) = 62 \qquad 9.98$$

From Fig. 9.20 $\sigma_{cr} = 20{,}000$ psi, $F_{cr} = (28{,}000)(0.332) = 9300$ lb and $E_tI/L = 1.48 \times 10^4$. In addition, $\beta = 1.38$; then from Fig. 9.21 $\alpha = 0.82$, which is so close to the assumed value that it will be used for the final determination of critical load. Thus

$$L_e/r = (0.82)(76.4) = 62.6$$

and from Fig. 9.20, $\sigma_{cr} = 27{,}000$ psi, from which

$$F_{cr} = (27{,}000)(0.332) = 8960 \text{ lb} \qquad 9.99$$

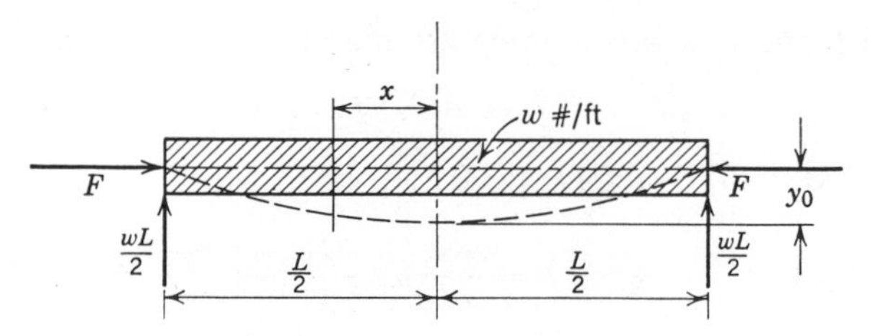

Fig. 9.22 Simple beam column.

Elementary beam-column theory

In the previous sections columns subjected to axial loads only were considered. Often, however, loads on columns may also be applied in a lateral direction so that the resultant is an ordinary beam also subjected to column loading. The simplest case of a beam column is shown in Fig. 9.22, where the beam deflects laterally under its own weight as well as by the assistance supplied from the centroid axis loading. At any section x the bending moment due to the weight w is

$$(M_x)_w = -\frac{wL}{2}\left(\frac{L}{2} - x\right) + \frac{w}{2}\left(\frac{L}{2} - x\right)^2 \qquad 9.100$$

or

$$(M_x)_w = -\frac{w}{2}\left(\frac{L^2}{4} - x^2\right) \qquad 9.101$$

The moment due to the axial force F is $-Fy$; therefore the total moment is

$$M_x = -\frac{w}{2}\left(\frac{L^2}{4} - x^2\right) - Fy \qquad 9.102$$

The beam deflection equation is

$$EI(d^2y/dx^2) = M_x \qquad 9.103$$

Therefore

$$M_x/EI = d^2y/dx^2 \qquad 9.104$$

or

$$\frac{M_x}{EI} = \frac{d^2}{dx^2}\left[-M_x - \frac{w}{2}\left(\frac{L^2}{4} - x^2\right)\right]\frac{1}{F} \qquad 9.105$$

from which

$$\frac{M_xF}{EI} = -\frac{d^2M_x}{dx^2} - \frac{d^2}{dx^2}\left[\frac{w}{2}\left(\frac{L^2}{4} - x^2\right)\right] \qquad 9.106$$

or

$$\frac{d^2M_x}{dx^2} + \frac{M_xF}{EI} - w = 0 \qquad 9.107$$

the solution for which is

$$M_x = A\sin kx + B\cos kx + w/k^2 \qquad 9.108$$

where A and B are constants of integration and

$$k^2 = F/EI \tag{9.109}$$

at $x = \pm L/2$, $M_x = 0$; therefore $A = 0$ and

$$B = \frac{-w}{k^2} \sec \frac{kL}{2} \tag{9.110}$$

so that

$$M_x = \frac{wEI}{F}\left(1 - \cos kx \sec \frac{kL}{2}\right) \tag{9.111}$$

The moment is maximum at midspan, and

$$M_{max} = \frac{wEI}{F}\left(1 - \sec \frac{kL}{2}\right) \tag{9.112}$$

For a pin-connected column the Euler critical load (Eq. 9.18) is

$$F_{cr} = \pi^2 EI/L^2 \tag{9.113}$$

and using this value in Eq. 9.112 gives the result

$$M_{max} = \left(\frac{F_{cr}}{F_{cr}}\right)\frac{wEI}{F}\left(1 - \sec \frac{kL}{2}\right) \tag{9.114}$$

or

$$M_{max} = \left(\frac{F_{cr}}{F}\right)\left(\frac{wL^2}{\pi^2}\right)\left(1 - \sec \frac{\pi}{2}\sqrt{F/F_{cr}}\right) \tag{9.115}$$

The maximum compression stress is then

$$\sigma_{max} = \frac{F}{A} + \frac{M_{max}}{Z} \tag{9.116}$$

An alternate method of solution for this problem is to assume that the elastic deflection curve can be represented by the equation

$$y = y_0 \cos(\pi x/L) \tag{9.117}$$

where y_0 is the midspan deflection. Substituting Eq. 9.117 into Eq. 9.102 and using Eq. 9.103,

$$EI\frac{d^2y}{dx^2} = -\frac{w}{2}\left(\frac{L^2}{4} - x^2\right) - F\left(y_0 \cos \frac{\pi x}{L}\right) \tag{9.118}$$

from which

$$-EIy_0\frac{\pi^2}{L^2}\cos\frac{\pi x}{L} = -\frac{w}{2}\left(\frac{L^2}{4} - x^2\right) - Fy_0 \cos\frac{\pi x}{L}$$

or

$$y = y_0 \cos\frac{\pi x}{L} = -\frac{w}{2}\left(\frac{L^2}{4} - x^2\right)\Big/(F - F_{cr}) \tag{9.119}$$

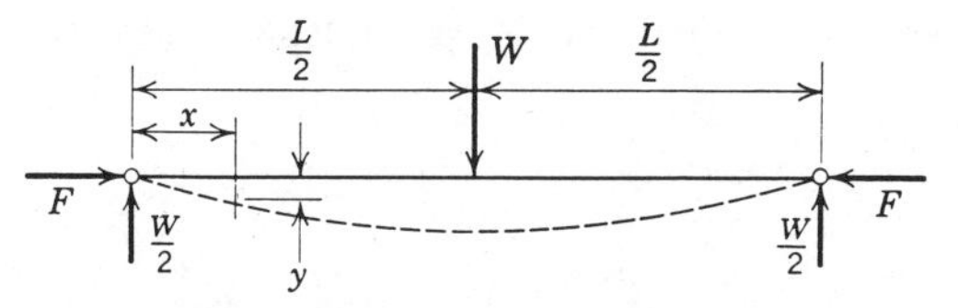

Fig. 9.23 Centrally loaded beam column.

Substituting Eq. 9.119 into Eq. 9.102 then gives

$$M_x = -\frac{w}{2}\left(\frac{L^2}{4} - x^2\right)\left(1 - \frac{F}{F - F_{cr}}\right) \qquad 9.120$$

The moment is maximum at midspan; therefore

$$M_{max} = \frac{wL^2}{8}\left(\frac{F_{cr}}{F_{cr} - F}\right) \qquad 9.121$$

which, as can be seen by computation, is practically equal to the value given by Eq. 9.115.

Consider now a beam column as shown in Fig. 9.23. From the beam equation

$$EI\frac{d^2y}{dx^2} = M = -\frac{Wx}{2} - Fy \qquad 9.122$$

or

$$\frac{d^2y}{dx^2} + \frac{F}{EI}y + \frac{Wx}{2EI} = 0 \qquad 9.123$$

the solution for which is

$$y = A \sin kx + B \cos kx - \frac{m^2 x}{k^2} \qquad 9.124$$

where A and B are constants of integration and

$$m^2 = W/2EI \qquad 9.125$$

At $x = 0$, $y = 0$, and $B = 0$; also, from the same conditions

$$A = \frac{W}{2Fk}\sec\frac{kL}{2} \qquad 9.126$$

so that

$$y = \frac{W}{2Fk}\left[\sin kx \sec\frac{kL}{2} - kx\right] \qquad 9.127$$

The maximum deflection occurs at midspan; therefore

$$y_{max} = \frac{W}{2Fk}\left(\tan\frac{kL}{2} - \frac{kL}{2}\right) \qquad 9.128$$

The maximum bending moment is also at midspan and is given by

$$M_{\max} = Fy_{\max} + \frac{WL}{4} = \frac{W}{2k}\tan\frac{kL}{2} \qquad 9.129$$

and the maximum compression stress is given by Eq. 9.116. In addition to Timoshenko and Gere (33) a summary paper (as of 1961) on the strength and design of metal beam columns is given by Austin (1). The plastic behavior of beam columns is reviewed by Muvdi and Sidebottom (20). Tapered beam columns are reviewed by Butler and Anderson (10).

9-2 BUCKLING OF COMPLEX STRUCTURAL ELEMENTS

Columns of variable cross section

Many structural elements are made in the form of variable cross section members; several such cases are considered in (4) and (33). Consider the variable cross section pin-connected column shown in Fig. 9.24, having two axes of symmetry. The column is symmetrical, and at $x = a$ the width is h_0 and the moment of inertia is I_0; other values are as shown. If $I_{L/2}$ is the midspan moment of inertia, then

$$I_x = I_{L/2}\left(\frac{h_x}{h_{L/2}}\right)^2 \qquad 9.130$$

From Eq. 9.2,

$$EI\frac{d^2y}{dx^2} = M = -Fy \qquad 9.131$$

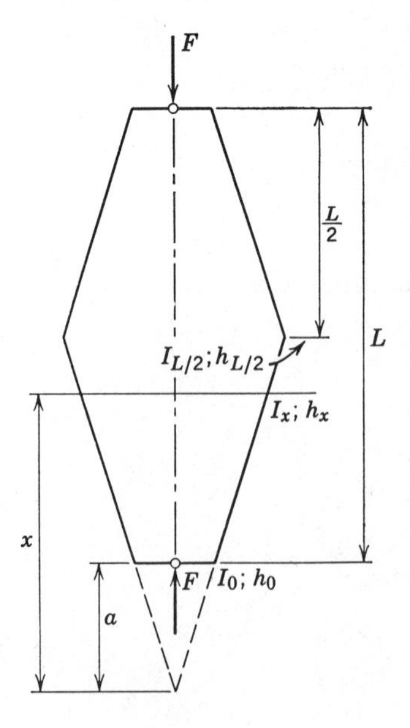

Fig. 9.24 Variable cross section pin-connected column.

which, together with Eq. 9.130, gives the result

$$\left(\frac{x^2}{a^2}\right)\frac{d^2y}{d(x/a)^2} + \alpha^2 y = 0 \qquad 9.132$$

where

$$\alpha^2 = \frac{Fa^2}{EI_{L/2}} \qquad 9.133$$

The solution for Eq. 9.132 is

$$y = \left(\frac{x}{a}\right)^{1/2}\left[A\sin\left(k\ln\frac{x}{a}\right) + B\cos\left(k\ln\frac{x}{a}\right)\right] \qquad 9.134$$

where

$$k = \sqrt{\alpha^2 - \tfrac{1}{4}} \qquad 9.135$$

At $y = 0$, $x/a = h_0/h_{L/2}$ and at $x = a$, $dy/d(x/a) = 0$; these conditions lead to two equations for A and B as follows:

$$A\sin\left(k\ln\frac{h_0}{h_{L/2}}\right) + B\cos\left(k\ln\frac{h_0}{h_{L/2}}\right) = 0 \qquad 9.136$$

and

$$Ak + B/2 = 0 \qquad 9.137$$

As Bleich (4) indicates, solutions for A and B different from zero exist only if

$$\tan\left(k\ln\frac{h_0}{h_{L/2}}\right) - 2k = 0 \qquad 9.138$$

in which the smallest root k defines the critical load F_{cr}, which is

$$F_{cr} = \frac{EI_{L/2}}{4a^2}(1 + 4k_1{}^2) \qquad 9.139$$

The Euler buckling load for a pin-connected column is given by Eq. 9.19; therefore Eq. 9.139 can be rewritten as

$$F_{cr} = \frac{\pi^2 EI_{L/2}}{L^2}\phi \qquad 9.140$$

where

$$\phi = \frac{1 + 4k_1{}^2}{\pi^2}\left(1 - \frac{h_0}{h_{L/2}}\right)^2 \qquad 9.141$$

which is plotted in Fig. 9.25. Suppose, for example, that the column shown in Fig. 9.24 is 8 ft long, 1 in. thick, and is 4 in. wide at the center and 2 in. wide at the ends. For an aluminum column with an elastic modulus of 10×10^6 psi the critical buckling load, from Eq. 9.140 is

$$F_{cr} = \frac{\pi^2(10 \times 10^6)(\frac{1}{12})(4)}{(96)^2}\phi \qquad 9.142$$

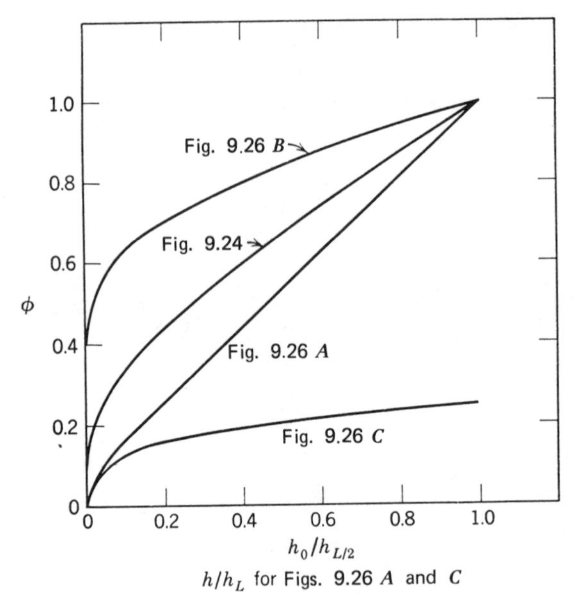

Fig. 9.25 Parameters for variable cross-section columns.

In this case $h_0/h_{L/2}$ is 0.5, so that from Fig. 9.25 ϕ is 0.68; the critical load is thus

$$F_{cr} = 3600(0.68) = 2450 \text{ lb}$$

For columns such as shown in Fig. 9.26, the critical buckling loads are obtained by an analysis similar to that just given, and the buckling load is given by Eq. 9.140 where the factor ϕ is given in Fig. 9.25. For Fig. 9.26*A*, I_L is substituted in Eq. 9.140 for $I_{L/2}$. A solid conical bar, Fig. 9.26*D*, has been solved by Timoshenko and Gere (33); their results are very close to those shown by the curve for Fig. 9.26*C* in Fig. 9.25.

Another form of variable cross section is shown in Fig. 9.27*A*; this column consists of a stacking of prismatic columns of different size. A simple way to solve this column problem is to assume the shape of the deflection curve and then apply the strain-energy method. One shape of the deflection curve* satisfying boundary conditions for the column shown in Fig. 9.27 is

$$y = \delta[1 - \cos(\pi x/2L)] \tag{9.143}$$

The strain energy is that due to bending and is calculated by use of Eq. 7.24, that is,

$$U = \int_0^{L_2} \frac{M_2^{\,2}\,dx}{2EI_2} + \int_{L_2}^{L} \frac{M_1^{\,2}\,dx}{2EI_1} \tag{9.144}$$

*In writing equations for assumed deflection curves it is necessary to check and see if the proposed equation meets the boundary conditions of the problem.

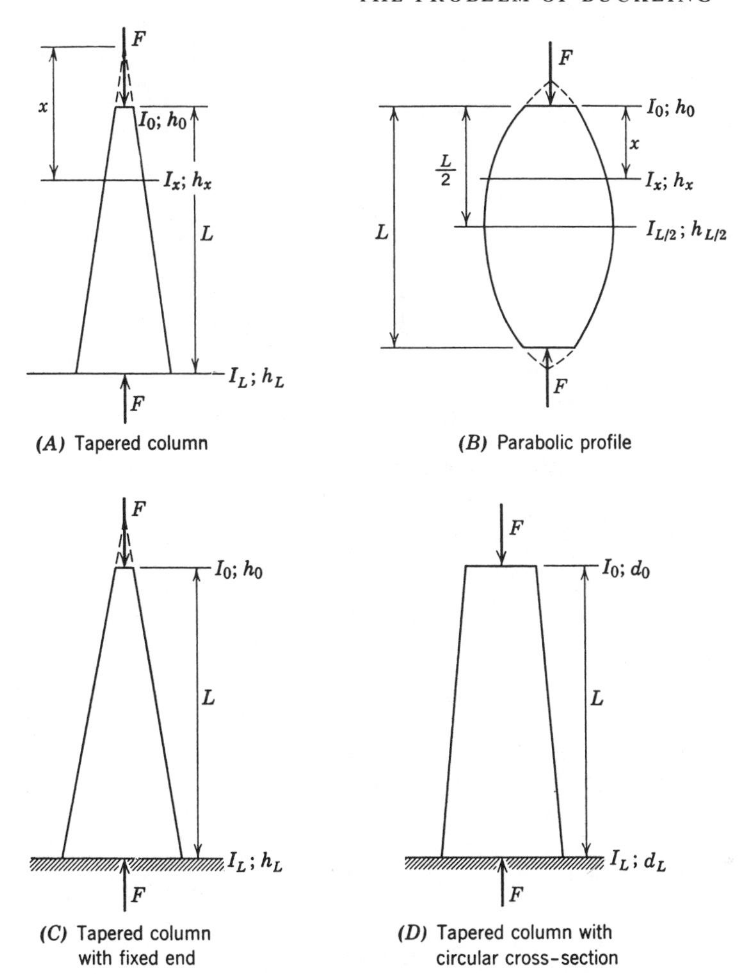

Fig. 9.26 Columns with variable cross section.

However,

$$M_1 = M_2 = F(\delta - y) = \delta \cos(\pi x/2L) \tag{9.145}$$

Therefore Eq. 9.144 becomes

$$U = \frac{F^2\delta^2}{2EI_2}\left[\frac{L_2}{2} + \frac{I_2}{I_1}\left(\frac{L_1}{2}\right) + \frac{L}{2\pi}\left(1 - \frac{I_2}{I_1}\right)\sin\frac{\pi L_2}{L}\right] \tag{9.146}$$

which represents the internal energy. The load F moves through a distance x, giving rise to work term of

$$W = Fx \tag{9.147}$$

where x has to be determined. Consider Fig. 9.28 where the increase in length of

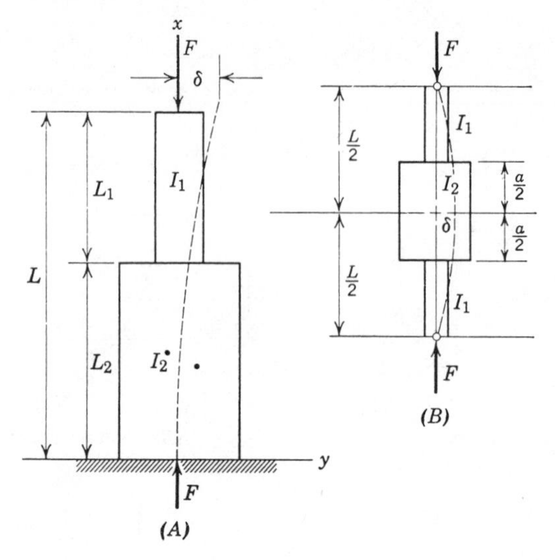

Fig. 9.27 Columns of variable cross section.

the deflected line is $ds - dx$. From the geometry indicated

$$ds = \sqrt{(dy)^2 + (dx)^2} \tag{9.148}$$

Therefore

$$ds - dx \approx \left(\frac{dy}{dx}\right)^2 \tfrac{1}{2} dx \tag{9.149}$$

and

$$x = \frac{1}{2} \int_0^L \left(\frac{dy}{dx}\right)^2 dx \tag{9.150}$$

Substituting Eq. 9.150 into Eq. 9.147 gives

$$W = \frac{F}{2} \int_0^L \left(\frac{dy}{dx}\right)^2 dx = \frac{\pi^2 F \delta^2}{16L} \tag{9.151}$$

By the principle of conservation of energy $U = W$; and, from Eqs. 9.146 and 9.151

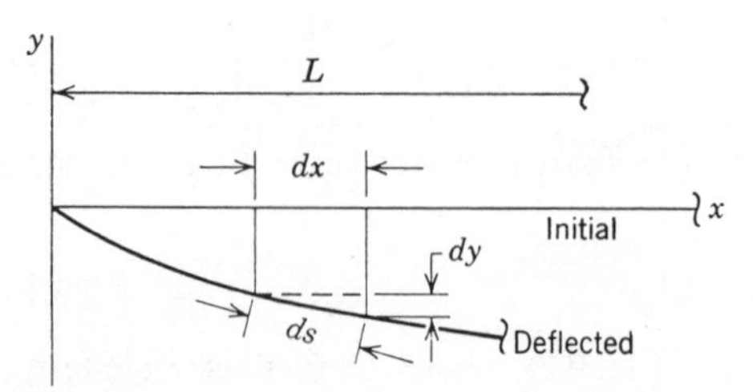

Fig. 9.28 Geometry of deflected line.

$$F_{cr}=\frac{\pi^2EI_2}{4L^2}\left[\frac{1}{L_2/L+(L_1/L)(I_2/I_1)-(1/\pi)(I_2/I_1-1)\sin(\pi L_2/L)}\right] \quad 9.152$$

For the pinned column shown in Fig. 9.27*B*, the equation for the deflection curve satisfying boundary conditions is

$$y=\delta\sin(\pi x/L) \quad 9.153$$

Thus

$$U=2\int_0^{(L/2)-a/2}\frac{M^2\,dx}{2EI_1}+2\int_{(L/2)-a/2}^{L/2}\frac{M^2\,dx}{2EI_2} \quad 9.154$$

and since $M=Fy=F\delta\sin(\pi x/L)$

$$U=\frac{F^2\delta^2}{EI_2}\left[\frac{I_2}{I_1}\left(\frac{L-a}{4}\right)-\sin\frac{\pi a}{L}\left(\frac{I_2}{I_1}-1\right)\frac{L}{4\pi}+\frac{a}{4}\right] \quad 9.155$$

However, from Eqs. 9.147 and 9.150,

$$W=\frac{F}{2}\int_0^L\left(\frac{dy}{dx}\right)^2dx=\frac{\pi^2F\delta^2}{4L} \quad 9.156$$

so that from the condition that $W=U$,

$$F_{cr}=\frac{\pi^2EI_2}{L^2}\left[\frac{1}{a/L+(I_2/I_1)[(L-a)/L]-\sin(\pi a/L)(I_2/I_1-1)1/\pi}\right] \quad 9.157$$

Columns with various loading

Uniformly distributed axial loads. The exact solution of this problem is due to Timoshenko and Gere (33) and involves use of Bessel functions. For an

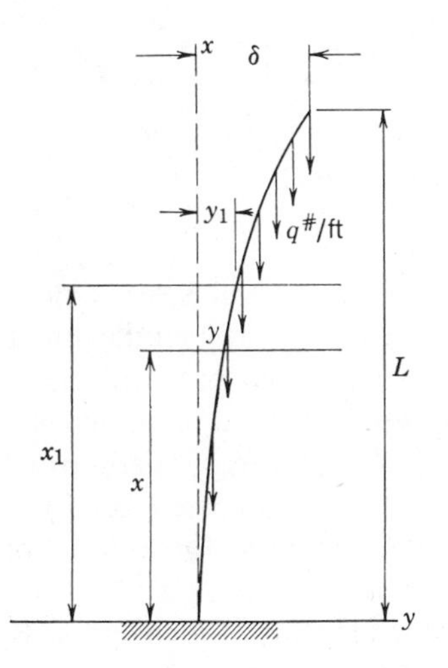

Fig. 9.29 Column with uniformly distributed axial load.

approximate solution, however, the strain-energy method may be used as previously demonstrated. Consider the column in Fig. 9.29 subjected to a uniformly distributed axial load. As before, it is assumed that the deflection curve is given by Eq. 9.143. At any section x the bending moment is

$$M_x = \int_x^L q(y_1 - y)\,dx_1 \tag{9.158}$$

The section above x can also be considered as a column having the deflection curve defined by Eq. 9.143; in this instance Eq. 9.143 becomes

$$y_1 = \delta[1 - \cos(\pi x_1/2L)] \tag{9.159}$$

From Eqs. 9.143 and 9.159, Eq. 9.158 integrates to

$$M_x = \delta q\left[(L - x)\cos\frac{\pi x}{2L} - \frac{2L}{\pi}\left(1 - \sin\frac{\pi x}{2L}\right)\right] \tag{9.160}$$

The strain energy in bending is given by Eq. 7.24 therefore, for the present case

$$U = \int_0^L \frac{M_x^{\,2}\,dx}{2EI} = \frac{\delta^2 q^2 L^3}{12EI\pi^3}(\pi^3 + 54\pi - 192) \tag{9.161}$$

The column moves vertically a distance x; therefore the work done is

$$W = q(L - x)x \tag{9.162}$$

However, since x is defined by Eq. 9.150,

$$W = \frac{q}{2}\int_0^L \left(\frac{dy}{dx}\right)^2 (L - x)\,dx \tag{9.163}$$

or

$$W = \frac{\pi^2\delta^2 q}{32\pi^2}(\pi^2 - 4) \tag{9.164}$$

Again, by the principle of conservation of energy, $W = U$, so that from Eqs. 9.161 and 9.164

$$(qL)_{\text{cr}} = \frac{3\pi^3(\pi^2 - 4)}{8(\pi^3 + 54\pi - 192)}\left(\frac{EI}{L^2}\right) = \frac{7.89EI}{L^2} \tag{9.165}$$

Bars on elastic foundations. In railway rails, for example, it is possible to set up high compressive forces that tend to buckle the rails laterally if the underlying support is essentially rigid. This is a very complicated problem and is treated in some detail in (4, 33). For present purposes consider a pin-connected beam on an elastic foundation as shown in Fig. 9.30; the energy method will be used to solve the problem. Elastic foundations were discussed briefly in Chapter 3 where it was pointed out that the reaction offered by the foundation to deflection is proportional to the deflections. The reaction per unit length of beam is Ky, where y is the deflection and K is the *modulus of foundation*. By definition, the load necessary to produce unit deformation is the *spring constant*;

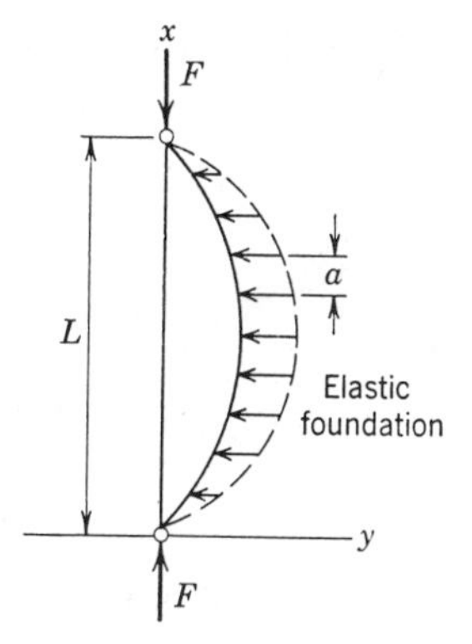

Fig. 9.30 Pin-connected column on an elastic foundation.

in the present case it is convenient to express the modulus of foundation (magnitude of foundation per unit length if deflection is unity) as

$$K = \alpha/a \tag{9.166}$$

where α is the spring constant and a is the distance between supports. The deflection may be represented by (33),

$$y = a_m \sin m(\pi x/L) \tag{9.167}$$

and from this the critical load [see (33)] is

$$F_{\text{cr}} = 2EI\sqrt{K/EI} = 2\sqrt{KEI} \tag{9.168}$$

Buckling of laced and batten-reinforced struts

Effect of shear force on critical load of columns. In many columns, particularly solid bars, the effect of shearing on the critical load can be neglected, as will be shown presently. For structures like latticed struts, however, treated in the following section, shear is very important, and the results obtained here will be useful in developing the strut problem.

When a bar buckles, a shear force acts in the cross section because the cross section is no longer perpendicular to the compressive force. In Fig. 9.4, for example, the bending moment at any cross section is

$$M_x = Fy \tag{9.169}$$

and the shear force (Eq. 2.58) is

$$V = \frac{d}{dx}M = F\frac{dy}{dx} \tag{9.170}$$

From Eqs. 7.24 and 7.27 the strain energy in the system is

$$U = \int_0^L \frac{M^2\,dx}{2EI} + \int_0^L \frac{V^2\,dx}{2GA} \tag{9.171}$$

Substituting Eqs. 9.169 and 9.170 into Eq. 9.171 and integrating gives the result

$$U = \int_0^L \frac{F^2 y^2\, dx}{2EI} + \frac{F^2}{2GA}\int_0^L \left(\frac{dy}{dx}\right)^2 dx \qquad 9.172$$

The deflection curve can be assumed to have the form

$$y = \delta \sin(\pi x/L) \qquad 9.173$$

Therefore, using the value of y from Eq. 9.173, Eq. 9.172 becomes

$$U = F^2\delta^2 L/4EI + \delta^2 F^2 \pi^2 L/4GAL^2 \qquad 9.174$$

The force F moves through a distance x defined by Eq. 9.150, which gives a work term of

$$W = \delta^2 F \pi^2/4L \qquad 9.175$$

By conservation of energy $U = W$; therefore from Eqs. 9.174 and 9.175

$$F_{cr} = \frac{\pi^2 EI}{L^2}\left[\frac{1}{1 + (EI/GA)(\pi^2/L^2)}\right] \qquad 9.176$$

For solid bars GA is large, and thus the effect of shear is minimum for such cases.

The laced strut. Having determined the influence of shear force on buckling now consider a structural latticed strut (Fig. 9.31). In this figure the dotted line represents a plain column and it is assumed that the laced structure deforms to the same shape and that the deflections can be expressed by Eq. 9.173. The strain-energy method will be used. For the structure in question the strain energy is that due to axial load, bending, and shear. For the axial load

$$U_{\text{axial}} = \frac{F^2 L}{2EA} + \sum \frac{F_d^{\,2} d}{2EA_d} + \sum \frac{F_b^{\,2} h}{2EA_b} \qquad 9.177$$

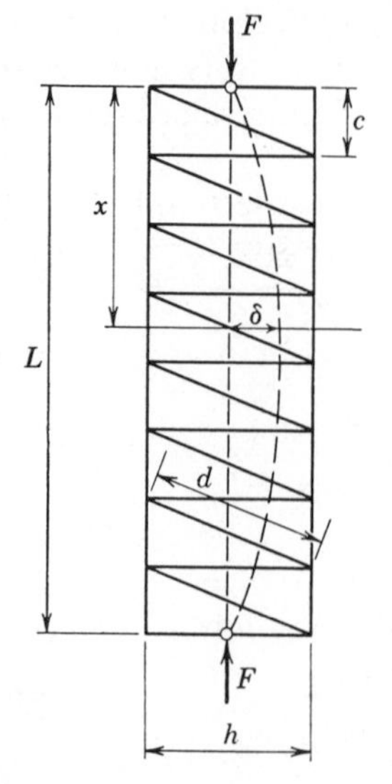

Fig. 9.31 Structural latticed strut.

where A is the cross-sectional area of the vertical cord, F_d is the force in the diagonals, and A_d its area; F_b is the force in the battens (transverse bars) and A_b the corresponding area. The bending moment at any section is

$$M_x = Fy = F\delta \sin(\pi x/L) \tag{9.178}$$

and the shear force is

$$V_x = (d/dx)M_x = F\delta(\pi/L)\cos(\pi x/L) \tag{9.179}$$

from which

$$F = M_x/h \tag{9.180}$$

$$F_d = (d/h)V_x \tag{9.181}$$

and

$$F_b = V_x \tag{9.182}$$

Use of Eq. 9.177 then gives

$$U = \frac{F^2\delta^2}{h^2}\sin^2\frac{\pi x}{L}\left(\frac{L}{2EA}\right) + \sum F^2\delta^2\frac{d^2}{h^2}\frac{\pi^2}{L^2}\cos^2\frac{\pi x}{L}\left(\frac{d}{2EA_d}\right) + \sum F^2\delta^2\frac{\pi^2}{L^2}\cos^2\frac{\pi x}{L}\left(\frac{h}{2EA_b}\right) \tag{9.183}$$

or

$$U = \frac{F^2\delta^2 L}{4}\left(\frac{1}{EI_0} + \frac{d^3\pi^2}{ch^2L^2EA_d} + \frac{h\pi^2}{cL^2EA_b}\right) \tag{9.184}$$

where

$$I_0 = Ah^2/2 \tag{9.185}$$

Finally, apply the principle of conservation of energy

$$F_{cr} = \frac{\pi^2 EI_0}{L^2}\left[\frac{1}{1 + (\pi^2 EI_0/L^2)(1/Ech^2)(d^3/A_d + h^3/A_b)}\right] \tag{9.186}$$

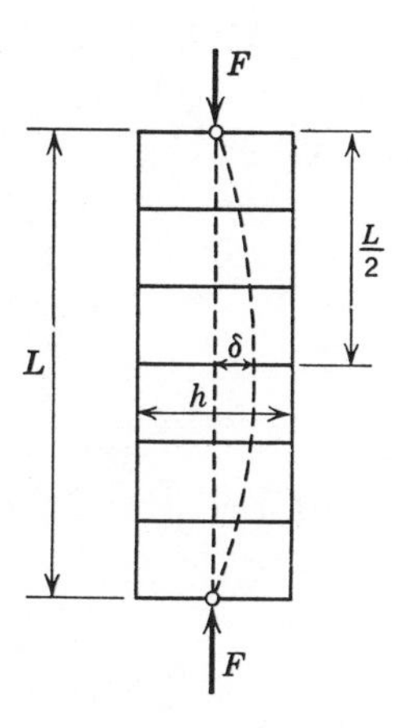

Fig. 9.32 Structural strut with batten plates only.

For a column with batten plates only (Fig. 9.32), the critical load is (18)

$$F_{cr} = \frac{\pi^2 EI_0}{L^2}\left[\frac{1}{1 + (\pi^2/24)(I_0/I_1)(c/L)^2 + \pi^2(EI_0/L^2)(ch/12EI_b)}\right] \quad 9.187$$

where $I_0 = Ah^2/2$ and I_1 and I_b are moments of inertia of the cord (vertical member) and battens respectively.

Buckling of laterally unsupported beams

Situations frequently occur where a beam or girder is not supported laterally at the same time critical load-lateral deflection occurs. This deflection involves a variable twisting of the cross section, and the action that takes place is called *lateral-torsional buckling* (16). Consider the beam of rectangular cross section shown in Fig. 9.33 which is simply supported, centrally loaded, and the ends are restrained against twisting. Since the deflections are considered small, the bending moment induced in the deflected position is

$$M = \frac{Fx \sin\theta}{2} = \frac{Fx\theta}{2} \quad 9.188$$

where θ is the small angle of twist involved. The strain energy of the system is thus made up of bending and torsional energy, or from Eqs. 7.25 and 7.37

$$U = 2\int_0^{L/2} \frac{EI(d^2y/dx^2)^2\,dx}{2} + C\int_0^{L/2}\left(\frac{d\theta}{dx}\right)^2 dx \quad 9.189$$

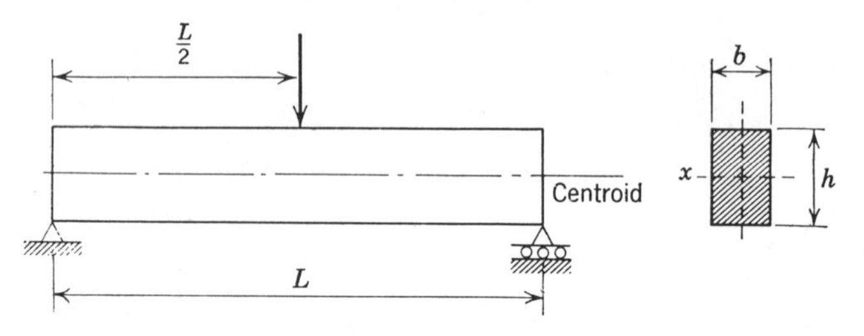

Simple supports—center loaded

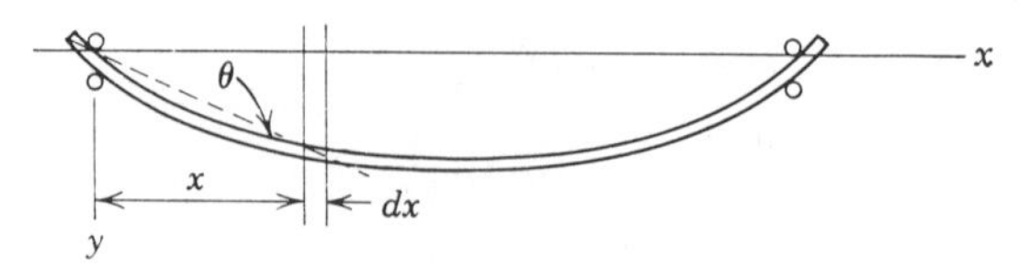

Sidewise deflection and rotation of above structure

Fig. 9.33 Laterally unsupported beam.

from which

$$U = \frac{F^2}{4EI}\int_0^{L/2} x^2\theta^2\,dx + C\int_0^{L/2} x^2\theta^2\,dx \qquad 9.190$$

The vertical deflection δ is given by Eq. 7.53; therefore from Eq. 9.190

$$\delta = \frac{\partial U}{\partial F} = \frac{F}{2EI}\int_0^{L/2} x^2\theta^2\,dx \qquad 9.191$$

The external work done is $F\delta$. From conservation of energy, $W = U$ and

$$F_{cr}\delta = \frac{F_{cr}^2}{4EI}\int_0^{L/2} x^2\theta^2\,dx + C\int_0^{L/2}\left(\frac{d\theta}{dx}\right)^2 dx \qquad 9.192$$

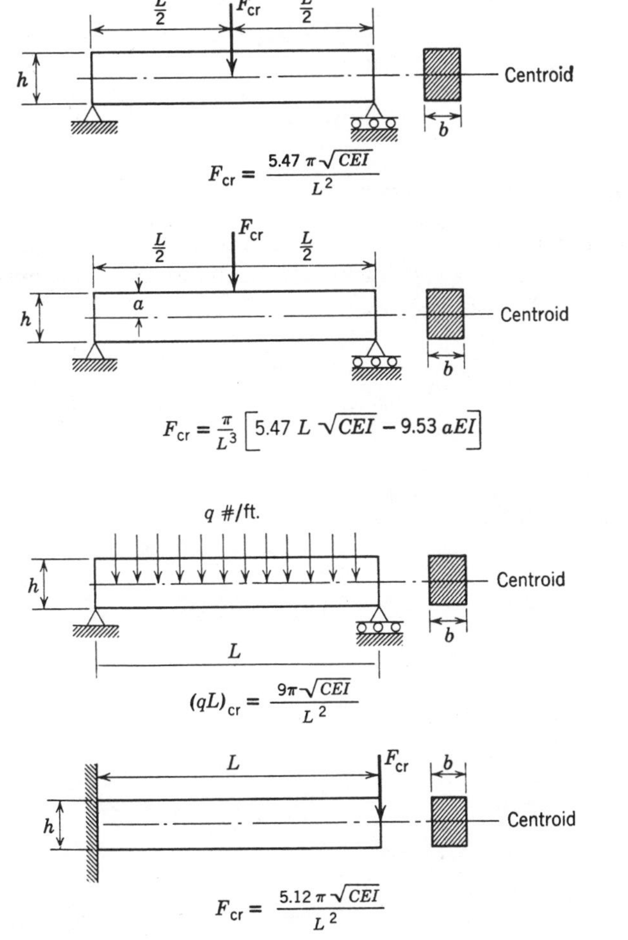

Fig. 9.34 Various beams without lateral support [*see 17, 31, and 33 for other solutions*].

from which, using Eq. 9.191,

$$\frac{F^2}{2EI}\int_0^{L/2} x^2\theta^2\,dx = \frac{F^2}{4EI}\int_0^{L/2} x^2\theta^2\,dx + C\int_0^{L/2}\left(\frac{d\theta}{dx}\right)^2 dx \qquad 9.193$$

or

$$F_{cr}^{\ 2} = \frac{4EIC\displaystyle\int_0^{L/2}\left(\frac{d\theta}{dx}\right)^2 dx}{\displaystyle\int_0^{L/2} x^2\theta^2\,dx} \qquad 9.194$$

In Eq. 9.194 the angle of twist θ is expressed (4) as a function of x or

$$\theta = f(x) = C_1 \sin \pi x/L \qquad 9.195$$

where C_1 is a constant which satisfies the boundary conditions of the problem. Substitution of Eq. 9.195 into Eq. 9.194 gives the result

$$F_{cr} = \frac{5.47\pi\sqrt{CEI}}{L^2} \qquad 9.196$$

A summary for various cases is shown in Fig. 9.34.

Twist buckling of a column

Columns with thin members in the cross section exhibit two modes of buckling. The column in Fig. 9.35 can buckle as a column and by a twisting motion that can be considered in two ways.

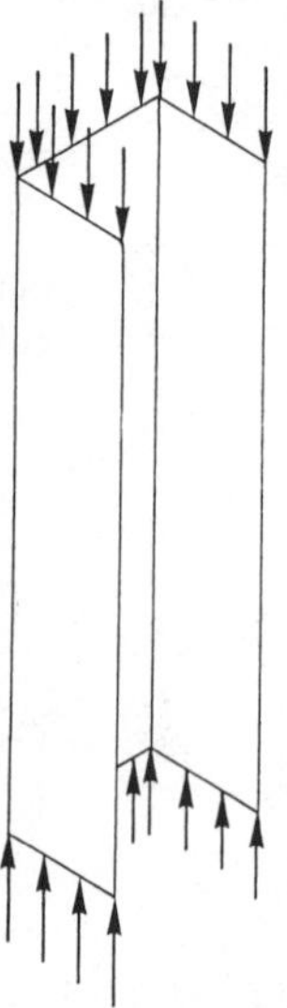

Fig. 9.35 Thin-wall column.

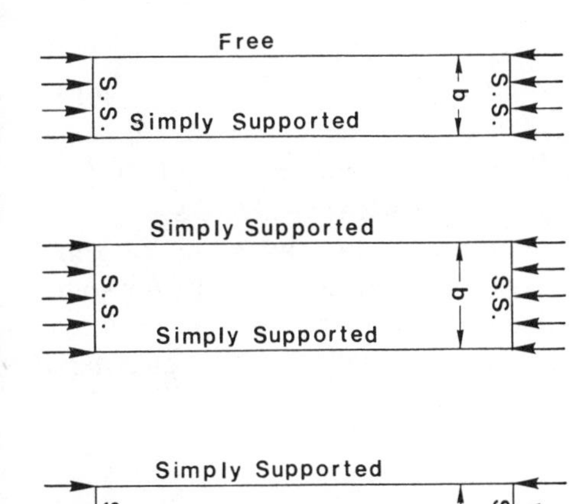

Fig. 9.36 Plate model of Fig. 9.35.

1. A beam section buckling about either principal axis or twisting about a vertical axis.
2. A beam section buckling about either principal axis made of plates with different end conditions that can buckle locally causing the column to collapse.

The motion of a beam section is considered by (4, 17, 19, 33). The section properties I_p (polar moment of inertia with respect to the shear center), GJ (torsional stiffness with respect to the centroid), and the principal moments of inertia are used to obtain buckling loads. The same results (29) can be obtained by considering the beam made up of plates as in Fig. 9.36. When the local buckling stress for one of the plates is exceeded, the column collapses. The buckling equation for a plate is

$$\sigma = K\frac{E}{1-\nu^2}\left(\frac{t}{b}\right)^2 \tag{9.197}$$

$K = 0.416$ for left member, Fig. 9.36. $K = 3.29$ for center member, Fig. 9.36.

The buckling stress can be found for each member of the section. Also, the σ value is selected to prevent buckling of the separate plates. The yield stress is selected since the buckling stress

$$\sigma_c = \frac{P_c}{A} = \frac{n\pi^2 EI}{AL^2} \tag{9.198}$$

is nearly always less than the yield stress σ_y. Therefore, in Eq. 9.197 if b_1 and b_2 are the same values, the left member is critical or

$$\sigma_y = 0.416\frac{E}{1-\nu^2}\left(\frac{t}{b}\right)^2$$

or

$$\frac{t}{b} = \left(\frac{1 - \nu^2}{0.416} \frac{\sigma_y}{E} \right)^{1/2} \qquad 9.199$$

For a steel column with 40 kpsi yield and b_1 of 4 in., the maximum t is 0.2135 in. Then for thinner sections the column fails at less than the yield stress.

Gerard and Becker (14) used the plate approach to solve many buckling problems in columns and stiffened plates. This approach and plots for some sections are presented in (7, 9, 22). McGuire (19) also presents an extensive treatment of the problem as well as that of lateral buckling. For a design criteria guide of the Column Research Council see (16).

Buckling of simple frames

A relatively simple type of frame structure containing compression members is shown in Fig. 9.37*A*; the situation is somewhat similar to that in a fixed-end column (Fig. 9.5) where each end is subjected to a compressive force and a restraining bending moment (Fig. 9.37*B*). The equation for the vertical member is given by

$$EI(d^2y/dx^2) = M = -Fy + M_0 \qquad 9.200$$

the solution to which is

$$y = C_1 \cos mx + C_2 \sin mx + M_0/F \qquad 9.201$$

where $m = \sqrt{F/EI}$ and C_1 and C_2 are constants of integration. At $y = 0$, $x = 0$ and the slope dy/dx is zero at $x = L/2$; therefore,

$$C_1 = -M_0/F \qquad 9.202$$

$$C_2 = M_0 L_1 / 2EI_1 m \qquad 9.203$$

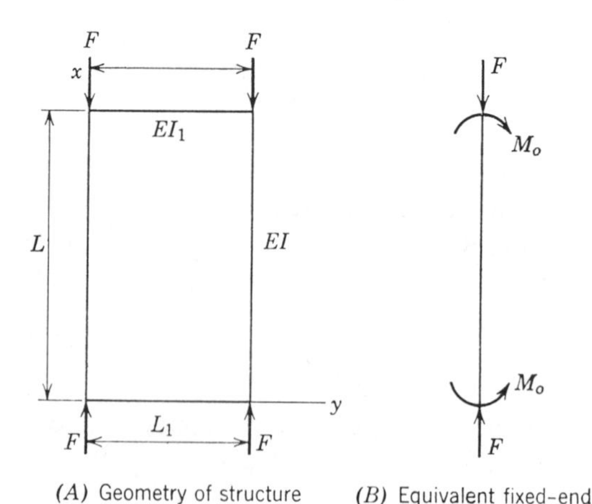

(*A*) Geometry of structure (*B*) Equivalent fixed-end column

Fig. 9.37 Buckling of a simple frame structure.

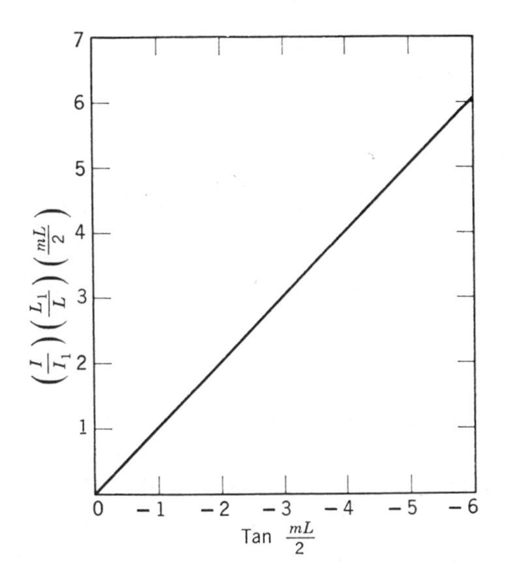

Fig. 9.38 Plot of tangent function.

and

$$m = \frac{-L_1 F}{2EI_1 \tan(mL/2)} = \sqrt{F/EI} \tag{9.204}$$

from which

$$\tan(mL/2) + (I/I_1)(L_1/L)(mL/2) = 0 \tag{9.205}$$

Equation 9.205 is plotted in Fig. 9.38, and the tangent function is plotted in

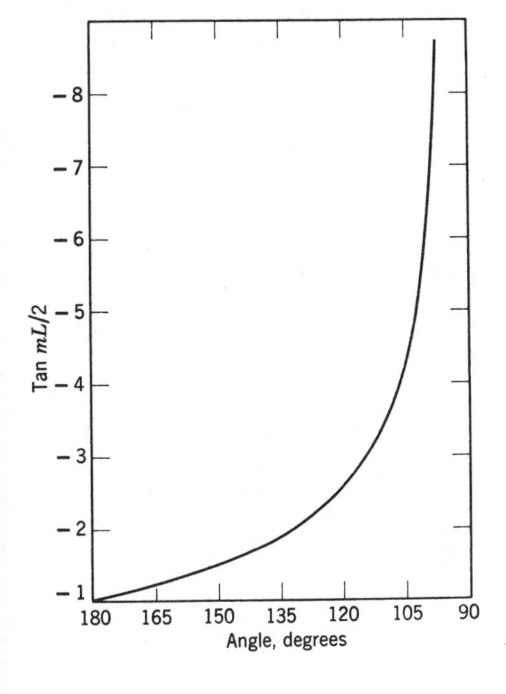

Fig. 9.39 Effect of angle on tangent function.

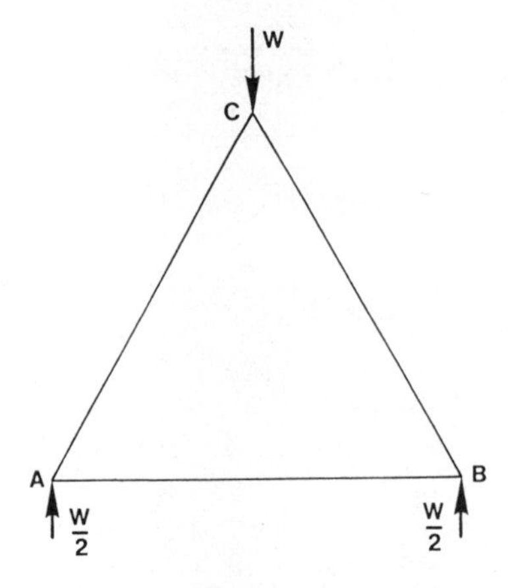

Fig. 9.40 Triangular frame.

Fig. 9.39. When the horizontal members have low resistance to buckling, that is, when the ordinate of Fig. 9.38 is large, the tangent function takes on progressively larger negative numbers which, as shown in Fig. 9.39, reaches a limit at $\pi/2$. Thus from Eq. 9.204

$$F_{cr} = \pi^2 EI/L^2 \qquad 9.206$$

which is the Euler equation for a pin-connected column. When the reverse is true, that is, when the horizontal bars resist buckling, the critical load is

$$F_{cr} = 4\pi^2 EI/L^2 \qquad 9.207$$

which is Euler's formula for a fixed-end column.

An equilateral triangular frame, Fig. 9.40, buckling in its own plane forms two buckling modes; one is a symmetric shape, Fig. 9.41*A* and the other an unsymmetrical mode, Fig. 9.41*B*. The members *AC* and *BC* have a compressive force

$$P = \frac{W}{\sqrt{3}}$$

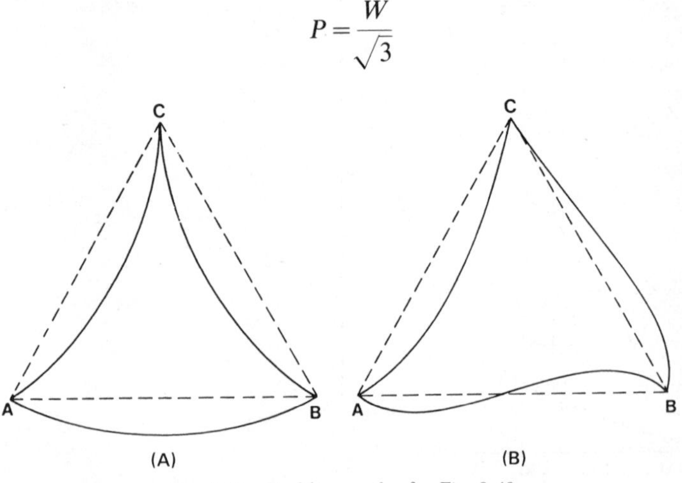

Fig. 9.41 Buckling modes for Fig. 9.40.

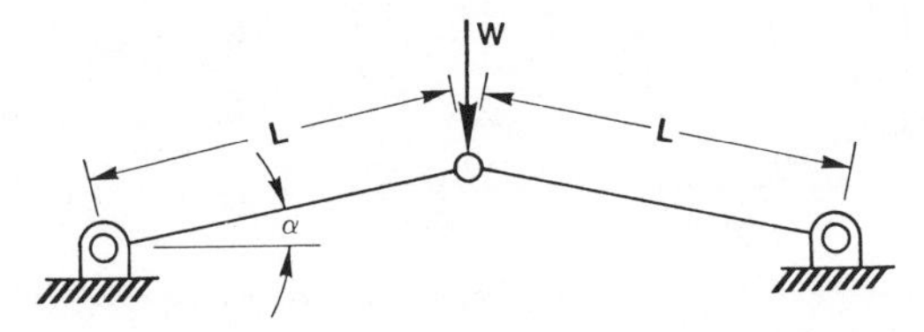

Fig. 9.42 Linkage buckling.

and member AB has a tension force of

$$\frac{W}{2\sqrt{3}}$$

Gregory (15) uses a column member with moments on both ends to allow for a moment transfer around the frame. Moments are applied at each end of the three beams to generate six slope equations. The two moments at A, B, and C are equal and opposite. Then, an Eigenvalue problem is solved for the mode in Fig. 9.41A

$$P = \frac{W}{\sqrt{3}} = 28.5\frac{EI}{L^2} \tag{9.208}$$

and for Fig. 9.41B

$$P = \frac{W}{\sqrt{3}} = 16\frac{EI}{L^2} \tag{9.209}$$

A linkage made of two pinned bars, Fig. 9.42, in a plane can buckle in two modes for small values of α. The first is an Euler buckling load

$$P_E = \frac{\pi^2 EI}{L^2}$$

The second is a load where the two linkages snap through. Britvec (5) maximizes the strain in the bars for the snap-through condition and finds that

$$\frac{P_s}{AE} = 1 - (\cos\alpha)^{2/3} \tag{9.210}$$

Dividing by P_E

$$\frac{P_s}{P_E} = \frac{AL^2}{I\pi^2}[1 - (\cos\alpha)^{2/3}] \tag{9.211}$$

If

$$\frac{P_s}{P_E} < 1 \tag{9.212}$$

the failure is by snapping through. However, if

$$\frac{P_s}{P_E} > 1 \tag{9.213}$$

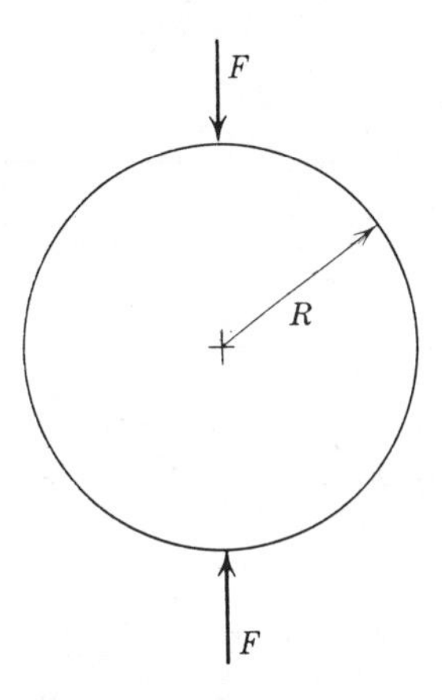

Fig. 9.43 Circular ring under compressive radial load.

the failure is by buckling of the inclined bars. Reference (5) also includes work on fixed-end bars and spring-loaded bars for elastic constraints.

Buckling of circular rings and arches

Consider the circular ring shown in Fig. 9.43, which is compressed by diametral forces F. The free body is shown in Fig. 9.44, where it is seen that at any section m the bending moment is

$$M = M_0 + (FR/2)(1 - \cos\theta) \qquad 9.214$$

where M_0 may be calculated by Castigliano's theorem (see Chapter 7). The strain energy of the system is

$$U = \int_0^{2\pi} \frac{M^2\,ds}{2EI} = \int_0^{2\pi} \frac{M^2\,R d\theta}{2EI} \qquad 9.215$$

However, there is no rotation at A; therefore from Eq. 7.55

$$0 = \frac{\partial U}{\partial M_0} = \int_0^{2\pi} \frac{M\,\partial M}{\partial M_0}\frac{R\,d\theta}{EI} \qquad 9.216$$

or

$$0 = \int_0^{\pi/2} \left[M_0 + \frac{FR}{2}(1 - \cos\theta) \right] d\theta \qquad 9.217$$

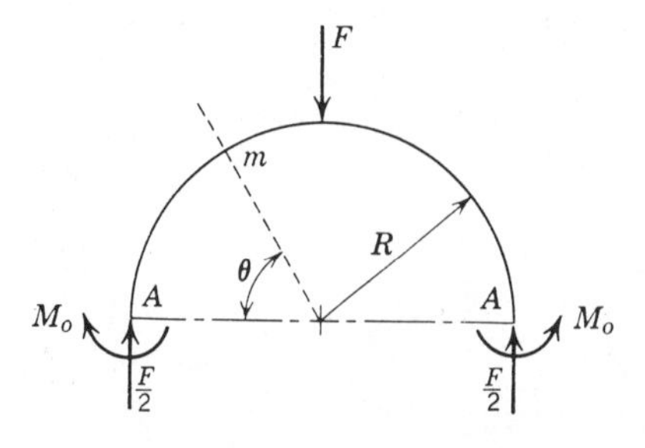

Fig. 9.44 Free body for ring in Fig. 9.43.

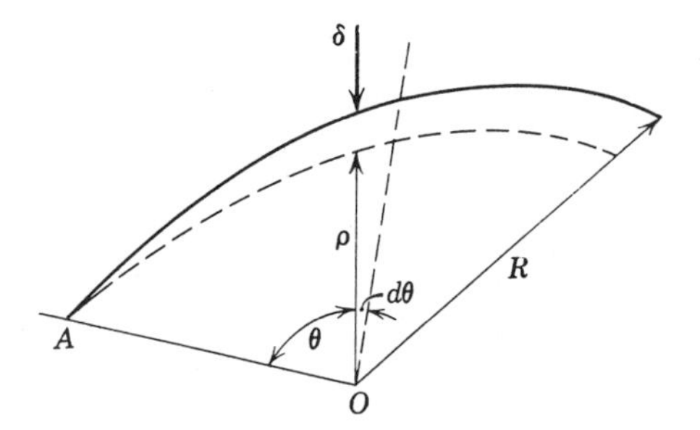

Fig. 9.45 Geometry of bent ring.

from which

$$M_0 = (FR/2\pi)(2 - \pi) \tag{9.218}$$

The moment given by Eq. 9.218 is then used in the differential equation for a bent curve,

$$d^2\delta/d\theta^2 + \delta = -MR^2/EI \tag{9.219}$$

where the symbols are defined in Fig. 9.45. The moment in Eq. 9.219 is given by Eq. 9.214; therefore

$$\frac{d^2\delta}{d\theta^2} + \delta = -\frac{M_0R^2}{EI} - \frac{FR^3(1 - \cos\theta)}{2EI} \tag{9.220}$$

the solution for which is

$$\delta = C_1 \sin\theta + C_2 \cos\theta - \frac{M_0R^2}{EI} - \frac{FR^3}{2EI} + \frac{FR^3\,\theta\sin\theta}{4EI} \tag{9.221}$$

where M_0 is defined by Eq. 9.218 and C_1 and C_2 are constants of integration. At $\theta = 0$ and $90°$, $d\delta/d\theta = 0$; therefore C_1 is zero and

$$C_2 = FR^3/4EI \tag{9.222}$$

Thus from these conditions and using Eq. 9.221

$$\delta = (FR^3/4EI)(\cos\theta + \theta\sin\theta - 4/\pi) \tag{9.223}$$

Therefore, under radial loading (Fig. 9.43), from Eq. 9.223 the increase in the horizontal diameter is

$$2\delta_{\theta=0} = (FR^3/2EI)\left(1 - \frac{4}{\pi}\right) \tag{9.224}$$

and the shortening of the vertical diameter is

$$2\delta_{\theta=90°} = (FR^3/4EI)(\pi - 8/\pi) \tag{9.225}$$

Such rings do not buckle but, if made of ductile material, will yield plastically at the critical section defined by Eq. 9.218 if the stress exceeds the yield limit for

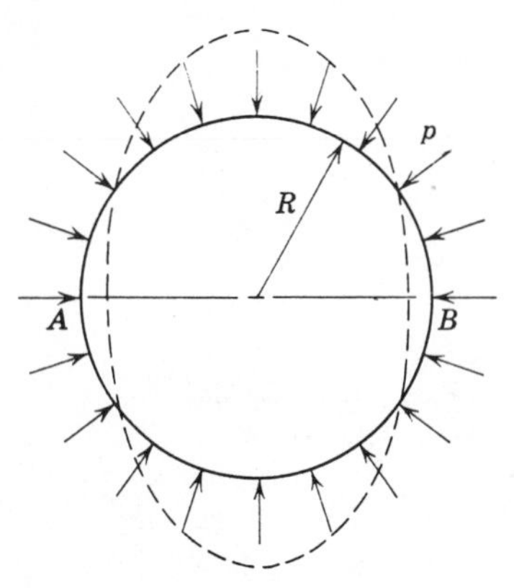

Fig. 9.46 Circular ring under uniform compressive radial loading.

the material. For example, in bending, the yield stress is

$$\sigma_y = M_y/Z \tag{9.226}$$

where σ_y is the yield strength of the material and M_y is replaced by M_0 in Eq. 9.218.

Consider now a circular ring that is uniformly loaded as shown in Fig. 9.46; for such thin rings the common form of buckling is to an elliptical form as shown by the dotted line. The free body is shown in Fig. 9.47. From conditions of equilibrium

$$S = p(R - \delta_0) \tag{9.227}$$

and the bending moment at any section defined by angle θ is, from the geometry indicated.

$$M = M_0 - pR(\delta_0 - \delta) \tag{9.228}$$

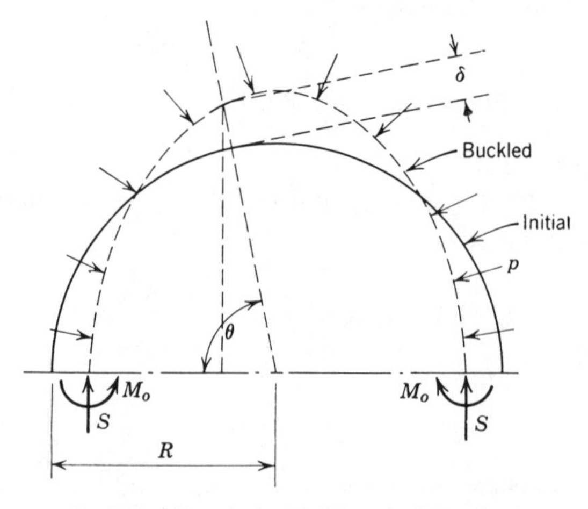

Fig. 9.47 Free body for the ring of Fig. 9.46.

Substituting Eq. 9.228 into Eq. 9.219 then gives

$$\frac{d^2\delta}{d\theta^2} + \delta = -\frac{R^2}{EI}[M_0 - pR(\delta_0 - \delta)] \qquad 9.229$$

the solution for which is (33)

$$\delta = C_1 \sin k\theta + C_2 \cos k\theta + \frac{pR^3\delta_0 - M_0R^2}{pR^3 + EI} \qquad 9.230$$

where

$$k = \sqrt{1 + pR^3/EI} \qquad 9.231$$

and C_1 and C_2 are constants of integration. At $\theta = 0$, $d\delta/d\theta = 0$, from which $C_1 = 0$ and at $\theta = 90^\circ$ $d\delta/d\theta$ is also zero and

$$\sin(k\pi/2) = 0 \qquad 9.232$$

The smallest root of Eq. 9.232 (other than zero) is $k\pi/2 = \pi$ and $k = 2$, or from Eq. 9.231

$$p_{cr} = 3EI/R^3 \qquad 9.233$$

Equation 9.233 represents the simplest buckling conditions; the other roots of Eq. 9.232 correspond to more complex forms.

A segment of the circular ring forms an arch structure which may also buckle under certain conditions. Consider, for example, the hinged arch shown in Fig. 9.48 subjected to uniform loading and buckling by the mode indicated by the dotted line. From Eq. 9.219

$$d^2\delta/d\theta^2 + \delta = -R^2(pR)\delta/EI \qquad 9.234$$

and the solution obtained by Hurlbrink and reported by Timoshenko and Gere (33) is

$$p_{cr} = (EI/R^3)(\pi^2/\alpha^2 - 1) \qquad 9.235$$

If the ends of the arch are fixed,

$$p_{cr} = \frac{EI}{R^3}(k^2 - 1) \qquad 9.236$$

where k is defined by Fig. 9.49 (33).

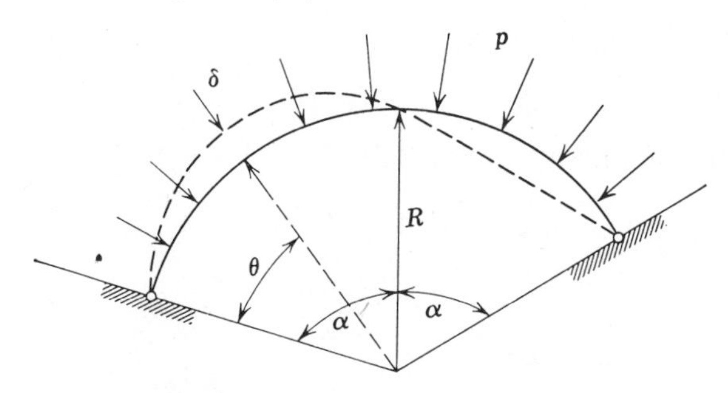

Fig. 9.48 Pin-connected arch under uniform loading.

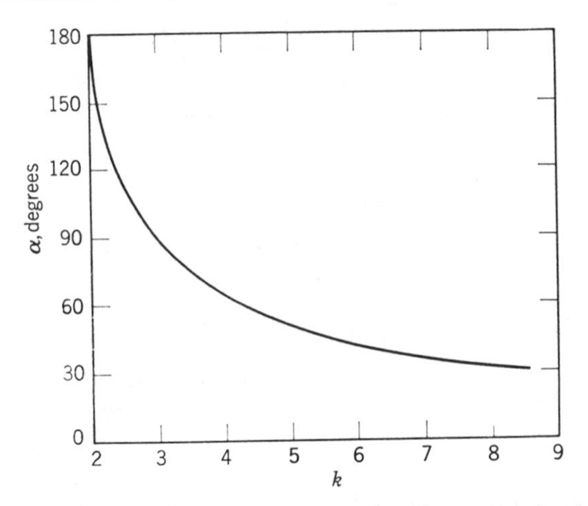

Fig. 9.49 Plot for determining value of parameter k.

9-3 BUCKLING OF TUBES, PLATES, AND SHELLS

Thin-walled vessels

Buckling by internal pressure. There are situations when tubes subjected to internal pressure can fail by buckling. One case is hydraulic tubing loaded by rams as shown in Fig. 9.50*A*. The force F due to the internal pressure is

$$F = \pi r^2 p \tag{9.237}$$

As the pressure is increased a critical value of F is reached at which buckling occurs. Consider, for example, Fig. 9.50*B*; because of the piston arrangement the tube walls do not support axial load and since

$$EI(d^2y/dx^2) = M \tag{9.238}$$

the equation defining buckling is

$$d^2y/dx^2 = (1/EI)(-Fy) \tag{9.239}$$

The Euler buckling load (Eq. 9.19) is

$$F_{cr} = \pi^2 EI/L^2 \tag{9.240}$$

Therefore the critical internal pressure, from Eq. 9.237, is

$$\pi r^2 p = \pi^2 EI/L^2 \tag{9.241}$$

or

$$p_{cr} = \pi EI/(Lr)^2 \tag{9.242}$$

However, since for a thin tube

$$I = \pi r^3 t$$

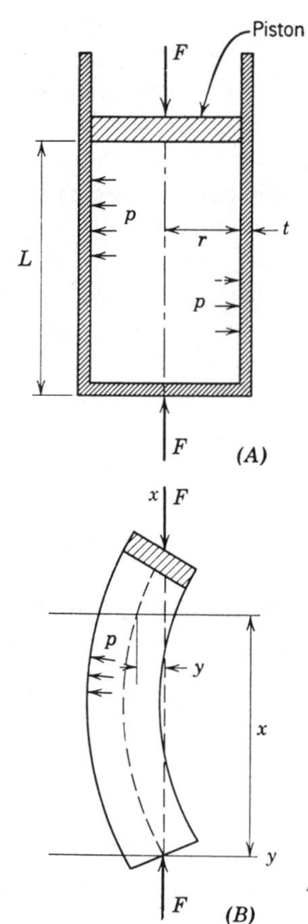

Fig. 9.50 Thin-walled cylinder under internal pressure from ram action.

Eq. 9.242 becomes

$$p_{cr} = \pi^2 Etr/L^2 \qquad 9.243$$

Various types of pressure vessel heads may collapse under the action of internal pressure due to the existence of high circumferential compression in the knuckle region. This situation has been reviewed by Langer (42) who reports Drucker's formula for a torispherical head,

$$p_{\text{collapse}} = \sigma_y\left[\left(0.33 + 5.5\frac{r}{d_m}\right)\frac{T}{L} + 28\left(1 - 2.2\frac{r}{d_m}\right)\left(\frac{T}{L}\right)^2 - 0.0006\right] \qquad 9.244$$

where σ_y = yield strength of the material, psi
r = knuckle radius, in.
d_m = nominal vessel diameter, in.
T = head thickness, in.
L = crown radius, in.

Ellipsoidal head collapse can be computed using Eq. 9.244 with values of r and L most closely approximating a true ellipse. Collapse internal pressure for toriconical heads formed by replacing the spherical cap by a tangent cone may also be found by use of the above formula.

Buckling by external pressure. The buckling of tubes by external pressure has been of considerable practical interest particularly for vacuum systems and structures like submarines. The critical buckling pressure for an externally pressurized tube is developed from the theory previously given for circular rings and in particular Eq. 9.233. In Eq. 9.233 the moment of inertia is given by

$$I = t^3/12 \qquad 9.245$$

where t is the shell thickness. In addition, from Eqs. 3.66 and 3.67 it is seen that for the combined stress case that E is replaced by $E(1 - \nu^2)$ so that Eq. 9.233 for a long tube under external pressure becomes

$$p_{cr} = \frac{Et^3}{4R^3(1 - \nu^2)} \qquad 9.246$$

or, in terms of stress, if it is assumed that the critical stress is defined by the hoop stress

$$\sigma_h = pd/2t = \sigma_{cr} \qquad 9.247$$

from which

$$\sigma_{cr} = \frac{E}{1 - \nu^2}\left(\frac{t}{2R}\right)^2 \qquad 9.248$$

where R is the midwall tube radius and t is the tube wall thickness. If the stress exceeds the proportional limit, then, as was the case with columns, E is replaced by the tangent modulus E_t so that Eq. 9.248 becomes

$$\sigma_{cr} = \frac{E_t}{1 - \nu^2}\left(\frac{t}{2R}\right)^2 \qquad 9.249$$

Equation 9.249 represents the special case of tube buckling when the shell is very long compared to the radius. For short shells that may buckle in different modes a more elaborate analysis is required (33).

If the tube is long and contains an initial eccentricity (out of roundness), of u on the radius, the approximate collapse pressure can be determined from the formula (31)

$$p = \frac{-B \pm (B^2 - 4C)^{1/2}}{2}$$

where

$$B = -\sigma_y \frac{t}{R}\left[1 + \frac{t(t + 6u)}{3\varepsilon R^2}\right]$$

$$C = \frac{\sigma_y^2}{3\varepsilon}\left(\frac{t}{R}\right)^4$$

and

$$\varepsilon = \frac{\text{yield strength of material}}{\text{modulus of elasticity}}$$

Equation 9.246 represents the special case of a thin-walled tube having a large L/R ratio. This equation results from the general equation [(33) in this chapter and (36) of Chapter 3] which is as follows:

$$p_{cr} = \frac{E}{12(1 - \nu^2)}\left[(n^2 - 1)\left(\frac{t}{R}\right)^3 + \frac{12\pi^4(1 - \nu^2)}{(L/R)^3(L/t)n^4(n^2 - 1)}\right] \qquad 9.250$$

In this equation when L/R is large and n, the number of lobes of buckling is 2, Eq. 9.246 is obtained. Both equations are valid for lateral pressure only. If the ratio L/R is small, Eq. 9.250 is used with the value of n determined from the equation

$$n = \sqrt[4]{\frac{6\pi^2(1 - \nu^2)^{1/2}}{(L/R)^2(t/R)}} \qquad 9.251$$

Thus for a short tube under external lateral pressure only, n, the number of lobes into which the tube will buckle, can be predicted and this predicted value is put into Eq. 9.250 for calculating the collapse pressure. Buckling resulting from both axial and lateral pressure is considered later.

For a spherical shell of midwall radius R and wall thickness t the quantities corresponding to Eqs. 9.246 and 9.249 are

$$(p_{cr})_{\text{sphere}} = \frac{2Et^2}{R^2\sqrt{3(1 - \nu^2)}} \qquad 9.252$$

and

$$(\sigma_{cr})_{\text{sphere}} = \frac{Et}{R\sqrt{3(1 - \nu^2)}} \qquad 9.253$$

These values are useful in designing reaction tubes acting in vacuum. For example, for the vacuum tube shown in Fig. 9.51 the following design equations may be used.

BUCKLING PRESSURE FOR SPHERICAL END CAP

$$p_{cr} = \frac{2Et^2}{R_m^2\sqrt{3(1 - \nu^2)}} \qquad 9.254$$

BUCKLING PRESSURE FOR THE CYLINDRICAL BODY

$$p_{cr} = \frac{E_t t^3}{4R_m{}^3(1 - \nu^2)} \qquad 9.255$$

LIMITING PRESSURE FOR THE CIRCULAR END PLATE

$$p_{cr} = \frac{4\sigma_y t_p{}^2}{3R_i{}^2} \quad \text{(plate clamped)} \qquad 9.256$$

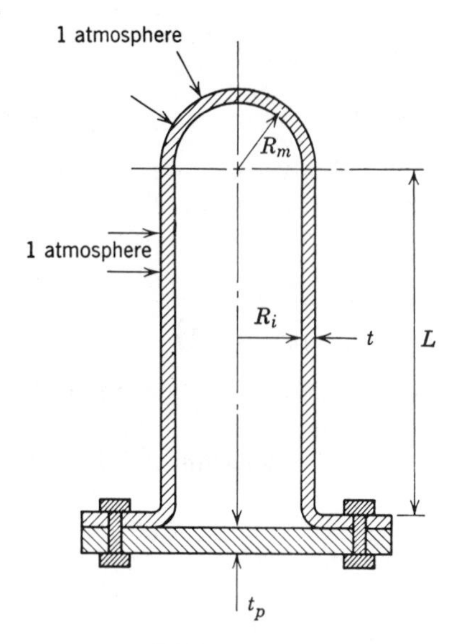

Fig. 9.51 A typical design of a vacuum tube.

The preceding theory and applications to the design of submarine hulls have been reviewed by Wenk (34) and the extension of the theory to include *sandwich-type* shells has been analyzed by Fulton (12).

Buckling by torsion loading. If a thin-walled tube is subjected to torsion loading it may fail by buckling if the applied twisting moment is too large. For long tubes the critical shear stress is (33)

$$\tau_{cr} = 0.60\,\phi E/(D/t)^{3/2} = T_{cr}/2\pi r^2 t \qquad 9.257$$

where T is the applied torque and ϕ represents a factor to account for inelastic effects and is defined as

$$\phi = \sqrt{E_t/E} \qquad 9.258$$

where E_t, the tangent modulus, is determined from a compression stress-strain curve at values of $\sigma = 2\tau$. As an alternative

$$\phi = G_s/G \qquad 9.259$$

where G_s is the secant modulus obtained from a shear stress-shear strain diagram of the material. For short tubes (2)

$$\tau_{cr} = \frac{1.26\phi E}{(D/t)^{5/4}(L/D)^{1/2}} = \frac{T_{cr}}{2\pi r^2 t} \qquad 9.260$$

where L is the tube length and

$$\frac{25}{D/t} < (L/D)^2 < D/t \qquad 9.261$$

when $(L/D)^2(D/t) < 1/2$ the cylinder wall may be treated as a flat shear web (25).

When a thin-walled tube has circular or square holes in the wall, a reduction in the critical stress can be expected. Starnes (28) found for a cylinder, R/t of 400, the following results from his tests for the applied torque T.

$$T/T_{\text{classic}} = 0.85 \quad \text{for} \quad 0 \le \bar{r} \le 0.6 \tag{9.262}$$

$$T/T_{\text{classic}} \quad \text{linear to 0.68 for } 0.6 \le \bar{r} \le 2.5 \tag{9.263}$$

The term $\bar{r}$ is a geometric parameter relating the hole size to the radius and thickness.

$$\bar{r} = \frac{r}{(Rt)^{1/2}} \tag{9.264}$$

If a 20-in. diameter tube with R/t of 400 is selected, then t is 0.025 in. and

$$\bar{r} = 20\frac{r}{R}$$

then for $\bar{r}$ of 2.5 the radius of the cut-out hole is 1.25 in. The buckling torque for a 20-in diameter tube with a 2.5-in. diameter hole in it is

$$T = 0.68\ T_{\text{classic}}$$

Compression buckling. Consider the thin-walled cylinder in Fig. 9.50 loaded only by the axial load F. At some critical load buckling will occur by one of various modes. In one mechanism the hoop direction offers support for the axial direction and the critical load is (lb/in.)

$$F_{\text{cr}} = \left[\frac{Et^2}{\sqrt{3(1-\nu^2)}}\right]\left(\frac{1}{R}\right) \tag{9.265}$$

where R is the radius and t is the wall thickness. The length of the "bellows-type" buckles formed is

$$L = \frac{\pi\sqrt{Rt}}{\sqrt[4]{12(1-\nu^2)}} \tag{9.266}$$

Starnes (28) also conducted tests for thin-walled tubes under compression buckling on tubes with a R/t of 400. These tests are conducted with round and square holes in the tubes. The results show a slight reduction in buckling load for square cutouts as compared to round cutouts. The results show

$$P/P_{\text{classic}} = 0.7 \qquad 0 \le \bar{r} \le 0.5 \tag{9.267}$$

$$P/P_{\text{classic}} \text{ linear to } 0.35 \qquad 0.5 \le \bar{r} \le 1.2 \tag{9.268}$$

Brush and Almruth (8) have extended Starnes' tests to include values for R/t of 100 and 200. It is found that the new test data decreases rapidly at respective values of around 6 and 12 for $\bar{r}$ from the P/P_{classic} line of 0.35.

If a 20-in. diameter tube with R/t of 400 is selected, it would have a 0.025-in.

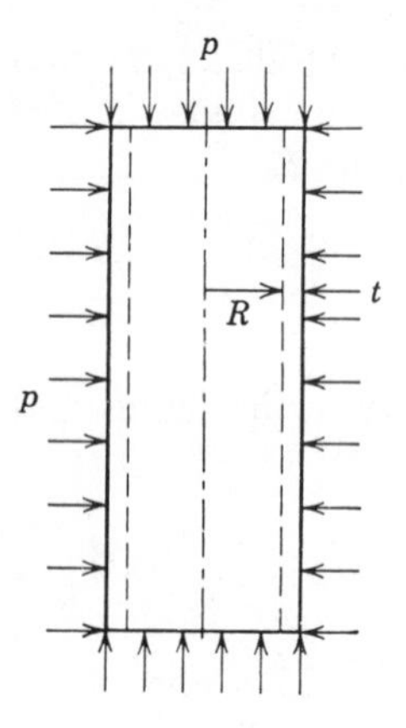

Fig. 9.52 Tube subjected to external axial and lateral pressure.

wall thickness. When $\bar{r}$ is 1.2 in Eq. 9.262

$$P = 0.35\,P_{\text{classic}}$$

The 20-in. diameter tube would have a 1.2-in. diameter hole in it.

For a cylinder with a composite wall under axial load,

$$F_{cr} = \left[\frac{2EAr}{\sqrt{1-\nu^2}}\right]\left(\frac{1}{R}\right) \tag{9.269}$$

where r is radius of gyration of the composite cross section. If the composite consists of a thin-faced material separated by a core, the moment of inertia is expressed by Eq. 4.259, modified for cylinders so that

$$I = (t_s/2)(t_s + t_c)^2 \tag{9.270}$$

and

$$F_{cr} = \left[\frac{2Et_s(t_s + t_c)}{\sqrt{1-\nu^2}}\right]\left(\frac{1}{R}\right) \tag{9.271}$$

where t_s and t_c are the skin and core thicknesses respectively.

For thin closed-end cylinders subjected to both axial and lateral pressure as shown in Fig. 9.52, the equations corresponding to Eqs. 9.246 and 9.248, for cylinders under lateral pressure only are as follows:

$$p_{cr} = \frac{2E(t/R)^2}{\sqrt{3(1-\nu^2)}} \tag{9.272*}$$

and

$$\sigma_{cr} = \frac{E(t/R)}{\sqrt{3(1-\nu^2)}} \tag{9.273}$$

These equations hold only for axisymmetric (bellows-type) buckling. For the

*This is identical to Eq. 9.265 when F_{cr} is expressed in terms of pressure on a closed-end tube.

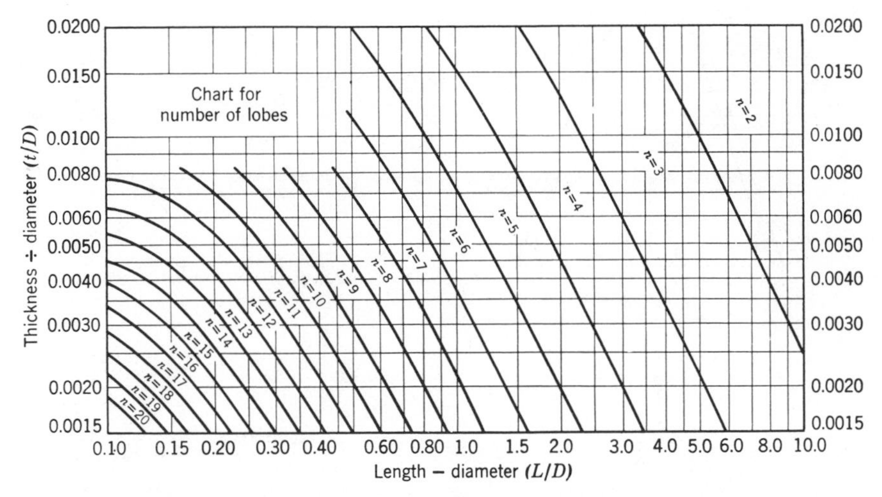

Fig. 9.53 Chart for determining number of lobes of buckling for tubes subjected to uniform simultaneous radial and axial pressure. [From Chapter 3 (36), courtesy of the American Society of Mechanical Engineers.]

usual lobar buckling the references previously cited show that

$$p_{cr} = \frac{1.345E}{(1-\nu^2)^{3/4}}\left[\frac{(t/R)^{5/2}}{1.57(L/R)-(t/R)^{1/2}}\right] \tag{9.274}$$

In any problem the lower value of p_{cr} from Eqs. 9.272 and 9.274 governs. The number of lobes of buckling is obtained from the curves shown in Fig. 9.53. For buckling of springs see (41).

Interaction equations. When design interaction equations are used for columns (43) and for circular cylinders (7), the equations determine if a failure will occur for torsion and compression buckling.

$$R_c + R_{st}^{\ 2} = 1 \tag{9.275}$$

$$R_c = \frac{\text{applied compression stress}}{\text{buckling compression shear stress}}$$

$$R_{st} = \frac{\text{applied torsional shear stress}}{\text{buckling torsional shear stress}}$$

for bending and torsion

$$R_b^{1.5} + R_{st}^{\ 2} = 1 \tag{9.276}$$

$$R_b = \frac{\text{applied bending stress}}{\text{buckling bending stress}}$$

Bruhn (7) lists other combinations of loadings.

Buckling of conical shells. This is a very complicated problem, but based on experiments the following empirical formula has been suggested for the

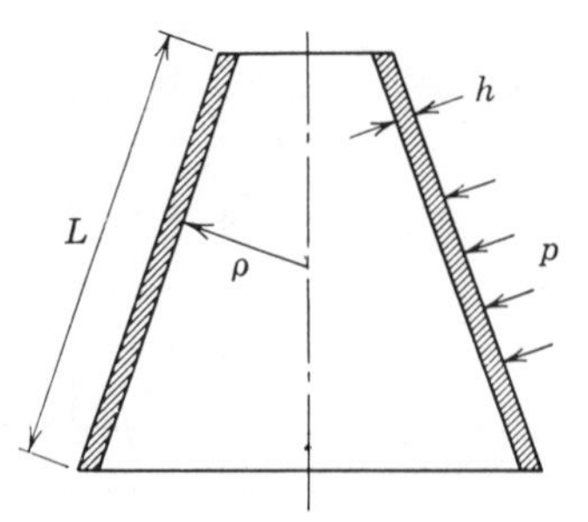

Fig. 9.54 Conical shell under external pressure.

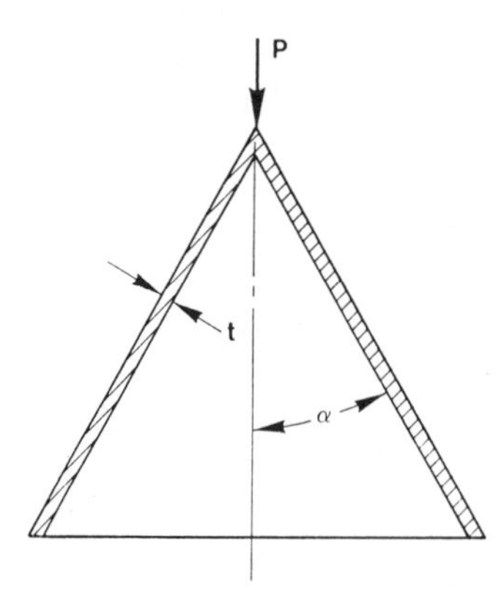

Fig. 9.55 Pointed conical shell.

geometry shown in Fig. 9.54 (26).

$$p_{cr} = \left(\frac{\pi^2}{14.45}\right)\frac{L^{1.1}Eh^{1.45}}{(1-\nu^2)^{0.725}\rho^{1.55}} \tag{9.277}$$

where ρ is the mean radius of curvature of the shell.

The buckling load for a pointed conical shell (17) in Fig. 9.55 is

$$P = \frac{2\pi E t^2 \cos^2\alpha}{\sqrt{3(1-\nu^2)}} \tag{9.278}$$

Plates and sheets

Plates are frequently used as structural elements to resist column-type loading and may be classified into at least two general types. One type, Fig. 9.56, is called a *wide column* – the edges are free and the load is applied to the pin-connected ends. The other type, Fig. 9.57, called a *panel*, has pin-connected ends where the load is applied but the edges are simply supported so that buckling is resisted in the lateral direction. For the wide column Euler's buckling load (25) is

$$W_{cr} = \frac{\pi^2 E t^3}{12(1-\nu^2)L_e^{\,2}} \tag{9.279}$$

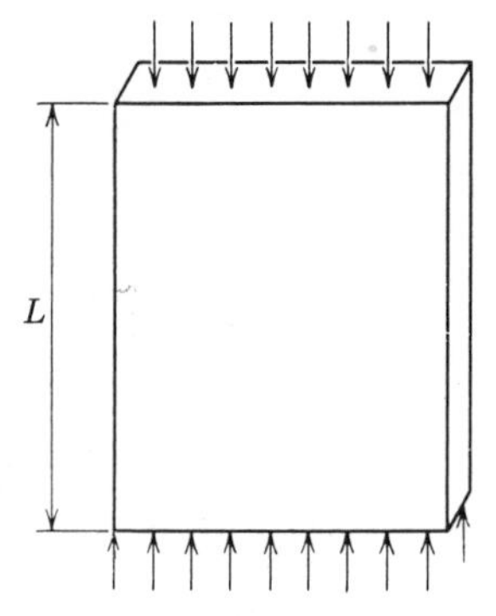

Fig. 9.56 Plate classified as a "wide" column.

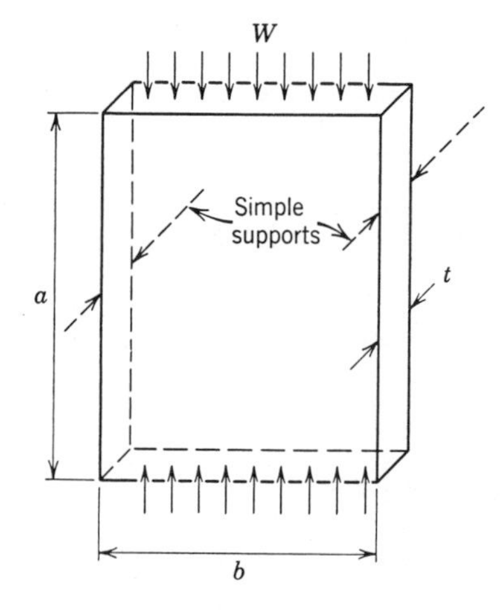

Fig. 9.57 Panel column.

where L_e is the equivalent column length ($L_e = L$ for a pin-connected column). Equation 9.279 in terms of stress is simply

$$\sigma_{cr} = \frac{\pi^2 E t^2}{12(1 - \nu^2) L_e^{\,2}} \qquad 9.280$$

or

$$\sigma_{cr} = \frac{\pi^2 E}{(1 - \nu^2)(L_e/r)^2} \qquad 9.281$$

For a panel where the unloaded edges are simply supported the critical buckling stress is given by

$$\sigma_{cr} = \frac{\pi^2}{12(1 - \nu^2)} \left(\frac{a}{mb} + \frac{mb}{a} \right)^2 E \left(\frac{t}{b} \right)^2 = \frac{\pi^2 k E}{12(1 - \nu^2)} \left(\frac{t}{b} \right)^2 \qquad 9.282$$

where

$$k = (a/mb + mb/a)^2 \qquad 9.283$$

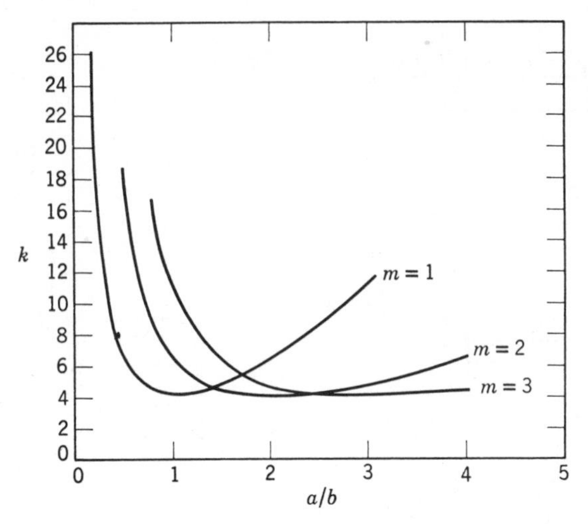

Fig. 9.58 Plot for determining parameter values for a panel column. [*After Saelman (23), courtesy of Rogers Publishing Co.*]

which is plotted in Fig. 9.58 (23). The quantity m is related to the buckle pattern and, for minimum buckling stress,

$$m = a/b \tag{9.284}$$

so that Eq. 9.282 becomes

$$\sigma_{cr} = \frac{4\pi^2 E}{12(1-\nu^2)}\left(\frac{t}{b}\right)^2 \tag{9.285}$$

Where m is not an integer the value of k (Eq. 9.283) corresponding to minimum buckling stress is determined from Fig. 9.58 by taking the smaller value of k corresponding to the a/b ratio. Saelman (23) emphasizes the fact that since the panel tends to buckle into square waves, the transverse supports that are spaced equal to or greater than the width of the panel will have no effect on the buckling stress—to increase the buckling stress supports must be placed closer than the width of the plate, as is evident from Fig. 9.58. For a long panel with complete edge restraint k is equal to 6.98, and by letting Poisson's ratio equal 0.3 (23), the critical buckling stress is

$$\sigma_{cr} = 6.3E(t/b)^2 \tag{9.286}$$

In many practical situations (for pin-connected structures), Eqs. 9.281 and 9.285 reduce to

$$\sigma_{cr} = E(t/L)^2 \tag{9.287}$$

and

$$\sigma_{cr} = 4E(t/b)^2 \tag{9.288}$$

With Eqs. 9.287 and 9.288 it is possible to compare the relative effectiveness of the two types of structures for the same stress application, thus

$$E(t/L)^2 = 4E(t/b)^2 \qquad 9.289$$

from which it is seen that $b = 2L$. This shows that, for the same stress, longitudinal supports can be twice as far apart as transverse supports, or the support structure is twice as efficient if it is assumed that the weight per unit length of the support structure is equal (23). To illustrate, Saelman gives the following examples.

1. For 50,000 psi stress, $\nu = 0.3$ and $E = 10 \times 10^6$ psi, find the required support spacing for transversely and longitudinally supported panels, pin-connected and 0.25 in. thick.

From Eq. 9.287,

$$L = \left[\frac{(10^7)(0.25)^2}{50{,}000}\right]^{1/2} = 3.5 \text{ in.}$$

and from Eq. 9.288,

$$b = \left[\frac{(4)(10^7)(0.25)^2}{50{,}000}\right] = 7.1 \text{ in.}$$

2. For an aluminum alloy beam 50 in. wide with maximum depth $d = 6.0$ in. and an estimated bending moment of 4×10^6 in. lb, determine the required thickness of the plate on the compression side assuming spanwise support beams.

The loading is

$$\frac{M}{0.9\,dw} = \frac{4 \times 10^6}{(0.9)(6)(50)} = 14{,}700 \text{ lb/in.}$$

and from Eq. 9.289

$$\frac{14{,}700}{t} = 4 \times 10^7 (t^2/b^2)$$

or

$$t^3 = \frac{(7.1)^2(14{,}700)}{4 \times 10^7} = 0.0185$$

from which

$$t = 0.265 \text{ in.}$$

In addition to those given by Saelman the following are added to illustrate the theory further.

3. A square tube is made of sheet steel 0.0625 in. thick; the outside width is 5 in. and the modulus of elasticity is 30×10^6 psi. What is the axial buckling load?

In this example the sides of the tube may be regarded as panels with simply supported edges with an average width of $5 - 0.0625 = 4.94$ in. Thus from Eq. 9.288

$$\sigma_{cr} = 4E\left(\frac{t}{b}\right)^2 = \frac{(4)(30 \times 10^6)(0.0625)^2}{(4.94)^2}$$

or

$$\sigma_{cr} = 19{,}300 \text{ psi}$$

The total load $W = \sigma_{cr} A$ or

$$W = (19{,}300)(4)(4.94)(0.0625) = 23{,}836 \text{ lb}$$

4. An aluminum alloy panel 4 in. wide and 0.1875 in. thick is end-loaded with the unloaded edges simply supported. The modulus in compression is 10,500,000 psi, and the secant modulus at yield is 0.7 E. Find the critical buckling stress. From Eq. 9.288

$$\sigma_{cr} = 4E\left(\frac{t}{b}\right)^2 = \frac{(4)(10{,}500{,}000)(0.1875)^2}{16} = 92{,}200 \text{ psi}$$

This value is beyond yield; therefore, using the secant value

$$\sigma_{cr} = \frac{(4)(7{,}350{,}000)(0.1875)^2}{16} = 64{,}500 \text{ psi}$$

The critical buckling stress for a simply supported square plate is also expressed by Eq. 9.285. Bulson (9) reports that when a hole (Fig. 9.59) is placed in the plate, the buckling stress decreases. A hole such that d/b is 0.333 will decrease k in Eq. 9.285 from 4 to 3.5, or a reduction of 12.5%. Further, when a simply supported square plate is tested with reinforced holes, the critical stress is equal to or greater than for a square plate without the hole. These

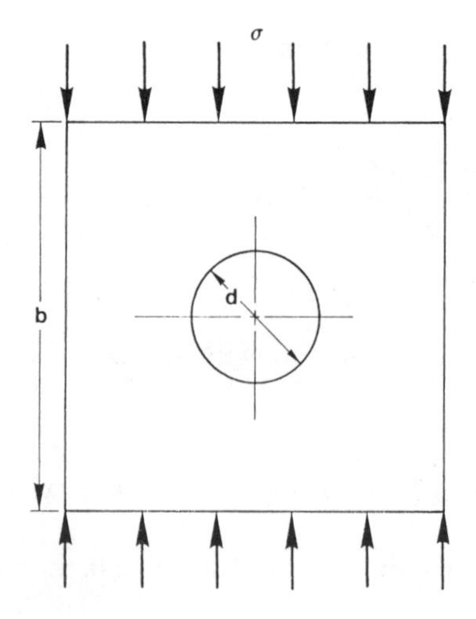

Fig. 9.59 Plate with a hole.

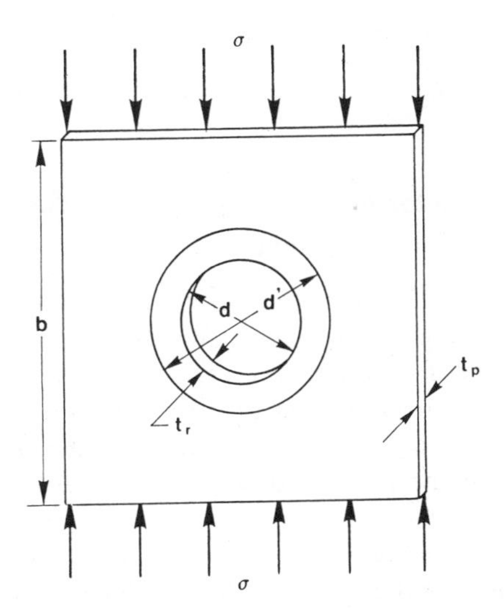

Fig. 9.60 Reinforced plate hole.

tests have been conducted for a d/b of 0.25 and the reinforcement values (Fig. 9.60) are

$$\frac{d'}{b} = \frac{1}{2} \qquad t_r = 2.0\, t_p \qquad k = 8.9$$

$$\frac{d'}{b} = \frac{3}{8} \qquad t_r = 3.4\, t_p \qquad k = 8.9$$

Shear panels, flat or curved (7), are a major factor in design of castings and beams. The critical shear stress for a simply supported rectangular panel is

$$\tau_{cr} = \frac{k\pi^2 E}{12(1 - v^2)}\left(\frac{t}{b}\right)^2 \qquad 9.290$$

The k for a square panel is 9.35 from theory (9) while tests show a 8.00 value from testing. The fixed-edge values are 14.6 from theory and 13.3 from testing. When a hole is placed in a square panel for d/b of 0.333, the k is reduced from 8 to 5 for the simply supported case and from 13.3 to 9.5 in the fixed-edge case. When a simply supported panel in shear with a reinforced hole as shown in Fig. 9.60 is tested, k is 19.1 for $t_r = 2.0\ t_p$, and k is 31.4 for $t_r = 3.4\ t_p$. Note the initial k value from theory is only 9.35. Holes can also be placed in rectangular panels to reduce weight. The critical stress for this case is presented in (9).

Interaction equations (13, 29) are sometimes used for plates acting under combined loading for a/b greater than 1.

COMBINED SHEAR AND AXIAL LOADING

$$R_c + R_s^2 = 1 \qquad 9.291$$

BEND AND SHEAR

$$R_b^2 + R_s^2 = 1 \qquad 9.292$$

BIAXIAL LOADING

$$R_x + R_y = 1 \qquad 9.293$$

The R's are obtained from

$$R_i = \frac{\text{actual loading}}{\text{critical loading acting alone}} \qquad 9.294$$

When three combined loads are used (13, 22, 29) interaction curves are available.

A simply supported panel can carry a higher buckling stress when reinforcement is applied to the panel. When rectangular bars (22) are applied to a panel under compression, k can be increased from 4 to 6. Further, a panel can be reinforced with an angle or a tee section to obtain the same results. In the work by Bulson (9), shear panels can be reinforced to carry larger stresses without buckling.

When solving sandwich plate problems, they are treated as plates on elastic foundations in (9). The problem of face wrinkling or dimpling is a local buckling problem in the plate. The entire plate then has a buckling load of its own sometimes called the primary buckling load.

*Shells with circumferential stiffeners**

Considerable savings in weight and materials can be made by the use of *reinforcing* or *stiffening rings* in vessels subjected to either internal or external pressure. When stiffeners are used, each stiffener is considered to resist the external load for a distance $L/2$ on either side of the ring, where L is the spacing between stiffening rings. For vessels under external pressure the load per unit length on the ring at collapse is equal to pL. Thus, if P is the load on the combined stiffener and shell per unit of circumferential length, use of Eq. 9.233 shows that

$$P = pL = 24EI/d^3 \qquad 9.295$$

where d is the midwall diameter. Since the maximum stress (hoop) is

$$\sigma_h = pd/2t \qquad 9.296$$

Eq. 9.295 can be rewritten as

$$I = d^2 Lt\sigma_h/12E \qquad 9.297$$

Furthermore, since stress (σ_h) equals strain (ε_h) times modulus (E), Eq. 9.297 can be written as

$$I = d^2 Lt\varepsilon_h/12 \qquad 9.298$$

The combined moment of inertia of the shell and stiffener may be considered as

*For a complete theoretical discussion of this subject, including ring-stiffened shells under both axial and lateral pressure, see References 21, 34, and 35.

equivalent to a single thicker shell, that is (33),

$$t_e = t + A_e/d_e = t + A_e/L \qquad 9.299$$

where t_e is the equivalent thickness of shell, A_e the cross-sectional area of a stiffener, and $L = d_e$ is the distance between stiffeners. Thus substituting Eq. 9.299 into Eq. 9.298 gives

$$I = \frac{d^2L}{12}(t + A_e/L)\varepsilon_h \qquad 9.300$$

where I is the required moment of inertia of the stiffening ring. In the ASME code the factor 12 in the denominator of Eq. 9.300 is replaced by the factor 14. The critical collapse pressure for a tube under external pressure is given by Eq. 9.246, which, with Poisson's ratio equal to 0.3, becomes

$$p_{cr} = 2.2E(t/d)^3 \qquad 9.301$$

In designing cylindrical vessels for external pressure a factor of 4 is frequently allowed, which, when applied to Eq. 9.301, gives

$$(p_{cr})_{allowable} = 0.55E(t/d)^3 \qquad 9.302$$

These equations hold for long tubes without stiffeners or where the stiffeners are spaced beyond the *critical length* which is (44)

$$L_{cr} = (4\pi\sqrt{6}/27)[\sqrt[4]{1 - v^2}]d\sqrt{d/t} \qquad 9.303$$

or for $v = 0.3$

$$L_{cr} = 1.11d\sqrt{d/t} \qquad 9.304$$

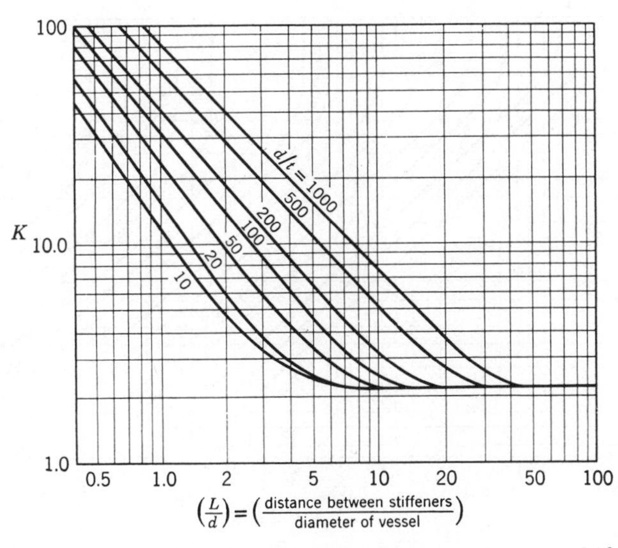

Fig. 9.61 Collapse coefficients for cylindrical shells under external pressure. [*After Brownell and Young (6), courtesy of John Wiley and Sons, Inc.*]

For vessels with stiffeners spaced at distances less than those given by Eq. 9.304 the design equations 9.301 and 9.302 are modified as follows:

$$p_{cr} = KE(t/d)^3 \qquad 9.305$$

and

$$(p_{cr})_{allowable} = (K/4)E(t/d)^3 \qquad 9.306$$

where K is a factor determined from Fig. 9.61. Finally, from Eqs. 9.296 and 9.305

$$(\sigma_h)_{collapse} = pd/2t = (dKE/2t)(t/d)^3 \qquad 9.307$$

from which

$$\sigma_h/E = \varepsilon_h = (K/2)(t/d)^2 \qquad 9.308$$

which is plotted in Fig. 9.62. As pointed out by Brownell and Young (6) the inflections in the parameters (Fig. 9.62) occur at the critical lengths given by Eq. 9.304. The vertical parameters of d/t above the inflections represent the region where the spacing between stiffeners exceeds the critical length and the collapse pressure is independent of the L/d ratio (Eq. 9.301 applies in this region). For the rest of the plots the stiffeners have an effect, and the collapse pressure is a function of the L/d ratio as expressed by K in Eq. 9.305. In order to use Fig. 9.62 to predict the L/d ratio for collapse it is necessary to know the appropriate value for ε_h (which is assumed equal to the uniaxial tensile value ε) for the material at the temperature under consideration; such values are obtained from stress-strain curves such as shown for several materials in

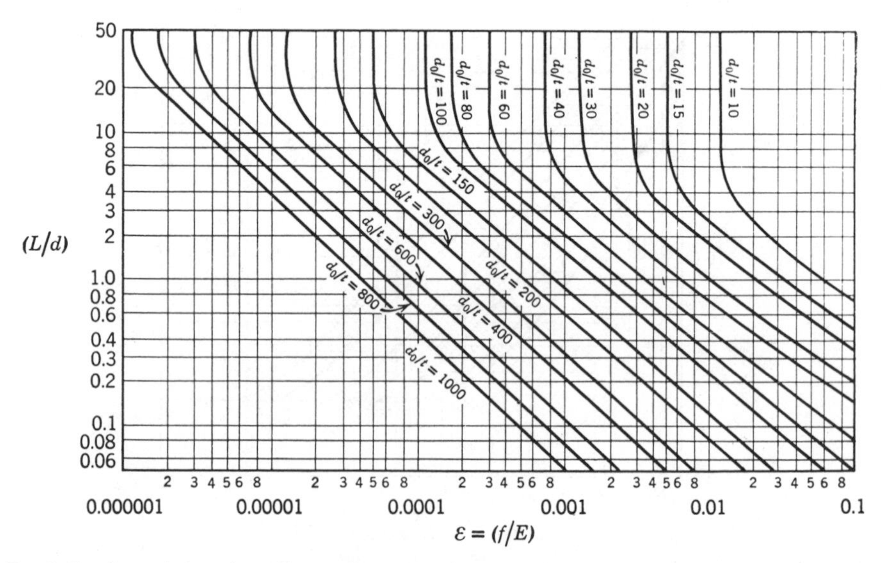

Fig. 9.62 General chart for collapse of vessels under external pressure. [After Brownell and Young (6) courtesy of John Wiley and Sons, Inc.]

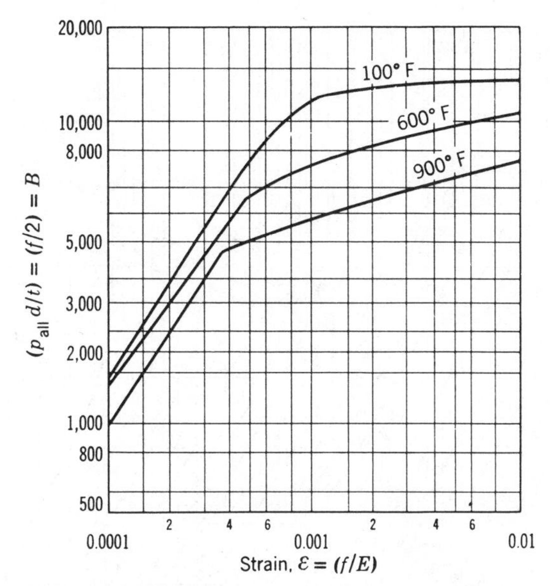

Fig. 9.63 Design chart for plain carbon steel showing the allowable external pressure from 100 to 900°F. [From (40), courtesy of the American Society of Mechanical Engineers.]

Chapter 1. By using a design factor of safety of 4,

$$p = 2\sigma_h t/d = 4p_{\text{allowable}} \tag{9.309}$$

or

$$p_{\text{allowable}} = B(t/d) \tag{9.310}$$

With Eq. 9.310 and stress-strain plots for the material in question, values of σ/E and B (which is equal to $\sigma/2$) can be obtained and plotted such as shown in Fig. 9.63, which, together with Fig. 9.62, determine safe working external pressures. The quantities L/d and d/t are first computed and the corresponding value of E is determined from Fig. 9.62, this value is used with Fig. 9.63 to determine the quantity $[(p)_{\text{allowable}}](d/t)$ from which $p_{\text{allowable}}$ is obtained. Since Figs. 9.62 and 9.63 have a common abscissa, they may be superposed as shown in Figs. 9.64 and 9.65 for carbon steel up to 900°F. The following design example is given (6).*

Design of a fractionating tower. A fractionating tower 14 ft inside diameter by 21 ft in length from tangent line to tangent line of the closures contains removable trays on a 39 in. tray spacing and is to operate under vacuum at 750°F.; the material of construction is SA-283 Grade B plain carbon steel which

*Courtesy of John Wiley and Sons, Inc. More detailed procedures are given in the ASME Boiler and Pressure Vessel Code. Figs. 9.64 and 9.65 were taken from the 1956 edition of the code. In the 1977 edition the format has been changed, but the results are essentially the same as previously given.

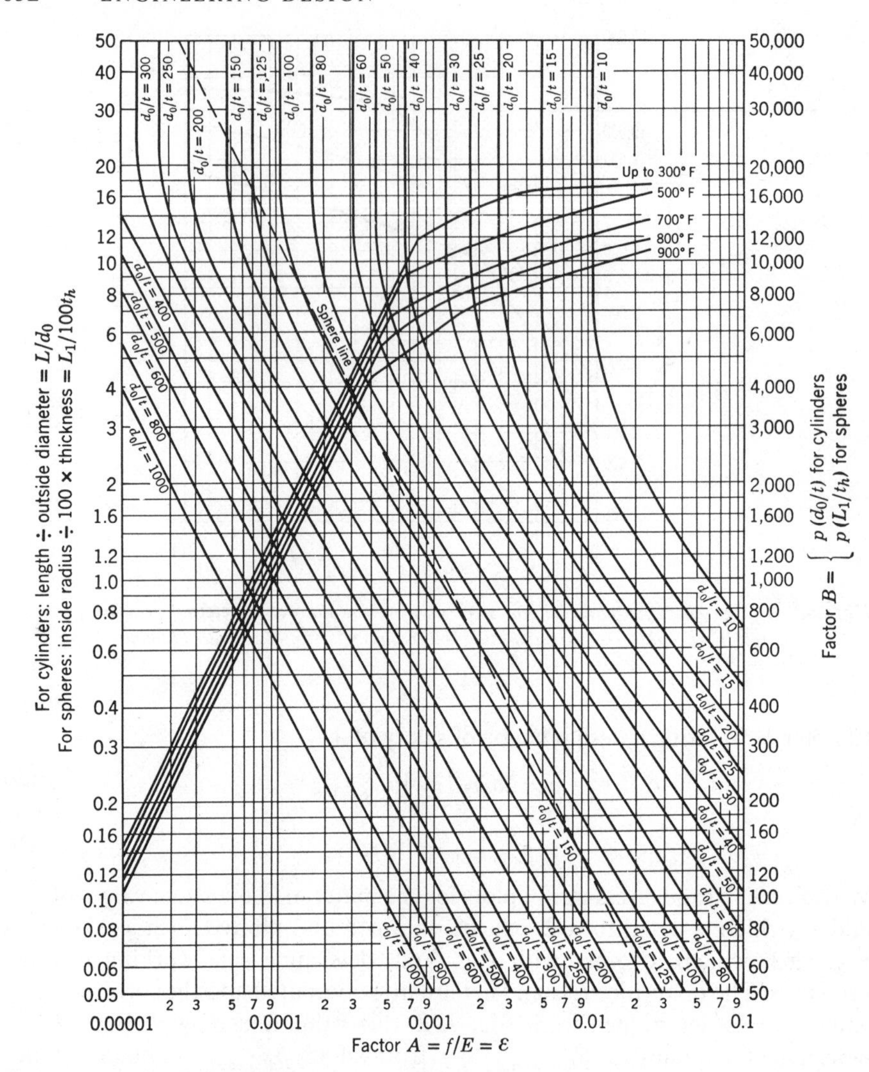

Fig. 9.64 Combined chart for determining thickness of carbon steel shells under external pressure for material yield strength of 24,000–30,000 psi. [From (40), courtesy of the American Society of Mechanical Engineers.]

has a yield strength of 27,000 psi. The required thickness of the shell with and without stiffeners at the tray positions is to be determined.

Shell thickness without stiffeners. The determination of the shell thickness is a successive approximation calculation. Assume a shell thickness of 5/8 in.

$$\frac{L}{d} = \frac{21 \times 12}{(14 \times 12) + 1.25} = 1.49$$

$$\frac{d}{t} = \frac{169.25}{0.625} = 271$$

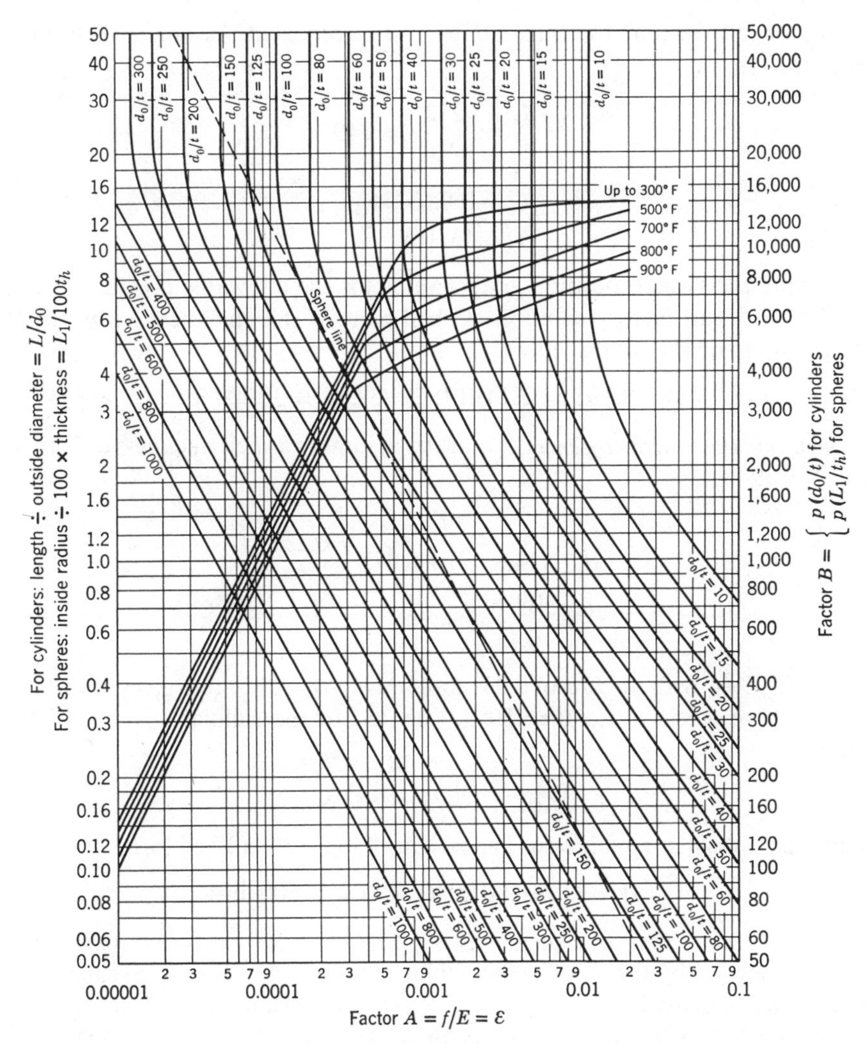

Fig. 9.65 Combined chart for determining thickness of carbon steel shells under external pressure for material yield strength of 30,000–38,000 psi. [*From* (40), *courtesy of the American Society of Mechanical Engineers.*]

Enter Fig. 9.64 with $L/d = 1.49$ and move horizontally to intersect the diagonal line for $d/t = 271$; this gives $\varepsilon = 0.0002$ in./in. Move vertically to the material line for 750° F and then move horizontally to the right-hand side of the chart and read $B = 2300$. The maximum allowable external pressure for the assumed shell thickness of $\frac{5}{8}$ in. is

$$p_{\text{allowable}} = \frac{B}{d/t} = \frac{2300}{271} = 8.48 \text{ psi}$$

Since this pressure is considerably lower than the desired external pressure of

15 psi for full vacuum, the calculation must be repeated. Assume a shell thickness of 13/16 in.

$$\frac{L}{d} = \frac{252}{169.63} = 1.49$$

$$\frac{\mathrm{d}}{t} = \frac{169.63}{(13/16)} = 208.5$$

Following the same procedure gives a value for the allowable pressure of 16.3 psi. This is a satisfactory solution. A shell plate of this thickness weighs 33.15 lb/ft^2; therefore the total shell weight is

$$W = \pi\gamma dL = (3.14)(14)(21)(33.15) = 30{,}700 \text{ lb}$$

Shell thickness with stiffeners. Start by assuming a shell thickness of 7/16 in. and $L = 39$ in.

$$\frac{L}{d} = \frac{39}{(14 \times 12) + 0.875} = 0.231$$

$$\frac{d}{t} = \frac{168.875}{0.4375} = 386$$

Enter Fig. 9.64 with an L/d of 0.231 and move $d/t = 386$; this gives $\varepsilon = 0.0012$ in./in. Move vertically to the 750°F. line and to the right to give $B = 6200$; then

$$p_{\text{allowable}} = 6200/386 = 16.1 \text{ psi}$$

which is a satisfactory solution. The weight of the shell is 17.85 lb/ft^2, therefore

$$W = (3.14)(14)(21)(17.85) = 16{,}500 \text{ lb}$$

This represents a saving in shell steel of 30,700 – 16,500 or 14,200 lb, which is offset somewhat by the weight of stiffening rings which will now be calculated.

Design of circumferential stiffeners. Start by assuming a 7-in. channel weighing 12.25 lb/ft and having moment of inertia $I = 24.1$ in.^{4} A_e is 3.58 sq in., which when substituted into Eq. 9.310

$$B = \frac{\sigma}{2} = \frac{pd}{4t} = \frac{(p_{\text{allowable}})d}{t + A_e/L} = \frac{(15)(168.875)}{0.4375 + 3.58/39} = 4790$$

Enter the right side of Fig. 9.64 with $B = 4790$ and move horizontally to the material line for 750°F; then move vertically to the bottom of the chart where $\varepsilon = 0.00045$; Substituting in Eq. 9.300 with 14 in the denominator instead of 12 gives

$$I = \frac{(169)^2 \times 39(0.4375 + 3.58/39)0.00045}{14} = 18.95 \text{ in.}^4$$

As the required moment of inertia is less than that provided by the assumed 7-in.

channel the design is satisfactory. The weight of five such stiffening rings is

$$W = (5)(3.14)(14)(12.25) = 2700 \text{ lb}$$

The weight of the shell is then 16,500 + 2700 = 19,200 lbs. Therefore the total weight saving by using the stiffened structure is 30,700 − 19,200 = 11,500 lb or nearly 40%.

The foregoing procedures have been largely developed by the Boiler and Pressure Vessel Committee of the American Society of Mechanical Engineers. The committee provides continuous updates on procedures and properties of materials so that the designer should always consult the latest edition of the Boiler and Pressure Vessel Code for current status. However, these procedures are mostly for common metals, and the designer frequently needs procedures to apply to a variety of materials. As a first-order approximation, the following procedures may be used for initial design of vessels subjected to external pressure:

FOR CYLINDERS. Compute 1.73 $(D/t)^{0.5}$. If the value obtained is less than the L/D value for the cylinder, then compute the allowable external pressure from

$$p_A = 2.2\frac{E}{F}\left(\frac{t}{D}\right)^3 \qquad 9.311$$

If the value obtained is greater than the L/D value for the cylinder, compute the allowable external pressure from

$$p_A = \frac{2.6(E/F)(t/D)^{2.5}}{(L/D) - 0.45\,(t/D)^{0.5}} \qquad 9.312$$

where

E = modulus of elasticity (psi)

t = wall thickness (in.)

D = outside diameter (in.)

L = vessel length (for stiffened cylinders the length between stiffeners)

F = factor of safety; generally, 3 to 4 but usually 5 or more for brittle materials and thermosetting plastics

Having computed p_A, then compute the hoop stress from

$$\sigma_h = \frac{p_A D}{2t} \qquad 9.313$$

The value of σ_h should not exceed one-half the minimum yield strength of the material (one half the tensile strength (compressive strength)) for materials with no definable yield strength.

FOR SPHERES. Compute the allowable external pressure from

$$p_A = 1.44\frac{E}{F}\left(\frac{t}{D}\right)^2 \qquad 9.314$$

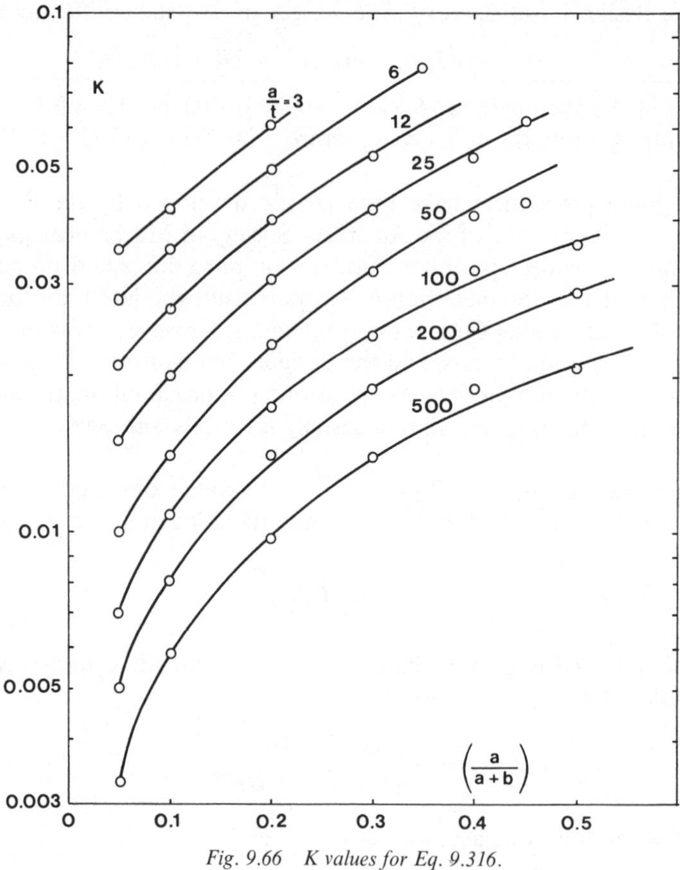

Fig. 9.66 K values for Eq. 9.316.

Compute the shell stress

$$\sigma_h = \frac{p_A D}{4t} \qquad 9.315$$

The value of σ_h should not exceed the limits given in section on cylinders above.

Buckling of toroidal shells. The stability of toroidal shells under external pressure is a very complex problem that has been studied by Sobel and Flügge (27). For approximate design purposes, their analysis shows that the allowable external pressure can be expressed by

$$p_A = \frac{K E t^2}{F a^2} \qquad 9.316$$

where K is shown in Fig. 9.66 and the geometry in Fig. 9.67. F is the factor of safety, and E is the material modulus.

Ring-stiffened cylinders under internal pressure. In a cylindrical vessel under

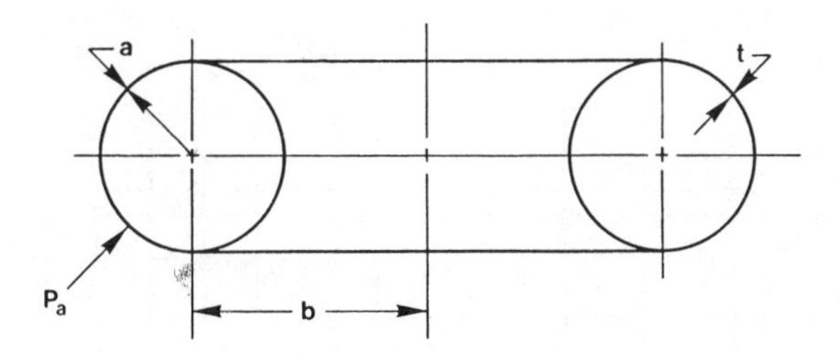

Fig. 9.67 Torus geometry for Eq. 9.316.

internal pressure the longitudinal bending stress due to the stiffener ring is [see (36), Chapter 3],

$$\sigma'_L = \frac{6M_0}{t^2} = \frac{(6)(0.85pR^2)(2D\lambda^2CE)}{Et^3\left[CE + 8D\lambda^3R\left(\dfrac{R_s^{\,2}+R^2}{R_s^{\,2}-R^2}+0.3\right)\right]} \qquad 9.317$$

where R is the midwall radius, C is the width of the stiffener, R_s is the outside radius of the stiffener, t is the shell thickness.

$$D = \frac{Et^3}{12(1-v^2)} \qquad 9.318$$

$$\lambda = \left[\frac{3(1-v^2)}{R^2t^2}\right]^{1/4} \qquad 9.319$$

The shell longitudinal stress due to internal pressure is

$$\sigma_L = pR/2t \qquad 9.320$$

so that the total stress at the stiffener is

$$\sigma = \sigma'_L + \sigma_L \qquad 9.321$$

The hoop stress in the unaffected parts of the shell is

$$\sigma_h = pR/t \qquad 9.322$$

so that the maximum stress ratio $(\sigma'_L + \sigma_L)/\sigma_h$ is

$$\text{stress ratio} = \frac{1}{2} + \frac{10.20}{6.61 + \dfrac{10.288}{(R/t^3)^{1/2}}\left(\dfrac{1.3K^2+0.7}{K^2-1}\right)\left(\dfrac{1}{C}\right)} \qquad 9.323$$

where $K = R_s/R$. The stress ratio at any distance x from the ring stiffener is determined by considering the discontinuity stress (see Chapter 3), so that

$$\text{stress ratio} = \frac{1}{2} + \frac{10.20C_2}{C_1} \qquad 9.324$$

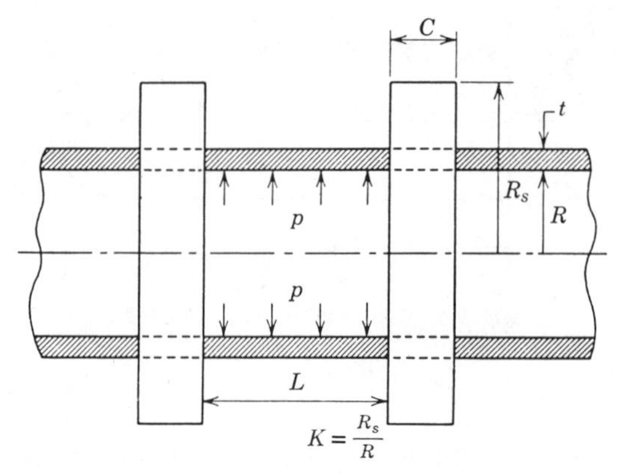

Fig. 9.68 Ring-stiffened cylinder under internal pressure.

where C_1 is the denominator in the second term of Eq. 9.323 and

$$C_2 = e^{-\beta x}(\cos \beta x + \sin \beta x) - 2e^{-\beta x} \sin \beta x = \phi_1 - 2\phi_4 \qquad 9.325$$

where ϕ_1 and ϕ_4 are defined by Eqs. 3.184 and 3.187. The maximum stress occurs at $x = 0$ and is higher than obtained by pressure stress alone. The stiffness of the vessel is increased considerably, however.

In many instances hoop bands can be used to strengthen a vessel. Consider Fig. 9.68. The normal hoop stress for the vessel shown is

$$\sigma_h = pR/t \qquad 9.326$$

or, for yielding, approximately, as shown in Chapter 3,

$$p_y = \sigma_y(t/R) \qquad 9.327$$

where σ_y is the yield strength of the material. If the bands are close enough and the edges rounded to minimize stress concentration, the vessel wall acts like a beam in shear so that

$$\tau_y = p_y L/2t = \sigma_y/2 \qquad 9.328$$

from which

$$p_y = \sigma_y(t/L) \qquad 9.329$$

Thus when L is smaller than R, an increase in the strength of the shell is realized. This same system can also be used to advantage in the design of heavy-walled vessels. In this case discontinuity stresses vanish and the rings can be spaced further apart. This system is particularly useful in building up a vessel where high strength is needed at the bore; by using the band design a relatively thin-walled, high-strength material can be used in conjunction with slip-on rings of low-strength material.

REFERENCES

1. W. J. Austin, *Proc. Am. Soc. Civ. Eng. (J. Struc. Div.)*, **87** (ST4), 1–32 (April 1961).
2. S. B. Batdorf, *A Simplified Method of Elastic-Stability Analysis for Thin Cylindrical Shells: I —Donnell's Equation*, National Advisory Committee for Aeronautics, Technical Note 1341 (1947).
3. L. S. Beedle, T. V. Galambos, and L. Tall, "Column Strength of Constructional Steels," in *New Concepts in Steel Design and Engineering*, pp. 33–50 United States Steel Corp., Pittsburgh, Pa. (1961).
4. F. Bleich, *Buckling Strength of Metal Structures*, McGraw-Hill Book Co., New York, N.Y. (1952).
5. S. J. Britvec, *The Stability of Elastic Systems*, Pergamon Press Inc., New York, N.Y. (1973).
6. L. E. Brownell and E. H. Young, *Process Equipment Design-Vessel Design*, John Wiley and Sons, Inc., New York, N.Y. (1959).
7. E. F. Bruhn, *Analysis and Design of Flight Vehicle Structures*, Tri-State offset Co., Cincinnati, Oh. (1965).
8. D. O. Brush and B. O. Almroth, *Buckling of Bars, Plates, and Shells*, McGraw-Hill Book Co., New York, N.Y. (1975).
9. P. S. Bulson, *The Stability of Flat Plates*, American Elsevier, New York, N.Y. (1971).
10. D. J. Butler and G. B. Anderson, *Weld. J. Res. Suppl.*, **42** (1), 29s–36s (January 1963).
11. J. O. Draffin, and W. L. Collins, *Statics and Strength of Materials*, Ronald Press, New York, N.Y. (1959).
12. R. E. Fulton, *Buckling Analysis and Optimum Proportions of Sandwich Cylindrical Shells under Hydrostatic Pressure*, Structural Research Series No. 199, Department of Civil Engineering, University of Illinois, Urbana, Ill. (June 1960).
13. G. Gerard, *Introduction to Structural Stability Theory*, McGraw Hill Book Co., New York, N.Y. (1962).
14. G. Gerard and H. Becker, *Handbook of Structural Stability*, NACA Technical Notes 3781–3786, NASA Technical Notes D-162 and D-163 (1957–1959).
15. M. Gregory, *Elastic Instability*, E & F. N. Spon Limited, London (1967).
16. B. G. Johnston, Ed., *Guide to Design Criteria for Metal Compression Members*, 2nd ed., Column Research Council, John Wiley and Sons, Inc. (1966).
17. C. Libove, "Elastic Stability," in, W. Flügge, Ed., *Handbook of Engineering Mechanics*, McGraw-Hill Book Co., New York, N.Y. (1962).
18. T. H. Lin, *J. Aero. Sci.*, **17** (3), 159–172 (March 1950).
19. W. McGuire, *Steel Structures*, Prentice-Hall Inc., Englewood Cliffs, N.J. (1967).
20. B. B. Muvdi and O. M. Sidebottom, *Inelastic Design of Load-Carrying Members—The Behavior of Beam Columns in the Inelastic Range*, WADD Tech. Report 60–580 (Part IV), (July 1961).
21. J. G. Pulos, and V. L. Salerno, *Axisymmetric Elastic Deformations and Stresses in a Ring Stiffened, Perfectly Circular Cylinder Shell under External Hydrostatic Pressure*, David Taylor Model Basin Report 1497, U.S. Navy Department (September 1961).
22. R. M. Rivello, *Theory and Analysis of Flight Structures*, McGraw-Hill Book Co., New York, N.Y. (1969).
23. B. Saelman, *Des. News*, **15** (4), 55–58 (July 4, 1960).
24. F. R. Shanley, *J. Aero. Sci.*, **13** (12), 678, (December 1946); also *Weight-Strength Analysis of Aircraft Structures*, Dover Publications, New York, N.Y. (1960).
25. F. R. Shanley, *Strength of Materials*, McGraw-Hill Book Co., New York, N.Y. (1957).
26. J. Singer, and A. Eckstein, *Research Council Israel, Bulletin*, **11C**, 97–112 (April 1962).
27. L. H. Sobel and W. Flügge, *AIAA Journal*, **5** (3), 425–431 (March 1967).

28. J. H. Starnes, Jr. "The Effects of Cutouts on the Buckling of Thin Shells," in Y. C. Fung and E. E. Sechler, Eds., *Thin-Shell Structures*, pp. 289–305, Prentice-Hall, Englewood (1974).

29. F. R. Steinbacher and G. Gerard, *Aircraft Structural Mechanics*, Pitman Publishing Corp., New York, N.Y. (1952).

30. S. Timoshenko, *Strength of Materials, Part I*, 3rd ed., D. Van Nostrand Co., Princeton, N.J. (1955).

31. S. Timoshenko, *Strength of Materials, Part II*, 3rd ed., D. Van Nostrand Co., Princeton, N.J. (1956).

32. S. Timoshenko, *History of Strength of Materials*, McGraw-Hill Book Co., New York, N.Y. (1953).

33. S. Timoshenko, and J. M. Gere, *Theory of Elastic Stability*, 2nd ed., McGraw-Hill Book Co., New York, N.Y. (1961).

34. E. Wenk, *Weld. J. Res. Suppl.*, **40** (6), 272s–288s (June 1961).

35. E. Wenk, R. C. Slankard, and W. A. Nash, *Proc. Soc. for Exp. Stress Anal.*, **12** (1), 163–180 (1954).

36. *Alcoa Structural Handbook*, Aluminum Company of America, Pittsburgh, Pa. (1955).

37. *Strength of Metal Aircraft Elements*, Military Handbook MIL-HDBK-5B, Government Printing Office, Washington D.C. (March 1971).

38. *Welding Research Council Bulletin Series*, Welding Research Council of the Engineering Center, 345 E. 47th St., New York, N.Y. (1959).

39. *Commentary on Plastic Design in Steel*, Welding Research Council and American Society of Civil Engineers, United Engineering Center, 345 E. 47th St., New York, N.Y. (1959).

40. *ASME Boiler and Pressure Vessel Code, Section VIII, Pressure Vessels*, The American Society of Mechanical Engineers, United Engineering Center, 345 E. 47th St., New York (1977).

41. *Prod. Eng.*, **31** (29), 62–66 (July 18, 1960).

42. *PVRC Interpretative Report of Pressure Vessel Research, Section I—Design Considerations*, Report 95, Welding Research Council Bulletin, Welding Research Council, New York, N.Y. (April 1964).

43. *Steel Construction Manual*, 7th ed., American Institute of Steel Construction, New York, N.Y. (1970).

44. American Society of Mechanical Engineers, "*Pressure Vessel and Piping Design*, 612–618, (1960).

PROBLEMS

9.1 In Problem 5.1 find the allowable L where buckling occurs. Take the worst case of P shifted to one side of the pipe and find its value. Consider that E also changes with respect to time.

9.2 A base plate for a pressure vessel support is to be designed. A vertical steel member acts as a column to support the 325 kips acting on the support. (See Fig. P9.2). Find the thickness of the members using (22).

9.3 Calculate the collapsing pressure on a brass condenser tube of 0.75-in. O.D. and a wall thickness of 0.030 in.

9.4 Find the end buckling load on a cantilever beam with a depth of 3 in., a width of 0.0625 in., and a beam length of 36 in. The material is wood with E of 1×10^6 lb/in.^{2} and G of 0.0625×10^6 lb/in.^{2}.

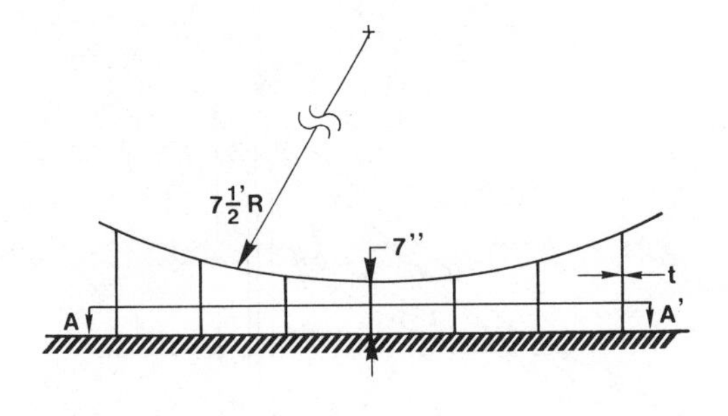

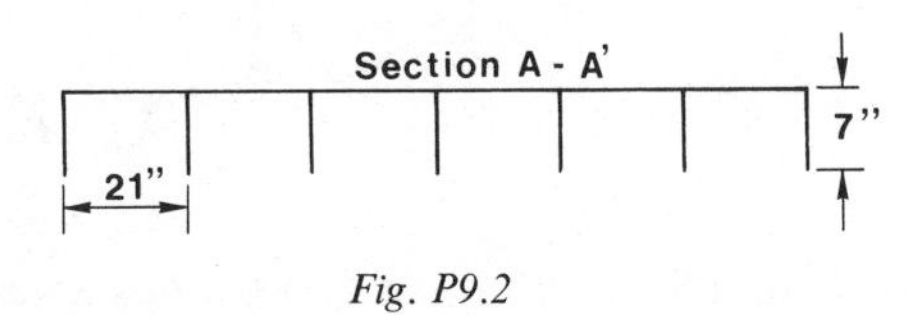

Fig. P9.2

9.5 Design a vacuum chamber which is to be 48 in. I.D. and 48 in. from the base plate to the inside of the hemispherical head. The base plate requires six 1-in. diameter holes for instrumentation. Size and show sketches for the chamber, base plate, flanges for a 12-in. square window, window glass, and other details.

9.6 The aluminum frame in Fig. P9.6 has a 500-lb load applied. Find the factor of safety for buckling.

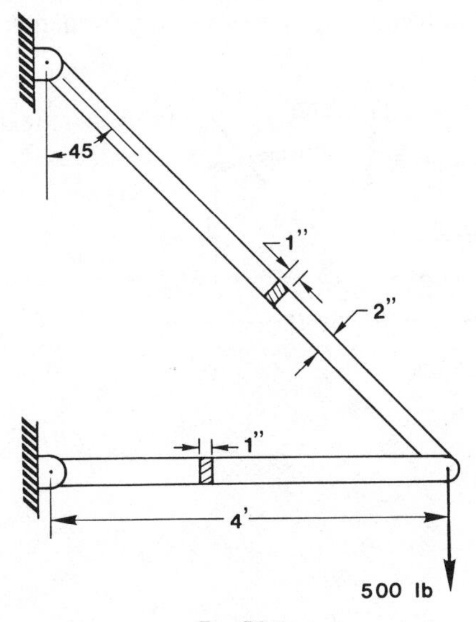

Fig. P9.6

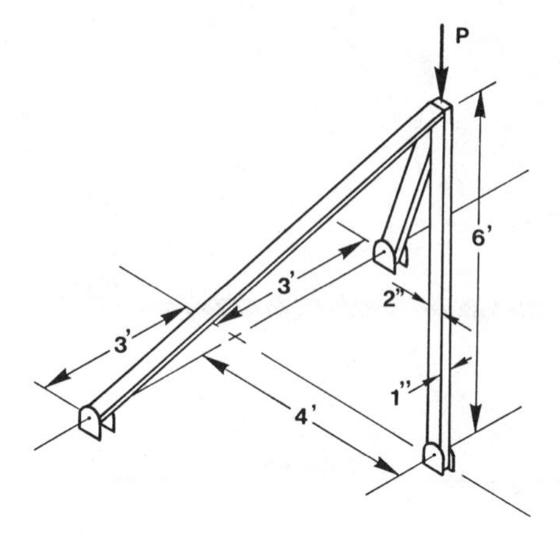

Fig. P9.7

9.7 A steel frame (Fig. P9.7) with σ_y of 42 kpsi has a vertical load P applied to it. For a factor of safety of 3, find the safe value for P.

9.8 In Problem 2.2 size and check the member OC for possible snap through or buckling.

9.9 Instruments are packaged in an aluminum suitcase-type container which has a 3.5 psi vacuum on it (Fig. P9.9). Find the buckling pressure and the required changes to meet the vacuum requirement. Consider a change of material, increased thickness, and stiffeners so that a proper design can be made.

9.10 Design a laced column with proper wire supports 1000 ft high to support a 100-lb antenna. Compare the cross section and weight of a laced column to a solid column meeting the same requirements.

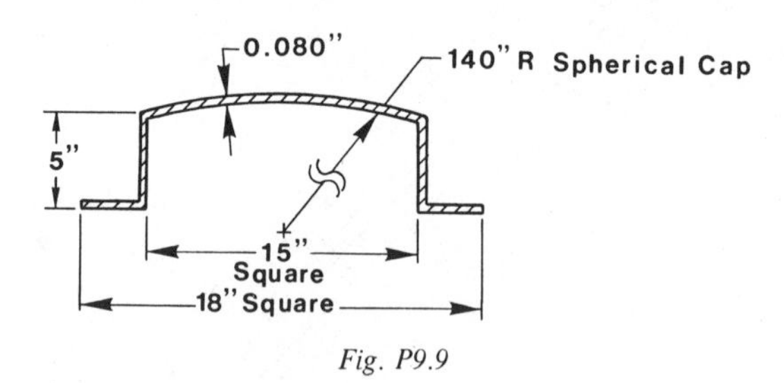

Fig. P9.9

chapter 10

SHOCK, IMPACT, AND INERTIA

Questions continually arise concerning the dynamic behavior of materials and structures. For example, it is important to know how the rate of load application influences the properties of materials; it is also necessary to know how to analyze a materials dynamics problem and to be able to apply the appropriate analysis, depending on whether the rate of loading is *slow*, *fast*, or *rapid* or if the condition of loading is one of *shock*, *impact*, or *impulsion*. These topics will be dealt with in this chapter, and, in addition, discussions will be presented dealing with the changes in properties of materials accompanying dynamic loading, the phenomenon of *delayed yielding* that occurs in some metals, and special problems such as projectile penetration of plates, scabbing, explosive internal pressure loading of cylinders, and inertia problems (8, 10, 15, 35, and 85).

10-1 MECHANICAL PROPERTIES AND DYNAMIC BEHAVIOR

Studies concerning the dynamic behavior of materials from a mechanical point of view have been largely limited to strain rate tension, bending and torsion tests, impact loading of bars and beams, explosive internal pressure tests of cylinders, elastic wave propagation, scabbing and projectile penetration of plates and studies concerning changes in properties due to explosive loading. As will be shown, design calculations are difficult even for simple cases; and, for complex situations only approximations can be made.

Classification of dynamic loadings

In general, the behavior of materials under dynamic conditions is vastly different from behavior under static conditions. If the time of load application is greater than about *three times the natural period of vibration* of a structure, the loading can be considered as applied *statically*, and the usual analyses of stress and strain can be made on a static basis. If the time of load application is less than about *half the natural period of vibration*, the structure is said to be loaded in *impact* or *shock* and calculations based on statics are often meaningless. This is because for short times of load applications the propagation, reflection, and interference of elastic and plastic waves predominate, and calculations of stress and strain must be based on a wave analysis. In most cases of this type the calculations are exceedingly complex, and often no exact analysis can be made at all because of the large number of unknowns. For example, in a punching operation, if the punch is slowly applied to the metal, calculations of punch pressure can be made on a static basis (see Chapter 6). On the other hand, if a bullet is fired at a plate, it is necessary to consider the mechanical properties of both target and projectile, velocity of the bullet, the attack angle, and the geometry of both target and projectile. A third type of loading called *fast or rapid loading* occurs approximately in the range *one-half to three times the* natural period of vibration. In this range wave phenomena are essentially inconsequential. However, because of dynamic effects mechanical properties of materials frequently change so that, even though calculations of stress and strain can be fairly well approximated by static analyses, the mechanical property changes have to be incorporated in the analysis. For example, in tension, the ultimate strength of some steels can be increased 50% by rapid loading, although the general behavior is the same as in slow or static loading.

Another common type of dynamic loading is called *fatigue*, in which a structure of some kind is alternately stressed and unstressed over a period of time. This is a complicated type of loading and considered later in Chapter 12. The final type of loading called *inertia loading* is concerned with rotating masses in which loads are imparted by centrifugal action. In general, the load is calculated from a consideration of geometry and velocity, then stresses and strains are computed by conventional static analyses.

Property changes under extreme conditions of loading

The difference in behavior of materials under static and dynamic conditions has been recognized for many years. Unfortunately, most of the reliable data for

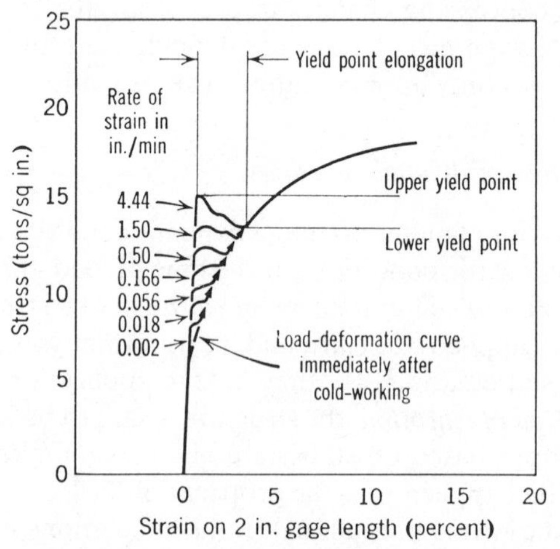

Fig. 10.1 Composite stress-strain curve illustrating the influence of rate of straining on the yield point elongation of annealed low-carbon steel. [After Winlock and Leiter as reported by Jevons (38), courtesy of John Wiley and Sons, Inc.]

dynamic loading conditions have been obtained from uniaxial tests; very few data exist for combined stress conditions. A very important finding concerning the behavior of materials under dynamic conditions is that the mechanical properties tend to improve with increasing rate of load application. This is particularly evident in tensile tests in the range of load application time between $0.5T$ and $3T$.* Tests in this time range can be made by a tension impact machine (for times around $0.5T$) and with a modified tensile machine (for times around $3T$). The standard Charpy impact test on a notched specimen often reveals brittle behavior for a material ordinarily ductile in tension; this type of behavior however, is related to a high degree of tensile triaxiality in the specimen which reduces plastic flow. Elastic strains in the impact test, however, may exceed those observed in static tests (97).

For the rapid loading type of test (time between $0.5T$ and $3T$), which is usually conducted in a modified tensile testing machine, results have been obtained similar to those shown in Fig. 10.1. The plots in this figure indicate an increase in yield strength of low-carbon steel with increasing rate of strain; it is not shown in Fig. 10.1, but the ductility also increases with increasing rate of loading (38, 90).

For tensile impact tests (loading time around $0.5T$ or less) it has been found that changes in yield strength are very difficult to record because of instrumentation limitations. However, significant changes do occur, and have been recorded, in both ultimate strength and ductility. These increases show up in tension impact tests at impact velocities up to 25 ft/sec and are little changed at higher speeds. Figure 10.2 shows, for example, the increase in ultimate strength that has been observed under tension impact conditions as a function of the

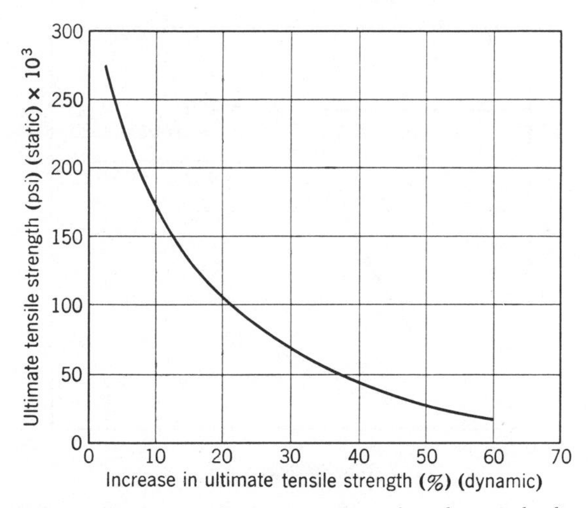

Fig. 10.2 Illustration of increase in strength in dynamic loading.

*T is the natural period of vibration.

static ultimate tensile strength. The data used in plotting Fig. 10.2 were obtained principally from (7) and include both ferrous and nonferrous metals and alloys. The significant feature of Fig. 10.2 is that under impact loading conditions the increases in ultimate strength are greatest for the lowest strength materials. This is particularly true for the annealed low-carbon steels exhibiting an upper yield point; that is, a considerable increase in ultimate (dynamic) strength can always be expected in such a steel, whereas, on the other hand, other materials or the same steel at the same hardness level (but not exhibiting an upper yield point) may occasionally show only a slight increase in ultimate strength. For this reason the curve in Fig. 10.2 should be considered approximate, with the realization that in perhaps 10 to 20% of the cases encountered the curve will not describe the material behavior as indicated.

It would appear that for metals it should be feasible to conduct a dynamic tension impact test, record the pertinent data, and then reconstruct from the stress-time plot the stress-strain diagram for the material. Unfortunately, this cannot be done because in impact loading the strain and strain rate vary continuously along the length of the bar. However, an interesting observation made in examining various stress-time plots is that the form of the curve apparently depends on the static yield-ultimate tensile strength ratio. This is schematically shown in Fig. 10.3 for three ranges of the yield-ultimate ratio (7).

Raising the temperature usually lowers strength properties and increases ductility values, although increases in both strength and ductility with increase in speed still persist (72), that is, temperature, per se, lowers strength and increases ductility, but increasing speed at any temperature increases both strength and ductility at that temperature; some typical strength data on pure copper are

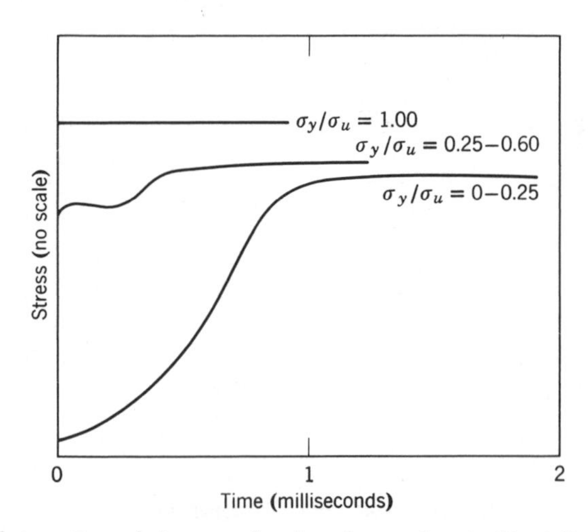

Fig. 10.3 Typical stress-time relations as a function of strength ratio. Note: Only the shape of the curve and its relation to the time axis is important; it is not implied that for the same material the order of the curves with respect to the stress axis would be as indicated.

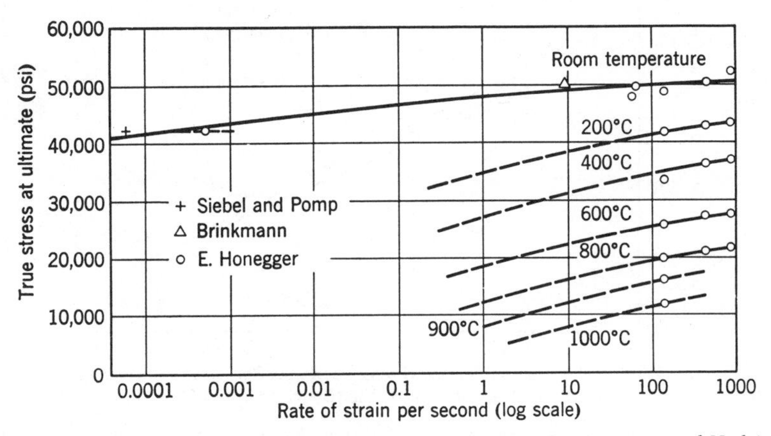

Fig. 10.4 Effect of rate of strain on the strength of pure copper. [*After Manjoine and Nadai (53), courtesy of the American Society for Testing and Materials.*]

shown in Fig. 10.4. Mild steel, because of the *blue brittleness* effect, is an exception.

Typical data for various metals are shown in Fig. 10.5 and in Fig. 1.23. Data for plastics are shown in Figs. 5.2 and 5.8, for wood in Fig. 10.6, for glass in Fig. 10.7, and for concrete in Fig. 10.8. For design purposes it appears necessary to conduct tests under the conditions expected to prevail when the materials are used. Attempts have been made, however, to obtain mathematical relationships between strength, temperature, and strain rate (5, 6, 11, 72). Typical design data

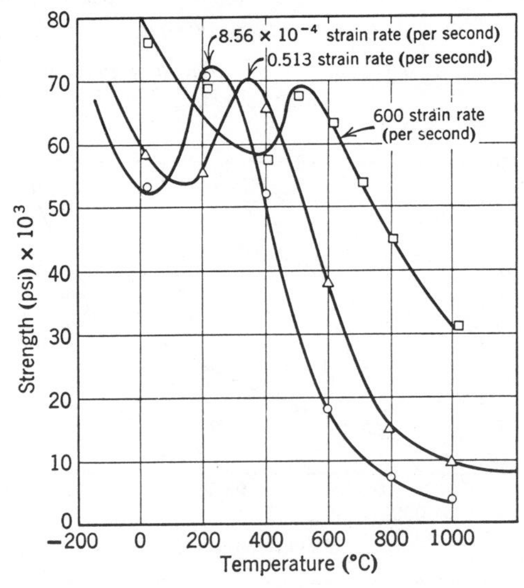

Fig. 10.5 Effect of temperature and strain rate on the tensile strength of steel. [*After Hollomon (33), courtesy of The American Welding Society.*]

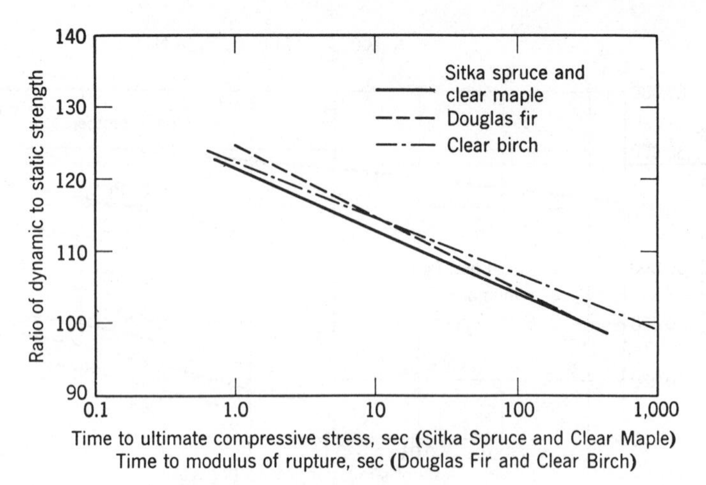

Fig. 10.6 Effect of rate of load application on the strength of wood. Note: curves are average data from a paper by Markwardt and Liska (99).

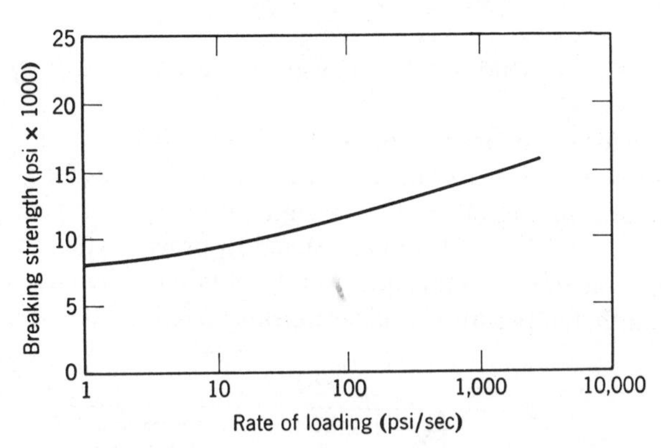

Fig. 10.7 Effect of rate of load application on the strength of glass [average of data supplied by Ritland (99)].

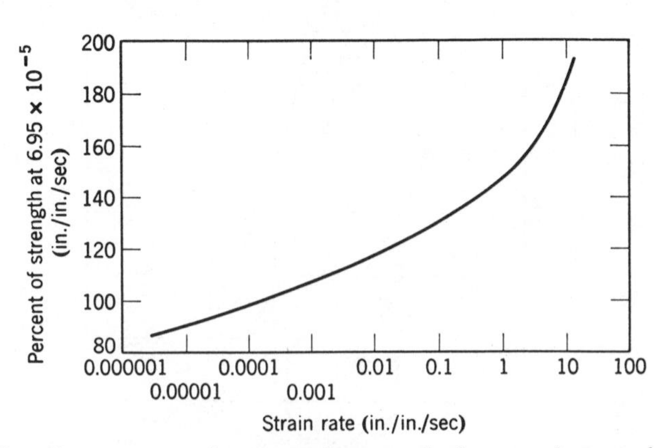

Fig. 10.8 Effect of strain rate on the compressive strength of concrete [average data of McHenry and Shideler (99).]

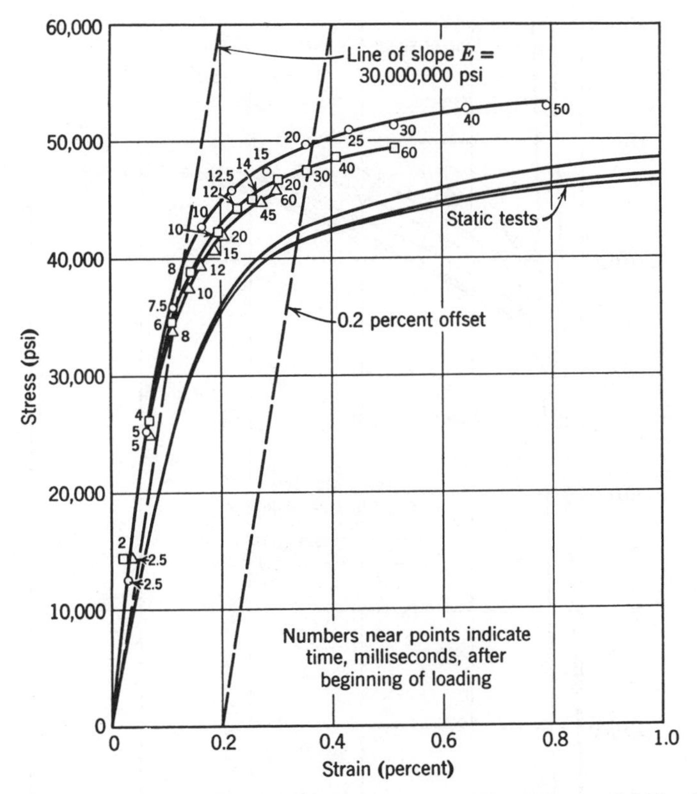

Fig. 10.9 Stress versus strain for three rapid load tests on type 302 stainless steel. [After Clark and Wood (9), courtesy of The American Society for Testing and Materials.]

for several materials showing variation of stress with time and time to reach a particular state of stress are shown in Figs. 10.9 to 10.17. The materials whose properties are shown in Figs. 10.9 to 10.15 do not exhibit a delay in yielding typical of low-carbon steel (to be discussed presently) and thus exhibit normal concave-downward stress-strain curves. The curves in this group are presented both as ordinary stress-strain curves, with times noted to reach particular stress and strain levels, and also as curves with a constant time parameter. For low-carbon steel it has been found that there is a definite time required for the initiation of plastic deformation when stresses above the yield point are applied dynamically. The time to initiate plastic strain in this material varies from about 5 msec at a tensile stress of about 50,000 psi, to about 6 secs at 27,000 psi, as shown in Fig. 10.16 for a steel having a static lower yield strength of about 30,000 psi. An explanation of these effects is given by Clark in his Campbell Memorial Lecture to the American Society For Metals (6).

The data discussed thus far have been for uniaxial loading only. Some information concerning mechanical properties of materials under dynamic combined stresses is given in (13, 23, 66, 71, 72, and 76), which deal with scabbing and

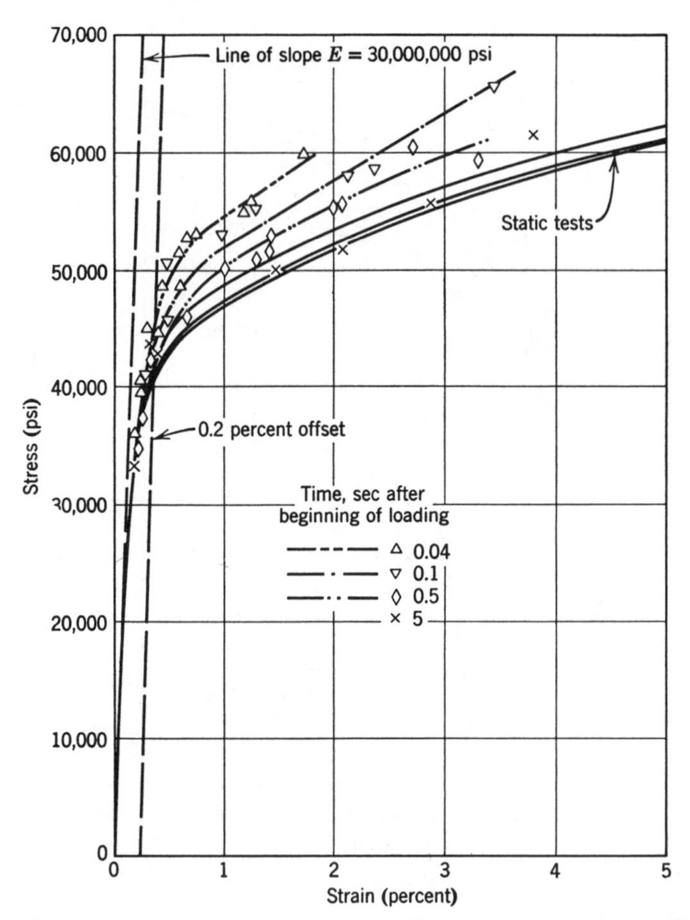

Fig. 10.10 Stress versus strain at various times after loading for type 302 stainless steel. [*After Clark and Wood (9), courtesy of The American Society for Testing and Materials.*]

comminution of solids. Data for dynamic torsion are given in (92). Of particular interest is a dynamic value called the *critical normal fracture strength*, which presumably should be identical with the dynamic ultimate tensile strength of a material. These two quantities, however, although related in some unknown way, differ considerably for various materials. Goldsmith (24) and Wood (91) also noticed this difference in dynamic ultimate and critical normal fracture strengths and concluded that the discrepancy was caused by restraints in dynamic combined loading which induced near balanced tensile triaxiality. Thus at the present time there is no way to estimate the fundamental fracture strength of a material under dynamic combined stresses except by experiment. Several data given for the critical normal fracture strength in (71) are given in Table 10.1.

A practical application of the effects of dynamic loading is the obtaining of

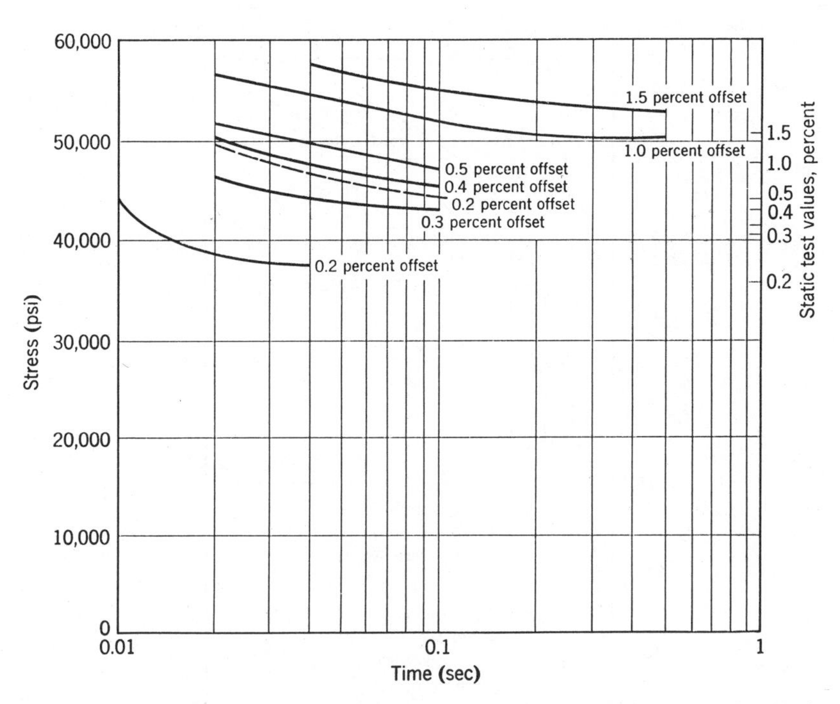

Fig. 10.11 Stress versus time at various constant strains for type 302 stainless steel. [After Clark and Wood (9), courtesy of The American Society for Testing and Materials.]

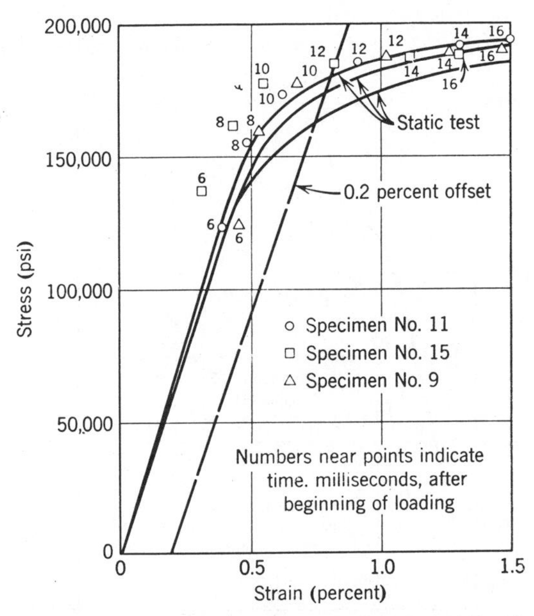

Fig. 10.12 Stress versus strain for three rapid load tests on A.I.S.I. 4130 steel. [After Clark and Wood (9), courtesy of The American Society for Testing and Materials.]

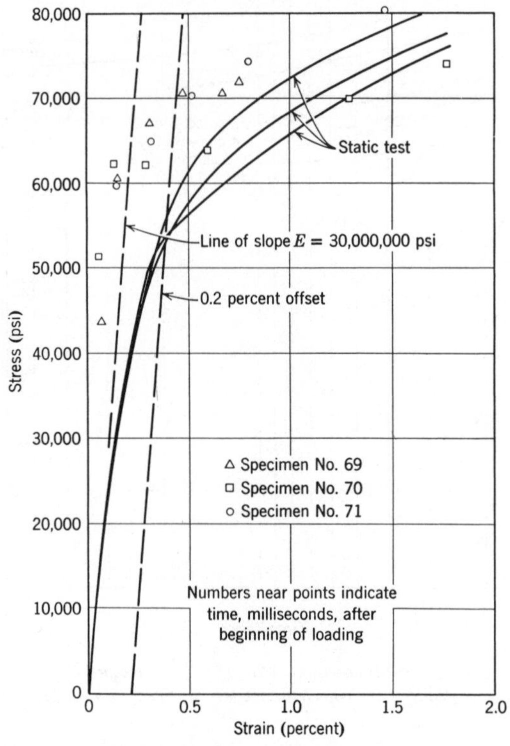

Fig. 10.13 Stress versus strain for three rapid load tests on normalized A.I.S.I. 4130 steel. [*After Clark and Wood (9), courtesy of The American Society for Testing and Materials.*]

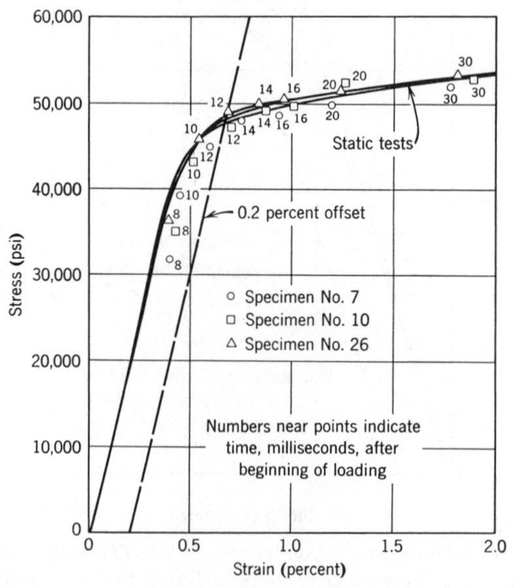

Fig. 10.14 Stress versus strain for three rapid load tests on 2024–T4 aluminum alloy. [*After Clark and Wood (9), courtesy of The American Society for Testing and Materials.*]

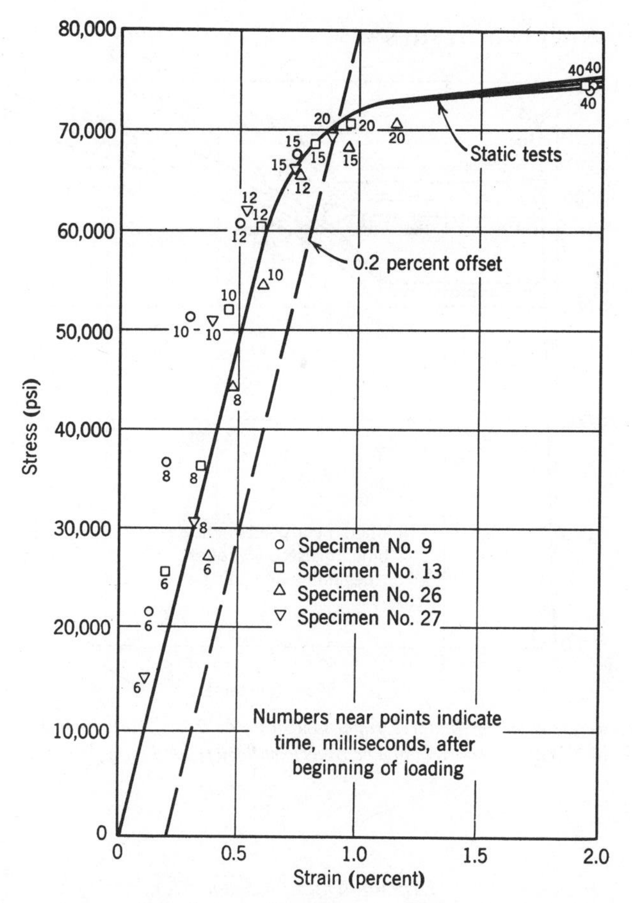

Fig. 10.15 *Stress versus strain for four rapid load tests on 7075–T6 aluminum alloy.* [*After Clark and Wood (9), courtesy of The American Society for Testing and Materials.*]

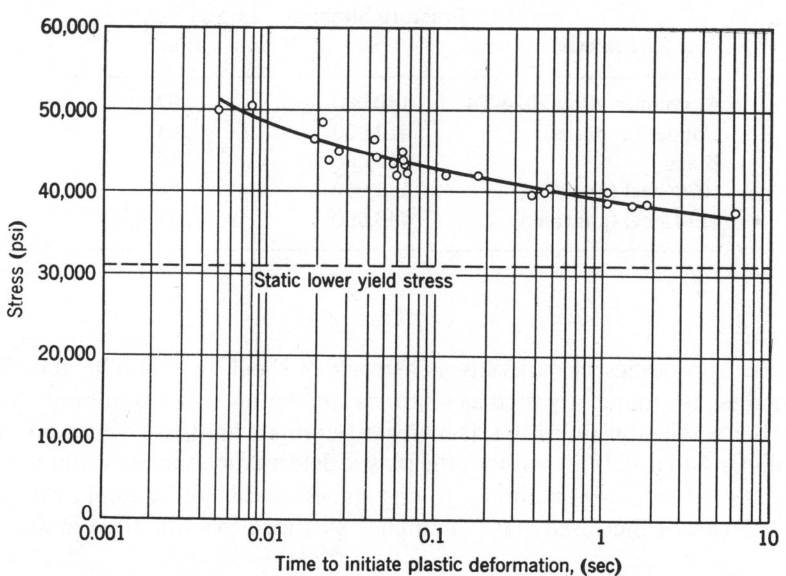

Fig. 10.16 *Stress versus time to initiate plastic deformation for 0.19% carbon steel, annealed.* [*After Clark and Wood (9), courtesy of The American Society for Testing and Materials.*]

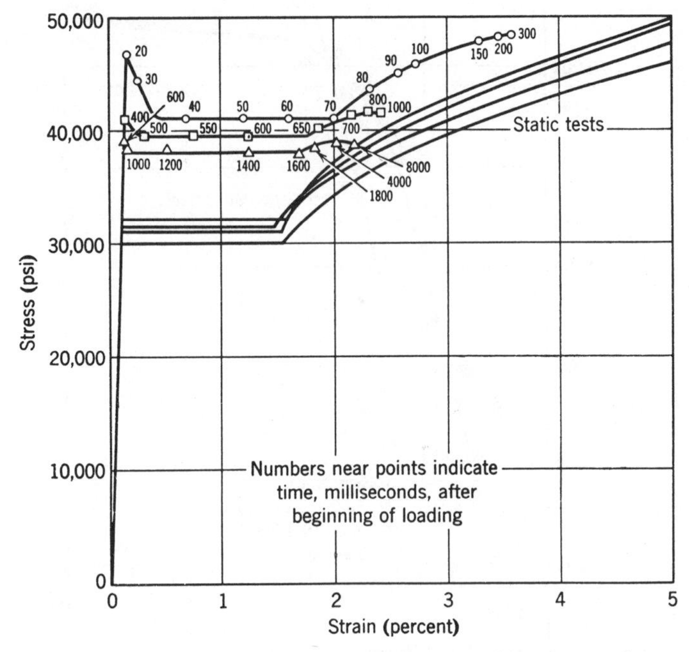

Fig. 10.17 Stress versus strain for three rapid load tests on 0.19% carbon steel, annealed. [After Clark and Wood (9), courtesy of The American Society for Testing and Materials.]

Table 10.1 Dynamic Data for Some Metals (71)

Material	Critical Normal Fracture Strength, psi	Associated Critical Impact Velocity, ft/sec
Aluminum alloy 2024-T4	140,000	202
Copper (annealed)	410,000	264
Brass	310,000	216
1020 steel (annealed)	180,000	84
4130 steel (annealed)	440,000	235

increased properties of materials as a result of shock loading (28, 100). This method is now frequently used as a fabrication method since it not only forms parts but at the same time imparts increased strength, hardness, wear resistance, and dimensional stability without the plastic deformation that accompanies the conventional working methods. In manganese steel, for example the yield strength can be increased 100% or higher by shock loading. See (76 and 86).

10-2 ANALYSES OF RATE-OF-LOADING AND INERTIA PROBLEMS

Wave propagation phenomena

The dynamic propagation of strain under elastic conditions has been known for some time; the elastic strain which propagates in a bar is proportional to the velocity of impact (82). When a bar, for example, is subjected to an impact load, the stress and strain in the bar are not constant but vary along the length and with time. Consequently, for calculation, methods used must involve stress and strain as functions of distance and time (56). The velocity of propagation of a particular strain

$$c = \sqrt{\frac{d\sigma}{d\varepsilon} \Big/ \gamma} \qquad 10.1$$

The impact velocity that will produce a plastic strain is given by the relation

$$v = \int_0^{\varepsilon} \sqrt{\left(\frac{d\sigma}{\gamma\, d\varepsilon}\right)}\, d\varepsilon \qquad 10.2$$

It is seen in Eq. 10.2 that the plastic strain increases with increasing impact velocity. However, the slope $d\sigma/d\varepsilon$ becomes zero when the strain reaches the value ε_m, corresponding to the ultimate tensile strength. Therefore the theory predicts that rupture will occur at the instant of impact at the impacted end of the bar when the impact velocity is

$$v_m = v \geq \int_0^{\varepsilon_m} \sqrt{\left(\frac{d\sigma}{\gamma\, d\varepsilon}\right)}\, d\varepsilon \qquad 10.3$$

Here v_m is called the *critical impact velocity*. Experiment [see (19)] confirms the preceding theory. In order to calculate the critical impact velocity for a metal, it is only necessary to have a stress-strain curve. From this curve a new plot can be made of ε versus $\sqrt{(d\sigma/d\varepsilon)/\gamma}$; the area under this new curve between 0 and ε_m then represents Eq. 10.3 and the critical velocity is obtained.

Penetration and scabbing of plates. The subject of penetration and scabbing of plates has been of special concern in ordnance work and in the design of barricades for high pressure or explosive work (59). The theory concerning both penetration and scabbing is too involved and specialized to go into here; however, some facts concerning the subject will be presented along with appropriate analytical techniques that can be used to obtain approximate results. When a projectile is directed towards a target, it may, depending on many factors, perforate the target on contact, partially pierce or dent the target, cause a scab to fly off the free of the target, or both penetration and scabbing may result (much like what happens when a BB shot from an air rifle strikes a window).

PLATE PENETRATION. Consider first a projectile striking a target. In general, the projectile will produce either a hole or a crater depending on the properties of the materials and the impact velocity. For steel plates a low velocity projectile

(<4000 ft/sec) produces a straight hole, although complete penetration may not occur. For thicker plates and velocities of 5 to 10,000 ft/sec, a large cuplike crater is produced (72). The principal factors to be considered in the impact are (*a*) projectile and target geometry and mechanical properties, (*b*) angle of impact, and (*c*) velocity.

The following formula (103) has been found to be effective in estimating the distance a projectile will penetrate into massive reinforced concrete provided that (*a*) the penetration distance is no more than about 20 times the diameter of the projectile, and (*b*) the target thickness is about twice the penetration distance (the case of full perforation of a target is considered later).

$$S = 5.423 \frac{Pv^{1.33}}{10^4 d^{1.8}} \qquad 10.4$$

where S = penetration distance into reinforced concrete (in.)
P = weight of projectile (lb)
v = striking velocity of projectile (ft/sec)
d = diameter of projectile (in.). When the projectile is not round, measure the area that strikes the target and convert that area to an equivalent diameter, $d = 1.126A^{1/2}$.

Equation 10.4 can be used for materials other than reinforced concrete on an approximate basis by assuming 1-ft thickness of reinforced concrete equivalent to 1 in. of steel, 4 in. of high-strength aluminum, and 2 ft of wood. Pieces of shrapnel penetrating near the center of a target may, if striking at the same velocity and angle, perforate the target near its edges. Also, a monolithic target is more effective in resisting perforation than a layered target of the same material and thickness; steel overlays on concrete walls are frequently used to prevent scabbing and spalling of the concrete and to reduce the thickness of concrete.

The value of v in Eq. 10.4 may not be known if the projectile originated at the source of an explosion such as a burst pressure vessel. However, the value of v can be estimated from the relationship (104)

$$v = 8000R^{1/2} \exp\left(\frac{-0.00169S_1}{P^{1/3}}\right) \qquad 10.5$$

where v = striking velocity, in feet/second of projectile weighing P lb at distance S_1 ft from the target.
exp = constant, 2.718.

$$R = \frac{(c/m)}{(1 + c/2m)} \quad \text{for a cylinder and}$$

$$= \frac{(c/m)}{\left(1 + \dfrac{3c}{5m}\right)} \quad \text{for a sphere}$$

where c = weight of explosive (pounds/unit length of a cylinder) in equivalent TNT pounds,

m = weight of case containing the explosive (per unit length of a cylinder), in pounds,

The TNT pounds equivalent can be expressed as:

a. *For a pressurized container:*

$$\text{TNT (lb)} = 6.313 \times 10^{-8} pV \ln\left(\frac{p}{15}\right)$$

where p = pressure in container; psi
V = volume of container (in.^{3})
ln = symbol for natural logarithms.

b. *For an exploded vessel:*

$$\text{TNT (lb)} = 0.0071 V p^{0.9}(p^{0.1} - 1.065)$$

where p = pressure (psi)
V = volume of vessel (ft^3).

In *b*, above, for a vessel of usual design as outlined by the Pressure Vessel Code (105), the pressure rise is generally of the order of eight times the nominal design pressure. Information on fragment size and distribution is inconclusive, but theories are presented in (106).

Whether or not a projectile perforates a target is of interest and has been studied by ordnance personnel (107). Approximately, the perforation limit is expressed as

$$t = 1.216S + 1.4d \qquad 10.6$$

where t = thickness of material perforated (in.)
S = distance penetrated into target (in.) (Eq. 10.4)
d = diameter of projectile (in.).

For example, consider a projectile 0.5-in. diameter weighing 0.2 lb, impacting a 30-in. thick target of reinforced concrete at 2500 ft/sec.

From Eq. 10.4

$$S = 5.423(0.2)\frac{(2500)^{1.33}}{(10^4)(0.5)^{1.8}} = 12.5 \text{ in.}$$

For this condition, using Eq. 10.6,

$$t = 1.216(12.5) + 1.4(0.5) = 15.9 \text{ in.}$$

Thus, the projectile will perforate about 16 in. of reinforced concrete, or, 16/12 = 1.33 in. of steel. In an approximate way, if the perforating characteristics of a specific projectile are known, this information can be used to estimate the behavior for some other projectile. For example, if a 75 mm projectile weighing 13.9 lb perforates steel 3.7 in. thick, then a similar projectile at the same velocity but weighing Y pounds and being X mm in diameter will perforate a target of the same material of thickness

$$t = 3.7\left(\frac{Y}{13.9}\right)\left(\frac{75}{X}\right)^2 \qquad 10.7$$

Table 10.2 Material Parameters for Eq. 10.8 [*Data after Wessman and Rose (88), courtesy of John Wiley and Sons, Inc.*]

Material	Parameters	
	a, psi	b, $\frac{\text{lb-sec}^2}{\text{ft}^2\text{-in.}^2}$
Sand gravel	620	0.0115
Clay-sand gravel	1,490	0.0048
Grassy earthwork	1,000	0.0056
Sand-clay earthwork	655	0.0037
Clay soil	490	0.0036
Damp clay	378	0.0028
Earthwork	432	0.0008
Good stonework	7,850	0.0109
Medium stonework	6,250	0.0087
Brickwork	4,490	0.0062
Limestone	17,100	0.0238
Oak, beech, ash	2,960	0.0055
Elm	2,280	0.0042
Fir, birch	1,650	0.0031
Poplar	1,550	0.0029

Table 10.3 *Comparison of Plate Perforation Data* (34, 78)

Projectile and Target	Test* Velocity (ft/sec)	Perforation per Eq. 10.6 (in. steel)
Mild steel blocks 0.627 in. × 0.627 in. × 0.28 in. shot at mild steel plate 0.08 to 0.33 in. thick	4830	0.36
Bullet 0.256 in. diameter × 1.3 in. long, (0.024 lb) shot at mild steel plate, 0.244 in. thick	2220	1.8
Same as above except plate = 0.532 in. thick	2220	1.8

*Perforation occurred in all cases.

In any practical case, of course, adequate factors of safety should be assigned.

When the projectile is pointed, a complete perforation can be achieved at about half the striking velocity of a flat-ended projectile, or to consider a needlelike projectile, when the ratio of the height of the pointed portion of the projectile to the diameter of the projectile is between about 1 and 2. At low velocities perforation may not occur, but the distance traveled into the plate by the projectile may be expressed by the relation (72)

$$D = (m/2bA)\ln[1 + (b/a)v^2] \qquad 10.8$$

where m is the mass of the projectile, A its cross-sectional area, v the striking velocity, and a and b are given in Table 10.2.

The preceding theory has been applied to data taken from (34 and 78), and the results are shown in Table 10.3. Additional studies relating to the mechanics of armor penetration by ogival projectiles are reported in (65, 95, and 96).

If actual penetration of the target occurs, it is often of interest to know the residual velocity of the ejected parts. In the simplest analysis, this is obtained by utilizing the principle of conservation of momentum. If m_1 is the mass of the projectile traveling at velocity v_1, and m_1 and v_r are the projectile mass and velocity after penetration of the target, and m_2 and v_r are the mass and velocity of the ejected portion of the target, then, for the same materials,

$$m_1 v_1 = m_1 v_r + m_2 v_r$$

or

$$v_r = Lv_1/(L + h) \qquad 10.9$$

where L is the projectile length and h the target thickness.

SCABBING. Scabbing, or fracturing of a material near a free surface, is the result of a transient compressive stress wave of high intensity impinging on that surface (44, 45, 72). A schematic representation of the process is shown in Fig. 10.18. Studies concerning the mechanism of scabbing have been reported

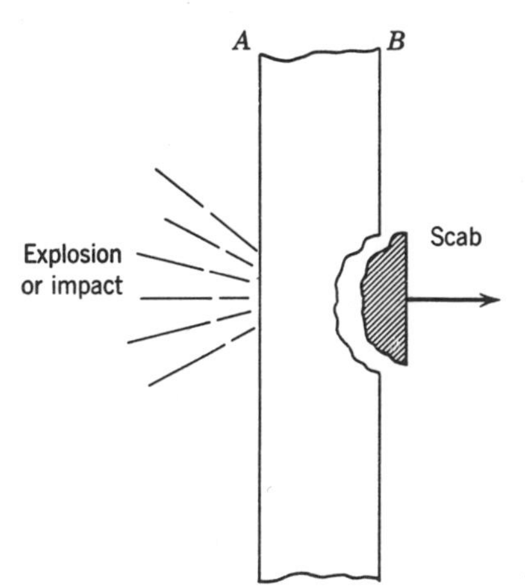

Fig. 10.18 Schematic representation of scabbing.

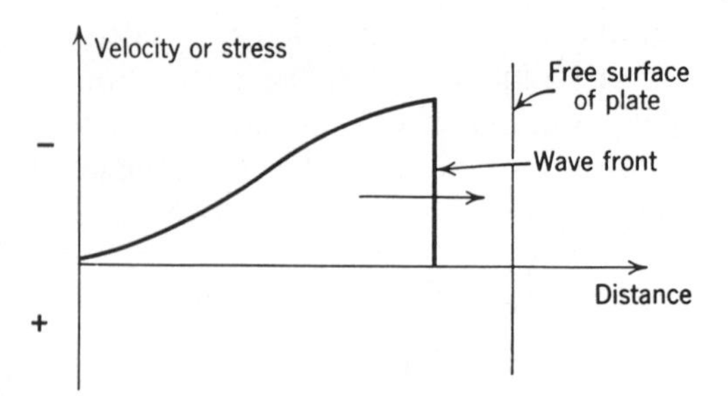

Fig. 10.19 Pressure pulse traveling through a plate.

by several authors (13, 39, 44, 45, 66, 71, 72); for the purposes of this book the following brief description is presented.

Consider Fig. 10.19 which illustrates a pressure pulse traveling through a plate (such as from A to B in Fig. 10.18). When the compression pulse strikes the free surface, it is reflected as a tension wave. As a result of the reflection an interference with the incident compression wave results, giving the situation shown in Fig. 10.20. The tension AB increases as the reflected wave moves to the left; at some point the stress reaches its fracture value and a scab flies off. The fracture value obtained is the *critical normal fracture stress*, mentioned earlier in the chapter, and the thickness δ of the scab is equal to one-half the distance within the wave that corresponds to a decrease in stress equal to the critical normal fracture stress (66). For a scab to form, the peak pressure in the incident pulse must be greater than the critical normal fracture stress σ_t, and $\sigma_c/\sigma_t > 1.0$ where σ_c is the compressive stress generated.

In addition to scabbing, other fractures that occur as a result of wave interference are shown in Fig. 10.21 (13). Note that all fractures occur in the thickest section rather than in the thinnest section as would happen in static loading. Rinehart explains this type of fracture as the result of highly localized stress concentration from the division and subsequent reflection of a single compressional wave into two tensile waves on striking two inclined free surfaces.

The thickness limit for scabbing may be determined from the relationship (107)

$$t = 1.333S + 2.3d \tag{10.10}$$

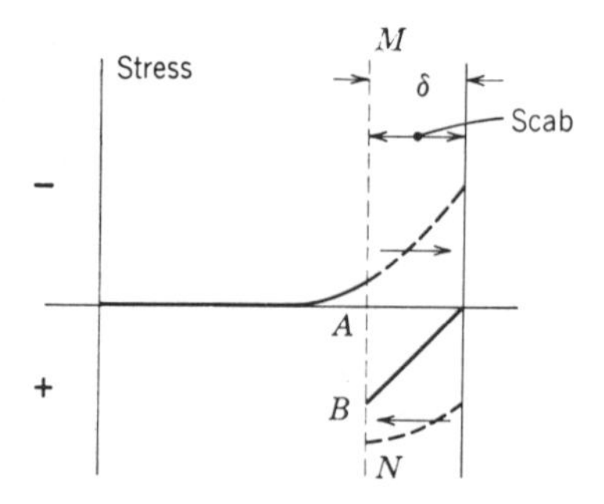

Fig. 10.20 Stress distribution near free surface of a plate just after reflection of pressure pulse.

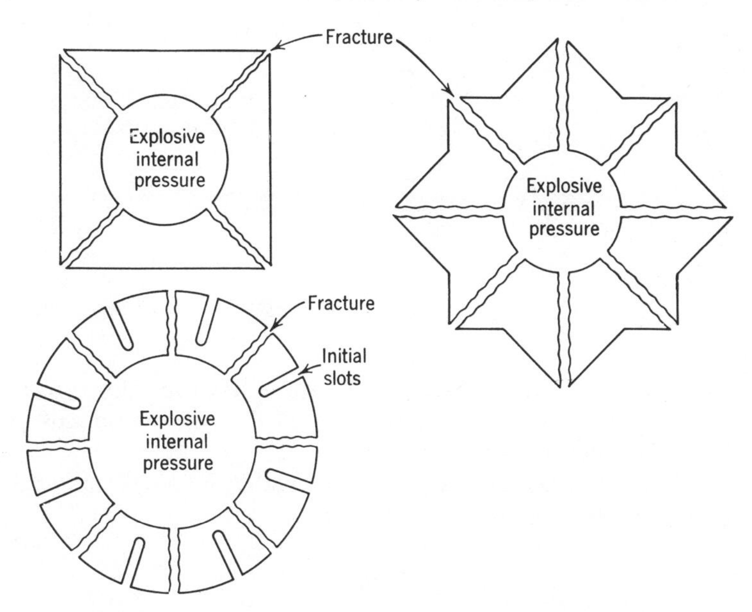

Fig. 10.21 Fracture modes of various cylinders subjected to explosive internal pressure.

where the terms are defined under Eq. 10.6. Thus, in the previous example,

$$t = 1.333(12.5) + 2.3(0.5) = 17.8 \text{ in.}$$

Thus, for these conditions the projectile will perforate about 16 in. of reinforced concrete, will cause a spall for concrete about 18 in. thick, or in terms of steel, 1.5 in. thick for spalling and 1.33 in. for perforation.

Vibration theory in dynamics analysis

In considering the dynamic behavior of materials it is important to keep in mind the factor of rate of loading so that the appropriate analysis can be made. If the natural period of vibration of an elastic structure can be ascertained, then design becomes a little easier since proper allowance can be made for the expected loads. For the simple case of impact of an elastic body the following analysis may be made.*

A single degree of freedom system is considered to be subjected to a pulse of force consisting of a half sine wave as illustrated by the full line in Fig. 10.22, such a simple system is illustrated in Fig. 10.23, but the equation can equally well represent the response of a single normal mode in a more complicated elastic system. The differential equation for the displacement x is

$$d^2x/dt^2 + \omega^2 x = F(t) sin(\pi t/T) \qquad 10.11$$

where T is the time of loading.

*From a private communication to J.H. Faupel through courtesy of Professor E. H. Lee, Brown University.

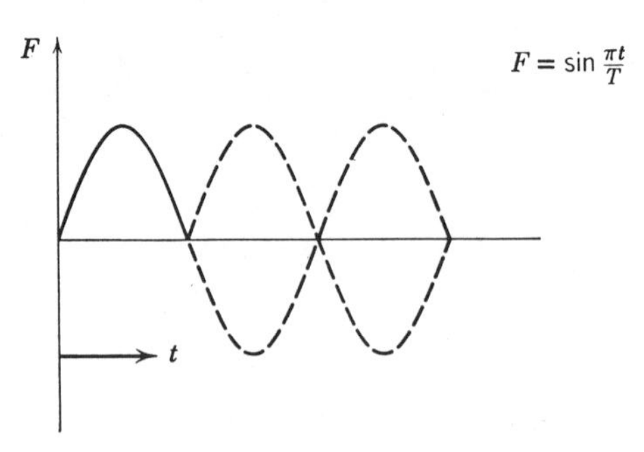

Fig. 10.22 Harmonic vibration.

It should be noted that the stress in the spring is proportional to the displacement x so that the maximum response in the following discussion can be interpreted directly as maximum stress.

The simplest way to analyze the problem of loading for a half sine wave pulse is to solve the problem for sine wave loading

$$F sin(\pi t/T) \qquad 10.12$$

for all t and then superpose the same solution displaced by an amount T on the t scale. The fact that superposition of these two waves produces the required loading is illustrated by the canceling of the broken lines in Fig. 10.22.

It is assumed that the system is initially undisturbed before the force acts; thus Eq. 10.11 is solved under the initial conditions

$$x(0) = (dx/dt)(0) = 0 \qquad 10.13$$

This equation can be solved in a variety of ways, either by adding a complementary function and a particular integral and finding two constants or directly using a Laplace transform. The solution is

$$x(t) = \frac{F}{\omega^2 - \pi^2/T^2}\left(\sin\frac{\pi t}{T} - \frac{\pi/T}{\omega}\sin\omega t\right) \qquad 10.14$$

In the present problem this solution applies directly in the region

$$0 < t < T \qquad 10.15$$

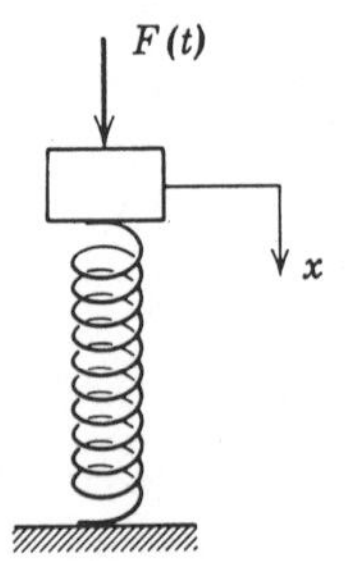

Fig. 10.23 A simple dynamical system.

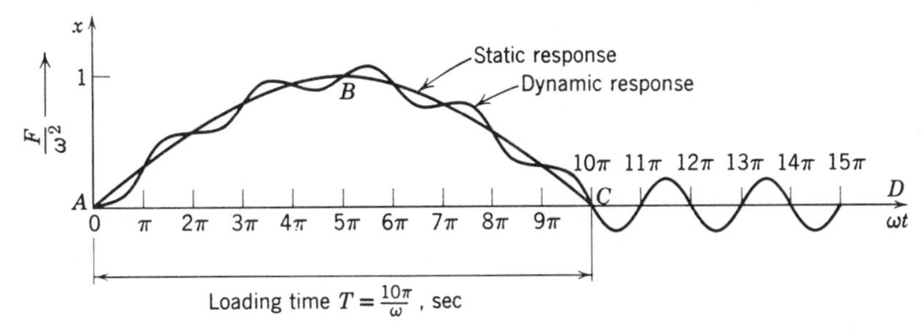

Fig. 10.24 Long-time loading.

For $t > T$ superposition is used and the displacement is given by

$$x(t) + x(t - T) \tag{10.16}$$

This solution assumes that resonance does not occur; in other words, that T is not equal to π/ω. In this special case a different form of a particular integral must be found and the response is given by

$$x(t) = (F/2\omega^2)(\sin \omega t - \omega t \cos \omega t) \tag{10.17}$$

If the system contained no mass, so that there was no dynamic effect, static response of the spring itself would occur. This is given by Eq. 10.11 with $d^2x/dt^2 = 0$, resulting in

$$(x)_{st} = \frac{F}{\omega^2} \sin(\pi t/T) \tag{10.18}$$

Using these solutions for different values of the loading time T, the influence of loading time on the displacements and stresses produced by the force pulse can be considered. In this way the duration of loading is varied, but the maximum force is kept constant, so that the maximum static response and the maximum static stress do not change.

To illustrate the influence of loading time, three cases are considered. For a loading time that is long compared with the natural period ($T = 10\pi/\omega$), the

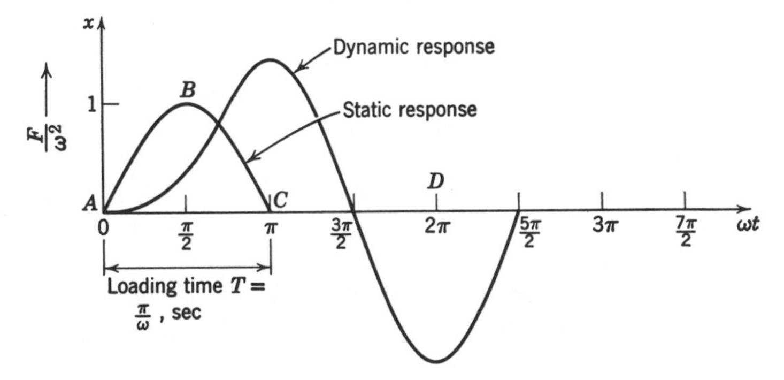

Fig. 10.25 Resonance loading.

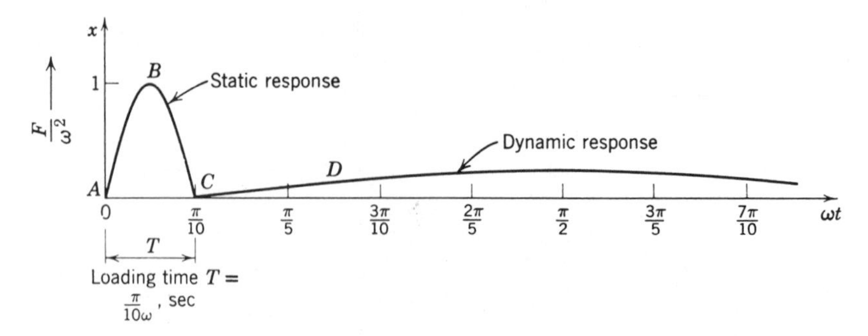

Fig. 10.26 Short-time loading.

response is illustrated in Fig. 10.24, in which both the static and dynamic responses are shown. Figure 10.25 shows a pulse with the frequency of the pulse equal to the natural frequency of the system, the resonance case. Figure 10.26 shows a short-time loading in which the pulse of force lasts for only one-twentieth of the natural period of the system.

In all three cases the static responses under load are given by the sine curves *ABC*, and after the load is removed there is no static response as shown by the lines *CD*. The dynamic response illustrated is obtained from Eqs. 10.14 or 10.17 for $t < T$ and Eq. 10.16 for $t > T$. For $t > T$, there is no load acting, and so the system simply vibrates at its natural frequency. It was therefore found simpler to add the two solutions analytically and to obtain a simple expression for the subsequent free vibration. The formulas used for all these cases are detailed in the following discussion. For Fig. 10.24, $x(t)$ is given by

$$x(t) = \frac{F}{\omega^2}\frac{100}{99}\left(\sin\frac{\omega t}{10} - \frac{1}{10}\sin\omega t\right) \qquad 10.19$$

and superposition gives the result

$$x(t) = (F/\omega^2)(100/99)(-\tfrac{2}{10}\sin\omega t) \qquad 10.20$$

Note for this case that the static response supplies a reasonable approximation to the dynamic response. In the dynamic case a comparatively high-frequency oscillation (at the natural frequency of the system) is superimposed on the static response, and this causes an increase in response and stress of about 10%. The factor 100/99 is so close to 1 that on the figure the low frequency part of the dynamic response is indistinguishable from the static response. After the load is removed, the displacement and stress associated with the resulting free vibration are considerably smaller than the value under load. This result is typical of what is to be expected for long-time loading. When the loading time is long compared with the natural period, essentially static response occurs, and a static solution for the stress will be satisfactory for design purposes.

Resonance loading is shown in Fig. 10.25. The displacement under load is

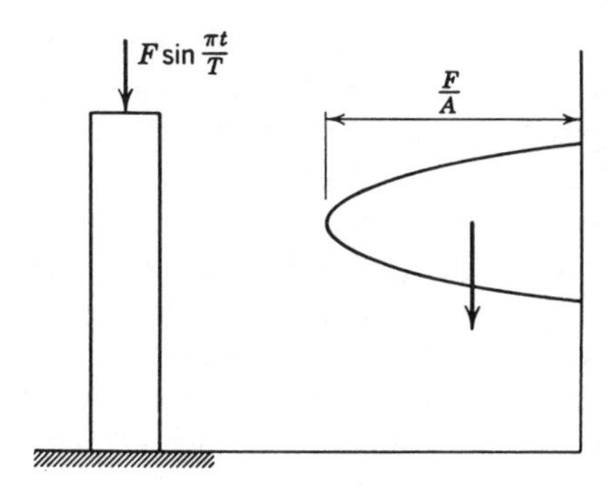

Fig. 10.27 Bar subjected to compression loading.

given by Eq. 10.17, and the superposition for $t > T$ gives the displacement

$$x(t) = -(F/\omega^2)(\pi/2)\cos \omega t \qquad 10.21$$

Here the dynamic response is quite different from the static response and results in a maximum stress which is about 60% greater than that in the static case. This stress is repeated both in tension and compression in the subsequent free oscillation of the system. Thus, when the duration of loading is equal to one-half of the natural period of the system, it is essential to analyze the dynamic case in order to obtain the stress variation. This is also necessary when these times are of the same order, although not exactly equal. The dynamic stress is again likely to be larger although not as large as in the resonance case. The nonresonance case, even when these times are of the same order, is given by Eqs. 10.14 and 10.16.

Figure 10.26 shows short-time loading. During loading the displacement is given by

$$x(t) = (F/99\omega^2)(10 \sin \omega t - \sin 10\omega t) \qquad 10.22$$

and after the load is removed by

$$x(t) = \frac{F}{\omega^2}\frac{10}{99}\left(\cos \frac{\pi}{20}\right)2 \sin\left(\omega t - \frac{\pi}{20}\right) \qquad 10.23$$

In this case it is seen that the rapid pulse of force supplies momentum to the system, but while the load is acting the displacement and stress are small. This momentum produces a free vibration which causes an increase in the displacement and stress to a maximum long after the force has ceased to act. Here the maximum stress, however, is only about 20% of the maximum static value. It *must not* be deduced from this result that short-time loads on mechanical systems in general give rise to stresses small compared with static values. This is because in the present case an idealized system with only one degree of freedom was considered. For an elastic body there will be an infinite sequence of normal modes with natural frequencies forming an infinite sequence above the fundamental frequency. Thus, although the period of loading may be short compared with the fundamental period, it may be in resonance for one of the higher modes, and excitation of such a mode may involve stresses larger than the static values. This can be demonstrated very simply by considering the system shown in Fig. 10.27 which consists of an elastic rod fixed at the lower end and subjected to the longitudinal compressive force $F(t)$ at the upper end. The normal modes of oscillation of the system are well known [see, for example, the discussion on page 296 of (83), (39)] and the stresses can be determined by computing the stresses for each mode and adding the results at each section of the rod. For short-time force pulses this normal mode method would not be very practical because of convergence difficulties. For short pulses the higher modes would be strongly excited and only low stress values would be associated with the lower modes. Thus the influence of the terms in the normal mode series would at first rise, and so many terms would have to be taken to obtain a satisfactory value for the stresses. However, in this particular simple case an alternative method of representing the solution is available by direct solution of the wave equation. This must give the same result as the normal mode method, but it forms a much more convenient means of analysis in pulse time that is short compared with the fundamental natural period of the rod. This solution is illustrated in Fig. 10.27. A stress wave reproducing the variation of force pulse travels down the rod, the maximum stress magnitude being F/A, where A is the cross section of the rod. When this strikes the free end, reflection occurs with the same sign for the stress, and the maximum value reached by the stress occurs at the fixed end and has magnitude $2F/A$. The pulse length in the rod is equal to $c_0 T$, where c_0 is the elastic wave velocity, and if this is less than twice the length of the rod, the doubling of the static stress will occur. For much longer pulses, reflection at the free end will cause a reduction below a value of two. For very long pulses many reflections have to be considered in order to determine the maximum stress, and then the normal mode solution if more convenient. This problem illustrates that for this system the *maximum stress is always twice the static stress* for sufficiently short force pulses. In a manner similar to the example illustrated in Fig. 10.27 the stress associated with the fundamental mode of vibration will be small in these cases, (since the pulse time is short as compared with the natural period) but nonetheless, the high modes will be sufficiently excited to produce this doubling of the maximum static stress. The stress distribution shown in Fig. 10.24 which consists of a short pulse of stress in one section of the rod and zero stress elsewhere illustrates why the normal mode attack does not provide satisfactory convergence. In effect, the normal mode solution supplies a Fourier series for this stress distribution, and a Fourier series expansion for such a pulse with high magnitude inside only a narrow region is well known to be an impractical method of representation.

The influence of the free vibration in modifying the static stress values in Fig. 10.24 is closely associated with the smoothness of application of the force. In this example the time derivative of the force changed discontinuously at A from 0 to a finite value. In practice the growth of force is likely to be smoother than this, which results in an oscillation of the system of much smaller magni-

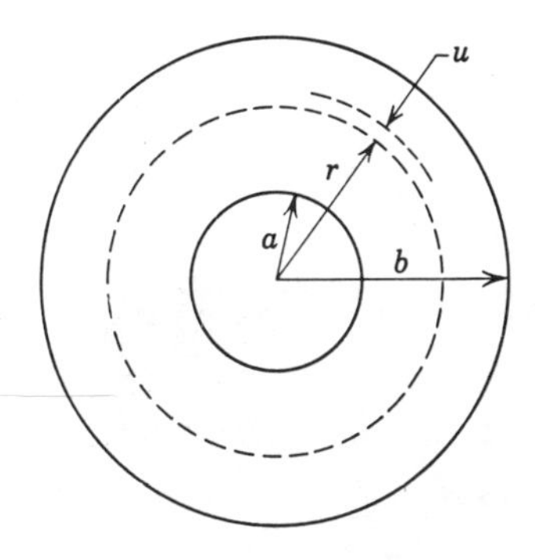

Fig. 10.28 Geometric model for estimation of the period of vibration of a heavy-walled cylinder by the Rayleigh energy balance method.

tude. Thus the overstress above the static stress is more closely represented by the particular integral associated with steady-state forced oscillation; in the example in Fig. 10.24 this corresponds to an overstress above the static of about 1%. It should be noted, however, that there is always some damping of a physical system, so that after several oscillations the amplitude is reduced, and this has an important influence on the stresses associated with the free vibration after the load has been removed. The application of vibration theory to dynamics problems has also been discussed by Murray (57).

Dynamic loading of heavy-walled cylinders. Sometimes it may not be possible to ascertain what the fundamental vibration characteristics of a structure are. Then it is possible to make an approximation of the period of vibration by using the Rayleigh energy balance method (83). Consider, for example, the heavy-walled cylinder shown in Fig. 10.28. From the geometry indicated and equating kinetic and potential energy

$$\gamma u^2 \omega^2 = E_0 u^2 / r^2 \tag{10.24}$$

Using Eq. 10.24 leads to an expression for the period of vibration*

$$T = \frac{2\pi r}{c_0} = \frac{\pi a(R+1)}{c_0} \tag{10.25}$$

where $R = b/a$. By using Eq. 10.25 it can be approximately deduced whether the period of vibration is equal to, less than, or greater than the time of load application; thus a decision can be made as to whether or not a dynamic loading problem exists. For the particular case of internal explosion the momentary pressure is so high that for all practical purposes the cylinder wall is subjected to an impact load, causing it to vibrate and transmit pressure waves. These waves, starting at the bore, radiate toward the outside where they are reflected, thus causing interference on their way back to the bore. Longitudinal waves are also set in motion; the resultant is a complexity of wave reflection and interference which, if the load is high enough, eventually leads to fracture.

*Vibrations of thick-walked shells are treated in detail in Reference 25.

Fig. 10.29 Typical appearance of cylinders shattered by explosive internal pressure. [*After Pearson and Rinehart (66); courtesy of The American Institute of Physics.*]

Some data on explosive loading of heavy-walled cylinders are presented in (66). The material tested was annealed AISI-1020 steel; cylinder sizes ranged from about 2 to 9 dia. ratio. The static bursting pressure for these cylinders determined by use of Eq. 6.377

$$p_b = \frac{2\sigma_y}{\sqrt{3}} \ln R(2 - \sigma_y/\sigma_u) \qquad 10.26$$

ranged from 42,500 psi for a cylinder of dia. ratio 2 to about 134,000 psi for a cylinder of dia. ratio 9. These pressures are considerably below the internal explosive pressure which was reported to be about 3,000,000 psi, acting for a period of about 30 msec; consequently, it might be expected that complete shattering of the cylinders would occur. Actually, the lower dia. ratio cylinder did shatter, as shown in Fig. 10.29, but the cylinder of dia. ratio 9 only cracked slightly within the wall as shown in Fig. 10.30.

If the critical normal fracture stress data for this steel given in Table 10.1 are used, then $\sigma_y = \sigma_u = 180{,}000$ psi, and the bursting pressures for cylinders of dia. ratio 2 and 9 would be 144,000 psi and 457,000 psi respectively. For these same cylinders it is also interesting to check on the period of vibration using Eq. 10.25. The results of these calculations are presented in Table 10.4. By comparing the loading transient and period of vibration data given in Table 10.4 it is seen that the cylinder of dia. ratio 2 was loaded in a time equal to $1.28T$,

Table 10.4 Dynamic Data for Heavy-Walled Cylinders

Diameter Ratio, R	Loading Transient, t (sec)	Period, T (sec)	Overload Factor
2.000	0.000030	0.0000233	0.194
5.875	0.000030	0.0000535	0.445
9.000	0.000030	0.0000778	0.648

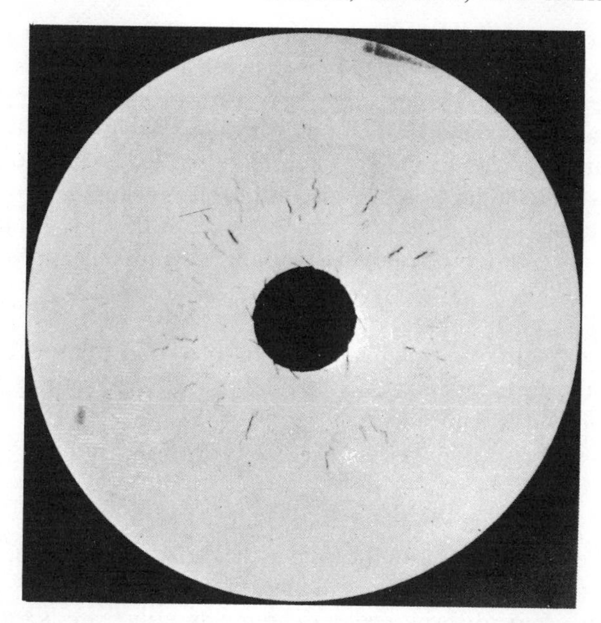

Fig. 10.30 Appearance of cross section of a cylinder of diameter ratio 9.00 after subjection to explosive internal pressure. [*After Pearson and Rinehart (66), courtesy of The American Institute of Physics.*]

whereas the cylinder of dia. ratio 9 was loaded in a time $0.38T$; the former would represent rapid loading, and the latter would be impact loading. Consequently, for rapid loading, since there is only a moderate increase in mechanical properties, it would be expected that the cylinder of dia. ratio 2 would burst under explosive pressure.

Overload factor. Closely associated with dynamic loading and vibration theory is the amount of overload permissible on a structure under dynamic loading. This is a very complicated problem, although attempts have been made to obtain an approximate empirical solution. Such a solution by Roop (73) proposes that the overload stress on a structure (apparently either uniaxial or combined loading) can be computed by dividing $T/4$ by the loading transient,

$$\sigma_{\text{overload}} = \sigma_{\text{static}}(1 + T/4t) \qquad 10.27$$

Applied to previous problem the overload factors for the cylinders are listed in Table 10.4. Thus, for the cylinder of dia. ratio 2 the maximum allowable pressure would be less than 100,000 psi. For the cylinder of dia. ratio 9 the overload factor is 0.648, giving a safe explosive internal pressure of nearly a million psi.

Dynamic loading of thin-walled cylinders. Typical problems that arise with the use of thin-walled tubing are sudden overloading by closing of valves,*

*A related problem "water hammer" is discussed in Reference 64.

detonations that may take place in processes carried out under conditions of high pressure and temperature, and in shells such as submarines, subjected to depth charge attack. In the first type of problem a solution can often be found by simply using an accumulator in the hydraulic system. However, this is not always required and the tube can be initially designed to handle the overload expected. A simple design solution to this problem defines a fast-closing valve as one that closes faster than a pressure wave can travel a distance equal to twice the length of the pipe carrying the fluid; this time is approximately (46)

$$T = 2L/c \tag{10.28}$$

where L is the length of the pipe and c is the speed of the pressure wave, which depends on the compressibility of the fluid, the strength of the pipe, and the speed of sound in the fluid; this is,

$$c = 68.1\sqrt{\frac{B}{\gamma[1 + (B/E)(D_0/t)]}} \tag{10.29}$$

where B is the bulk modulus of the fluid, γ the density of the fluid, D_0 the inside diameter of the tube, and t its wall thickness. Equation 10.29 is plotted in Fig. 10.31 for a pipe carrying water. The pressure increase is found from the relation

$$\Delta p = \frac{0.0395\gamma c Q}{D_0^{\,2}} \tag{10.30}$$

where Q is the fluid flow rate. As an example illustrating the foregoing theory consider a steel pipe of 0.0937 in. wall thickness and 0.75 in. inside diameter carrying water at 0.15 ft^3/sec at 2000 psi in a pipe 600 ft long. Suppose a valve in the line is shut off in 0.10 sec; the density of water is 62.4 lb/ft^3 and its bulk

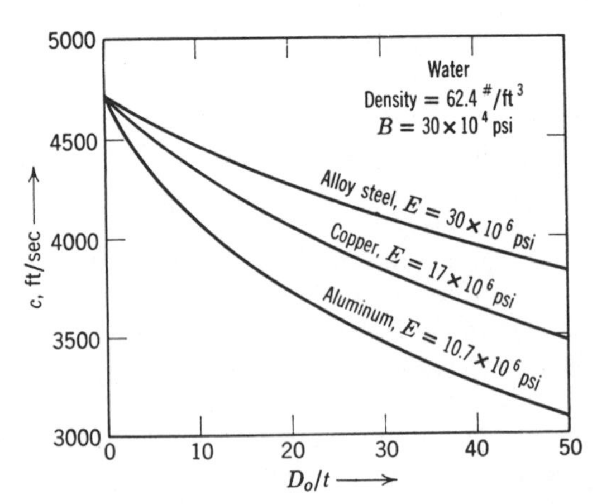

Fig. 10.31 Plot of Eq. 10.29 for wave velocity in a hydraulic pipe. [After Lapera (46), courtesy of The McGraw-Hill Publishing Co.]

modulus is 300,000 psi. Therefore, from Fig. 10.31, c is 4500 ft/sec and the time from Eq. 10.28 is

$$T = \frac{2(600)}{4500} = 0.267 \text{ sec}$$

The pressure increase, from Eq. 10.30 is

$$\Delta p = \frac{(0.0395)(62.4)(4500)(0.15)}{(0.84)^2} = 2950 \text{ psi}$$

The hoop stress for a thin-walled cylinder is given by Eq. 3.371; therefore the total stress on valve cutoff is

$$\sigma_h = \frac{pR}{t} = \frac{pd}{2t} = \frac{(2000 + 2950)(0.75)}{(2)(0.0937)} = 19{,}800 \text{ psi}$$

which is well within the allowable stress limit for the material.

The problem of thin tubes subjected to explosive pressure has been considered by Hanita (27), Luker and Mosier (51), and Randall and Ginsburgh (69). Hanita considered the problem of explosive *external pressure* such as might be experienced by a submarine under depth charge attack; his theory has been simplified here, however, for design convenience in obtaining approximate answers. For a thin-walled tube the wall acts as a spring of stiffness K and the radial motion (see Chapter 3) is

$$u = \varepsilon_h R = R\left[\frac{1}{E}(\sigma_h - \nu\sigma_r)\right] = \frac{pR}{E}\left(\frac{R}{t} - \nu\right) \qquad 10.31$$

If the radial stress is considered small compared with the hoop stress, then $\sigma_r \sim 0$ and Eq. 10.31 becomes

$$u = pR^2/Et \qquad 10.32$$

and

$$K = p/u = Et/R^2 \qquad 10.33$$

The natural period of vibration of the system is

$$T = 2\pi\sqrt{m/K} = 2\pi R\sqrt{\gamma/gE} \qquad 10.34$$

Equation 10.34, with appropriate values can then be substituted into Eq. 10.27 to obtain the dynamic overload factor.

The same analysis can be applied to explosive internal pressure (69).

Thrust from pressure relief. In many instances where sudden release of pressure may be a problem it is desirable to have means for computing the resultant thrust that equipment may have to sustain. Such problems exist in reactors subject to runaway reactions or in pressure vessel systems equipped with safety valves. This is a very complicated problem since the thrust developed depends on the pressure generated within the vessel, ambient pressure, areas of vessel cross section and exit nozzle, and physical and chemical properties

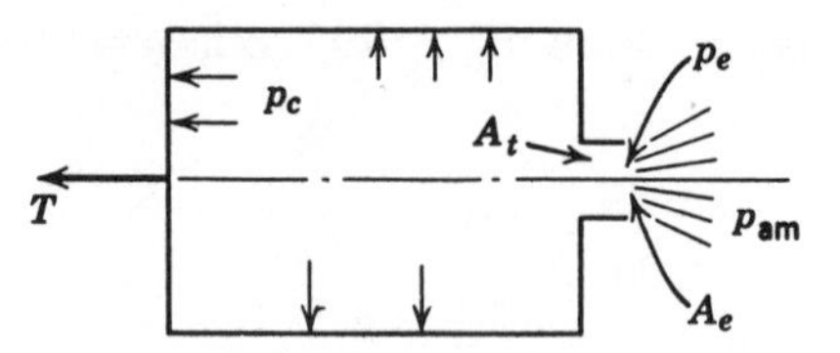

***(A)* Approximation of a converging rocket nozzle.**

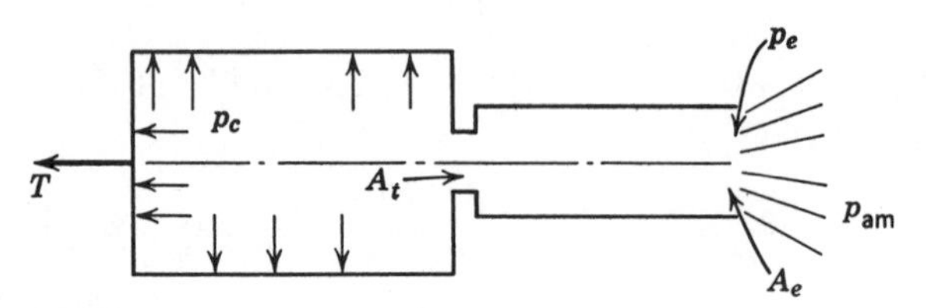

***(B)* Approximation of a DeLaval-type, converging-diverging rocket nozzle.**

Fig. 10.32 Illustration of thrust from nozzles.

of the materials involved, and many others. The nature of the thrust is also of interest, that is, whether the thrust is continuous, like in a chemical rocket, or whether the thrust is initially high and rapidly dies out. For present purposes the concern is in the initial thrust which would be the basis of design for a structure subject to such sudden impulses of force. In order to calculate thrusts, information developed in rocket technology will be utilized. Briefly, it has been found that (101) (Fig. 10.32),

$$\text{thrust} = F = C_F \times A_t \times p_c \tag{10.35}$$

where A_t = cross-sectional area of nozzle throat, in.^{2}
p_c = pressure in the vessel, psi
C_F = thrust coefficient, having the value

$$C_F = \left\{ \frac{2k^2}{k-1} \left(\frac{2}{k+1} \right)^{(k+1)/(k-1)} \left[1 - \left(\frac{p_e}{p_c} \right)^{(k-1)/k} \right] \right\}^{1/2} + \frac{p_e - p_{am}}{p_c} \left(\frac{A_e}{A_t} \right) \tag{10.36}$$

where k = ratio of specific heats
p_e = pressure at nozzle exit, psi
p_{am} = ambient pressure, psi
A_e = cross-sectional area of nozzle exit, in.^{2}

In an optimally designed nozzle $p_e = p_{am}$ so that the last term in Eq. 10.36 is zero. Then, for practical use, Eq. 10.35 with appropriate values for C_F can be used in situations where the gas is ejected through a De Laval-type nozzle, which is converging-diverging, or for a vessel where the gas is vented into a tailpipe whose cross section is larger than the vent opening. For most practical cases of this type the thrust can be given as

$$T = 2.5p_c \times A_t \tag{10.37}$$

For vessels having a nozzle that vents directly into the atmosphere the situation resembles a simple converging nozzle and the thrust may be taken as

$$T = 1.5p_c \times A_t \qquad 10.38$$

Wind load stresses. Structures are often subjected to dynamic loading caused by gusting wind (102). Wind loading can also be a major factor in setting up damaging vibrations. For example, the first Tacoma Narrows suspension bridge failed as a result of resonance between vortex action and torsional vibration. Transmission lines have also been known to fail as a result of fatigue action brought about by high frequency vibrations (singing transmission lines). When wind loading is apt to be a problem, careful analyses must be made; suggestions for design procedure are outlined in (102). In some cases the wind merely acts to produce a static force and the problem is to determine the loading in terms of the wind velocity. This can be done by using the following equation:

$$F = 0.003V^2\left(\frac{2 \sin A}{1 + \sin^2 A}\right) \qquad 10.39$$

where V = wind velocity, miles per hour
F = normal pressure on surface, pounds per square foot
A = angle of surface from horizontal, degrees

Vortex shedding (18, 30) develops when an object whose projected cross section D faces into a constant velocity wind. This condition causes a forcing perpendicular to the wind direction. If the forcing frequency is near the natural frequency of the structure or object, an increasing response will be noticed or destruction of the structure. The forcing frequency is found from the Strouhal number or

$$\frac{fD}{V} = 0.22 \qquad 10.40$$

The wind can also cause a combined bending and torsional vibration (2, 18) in a structural member if it is a flat plate or air-foil shaped. A lift force is developed away from the elastic axis, hence, causing a torque. The member twists and bends until the stiffness of the member overcomes the force of the wind setting up a vibration called *flutter*. When the force of the wind velocity overcomes the stiffness of the member, it soon fails.

Water impact loads. Considerations should be given to a design impacting water. Kornhauser (43) discusses not only low velocity impact but also hypervelocity water impacts. For water impact of 200 fps (137 mph) or less, the expression is

$$\frac{p}{\frac{1}{2}\rho_0 v^2} = 2\left(\frac{C_0}{v}\right) \qquad 10.41$$

where $g\rho_0$ is the water density (62.4 lb/ft^3) and C_0 is 5000 fps. The left side of

Eq. 10.41 is the ratio of the impacting pressure to the stagnation pressure on the body. When the surface area of the water impact is known, the total applied force can be calculated.

Earthquake loads. Earthquake records (29, Vol. III); (55, 60, 61) indicate maximum acceleration levels of 0.2–0.5 *g*'s. However, these levels are not constant but resemble sawtooth shaped high frequency signals.

When designing, some criteria must be adapted. If these were vibration loads, a static design load would be

$$L_D = W(0.5)(10)$$

where a magnification factor of 10 with 5% critical damping would be used from the forced vibrations of a spring-mass system.

Another method is to take a design load V where

$$V = KCW \tag{10.42}$$

Then, K is 1.50 for mechanical designs and

$$C = \frac{0.05}{(T)^{1/3}} \tag{10.43}$$

T is the period (seconds) of the natural frequency if it is known. When only the size of a design is known

$$T = \frac{0.05H}{\sqrt{D}} \tag{10.44}$$

H is the height of the structure or design above the base in feet, and D is the depth of the design parallel to the applied force. Equations 10.42, 10.43, and 10.44 size the design for earthquake loads. It still must be checked by applying actual earthquake acceleration records to the base of a computer model to verify the design.

Human loading or limitations. The human body (30, 43) can stand 20 *g*'s for velocity changes of 80 fps over 0.1-sec and impact up to 10,000 fps velocity changes over 10-sec impact duration. However, at a velocity change of 80 fps, the impact durations decrease to 0.01 sec. for 100 *g*'s. These values represent extreme limits for healthy individuals. Therefore, in any situation the average population should never be exposed to design loads of more than 10 *g*'s.

Buckling of tubes. Tube selection by buckling and frequency criteria has been treated by Van Meter (84), who points out that the choice of tubular structural elements for dynamic applications is often made on the basis of critical buckling stress and the lowest natural frequency. The theory of simple columns was given in Chapter 9 where it was shown that the Euler critical stress was

$$\sigma_{cr} = \frac{F_{cr}}{A} = \frac{K\pi^2 E}{(L/r)^2}$$

where K is a constant depending on the end conditions of the column. For tubes, the moment of inertia I is Ar^2 and for convenience two constants are defined as

$$C_1 = 1/\pi^2 E \qquad 10.45$$

and

$$C_2 = 1/K \qquad 10.46$$

If a factor of safety n is applied ($F_{cr} = nF$), the moment of inertia can be expressed through use of the foregoing equations as

$$I = C_1 C_2 n F L^2 \qquad 10.47$$

The first mode natural frequency for lateral beam vibration is (83)

$$f = K_f \sqrt{(gE/\gamma L^4)(1/A)(1 - F/F_{cr})}$$

from which

$$I/A = \frac{f^2 \gamma L^4}{K_f^2 g E}\left(\frac{n}{n-1}\right) \qquad 10.48$$

where K_f is a constant depending on the end conditions of the beam. To illustrate this theory Van Meter gives the following example for a pin-connected aluminum tube with a 30-in. span carrying an axial load of 175 lb. The minimum natural frequency is to be 40 cps, and a factor of safety of 1.5 is applied. For these conditions the optimum tube can be found as follows. Since aluminum has a modulus of elasticity of 10,000,000 psi, $C_1 = 10.15 \times 10^{-9}$. For a pin-connected column K equals 1.0; therefore $C_2 = 1.0$, and from Eq. 10.47 the moment of inertia is about 2.3×10^{-3} in.^{4} Calculation of I/A gives a value of about 0.039. Then, having values for I and I/A, the optimum tube is the one that satisfies the given requirements. For the present case a tube 0.625 in. outside and 0.028 in. wall thickness is the minimum weight tube satisfying the conditions of the problem. Further analyses of the buckling of struts under dynamic loading are given by Davidson (14). The design of foundations for forging hammers is discussed by Heaviside and Wallace (31).

Hertz theory of impact. Hertz (see 82), assuming that the time of loading was long compared with the natural period of vibration, developed a theory that allows the construction of stress-time curves as shown schematically in Fig. 10.33. Details of the theory may be found in (82); for the present the

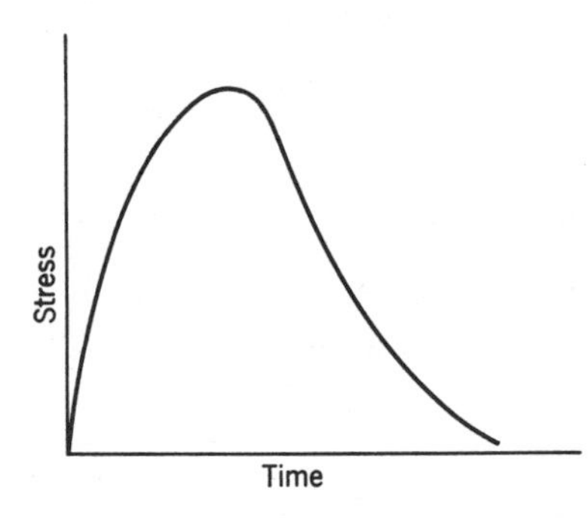

Fig. 10.33 Schematic stress-time curve from the Hertz theory of impact.

following will suffice for design purposes. For spherical contact

$$\alpha = (k_1 + k_2)q_0\pi^2 a/2 \tag{10.49}$$

and

$$a = (k_1 + k_2)q_0\pi^2/4\phi \tag{10.50}$$

where α is displacement of the spherical surface of contact of radius a, q_0 the contact pressure, and

$$k_1 = \frac{1 - \nu_1^{\ 2}}{\pi E_1} \tag{10.51}$$

$$k_2 = \frac{1 - \nu_2^{\ 2}}{\pi E_2} \tag{10.52}$$

and

$$\phi = \frac{R_1 + R_2}{2R_1 R_2} \tag{10.53}$$

where the subscripts 1 and 2 denote different materials if the impact is between two different materials and R_1 and R_2 are the radii of curvature of the objects. The maximum pressure q_0 is obtained by equating the sum of the pressures over the contact area to the compressive force F; then for the hemispherical pressure distribution

$$q_o = 3F/2\pi a^2 \tag{10.54}$$

For two spheres of radius R_1 and R_2, respectively, in contact, from Eqs. 10.49 to 10.53

$$a = \sqrt[3]{\frac{3\pi}{4} \frac{F(k_1 + k_2)R_1 R_2}{R_1 + R_2}} \tag{10.55}$$

and

$$\alpha = \sqrt[3]{\frac{9\pi^2}{16} \frac{F^2(k_1 + k_2)^2(R_1 + R_2)}{R_1 R_2}} \tag{10.56}$$

It is assumed that the two spheres are approaching each other with velocities v_1 and v_2 respectively. At the instant of contact a force F is induced between the spheres and the rates of change during impact are

$$m_1(dv_1/dt) = -F \tag{10.57}$$

$$m_2(dv_2/dt) = -F \tag{10.58}$$

where m_1 and m_2 are the masses of the spheres. If α is the distance the two spheres approach each other as a result of compression, the velocity of this approach is

$$d\alpha/dt = v_1 + v_2 \tag{10.59}$$

and from Eqs. 10.57 and 10.58

$$\frac{d^2\alpha}{dt^2} = (-F)\frac{m_1 + m_2}{m_1 m_2} \tag{10.60}$$

If now it is noted by definition that

$$n = \sqrt{\frac{16}{9\pi^2}\frac{R_1 R_2}{(k_1 + k_2)^2(R_1 + R_2)}} \tag{10.61}$$

and

$$n_1 = \frac{m_1 + m_2}{m_1 m_2} \tag{10.62}$$

it can be shown, using Eqs. 10.55 and 10.56, that

$$F = n\alpha^{3/2} \tag{10.63}$$

and Eq. 10.60 becomes

$$\frac{d^2\alpha}{dt^2} = -(n)n_1\alpha^{3/2} \tag{10.64}$$

Integration of Eq. 10.64 leads to the expression

$$\tfrac{1}{2}[(d\alpha/dt)^2 - v^2] = -\tfrac{2}{5}(n)n_1\,\alpha^{5/2} \tag{10.65}$$

Finally, for calculating the duration of impact, Eq. 10.65 becomes

$$dt = \frac{d\alpha}{\sqrt{v^2 - \frac{4}{5}(n)n_1\alpha^{5/2}}} \tag{10.66}$$

Equation 10.66 can be used to determine various time values associated with various stresses and a plot such as shown in Fig. 10.33 computed. The specific time values are obtained by plotting t against the integral of the right-hand side of Eq. 10.66 and then using this plot for scaling-off the time values. For the two spheres impacting on each other, using Eq. 10.66 (82),

$$t = 2.94\left(\frac{5}{4}\frac{v^2}{nn_1}\right)^{2/5} v \tag{10.67}$$

and for the particular case of two equal spheres Eq. 10.67 becomes

$$t = 2.94\left[\frac{5\sqrt{2}\pi\gamma}{4}\left(\frac{1 - v^2}{E}\right)\right]^{2/5}\frac{R}{v^{1/5}} \tag{10.68}$$

Dynamic behavior of viscoelastic materials. The foregoing analyses all assume Hookean behavior, that is, instantaneous linear elastic response to loading and unloading. Viscoelastic materials, however, are particularly rate sensitive, and the laws of elastic behavior are usually inadequate, *per se*, to describe their mechanical behavior. This was discussed in Chapter 5. Assuming sinusoidal variation of mechanical response with time (Fig. 10.34), it is seen that

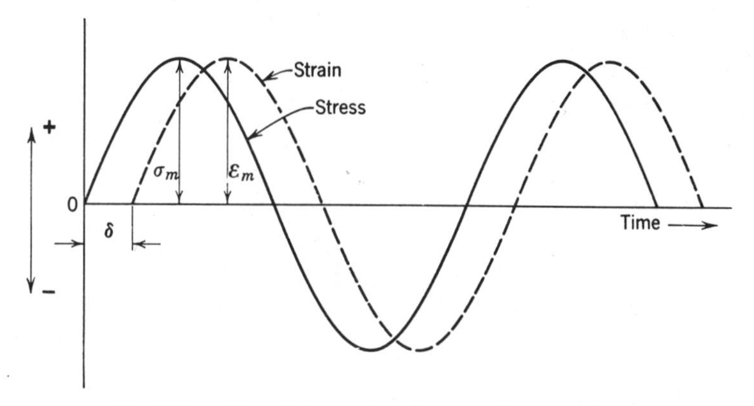

Fig. 10.34 Sinusoidal variation of stress and strain with time.

the strain response lags the stress by a phase angle δ; the stress response can thus be split into two components as shown in Fig. 10.35, one in phase with the strain response and one 90° ($\pi/2$ radians) out-of-phase. The in-phase component represents elastic response; the maximum value is $(\sigma_m)_e$, whereas the maximum out-of-phase component is $(\sigma_m)_v$. The over-all observed *apparent* stiffness (modulus) of the system is expressed by the *dynamic modulus* E_d; thus from Fig. 10.34

$$E_d = \sigma_m/\varepsilon_m \tag{10.69}$$

The modulus E_d, however, by virtue of the in- and out-of-phase characteristics just mentioned is composed of two parts. One part represents *elastic* behavior and is called the *storage modulus* E_s; it is much like the usual Young's elastic modulus and is expressed (Fig. 10.35) by

$$E_s = (\sigma_m)_e/\varepsilon_m \tag{10.70}$$

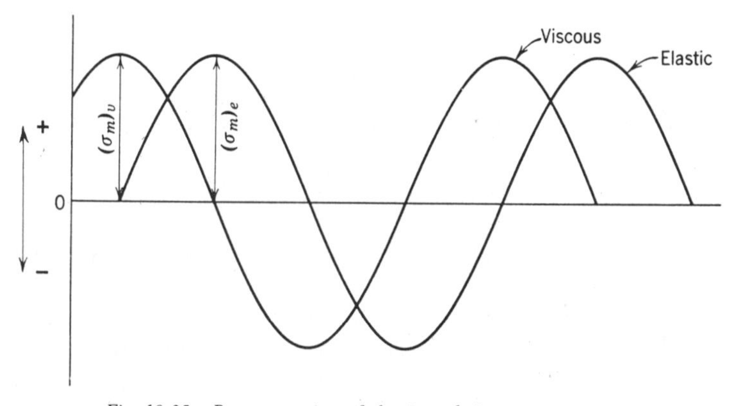

Fig. 10.35 Representation of elastic and viscous components.

The other component represents viscous behavior and represents the dissipation per cycle (heat loss) and is expressed as

$$E_e = (\sigma_m)_v / \varepsilon_m \qquad 10.71$$

From Eqs. 10.70 and 10.71

$$(\sigma_m)_e/(\sigma_m)_v = E_s\varepsilon_m / E_e\varepsilon_m = E_s/E_e = \tan\delta \qquad 10.72$$

where $\tan\delta$ is called the *loss tangent*. It is also evident that

$$\sigma_m = \sqrt{(\sigma_m)_c^{\,2} + (\sigma_m)_v^{\,2}} \qquad 10.73$$

from which

$$\pm E_d = \sqrt{E_s^{\,2} + E_e^{\,2}} \qquad 10.74$$

The obtaining of the various property values to use in design problems is difficult, requiring specialized apparatus and methods of interpretation. Information concerning these problems may be obtained in the references cited in Chapter 5 as well as in (40, 49, 50, and 54) at the end of this chapter. One procedure for obtaining the data is to use a rotating beam of circular cross section fixed in a chuck at one end and weighted at the other end. The modulus of the material is then calculated from the deflection of the bar and the applied force. The combination of force and rotation causes the bar not only to deflect vertically but also laterally, depending on the loss factor, Mathematically, the modulus is

$$E_d = FL^3/yR^4 \qquad 10.75$$

and the loss tangent is $\tan\delta$, the angle between the vertical load F and the observed sidewise deflection observed by viewing the end of the bar. Having E_d and $\tan\delta$ the *storage modulus* can be obtained from Eqs. 10.72 and 10.74, this is the modulus value used in design. For springs see (52).

Inertia effects

Rotating bars. Solid objects rotating about a center of spin generate high internal stresses produced by the centrifugal action of the mass of material away from the center of spin. Consider, for example, the bar shown in Fig. 10.36 which has a circular cross section and rotates at angular velocity ω. For an element of mass dm distant r from the center of rotation the centrifugal force exerted is

$$dF = \omega^2 r\,dm \qquad 10.76$$

However,

$$dm = \gamma A\,dr/g \qquad 10.77$$

where γ is the density of the material and A its cross-sectional area. Therefore

$$dF = \gamma A\omega^2 r\,dr/g \qquad 10.78$$

The stress on section r is caused by the mass of material from r to L, therefore

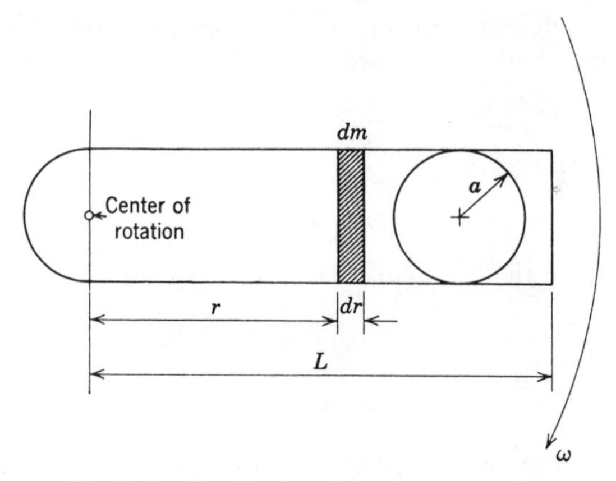

Fig. 10.36 Rotating bar of circular cross section.

on any section the total force is

$$F = \int dF = \int_r^L \frac{\gamma A \omega^2 r}{g}\, dr \qquad 10.79$$

from which

$$F_r = (\gamma A \omega^2/2g)(L^2 - r^2) \qquad 10.80$$

The stress is consequently

$$\sigma_r = F_r/A = (\gamma \omega^2/2g)(L^2 - r^2) \qquad 10.81$$

The maximum stress is at $r = 0$ or

$$(\sigma_r)_{max} = \gamma \omega^2 L^2/2g \qquad 10.82$$

The total elongation of the bar is

$$\mathscr{E} = \int_r^L d\mathscr{E} = \int_r^L \frac{\sigma_r}{E}\, dr \qquad 10.83$$

Substituting Eq. 10.81 into Eq. 10.83 gives

$$\mathscr{E} = \int_r^L \frac{\gamma \omega^2}{2gE}(L^2 - r^2)\, dr \qquad 10.84$$

from which

$$\mathscr{E} = \frac{\gamma \omega^2}{2gE}\left(\frac{2}{3}L^3 - L^2 r + \frac{r^3}{3}\right) \qquad 10.85$$

and with $r = 0$, Eq. 10.85 becomes

$$\mathscr{E} = \gamma \omega^2 L^3/3gE \qquad 10.86$$

Suppose the rotating bar has a conical profile and circular cross section as

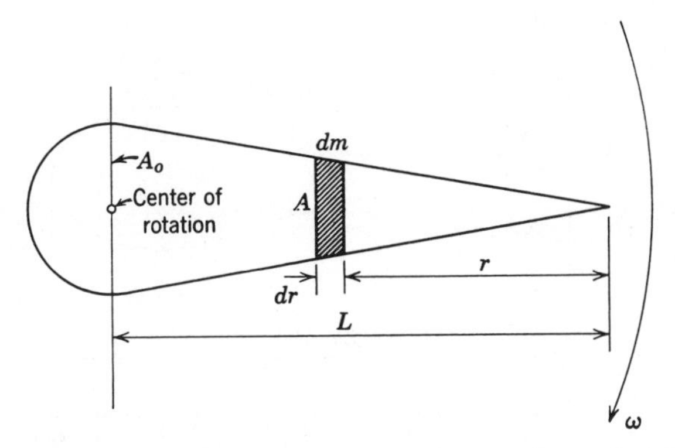

Fig. 10.37 Rotating bar of circular cross section and conical profile.

shown in Fig. 10.37. Then

$$dm = (\gamma/g)(A)\,dr \tag{10.87}$$

However, for a cone

$$A_0/L^2 = A/r^2 \tag{10.88}$$

where A_0 is the cross-sectional area at the center of rotation. Therefore from Eq. 10.88

$$A = A_0 r^2/L^2 \tag{10.89}$$

and Eq. 10.87 becomes

$$dm = (\gamma/g)(A_0/L^2)r^2\,dr \tag{10.90}$$

and consequently

$$dF = (\gamma\omega^2)(L-r)(A_0/gL^2)r^2\,dr \tag{10.91}$$

and the total force is

$$F = \int dF = \frac{\gamma\omega^2}{g}\frac{A_0}{L^2}\int_0^r (L-r)r^2\,dr \tag{10.92}$$

from which

$$F = \frac{\gamma\omega^2 A_0}{gL^2}\left(\frac{Lr^3}{3} - \frac{r^4}{4}\right) \tag{10.93}$$

and the maximum stress (at $r = L$) is

$$(\sigma_r)_{\max} = \frac{\gamma\omega^2}{gL^2}\left(\frac{L^4}{3} - \frac{L^4}{4}\right) = \frac{\gamma\omega^2 L^2}{12g} \tag{10.94}$$

The total elongation is

$$\mathscr{E} = \int d\mathscr{E} = \int_0^r \frac{\sigma_r}{E}\,dr = \frac{\gamma\omega^2 A_0}{gL^2 E A_0(r^2/L^2)}\int_0^r \left(\frac{Lr^3}{3} - \frac{r^4}{4}\right)dr \tag{10.95}$$

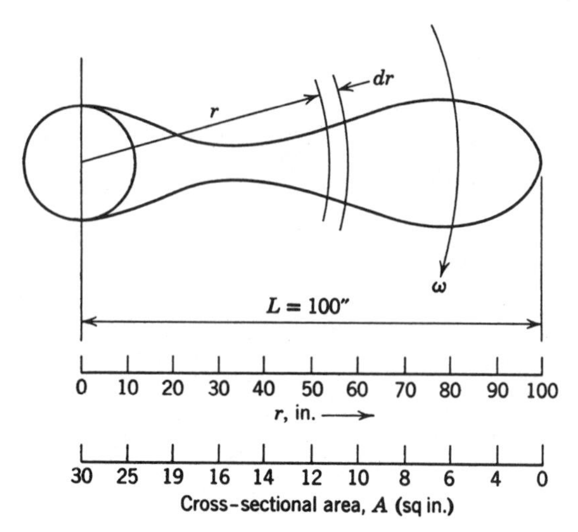

Fig. 10.38 Rotating tapered aluminum blade.

from which

$$\mathscr{E} = \frac{\gamma\omega^2}{gE}\left(\frac{Lr^2}{6} - \frac{r^3}{12}\right) \tag{10.96}$$

Letting $r = L$ in Eq. 10.96 then gives the result

$$\mathscr{E} = \gamma\omega^2 L^3/12gE \tag{10.97}$$

If the profile of the rotating bar is not easily expressible by an equation, a graphical procedure can be used as follows. Suppose the bar has the shape shown in Fig. 10.38, where the cross-sectional area at different radial distances is also shown. The bar is a tapered aluminum blade rotating at 1500 rpm. The density γ of the material is 165 lb/ft.³

At any radius r

$$dw = A\,dr\,\gamma \tag{10.98}$$

and the centrifugal force is

$$dF = dw\omega^2 r/g \tag{10.99}$$

Substituting Eq. 10.98 into Eq. 10.99 gives

$$dF = A\,dr\gamma\omega^2 r/g \tag{10.100}$$

At radius r the centrifugal force causes a tension given by

$$F_r = \int_r^L A\,dr\gamma\omega^2 r/g \tag{10.101}$$

or

$$F_r = \frac{\gamma\omega^2}{g}\int_r^L Ar\,dr \tag{10.102}$$

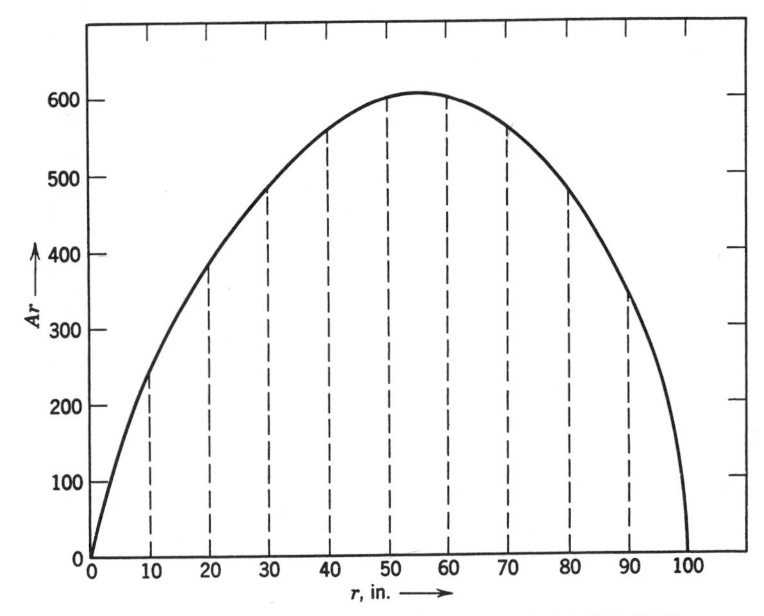

Fig. 10.39 Plot of Ar versus r for the rotating blade in Fig. 10.38.

Placing the numerical values given below into Eq. 10.102 leads to F_r.

$A =$ area, sq in.

$\gamma =$ density $= (165/1728)$ lb/in.^{3}

$$\omega = \text{angular velocity} = \frac{1500}{60} 2\pi = 157 \text{ rad/sec}$$

$g = (32.2)(12) = 386.4$ in./sec^2

$L = 100$ in.

With these values

$$F_r = \int_r^{100} \frac{(A\,dr)(165/1728)(157)^2 r}{386.4}$$

or

$$F_r = \int_r^{100} 6.1 Ar\,dr \qquad 10.103$$

which is integrated graphically as follows. A plot is made of Ar versus r as shown in Fig. 10.39 (the accompanying data are listed in Table 10.5). The tension at any particular distance r is the area under the curve within the limits of r and 100 multiplied by the factor 6.1. The stress is then simply F_r/A_r. To find the elongation the stress is plotted against r in Fig. 10.40 and graphically inte-

Table 10.5 Data for Calculating Stress in Rotating Blade of Fig. 10.38

r, in.	Ar, in.^{3}	$\int_r^{100} Ar\,dr$, lb	$6.1 \int_r^{100} Ar\,dr$, lb	Stress, σ (psi)
0	0	43,700	268,000	8,925
10	250	42,300	258,000	10,300
20	380	39,050	238,000	12,500
30	480	34,700	212,000	14,600
40	560	29,500	180,000	12,850
50	600	23,650	144,000	12,000
60	600	17,600	107,200	10,720
70	560	11,750	71,600	8,975
80	480	6,500	39,600	6,600
90	360	2,300	14,050	3,520
100	0	0	0	0

grated. The elongation is

$$\mathscr{E} = \int_r^L \frac{\sigma_r}{E}\,dr \tag{10.104}$$

or

$$\mathscr{E} = \frac{1}{E}\int_0^{100} \sigma\,dr = \frac{960{,}000}{10{,}000{,}000} = 0.096 \text{ in.}$$

Rotating cylinders and disks of constant thickness. Consider first a circular disk of uniform thickness rotating about its polar axis of symmetry as shown

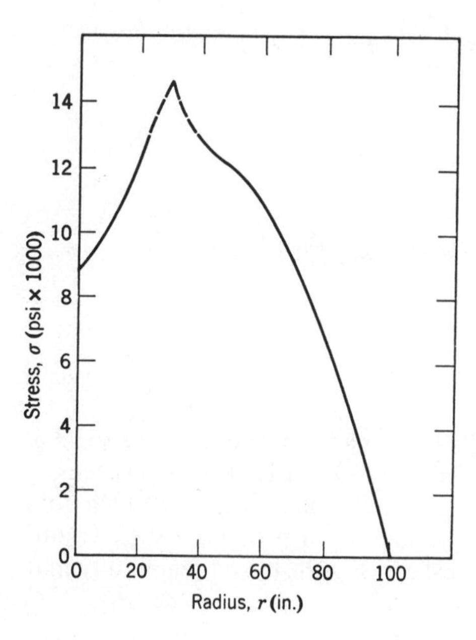

Fig. 10.40 Plot of stress versus r for the rotating blade in Fig. 10.38.

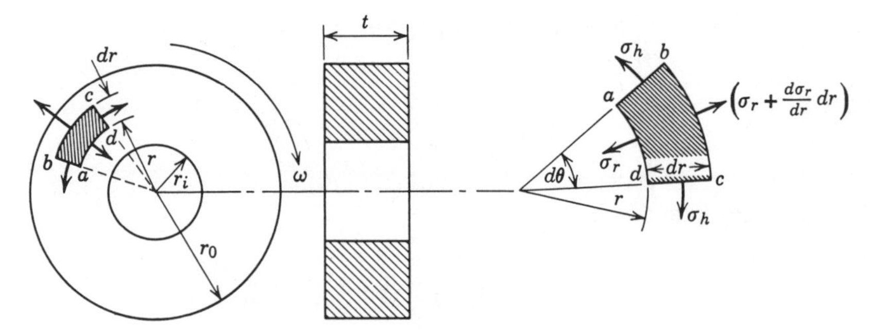

Fig. 10.41 Geometry and equilibrium of rotating disk.

in Fig. 10.41. It is assumed that the stresses are uniform through the thickness of the disk and are a function of radial distance and speed only. For the element *abcd* in Fig. 10.41 the inertia force due to rotation at angular velocity ω is

$$F_{\text{inertia}} = (\gamma\omega^2 r^2/g)\, dr\, d\theta \qquad 10.105$$

The load given by Eq. 10.105 is added to that given by Eq. 3.452 so that

$$\sigma_h - \sigma_r - r\, d\sigma_r/dr = \gamma\omega^2 r^2/g = 0 \qquad 10.106$$

The stresses (σ_h and σ_r) in Eq. 10.106 are replaced by Eqs. 3.460 and 3.461 so that

$$\frac{d^2u}{dr^2} + \frac{1}{r}\frac{du}{dr} - \frac{u}{r^2} + (1 - v^2)\frac{\gamma\omega^2 r}{gE} = 0 \qquad 10.107$$

the solution for which is

$$u = -\left[(1 - v^2)\frac{\gamma\omega^2}{gE}\right]\frac{r^3}{8} + C_1 r + \frac{C_2}{r} \qquad 10.108$$

where C_1 and C_2 are constants of integration. At the bore and outside surface of the disk $\sigma_r = 0$, and

$$\sigma_r = \frac{E}{1 - v^2}\left[-\frac{3 + v}{8} N r^2 + (1 + v)C_1 - (1 - v)\frac{C_2}{r^2}\right] \qquad 10.109$$

where

$$N = (1 - v^2)(\gamma\omega^2/gE) \qquad 10.110$$

In Eq. 10.109 $\sigma_r = 0$ at $r = r_0$ and $r = r_i$; therefore from this condition and Eqs. 10.109 and 10.110

$$C_1 = \frac{3 + v}{8(1 + v)}(r_0{}^2 + r_i{}^2)N \qquad 10.111$$

$$C_2 = \frac{3 + v}{8(1 - v)}(r_0{}^2 r_i{}^2 N) \qquad 10.112$$

$$\sigma_h = \frac{3+\nu}{8(1-\nu^2)} EN\left(r_i^{\,2} + r_0^{\,2} - \frac{1+3\nu}{3+\nu} r^2 + \frac{r_0^{\,2} r_i^{\,2}}{r^2}\right) \qquad 10.113$$

and

$$\sigma_r = \frac{3+\nu}{8(1-\nu^2)} EN\left(r_i^{\,2} + r_0^{\,2} - r^2 - \frac{r_i^{\,2} r_0^{\,2}}{r^2}\right) \qquad 10.114$$

The maximum values are

$$(\sigma_h)_{\max} = \frac{\gamma\omega^2 r_0^{\,2}}{g}\left(\frac{3+\nu}{4}\right)\left(1 + \frac{1-\nu}{3+\nu}\frac{r_i^{\,2}}{r_0^{\,2}}\right) \qquad 10.115$$

and

$$(\sigma_r)_{\max} = \frac{\gamma\omega^2 r_0^{\,2}}{g}\left(\frac{3+\nu}{8}\right)\left(1 - \frac{r_i}{r_0}\right)^2 \qquad 10.116$$

From Eqs. 10.115 and 10.116 it is seen that σ_h is always greater than σ_r.

The solution for a rotating *thin ring* is obtained by letting r_i/r_0 approach unity in Eq. 10.115, to give

$$(\sigma_h)_{\max} = \gamma\omega^2 r_0^{\,2}/g \qquad 10.117$$

To investigate the effect of decreasing the hole size two solutions are required. If r_i/r_0 approaches zero in Eq. 10.115

$$(\sigma_h)_{\max} = (\gamma\omega^2 r_0^{\,2}/g)\left(\frac{3+\nu}{4}\right) \qquad 10.118$$

For a *solid disk*, however, $u = 0$ for $r = 0$; hence C_2 is zero in Eq. 10.108. Again, at $r = r_0, \sigma_r = 0$; therefore

$$(C_1)_{\text{solid}} = \frac{3+\nu}{8(1+\nu)} N r_0^{\,2} \qquad 10.119$$

$$\sigma_h = \frac{\gamma r_0^{\,2}\omega^2}{g}\left(\frac{3+\nu}{8}\right)\left(1 - \frac{1+3\nu}{3+\nu}\frac{r^2}{r_0^{\,2}}\right) \qquad 10.120$$

and

$$\sigma_r = \frac{\gamma r_0^{\,2}\omega^2}{g}\left(\frac{3+\nu}{8}\right)\left(1 - \frac{r^2}{r_0^{\,2}}\right) \qquad 10.121$$

from which

$$(\sigma_h)_{\max} = (\sigma_r)_{\max} = \frac{\gamma\omega^2 r_0^{\,2}}{g}\left(\frac{3+\nu}{8}\right) \qquad 10.122$$

From a design point of view it is important to note that in large forgings there may be defects at the center that would act like small holes or stress concentrators in a rotating body assumed to be acting as a solid. This condition can be improved somewhat by drilling a larger hole in the disk if possible. The following example is taken from (80):*

*Courtesy of D. Van Nostrand Co.

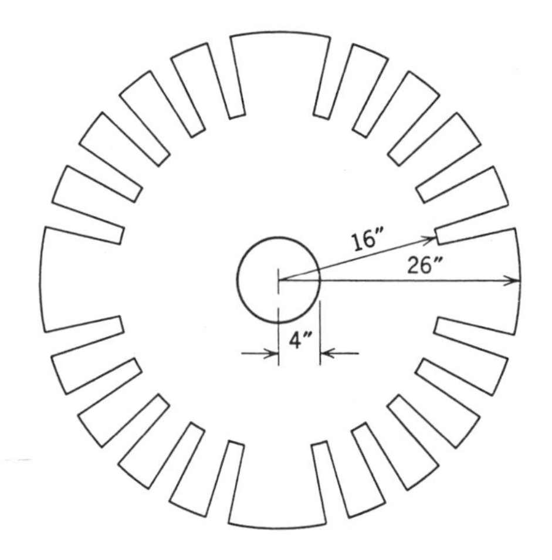

Fig. 10.42 Geometry of a slotted rotor.

Consider Fig. 10.42 which shows a slotted rotor made of steel and rotating at a speed of 1800 rpm. It is assumed that the weight of the winding in the slots is the same as that of the material removed. Because of the radial slots the part of the rotor between the outer and the 16-in. radii can support no tensile hoop stress. The centrifugal force due to this rotating ring is transmitted as a radial tensile stress across the surface of the cylinder of 16 in. radius; the magnitude of this stress is

$$(\sigma_r)_{16\text{ in.}} = \frac{1}{2\pi(16)} \int_{r=16}^{r=26} (\gamma/g)\omega^2 r \, dV$$

Taking $\gamma = 0.284$ lb/in.^{3} this equation becomes

$$(\sigma_r)_{16\text{ in.}} = \left(\frac{2\pi}{32\pi}\right)\left(\frac{\gamma\omega^2}{g}\right) \int_{16}^{26} r^2 \, dr = 7334 \text{ psi}$$

Then from Eq. 3.466, the hoop stress at the inner edge caused by the external pressure $(\sigma_r)_{16}$ is

$$\sigma_h = -\frac{2p_0 c^2}{c^2 - a^2} = \frac{2(\sigma_r)_{16}(16)^2}{16^2 - 4^2} = 15{,}700 \text{ psi}$$

The maximum hoop stress at the inner edge due to the mass between the 4- and 16-in. radii calculated as a rotating disk (Eq. 10.115) is 5580 psi. Thus the total hoop stress at the inner edge is $15{,}700 + 5580 = 21{,}280 \approx 21{,}300$ psi.

The foregoing theory applied to disks. The corresponding equations for rotating cylinders are as follows (67). See also (93, 94).

SOLID CYLINDER

$$\sigma_r = \frac{3 - 2\nu}{8(1 - \nu)}\left(\frac{\gamma\omega^2}{g}\right)(r_0{}^2 - r^2) \tag{10.123}$$

$$\sigma_h = \frac{\gamma\omega^2}{8g(1 - \nu)}[r_0{}^2(3 - 2\nu) - r^2(1 - 2\nu)] \tag{10.124}$$

$$\sigma_z = \frac{\nu\gamma\omega^2}{4g(1-\nu)}(r_0{}^2 - 2r^2) \qquad 10.125$$

HOLLOW CYLINDER

$$\sigma_r = \frac{3-2\nu}{8(1-\nu)}\left(\frac{\gamma\omega^2}{g}\right)\left[r_0{}^2 - r_i{}^2 - r^2 - \left(\frac{r_0 r_i}{r}\right)^2\right] \qquad 10.126$$

$$\sigma_h = \frac{3-2\nu}{8(1-\nu)}\left(\frac{\gamma\omega^2}{g}\right)\left[r_0{}^2 + r_i{}^2 + \left(\frac{r_i r_0}{r}\right)^2 - r^2\left(\frac{1-2\nu}{3-2\nu}\right)\right] \qquad 10.127$$

$$\sigma_z = \frac{\nu}{4(1-\nu)}\left(\frac{\gamma\omega^2}{g}\right)(r_0{}^2 + r_i{}^2 - 2r^2) \qquad 10.128$$

Allowable speeds for rotating cylinders and disks of constant thickness. Rotating disks and cylinders are frequently used in machinery to provide stability as in flywheels, a large source of energy as in cutting and shearing machines, and as components of turbines and various electrical machines using rotors. These, and associated machine design problems, are considered in details in (98) and will not be repeated here. In the following paragraphs the discussion will be concerned with the problem of calculating allowable speeds for rotating masses to avoid plastic flow or sometimes bursting.

By using the distortion energy theory expressed by Eq. 3.523, with the stresses given by the equations in the previous section for rotating disks and cylinders, it is a simple matter of substitution to show that for a *solid disk* the allowable speed is

$$(\omega)_{\text{solid}} = \frac{1}{r_0}\sqrt{\frac{8g\sigma_w}{(3+\nu)\gamma}} \qquad 10.129$$

where σ_w is the working stress (a fraction of σ_y, the material yield strength in Eq. 3.523). Equation 10.129 is plotted in Fig. 10.43. In a *hollow disk* the critical location is at the bore and application of Eq. 3.523 shows that the allowable speed is

$$(\omega)_{\text{hollow}} = \left[\frac{8g\sigma_w}{(3+\nu)\gamma r_0}\left(\frac{1}{2+(r_i/r_0)^2(1-(1+3\nu)/(3+\nu))}\right)\right]^{1/2} \qquad 10.130$$

which is plotted in Fig. 10.44. Use of Eq. 10.130 shows that, for all practical purposes, all ratios of r_i/r_0 are covered by the single plot in Fig. 10.44. It is also evident that the solid disk is considerably stronger. There is, however, considerable interest in disk rotation beyond the elastic limit in order to obtain a favorable residual stress distribution and so obtain added elastic strength for the same material weight. This subject is discussed in detail in Chapter 11. To conclude the present discussion on allowable speed some of the results obtained in Chapter 11 will be used to assist in the calculation of the *bursting speed* of disks. The burst speed is an important design consideration and is required in order to assign a realistic factor of safety to the over-all structure.

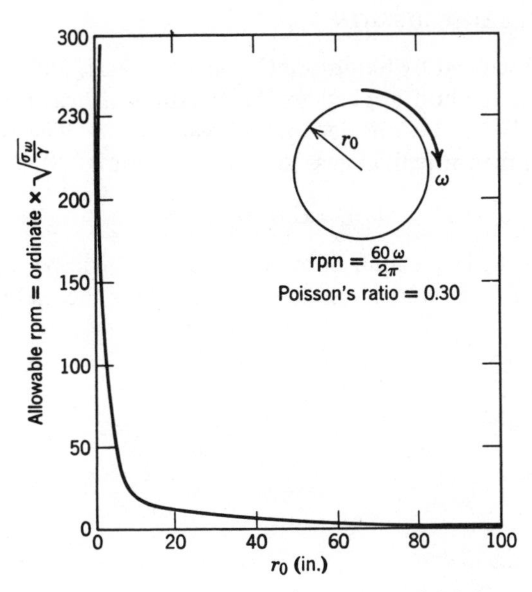

Fig. 10.43 Design curve for rotating solid disk.

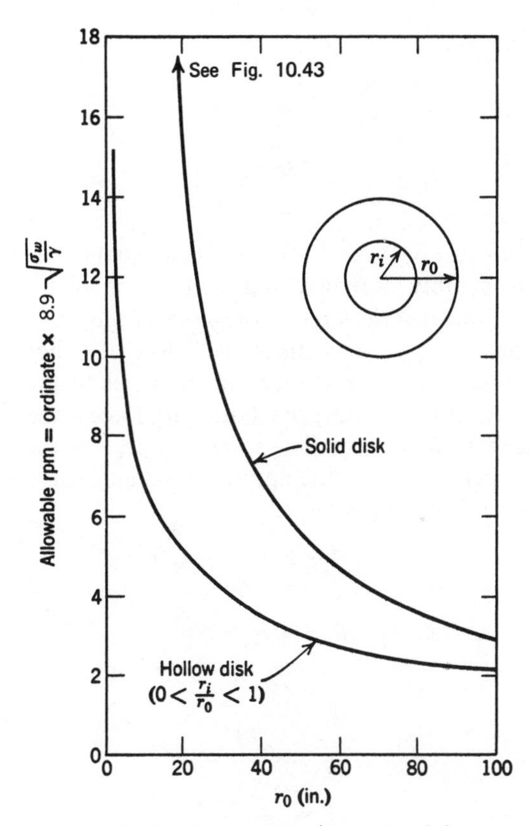

Fig. 10.44 Design curves for rotating disks.

This subject is discussed by Skidmore (77) and by Weiss and Prager (87). As a simplification the method applied to the bursting of heavy-walled cylinders (Chapter 6) will be used. Consider first a *solid cylinder*. The rotational speed required to cause the entire thickness to become plastic is (58)

$$(\omega)_y = (2/r_0)\sqrt{\sigma_y g/\gamma} \qquad 10.131$$

which represents the lower limit of bursting speed. The upper limit is obtained by substituting the material ultimate strength σ_u for the yield strength σ_y in Eq. 10.131 thus

$$(\omega)_u = (2/r_0)\sqrt{\sigma_u g/\gamma} \qquad 10.132$$

Then by limit analysis

$$(\omega)_{\text{burst}} = \frac{(\omega)_y + (\omega)_u}{2} = \frac{\sqrt{g/\gamma}(\sqrt{\sigma_y} + \sqrt{\sigma_u})}{r_0} \qquad 10.133$$

Similarly, for a *solid disk*

$$(\omega)_y = \frac{\sqrt{3}}{r_0}\sqrt{\sigma_y g/\gamma} \qquad 10.134$$

$$(\omega)_u = \frac{\sqrt{3}}{r_0}\sqrt{\sigma_u g/\gamma} \qquad 10.135$$

and

$$(\omega)_{\text{burst}} = \frac{\sqrt{3g/\gamma}}{2r_0}(\sqrt{\sigma_y} + \sqrt{\sigma_u}) \qquad 10.136$$

Similar calculations can be made for hollow disks and cylinders, but the computations are considerably more involved than for the solid bodies.

Rotating disks of constant thickness with boundary forces. In the foregoing discussions of constant thickness disks the effect of a boundary force was neglected; such forces, however, frequently occur, as in the cups on turbine blades, and so on. Another example of a boundary force is the press-fit pressure of a cylinder on a shaft. When boundary forces are present, the resultant stresses are obtained by superposition of the applied boundary stresses and the inertia stresses. The analysis follows (80):

$$\sigma_r = A + B/r^2 - \beta_1\omega^2 r^2 \qquad 10.137$$

and

$$\sigma_h = A - B/r^2 - \beta\omega^2 r^2 \qquad 10.138$$

where

$$\beta_1 = \frac{(3 + \nu)\gamma}{8g} \qquad 10.139$$

and

$$\beta = \frac{(1 + 3\nu)\gamma}{8g} \qquad 10.140$$

A and B are constants determined by the boundary conditions. For example, in a hollow disk shrunk on a shaft, the radial stress at $r = a$ is

$$(\sigma_r)_{r=a} = -p \tag{10.141}$$

and at the outer edge

$$(\sigma_r)_{r=b} = 0 \tag{10.142}$$

so that for this case

$$A = \frac{\beta_1 \omega^2 (b^4 - a^4) + a^2 p}{b^2 - a^2} \tag{10.143}$$

and

$$B = a^2 b^2 \left(\frac{p}{a^2 - b^2} - \beta_1 \omega^2 \right) \tag{10.144}$$

The problem can then be solved by use of the distortion energy theory (Eq. 3.523) and the preceding expressions for σ_r, σ_h, A, and B. Since the critical element is at the bore ($r = a$), it is seen that

$$(\sigma_r)_{\max} = \frac{\beta_1 \omega^2 (b^4 - a^4) + a^2 p}{b^2 - a^2} + b^2 \left(\frac{p}{a^2 - b^2} - \beta_1 \omega^2 \right) - \beta_1 \omega^2 a^2 \tag{10.145}$$

and

$$(\sigma_h)_{\max} = \frac{\beta_1 \omega^2 (b^4 - a^4) + a^2 p}{b^2 - a^2} - b^2 \left(\frac{p}{a^2 - b^2} - \beta_1 \omega^2 \right) - \beta \omega^2 a^2 \tag{10.146}$$

Equations 10.145 and 10.146 substituted into Eq. 3.523 give the allowable speed ω. The problem of shaft loosening during high-speed rotation is discussed in Chapter 11.

To provide more flexibility in the method so that it can be applied to a variety to problems and in particular the problem of variable thickness disks, it is convenient to express the equations for σ_r and σ_h in another form suggested by Grammel and reported as follows (80):
Let

$$s = \sigma_r + \beta_1 \omega^2 r^2 \tag{10.147}$$

$$t = \sigma_h + \beta \omega^2 r^2 \tag{10.148}$$

and

$$w = 1/r^2 \tag{10.149}$$

Then

$$s = A + Bw \tag{10.150}$$

and

$$t = A - Bw \tag{10.151}$$

which are plotted in Fig. 10.45. In Fig. 10.45 A is the common ordinate, and values for s and t are obtained from the two lines of equal and opposite slope B.

Rotating disks of variable cross section. The problem of a rotating disk of variable cross section has been considered by several investigators (17, 22, 32,

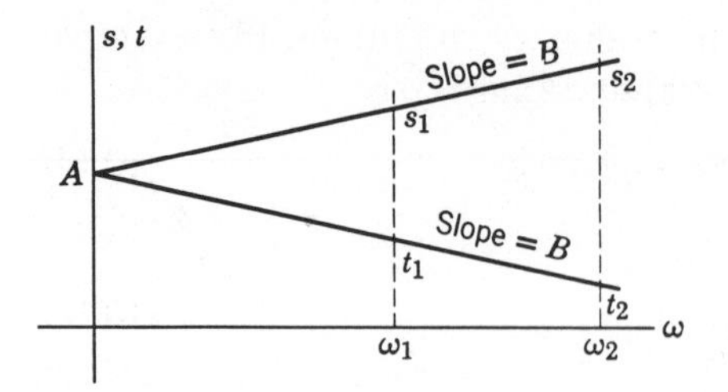

Fig. 10.45 Plot of Eqs. 10.150 and 10.151.

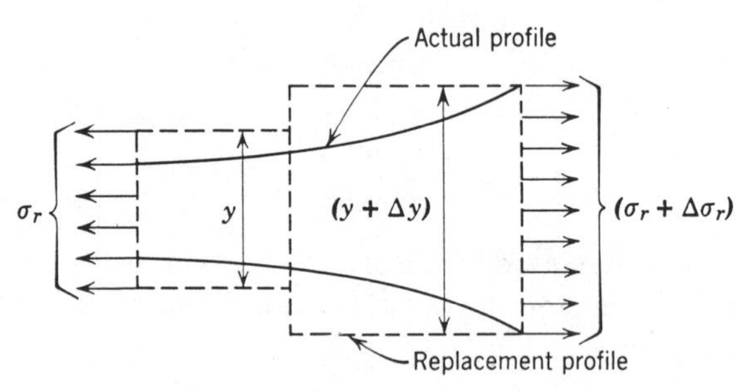

Fig. 10.46 Rotating disk of variable cross section and its replacement equivalent.

80). From a practical viewpoint the most popular methods use approximate procedures and, in general, follow the analysis given next which was formulated by Timoshenko (80) based on the analyses and results of, principally, Stodola, Martin, Biezeno and Grammel, and Donath.

In the approximate method the actual disk is replaced by one of a system of disks of uniform thickness, in which the stresses are calculated by the methods used for constant thickness disks previously described. In this procedure one of the problems that arises is the calculation of the stress at the interface zone of two constant thickness disks where the geometry (and stress) changes abruptly. For such locations, from conditions of equilibrium (Fig. 10.46)

$$\sigma_r y = (\sigma_r + \Delta\sigma_r)(y + \Delta y) \qquad 10.152$$

where y is the thickness of a disk and $y + \Delta y$ is the thickness of an adjacent section. From Eq. 10.152

$$\Delta\sigma_r = -\left(\frac{\Delta y}{y + \Delta y}\right)\sigma_r \qquad 10.153$$

which represents the change in radial stress in going from one section to another section. The hoop stress is found from the condition of compatability which requires the strain to be the same at the interface for both sections; that is,

$$\varepsilon_h = (1/E)(\sigma_h - v\sigma_r) = (1/E)[\sigma_h + \Delta\sigma_h - v(\sigma_r + \Delta\sigma_r)] \qquad 10.154$$

from which

$$\Delta\sigma_h = v\Delta\sigma_r \qquad 10.155$$

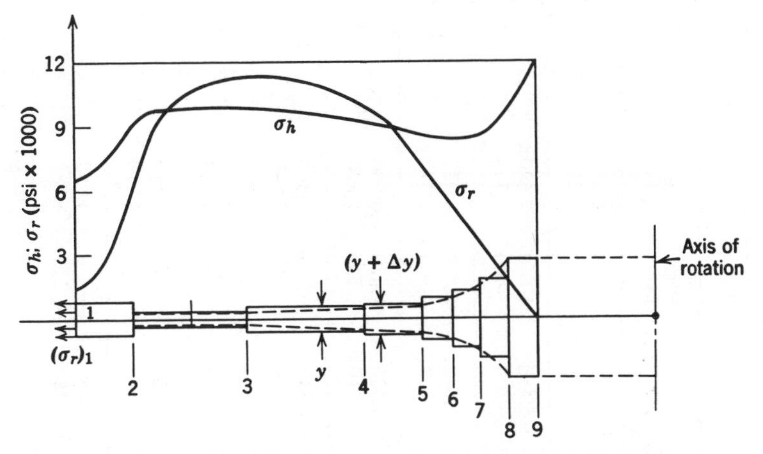

Fig. 10.47 Stress distribution in a rotating disk of variable profile.

For generality, to include cases where boundary loads are applied in addition to the inertia loads, the equations for s and t previously given will be used. Thus from Eqs. 10.150 and 10.151,

$$\Delta s = \Delta\sigma_r = -\frac{\Delta y \sigma_r}{y + \Delta y} \tag{10.156}$$

and

$$\Delta t = \Delta\sigma_t = \nu \, \Delta s \tag{10.157}$$

Equations 10.139, 10.140, 10.150, 10.151, and 10.157 in conjunction with Fig. 10.45 can be used to calculate the stresses in a rotating disk of variable thickness. Consider, for example, the disk shown in Fig. 10.47 for which the following data are supplied:

rotational speed ω	3,000 rpm
Poisson's ratio ν	0.3
density γ	0.283 lb/in.^{3}
$\beta_1\omega^2$	30.00 lb/in.^{4}
$\beta\omega^2$	17.30 lb/in.^{4}

Additional data required for the calculations are presented in Table 10.6. It is assumed that a boundary force is applied at the outside edge of the disk such that

$$(\sigma_r)_1 = 1420 \text{ psi}$$

The corresponding value of the hoop stress at the outside edge $(\sigma_h)_1$ is unknown and has to be assigned an assumed value. The simplest procedure is to assign a value of $(\sigma_h)_1$ such that s and t in Eqs. 10.147 and 10.148 will be equal. Thus

$$(\sigma_h)_1 = (\sigma_r)_1 + \beta_1\omega^2 r_1{}^2 - \beta\omega^2 r_1{}^2$$

Table 10.6 Data for Rotating Disk Shown in Fig. 10.47

1	2	3	4	5	6	7	8	9	10	11	12	13	14
Cross section	r, in.	r^2, in.^{2}	$10^3 w$, in.$^{-2}$	$\beta_1 \omega^2 r^2$, psi	$\beta \omega^2 r^2$, psi	y, in.	$-\dfrac{\Delta y}{y + \Delta y}$	σ_r, psi	Δs, psi	Δt, psi	σ_h, psi	$(\sigma_r)_{avg}$, psi	$(\sigma_h)_{avg}$, psi
1	19.7	388	2.58	11,620	6,680			1,420 0			6,360 710	1,420	6,800
						0.985							
2	17.7	314	3.18	9,410	5,420		+1.50	3,680 −80	5,450 −130	1,640 −40	7,620 795	6,250	8,900
						0.394							
3	13.8	190	5.26	5,700	3,290		−0.33	14,000 −510	−4,670 170	−1,390 60	10,020 1,070	11,400	10,000
						0.591							
4	9.85	97	10.31	2,910	1,680		−0.40	13,500 −1,020	−5,400 410	−1,620 130	8,900 1,820	10,300	9,300
						0.985							
5	7.87	62	16.1	1,860	1,070		−0.286	9,680 −1,340	−2,740 380	−825 110	7,300 2,640	7,600	8,600
						1.38							
6	6.89	47.4	21.1	1,420	820		−0.263	7,550 −1,550	−597 410	−597 130	6,530 3,300	5,700	8,400
						1.87							
7	5.91	34.9	28.6	1,050	603		−0.321	5,970 −1,990	−1,920 640	−570 200	6,110 4,270	3,980	8,600
						2.76							
8	4.92	24.2	41.3	726	418		−0.30	4,280 −2,590	−1,250 780	−370 230	5,960 5,720	2,260	9,500
						3.94							
9	3.94	15.5	64.5	465	268			2,530 −4,000			6,350 8,100	0	11,500

which, with the values given in Columns 5 and 6 of Table 10.6, give the result

$$(\sigma_h)_1 = 1420 + 11{,}620 - 6680 = 6360 \text{ psi}$$

Then from Eqs. 10.147 and 10.148

$$s_1 = (\sigma_r)_1 + \beta_1\omega^2 r_1{}^2 = 1420 + 11{,}620 = 13{,}040 \text{ psi}$$

and

$$t_1 = (\sigma_h)_1 + \beta\omega^2 r_1{}^2 = 6360 + 6680 = 13{,}040 \text{ psi}$$

Since s_1 and t_1 were taken equal, the s and t lines coincide in Fig. 10.45; from this condition and using Column 4 of Table 10.6 it is seen that for Section 2, Fig. 10.47,

$$s_2 = t_2 = 13{,}040 \text{ psi}$$

and again from Eqs. 10.147 and 10.148

$$(\sigma_r)_2 = s_2 - \beta_1\omega^2 r_2{}^2 = 13{,}040 - 9{,}410 = 3630 \text{ psi}$$

and

$$(\sigma_h)_2 = t_2 - \beta\omega^2 r_2{}^2 = 13{,}040 - 5420 = 7620 \text{ psi}$$

At Section 2, however, there is an abrupt change in cross section. To take this into account, Eqs. 10.156 and 10.157 together with the data in Column 8 of Table 10.6 are applied; thus

$$(\Delta s)_2 = (\Delta\sigma_r)_2 = \left(-\frac{\Delta y}{y + \Delta y}\right)\sigma_r = (1.50)(3630) = 5450 \text{ psi}$$

and

$$(\Delta t)_2 = (\Delta\sigma_h)_2 = \nu(\Delta s)_2 = (0.3)(5450) = 1640 \text{ psi}$$

These Δ values are then added to the values of s and t and the process repeated until the values given by the upper lines in Columns 9 to 12 of Table 10.6 are obtained. In the beginning of the problem the value of $(\sigma_h)_1$ at the outer edge was assumed; therefore at the inner edge the boundary conditions may not be satisfied and the computed value of $(\sigma_r)_9$ will be incorrect. In order to correct this condition assume that $(\sigma_r)_1 = 0$, $\omega = 0$, and assume an arbitrary value for $(\sigma_h)_1$ [in the example given $(\sigma_h)_1$ is taken as 710 psi]. Then the process is repeated as described. For this case $s = \sigma_r$ and $t = \sigma_h$. The results of these calculations are given in the lower lines of Columns 9 to 12 in Table 10.6. The solution satisfying the true condition at the inner edge is then obtained by superposing on the first stress distribution the stresses of the second distribution multiplied by

$$n = \frac{(\sigma_r)_9^* - (\sigma_r)_9}{(\sigma_r)_9'}$$

and the average stresses at the interface reactions are

$$(\sigma_r)^* = (\sigma_r + \Delta s/2) + n(\sigma_r' + \Delta s'/2)$$

and

$$(\sigma_h)^* = (\sigma_h + \Delta t/2) + n(\sigma_h' + \Delta t'/2)$$

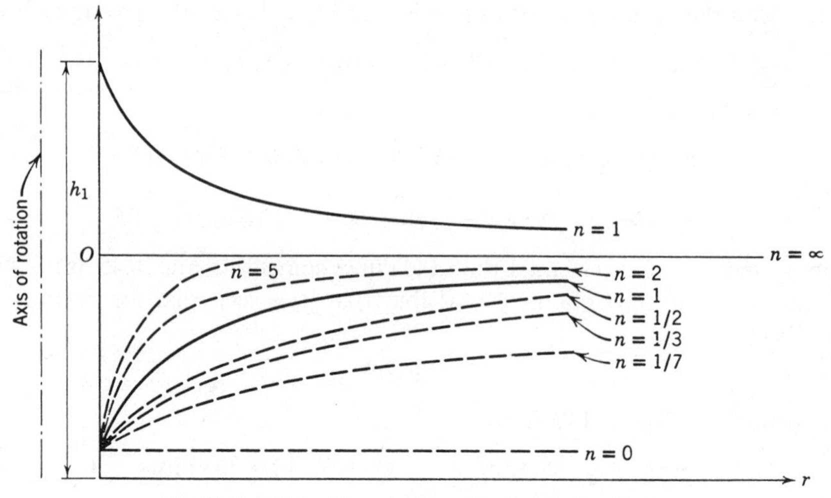

Fig. 10.48 Disk of hyperbolic profile given by Eq. 10.163.

where* means the actual value and the prime mark denotes the values calculated in the second distribution. The final results are shown in Columns 13 and 14 of Table 10.6 and by the stress curves in Fig. 10.47. Additional numerical examples are given in (17, 80).

In the exact analysis of disks of variable thickness† consider Fig. 10.48 which shows an element of a rotating variable thickness disk. For equilibrium

$$(d/dr)(\sigma_r hr) + \gamma hr^2\omega^2/g = \sigma_h h \qquad 10.158$$

Equation 10.158 contains two unknowns; therefore it is necessary to obtain another equation in order to solve for both σ_r and σ_h. This is done by examining the deformation that takes place, which is the same as for the heavy-walled cylinder discussed in Chapter 3. From Eq. 3.456

$$\varepsilon_h = u/r = (1/E)(\sigma_h - v\sigma_r) \qquad 10.159$$

and from Eq. 3.455

$$\varepsilon_r = du/dr = (1/E)(\sigma_r - v\sigma_h) \qquad 10.160$$

In Eqs. 10.159 and 10.160 the thickness is not a factor; therefore by differentiating Eq. 10.159 with respect to r an expression for du/dr is obtained which can be equated to Eq. 10.160. This operation gives

$$\frac{d}{dr}\sigma_h - v\frac{d}{dr}\sigma_r + \frac{1+v}{r}(\sigma_h - \sigma_r) = 0 \qquad 10.161$$

Equations 10.158 and 10.161 are now sufficient for a solution to the problem. First, σ_h is eliminated between these two equations giving an expression for σ_r;

†Applies only to disks with a central hole.

then from Eq. 10.158 an expression for $d\sigma_h/dr$ is obtained. When both of these results are substituted into Eq. 10.161

$$r^2 \frac{d^2}{dr^2}(hr\sigma_r) + r\frac{d}{dr}(hr\sigma_r) - hr\sigma_r - \frac{r}{h}$$

$$\times \frac{dh}{dr}\left[r\frac{d}{dr}(hr\sigma_r) - \nu hr\sigma_r\right] + (3+\nu)\frac{\gamma}{g}\omega^2 r^2 h = 0 \qquad 10.162$$

Equation 10.162 can be integrated if

$$h = h_1/r^n \qquad 10.163$$

which is Stodola's hyperbolic relationship for profiles of order n as shown in Fig. 10.48. It is seen that when $n = 0$ a disk of constant thickness is obtained. It is also seen that all other values of n give infinite thickness at the origin so that the theory is useful only for disks with a central hole. The following is a solution to Eq. 10.162. First Eq. 10.163 is placed in Eq. 10.162 giving (17)

$$r^2 \frac{d^2}{dr^2}(hr\sigma_r) + r(1+n)\frac{d}{dr}(hr\sigma_r) - (1+\nu n)(hr\sigma_r) + (3+\nu)\frac{\gamma}{g}\omega^2 r^2 h = 0 \qquad 10.164$$

the solution to which is

$$\sigma_r = \frac{1}{hr}\left[C_1 r^{m_1} + C_2 r^{m_2} - \frac{(3+\nu)\gamma\omega^2}{g[8-(3+\nu)n]} r^3 h\right] \qquad 10.165$$

where h and n are defined by Eq. 10.163 and

$$m_1 = -n/2 + \sqrt{n^2/4 + \nu n + 1} \qquad 10.166$$

and

$$m_2 = -n/2 - \sqrt{n^2/4 + \nu n + 1} \qquad 10.167$$

Equation 10.165 defines the radial stress at any point in terms of the boundary conditions that define C_1 and C_2. The hoop stress can then be obtained from Eq. 10.158 and the displacement from Eq. 10.159. The following example is given to illustrate the use of the foregoing theory. Suppose the disk is to be 4 in. thick at the bore location and $\frac{1}{2}$ in. thick at the periphery; the disk has a bore radius of 1 in. and a peripheral radius of 8 in. From Eq. 10.163,

$$\frac{h_{\text{bore}}}{h_{\text{outside}}} = \left(\frac{r_{\text{outside}}}{r_{\text{bore}}}\right)^n$$

or

$$4/\tfrac{1}{2} = 8^n$$

from which

$$n = 1$$

Then from Eqs. 10.166 and 10.167 assuming $\nu = 0.30$,

$$m_1, m_2 = -\tfrac{1}{2} \pm \sqrt{\tfrac{1}{4} + 0.3 + 1} = -0.50 \pm 1.25$$

from which

$$m_1 = 0.75 \qquad \text{and} \quad m_2 = -1.75$$

Since $n = 1$, Eq. 10.163 shows that $hr = h_1 =$ constant; therefore from Eq. 10.165

$$\sigma_r = C_1 r^{0.75} + C_2 r^{-1.75} - 0.70(\gamma\omega^2/g)r^2$$

At the bore and periphery $\sigma_r = 0$; therefore

$$0 = C_1(1)^{0.75} + C_2(1)^{-1.75} - 0.70(\gamma\omega^2/g)(1)$$

and

$$0 = C_1(8)^{0.75} + C_2(8)^{-1.75} - 0.70(\gamma\omega^2/g)64$$

from which

$$C_1 = 9.42\gamma\omega^2/g$$

and

$$C_2 = -8.72\gamma\omega^2/g$$

Therefore

$$\sigma_r = (9.42r^{0.75} - 8.72r^{-1.75} - 0.70r^2)\gamma\omega^2/g$$

which is plotted in Fig. 10.49. Finally, from Eq. 10.158

$$\sigma_h = r(d\sigma_r/dr) + (\gamma\omega^2/g)r^2$$

from which $d\sigma_r/dr$ is obtained from this equation for σ_r; thus

$$\sigma_h = r(7.06r^{-0.25} + 15.2r^{-2.75} - 1.40r)\gamma\omega^2/g + \gamma\omega^2 r^2/g$$

or

$$\sigma_h = (7.06r^{0.75} + 15.2r^{-1.75} - 1.40r^2 + r^2)\gamma\omega^2/g$$

from which

$$\sigma_h = (7.06r^{0.75} + 15.2r^{-1.75} - 0.40r^2)\gamma\omega^2/g$$

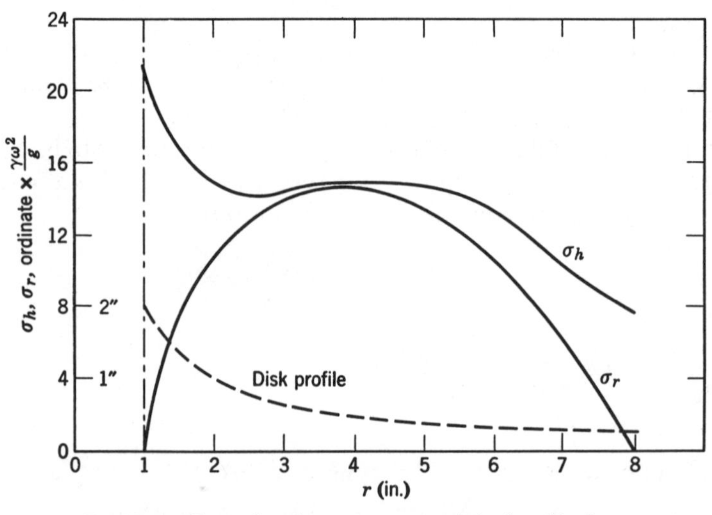

Fig. 10.49 Stress distribution in rotating disk of profile shown.

which is also plotted in Fig. 10.49. The problem of shaft loosening is discussed in Chapter 11.

Thus far consideration has been given to disks of both constant and variable sections; it is now convenient to discuss the *optimum* disk or *disk of constant strength*. The basic idea behind the constant strength disk is to save material weight by removing it from locations that increase the centrifugal force but do not add anything to strength *per se*. For constant strength, σ_r equals σ_h in Eq. 10.158 and if both stresses are set equal to σ, then

$$\sigma[h(dr/dr) + r(dh/dr)] + \gamma(hr^2\omega^2/g) - \sigma h = 0 \qquad 10.168$$

from which

$$dh/dr + \gamma(hr\omega^2/g\sigma) = 0 \qquad 10.169$$

the solution to which is

$$h = Ce^{-\gamma\omega^2 r^2/2\sigma} \qquad 10.170$$

where C is a constant of integration. At $r = 0$, $h = h_0$; therefore from Eq. 10.170

$$\ln h = -(\gamma\omega^2 r^2/2\sigma) + \ln C \qquad 10.171$$

or

$$\ln h_0 = \ln C \qquad 10.172$$

from which

$$C = h_0 \qquad 10.173$$

Equation 10.170 then becomes

$$h = h_0 e^{-\gamma\omega^2 r^2/2\sigma} \qquad 10.174$$

where h_0 is the central thickness of the disk. Note that since the stress is constant the solution applies only for a disk with no central hole. If there were a central

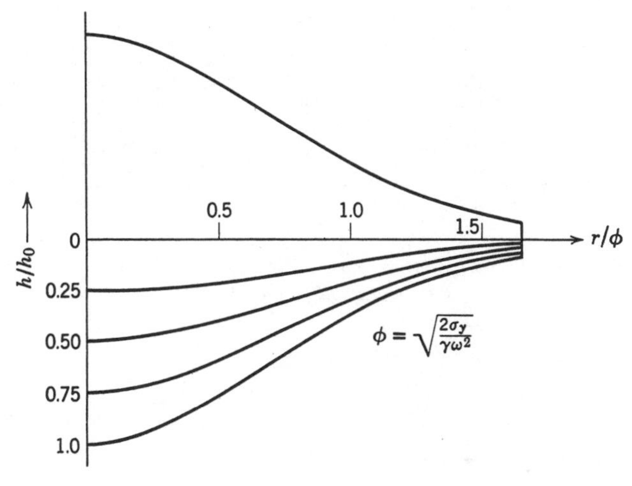

Fig. 10.50 Various profiles for rotating disks of constant strength.

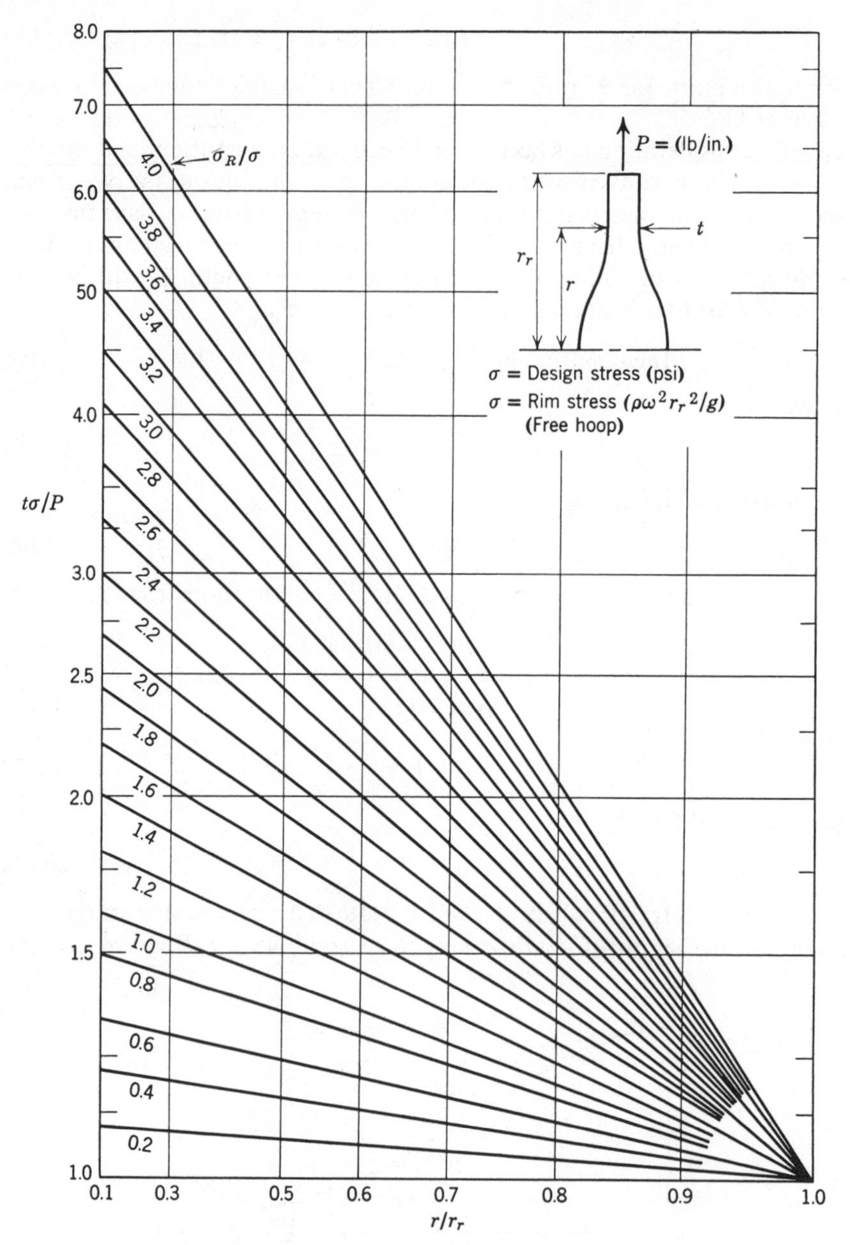

Fig. 10.51 Design chart for constant stress rotating disk. [After Rowe (74), courtesy of The McGraw-Hill Publishing Co.]

hole, the radial stress would be zero and the conditions of the problem would be violated. Typical profiles obtained from Eq. 10.174 are shown in Fig. 10.50. For design convenience Rowe (74) considers a constant strength disk subjected to peripheral loading and gives the design charts shown in Figs. 10.51 and 10.52. For example, consider a disk with an allowable stress of 20,000 psi with an outside radius of 12 in. which is to rotate at 600 rad/min with a rim load of 5000 lb/in. of circumference. From Fig. 10.52 for $\omega = 600$ and $r = 12$, σ_R (the *free hoop stress* given by the equation $\sigma_R = \gamma r_{\text{rim}} \omega^2 / g$, and being an imagi-

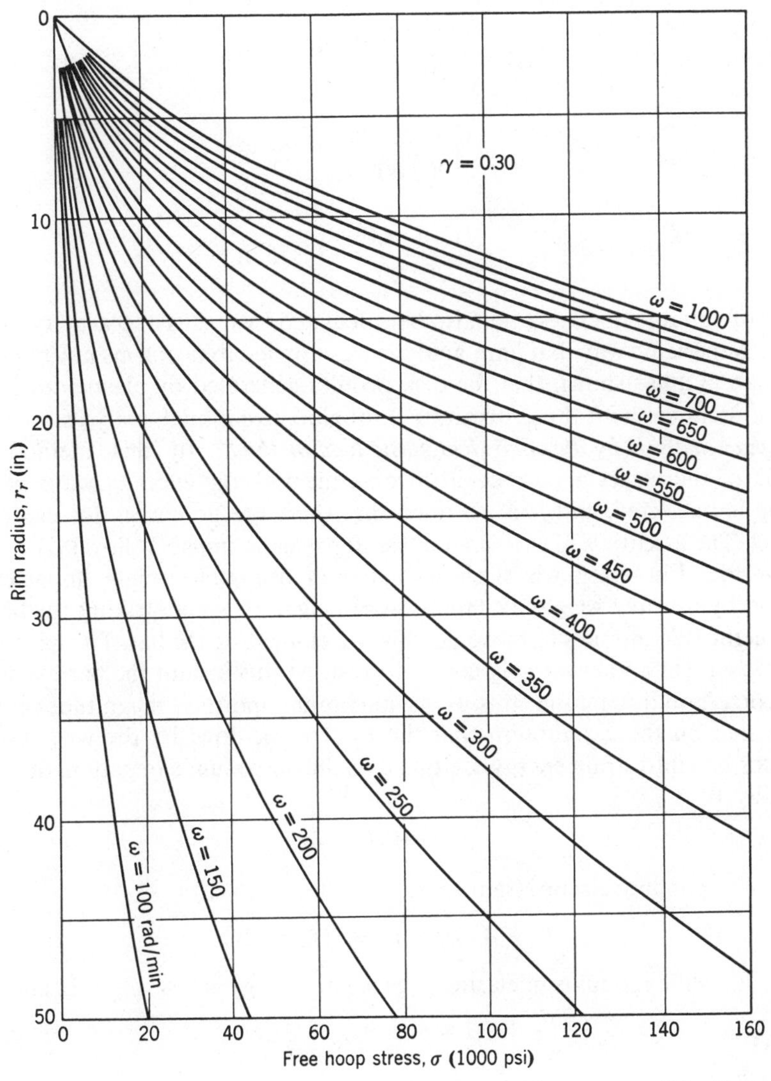

Fig. 10.52 Free hoop stress in rotating disks. [After Rowe (74), courtesy of The McGraw-Hill Publishing Co.]

nary stress which would exist in a ring rotating at the specified speed and having the same diameter as the rim) is 40,000 psi. Thus

$$\sigma_R/\sigma = 40{,}000/20{,}000 = 2$$

To find the various section thicknesses enter Fig. 10.51 at the appropriate r/r_{rim} value and read $h\sigma/P$ on the ordinate for the line $\sigma_R/\sigma = 2$. The actual thickness is then

$$h = (h\sigma/P)(P/\sigma) \tag{10.175}$$

For example, at $r = 3$ in.

$$t_{3\text{ in.}} = \frac{(2.5)(5000)}{20{,}000} = 0.625 \text{ in.}$$

Also

$$t_{6\text{ in.}} = \frac{(2.1)(5000)}{20{,}000} = 0.525 \text{ in.}$$

and

$$t_{9\text{ in.}} = \frac{(1.5)(5000)}{20{,}000} = 0.375 \text{ in.}$$

Disks of variable thickness can also be calculated by means of plasticity theory in much the same way that limit analysis was applied to problems in Chapter 6. Heyman (32) has shown that the disk profiles generated by plastic design are almost identical with those obtained through conventional analyses.

Impact analysis by the equivalent static load method. In this method the inertia of the target is considered inconsequential, the effect of wave propagation is neglected, and it is assumed that there are no energy losses during impact. The essentials of such an analysis for tension impact follow (81).

Consider Fig. 10.53 which shows a bar of negligible weight impacted in tension by a heavy weight W falling from height H. After striking the flange, the weight W continues to move downward, elongating the bar. The bar resists this movement and eventually comes to rest. At this instant the bar extension and corresponding tensile stresses are maximum and their magnitudes can be calculated on the assumption that the total work done by the weight W is transformed into strain energy. Letting δ be the maximum elongation, the work done by W is

$$U_w = W(H + \delta) \tag{10.176}$$

The corresponding elastic strain energy from Eq. 7.16 is

$$U = W^2L/2AE_0 = AE_0\delta^2/2L \tag{10.177}$$

where E_0 is the modulus of elasticity. Since no energy is lost, $U_w = U$ and

$$W(H + \delta) = AE_0\delta^2/2L \tag{10.178}$$

from which

$$\delta = \delta_{\text{static}} + \sqrt{(\delta_{\text{static}})^2 + [(\delta_{\text{static}})v^2]/g} \tag{10.179}$$

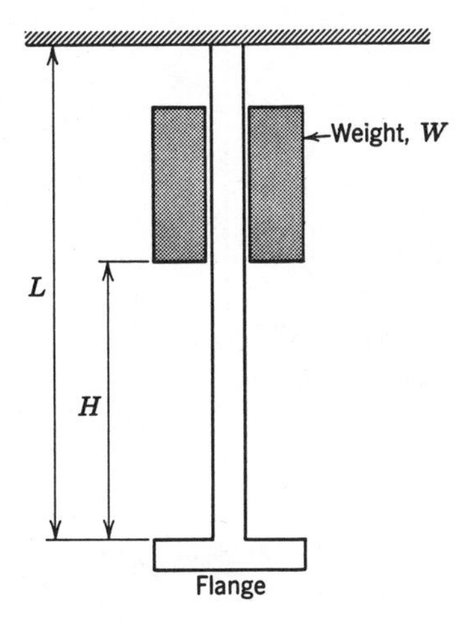

Fig. 10.53 Bar subjected to tension impact.

where

$$\delta_{\text{static}} = WL/AE_0 \tag{10.180}$$

and

$$\upsilon = \sqrt{2gH} \tag{10.181}$$

where υ is the velocity of the falling body at the moment it strikes the flange. If H is large compared with δ_{static},

$$\delta = \sqrt{[(\delta_{\text{static}})\upsilon^2]/g} \tag{10.182}$$

and the corresponding tensile stress is

$$\sigma = \delta E_0/L = (E_0/L)\sqrt{[(\delta_{\text{static}})\upsilon^2]/g} = \sqrt{(2E_0/AL)(W\upsilon^2/2g)} \tag{10.183}$$

In Eq. 10.183 the expression under the radical is directly proportional to the kinetic energy of the falling body, to the modulus of elasticity of the material, and is inversely proportional to the volume of the bar. Thus the stress can be diminished not only by an increase in cross-sectional area but also by an increase in length or by a decrease in modulus.

By substituting the working stress σ_w for σ in Eq. 10.183, the following formula is obtained for proportioning a bar subjected to axial impact:

$$AL = [E_0/(\sigma_w)^2](W\upsilon^2/g) \tag{10.184}$$

Thus it is seen that for a given material the volume of the bar must be proportional to the kinetic energy of the falling body in order to keep the maximum stress constant.

If in the preceding example the height $H = 0$ and the velocity is zero (sudden load application), Eq. 10.179 becomes

$$\delta = 2\delta_{\text{static}} \tag{10.185}$$

and

$$\sigma = 2W/A \tag{10.186}$$

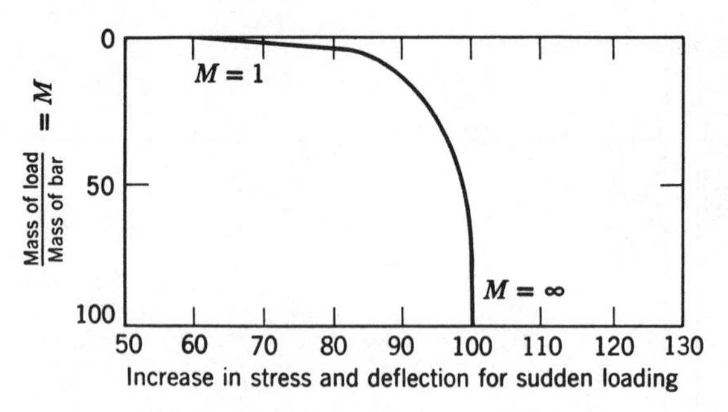

Fig. 10.54 Mass effect in sudden loading.

which is the same result predicted by vibration theory for short-time loading. Figure 10.54 shows the relationship between ratios of masses of impacted to impacting body versus stress and deflection. It can be seen that the assumption that sudden loading causes twice as much stress and deflection as static loading is always on the safe side.

If the mass of the bar shown in Fig. 10.53 is included in the calculation, then instead of Eq. 10.179 it can be shown that the maximum stress is given by Eq. 10.183 with

$$\delta = \delta_{\text{static}} + \left[(\delta_{\text{static}})^2 + \frac{(\delta_{\text{static}})v^2}{g} \left(\frac{1}{1 + qL/3W} \right) \right]^{1/2} \qquad 10.187$$

where q is the weight of the bar per unit length. Following the same procedure impact analysis of bars in bending or torsion can be made; the solutions to a few typical problems, including cases of combined stress, are given in Table 10.7.

A final item to be considered in this section is the comparison of effects of static and dynamic loading for simple structures. For example, in Fig. 10.55 are three bars, one of constant and two of variable diameter. By neglecting stress concentration effects at the locations of change in section, the maximum static stress for a given load F is the same in all three bars, $\sigma = F/A$. For dynamic loading, however, energy absorption is critical and varies in the three bars. For tension, from Eq. 10.177 and Fig. 10.55,

$$U_A = \frac{2F^2L}{2A_1E_0} \qquad 10.188$$

$$U_B = \frac{F^2L}{2A_1E_0} + \frac{F^2L}{2A_2E_0} = \frac{F^2L}{2E_0}\left(\frac{1}{A_1} + \frac{1}{A_2}\right) \qquad 10.189$$

$$U_C = \frac{3F^2L}{4A_2E_0} + \frac{F^2L}{4A_1E_0} = \frac{F^2L}{E_0}\left(\frac{3}{4A_2} + \frac{1}{4A_1}\right) \qquad 10.190$$

From Eqs. 10.188 and 10.189,

$$U_B = \tfrac{5}{8}U_A \qquad 10.191$$

Table 10.7 Results of Some Impact Analysis by the Equivalent Static Load Method

Geometry	Stress, σ		Strain, δ	
	Weight of Member Neglected	Weight of Member Considered	Weight of Member Neglected	Weight of Member Considered
q, W, L, Tension bar, H, Area, δ, q = Lb./unit length	$\frac{W}{A} + \left[\left(\frac{W}{A}\right)^2 + \frac{2E_0WH}{AL}\right]^{1/2}$	$\frac{W}{A} + \left[\left(\frac{W}{A}\right)^2 + \frac{2E_0WH}{AL}\left(\frac{1}{1 + qL/3W}\right)\right]^{1/2}$	$\frac{WL}{AE_0} + \left[\left(\frac{WL}{AE_0}\right)^2 + \frac{2WLH}{AE_0}\right]^{1/2}$	$\frac{WL}{AE_0} + \left[\left(\frac{WL}{AE_0}\right)^2 + \frac{2WLH}{AE_0}\left(\frac{1}{1 + qL/3W}\right)\right]^{1/2}$
W, H, h, b, δ, $\frac{L}{2}$, $\frac{L}{2}$, Beam bending	$\frac{3WL}{2Ah} + \left[\left(\frac{3WL}{2Ah}\right) + \frac{18WHE_0}{LA}\right]^{1/2}$	$\frac{3WL}{2Ah} + \left[\left(\frac{3WL}{2Ah}\right)^2 + \frac{18WHE_0}{LA}\left(\frac{1}{1 + \frac{17}{35}qL/W}\right)\right]^{1/2}$	$\frac{WL^3}{48E_0I} + \left[\left(\frac{WL^3}{48E_0I}\right)^2 + \frac{WHL^3}{24E_0I}\right]^{1/2}$	$\frac{WL^3}{48E_0I} + \left[\left(\frac{WL^3}{48E_0I}\right)^2 + \frac{WHL^3}{24E_0I}\left(\frac{1}{1 + \frac{17}{35}qL/W}\right)\right]^{1/2}$
L, r, W, W, H, $\frac{h}{2}$, θ, Torsion	Shear Stress $\tau = \left[\frac{8Gr^2}{AL}(WH)\right]^{1/2}$		Angle of Twist $\theta = \frac{TL}{GJ}\left\{1 + \left[1 + \frac{4H}{(TL/GJ)h}\right]^{1/2}\right\}$	

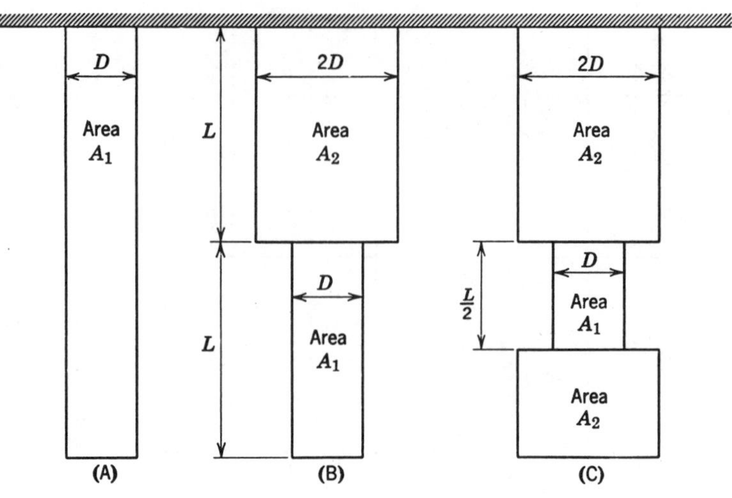

Fig. 10.55 Geometry of impact bars.

and from Eqs. 10.188 and 10.190,

$$U_C = \tfrac{7}{16} U_A \tag{10.192}$$

It is thus apparent that the bars of variable section have less stored strain energy than the uniform bar. Physically this means, for the dimensions given, that the variable section bars have less ability to strain and absorb energy; consequently, at high stress levels danger of brittle failure is present. This has been recognized for years, and when energy loads are expected in bolts, for example, the bolts are made longer to increase the energy absorption. In

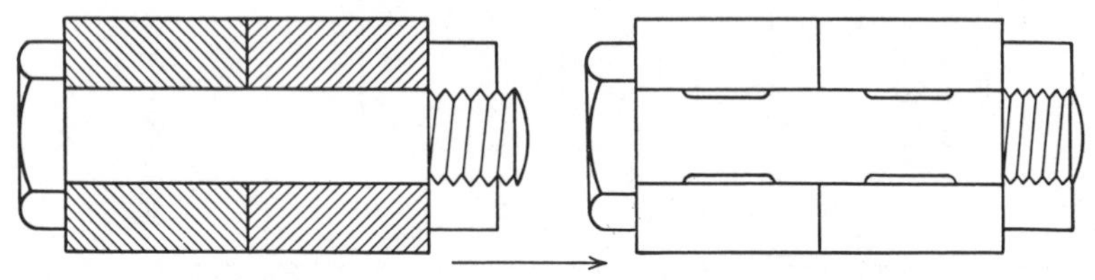

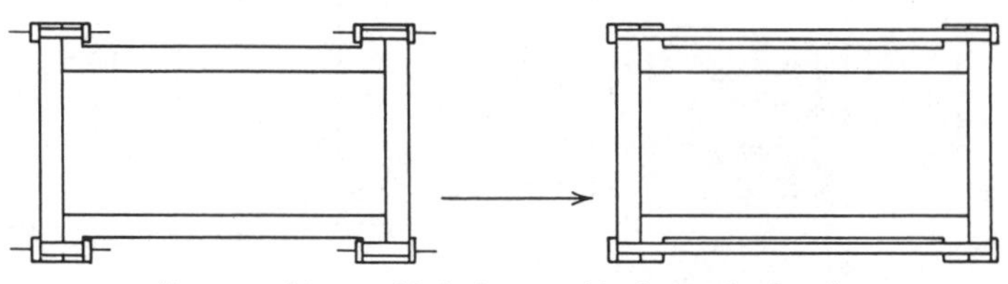

Fig. 10.56 Bolt design for energy absorption.

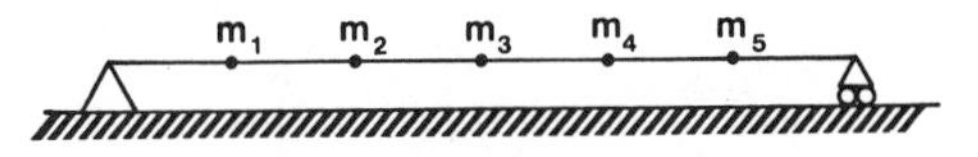

Fig. 10.57 Uncoupled vibrations.

addition, in various threaded parts the bolt is often reduced in diameter so the energy absorption will not be concentrated at the roots of threads at the end of the bolt (see Fig. 10.56).

Vibration effects. When vibrations are observed or measured it is often difficult to determine exactly what is happening. The motions can be coupled or decoupled vibrations. In uncoupled vibrations (Fig. 10.57), the lumped masses (18, 83) are located on the elastic axis, which is the neutral axis or shear center of a beam-like structure. Vibrations in the axial, torsional, and bending motions can be excited separately.

Coupled vibrations in Figs. 10.58 and 10.59 (2, 18), on the other hand, will allow motion or excitation from one direction to the other. In Fig. 10.58, when the torque T is applied, m_1 is displaced y_1, and the plate (represented by I_2) is rotated by θ_2 or rotating and translating coupled motion. Also, in Fig. 10.59 when F is applied to m_1, the rest of the masses go through bending, translating, and rotating motion.

When vibrations problems are complicated, such as plates and beams in

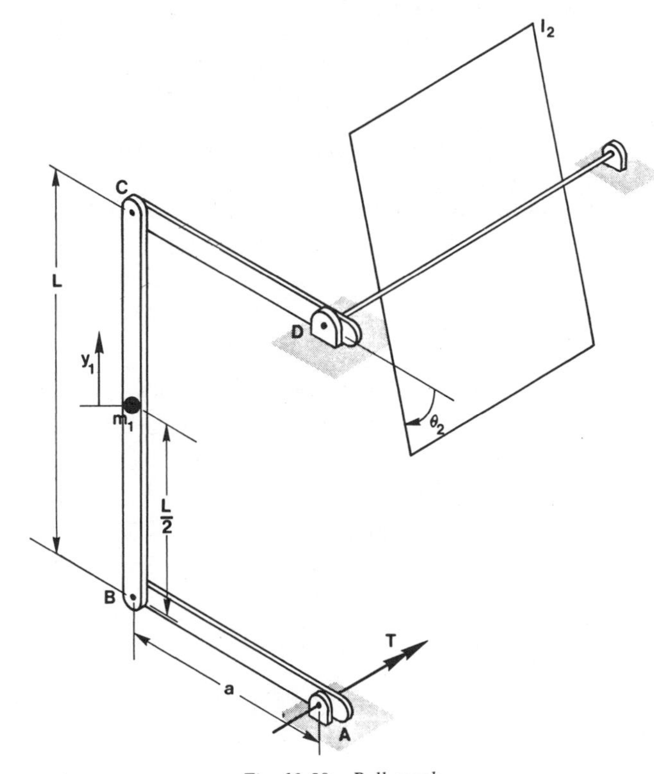

Fig. 10.58 Bell crank.

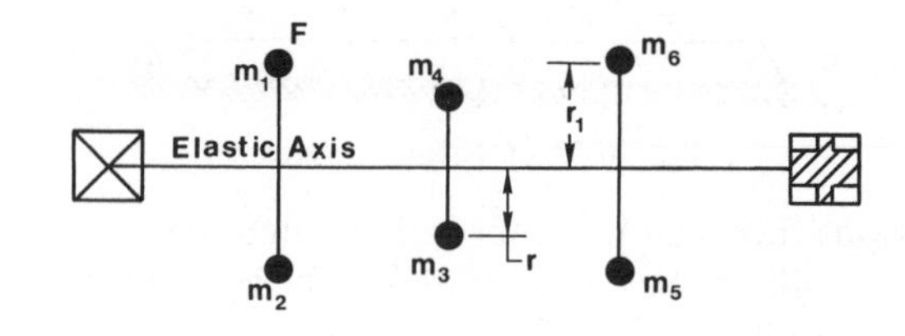

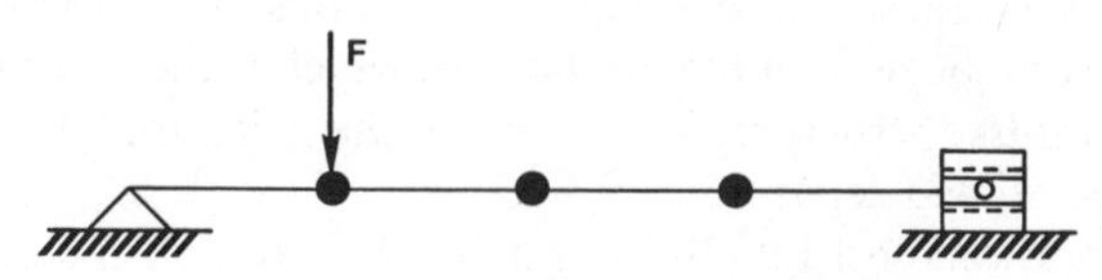

Fig. 10.59 Torsion-bending model.

castings and welded structures, normal modes (37, 68, 83) and finite element modeling are used to obtain uncoupled modes of vibration. In Chapters 7 and 8 methods are developed to find deflections and stresses in a static loading condition. Inertial terms for vibration can be substituted into Eqs. 7.330 and 7.325 for the stiffness and flexibility influence coefficients.

$$R_i = K_{ij} r_j \qquad 10.193$$

$$r_i = C_{ij} R_j \qquad 10.194$$

R_i and R_j are the same terms and represent either linear or rotational inertia terms of the form $-m\ddot{x}$ or $-I\ddot{\theta}$.

When only linear terms are present (37, 63), Eqs. 10.193 and 10.194 reduce to equations in one direction. The development for normal modes begins with

$$-m_i \ddot{x}_i = K_{ij} x_j \qquad 10.195$$

$$x_i = C_{ij}(-m_j \ddot{x}_j) \qquad 10.196$$

Each total displacement x_i is a product of a mode shape u_{ia} and a time term q_a. In order to obtain solutions for these terms the following equations are used.

$$x_i = u_i q \qquad 10.197$$

where

$$q = \sin(\omega t + \beta) \qquad 10.198$$

Therefore,

$$\ddot{x}_i = -u_i \omega^2 q \qquad 10.199$$

Substituting Eqs. 10.197 and 10.199 into Eqs. 10.195 and 10.196 yields

$$-m_i(-u_i \omega^2 q) = K_{ij}(u_j q)$$

And for Eq. 10.196

$$u_i q = -C_{ij} m_j(-u_i \omega^2 q)$$

Rearranging

$$q(K_{ij}u_j - \omega^2 m_i u_i) = 0 \qquad 10.200$$

$$q(\omega^2 C_{ij} m_j u_j - u_i) = 0 \qquad 10.201$$

The term q is Eq. 10.198 and is not zero and can be divided out of Eqs. 10.200 and 10.201. Equation 10.197 acts as $e^{-\lambda t}$ in a differential equation where solutions for ω are obtained intead of λ values. Equations 10.200 and 10.201 represent linear vibrations using the stiffness and flexibility influence coefficients. They can also be written as

$$[K_{ij} - \omega^2 m_i][u_i] = 0 \qquad 10.202$$

$$[\omega^2 C_{ij} m_j - 1][u_j] = 0 \qquad 10.203$$

Again, the u's are not zero so the frequency determinants are

$$[K_{ij} - \omega^2 m_i] = 0 \qquad 10.204$$

$$[\omega^2 C_{ij} m_j - 1] = 0 \qquad 10.205$$

Expanding 10.204 and 10.205 for three coordinates

$$\begin{bmatrix} (K_{11} - \omega_2 m_1) & K_{12} & K_{13} \\ K_{21} & (K_{22} - \omega^2 m_2) & K_{23} \\ K_{31} & K_{32} & (K_{33} - \omega^2 m_3) \end{bmatrix} = 0 \qquad 10.206$$

$$\begin{bmatrix} (\omega^2 C_{11} m_1 - 1) & C_{12} m_2 & C_{13} m_3 \\ C_{21} m_1 & (\omega^2 C_{22} m_2 - 1) & C_{23} m_3 \\ C_{31} m_1 & C_{32} m_2 & (\omega^2 C_{33} m_3 - 1) \end{bmatrix} = 0 \qquad 10.207$$

Three ω values, ω_1, ω_2, and ω_3 can be obtained from each of Eqs. 10.206 and 10.207, and they will be the same values. The first two ω's will be good values; however, the third will be suspect. A good rule is always have more than twice the number of coordinates as reliable frequencies desired. For Eqs. 10.206 and 10.207, six or more coordinates are needed to obtain three reliable frequencies. However, in finite element models this seldom appears as a problem since most of the elements are small.

When values for the ω's are obtained from Eqs. 10.206 or 10.207, they are substituted into Eqs. 10.202 or 10.203 to obtain the mode shapes or deflection pattern for each of the natural frequencies ω_1, ω_2, and ω_3. The ω's and u_i's can be iterated from Eqs. 10.202 and 10.203; however, sweeping matrices (37, 83) must be used. When more than three frequencies are desired, the effort mounts to hand calculate the C_{ij}'s or K_{ij}'s as well as to obtain mode shapes and frequencies.

When mode shapes and frequencies are obtained, then Eq. 10.197 becomes

$$x_i = \sum_a u_{ia} q_a \qquad 10.208$$

With frequencies, ω_a for each q_a and u_{ia} is normalized. A generalized mass and stiffness can be developed (2, 37, 63) from orthogonality and the known mode shapes

$$M_a = \sum_a m_i u_{ia}^2 \qquad 10.209$$

The sum of all the generalized masses is the total mass in that direction. The generalized stiffness is

$$K_{ka} = \omega_a^2 M_a \qquad 10.210$$

The basic equation for q_a is then

$$M_a \ddot{q}_a + \omega_a^2 M_a q_a = 0 \qquad 10.211$$

which is a spring mass system. The normal coordinate approach has reduced a system, coupled or uncoupled, into a simpler system of independent spring mass oscillators. These can be forced independently (2, 37, 68) through the mode shapes as a force is applied to each coordinate with an x_i. When many coordinates are involved, only a computer can handle the calculations.

One form of forcing commonly used is data in the form of power spectral density plots. These are used as a means (29, Vol. III) to place an envelope around all known test data and, hence, insure a safe new design. The response of a lightly damped single spring mass system to a white noise power spectral density can be represented in three ways from the force, displacement, and acceleration.

Force Φ_{ff}(kip^2 – sec/rad) is (20):

$$|Y(t)|^2 = \frac{\omega \Phi_{ff}(\omega)}{2\zeta k^2} \qquad 10.212$$

where

ω = natural frequency (rad/sec)
k = spring stiffness (kip/in.)
ζ = critical damping (%)

Displacement f (in.^{2}-sec/rad) is (37):

$$|Y(t)|^2 = \frac{\omega f(\omega)}{8\zeta} \qquad 10.213$$

Acceleration W (g^2-sec/rad) is (12, 29, Vol. I):

$$|Y(t)|^2 = \frac{(386)^2 \pi W(\omega)}{4\zeta\omega^3} = \frac{(386)^2 W(f)}{8\zeta(2\pi f)^3} \qquad 10.214$$

where

f = natural frequency (cps or hertz)

The displacement $|Y(t)|^2$ is the mean square response and the peak response (70)

$$C = \sqrt{2|Y(t)|^2} \qquad 10.215$$

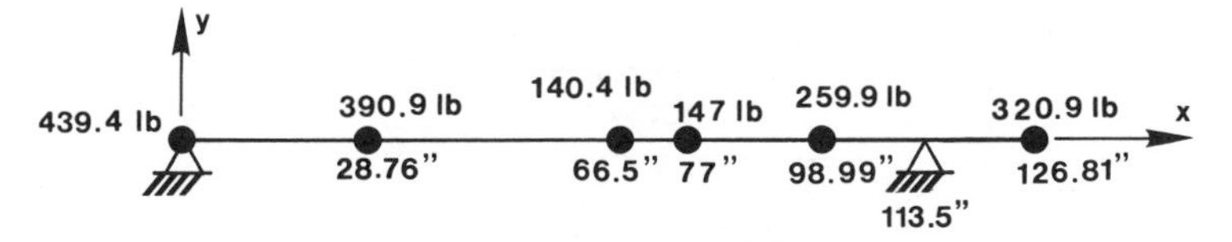

Fig. 10.60 Lumped mass beam.

Equations 10.212–10.215 can be used to estimate a single modal response to known power spectral density data provided the natural frequencies are not too close together. For example, in Eq. 10.211 for ω_1, k is $\omega_1{}^2 M_1$, and ζ is the percent of critical damping in the mode or the system damping for estimates. The value $|Y(t)|^2$ represents $|q_1(t)|^2$.

In Fig. 10.60 a lumped mass system is shown, and Eq. 10.196 with the calculated C_{ij}'s and the m_i matrix is presented. The mode shapes and frequencies are to be found for bending and in the longitudinal direction. The bending equation is

$$\begin{bmatrix} y_1 \\ y_2 \\ y_3 \\ y_4 \\ y_5 \end{bmatrix} = \frac{\omega^2 \times 10^6}{3EI(113.5)} \begin{bmatrix} 5.940 & 6.655 & 5.628 & 2.926 & -2.308 \\ 6.655 & 9.769 & 8.893 & 5.127 & -3.745 \\ 5.628 & 8.893 & 7.899 & 4.982 & -3.568 \\ 2.926 & 5.127 & 4.982 & 1.962 & -2.031 \\ -2.308 & -3.745 & -3.568 & -2.031 & 2.584 \end{bmatrix} \begin{bmatrix} 390.0 & 0 & 0 & 0 & 0 \\ 0 & 140.4 & 0 & 0 & 0 \\ 0 & 0 & 147 & 0 & 0 \\ 0 & 0 & 0 & 259.9 & 0 \\ 0 & 0 & 0 & 0 & 320.9 \end{bmatrix} \begin{bmatrix} y_1 \\ y_2 \\ y_3 \\ y_4 \\ y_5 \end{bmatrix} \tag{10.216}$$

where EI is 50×10^8 lb in^2. The bending mode shapes and frequencies are shown in Fig. 10.61. The modal weights and percentage in each mode in the y direction are

$$W_1 = 529 \text{ lb } (42\%) \quad W_2 = 204 \text{ lb } (16\%)$$
$$W_3 = 440 \text{ lb } (35\%) \quad \sum W_a = 1259.1 \text{ lb}$$

$$\begin{bmatrix} x_1 \\ x_2 \\ x_3 \\ x_4 \\ x_5 \end{bmatrix} = \frac{\omega^2 \times 10^2}{EA(113.5)} \begin{bmatrix} 24.37 & 13.52 & 10.50 & 4.17 & 0 \\ 13.52 & 31.26 & 24.27 & 9.64 & 0 \\ 10.50 & 24.27 & 28.11 & 11.17 & 0 \\ 4.17 & 9.64 & 11.17 & 14.36 & 0 \\ 0 & 0 & 0 & 0 & 21.00 \end{bmatrix} \begin{bmatrix} 390.9 & 0 & 0 & 0 & 0 \\ 0 & 140.4 & 0 & 0 & 0 \\ 0 & 0 & 147 & 0 & 0 \\ 0 & 0 & 0 & 259.9 & 0 \\ 0 & 0 & 0 & 0 & 320.9 \end{bmatrix} \begin{bmatrix} x_1 \\ x_2 \\ x_3 \\ x_4 \\ x_5 \end{bmatrix} \tag{10.217}$$

where EA is 10^8 lb. The longitudinal mode shapes and frequencies are shown in Fig. 10.62. The modal weights and percentage in each mode in the x direction are

$$W_1 = 870 \text{ lb } (69\%)$$
$$W_2 = 320.9 \text{ lb } (25\%)$$
$$\sum W_a = 1259.1 \text{ lb}$$

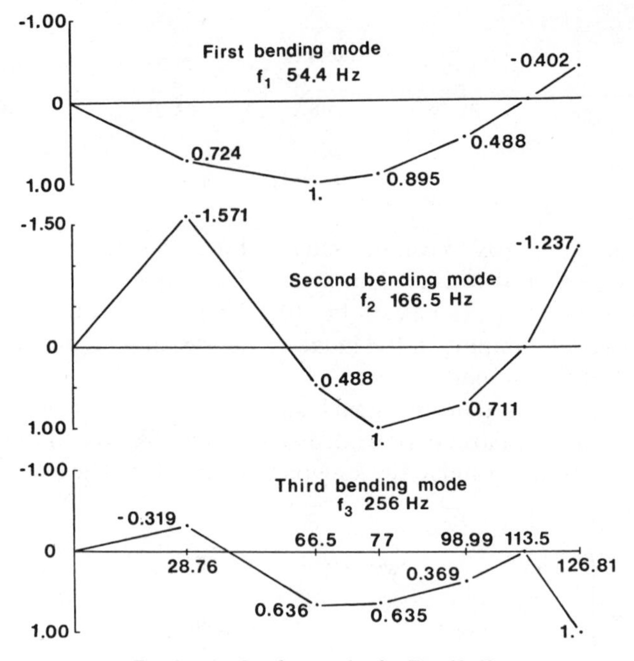

Fig. 10.61 Bending modes for Fig. 10.60.

When designing a system there are some guidelines to follow which will prevent vibration problems in the actual operating environment.

1. Draw free-body diagrams of the load path of the heaviest members from their mounting brackets through the supporting members down to whatever the system is attached to.

2. Make numerical estimates of GJ, GA, EI, EA, and make sure the values

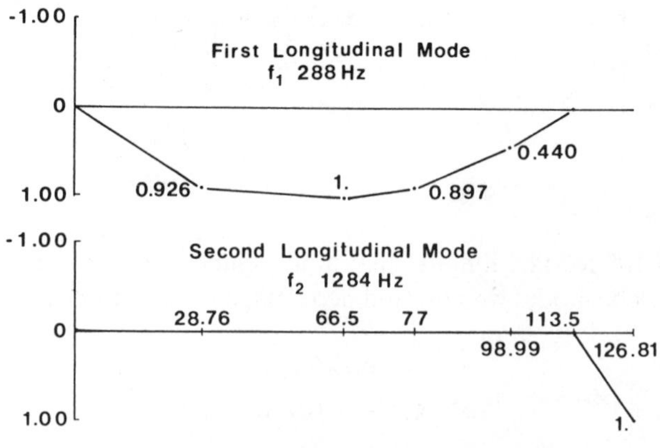

Fig. 10.62 Longitudinal modes for Fig. 10.60.

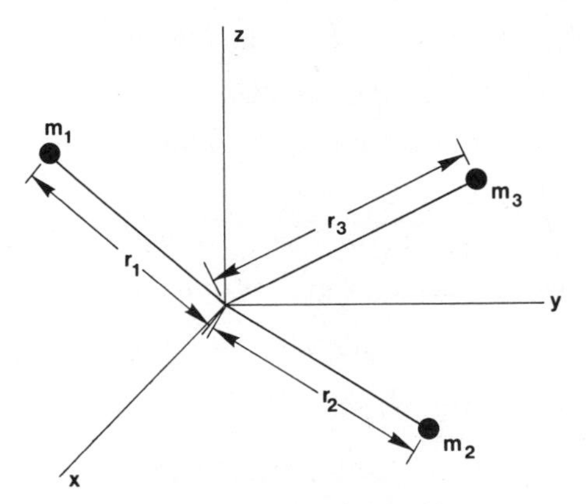

Fig. 10.63 Lumped mass rigid body.

do not suddenly drop. Also, be aware of the E and G available from different materials.

3. Whenever possible, do not place equipment that has unbalanced moving parts on cantilever structures or brackets.

4. Whenever possible, select closed cross sections as supporting members and never expect a slitted cross section to carry the same loads as a closed one.

5. Know the environment a design is to experience – say 200 Hz and less. While natural frequencies are difficult to predict for a yet-to-be designed system, much information is available on frequencies of plates, beams, and so on. If a system frequency of 200 Hz is desired, make the beams, plates, and other members have individual frequencies of 400 Hz or better. Then when the components are assembled, the lowest natural frequency will fall well below 400 Hz and, hopefully, not below 200 Hz.

6. In a detailed model, physical members then can be made thinner more easily than material can be added.

Information on component frequencies can be found in Timoshenko, Young, and Weaver (83) and Den Hartog (18). Harris and Crede (29, 30) discuss component frequencies and forcing inputs. Nowacki (62) gives overall discussion to shells, plates, frames, and continuous beams.

Rigid body effects. Rigid body motion (26, 79) introduces problems such as forcing and unbalance to a system. A system of bodies and particles making up a rigid body (Fig. 10.63) has a coordinate system placed at its center of gravity or a point fixed in space on the body. The coordinates move with the rigid body. The moment equations for the x, y, z body coordinate system are

$$M_x = I_{xx}\dot{\omega}_x + I_{xy}(\dot{\omega}_y - \omega_x\omega_z) + I_{xz}(\dot{\omega}_z + \omega_x\omega_y) + (I_{zz} - I_{yy})\omega_y\omega_z + I_{yz}(\omega_y^2 - \omega_z^2) \qquad 10.218$$

$$M_y = I_{xy}(\dot{\omega}_x + \omega_y\omega_z) + I_{yy}\dot{\omega}_y + I_{yz}(\dot{\omega}_z - \omega_x\omega_y) + (I_{xx} - I_{zz})\omega_x\omega_z + I_{xz}(\omega_z^2 - \omega_x^2) \qquad 10.219$$

$$M_z = I_{xz}(\dot{\omega}_x - \omega_y\omega_z) + I_{yz}(\dot{\omega}_y + \omega_x\omega_z) + I_{zz}\dot{\omega}_z + (I_{yy} - I_{xx})\omega_x\omega_y + I_{xy}(\omega_x^2 - \omega_y^2) \qquad 10.220$$

The equations simplify if the system is balanced to a set of desired coordinates to give principal axis at the center of gravity or if the principal axis is selected.

$$M_x = I_{xx}\dot{\omega}_x + (I_{zz} - I_{yy})\omega_y\omega_z \qquad 10.221$$

$$M_y = I_{yy}\dot{\omega}_x + (I_{xx} - I_{zz})\omega_z\omega_x \qquad 10.222$$

$$M_{zz} = I_{zz}\dot{\omega}_z + (I_{yy} - I_{xx})\omega_x\omega_y \qquad 10.223$$

In addition to Eqs. 10.221–10.223 forces are applied at the mass center to maintain equilibrium. The moments of inertia for a volume and point masses are defined as

$$I_{xx} = \int (y^2 + z)^2 dm = \sum_i (y_i^2 + z_i^2)m_i \qquad 10.224$$

$$I_{yy} = \int (x^2 + z^2)dm = \sum_i (x_i^2 + z_i^2)m_i \qquad 10.225$$

$$I_{zz} = \int (x^2 + y^2)dm = \sum_i (x_i^2 + y_i^2)m_i \qquad 10.226$$

$$I_{xy} = I_{yx} = -\int xydm = -\sum x_iy_im_i \qquad 10.227$$

$$I_{xz} = I_{zx} = -\int xzdm = -\sum x_iz_im_i \qquad 10.228$$

$$I_{yz} = I_{zy} = -\int yzdm = -\sum y_iz_im_i \qquad 10.229$$

The parallel axis theorem also applies to rigid bodies. If a rigid body is part of a system, its body inertia could be a significant portion of the transformed inertia.

Greenwood (26) develops ω_x, ω_y, and ω_z in terms of pitch, roll, and yaw Eulerian angles. These angles describe the physical behavior of many vehicles and systems that can be observed and experienced while operating them. Eulerian angles are a difficult concept to comprehend, and each system studied should be compared to examples in (26, 79).

Den Hartog (18) discusses the effects of cantilever mounted flywheels and motors while Timoshenko (83, 3rd. ed.) discusses the effects of disks on the critical speed of shafts. For the effects of disks and rigid bodies on bearing loads and system behavior see the discussions in (1, 75).

REFERENCES

1. R. N. Arnold and L. Maunder, *Gyrodynamics*, Academic Press, New York, N.Y. (1961).
2. R. L. Bisplinghoff, H. Ashley, and R. L. Halfman, *Aeroelasticity*, Addison-Wesley, Reading, Mass. (1955).
3. S. R. Bodner and P. S. Symonds, "Plastic Deformations in Impact and Impulsive Loading of Beams," in *Plasticity*, 488–500, Pergamon Press, New York, N.Y. (1960).
4. W. R. Campbell, *Proc. Soc. Exp. Stress Anal.*, **10** (1), 113–124 (1952).
5. D. S. Clark, D. H. Hyers, D. S. Wood, and P. E. Duwez, *Mechanics of the Dynamic Performance of Metals*, Report 36, NRC Project 82, National Defense Research Committee, Division 18, California Institute of Technology (September 1, 1944).
6. D. S. Clark, *Trans. Am. Soc. Met.*, **46**, 34–62 (1954).
7. D. S. Clark and D. S. Wood, *Trans. Am. Soc. Met.*, **42**, 45–74 (1950).
8. D. S. Clark and D. S. Wood, *Proc. Am. Soc. Test. Mater.*, **50**, 1–9 (1950).
9. D. S. Clark and D. S. Wood, *Proc. Am. Soc. Test. Mater.*, **49**, 717–737 (1949).
10. J. R. Cotner and J. Weertman, *Bibliography on High Speed Deformation of Materials*, Report AD-261376, Armed Services Technical Information Agency (May 15, 1961).
11. A. H. Cottrell, *Chart. Mech. Eng.*, **4** (9) 448–461 (November 1957).
12. S. H. Crandall and W. D. Mark, *Random Vibrations in Mechanical Systems*, Academic Press, New York, N.Y. (1963).
13. N. Davids, (Ed.), *Stress Wave Propagation in Materials*, Interscience Publishers, New York, N.Y. (1960).
14. J. F. Davidson, *J. Mech. Phys. Solids*, **2** (1), 54–66, October (1953).
15. R. M. Davies, *Appl. Mech. Rev.*, **6** (1), 1–3, (January 1953).
16. K. J. De Juhasz, *J. Franklin Inst.*, **248** (1), 15–48 (July 1949).
17. J. P. Den Hartog, *Advanced Strength of Materials*, 1st ed., McGraw-Hill Book Co., New York, N.Y. (1952).
18. J. P. Den Hartog, *Mechanical Vibrations*, McGraw-Hill New York, N.Y. (1956).
19. P. E. Duwez, and D. S. Clark, *Proc. Am. Soc. Test. Mater.*, **47**, 502–532 (1947).
20. D. G. Fertis, *Dynamics and Vibrations of Structures*, John Wiley and Sons, Inc., New York, N.Y. (1973).
21. F. E. Fisher, *The Motion of a Slender Beam Embedded in a Rotating Coordinate System*, 70–11, 635, University Microfilms, Ann Arbor, Mich. (1970).
22. T. H. Gawain, *Prod. Eng.*, **22** (7), 152–155 (July 1951).
23. G. Gerard and R. Papirno, *Trans. Am. Soc. Met.*, **49**, 132–148 (1957).
24. W. Goldsmith, *J. Appl. Mech.*, **20** (2), 307–308 (June 1953).
25. J. E. Greenspon, *J. Acoust. Soc. Am.*, **32** (5), 571–578 (May 1960).
26. D. T. Greenwood, *Principles of Dynamics*, Prentice-Hall, Englewood Cliffs, N.J. (1965).
27. K. Hanita, *Jap. Sci. Rev.*, **2** (1), 27–48 (1951).
28. W. A. Harper, *Prod. Eng.*, **30** (15), 62–63 (April 13, 1959). See also W. W. Wood, *Prod. Eng.*, **34** (20), 58–69 (September 30, 1963).
29. C. M. Harris and C. E. Crede (Eds.), *Shock and Vibration Handbook*, *Vols. I, II, III*, McGraw-Hill Book Co., New York, N.Y. (1961).
30. C. M. Harris and C. E. Crede (Eds.), *Shock and Vibration Handbook*, 2nd ed., McGraw-Hill Book Co., New York (1976).
31. P. R. Heaviside and J. F. Wallace, *Met. Treat. Drop Forg.*, **27** (178), 283–288 (July 1960).
32. J. Heyman, *Proc. Third U.S. Natl. Congr. Appl. Mech.*, 551–556, American Society of Mechanical Engineers, New York, N.Y. (1958).

33. J. H. Holloman, *The Problem of Fracture*, American Welding Society, New York, N.Y. (1946).

34. K. Honda, G. Takemae, and T. Watanabe, *Tohoku Imp. Univ. Sci. Rep.*, **19** (6), 703–725 (December 1930).

35. H. G. Hopkins, *Appl. Mech. Rev.*, **14** (6), 417–431 (June 1961).

36. W. H. Hoppmann, *Proc. Soc. Exp. Stress Anal.*, **10** (1), 157–164 (1952).

37. W. C. Hurty and M. F. Rubinstein, *Dynamics of Structures*, Prentice-Hall, Englewood Cliffs, N.J. (1964).

38. J. D. Jevons, *The Metallurgy of Deep Drawing and Pressing*, 2nd ed. John Wiley and Sons, Inc., New York, N.Y. (1942).

39. W. Johnson, *Impact Strength of Materials*, Edward Arnold, London (1972).

40. D. H. Kaelble, *J. Soc. Plast. Eng.*, **15** (12), 1071–1077 (December 1959).

41. H. F. Kohlbacker, *Ordnance*, **35** (186), 632–635 (May–June 1951).

42. H. Kolsky, *Stress Waves in Solids*, 2nd ed., Dover, New York, N.Y. (1963).

43. M. Kornhauser, *Structural Effects of Impact*, Spartan Books, Baltimore, Md. (1964).

44. S. Kumar and N. Davids, *J. Franklin Inst.*, **265** (5), 371–383 (May 1958).

45. S. Kumar and N. Davids, *J. Franklin Inst.*, **263** (4), 295–302 (April 1957).

46. D. J. Lapera, *Prod. Eng.*, **33** (7), 105–107 (April 2, 1962).

47. E. H. Lee and H. Wolf, *J. Appl. Mech.*, **18** (4), 379–386 (December 1951).

48. E. H. Lee, *Q. Appl. Math.*, **10** (34), 335–346, (January 1953).

49. E. H. Lee and D. R. Bland, *J. Soc. Plast. Eng.*, **11** (7), 28–35 (September 1955).

50. T. C. Lee, *J. Appl. Mech.*, **19** (3), 263–266 (September 1952).

51. J. A. Luker and S. A. Moiser, *Ind. Eng. Chem.*, **51** (4), 589–594 (April 1959).

52. K. W. Maier, *Prod. Eng.*, **25** (1), 162–167 (January 1954); **26** (3), 162–174 (March 1955); **28** (8), 167–174 (August 1957).

53. M. Manjoine and A. Nadai, *Proc. Am. Soc. Test. Mater.*, **40**, 822–837 (1940).

54. B. Maxwell and C. Guimon, *J. Appl. Polym. Sci.*, **6** (19), 83–93 (January–February 1962).

55. W. McGuire, *Steel Structures*, Prentice-Hall, Englewood Cliffs, N.J. (1968).

56. J. Miklowitz, *Appl. Mech. Rev.*, **13** (12), 865–878 (December 1960).

57. W. M. Murray, "Effects of Shock Loading," in *Design Work Sheets*, 12th series, X-65, 152–156, compiled by Product Engineering, McGraw-Hill Publishing Co., New York, N.Y.

58. A. Nadai, *Theory of Flow and Fracture of Solids*, McGraw-Hill Book Co., New York, N.Y. (1950).

59. N. M. Newmark, *Proc. Am. Soc. Civ. Eng.*, Paper No. 2786, **121**, pp. 45–64 (1956).

60. N. M. Newmark and E. Rosenblueth, *Fundamentals of Earthquake Engineering*, Prentice-Hall, Englewood Cliffs, N.J. (1971).

61. C. H. Norris et. al., *Structural Design for Dynamic Loads*, McGraw-Hill Book Co., New York, N.Y. (1959).

62. W. Nowacki, *Dynamics of Elastic Systems*, John Wiley and Sons, Inc., New York, N.Y. (1963).

63. G. J. O'Hara and P. F. Cunniff, *Elements of Normal Mode Theory*, NRL Report 6002, Naval Research Laboratory, Washington, D.C. (1963).

64. J. Parmakian, *Waterhammer Analysis*, Dover Publications, New York, N.Y. (1955).

65. B. Paul and M. Zaid, *J. Franklin Inst.*, **265** (4), 317–335 (April 1958).

66. J. Pearson and J. S. Rinehart, *J. Appl. Phys.*, **23** (4), 434–441 (April 1952).

67. J. Prescott, *Applied Elasticity*, Dover Publications, New York, N. Y. (1946).

68. J. S. Przemieniecki, *Theory of Matrix Structural Analysis*, McGraw-Hill Book Co., New York, N.Y. (1968).

69. P. N. Randall and I. Ginsburgh, *Trans. Am. Soc. Mech. Eng. (J. Basic Eng.)*, **83** (4), Series D, 519–528 (December 1961).
70. S. O. Rice, "Mathematical Analysis of Random Noise," in N. Wax, Ed., *Noise and Stochastic Processes*, Dover Publications, New York, N.Y. (1954).
71. J. S. Rinehart, *J. Appl. Phys.*, **22** (5), 550–560 (May 1951).
72. J. S. Rinehart and J. Pearson, *Behavior of Metals Under Impulsive Loads*, American Society for Metals, Novelty, Oh. (1954).
73. W. P. Roop, "Design for Energy Absorption," in *Properties of Metals in Materials Engineering*, 140–170, American Society for Metals, Novelty, Oh. (1949).
74. J. H. Rowe, *Prod. Eng.*, **28** (4), 211–215 (April 1957).
75. J. B. Scarborough, *The Gyroscope*, Interscience, New York, N.Y. (1958).
76. P. G. Skewmon and V. F. Zackay, Eds., *Response of Metals to High Velocity Deformation*, Vol. 9, Metallurgical Society Conferences, Interscience Publishers, New York, N.Y. (1961).
77. W. E. Skidmore, *Proc. Soc. Exp. Stress Anal.*, **8** (2), 29–48 (1951).
78. K. E. Spells, *Proc. Phys. Soc. (London)*, **64** (B), Part 3, 212–218 (March 1951).
79. S. Timoshenko and D. H. Young, *Advanced Dynamics*, McGraw-Hill Book Co., New York, N.Y. (1948).
80. S. Timoshenko, *Strength of Materials*, Part II, 3rd ed. D. Van Nostrand Co., Princeton, N.J. (1956).
81. S. Timoshenko, *Strength of Materials*, Part I, 2nd ed., D. Van Nostrand Co., Princeton, N.J. (1940).
82. S. Timoshenko and J. N. Goodier, *Theory of Elasticity*, 3rd ed., McGraw-Hill Book Co., New York, N.Y. (1970).
83. S. Timoshenko, D. H. Young, and W. Weaver, Jr., *Vibration Problems in Engineering*, 4th ed., D. Van Nostrand Co., Princeton, N.J. (1974).
84. P. W. Van Meter, *Des. News*, **17** (5), 60–61 (March 7, 1962).
85. E. Volterra, R. A. Eubanks, and D. Muster, *Proc. Soc. Exp. Stress Anal.*, **13** (1), 85–96 (1955).
86. J. M. Walsh, M. H. Rice, R. G. McQueen, and F. L. Yarger, *Phys. Rev.*, **108** (2), 196–216 (October 15, 1957).
87. H. J. Weiss and W. Prager, *J. Aeronaut. Sci.*, **21** (3), 196–200 (March 1954).
88. H. E. Wessman, and W. A. Rose, *Aerial Bombardment Protection*, John Wiley and Sons, Inc., New York, N.Y. (1942).
89. M. P. White, *J. Appl. Mech.*, **16** (1), 39–52 (March 1949).
90. J. Winlock, *Trans. Am. Inst. Min. Metal. Eng. (J. Met.)*, **197**, 797–803 (June 1953).
91. D. S. Wood, *J. Appl. Mech.*, **19** (4), 521–525 (December 1952).
92. C. E. Work and T. J. Dolan, *The Influence of Temperature and Strain Rate on the Properties of Metals in Torsion*, University of Illinois Engineering Experimental Station Bulletin Series 420, **51** (24) (November 1953).
93. M. Zaid, *J. Aeronaut. Sci.*, **19** (10), 697–704 (October 1952).
94. M. Zaid, *J. Aeronaut. Sci.*, **20** (6), 369–377 (June 1953).
95. M. Zaid and B. Paul, *J. Franklin Inst.*, **268** (1), 24–45 (July 1959).
96. M. Zaid and B. Paul, *J. Franklin Inst.*, **263** (4), 117–127 (April 1957).
97. *Symposium on Impact Testing*, Special Tech. Pub. No. 176, American Society Testing and Materials, Philadelphia, Pa. (1956).
98. *Machinery's Handbook*, 20th ed., The Industrial Press, New York, N.Y. (1975).
99. *Symposium on Speed of Testing of Non-Metallic Materials*, Special Tech. Pub. No. 185, American Society Testing and Materials, Philadephia, Pa. (1956).
100. *Steel*, **151** (6), 64–70 (August 6, 1962).

101. *Rocket Encyclopedia*, Aero. Publishers, Los Angeles, Cal. (1959).

102. *Structural Design for Dynamic Loads*, McGraw-Hill Book Co., New York, N.Y. (1959).

103. "Final Report on Concrete Penetration," *OSRD Report 6459* (March 1946).

104. "'The Initial Velocities of Fragments from Bombs, Shells and Grenades," Report 405, Ballistics Research Laboratory, Aberdeen, Md. (September 1943).

105. ASME Boiler and Pressure Vessel Code, Section VIII, American Society of Mechanical Engineers, New York, N.Y. (1977).

106. "Structures to Resist the Effects of Accidental Explosions," Departments of the Army, Navy, and Air Force, TM 5–1300 (June 1969).

107. "Fundamentals of Protective Design," U.S. Army Corps of Engineers (1946).

PROBLEMS

10.1 A $\frac{1}{2}$ in. thick steel disk with a 24-in. O. D. and a 1-in. I. D. is made from steel with a yield of 270 kpsi. Find the maximum angular velocity the disk can withstand and not yield. If the speed is increased by 50% for the same outside diameter and material, what modifications must be made to not exceed the yield point?

10.2 A cantilever 1.25-in. diameter steel rod has a weight W dropped from a height h on to the tip of the rod (Fig. P10.2). If the yield is 50 kpsi, what combinations of W and h are needed to yield the rod?

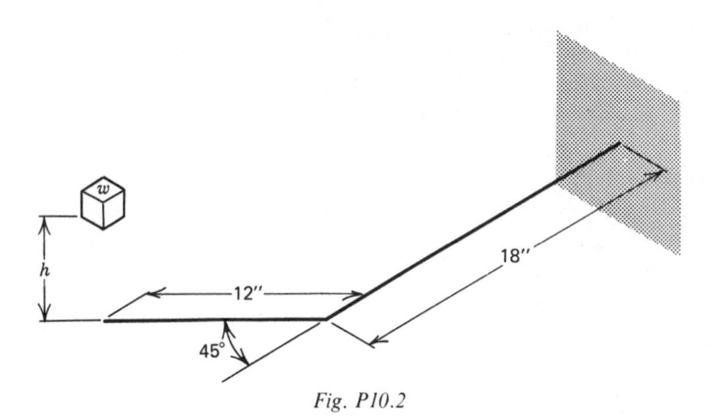

Fig. P10.2

10.3 A pair of 500-lb loads are applied at the top and bottom of the 0.25-in. diameter S hook. (See Fig. P10.3.) The loads are applied through rings that have 0.010-in. clearance with respect to the S hook. If the yield strength is 240 kpsi and the utimate strength 270 kspi, find the factor of safety before yielding.

10.4 A $\frac{1}{2}$-in. diameter steel rod supports two 40-lb weights (Fig. P10.4). Neglecting the weight of the rod, find the natural frequency and mode shapes in bending and in the longitudinal direction.

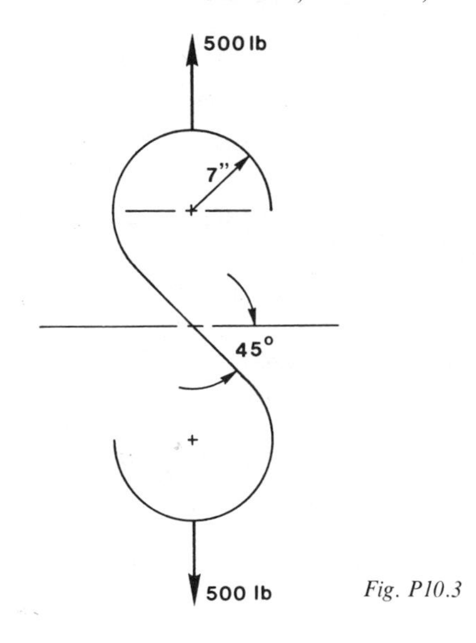

Fig. P10.3

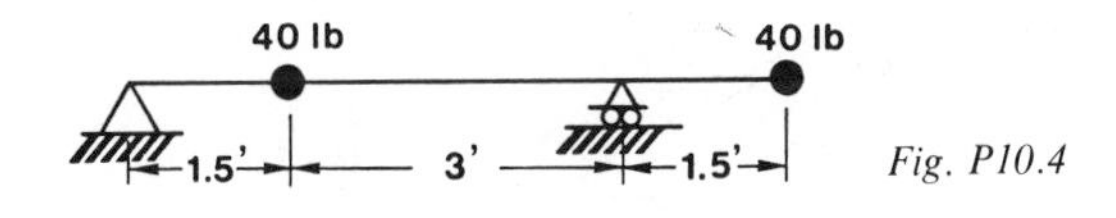

Fig. P10.4

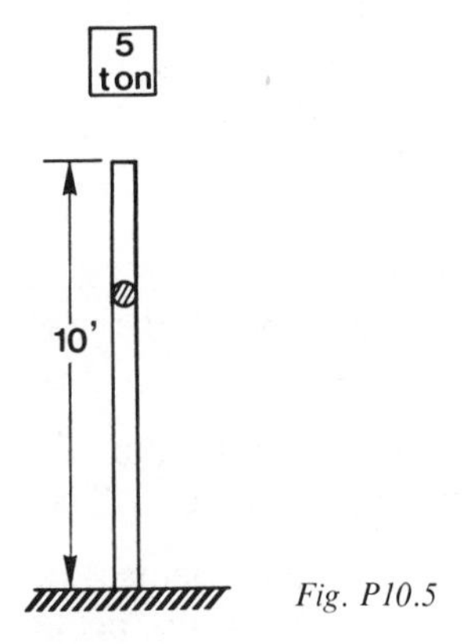

Fig. P10.5

10.5 A steel hammer weighing 5 tons strikes a round aluminum bar (Fig. P10.5). If the striking velocity is 5 fps, find the maximum stress and compare it with the approximate stress. Is there a limiting velocity from the physical set up? E is 10×10^6 lb/in.^{2}, and γ is 0.101 lb/in.^{3} for the aluminum.

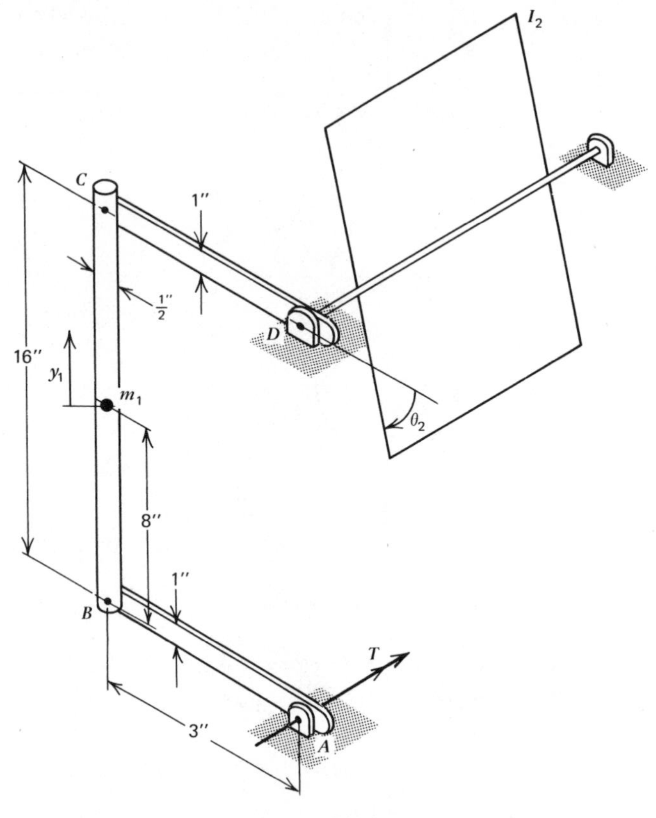

Fig. P10.8

10.6 A 1020 hollow steel shaft $2\frac{1}{2}$ in. O. D. and 2 in. I. D. is being considered for a 500-HP, 10,000 rpm-motor. If the shaft is 5 ft long, what is the factor of safety?

10.7 In Problem 10.5 the 5-ton load is connected to the column and receives a 3 g horizontal acceleration. What will be the resulting stresses?

10.8 Using flexibility influence coefficients (Eq. 10.194), find the two natural frequencies and modal shapes of the linkage system in Fig. P10.8 when input pin A is fixed. The moment of inertia I_2 for the flat plate is 0.080 in.-lbs.-sec.^{2} Links CD and AB are 0.625 in. thick. Lump all the mass for the aluminum linkage AB, BC, and CD at y_1.

10.9 A steel, 0.250-in. diameter beam AB is 48 in. long, and all the arms to the weights are 1/16-in. diameter steel (Fig. P10.9). Using Eq. 10.194, find the natural frequencies and mode shapes. For torsion consider pin A fixed and pin B free.

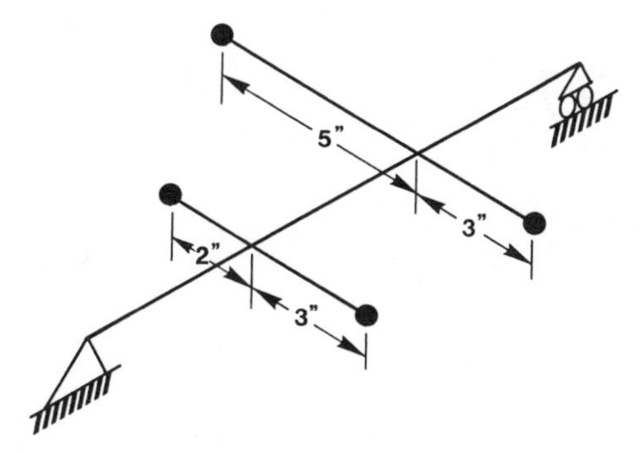

Fig. P10.9

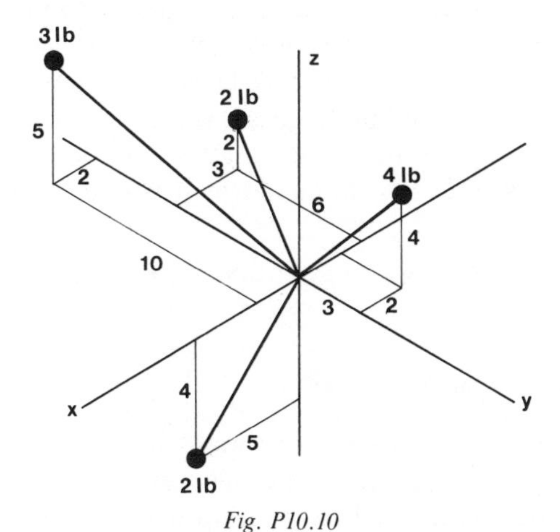

Fig. P10.10

10.10 A group of weights with a desired coordinate system is to be balanced (Fig. P10.10). First add weights to balance the moments about the y axis and force the center of gravity to be at the origin. Second, add weights to make the products of inertia zero. Compare the weights in the first and second case.

chapter 11

PRESTRESSING FOR STRENGTH

In this chapter the main interest is in stresses inherently present in a body regardless of conditions outside the boundaries of the body, and in stresses generated within a body that do not disappear on removal of the external load. The first kind of stresses are called *initial stresses* and are found in bodies stressed by their own weight, in rings that have an initial open gap that is closed by welding the two ends together, in shrink-fit construction, and in multiphase systems when an internal phase undergoes a volume change. In all these examples the material is always in the elastic range. The second type of stress, called *residual stress*, is always due to nonuniform plastic deformation (38).

11-1 THE NATURE AND SIGNIFICANCE OF INITIAL AND RESIDUAL STRESS

The origin of residual stresses

Residual stresses are induced whenever a material is nonuniformly plastically deformed. In such operations the permanent strain produced prevents the elastic component from completely recovering, and as a result a stress is generated within the material. Such deformation may be induced in a material by plastic bending (15, 26), as shown schematically in Fig. 11.1, by shot peening (1, 15), the effects of which are illustrated in Fig. 11.2, by autofrettage (7, 13), overstrain of rotating parts (6, 43), rolling and forming operations (10, 40, 43), and grinding (10, 12, 18). Phase transformations in the solid state may also induce plastic deformation and residual stresses (3, 14, 43).

Significance of residual stress

The significance of residual stress is that such stresses may be exceedingly harmful or, if properly controlled, may result in substantial benefits (7, 8, 35). Of the detrimental effects of residual stress the following items may be listed as typical.

1. The warping of parts subjected to machining operations. The magnitude of warping depends on the magnitude of the residual stresses present and their

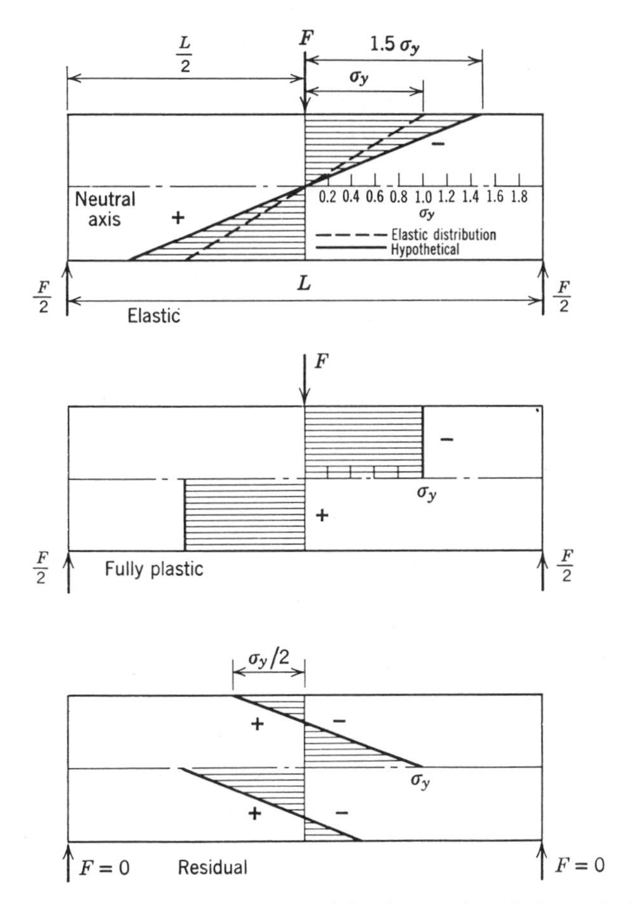

Fig. 11.1 Example of the development and distribution of residual stress in bending.

distribution. When the residual stresses are symmetrically distributed, and machining is uniform, warping may not take place. However, if nonuniform machining of such a part takes place or if nonuniformly distributed residual stress occurs, considerable warpage is sure to result.

2. Season cracking of drawn products (30, 43). Yellow brass is very susceptible to this type of failure, which is attributed to the corrosive action of the ammonia content of the atmosphere on the strained metal. Cracking of magnesium and aluminum base alloys, stainless steel, and in various carbon and alloy steels has also been observed but not to the extent that brass has been reported. Caustic embrittlement around rivets in boiler plate, cracking of gun shells, cracking of steel castings during pickling, and fire cracking of nickel, silver, bronze, and white gold alloys when rapidly heated are other examples. Brittle fracture has long been thought due to the presence of residual stress in a structure, but this may not be entirely true if the results obtained by Mylonas (25) are generally applicable. Mylonas found that loss of ductility in steel and attendant

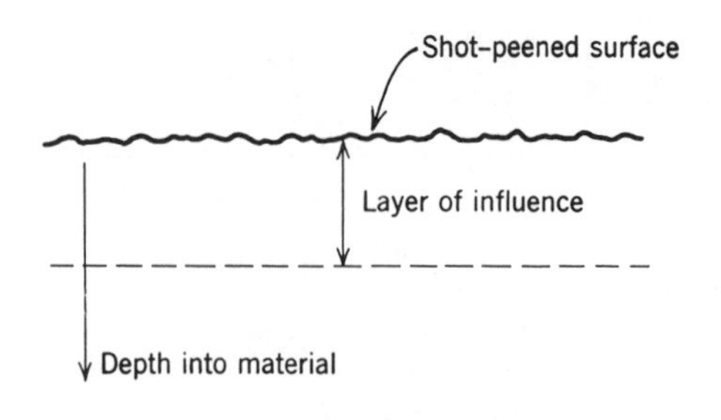

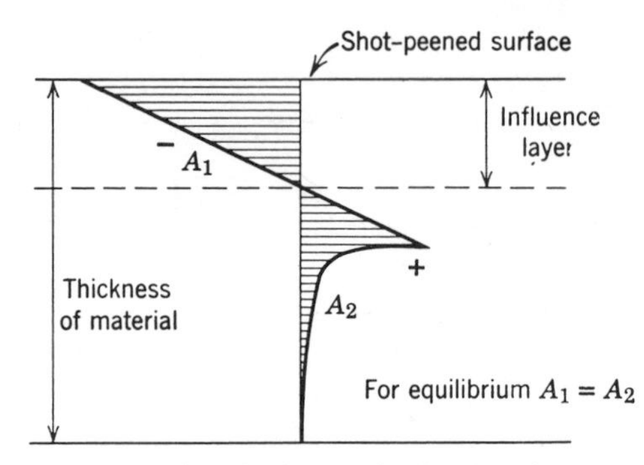

Fig. 11.2 Representation of residual stress developed in shot-peening a surface.

fracture could be imparted by prestrain alone; he found, for example, that notched plates with residual stress but no prestrain did not fracture in a brittle manner.

3. Materials that are quenched for the purpose of inducing hardness often crack following the quench (30, 43).

Fortunately, not all the effects of residual stress are deleterious. Following are typical examples of the beneficial effects of residual stresses, when these stresses are a result of controlled conditions.

1. Residual stresses induced in a material by plastic expansion are utilized in increasing the range of elastic operation of gun tubes and pressure vessels (7, 13, 43). The process involved in this application is called *autofrettage*. The object of such an operation is to induce residual stresses into the material which act in a direction opposite to that of the stresses set up by service loads. The elastic strength of rotating disks can also be enhanced by previous overloading which causes plastic deformation and results in beneficial residual stress.

2. Residual stresses induced by shrinking a collar on a cylinder (8, 20, 21) are utilized in improving the strength of guns, hydraulic cylinders, and extrusion dies. Many structural applications also take advantage of prestressed concrete. Liquid oxygen at − 197° F, for example, was successfully stored in prestressed concrete tanks for nearly ten years (46).

3. Residual stresses derived from such operations as shot peening (35, 43), surface rolling, etc., are reported to increase the fatigue life of machine parts.

4. Prestressed concrete is used extensively in building construction (19,45, 47). Prestressed concrete beams, for example, develop final compressive strengths around 4000 psi. Some beams are made by stringing concrete blocks on steel cables and then tensioning the cables to force out excess joint mortar. The beam is cured under tension for several days until the mortar reaches a compressive strength of around 2000 psi when final force is applied to the cables.

Control of residual stresses

The state of residual stress in a metal may be controlled by either mechanical or metallurgical means. In alloys susceptible to precipitation hardening there is opportunity to control residual stress to some extent by prestraining the metal before the precipitation-hardening treatment (43). Somewhat better control of residual stress is reported when the cold-working operation occurs between the quench and aging treatments. Quenching in oil instead of cold water also controls the magnitude of residual stress. In steel, considerable control of the magnitude and distribution of residual stress may be effected by quenching cylinders at the bore (9) rather than at the outside surface by combinations of surface-bore quenches, by controlling composition (4), or possibly by austempering or martempering (14). See also (23, 24, 32, 36, and 42).

Removal of residual stress

Residual stresses in a metal are usually removed or reduced in magnitude by various stress-relief treatments (15, 27). These treatments include both stress relief by annealing or relief of stress by deformation of the metal. Usually, an anneal at a high temperature followed by extremely slow cooling results in almost complete relief of residual stress. However, in practice, it is often necessary to reduce the magnitude of residual stress without seriously impairing the mechanical properties or resistance to corrosion of the material. Then, particularly in alloys hardened by precipitation, it is usually necessary to compromise between the mechanical properties affected by age hardening and the amount of stress relief by annealing.

11-2 MEASUREMENT OF RESIDUAL STRESS

X-ray diffraction method

The basic principle of the X-ray method is in the ability of X-rays to measure interatomic spacings. The atomic plane spacing acts as a gauge length, from which strain can be computed. These measured strains, at various angles, together with the principles of stress analysis developed in the theory of elasticity can be combined in a form to yield the magnitude and direction of the principal stresses. Uniaxial or biaxial states of stress only can be determined by this method, except where amplified by mechanical methods (11).

Mechanical methods

The various mechanical methods of residual stress analysis can be classified, generally, into two groups (41, 44).

1. Those in which a measurement is made of the amount of bending of a strip when it is cut from an object.

2. Those in which a measurement is made of the changes in dimensions of a body when material is removed from it.

For discussions and methods of analysis see (2, 17, 29–31, 41, 43, and 44).

11-3 PRACTICAL APPLICATION OF INITIAL AND RESIDUAL STRESSES

Shrink-fit construction

The fundamental basis of shrink-fit construction as in all types of applications where residual or initial stresses are purposely induced in the material is to utilize favorably distributed internal stresses rather than increased weight of material to resist service loads. In the following discussion two-shell construction will be treated first since it is widely used; following this, multishell construction will be examined.

In shrink-fit construction the intent may be to increase the elastic range of operation of a pressure vessel, to increase the fatigue life of a wheel and axle assembly, or perhaps to strengthen an extrusion die. In many instances a low-cost material, weak from the standpoint of inherent strength, can be used in a prestressed condition to accomplish the same results as a more expensive, higher strength material (8, 20, 21).*

Shrink-fit constructions are well suited to many different design situations, such as replacable liners in pressure vessels or strengthening of a liner by a shrunk-on shell that helps retard crack propagation. Its versatility is also demonstrated in assemblies composed of materials having different properties or degrees of corrosion resistance, such as a shrunk-on steel shell over a copper or aluminum liner, where maximum utilization is made of the properties of both materials.

In this discussion, equations for calculation of principal stresses and shrinkage allowance, in which the geometry of the assembly is arbitrarily determined and stresses calculated on the basis of this geometry and the properties of the

*The special case of finite interval of contact between tubes is considered in Reference 34.

materials, will be reviewed. In addition, however, methods will be given for determining the best (optimum) geometry so that maximum performance may be expected with minimum weight of material. Examples are also given of such design methods applied to two-shell, shrink-fit-type pressure vessels.

Basic theory. Fundamentals of shrink-fit theory are covered by analyses of heavy-walled cylinders subject to internal and external pressure. Fabrication procedures involved, lubrication of shrink-fit assemblies, preparation of contact surfaces, and similar manufacturing problems will not be considered here.

Cross sections of various types of shrink-fit construction, Fig. 11.3, are (*A*) hub and solid shaft under shrink-fit pressure only, (*B*) hub and hollow shaft under shrink-fit pressure only, (*C*) two-shell vessel with internal pressure and shrink-fit pressure, and (*D*) hub and shaft subjected to shrink-fit pressure and centrifugal forces. In all cases the component parts are not necessarily of the same material.

In producing a shrink-fit type of construction, it is common to heat the outer component in order to expand it beyond the interference on the diameter, and then slip it over the inner component while expanded. The temperature differential through which the outer part must be heated in order to obtain the required expansion is

$$t_f - t_i = \frac{D_f - D_i}{\alpha D_i} \tag{11.1}$$

where t_i and t_f are initial and final temperatures, D_i and D_f are initial and final diameters of the cylinder, and α is the coefficient of thermal expansion. To

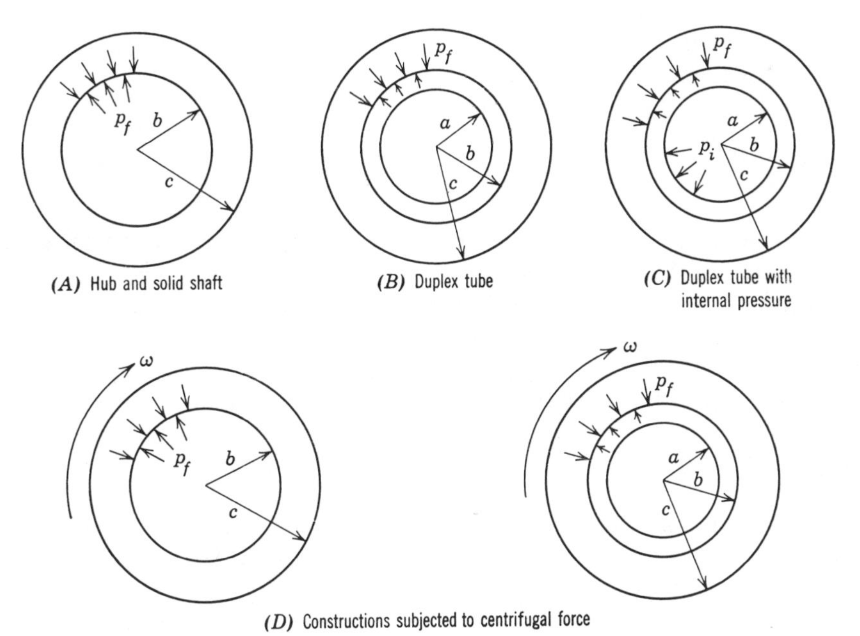

Fig. 11.3 Various types of shrink-fit construction.

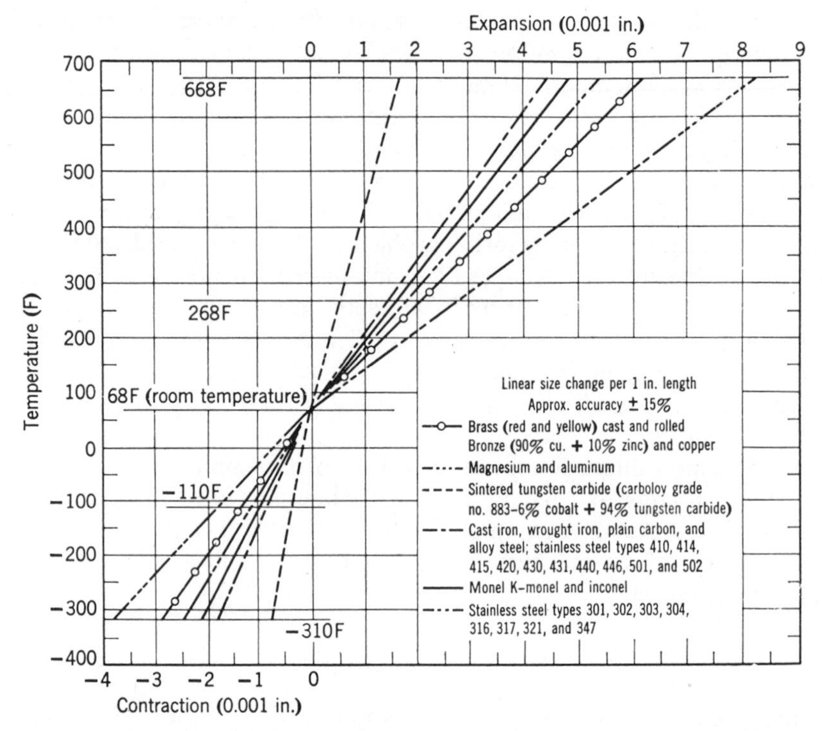

Fig. 11.4 Size change due to heating or cooling. [*Courtesy of the American Society of Tool Engineers; see (8).*]

increase the temperature differential or decrease the high temperature of the outer component, the inner component of the assembly may also be cooled so that it contracts. If cooling only is used, then instead of Eq. 11.1

$$-(t_i - t_f) = \frac{D_f - D_i}{\alpha D_i} \qquad 11.2$$

The approximate size change, per inch of length or diameter, which occurs when parts are heated or cooled is shown on Fig. 11.4. To determine size change for a given material on heating or cooling, start on Fig. 11.4 at the temperature specified, follow across the graph horizontally to the material in question, then follow the vertical lines to read the size change per inch of dimension. The size change per inch multiplied by the length of diameter of the part gives the total size change. For example, suppose a 6 in. diameter steel shaft is to be cooled from room temperature to $-310°F$. Start at $-310°F$, follow horizontally across to the line for steel, then down to read approximately -0.0018 in./in. Multiply this size change per inch by the diameter of the shaft. The result, $-0.0018 \times 6 = -0.0108$ in., which is the total size change cooled from room temperature. To determine the diameter of this shaft at $-310°F$, subtract the total size change from the diameter at room temperature. The result, $6.0 - 0.0108 = 5.9892$ in., is the diameter at the assembly temperature.

Figure 11.4 may also be used to determine the temperature to which a part must be heated or cooled to effect a specified size change. For example, to shrink a 5 in. diameter steel shaft 0.005 in. for assembly, divide the total shrink required by the diameter; the shrinkage must be $0.005/5 = 0.001$ in./in. of diameter. Using the curve in Fig. 11.4, start at -0.001 in., and follow up to the line for steel, then across to the temperature scale. The chart shows that the part must be cooled from room temperature to approximately $-150°$F.

For those materials not covered in Fig. 11.4, Eqs. 11.1 and 11.2 may be used. For example, assume a ring of 10 in. inside diameter of a material having a linear coefficient of thermal expansion of 0.000007 in./in./deg F. This ring must be heated from room temperature (68°F) to 500°F to make a shrink assembly. By Eq. 11.1,

$$
\begin{aligned}
D_f &= \alpha D_i(t_f - t_i) + D_i \\
&= 0.000007(10)(500 - 68) + 10 \\
&= 10.03024 \text{ in.}
\end{aligned}
$$

which is the inside diameter of the ring when heated.

Some physical and mechanical properties of common materials that may be encountered in making shrink-fits are listed in Table 1.3. In this table properties are included for annealed material, where it is assumed that no residual stresses are present that would influence shrink-fit calculations. Steels tempered to 1000°F or higher usually contain fairly low heat-treatment residual stresses. However, caution must be exercised in noting mechanical properties because, as shown on Fig. 11.5, properties vary with section size.* Consequently, for

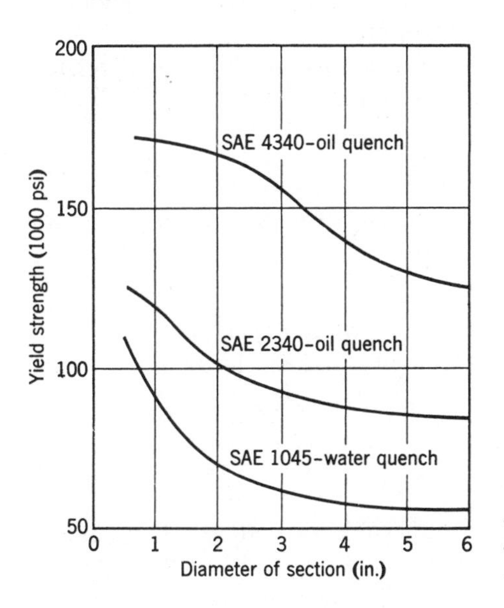

Fig. 11.5 Variation in yield strength with section size of quenched steels tempered at 1000°F.

*See also the discussion of this subject in Chapter 1.

heat-treated materials particularly, the properties noted should be in reference to the size of the part involved.

As a result of the shrinking operation, stresses are induced in the component parts from the shrink-fit pressure generated at the interface. As shown in Fig. 11.3, the effect is such that the outer member is subjected to internal pressure and the inner member to external pressure. The principal stresses for a cylinder under internal and external pressure are given by Eqs. 3.466 and 3.467.

$$\sigma_h = \frac{a^2 p_i - c^2 p_0}{c^2 - a^2} + \frac{(p_i - p_0) a^2 c^2}{r^2 (c^2 - a^2)} \qquad 11.3$$

and

$$\sigma_r = \frac{a^2 p_i - c^2 p_0}{c^2 - a^2} - \frac{(p_i - p_0) a^2 c^2}{r^2 (c^2 - a^2)} \qquad 11.4$$

For the inner component (identified by the subscript i), which is under external pressure, the principal stresses by Eqs. 11.3 and 11.4 become

$$\sigma_{hi} = \frac{-p_f b^2}{b^2 - a^2}\left(1 + \frac{a^2}{r^2}\right) \qquad 11.5$$

$$\sigma_{ri} = \frac{-p_f b^2}{b^2 - a^2}\left(1 - \frac{a^2}{r^2}\right) \qquad 11.6$$

If the inner component is a solid shaft, $a = 0$ and

$$\sigma_{hi} = \sigma_{ri} = -p_f \qquad 11.7$$

that is, the hoop and radial stresses are uniformly compressive throughout the shaft and equal in magnitude to the shrink-fit pressure.

The outer component (identified by the subscript o) is under internal pressure, and the principal stresses according to Eqs. 11.3 and 11.4 are

$$\sigma_{ho} = \frac{p_f b^2}{c^2 - b^2}\left(1 + \frac{c^2}{r^2}\right) \qquad 11.8$$

$$\sigma_{ro} = \frac{p_f b^2}{c^2 - b^2}\left(1 - \frac{c^2}{r^2}\right) \qquad 11.9$$

The various stress distributions obtained are shown in Fig. 11.6.

In order to calculate the shrinkage, the amount of radial deformation involved must be known, which is

$$u = \frac{1 - \nu}{E}\left(\frac{a^2 p_i - c^2 p_0}{c^2 - a^2}\right) r + \frac{1 + \nu}{E}\left(\frac{a^2 c^2}{r}\right)\left(\frac{p_i - p_0}{c^2 - a^2}\right) \qquad 11.10$$

For a composite cylinder, Eq. 11.10 is applicable to both parts; p_i and p_0 represent the shrink-fit pressure, for example, and b would be substituted for c for the inner component, and so on. The shrinkage allowance is the total deformation involved and is equal to the absolute sum of the individual de-

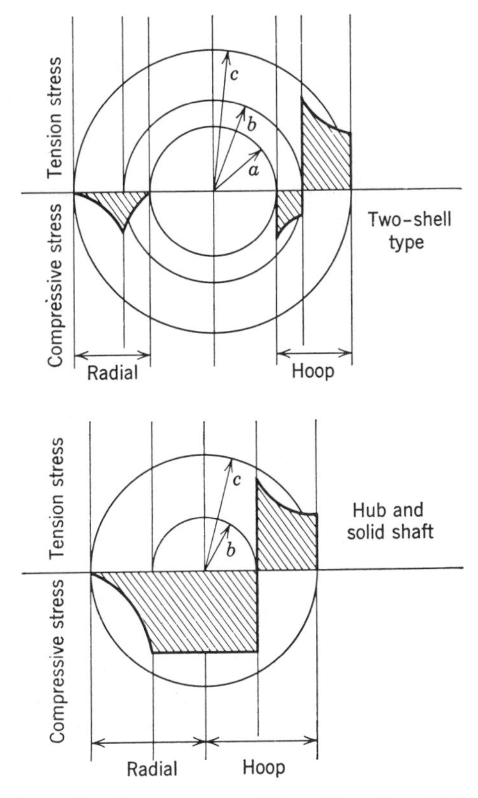

Fig. 11.6 Stress distributions in assemblies subjected to shrink-fit pressure.

formations of the inner and outer parts, that is, the radial shrinkage allowance Δ is

$$\Delta = |u_i + u_0| \tag{11.11}$$

Before considering practical design cases and calculations, a few cautions in using the shrink-fit type of construction should be mentioned. First, it is necessary to know exactly the initial stresses in the components before assembly. If high residual stresses due to heat treatment are present, for example, they must be taken into account in the design equations; otherwise the stresses and deformations obtained will not be in accordance with those calculated. Shrinkage temperatures over 500°F may have an annealing effect on certain steels tempered to a high hardness level. High thermal gradients at the instant of contact of the components may induce high thermal stress and accompanying plastic flow if the stresses are above the yield strength of the material. Freezing of shells on liners or shafts before complete assembly is also a frequently encountered inconvenience. Dimensional tolerances must be accurately controlled because stresses set up in shrink fitting are directly dependent on the accuracy of machining.

Design for shrink-fit pressure only. The simplest case to be considered is when shrink-fit pressure only is taken into account; the effects of internal pressure and centrifugal forces are not present. Four general situations will be discussed.

DIFFERENT MATERIALS—HOLLOW SHAFT. In the general case, the assembly is a two-part system, each part having different properties. The assembly is held together by a shrink-fit pressure, p_f. The outer component has a Poisson's ratio of ν_0 and an elastic modulus of E_0. Such an assembly might, for example, be a steel shell on a brass liner.

In accordance with Eqs. 11.3 and 11.4 the hoop and radial stresses can be computed. These equations, however, are not sufficient to define the value of the shrink-fit pressure; consequently, the deformations involved must be considered. By Eq. 11.10 the radial deformation for the inner cylinder is

$$u_{rb} = u_i = \frac{-p_f}{E_i}\left(\frac{a^2+b^2}{b^2-a^2} - \nu_i\right)b \qquad 11.12$$

Deformation for the outer cylinder is

$$u_{rb} = u_0 = \frac{p_f}{E_0}\left(\frac{b^2+c^2}{c^2-b^2} + \nu_0\right)b \qquad 11.13$$

The total shrinkage allowance according to Eq. 11.11 is the absolute sum of the individual deformations; thus, summing Eqs. 11.12 and 11.13,

$$\Delta = p_f b\left[\frac{1}{E_0}\left(\frac{b^2+c^2}{c^2-b^2} + \nu_0\right) + \frac{1}{E_i}\left(\frac{a^2+b^2}{b^2-a^2} - \nu_i\right)\right] = p_f bK \qquad 11.14$$

From Eq. 11.14 the shrink-fit pressure is

$$p_f = \Delta/bK \qquad 11.15$$

where

$$K = \frac{1}{E_0}\left(\frac{b^2+c^2}{c^2-b^2} + \nu_0\right) + \frac{1}{E_i}\left(\frac{a^2+b^2}{b^2-a^2} - \nu_i\right) \qquad 11.16$$

SAME MATERIALS—HOLLOW SHAFT. For the special case where the two components are of identical materials, Eqs. 11.14 and 11.15 become

$$\Delta = \frac{2p_f b^3(c^2-a^2)}{E(c^2-b^2)(b^2-a^2)} \qquad 11.17$$

and

$$p_f = \frac{\Delta E(c^2-b^2)(b^2-a^2)}{2b^3(c^2-a^2)} \qquad 11.18$$

DIFFERENT MATERIALS—SOLID SHAFT. If the inner component is a solid shaft, $a = 0$ and Eqs. 11.14 and 11.15 become

$$\Delta = p_f b\left[\frac{1}{E_0}\left(\frac{c^2+b^2}{c^2-b^2} + \nu_0\right) + \frac{1-\nu_i}{E_i}\right] \qquad 11.19$$

and

$$p_f = \frac{\Delta}{b\left[\frac{1}{E_0}\left(\frac{c^2+b^2}{c^2-b^2}+\nu_0\right)+\frac{1-\nu_i}{E_i}\right]} \qquad 11.20$$

SAME MATERIAL—SOLID SHAFT. If the solid shaft and hub are made of the same material Eqs. 11.17 and 11.18 become

$$\Delta = \frac{2p_f bc^2}{(c^2-b^2)E} \qquad 11.21$$

and

$$p_f = \frac{\Delta E(c^2-b^2)}{2bc^2} \qquad 11.22$$

Effect of centrifugal forces. Shrink-fit assemblies are often used in applications where high speeds are encountered; consequently, it is necessary in design to take this effect into account in shrink-fit calculations. The effect of rotation is to develop high centrifugal forces that tend to change the shrinkage between the two parts and thus influence the interface pressure; four cases follow.

SAME MATERIAL—HOLLOW SHAFT. The compound cylinder can be considered or composed of two parts, an inner component that develops a radial deformation at b of u_i and an outer component that develops a radial deformation at b of u_0. From Eqs. 10.108, 10.110, 10.111, and 10.112 these deformations are

$$u_i = \frac{N}{8}\left[-b^3 + b(b^2+a^2)\left(\frac{3+\nu}{1+\nu}\right) + a^2 b\left(\frac{3+\nu}{1-\nu}\right)\right] \qquad 11.23$$

and

$$u_0 = \frac{Nb}{4(1-\nu^2)}\left[b^2(1-\nu)+c^2(3+\nu)\right] \qquad 11.24$$

where

$$N = (1-\nu^2)(\gamma\omega^2/gE) \qquad 11.25$$

The change in shrinkage Δ' because of centrifugal action is

$$\Delta' = u_0 - u_i \qquad 11.26$$

which gives

$$\Delta' = \frac{b\gamma V^2}{4gE}(3+\nu)\left(1-\frac{a^2}{c^2}\right) \qquad 11.27$$

where V is the linear velocity at $r = c$, that is,

$$V = c\omega \qquad 11.28$$

DIFFERENT MATERIALS—HOLLOW SHAFT. When the elastic constants are different in the two parts,

$$\Delta' = \frac{V^2 b}{4g}\left\{\frac{\gamma_0}{E_0}\left[(3+\nu_0)+(1-\nu_0)\frac{b^2}{c^2}\right] - \frac{\gamma_i}{E_i}\left[(3+\nu_i)\frac{a^2}{c^2}+(1-\nu_i)\frac{b^2}{c^2}\right]\right\} \qquad 11.29$$

SAME MATERIAL—SOLID SHAFT. For the special case of a solid shaft, $a = 0$ and

$$\Delta' = \frac{V^2 b\gamma(3 + \nu)}{4gE} \qquad 11.30$$

DIFFERENT MATERIALS—SOLID SHAFT. For this type of construction

$$\Delta' = \frac{V^2 b}{4g}\left\{\frac{\gamma_0}{E_0}\left[(3 + \nu_0) + (1 - \nu_0)\frac{b^2}{c^2}\right] - \frac{\gamma_i}{E_i}(1 - \nu_i)\frac{b^2}{c^2}\right\} \qquad 11.31$$

Loosening of shrink-fit construction. A shrink-fit assembly made up of two hollow disks and subjected to centrifugal action is shown in Fig. 11.3*D*. In order to determine the conditions under which the two components would separate, the deformations involved must be considered. The radial deformations at the interface due to centrifugal action are given by Eqs. 11.23 and 11.24. The same deformations due to the shrink-fit pressure are given by Eqs. 11.12 and 11.13. The condition for loosening is thus

$$\text{Eqs. } 11.23 + 11.12 = \text{Eqs. } 11.24 + 11.13 \qquad 11.32$$

Axial holding ability. The axial force that a shrink-fit assembly can withstand without separation of the parts is

$$F = 2\pi L b p_f f \qquad 11.33$$

where F is the force, L the length of the assembly, and f is the coefficient of friction. For determining various degrees of axial holding ability (with f known) the value of the shrink-fit pressure in Eq. 11.33 can be replaced by its value as given by previous equations.

Torsional holding ability. The torque which a shrink-fit assembly can withstand without relative movement of the parts is

$$T = 2\pi L b^2 p_f f^1 \qquad 11.34$$

where f^1, the coefficient of friction in torsion, must usually be found experimentally (8, 39).

Elastic-breakdown of two-shell shrink-fit vessels. As a check on certain portions of the foregoing theory some tests were carried out at the Du Pont Company (8). Cylinders were fabricated having ratios of outside to bore radius of 2.75 and 4.00. Two cylinders were of monobloc-type construction and the other two were of the two-shell shrink-fit type. Mechanical properties and pertinent geometrical data concerning these cylinders are presented in Table 11.1. In testing the assemblies the ends were sealed, and curves of external surface hoop strain versus internal pressure were plotted. The initial portion of the pressure-strain curve is linear in accordance with elastic theory, but at the beginning of yielding of the cylinder, deviation from linearity occurs; this pressure is called the "elastic-breakdown pressure." This value can be predicted with fair accuracy both for monobloc and shrink-fit-type cylinders as shown in

Table 11.1 Properties and Geometry of Alloy Steel Test Cylinders

Property or Dimension	Monobloc 1	Monobloc 2	Shrink-Fit 3	Shrink-Fit 4
Inside diameter, in.	1.501	1.377	1.498	1.374
Outside diameter, in.	4.128	5.502	4.128	5.502
Interface diameter, in.	—	—	2.500	2.750
Over-all diameter ratio	2.75	4.00	2.756	4.010
Liner diameter ratio	—	—	1.670	2.000
Shrinkage, in./in. (radial)	—	—	0.00284	0.00349
Yield strength, shell, psi	125,000	113,000	125,000	125,000
Yield strength, liner, psi	—	—	103,000	103,000

NOTE: Cr–Ni–Mo–V steel quenched and tempered at 1100°F (see Reference 9).

Table 11.2. For the monobloc cylinders (Chapter 3),

$$p_i = \frac{\sigma_y}{\sqrt{3}}\left(\frac{R^2 - 1}{R^2}\right)$$

For the shrink-fit cylinders, as shown in Table 11.1, the shells and liners had different mechanical properties; consequently, the optimum shrink-fit interference was calculated. Having the required interference, the temperature differential required in order to expand the shell by the amount of the interference was calculated using Eq. 11.1. The pressure at elastic failure can be calculated.

Through utilization of the previous concepts design characteristics of the two types of construction can easily be compared. Characteristics of two-shell shrink-fit vessels under internal pressure when the materials and properties are the same for both parts will be used as one case, monobloc construction as the other. The elastic response of both monobloc and two-shell vessels under internal pressure is shown in Fig. 11.7. The ordinate can be considered as representing the elastic-breakdown pressure, whereas the abscissa represents the ratio of outside to bore radius (diameter ratio). The curve for the monobloc

Table 11.2 Comparison of Theoretical and Experimental Results for Test Cylinders

Construction	Elastic-Breakdown Pressure, psi		Deviation, %
	Observed	Calculated	
Monobloc 1	65,000	63,000	−4
Monobloc 2	65,000	71,000	+9
Shrink-fit 3	76,000	82,000	+8
Shrink-fit 4	85,000	96,000	+13

See Reference 8.

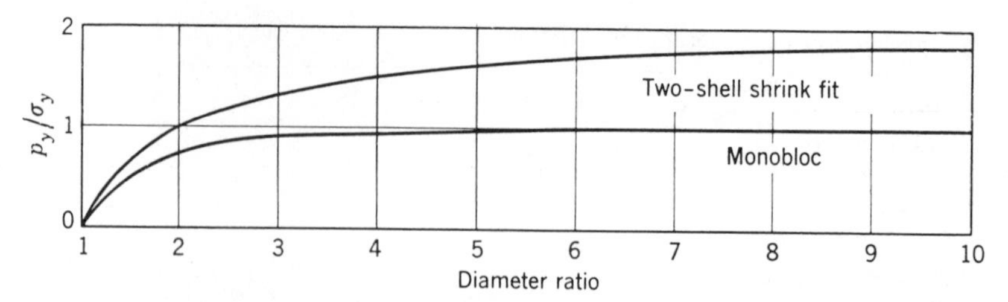

Fig. 11.7 Yield characteristics of heavy-walled cylinders.

cylinder shows that little, if any, gain can be expected in elastic-breakdown pressure by increasing the diameter ratio beyond about 3.50 or 4.00. By shrink-fitting, on the other hand, considerable advantage is gained, and for infinite diameter ratio the elastic-breakdown pressure can be doubled. More will be said about this later in the discussion concerning plastic-radial expansion (autofrettage). Thus it is seen that shrink-fit construction is always better than the monobloc and becomes twice as good at infinite diameter ratio. In order to determine the equivalence of the two types of construction, Fig. 11.8 shows the diameter ratios required for both types of vessels to give the same elastic-breakdown pressure. The latter curve shows, for example, that a monobloc vessel at diameter ratio 6.00 can be replaced by a two-shell shrink-fit type of diameter ratio 1.95. Similar curves can be constructed either for the case the materials are different or the mechanical properties of the two parts are different. For example, the elastic-breakdown pressure can be increased over one-third by increasing the strength of the outside shell from $0.5\sigma_i$ to $1.25\sigma_i$.

STRESSES IN PRESS-FIT BUSHINGS. The foregoing theory is readily adaptable to the special case of thin rings or bushings that are press or shrink-fitted into some kind of a holder. In Fig. 11.9, for example, the maximum tensile stress in the collar at interface b is obtained from Eqs. 11.8 and 11.13,

$$(\sigma_h)_b = E_0 \frac{\Delta}{b}\left(\frac{b^2+c^2}{c^2-b^2}\right)\left[\frac{1}{\dfrac{b^2+c^2}{c^2-b^2}+\nu_0+\dfrac{E_0}{E_i}\left(\dfrac{a^2+b^2}{b^2-a^2}-\nu_i\right)}\right] \tag{11.35}$$

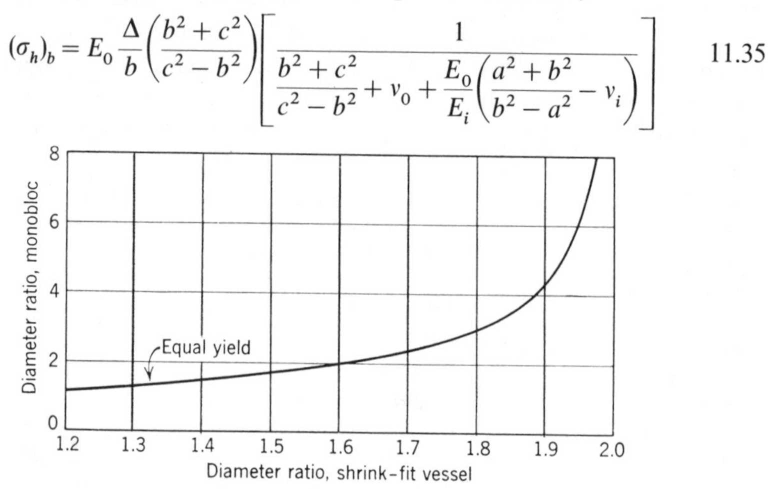

Fig. 11.8 Equivalence of monobloc and shrink-fit pressure vessels.

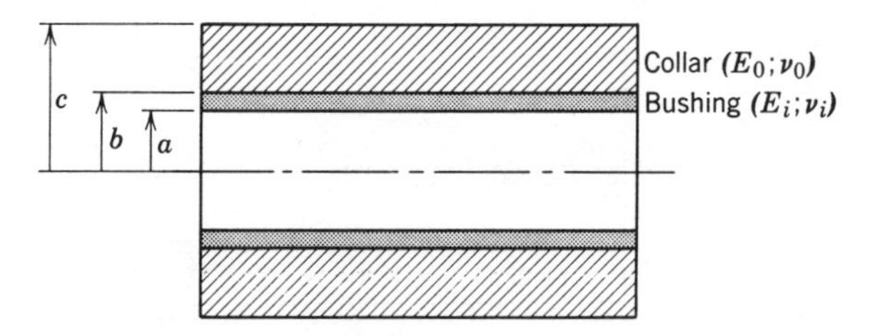

Fig. 11.9 Thin bushing with shrunk-on collar.

For more complicated situations such as shown in Figs. 11.10 and 11.11 similar design formulas have been worked out by Schuette (33) which are as follows.

For the construction shown in Fig. 11.10, the average effective hoop stress in the collar at interface b is

$$(\sigma_h)_b = \frac{\dfrac{E_0 \Delta}{b}\left[\dfrac{b^2}{c^2 - b^2} + \dfrac{(c+x)^2}{(c+x)^2 - b^2}\right]}{\dfrac{E_0}{E_i}\left(\dfrac{b^2 + a^2}{b^2 - a^2} - \nu_i\right) + \nu_0 + \dfrac{b^2}{c^2 - b^2} + \dfrac{(c+x)^2}{(c+x)^2 - b^2}} \qquad 11.36$$

where

$$x = \left(\frac{b + 2a}{3}\right)\left(\frac{d - c}{d - c + b}\right)\left(\frac{H}{H + h}\right)^{(b + 2H)/b}$$

For the construction shown in Fig. 11.11, the average effective hoop stress in the collar at interface b is

$$(\sigma_h)_b = \frac{\dfrac{E_0 \Delta}{b}\left[\dfrac{b^2}{(c+y)^2 - b^2} + \dfrac{(c+y)^2}{(c+y)^2 - b^2}\right]}{\dfrac{E_0}{E_i}\left(\dfrac{b^2 + a^2}{b^2 - a^2} - \nu_i\right) + \dfrac{b^2}{(c+y)^2 - b^2} + \dfrac{(c+y)^2}{(c+y)^2 - b^2} + \nu_0} \qquad 11.37$$

where

$$y = \tfrac{1}{2}(d - c)\left[1 - \frac{(H/b)^2}{(H/b)^2 + 2}\right]$$

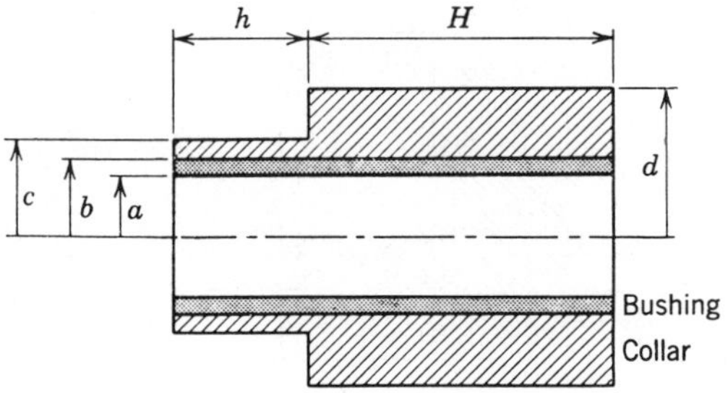

Fig. 11.10 Step-down shrink-fit construction.

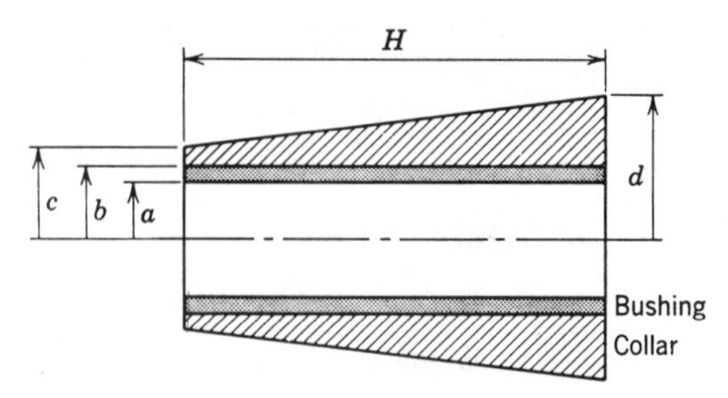

Fig. 11.11 Tapered shrink-fit construction.

Multishell shrink-fit construction. Consider the multilayer construction shown in Fig. 11.12 made up of several shells having different elastic properties. For any particular shell the diameter ratio is

$$\text{diameter ratio} = \frac{R_n}{R_{n-1}}; \quad \frac{R_{n+1}}{R_n}, \text{ etc.} \tag{11.38}$$

In practice it is common to assemble first the outer shells and then shrink the outside assembly on the inner shell or liner. The analysis for this assembly of shells follows that previously given; for example, for the nth shell, by Eqs. 11.3 and 11.4,

$$(\sigma_h)_n = \frac{1}{\left(\dfrac{R_n}{R_{n-1}}\right)^2 - 1}\left[p_{n-1} - p_n\left(\frac{R_n}{R_{n-1}}\right)^2 + (p_{n-1} - p_n)\left(\frac{R_n}{R}\right)^2\right] \tag{11.39}$$

and

$$(\sigma_r)_n = \frac{1}{\left(\dfrac{R_n}{R_{n-1}}\right)^2 - 1}\left[p_{n-1} - p_n\left(\frac{R_n}{R_{n-1}}\right)^2 - (p_{n-1} - p_n)\left(\frac{R_n}{R}\right)^2\right] \tag{11.40}$$

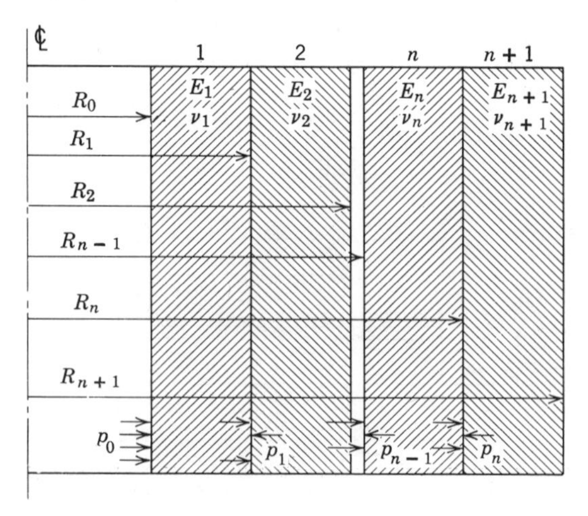

Fig. 11.12 Multilayer shrink-fit construction.

Between any two shells the pressure is $-\sigma_r$; therefore three pressures are present, the operating pressure, p_0, the radial stress at the interface, and the original shrink-fit pressure.

Wire and ribbon wound cylinders

There are several modifications of this type of vessel, the more prominent being the ribbon or wire-wrapped vessel and the interlocking ribbon wound vessel (also called *Wickelofen* wrapping). Both types of vessels provide a unique way of introducing favorable initial stresses into a vessel by building up on a base cylinder a shroud of winding that is relatively simple to apply. One of the principal advantages of a wound vessel is that very high strength filaments can be obtained in wire form, whereas the materials may not be available in massive form as would be required for shrink fitting. In addition, because of the laminar effect produced, crack propagation is arrested. On the other hand, there have been difficulties with such vessels in providing satisfactory end closures; also they do not support axial load. In the Wickelofen vessel the outside layers are built up of spiral-wound interlocking ribbons of material so that the axial load can be supported. Details of manufacture may be found in (6) of Chapter 9. For the purposes of this book some of the salient features of the analytical procedure applied to such vessels will be considered. Wound thin shells were discussed in Chapter 4; the following discussion relates to wound heavy-walled cylinders.

Consider Fig. 11.13 which shows a heavy-walled cylinder wound with wire at a constant wire tension T. As the wire is applied a compression is induced in the solid cylinder by an equivalent external pressure p_w, which varies depending on the amount of wrapping. Assuming a cylinder of radius b^1 the stresses

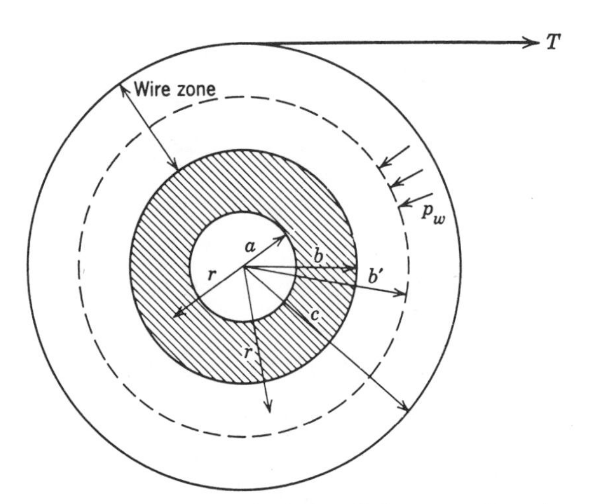

Fig. 11.13 Geometry of a thick-walled wire-wound vessel.

induced, by Eqs. 3.466 and 3.467 (with $p_i = 0$ and $p_0 = p_w$), are

$$\sigma_h = \frac{-p_w(b^1)^2}{a^2 - (b^1)^2}[(a/r)^2 + 1] \tag{11.41}$$

and

$$\sigma_r = \frac{p_w(b^1)^2}{a^2 - (b^1)^2}[(a/r)^2 - 1] \tag{11.42}$$

As each layer is applied the tension T in the underlying layer is reduced by the compressive force induced, thus

$$(\sigma_h)_{\text{wire zone}} = T + \sigma_r\left(\frac{r^2 + a^2}{r^2 - a^2}\right) \tag{11.43}$$

For equilibrium, from Eq. 3.453

$$\sigma_h - \sigma_r - r\, d\sigma_r/dr = 0 \tag{11.44}$$

Therefore from Eqs. 11.43 and 11.44

$$T + \sigma_r\left(\frac{r^2 + a^2}{r^2 - a^2}\right) - \sigma_r - \frac{r\, d\sigma_r}{dr} = 0 \tag{11.45}$$

or

$$d\sigma_r - 2a^2\left[\frac{\sigma_r\, dr}{r(r^2 - a^2)}\right] = \left(\frac{T}{r}\right)dr \tag{11.46}$$

The radial stress is constant if multiplied by $[r^2/(a^2 - r^2)]$; therefore,

$$-d\sigma_r\left(\frac{r^2}{r^2 - a^2}\right) + 2a^2\left[\frac{\sigma_r r^2\, dr}{r(r^2 - a^2)^2}\right] = -\frac{Tr^2\, dr}{r(r^2 - a^2)} \tag{11.47}$$

so that

$$d\left(\frac{r^2}{r^2 - a^2}\right)\sigma_r = \frac{Tr\, dr}{r^2 - a^2} \tag{11.48}$$

which on integrating gives

$$\sigma_r\left(\frac{r^2}{r^2 - a^2}\right) = \frac{T}{2}\ln(r^2 - a^2) + C \tag{11.49}$$

where C is a constant of integration determined by the boundary conditions. At $r = c$, $\sigma_r = 0$ and,

$$C = -(T/2)\ln(c^2 - a^2) \tag{11.50}$$

Substituting Eq. 11.50 into Eq. 11.49 then gives

$$\sigma_r = -\frac{T(r^2 - a^2)}{2r^2}\left(\ln\frac{c^2 - a^2}{r^2 - a^2}\right) \tag{11.51}$$

Then, for the wire zone, from Eqs. 11.43 and 11.51

$$(\sigma_h)_{\text{wire}} = T\left[1 - \left(\frac{r^2 + a^2}{2r^2}\right)\left(\ln\frac{c^2 - a^2}{r^2 - a^2}\right)\right] \tag{11.52}$$

At the interface $r = b$,

$$(\sigma_r)_b = -p_w = -\frac{T(b^2 - a^2)}{2b^2}\left(\ln\frac{c^2 - a^2}{b^2 - a^2}\right) \tag{11.53}$$

and

$$p_w = T\left(\frac{b^2 - a^2}{2b^2}\right)\left(\ln\frac{c^2 - a^2}{b^2 - a^2}\right) \tag{11.54}$$

As an example of the use of the preceding theory consider a cylinder of inner radius 6 in. operating at 20,000 psi internal pressure; the allowable stress in the core material is 25,000 psi, and the shell is wire wrapped to an outside radius of 10 in.; the required shell thickness and winding tension will be determined.

From Eq. 3.469 the axial stress for a closed-end cylinder is

$$\sigma_z = \frac{p_i a^2}{b^2 - a^2} \tag{11.55}$$

Therefore

$$25{,}000 = \frac{20{,}000(6)^2}{b^2 - (6)^2}$$

or $b = 8.05$ in. The wrapping thickness is therefore $10.00 - 8.05 = 1.95$ in., and the core thickness is $8.05 - 6 = 2.05$ in. When internal pressure is applied to the vessel, by Eq. 3.466

$$\sigma_h = \frac{a^2 p_i}{b^2 - a^2} + \frac{b^2 a^2}{r^2} p_i \left(\frac{1}{b^2 - a^2}\right) \tag{11.56}$$

or

$$\sigma_h = \frac{(20{,}000)(36)}{100 - 36} + \frac{(100)(36)(20{,}000)}{(100 - 36)r^2}$$

The stress is maximum at $r = a$; therefore at the bore

$$\sigma_h = 42{,}500 \text{ psi}$$

This stress is in excess of the allowable stress, and the required residual stress is

$$(\sigma_h)_{\text{residual}} = 42{,}500 - 25{,}000 = 17{,}500 \text{ psi}$$

which has to be negative in order to bring the bore hoop stress to the maximum allowable value. Thus considering the core as being under external pressure only,

$$-17{,}500 = -\frac{p_w b^2}{b^2 - a^2} + \frac{b^2 a^2}{r^2}\left(\frac{-p_w}{b^2 - a^2}\right)$$

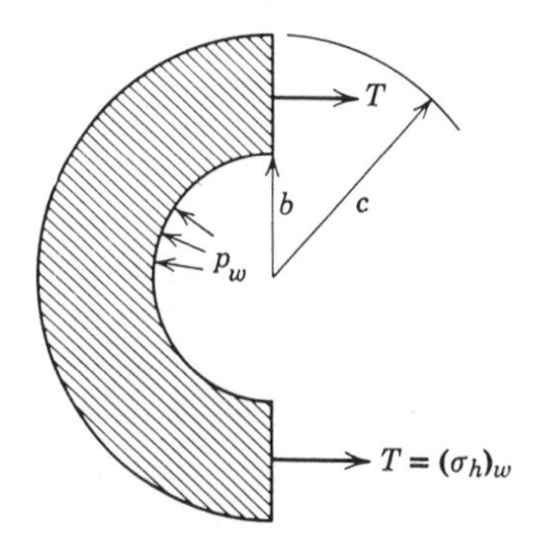

Fig. 11.14 Free-body element of wire-wrapped vessel.

from which

$$p_w = -3890 \text{ psi}$$

Then by Eq. 11.54 the required winding tension is

$$T = \frac{3890}{\left[\dfrac{(8.05)^2 - (6)^2}{2(8.05)^2}\right] \ln \left[\dfrac{(10)^2 - (6)^2}{(8.05)^2 - (6)^2}\right]} = 22{,}000 \text{ psi}$$

The value of 22,000 psi substituted into Eq. 11.52 then gives the initial stress value in the wire at any value of r.

For optimum design the tension T is varied from layer to layer in the built-up shell in such a manner that when the completed vessel is subjected to internal pressure each layer of wire will be under the same tensile stress. Assume, for example, that the uniform tension in the wire is $(\sigma_h)_w$; then, with an internal pressure p_i the hoop stress in the core is

$$\sigma_h = \frac{1}{b^2 - a^2}\left[a^2 p_i - b^2 p_w + \left(\frac{ab}{r}\right)^2 (p_i - p_w)\right] \tag{11.57}$$

From Fig. 11.14, for constant σ_h in the wire, equilibrium requires that

$$2T(c - b) = p_w(2b) \tag{11.58}$$

or

$$p_w = \frac{T(c - b)}{b} \tag{11.59}$$

Substituting Eq. 11.59 into Eq. 11.57 gives

$$(\sigma_h)_{\text{core}} = \frac{1}{b^2 - a^2}\left\{a^2 p_i - Tb(c - b) + \left[p_i - T\left(\frac{c - b}{b}\right)\right]\left(\frac{ab}{r}\right)^2\right\} \tag{11.60}$$

and the total distribution is

$$(\sigma_h)_{\text{core + wire}} = \frac{p_i a^2}{c^2 - a^2}\left(1 + \frac{c^2}{r^2}\right) \tag{11.61}$$

When the internal pressure is zero,

$$\sigma_h = -\frac{p_w}{b^2 - a^2}\left[\left(\frac{ab}{r}\right)^2 + b^2\right] \qquad 11.62$$

and the tension in the wire wrapping is

$$(\sigma_h)_{\text{wire}} = T - \frac{p_i a^2}{c^2 - a^2}\left(1 + \frac{c^2}{r^2}\right) \qquad 11.63$$

That is, the wire stress equals the uniform tension T minus the tension due to p_i if p_i were applied. From Eq. 11.41

$$(\sigma_h)_{\text{wire}} = -p_{\text{w}}\left(\frac{r^2 + a^2}{r^2 - a^2}\right) \qquad 11.64$$

The compression given by Eq. 11.64 must be subtracted from Eq. 11.63; therefore

$$p_{\text{w}} = \frac{T(c - r)}{r} - \frac{p_i a^2}{c^2 - a^2}\left(1 - \frac{c^2}{r^2}\right) \qquad 11.65$$

or

$$p_w = \frac{c - r}{r}\left[T - \frac{p_i a^2}{c^2 - a^2}\left(1 + \frac{c}{r}\right)\right] \qquad 11.66$$

Thus combining Eqs. 11.63, 11.64, and 11.66 gives the required tension for application T_{app} to give a constant tension T in each layer when p_i is applied,

$$T_{\text{app}} = T\left[\frac{c(r^2 + a^2) - 2ra^2}{r(r^2 - a^2)}\right] - \left(\frac{2p_i a^2}{r^2 - a^2}\right) \qquad 11.67$$

Plastic radial expansion—autofrettage

Several cases of residual stresses in materials due to nonuniform plastic flow will be discussed in the following paragraphs. First, consideration will be given to an old but extremely useful process called *autofrettage* (self-hooping) which was originally applied to artillery pieces such as cannons but is now used in a wide variety of industrial applications; it is applied principally to heavy-walled cylinders* and a typical sequence of hoop stress distributions is shown in Fig. 11.15. In the elastic range application of pressure p gives the elastic distribution calculated from Eq. 3.471.

$$(\sigma_h)_{\text{elastic}} = \frac{p_1}{R^2 - 1}\left(1 + \frac{c^2}{r^2}\right) \qquad 11.68$$

At a stress σ_0, defined by Eq. 3.523, yielding at the bore commences and when the wall is completely plastic the hoop stress distribution caused by pressure p_2

*Residual stresses in heavy-walled spheres are obtained by the same method.

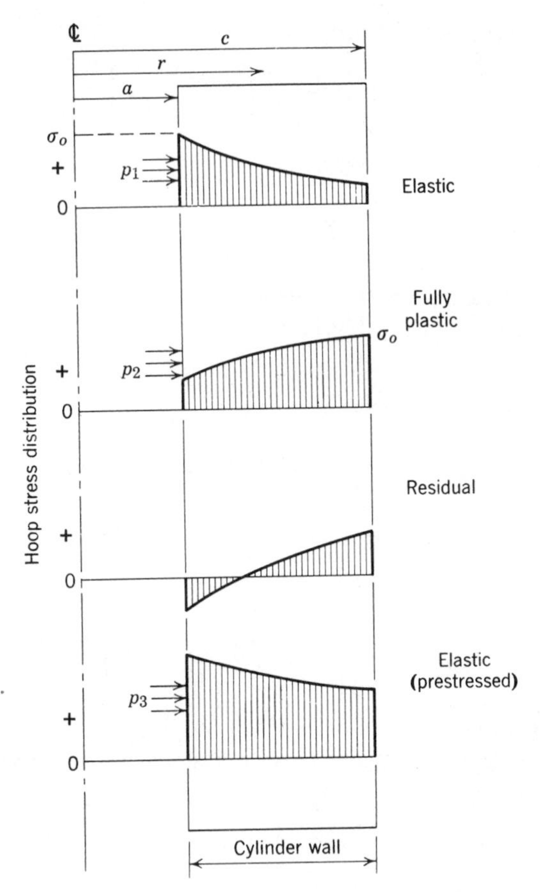

Fig. 11.15 Development of stress distributions in a cylindrical vessel prestressed by autofrettage. See also (22).

is given by Eq. 6.366

$$(\sigma_h)_{\text{plastic}} = \frac{2\sigma_0}{\sqrt{3}}\left(1 + \ln\frac{r}{c}\right) \tag{11.69}$$

If the internal pressure is now released a residual hoop stress is left in the cylinder wall which is given by the difference of Eqs. 11.68 and 11.69,

$$(\sigma_h)_{\text{residual}} = \frac{2\sigma_0}{\sqrt{3}}\left(1 + \ln\frac{r}{c}\right) - \frac{p_2}{R^2 - 1}\left(1 + \frac{c^2}{r^2}\right) \tag{11.70}$$

where p_2 replaces p_1. The second term in Eq. 11.70 is the "elastic" stress that would prevail if the cylinder were elastic under the action of pressure p_2. Finally, if a new service pressure p_3 is applied, the hoop stress distribution is given by the sum of Eqs. 11.68 and 11.70,

$$(\sigma_h)_{\text{service}} = \frac{p_3}{R^2 - 1}\left(1 + \frac{c^2}{r^2}\right) + \left[\frac{2\sigma_0}{\sqrt{3}}\left(1 + \ln\frac{r}{c}\right) - \frac{p_2}{R^2 - 1}\left(1 + \frac{c^2}{r^2}\right)\right] \tag{11.71}$$

In a similar manner the radial and longitudinal stress distributions may be computed. In the autofrettage process (7, 13, 43) the amount of residual stress left at the bore on release of pressure is dependent on the pressure, material, and diameter ratio (O.D. ÷ I.D.). Full autofrettage is obtained when the theoretical maximum residual stresses are left at the bore; this condition may be called 100% autofrettage. If only half the maximum possible residual stresses are left at the bore, the cylinder would be 50% autofrettaged. These percentage values are not to be confused with percent overstrain of the cylinder wall. Depending on the cylinder, various degrees of overstrain are required to induce various amounts of autofrettage. Full overstrain of the cylinder wall is called 100% overstrain, whereas, if the plastic zone has penetrated only half the wall thickness, the cylinder would be 50% overstrained. It will be shown, based on these concepts, that in some instances 100% autofrettage can be accomplished by only 10% overstrain of the cylinder wall.

In many instances a low cost unalloyed steel that is weak from the standpoint of strength of material can be used in a prestressed cylinder to accomplish the same result as an expensive high strength alloy steel. In addition, aside from mechanical prestressing, it is often possible to increase the strength of an otherwise low strength material by suitable metallurgical treatment that gives rise to a *strain-aging* effect; this will be discussed later. Depending on the material, the yield strength can be increased by 5 to 50% or more by a suitable strain-aging treatment. Also, both mechanical and metallurgical effects may combine to give improved performance of a vessel.

Elastic-breakdown criterion. When a cylinder containing no residual stresses due to heat treatment, fabrication, or prestress is subjected to a high enough pressure, plastic flow begins at the bore as explained in Section 6-3.

If residual stresses are present in the cylinder wall before application of service pressure, they will cause either an increase or a decrease in the elastic-breakdown pressure depending on whether they are tension or compression. Residual stresses, if present, must be allowed for in the calculation as additions to the stress terms in Eqs. 3.466 to 3.470. For internal pressure only in a closed-end cylinder Faupel and Furbeck (9) have shown that

$$p_y^2\left(\frac{R^2}{R^2-1}\right)^2 + p_y\left(\frac{R^2}{R^2-1}\right)\sigma_h^1 + \tfrac{1}{3}[(\sigma_h^1)^2 + (\sigma_z^1)^2 + \sigma_h^1\sigma_z^1 - \sigma_y^2] = 0 \qquad 11.72^*$$

where p_y is the yield pressure.

The residual stresses, σ_h^1 and σ_z^1, if introduced by means of fabrication or heat treatment, will generally not be known and will have to be determined experimentally by a boring-out method. For prestressed cylinders fabricated with autofrettage or shrink-fit construction, for example, values of σ_h^1 and σ_z^1 can be obtained analytically.

*In the subsequent calculations the yield strength σ_y will be used rather than the symbol σ_0.

Residual stresses due to partial overstrain and no reverse yielding. Under certain special circumstances, which will be discussed later, complications in analysis result if the autofrettage pressure is too high, causing reverse yielding on unloading; in this section the case (normal) of ordinary elastic recovery will be considered.

If the pressure is released when the plastic-elastic interface has been extended to radius b, residual stresses and strains are induced in the cylinder wall as a result of the permanent expansion involved during plastic radial expansion. The theoretical residual stresses and strains are obtained by calculating the stresses and strains under load and subtracting from these values the elastic stress or strain distribution that would prevail if the material were elastic at the autofrettage pressure. For partial overstrain Eqs. 6.346 to 6.349 and 6.359 to 6.362 and Eqs. 3.471 to 3.475 are involved, for example, in calculating the following residual stresses.

$$(\sigma_h{}^1)_\varepsilon = \frac{\sigma_y}{\sqrt{3}}\left[1+\left(\frac{c}{r}\right)^2\right]\left\{\left(\frac{b}{c}\right)^2 - \frac{1}{R^2-1}\left[1-\left(\frac{b}{c}\right)^2 + 2\ln\frac{b}{a}\right]\right\} \quad 11.73$$

$$(\sigma_r{}^1)_\varepsilon = \frac{\sigma_y}{\sqrt{3}}\left[1-\left(\frac{c}{r}\right)^2\right]\left\{\left(\frac{b}{c}\right)^2 - \frac{1}{R^2-1}\left[1-\left(\frac{b}{c}\right)^2 + 2\ln\frac{b}{a}\right]\right\} \quad 11.74$$

$$(\sigma_z{}^1)_{0\varepsilon} = 0 \quad 11.75$$

$$(\sigma_z{}^1)_{c\varepsilon} = \frac{\sigma_y}{\sqrt{3}}\left\{\left(\frac{b}{c}\right)^2 - \frac{1}{R^2-1}\left[1-\left(\frac{b}{c}\right)^2 + 2\ln\frac{b}{a}\right]\right\} \quad 11.76$$

$$(\sigma_h{}^1)_p = \frac{\sigma_y}{\sqrt{3}}\left\{1+\left(\frac{b}{c}\right)^2 + 2\ln\frac{r}{b} - \frac{1}{R^2-1}\left[1+\left(\frac{c}{r}\right)^2\right]\left[1-\left(\frac{b}{c}\right)^2 + 2\ln\frac{b}{a}\right]\right\} \quad 11.77$$

$$(\sigma_r{}^1)_p = \frac{\sigma_y}{\sqrt{3}}\left\{\left(\frac{b}{c}\right)^2 - 1 + 2\ln\frac{r}{b} - \frac{1}{R^2-1}\left[1-\left(\frac{c}{r}\right)^2\right]\left[1-\left(\frac{b}{c}\right)^2 + 2\ln\frac{b}{a}\right]\right\} \quad 11.78$$

$$(\sigma_z{}^1)_{cp} = \frac{\sigma_y}{\sqrt{3}}\left\{\left(\frac{b}{c}\right)^2 + 2\ln\frac{r}{b} - \frac{1}{R^2-1}\left[1-\left(\frac{b}{c}\right)^2 + 2\ln\frac{b}{a}\right]\right\} \quad 11.79$$

A typical residual-stress distribution plot for a closed-end cylinder is shown in Fig. 11.16.

The cylinder can be expanded by internal pressure until the entire wall is in the plastic range. This procedure, however, is not always necessary in order to arrive at the theoretical maximum degree of residual stress at the bore of the cylinder. The theoretical maximum amount of residual stress that can be supported by the cylinder wall is determined by Eq. 3.523. Since the radial residual stress at the bore is zero, the highest stress system that can be accommodated is

$$(\bar{\sigma})_c{}^2 = (\sigma_h{}^1)^2 + (\sigma_z{}^1)^2 - \sigma_h{}^1\sigma_z{}^1 \quad 11.80$$

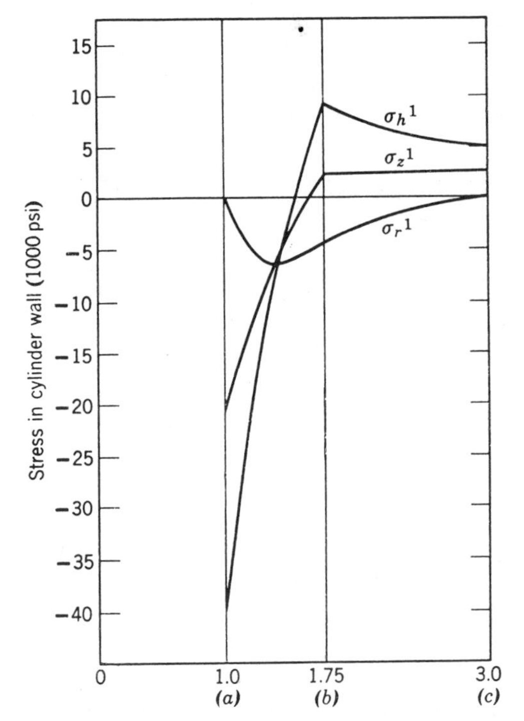

Fig. 11.16 Residual stress distributions in wall of closed-end, heavy-walled cylinder unloaded from the critical pressure.

for a closed-end cylinder, and

$$(\bar{\sigma})_0{}^2 = (\sigma_h{}^1)^2 \qquad 11.81$$

for an open-end cylinder. In other words, when the equivalent residual stress system exceeds the compressive yield strength of the material, yielding occurs on unloading as well as on original pressurizing. More will be said about this later in the discussion of reyielding.

Complete overstrain. For full overstrain of the cylinder wall the values of b and c become equal in the foregoing equations, and residual stresses are determined by use of Eqs. 6.365 to 6.369 and Eqs. 3.471 to 3.475 as explained previously for partial overstrain.

Reverse yielding. It is a simple matter to deduce from Eq. 11.80 that the limiting condition for no reverse yielding is a fully overstrained closed-end cylinder having a diameter ratio of 2.22. For the open-end cylinder (Eq. 11.81) this limiting ratio is 1.94. For diameter ratios greater than these, reverse yielding will occur on release of autofrettage pressure (Eq. 6.365) if the pressure is greater than twice the elastic-breakdown pressure for a closed-end cylinder or 1.83 times the elastic-breakdown pressure for an open-end cylinder. The theoretical optimum autofrettage pressure is thus, for most cases, about double the initial yield pressure for cylinders having diameter ratios in excess of 2.0

and the overstrain pressure (Eq. 6.365 of Chapter 6) for cylinders of lesser diameter ratio. Since the conditions of both open and closed ends yield about the same results, the reverse-yield analysis to be given will be assumed to cover all cases. The following analysis is due to Prager and Hodge (28). Equation 6.344, valid for ordinary loading, becomes for reverse yielding

$$p_r = \frac{2\sigma_y}{\sqrt{3}}\left[1 - \left(\frac{b^1}{c}\right)^2 + 2\ln\frac{b^1}{a}\right] \tag{11.82}$$

where b^1 is the reverse-yield boundary. Then, for the elastic portion of the wall, the stress reductions due to reverse-yielding are

$$(\Delta\sigma_h)_\varepsilon = \frac{-2\sigma_y}{\sqrt{3}}\left(\frac{b^1}{c}\right)^2\left(1 + \frac{c}{r}\right)^2 \tag{11.83}$$

$$(\Delta\sigma_r)_\varepsilon = \frac{-2\sigma_y}{\sqrt{3}}\left(\frac{b^1}{c}\right)^2\left(1 - \frac{c}{r}\right)^2 \tag{11.84}$$

$$(\Delta\sigma_z)_\varepsilon = \frac{-2\sigma_y}{\sqrt{3}}\left(\frac{b^1}{c}\right)^2 \tag{11.85*}$$

In the plastic portion of the cylinder wall

$$(\Delta\sigma_h)_p = \frac{-2\sigma_y}{\sqrt{3}}\left[2\ln\frac{r}{b^1} + 1 + \left(\frac{b^1}{c}\right)^2\right] \tag{11.86}$$

$$(\Delta\sigma_r)_p = \frac{-2\sigma_y}{\sqrt{3}}\left[2\ln\frac{r}{b^1} - 1 + \left(\frac{b^1}{c}\right)^2\right] \tag{11.87}$$

$$(\Delta\sigma_z)_p = \frac{-2\sigma_y}{\sqrt{3}}\left[2\ln\frac{r}{b^1} + \left(\frac{b^1}{c}\right)^2\right] \tag{11.88}$$

The residual-stress distribution is obtained by adding the stresses under pressure to the stress reduction given before; that is,

$$\sigma' = \sigma + \Delta\sigma \tag{11.89}$$

In order to illustrate the reverse-yield effect, stress distribution plots have been made for a cylinder of diameter ratio 3 and having a tensile yield strength σ_y of about 35,000 psi. Since the cylinder has a diameter ratio greater than 2.22, reverse yielding can be induced by pressurizing the cylinder at a pressure greater than twice the elastic-breakdown pressure. In this example, complete overstrain of the cylinder wall is assumed, and the distributions so obtained are shown in Fig. 11.17. When the pressure is released, reverse yielding starts at a location defined by Eq. 11.82; the final residual stresses are calculated using Eqs. 11.83 to 11.89 and are plotted in Fig. 11.18. It can be seen that there is an increase in compressive stress that occurs as the distance from the bore

*For the open-end cylinder any σ_z term would always be zero.

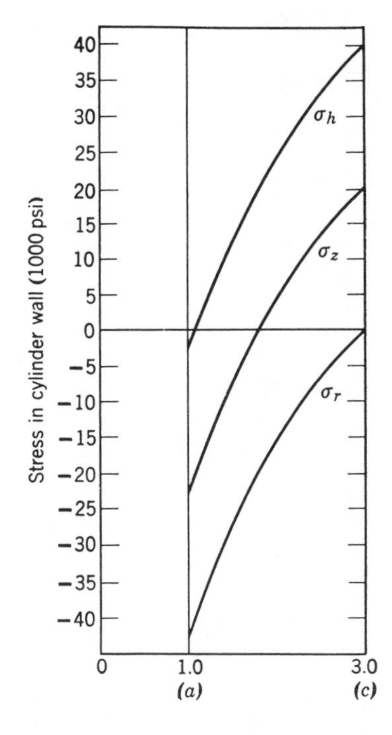

Fig. 11.17 Stress distributions in wall of a closed-end, heavy-walled cylinder at the over-strain pressure (43,800 psi).

increases. This condition does not necessarily mean that increased benefit is obtained by allowing reverse yielding to occur; this is shown by the fact that the equivalent stress is constant through the reverse-yield zone. There may be some benefit in having a zone of high equivalent residual compressive stress, but this has not been confirmed. Along with the possible benefit is a counter-effect; the equivalent tension stress is increased at the outside surface of the cylinder and continues to increase as the reyield boundary moves farther and farther from the bore. Thus, eventually, failure could start at the outside of the cylinder, particularly under fluctuating stresses or in the presence of serious stress concentrators. The reverse-yield phenomenon will probably be most useful for autofrettaged cylinders in which it is necessary to machine the bore following the autofrettage treatment.

Autofrettage design criteria. Experiment shows that, at least for steel, hardness and diameter ratio determine the ability of a cylinder to be satisfactorily autofrettaged, as shown in Fig. 11.19 (7). This figure shows, for various diameter ratios, the required yield strengths of the cylinder material in order that a fully autofrettaged ($p = 2p_y$) cylinder will operate linearly on subsequent pressure cycles to the autofrettage pressure. For the lowest strength material the minimum required diameter ratio is about 3.75, and for the strongest material the diameter ratio is about 2.5. This is a narrow range of diameter ratios, but within this range designs can be accomplished to almost any specification on elastic range operation and burst pressure (Eq. 6.377). The numbers

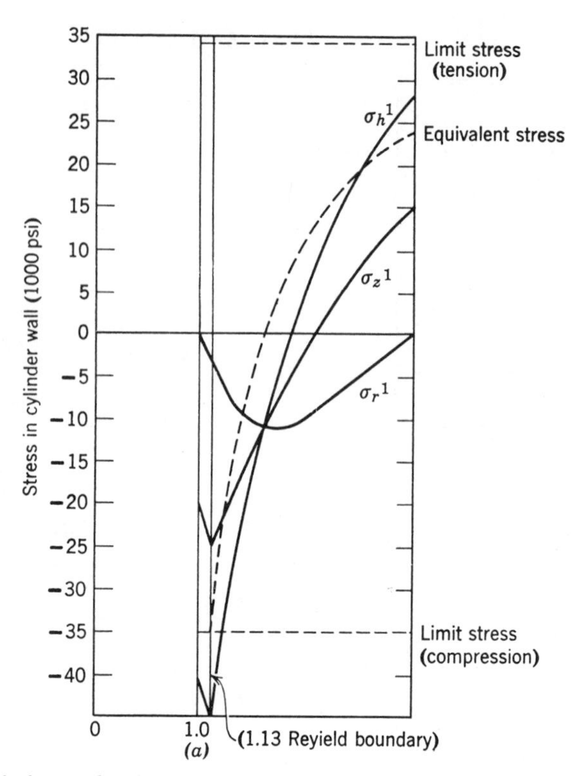

Fig. 11.18 Residual stress distributions in wall of a closed-end, heavy-walled cylinder unloaded from above the critical pressure.

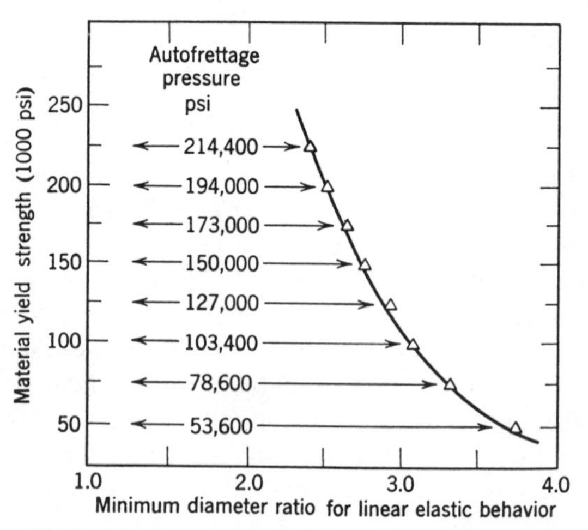

Fig. 11.19 Design curve for the autofrettage process.

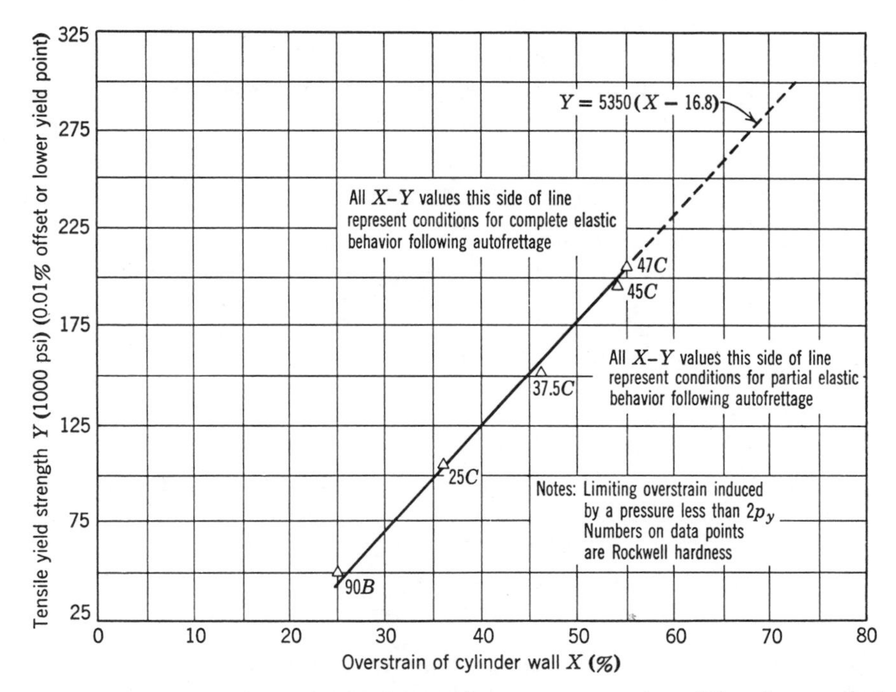

Fig. 11.20 Limiting conditions for autofrettage design, pressure-strain stability of overstrained cylinders.

on the left-hand side of the curve in Fig. 11.19 represent the range of elastic linear operation available for materials of different strengths fabricated to the required minimum diameter ratio and pressurized to give 100% autofrettage. For example, a cylinder with a diameter ratio of 3.75 and made of material having a yield strength of 50,000 psi has an elastic-breakdown pressure p_y of 26,800 psi and hence a total linear elastic range of $2p_y$, or 53,600 psi internal pressure after 100% autofrettage. By various combinations of material, diameter ratio, and elastic requirement, cylinders to almost any pressure-strain specification can be designed.

Because of size or material restrictions it may not always be possible to obtain 100% autofrettage. In such cases it becomes important to know, for material strength available, and the cylinder specified by design considerations, what range of linear operation can be expected and how to autofrettage the cylinder in order to obtain this range of linear behavior. For autofrettage under these circumstances, use is made of Fig. 11.20, which relates the permissible overstrain to the yield strength of the material. For example, for a material having a yield strength of about 50,000 psi, 25% overstrain is the maximum that can be tolerated by the cylinder in order for the cylinder to behave linearly on recycling to the autofrettage pressure. Figure 11.20 is applicable only when the autofrettage pressure is not high enough to cause reverse yielding on release of pressure. For example, using a material having a yield strength of 50,000 psi, the limiting overstrain is 25%. This 25% overstrain, however, must be caused by a pressure not exceeding that which will give 100% autofrettage. This means that for cylinders of diameter ratio less than 2.22, it is necessary to cause the overstrain by a pressure less than twice the initial yield pressure. In using Fig. 11.20, it is convenient to note that

$$Y = 5350(X - 16.8) \tag{11.90}$$

EXAMPLE. Calculate, for completely linear elastic behavior, the percent overstrain of the cylinder wall, the autofrettage pressure, and percent autofrettage for a cylinder of diameter ratio $R = 2.25$, made of a material having a yield strength of 125,000 psi.

From Eq. 11.90, using $Y = 125{,}000$, the amount of overstrain is found to be 40%. In terms of the cylinder dimensions, 40% overstrain is defined as

$$0.4 = \frac{b-a}{c-a} = \frac{R_1 - 1}{R - 1}, \qquad \text{from which } R_1 = 1.50$$

Use of Eq. 6.344 now reveals that the autofrettage pressure is 98,600 psi. The initial yield pressure (Chapter 3) is 58,000 psi. For 100% autofrettage, the pressure would have to be $2p_y$ or 116,000 psi; therefore the amount of autofrettage accomplished is

$$\frac{p_r - p_y}{p_y} 100 = 70\%$$

When performing the autofrettage treatment, it is well to keep in mind the effect of the end condition; that is whether the cylinder has open or closed ends. Often, especially with large vessels, it is usually more convenient to autofrettage with a filler bar equipped with "O" ring seals, giving an open-end cylinder. In either case, any side holes in the cylinder are also prestressed, and allowance should be made for stress concentration; this is discussed in Chapter 13. For the autofrettage of a closed-end cylinder, special care must be exercised since tangential holes in the cylinder wall (for valves, etc.) produce elliptical holes at the bore that give rise to high stress concentration effects at the end of the major axis of the ellipse. In order that the longitudinal stress in the cylinder does not induce an unduly high stress at an elliptical side hole, the ellipse geometry has to be controlled within rather narrow limits. For example, for a cylinder of diameter ratio 2.5, the limiting ellipse axis ratio to avoid excessive stress and possible crack initiation is about 4 [see (7)].

Autofrettage of steel cylinders is frequently accompanied by a *stabilization heat treatment* at around 600 to 700°F. The exact nature of what this treatment does is not known, but its effects can be far reaching. First, if residual stresses are present that are greater than the yield strength of the material at temperature, stress relief will occur. Under certain circumstances, apparently largely dependent on the steel composition, strain aging might also occur, and this has the effect of increasing the strength of the material and offsetting any loss due to stress relief. Some tests reported in the literature seem to indicate that the benefits of a thermal treatment following autofrettage are of metallurgical nature, such as strain aging with attendant increase in yield strength of the aged material.

In order to find the optimum stabilization temperature (if one exists), tensile specimens are prestrained an amount equivalent to that at the bore using Eq. 6.29, thus,

$$\varepsilon_t = \frac{u}{a\sqrt{3}} \tag{11.91}$$

where ε_t is the tensile strain equivalent to a cylinder bore diametral deformation u. Specimens thus prestrained are aged 1 hr per inch thickness at various proposed stabilization temperatures, following which they are tested in tension at room temperature. The room temperature, stress-strain curves then indicate any strain-aging effect and under what conditions it is maximum. Companion tensile tests are also conducted at the proposed stabilization temperatures in order to indicate how the yield strength changes with temperature. With a decrease in yield strength, stress-relief will occur if the residual bore stress exceeds the equivalent yield strength of the material at the stabilization temperature.

Overspinning of Disks and Cylinders

Turbine rotors and similar equipment are frequently prestressed by overspinning to induce the same effects as autofrettage of static cylinders. The elastic analysis for such members was given in Chapter 10 (see Eqs. 10.113 to 10.116 and 10.120 to 10.122, for example). In order to determine residual stresses in rotating disks and cylinders, it is first necessary to develop the equations for stress in the plastic range; for convenience, several cases will be considered.

Circular hollow disk. Consider Fig. 11.21 which shows a rotating hollow disk, where the "elastic" stresses are given by Eqs. 10.113 to 10.116. When the angular velocity reaches the value defined by Eq. 10.130, yielding starts at the inner rim in accordance with Eq. 10.115. The condition for equilibrium is given by Eq. 10.106 where it is seen that $\sigma_h = \sigma_r$ at $r = 0$; consequently, from Eq. 3.523

$$\sigma_h = \sigma_y \tag{11.92}$$

a condition that experience has shown can be applied to the entire disk, that is,

$$\frac{d\sigma_r}{dr} + \frac{\sigma_r - \sigma_y}{r} + \frac{\gamma\omega^2 r}{g} = 0 \tag{11.93}$$

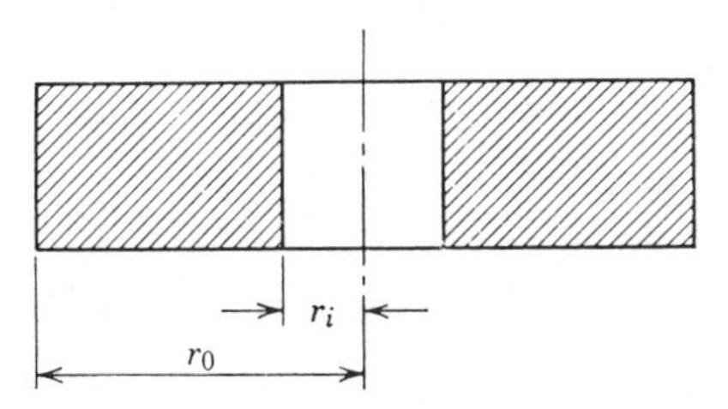

Fig. 11.21 Geometry of circular hollow disk.

the solution to which is

$$\sigma_r = \sigma_y - \frac{\gamma}{g}\frac{\omega^2 r^2}{3} + \frac{C}{r} \qquad 11.94$$

where C is a constant of integration determinable from the boundary condition. The radial stress σ_r is zero at the inner and outer edges of the rotating disk; therefore

$$\sigma_y + \frac{C}{r_i} - \frac{\gamma\omega^2 r_i^{\,2}}{3g} = 0 \qquad 11.95$$

and

$$\sigma_y + \frac{C}{r_0} - \frac{\gamma\omega^2 r_0^{\,2}}{3g} = 0 \qquad 11.96$$

from which

$$C = \sigma_y r_i \left[\frac{(r_i/r_0)^2 - 1}{1 - (r_i/r_0)^3}\right] \qquad 11.97$$

The speed for full plasticity of the disk is thus

$$\omega = \sqrt{\frac{3\sigma_y g(r_0 - r_i)}{\gamma(r_0^{\,3} - r_i^{\,3})}} \qquad 11.98$$

With these results it is seen that in the plastic range

$$\sigma_r = \sigma_y \left\{1 - \frac{r_i}{r}\left[\frac{1 - (r_i/r_0)^2}{1 - (r_i/r_0)^3}\right] - \left(\frac{r}{r_0}\right)^2 \left[\frac{1 - r_i/r_0}{1 - (r_i/r_0)^3}\right]\right\} \qquad 11.99$$

and

$$\sigma_h = \sigma_y \qquad \text{(constant across the disk)} \qquad 11.100$$

The residual stresses can now be calculated by subtracting from Eqs. 11.99 to 11.100 the stresses that would prevail if the disk were elastic at the speed given by Eq. 11.98.

Circular solid disk. Here, yielding starts at the center of the disk at the speed defined by Eq. 10.129. The analysis is the same as for the hollow disk except that the constant C must vanish in order to avoid infinite stress at $r = 0$. Therefore

$$\sigma_r = \sigma_y - \gamma\omega^2 r^2/3g \qquad 11.101$$

The radial stress σ_r is zero at the outside rim; therefore for full plasticity, from Eq. 11.101

$$\omega = (1/r_0)\sqrt{3\sigma_y g/\gamma} \qquad 11.102$$

$$\sigma_r = \sigma_y[1 - (r/r_0)^2] \qquad 11.103$$

and

$$\sigma_h = \sigma_y \qquad \text{(constant across disk)} \qquad 11.104$$

The residual stresses are then computed in the same way as mentioned for the hollow disk.

Rotating cylinders. The analyses for cylinders are quite complex and may be found in references on plasticity (see Chapter 6). Some final results will be given here, however. For the *solid* cylinder, the speed for full plasticity is given by Eq. 10.131, and the elastic stress distribution by Eqs. 10.123 to 10.125. In the plastic range

$$\sigma_h = \sigma_r = 2\sigma_y[1 - (r/r_0)^2] \tag{11.105}$$

and

$$\sigma_z = 2\sigma_y[\tfrac{1}{2} - (r/r_0)^2] \tag{11.106}$$

Residual stresses are determined by subtracting from Eqs. 11.105 to 11.106 the elastic stresses given by Eqs. 10.123 to 10.125 with the speed defined by Eq. 10.131 For the *hollow* cylinder, the speed for full plasticity is

$$\omega = \sqrt{(4g\sigma_y/\gamma\sqrt{3})\ln(r_i/r_0)[1/(r_i^2 - r_0^2)]} \tag{11.107}$$

and in the plastic range

$$\sigma_r = \frac{2\sigma_y}{\sqrt{3}}\ln\frac{r}{r_0} - \frac{\gamma\omega^2}{2g}(r^2 - r_0^2) \tag{11.108}$$

Assuming in the plastic range that ε_z is zero, it is seen from Eq. 6.21 that

$$\sigma_z = \frac{\sigma_h + \sigma_r}{2} \tag{11.109}$$

The hoop stress is obtained from Eq. 3.523,

$$\sigma_h = \frac{\frac{3}{2}\sigma_r \pm \sqrt{\frac{9}{4}\sigma_r^2 - 3(\frac{3}{4}\sigma_r^2 - \sigma_y^2)}}{\frac{3}{2}} \tag{11.110}$$

Again, residual stresses are obtained as outlined for the disks and solid cylinder.

Residual machining stresses

All machining operations, including fine grinding and lapping (18), introduce residual stresses in the workpiece; the magnitude and distribution of these stresses depend on the material being machined, the tool geometry and sharpness, amount of feed, speed and depth of cut, and other similar variables. In general, these stresses are concentrated in a shallow layer immediately below the machined surface; for example, in carbon and low-alloy steels, the penetration extends only a few thousandths of an inch below the surface. On the other hand, in austenitic stainless steel, successive cuts act cumulatively to build up high stresses extending to a depth as great as $\frac{1}{4}$ in. below the surface. Henriksen (12) determined the residual machining stresses imparted to rectangular bars of various carbon, alloy, and stainless steels using various shaper and planer speeds, feeds, depths of cut, and tool geometries. He found that (*a*) ductile materials sustain residual surface tensile stresses, (*b*) the induced-residual stresses are related to the work-hardening characteristics of the material and appear to be a function of the mechanical action of the tool rather than of thermal effects, and (*c*) the induced stresses depend on the geometry of the tool. Colwell, Sinnott, and Tobin (5), investigated the residual surface stresses induced by grinding a hardened SAE 4340 steel and found that the depth of penetration of the residual stresses increased as the severity of the grinding increased, although the depth of penetration did not exceed 0.006 in. They also observed that the harder the steel, the greater is the absolute value of residual stress (100,000 psi tension to 135,000 psi compression) and that severe grinding produced predominantly tensile surface stresses, whereas light grinding also produced surface compression stresses. Somewhat similar observations were made by Letner (18).

REFERENCES

1. J. O. Almen, *Prod. Eng.*, **17** (8), 81–85 (August 1946).
2. W. M. Baldwin, *Proc. Am. Soc. Test. Mater.*, **49**, 1–45 (1949).
3. H. Bühler, H. Buchholtz, and E. Schulz, *Arch. für das Eisenhütt.*, **5** (8), 413–418 (February 1932).
4. H. Bühlër and E. Scheil, *Arch. für das Eisenhütt.*, **6** (7), 283–288 (January 1933).
5. L. V. Colwell, M. J. Sinnot, and J. C. Tobin, *Trans. Am. Soc. Mech. Eng.*, **77** (7), 1099–1105 (October 1955).
6. H. Dobkin, *Steel*, **118** (22), 104–106; 150 (June 3, 1946).
7. J. H. Faupel, *J. Franklin Inst.*, **259** (5), 405–419 (May 1955); **269** (6), 474–489 (June 1960).
8. J. H. Faupel, *Mach. Des.*, **26** (11), 114–124 (January 1954).
9. J. H. Faupel, and A. R. Furbeck, *Trans. Am. Soc. Mech. Eng.*, **75** (3), 345–359 (April 1953).
10. J. Frisch, and E. G. Thomsen, *Trans. Am. Soc. Mech. Eng.*, **73** (3), 337–346 (April 1951).
11. K. Heidlhofer, *Evaluation of Residual Stress*, McGraw-Hill Book Co., New York, N.Y. (1948).
12. E. K. Henriksen, *Trans. Am. Soc. Mech. Eng.*, **73** (1), 69–76 (January 1951).
13. R. Hill, *The Mathematical Theory of Plasticity*, Oxford, The Clarendon Press (1950).
14. J. H. Hollomon and L. D. Jaffe, *Ferrous Metallurgical Design*, John Wiley and Sons, Inc., New York, N.Y. (1947).
15. O. J. Horger, "Residual Stresses," in M. Hetenyi, Ed., *Handbook of Experimental Stress Analysis*, 459–578, John Wiley and Sons, Inc., New York, N.Y. (1950).
16. R. A. Jones, and A. I. Andrews, *J. Am. Ceram. Soc.*, **31** (10), 274–279 (October 1948).
17. R. L. Ketter, *The Influence of Residual Stress on the Strength of Structural Members*, Report No. 44, Bulletin Series, Welding Research Council, New York, N.Y. (November 1958).
18. H. R. Letner, *Proc. Soc. Experimental Stress Analysis*, **10** (2), 23–26 (1952). *Trans. Am. Soc. Mech. Eng.*, **77** (7), 1089–1098 (October 1955).
19. T. Y. Lin, *Design of Prestressed Concrete Structures*, 2nd ed., John Wiley and Sons, Inc., New York, N.Y. (1963).
20. C. W. MacGregor and L. F. Coffin, *J. Franklin Inst.*, **243** (5), 391–421 (May 1947).
21. W.R.D. Manning, *Engineering*, **163** (4240), 349–352 (May 2, 1947).
22. W. R. D. Manning, *Engineering*, **169** (4396), 479–481 (April 28, 1950); (4397), 509–511 (May 5, 1950); (4399), 562–563 (May 19, 1950).
23. J. L. Meriman, et al., *Weld. J. Res. Suppl.* (J. Am. Weld. Soc.), **11** (6), 340s–343s (June 1946).
24. G. P. Michailov, *Weld. J. Res. Suppl.* (J. Am. Weld. Soc.), **9** (5), 255s (May 1944).
25. C., Mylonas, *Weld. J. Res. Suppl.* (J. Am. Weld. Soc.), **26** (11), 516s–520s (November 1961).
26. A. Nadai, "Inherent and Residual Stresses," in *Plasticity*, 258–269, McGraw-Hill Book Co., New York, N.Y. (1931).

27. W. R. Osgood, Ed., *Residual Stresses in Metals and Metal Construction*, Reinhold Publishing Corp., New York, N.Y. (1954).

28. W. Prager, and P. G. Hodge, *Theory of Perfectly Plastic Solids*, John Wiley and Sons, Inc., New York, N.Y. (1951).

29. D. G. Richards, *Proc. Soc. Exp. Stress Anal.*, **3** (1), 40–61 (1945).

30. G. Sachs, "Residual Stresses, Their Measurement, and Their Effects on Structural Parts," in *Symposium on the Failure of Metals by Fatigue*, 237–247 Melbourne University, Australia (1946).

31. G. Sachs and G. Espey, *Met. Tech.*, **8** (T.P. 1384), 1–11 (October 1941).

32. G. Sachs and G. Espey, *Iron Age*, **148** (12), 63–71 (September 18, 1941); (13), 36–42 (September 25, 1941).

33. E. H. Schuette, *Prod. Eng.*, **29** (35), 51–53 (September 1, 1958).

34. R. T. Severn, *Q. J. Mech. and Math.*, **12** (Part 1), 82–88 (1959).

35. J. C. Straub, *Prod. Eng.*, **31** (38), 210–212 (mid-September 1960).

36. D. E. Thomas, *J. Appl. Phys.*, **19** (2), 190–193 (February 1948).

37. S. Timoshenko and J. N. Goodier, *Theory of Elasticity*, 3rd ed., McGraw-Hill Book Co., New York, N.Y. (1970).

38. S. Timoshenko, *Strength of Materials, Part II*, 3rd ed., D. Van Nostrand Co., Princeton, N.J. (1956).

39. R. Wadler, *Des. News*, **16** (15), 74–75 (July 21, 1961).

40. W. M. Wilson and C. C. Hao, *Residual Stresses in Welded Structures*, University of Illinois Bulletin No. 40, Vol. 43 (February 26, 1946).

41. Symposium on Residual Stress, *Proc. Soc. Exp. Stress Anal.*, **2** (1), 147–225 (1944).

42. "Behavior of Residual Stress Under External Load and Their Effect on Strength of Welded Structures" *Weld. J. Res. Suppl.* (J. Am. Weld. Soc.), **9** (9) 473s–480s (September 1944).

43. *Symposium on Internal Stresses in Metals and Alloys*, Institute of Metals and Alloys, Institute of Metals, London (1948).

44. *Residual Stress Measurements*, American Society for Metals, Novelty, Oh. (1952).

45. "Current Work on Prestressed Concrete," *Engineering*, **193** (5018), 811–812 (June 22, 1962).

46. *Chemical Process.*, **24** (10), 126–127 (October 1961).

47. *Proceedings World Conference on Prestressed Concrete*, World Conference on Prestressed Concrete, Inc., San Francisco, Cal. (July, 1957).

PROBLEMS

11.1 A thick-walled cylinder with an inside radius of 6 in. and internal pressure of 30 kpsi is to be designed from a steel with an allowable stress of 50 kpsi. Find (*a*) the thickness of a solid wall, (*b*) the thickness using a wall made of two cylinders, and (*c*) the thickness using a wall made of three cylinders.

11.2 A thick-walled, closed-end cylinder has an I.D. of 2 in. and an O.D. of 6 in. Assuming the material is perfectly plastic and has a yield stress of 100 kpsi, find the pressure for yielding midway through the wall. Find the resulting residual stress at the inner radius.

11.3 A pressure vessel supports an internal pressure of 12 kpsi (Fig. P11.3). The yield strength of the wall material is 100 kpsi. What is the optimum wall thickness for (*a*) monobloc construction, (*b*) two-ring shrink fit

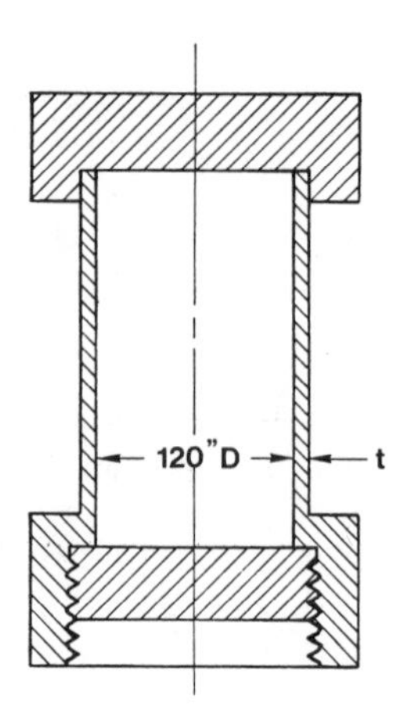

Fig. P11.3

construction, and (*c*) three-ring shrink fit construction? Use a safety factor of 1.5. Sketch the respective constructions, show the stress distribution, and dimension the individual layers.

11.4 A 0.500-in. round shaft 18 in. between bearings is bent a maximum of 0.015 in. out from its center line from permanent set. Find its present residual stress pattern and the value of the concentrated force to spring the shaft back to its initial center-line. Is there any residual stress in the shaft after it is straightened?

11.5 A 175-lb, 24-in. diameter steel gear is pressed onto a 3-in. diameter steel shaft. The diametrical press is 0.005 in. Plot the transmitted torque versus rpm. Obtain the limit on the shaft and the gear for horsepower or rpm. What must the minimum yield strengths of the shaft and the gear be?

11.6 A smooth bore cannon 5.5-in. O.D. fires a 2-in. diameter steel ball. The cannon material yield strength is 50 kpsi with an allowable firing stress of 10 kpsi, and the barrel is 6 ft long. It is proposed to increase the range by wrapping the barrel with five layers of 0.0625-in. diameter wire with a tensile strength of 293.8 kpsi and an allowable stress of 73.5 kpsi. Find the ratio of the increase in range.

11.7 In a thick slotted ring r_i is 4 in. and r_0 is 5 in., and the material is steel with a yield strength of 60 kpsi. Find the minimum gap to avoid failure and the uniform pressure on the outside to keep the gap closed.

11.8 What is the increase in speed when the gear in Problem 11.5 is changed to a solid gear with flanges that are driven by the shaft? How much will the transmitted horsepower increase?

chapter 12

FATIGUE

12-1 FATIGUE OF MATERIALS AND STRUCTURES

Fatigue of a material or structure is a term applied to denote a failure by fracture brought about by repeated reversal or removal of the applied load. In this chapter only mechanical fatigue is considered; thermal fatigue, brought about by repeated thermal expansions and contractions, is discussed in Chapter 15. The essential feature of fatigue is the loading-unloading nature of the process, and for this reason time is not always the critical factor. For example, one structure might retain its integrity under millions of stress or strain reversals whereas another could fail in a few cycles. The former type of behavior is the conventional description of fatigue that consists of loading and unloading a part and plotting applied stress by the number of cycles to failure (S-N diagram). The latter behavior, called low-cycle fatigue, is associated with rather large stresses causing plastic deformation and subsequent hysteresis effects that change from cycle to cycle.

The problem of fatigue has been of major concern in engineering for over 100 years, and voluminous literature on the subject has accumulated. Much of the work to date has been of a general engineering nature consisting of service tests of full-size structures and attempts to correlate large-scale behavior with laboratory small-scale tests. The fundamental work on fatigue may be classified into two general categories; first, since fatigue is associated with plastic flow and fracture, the hypotheses and laws governing the mechanisms of flow and fracture may be said to also apply to fatigue failures. Secondly, studies of fatigue have been along statistical-analytical lines with attempts to determine from multitudes of tests a more rational measure of fatigue life. Many difficulties are experienced in studying fatigue in relation to service behavior since fatigue is markedly influenced by such factors as size, shape, temperature, state-of-stress, type of testing machine, and specimen. It has been said that 90% of all service failures are caused by fatigue and that 90% of all fatigue failures are caused by improper design. For example, in Fig. 12.1 the crack shown is the result of the sharp corner that should have been ground out to a smooth radius before the part was put in service. In this chapter the subject of mechanical fatigue will be discussed primarily from the point of view of design and engineer-

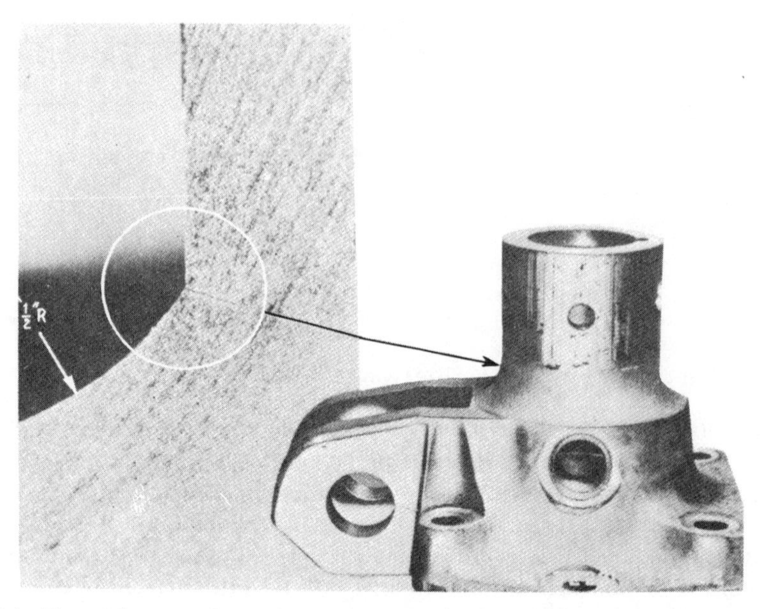

Fig. 12.1 Typical fatigue with origin at a point of stress concentration. [*After Grover, Gordon, and Jackson (7), courtesy of The Bureau of Naval Weapons, U.S. Department of the Navy.*]

ing application. For additional information see (7, 26, 28, and 35). The following paragraphs describe fatigue behavior and many of its effects in the traditional manner. Prediction of fatigue behavior on the basis of fatigue fracture mechanics concepts has taken on significant importance and is discussed, along with examples, in Chapter 14.

The nature of fatigue

Although the precise physical mechanism of fatigue is not known, it is believed that fatigue is very closely associated with flow and fracture so that considerable information can be gained from an examination of fatigue fractures. For example, in a piece of mild steel fractured statically, considerable plastic flow precedes fracture, and the surfaces at the fractured section show a silky fibrous structure due to stretching of the grains. In fatigue, the crack usually starts at some discontinuity, stress-raiser, inclusion in the material, or the like. Once the crack is initiated it propagates under the action of load application and removal until the cross section is reduced to the point where the remaining area can no longer sustain the load, and this remaining section cracks suddenly. Two zones can usually be distinguished in a fatigue crack, one due to the gradual development of the crack and the other due to sudden fracture. The latter resembles the fracture of a tensile test specimen of a brittle material so that the fracture appears brittle even though, by usual standards, the material is ductile.

The entire basis for fatigue analysis lies in the notion that there is some limiting

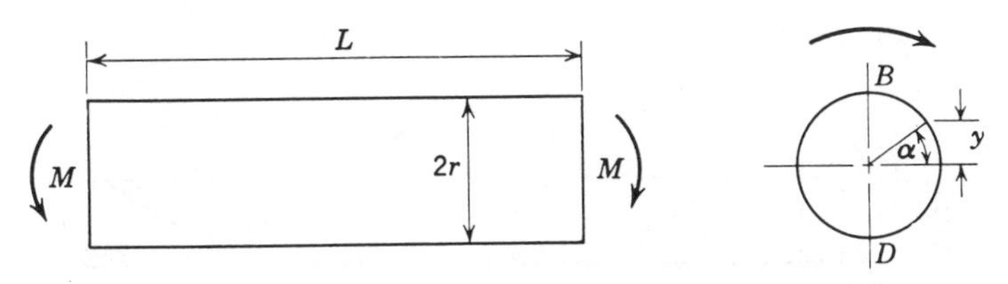

Fig. 12.2 Bar subjected to pure bending fatigue stress.

stress value that can be imposed on a material dynamically that will not cause failure in an infinite number of stress 'fluctuations'. The number and effect of variables involved are so great that the use of tests to predict service behavior in fatigue is almost a hopeless venture. However, many interesting analytical relationships have been developed that aid the designer in selection of materials and dimensions for machine and structural parts subjected to fluctuating or fatigue stresses (35).

A considerable volume of fatigue testing is done under conditions of sinusoidal loading in pure bending as shown in Fig. 12.2. The specimen of length L is suitably held in a machine that rotates it while under a condition of pure bending; this gives a complete reversal of stress, and the specimen is usually of circular cross section. Details of various fatigue-testing machines and specimens may be found in the references cited.

Analytically, from Fig. 12.2, the bending stress imposed on the specimen is

$$\sigma = M/Z = My/I = (M/I)(r \sin \alpha) = \sigma_{max} \sin \alpha \qquad 12.1$$

with a constant rotational speed and a stress variation as shown in Fig. 12.3. Stress fluctuations can be imposed by various types of machines. In this example the machine imposes pure bending with the specimen fibers alternately passing through +, 0, and − stress values. Other machines use cantilever-type specimens, tension-compression specimens, or torsion specimens. The loads may be applied so as to vary from a maximum plus value to a maximum minus value of the same magnitude. This produces a condition of *complete reversal of stress* where the mean stress equals zero. Loading may also be applied from 0 to a plus or minus value; this gives a condition of *repeated loading* with the mean stress equal to $\pm\sigma_{max}/2$. The third type of loading is either repeated or complete

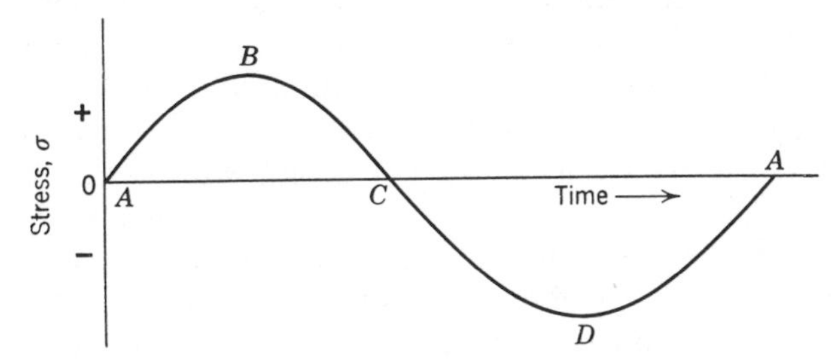

Fig. 12.3 Sinusoidal variation of stress with time.

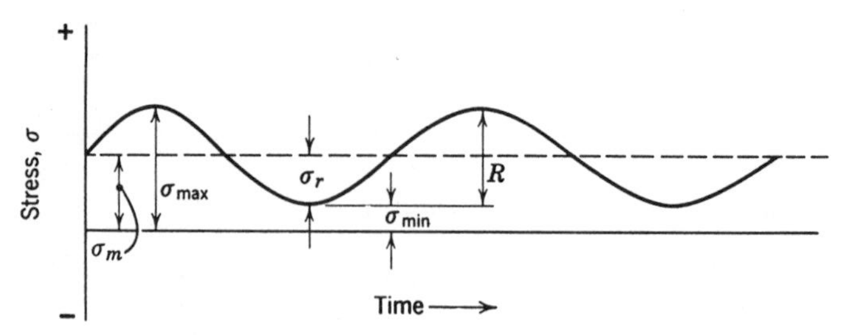

Fig. 12.4 Typical sinusoidal fatigue loading with a mean stress.

reversal of load applied to a specimen statically loaded to some plus or minus value. A very common type of loading is shown in Fig. 12.4 where σ_m is the mean stress, σ_{max} the maximum stress, σ_{min} the minimum stress, σ_r the variable stress, and R the range of stress. Thus for any particular system

$$\sigma = \sigma_m + \sigma_r \sin (2\pi t/T) \qquad 12.2$$

where T is the time to complete one cycle.

$$\sigma_m = (\sigma_{max} + \sigma_{min})/2 \qquad 12.3$$

$$\sigma_r = (\sigma_{max} - \sigma_{min})/2 \qquad 12.4$$

$$R = 2\sigma_r \qquad 12.5$$

$$\sigma_{max} = \sigma_m + R/2 \qquad 12.6$$

and

$$\sigma_{min} = \sigma_m - R/2 \qquad 12.7$$

Figure 12.4 is a plot of stress versus time, which is the response of a part or system to some known force or pressure variation. It does not include the effects of vibration. The element sizes from a fatigue analysis must be used to find natural frequencies and mode shapes to obtain these effects. Techniques to allow for the effects of shock and vibration will be discussed in Section 12-2. In a typical fatigue test the specimen is stressed to some value and the number of stress fluctuations to fracture noted; then the applied stress is plotted against the number of cycles to failure. For many ferrous metals and thermosetting plastics the *S-N* (stress versus cycles) curve approaches an asymptotic value at some stress level called the *endurance or fatigue limit*. For example, the curve shown in Fig. 12.5 for a mild steel has an endurance limit of about 27,000 psi. For most nonferrous metals, wood, and thermoplastic materials there is no definite endurance limit, and a value is arbitrarily selected at some predetermined number of cycles. This is shown schematically in Fig. 12.6. The curves in Figs. 12.5 and 12.6 are for uniaxial tests like bending or push-pull tension-compression loading. For combined stresses, for example, a tube subjected to pulsating internal pressure or a bar subjected to repeated torsion, the ordinate would be principal stress and the abscissa the number of cycles of failure.

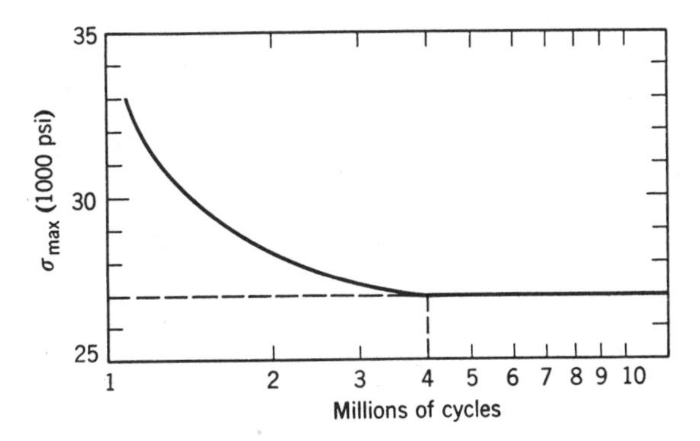

Fig. 12.5 Fatigue data for a mild steel subjected to completely reversed stresses.

Theories of failure. In the interpretation of fatigue results a number of procedures have been suggested (7). Experience has shown, however, that the most reliable methods are those that combine a modification of the *Goodman's law* or *Soderberg's law* with the distortion energy theory discussed in Chapter 3; this will be explained presently. Goodman's law is an empirical representation of fatigue behavior that takes into account the effect of mean stress. Goodman's original law was based on the ultimate tensile strength of the material, whereas the modified Goodman law, or Soderberg law is based on the yield strength of the material. Considerable support exists for both methods (7) although Soderberg's law is more conservative and hence somewhat safer to use. Mathematically, the empirical Goodman's law is represented by the equation

$$y = (1 - x) \tag{12.8}$$

which is plotted in Fig. 12.7 along with Soderberg's law. In terms of the fatigue variables mentioned earlier Goodman's law becomes

$$\sigma_r = \sigma_e - \sigma_m(\sigma_e/\sigma_u) \tag{12.9}$$

and Soderberg's law becomes

$$\sigma_r = \sigma_e - \sigma_m(\sigma_e/\sigma_y) \tag{12.10}$$

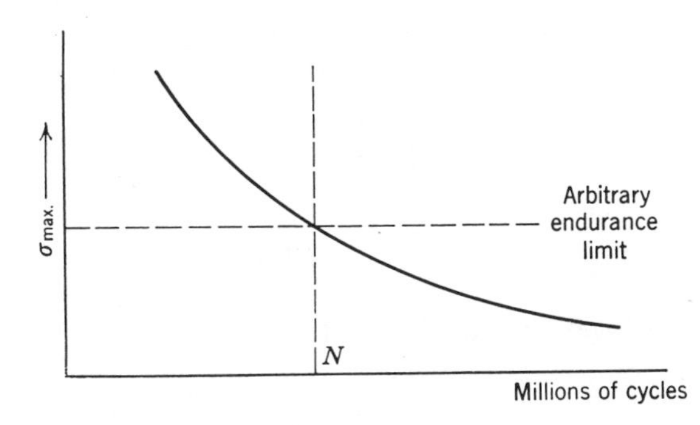

Fig. 12.6 Endurance limit determination for nonferrous metals.

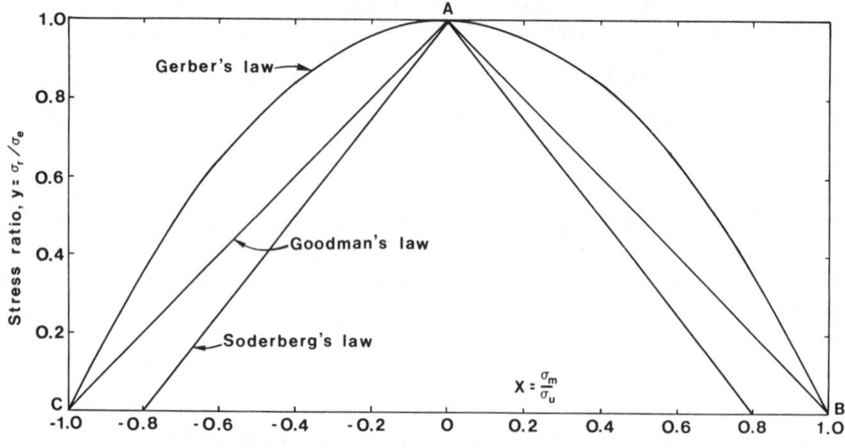

Fig. 12.7 Graphical representation of three fatigue laws.

where σ_r is the variable stress, σ_e the endurance limit, σ_u the ultimate tensile strength, σ_y the yield strength, and σ_m the mean stress. In terms of maximum and minimum stress, from Eqs. 12.6 and 12.7

Goodman's law
$$\sigma_{max} = \sigma_e + \sigma_m(1 - \sigma_e/\sigma_u) \qquad 12.11$$
$$\sigma_{min} = -\sigma_e + \sigma_m(1 + \sigma_e/\sigma_u) \qquad 12.12$$

Soderberg's law
$$\sigma_{max} = \sigma_e + \sigma_m(1 - \sigma_e/\sigma_y) \qquad 12.13$$
$$\sigma_{min} = -\sigma_e + \sigma_m(1 + \sigma_e/\sigma_y) \qquad 12.14$$

Equations 12.11 to 12.14 together with Eq. 3.523 with the endurance limit σ_e substituted for σ_0 form the basis for practical fatigue calculations. For uniaxial loading Eqs. 12.11 to 12.14 are used directly. For combined stresses the maximum and mean stresses are expressed in terms of distortion energy as follows.

The distortion energy for uniaxial loading is given by Eq. 3.522,

$$U = \left(\frac{1+v}{3E}\right)\sigma_0^{\,2} \qquad 12.15$$

In terms of fluctuating stress the distortion energy becomes

$$U_{max} = \left(\frac{1+v}{3E}\right)(\sigma_{max})^2 \qquad 12.16$$

and

$$U_{mean} = \left(\frac{1+v}{3E}\right)(\sigma_{mean})^2 \qquad 12.17$$

These relationships together with Eqs. 12.13 and 12.14 give the result

$$\sqrt{U_{max}} = (1 - \sigma_e/\sigma_y)\sqrt{U_{mean}} + \left(\sqrt{\frac{1+v}{3E}}\right)(\sigma_e) \qquad 12.18$$

Similarly, for combined stresses, from Eq. 3.521

$$U_{max} = \left(\frac{1+v}{3E}\right)[(\sigma_1)^2_{max} + (\sigma_2)^2_{max} + (\sigma_3)^2_{max} - (\sigma_1)_{max}(\sigma_2)_{max} - (\sigma_2)_{max}(\sigma_3)_{max} - (\sigma_3)_{max}(\sigma_1)_{max}] \quad 12.19$$

and

$$U_{mean} = \left(\frac{1+v}{3E}\right)[(\sigma_1)^2_{mean} + (\sigma_2)^2_{mean} + \ldots] \quad 12.20$$

Substituting Eqs. 12.19 and 12.20 into Eq. 12.18 gives

$$\sigma_e = [(\sigma_1)^2_{max} + (\sigma_2)^2_{max} + (\sigma_3)^2_{max} - (\sigma_1)_{max}(\sigma_2)_{max} - (\sigma_2)_{max}(\sigma_3)_{max} - (\sigma_3)_{max}(\sigma_1)_{max}]^{1/2} - (1 - \sigma_e/\sigma_y)[(\sigma_1)^2_{mean} + (\sigma_2)^2_{mean} + \ldots]^{1/2} \quad 12.21$$

Equation 12.21 is expressed in terms of *principal stresses*. For the two-dimensional case, from Eqs. 3.18 to 3.20, in terms of *component stresses*, Eq. 12.21 becomes

$$\sigma_e = [(\sigma_x)^2_{max} - (\sigma_x)_{max}(\sigma_y)_{max} + (\sigma_y)^2_{max} + 3(\tau_{xy})^2_{max}]^{1/2} - (1 - \sigma_e/\sigma_y) \times [(\sigma_x)^2_{mean} - (\sigma_x)_{mean}(\sigma_y)_{mean} + (\sigma_y)^2_{mean} + 3(\tau_{xy})^2_{mean}]^{1/2} \quad 12.22$$

In Fig. 12.7 the equations for the curves are

Soderberg's law $$\frac{\sigma_m}{\sigma_y} + \frac{\sigma_r}{\sigma_e} = 1 \quad 12.23$$

Goodman's law $$\frac{\sigma_m}{\sigma_u} + \frac{\sigma_r}{\sigma_e} = 1 \quad 12.24$$

Gerber's law $$\left(\frac{\sigma_m}{\sigma_u}\right)^2 + \frac{\sigma_r}{\sigma_e} = 1 \quad 12.25$$

If the Soderberg curve, Eq. 12.23, for a simple stress is examined (3)

$$\frac{K_1\sigma_m}{\sigma_y} + \frac{K_2\sigma_r}{\sigma_e} = \frac{\sigma_m}{\sigma_y/K_1} + \frac{\sigma_r}{\sigma_e/K_2} = 1 \quad 12.26$$

For all three equations (Eqs. 12.23–12.25), the factors influencing fatigue can be applied either to σ_m and σ_r or σ_y and σ_e.

When stresses are complex, σ_m and σ_a can be treated using Eq. 3.526 where for plane stress

$$\sigma_m = \sqrt{\sigma_{xm}^2 - \sigma_{xm}\sigma_{ym} + \sigma_{ym}^2 - 3\tau_{xym}^2} \quad 12.27$$

$$\sigma_r = \sqrt{\sigma_{xr}^2 - \sigma_{xr}\sigma_{yr} + \sigma_{yr}^2 - 3\tau_{xyr}^2} \quad 12.28$$

The factors influencing fatigue are then applied according to what material data are available. Some data (30, 34) are available giving σ_y, σ_u, and σ_e for specific sizes of test samples and a $\sigma_r - \sigma_m$ plot must be constructed, while other material data are in the form of $\sigma_r - \sigma_m$ plots (37) for a standard test sample. This means for some calculations the correction terms in Eq. 12.29, and Section 12-2 will be applied to either the stresses or the data—but not both. In order to generate a design curve from Fig. 12.7, σ_e is one of the important factors formulated by Marin and presented by Shigley (23) where

$$\sigma_e = k_a k_b k_c k_d k_e k_f \ldots k_1 \sigma_e^1 \tag{12.29}$$

σ_e^1 represents data from a smooth polished rotating beam specimen. The k values can be applied to the stresses or to correct σ_e. Material data can have some k values incorporated in the test or no k values at all. When developing a design curve for combined stresses, it is better to place the k values with the individual stresses where possible. The factors, k values, influencing fatigue behavior will be discussed where most of the corrections are to σ_e or σ_r. σ_e^1 will be discussed in Section 12-3.

12-2 SOME FACTORS INFLUENCING FATIGUE BEHAVIOR

The number of variables and combinations of variables that have an influence on the fatigue behavior of parts and structures is discouragingly large, and a thorough discussion concerning this subject is virtually impossible. At best, the designer can make rough estimates and predictions, but even to do this requires some knowledge of at least the various principal factors involved. In the following discussion some high-spot information is presented with the caution that fatigue behavior is extremely complicated and any data or methods of utilizing the data should be viewed in a most critical way.

Surface condition, k_a

By surface condition is meant the degree of smoothness of the part and the presence or absence of corrosive effects. In general, a highly polished surface gives the highest fatigue life, although there is evidence suggesting that the *uniformity* of finish is more important than the finish itself. For example, a single scratch on a highly polished surface would probably lead to a fatigue life somewhat lower than for a surface containing an even distribution of scratches. Typical trend data of Karpov and reported by Landau (14) are shown in Fig. 12.8 for steel. Reference (15) also shows data for forgings that are similar to the k_a for tap water. The *Machinery Handbook* (33) shows a detailed breakdown of surface roughness versus machining or casting processes. This information can be used for steels to find the k_a from a theoretical model developed by Johnson (10) in Fig. 12.9.

Size and shape, k_b

The subject of size and shape effects in design is discussed in Chapter 1; the same general conclusions and methods presented there also apply to fatigue loading. For example, referring to Fig. 1.49, it is seen that the small bar has less

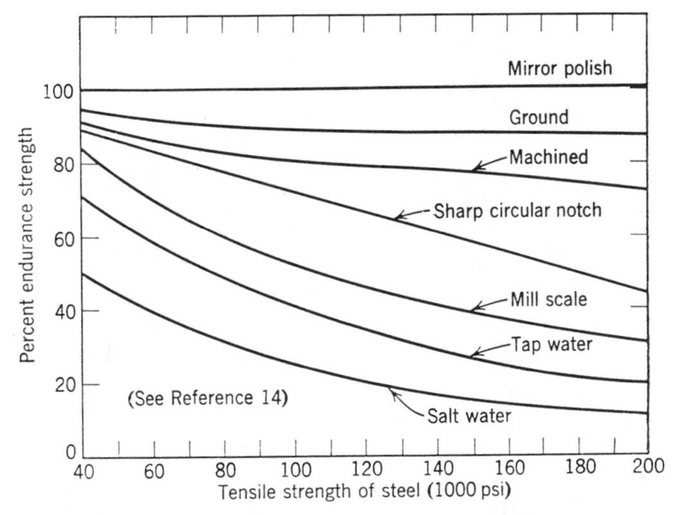

Fig. 12.8 Effect of surface condition on fatigue of steel.

volume of material exposed to a high stress condition for a given loading and consequently should exhibit a higher fatigue life than the larger bar. Some data illustrating this effect are shown in Fig. 12.10 (7). Shape (moment of inertia) also has an effect as shown in Fig. 12.11 (7). In design it is important to consider effects of size and shape, but by proper attention to these factors a part several inches in diameter can be designed on the basis of fatigue data obtained on small specimens. A rough guide presented by Castleberry, Juvinall, and Shigley is

$$k_b = \begin{bmatrix} 1 & \text{for } d \le 0.030 \text{ in. (11, 23)} \\ 0.85 & 0.3 \le d \le 2 \text{ in. (11, 23)} \\ 1 - \dfrac{(d-0.03)}{15} & 2 \le d \le 9 \text{ in. (1)} \end{bmatrix} \qquad 12.30$$

The k_b, greater than 0.5, is for steel and only serves as a guide to other materials. When $d \le 0.30$ many materials fall into the range of spring diameters where ultimate and endurance limit strengths (3) are stated as a function of wire diameter and the k_b is greater than one.

Reliability, k_c

The k_c value corrects σ_e^1 for an 8% standard deviation when no other data are available. In Fig. 12.12 the $\sigma_r - \sigma_m$ curve can be developed; however, the solid line is the average or mean of all data. The reliability of a design using any point on the solid line is 0.50. Tests have been conducted (12) where the dotted lines A and B represent data spread of $\pm 3\sigma$ derived from several tests along the curve.

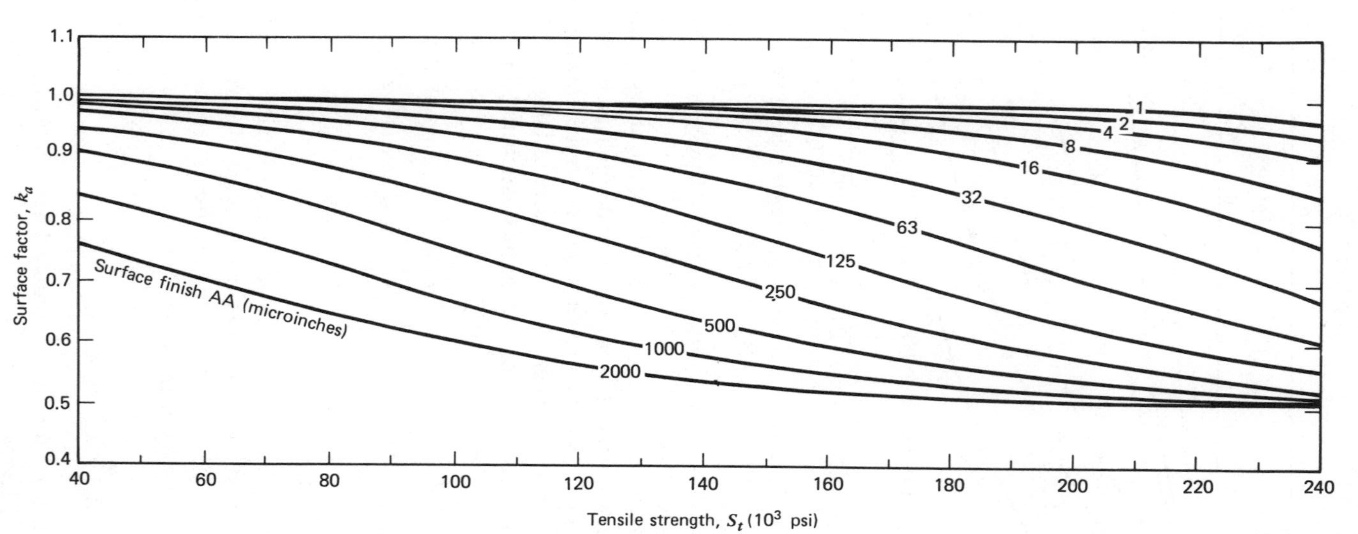

Fig. 12.9 k_a *versus surface roughness and tensile strength. [After Johnson (10), courtesy of Machine Design.]*

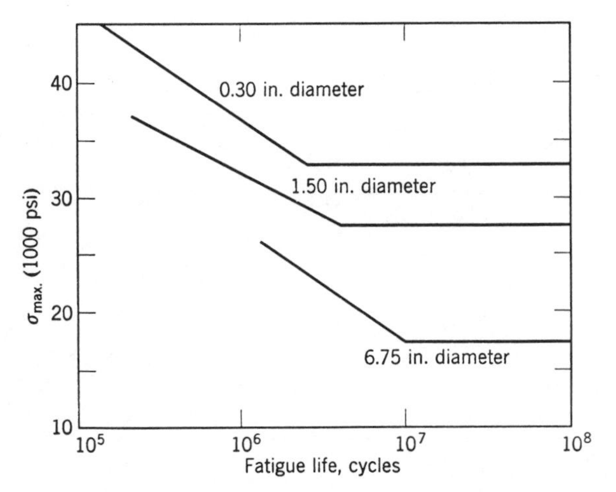

Fig. 12.10 Effect of size on the fatigue strength of steel. [After Grover, Gordon, and Jackson (7), courtesy of The Bureau of Naval Weapons, U.S. Department of the Navy.]

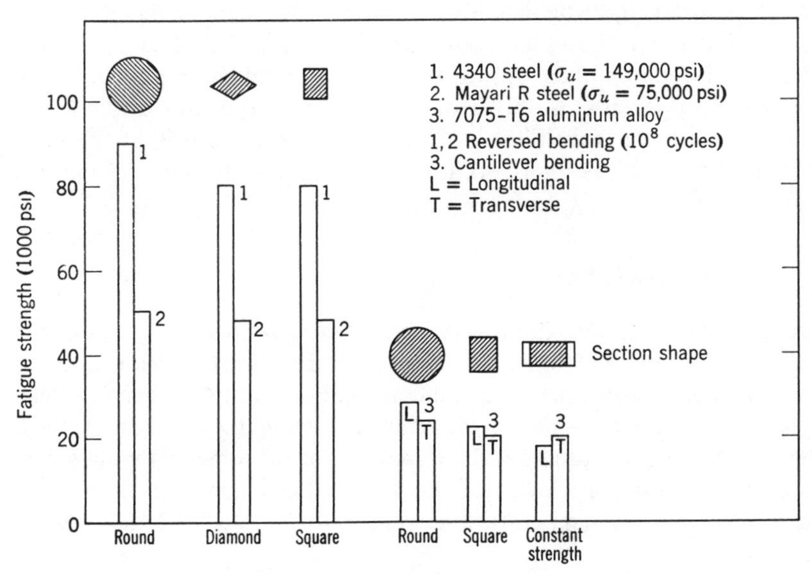

Fig. 12.11 Effect of section shape on fatigue strength. [After Grover, Gordon, and Jackson (7), courtesy of The Bureau of Naval Weapons, U.S. Department of the Navy.]

The k_c values (23) are as follows:

$$R(.50) = 1.00 \tag{12.31}$$

$$R(.90) = 0.897 \tag{12.32}$$

$$R(0.95) = 0.868 \tag{12.33}$$

$$R(0.99) = 0.814 \tag{12.34}$$

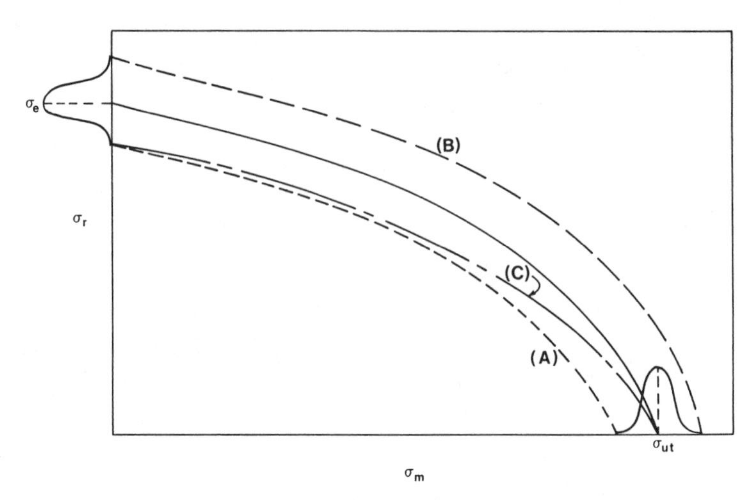

Fig. 12.12 $\sigma_r - \sigma_m$ *with material property variations.*

The values correct σ_e or increase the amplitude stress depending on the calculation. If a curve is developed and line A in Fig. 12.12 is to be drawn, some knowledge about the spread of the data about the σ_{ut} mean should be obtained. However, if k_c is only used on $\sigma_e{}^1$, it should be noted that a line c is generated where the reliability is 0.99 at σ_e and slips to 0.50 at the ultimate, σ_{ut}. The curve would be more accurate if k_c is applied to both σ_{ut} and σ_e until material data are available.

Temperature, k_d

In general, the endurance limit increases as temperature decreases, but specific data should be obtained for any anticipated temperature condition since factors other than temperature, *per se*, could control. For example, for many steels the range of temperature associated with transition from ductile to brittle behavior (see Chapter 14) has to be allowed for. In addition, for some materials, structural phase changes occur at elevated temperature that might tend to increase the fatigue life. The low temperature k_d values (4) for $-186°C$ to $-196°C$ are approximately:

Carbon steels	2.57
Alloy steels	1.61
Stainless steels	1.54
Aluminum alloys	1.14
Titanium alloys	1.40

The values decrease linearly with temperature to the room temperature value of one. These k_d values increase σ_e and decrease σ_m and σ_r.

Unlike low temperature values, k_d is not linear above room temperatures for metals. Typical k_d values are:

Magnesium	(572° F)	0.4
Aluminum	(662° F)	0.24
Cast alloys	(500° F)	0.55
Titanium	(752° F)	0.70
Heat resistant steel	(1382° F)	0.63
Nickel alloys	(1382° F)	0.70

Many k_d values for specific alloys and temperatures can be found in (4, 30, 37).

Stress concentration, k_e

The subject of stress concentration is considered separately in Chapter 13 and the discussion concerning the effect of mechanical stress concentrators like grooves, notches, and so on, on fatigue behavior is included as part of Chapter 13. Later in this chapter the effect of stress concentrations like inclusions in the material is considered. In general, the presence of any kind of a stress raiser lowers the fatigue life of a part or structure.

Stress concentrations are introduced in two ways:

1. The geometry of a design and loading creates stress concentrations (Chapter 13). This is introduced into the design calculations by

$$q = \frac{K_f - 1}{K_t - 1} \qquad 12.35$$

where q = the notch sensitivity factor (3, 23, 24)
K_t = theoretical factors (3, 23, 24)
Often to find q a notch radius, r, is required which is generally not known until the design is completed. Therefore, to start a design

$$K_f = K_t \qquad 12.36$$

K_f is used in Eq. 12.29 to correct σ_e for a single state of stress as

$$k_e = \frac{1}{K_f} \qquad 12.37$$

Otherwise, for combined states of stress K_f is used in Eq. 12.28 for the design of ductile materials and in Eqs. 12.27 and 12.28 for the design of brittle materials. For example: for a ductile material with a bending stress and an axial load.

$$\sigma_{xa} = K_{fb}\frac{M_a c}{I} + K_{ft}\frac{P_a}{A} \qquad 12.38$$

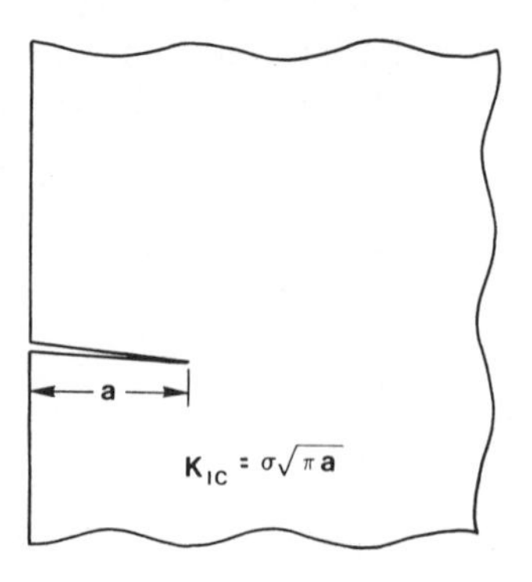

Fig. 12.13 Surface crack on a large part.

where K_{fb} and K_{ft} stand for the stress concentration factors for bending and tension.

2. Stress concentrations also develop as cracks in a material whether caused by machining, heat treatment, or a flaw in the material. Therefore, some consideration should be given to crack size. In Fig. 12.13 grooves or cracks in polished samples (4) for 1–10 rms finish are less than 0.001 mm (3.95×10^{-5} in.) while in a rough turn, 190–1500 rms, the cracks are 0.025–0.050 mm (0.001–0.002 in.) long. The minimum detectable crack (16) with X-ray or fluoroscope is about 0.16 mm (0.006 in.). In Fig. 12.14 the inherent flaws (5) in steel, $2a$ length, under the surface run from 0.001 to 0.004 in. decreasing with strength. Aluminum and magnesium alloys vary from 0.003–0.004 in. while copper alloy has, $2a$ crack lengths, of up to 0.007 in.

In Chapter 14 cracks are discussed as well as what K_{IC} and K_{th} mean in fatigue. When cracks grow (8) K_{th} is exceeded, and when cracks split a part into pieces K_{IC} has been exceeded. In Eq. 14.28

$$\frac{1}{\pi}\left(1.5\frac{E}{\sigma} \times 10^{-4}\right)^2 \leq a_{th} \leq \frac{1}{\pi}\left(1.8\frac{E}{\sigma} \times 10^{-4}\right)^2 \qquad 12.39$$

where E is the modulus and σ is the highest stress. Work by Siebel and Gaier presented by Forrest (4) on machining grooves will be compared with a_{th} in Table 12.1. An operating stress of 30 kpsi is selected for the illustration. The representative rms values are from (9 and 33).

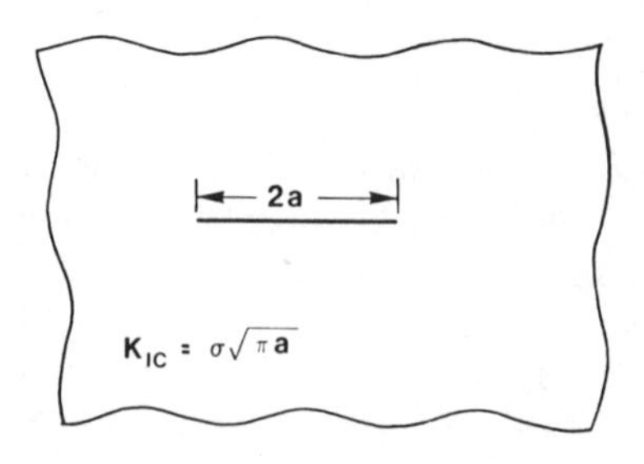

Fig. 12.14 Inherent flaw in a large part.

Table 12.1. a_{th} Compared to Machined Grooves

	rms	Groove Depth (Inches × 10^{-3})	a_{th} (30 kpsi) (Inches × 10^{-3})
Polish	8	0.04	Steel (7–10)
Fine grind	10	0.08	Aluminum (0.8–1.2)
Rough grind	70	0.2–0.4	Magnesium (0.3–0.5)
Fine turn	10–90	0.4–0.8	Titanium (2–3)
Rough turn	90–500	0.8–2	
Very rough turn	>500	>2	

The surface crack in Table 12.1 and Fig. 12.13 is accounted for in k_a, surface conditions, and its effects are further reduced by residual stresses k_f, surface treatment k_h, and discussed in fretting k_i. However, the inherent flaw (Fig. 12.14) must be detected by nondestructive testing such as X-rays. Then, the part is either scrapped or repaired.

Residual stress, k_f

The subject of residual stress is considered separately in Chapter 11 which may be referred to for more details. For present purposes it is to be noted that, in general, a favorable residual stress distribution in a part leads to an increased fatigue life; typical applications are shot peening or surface rolling of shafts and autofrettage of cylinders.

Shot peening on any part surface—whether it be machined, surface hardened, or plated—will generally increase endurance strength. The shot peening residual stress is compressive and generally half of the yield strength and with a depth of 0.020 to 0.040 in. The shot peening effect (3) disappears for steel above 500°F and for aluminum above 250° F. The correction to the endurance strength for shot peening is

$$k_f = (1 + Y) \qquad 12.40$$

where Y is the improvement.

Typical values for steel are:

Surface	Y
Polished	0.04–0.22
Machined	0.25
Rolled	0.25–0.5
Forged	1–2

The roughest surface will realize the largest values of Y improvement. However, the overall net effect of shot peening is to increase σ_e so that

$$0.70\,\sigma_e^{\,1} \leq \sigma_e \leq 0.90\,\sigma_e^{\,1}$$

where $\sigma_e^{\,1}$ is the endurance strength of a mirror-polished test sample, and σ_e is calculated from

$$\sigma_e = k_a k_f \sigma_e^{\,1}$$

Surface rolling induces a deeper layer than shot peening (0.040 to 0.5 in.). The Y improvements are:

Surface or Material	Y
Straight steel shafts	0.2–0.8
Polished or machined steel parts	0.06–0.5
Magnesium	0.5
Aluminum	0.2–0.3
Cast iron	0.2–1.93
Any condition	0.1–0.9

Actual cases with discussion are presented by Frost (5), Forrest (4), Faires (3), Lipson and Juvinall (15) as well as (24; 34 Vol. II; 2).

Cold-working of axles also imparts compressive residual stresses that tend to increase the fatigue life. If, however, collars are press-fitted on shafts, an effective stress raiser is formed at the interface (Fig. 12.15) which offsets any beneficial effect of the compressive residual stress and usually results in a lower fatigue life. This difficulty may be overcome to a large extent by the modified arrangements of the collar shown.

Internal structure, k_g

For the purposes of this book the only internal structural aspects of fatigue behavior of materials of interest are inclusions that act as stress concentrators and (probably related to inclusions) directional effects giving rise to different fatigue properties in the longitudinal and transverse directions of fabricated

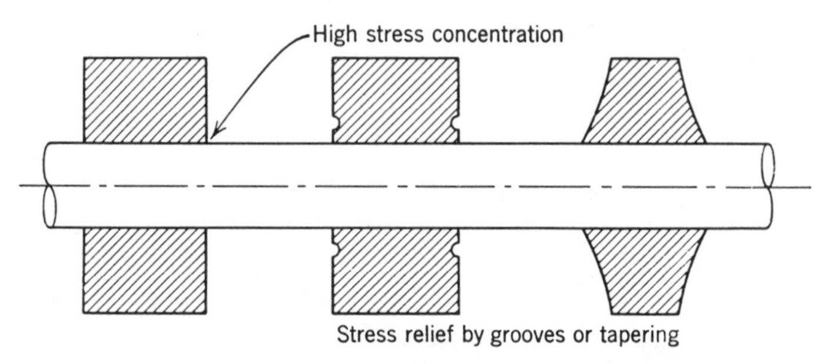

Fig. 12.15 Various assemblies of collars shrunk on a shaft

materials. By longitudinal is meant the axis of rolling direction in sheet, for example. More will be said about this later in the design application examples where it is pointed out that the transverse fatigue properties of many steels, for example, are distinctly lower than the longitudinal fatigue properties.

Environment, k_h

The effects of tap and salt water on steel are shown in Figure 12.8. The same effects for nonferrous metal (4, 20) for all tensile strengths are

$$0.40 \leq k_h \leq 0.64 \qquad 12.41$$

Two exceptions are electrolytic copper and copper-nickel alloys for which

$$0.85 \leq k_h \leq 1.06 \qquad 12.42$$

and nickel-copper alloys for which

$$0.64 \leq k_h \leq 0.86 \qquad 12.43$$

These results are from tests conducted from 1930–1950; therefore, care should be taken with newer alloys.

The effect of steam on steel under pressure is

$$0.70 \leq k_h \leq 0.94 \qquad 12.44$$

However, for a jet of steam acting in air on steel the values are one-half of Eq. 12.44.

A corrosive environment on anodized aluminum and magnesium yields

$$0.76 \leq k_h \leq 1 \qquad 12.45$$

while for nitrided steel

$$0.68 \leq k_h \leq 0.80 \qquad 12.46$$

Surface treatment and hardening, k_i

Surface treatment protects the surface from gross corrosion. Results (5) for chrome plating of steel are represented by the reduction of the fatigue strength

$$Y = 0.3667 - 9.193 \times 10^{-3} \sigma_e{}^1 \qquad 12.47$$

where $\sigma_e{}^1$ is the fatigue strength of the base material. Then k_i is found by using Eq. 12.47.

$$k_i = 1 + Y \qquad 12.48$$

For nickel plating (24), Y is -0.99 in 1008 steel and -0.33 in 1063 steel. If shot peening (3) is performed after nickel and chrome plating, the fatigue strength can be increased above that of the base metal.

The endurance strength of anodized aluminum in general is not affected. Osgood (20) and Forrest (4) present effects of other surface treatments.

Table 12.2. *Surface Hardening—Y Increases and Layer Thickness*

	Flame and Induction Hardening	Carburizing	Nitriding
Layer thickness (inches)	0.125–0.500 (induction) ≈ 0.125 (flame)	0.03–0.1	0.004–0.02
Steel	0.66–0.80	0.62–0.85	
Alloy steel	0.06–0.64	0.02–0.36	0.30–1.00

Surface hardening of steel produces a hardened layer to resist wear and cracking. Table 12.2 compares typical layer thicknesses and increases in Y from (2, 3, 4, 5, 34 Vol. II). Rotating beam samples exhibit Y values of 0.20–1.05 for carburizing or nitriding. The soft layer under the surface hardening should always be checked to see if its endurance strength is adequate. In the soft layer the tension residual stresses will balance the compression stresses in the outer layer. These effects are also discussed in Chapter 11.

Fretting, k_j

Fretting can occur in parts where motions of 0.0001–0.004 in maximum take place between two surfaces. The surface pairs exist as

1. Tapered cone and shaft assemblies
2. Pin, bolted, or riveted joints
3. Leaf springs
4. Ball and bearing race
5. Mechanical slides under vibration
6. Spline connections
7. Spring connections
8. Keyed shafts and joints

These surfaces, under pressure, work against each other producing pits and metal particles. Extensive action can result in cracks and finally failure. The action can be recognized in disassembled parts by a rust color residue in ferrous parts and by a black residue for aluminum and magnesium parts. Desirable surface pairs (4, 11, 20) can be selected to reduce fretting. Like steel surfaces react well while cast iron must be lubricated to obtain the same performance.

Sors (24) concludes k_j is 0.70–0.8 in general and 0.95 for good surface matches. Frost (5) reports similar values with some surface pairs lower.

Fretting may be reduced (2, 3, 7) by constraining the motion or by closer fits, lubricated surfaces or gaskets, and residual stresses imparted to the surfaces. A constraining example: a flywheel on a shaft with a keyed cone fit held in place with a loading nut. Surface lubrication with molybdenum disulfide, MoS_2 as well as other inhibitors extends the life of a surface before fretting. Further, reductions can be obtained if the surfaces are shot peened or surface rolled prior to assembly.

Shock or vibration loading, k_k

These effects are used to increase the stresses σ_r and σ_m. The effects of shock loading applied to the particular stress vary from

$$1 \le k_k \le 3 \qquad 12.49$$

One is for smooth operation, and 3 is found by rearranging Eq. 10.200 into

$$\delta = \delta_{st}\left[1 + \sqrt{1 + \frac{2H}{\delta_{st}}}\right]$$

If

$$H = 1.5\delta_{st}$$

then

$$\delta = 3\delta_{st}$$

This means during a varying load situation if a clearance of $1.5\delta_{st}$ is allowed, then triple the stress will be developed. It should be noted that δ_{st} is the static deflection at the point of load application due to the load. δ_{st} should always be checked since a very stiff system could inadvertently be under heavy shock loading. Rapid loading for some materials (Chapter 10) can increase the yield strength of materials giving a built-in safety factor.

Many parts are subjected to varying degrees of vibration. If the accelerations are known, the weights or applied forces can be multiplied by 10 (see Chapter 10). This is the response of a spring mass with 5% structural damping to a $1g$ input. Extreme care should be taken to see if the design loads take into account the effects for shock and vibration.

Radiation, k_l

Radiation tends to increase tensile strength and decrease ductility. The effect is discussed in more detail in (20).

Speed

For most metallic materials and other materials that do not have viscoelastic properties, stress frequencies in the range from about 200 to 7000 cycles per minute have little or no effect on fatigue life. The fatigue life could be affected, however, if during the rapid stress fluctuations, the temperature increased appreciably. For speeds over 7000 cycles per minute there is some evidence that the fatigue life increases a small amount.

For viscoelastic materials (plastics), considerably more caution must be exercised in interpreting fatigue data. Normally, fatigue tests are conducted at as rapid a frequency of stressing as possible, with due consideration for temperature rise. However, plastics will exhibit different fatigue characteristics depending on the stress frequency, which, depending on the material, will yield different results in ranges of high and low loss factors. In general, in applications involving fatigue loading it is best to use materials that exhibit low loss factor

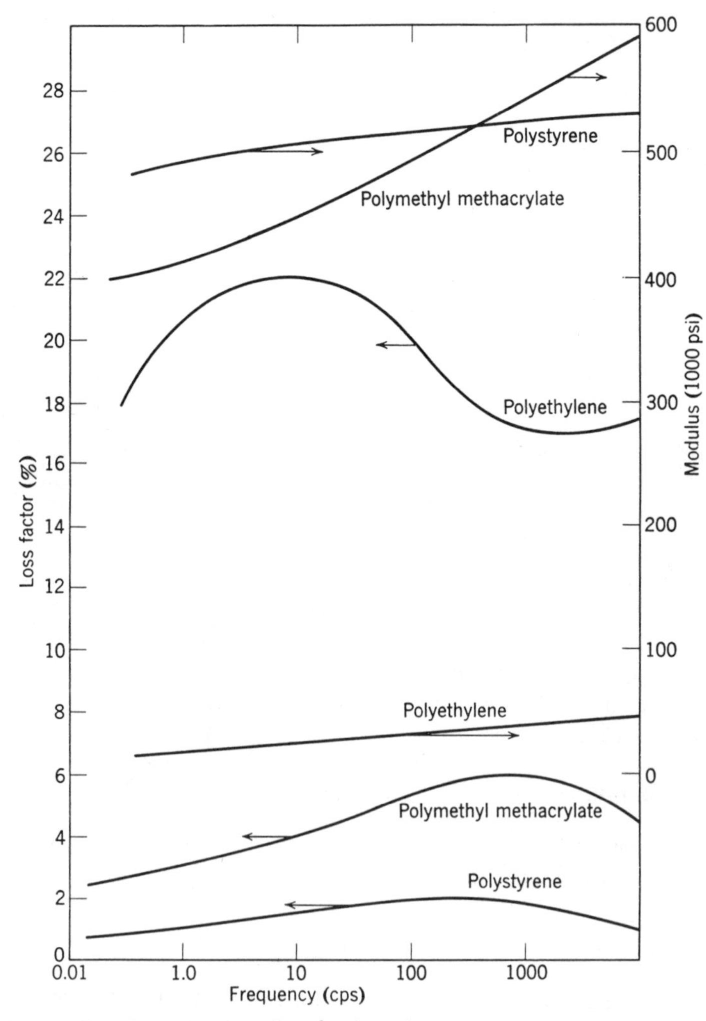

Fig. 12.16 Dynamic data for three plastics at room temperature.

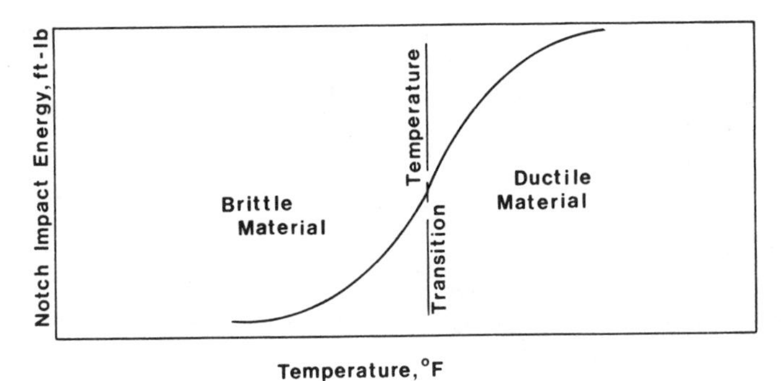

Fig. 12.17 Typical Charpy or Izod test of steel.

under the conditions expected to persist in the application; if the part is used as a damper or energy absorber, it should be used at a frequency characterized by a high loss factor. For example, a vibrating part made of polymethyl methacrylate at room temperature should not be used at a frequency of 600 cycles per minute since this is where the loss factor is maximum (Fig. 12.16). Thus, in evaluating fatigue data for plastics, curves such as shown in Fig. 12.16 should be available.

Mean stress

A structure stress-cycled about some mean stress other than zero has different fatigue characteristics than one cycled about zero mean stress. The precise reason for this is unknown, but it is believed due to hysteresis effects caused by plastic flow that changes the fatigue characteristics on each cycle. The effect of mean stress has been included empirically in the design examples discussed later in this section.

The mean stress, Eq. 12.27, for brittle metals requires the application of K_f values as shown in Eq. 12.38. For some steels, for example, the criterion for brittleness can be found approximately from Charpy or Izod test data shown in Fig. 12.17. Above the transition temperature the metal acts in a ductile manner while below the transition temperature the metal acts in a brittle fashion. Further discussion of this concept is in Chapter 14 dealing with fracture of materials.

12-3 FATIGUE PROPERTIES OF MATERIALS

Fatigue life of a material is not a property like modulus, which, under normal conditions, is a material constant. The endurance limit of a material is influenced by the type of test used and numerous other variables, of which many have already been mentioned. Therefore for any particular material it is necessary to examine fatigue data with respect to the end use conditions. Nevertheless, as a guide, a few properties are presented here to assist in

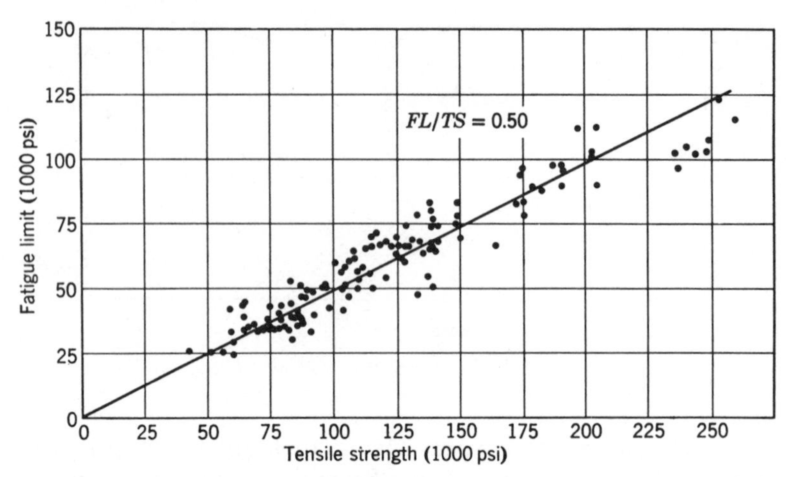

Fig. 12.18 Rotating-bending fatigue limits of cast and wrought steels. [After Grover, Gordon, and Jackson (7), courtesy of The Bureau of Naval Weapons, U.S. Department of the Navy.]

the selection of materials for particular applications. More detailed property information on specific materials should be obtained from the fabricator or vendor. Information can also be obtained from (30, 34, 36, 37, 38) as well as from publications like *Machine Design*, *Modern Plastics*, and *Materials Engineering*, which publish data yearly. Most organizations that publish standards and mechanical strength properties such as SAE, ASME, AISC, AITC, and others are listed in (31).

As previously mentioned most *ferrous materials* are characterized by a more or less definite endurance limit which is of the order of half the ultimate tensile strength of the material. Typical data for a wide range of ferrous materials are

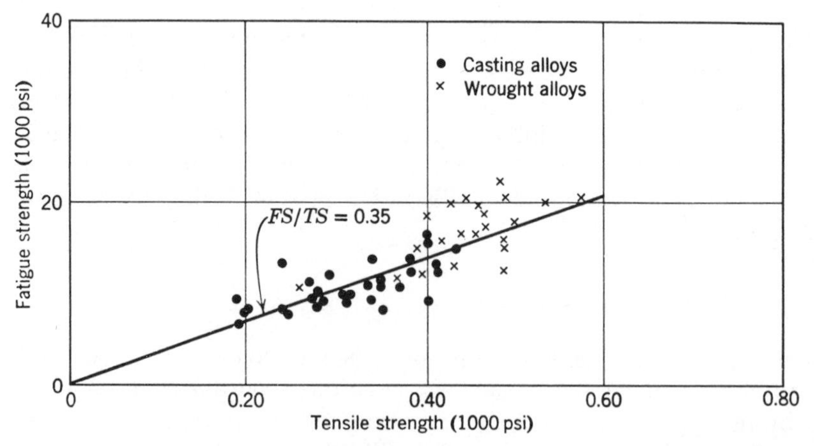

Fig. 12.19 Rotating-bending fatigue strengths at 100 million cycles of magnesium alloys. [After Grover, Gordon, and Jackson (7), courtesy of The Bureau of Naval Weapons, U.S. Department of the Navy.]

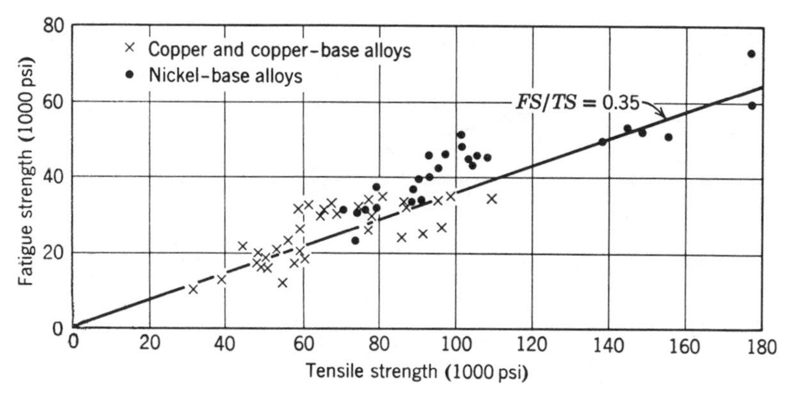

Fig. 12.20 Rotating-bending fatigue strengths at 100 million cycles of some nonferrous alloys. [After Grover, Gordon, and Jackson (7), courtesy of The Bureau of Naval Weapons, U.S. Department of the Navy.]

shown in Fig. 12.18. In using such data it is necessary to consider the fact that most steels exhibit anisotropy of fatigue properties and that the values reported in the curves (like Fig. 12.18) are probably from tests of specimens cut in the longitudinal direction. The annealed austenitic stainless steels have very good fatigue-corrosion resistance and are not as notch-sensitive as other steels; however, in the cold-worked condition, their fatigue properties are about the same as those exhibited by other steels. Typical fatigue data for a variety of other materials are shown in Figs. 12.19 to 12.23.

Titanium alloys (4) behave as cast and wrought steel in Fig. 12.18. Wood, plastics with low modulus filler, and plywood (4, 7, 24) for 10^7 cycles have

$$0.20 \leq \frac{FS}{TS} \leq 0.40 \qquad 12.50$$

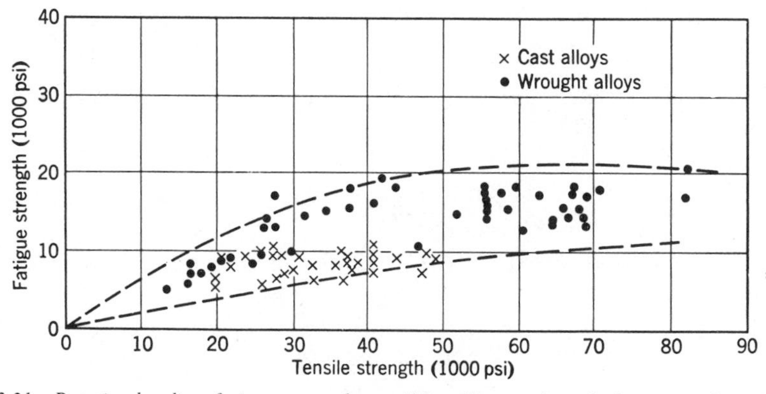

Fig. 12.21 Rotating-bending fatigue strengths at 500 million cycles of aluminum alloys. [After Grover, Gordon, and Jackson (7), courtesy of The Bureau of Naval Weapons, U.S. Department of the Navy.]

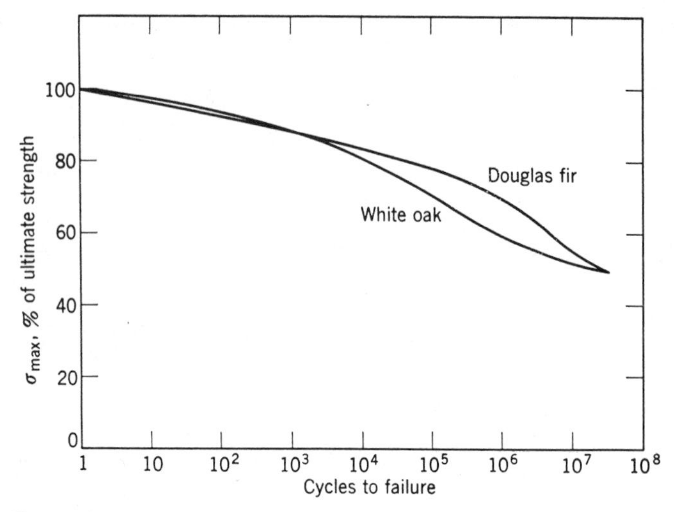

Fig. 12.22 Typical fatigue test data for wood. Repeated tension parallel to grain at 900 cycles per minute [see (17).]

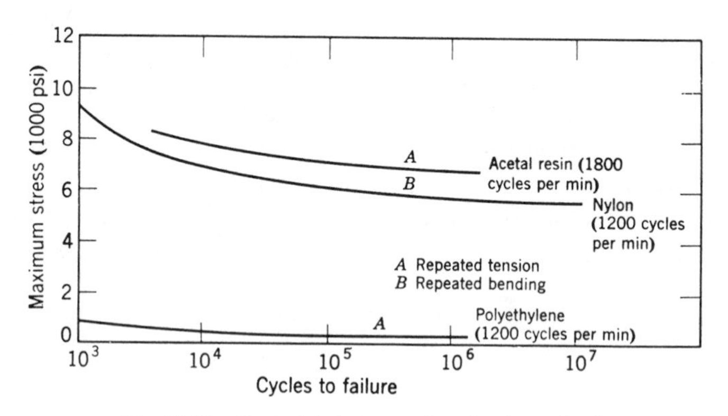

Fig. 12.23 Typical fatigue test data for three plastics.

Unidirectional nonmetallics with high modulus fibers with tension fatigue loads for 10^7 cycles (13, 21) have

$$0.70 \leq \frac{FS}{TS} \leq 0.95 \qquad 12.51$$

However, for fully reversed stress (21, 22) for unidirectional and cross plied laminates

$$0.30 \leq \frac{FS}{TS} \leq 0.60 \qquad 12.52$$

Surface endurance strengths (15) such as in gear teeth with contact stresses (Chapter 13) are generally 2.5–5.0 times flexural endurance limits.

12-4 APPLICATION OF FATIGUE DATA TO DESIGN

Interpretation of data

One of the principal problems encountered in the handling of fatigue data* is the often large scatter of results, making it difficult to plot meaningful *S-N* curves. Sometimes the data are simply bounded by lines that just enclose the data of widest scatter and the results are reported as bands or ranges of results. Sometimes an average curve is attempted. Many methods of analyzing or grouping data for interpretation have been reported (7), but for present purposes a statistical method suggested by Peterson will be discussed. Peterson's method involves a statistical analysis of data as do other methods, but it has the advantage of being applicable to data recorded from tests to which no statistical planning was applied. It should be noted that the various methods available provide only a reasonable basis for establishing an *S-N* curve that can then be used as the basis for further predictions.

Peterson's method is based on the assumptions that the population is normally dispersed with respect to stress and that the results obtained at one lifetime can be extrapolated to another lifetime. If X represents stress, the proportional number of values y that are X units from the mean value is

$$y = \frac{1}{\sigma\sqrt{2\pi}} e^{-X^2/2\sigma^2} \qquad 12.53$$

where σ is the standard deviation (not stress as σ usually signifies) defined as

$$\sigma = \sqrt{[\sum X_i^2 - m(\bar{X}_i)^2]/m} \qquad 12.54$$

where

$$\bar{X}_i = \frac{1}{m} \sum_i X_i \qquad 12.55$$

and m is the number of samples tested. The data obtained from the fatigue tests are plotted as *S-N* curves in such a way that there is a balance between points above and points below the curve. Let N_i denote an observed lifetime for a specimen tested at stress σ_i and let σ_{ai} denote the stress on the average curve at lifetime N_i; then

$$\Delta\sigma_i = \sigma_i - \sigma_{ai} \qquad 12.56$$

Values of $\Delta\sigma_i/\sigma_{ai}$ should add up to zero; if they do not, an adjustment in the average curve is required. Next a lifetime N_e is selected for calculation; to this will correspond an average stress σ_e on the curve and by linear extrapolation

$$\sigma_i = \sigma_e(\Delta\sigma_i/\sigma_{ai} + 1) \qquad 12.57$$

*Most fatigue data are for completely reversed stress in pure bending. For correlation purposes it has been found that tension-compression or push-pull tests give an endurance limit approximately 75% of that obtained in pure bending fatigue tests.

Table 12.3 Experimental Data and Illustration of Calculation Using Peterson's Method [After Grover, Gordon, and Jackson (7), courtesy of The Bureau of Naval Weapons, Department of the Navy.]

Specimen	σ_i, 1000 psi	N_i, Cycles	σ_{ai}, 1000 psi	$\Delta\sigma_i$, 1000 psi	$(\Delta\sigma_i/\sigma_{ai})^2$
1	50	13,000	54.5	−4.5	0.0068
2	50	13,100	54.4	−4.4	0.0065
3	50	24,000	49.5	+0.5	0.0001
4	50	28,000	48.3	+1.7	0.0012
5	50	40,000	45.3	+4.7	0.0107
6	40	42,000	45.0	−5.0	0.0123
7	40	53,000	43.0	−3.0	0.0048
8	40	101,000	37.6	+2.4	0.0040
9	40	111,000	37.0	+3.0	0.0065
10	30	211,000	31.5	−1.5	0.0022
11	30	226,000	31.0	−1.0	0.0010
12	30	255,000	30.1	−0.1	0
13	30	305,700	29.0	+1.0	0.0011
14	30	500,000	27.0	+3.0	0.0123
15	25	1,169,000	25.1	−0.1	0
16	25	1,728,900	24.6	+0.4	0.0002
17	25	2,116,000	24.4	+0.6	0.0006
18	25	2,289,300	24.3	+0.7	0.0008
19	25	4,374,000	23.7	+1.3	0.0030
20	25	4,550,000	23.6	+1.4	0.0035
21	25	7,280,000	23.3	+1.7	0.0053
22	24	993,000	25.3	−1.3	0.0026
23	24	2,579,000	24.2	−0.2	0
24	24	7,913,700	23.3	+0.7	0.0009
25	24	8,230,000	23.2	+0.8	0.0011
26	24	8,389,000	23.2	+0.8	0.0011
27	23	1,365,000	24.9	−1.9	0.0058
28	23	1,940,000	24.5	−1.5	0.0037
29	23	9,390,000	23.1	−0.1	0
30	23	12,131,000	23.0	0	0
31	23	13,260,000	22.9	+0.1	0
32	22	9,903,000	23.1	−1.1	0.0022
33	22	14,488,000	22.8	−0.8	0.0012
34	22	20,524,000	22.7	−0.7	0.0009
35	22	21,305,000	22.7	−0.7	0.0009
36	22	32,490,000	22.6	−0.6	0.0007
					0.1054

Table 12.4 Population Density in a Normal Distribution [From (7), courtesy of Bureau of Naval Weapons, Department of the Navy]

K	Values within $\pm K\sigma$ of Mean, %
1.0	68.27
1.645	90.00
1.96	95.00
2.0	95.44
2.575	99.00
3.0	99.74
4.0	99.99

and

$$\sigma' = \sigma_e \sqrt{\frac{\sum(\Delta\sigma_i/\sigma_{ai})^2}{m}} \qquad 12.58$$

For example, some actual data are presented in Table 12.3 for a selected stress σ_e of 23,000 psi. Applying Eq. 12.58 gives

$$\sigma' = 23{,}000\sqrt{0.1054/36} = 1250 \text{ psi}$$

which is a measure of standard deviation in stress at lifetime N_e of the population from which m samples were drawn. The percentage of values with $\pm K\sigma$ from the mean is given in Table 12.4.

Cumulative damage. Many structures and machine parts are cyclically stressed over a lifetime at several amplitudes, and this situation makes it somewhat difficult for the designer trying to predict what kind of a random amplitude the part is likely to sustain. An approximate method for handling such a situation was suggested by Miner (6). This method must be used with considerable caution since there is no allowance made for other variables that could affect the results. However, if due regard is given to the limitations and the data are carefully selected, a fairly good approximation can be obtained. Miner's suggestion was that if a specimen is cycled for n cycles at stress σ_n with an associated life of N_n cycles, m cycles at σ_m, p cycles at σ_p, and so on, then the cumulative cycle ratio would be unity; that is,

$$\frac{n}{N_n} + \frac{m}{N_m} + \frac{p}{N_p} + \ldots = 1 \qquad 12.59$$

which is illustrated in Fig. 12.24. A practical example of the use of this concept is given by Graham (6)* for a steel gear and pinion application. First the *S-N* curve for the material is expressed mathematically as

$$N = C\sigma^{-1/a} \qquad 12.60$$

*Courtesy of The Society of Automotive Engineers.

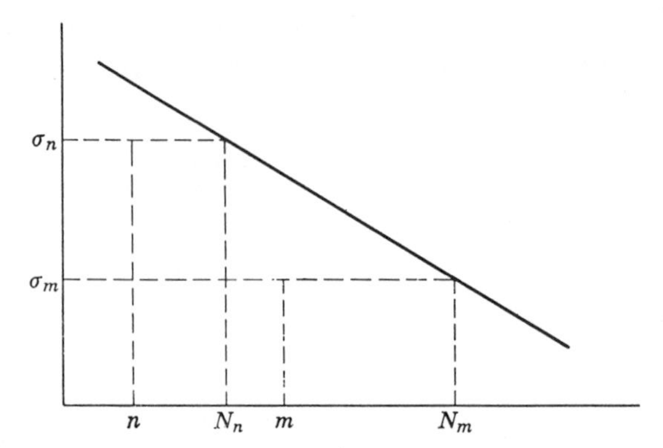

Fig. 12.24 Miner's theory of cumulative damage.

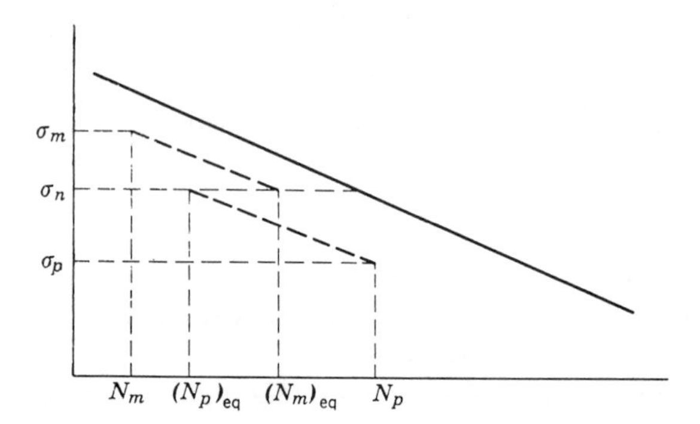

Fig. 12.25 Graphical illustration of Eq. 12.61

where C is a constant and a is the slope of the curve. Taking the ratio of cycles at σ_n to cycles at σ_m, Eq. 12.60 becomes

$$N_n = N_m(\sigma_m/\sigma_n)^{1/a} \qquad 12.61$$

which relates cycles N_m at σ_m to an equivalent N_n cycles at σ_n as shown in Fig. 12.25. Essentially, the cycles required at σ_n to accomplish the same damage at other stress levels is obtained. In Fig. 12.25, $(N_m)_{eq}$ cycles are required at σ_n to do the same amount of damage as N_m cycles at σ_m. The *S-N* data are shown in Fig. 12.26 and the results of calculation are shown in Table 12.5. The load level chosen for σ_n was 58,750 in.-lb torque and the values in the second column of Table 12.5 show how many cycles at 58,750 in.-lb would be required to do the same amount of damage as one cycle at the torque in the first column. For example, one cycle between 90,000 to 100,000 in.-lb would do the same damage to the gear as 39.5 cycles at a torque of 58,750 in.-lb. The equivalent cycles per hour is

$$\text{cycles/hr} = (\text{rpm})(60)(T_t/T_d)^{1/a} \qquad 12.62$$

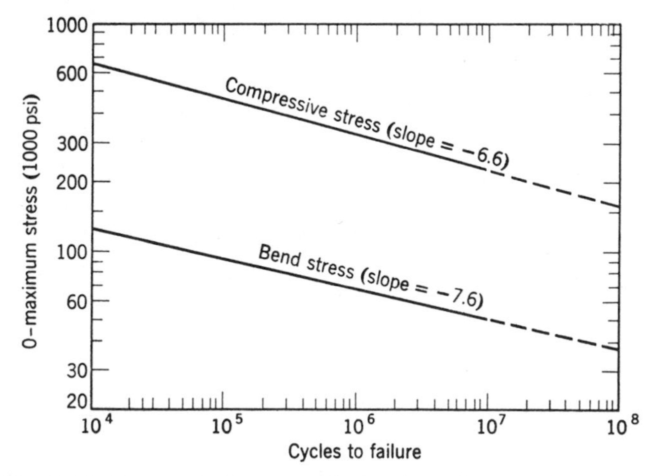

Fig. 12.26 Spur gear fatigue curves for bending and compression stresses.

Table 12.5. *Analysis of Fatigue Data*

Axle Torque Levels, 1000 in.-lb.	$\left(\frac{T}{58,750}\right)^{7.6}$ Bending	Observed Cycles per Hour	Equivalent Revolutions per Hour
90–100	39.5	1.8	71
80–90	17.0	8.1	138
70–80	6.5	20.5	133
60–70	2.2	33.4	73
50–60	0.6	53.1	32
40–50	0.13	76.0	10
30–40	0.0195	107.0	2
20–30	0.0015	130.0	0
Total equivalent revolutions per hour			459
Total equivalent pinion revolutions per hour			1,940
Stress in gear at 58,750 in.-lb torque			78,000 psi
Stress in pinion at 58,750 in.-lb torque			58,300 psi
Expected cycles to failure (Fig. 12.26)			
	Gear		320,000
	Pinion		3,000,000
Expected life under each condition			
	Gear		700 hr
	Pinion		1,540 hr

Table 12.6. *Miner's Constant C*

Reference	Value	Limitations
Sors (24)	1.75–2.35 Mild steel	Increasing Stress
	1.46–1.56 Ni Cr Mo steel	Increasing stress
	0.75–1.1 Mild steel	Decreasing stress
	0.91–1.01 Ni Cr Mo steel	Decreasing stress
Juvinall (11)	0.18–23	
Frost (5)	0.3–3.0	
Freudenthal (5)	0.1–0.6	Randomized six-level block test on high strength aluminum
Miner's tests (11)	0.61–1.45	
	0.7–2.2	
Forrest (4)	0.3–1.64	Spectrum test on aluminum
	0.14–22.8	High stress to low on aluminum (two-step tests)
	0.74–2.27	Low stress to high stress aluminum (two-step tests)
	0.65–1.7	H-L (two-step tests) on steels
	0.65–2.35	L-H (two-step tests) on steels

where T_t is the axle torque and T_d the design torque of 58,750 in.-lb. Note that most of the damage occurs at high load levels. The summation of Column 4 gives the total damage in terms of revolutions per hour of the axle at a torque of 58,750 in.-lb (459 rph). The cycles to failure of the gear at this torque, from the *S-N* curve are 320,000 cycles. By dividing the cycles to failure by the equivalent damage cycles per hour the life of the gear is predicted at 700 hr. By the same procedure the pinion life was predicted at 1540 hr.

Miner's rule is

$$\sum \frac{n_i}{N_i} = C \qquad 12.63$$

The constant C is reported as one; however, in Table 12.6 the values vary. Osgood (20) thoroughly discusses the various methods using Miner's rule in cumulative damage.

Low cycle fatigue. In general, any cyclic loading where there are less than about 10,000 cycles of stress beyond the elastic limit is considered low cycle fatigue. In usual mechanical testing a tensile test may be regarded as a one-half cycle test. For design purposes Tavernelli and Coffin (25) have found that the following formula agrees quite well with experimental data on steel, copper, aluminum, nickel, stainless steel, and titanium.

$$\sigma = (Ec/2N^{1/2}) + \sigma_e \qquad 12.64$$

Fig. 12.27 Prismatic bar subjected to alternating axial fatigue stress.

where σ is the stress amplitude (pseudoelastic stress), σ_e the endurance limit, E the modulus of elasticity, N the cycles to failure and

$$c = \tfrac{1}{2} \ln \left(\frac{100}{100 - \%\text{R. A.}} \right) \tag{12.65}$$

where R. A. is the reduction of area in a tensile test of the material. For example, suppose that a steel has an endurance limit of 20,000 psi at 10^8 cycles and a c value of 0.525. Then if a pseudoelastic stress of 50,000 psi is applied, N would be about 69,000 cycles. A more detailed review of low cycle fatigue of metals is given in (28). Fatigue analyses of pressure vessels and pressure vessel parts are made using the above procedures, and the detailed procedure is outlined in the ASME Boiler and Pressure Vessel Code (32).

Some examples illustrating the use of the preceding analyses follow.

EXAMPLE 1. PRISMATIC BAR SUBJECTED TO ALTERNATING AXIAL STRESS (FIG. 12.27). *Given:* F_{max}; F_{min}; σ_e; σ_y.

In order to compute the required cross-sectional area A of the bar shown in Fig. 12.27 the maximum stress is calculated from Eq. 12.11

$$\sigma_{max} = \sigma_e + \sigma_m \left(1 - \frac{\sigma_e}{\sigma_y} \right)$$

However,

$$\sigma_{max} = \frac{F_{max}}{A}$$

and the mean stress is

$$\sigma_m = \frac{F_{max} + F_{min}}{2A}$$

From these relationships

$$A = \frac{1}{\sigma_e} \left[F_{max} - \frac{F_{max} + F_{min}}{2} \left(1 - \frac{\sigma_e}{\sigma_y} \right) \right]$$

EXAMPLE 2. PRISMATIC BAR SUBJECTED TO FLUCTUATING BENDING STRESS (FIG. 12.28). *Given:* M_{max}; M_{min}; σ_e; σ_y.

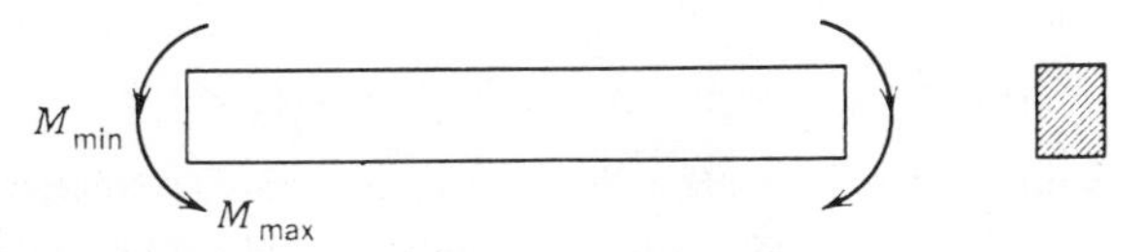

Fig. 12.28 Prismatic bar subjected to alternating bending fatigue stress.

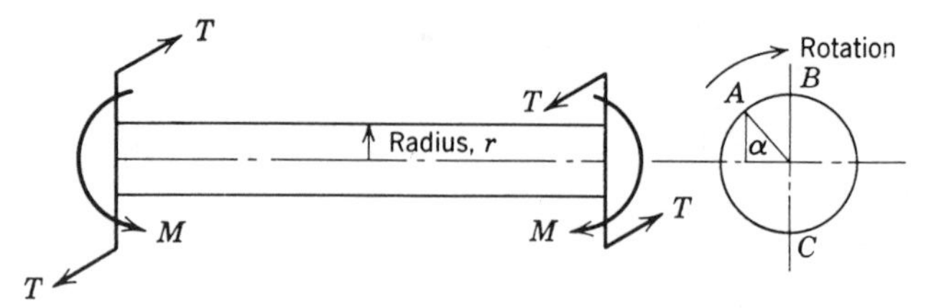

Fig. 12.29 Prismatic bar subjected to static pure bending and static torsion while being rotated.

In this example the required section modulus I/c is computed.

$$\sigma_{\max} = \sigma_e + \sigma_m\left(1 - \frac{\sigma_e}{\sigma_y}\right) = M_{\max}(c/I)$$

$$\sigma_m = \left(\frac{M_{\max} + M_{\min}}{2}\right)\left(\frac{c}{I}\right)$$

from which

$$\left(\frac{I}{c}\right) = \frac{1}{\sigma_e}\left[M_{\max} - \left(\frac{M_{\max} + M_{\min}}{2}\right)\right]$$

EXAMPLE 3. ROTATING SHAFT SUBJECTED TO PURE BENDING AND STATIC TORQUE (FIG. 12.29). For this situation there is a constantly applied torque T giving a shear stress of

$$\tau_{xy} = Tr/J$$

Because of the bending, an axial stress is also generated, which by Eq. 12.1 is

$$\sigma_x = \frac{Mr \sin \alpha}{I}$$

As the shaft rotates, the axial stress σ_x is maximum at $\alpha = \pi/2$, becomes zero at $\alpha = 0$, and is minimum $(-)$ at $\alpha = 3\pi/2$. The mean axial stress is thus at $\alpha = 0$ and π and is zero. The shear is always applied and its mean value (as well as its maximum) is given by

$$\tau_{xy} = Tr/J$$

From Eq. 12.22, where σ_y is zero, σ_x is given by Eq. 12.13 and τ_{xy} is given previously.

$$\sigma_e = \sqrt{\sigma_x^{\,2} + 3\tau_{xy}^{\,2}} - (1 - \sigma_e/\sigma_y)\sqrt{3\tau_{xy}^{\,2}}$$

or

$$r = \sqrt[3]{\frac{4T}{\pi(\sigma_e/\sigma_y)}}\left\{\left[\sqrt{4\left(\frac{M}{T}\right)^2 + 3}\right] - \left[\sqrt{3}\left(1 - \frac{\sigma_e}{\sigma_y}\right)\right]\right\}$$

EXAMPLE 4. FATIGUE STRENGTH OF ANISOTROPIC MATERIALS. In the following example several cases will be examined where the material is anisotropic and, sometimes, where stress raisers are present. The general case of anisotropic material is discussed in Chapter 4, where it is shown that

$$2(\bar{\sigma})^2 = L(\sigma_1 - \sigma_2)^2 + M(\sigma_2 - \sigma_3)^2 + N(\sigma_3 - \sigma_1)^2$$

For fatigue application of this relationship the equivalent stress $\bar{\sigma}$ is replaced by the equivalent endurance limit of the material σ_e. If, for example, the endurance limit data are for completely reversed pure bending, the correlation value would be 75% of this figure as

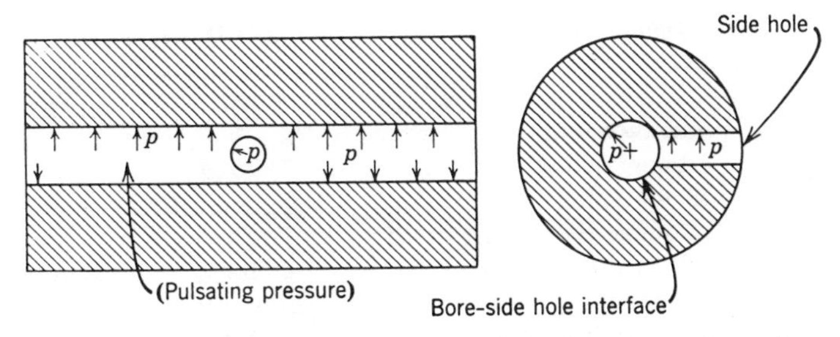

Fig. 12.30 Heavy-walled cylinder with side hole subjected to pulsating internal pressure.

mentioned earlier. The anisotropy factors are defined as

$$L = 2(\sigma_e/\sigma_{01})^2 - (\sigma_e/\sigma_{03})^2[\beta^2 - (\alpha/\beta)^2 + 1]$$
$$M = 2(\sigma_e/\sigma_{02})^2 - 2(\sigma_e/\sigma_{01})^2 + (\sigma_e/\sigma_{03})^2[\beta^2 - (\alpha/\beta)^2 + 1]$$

and

$$N = (\sigma_e/\sigma_{03})^2[\beta^2 - (\alpha/\beta)^2 + 1]$$

where σ_{01}, σ_{02}, and σ_{03} are endurance limits in the 1, 2, and 3 directions respectively and

$$\alpha = \sigma_{02}/\sigma_{01} \quad \text{and} \quad \beta = \sigma_{03}/\sigma_{01}$$

Consider, for example, a heavy-walled steel cylinder subjected to pulsating internal pressure.* The material is anisotropic, that is, the endurance limits are different in the longitudinal and transverse directions. Furthermore, the cylinder has a side hole (Fig. 12.30) which causes a mechanical stress concentration effect.† For this situation the critical location is at the bore-side hole interface where the stresses are as follows:

$$\sigma_1 = \sigma_h = Kp_e\left(\frac{R^2+1}{R^2-1}\right) \qquad \text{(hoop stress)}$$

$$\sigma_2 = \sigma_r = -p_e \qquad \text{(radial stress)}$$

and

$$\sigma_3 = \sigma_z = -p_e \qquad \text{(longitudinal stress)}$$

where p_e is the endurance internal pressure, K a stress concentration factor for the hoop stress at the bore-side hole interface, and R the ratio of outside to bore diameter of the cylinder. In addition.

$$(\sigma_1 - \sigma_2)^2 = (p_e)^2\left[K\left(\frac{R^2+1}{R^2-1}\right) + 1\right]^2$$

$$(\sigma_2 - \sigma_3)^2 = 0$$

and

$$(\sigma_3 - \sigma_1)^2 = (p_e)^2\left[K\left(\frac{R^2+1}{R^2-1}\right) + 1\right]^2$$

*Cylinders under pulsating internal pressure have also been considered by Narduzzi and Welter (19) and by Morrison, Crossland, and Parry (18).

†See Chapter 13 for a discussion of this problem.

Table 12.7 Fatigue Data for Anisotropic Steel

Steel Hardness, Rockwell C	Endurance Limit of Longitudinal Samples for Completely Reversed Pure Bending, psi	Correlation Endurance Limit, σ_e (75% of Reversed Bending Strength), psi	Ratio of Longitudinal to Transverse Endurance Limit for Pure Bending	Anisotropy Factors		
				L	M	N
25	116,250	87,500	1.588	1.92	0	3.124
37	148,400	111,500	1.628	2.03	0	3.270
47	151,130	113,000	1.750	2.38	0	3.720

With these quantities

$$2(\sigma_e)^2 = L(p_e)^2\left[K\left(\frac{R^2+1}{R^2-1}\right)+1\right]^2 + N(p_e)^2\left[K\left(\frac{R^2+1}{R^2-1}\right)+1\right]^2$$

which reduces to

$$p_e = \frac{\sqrt{2}\sigma_e}{\left[K\left(\frac{R^2+1}{R^2-1}\right)+1\right]\sqrt{L+N}}$$

It is emphasized again that the correlation value of σ_e must represent tension-compression, push-pull data for the material, which is approximately 75% of the endurance limit obtained in reversed bend tests. The preceding expression for endurance pressure has been applied to some fatigue tests carried out in the Du Pont Company with cylinders having a diameter ratio R of 2.5. It is shown in Chapter 13 that side holes in pressurized cylinders intensify the hoop stress by a factor of about 2.5; therefore for this case the preceding formula for endurance pressure becomes

$$p_e = \frac{\sigma_e}{3.16\sqrt{L+N}}$$

where σ_e, L, and N have to be calculated. The pertinent data are shown in Table 12.7 and the results of the calculations in Table 12.8. To illustrate, for the steel having a Rockwell

Table 12.8 Fatigue Data for Cylinders with Side Holes Subjected to Pulsating Internal Pressure

Steel Hardness, Rockwell C	Endurance Pressure, p_e (psi)		Difference, %
	Observed*	Calculated	
25	12,500	12,350	0
37	14,000	15,500	11
47	16,000	14,600	−9

* 250 pressure fluctuations per minute.

Table 12.9 Fatigue Data for Plain Cylinders Subjected to Pulsating Internal Pressure

Steel Hardness, Rockwell C	Endurance Pressure, p_e (psi)		Difference, %
	Observed*	Calculated	
25	26,000	23,200	−11
37	30,000	29,200	−3
47	35,000	27,600	−21

* 250 pressure fluctuations per minute.

hardness of 25 C,

$$\sigma_e = (116{,}250)(0.75) = 87{,}500 \text{ psi}$$

$$L = 2\left(\frac{\sigma_e}{\sigma_{01}}\right)^2 - \left(\frac{\sigma_e}{\sigma_{03}}\right)^2\left[\left(\frac{\sigma_{03}}{\sigma_{01}}\right)^2 - \left(\frac{\sigma_{02}/\sigma_{01}}{\sigma_{03}/\sigma_{01}}\right)^2 + 1\right]$$

or

$$L = 2(1.588)^2 - [(1.588)^2 - (0.63)^2 + 1] = 1.92$$

and

$$N = \left(\frac{\sigma_e}{\sigma_{03}}\right)^2\left[\left(\frac{\sigma_{03}}{\sigma_{01}}\right)^2 - \left(\frac{\sigma_{02}}{\sigma_{03}}\right)^2 + 1\right] = 3.124$$

$$p_e = \frac{87{,}500}{3.16\sqrt{1.92 + 3.124}} = 12{,}350 \text{ psi}$$

If the cylinders in question did not have the side holes, there would be no mechanical stress raiser and K would be unity, giving for the same size cylinders as above ($R = 2.5$)

$$p_e = \frac{\sigma_e}{1.68\sqrt{L+N}}$$

Some results obtained through the use of this formula are shown in Table 12.9.

In many instances the bore of the cylinder is protected against corrosion by a thin liner of some material which does not affect the mechanical strength of the cylinder but simply keeps the hydraulic fluid from contacting the bore of the cylinder. The inclusions are thus covered up and at the critical stress area there is no longitudinal stress so that

$$\sigma_1 = \sigma_h = p_e\left(\frac{R^2+1}{R^2-1}\right)$$

$$\sigma_2 = \sigma_r = -p_e$$

and

$$\sigma_3 = \sigma_z = 0$$

Consequently,

$$(\sigma_1 - \sigma_2)^2 = (p_e)^2\left(1 + \frac{R^2+1}{R^2-1}\right)^2$$

$$(\sigma_2 - \sigma_3)^2 = (p_e)^2$$

$$(\sigma_3 - \sigma_1)^2 = (p_e)^2\left(\frac{R^2+1}{R^2-1}\right)^2$$

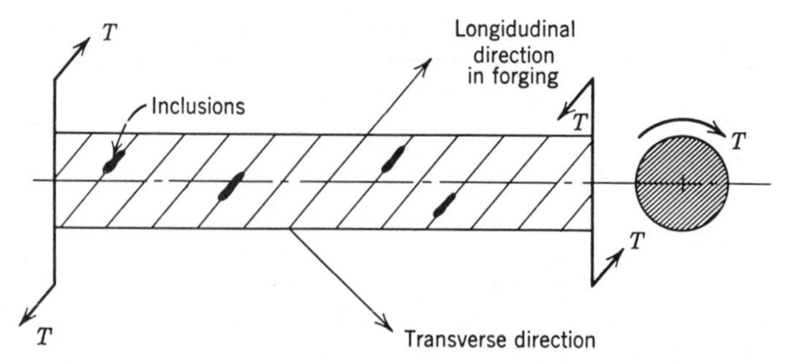

Fig. 12.31 Anisotropic bar of steel with inclusions subjected to torsional fatigue stress.

and

$$p_e = \frac{\sqrt{2}\,\sigma_e}{\left\{\left[\dfrac{4R^4}{(R^2-1)^2}\right]L + M + N\left(\dfrac{R^2+1}{R^2-1}\right)^2\right\}^{1/2}}$$

The preceding formula applied to steel having a hardness of 25 Rockwell C indicated an endurance pressure of 29,000 psi; the experimental value was 35,000 psi.

As another example consider Fig. 12.31 which shows a bar subjected to torsion in which the inclusions (longitudinal direction of the bar from which the fatigue specimens were taken) are oriented at 45° to the axis of the specimen. Application of torque thus produces only a stress σ_1 at the tip of the inclusion; at the tip boundary the other principal stress is zero since there can be no stress at a free boundary. Therefore $\sigma_2 = \sigma_3 = 0$ and

$$(\sigma_1 - \sigma_2) = \sigma_1{}^2$$
$$(\sigma_2 - \sigma_3)^2 = 0$$
$$(\sigma_3 - \sigma_1)^2 = \sigma_1{}^2$$

and

$$2(\sigma_e)^2 = L\sigma_1{}^2 + N\sigma_1{}^2 = \sigma_1{}^2(L+N)$$

from which

$$\sigma_1 = \frac{\sqrt{2}\,\sigma_e}{\sqrt{L+N}}$$

Some results obtained through the use of this formula are compared in Table 12.10 with experimental data.

Table 12.10. *Fatigue Data for Anisotropic Bars Subjected to Reversed Torsion Stress*

Steel Hardness, Rockwell C	Endurance Limit for Longitudinal Samples in Pure Bending (psi)	Correlation Endurance Limit (see Table 12.7) (psi)	Endurance Limit in Torsion, σ_1, (psi)		Difference, %
			Observed	Calculated	
25	116,250	87,500	75,830	55,000	− 27
37	148,400	111,500	98,750	66,500	− 33
47	151,130	113,000	108,860	65,250	− 40

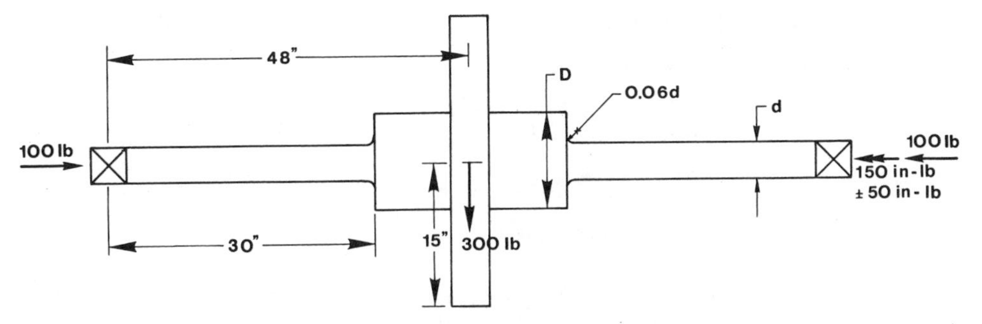

Fig. 12.32 Bull gear and shafting.

EXAMPLE 5. COMBINED FATIGUE STRESSES.

A large 300-lb gear, Fig. 12.32, transmits torque through a shaft whose bearings are preloaded with a 100-lb force so that the load in the shaft varies from 0–200 lb. The critical area is the change in shaft diameter which is made of SAE 4140 steel. The two shaft diameters are required. Assume $D/d = 2$. Reactions from gear contact are neglected.

$$T_m = 150 \text{ in.-lb}$$
$$T_r = \pm 50 \text{ in.-lb}$$
$$M_r = 150 \text{ lb } (30 \text{ in.})$$
$$F_m = 100 \text{ lb}$$
$$F_r = 100 \text{ lb}$$

$$\tau_m = \frac{T_m r}{J} = \frac{16T_m}{\pi d^3}; \qquad \tau_r = K_{tT}\frac{16T_r}{\pi d^3}; \qquad \sigma_{xr} = K_{tB}\frac{M_r c}{I} = K_{tB}\frac{32M_r}{\pi d^3};$$

$$\sigma_{xm} = \frac{4F_m}{\pi d^2}; \qquad \sigma_{xr} = K_{tF}\frac{4F_r}{\pi d^2}$$

From (3, 23, 24)

$$K_{tF} = 2.4 \qquad K_{tB} = 1.85 \qquad K_{tT} = 1.63$$

The stress concentration will be applied only to Eq. 12.28 since the system should operate above its transition temperature. The $\sigma_r - \sigma_m$ curve will be drawn and allowable stress levels obtained (Fig. 12.33).

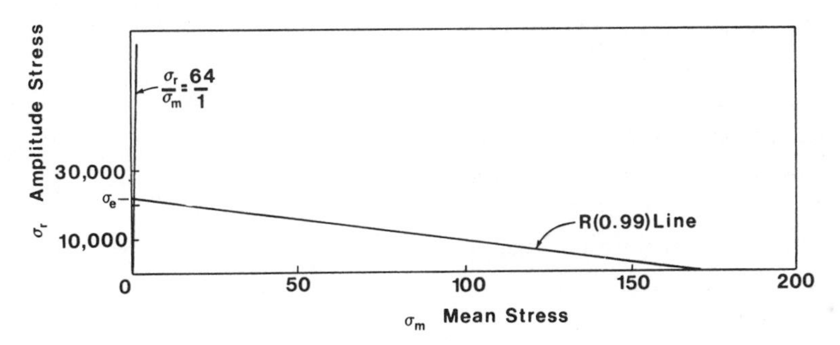

Fig. 12.33 $\sigma_r - \sigma_m$ curve for Fig. 12.32 example.

For the shoulder the combined stress are

$$\sigma_r = \sqrt{\sigma_{xr}^{\ 2} + 3\tau_{xyr}^{\ 2}}$$

$$\sigma_{xr} = K_{tB}\frac{32M_r}{\pi d^3} + K_{tF}\frac{4F_r}{\pi d^2} = \frac{16}{\pi d^3}(16{,}650 + 60d)$$

$$\tau_{xyr} = K_{tT}\frac{16T_r}{\pi d^3} = \frac{16}{\pi d^3}(81.5)$$

substituting

$$\sigma_r = \frac{16}{\pi d^3}\sqrt{(16{,}650 + 60d)^2 + 3(81.5)^2}$$

$$\sigma_m = \sqrt{\sigma_{xm}^{\ 2} + 3\tau_{xym}^{\ 2}}$$

$$\sigma_{xm} = \frac{4F_m}{\pi d^2} = (25)\frac{16}{\pi d^3}$$

$$\sigma_{xym} = \frac{16T_m}{\pi d^3} = \frac{16}{\pi d^3}\left[150\right]$$

substituting

$$\sigma_m = \frac{16}{\pi d^3}\sqrt{(25d)^2 + 3(150)^2}$$

The $\sigma_r - \sigma_m$ curve will be entered on the slope

$$\frac{\sigma_r}{\sigma_m} = \frac{\sqrt{(16{,}650 + 60d)^2 + 3(81.5)^2}}{\sqrt{(25d)^2 + 3(150)^2}}$$

If the d term representing the axial force is dropped

$$\frac{\sigma_r}{\sigma_m} = 64.09$$

This slope on the $\sigma_r - \sigma_m$ curve is very near the value for σ_e

4140 steel $\frac{1}{2}$-in. diameter $\sigma_{ut} = 210$ kpsi heat treated and drawn to 800°F

$$\sigma_e = k_a k_b k_c \dots k_1 \sigma_e^{\ 1}$$

$\sigma_e^{\ 1}$ $\sigma_{ut/2}$ but never greater than 100 kpsi
k_a Fig. 12.9 (8rms) 0.9
k_b 0.85 diameter d assumed less than 2 in.
k_c 0.814 for both σ_{ut} and σ_e Eq. 12.34 $R(0.99)$
k_d 1 since oils operate at 180°F
k_e included in the stresses, 1.
k_f shot peening of the corner polished 1.04
k_g 1
k_h oil environment corrosion small, 1
k_i no surface treatment or hardening 1
k_j no fretting, 1
k_k must know operating speed to compare to critical speed, impact, 1/3.

$$\sigma_e = (0.9)(0.85)(0.814)(1.04)(\tfrac{1}{3})100{,}000 \text{ psi}$$

$$\sigma_e = 21{,}587$$

The factor of safety is selected as 1.5

$$\sigma_r = \frac{\sigma_e}{1.5} = \frac{16}{\pi d^3}\sqrt{(16{,}650)^2 + 3(81.5)^2}$$

$$d = \left[\frac{16(1.5)}{\pi(21{,}587)}\sqrt{(16{,}650)^2 + 3(81.5)^2}\right]^{1/3} = 1.81 \text{ in.}$$

$$D = 3.61 \text{ in.} \qquad r = 0.108 \text{ in.}$$

The next steps are to

1. check the critical speed
2. With $r = 0.108$ $q \approx 1$ so $k_f \approx k_t$
3. Include the axial force and calculate the actual factor of safety.

12-5 FATIGUE CONSIDERATIONS IN DESIGN CODES

National codes and standards frequently provide methods of analysis that address the problem of fatigue, or cyclic loading. In applications where use of such codes is mandatory, the rules must be followed in detail. A notable example of this is provided by the ASME Boiler and Pressure Vessel Code (32) which contains detailed procedures for evaluating the fatigue behavior of pressure vessels and pressure vessel parts. This is an excellent treatment of the subject and is recommended to the reader for study and use [see also (27)]. Information on the American Institute of Steel Construction Code (AISC) is presented in (23, 36). An excellent source of data is also presented in (37).

In the example in Fig. 12.32, an allowable stress can be developed for comparison. If the impact constant k_k is placed in the stress calculations, the allowable stress is 0.31 σ_{ut}.

12-6 SUMMARY

1. When designing a system or a component, the critical frequencies, deflections, shock and vibration levels, operating temperatures, and surrounding environment must be known. The system must meet the overall requirements while often components must exceed them, as in deflections. Components with maximum deflections A when assembled will always have larger maximum deflections, as do springs in series.

2. Determine the component critical parameters such as deflections, stresses, frequency, or failure modes. Attempt to assign a value to these parameters even though it is understood that they will often change.

3. When the components are designed, the maximum and minimum static loads should be corrected to include the shock and vibration effects in the

design. This does not mean that a vibration analysis can be neglected. Also, fracture mechanics (Chapter 14) should be checked to include important parameters.

4. Free-body diagrams when drawn will determine end conditions at mating parts. Such items as self-aligning bearings making simply supported ends while double bearings approach fixed-end conditions are important.

5. The determination of the effects of k values, Eq. 12.29, on endurance limits will force the designer to select the materials and/or the manufacturing process such as machined or cast parts, heat treatment, and surface coating or treatment.

6. When the $\sigma_r - \sigma_m$ curve is selected or drawn, the proper safety factor should be determined.

7. When complex systems or parts are designed, a finite element check for frequencies, deflections, and loads from the known operating conditions will allow a check on a system design before parts are fabricated. The parts should further be checked by fracture mechanics (Chapter 14) for acceptability.

8. When a part or a system is manufactured, finished products should be tested to failure to verify the designing and manufacturing. However, there are cases where only one item is produced and used. Then, some form of instrumentation recordings or inspections at servicing must be performed to maintain a check on the system.

9. Keep a record of successful designs and failures so that design criteria can be modified for a particular class of designs. This is how design criteria and future design codes will be developed.

REFERENCES

1. G. Castleberry, *Mach. Des.*, **50** (4), 108–110 (February 23, 1978).
2. A. D. Deutschman, W. J. Michels, and C. E. Wilson, *Machine Design*, MacMillan Publishing Co., New York, N.Y. (1975).
3. V. M. Faires, *Design of Machine Elements*, 4th ed., MacMillan Co., New York, N.Y. (1965).
4. P. G. Forrest, *Fatigue of Metals*, Addison-Wesley, Reading, Mass. (1962).
5. N.E. Frost, K. J. Marsh, and L. P. Pook, *Metal Fatigue*, Oxford University Press, London (1974).
6. J. A. Graham, *A Direct Approach to Designing Load Carrying Members*, Div. IV Meeting of the Society of Automotive Engineers, Iron and Steel Technical Committee, Detroit, Mich. (January 10, 1962).
7. H. J. Grover, S. A. Gordon, and L. R. Jackson, *Fatigue of Metals and Structures*, NAVWEPS Report 00–25–534, Bureau of Naval Weapons, Department of the Navy, Washington, D.C. (1960).
8. H. C. Hagendorf, and F. A. Pall, *A Rational Theory of Fatigue Crack Growth*, NA–74–278, Rockwell International, Los Angeles, Cal. (1974).
9. C. R. Hine, *Machine Tools and Processes for Engineers*, McGraw-Hill Book Co., New York, N.Y. (1971).

10. R. C. Johnson, *Mach. Des.*, **45** (11), 108 (May 3, 1973).

11. R. C. Juvinall, *Stress, Strain, and Strength*, McGraw-Hill Book Co., New York, N.Y. (1967).

12. D. B. Kececioglu and L. B. Chester, *Trans. Soc. Mech. Eng. (J. Eng. Ind.)*, **98** (1), Series B, 153–160 (February 1976).

13. H. S. Kliger, *Plast. Des. Forum*, **2** (3), 36–40 (May/June 1977).

14. D. Landau, *Fatigue of Metals—Some Facts for the Designing Engineer*, 2nd. ed., the Nitralloy Corp., New York, N.Y. (1942).

15. C. Lipson and R. C. Juvinall, *Stress and Strength*, MacMillan Co., New York, N. Y. (1963).

16. R. C. McMaster, *Non-Destructive Testing Handbook*, Vol. I, Ronald Press, New York, N.Y. (1959).

17. D. F. Miner and J. B. Seastone, *Handbook of Engineering Materials*, John Wiley and Sons, Inc., New York, N.Y. (1955).

18. J. L. M. Morrison, B. Crossland, and J. S. C. Parry, *Proc. Inst. Mech. Eng. (London)*, **174** (2), 95–117 (1960).

19. E. D. Narduzzi and G. Welter, *Weld. J. Res. Suppl.*, **19** (5), 230s–238s, (May 1954).

20. C. C. Osgood, *Fatigue Design*, Wiley-Interscience, New York, N.Y. (1970).

21. M. J. Owen, "Fatigue of Carbon-Fiber-Reinforced Plastics," in L. J. Broutman and R. H. Krock, Eds., *Composite Materials*, Vol. 5, Academic Press, New York, N.Y. (1974).

22. M. J. Salkind, "Fatigue of Composites," *Composite Materials*, STP 497, ASTM, Philadelphia, Pa. (1971).

23. J. E. Shigley, *Mechanical Engineering Design*, 3rd. ed., McGraw-Hill Book Co., New York, N.Y. (1977).

24. L. Sors, *Fatigue Design of Machine Components*, Pergamon Press, Oxford (1971).

25. J. F. Tavernelli and L. F. Coffin, *Experimental Support for Generalized Equation Predicting Low Cycle Fatigue*, Instron Engineering Corp., Application Series M-3 (1959).

26. S. Timoshenko, *Strength of Materials*, Part II, 3rd. ed., D. Van Nostrand Co., Princeton, N.J. (1956).

27. P. H. Wirsching, and J. E. Kempert, *Mach. Des.*, **48** (21), 108–113 (September 23, 1976).

28. J. T. P. Yao and W. H. Munse, *Weld. J. Res. Suppl.*, **41** (4), 182s–192s (April 1962).

29. *Advanced Approaches to Fatigue Evaluation*, NASA SP-309, Sixth ICAF Symposium (May 1971).

30. *Aerospace Structural Metals Handbook*, AFML-TR-68-115, Mechanical Properties Data Center, 13919 West Bay Shore Drive, Traverse City, Mich. (1977).

31. *An Index of U.S. Voluntary Engineering Standards*, W. J. Slattery, Ed., NBS 329 plus Supplements 1 and 2, U. S. Government Printing Office, Washington, D.C. (1971).

32. *ASME Boiler and Pressure Vessel Code*, The American Society of Mechanical Engineers, United Engineering Center, 345 E. 47th St., New York, N.Y. (1977).

33. *Machinery's Handbook*, 20th ed., Industrial Press Inc., New York, N.Y. (1975).

34. *Metals Handbook Vol. I–V*, 8th ed., American Society for Metals, Metals Park, Oh.

35. *Prevention of the Failure of Metals Under Repeated Stress*, John Wiley and Sons, Inc., New York, N.Y. (1941).

36. *Steel Construction*, 7th ed., AISC Manual, American Institute of Steel Construction, 101 Park Ave., New York, N.Y. (1970).

37. *Strength of Metal Aircraft Elements*, Military Handbook MIL-HDBK-5B, Washington D.C. (1971).

38. *Timber Construction Manual*, 2nd ed. AITC, John Wiley and Sons, Inc., New York, N.Y. (1974).

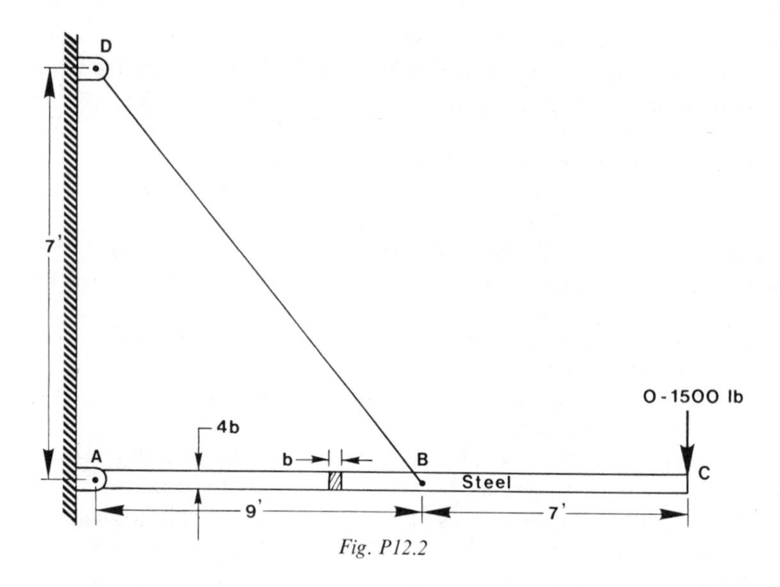

Fig. P12.2

PROBLEMS

12.1 In Problem 10.3 find the modification to the clearance and load if the S hook is to be used for 0–500-lb repeated loads with a factor of safety of 1.25.

12.2 The beam is repeatedly loaded as shown in Fig. P12.2. Select material and size the members for a factor of safety of 1.5.

12.3 A closed-end, thick-wall cylinder supports an internal pressure of 50 kpsi $\pm$ 10 kpsi. The inside diameter is 10 in., and the cylinder has two openings for 0.250-in. I.D. thick-wall tubes. Design the cylinder for a factor of safety of 1.5 using SAE 1045 steel.

12.4 In Problem 2.2 size and select material for all of the elements in the lifting tongs. Include the effect of three mph at lift.

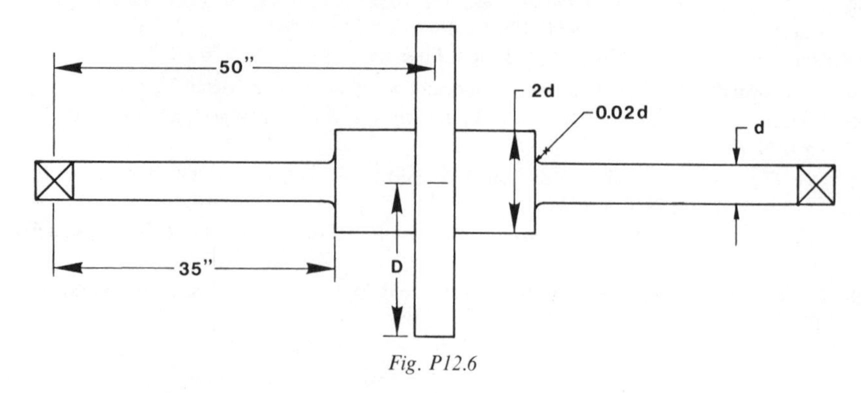

Fig. P12.6

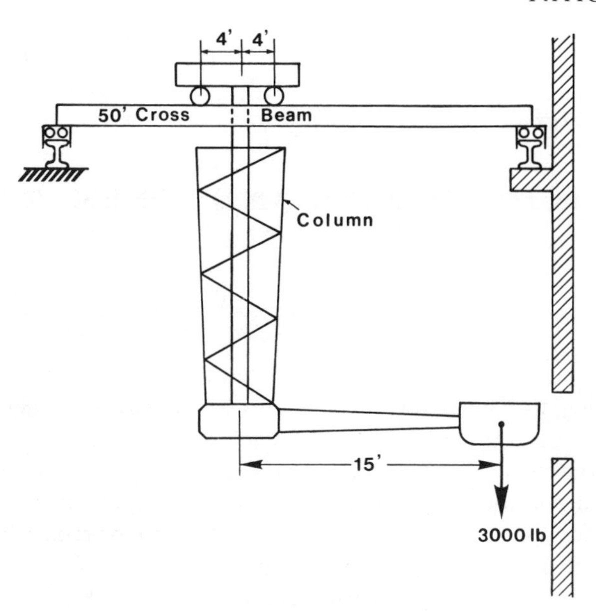

Fig. P12.7

12.5 In Problem 2.3 find the factor of safety for infinite life for a SAE 4130 hot-rolled and heat-treated shaft using only the loads from the chain drive.

12.6 A 600-lb disk loaded as in Fig. 12.32 is mounted to a shaft by means of a key way (Fig. P12.6). If the disk is 2 in. thick and the shaft rotates 3600 rpm, what will be the value of d for a shaft material of SAE 4340 steel and σ_{ut} of 260 kpsi?

12.7 A furnace loader is shown (Fig. P12.7). First find the required column weight to keep the loader from tipping. Then, select a standard cross beam if one is available.

12.8 In Fig. 13.10 of the text $F = 20$ tons maximum. Determine the sizes w, a, h.

12.9 In Fig. 13.39 if $p = 500$ psi $\pm$ 100 psi and the pressure vessel is constructed of steel with $\sigma_y = 80{,}000$ psi, $\sigma_u = 95{,}000$ psi, size the wall thickness for infinite life.

12.10 In Problem 12.9 a pipe, 1 in. O.D. exists from the side of the vessel some distance away from the end. Find the fatigue life considering only the pressure stresses around the 1-in. O.D. pipe. Use the same vessel materials.

chapter 13

NOTCHES, HOLES, AND STRESS RAISERS

Stress concentration* effects were first related to theoretical concepts early in the last century in connection with failures of railway axles (89, 90); in this situation the stress was intensified at the wheel-axle junction which was subjected to an alternating bending stress. Since those early days the theory of elasticity has developed considerably, and this fact plus simultaneous development of reliable theories of mechanical failure of materials, and experimental techniques to ascertain the magnitude of stresses induced by various factors, has led to a presently large volume of information on the subject (87). In this chapter the subject of stress concentration in relation to design technique will be discussed at some length; the role of stress concentration in relation to brittleness of materials and crack propagation is discussed in Chapter 14.

Stress intensification effects in machine parts and structures range from purely internal phenomena associated with the brittle behavior of materials to effects commonly observed when various types of external notches, fillets, or holes are made part of the operating system. For this reason it is difficult to present a unified picture of the subject, and for this reason it is necessary to approach its various aspects in the light of information presented in other chapters of this book. For example, in order to understand certain brittle behavior problems discussed in Chapter 14 it is convenient to refer to those parts of this chapter having to do with the stress intensifying effect of notches; this is the basis of Griffith's theory of the brittle behavior of glass. Also, in order to fully appreciate fatigue failures it is necessary to understand Griffith's theory and also the behavior of parts that are externally notched. To this end, appropriate cross referencing is made between this chapter and Chapters 14 and 12; the composite picture can be obtained only by studying the particular sections of these three chapters which are interrelated. For the purposes of this chapter, stress concentration is discussed under three headings: first, some experimental techniques are presented which often provide the only measure of how stress changes in

*For example, in tension, if

$$\sigma = \text{stress} = K\frac{F}{A} = K\frac{\text{load}}{\text{area}} \qquad 13.1$$

K is a stress concentration factor.

various parts containing fillets, notches, and holes; second, several theoretical developments and their application to design are reviewed; and finally, a collection of design charts and curves for a variety of conditions is given. Many important references are cited at the end of the chapter and may be consulted for more details.

13-1 EXPERIMENTAL STRESS ANALYSIS TECHNIQUES

Experimental stress analysis techniques have been of great value in stress analysis and design work. Some particularly useful techniques are: (a) Brittle coating method (39, 54); (b) Stresscoat (11, 38, 54); (c) Photoelasticity (5, 12, 16, 21, 28, 54) and (d) Photostress (100). Various mechanical and electrical techniques (extensometer, strain gages, etc.) are commonplace and are often used in conjunction with the other methods mentioned above (37).

13-2 ANALYTICAL TREATMENT OF STRESS CONCENTRATION PROBLEMS

In the following section mathematical procedures are presented for stress concentration problems amenable to analytical solution through use of the theories of elasticity and plasticity.† The last section of the chapter contains additional results of studies by various workers and includes both mathematically and experimentally obtained results. For convenience all the results of stress concentration studies are grouped together in the figures and graphs in the last section of this chapter. It is important to note from the outset that stress concentration effects are manifested for all types of stressing. For uniaxial problems, like a bar in tension or bending, the stress concentration factor is applied to the tensile or compressive stress involved; in shear and torsion the stress concentration factor is applied to the shear stress and, for combined stresses, each principal stress may be modified by a stress concentration factor; this will become evident in the subsequent analyses. Many problems can only be solved experimentally, but Hetenyi and Liu (40) have developed an approximate analytical method that may be applied, for example, to fillets.

Circular hole in a plate

Stress concentrations associated with holes in plates have been the subject of considerable analysis. As an example the analytical treatment of a small circular hole in a thin plate will be presented here.*

† Notch stress theory is too complex to include here; see treatment by Neuber (67). Effects of notches are shown in Section 13-3.

*Analysis of stress and deflection of various perforated plates is discussed in (43, 59, 102) Mansfield (61) has developed formulas for analyzing "neutral holes," that is, reinforced holes in a plane sheet under any particular loading system which do not alter the stress distribution in the main body of the sheet. These "neutral holes" have the same stiffness and strength as the portion of the sheet that has been cut out. Conte (6) has considered a circular plate with an eccentric hole.

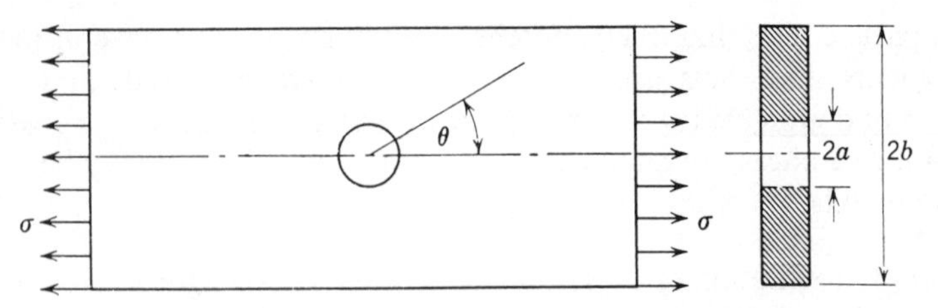

Fig. 13.1 Thin plate with small central circular hole subjected to uniform tension stress.

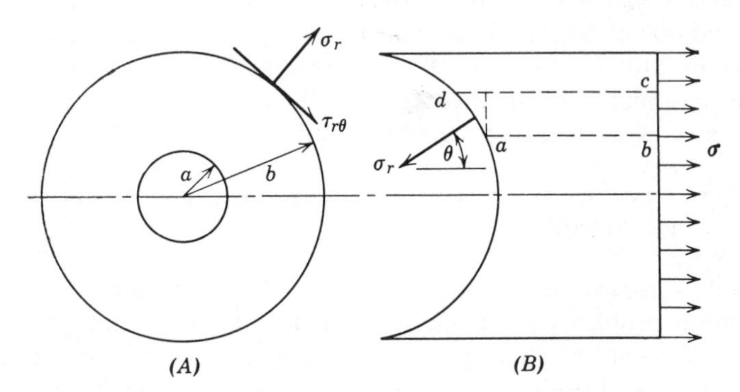

Fig. 13.2 Free bodies derived from hole location in Fig. 13.1.

Consider the plate shown in Fig. 13.1 of unit thickness and containing a small circular hole at its center and subjected to axial tension σ. It is assumed that the radius of the hole is small compared with the width of the plate. The first step in the solution of this problem is to set up the free bodies and examine equilibrium conditions. Thus if the hole is "removed" from the plate of Fig. 13.1, the free bodies are as shown in Fig. 13.2.

For the element *abcd* of differential area dA in Fig. 13.2*B*, the equilibrium force diagram is shown in Fig. 13.3. For equilibrium, a force $A\sigma$ is acting which may be resolved into components $A\sigma_r$, a normal force, and $A\tau_{r\theta}$, a shear force, where A represents the area of the stressed portion. The radial summation of forces gives

$$\sigma_r(1/\cos\theta) = \sigma\cos\theta \tag{13.2}$$

from which

$$\sigma_r = \sigma\cos^2\theta \tag{13.3}$$

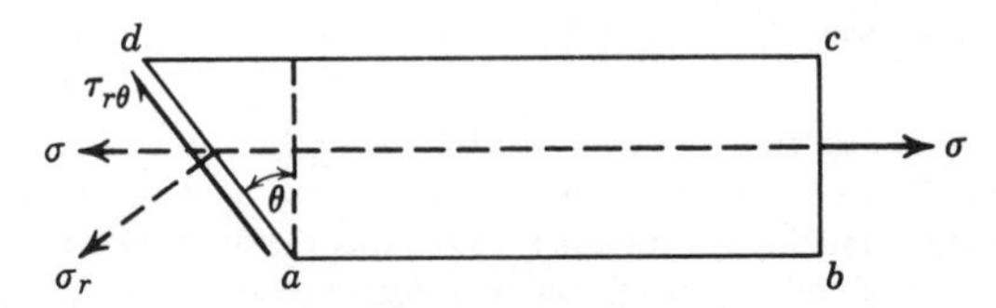

Fig. 13.3 Equilibrium force diagram of section abcd.

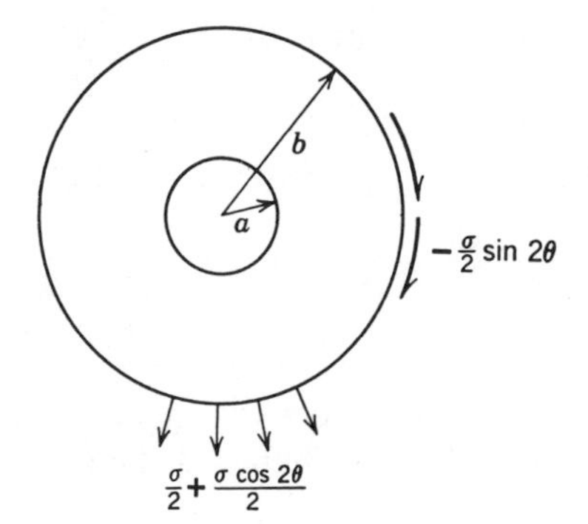

Fig. 13.4 State of stress around hole in plate.

The hoop-direction force summation gives

$$-\tau_{r\theta}(1/\cos\theta) = \sigma\sin\theta \qquad 13.4$$

from which

$$\tau_{r\theta} = -(\sigma/2)\sin 2\theta \qquad 13.5$$

Thus at radius $r = b$,

$$\sigma_r = \sigma\cos^2\theta = (\sigma/2)(1+\cos 2\theta) = \sigma/2 + \sigma\cos 2\theta/2 \qquad 13.6$$

and

$$\tau_{r\theta} = -(\sigma/2)\sin 2\theta \qquad 13.7$$

which is equivalent to the state of stress around the hole as shown in Fig. 13.4. For convenience the stress distribution in Fig. 13.4 is split into two distributions as shown in Fig. 13.5. Thus in Fig. 13.5, two cases can be analyzed and superposed. In Fig. 13.5*A* the problem reduces to the analysis of a heavy-walled cylinder with zero internal pressure and an external pressure $\sigma/2$. Thus from Eqs. 3.466 and 3.467

$$(\sigma_h)_A = \frac{\sigma}{2}\left(\frac{b^2}{b^2-a^2}\right)\left(1+\frac{a^2}{r^2}\right) \qquad 13.8$$

and

$$(\sigma_r)_A = \frac{\sigma}{2}\left(\frac{b^2}{b^2-a^2}\right)\left(1-\frac{a^2}{r^2}\right) \qquad 13.9$$

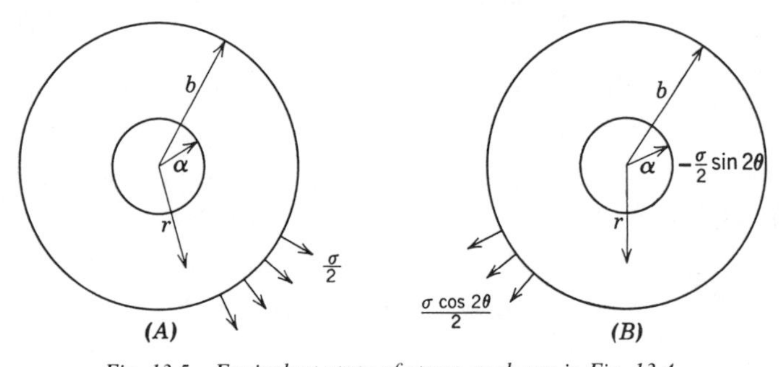

Fig. 13.5 Equivalent state of stress as shown in Fig. 13.4.

However, since it is assumed that a is small compared with b, the factor

$$b^2/(b^2 - a^2)$$

becomes unity in Eqs. 13.8 and 13.9 and the resulting stresses are as follows:

$$(\sigma_h)_A = (\sigma/2)(1 + a^2/r^2) \tag{13.10}$$

and

$$(\sigma_r)_A = (\sigma/2)(1 - a^2/r^2) \tag{13.11}$$

In analyzing the stress distribution in Fig. 13.5B it is convenient to use the theory of elasticity and stress functions as described in Chapter 3. Thus using the compatibility relationship of Eq. 3.111,

$$\left(\frac{\partial^2}{\partial r^2} + \frac{1}{r}\frac{\partial}{\partial r} + \frac{1}{r^2}\frac{\partial^2}{\partial \theta^2}\right)\left(\frac{\partial^2 \phi}{\partial r^2} + \frac{1}{r}\frac{\partial \phi}{\partial r} + \frac{1}{r^2}\frac{\partial^2 \phi}{\partial \theta^2}\right) = 0 \tag{13.12}$$

where ϕ is a stress function defined for the present case by the expression (88)

$$\phi = f(r) \cos 2\theta \tag{13.13}$$

Substituting Eq. 13.13 into Eq. 13.12 gives the following differential equation:

$$\left(\frac{d^2}{dr^2} + \frac{1}{r}\frac{d}{dr} - \frac{4}{r^2}\right)\left(\frac{d^2 f}{dr^2} + \frac{1}{r}\frac{df}{dr} - \frac{4f}{r^2}\right) = 0 \tag{13.14}$$

the solution for which is

$$f(r) = C_1 r^2 + C_2 r^4 + C_3(1/r^2) + C_4 \tag{13.15}$$

where C_1, C_2, C_3, and C_4 are constants of integration. Combining Eqs. 13.13 and 13.15 gives

$$\phi = [C_1 r^2 + C_2 r^4 + C_3(1/r^2) + C_4] \cos 2\theta \tag{13.16}$$

Having an expression for the stress function ϕ, Eqs. 3.97 to 3.99 can then be used to determine the stresses; thus

$$(\sigma_h)_B = \frac{\partial^2 \phi}{\partial r^2} = \left(2C_1 + 12C_2 r^2 + \frac{6C_3}{r^4}\right) \cos 2\theta \tag{13.17}$$

$$(\sigma_r)_B = \frac{1}{r}\frac{\partial \phi}{\partial r} + \frac{1}{r^2}\frac{\partial^2 \phi}{\partial \theta^2} = -\left(2C_1 + \frac{6C_3}{r^4} + \frac{4C_4}{r^2}\right) \cos 2\theta \tag{13.18}$$

and

$$(\tau_{r\theta})_B = \frac{-\partial}{\partial r}\left(\frac{1}{r}\frac{\partial \phi}{\partial \theta}\right) = \left(2C_1 + 6C_2 r^2 - \frac{6C_3}{r^4} - \frac{2C_4}{r^2}\right) \sin 2\theta \tag{13.19}$$

The constants C_1, C_2, C_3, and C_4 are evaluated through consideration of the boundary conditions of the problem which are as follows:

$$(\sigma_r)_B = 0 \qquad \text{at } r = a \tag{13.20}$$

$$(\sigma_r)_B = (\sigma/2) \cos 2\theta \qquad \text{at } r = b \tag{13.21}$$

$$(\tau_{r\theta})_B = 0 \qquad \text{at } r = a \tag{13.22}$$

and

$$(\tau_{r\theta})_B = (-\sigma/2)\sin 2\theta \qquad \text{at } r = b \tag{13.23}$$

Thus through application of Eqs. 13.17 to 13.23, the constants are obtained and their values are

$$C_1 = -\sigma/4 \tag{13.24}$$

$$C_2 = 0 \tag{13.25}$$

$$C_3 = -\sigma a^4/4 \tag{13.26}$$

and

$$C_4 = \sigma a^2/2 \tag{13.27}$$

Substituting Eqs. 13.24 to 13.27 into Eqs. 13.17 to 13.19 gives

$$(\sigma_h)_B = -\frac{\sigma}{2}\left(1 + \frac{3a^4}{r^4}\right)\cos 2\theta \tag{13.28}$$

$$(\sigma_r)_B = \frac{\sigma}{2}\left(1 + \frac{3a^4}{r^4} - \frac{4a^2}{r^2}\right)\cos 2\theta \tag{13.29}$$

and

$$(\tau_{r\theta})_B = -\frac{\sigma}{2}\left(1 - \frac{3a^4}{r^4} + \frac{2a^2}{r^2}\right)\sin 2\theta \tag{13.30}$$

The total stress distribution is the sum of the distributions given by subscripts A and B; therefore combining Eqs. 13.8, 13.9, 13.28, 13.29 and 13.30 gives

$$\sigma_h = (\sigma_h)_A + (\sigma_h)_B = \frac{\sigma}{2}\left(1 + \frac{a^2}{r^2}\right) - \frac{\sigma}{2}\left(1 + \frac{3a^4}{r^4}\right)\cos 2\theta \tag{13.31}$$

$$\sigma_r = (\sigma_r)_A + (\sigma_r)_B = \frac{\sigma}{2}\left(1 - \frac{a^2}{r^2}\right) + \frac{\sigma}{2}\left(1 + \frac{3a^4}{r^4} - \frac{4a^2}{r^2}\right)\cos 2\theta \tag{13.32}$$

and

$$\tau_{r\theta} = (\tau_{r\theta})_A + (\tau_{r\theta})_B = -\frac{\sigma}{2}\left(1 - \frac{3a^4}{r^4} + \frac{2a^2}{r^2}\right)\sin 2\theta \tag{13.33}$$

Having the above relationships the stress distribution can be computed for various values of r and θ. Taking, for example, the hoop stress (Eq. 13.31), if $\theta = 90°$,

$$(\sigma_h)_{90°} = \frac{\sigma}{2}\left[2 + \frac{a^2}{r^2}\left(1 + \frac{3a^2}{r^2}\right)\right] \tag{13.34}$$

and the stress distribution shown in Fig. 13.6 is obtained. It is seen that in this case the stress at the edge of the hole is *three times* the "normal" stress; that is, the stress concentration factor is 3.0. By the same procedure, the stress at the edge of the hole for $\theta = 0°$ is $-\sigma$. If a biaxial state of stress is present the same procedure as outlined above is used and the results are superposed, several examples are shown in Fig. 13.7.

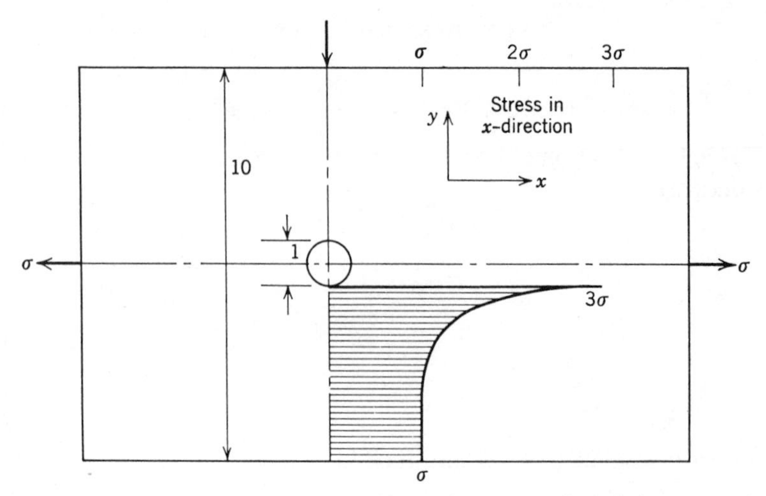

Fig. 13.6 *Stress distribution in direction of applied load for a plate with a hole diameter to plate width ratio of $\frac{1}{10}$.*

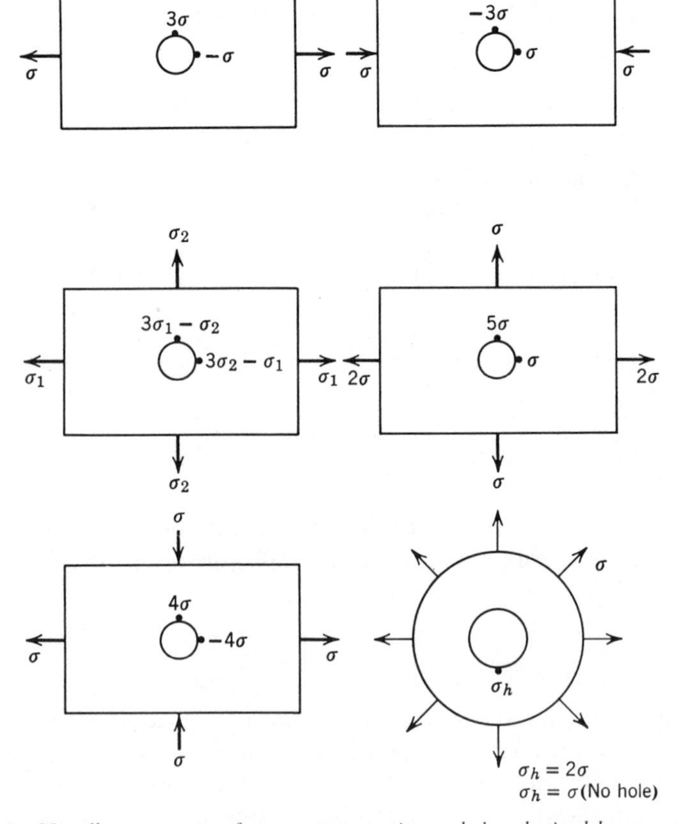

Fig. 13.7 *Miscellaneous cases of stress concentration at holes obtained by superposition.*

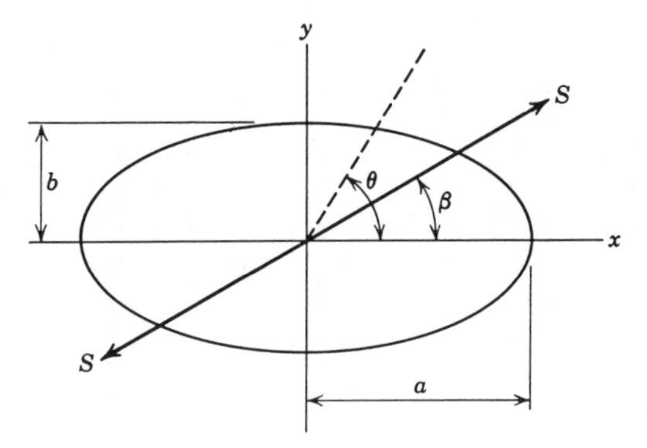

Fig. 13.8 Geometry of an elliptic hole in an infinite plate subjected to a stress S.

Elliptical hole in a plate

An analysis for the stress concentrating effect of an elliptical hole in a plate has been given by Inglis (45), Conway (7), and Wang (96). Consider Fig. 13.8 which shows an elliptical hole in an infinite plate subjected to a unit stress S. For the geometry given,

$$\sigma_x + \sigma_y = S[1 - m^2 - 2\cos 2(\beta - \theta) + 2m\cos 2\theta \cos 2(\beta - \theta) - 2m \sin 2\theta \sin 2(\beta - \theta)]/(1 + m^2 - 2m\cos 2\theta) \qquad 13.35$$

where σ_x and σ_y are stresses in the cartesian directions, S is unit stress, $m = (a - b)/(a + b)$, β is the angle between the applied load direction and the major axis, and θ is an angle defining any point on the perimeter of the ellipse with respect to the major axis.

When the loading direction is along the x axis ($\beta = 0$), the stresses at the ends of the ellipse axes are found as follows. At the ends of the minor axis $\theta = \pi/2$, $\sigma_y = 0$, and from Eq. 13.35

$$\sigma_x = S_x(1 + 2b/a) \qquad 13.36$$

At the ends of the major axis, σ_x, β, and θ are all equal to zero and by Eq. 13.65

$$\sigma_y = -S_x \qquad 13.37$$

When the loading is along the y axis, the stress at the ends of the minor axis is found by letting β and θ equal $\pi/2$, with σ_y equal to zero in Eq. 13.35, thus

$$\sigma_x = -S_y \qquad 13.38$$

and at the ends of the major axis, σ_x and θ equal zero and β equals $\pi/2$ in Eq. 13.35; thus

$$\sigma_y = S_y[1 + 2a/b] \qquad 13.39$$

Thus with appropriate values of σ_x, σ_y, β, and θ substituted in Eq. 13.35, the state of stress for any biaxial loading condition can be determined for any

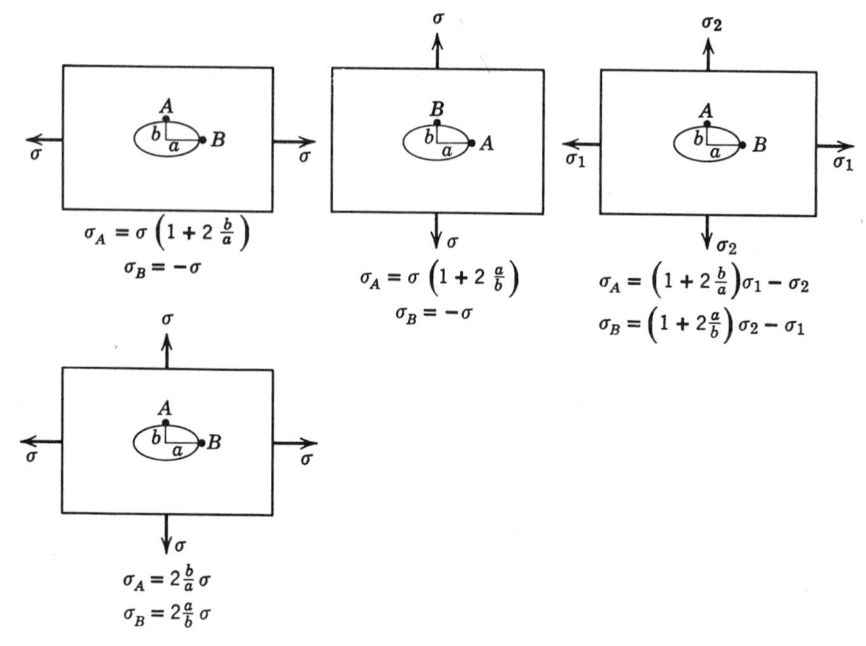

Fig. 13.9 Stress concentrations around elliptic holes obtained by superposition.

point on the perimeter of the ellipse. Equation 13.35, for an elliptic hole in an infinite plate represents the general case; the circular hole is a special case of an ellipse with equal axes. Therefore, if the state of stress is required for a circular hole in a plate under uniaxial or biaxial loading, Eq. 13.35 is modified by letting $a = b$. For a circular hole in a plate under uniaxial load, σ_x and σ_y in Eqs. 13.36 and 13.39 become equal to $3S_x$ and $3S_y$, respectively – or, a stress concentration factor of 3 is obtained. If the hole is not small with respect to the plate dimen-

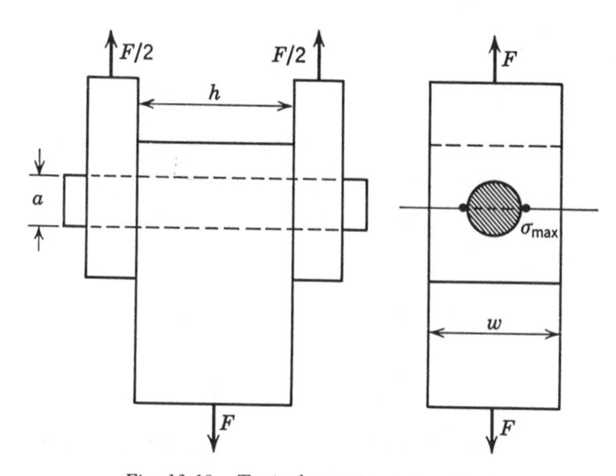

Fig. 13.10 Typical trunnion connection.

sions, other modifications are required; this is discussed later when the analysis is applied to the case of a side hole in a circular cylinder. Several cases are summarized in Fig. 13.9.

Trunnion and pin connections

In many applications load is transmitted from one member to another by pin or trunnion* connections through circular holes as shown in Fig. 13.10. In this case the inner surface of the hole is subjected to a high bearing stress which results in a stress concentration effect.

Contact stresses†

In applications involving balls and rollers, parts are subjected to high local stresses by virtue of the small area of contact between the parts in the assembly. Consider, for example, the contact of two spheres as shown in Fig. 13.11. When there is no load applied, the contact is at point O, and the distances y_1 and y_2 at location x, for small values of x, are

$$y_1 = x^2/2R_1 \qquad 13.40$$

and

$$y_2 = x^2/2R_2$$

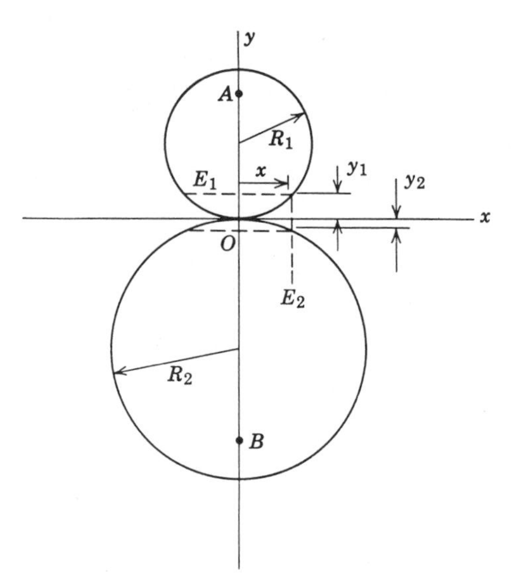

Fig. 13.11 Two spheres of different materials in contact.

*For stresses in the pins or linkages see Chapter 3.

† A review of this subject including friction effects has been presented by Barwell (2). Contact stress and deformation in a thin elastic layer are discussed in (33, 48).

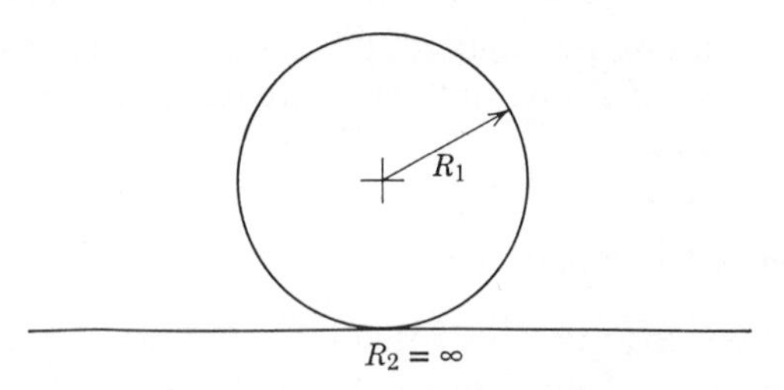

Fig. 13.12 Sphere resting on a plane.

from which

$$\delta = y_1 + y_2 = x^2(1/2R_1 + 1/2R_2) \qquad 13.41$$

For a ball on a plane as shown in Fig. 13.12, R_2 is finite and Eq.13.41 becomes

$$\delta = x^2/2R_1 \qquad 13.42$$

For a ball in a spherical seat, as shown in Fig. 13.13,

$$\delta = y_1 - y_2 = x^2(1/2R_1 - 1/2R_2) \qquad 13.43$$

It is assumed that the point (area) of contact is fixed and that the two balls approach each other; the ball of radius R_1 deforms an amount w_1 and the ball of radius R_2 deforms an amount w_2. Then, for points on the y axis at large distances from O, the points (such as A and B, Fig. 13.11) approach each other an amount Δ and points near O approach each other an amount $\Delta - (w_1 + w_2)$; if the points near O are within the surface of contact,

$$y_1 + y_2 = \Delta - (w_1 + w_2) \qquad 13.44$$

and it is shown in the theory of elasticity (88) that the displacements w_1 and w_2 are

$$w_1 = \frac{1 - \nu_1{}^2}{\pi E_1} \int\int q \, ds \, d\phi \qquad 13.45$$

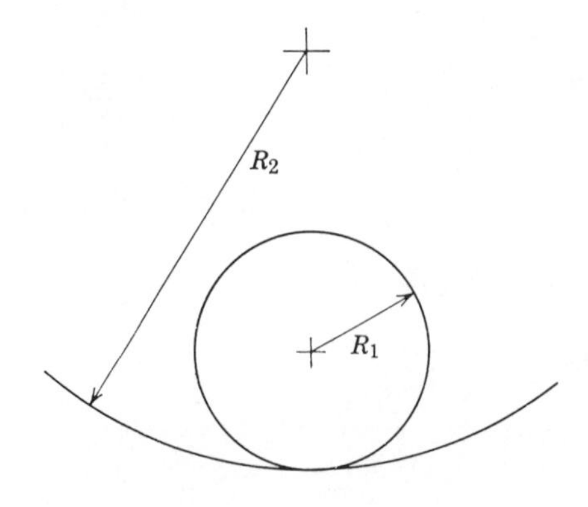

Fig. 13.13 Ball in a spherical seat.

and

$$w_2 = \frac{1 - v_2{}^2}{\pi E_2} \iint q \, ds \, d\phi \tag{13.46}$$

where v and E are Poisson's ratio and the modulus of elasticity, respectively, q is the contact pressure, ds is an element of a chord of the circle of contact, and ϕ is any angle of the chord from a diameter of the circle of contact. Thus from Eqs. 13.44 to 13.46

$$\Delta - \beta x^2 = \frac{1}{\pi}\left(\frac{1 - v_1{}^2}{E_1} + \frac{1 - v_2{}^2}{E_2}\right) \iint q \, ds \, d\phi \tag{13.47}$$

where β is a constant dependent on the conditions as expressed by Eqs. 13.41 to 13.43. Equation 13.47 is satisfied if

$$\int q \, ds = q_0 \frac{A}{a} \tag{13.48}$$

where q_0 is the pressure at the surface of contact O, a is the radius of the surface of contact, and

$$A = (\pi/2)(a^2 - x^2 \sin^2 \phi) \tag{13.49}$$

Substituting Eqs. 13.48 and 13.49 into Eq. 13.47 gives

$$q_0 \left(\frac{1 - v_1{}^2}{E_1} + \frac{1 - v_2{}^2}{E_2}\right) \frac{\pi}{4a} (2a^2 - x^2) = \Delta - \beta x^2 \tag{13.50}$$

$$\Delta = q_0 \frac{\pi a}{2} \left(\frac{1 - v_1{}^2}{E_1} + \frac{1 - v_2{}^2}{E_2}\right) \tag{13.51}$$

and

$$a = \frac{q_0 \pi}{4\beta} \left(\frac{1 - v_1{}^2}{E_1} + \frac{1 - v_2{}^2}{E_2}\right) \tag{13.52}$$

Finally, from equilibrium conditions

$$F = (q_0/a)(\tfrac{2}{3}\pi a^3) \tag{13.53}$$

from which

$$q_0 = 3F/2\pi a^2 \tag{13.54}$$

Thus, for two balls in contact and pressed together by a force F, the radius of the contact area is, from Eqs. 13.41 and 13.52

$$a = \sqrt[3]{\frac{3}{4}\left(\frac{F R_1 R_2}{R_1 + R_2}\right)\left(\frac{1 - v_1{}^2}{E_1} + \frac{1 - v_2{}^2}{E_2}\right)} \tag{13.55}$$

and the centers of the balls approach by an amount

$$\Delta = \sqrt[3]{\frac{9\pi}{16}\left(\frac{F^2}{R_1 R_2}\right)(R_1 + R_2)\left(\frac{1 - v_1{}^2}{E_1} + \frac{1 - v_2{}^2}{E_2}\right)} \tag{13.56}$$

The maximum pressure is obtained by substituting Eq. 13.55 into Eq. 13.54. Numerous variations of the preceding cases are obtained by selecting various values for R_1, R_2, v, and E. The corresponding stresses (88) are

$$\sigma_y = q\left[\frac{y^3}{(a^2+y^2)^{3/2}} - 1\right] \qquad 13.57$$

$$\sigma_r = \sigma_h = \frac{q}{2}\left\{\frac{2y(1+v)}{(a^2+y^2)^{1/2}} - (1+2v) - \left[\frac{y}{(a^2+y^2)^{1/2}}\right]^3\right\} \qquad 13.58$$

At the center of the contact area O

$$\sigma_y = -q \qquad 13.59$$

and

$$\sigma_r = \sigma_h = -q\left(\frac{1+2v}{2}\right) \qquad 13.60$$

In order to find the critical failure location it is convenient to use the shear theory (see Chapter 3), which from Eqs. 13.57 and 13.58 shows that

$$\tau_{max} = \left(\frac{\sigma_h - \sigma_y}{2}\right) = \frac{q}{2}\left\{\frac{1-2v}{2} + \tfrac{2}{9}(1+v)[2(1+v)]^{1/2}\right\} \qquad 13.61$$

which corresponds to a location $y = 0.638a$. For brittle materials (using the maximum stress theory, Chapter 3), the maximum value is not at the surface where $y = 0$, but at the periphery of the area of contact (88) where

$$\sigma_r = (q_0/3)(1-2v) \qquad 13.62$$

and

$$\sigma_h = -(q_0/3)(1-2v) \qquad 13.63$$

By the stress theory, failure is governed by the maximum tensile stress which is expressed by Eq. 13.62.

In the general case of contact, like a ball on a cylindrical surface, the area of contact is elliptical and expressions for deformation and stress are obtained in a manner similar to that described for circular areas of contact. The detailed analyses may be found in standard texts on the mathematical theory of elasticity; for present purposes only a few final results will be given (88). The maximum pressure q_0 is

$$q_0 = \tfrac{3}{2}(F/\pi ab) \qquad 13.64$$

where a is the semimajor axis and b is the semiminor axis of the ellipse of contact. The values of a and b are determined from the equations

$$a = m\sqrt[3]{\frac{3F}{4(A+B)}\left(\frac{1-v_1{}^2}{E_1} + \frac{1-v_2{}^2}{E_2}\right)} \qquad 13.65$$

and

$$b = n\sqrt[3]{\frac{3F}{4(A+B)}\left(\frac{1-v_1{}^2}{E_1} + \frac{1-v_2{}^2}{E_2}\right)} \qquad 13.66$$

where

$$A+B=\frac{1}{2}\left(\frac{1}{R_1}+\frac{1}{R_1{}^1}+\frac{1}{R_2}+\frac{1}{R_2{}^1}\right) \qquad 13.67$$

and R_1 and $R_1{}^1$ and R_2 and $R_2{}^1$ are principal radii of curvature at the point of contact of the bodies. To determine A and B the following additional relationship is required:

$$B-A=\frac{1}{2}\left[\left(\frac{1}{R_1}-\frac{1}{R_1{}^1}\right)^2+\left(\frac{1}{R_2}-\frac{1}{R_2{}^1}\right)^2 + 2\left(\frac{1}{R_1}-\frac{1}{R_1{}^1}\right)\left(\frac{1}{R_2}-\frac{1}{R_2{}^1}\right)\cos 2\theta\right]^{1/2} \qquad 13.68$$

where θ is the angle between the normal planes containing the curvatures $1/R_1$ and $1/R_2$. To facilitate calculation the relation

$$\cos\theta=\frac{B-A}{A+B} \qquad 13.69$$

is introduced which is useful for obtaining the various m and n values in Eqs. 13.65 and 13.66 as shown in Fig. 13.14. On the surface of contact the stresses (see Fig. 13.15) are

$$\sigma_y=-q_0 \qquad 13.70$$

$$\sigma_x=-2\nu q_0-(1-2\nu)q_0\left(\frac{b}{a+b}\right) \qquad 13.71$$

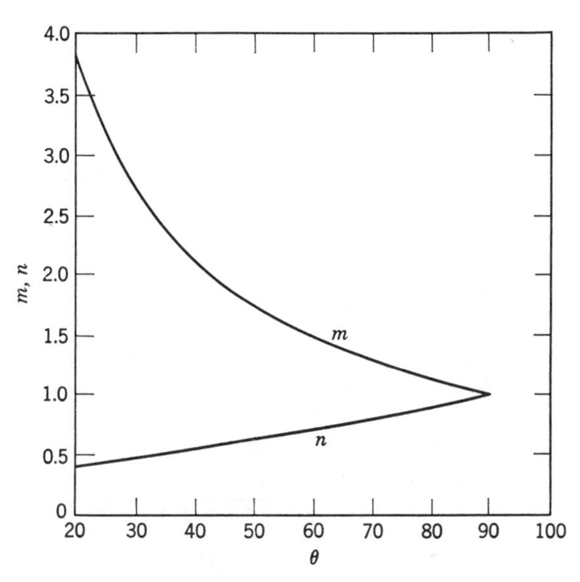

Fig. 13.14 Graph for determining m and n as a function of θ.

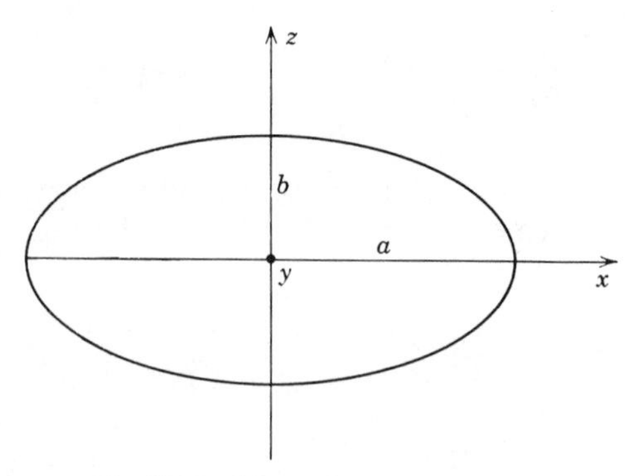

Fig. 13.15 Elliptical surface of contact.

$$\sigma_z = -2\nu q_0 - (1-2\nu)q_0\left(\frac{a}{a+b}\right) \quad 13.72$$

$$(\tau)_{x=\pm a} = (1-2\nu)q_0\frac{\beta}{e^2}\left(\frac{1}{e}\operatorname{arctanh} e - 1\right) \quad 13.73$$

$$(\tau)_{x=\pm b} = (1-2\nu)q_0\frac{\beta}{e^2}\left(1 - \frac{\beta}{e}\arctan\frac{e}{\beta}\right) \quad 13.74$$

where

$$\beta = b/a \quad 13.75$$

and

$$e = (1/a)\sqrt{a^2 - b^2} \quad 13.76$$

when the ratio a/b approaches infinity, the problem of two cylindrical rollers in contact is obtained. Here the contact area is a rectangular strip of width b; thus, for the case shown in Fig. 13.16,

$$(F/L) = (\pi b/2)q_0 \quad 13.77$$

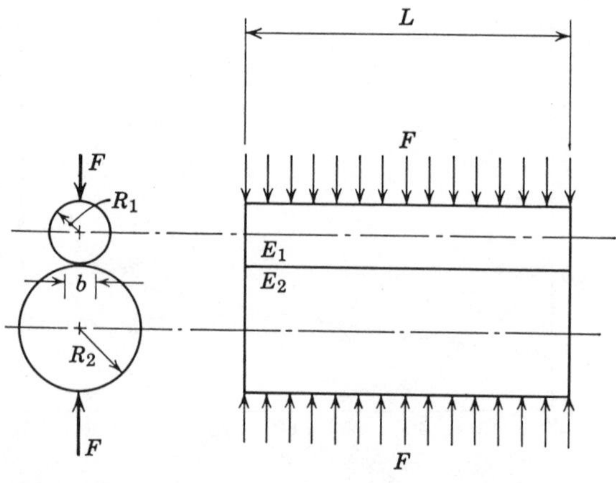

Fig. 13.16 Two cylindrical rollers in contact.

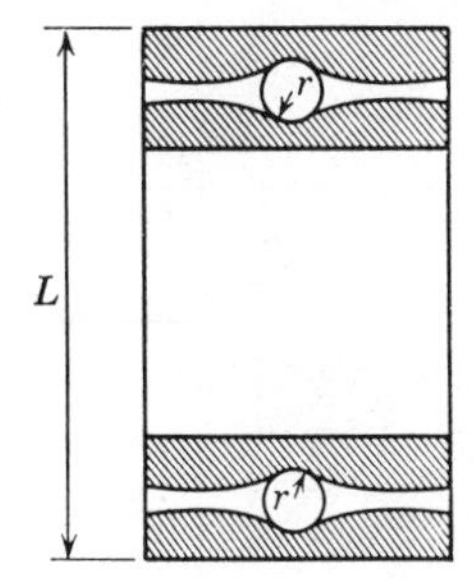

Fig. 13.17 Typical ball bearing assembly.

and

$$b = \sqrt{\frac{4FR_1R_2}{L\pi(R_1 + R_2)}\left(\frac{1 - \nu_1{}^2}{E_1} + \frac{1 - \nu_2{}^2}{E_2}\right)} \tag{13.78}$$

The preceding analysis is useful in designing ball bearing assemblies as illustrated in the following example. Consider the system shown in Fig. 13.17. In this case $R_1 = R_1{}^1$, $R_2 = -r$, and $R_2{}^1 = -L/2$. Having these quantities the other quantities can be calculated and the maximum pressure determined. For design convenience Isakower (47) has reduced these formulas to simple equations and graphs applicable to metal systems as shown in Figs. 13.18 and 13.19. The materials factors to be used are listed in Table 13.1. For example, consider an aluminum ball having a diameter of 1 in. riding in a copper socket having a $1\frac{1}{2}$ in. diameter. What is the maximum compressive stress developed by a

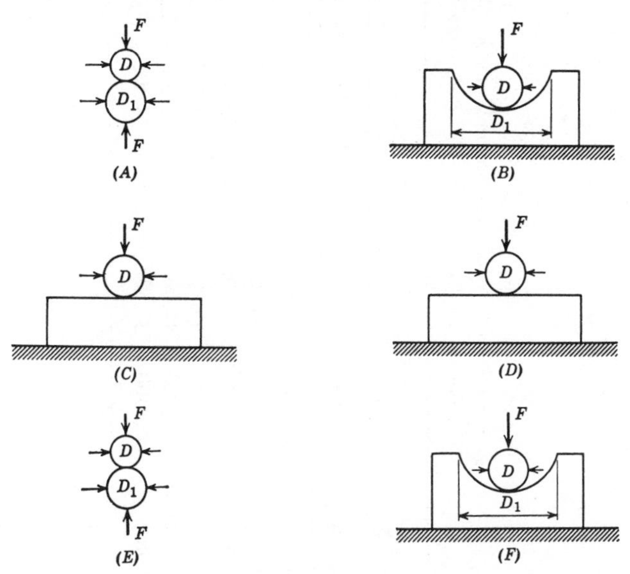

Fig. 13.18 Formulas for contact stresses in various systems. (a) Cylinders, $\sigma_c = K_1 K_2(F/D^2)^{1/2}$. (b) Cylinder, $\sigma_c = K_1 K_3(F/D^2)^{1/2}$. (c) Cylinder, $\sigma_c = K_1(F/D^2)^{1/2}$. (d) Sphere, $\sigma_c = K_4(F/D^2)^{1/3}$. (e) Spheres, $\sigma_c = K_4 K_5(F/D^2)^{1/3}$. (f) Sphere in spherical seat, $\sigma_c = K_4 K_6(F/D^2)^{1/3}$. (After Isakower, Reference 47, courtesy of The McGraw-Hill Publishing Co.)

Table 13.1 Materials Factor K_1 and K_4 for Different Systems (Fig. 13.18). [After Isakower (47), courtesy of The McGraw-Hill Publishing Co.]

	Steel		Titanium		Copper		Brass		Gray Cast Iron		Aluminum	
	K_1	K_4	K_1	K_4	K_1	K_4	K_1	K_4	K_1	K_4	K_1	K_4
Steel	3,195	58,500										
Titanium	2,770	48,300	2,485	41,700								
Copper	2,700	46,600	2,440	40,800	2,390	39,400						
Brass	2,665	45,900	2,405	39,900	2,360	39,100	2,335	38,300				
Gray Cast Iron	2,620	44,600	2,375	39,200	2,330	38,100	2,303	37,600	2,275	37,000		
Aluminum	2,300	37,500	2,130	34,000	2,100	33,000	2,080	32,800	2,055	32,400	1,890	28,700

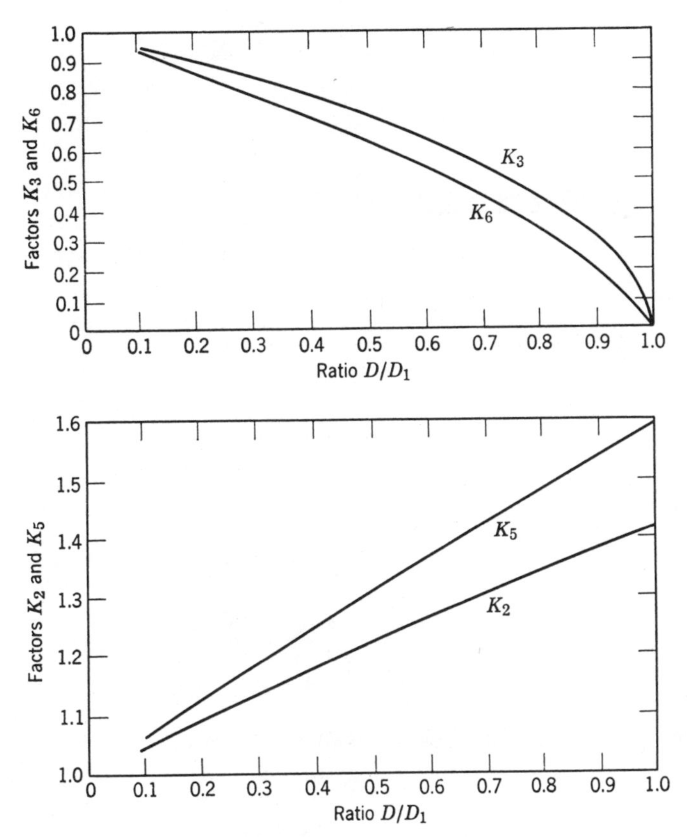

Fig. 13.19 Curves for determining dimensional factors. [After Isakower (47), courtesy of The McGraw-Hill Publishing Co.]

load of 230 lb? The problem may be solved by reference to Isakower's method (see Fig. 13.18*F*) or Eqs. 13.54 and 13.55. Both methods show that for the materials involved the maximum compressive stress is approximately 96,000 psi.

Stresses due to concentrated forces

Load on a flat infinite surface. A concentrated line load on an infinite surface is shown in Fig. 13.20; this is called a *simple radial stress distribution* and is most conveniently analyzed by the use of stress functions (83, 88). In the present case the stress function is

$$\phi = -(Fr\theta/\pi)\sin\theta \qquad 13.79$$

and the stress distribution is given by Eqs. 3.97 to 3.99; that is, with zero body forces

$$\sigma_\theta = \partial^2\phi/\partial r^2 \qquad 13.80$$

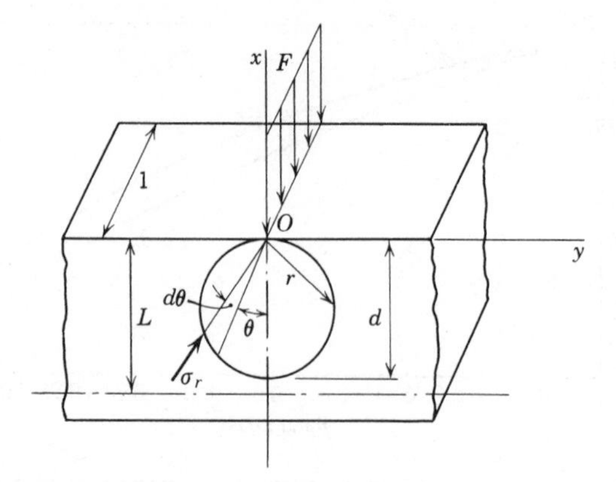

Fig. 13.20 Concentrated line load on an infinite surface.

$$\sigma_r = \frac{1}{r}\frac{\partial \phi}{\partial r} + \frac{1}{r^2}\frac{\partial^2 \phi}{\partial \theta^2} \qquad 13.81$$

and

$$\tau_{r\theta} = -\frac{\partial}{\partial r}\left(\frac{1}{r}\right)\frac{\partial \phi}{\partial \theta} \qquad 13.82$$

Substituting Eq. 13.79 into Eqs. 13.80 to 13.82 gives

$$\sigma_\theta = 0 \qquad 13.83$$

$$\sigma_r = -\frac{2F}{\pi}\frac{\cos\theta}{r} \qquad 13.84$$

and

$$\tau_{r\theta} = 0 \qquad 13.85$$

For any point on the circle shown

$$r = d\cos\theta \qquad 13.86$$

Therefore, combining Eqs. 13.84 and 13.86 gives

$$\sigma_r = -2F/\pi d \qquad 13.87$$

and it is seen that, except for the point of load application, the stress is the same at all points on the circle. That the boundary condition is satisfied is shown by noting that the resultant of forces on a surface r has to balance the load F; thus,

$$2\int_0^{\pi/2} r\sigma_r \cos\theta \, d\theta = -\frac{4F}{\pi}\int_0^{\pi/2} \cos^2\theta \, d\theta = -F$$

Suppose it is desired to determine the stresses on a plane at distance L from the surface (Fig. 13.20). First, the radial stress is resolved into the stress components

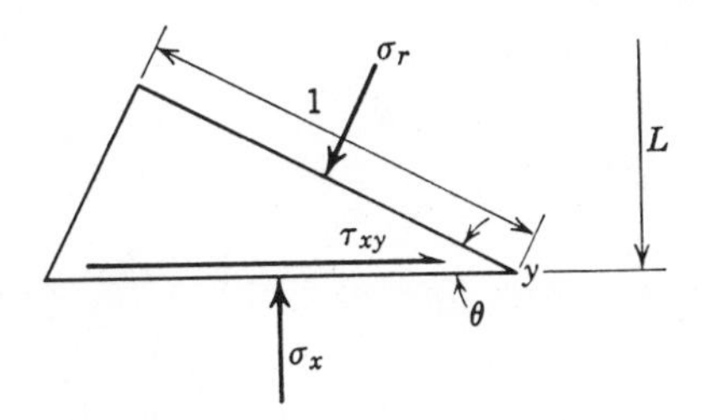

Fig. 13.21 Free body for force in x direction.

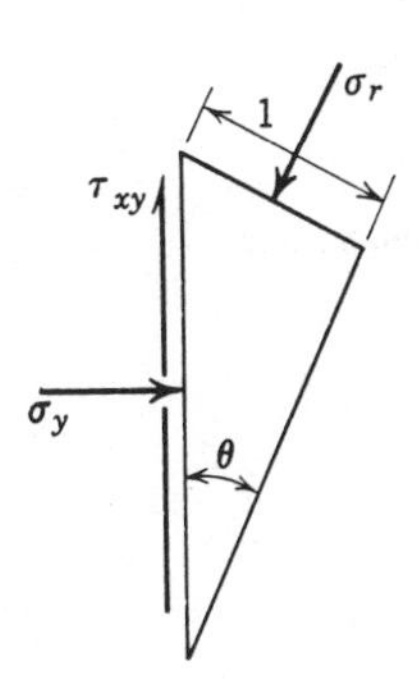

Fig. 13.22 Free body for force in y direction.

σ_x and τ_{xy}; summation of forces in the x direction then gives (Fig. 13.21)

$$\sigma_x(1/\cos\theta) = \sigma_r \cos\theta$$

$$\sigma_x = \sigma_r \cos^2\theta \qquad 13.88$$

from which

$$\sigma_x = -(2F/\pi r)\cos^3\theta$$

But $\cos\theta = L/r$; therefore

$$\sigma_x = -(2F/\pi L)\cos^4\theta$$

Similarly, summing force in the y direction (Fig. 13.22) gives

$$\sigma_y = -(2F/\pi L)\sin^2\theta\cos^2\theta$$

and finally

$$\tau_{xy} = -(2F/\pi L)\sin\theta\cos^3\theta$$

Having the component stresses, the principal stresses are calculated by substituting into Eqs. 3.18 to 3.20.

Knife edge or pivot. The geometry for this case is shown in Fig. 13.23. Following the same method as given and replacing Fr/π by C in the stress function,

$$\phi = -Cr\theta\sin\theta \qquad 13.89$$

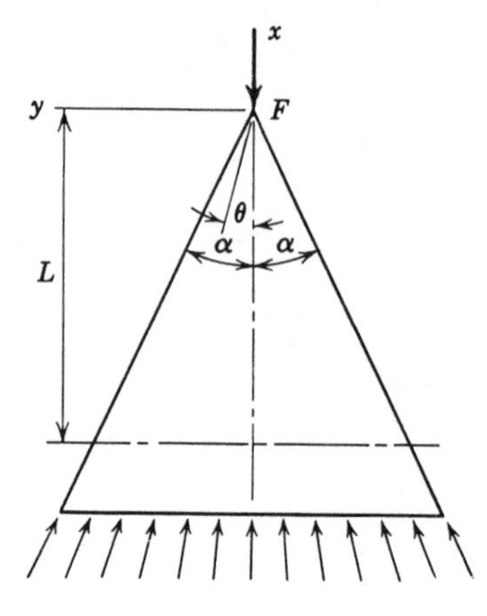

Fig. 13.23 Geometry of a knife edge or pivot.

where C is a constant, the stresses are as follows:

$$\sigma_r = -(2C\cos\theta)/r \tag{13.90}$$

and

$$\sigma_\theta = \tau_{r\theta} = 0 \tag{13.91}$$

The constant C is determined from the condition of force equilibrium in the x direction, thus

$$F = 2\int_{-\alpha}^{\alpha} \sigma_r r\cos\theta\, d\theta = 2C\int_0^{\alpha} \frac{2\cos^2\theta}{r}\, r\, d\theta \tag{13.92}$$

from which

$$C = \frac{F}{2(\alpha + \frac{1}{2}\sin 2\alpha)} \tag{13.93}$$

Substituting Eq. 13.93 into Eq. 13.90 then gives

$$\sigma_r = -\frac{F}{\alpha + \frac{1}{2}\sin 2\alpha}\frac{\cos\theta}{r} \tag{13.94}$$

The component stresses on a plane distant L from the loading point are as follows:

$$\sigma_x = -\frac{F\cos^4\theta}{L(\alpha + \frac{1}{2}\sin 2\alpha)} \tag{13.95}$$

$$\sigma_y = -\frac{F\sin^2\theta\cos^2\theta}{L(\alpha + \frac{1}{2}\sin 2\alpha)} \tag{13.96}$$

and

$$\tau_{xy} = -\frac{F\sin\theta\cos^3\theta}{L(\alpha + \frac{1}{2}\sin^2\alpha)} \tag{13.97}$$

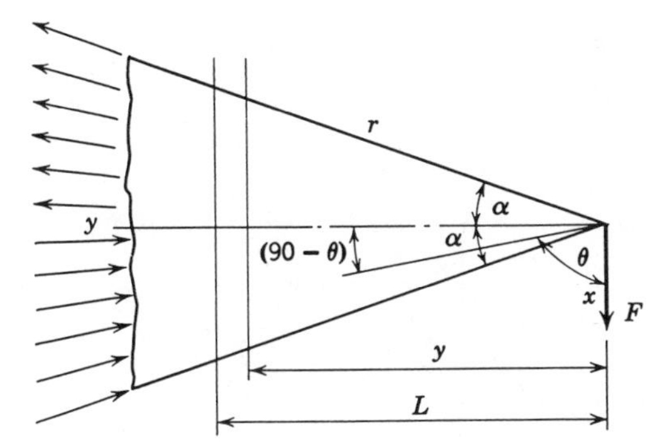

Fig. 13.24 Geometry of a wedge cantilever.

which when substituted into Eqs. 3.18 to 3.20 give values for the principal stresses.

Wedge cantilever. For the wedge cantilever shown in Fig. 13.24 the stress function is

$$\phi = -Cr\theta \sin\theta \tag{13.98}$$

from which

$$\sigma_r = -(2C/r)\cos\theta \tag{13.99}$$

where C is a constant determined from the relationship

$$F = \int_{-\alpha}^{\alpha} r\sigma_r \cos\theta \, dr \tag{13.100}$$

Substituting Eq. 13.99 into Eq. 13.100 thus gives

$$C = \frac{F}{2\alpha - \sin 2\alpha} \tag{13.101}$$

so that

$$\sigma_r = \frac{F\cos\theta}{r(\alpha - \frac{1}{2}\sin 2\alpha)} \tag{13.102}$$

and

$$\sigma_\theta = \tau_{r\theta} = 0 \tag{13.103}$$

For the component stresses at distance L from the load

$$\sigma_x = \sigma_r \sin^2(90^\circ - \theta) = \sigma_r \cos^2\theta \tag{13.104}$$

or

$$\sigma_x = -\frac{F\cos^3\theta}{r(\alpha - \frac{1}{2}\sin 2\alpha)} \tag{13.105}$$

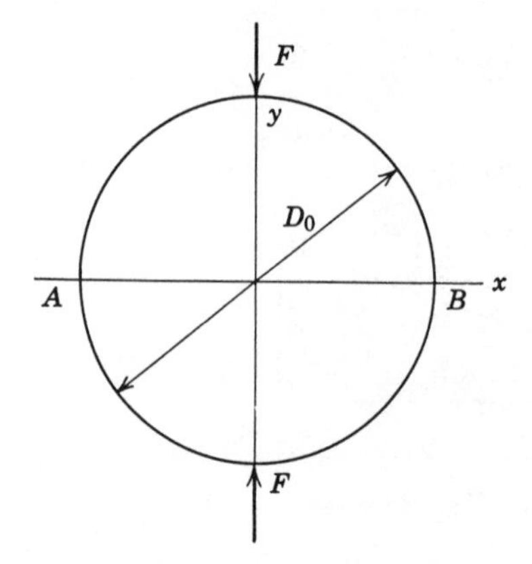

Fig. 13.25 Roller subjected to dimetral forces.

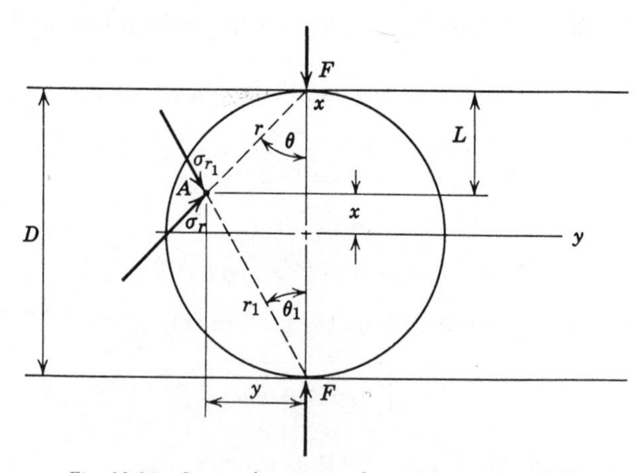

Fig. 13.26 Stress description for roller in Fig. 13.25.

But $\cos(90° - \theta) = y/r = \sin\theta$; therefore,

$$\sigma_x = -\frac{F\cos^3\theta\sin\theta}{L(\alpha - \frac{1}{2}\sin 2\alpha)} \qquad 13.106$$

Similarly,

$$\sigma_y = -\frac{Fx\sin^4\theta}{L^2(\alpha - \frac{1}{2}\sin 2\alpha)} \qquad 13.107$$

and

$$\tau_{xy} = -\frac{Fx^2\sin^4\theta}{L^3(\alpha - \frac{1}{2}\sin 2\alpha)} \qquad 13.108$$

Stress in a roller. Consider a roller acted on by two forces F as shown in Fig. 13.25. To solve this problem consider the solution to the semi-infinite plate shown in Fig. 13.20. In terms of the present problem the geometry involved

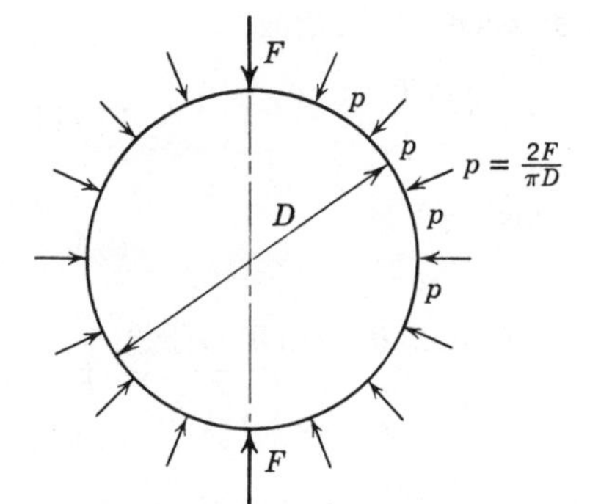

Fig. 13.27 Disk subjected to dimetral loading plus external pressure.

is shown in Fig. 13.26. For point A, as already shown, there is a radial stress given by Eq. 13.84

$$\sigma_r = -\frac{2F}{r\pi}\cos\theta \qquad 13.84$$

Similarly, for the lower force F

$$\sigma_{r_1} = -(2F\cos\theta_1)/\pi r_1 \qquad 13.109$$

so that the element is subjected to two compressive stresses orthogonal to each other. On the boundary where

$$\cos\theta/r = \cos\theta_1/r_1 = 1/D \qquad 13.110$$

there are orthogonal stresses each equal to $-2F/\pi D$. The element of diameter D can be considered as a disk subjected to uniform external pressure and radial concentrated loads as shown in Fig. 13.27. The boundary of the disk must be free of stress; therefore by superposition the stresses shown in Fig. 13.28 are obtained. The problem is thus solved by superposing the solutions obtained for a concentrated load on a semi-infinite plate and for uniform external tension

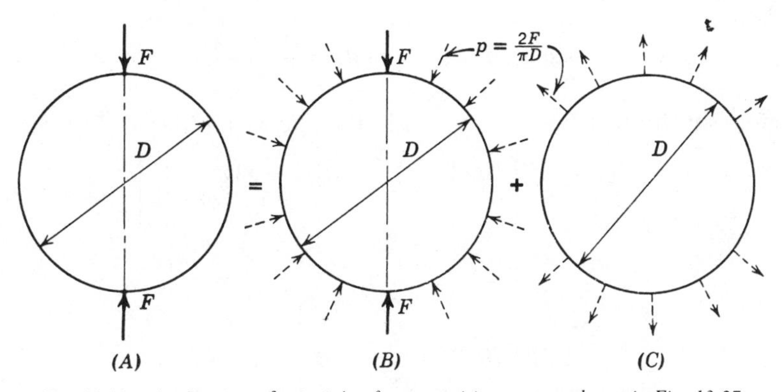

Fig. 13.28 Application of principle of superposition to case shown in Fig. 13.27.

on a disk. The solution for the plate has already been given; thus

$$\sigma_y = \sigma_r \sin^2\theta + \sigma_{r_1} \sin^2\theta_1 = -\frac{2F}{\pi}\left(\frac{\cos\theta}{r}\sin^2\theta + \frac{\cos\theta_1 \sin^2\theta_1}{r_1}\right) \qquad 13.111$$

$$\sigma_x = \sigma_r \cos^2\theta + \sigma_{r_1} \cos^2\theta_1 = -\frac{2F}{\pi}\left(\frac{\cos^3\theta}{r} + \frac{\cos^3\theta_1}{r_1}\right) \qquad 13.112$$

$$\tau_{xy} = \sigma_r \sin\theta\cos\theta - \sigma_{r_1} \sin\theta_1 \cos\theta_1 = -\frac{2F}{\pi}\left(\frac{\sin\theta\cos^2\theta}{r} - \frac{\sin\theta_1\cos^2\theta_1}{r_1}\right) \qquad 13.113$$

For a uniform disk under external radial loading the solution is given by Eqs. 3.466 and 3.467 with p_i and a equal to zero; thus, for the solid disk under uniform tension the stress is everywhere

$$\sigma_h = \sigma_r = p_0 \qquad 13.114$$

so that for this case

$$p_0 = 2F/\pi D \qquad 13.115$$

Thus at any point in the disk the components of σ_h and σ_r are added to Eqs. 13.111 to 13.113, thus

$$\sigma_y = -\frac{2F}{\pi}\left(\frac{\cos\theta\sin^2\theta}{r} + \frac{\cos\theta_1\sin^2\theta_1}{r_1} - \frac{1}{D}\right) \qquad 13.116$$

$$\sigma_x = -\frac{2F}{\pi}\left(\frac{\cos^3\theta}{r} + \frac{\cos^3\theta_1}{r_1} - \frac{1}{D}\right) \qquad 13.117$$

and

$$\tau_{xy} = -\frac{2F}{\pi}\left(\frac{\sin\theta\cos^2\theta}{r} - \frac{\sin\theta_1\cos^2\theta_1}{r_1}\right) \qquad 13.118$$

From Fig. 13.26

$$\sin\theta = \frac{y}{r}; \qquad \cos\theta = \frac{D/2 - x}{r} \qquad 13.119$$

$$\sin\theta_1 = \frac{y}{r_1}; \qquad \cos\theta_1 = \frac{D/2 + x}{r_1} \qquad 13.120$$

Therefore substituting Eqs. 13.119 and 13.120 into Eqs. 13.116 to 13.118,

$$\sigma_y = -\frac{2F}{\pi}\left[\frac{(D/2 - x)y^2}{r^4} + \frac{(D/2 + x)y^2}{r_1{}^4} - \frac{1}{D}\right] \qquad 13.121$$

$$\sigma_x = -\frac{2F}{\pi}\left[\frac{(D/2 - x)^3}{r^4} + \frac{(D/2 + x)^3}{r_1{}^4} - \frac{1}{D}\right] \qquad 13.122$$

and

$$\tau_{xy} = -\frac{2F}{\pi}\left[\frac{(D/2 - x)^2 y}{r^4} - \frac{(D/2 + x)^2 y}{r_1{}^4}\right] \qquad 13.123$$

where

$$r^2 = y^2 + (D/2 - x)^2$$

and

$$r_1{}^2 = y^2 + (D/2 + x)^2$$

Thus for points on the diameter normal to the radial load F, for example, $x = 0$ and

$$r = r_1 = [y^2 + (D/2)^2]^{1/2}$$

so that

$$\sigma_y = -\frac{2F}{\pi D}\left[\frac{16D^2y^2}{(D^2 + 4y^2)^2} - 1\right] \tag{13.124}$$

$$\sigma_x = -\frac{2F}{\pi D}\left[\frac{4D^4}{(D^2 + 4y^2)^2} - 1\right] \tag{13.125}$$

and

$$\tau_{zy} = 0$$

The maximum value is at the center of the disk where $x = 0$, thus

$$(\sigma_y)_{\max} = 2F/\pi D \tag{13.126}$$

and

$$(\sigma_x)_{\max} = -6F/\pi D \tag{13.127}$$

Similarly, along the x axis, where $y = 0$

$$r = (D/2) - x$$

and

$$r_1 = (D/2) + x$$

so that

$$\sigma_y = 2F/\pi D \tag{13.128}$$

$$\sigma_x = \frac{2F}{\pi}\left(\frac{4D}{D^2 - 4x^2} - \frac{1}{D}\right) \tag{13.129}$$

and

$$\tau_{xy} = 0$$

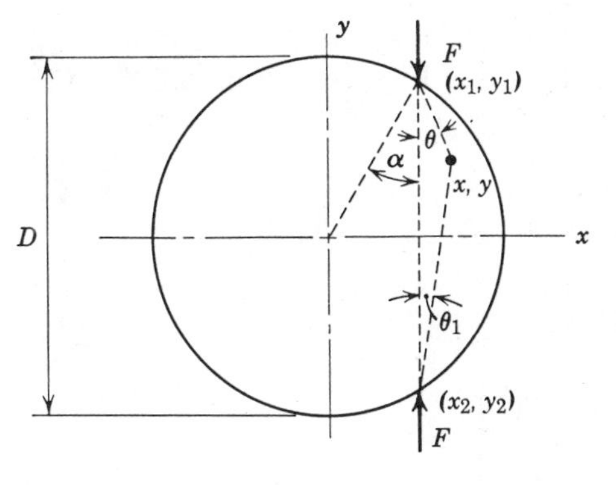

Fig. 13.29 Eccentrically loaded disk.

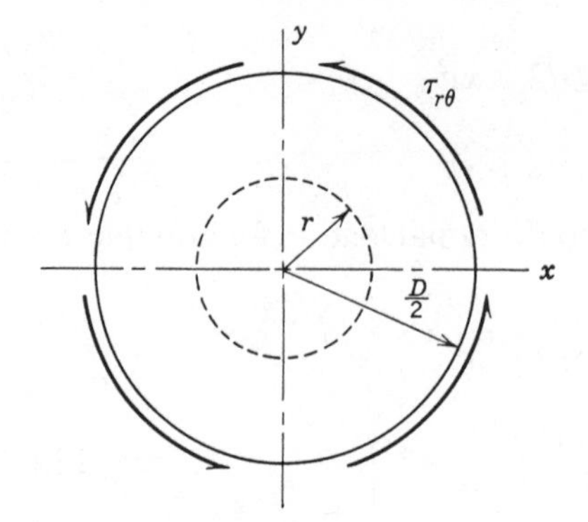

Fig. 13.30 Disk subjected to pure shear.

By a similar analysis the stress at any point x, y within an eccentrically loaded disk can be calculated. For the geometry shown in Fig. 13.29, for example,

$$\sigma_x = -\frac{2F}{\pi}\left[\frac{(y_1 - y)(x - x_1)^2}{r^4} + \frac{(y_1 + y)(x - x_1)^2}{r_1^{\,4}} - \frac{\sin(\pi/2 + \alpha)}{D}\right] \qquad 13.130$$

$$\sigma_y = -\frac{2F}{\pi}\left[\frac{(y_1 - y)^3}{r^4} + \frac{(y_1 + y)^3}{r_1^{\,4}} - \frac{\sin(\pi/2 + \alpha)}{D}\right] \qquad 13.131$$

and

$$\tau_{xy} = \frac{2F}{\pi}\left[\frac{(y_1 - y)^2(x - x_1)}{r^4} - \frac{(y_1 + y)^2(x - x_1)}{r_1^{\,4}}\right] \qquad 13.132$$

where

$$r^2 = (x - x_1)^2 + (y_1 - y)^2 \qquad 13.133$$

and

$$r_1^{\,2} = (x - x_1)^2 + (y + y_1)^2 \qquad 13.134$$

For pure shear, as shown in Fig. 13.30,

$$\sigma_r = \sigma_h = 0 \qquad 13.135$$

and

$$\tau_{r\theta} = T/2\pi r^2 \qquad 13.136$$

where T is the applied torque.

Local stresses in a beam. This problem was referred to in Chapter 2 in connection with testing of brittle materials (see Fig. 2.60). For this situation the stresses and deflection are as follows:

$$\sigma_x = \frac{F}{h}\left[\frac{3y}{h}\left(\frac{L}{h} - \frac{2}{\pi}\right) + \frac{1}{\pi} + \frac{2y}{\pi h}\left(\frac{4y^2}{h^2} - \frac{3}{5}\right)\right] \qquad 13.137$$

$$\sigma_y = \frac{F}{\pi}\left[\frac{1}{h} + \frac{2y}{h^2}\left(\frac{3}{2} - \frac{2y^2}{h^2}\right) - \frac{4}{h + 2y}\right] \qquad 13.138$$

$$\delta_B = \frac{FL^3}{48EI}\left[1 + 3.12\left(\frac{h}{L}\right)^2\right] \qquad 13.139$$

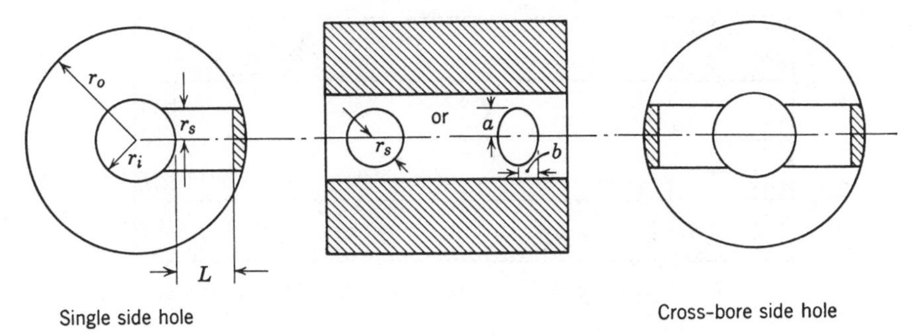

Fig. 13.31 Geometry of cylinder containing side holes.

Heavy-walled cylinder* with side holes. The following discussion (20) concerns the effect of adding either elliptic or circular side holes to a heavy-walled cylinder subjected to internal pressure (Fig. 13.31).

A knowledge of the stresses at the cylinder bore-side hole interface is important because many heavy-walled vessels contain oil holes for lubrication and ports for valves. In particular, for high-pressure applications, a realistic picture of the state of stress in a vessel with side ports is needed because fatigue life is very critical and present-day limitations of strength and ductility in commercial pressure vessel materials prevent high factors of safety.

In arriving at theoretical values of stress concentration factors at the bore-side hole interface, the cylinder bore is treated like a plate with a hole (Eq. 13.37). When the hole also has pressure in it, one would intuitively expect K factors (stress-concentration factors) in excess of 3 and possibly reaching values as high as 5 or 6. Experimental data and theoretical analysis do not support this reasoning. In the following analysis the plate theory is used; first, however, it is necessary to determine how to apply plate theory in a circular cylinder. The distribution of stress in a cylinder containing side holes, where both cylinder bore and side holes are subjected to internal pressure, is complex, and thus it is necessary to reduce the situation to a simpler one that is amenable to exact analysis. This is accomplished by noting that the addition of a hydrostatic stress on a system does not alter its fundamental behavior.

Figure 13.32 shows a closed-end cylinder containing an elliptic side hole subjected to internal pressure; to this picture is added hydrostatic tension (Fig. 13.33) equal in magnitude to the internal pressure. Superposition then gives the situation shown in Fig. 13.34, which is amenable to exact analysis by the equations for a cylinder subjected to external pressure. Thus, for the cylinder in Fig. 13.33 from Eqs. 3.466, 3.467, and 3.469,

$$\sigma_h = \frac{p_0 r_0^2}{r_0^2 - r_i^2}\left(1 + \frac{r_i^2}{r^2}\right) \qquad 13.140$$

*The analysis of multibore cylinders under pressure is given by Kraus (53). See also Section 13.3. The stresses in a heavy-walled cylinder having a square bore under concentrated loading are given by Seika (81). For spherical caps see (42).

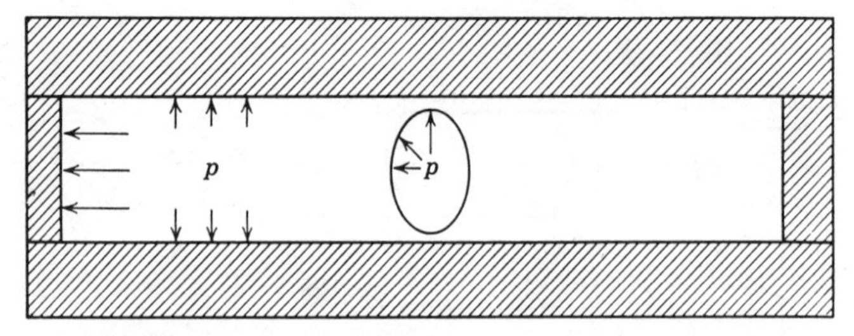

Fig. 13.32 Closed-end cylinder with side hole subjected to internal pressure.

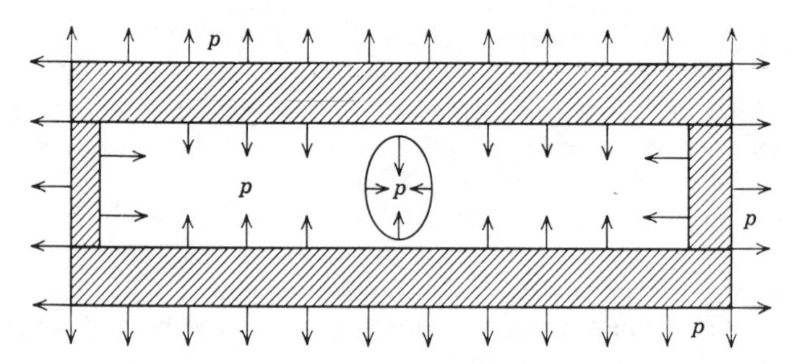

Fig. 13.33 Cylinder subjected to hydrostatic tension.

$$\sigma_r = \frac{p_0 r_0^{\,2}}{r_0^{\,2} - r_i^{\,2}}\left(1 - \frac{r_i^{\,2}}{r^2}\right) \qquad 13.141$$

and

$$\sigma_z = \frac{p_0 r_0^{\,2}}{r_0^{\,2} - r_i^{\,2}} \qquad 13.142$$

where p_0 is the external pressure.

The location of maximum stress concentration is at A on the bore surface, Fig. 13.34; therefore, in Eqs. 13.140 and 13.141, $r = r_i$ and

$$\sigma_{h(\max)} = \frac{2p_0 R^2}{R^2 - 1} \qquad 13.143$$

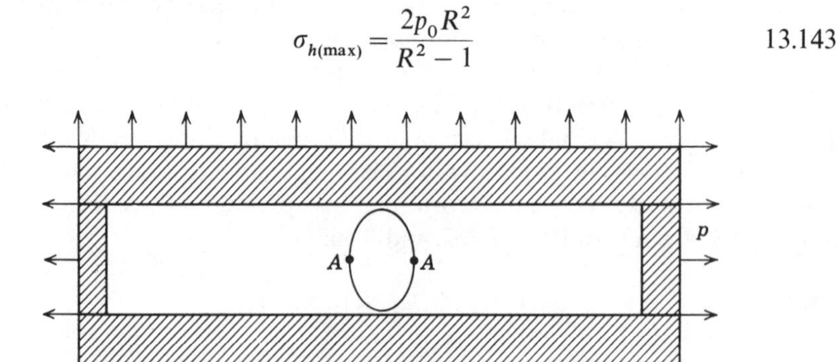

Fig. 13.34 Superposition of Figs. 13.32 and 13.33.

$$\sigma_r = 0 \tag{13.144}$$

$$\sigma_z = \frac{p_0 R^2}{R^2 - 1} \tag{13.145}$$

where $R = r_0/r_i$.

Elastic plate analysis can now be applied to the cylinder problem. The hoop stress generated at the bore of the cylinder (Eq. 13.143) is interpreted as the unit stress S_x in Eq. 13.36; thus

$$\sigma_x = \frac{2p_0 R^2}{R^2 - 1}\left(1 + 2\frac{b}{a}\right) \tag{13.146}$$

Because the cylinder has closed ends, a longitudinal stress is generated (Eq. 13.145) which is interpreted as unit stress S_y in Eq. 13.38, so that

$$\sigma_x = \frac{p_0 R^2}{R^2 - 1} \tag{13.147}$$

Stresses σ_x given by Eqs. 13.146 and 13.147 are both generated at location A, Fig. 13.34, and are additive; thus the total effective hoop stress in the cylinder at A is

$$(\sigma_h)_A = \frac{2p_0 R^2}{R^2 - 1}\left(1 + 2\frac{b}{a}\right) - \frac{p_0 R^2}{R^2 - 1} \tag{13.148}$$

The normal hoop stress at A is given in Eq. 13.143; therefore at this location the stress concentration factor K is

$$K = \frac{[p_0 R^2/(R^2 - 1)](1 + 4b/a)}{2p_0 R^2/(R^2 - 1)} = \frac{1 + 4b/a}{2} \tag{13.149}$$

Use of Eq. 13.149 now permits the calculation of specific stress concentration factors depending on the geometry of the side holes. If the side hole is small and circular, $a = b$ and $K = 2.5$. For an elliptic hole of geometry $a/b = 2$, $K = 1.50$. These K values indicate the influence of the longitudinal stress in a cylinder on stress concentration at side holes—for example, with no longitudinal stress and a circular side hole K would be 3.0, the same as in a plate under tension—the presence of the longitudinal stress decreases this value to 2.5 or a decrease of 16.7%.

The K factor is determined after the application of hydrostatic tension to the cylinder (Fig. 13.34). Therefore in using the preceding theory to solve practical problems, one must keep this fact in mind and make calculations accordingly. For example, suppose the cylinder has open rather than closed ends; then, the effect of superposition of hydrostatic tension would give the result in Fig. 13.34 minus the tensile force distributed over the bore area, and the resultant longitudinal stress would be p_0 rather than the stress expressed by Eq. 13.145. Thus zero longitudinal stress could result only by the application of a compression force to the end of the cylinder, which would be canceled by the hydrostatic tension. For an open-end cylinder the K factor for a small circular hole would be not 3.0 but $3.0 - [(R^2 - 1)/2R^2]$; in other words, for an open-end cylinder the K factor depends on the diameter ratio R.

The foregoing portion of the analysis is a basis on which to calculate other values. For example, when the side hole is not small or when there are two side holes diametrically opposed, the intensification factors for small single holes are not valid. To account for the size of the side hole, the longitudinal stress σ_z is based on the net section of the cylinder cross section supporting the load; this automatically takes into account the effect of changing the side-hole geometry.

Consider Fig. 13.31 and take the pressure length of the side hole as three fourths of the hole length (this was the case in the experimental work). Thus

$$L = \tfrac{3}{4}(r_0 - r_i) \qquad 13.150$$

Now, because stress is equal to load divided by area,

$$\sigma_z = \frac{\text{force}}{\text{net area}} = \frac{p_0 \pi r_0^{\,2}}{\pi(r_0^{\,2} - r_i^{\,2}) - 2nr_s L} \qquad 13.151$$

where $n = 1$ for one side hole and 2 for two diametrically opposite holes. Thus, from Eq. 13.151, for a single side hole (elliptic or circular)

$$\sigma_z = \frac{p_0 \pi R^2 R_s}{(R-1)[\pi R_s(R+1) - 1.5]} \qquad 13.152$$

and for two diametrically opposite side holes

$$\sigma_z = \frac{p_0 \pi R^2 R_s}{(R-1)[\pi R_s(R+1) - 3.0]} \qquad 13.153$$

where R_s equals side-hole ratio, r_i/r_s for a circular hole and r_i/a for an elliptic hole.

CIRCULAR SIDE HOLES Equation 13.151 gives an expression for the longitudinal stress as a function of the geometry of the cylinder. To assign the proper stress intensification values to σ_h and σ_z for calculating the K factor for large side-hole sizes, reference is made to the compilation of data by Peterson (71). Pertinent data from Peterson (Fig. 13.35) come from a consideration of two circular holes in a plate subjected to axial loading. These curves show the effect on stress concentration of the proximity of holes; for a cylinder with two side holes diametrically opposed the proximity factor is interpreted as the side-hole ratio r_i/r_s. Curve A is used to arrive at intensification values for the hoop stress at location A, Fig. 13.34. σ on the plot represents the hoop stress σ_h in the cylinder which is intensified to σ_{max} as shown. Thus for a circular cross-bore side hole, if R_s is equal to 7, σ_h is intensified by a factor equal to about 2.98.

The intensification factors for σ_z are obtained from curve B. For this case σ represents the longitudinal stress σ_z in the cylinder with a maximum value σ_{max}, as shown. This maximum longitudinal stress is positive, but according

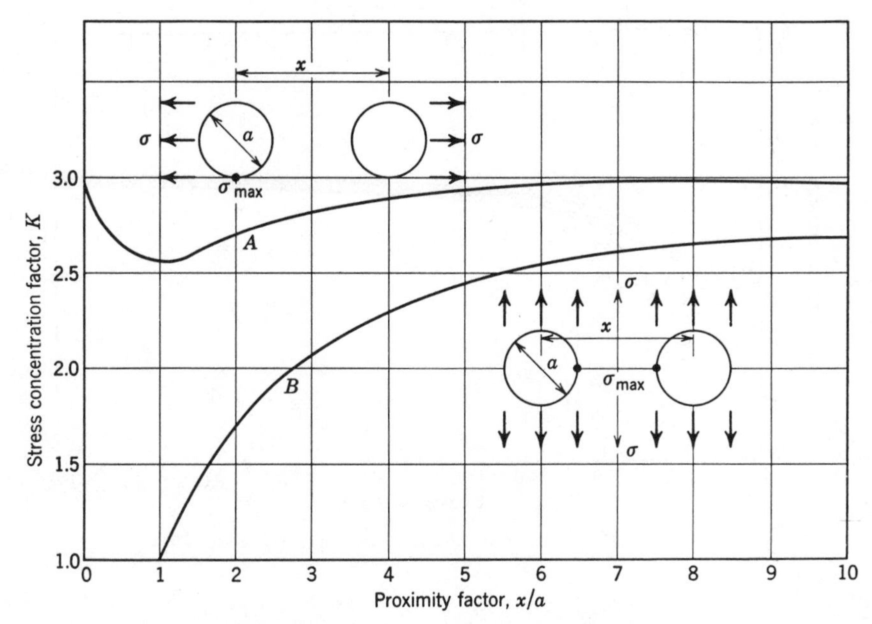

Fig. 13.35 Stress intensification factors (71).

to plate theory it would induce a negative stress at a location 90° from σ_{max} (the σ_h direction in the cylinder). Thus, as an approximation, if σ_{max} were 3σ, the longitudinal stress would be reflected in the σ_h direction as $-\sigma_z$; if σ_{max} were 1.5σ, the longitudinal stress would be reflected as 50% of -1.0 or $-0.5\sigma_z$. Thus, if the side-hole ratio were 7, σ_h would be intensified by a factor equal to about 2.98, and to this would be added the effect of the longitudinal stress $-0.86\sigma_z$. Intensification factors for circular side holes have been listed in Table 13.2 for a series of cylinders. With these factors the stress concentration

Table 13.2 Stress Intensification Factors for a Cylinder

Side Hole Ratio, R_s	Factor for σ_h (α)	Factor for σ_z, (γ)
1	2.57	−0.33
2	2.71	−0.58
3	2.82	−0.70
4	2.88	−0.77
5	2.92	−0.81
6	2.95	−0.84
7	2.96	−0.86
8	3	−0.88
9	3	−0.90
10	3	−0.92

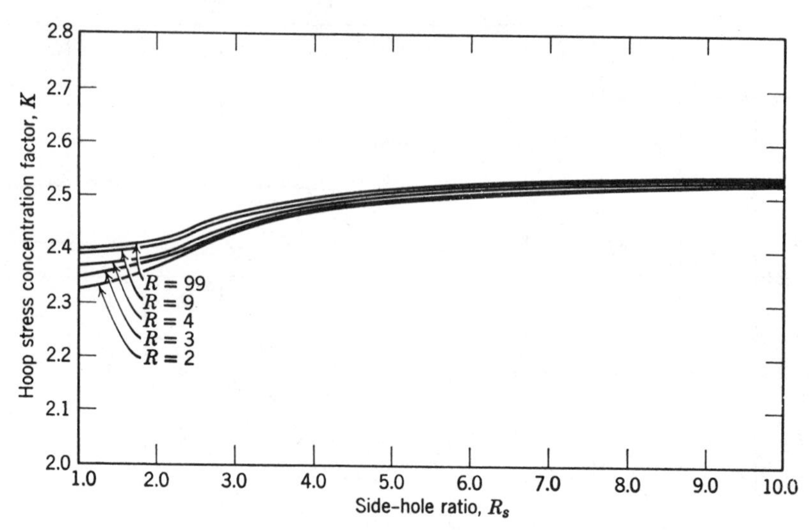

Fig. 13.36 K values for various cylinder geometries (20).

factor K for the hoop stress at the bore-side hole interface can be calculated as follows:

$$K = \frac{\alpha\sigma_h + \gamma\sigma_z}{\sigma_h} \qquad 13.154$$

where α and γ are intensification factors determined from Fig. 13.35, σ_h is calculated from Eq. 13.143, and σ_z from Eq. 13.151. Figure 13.36 shows a design curve for cylinders with side holes.

ELLIPTIC SIDE HOLES. In analyzing elliptic side holes the procedure used to determine the effect of circular side holes is applied, except that different intensification factors are applied to σ_h and σ_z. For example, for a circular cylinder having a diameter ratio r_o/r_i of 2.5 containing a single elliptic side hole of axis ratio a/b equal to 2.0,

$$(\sigma_h)_{\max} = \frac{4p_0R^2}{R^2 - 1} \qquad 13.155$$

$$\sigma_z = \frac{p_0\pi R^2 R_s}{(R-1)[\pi R_s(R+1) - 1.5]} \qquad 13.156$$

and

$$K = 2 - \frac{3.5\pi}{2\pi(3.5) - 1.5} = 1.46 \qquad 13.157$$

Because data are not available for multiple elliptical holes similar to those in Fig. 13.35 for multiple circular holes, this case cannot be analyzed.

Application of this theory for side holes in cylinders is illustrated in three design examples.

EXAMPLE 1. DESIGN OF PLAIN MONOBLOC CYLINDERS WITH CIRCULAR SIDE HOLES. If K values are available, it is possible to estimate the elastic limit pressure for a cylinder containing side holes. Assuming that the cylinder is initially stress free, the elastic-limit pressure is obtained by substituting maximum principal stress values for a cylinder, corrected for stress concentration, into the distortion energy criterion for failure; thus, for a closed-end cylinder (Fig. 13.34),

$$(\sigma_h)_{\max} = Kp\frac{R^2+1}{R^2-1} \tag{13.158}$$

$$\sigma_{r(\max)} = -p \tag{13.159}$$

and

$$(\sigma_z)_{\max} = -p \tag{13.160}$$

These stresses are related by the distortion energy criterion to the yield strength of the material in tension in accordance with Eq. 3.524; thus

$$\sigma_0 = (\sigma_h{}^2 + \sigma_r{}^2 + \sigma_z{}^2 - \sigma_h\sigma_r - \sigma_r\sigma_z - \sigma_z\sigma_h)^{1/2} \tag{13.161}$$

Substitution of Eqs. 13.158 to 13.160 into Eq. 13.161 thus gives an expression for the elastic limit pressure p_y of a cylinder:

$$p_y = \sigma_0(R^2-1)[R^4(K^2+2K+1)+2R^2(K^2-1)+(K^2-2K+1)]^{-1/2} \tag{13.162}$$

If $R = 2$ and $R_s = 4$, from Fig. 13.36, $K = 2.47$ and by Eq. 13.162

$$p_y = 0.195\sigma_0 \tag{13.163}$$

If there were no side hole,

$$p_y = \sigma_0(3)^{-1/2}\left(\frac{R^2-1}{R^2}\right) = 0.432\sigma_0 \tag{13.164}$$

The cylinder with the side holes will yield (locally) at a pressure 55% lower than the cylinder without side holes.

If R is increased to 4 in a cylinder with no side holes, then by Eq. 13.164

$$p_y = 0.54\,\sigma_0 \tag{13.165}$$

which is an elastic-limit pressure increase of 25% over that obtained for the cylinder with a wall ratio of 2. However, for $R = 4$ and $R_s = 4$, $K = 2.485$, and by Eq. 13.162

$$p_y = 0.258\sigma_0 \tag{13.166}$$

which is an increase in elastic limit pressure of about 32%. Thus benefit can be obtained, elastically, in increasing the wall ratio; even though the K values increase with increasing wall ratio, the elastic limit properties increase at a more rapid rate (due to increase in wall ratio) than the decrease in elastic limit properties caused by increasing K. For each case, however, the situation should be examined in order that realistic values may be assigned to specific designs.

On the other hand, difficulty could be encountered in a particular case, that is designed to operate within the elastic range, if additional side holes of different R_s values were used. For example, if $R = 2$ and $R_s = 1$, then by Eq. 13.162 and Fig. 13.36,

$$p_y = 0.218\sigma_0 \tag{13.167}$$

Now if to this cylinder an additional set of side holes is introduced having an R_s value

Table 13.3 Elastic Limit Pressures for Closed-End Cylinders

Diameter Ratio, R	Relative Elastic Limit for Side-Hole Ratio of			
	∞	5	1	Variable ($K = 6$)
2	1.00	0.447	0.473	0.210
3	1.185	0.560	0.575	0.285
4	1.250	0.605	0.625	0.296
9	1.315	0.645	0.670	0.322

of, say, 10 (small oil hole), then

$$p_y = 0.151\,\sigma_0 \qquad 13.168$$

or a decrease in elastic limit pressure of 31%.

EXAMPLE 2. COMPARATIVE DESIGN OF CYLINDERS WITH CIRCULAR SIDE HOLES. In this example some comparisons are made relative to the elastic limit pressures of cylinders of various R values containing circular side holes of various R_s values. The comparison includes a conventional design where the stress concentration factor is assumed to be 6.00.

As in Example 1, Eqs. 13.158 to 13.160 define the state of stress in the cylinders, which when put into Eq. 13.161 gives the elastic limit pressure defined by Eq. 13.162, where K is determined from Fig. 13.36. The results of some calculations are shown in Table 13.3, where relative elastic limits for cylinders of various geometries are listed. When the side hole ratio R_s is infinity, it is assumed that the cylinder acts as a plain monobloc with a relative elastic limit of 1.00 for a diameter ratio of 2.0 and 1.315 for a cylinder where $R = 9.0$. This is another way of saying that if the ratio of a cylinder is increased from 2 to 9, the elastic limit pressure increases 31.5%. Furthermore, still referring to Table 13.3, a cylinder having diameter ratio of 9 and a sidehole ratio of 5 exhibits an elastic limit pressure equal to 64.5% of that exhibited by a plain monobloc cylinder having a diameter ratio of 2.0.

The real value of having more precise values for stress concentration factors in cylinders with side holes can be illustrated by noting the relative elastic limit values in Table 13.3 corresponding to the last column marked "Variable ($K = 6$)." In many conventional designs it has been customary to use K values as high as 6.0, regardless of the side hole size, simply because the true value has not been known. When such a large factor is used, for example, the calculated elastic limit pressure for a cylinder having a diameter ratio of 4.0 is only 29.6% of that exhibited by a plain monobloc cylinder having a diameter ratio of 2.0, regardless of the hole size (R_s value). Now, if R_s is taken into account, Table 13.3 shows for the same cylinder having a diameter ratio of 4.0, that for an R_s of 5.0, the elastic limit pressure is 60.5% of that exhibited by the monobloc of diameter ratio $R = 2.0$. In other words, a factor of about 100% is involved, and when the true value of K is used overdesign is avoided. Safety factors are always used; however, if the safety factor is 2, the equipment should be proportioned to give this factor of safety and not a factor of 4, which may occur if the proper values of K are not used.

EXAMPLE 3. USE OF ELLIPTIC SIDE HOLES. When a circular side hole is placed in a cylinder, the maximum stress concentration occurs in the hoop direction at the side-hole bore interface. This factor can be reduced by making the side hole elliptic in shape; however, it is important not to introduce a K factor at the ends of the major axis of the ellipse which, when applied to the longitudinal stress, would create a situation worse than that obtained in the hoop direction for a circular side hole.

The limiting case for a single small elliptic side hole will now be considered for cylinders with both open- and closed-ends. In the closed-end cylinder, the total effective hoop stress at the ends of the minor ellipse axis is given by the sum of Eqs. 13.36 and 13.38,

$$\sigma_h = \frac{2p_0R^2}{R^2-1}\left(1+2\frac{b}{a}\right) - \frac{p_0R^2}{R^2-1} \tag{13.169}$$

The total effective longitudinal stress at the ends of the major ellipse axis is given by the sum of Eqs. 13.37 and 13.39,

$$\sigma_z = \frac{p_0R^2}{R^2-1}\left(1+2\frac{a}{b}\right) - \frac{2p_0R^2}{R^2-1} \tag{13.170}$$

From Eqs. 13.143 and 13.169, the stress concentration factor at the end of the minor axis is calculated to be

$$K_b = \frac{1+4(b/a)}{2} \tag{13.171}$$

and by Eqs. 13.143 and 13.170, the stress concentration factor at the end of the major axis is

$$K_a = 2\frac{a}{b} - 1 \tag{13.172}$$

To determine the limiting ellipse axis ratio it is necessary to equate the equivalent stresses, as given by Eq. 13.161, at the ends of the two axes; thus

$$(\sigma_0)_a{}^2 = (\sigma_0)_b{}^2 = (K_b\sigma_h)^2 + 2p(K_b\sigma_h) = (K_a\sigma_z)^2 + 2p(K_a\sigma_z) \tag{13.173}$$

For a closed-end cylinder, $\sigma_h = \sigma_z(R^2+1)$; therefore, by substituting this value of σ_h in Eq. 13.173 along with values of K_b and K_a given by Eqs. 13.171 and 13.172, the limiting axis ratio a/b may be determined as a function of the diameter ratio, R, as shown in Table 13.4.

If the ends of the cylinder are open, the longitudinal stress (Eq. 13.145) becomes

$$\sigma_z = p_0 \tag{13.174}$$

and Eqs. 13.169 and 13.170 become, respectively,

$$\sigma_h = \frac{2p_0R^2}{R^2-1}\left(1+2\frac{b}{a}\right) - p_0 \tag{13.175}$$

Table 13.4 Limiting Values of Axis Ratio for Elliptic Side Hole in Closed-End Cylinder

Diameter Ratio, R	Axis Ratio, a/b
1.5	2.57
2.0	3.28
2.5	4.09
3.0	5.00
3.5	6.02
4.0	7.13
5.0	9.68
10.0	29.21

and

$$\sigma_z = p_0\left(1 + 2\frac{a}{b}\right) - \frac{2p_0R^2}{R^2 - 1} \qquad 13.176$$

Similarly, Eqs. 13.171 and 13.172 become, respectively,

$$K_b = \left(1 + 2\frac{b}{a}\right) - \frac{R^2 - 1}{2R^2} \qquad 13.177$$

and

$$K_a = \left(1 + 2\frac{a}{b}\right) - \frac{2R^2}{R^2 - 1} \qquad 13.178$$

Now, as before, by equating the equivalent stress $(\sigma_0)_a$ and $(\sigma_0)_b$, it can be shown that there is no limiting axis ratio; in other words, for an open-end cylinder under internal pressure yielding will always initiate at the ends of the minor axis. The end condition of the cylinder has a large effect on the stress-concentrating effect of side holes, whereas in plain cylinders without side holes the effect is almost negligible.

Intermediate axis ratios between 1 and the critical values can be used to reduce stress concentration effects in cylinders. For many cylinders such ellipses can be obtained by tangential drilling of the side hole, using a cylindrical drill.

Thin-walled cylinders with thick ends. Often a thin-walled cylinder is fitted with a relatively thick end like a flange or cover plate; this end has a pronounced effect on the cylinder stresses in the vicinity of the closure – the effects are calculated using the theory of discontinuity stresses discussed in Chapter 3. For the present case, where $R/h > 10$, design curves have been calculated (46) and are presented in Fig. 13.37 for cylinders subjected only to axial loading (Fig. 13.38), and to internal pressure (Fig. 13.39). The curves are used as follows.

CYLINDER SUBJECTED TO AXIAL LOAD (FIG. 13.40)*

$$\text{Axial stress} = \sigma_z = (F/h)(1 \pm K_2) \qquad 13.179$$

$$\text{Hoop stress} = \sigma_h = (vF/h)(2 \pm K_2) \qquad 13.180$$

where F is the force in pounds per inch, v is Poisson's ratio, and K_2 is a factor defined by Fig. 13.37. For example, in Fig. 13.38, if F is 1000 lb/in., $R = 12$ in., $h = 0.4$ in., and $v = 0.30$, K_2 is found to be 0.545; then

$$(\sigma_z)_{\text{inner fiber}} = \left(\frac{1000}{0.4}\right)(1 - 0.545) = 1138 \text{ psi}$$

$$(\sigma_z)_{\text{outer fiber}} = \left(\frac{1000}{0.4}\right)(1 + 0.545) = 3863 \text{ psi}$$

$$(\sigma_h)_{\text{inner fiber}} = \frac{(0.3)(1000)}{0.4}(2 - 0.545) = 1093 \text{ psi}$$

$$(\sigma_h)_{\text{outer fiber}} = \frac{(0.3)(1000)}{0.4}(2 + 0.545) = 1909 \text{ psi}$$

*See also the discussion on discontinuity stresses in Chapter 3.

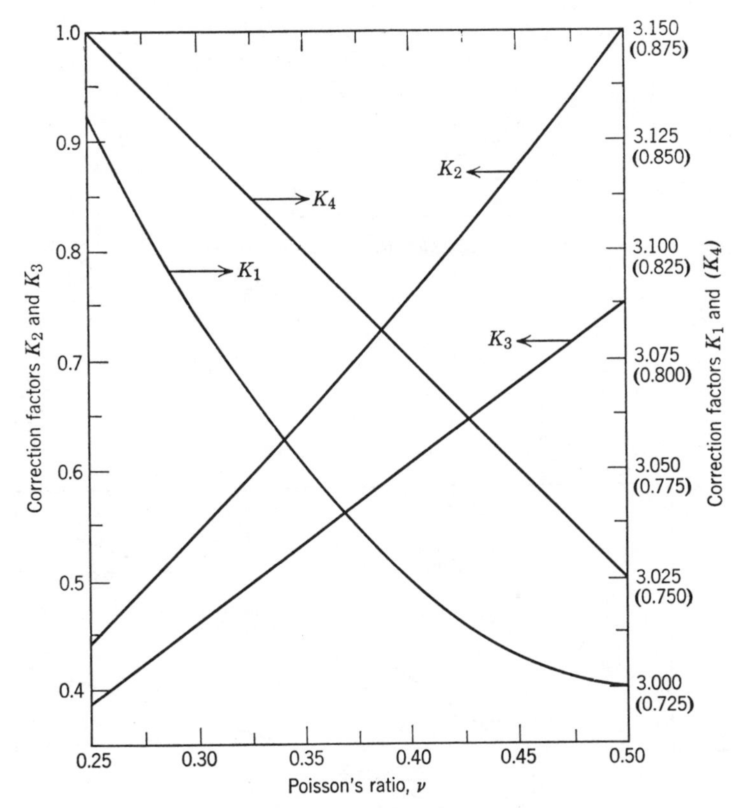

Fig. 13.37 Correction factors for thick-end cylinders. [After Isakower (46), courtesy of The McGraw-Hill Publishing Co.]

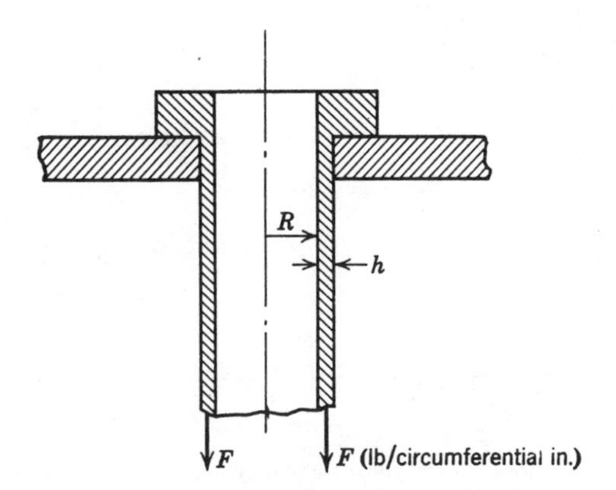

Fig. 13.38 Cylinder subjected to axial loading.

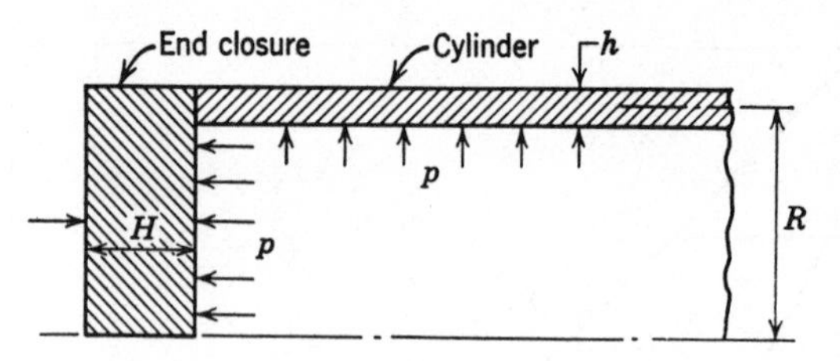

Fig. 13.39 Cylinder subjected to internal pressure.

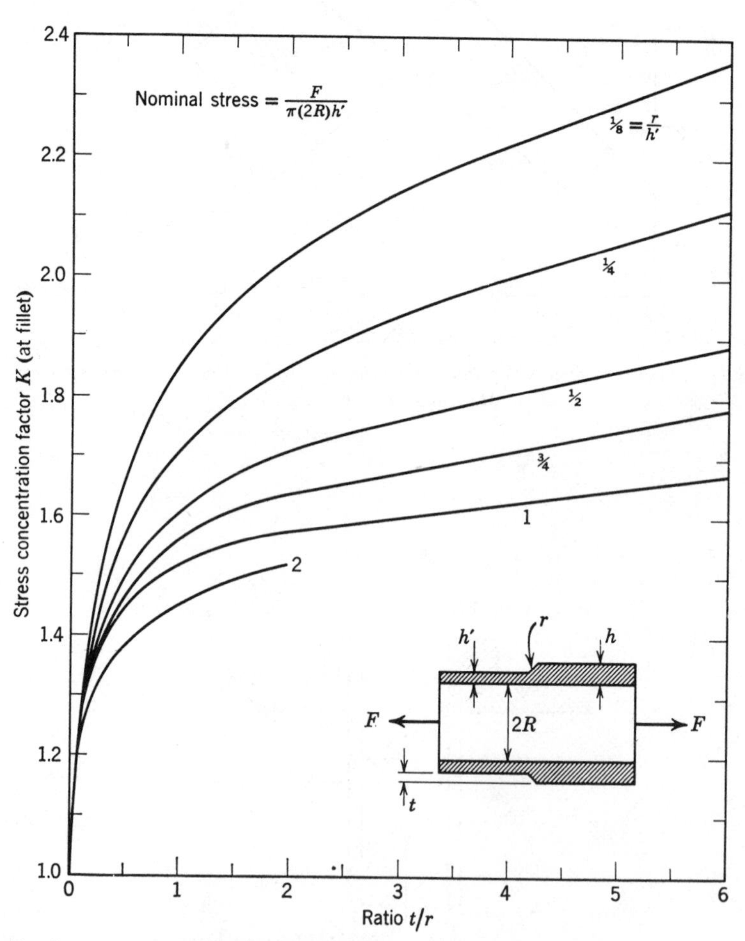

Fig. 13.40 Stress concentration factors for a stepped cylinder with shoulder fillets. [After Lee and Ades (55), courtesy of The Society for Experimental Stress Analysis.]

Without the correction factors the conventional stresses are given by Eqs. 13.179 and 13.180 with K_2 equal to zero and the 2 in Eq. 13.180 replaced by 1, thus,

$$\sigma_z = F/h = 1000/0.4 = 2500 \text{ psi}$$

and

$$\sigma_h = \nu F/h = 0.3(2500) = 750 \text{ psi}$$

Thus, at the thick end the stress-concentration factors are as follows:

	Stress Concentration Factors
Axial stress	3863/2500 = 1.55
Hoop stress	1090/751 = 2.54

CYLINDER SUBJECTED TO INTERNAL PRESSURE (FIG. 13.39)

$$\text{Axial stress} = \sigma_z = (pR/2h)(1 \pm K_1) \qquad 13.181$$

$$\text{Hoop stress} = \sigma_h = (pR/h)(1 - K_4 \pm K_3) \qquad 13.182$$

where p is the internal pressure. For example, in Fig. 13.39, if $R = 10$ in., $h = 0.3$ in., $p = 500$ psi, and $\nu = 0.3$, then since the cover plate is very thick and can be assumed rigid, the cylinder stresses at the end closure are as follows, using Fig. 13.39:

$$(\sigma_z)_{\text{inner fibers}} = \frac{(500)(10)}{(2)(0.3)}(1 + 3.086) = 34{,}100 \text{ psi}$$

$$(\sigma_z)_{\text{outer fibers}} = \frac{(500)(10)}{(2)(0.3)}(1 - 3.086) = -17{,}150 \text{ psi}$$

$$(\sigma_h)_{\text{inner fibers}} = \frac{(500)(10)}{0.3}(1 - 0.850 + 0.4625) = 10{,}210 \text{ psi}$$

$$(\sigma_h)_{\text{outer fibers}} = \frac{(500)(10)}{3.0}(1 - 0.850 - 0.4625) = -5208 \text{ psi}$$

The conventional stresses are found by letting the K values equal zero in Eqs. 13.181 and 13.182, thus

$$\sigma_z = \frac{pR}{2h} = \frac{(500)(10)}{(2)(0.3)} = 8333 \text{ psi}$$

and

$$\sigma_h = \frac{pR}{h} = \frac{(500)(10)}{0.3} = 16{,}666 \text{ psi}$$

In this instance the axial stress is intensified by a factor of 34,100/8333 = 4.1, while the hoop stress is intensified (negatively) by 10,210/16,666 = 0.6.

Thin-walled vessels with holes and nozzles. As in heavy-walled cylinders with side holes, serious stress intensification effects occur in various types of pressure vessels containing holes, nozzles, branch pipes, and the like*. In thin-walled vessels, much research has been carried out for a number of years in an attempt to resolve some of the problems involved (63, 97, 101). The ASME pressure vessel code provides general guidance for the reinforcement of openings in vessels and the following example is taken from the 1977 code (101):† An $11\frac{3}{4}$ in. I.D., $\frac{1}{2}$ in. wall, nozzle conforming to Specification SA-105, is attached by welding to a vessel that has an inside diameter of 60 in. The shell thickness is $\frac{3}{4}$ in. The shell plate is to conform to Specification SA-414, Grade D, and the reinforcing element to conform to Specification SA-285, Grade C. The vessel is to operate at 250 psi and 700°F with no allowance for corrosion. Check the adequacy of the reinforcing element and the attachment welds shown in Fig. 13.41.

WALL THICKNESS REQUIRED

Shell
$$t_r = \frac{PR}{SE - 0.6P} = \frac{(250)(30)}{(14{,}300)(1) - 0.6(250)} = 0.530 \text{ in.}$$

where t_r = shell thickness, in.
P = internal pressure, psi
R = inside radius of shell, in.
S = allowable stress, psi
E = 1 (from code requirements)

Nozzle
$$t_{rn} = \frac{PR_n}{SE - 0.6P} = \frac{(250)(5.875)}{(16{,}600)(1) - 0.6(250)} = 0.089 \text{ in.}$$

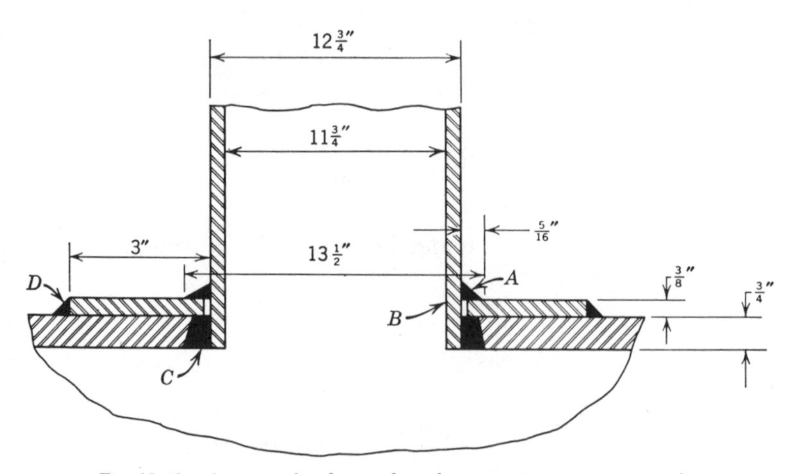

Fig. 13.41 An example of a reinforced opening in a pressure vessel.

*The discussion here is limited to effects of pressure only. See also Reference 31 of Chapter 3.
†Courtesy of The American Society of Mechanical Engineers.

SIZE OF WELD REQUIRED

Inner reinforcing fillet weld

$$t_w = 0.7t_{\min} = (0.7)(0.375) = 0.263 \text{ in.}$$

$$t_w = 0.7 \text{ (weld size)} = \frac{0.7(13.5 - 12.75)}{2} = 0.263 \text{ in} \quad \text{(existing)}$$

Outer reinforcing fillet weld

$$t = \tfrac{1}{2}t_{\min} = 0.5(0.375) = 0.1875 \text{ in.}$$

$$t = 0.7 \text{ (weld size)} = 0.7(0.3125) = 0.219 \text{ in.} \quad \text{(existing)}$$

Therefore the weld sizes are satisfactory.

AREA OF REINFORCEMENT REQUIRED

$$A = XF\,dt_r = (1)(11.75)(0.530) = 6.228 \text{ in.}^2$$

AREA OF REINFORCEMENT AVAILABLE

$$A_1 = \text{larger of the following:}$$

$$A_1 = (E_1 t - Ft_r)\,d = [(1)(0.75) - (1)(0.530)](11.75) = 2.585 \text{ in.}^2$$

or

$$A_1 = 2(E_1 t - Ft_r)(t_n + t) = 2[(1)(0.75) - (1)(0.530)](0.5 + 0.75) = 0.550 \text{ in.}^2$$

$$A_2 = \text{smaller of the following:}$$

$$A_2 = (t_n - t_{rn})(5t) = [0.5 - (0.089)(5)(0.75)] = 1.541 \text{ in.}^2$$

or

$$A_2 = 2(t_n - t_{rn})(2.5t_n + t_e) = 2\left\{(0.5 - 0.089)\left[(2.5)(0.5) + 0.375 \times \left(\frac{13{,}200}{14{,}300}\right)\right]\right\} = 1.312 \text{ in.}^2$$

$$A_4 = (2)(0.5)[(0.3125)^2 + (0.375)^2] = 0.238 \text{ in.}^2$$

The area provided by A_1, A_2, and A_4 is thus 4.135 in.^{2}

$$A_5 = (D_p - d - 2t_n)t_e = (18.75 - 11.75 - 1.0)(0.375)\left(\frac{13{,}200}{14{,}300}\right) = 2.077 \text{ in.}^2$$

The total area available is thus 6.212 in.^{2}. In the preceding equations,

A_1 = area in excess thickness in vessel wall available for reinforcement (in.^{2}).

A_2 = area in excess thickness in nozzle wall available for reinforcement (in.^{2})

A_4 = cross-sectional area of welds available for reinforcement (in.^{2})

A_5 = cross-sectional area of material added as reinforcement (in.^{2})
D_p = outside diameter of reinforcing element (in.)
d = diameter in the plane under consideration of the finished opening (in.)
$F = 1$ (From code requirement)

The area available is 6.212 in.^{2}, but the area required is 6.228 in.^{2}; therefore the opening is not adequately reinforced, and the size of the reinforcing element must be increased. This is done by increasing the O.D. of the reinforcing element by $\frac{1}{4}$ in.

LOAD TO BE CARRIED BY WELDS

$$W = (A - A_1)S = (6.228 - 2.585)(14{,}300) = 52{,}095 \text{ lb}$$

The unit stress in the outer fillet weld in shear is (0.49) (13,200) = 6468 psi. The inner fillet weld shear is (0.49) (13,200) = 6468 psi. The groove weld tension is (0.74)(14,300) = 10,582 psi, and the nozzle wall shear is (0.70)(14,300) = 10,010 psi.

STRENGTH OF CONNECTION ELEMENTS

1. Inner fillet weld shear = $(\pi/2)$ (O.D.) (weld leg) (6468) = 48,552 lb
2. Nozzle wall shear = $(\pi/2)$ (mean diameter) (t_n) (10,010) = 96,259 lb
3. Groove weld tension = $(\pi/2)$ (O.D.) (t) (10,582) = 158,869 lb
4. Outer fillet weld shear = $(\pi/2)$ (reinforcing O. D.) (weld leg) (6468) = 60,197 lb

The possible paths of failure are (*a*) through *B* and *D*, 96,259 + 60,197 = 156,456 lb; (*b*) through *A* and *C*, 48,552 + 158,869 = 207,421 lb; and (*c*) through *C* and *D*, 158,869 + 60,197 = 219,066 lb. Thus all paths are stronger than the required strength of 52,095 lb. The design strength of the outer fillet weld attaching the reinforcing element to the shell is 60,197 or greater than the reinforcing element strength of (2.163)(13,200) = 28,552 lb. Additional examples are given in (63, 98, 101).*

Following the same line of reasoning outlined previously for side holes in heavy-walled cylinders various stress intensification factors can be calculated for thin-walled vessels. First, consider a simple closed-end, thin-walled cylinder containing a side hole and pipe as shown in Fig. 13.42. In this example discontinuity stresses are not included as they are in a later example. It is assumed that the wall thickness of the pipe branch is very small in comparison with the wall thickness of the main cylinder and that the branch is subjected only to hoop stress. The location of maximum stress intensification is at *A*, where

$$\sigma_A = \frac{3pR}{t} - \frac{pR}{2t} = \frac{5pR}{2t} \qquad 13.183$$

*Reinforcement of a hole in a concrete slab is discussed in (41).

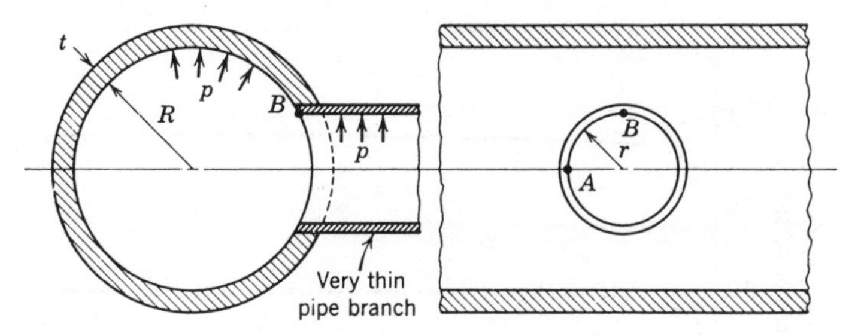

Fig. 13.42 Thin-walled cylinder with side hole subjected to internal pressure.

The normal hoop stress at A in the absence of the hole is

$$\sigma_A = pR/t \tag{13.184}$$

In Eq. 13.183 the stress given by Eq. 13.184 is intensified by a factor of three and is also subjected to minus one times the longitudinal stress since this is a case of biaxial loading (as in Fig. 13.7). Also at location A are longitudinal and radial stresses

$$\sigma_z = \sigma_r = -p \tag{13.185}$$

The equivalent stress at A, from Eq. 3.523, is

$$(\sigma_0)_A{}^2 = \left(\frac{5pR}{2t}\right)^2 + (-p)^2 + (-p)^2 - \left(\frac{5}{2}\frac{pR}{2t}\right)(-p) - (-p)(-p) + \left(\frac{5}{2}\frac{p^2R}{2t}\right) \tag{13.186}$$

from which

$$(\sigma_0)_A = \frac{p}{2t}\sqrt{25R^2 + 10Rt + 4t^2} \tag{13.187}$$

Without the hole, the cylinder stresses are given by Eqs. 3.371 to 3.372. In addition to these stresses the average radial stress is taken as

$$\sigma_r \approx -p/2 \tag{13.188}$$

so that application of Eq. 3.523 gives

$$\sigma_0 = (p/2t)\sqrt{3R^2 + 6Rt + t^2} \tag{13.189}$$

The combined stress-stress concentration factor is thus

$$K = \frac{\text{Eq. 11.187}}{\text{Eq. 11.189}} = \sqrt{\frac{25\alpha^2 + 10\alpha + 4}{3\alpha^2 + 6\alpha + 1}} \tag{13.190}$$

where

$$\alpha = R/t \tag{13.191}$$

Various values are plotted in Fig. 13.43.

For a thin-walled sphere under internal pressure and with a small side hole

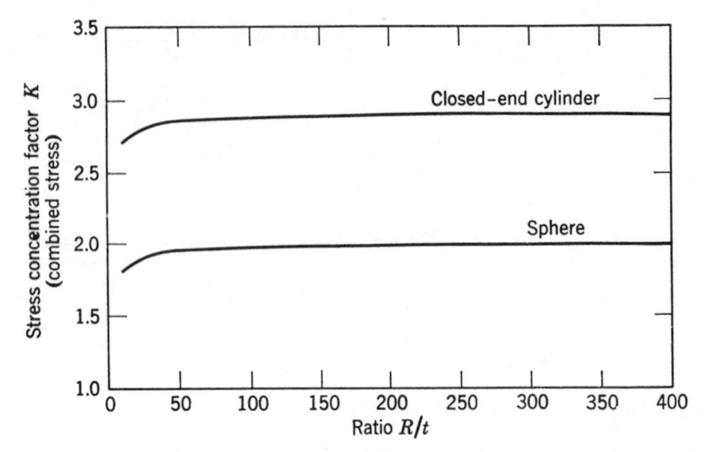

Fig. 13.43 Stress concentration factors for thin-walled cylinders and spheres with side holes under internal pressure.

the hoop stress at the hole boundary is everywhere

$$\sigma_h = \frac{3pR}{2t} - \frac{pR}{2t} = \frac{pR}{t} \qquad 13.192$$

The radial stress is $-p/2$; therefore, from Eq. 3.523,

$$\sigma_0 = (p/2t)\sqrt{(2R+t)^2} \qquad 13.193$$

Without the hole

$$\sigma_0 = (p/2t)\sqrt{R^2 + 4Rt + t^2} \qquad 13.194$$

so that

$$K = \sqrt{\frac{4\alpha^2 + 4\alpha + 1}{\alpha^2 + 4\alpha + 1}} \qquad 13.195$$

which is also plotted in Fig. 13.43.

A somewhat more complicated situation exists when the branch pipe has a wall thickness of the same order as that of the main cylinder, as shown in Fig. 13.44. Consider first the stresses at the bore of the cylinder. At location A, the hoop directions of the cylinder and pipe branch coincide; as an approximation it is considered that the hoop stress at A is composed of three parts, a hoop stress due to pressure in the main cylinder,

$$(\sigma_{h1})_A = 3pR/T \qquad 13.196$$

where 3 is the usual stress concentration factor for a hole in a plate, a hoop stress due to the effect of the longitudinal stress in the main cylinder,

$$(\sigma_{h2})_A = -pR/2T \qquad 13.197$$

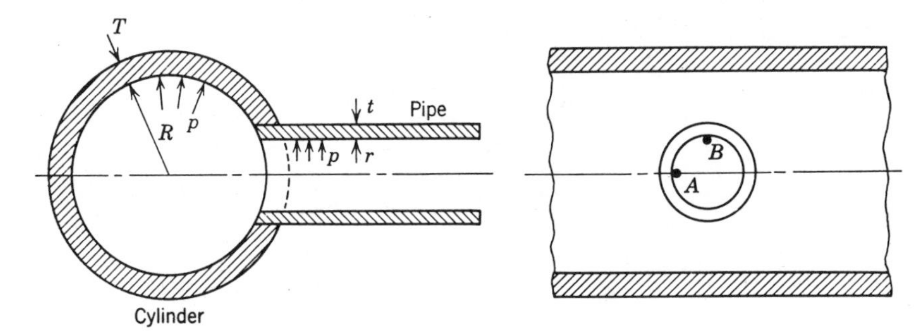

Fig. 13.44 Thin-walled cylinder with side hole under internal pressure.

and a discontinuity stress given as

$$(\sigma_{h3})_A = \frac{Kpr}{t} \tag{13.198}$$

The stress given by Eq. 13.198 is approximated by considering the longitudinal direction of the main cylinder as a support ring and on this basis by calculating the discontinuity stress from the ring-stiffened cylinder theory given in Chapter 9. Thus, from Eq. 9.323

$$\text{stress ratio} = \frac{1}{2} + \frac{10.20(1-0)}{6.61 + \dfrac{10.288}{(r/t^3)^{1/2}}\left[\dfrac{1.3K^2+0.7}{K^2-1}\right]\dfrac{1}{T}} \tag{13.199}$$

where K is defined as shown in Fig. 9.68. In the present case K is a large number so that the bracket term reduces to 1.3. The stress ratio defined by Eq. 13.199 is then equal to the stress concentration factor K in Eq. 13.198. Therefore, from Eqs. 13.196 to 13.198,

$$(\sigma_h)_A = p(2.5R/T + Kr/t) \tag{13.200}$$

If $R = r$ and $T = t$, then

$$(\sigma_h)_A = (pR/T)[2.5 + K] \tag{13.201}$$

More specifically, if $R/T = r/t = 10$, then K from Eq. 13.199 is 1.44 and

$$(\sigma_h)_A = (pR/T)(3.94) \tag{13.202}$$

Thus, for this case, the total stress concentration factor at A is 3.94, which is considerably higher than experienced in heavy-walled piping.

Applying this information to the code example previously given, r is 5.875 in., t is 0.50 in., T is 1.125 in., and R is 30. Then application of Eqs. 13.199 and 13.200 gives an allowable pressure, at 700°F, of 165 psi, using an allowable stress value of 14,350 psi. If the yield strength of the material is used as the allowable stress value, then the allowable pressure is calculated as 356 psi.

Nozzles and pipe branches on vessels frequently lead to high stresses in the vessel wall due to radial loads and bending moments; these may be applied

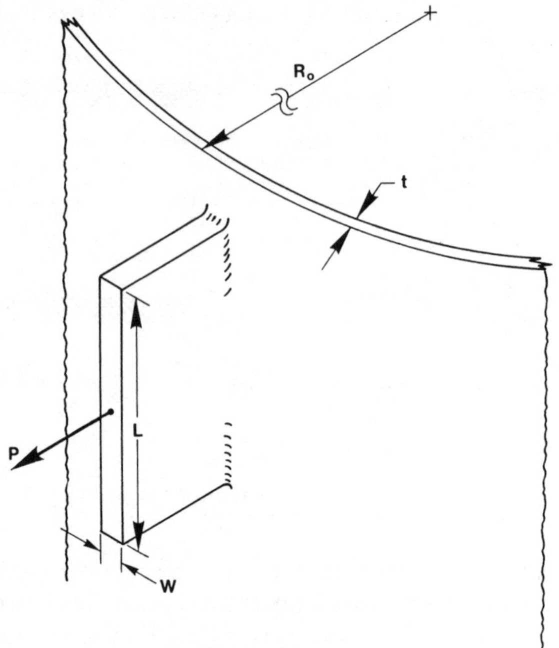

Fig. 13.45 Clip geometry.

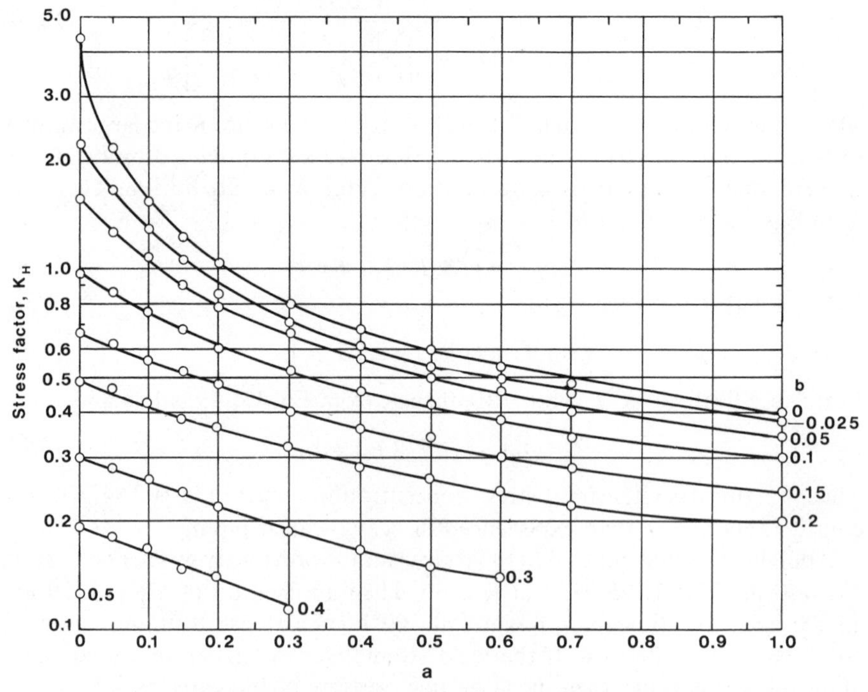

Fig. 13.46 K_H for Eq. 13.205. [Data from Roark (76).]

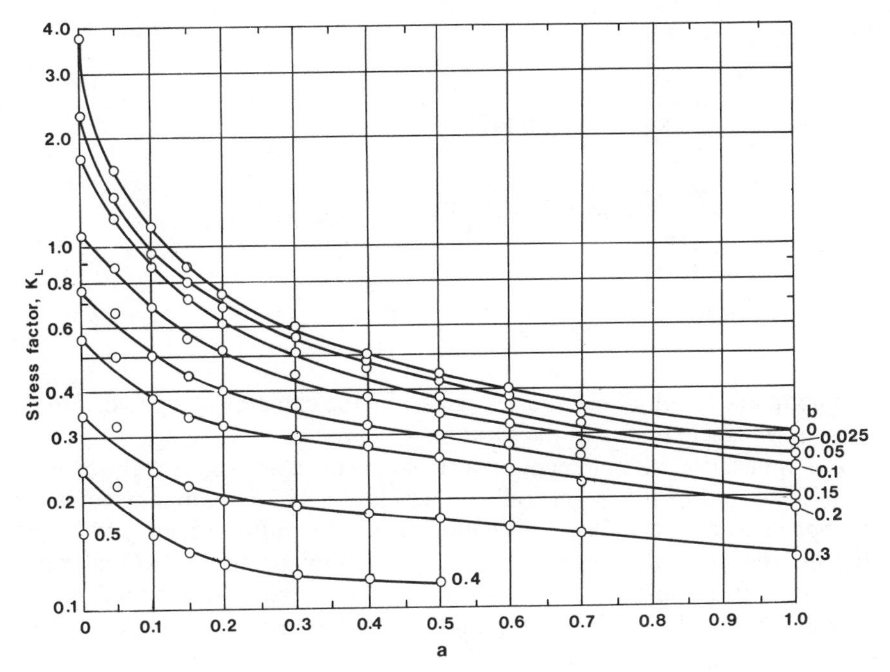

Fig. 13.47 K_L for Eq. 13.206. [Data from Roark (76).]

directly or be the result of expansions in heated or cooled piping systems. The determination of such stresses is illustrated in an example of a piping flexibility calculation in Chapter 15.

Thin-walled vessels with clips. Pressure vessels such as storage tanks and distillation towers have ladders, pipes, and platforms supported by the vessel walls. These items represent concentrated forces acting on the thin vessel wall. Roark (76) performed experiments on thin-wall cylinders where loads were applied through clips or attachments. The area of load application (Fig. 13.45) was varied to give typical ranges as in common practice. The clip width in the hoop direction is W, the length in the longitudinal direction is L, and the outside radius of the vessel is R_0. In Figs. 13.46 and 13.47, two parameters a and b determine K_H and K_L:

$$a = \frac{L}{2R_0} \qquad 13.203$$

$$b = \frac{W}{2R_0} \qquad 13.204$$

When these are determined, K_H and K_L can be selected from Figs. 13.46 and 13.47. Then

$$\sigma_H = K_H \frac{P}{2t^2} \qquad 13.205$$

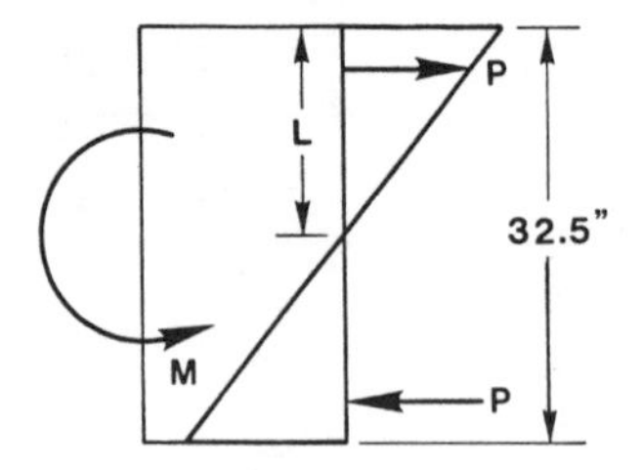

Fig. 13.48 Moment resolved into clip forces.

$$\sigma_L = K_L \frac{P}{2t^2} \tag{13.206}$$

The hoop and longitudinal stresses caused by the clip are superposed with the remaining vessel stresses at the location.

The application of a moment on a thin-walled vessel can be analyzed using Eqs. 13.205 and 13.206. A vertical load develops a 42,520 in.-lb moment which is applied to a thin-walled vessel through a clip width, 0.5 in., and length, 32.5 in. The vessel R_0 is 93.875 in., and the wall thickness t is 0.875 in., which includes a 0.1875-in. corrosion allowance. The stresses are required so that they can be combined with other known stresses to determine if the wall is overstressed.

The 42,520 in.-lb moment is reacted at the wall (Fig. 13.48) in a beam stress pattern. The resulting force couple is found from the triangular stress pattern.

$$P\left(\frac{32.5}{2}\frac{2}{3}\right)2 = 42{,}520 \text{ in.-lb}$$

$$P = 1962.5 \text{ lb}$$

This force is considered to act on the top half or bottom half of the clip. Equations 13.203 and 13.204 reduce to

$$a = \frac{(32.5/2)}{2(93.875)} = 0.08655$$

$$b = \frac{(0.5/2)}{2(93.875)} = 0.00133$$

From Figs. 13.59 and 13.60

$$K_H = 1.668 \qquad K_L = 1.30$$

The thickness used for calculating stresses is the actual thickness minus the corrosion allowance, or

$$t = 0.875 - 0.1875 = 0.6875 \text{ in.}$$

All of the parameters are substituted into Eqs. 13.205 and 13.206 yielding

$$\sigma_H = (1.668)\frac{(1962.5)}{2(0.6875)^2} = 3463 \text{ psi}$$

$$\sigma_L = (1.30)\frac{1962.5}{2(0.6875)^2} = 2699 \text{ psi}$$

The hoop stress will be summed with the additional hoop stress created by the operating pressure. The longitudinal stress is also summed with the longitudinal stress caused by the operating pressure as well as bending stresses caused by wind loads, earthquake loads, and other design conditions. The stresses are checked by the ASME Pressure Vessel Code (100) or by checking Eq. 3.526 to determine if the allowable stresses have been exceeded.

Stress concentration in filamentary structures. Theoretical analyses of the stress distributions in a sheet of parallel filaments carrying normal loads and imbedded in a matrix carrying only shear have been made by Hedgepeth (36). The application of such analyses to structures like filament-wound rocket motor cases, for example, is becoming more and more prevalent.* Hedgepeth found that one broken filament resulted in a stress concentration factor of 1.33 and that, in general, for any number of broken filaments the stress concentration factor could be expressed by

$$K_r = \frac{(4)(6)(8)\ldots(2r+2)}{(3)(5)(7)\ldots(2r+1)} \qquad 13.207$$

where K_r is the stress concentration factor for r broken filaments.

Further results were obtained by Hedgepeth and Van Dyke in 1967 as discussed by Cooper (8). They found the dynamic effects were 15% greater than Eq. 13.207 for one filament broken up to 27% larger as the number of filaments grew large. Cooper further discusses other problems in this area.

Stress concentration in the plastic range. Factors of stress concentration in the plastic range have been determined by Hardrath and Ohman (35), who conducted a series of tests on flat plates containing notches and fillets. It was found that the following formula could be used to determine the plastic range stress concentration factor:

$$K_p = 1 + [(K_e - 1)E_s]/E_\infty \qquad 13.208$$

where K_e is the elastic stress concentration factor for any particular geometry, E_s is the secant modulus (see Chapter 1) of the material at the point of maximum stress, and E_∞ is the secant modulus of the material at points far removed from the stress concentrator. The plane elastic-plastic problem has been analyzed by Galin (29) and Wu (99).

Stress concentration and fatigue. It has been customary practice to introduce the subject of *notch sensitivity* in discussions concerning stress concentration and fatigue (10, 25, 69, 70, 86); in this book notch sensitivity was discussed in Chapter 1 in connection with the problems of brittleness and ductility of materials. It is well known that the effect of a notch on the fatigue properties of a material varies with material and notch geometry and that the observed decrease

*See also Chapter 4 for a discussion of composite materials. A discussion of stress concentration in fabric structures is given in Reference (94)

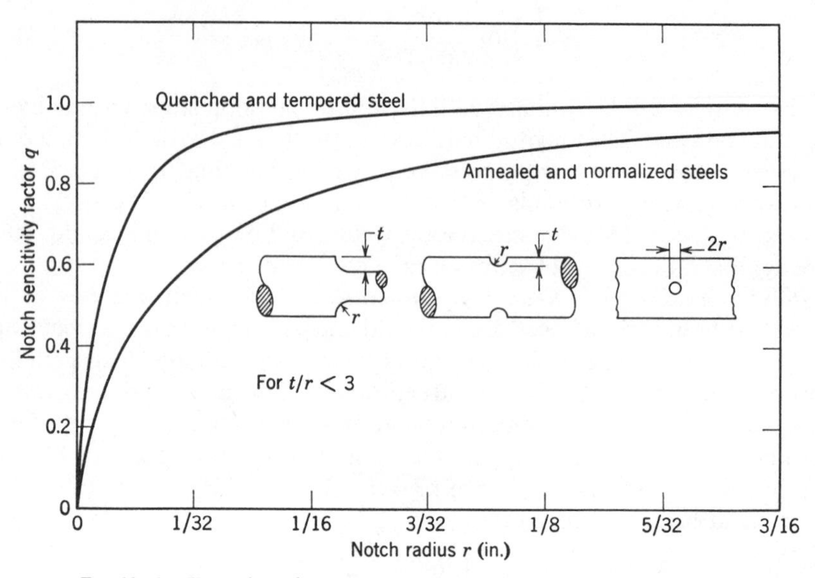

Fig. 13.49 Typical notch sensitivity data for steel. [*Data by Peterson (71).*]

in fatigue life is less than would be predicted from considering the theoretical static stress concentration factor; thus notch sensitivity has been defined as the degree to which the theoretical value is obtained, that is (see Fig. 13.49),

$$q = \frac{K_f - 1}{K_t - 1} \tag{13.209}$$

where q is the notch sensitivity factor, K_f is the fatigue notch factor (ratio of fatigue life of unnotched to notched specimens), and K_t is the theoretical factor. Eq. 13.209 may be written for all states of stress. The determination of q is difficult since many specimens containing the notch geometry of interest, as well as unnotched specimens, have to be tested; thus the theoretical factor is frequently used directly, which, from a design point of view, is satisfactory since the results of such an analysis will be on the safe side. Typical fatigue calculations were presented in Chapter 12 and, in particular, examples were given where stress concentration factors were used. As an additional example, suppose that an analysis is desired concerning the fatigue behavior of a structural member subjected to pulsating bending as shown in Fig. 12.28. Consider further that the bar has a shoulder fillet that gives rise to a stress concentration effect. Thus from Eq. 12.13 and Eq. 13.209 since the mean stress is zero,

$$\sigma_e = K\sigma_{\max} \tag{13.210}$$

or

$$\sigma_e = [q(K_t - 1) + 1]\sigma_{\max} \tag{13.211}$$

if q is known from experimental data. If a design factor of safety F is applied,

Eq. 13.211 becomes

$$\sigma_e = F\sigma_{\max}[q(K_t - 1) + 1] \tag{13.212}$$

Similar analyses would be made when the mean stress is not zero. Cases involving combined stress have already been given in Chapter 12.

Brittle materials. The maximum normal stress theory has long been used as a design criterion when brittle materials are involved. It is now clear, however, that "brittle" materials may be brittle by virtue of internal stress concentrations which make it possible for the material to be stressed locally to its ultimate strength long before this stress would be reached statically in the absence of stress concentrators. A discussion of this condition was given in Chapter 1. Additional support for this view is given by the data presented in Chapter 12 for the fatigue strength of cylinders containing stress concentrators and also by analysis of data for cast iron cylinders under internal pressure given by Fisher (23). Fisher observed that the cast iron contained graphite inclusions of elliptical form that gave rise to a stress concentration factor of about 3.2, which, when applied to the hoop stress in Eq. 13.161, resulted in agreement with predictions made by use of the distortion energy theory of failure, once thought applicable only to "ductile" materials. Fisher's analysis was later substantiated by Clough and Shank (4). Thus, fundamentally, all materials can probably be characterized by the distortion energy criterion, provided that the stress terms are properly adjusted for any stress concentration present. Other cases of stress concentration due to inclusions are described in (14, 15, 19, 70, 77). The relation of inclusions to fatigue properties is discussed in (13).

13-3 DESIGN DATA FOR STRESS CONCENTRATION PROBLEMS

Various procedures, charts, and curves relative to problems involving stress concentration abound in the literature. For additional information the references cited may be consulted, in particular, Peterson (71), who has provided various design approaches in handling stress concentration problems. Peterson also includes a bibliography of 174 references. Other notable references include Neuber (67), Cox (9), and Lipson, Noll, and Clock (60). In (60) 270 references are cited, and, in addition, nearly 100 charts containing notch sensitivity data are presented. Information on the effect of holes on the twisting of strips is given by Tamate (84); on the effect of holes in web plates by Edmunds (18); on tapered circular reinforcement of holes by Kaufman (49); on circular reinforcements of rectangle cross section by Kaufman (49); on reinforcement of openings in pressure vessels by Waters (97) and Mershon (63); on nozzle connections on pressure vessels by Hardenbergh (34) and Schoessow (78); on stress concentration in nonlinear theory by Neuber (65); and on flexure of circular plates with rings of holes by Kraus (52) and Radkowski (73). Rotating disks are considered in (1).

REFERENCES

1. K. E. Barnhart, A. L. Hale, and J. L. Meriam, *Proc. Soc. Exp. Stress Anal.*, **9** (1), 35–52 (1951).
2. F. T. Barwell, *Proc. Inst. Mech. Eng. (London)*, **175** (17), 853–879 (1961).
3. D. C. Berkey, *Proc. Soc. Exp. Stress Anal.*, **1** (2), 56–60 (1943).
4. W. R. Clough and M. E. Shank, *Trans. Am. Soc. Met.*, **49**, 241–262 (1957).
5. E. G. Coker and L. N. G. Filon, *A Treatise on Photoelasticity*, Cambridge University Press, London (1931).
6. S. D. Conte, *Q. Appl. Math.*, **9** (4), 435–440 (January 1952).
7. H. D. Conway, *J. Appl. Mech.*, **21** (1), 42–44 (March 1954).
8. G. A. Cooper, "Micromechanics Aspects of Fracture and Toughness," in L. J. Broutman, Ed., *Composite Materials*, Vol. 5, Academic Press, New York, N.Y. (1974).
9. H. L. Cox, *Four Studies in the Theory of Stress Concentration*, Aeronautical and Research Council Monograph No. 2704, Ministry of Supply, H. M. Stationery Office, London (1953).
10. H. L. Cox, *Proc. Inst. Mech. Eng.* (International Conference on Fatigue of Metals), Paper No. 16, 12 pp. (1956).
11. N. A. Crites, *Prod. Eng.*, **32** (42), 63–72 (November 27, 1961).
12. N. A. Crites and A. R. Hunter, *Prod. Eng.*, **33** (18), 57–69 (September 3, 1962).
13. H. N. Cummings, F. B. Stulen, and W. C. Schulte, *Trans. Am. Soc. Met.*, **49**, 482–516 (1957).
14. S. C. Das, *J. Appl. Mech.*, **21** (1), 83–87 (March 1954).
15. S. C. Das, *Zeit. Angew. Math. u. Phys.*, **5** (5), 389–399 (September 15, 1954).
16. T. J. Dolan, W. M. Murray, and D. C. Drucker. "Photoelasticity." Chapter 17, pp. 828–976, of *Handbook of Experimental Stress Analysis*, John Wiley and Sons, New York, N.Y. (1950).
17. A. J. Durelli, R. L. Lake, and E. Phillips, *Mach. Des.*, **23** (12), 165–167 (December 1951).
18. H. G. Edmunds, *Engineering*, **183** (4739), 18–19 (January 4, 1957).
19. R. A. Eubanks, *J. Appl. Mech.*, **21** (1), 57–62 (March 1954).
20. J. H. Faupel and D. B. Harris, *Ind. and Eng. Chem.*, **49** (12), 1979–1986 (December 1957).
21. H. Fessler and B. H. Lewin, *Brit. J. Appl. Phys.*, **7** (2), 76–79 (February 1956).
22. H. Fessler and R. T. Rose, *Brit. J. Appl. Phys.*, **4** (3), 76–79 (March 1953).
23. J. C. Fisher, *Bull. Am. Soc. Test. Mater.* (TP-76), 74–75 (April 1952).
24. L. Föppl, *Zeit. Deut. Ing.*, **76**, 505 (January-June 1932).
25. O. Föppl, *The Engineer*, **185** (4801), 114–115 (June 30, 1948).
26. M. M. Frocht, *J. Appl. Mech.*, **2** (1), A67–A68 (March 1935).
27. M. M. Frocht and D. Landsberg, *Proc. Soc. Exp. Stress Anal.*, **8** (2), 149–170 (1951).
28. M. M. Frocht, *Photoelasticity*, Vol. 1 (1941) and Vol. 2, John Wiley and Sons, New York, N.Y. (1948).
29. L. A. Galin, *Prik. Matern. i Mek.*, **10**, 365–386 (1946). (*Plastic Regions Around Circular Holes in Plates and Beams.* Translated by Brown University, Report All-T1-34, December 1947).
30. J. N. Goodier, *Trans. Am. Soc. Mech. Eng.*, **55**, APM-55-7, 39–44 (1933).
31. J. N. Goodier, *Phil. Mag.*, **22** (Series 7), 69–80 (1936).
32. W. Griffel, *Prod. Eng.*, **34** (19), 98–104 (September 16, 1963); **34** (23), 104–113 (November 11, 1963).
33. M. Hannah, *Q. J. Mech. and Appl. Math.*, **4** (3), 94–105 (March 1951).
34. D. E. Hardenbergh, *Exp. Mech.*, **1** (5), 152–158 (May 1961).
35. H. F. Hardrath and L. Ohman, *A Study of Elastic and Plastic Stress Concentration Factors*

Due to Notches and Fillets in Flat Plates, NACA Tech. Note No. 2566, Washington, D.C. (December 1951).

36. J. M. Hedgepeth, *Stress Concentrations in Filamentary Structures*, NASA TN-D882, Washington, D.C. (May 1961).

37. M. Hetenyi, *J. Appl. Mech.*, **6** (4), A151–A155 (December 1939).

38. M. Hetenyi, Ed., *Handbook of Experimental Stress Analysis*, John Wiley and Sons, New York, N.Y. (1950).

39. M. Hetenyi, "Brittle Models and Brittle Coatings," in *Handbook of Experimental Stress Analysis*, John Wiley and Sons, New York, N.Y. (1950).

40. M. Hetenyi and T. D. Liu, *J. Appl. Mech.* (Paper No. 55–A–81).

41. R. Hicks, *Engineering*, **176** (4587), 808–809 (December 1953).

42. P. G. Hodge and C. Lakshmikantham, *Yield Point Loads of Spherical Caps with Cutouts*, U.S. Dept. of Commerce Report AD 272–550 (Office of Technical Services) (December 1961).

43. R. C. J. Howland, *Phil. Trans. R. Soc. (London)*, **229** (Series A), 49–86 (January 6, 1930). (Also *Proc. R. Soc. (London)*, **148** (Series A), 471–491 (1935).

44. A. Hütter, *Z. Angew. Math. u. mech.*, **22** (6), 322–335 (December 1942).

45. C. E. Inglis, *Engineering*, **95**, 415 (March 28, 1913).

46. R. I. Isakower, *Prod. Eng.*, **30** (21), 65–67 (May 25, 1959).

47. R. I. Isakower, *Prod. Eng.*, **32** (25), 65–66 (June 19, 1961).

48. K. L. Johnson, *Proc. Inst. Mech. Engrs. (London)*, **173** (34), 795–810 (1959).

49. A. Kaufman, P. T. Bizon, and W. C. Morgan, *Investigation of Tapered Circular Reinforcements around Central Holes in Flat Sheets under Biaxial Loads in the Elastic Range*, NASA TN-D1101, Washington, D.C. (February 1962).

50. A. Kaufman, P. T. Bizon, and W. C. Morgan, *Investigation of Circular Reinforcements of Rectangular Cross Section around Central Holes in Flat Sheets under Biaxial Loads in the Elastic Range*, NASA TN-D1195, Washington, D.C. (February 1962).

51. W. T. Koiter, *Q. Appl. Math.*, **15** (3), 303–309 (October 1957).

52. H. Kraus, *J. Appl. Mech.*, **29** (3), Series E, 489–496 (September 1962).

53. H. Kraus, *Int. J. Mech. Sci.*, **4**, 187–194 (1962).

54. G. H. Lee, *An Introduction to Experimental Stress Analysis*, John Wiley and Sons, New York, N.Y. (1950).

55. L. H. N. Lee and C. S. Ades, *Proc. Soc. Exp. Stress Anal.*, **14** (1), 99–108 (1956).

56. M. M. Leven, *Proc. Soc. for Exp. Stress Anal.*, **7** (2), 141–154 (1949).

57. M. M. Leven and M. M. Frocht, *J. Appl. Mech.*, **19** (4), 560–561 (December 1952).

58. M. M. Leven and J. B. Hartman, *Proc. Soc. for Exp. Stress Anal.*, **9** (1), 53–62 (1951).

59. C. B. Ling, *J. Appl. Mech*, **24** (3) 365–375 (September 1957). (Also *J. Appl. Phys.*, **19** (1), 77–83 (1948).)

60. C. Lipson, G. S. Noll, and L. S. Clock, *Stress and Strength of Manufactured Parts*, McGraw-Hill Book Co., New York, N.Y. (1950).

61. E. H. Mansfield, *Q. J. Mech. and Appl. Math.*, **6** (Part 3), 370–378 (1953).

62. J. B. Mantle and T. J. Dolan, *Proc. Soc. Exp. Stress Anal.*, **6** (1), 66–73 (1948).

63. J. L. Mershon, *PVRC Research on Reinforcement of Openings in Pressure Vessels*, Report No. 77, Welding Research Council, New York, N.Y. (May 1962).

64. P. M. Naghdi, *J. Appl. Mech.*, **22** (1), 89–93 (March 1955).

65. H. Neuber, *J. Appl. Mech.*, **28** (4), 544–550 (December 1961).

66. H., Neuber, *Research on the Distributions of Tension in Notched Construction Parts*, WADD Tech. Report 60–906, Wright-Patterson Air Force Base, Oh. (January 1961).

67. H. Neuber, *Theory of Notch Stresses*, J. W. Edwards Co., Ann Arbor, Mich. (1946).

68. G. H. Neugebauer, *Prod. Eng.*, **14** (2), 82–87 (February 1943).

69. E. Orowan, *Weld. J. Res. Suppl.*, **27** (6), 273s–282s (June 1952). Discussion, **27** (10), 502s–504s (October 1952).

70. R. E. Peterson, *Proc. Inst. Mech., Eng.* (International Conference on Fatigue of Metals), Paper No. 4, 10 pp. (1956).

71. R. E. Peterson, *Stress Concentration Design Factors*, John Wiley and Sons, New York, N.Y. (1974).

72. R. E. Peterson and A. M. Wahl, *J. Appl. Mech.*, **3** (1), A15–A22 (March 1936).

73. P. P. Radkowski, *Proc. 2nd U.S. National Congress of Appl. Mech.*, 277–282, Am. Soc. of Mech. Eng., New York, N.Y. (1955).

74. R. J. Roark, R. S. Hartenberg, and R. Z. Williams, *Influence of Form and Scale on Strength*, Eng. Exp. Station Bulletin, University of Wisconsin (1938).

75. R. J. Roark, W. C. Young, *Formulas for Stress and Strain*, 5th ed. McGraw-Hill Book Co., New York, N.Y. (1975).

76. R. J. Roark, *Stresses and Deflections in Thin Shells and Curved Plates Due to Concentrated and Variously Distributed Loading*, NACA TN 806, Washington, D.C. (May 1941).

77. M. A. Sadowsky and E. Sternberg, *J. Appl. Mech.* **16**(2), 149–157 (June 1949).

78. G. J. Schoessow and L. F. Kooistra, *J. Appl. Mech.*, **12** (2), A107–A112 (June 1945).

79. S. Sjöström, *On the Stresses at the Edge of an Eccentrically Loaded Circular Hole in a Strip Under Tension*, Report 36, Aeronautical Research Institute of Sweden (Stockholm) (1950).

80. F. B. Seely and T. J. Dolan, *Stress Concentration at Fillets, Holes and Keyways as Found by the Plaster Model Method*, Eng. Exp. Station Bulletin 276, University of Illinois (1935).

81. M. Seika, *J. Appl. Mech.*, **25** (4), 571–574, December (1958).

82. R. Sonntag, *Z. Angew. Math. u. Mech.*, **9** (1), 1–22 (February 1929).

83. P. S. Symonds, *J. Appl. Mech.*, **13** (3), A183–A197 (September 1946).

84. O. Tamate, *J. Appl. Mech.*, **24** (1), 115–121 (March 1957).

85. S. Timoshenko, *J. Franklin Inst.*, **197** (4), 506–516 (April 1924).

86. S. Timoshenko, *Proc. Inst. Mech. Eng. (Applied Mechanics—War Emergency Issue)*, **157** (28), 163–169, London (1947).

87. S. Timoshenko, *Strength of Materials, Part II*, 3rd ed., D. Van Nostrand Co., Princeton, N.J., (1956).

88. S. Timoshenko and J. N. Goodier, *Theory of Elasticity*, 2nd ed., McGraw-Hill Book Co., New York, N.Y. (1970).

89. S. Timoshenko, *History of the Strength of Materials*, McGraw-Hill Book Co., New York, N.Y. (1953).

90. S. Timoshenko, *Proc. Soc. Exp. Stress Anal.*, **12** (1), 1–12 (1954).

91. S. Timoshenko and W. Dietz, *Trans. Am. Soc. Mech. Eng.*, **47** (Paper No. 1958), 199–237 (1925).

92. S. Timoshenko, *Phil. Mag.*, **44** (Series 6), 1014–1019 (1922).

93. S. Timoshenko and G. H. McCullough, *Elements of Strength of Materials*, 2nd ed., D. Van Nostrand Co., Princeton, N.J. (1940).

94. A. D. Topping, *Aerosp. Eng.*, **20**, 18–19; 53–58 (April 1961).

95. A. M. Wahl and R. Beeuwkes, *Trans. Am. Soc. of Mech. Eng.*, **56**, 617–636 (1934).

96. C. T. Wang, *Applied Elasticity*, McGraw-Hill Book Co., New York, N.Y. (1953).

97. O. Waters, *Weld. J. Res. Suppl.*, **23** (6), 277s–288s (June 1948).

98. H. J. Weiss and W. Prager, *J. Appl. Mech.*, **19** (3), 397–401 (September 1952).

99. M. H. L. Wu, *Linearized Solution and General Plastic Behavior of Thin Plates With Circular Hole in the Strain Hardening Range*, NACA Tech. Note No. 2301, Washington, D. C. (March 1951).

100. F. Zandman, *Prod. Eng.*, **30** (9), 43–46 (March 2, 1959).

101. *Section VIII, Unified Pressure Vessels*, ASME Boiler and Pressure Vessel Code, Am. Soc. Mech. Eng., New York (1977). See also *Section III, Nuclear Vessels* (1977).

102. *Welding Research Council Bulletin* No. 80, (August 1962), J. B. Mahoney, V. L. Salerno, and M. A. Goldberg, "Analysis of a Perforated Circular Plate Containing A Rectangular Array of Holes"; and D. Bynum and M. M. Lemcoe, "Stresses and Deflections in Laterally Loaded Perforated Plates."

PROBLEMS

13.1 A distillation column supports a repair platform, 4×6 ft, loaded 100 lb/ft^2 and is supported as shown at both ends (Fig. P13.1). The corrosion allowance is 0.15 in. with an allowable stress of 20 kpsi and a stress intensity value of 26.7 kpsi. The operating pressure is 50 psig, and the bending moment in this location is 10^4 in.-lb. Find the length of the 1-in. thick clips to support the platform.

13.2 A 0.060-in. thick metal container with lugs is mounted to support points (Fig. P13.2). A locating notch is placed in the side. An effective beam depth of $30t$ is used. Find the tensile strength, σ_{ut}, using an allowable stress of $\sigma_{ut/4}$.

13.3 In Problem 13.1 a thick steel cover is used at the top of the distillation column. Find the factor of safety for the internal pressure.

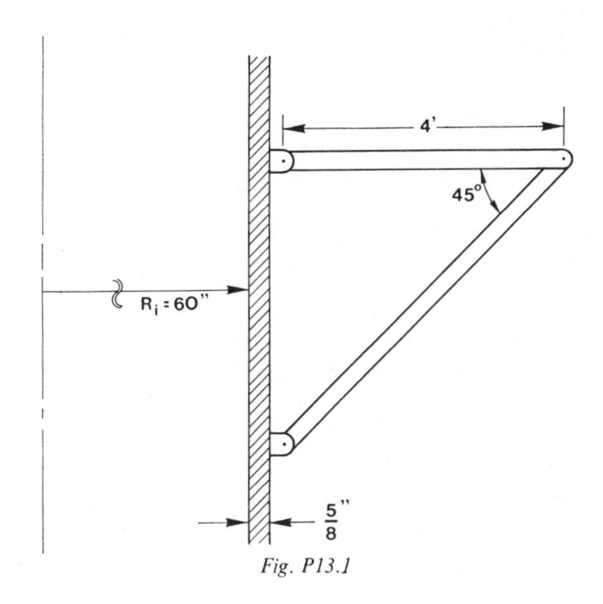

Fig. P13.1

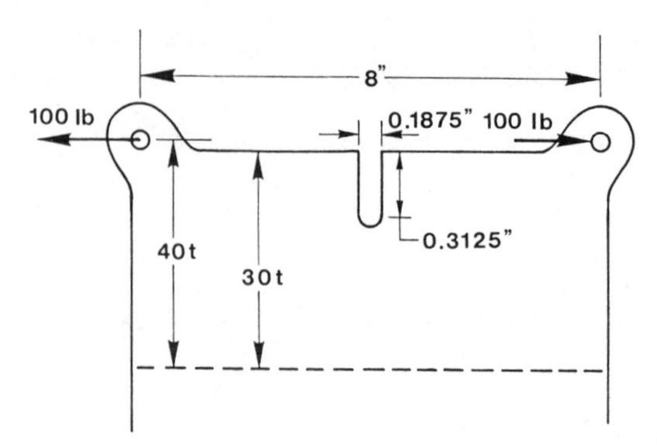

Fig. P13.2

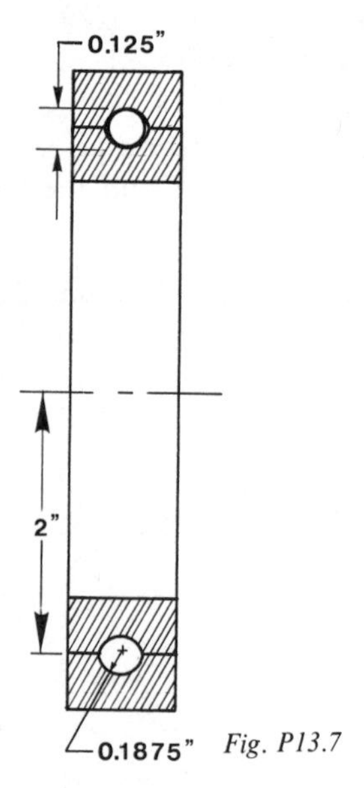

Fig. P13.7

13.4 In Problem 13.1 a 15-in. diameter line enters the distillation column. Size the line thickness and reinforcement as the line enters the column.

13.5 A trunnion connection (Fig. 13.10) carries 5 tons. Size the parts and select the material. Do not neglect fatigue and fretting.

13.6 A 3-in. diameter steel cyclinder 2 in. long rests on a flat cast iron plate. A 2 kip load is applied to the top of the cylinder. Find the failure stress using Eq. 3.524 and compare Eqs. 13.70–13.74 with 13.116–13.118.

13.7 A steel sphere in a bearing carries a 30-lb radial load (Fig. P13.7). Find the failure stress using Eq. 3.524.

13.8 A steel pressure vessel with 5000 psig internal pressure and 60-in. inside radius has a 2-in. inside diameter outlet. Use the same properties as in Problem 13.1 and find the outside dimensions of the pressure vessel and outlet.

chapter 14

BRITTLE FRACTURE AND DUCTILITY

In Chapter 1 some general principles and design criteria were presented relative to brittle fracture and ductility of materials. In this chapter a more in-depth treatment of the subject will be given.

14-1 BRITTLE FAILURE OF DUCTILE MATERIALS

Many cases are on record where failures* of parts presumed adequately designed made of ductile materials occurred resulting in great capital loss and sometimes in the death of personnel in the vicinity of the failure. In every instance the ductility thought to be inherently present in the material was mysteriously absent, resulting in brittle glasslike failures at loads far below those considered safe by conventional design criteria. What went wrong? An attempt will be made to unravel some of the mystery.

Common experience indicates that materials like glass and ceramics are hard and brittle, whereas materials like aluminum and structural steel are soft and ductile. Common experience also indicates that the relation between "hard and brittle" and between "soft and ductile" is not constant. Bismuth, for example, is soft yet brittle and piano wire is very hard, yet ductile in small sizes.

Further reflection also reveals a number of anomalies that conflict with common experience. For example, if one were to take a good grade steel file and support it as a cantilever beam and apply either a sudden or gradually increasing load at the free end, the file would break in a brittle fashion. On the other hand, the surface of the file provides a very tough rubbing surface when the file is hand operated. The material, however, is still not ductile as evidenced by the fact that teeth can be chipped from the surface from time to time. The material appears to be both tough and brittle at the same time. There are many other anomalous behaviors. Bismuth metal, ordinarily considered brittle, is easily extruded; porcelain and glass rods can be twisted and even marble can be made to flow plastically at room temperature provided it is subjected to high lateral pressure. Some polymeric materials are very ductile

*In steel cleavage cracks propagate at speeds of about 5000 ft/sec.

Fig. 14.1 Photograph of a T-2 tanker that failed at pier. [*After Parker (23), courtesy of John Wiley and Sons.*]

under high-level combined stresses but exhibit practically no ductility under low-level combined stresses [see (2, 3) of Chapter 5].

Most of the serious cases involving these strange anomalous behaviors have been associated with large, costly structures. This has been such a serious problem that large-scale investigations have been sponsored by the Government and private agencies for the purposes of analyzing the failures and developing remedial measures. The results of many of the investigations may be found in (23, 24, 26).

There are numerous examples of spectacular failures among which are (*a*) the sudden bursting of a multimillion-gallon storage tank in Boston in 1919, (*b*) sudden rupture of a hydrogen storage tank in Schenectady in 1943, (*c*) liberty ships that fell apart on the high seas or even when lying quietly at a pier (Fig. 14.1), and (*d*) an airplane that fell apart in flight as a result of a sudden failure of a main spar of an internally braced cantilever wing (21). Details on all of these, and other failures, may be found in the literature cited.

Some of the reasons for the failures are now known. The exact causes cannot always be determined, but it is definitely known that all such failures are initiated by some critical combination of two or more of the following: structure and composition (purity), size and shape, temperature, rate of loading, and state of stress, including effect of flaws.

Structure, composition, and purity of material are known to be important factors because pure single crystals of many materials show high ductility, whereas the commercial polycrystalline forms would be classified as "brittle."

Size and shape are important because of strain-energy considerations as already mentioned and because it is known that serious metallurgical problems arise when the properties have to be generated by heat treatment; if section sizes are too large, the heat cannot be removed efficiently on quenching, and variable structures and properties result.

Finally, the tendency for a material to be brittle increases as temperature decreases, as rate of loading increases, and as the state of stress tends toward triaxial tension.

Careful consideration to these few factors can reduce failures considerably. In addition, when considering the relative toughness of a material, its elastic range strength should be carefully noted. It was shown previously that toughness could be measured by the area under a stress-strain curve to the ultimate stress. On this basis, depending on the properties of a material as shown in Fig. 14.2, a "brittle" material could appear to be tougher than a "ductile" material.

Further insight into this problem is gained by considering the Griffith theory. It was indicated that Eq. 1.43 adequately predicted the behavior of brittle materials like glass. For normally ductile materials, however, the ϕ term has to include the energy of plastic deformation; the term 4ϕ in the equation then becomes a quantity commonly referred to as fracture toughness so that the critical fracture stress is

$$\sigma = \sqrt{\frac{E(4\phi)}{\pi L}} = \sqrt{\frac{EK}{\pi L}} \qquad 14.1$$

where K, the fracture toughness, is determined pseudoexperimentally. Basically, fracture toughness is defined as the component of work irreversibly absorbed in local plastic flow and cleavage surface tension to create a unit area of fracture; consequently, the fracture criterion is that K, the fracture toughness, be equal to K', the strain-energy release rate, that is, the quantity of stored elastic strain

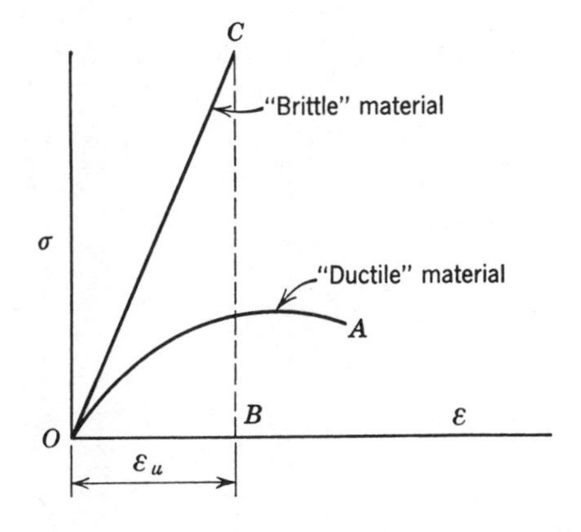

Fig. 14.2 Stress-strain curves for typical "brittle" and "ductile" materials.

Table 14.1 Elastic Energy Release Rate for Brittle Fracture—(Release Rate = Fracture Toughness)^2 / E – See Also (28, 30, 34)

Geometry	Type	Release Rate, K'
Infinite plate (Crack, L, σ)	Plane stress (thin sheet, through crack)	$\frac{\pi\sigma^2 L}{2E}$
	Plane strain (thick plate, internal crack)	$\frac{\pi\sigma^2 L(1-\nu^2)}{E}$
Infinite plate. ($2L$, σ)	Plane stress	$\frac{\pi\sigma^2 L}{2E}$
	Plane strain	$\frac{\pi\sigma^2 L(1-\nu^2)}{E}$
Semi-infinite plate (L, σ)	Plane stress	$\frac{\pi\sigma^2 L}{2E}$
	Plane strain	$\frac{\pi\sigma^2 L(1-\nu^2)}{E}$

Table 14.1 Contd.

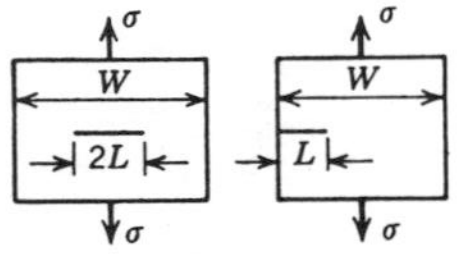 Finite plate (W, $2L$; W, L; σ)	Plane stress	$$\frac{\sigma^2 W}{E} \tan\left(\frac{\pi L}{W}\right)$$
	Plane strain	$$\frac{\sigma^2 W(1 - \nu^2)}{E} \tan\left(\frac{\pi L}{W}\right)$$
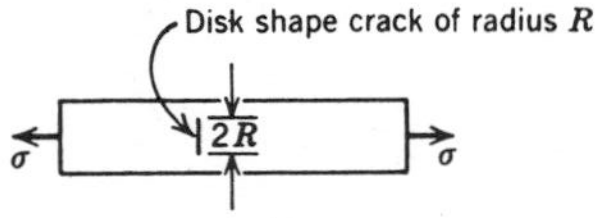 Disk shape crack of radius R ($2R$; σ) Cylindrical bar	Tension bar	$$\frac{4R\sigma^2(1 - \nu^2)}{\pi E}$$
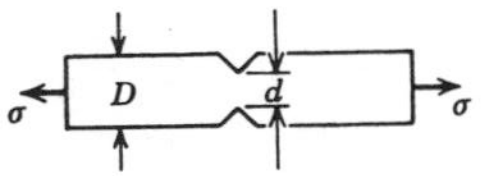 Notched cylinder (D, d; σ)	Tension bar	$$\frac{\sigma^2 d(1 - \nu^2)}{E}\left[8\pi\left(1 - \frac{d^2}{D^2}\right)\right]\left[\frac{1}{5 + 3\left(1 - \left(\frac{d}{D}\right)^2\right)}\right]^2$$

energy released from a cracking material as a result of extension by a unit area of the advancing crack. It is assumed that a unit area of cleavage fracture of a given material at a given temperature absorbs the same amount of energy; thus the fracture criterion is (10, 11, 28)

$$K = K' \tag{14.2}$$

and K' can be determined experimentally. K and K' are for either the elastic (plane strain) or elastic-plastic (plane stress) conditions. Various formulae for K' have been published, a few of which are given in Table 14.1. In the experimental determination of K' a specimen of suitable geometry is loaded until rapid cracking propagation occurs. In the formulas, L or $L/2$ is the crack length or "depth" of a scratch. Having experimentally determined K', the value of K is determined, which is inserted in Eq. 14.1 to give the size of flaw L at a stress level σ to produce brittle failure. Such analyses have been applied to pressure vessels, missile cases, and power-generating equipment to determine the probable effect of known flaws and cracks in the material. Some values for K are given in Table 14.2. The following example illustrates the use of the theory.

EXAMPLE A sheet of SAE 4340 steel has a fracture toughness value of 100 in.-lb/in.^{2} What depth of surface scratch would cause the material to fail at its yield strength of 150,000 psi? From Table 14.2 for plane strain

$$K' = \frac{\pi\sigma^2(1 - v^2)L}{E} = 100$$

Table 14.2. *Fracture Toughness of Some Metals*

Material	Tensile Yield Strength (σ, kpsi)	Fracture Toughness K (in.-lb/in.^{2})	K_I (kpsi$\sqrt{\text{in.}}$)	K_{IC} (kpsi$\sqrt{\text{in.}}$)
Aluminum Alloys				
2024-T3	50	300	58	20–26
2024-T4	50	300	58	31
7075-T6	72	115	36	25–27
7079-T651	67			27–30
Titanium Alloys				
B-210 VAC	149	274	69	34–60
Ti-6Al-6V-2Sn	145			57–62
	160			30–32
Steels				
SAE 4340	230	125	65	40–63
Cr-Mo-Va	101	362	110	
Ni-Mo-Va	89	245	91	
Type 304 stainless	30	450	123	100
9 Ni-4Co-0.20C	180			130–174

from which

$$L = \frac{(100)(30 \times 10^6)}{(150 \times 10^3)^2(\pi)(0.91)} = 0.047 \text{ in.}$$

14-2 BASIS OF FRACTURE

The term fracture toughness is developed from the energy in a crack (18, 14). In Fig. 14.3 there are three displacement modes for a crack. The total energy release rate (18)

$$G = G_{\text{I}} + G_{\text{II}} + G_{\text{III}} \qquad 14.3$$

Mode I for plane stress

$$G_{\text{I}} = \frac{K_{\text{I}}^2}{E} \qquad 14.4$$

and plain strain

$$G_{\text{I}} = (1 - v^2)\frac{K_{\text{I}}^2}{E} \qquad 14.5$$

Mode II for plane stress

$$G_{\text{II}} = \frac{K_{\text{II}}^2}{E} \qquad 14.6$$

and for plane strain

$$G_{\text{II}} = (1 - v^2)\frac{K_{\text{II}}^2}{E} \qquad 14.7$$

And mode III for plane strain and plane stress

$$G_{\text{III}} = (1 + v)\frac{K_{\text{III}}^2}{E} \qquad 14.8$$

There is much work devoted to the energy stored in cracks, the shape of cracks and the resulting stresses, testing of materials with cracks, and the study of failures (fractography). The seven volume treatise on fracture (20) edited by Liebowitz and the five volumes (27) edited by Sih represent examples of the extent of the work in fracture mechanics. Many books such as Broek (3), Hertzberg (9), Knott (14), Rolfe and Barsom (25), Tetelman and McEvily (31) cover not only the background of fracture but also interpretations of the physical problems.

The fracture mechanics studies have concentrated on plane strain ($\varepsilon_Z = 0$ in Fig. 14.3) where the energy release is G (Eq. 14.3), and a material failure parameter K_{IC} is developed from testing of material samples. When testing samples an important plastic zone ahead of the material sample crack can be observed. A slight depression will develop as thinner samples are tested. In

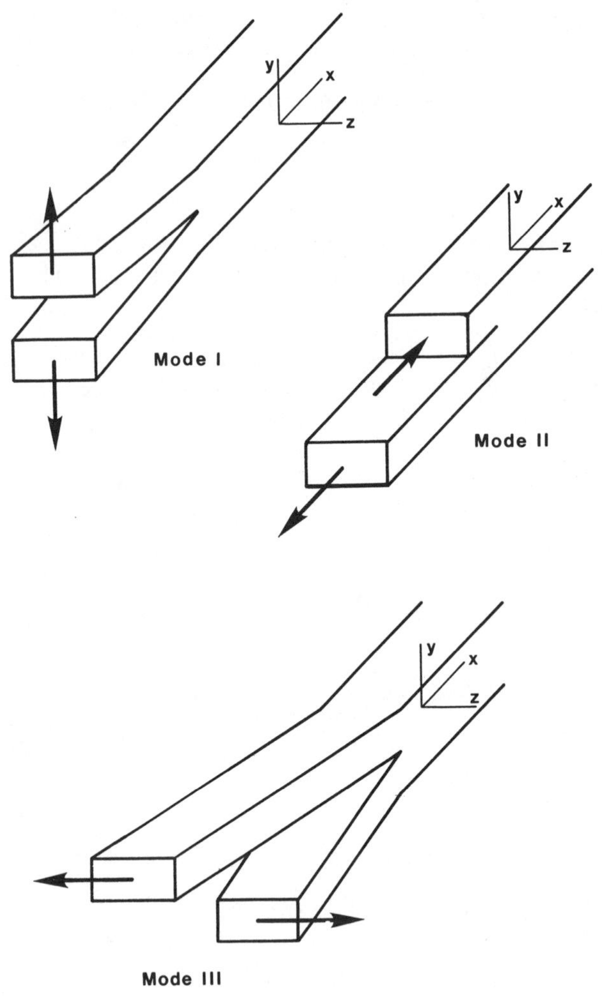

Fig. 14.3 Displacement modes for crack surfaces.

plane strain, testing with a specified thickness minimizes this plastic zone and yields a *linear elastic fracture mechanics* K_{IC}. When the sample thickness is decreased, the plastic zone increases and the energy release J for an elastic-plastic condition must be formulated. However, when K_{IC} data are obtained

$$J = G \qquad 14.9$$

This is true because the energy in the plastic zone is very small compared to the elastic energy, G. When thin samples are tested in the plane-stress condition ($\sigma_Z = 0$ in Fig. 14.3) R curves (38) or K_R values are developed. The thin samples are subjected to plane stress when the plastic zone radius r_y and the sample thickness t are related by

$$r_y \approx \frac{1}{2\pi} \frac{K_c^2}{\sigma_{ys}^2} \approx t \tag{14.10}$$

The plane-strain condition

$$r_y \approx \frac{1}{6\pi} \frac{K_c^2}{\sigma_{ys}^2} \approx 0.1t \tag{14.11}$$

The other samples with thicknesses between these limits are called the mixed mode. The material discussed in this chapter will be confined to the linear elastic fracture mechanic (LEFM) or plane-strain condition where most of the test data have been developed. The plane-stress versus plane-strain cases are discussed in more detail by Hertzberg (9) and by Rolfe and Barsom (25).

The values K_{I}, K_{II}, and K_{III} are often called stress-intensity factors or material toughness. Examining Eqs. 14.4–14.8 and the K equations in Table 14.1 the stress σ, crack length L, and K can be obtained from the test values of K_{I}, K_{II}, K_{III}.

The mode I (Fig. 14.3) crack parameter K_{IC} has received much study and use. The plane-strain test (Fig. 14.4) (ASTM designation: E-399-74) (4) has the following limitations on the test specimen:

1. Thickness B is greater than

$$2.5\left(\frac{K_{\mathrm{IC}}}{\sigma_{ys}}\right)^2$$

2. The crack length a is greater than

$$2.5\left(\frac{K_{\mathrm{IC}}}{\sigma_{ys}}\right)^2$$

3. The length W is greater than

$$5.0\left(\frac{K_{\mathrm{IC}}}{\sigma_{ys}}\right)^2$$

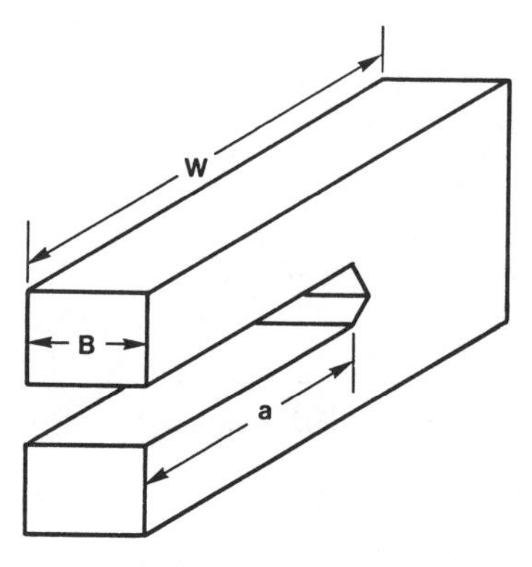

Fig. 14.4 Plane-strain test sample.

A value K_{IC} is derived from the test, and some typical values are shown in Table 14.2 (19). The K_{IC} is valid for a thickness

$$B \geq 2.5\left(\frac{K_{IC}}{\sigma_{ys}}\right)^2 \tag{14.12}$$

For a 304 stainless steel in Table 14.2

$$B \geq 2.5\left(\frac{100 \times 10^3}{30 \times 10^3}\right)^2$$

$$B \geq 27.78 \text{ in.}$$

While for SAE 4340 steel

$$B \geq 2.5\left(\frac{52 \times 10^3}{230 \times 10^3}\right)^2$$

$$B \geq 0.127 \text{ in.}$$

Therefore, for plane strain the thickness varies for different materials. When thinner samples are tested, the K_C values (Fig. 14.5) will tend to increase through the mixed mode condition until a maximum is obtained for plane stress. The condition for plane stress is from Eq. 14.10, and the increase in K_C (25) is 2–10 times K_{IC}. Broek (3) reports the thickness B_o where the maximum K_C is obtained as

$$B_o \approx \frac{K_{IC}}{3\pi \sigma_{ys}{}^2} \tag{14.13}$$

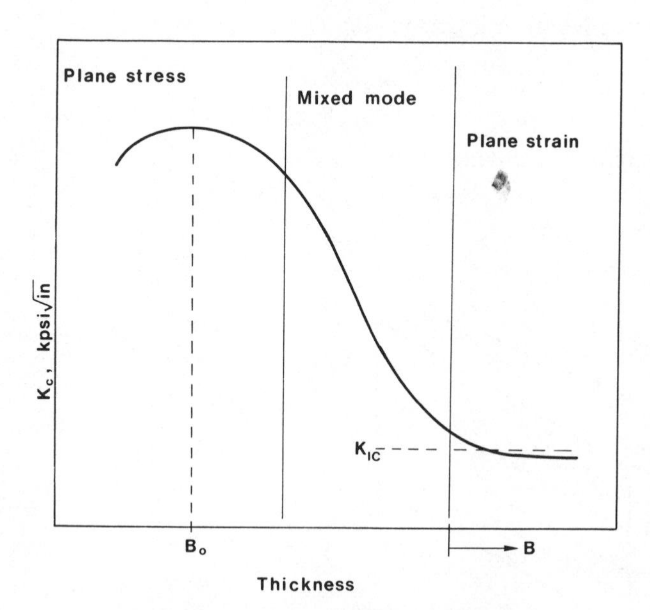

Fig. 14.5 Fracture toughness versus thickness.

Table 14.3. *Fig*. 14.5 *Parameters*

	B_o(in.)	B(in.)	K_C/K_{IC}Max
Al 7075-T7351	0.2	> 1.0	2.66
7075-T6	0.10	> 1.0	2.11
Ti-4Al-3Mo-1V	0.07–0.1	> 0.14	2.07
Maraging steel 300 grade	0.12	> 0.20	2.90
Ti-6Al-14V annealed	0.05	> 1.0	2.28
Ti-6Al-16V2-Sn annealed	0.20	> 1.0	1.76
18 Ni-Co-Mo	0.10	> 1.0	2.75

Typical values (19) are tabulated from data and placed in Table 14.3. Further work (38) in plane stress obtaining R curves will yield more data to determine behavior of the K_c curve (Fig. 14.5) below B_o.

The values in Tables 14.2 and 14.3 are for flat plate specimens tested at room temperature. As the temperature lowers so do the K_{IC} values. Relationships (7; 27 Vol. 3) for shells and cylinders have been developed as curvature has an effect in calculations.

The high cost of obtaining K_{IC} data (16), has developed interest in relating K_{IC} values to a simpler test, such as the Charpy impact test or others. Attempts to develop K_{IC} values for temperature changes from Charpy V notch tests are suggested by Begley and Logsdon as discussed by Rolfe and Barsom (25). Hertzberg (9) reviews similar methods. For steels at 100% brittle fracture near the nil ductility temperature in Fig. 14.6, using the σ_{ys} at the tested temperature

$$K_{IC} = 0.45\,\sigma_{ys}\sqrt{\text{in.}} \tag{14.14}$$

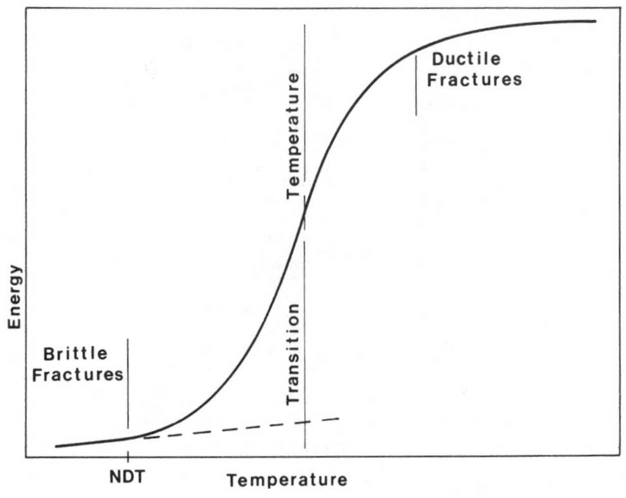

Fig. 14.6 Typical Charpy V notch test results for steel.

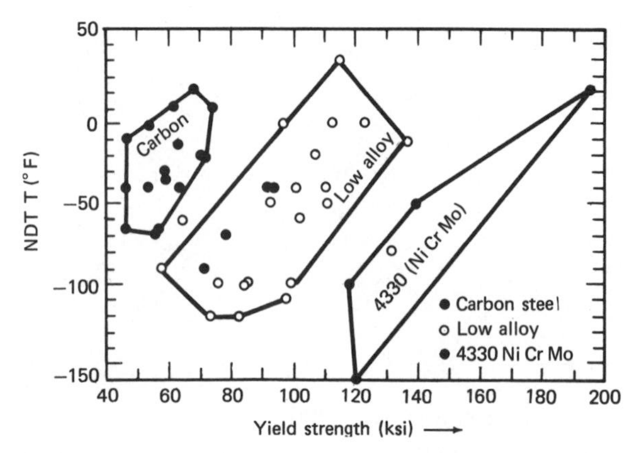

Fig. 14.7 The NDT temperature and yield strength of quenched and tempered commercial cast steels (1.5-in. sections). [(40), *courtesy of Machine Design.*]

and for 100% ductile fracture

$$\left(\frac{K_{IC}}{\sigma_{ys}}\right)^2 = 5\left(\frac{CVN}{\sigma_{ys}}\right) - 0.05 \qquad 14.15$$

In the absence of data, the room temperature σ_{ys} would yield rough approximations for designing. In Fig. 14.6 the nil ductility temperature (NDT) is approximately shown. Typical NDT values (40) in Fig. 14.7 and 14.8 also give indications of the transition temperatures.

Critical crack sizes, L_c or a_c, can also be found from the use of K_{IC} in Tables 14.1 and 14.2. The lengths are the size flaws in a material where a crack will propagate without stopping. The detection of these cracks is important. Nondestructive testing to detect cracks in 7075 aluminum and 4000 series

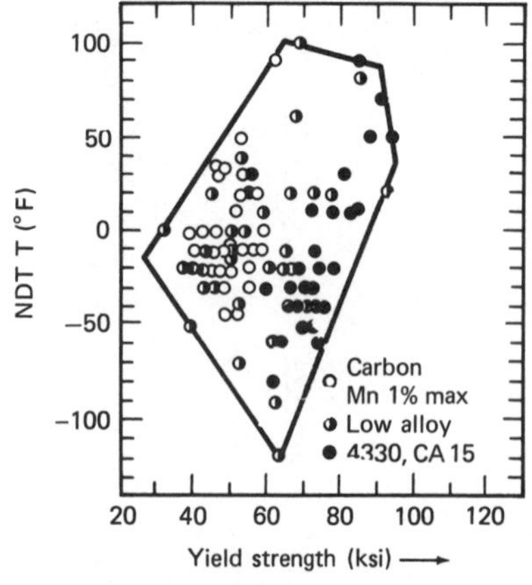

Fig. 14.8 The NDT temperature and yield strength of normalized and tempered commercial cast steel (1.5-in. section). [(40), *courtesy of Machine Design.*]

steels is discussed in (2, 19). Some cracks of 0.01–0.05 in. are detected, and only in crack sizes above 0.20 in. is the detection accuracy 70% and above. The methods tested included ultrasonics, dye penetrant, X-rays, and eddy current. The methods indicate that the critical crack size must be maintained at lengths where detection or visual inspection can forecast possible failure.

Some codes and standards (33) call for hydrostatic testing under certain applicable conditions to cause a crack to propagate and cause failure during testing to avoid growth after installation and failure during operation. Reference (33) applies the criteria to heavy section steel structures and shows the effects of cracks during a hydrostatic test.

The determination of which crack configuration can cause a failure in a design is important. In Table 14.4 some of the common crack shapes and the simplest to calculate are shown. These common crack shapes can be selected by examining the different design failures (5; 39, Vols. 9, 10). The proper calculation for the crack shapes are compiled in (20, Vol. II; 27, Vol. I; 30). Fracture mechanics data can be found in (19, 35–37). Several techniques (15) are available to determine stresses in cracks.

The crack shape in Table 14.4 (A) (20, Vol. I) can also be used for drilled holes in plates. Paris and Sih (41) present data where

$$K_{\mathrm{I}} = \sigma\sqrt{L\pi}\, F\left(\frac{L}{r}\right) \tag{14.16}$$

Only data for two values of L/r will be presented here:

	$F(L/r)$ (One Crack)		$F(L/r)$ (Two Cracks)	
L/r	Uniaxial Stress	Biaxial Stress	Uniaxial Stress	Biaxial Stress
0.10	2.73	1.98	2.73	1.98
10	0.75	0.75	1.03	1.03

The lengths of cracks are difficult to estimate. However, in Chapter 12 surface cracks during machining are discussed. An estimate of crack growth can be made with a linear representation of Eq. 14.16.

Elliptical crack shapes can be used to gain information for the crack shapes (1; 5; 39, Vol. 9; 10) in Table 14.4 (B) and (D). A summary (3) of work for an elliptic crack on the surface yields the maximum intensity

$$K_{\mathrm{I}\max} = 1.1\,\sigma\sqrt{\frac{\pi a}{Q}} \tag{14.17}$$

where:

$$Q = \sqrt{\Phi^2 - \frac{0.212\,\sigma^2}{\sigma_{ys}^{\;2}}}$$

and

$$\Phi = \frac{3\pi}{8} + \frac{\pi}{8}\frac{a^2}{c^2}$$

Table 14.4 Common Crack Shapes

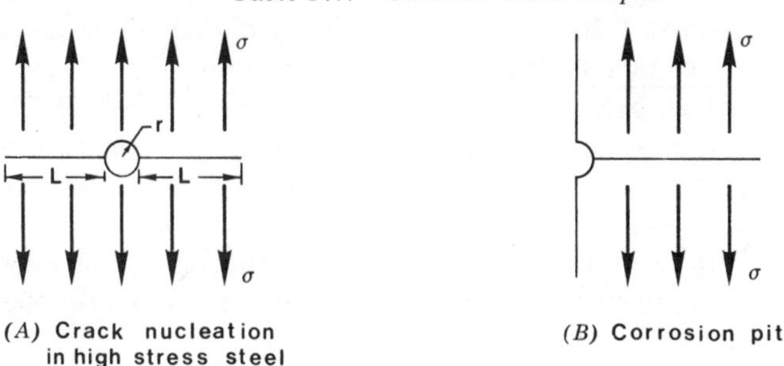

(A) Crack nucleation in high stress steel

(B) Corrosion pit

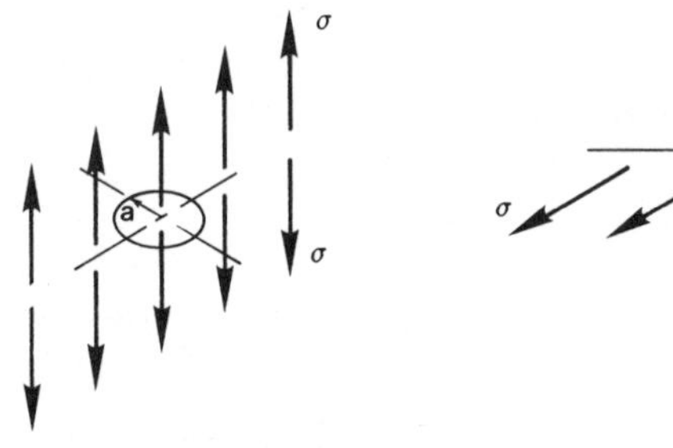

(C) Disc flow under tension

(D) Partial disc

When various values of σ/σ_{ys} are used and plotted versus $a/2c$, Fig. 14.9 results. An example is presented using these data. When the crack shape is buried,

$$K_{\mathrm{Imax}} = \sigma\sqrt{\frac{\pi a}{Q}} \qquad 14.18$$

The material presented in this chapter is derived for flat plates. It can be used for cases of cylinders and spheres made from flat plates. The answers are close approximations as curvature affects the K values. However, where approximations as to the crack shape are made for cases of cylinders or spheres, a flat plate analysis will be accurate enough for estimating values. Precise values will require including the effect of curvature.

When modes (Fig. 14.3) are combined Broek (3) discusses mode I and II. The fracture condition is

$$G = G_{\mathrm{I}} + G_{\mathrm{II}}$$

$$G = \frac{K_{\mathrm{I}}^2}{E} + \frac{K_{\mathrm{II}}^2}{E}$$

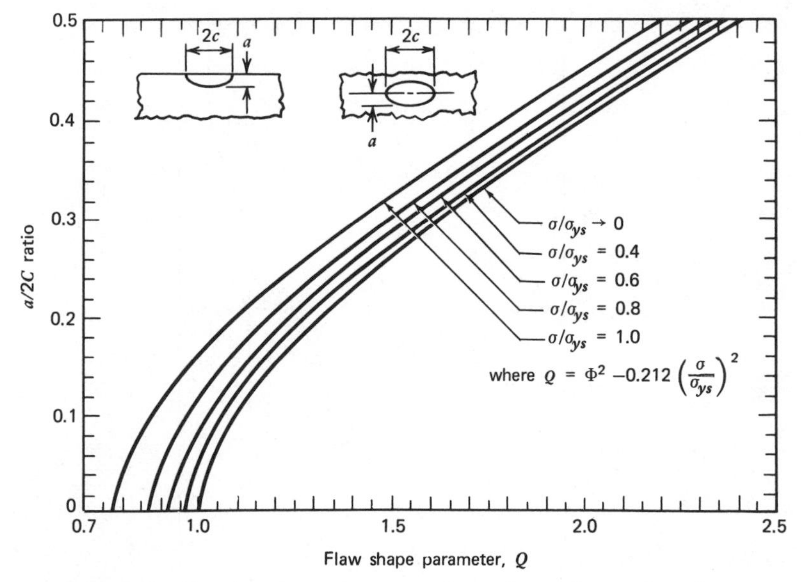

Fig. 14.9 Flaw-shape parameter curves for surface and internal cracks.

A relation has been developed

$$\left(\frac{K_{\rm I}}{K_{\rm IC}}\right)^2 + \left(\frac{K_{\rm II}}{K_{\rm IIC}}\right)^2 = 1 \qquad 14.19$$

with $K_{\rm IIC} \approx 0.75\ K_{\rm IC}$. A future examination of plane-strain energy suggests a fracture criteria for modes I, II, and III of

$$\left(\frac{K_{\rm I}}{K_{\rm IC}}\right)^2 + \left(\frac{K_{\rm II}}{K_{\rm IIC}}\right)^2 + \frac{1}{1-v}\left(\frac{K_{\rm III}}{K_{\rm IIIC}}\right)^2 = 1 \qquad 14.20$$

Reference (30) also performs calculations for combined modes.

Crack propagation is an area of interest and is represented by

$$\frac{da}{dn} = C(\Delta K)^n \qquad 14.21$$

where: C,n = material constants found from testing

$\Delta K = K_{\rm max} - K_{\rm min}$

da/dn = the crack growth, inches per cycle

Another form of Eq. 14.21 is

$$\frac{da}{dn} = C_1\left(\frac{\Delta K}{E}\right)^m \qquad 14.22$$

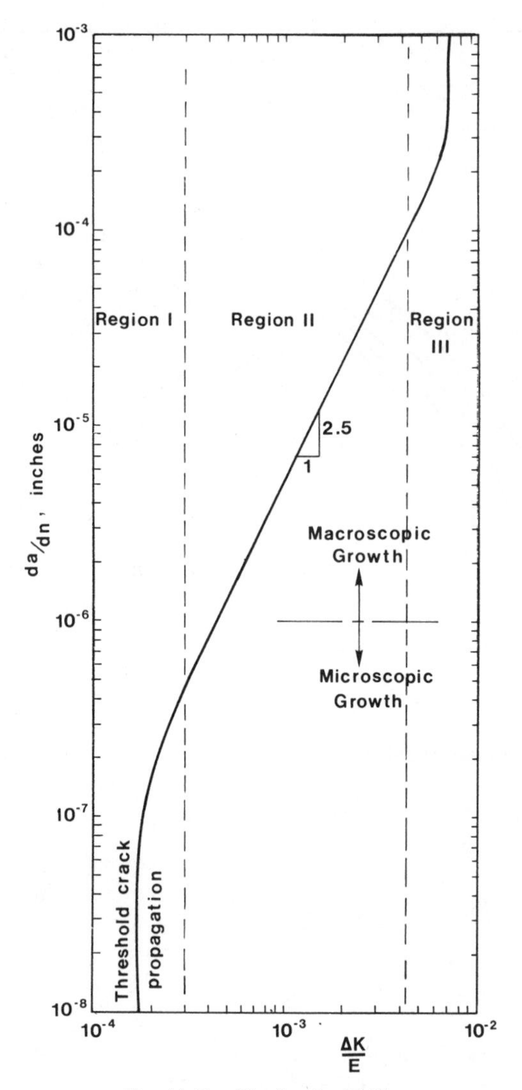

Fig. 14.10 Plot for Eq. 14.22.

where: E = the modulus of the material

$\Delta K = K_{max} - K_{min}$

C_1 = a material constant ($\approx 1.63 \times 10^4$)

$m = 3.2\text{–}3.6$

Equation 14.22 is plotted in Fig. 14.10, and for many materials (6, 9) made of steel, aluminum, and titanium the *da/dn* curve reduces to a single band of data. Broek (3), however, points out that some alloys of the same materials

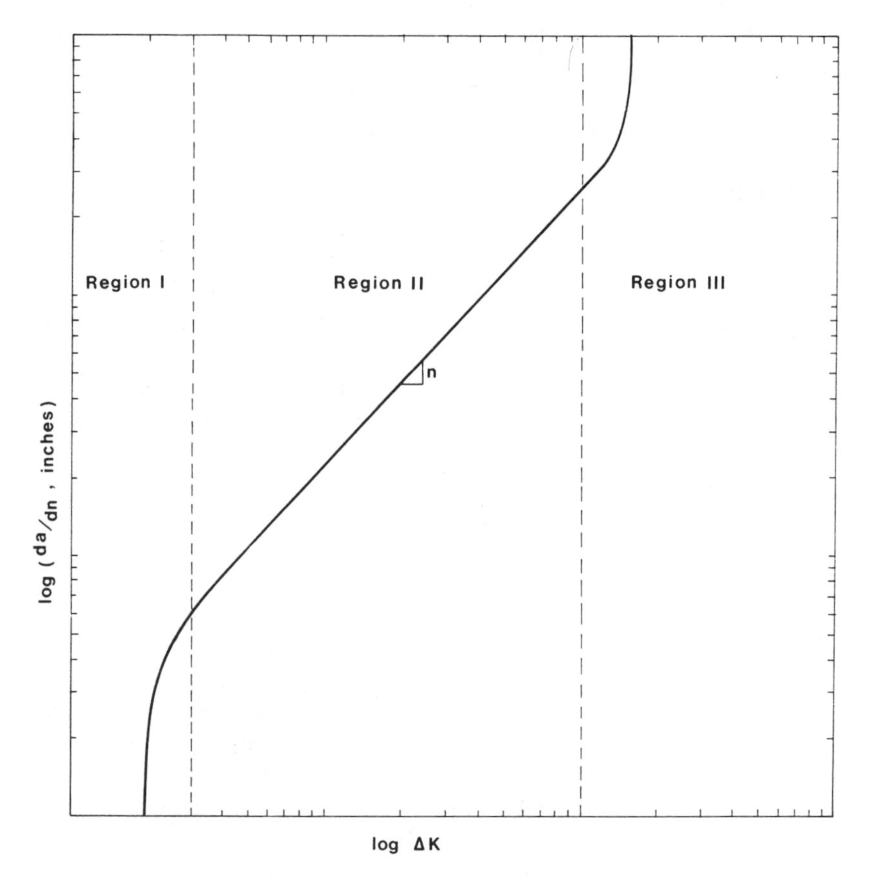

Fig. 14.11 Plot for Eq. 14.21.

in Fig. 14.10 do not conform to Eq. 14.22. Then data must be plotted (Fig. 14.11) according to Eq. 14.21 for each individual material. In Fig. 14.11 the power term n holds in Region II while Region III represents pending failure. The steels reduce to basic equations (25, 29). Martensitic steels:

$$\frac{da}{dn} = 0.66 \times 10^{-8}(\Delta K)^{2.25} \qquad 14.23$$

Ferrite-Pearlite steels:

$$\frac{da}{dn} = 3.6 \times 10^{-10}(\Delta K)^{3.0} \qquad 14.24$$

Austenitic stainless steels (304, 316):

$$\frac{da}{dn} = 3.0 \times 10^{-10}(\Delta K)^{3.25} \qquad 14.25$$

Typical high-strength aluminum (6, 36):

$$\frac{da}{dn} = 5.4 \times 10^{-19}(\Delta K)^{3.31} \tag{14.26}$$

The n values range from 2.7–4.8 for low- to high- strength alloys. A typical high strength titanium:

$$\frac{da}{dn} = 1.96 \times 10^{-25}(\Delta K)^{4.45} \tag{14.27}$$

The n values for titanium range from 3.5–4.4 depending on the alloy. Data for other alloys can be found in (35, 37).

In Fig. 14.10 it has been determined by Harrison and discussed further by Rolfe and Barsom (25) that in Region I the threshold for crack growth is

$$1.5 \times 10^{-4}\sqrt{\text{in.}} \leq \frac{K_{th}}{E} \leq 1.8 \times 10^{-4}\sqrt{\text{in.}} \tag{14.28}$$

This relationship appears to hold (8) regardless of material.

14-3 DESIGNING FOR FRACTURE

Some success using various notched tensile specimens has been obtained (32), but such data are difficult to interpret. Qualitatively, however, it has been found that various notch-toughness tests on steel, for example, produce results typified in Fig. 14.12. Such curves indicate a transition temperature, roughly at about the temperature a material's behavior changes from ductile to brittle, that is, where it becomes particularly notch sensitive. Most metals with body-centered cubic structures (like steel) and also some metals with hexagonal lattice structures show a sharp transition temperature and are brittle at low

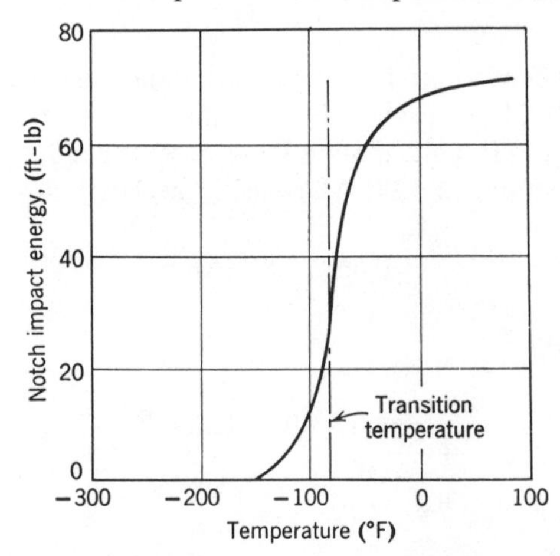

Fig. 14.12 Brittle transition temperature for 0.5% Mo steel. [After Jastryzebski (12), courtesy of John Wiley and Sons.]

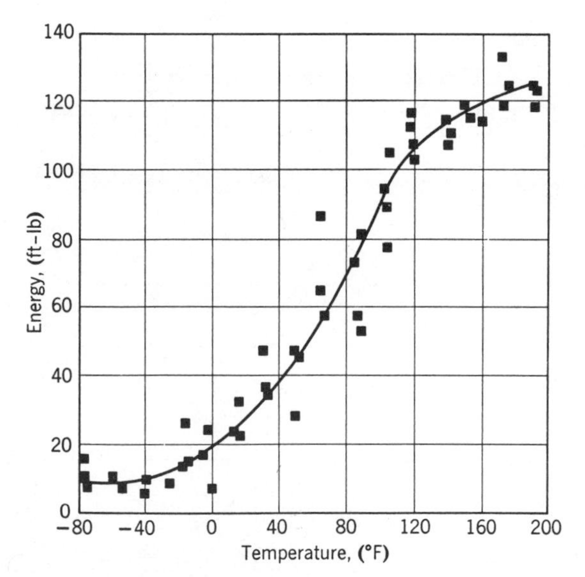

Fig. 14.13 Standard V notch Charpy impact data for low-carbon, low-alloy, hot-rolled steel. [After Parker (23), courtesy of John Wiley and Sons.]

temperatures. Other materials like aluminum and copper, which have a face-centered cubic lattice structures retain ductility to quite low temperatures.

Referring to Fig. 14.1, it is possible to rationalize the failure on the basis of transition temperature data shown in Fig. 14.13 which are representative of the kind of material used in the fabrication of the body of the ship. Note that the material has a brittle transition temperature around 40–60°F, a temperature not uncommon for sea water; this isolated fact does not explain the entire process of failure—it merely indicates that the material could be brittle at a particular temperature.

Brittleness is also observed in static loading of bars of identical material, Figs. 14.14A and B, but with B containing a notch. The toughness of the material as measured by the two specimens is different as mentioned earlier; this effect is called notch sensitivity. Many failures of supposedly ductile materials have failed in a brittle fashion, which has caused great concern, for it would be desirable to have a structure perform with the ductility indicated to be present by an ordinary tension test. As design guides, methods have been proposed by McClintock (21) and by Pellini (24) which are briefly described below.

In the McClintock analysis, consider a member such as shown in Fig. 14.14C having a continuous notch. For this condition of brittleness, geometry and properties combine such that

$$F_L\left(\frac{\sigma_u}{\sigma_y}\right) > \frac{\sqrt{3}}{2}\left(\frac{W}{w}\right) \qquad 14.29$$

where F_L, the load factor, is the ratio of the actual load sustained to the load

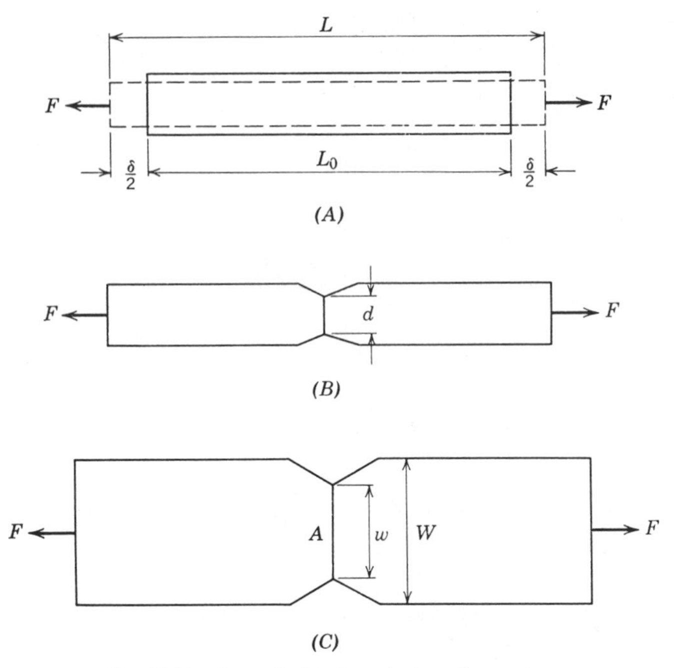

Fig. 14.14 Unnotched and notched tensile members.

calculated by multiplying σ_u by the cross-sectional area A at the notch; σ_u is the ultimate tensile strength; and σ_y is the yield strength. F_L nearly always varies between zero and unity. Now, with a geometry specified an estimate of what properties are needed can be made. For example, suppose W/w equals 2.0; then to avoid brittle failure,

$$F_L\left(\frac{\sigma_u}{\sigma_y}\right) > \sqrt{3} \tag{14.30}$$

If σ_u/σ_y is unity, brittle failure occurs because Eq. 14.30 cannot be satisfied. If $\sigma_u/\sigma_y = 1.5$, brittle failure still persists. Thus, for the geometry indicated it would be necessary to have σ_u/σ_y greater than 1.732 since F_L can also be less than unity.

Pellini and his associates, from data gathered over many years, have developed an analysis applicable only to steel but providing relatively "fracture-safe design" for engineering structures. The procedure is applicable to all steels having distinct transition features (Figs. 14.12 and 14.13) and is based on the concept of the *fracture analysis diagram*, which represents a consolidation of knowledge of flaw size, stress, and temperature for initiation and propagation of brittle fractures.

Consider Fig. 14.15A. In a flaw-free steel the tensile and yield strengths increase with decreasing temperature and meet at a common point where ductility is essentially zero. At this point on the diagram the temperature is indicated NDT (no flaw); this is the *nil ductility temperature* for a flawless steel. Some typical NDT values are shown in Figs. 14.7 and 14.8. If the steel contains flaws, the fracture stress decreases sharply as shown by the dotted curves. The highest temperature at which the decreasing fracture stress for fracture initiation as a result becomes contiguous with the yield strength curve is called the *nil ductility temperature* for a steel with flaws, and is the temperature provided by the common Charpy test (Figs. 14.12 and 14.13). Below this NDT tempera-

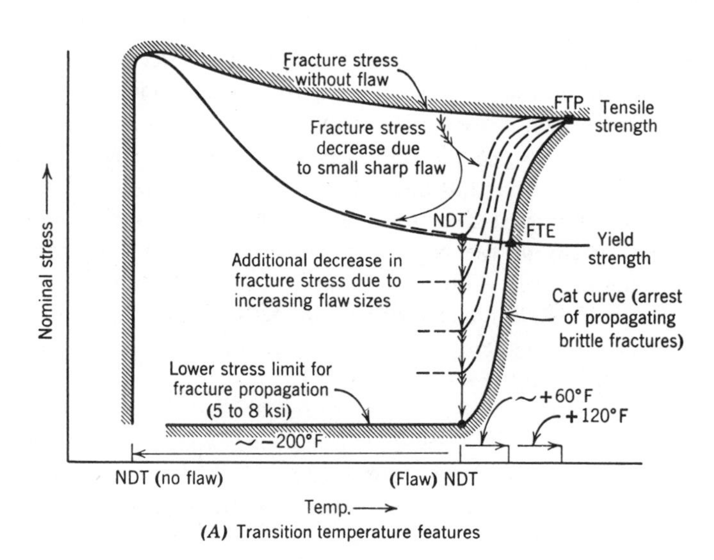

(A) Transition temperature features

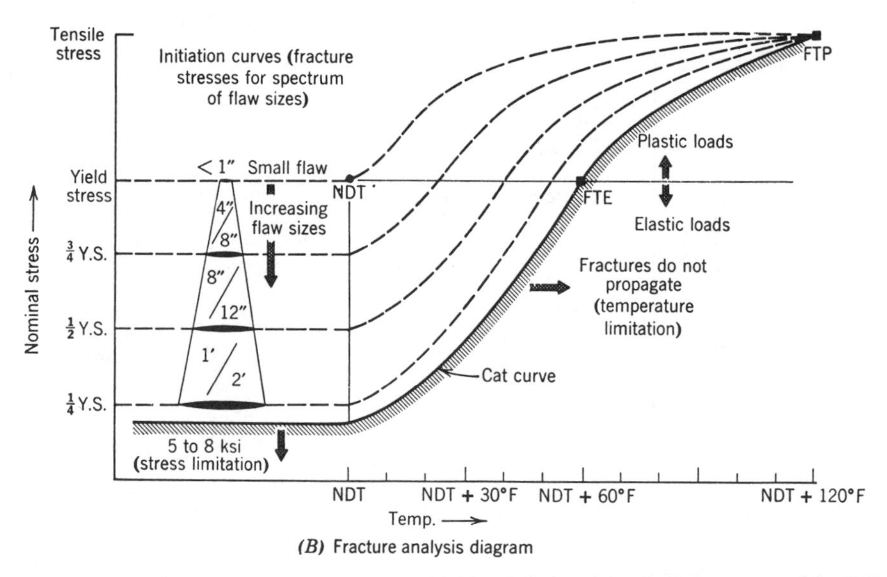

(B) Fracture analysis diagram

Fig. 14.15 Fracture analysis diagrams for steel. [After Pellini and Puzak (24) courtesy of the U.S. Naval Research Laboratory.]

ture the fracture stress is approximately inversely proportional to the square root of the flaw size and follows a pattern as shown on Fig. 14.15B.

To the extreme right of the curves in Fig. 14.15 is a line called CAT, the *crack arrest temperature*. The CAT curve represents the temperature of arrest of a propagating brittle fracture for various levels of applied stress. At the yield strength the CAT curve intersects a point called FTE, *fracture transition elastic*, the highest temperature of fracture propagation for elastic loads. Similarly at the ultimate strength the CAT curve intersects at a point called FTP, *fracture transition plastic*, which is the temperature above which fractures are entirely shear and the stress is equal to the tensile strength of the steel. At the bottom of the curves is a line marked 5–8 ksi; this is the stress level (5000–8000 psi) below which fractures do not ordinarily propagate.

The curves in Fig. 14.15 provide the basis for a practical fracture analysis of steel engineering structures that has shown surprising accuracy. The curves predict, for example, that for a given level of stress larger flaw sizes are required for fracture initiation above the NDT temperature. In design, a temperature above NDT is generally used. Thus, for example, if the NDT temperature for a particular steel is 10°F the design criterion might be NDT plus 30°F, or a design temperature of 40°F. At this temperature the steel could be stressed to half its yield strength with flaws in the size range of 8–12 in.

In actual practice, Pellini analyzed the case of a 7.5-ft diameter, 3-in. thick pressure vessel fabricated from 0.19% carbon steel having a yield strength of about 63,000 psi and a NDT temperature of 100°F. At hydrostatic test pressure at 55°F the shell stress reached about 38,000 psi, and fracture occurred. Testing

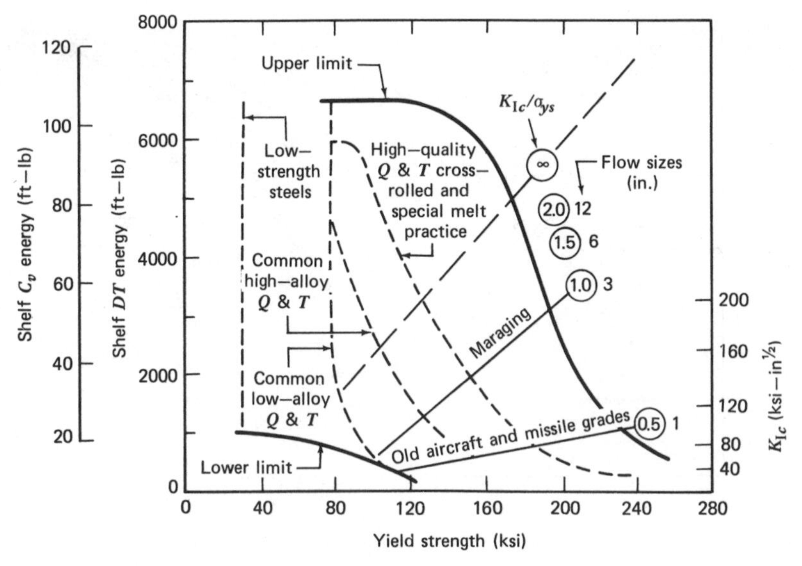

Fig. 14.16 Ratio analysis diagram for steel. [*After Lange (17), courtesy of the U.S. Naval Research Laboratory.*]

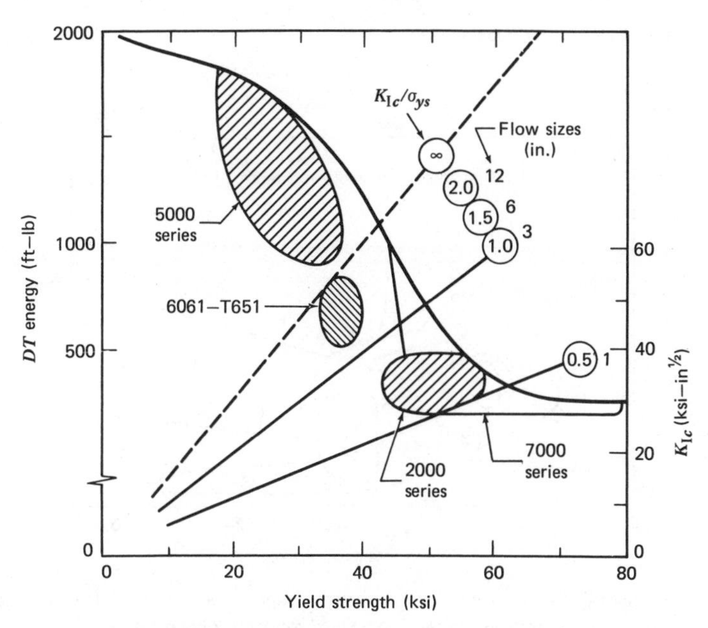

Fig. 14.17 Ratio analysis diagram for high-strength aluminum alloys. [After Lange (17), courtesy of the U.S. Naval Research Laboratory.]

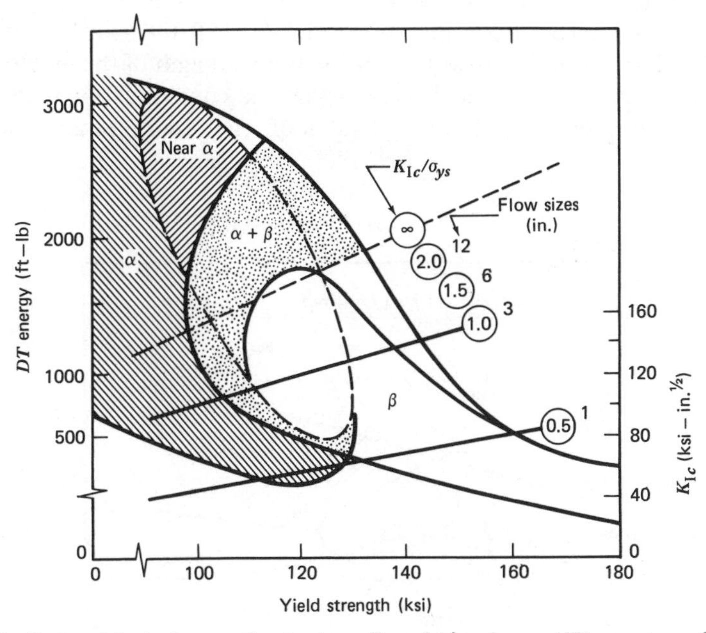

Fig. 14.18 Ratio analysis diagram for titanium alloys. [After Lange (17), courtesy of the U.S. Naval Research Laboratory.]

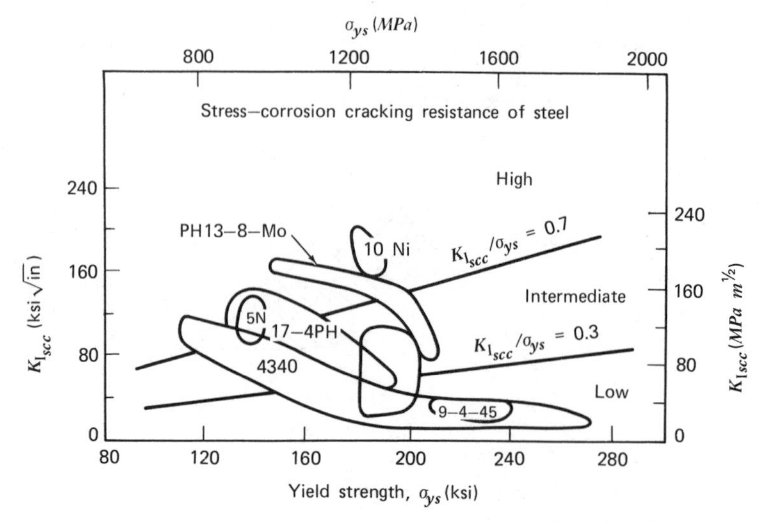

Fig. 14.19 Summary of SCC characteristics of selected high-strength steels. [After Judy and Goode (13), courtesy of the U.S. Naval Research Laboratory.]

disclosed that the material had been inadvertently annealed which lowered its yield strength to about 36,000 psi, the approximate stress developed during hydrostatic testing; this condition plus the finding of weld cracks of the order of $1\frac{1}{2}$ in. explain the behavior. Referring to Fig. 14.15B, the temperature was 15° below NDT and was the stress level at the yield strength of the material in the presence of small flaws. Thus it is seen that the conditions for fracture were present—and fracture occurred. Several additional examples are analyzed in (24).

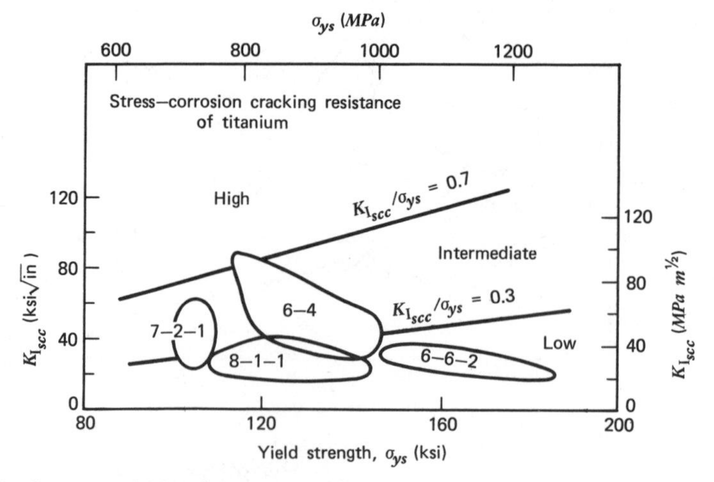

Fig. 14.20 Summary of SCC characteristics of selected high-strength titanium alloys. [After Judy and Goode (13), courtesy of the U.S. Naval Research Laboratory]

Ratio analysis diagrams (17) have been developed for use in designing such as Figs. 14.16, 14.17, and 14.18. Stress corrosion values are shown in Figs. 14.19 and 14.20. These represent values to be used in place of K_{IC} for a corrosive environment.

14-4 USE OF FRACTURE IN DESIGN

An important aspect of fracture mechanics as applied to engineering structures and component parts is the effect that cracks and/or defects have on the basic integrity of the part; in particular, for static loading, how deep can be flaw be without triggering spontaneous failure? Or, for cyclic loading, how many cycles of stress can be imposed on the structure, or part, before the flaw grows and propagates, eventually causing catastrophic failure?

The basic stress analyses are documented in many places (see references) and are too extensive to detail here. However, the practical elements can be extracted and applied to real problems that exist in large numbers in structural designs and component parts.

The simplest criterion for critical flaw size in a part is given by Eq. 14.17

$$a_{\text{cr.}} = \frac{K^2 Q}{1.21\pi\sigma^2} \qquad 14.31$$

where $a_{\text{cr.}}$ = critical depth of flaw
K = critical stress intensity (fracture toughness)
Q = flaw shape factor (see Fig. 14.9)
σ = applied stress

The geometrical aspects of the flaw are shown in Fig. 14.9 (42, 37). Thus, knowing certain data and geometry the integrity of a part can be determined by use of Eq. 14.31. For example, suppose that there is a flaw on the bore of a cylinder subject to internal pressure; the following data are given:

Cylinder I.D. = 20 in.
Cylinder O.D. = 21 in.
Length of flaw along bore = 0.2 in.
Depth of flaw = 0.08 in.

Material: Steel having a yield strength of 165,000 psi and a K value of $65 \text{ ksi}\sqrt{\text{in.}}$

Design conditions: 4000 psig at 75°F.

The maximum stress is

$$\sigma_n = p\frac{R}{t} = \frac{4000(10)}{0.5} = 80{,}000 \text{ psi}$$

From Fig. 14.9, $\sigma/\sigma_{ys} = 80{,}000/165{,}000 = 0.485$. Also, $a/2c = 0.08/0.2 = 0.40$ from which $Q = 2.1$. Then, using Eq. 14.31

$$a_{cr.} = \frac{(65{,}000)^2(2.1)}{1.21\pi(80{,}000)^2} = 0.365 \text{ in.}$$

The vessel wall thickness is 0.5 in.; therefore, if a 4 : 1 ratio flaw reaches a depth of 0.365 in. a failure could be triggered by one pressure application to 4000 psig.

The second aspect of the problem has to do with cyclic loading. Assuming the same cylinder as above but with internal pressure cycling from zero to 4000 psig, how many cycles can be sustained before failure occurs?

The fatigue aspect of fracture mechanics is also well documented (25, 37, 43) and the principles can be applied here to solve a practical problem.

The number of cycles that can be sustained by a part containing a flaw is, in the simplest form, expressed as

$$\frac{da}{dN} = cK_1{}^n \qquad 14.32$$

where da/dN is the growth per cycle of a flaw, a is the depth of the flaw, N is the number of cycles sustained, c and n are material constants, and K_1 is the stress range expressed as

$$K_1 = \sigma\sqrt{\frac{1.21\pi a}{Q}} \qquad 14.33$$

where σ is the maximum applied stress. The material constants c and n are obtained experimentally from plots such as shown in Fig. 14.11. The ΔK abscissa is a critical part of the plot and is obtained by use of special specimens, each specimen type being applicable to a particular geometry (4, 15, 25, 37). For the example at hand $c = 3.6 \times 10^{-10}$ and $n = 3$.

$$K_1 = \frac{4000(10)}{0.5}\sqrt{\frac{1.21\pi a}{2.1}} = 107634\sqrt{a}$$

In order to determine cycle life, Eq. 14.32 is integrated as follows:

$$N = \int_{t_i}^{t_c} \frac{da}{c\left[\dfrac{107634\sqrt{a}}{1000}\right]^n} \qquad 14.34$$

In Eq. 14.34 a factor of 1000 is applied to convert the pressure in psi to ksi to conform with the c factor which is based on ksi. The integral limits are t_c, the critical flaw depth, and t_i, the initial flaw depth. Thus,

$$N = \int_{0.08}^{0.365} \frac{da}{(3.6 \times 10^{-10})(107.634)a^{1.5}}$$

$$N = \frac{2228}{-0.5}\left[\frac{1}{0.365^{0.5}} - \frac{1}{0.08^{0.5}}\right] = 8381 \text{ cycles}$$

This indicates, theoretically, that starting with a flaw 0.08 in. deep the cylinder could sustain 8381 pressure cycles to 4000 psig at which time the flaw would have grown to a depth of 0.365 in., the critical depth at which one additional pressure cycle would trigger failure. In a real case a suitable factor of safety would be applied.

The above represents a relatively simple, but useful, procedure to assess the effect of flaws in structures and parts. As with other technologies, the field is expanding rapidly, and the user must keep up to date on methods available for solving practical problems. In particular, the techniques of finite element analysis are being applied to crack growth and general areas of fracture mechanics (45).

REFERENCES

1. G. M. Boyd, Ed., *Brittle Fracture in Steel Structures*, Butterworth and Co., London (1970).
2. J. Branger and F. Berger, Eds., *Problems with Fatigue in Aircraft*, Proceedings of Eighth ICAF Symposium, ICAF Doc. No. 801, Lausanne, Switzerland (1975).
3. D. Broek, *Elementary Engineering Fracture Mechanics*, Noordhoff International Publishing, Leyden, The Netherlands (1974).
4. W. F. Brown, Jr. and J. E. Srawley, *Plane Strain Crack Toughness Testing of High Strength Metallic Materials*, ASTM STP 410, ASTM, Philadelphia, Pa. (1966).
5. V. J. Colangelo and F. A. Heiser, *Analysis of Metallurgical Failures*, John Wiley and Sons, Inc., New York, N.Y. (1974).
6. N. E. Frost, K. J. Marsh, and L. P. Pook, *Metal Fatigue*, Oxford University Press, London (1974).
7. Y. C. Fung and E. E. Sechler Eds., *Thin-Shell Structures*, Prentice-Hall, Inc., Englewood Cliffs, N.J. (1974).
8. H. C. Hagendorf and F. A. Pall, *A Rational Theory of Fatigue Crack Growth*, NA-74-278, North American Rockwell, Los Angles, Cal. (1974).
9. R. W. Hertzberg, *Deformation and Fracture Mechanics of Engineering Materials*, John Wiley and Sons, Inc., New York, N.Y. (1976).
10. G. R. Irwin, *Weld. J. Res. Suppl.*, **41** (11), 519s–528s (November 1962).
11. G. R. Irwin and J. A. Kies, *Weld. J. Res. Suppl.* **33** (4), 193s–198s (April 1954).
12. Z. B. Jastryzebski, *Nature and Properties of Engineering Materials*, John Wiley and Sons, Inc., New York, N.Y. (1976).
13. R. W. Judy, Jr. and R. J. Goode, *Prevention and Control of Subcritical Crack Growth in High-Strength Metals*, NRL Report 7780, U.S. Naval Research Laboratory, Washington, D.C. (August 2, 1974).
14. J. F. Knott, *Fracture Mechanics*, Butterworth and Co., London (1973).
15. A. S. Kobayashi, Ed., *Experimental Techniques in Fracture Mechanics*, Vol. 1 and 2, Society for Experimental Stress Analysis, Westport, Conn. (1973) (1975).
16. E. A. Lange, *Mach. Des.*, **44** (22), 105–111 (September 7, 1972).
17. E. A. Lange, *Fracture Toughness of Structural Metals*, NRL Report 7046 (May 4, 1970).
18. B. R. Lawn and T. R. Wilshaw, *Fracture of Brittle Solids*, Cambridge University Press, Cambridge (1975).
19. H. Liebowitz, Ed., *Fracture Mechanics of Aircraft Structures*, AGARD-AG-176, National Technical Information Service, Springfield, Va. (1974).

20. H. Liebowitz, Ed., *Fracture*, Vols. 1–7, Academic Press, New York, N.Y. (1968–1972).

21. F. A. McClintock, *Weld. J. Res. Suppl.*, **40** (5), 202s–209s (May 1961).

22. C. C. Osgood, *A Basic Course in Fracture Mechanics*, Penton Education Division, Cleveland, Oh. (1971).

23. E. R. Parker, *Brittle Behavior of Engineering Structures*, John Wiley and Sons, Inc., New York, N.Y. (1957).

24. W. S. Pellini and P. P. Puzak, *Fracture Analysis Diagram Procedures for the Fracture-Safe Engineering Design of Steel Structures*, Report 5920, U.S. Naval Research Laboratory, Washington, D.C. (March 15, 1963). Also published as *Welding Research Council Bulletin* No. 88, American Welding Society, New York. See also *Practical Considerations in Applying Laboratory Fracture Test Criteria to the Fracture-Safe Design of Pressure Vessels*, Report 6030, U.S. Naval Research Laboratory, Washington, D.C. (November 5, 1963).

25. S. T. Rolfe and J. M. Barsom, *Fracture and Fatigue Control in Structures*, Prentice-Hall, Inc., Englewood Cliffs, N.J. (1977).

26. M. E. Shank Ed., *Control of Steel Construction to Avoid Brittle Failure*, Welding Research Council, New York, N.Y. (1957).

27. G. C. Sih, Ed., *Mechanics of Fracture*, Vol. 1: *Methods of Analysis and Solutions of Crack Problems* (1973); Vol. 2: *Three-Dimensional Crack Problems* (1975); Vol. 3: *Plates and Shells with Cracks* (1976); Vol. 4: *Elastodynamic Crack Problems* (1976); Vol. 5: *Stress Analysis of Notch Problems* (1977), Noordhoff International Publishing, Leyden, the Netherlands.

28. J. W. Spretnak, *A Summary of the Theory of Fracture in Metals*, Defense Metals Information Center Report DMIC-157, Battelle Memorial Institute, Columbus, Oh. (August 7, 1961).

29. R. I. Stephens, *Linear Elastic Fracture Mechanics and Its Application to Fatigue*, SAE Paper No. 740220, Automotive Engineering Congress, Detroit, Mich. (February 1974).

30. H. Tada, P. C. Paris, and G. R. Irwin, *The Stress Analysis of Cracks Handbook*, Del Research Corp., Hellertown, Pa. (1973).

31. A. S. Tetelman and A. J. McEvily, Jr., *Fracture of Structural Materials*, John Wiley and Sons, Inc., New York, N.Y. (1967).

32. C. F. Tipper, *The Brittle Fracture Story*, Cambridge University Press, New York, N.Y. (1962).

33. E. T. Wessel, W. G. Clark, Jr., and W. H. Pryle, *Fracture Mechanics Technology Applied to Heavy Section Steel Structures*, Paper 72, Fracture 1969, P. L. Pratt, Ed., Chapman and Hall Ltd., London (1969).

34. D. H. Winnie and B. M. Wundt, *Trans. Am. Soc. Mech. Eng.*, **80** (8), 1643–1658 (November 1958).

35. *Aerospace Structural Metals Handbook*, AFML-TR-68-115, Mechanical Properties Data Center, 13919 West Bay Shore Drive, Traverse City, Mich. (1977).

36. *Alloy Design for Fatigue and Fracture Resistance*, AGARD-CP-185, National Technical Information Service, Springfield, Va. (1975).

37. *Damage Tolerant Design Handbook, A Compilation of Fracture and Crack Growth Data for High Strength Alloys*, MCIC-HB-01, Metals and Ceramics Information Center, Battelle Memorial Institute, Columbus, Oh.

38. *Fracture Toughness Evaluation by R Curve Methods*, ASTM STP 527, American Society for Testing and Materials, Philadelphia, Pa. (1973).

39. *Metals Handbook*, American Society for Metals, Vols. 1–10, 8th ed. Metals Park, Oh.

40. J. D. McNaughton and P. F. Wieser, Metals Reference Issue, *Mach. Des.*, 8–11 (February 14, 1974).

41. P. C. Paris, and G. C. Sih, "Stress Analysis of Cracks," ASTM STP 381, 30–83, American Society of Testing and Materials (1965).

42. C. F. Tiffany and J. N. Masters, "Applied Fracture Mechanics," ASTM STP 381, 249–278, American Society of Testing and Materials (1965).

43. J. M. Barson, "Fatigue Behavior of Pressure Vessel Steels," WRL Bulletin No. 194, Welding Research Council, New York, N.Y. (May 1974).

44. PVRC Recommendations on Toughness Requirements for Ferritic Materials, WRC Bulletin No. 175, Welding Research Council, New York, N.Y. (August 1972).

45. T. P. Rich and D. J. Cartwright, Eds., *Case Studies in Fracture Mechanics*, AMMRC MS 77–5, Army Materials and Mechanics Research Center Watertown, Mass. (June 1977).

PROBLEMS

14.1 A concrete beam with a compression σ_{uc} of 10 kpsi and τ_u of 2 kpsi is loaded so the tensile stress is 1 kpsi. The K_{IC} value (9) is 0.21–1.30. Find the approximate surface finish using the related information in Chapter 12 when E is 4.5×10^6 psi.

14.2 A 155 mm gun bore with D_o/D_i of 2 and gas pressure of 40 kpsi is found to have a 0.060-in. radial crack. If the σ_{ys} is 180 kpsi and K_{IC} is 100 kpsi $\sqrt{\text{in.}}$, estimate the rupture pressure and the factor of safety. Is the wall thickness greater than B? If the material is steel, find the pressure to start the crack propagating from K_{th}.

14.3 Annealed 18 Ni maraging steel (81, Chapter 15) exhibits the following yield properties during low temperature testing:

80°F	175 kpsi	– 200°F	216 kpsi
– 100°F	200 kpsi	– 320°F	250 kpsi
– 423°F	283 kpsi		

Find B required for testing and plot a_c versus temperature for a design stress of one fourth the room temperature yield.

14.4 A large 30-in. diameter pipe with a 0.375-in. wall ruptures during testing with a partial crack 8.82 in. long and 0.346 in. deep. The hoop stress at failure is 12.2 kpsi. Using through crack failure, estimate K_{IC}. If the yield is 61 kpsi, what is B?

14.5 In Fig. 14.15, cracks do not grow below 5 – 8 kpsi for steel. Eq. 14.28 is another criterion for crack growth. Take a typical crack shape and compare the stress in both criteria.

14.6 A 0.250-in. hole is drilled in 0.1875-in. SAE 4340 steel plate. Using Chapter 12 for crack length estimates, find the maximum stress allowed for uniaxial stress. If the same hole is drilled in 0.1875-in. 304 stainless, find the maximum allowable stress. What is the stress allowed for no crack growth in both cases? Should a corrosive condition exist, is there any change in the stress values?

14.7 A small circular flaw exists in a 1-in. diameter bar of 7075-T6 aluminum with a uniaxial stress of 40 kpsi. What must the radius be for fracture?

14.8 In Problem 14.2 the 0.060-in. radial crack is found to be 0.040 in. long. For a stress of 0–40 kpsi, find the number of cycles before failure.

14.9 Using 0.75 σ_{max} for the SAE 4340 steel plate in Problem 14.6, find the number of cycles before failure.

14.10 Using one half the flaw radius in Problem 14.7, find the number of cycles before failure.

chapter 15

THERMAL STRESS, CREEP, AND STRESS RUPTURE

In this chapter consideration is given to some of the more or less ordinary effects that temperature has on the mechanical properties and behavior of materials. Most of the discussion is focused on structural applications under different elastic thermal conditions and on thermal problems associated with creep, fatigue, dilation, and rupture. The behavior of polymeric materials is not considered here, however, because of the special analyses required; this class of materials is treated separately in Chapter 5. See also (79).

15-1 THE NATURE OF THERMOMECHANICAL BEHAVIOR

The mechanical behavior of materials when subjected to temperature effects is one of the most important areas in the mechanics of materials; its effects are far-reaching, for few, if any, practical engineering considerations can be isolated from some effect of temperature or a change in temperature. Some everyday examples are simple dilation effects like the lengthening of bridges on a hot day or the bursting of water pipes in freezing weather, distortions set up in structures by thermal gradients, the sometimes drastic changes in the properties of materials (creep, fatigue, tensile strength, dimensional stability, ductility, etc.) with change in temperature, oxidation, surface deterioration, and breakdown of lubricants, to name a few (14).

Even defining *high or low temperature* is difficult because it is relative (25); *room temperature* is *hot* for gallium whose melting point is around 30°C, *warm* to lead with a melting point near 300°C, and *cold* to iron melting around 1700°C. In a very approximate way *high temperature* might be considered as any temperature in excess of about one-fourth the melting point of a material, and the maximum temperature for design purposes about half the melting point. Cryogenic temperatures extend from about 0 to -269°C, the temperature of liquid helium. Typical data are given in Table 15.1.

The data for metals in Table 15.1 can be presented in a more fundamental way by showing them graphically in order of increasing melting temperature. In such a plot lines are drawn horizontally from an ordinate; the end of the line represents the melting temperature of the material in degrees absolute, plotted

*Table 15.1 Approximate Melting Temperatures for Some Materials**

Range, °C	Material
3890	Hafnium carbide-tantalum carbide complex
3887	Hafnium carbide
3300–3700	Zirconium carbide (3350); tantalum carbide (3400); graphite (sublimes); tungsten (3370).
2750–3300	Magnesia (2800); tungsten carbide (2867); tantalum (2870); boron nitride (3000); zirconium boride (2995); titanium carbide (3140); fused thoria (3300); zirconia (2760); rhenium (3180).
2200–2750	Silicon carbide (2400); columbium (2415); boron carbide (2450); molybdenum (2620); beryllium oxide (2550); boron (2300); titanium boride (2600).
1650–2200	Alumina (2100); silica (1728); platinum (1750); rhodium (1865); silicon nitride (1900); titanium (1795); chromium (1650); hafnium (1870); zirconium (1850); vanadium (1710).
1100–1650	Manganese (1260); cobalt (1485); nickel (1450); iron (1540); nickel alloys (1460); nickel silver (1110); stainless steel (austenitic) (1450); beryllium (1285); cupro nickel (1240).
700–1100	Gold (1065); silver (960); copper (1085); brass (1035); bronze (1040); arsenic (815); silicon bronze (1090); aluminum bronze (1085).
0–700	Aluminum alloys (660); magnesium alloys (650); pewter (296); lead alloys (320); babbitt (282); antimony (630); tin (231); selenium; zinc (420); cadmium (320); bismuth (271); tellurium (450); lithium (186).

* Numbers in parentheses are specific temperatures.

NOTE: For temperature conversion [C = Centigrade (Celsius); F = Fahrenheit; K = Kelvin, absolute; R = Rankine, absolute; R′ = Réaumer].

$$°C = \frac{°F - 32}{1.8}$$

$$°F = 1.8°C + 32$$

$$°K = °C + 273.16$$

$$°R = °F + 460$$

$$°R' = \frac{°F - 32}{2.25} = \frac{°C}{1.25}$$

on the abscissa. For each material this terminal point is marked 1.0. Such a plot provides a homologous description of the metal (25). It has been observed that most metals behave similarly at the same homologous temperature. For example, in cold-rolled nickel and copper, softening takes place at a homologous temperature of about 0.35. That is, for both materials softening takes place at about 0.35 of the distance from the ordinate to the melting temperature (1.0).

If an object that is heated or cooled is not free to dilate but is held in place, thermal stresses are induced that are proportional to the coefficient of thermal expansion α, the modulus of elasticity E, and the temperature change (ΔT). Thus

Table 15.2 Unit Thermal Stress Values

Material	Modulus of Elasticity at 100°C (psi × 10^{-6})	Coefficient of Thermal Expansion to 100°C (in./in./°C × 10^{6})	Stress for 1°C Temperature Change (psi)
Alumina	45	8.1	362
Aluminum	10	28	280
Carbon steel	30	12	400
304 Stainless	28	19.1	535
Copper	19	18.2	346
Haynes 25	35	14.4	500
Haynes cermet	39	8.5	330
Inconel X-750	31	16.5	510
Molybdenum	50	5.1	255
Magnesium	6.5	30	195
Silicon carbide	58	4.7	272
Titanium	16	9.5	153
Tungsten	59	4.6	272
Udimet 500	31	14.9	465

within the elastic range the order of magnitude of thermal stress in a restrained object is roughly $E\alpha(\Delta T)$. The unit thermal stress obtained in several materials for a 1°C increase in temperature around 100°C and below is shown in Table 15.2. For steel it is thus seen that by simply placing a restrained bar of the material in boiling water a thermal stress of nearly 30,000 psi is obtained—enough to cause initial yielding for some steels.

Types of problems

The mechanics of design at high or low temperature requires considerable departure from conventional procedure at room temperature (10, 12, 45). Ordinary calculations involving changes in properties as a result of a change in temperature or problems involving elastic constants like the modulus or coefficient of expansion, for example, can be made in a great many practical situations with only theory as the guide. In specialized or complicated cases experiments are usually required in support of theory or to furnish direct data. At high temperature, one of the principal difficulties is that the behavior of a design becomes time dependent; at low temperature brittle behavior often becomes a problem as discussed in Chapter 1.

In general, thermomechanical design problems can be identified in one, or in a combination of two or more, of the following areas:

1. That in which there is only a comparatively small uniform change in temperature.

2. That in which there is a large variation in temperature and thermal dilations.

3. That in which time-dependent flow and fracture occur in addition to the usual elastic effects.

4. The field of *brittle* materials that requires somewhat special handling [see also Chapters 1, 14, and (73)].

In many practical problems the situation can often be reduced to the establishment of a limiting stress and a limiting deformation, both as functions of time; a compromise between these limits is usually required. In a turbine application, for example, deformation may be the more critical factor, whereas in a pressure vessel rupture might be the critical factor. If other mechanical factors are superposed, such as mechanical cyclic loading, or even if thermal cycling is imposed, rough calculations to determine the working stresses and strains can be made, but often the solution to the problem is obtained by estimates based on either model tests or past experience. An example of this type of situation is given by the plot shown in Fig. 15.1 for various temperature effects on nickel alloy gas turbine blades (25). In normal service, 10^8 cycles are reached in about 1000 hr;

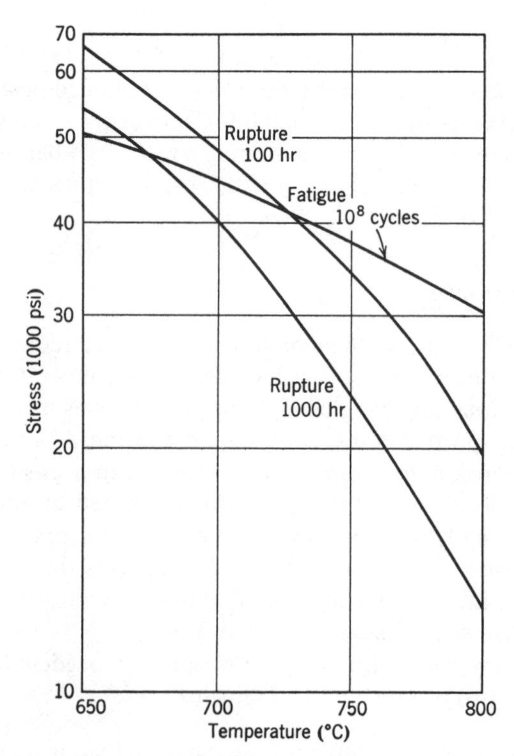

Fig. 15.1 Stress-rupture and fatigue failure of gas turbine blading alloy Nimonic 80A. [After Guard (25), courtesy of The McGraw-Hill Publishing Co.]

this is the basis for comparison. At 665°C, fatigue failure is the governing condition, that is, 50,000 psi stress results in a fatigue failure at 10^8 cycles or 1000 hr, whereas 55,000 psi stress would be required to produce a stress-rupture failure in the same time. At higher temperatures the stress-rupture characteristics govern. Additional practical problems of this type associated with thermal effects have been considered by Blaser (6).

Another interesting, as well as practical, problem is that of friction and wear at high temperature. Depending on the problem, these factors may be very important, even governing. Kingsbury (35), for example, conducted friction and wear tests at temperatures up to 1000°C on copper, zinc, titanium, and ordinary constructional steel and found that as temperatures were raised to 400°C the friction and wear rate of steel reached a maximum at about 100°C and then diminished, whereas for the other metals friction remained constant and wear rate increased. On the basis of these tests, metals with low W/p ratios should have favorable friction properties at high temperature (W is the surface energy and p the hardness of the metal).

Materials and properties

Elevated temperature. The primary requirements for any high temperature material are (a) high melting point, (b) low vapor pressure, (c) resistance to oxidizing and reducing atmospheres, (d) resistance to attack by various chemicals, and (e) physical strength and ability to absorb deformation as near to the melting point as possible. Every material presently known that could conceivably be called a material of construction has a melting point less than 4000°C, a low level indeed when compared with available plasma and thermonuclear temperatures ranging from 10^4 to 10^8°C, respectively. A comparison of melting temperatures for several materials is given in Table 15.1.

Some of the phenolic plastics reinforced with asbestos or glass fibers have been used, for short periods of time, at temperatures to 15,000°C in low-stress applications. The ability of such materials to withstand such high temperatures comes about through a series of convenient circumstances which, when combined, tend to delay structural failure if the part is exposed for only a short time (a few minutes). The heat capacity of these materials is about 16 times that of copper, and this condition, together with the fact that the thermal conductivity is only a thousandth that of copper, limits heat effects to the surface. At the same time, their capacity to generate surface gas, which acts as a protective blanket, is about 13 times that of copper. Thus these materials can operate under very severe conditions. On the other hand, however, their load-carrying ability is low (48). For sustained loads a *superalloy* or special material coated with molybdenum or a refractory nonmetallic is commonly used (26).

Mechanical property data on high-temperature materials are rather difficult to assess since much of the published information gives no details about test conditions, size or shape of specimens, strain rates, and so on. In the range 20°C to about 750°C, considerable creep and stress-rupture data are available for

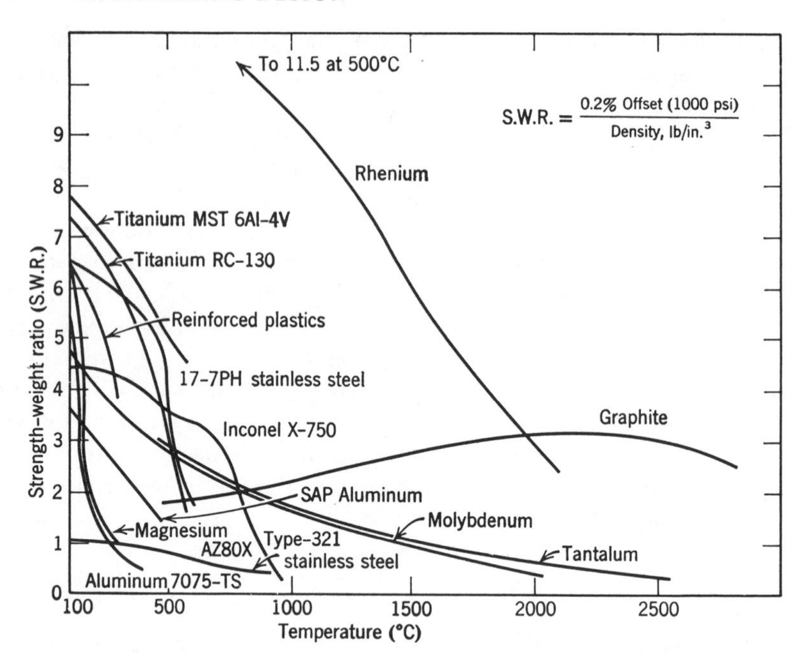

Fig. 15.2 Relative strength-weight ratio for some materials at elevated temperatures.

materials in tension but very few data for fatigue, impact, or combined stress behavior. Between about 750 and 1700° C data are available for many refractories and ceramics, particularly on bending and thermal shock strength. Beyond about 1700° C data are virtually nonexistent. A few typical data have been assembled in Fig. 15.2 which can serve as guides in the selection of materials for high-temperature service.* Figure 15.2 shows, for example, that near room temperature aluminum and magnesium alloys are at least six times as efficient as a common stainless steel, but no better than titanium alloys or percipitation-hardened stainless steel. As temperature increases the light alloys tend to lose their advantage and at the highest temperatures the refractory metals and graphite take over. Curves such as shown should be used with caution and with due regard for cost, corrosion, state of stress, and other factors. Where load-carrying ability is of primary importance, the best nickel and cobalt alloys are limited to service at a maximum temperature of about 1000° C.

Graphite is an important high-temperature material, but since it oxidizes at relatively low temperatures it has to be protected by a skin of silicon carbide or some other refractory nonmetallic material. *Pyrolytic graphite*, however, is one of the best high-temperature materials available and is used extensively in rocket nozzles (13).

Cermets have been developed to combine the ductility and thermal shock

*See (17, 47) for discussion of materials requirements for missiles, rockets, and space applications.

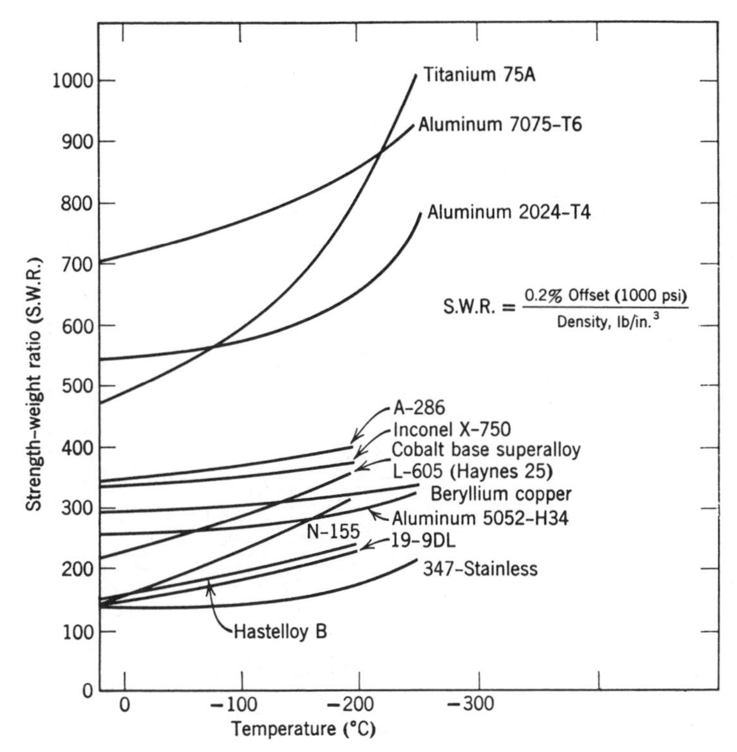

Fig. 15.3 Strength-weight ratio for some materials at cryogenic temperatures.

resistance of metals with the high-temperature strength of ceramics; these have been quite successful in high-temperature atmospheres for protective tubes in liquid metals, bearings, and reaction chambers. Their major use is in the range of about 800 to 1500° C.

Dispersion-hardened (SAP-type alloys) have also been developed in an attempt to gain more strength closer to the melting point of the material. With rare exception, however, about half the melting point of a material is the mechanical limit for service under most practical conditions. An exception to this is TD (thoria dispersed) nickel which has outstanding properties between 900 to 1300° C. At temperatures over about 1100° C it is stronger than the superalloys.

Many of the ceramic materials have high melting points, but for continuous use practically all the commercial refractories are limited to service at a maximum temperature of about 2000° C.

Cryogenics. Both materials and design problems for cryogenic temperatures are assuming increased proportions, particularly since the advent of space vehicles. Although it is probably true that the requirements for rockets and space vehicles have spurred development in this field, many other applications of cryogenics have emerged as byproducts. Along with this, problems have arisen in materials engineering and stress analysis that are of interest within the scope of this book.

Table 15.3 Cryogenic Data for Some Materials

	Yield Strength, psi			Tensile Strength, psi			Modulus of Elasticity, psi		
Material	20°C	Low Temp., °C		20°C	Low Temp., °C		20°C	Low Temp., °C	
Aluminum bronze	27,000	−40 −80 −120 −180	27,000 27,000 27,000 29,000	77,000	−40 −80 −120 −120	80,000 83,000 88,000 96,000			
Copper (cold rolled)	24,000	−40 −69 −196	25,000 27,000 27,000	32,000	−40 −69 −196	37,000 39,000 53,000	17 × 10^6	−40 −69 −196	18.4 × 10^6 19.2 × 10^6 17.3 × 10^6
Copper (annealed)				20,000	−250	60,000			
Phosphor bronze "C"	112,000	−196 −253	138,000 154,000	117,000	−196 −253	151,000 167,000			
Muntz metal				50,000	−200	65,000			
Austenitic stainless	35,000	−200 −250	70,000 90,000	90,000	−200 −250	250,000 180,000			
Low carbon steel				60,000	−40	67,000			
Low alloy steel				65,000					
Inconel 600 (annealed)	36,800	−79	42,400	93,800	−79	106,400			
17-7PH stainless (TH-1050)	183,000	−78 −250	195,000 256,000	194,000	−78 −250	216,000 263,000	27.5 × 10^6	−78 −250	28.9 × 10^6 31 × 10^6
Aluminum alloy (7079-T6)	60,000	−78 −250	65,000 82,000	72,000	−78 −250	75,000 93,000	10 × 10^6	−78 −250	10.1 × 10^6 10.6 × 10^6
Magnesium alloy (ZK60A-T5)	38,000	−78 −250	52,000 58,000	46,000	−78 −250	59,000 69,000	5.9 × 10^6	−78 −250	6.6 × 10^6 6.5 × 10^6
Titanium alloy (6AL-4V) (annealed)	129,000	−78 −250	156,000 260,000	141,000	−78 −250	167,000 267,000	15.5 × 10^6	−78 −250	15.9 × 10^6 19.4 × 10^6

At cryogenic temperatures near absolute zero (− 273° C), lead, for example, exhibits the resiliency of steel, bearings can become virtually frictionless, and some stainless steels cryogenically formed exhibit working strengths of over 200,000 psi (72).

The selection of materials for use at cryogenic temperatures involves many of the factors commonly associated with materials selection for other temperature ranges; for example, the designer must consider such things as strength, ductility, toughness, flexibility, and its notch sensitivity. Generally, as the temperature goes down, the strength of most materials goes up; this includes both metals and nonmetals, including glass-reinforced plastics. Some representative properties are shown in Fig. 15.3 and Table 15.3. It is also to be noted that as the strength increases, the ductility decreases; this in turn increases the notch sensitivity of materials.

A large portion of cryogenic equipment in the form of vessels and structures is made of the austenitic stainless steels, 5000-series aluminum alloys, nickel alloy steels and copper. These materials remain fairly ductile (81) even to liquid helium temperature. Martensitic carbon steels are not used for such applications because of their brittleness at cryogenic temperatures (15, 30, 42, 56, 80).*

*One of the main problems is brittle fracture (see section on brittleness and ductility in Chapters 1, 14).

15-2 TIME-INDEPENDENT THERMOMECHANICS

Simple dilation

When an unrestrained material is heated or cooled, it dilates in accordance with its characteristic coefficient of thermal expansion. For many materials the coefficient of expansion increases with increase in temperature and for practical situations an average value is frequently used. The coefficients characteristic of a few materials are listed in Table 15.4. The change in length ΔL of a bar of material of length L_0 and coefficient of thermal expansion α subjected to a temperature change ΔT is (61),

$$\Delta L = \alpha L_0(\Delta T) \tag{15.1}$$

Thus, for example, a bar of copper 1 in. long heated from 100 to 300° C would experience a change in length of 0.00406 in. The change in volume ΔV of a

Table 15.4 Coefficients of Thermal Expansion for Some Materials [See Also Fig. 11.10 and (70).]

Material	Temperature or Temperature Range	Coefficient, in./in./°C × 10⁻⁶
Acetal resin	−30 to + 30°C	83.0
Aluminum	+20 to + 600°C	28.7
Aluminum oxide	0 to + 1,400°C	8.0
Beryllium	+20 to + 700°C	16.8
Carborundum	0 to + 1,800°C	9.2
Copper	0 to + 1,000°C	20.3
Graphite	+20 to + 800°C	3.5
Lead	+20 to + 300°C	31.3
Molybdenum	+27 to + 2,127°C	7.2
Nickel	+25 to + 900°C	16.3
Platinum	0 to + 1,000°C	10.2
Polyamide resin	0 to + 100°C	72.0
Polyethylene resin	0 to + 60°C	140.0
Silicon	1,000°C	3.3
Silicon carbide	0 to + 1,400°C	4.4
Silver	0 to + 900°C	22.4
Tantalum	27 to + 2,400°C	7.8
Titanium	20 to + 800°C	10.1
Tungsten	27 to + 2,400°C	5.8
Tungsten carbide	20 to + 400°C	5.6
Zirconium	20 to + 700°C	7.1

	Coefficient of Thermal Expansion, in./in./°F From 70°F and (× 10⁻⁶)														
	−100	100	200	300	400	500	600	700	800	900	1000	1100	1200	1300	1400
Aluminum	12.1	12.7	13.0	13.3	13.6	13.9	14.2								
Admiralty							11.2								
Austenitic stainless	8.9	9.2	9.3	9.5	9.6	9.7	9.8	10.0	10.1	10.2	10.3	10.4	10.5	10.6	10.6
Beryllium			6.5	7.0		7.5	8.0				9.0				9.6
Brass 66/34	9.3	9.6	9.7	10.0	10.2	10.5	10.7	10.9	11.2	11.4	11.6	11.9	12.1		
Bronze			10.0	10.1	10.2	10.3	10.4	10.5	10.6	10.7	10.8	10.9	11.0		
Steel	5.8	6.2	6.4	6.6	6.8	7.0	7.2	7.4	7.7	7.9	8.0	8.1	8.2	8.3	8.4
Copper	9.0	9.4	9.6	9.7	9.8	9.9	10.1	10.2	10.3	10.4	10.5				
Cupro-nickel			8.5	8.7	8.9										
Monel	6.5	7.2	7.4	7.6	7.8	8.0	8.1	8.3	8.4	8.5	8.6	8.7			
Inconel	6.1	6.4	6.6	6.8	7.0	7.3	7.5	7.7	7.9	8.1	8.3.	8.5	8.6	8.7	8.9
Nickel	6.5	7.1	7.2	7.5	7.7	7.8	8.0	8.2	8.3	8.5	8.6	8.7	8.8	8.9	8.9
Cast iron		5.6	5.8	6.0	6.1	6.3	6.5	6.7	6.8	7.0	7.2				

material of original volume V_0, coefficient α, subjected to a temperature change ΔT is

$$\Delta V = 3\alpha V_0(\Delta T) \tag{15.2}$$

Thus a sphere of copper, 1 in. in diameter heated from 100 to 300° C, would experience a change in volume of 0.0064 cu in.

These simple dilation effects are very important as shown by several practical examples. In machine work, for example, careful cooling of the workpiece during machining is essential since the heat generated during machining results in a dilation and a measurement made under these conditions is subject to change when the part has cooled down. Bridges, piping systems, and railway rails have expansion joints to accommodate length changes that take place with changes in temperature; without such provision, severe stresses could be set up and buckling or fracture of the part could occur. Another very practical example is found in the fabrication of shrink-fit vessels. Here an outer ring or collar is heated so that it expands beyond the interference originally provided between it and the inner cylinder or liner; the collar is then slipped on over the liner and on subsequent cooling large compressive stresses are set up in the liner that oppose the tensile stresses generated by internal pressure when the vessel is in service.

It was shown by Eq. 15.1 that on heating, for example, a length change ΔL was involved, its value being $\alpha L_0(\Delta T)$. If the part is not free to dilate, ΔL is zero and an internal stress is induced whose value is related to the ΔL that would have occurred had the part been free to expand. In accordance with Hooke's law (Chapter 1), for a simple bar in tension or compression, the stress σ, strain ε, and modulus E are related as follows:

$$\sigma = E\varepsilon \tag{15.3}$$

The strain ε in Eq. 15.3 now substitutes for the ΔL in the present problem and the thermal stress induced in the restrained bar by a temperature change ΔT is

$$\sigma = E\frac{\Delta L}{L_0} = (E/L_0)(\alpha L_0)(\Delta T) = E\alpha(\Delta T) \tag{15.4}$$

where the stress σ is tension or compression depending on the sign of ΔT. Note that if the bar is heated, ΔT is *positive*, which means that the resulting thermal

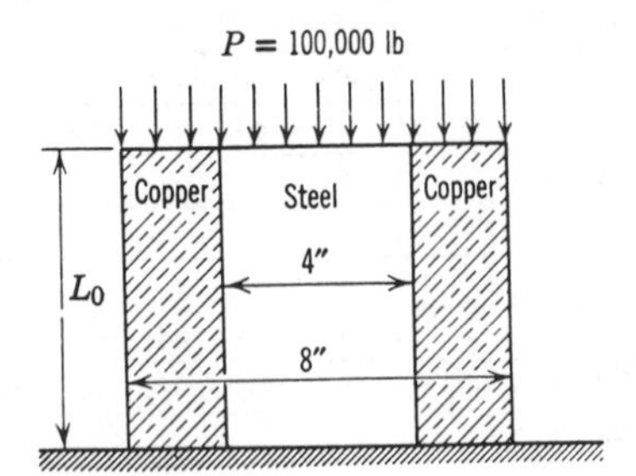

Fig. 15.4 Composite structure.

stress is *compression*. It is assumed that the length of the bar is such that buckling will not occur. It is also important that the correct values of E and α are used since they usually vary with temperature.

Consider the structure shown in Fig. 15.4. It consists of a core of steel surrounded by a hollow cylinder of copper to form a platform holding a dead weight of 100,000 lb. In order to calculate the ΔT for the assembly which will result in equal stress in the steel and copper components, the following data are given:

modulus of copper $= E_c = 19 \times 10^6$ psi

modulus of steel $= E_s = 30 \times 10^6$ psi

coefficient of expansion for copper $= \alpha_c = 18.2 \times 10^{-6}$ in./in./°C

coefficient of expansion for steel $= \alpha_s = 13.0 \times 10^{-6}$ in./in./°C

diameter of steel core = 4 in.

diameter of copper cylinder = 8 in.

From Hooke's law the stresses in the two parts are

$$\sigma_c = E_c\varepsilon = (19 \times 10^6)\varepsilon \tag{15.5}$$

$$\sigma_s = E_s\varepsilon = (30 \times 10^6)\varepsilon \tag{15.6}$$

Since the strain ε is identical for both parts,

$$\sigma_s = (30/19)\sigma_c \tag{15.7}$$

From statics

$$P = A\sigma$$

or

$$100{,}000 = (\tfrac{30}{19})(\sigma_c)(12.57) + \sigma_c(37.7) = 57.50\sigma_c$$

from which

$$\sigma_c = 1740 \text{ psi}$$

$$\sigma_s = 2750 \text{ psi}$$

and

$$\varepsilon = \sigma_c/E_c = 1740/(19 \times 10^6) = 0.0000915 \text{ in./in.}$$

For equal stresses

$$(\Delta\alpha)(\Delta T) = \Delta\sigma/E_c \tag{15.8}$$

or

$$(5.2 \times 10^{-6})(\Delta T) = 1010/(19 \times 10^6)$$

and

$$\Delta T = 10.21°\text{C}$$

As another example consider Fig. 15.5 which shows a steel-copper composite where the diameter ratio is 2 for each part. The ΔT required to give the same stress distribution in the parts as a 0.005-in. shrink on the interface diameter will be calculated.

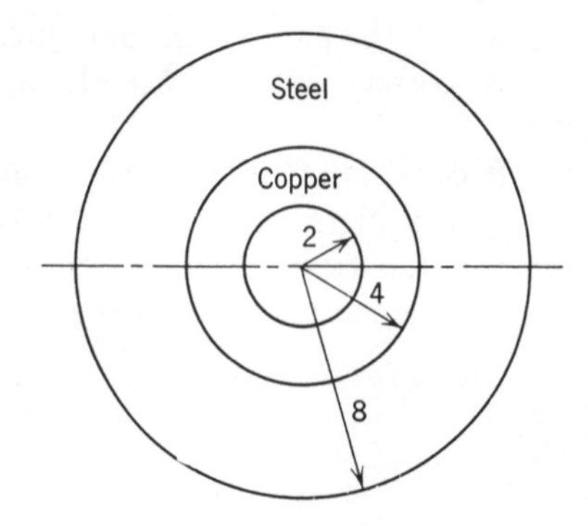

Fig. 15.5 Composite cylinder.

The change in diameter at the interface due to thermal expansion is

$$\Delta D = D(\Delta T)(\alpha_c - \alpha_s) = 0.005 \tag{15.9}$$

from which

$$\Delta T = 240°\text{C}$$

The temperature should be checked against the properties of the materials to insure that a 240°C increase is not high enough to result in yielding of the parts.

In many practical situations the restraint involved is not absolutely rigid, and then there is some reduction in thermal stress. For such cases, Eq. 15.4 is modified by a correction term K

$$\sigma = KE\alpha(\Delta T) \tag{15.10}$$

The K-factor in Eq. 15.10 is called the *coefficient of restraint*; its application is illustrated in the following example* which deals with two bars held together by a rivet, as shown in Fig. 15.6 (20). The bars are heated to different uniform temperatures involving changes in length. From Eqs. 15.1 and 15.4

$$\alpha L_0(\Delta T) + \sigma L_0/E = \alpha' L_0(\Delta T)' + \sigma' L_0/E' \tag{15.11}$$

and

$$\sigma A = -\sigma' A' \tag{15.12}$$

where A is the cross-sectional area. From the last two equations

$$\sigma = \frac{-E\alpha[(\Delta T) - (\alpha'/\alpha)\Delta T']}{1 + AE/A'E'} \tag{15.13}$$

and

$$\sigma' = -\sigma(A/A') \tag{15.14}$$

Rewriting Eq. 15.13 gives

$$\sigma = -K\alpha E(\Delta T) \tag{12.15}$$

in which

$$K = \frac{1 - [\alpha'(\Delta T)'/\alpha(\Delta T)]}{1 + AE/A'E'} \tag{15.16}$$

*Courtesy of The McGraw-Hill Publishing Co.

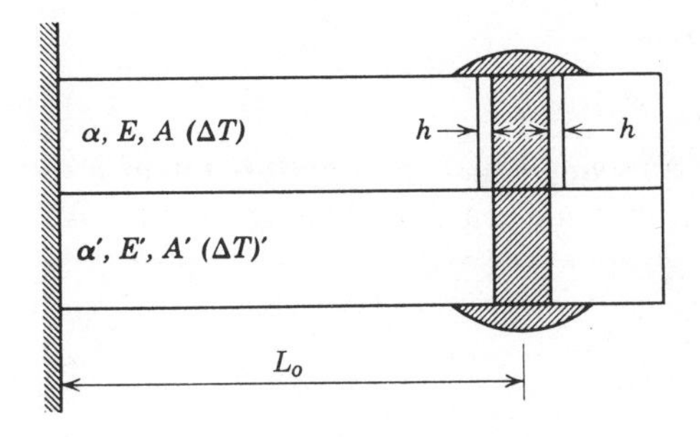

Fig. 15.6 System with partial restraint.

For values of $K > 0$, σ is compressive and σ' is tension; the reverse is true for $K < 0$. If a gap h is provided in one of the bars to allow for some movement before restraint sets in, then from Eq. 15.16

$$K = \frac{1 - [\alpha'(\Delta T)'/\alpha(\Delta T)] - [h/\alpha(\Delta T)]1/L}{1 + AE/A'E'} \qquad 15.17$$

For the condition of zero stress, $K = 0$ and the gap has a value of

$$h = L_0[\alpha(\Delta T) - \alpha'(\Delta T)'] \qquad 15.18$$

indicating that as L_0 increases, the gap h has to increase also to be effective. A steel bolt with a $(\Delta T)'$ of zero can restrain an aluminum block with $(\Delta T) = 200°C$ if the gap is $0.0043\ L_0$.

Analysis of bimetallics

One of the practical uses of thermal effects in metals is seen in thermostatic devices which, when subjected to a temperature change, register measurable deformations that can be interpreted in terms of the geometry and elastic constants of the materials making up the bimetallic strip (59).

A common type of thermostatic element is shown in Fig. 15.7. It consists of two strips of metal of equal thickness firmly joined together. Strip B is assumed to have a greater coefficient of expansion than strip A so that when the element is heated a curvature will be obtained. Strip B has a modulus E_B and strip A a modulus E_A. Corresponding cross-sectional areas are A_B and A_A.

When the element is subjected to a temperature change ΔT, the strips elongate

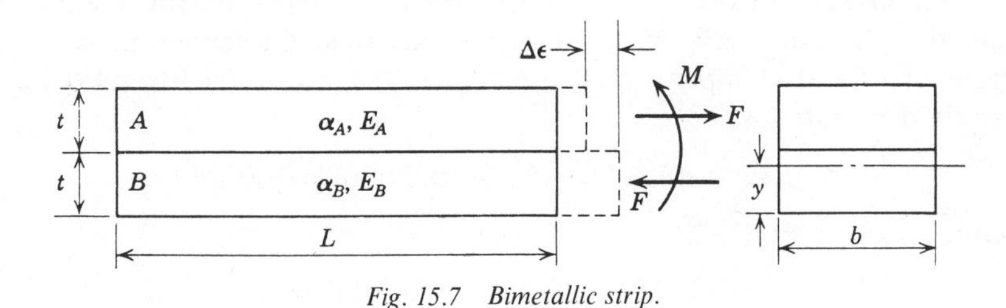

Fig. 15.7 Bimetallic strip.

different amounts giving a strain difference

$$\Delta\varepsilon = (\alpha_B - \alpha_A)(\Delta T) \tag{15.19}$$

This strain difference is eliminated by applying equal and opposite forces F to the strips, which at the same time introduces a reactive bending moment M. Then for each strip

$$\varepsilon_A = \frac{\sigma_A}{E_A} = \frac{F}{btE_A} = \frac{F}{A_A E_A} \tag{15.20}$$

and

$$\varepsilon_B = \frac{-\sigma_B}{E_B} = \frac{-F}{btE_B} = \frac{-F}{A_B E_B} \tag{15.21}$$

For equilibrium the difference of Eqs. 15.20 and 15.21 must equal Eq. 15.19; therefore

$$\frac{F}{bt}\left(\frac{1}{E_A} + \frac{1}{E_B}\right) = (\Delta T)(\alpha_B - \alpha_A) \tag{15.22}$$

or

$$F = \frac{(\alpha_B - \alpha_A)(\Delta T)(A_A E_A)}{(1 + A_A E_A / A_B E_B)} \tag{15.23}$$

where F is the increase in force due to the change in temperature. If, for example, the strip is not free to bend, then the force F is calculated from Eq. 15.23 and substituted into Eqs. 15.20 or 15.21 to obtain the strain resulting from the increase in force. Having ε_A and ε_B the stresses in each part are then

$$\sigma_B = \varepsilon_B E_B \tag{15.24}$$

and

$$\sigma_A = \varepsilon_A E_A \tag{15.25}$$

The total axial dilation of the laterally restrained composite is then either

$$\mathscr{E} = [\alpha_A(\Delta T) + \varepsilon_A]L \tag{15.26}$$

or

$$\mathscr{E} = [\alpha_B(\Delta T) - \varepsilon_B]L \tag{15.27}$$

As examples of the application of the theory to restrained composites consider Fig. 15.4 and 15.7. In the composite of steel and copper shown in Fig. 15.4 assume that the length is 30 in., the materials are welded together, there are no external forces, and a temperature differential of 50°C is applied. From Eq. 15.19, the strain differential is

$$\Delta\varepsilon = (\alpha_c - \alpha_s)(\Delta T) = (5.2 \times 10^{-6})(50) = 0.00026$$

which results in a total axial movement of

$$\mathscr{E} = \varepsilon L = (0.00026)(30) = 0.007 \text{ in.}$$

The increase in force F is given by Eq. 15.23

$$F = \frac{(5.2 \times 10^{-6})(50)(4\pi)(30 \times 10^6)}{1 + \dfrac{4\pi(30 \times 10^6)}{(\pi/4)(48)(19 \times 10^6)}} = 64{,}500 \text{ lb}$$

From Eqs. 15.20, 15.21, 15.24, 15.25, and 15.26

$$\sigma_s = \varepsilon_s E_s = \frac{FE_s}{A_s E_s} = \frac{F}{A_s} = \frac{64{,}500}{\pi(16)/4} = 5140 \text{ psi}$$

$$\sigma_c = \varepsilon_c E_c = -\frac{FE_c}{A_c E_c} = -\frac{F}{A_c} = -\frac{64{,}500}{(\pi/4)(48)} = -1715 \text{ psi}$$

Suppose the structure shown in Fig. 15.7 is 100 in. long and made of the same materials as in the previous example. It is assumed here that the bar is laterally restrained, no external forces act, and the bottom material (copper) is 0.1 in. thick while the steel is 0.3 in. thick. The temperature differential is 50°C.

From Eq. 15.19, the strain differential will be

$$\Delta\varepsilon = (5.2 \times 10^{-6})(50) = 0.00026$$

From Eq. 15.23, the increase in force is

$$F = \frac{(5.2 \times 10^{-6})(50)(0.3)(30 \times 10^6)}{1 + \dfrac{(0.3)(30 \times 10^6)}{(0.1)(19 \times 10^6)}} = 407 \text{ lb}$$

and the stresses are

$$\sigma_c = \frac{-F}{A_c} = \frac{-407}{0.10} = -4070 \text{ psi}$$

and

$$\sigma_s = \frac{F}{A_s} = \frac{407}{0.3} = 1357 \text{ psi}$$

When the composite is not restrained, it bends in proportion to the elastic constants of the materials in the composite. For convenience, since the moduli of the two strips are different, the element is analyzed as a composite structure in which an equivalent cross section, Fig. 15.8, is determined based on the ratio

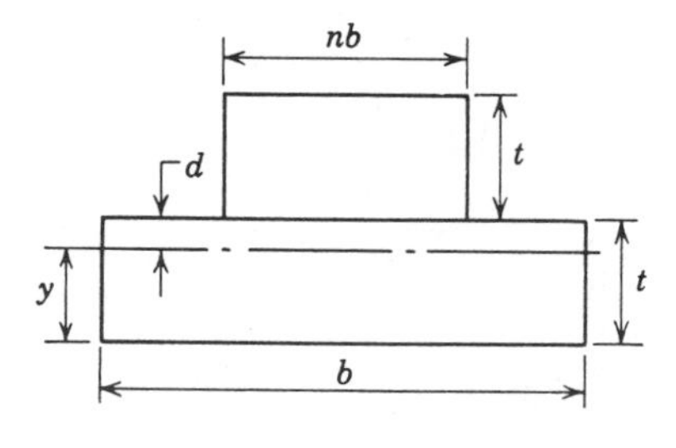

Fig. 15.8 Modulus corrected cross-section of element in Fig. 15.7.

of the moduli; thus defining

$$n = E_B/E_A \tag{15.28}$$

Eq. 15.23 can be rewritten as

$$F = \frac{(\Delta T)bt(\alpha_B - \alpha_A)}{1/E_A + 1/(nE_A)} = \frac{btnE_A(\Delta T)(\alpha_B - \alpha_A)}{n+1} \tag{15.29}$$

The neutral axis is found from the relation (see Appendix B)

$$\bar{y} = \int \sigma \frac{dy}{A} \tag{15.30}$$

which, referring to Fig. 15.8, becomes

$$\bar{y} = \frac{t}{2}\left(\frac{3n+1}{n+1}\right) \tag{15.31}$$

The foregoing theory has also been utilized in the stress analysis of tube sheets and heat exchangers (55, 74). The moment of inertia of this equivalent section about the interface axis is (see Appendix B)

$$I = I_c + Ad^2 \tag{15.32}$$

where I_c is the moment of inertia about the centroid and d is the distance from the interface to the centroid axis; thus

$$I = b\left\{\frac{t^3}{12}(n+1) + nt\left[\left(\frac{3}{2}t - \bar{y}\right)^2\right] + t\left(\bar{y} - \frac{t}{2}\right)^2\right\} \tag{15.33}$$

But since

$$M = EI/r \tag{15.34}$$

where $1/r$ is the curvature

$$\frac{1}{r} = \frac{(\Delta T)nt^2(\alpha_B - \alpha_A)}{(n+1)\{(t^3/12)(n+1) + nt[(\frac{3}{2}t - \bar{y})^2] + t(\bar{y} - t/2)^2\}} \tag{15.35}$$

or

$$1/r = K(\Delta T)(\alpha_B - \alpha_A)/t = K\phi \tag{15.36}$$

where

$$K = \frac{12n}{(n+1)\left\{n+1+3n\left[9-6\left(\frac{3n+1}{n+1}\right)+\left(\frac{3n+1}{n+1}\right)^2\right] + 3\left[\left(\frac{3n+1}{n+1}\right)^2 - 2\left(\frac{3n+1}{n+1}\right)+1\right]\right\}} \tag{15.37}$$

and

$$\phi = (\Delta T)(\alpha_B - \alpha_A)/t \tag{15.38}$$

In many instances, particularly in commercial thermostatic elements, the

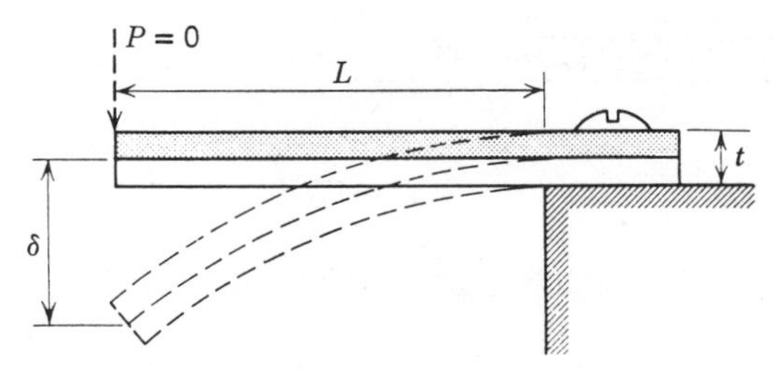

Fig. 15.9 Cantilever bimetallic element.

value of n is assumed to be unity since there is often little difference in the moduli of the materials making up the strip. With $n = 1$ the calculations are simplified and Eq. 15.36 becomes

$$\frac{1}{r} = \frac{3(\Delta T)(\alpha_B - \alpha_A)}{4t} \qquad 15.39$$

As an example of the use of this analysis, consider the element shown in Fig. 15.9. The end deflection is easily calculated using Castigliano's theorem (Chapter 7),

$$\delta = \partial U / \partial P \qquad 15.40$$

where U is the elastic strain energy and P is a *phantom load* (equal to zero) in the direction of the deflection δ. Thus from Eqs. 15.34 and 15.40

$$\delta = \frac{\partial}{\partial P} \int_0^L \frac{M^2\,dx}{2EI} = \int_0^L \frac{M \partial M}{\partial P} \frac{dx}{EI} = \int_0^L \left(\frac{EI}{r} + Px \right)(x) \frac{dx}{EI} \qquad 15.41$$

from which

$$\delta = L^2 / r \qquad 15.42$$

where r is obtained from Eq. 15.39. Deflection formulas for three types of bimetallic strips are shown in Fig. 15.10; additional formulas for various coil and disk-type thermostats are given in Reference 77.

In arriving at Eq. 15.39 it was noted that many bimetallic elements contain materials whose moduli are quite close. It turns out that this is the ideal case as shown by the curve in Fig. 15.11.

Thermal stresses in plates

Only the elementary theory which assumes that a thin plate subjected to a temperature differential assumes a spherical contour will be considered here. Such a condition is shown in Fig. 15.12.

A thin unrestrained plate subjected to ΔT will show fiber strains of

$$\varepsilon = \alpha(\Delta T) \qquad 15.43$$

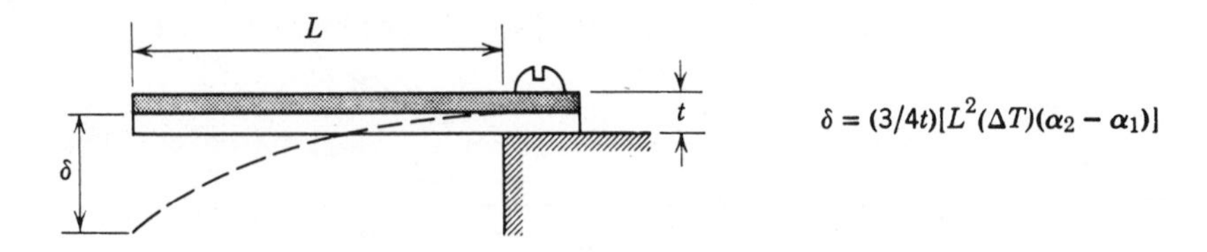

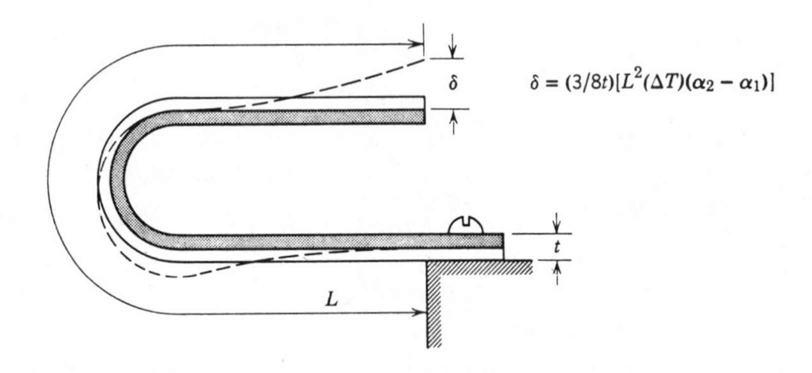

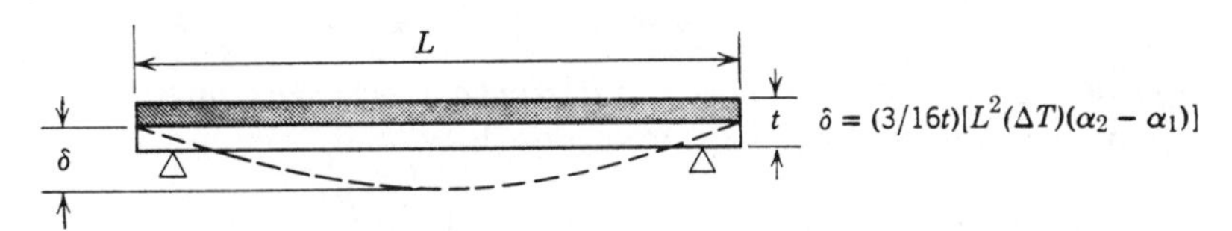

Fig. 15.10 Various bimetallic elements. [Adapted from (77), courtesy of The Metals Control Corp.]

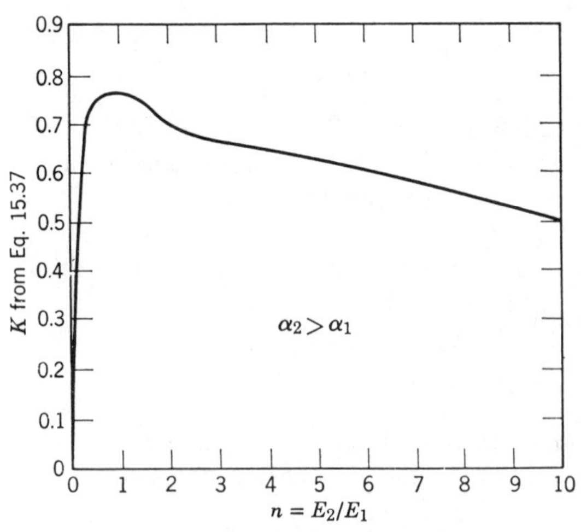

Fig. 15.11 Design curve for a bimetallic strip.

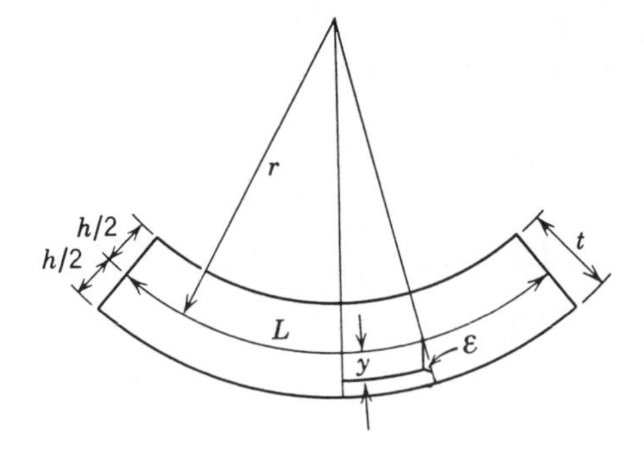

Fig. 15.12 Spherical contour of thin plate due to a temperature differential.

and from the geometry

$$1/r = \varepsilon/t = \alpha(\Delta T)/t \tag{15.44}$$

The deflection for such a case is

$$\delta = L^2/8r \tag{15.45}$$

and there is no stress since the plate is free to dilate.

Consider now any flat plate (Fig. 15.13) subjected to a linear thermal gradient through the thickness where $T_1 > T_2$. From Chapter 3 for pure bending of plates it was shown that the uniform moment developed in bending is

$$M = \frac{D(1+\nu)}{r} \tag{15.46}$$

where the flexural rigidity D is given by

$$D = \frac{Et^3}{12(1-\nu^2)} \tag{15.47}$$

If free expansion is restrained, internal stresses are introduced by reactive moments obtained from Eqs. 15.44 and 15.46, thus

$$M = \frac{\alpha(\Delta T)(1+\nu)D}{t} \tag{15.48}$$

and the stress is found by applying the bending stress formula

$$\sigma = \frac{M}{Z} \tag{15.49}$$

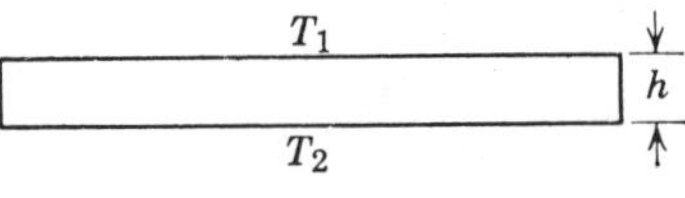

Fig. 15.13 Any plate of uniform thickness subjected to a thermal differential.

Table 15.5 Thermal Stress in Various Plates

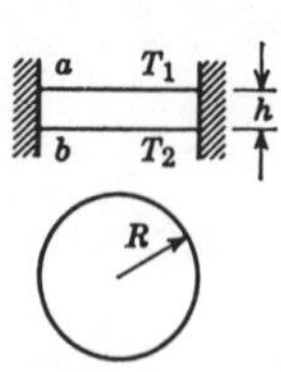

Circular Plate, Clamped at Edge

$$(\sigma)_a = \frac{-\alpha(\Delta T)E}{2(1-\nu)}$$

$$(\sigma)_b = \frac{\alpha(\Delta T)E}{2(1-\nu)}$$

$$\delta = 0$$

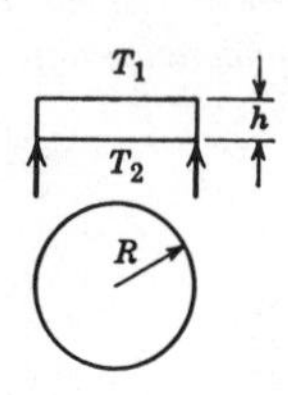

Circular Plate, Simply Supported

$$\sigma = 0$$

$$\delta = \frac{4\alpha(\Delta T)R^2}{8h}$$

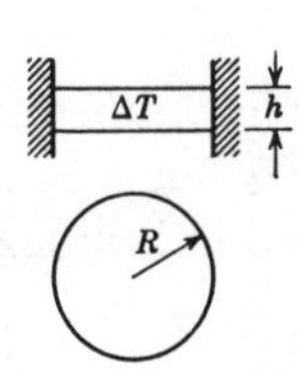

Circular Plate, Clamped at Edge

$$\sigma = \frac{-\alpha E(\Delta T)}{1-\nu}$$

$$\delta = 0$$

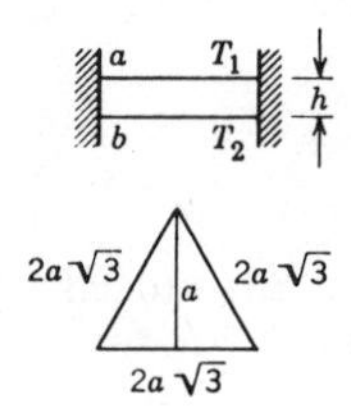

Triangular Plate, Clamped at Edge

$$(\sigma_{\max})_a = \frac{-3\alpha(\Delta T)E}{4} \text{ at corners}$$

$$(\sigma_{\max})_b = \frac{3\alpha(\Delta T)E}{4} \text{ at corners}$$

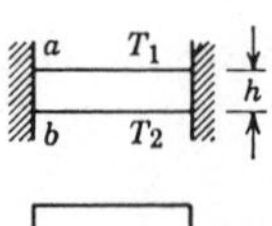

Square Plate, Clamped at Edge

$$(\sigma_{\max})_a = \frac{-\alpha E(\Delta T)}{2}$$

$$(\sigma_{\max})_b = \frac{\alpha E(\Delta T)}{2}$$

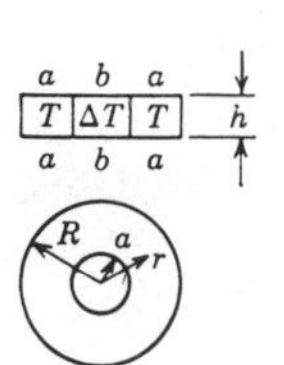

Circular Plate with Uniform Circular Area of ΔT at Center

$$(\sigma_r)_a = \frac{-\alpha E(\Delta T)}{2}\left(\frac{a}{r}\right)^2$$

$$(\sigma_h)_a = \frac{\alpha E(\Delta T)}{2}\left(\frac{a}{r}\right)^2$$

$$(\sigma_r)_b = (\sigma_h)_b = \frac{-\alpha E(\Delta T)}{2}$$

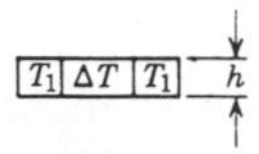

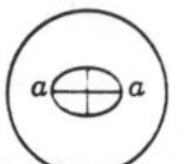

Circular Plate with Uniform Elliptical Area of ΔT at Center

$$(\sigma_h)\text{max} = (\sigma_h)_a = \frac{\alpha E(\Delta T)}{(1 + b/a)}$$

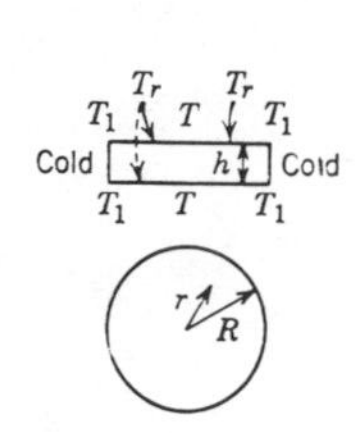

Circular Plate with Uniform Radial Temperature Change

$$(\sigma_r) = \alpha E\left[\frac{1}{R^2}\int_0^R (Tr - T_1)r\,dr - \frac{1}{r^2}\int_0^r (Tr - T_1)r\,dr\right]$$

$$(\sigma_h) = \alpha E\left[-(Tr - T_1) + \frac{1}{R^2}\int_0^R (Tr - T_1)r\,dr + \frac{1}{r^2}\int_0^r (Tr - T_1)r\,dr\right]$$

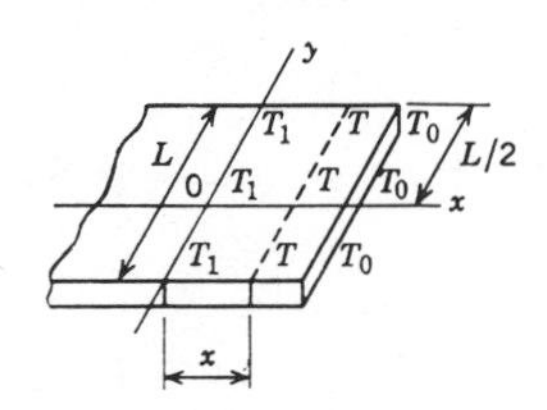

Rectangular Plate with Uniform Longitudinal Temperature Change

$$(\sigma_x)_{y=L/2} = E\alpha(T - T_0);\ \max x = 0$$

$$(\sigma_y)_{y=0} = -E\alpha(T - T_0)$$

$$(\sigma_y)_{y=L/2} = 0$$

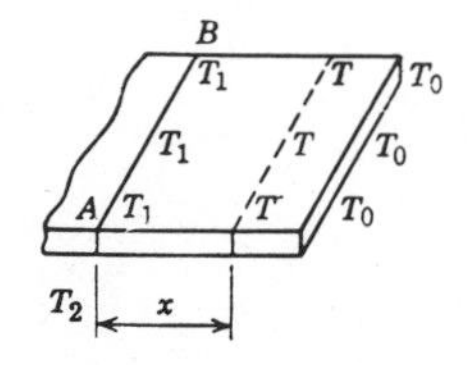

Rectangular Plate with Uniform Temperature Change Longitudinally and through Thickness

$$(\sigma_x)_{\max A,\,B,} = \frac{E\alpha}{2}\left[T_1 + T_2 - 2T_0 + \frac{1 - \nu}{(3 + \nu)}(T_1 - T_2)\right]$$

Formulas after Goodier, Reference 23, courtesy of The American Society of Mechanical Engineers, and Roark, Reference 49, courtesy of The McGraw-Hill Book Co.

The maximum stress for a rectangular element of unit width is

$$\sigma = \frac{6M}{t^2} = \frac{6\alpha(\Delta T)(1+\nu)D}{t^3} = \frac{\alpha(\Delta T)E}{2(1-\nu)} \tag{15.50}$$

Compression stresses are induced on the hot face and tension stresses on the cold face. Additional formulas for thermal stresses in various plates, taken largely from (23, 49), are shown in Table 15.5. An analysis of thermal stresses in perforated plates is given in (29).

Thermal stresses in thin cylindrical shells

Consider a long thin cylinder subjected to a uniform temperature change ΔT and restrained along the edge (Fig. 15.14). The built-in edges are analyzed in accordance with methods developed for beams on elastic foundations given in Chapter 3. For the present case the reactive moments and forces are determined from Eqs. 3.169 and 3.193 with the value of the deflection δ being given by the relationship

$$\delta = r\alpha(\Delta T) \tag{15.51}$$

Thus, the shear and moment equations become

$$S_0 = 4\delta\beta^3 D = 4r\alpha(\Delta T)\beta^3 D \tag{15.52}$$

and

$$M_0 = 2\delta\beta^2 D = 2r\alpha(\Delta T)\beta^2 D \tag{15.53}$$

Specific values of deflection and moment may then be computed for various positions along the length of the cylinder,

$$\delta = -\frac{1}{2\beta^3 D}[\beta M_0 \phi_2(\beta x) + S_0 \phi_3(\beta x)] \tag{15.54}$$

and

$$M = \frac{1}{2\beta^2 D}[2\beta M_0 \phi_4(\beta x) + S_0 \phi_1(\beta x)] \tag{15.55}$$

where ϕ values are given in Table 3.2 and

$$\beta = \sqrt[4]{Et/4r^2 D} = \sqrt[4]{[3(1-\nu^2)]/r^2 t^2}$$

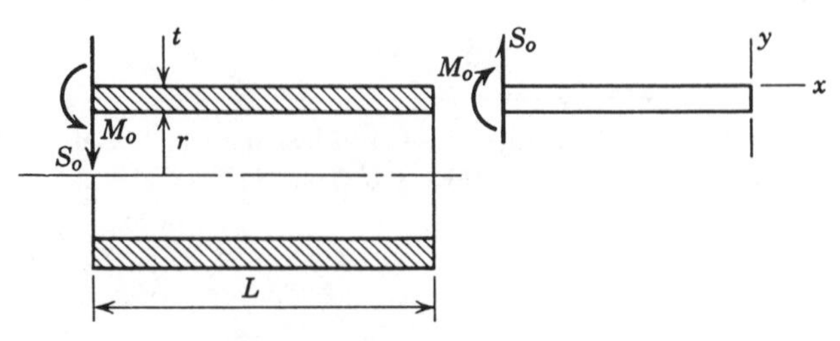

Fig. 15.14 Restrained thin-walled tube under thermal stress.

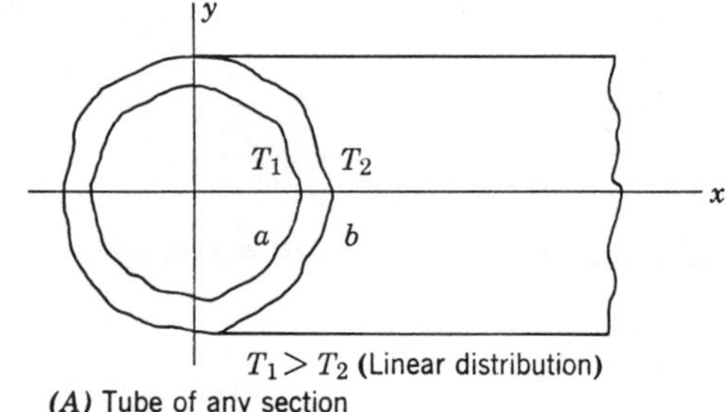

(A) Tube of any section

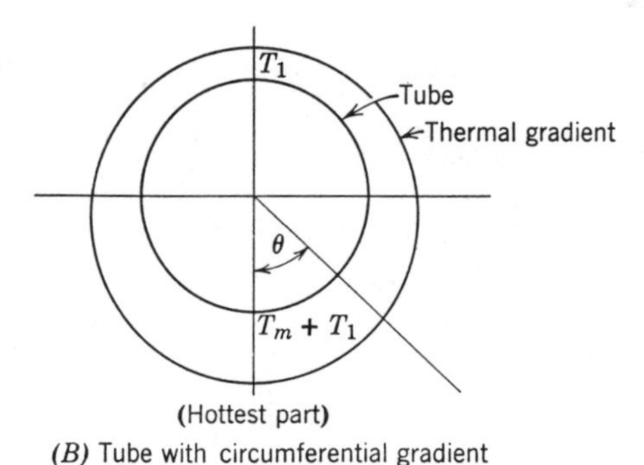

(B) Tube with circumferential gradient

Fig. 15.15 Thin-walled cylindrical tubes of uniform thickness subjected to thermal gradients.

Radial temperature gradient. Consider the tube in Fig. 15.15*A*, assumed to be cylindrical with a linear temperature gradient across the wall. Assuming the absence of any bending stresses, which would be at points far removed from the ends of the shell, the thermal stress is the same as in a restrained plate and is given by Eq. 15.50. Since T_1 is assumed greater than T_2, the outer surface will be in tension and the inner surface in compression, thus

$$\sigma_z = \sigma_h = \frac{E\alpha(\Delta T)}{2(1-\nu)} \qquad 15.56$$

As shown in Fig. 15.14 the longitudinal stress near the end of the cylinder must be balanced by a moment, thus

$$\sigma_z = -M\frac{c}{I} = -\frac{6M}{t^2}$$

or

$$M = -\frac{\sigma_z t^2}{6} = -\frac{E\alpha(\Delta T)t^2}{12(1-\nu)} \qquad 15.57$$

In order to obtain a free edge a moment equal to but opposite in sign to that given by Eq. 15.57 must be applied. The stresses obtained from such a moment are calculated as follows. First, at $x = 0$, the axial moment is the negative of Eq. 15.57 or

$$(M_z)_{x=0} = \frac{E\alpha(\Delta T)t^2}{12(1-\nu)}$$

In the hoop direction, from Eqs. 3.315 and 3.316,

$$(M_h)_{x=0} = \nu M_x$$

from which

$$(M_h)_{x=0} = \frac{\nu E\alpha(\Delta T)t^2}{12(1-\nu)} \quad 15.58$$

There is also a compressive force in the hoop direction given by the relationship

$$F_h = -A\sigma_h = -tE\varepsilon_h = -\frac{tE(\delta_h)_{x=0}}{r} \quad 15.59$$

The term $(\delta_h)_{x=0}$ is evaluated through use of Eq. 3.417; thus, since S_0 is zero,

$$(\delta_h)_{x=0} = \frac{e^{-\beta x}}{2\beta^3 D}[\beta M_0(\sin\beta x - \cos\beta x)]$$

or

$$(\delta_h)_{x=0} = \frac{1}{2\beta^3 D}(-\beta M_0) = -\frac{M_0}{2\beta^2 D} \quad 15.60$$

which when substituted into Eq. 15.59 gives

$$F_h = \frac{M_0 tE}{2\beta^2 Dr} = \frac{Et\alpha(\Delta T)\sqrt{1-\nu^2}}{2\sqrt{3}(1-\nu)} \quad 15.61$$

Therefore, from Eqs. 15.56, 15.58 and 15.61 the total thermal hoop stress maximum at edge is (see also Reference 62),

$$\sigma_h = \frac{E\alpha(\Delta T)}{2\sqrt{3}(1-\nu)}[\sqrt{3}(1-\nu) + \sqrt{1-\nu^2}] \quad 15.62$$

The stress given by Eq. 15.62 is 20 to 25% larger than the stress at points on the cylinder far removed from the end and explains why many thermal cracks start at the free end of tubes. If the temperature gradient is in the axial direction with the temperature being lowest at the ends of the cylinder, then (63)

$$(\sigma_z)_{\max} = \frac{6\beta D\alpha r}{2b}(T_0 - T_i) = \frac{6\beta D\alpha r(\Delta T)}{2b} \quad 15.63$$

where ΔT is positive and b is the distance from the end of the cylinder where $T = T_i$ to a location along the cylinder length where $T = T_0$. The value of σ_z calculated from Eq. 15.63 will always be on the high side except when b is large. The problem of thermal stresses in cylinders with both axial and radial heat flow has been solved by Weil (66) and Bijlaard (4*A*).

Circumferential temperature gradient. Refer to Fig. 15.15*B*. In this case the temperature varies around the circumference and through the wall of the tube but not along the length (23). For the case shown T_1 is the uniform inside temperature and T_m is the maximum temperature difference. Then the outside temperature is

$$T_2 = T_1 + \tfrac{1}{2}T_m(1 + \cos\theta) \qquad 15.64$$

At the hottest part of the tube the thermal stresses are as follows:

$$(\sigma_h)_{\max} = \pm\frac{E\alpha T_m}{2(1 - \nu)} \qquad 15.65$$

positive at the inside and negative at the outside.

$$(\sigma_z)_{\max} = \pm\frac{3E\alpha T_m}{4} \qquad 15.66$$

positive at the outside and negative at the inside.

Composite cylinders. For a uniform ΔT in a two-shell system the composite may be analyzed as a pair of shells with interface pressure. An analysis will now be presented that is more general and that will accommodate multishell constructions with both uniform ΔT and variable ΔT in each shell of the assembly. In general, the analysis follows that proposed by L. H. Abraham (1).

If a unit circle of material increases in temperature, it will undergo a free expansion of

$$\varepsilon_h = \alpha(\Delta T) \qquad 15.67$$

If, however, the unit circle of material is contained in a larger body of material that does not experience the same change in temperature, then free motion is restricted and, in a uniform ΔT the strain is

$$\varepsilon_h{}^1 = -\sigma_h/E \qquad 15.68$$

The surrounding material also expands so that the final dilation is

$$\varepsilon_h{}^1 = \alpha(\Delta T) + \sigma_h/E \qquad 15.69$$

or

$$\sigma_h = E[-\alpha(\Delta T) + \varepsilon_h{}^1] \qquad 15.70$$

If, however, there is a thermal gradient present, Eq. 15.68 becomes

$$\varepsilon_h{}^1 = (1/E)(\sigma_h - \nu\sigma_z) \qquad 15.71$$

or, since (as shown previously) $\sigma_h = \sigma_z$, Eq. 15.71 becomes

$$\varepsilon_h{}^1 = -\frac{\sigma}{E}(1 - \nu) \qquad 15.72$$

from which

$$\sigma = \frac{E}{1 - \nu}[-\alpha(\Delta T) + \varepsilon_h{}^1] \qquad 15.73$$

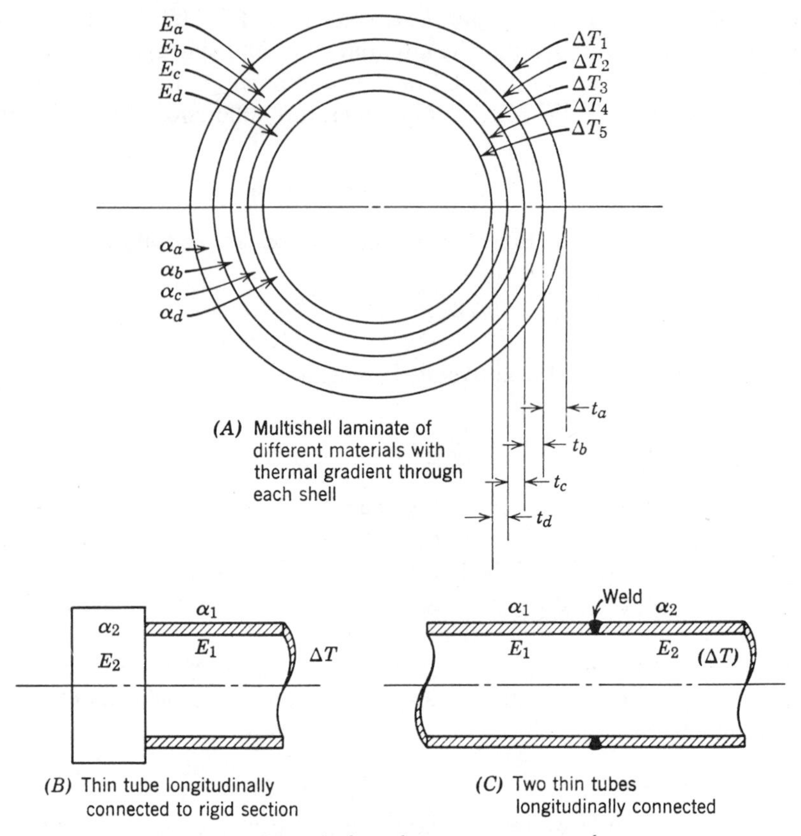

Fig. 15.16 Bimetal tubes subject to temperature change.

Consider a four-shell laminate consisting of different materials and a different thermal gradient through each shell (Fig. 15.16*A*). The variable temperature gradients are $\Delta T_1, \Delta T_2, \Delta T_3, \Delta T_4$, and ΔT_5, the thicknesses t_a, etc., the elastic moduli E_a, etc., and Poisson's ratio ν_a, etc. The average ΔT for each layer are $(\Delta T)_a$, $(\Delta T)_b$, $(\Delta T)_c$, and $(\Delta T)_d$, assumed to occur at midwall. Then, from Eq. 15.73,

$$(\sigma_a)_{\Delta T_1} = \frac{E_a}{1-\nu_a}[-\alpha_a(\Delta T_1) + \varepsilon_h{}^1] \tag{15.74}$$

$$(\sigma_a)_{\Delta T_2} = \frac{E_a}{1-\nu_a}[-\alpha_a(\Delta T_2) + \varepsilon_h{}^1] \tag{15.75}$$

$$(\sigma_b)_{\Delta T_2} = \frac{E_b}{1-\nu_b}[-\alpha_b(\Delta T_2) + \varepsilon_h{}^1] \tag{15.76}$$

$$(\sigma_b)_{\Delta T_3} = \frac{E_b}{1-\nu_b}[-\alpha_b(\Delta T_3) + \varepsilon_h{}^1] \tag{15.77}$$

$$(\sigma_c)_{\Delta T_3} = \frac{E_c}{1 - \nu_c}[-\alpha_c(\Delta T_3) + \varepsilon_h{}^1] \qquad 15.78$$

$$(\sigma_c)_{\Delta T_4} = \frac{E_c}{1 - \nu_c}[-\alpha_c(\Delta T_4) + \varepsilon_h{}^1] \qquad 15.79$$

$$(\sigma_d)_{\Delta T_4} = \frac{E_c}{1 - \nu_d}[-\alpha_d(\Delta T_4) + \varepsilon_h{}^1] \qquad 15.80$$

and

$$(\sigma_d)_{\Delta T_5} = \frac{E_d}{1 - \nu_d}[-\alpha_d(\Delta T_5) + \varepsilon_h{}^1] \qquad 15.81$$

The corresponding axial force in each layer is

$$F_a = \frac{(\sigma_a)_{\Delta T1} + (\sigma_a)_{\Delta T2}}{2} A_a \qquad 15.82$$

$$F_b = \frac{(\sigma_b)_{\Delta T2} + (\sigma_b)_{\Delta T3}}{2} A_b \qquad 15.83$$

$$F_c = \frac{(\sigma_c)_{\Delta T3} + (\sigma_c)_{\Delta T4}}{2} A_c \qquad 15.84$$

and

$$F_d = \frac{(\sigma_d)_{\Delta T4} + (\sigma_d)_{\Delta T5}}{2} A_d \qquad 15.85$$

However, for equilibrium the summation of axial forces must be zero, that is,

$$F_a + F_b + F_c + F_d = 0 \qquad 15.86$$

Thus, by substituting Eqs. 15.74 to 15.81 into Eqs. 15.82 to 15.85 and applying Eq. 15.86,

$$\varepsilon_h{}^1 = \frac{A_a E_a \alpha_a (\Delta T)_a + A_b E_b \alpha_b (\Delta T)_b \cdots}{A_a E_a + A_b E_b \cdots} \qquad 15.87$$

and the stresses are found by applying Eqs. 15.74 to 15.81. For example, consider a two-shell system as shown in Fig. 15.17, consisting of an inner cylinder of

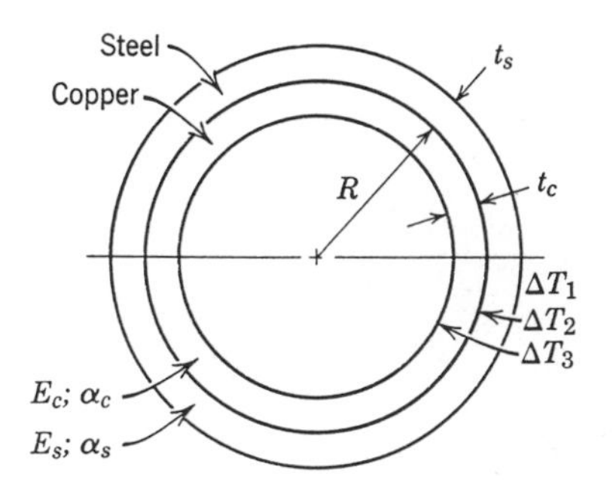

Fig. 15.17 Two-shell laminate with thermal gradient in each part.

copper and an outer cylinder of steel. From Eq. 15.87

$$\varepsilon_h^{\,1} = \frac{t_s E_s \alpha_s (\Delta T)_s + t_c E_c \alpha_c (\Delta T)_c}{t_s E_s + t_c E_c} \tag{15.88}$$

or

$$\varepsilon_h^{\,1} = \frac{t_s E_s \alpha_s \left(\dfrac{\Delta T_1 + \Delta T_2}{2}\right) + t_c E_c \alpha_c \left(\dfrac{\Delta T_2 + \Delta T_3}{2}\right)}{t_s E_s + t_c E_c} \tag{15.89}$$

where ΔT_1, ΔT_2, and ΔT_3 are temperature differentials at the boundaries. Then the stresses in the steel ring at the outside and inside surfaces are

$$(\sigma_s)_{\text{O.D.}} = \frac{E_s}{1 - v_s}[\varepsilon_h^{\,1} - \alpha_s (\Delta T)_1] \tag{15.90}$$

$$(\sigma_s)_{\text{I.D.}} = \frac{E_s}{1 - v_s}[\varepsilon_h^{\,1} - \alpha_s (\Delta T)_2] \tag{15.91}$$

Similarly, for the copper cylinder,

$$(\sigma_c)_{\text{O.D.}} = \frac{E_c}{1 - v_c}[\varepsilon_h^{\,1} - \alpha_c (\Delta T_2)] \tag{15.92}$$

and

$$(\sigma_c)_{\text{I.D}} = \frac{E_c}{1 - v_c}[\varepsilon_h^{\,1} - \alpha_c (\Delta T_3)] \tag{15.93}$$

Suppose that the following numerical data are given:

$$\Delta T_1 = 250°\text{C}$$
$$\Delta T_2 = 125°\text{C}$$
$$\Delta T_3 = 50°\text{C}$$
$$E_s = 30 \times 10^6 \text{ psi}$$
$$E_c = 15 \times 10^6 \text{ psi}$$
$$\alpha_s = 14.4 \times 10^{-6} \text{ in./in./°C}$$
$$\alpha_c = 17.8 \times 10^{-6} \text{ in./in./°C}$$
$$t_s = 0.10 \text{ in.}$$
$$t_c = 0.25 \text{ in.}$$
$$R = 10 \text{ in.}$$
$$v = 0.3 \text{ (for both parts)}$$

From Eq. 15.89,

$$\varepsilon_h^{\,1} = \frac{(0.10)(14.4 \times 10^{-6})(30 \times 10^6)\left(\dfrac{250 + 125}{2}\right) + (0.25)(17.8 \times 10^{-6})(15 \times 10^6) \times \left(\dfrac{125 + 50}{2}\right)}{(0.10)(30 \times 10^6) + (0.25)(15 \times 10^6)}$$

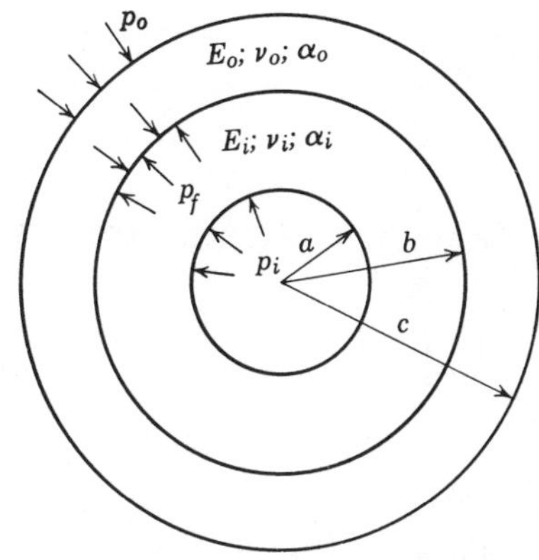

Fig. 15.18 Two-shell system subjected to surface and interface pressures.

or

$$\varepsilon_h^{\,1} = 0.0021 \text{ in./in.}$$

The change in radius R is then

$$\Delta R = \varepsilon_h^{\,1} R = 0.0021(10) = 0.021 \text{ in.}$$

From Eqs. 15.90 to 15.93 the stresses are as follows:

$$(\sigma_s)_{\text{O.D.}} = \frac{30 \times 10^6}{1 - 0.3}[0.0021 - 14.4 \times 10^{-6}(250)] = -64{,}300 \text{ psi}$$

$$(\sigma_s)_{\text{I.D.}} = \frac{30 \times 10^6}{1 - 0.3}[0.0021 - 14.4 \times 10^{-6}(125)] = +12{,}900 \text{ psi}$$

$$(\sigma_c)_{\text{O.D.}} = \frac{15 \times 10^6}{1 - 0.3}[0.0021 - 17.8 \times 10^{-6}(125)] = -2580 \text{ psi}$$

$$(\sigma_c)_{\text{I.D.}} = \frac{15 \times 10^6}{1 - 0.3}[0.0021 - 17.8 \times 10^{6}(50)] = +25{,}900 \text{ psi}$$

As another example suppose that the steel and copper shells in Fig. 15.17 are reversed and the assembly is heated to some temperature where the parts just slip together (Fig. 15.18).* If the assembly is now allowed to cool down, thermal stresses will be induced in both parts. It will be assumed in this example that there is no thermal gradient in the shells.

In this case the shells are contracting; therefore Eq. 15.67 becomes

$$\varepsilon_h = -\alpha(\Delta T) \qquad 15.94$$

Equation 15.68 becomes

$$\varepsilon_h^{\,1} = \sigma_h/E \qquad 15.95$$

Equation 15.69 becomes

$$\varepsilon_h^{\,1} = -\alpha(\Delta T) - \sigma_h/E \qquad 15.96$$

*With $p_i = p_0 = 0$.

and Eq. 15.70 becomes

$$\sigma_h = -E[\alpha(\Delta T) + \varepsilon_h{}^1] \tag{15.97}$$

Thus, applying Eqs. 15.89 and 15.96,

$$\varepsilon_h{}^1 = -\frac{t_s E_s \alpha_s(\Delta T) - t_c E_c \alpha_c(\Delta T)}{t_s E_s + t_c E_c} \tag{15.98}$$

or

$$\varepsilon_h{}^1 = -\frac{(\Delta T)[t_s E_s \alpha_s + t_c E_c \alpha_c]}{t_s E_s + t_c E_c} \tag{15.99}$$

Then from Eq. 15.97

$$\sigma_c = -E_c[\varepsilon_h{}^1 + \alpha_c(\Delta T)] = -E_c\left[\frac{-(\Delta T)\left(\alpha_c + \frac{t_s E_s}{t_c E_c}\alpha_s\right)}{1 + t_s E_s/t_c E_c} + \alpha_c(\Delta T)\right]$$

or

$$\sigma_c = -\frac{E_c(\Delta T)(\alpha_c - \alpha_s)}{1 + t_c E_c/t_s E_s} \tag{15.100}$$

where ΔT is negative (cooling); therefore the copper tube is in tension. By similar analysis it is easily shown that the steel tube is in compression.

Composite bimetallic shells may also be in forms as shown in Figs. 15.16*B* and 15.16*C*, where the different materials are connected longitudinally; both of these situations may be analyzed by reference to the theory of discontinuity stress discussed in Chapter 3. In Fig. 15.16*B* a thin shell of one material is longitudinally connected to a relatively heavy section of another material which may be assumed to be rigid. In this case, the differential expansion at the interface as a result of a change in temperature is $R(\Delta T)(\alpha_2 - \alpha_1)$ and Eq. 3.429 becomes $M = 2\delta\beta^2 D$, which, when substituted into Eq. 3.430, gives the result

$$\sigma = 1.82E_1(\Delta T)(\alpha_2 - \alpha_1)$$

where E_1 is the modulus of elasticity, at temperature, of the material for the thin part of the shell.

In Fig. 15.16*C* the two sections of shell are of the same order of thickness so that Eq. 3.429 can be used directly. Thus, following the same procedure as above, the maximum stress in the vicinity of the juncture of the two shells is

$$\sigma_{\alpha_1} = 0.293E_1(\Delta T)(\alpha_2 - \alpha_1)$$

and

$$\sigma_{\alpha_2} = 0.293E_2(\Delta T)(\alpha_2 - \alpha_1)$$

Thermal stresses in thick cylindrical shells

When a thick-walled cylindrical body is subjected to a thermal gradient, non-uniform deformation is induced and thermal stresses are developed. In the following discussion several cases will be considered.

Long hollow cylinder. Assuming that the ends of the cylinder are unrestrained, the longitudinal strain developed as a result of the stress is uniform and constant. The radial and hoop strains are given by Eqs. 3.455 and 3.456; the total strain is made up of a strain dependent on the induced stresses and a strain due to free thermal expansion; thus

$$\varepsilon_h = \frac{1}{E}[\sigma_h - \nu(\sigma_r + \sigma_z)] + \alpha(\Delta T) = \frac{u}{r} \qquad 15.101$$

$$\varepsilon_r = \frac{1}{E}[\sigma_r - \nu(\sigma_h + \sigma_z)] + \alpha(\Delta T) = \frac{du}{dr} \qquad 15.102$$

$$\varepsilon_z = \frac{1}{E}[\sigma_z - \nu(\sigma_h - \sigma_r)] + \alpha(\Delta T) = 0 \qquad 15.103$$

Also, from Chapter 3, equilibrium of an element of the cylinder is given by Eq. 3.453,

$$r\frac{d\sigma_r}{dr} + \sigma_r - \sigma_h = 0 \qquad 15.104$$

From Eqs. 15.101 and 15.102

$$\varepsilon_r = \frac{d}{dr}(r\varepsilon_h) \qquad 15.105$$

from which

$$r\frac{d}{dr}\varepsilon_h + \varepsilon_h - \varepsilon_r = 0 \qquad 15.106$$

Substituting Eqs. 15.101 and 15.102 into Eq. 15.106 and using Eq. 15.104 gives the result

$$\frac{d^2}{dr^2}\sigma_r + \frac{3}{r}\frac{d}{dr}\sigma_r = \frac{1}{r^3}\frac{d}{dr}\left(r^3\frac{d}{dr}\sigma_r\right) = \left(-\frac{\alpha E}{1-\nu}\right)\left(\frac{1}{r}\right)\left(\frac{dT}{dr}\right) \qquad 15.107$$

The solution of Eq. 15.107, with the boundary condition $\sigma_r = 0$ at the outside and inside surfaces, $r = r_0$ and $r = r_i$, for a hollow cylinder gives

$$\sigma_h = \frac{\alpha E}{1-\nu}\left(\frac{1}{r^2}\right)\left(\frac{r^2 + r_i^2}{r_0^2 - r_i^2}\int_{ri}^{ro} Tr\,dr + \int_{ri}^{r} Tr\,dr - Tr^2\right) \qquad 15.108$$

and

$$\sigma_r = \frac{\alpha E}{1-\nu}\left(\frac{1}{r^2}\right)\left(\frac{r^2 - r_i^2}{r_0^2 - r_i^2}\int_{ri}^{ro} Tr\,dr - \int_{ri}^{r} Tr\,dr\right) \qquad 15.109$$

There is also a longitudinal stress, which from Eq. 15.103, is

$$\sigma_z = \frac{\alpha E}{1-\nu}\left(\frac{2}{r_0^2 - r_i^2}\int_{ri}^{ro} Tr\,dr - T\right) \qquad 15.110$$

The solution of Eqs. 15.108, 15.109, and 15.110 depends on the thermal condition, that is, what the value of T is. In some instances a linear temperature distribution may be assumed; thus if the temperatures of the inner and outer surfaces are t_i and zero respectively, the temperature T at radius r is

$$T = t_i\left(\frac{r_0 - r}{r_0 - r_i}\right) \qquad 15.111$$

Substituting Eq. 15.111 into Eqs. 15.108 to 15.110 then gives

$$\sigma_h = \frac{E\alpha t_i}{3(1-\nu)(r_0 - r_i)}\left[2r + \frac{r_i^3}{r^2} - \left(1 + \frac{r_i^2}{r^2}\right)\left(\frac{r_0^3 - r_i^3}{r_0^2 - r_i^2}\right)\right] \qquad 15.112$$

$$\sigma_r = \frac{E\alpha t_i}{3(1-\nu)(r_0 - r_i)}\left[r - \frac{r_i^3}{r^2} - \left(1 - \frac{r_i^2}{r^2}\right)\left(\frac{r_0^3 - r_i^3}{r_0^2 - r_i^2}\right)\right] \qquad 15.113$$

and

$$\sigma_z = \frac{E\alpha t_i}{3(1-\nu)(r_0 - r_i)}\left[3r - 2\left(\frac{r_0^3 - r_i^3}{r_0^2 - r_i^2}\right)\right] \qquad 15.114$$

If $r_0/r_i = 1 + m$, the values of σ_h in Eq. 15.112 become

$$(\sigma_h)_{ri} = -\frac{E\alpha t_i}{2(1-\nu)}\left(1 + \frac{m}{6+3m}\right) \qquad 15.115$$

and

$$(\sigma_h)_{ro} = \frac{E\alpha t_i}{2(1-\nu)}\left(1 - \frac{m}{6+3m}\right) \qquad 15.116$$

For a thin-walled cylinder m is small and the hoop stress becomes

$$\sigma_h = \mp\frac{E\alpha t_i}{2(1-\nu)} = \mp\frac{E\alpha(\Delta T)}{2(1-\nu)} \qquad 15.117$$

where σ_h is negative at the inside surface, the same result as given earlier by Eq. 15.57. The value of σ_z from Eq. 15.114 also has the same value as given by Eq. 15.117.

The linear temperature distribution is unsuitable for thick tubes and a logarithmic distribution is usually assumed that has the form

$$T = \frac{t_i}{\ln(r_0/r_i)}\ln(r_0/r) \qquad 15.118$$

With this type of temperature distribution substituted into Eqs. 15.108 to 15.110, the following expressions are obtained:

$$\sigma_h = \frac{E\alpha t_i}{2(1-\nu)\ln(r_0/r_i)}\left[1 - \ln\frac{r_0}{r} - \frac{r_i^2}{r_0^2 - r_i^2}\left(1 + \frac{r_0^2}{r^2}\right)\ln\frac{r_0}{r_i}\right] \qquad 15.119$$

$$\sigma_r = \frac{E\alpha t_i}{2(1-\nu)\ln(r_0/r_i)}\left[-\ln\frac{r_0}{r} - \frac{r_i^2}{r_0^2 - r_i^2}\left(1 - \frac{r_0^2}{r^2}\right)\ln\frac{r_0}{r_i}\right] \qquad 15.120$$

and

$$\sigma_z = \frac{E\alpha t_i}{2(1-\nu)\ln(r_0/r_i)}\left(1 - 2\ln\frac{r_0}{r} - \frac{2r_i^2}{r_0^2 - r_i^2}\ln\frac{r_0}{r_i}\right) \qquad 15.121$$

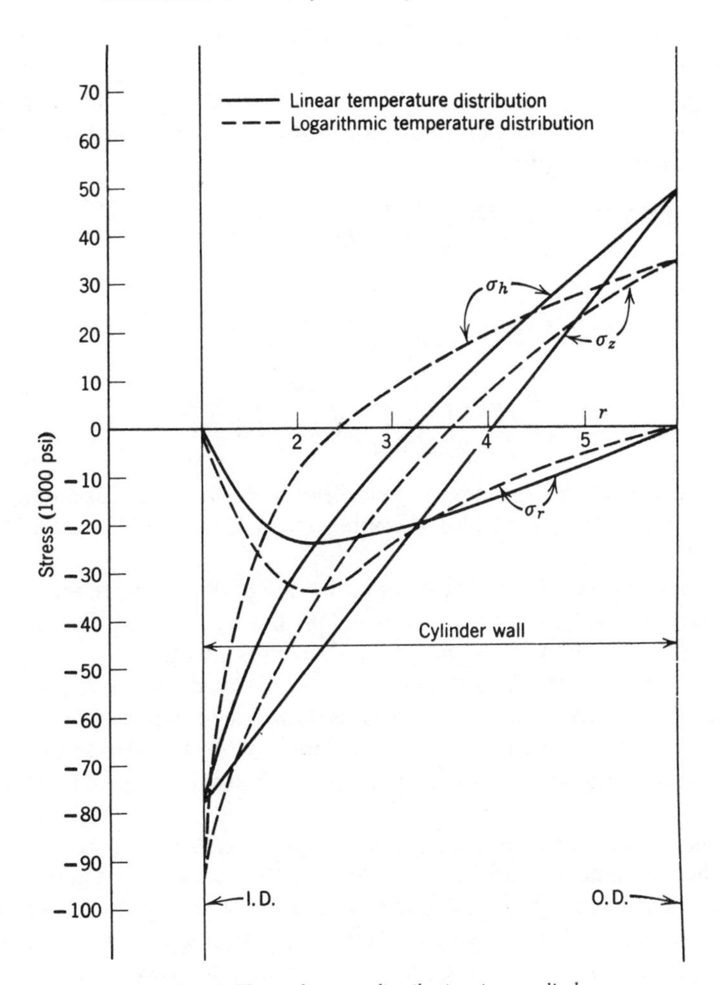

Fig. 15.19 Thermal stress distribution in a cylinder.

Again, for a thin cylinder Eqs. 15.119 and 15.121 reduce to Eq. 15.117.

As an example of the use of the preceding equations consider the following cylinder.

$$E = 30 \times 10^6 \text{ psi}$$
$$\alpha = 14.4 \times 10^{-6} \text{ in./in./}^\circ\text{C}$$
$$\nu = 0.3$$
$$r_0 = 6 \text{ in.}$$
$$r_i = 1 \text{ in.}$$
$$t_i = 200^\circ\text{C}$$
$$t_0 = 0^\circ\text{C}$$

The results are plotted in Fig. 15.19 for both a linear and logarithmic thermal gradient.

Solid cylinder. The solution of Eq. 15.107 with the boundary conditions $\sigma_r = 0$ at $r_0 = 0$ and zero deformation at $r = 0$, together with Eqs. 15.104 and 15.103 (with $\varepsilon_z = 0$), gives

$$\sigma_h = \frac{\alpha E}{1-\nu}\left(\frac{1}{r_0^{\,2}}\int_0^{r_0} Tr\,dr + \frac{1}{r^2}\int_0^{r} Tr\,dr - T\right) \qquad 15.122$$

$$\sigma_r = \frac{aE}{1-\nu}\left(\frac{1}{r_0^{\,2}}\int_0^{r_0} Tr\,dr - \frac{1}{r^2}\int_0^{r} Tr\,dr\right) \qquad 15.123$$

and

$$\sigma_z = \frac{\alpha E}{1-\nu}\left(\frac{2}{r_0^{\,2}}\int_0^{r_0} Tr\,dr - T\right) \qquad 15.124$$

Specific values of the stresses, as before, may be found by assuming a temperature distribution T so that the integrals in Eqs. 15.122 to 15.124 can be evaluated.

The application of thermal stresses in the design of nuclear reactors has been considered in some detail by Thompson (58) and by Freudenthal (18). Thompson, for example, considers the thermal stresses in cylinders resulting from flow of power where the density of the power generated has a given distribution. He indicates that to minimize stresses it is desirable to choose a material with a low $kE\alpha/(1-\nu)$ factor (k is the thermal conductivity), and also to use relatively small diameter tubes and to develop the power in a thin layer of material at the surface.

Cylinders with both temperature and pressure gradients. There are many cases where cylinders, in addition to having a thermal gradient across the wall, are also subject to internal and external pressure, which give rise to a pressure gradient across the wall.* Typical examples are in nuclear engineering and in the nozzle section of rockets. Consider the cylinder shown in Fig. 15.18 consisting of two shells of different materials subjected to internal and external pressure. Since the materials are different, an interface pressure develops at radius b (see Chapter 4) and the interface radial displacement is expressed by the relationship

$$p_i\frac{b}{E_i}\left(\frac{2a^2}{b^2-a^2}\right) - p_f\frac{b}{E_i}\left(\frac{b^2+a^2}{b^2-a^2} - \nu_i\right) + \frac{2\alpha_i b}{b^2-a^2}\int_a^b Tr\,dr$$
$$= p_f\frac{b}{E_0}\left(\frac{c^2+b^2}{c^2-b^2} + \nu_0\right) - \frac{p_0 b}{E_0}\left(\frac{2c^2}{c^2-b^2}\right) + \frac{2\alpha_0 b}{c^2-b^2}\int_b^c Tr\,dr \qquad 15.125$$

from which p_f may be solved. It is assumed that the conditions are such that a positive pressure p_f exists; if p_f is negative, then the cylinder parts have lost

*An associated problem is cyclic thermal stressing with an attendant "thermal ratchet effect" which is discussed at the end of this chapter in the section on thermal fatigue.

contact. Having the value of p_f it is a simple matter to calculate stress distributions in the shells. For the cylinder parts the thermal stress equations are given by Eqs. 15.108 to 15.110; stresses due to pressure are given by Eqs. 3.466 and 3.467; thus by combining stress terms it is seen that for the inner cylinder

$$\sigma_h = \frac{E_i \alpha_i}{1-\nu_i}\left(\frac{1}{r^2}\right)\left(\frac{r^2+a^2}{b^2-a^2}\int_a^b Tr\,dr + \int_a^r Tr\,dr - Tr^2\right) + \frac{a^2 p_i - b^2 p_f}{b^2-a^2} + \frac{a^2 b^2 (p_i - p_f)}{r^2(b^2-a^2)} \qquad 15.126$$

$$\sigma_r = \frac{E_i \alpha_i}{1-\nu_i}\left(\frac{1}{r^2}\right)\left(\frac{r^2-a^2}{b^2-a^2}\int_a^b Tr\,dr - \int_a^r Tr\,dr\right) + \frac{a^2 p_i - b^2 p_f}{b^2-a^2} - \frac{a^2 b^2 (p_i - p_f)}{r^2(b^2-a^2)} \qquad 15.127$$

$$\sigma_z = \frac{E_i \alpha_i}{1-\nu_i}\left(\frac{2}{b^2-a^2}\int_a^b Tr\,dr - T\right) \qquad 15.128$$

and for the outer cylinder

$$\sigma_h = \frac{E_0 \alpha_0}{1-\nu_0}\left(\frac{1}{r^2}\right)\left(\frac{r^2+b^2}{c^2-a^2}\int_b^c Tr\,dr + \int_b^r Tr\,dr - Tr^2\right) + \frac{b^2 p_f - c^2 p_0}{c^2-b^2} + \frac{b^2 c^2 (p_f - p_0)}{r^2(c^2-b^2)} \qquad 15.129$$

$$\sigma_r = \frac{E_0 \alpha_0}{1-\nu_0}\left(\frac{1}{r^2}\right)\left(\frac{r^2-b^2}{c^2-b^2}\int_b^c Tr\,dr - \int_b^r Tr\,dr\right) + \frac{b^2 p_f - c^2 p_0}{c^2-b^2} - \frac{b^2 c^2 (p_f - p_0)}{r^2(c^2-b^2)} \qquad 15.130$$

and

$$\sigma_z = \frac{E_0 \alpha_0}{1-\nu_0}\left(\frac{2}{c^2-b^2}\int_b^c Tr\,dr - T\right) \qquad 15.131$$

The corresponding radial displacements are found by use of the relationship

$$\text{radial displacement} = u = r\varepsilon_h \qquad 15.132$$

Therefore, for example, the interface deformation at $r = b$ is

$$u_b = \frac{p_i b}{E_i}\left(\frac{2a}{b^2-a^2}\right) - \frac{p_f b}{E_i}\left(\frac{b^2+a^2}{b^2-a^2} - \nu_i\right) + \frac{2ab}{b^2-a^2}\int_a^b Tr\,dr \qquad 15.133$$

The preceding problem has also been considered in some detail by Chang (8) and Gatewood (19). Thermal stresses in piping systems and tube sheets in heat exchangers are discussed in (55, 74–76). The elastic-plastic behavior of thick-walled cylinders subjected to internal and external pressure and temperature gradients is discussed in (5).

Thermal stresses in spherical shells

In the following discussion it is assumed that the temperature varies only through the wall of the sphere. First, a thick sphere will be considered and then from these results a thin-walled sphere will be considered.*

In solving this problem the same equations expressing equilibrium and compatability for a cylinder are utilized. In this symmetrical case the shell is subjected to a state of equal biaxial stress plus a radial stress; therefore, for a sphere, Eqs. 15.101 and 15.102 become

$$\varepsilon_h = \frac{1}{E}[\sigma_h - \nu(\sigma_r + \sigma_h)] + \alpha(\Delta T) = \frac{u}{r} \tag{15.134}$$

and

$$\varepsilon_r = \frac{1}{E}(\sigma_r - 2\nu\sigma_h) + \alpha(\Delta T) = \frac{du}{dr} \tag{15.135}$$

Substituting Eqs. 15.134 and 15.135 in Eq. 15.106 and using Eq. 15.104 the equation for radial stress becomes

$$\frac{d}{dr^2}\sigma_r + \frac{4}{r}\frac{d}{dr}\sigma_r = \frac{1}{r^4}\frac{d}{dr}\left(r^4\frac{d}{dr}\sigma_r\right) = -\frac{2\alpha E}{1-\nu}\left(\frac{1}{r}\right)\frac{dT}{dr} \tag{15.136}$$

For a hollow sphere, $\sigma_r = 0$ at the surfaces; therefore the solution of Eq. 15.136 with these boundary conditions gives

$$\sigma_h = \frac{2\alpha E}{1-\nu}\left[\frac{2r^3 + r_i^3}{2(r_o^3 - r_i^3)}\left(\frac{1}{r^3}\right)\int_{ri}^{ro} Tr^2\,dr + \frac{1}{2r^3}\int_{ri}^{r} Tr^2\,dr - \frac{T}{2}\right] \tag{15.137}$$

and

$$\sigma_r = \frac{2\alpha E}{1-\nu}\left[\frac{r^3 - r_i^3}{r_o^3 - r_i^3}\left(\frac{1}{r^3}\right)\int_{ri}^{ro} Tr^2\,dr - \frac{1}{r^3}\int_{ri}^{r} Tr^2\,dr\right] \tag{15.138}$$

For the sphere it is common to express T as a steady heat flow function

$$T = t_i\left(\frac{r_i}{r}\right)\left(\frac{r_o - r}{r_o - r_i}\right) \tag{15.139}$$

which when substituted into Eqs. 15.137 and 15.138 gives

$$\sigma_h = \frac{\alpha E t_i}{1-\nu}\left(\frac{r_o r_i}{r_o^3 - r_i^3}\right)\left(r_i + r_o - \frac{1}{2r}(r_o^2 + r_o r_i + r_i^2 + r_i^2) - \frac{r_i^2 r_o^2}{2r^3}\right) \tag{15.140}$$

and

$$\sigma_r = \frac{\alpha E t_i}{1-\nu}\left(\frac{r_o r_i}{r_o^3 - r_i^3}\right)\left(r_i + r_o - \frac{1}{r}(r_o^2 + r_o r_i + r_i^2) + \frac{r_i^2 r_o^2}{r^3}\right) \tag{15.141}$$

*The elastic-plastic behavior of thick-walled spheres subject to thermal gradients is discussed in Reference 33.

In the preceding equations σ_h reaches a maximum at the outside surface of the sphere, whereas the radial stress is maximum at $3r_0{}^2r_i{}^2$. Again, by letting $r_0/r_i = 1 + m$, as in the cylinder the hoop stresses are as follows:

$$(\sigma_h)_{ri} = -\frac{\alpha E t_i}{2(1-\nu)}\left(1 + \frac{2m}{3}\right) \qquad 15.142$$

and

$$(\sigma_h)_{ro} = \frac{\alpha E t_i}{2(1-\nu)}\left(1 - \frac{2m}{3}\right) \qquad 15.143$$

Thus, for a thin shell, m approaches zero and

$$\sigma_h = \mp \frac{\alpha E t_i}{2(1-\nu)} \qquad 15.144$$

where the stress is negative at the inside surface.

For a *solid sphere,* σ_r is zero at the outside surface and the deformation is zero at the center; therefore from Eq. 15.136

$$\sigma_h = \frac{\alpha E}{1-\nu}\left(\frac{2}{r_0{}^3}\int_0^{ro} Tr^2\,dr + \frac{1}{r^3}\int_0^{r} Tr^2\,dr - T\right) \qquad 15.145$$

and

$$\sigma_r = \frac{2\alpha E}{1-\nu}\left(\frac{1}{r_0{}^3}\int_0^{ro} Tr^2\,dr - \frac{1}{r^3}\int_0^{r} Tr^2\,dr\right) \qquad 15.146$$

Thermal stresses in structural shapes

The analysis of thermal stresses has been worked out by Brooks (78) for angles, T-, H-, and I-sections and channels.

Thermal stresses in a beam with thermal gradient

The problem considered here is a beam of rectangular cross section in which the temperature varies only through the thickness of the member (Fig. 15.20). Shear stresses are zero so that the equilibrium conditions expressed by Eqs. 3.76 and 3.77 are satisfied identically and the compatibility relationship, Eq. 3.85

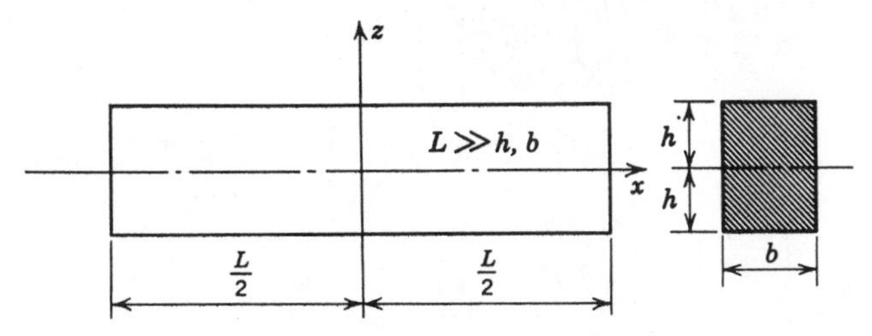

Fig. 15.20 Rectangular beam section.

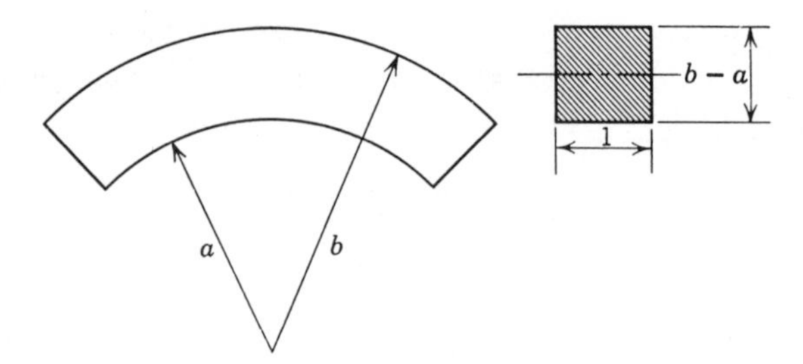

Fig. 15.21 Curved beam section.

becomes

$$\frac{\partial^2}{\partial z^2}(\sigma_x + \alpha ET) = 0 \qquad 15.147$$

the solution to which (with the boundary conditions that the summation of forces and moments are zero) is

$$\sigma_x = \alpha E\left(-T + \frac{1}{2h}\int_{-h}^{h} T\,dz + \frac{3z}{2h^3}\int_{-h}^{h} Tz\,dz\right) \qquad 15.148$$

and the beam deflects to the arc of a circle of radius $h/(\alpha\Delta T)$, where h is the depth of the beam.

If the beam is initially bent to the arc of a circle (Fig. 15.21), the hoop thermal stress is (7)

$$\sigma_h = -\frac{A_1}{r^2} + \frac{A_2}{a^2}\left(2\ln\frac{r}{a} + 3\right) + \frac{2A_3}{a^2} - E\alpha\left(T - \frac{1}{r^2}\int_a^r Tr\,dr\right) \qquad 15.149$$

where

$$A_1 = \frac{E\alpha}{N}\left\{\left[2\left(\frac{b}{a}\right)^2 \ln\frac{b}{a}\left(2\ln\frac{b}{a} - 1\right) + \left(\frac{b}{a}\right)^2 - 1\right]\int_a^b Tr\,dr - 4\left(\frac{b}{a}\right)^2 \ln\frac{b}{a}\int_a^b Tr\ln\frac{r}{a}\,dr\right\}$$

$$A_2 = \frac{E\alpha}{N}\left\{\left[2\left(\frac{b}{a}\right)^2 \ln\frac{b}{a} - \left(\frac{b}{a}\right)^2 + 1\right]\int_a^b Tr\,dr - 2\left[\left(\frac{b}{a}\right)^2 - 1\right]\int_a^b Tr\ln\frac{r}{a}\,dr\right\}$$

$$A_3 = -A_4 = \frac{E\alpha}{N}\left\{-2\left(\frac{b}{a}\right)^2\left(\ln\frac{b}{a}\right)^2\int_a^b Tr\,dr + \left[2\left(\frac{b}{a}\right)^2 \ln\frac{b}{a} + \left(\frac{b}{a}\right)^2 - 1\right]\int_a^b Tr\ln\frac{r}{a}\,dr\right\}$$

$$N = 4\left(\frac{b}{a}\right)^2\left(\ln\frac{b}{a}\right)^2 - \left[\left(\frac{b}{a}\right)^2 - 1\right]^2$$

Some results using the preceding equation are shown in Fig. 15.22.

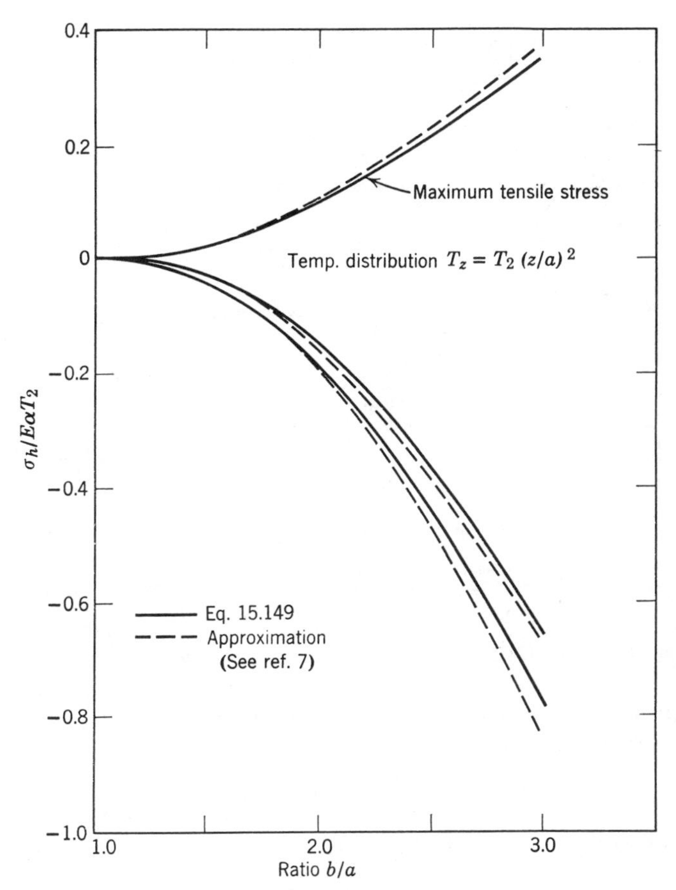

Fig. 15.22 Thermal stress distribution in a curved beam. [After Boley and Weiner (7), courtesy of John Wiley and Sons, Inc.]

Thermal buckling

The fundamental physical parameter in elastic buckling is the modulus of elasticity which usually decreases with increase in temperature as shown in Fig. 15.23. It is important then in any situation involving temperature or thermal gradients to control conditions and geometry so that buckling will not occur. In this section only buckling resulting from thermal stresses and not buckling from a simple reduction in the modulus because of increasing temperature is considered.

Restrained beam. A very simple example of buckling is shown by restrained beams of fairly long length subjected to a change in temperature. Such a beam is shown in Fig. 15.24. When the temperature is increased, there is a compressive stress induced in the member equal to

$$\sigma = E\alpha(\Delta T) \tag{15.150}$$

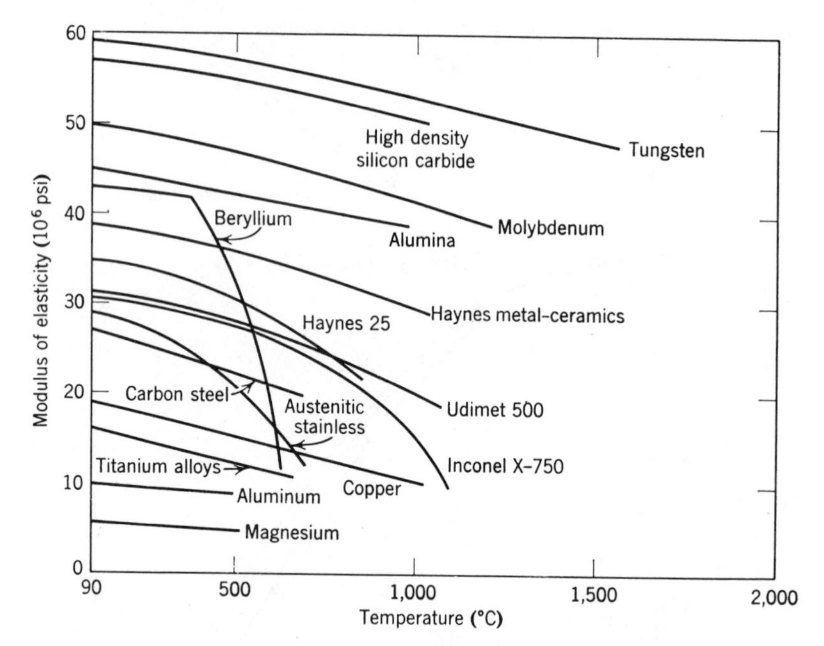

Fig. 15.23 Variation of modulus with temperature.

As a result of the temperature change the modulus E also changes to E'; then the equivalent end load on the beam is

$$P = A\sigma = AE'\alpha(\Delta T) \qquad 15.151$$

The critical load for buckling, in accordance with Eulers' theory, is given by Eq. 9.18,

$$P_{cr} = 4\pi^2 E' I / L^2 \qquad 15.152$$

Therefore from Eqs. 15.151 and 15.152 the ΔT for this particular case required to initiate buckling is

$$\Delta T = 4\pi^2 I / A\alpha L^2 \qquad 15.153$$

Thermostatic element. A bimetallic strip may buckle if it is heated excessively. The center deflection of a strip, Eq. 15.45, is

$$\delta = L^2/8r \qquad 15.154$$

and for a simple element the deflection picture would correspond roughly to

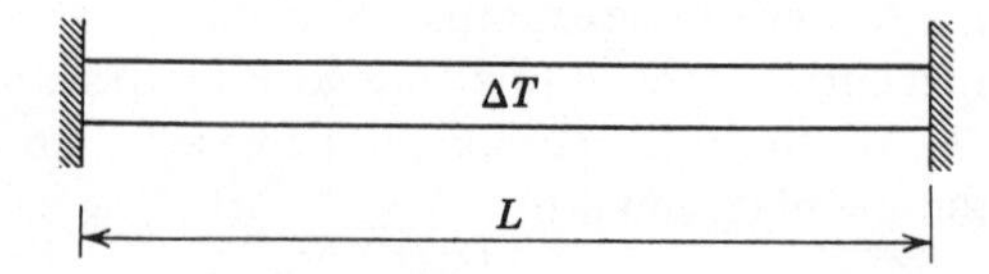

Fig. 15.24 Restrained beam subjected to a differential temperature.

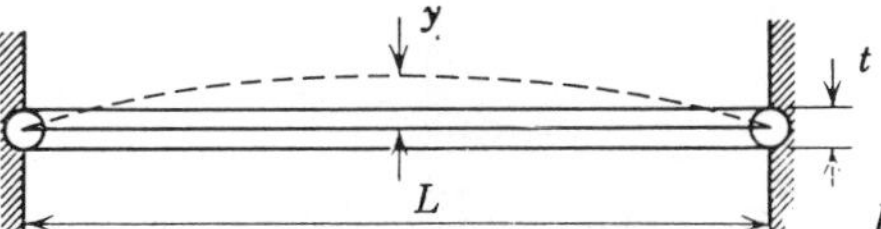

Fig. 15.25 Initial curvature of a bimetallic strip.

that shown in Fig. 15.25. For the purposes of the analysis here it is assumed that the strip has hinged ends. If the strip is initially bent to a slight curvature, a temperature can be reached that produces enough thrust to initiate a buckle. On cooling, the process can reverse and result in a second buckle. This buckling has been described (60) as similar to that occurring in the buckling of arches, the critical condition for which can be related to Eq. 15.154 as follows:

$$L^2/8ry = 1 + \sqrt{4(1-m)^3/27m^2} \tag{15.155}$$

where

$$m = 4I/Ay^2$$

y being the initial departure from flatness before heating. As a rough approximation to determine the temperature of buckling the expression for curvature given by Eq. 15.39, may be used; thus

$$3L^2(\Delta T)(\alpha_B - \alpha_A)/16yt = 1 + \sqrt{4(1-m)^3/27m^2} \tag{15.156}$$

or

$$\Delta T = (1 + C)(16yt)/[3L^2(\alpha_B - \alpha_A)] \tag{15.157}$$

where

$$C = \sqrt{4(1-m)^3/27m^2}$$

Equation 15.157 shows that as the initial curvature increases a larger ΔT is required to initiate an elastic thermal buckle.

Flat plates. Thermal buckling of plates has been described by Gossard (24) and Hoff (28). Gossard and his associates conducted experiments to test their theories and found good agreement between theory and experiment for deflections measured at the center of a plate subjected to "a tentlike" temperature distribution. The test specimen geometry is shown in Fig. 15.26, and results

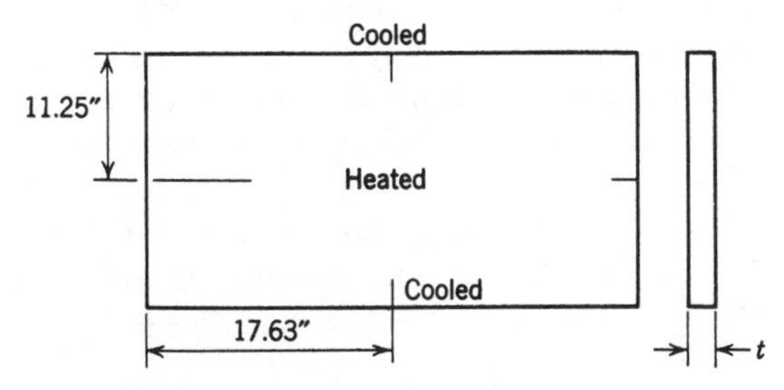

Fig. 15.26 Geometry of plate subjected to thermal buckling. [After Gossard, Seide, and Roberts (24), courtesy of The National Aeronautics and Space Administration.]

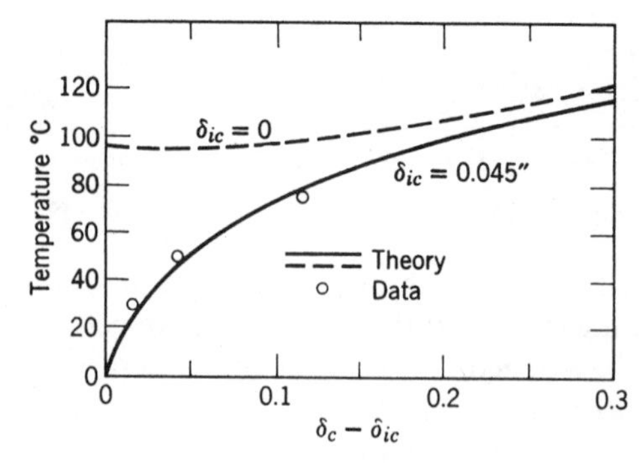

Fig. 15.27 Calculated and experimental deflections of plate in Fig. 15.26. [After Gossard, Seide, and Roberts (24), courtesy of The National Aeronautics and Space Administration.]

are plotted in Fig. 15.27. The theoretical curve is given by Gossard's equation

$$b^2 E\alpha T_0 t/\pi D^2 = 5.39(1 - \delta_{ic}/\delta_c) + 1.12(1 - \nu^2)(\delta_c^{\,2} - \delta_{ic}^{\,2})/t^2 \qquad 15.158$$

where $b = 11.25$ in.

$a = 17.63$ in.

$D = Et^3/[12(1 - \nu^2)]$

$a/b =$ aspect ratio $= 1.57$

$\nu =$ Poisson's Ratio $= 0.33$

$\alpha = 0.228 \times 10^{-4}$ in./in./°C

$t = 0.25$ in.

$\delta_{ic} =$ initial plate deflection at center $= 0.045$ in.

$\delta_c =$ center deflection of plate

$T_0 =$ temperature differential between center and edge of plate

The critical temperature T_c for buckling for the geometry and material involved is

$$T_c = 5.39\pi^2 D/b^2 E\alpha t \qquad 15.159$$

Creep buckling of columns and tubes. A structure may be designed to withstand ordinary static buckling by the methods described in Chapter 9; allowance for short-time temperature increases may also be allowed for in order that the structure retain its integrity. If, however, prolonged heating persists, the structure may be subjected to creep behavior (discussed later in this chapter) so that the dimensions change enough to induce instability. The time required to reach this stage is called the *critical time*, and the resulting collapse of the part is called *creep buckling*. For a pin-connected *H*-column, for example, it has been shown (7) that the critical creep buckling time is

$$t_{cr} = \frac{1}{24}\left(\frac{\pi h}{L}\right)^2\left(\frac{KA}{F_0}\right)^n \ln\left(1 + \frac{4}{a_0^{\,2}}\right) \qquad 15.160$$

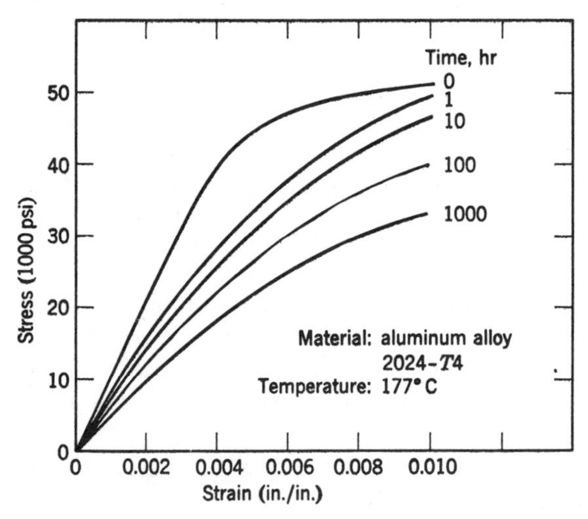

Fig. 15.28 Isochronous stress-strain curves. [After Goldin (22), courtesy of The Institute of Aeronautical Sciences.]

where K and n are defined by Eq. 15.197.

$$\frac{A}{2} = \text{area of one flange of section}$$

$h =$ separation of the flanges (from centroid of each part)

$F_0 =$ load

$a_0 =$ initial deflection at time $t = 0$

$L =$ length

For other than H-columns creep buckling analyses become extremely complicated, and solutions are not available. However, an approximation has been devised. In the approximate method isochronous stress-strain curves,* such as shown in Fig. 15.28, are used. At any given time the column in question is analyzed as an inelastic member with the stress-strain curve as given for the particular time as shown in Fig. 15.28. This procedure results in an overestimation of strain in a location where stress is increasing and underestimates the strain for decreasing stress.

Another approximate method is known as the *tangent-modulus* method and utilizes the tangent to an isochronous curve in a manner similar to the tangent

*At elevated temperature strain depends on time as well as on stress; therefore the usual stress-strain curves which are independent of time cannot be obtained at high temperature. To approximate the behavior, cross-plots of data obtained under particular stresses are used to obtain families of curves known as isochronous stress-strain curves. These are only "apparent" stress-strain curves, however, and must be used with caution. Such curves are questionable because of an implied existence of a mechanical equation of state.

modulus in plastic column theory (see Chapter 9). For example, starting with Eq. 9.90

$$F_{cr} = \pi^2 E_t I / L^2 \qquad 15.161$$

and dividing both sides by the cross-sectional area of the column

$$\frac{F_{cr}}{A} = \sigma_{cr} = \frac{\pi^2 E_t I}{L^2 A} = \left(\frac{\pi r}{L}\right)^2 \frac{d\sigma}{\sigma\varepsilon} \qquad 15.162$$

where r is the radius of gyration of the cross section and E_t is written as $d\sigma/d\varepsilon$. Next, creep data in tension are required so that the term $d\sigma/d\varepsilon$ can be evaluated and substituted into Eq. 15.162; knowing the geometry of the cross section, all terms in the equation are known and the critical buckling time corresponding to the particular stress can be found.

An analysis of creep buckling of thin cylindrical shells has been proposed by Wah (64); this analysis considers the collapse of cylindrical shells under steady-state creep and temperatures of the order of 250°C. The final result of the analyses is an equation defining the critical time for failure

$$t_{cr} = C\left(\frac{p_{cr}}{p} - 1\right)\left(\frac{\log \coth 3 \frac{\bar{\sigma}_0}{\sigma_0} f}{\cosh \sqrt{3}(\bar{\sigma}_0/\sigma_0)}\right) \qquad 15.163$$

where $C = \dfrac{2.303(1+v)\sigma_0 e^{\Delta H/RT}}{3EcT}$

p_{cr} = elastic external buckling pressure $= \dfrac{n^2 E h^3}{12a^3(1+v)}$

p = applied pressure
$\bar{\sigma}_0 = p(a/b)$
σ_0 = constant in creep law
a = mean radius of cylinder
b = wall thickness
n = number of lobes in buckling
c = constant
T = temperature
$f_0 = \dfrac{w}{b(1 - p/p_{cr})}$
w = initial deflection before load application
e = base of natural logarithms
ΔH = activation energy
R = gas constant.

Table 15.6 Results of Wah on Creep Buckling of Aluminum Tubes (64)

Cylinder length, in.	a, in.	b, in.	w, in.	Temperature, °C	p, psi	p_{cr}, psi	n	t_{cr}, Hour Theory	Exp.
10	2.9966	0.08025	0.0025	260	196	422.5	5	0.73	0.23
12	2.9966	0.08139	0.0136	260	130	263	4	0.23	0.00
13	2.9984	0.08140	0.0026	246	194	334.5	4	1.03	2.90
18	2.9958	0.08158	0.0121	246	70.4	164	3	31	28.3
25	2.9914	0.08121	0.0023	218	138	178.5	3	11.7	13.0
48	2.9916	0.08113	0.0085	204	53.9	58.4	2	0.53	1.2

Data for 6061-T6 Aluminum

Temperature, °C	E, psi	C
204	9.00×10^6	10,826
218	8.81×10^6	3,403
246	8.60×10^6	1,140
260	8.00×10^6	155.6

$\sigma_0 = 2{,}500$ psi
$c = 28.7 \times 10^6$/hr/°K
$\Delta H = 39{,}000$ cal/mol
$R = 2$ cal/mol/°K
ν = Poisson ratio = 0.30

Wah applied this method to 6 in. diameter 6061-T6 aluminum tubes and obtained the results shown in Table 15.6.

Thermal stress in sandwich structures

Consider the sandwich structure shown in Fig. 15.29. When a thermal differential is applied, strain compatability requires that

$$\varepsilon = \frac{\sigma_1}{E_1} + \alpha(\Delta T_1) = \frac{\sigma_2}{E_2} + \alpha(\Delta T_2) \qquad 15.164$$

Equilibrium requires that the force summation be zero, that is,

$$\sigma_1 t_1 + \sigma_2 t_2 = 0 \qquad 15.165$$

It is assumed that the core material sustains no load; therefore, if the facing material has a modulus E, Eqs. 15.164 and 15.165 show that

$$\sigma_1 = -\frac{E\alpha(\Delta T)}{1 + t_1/t_2} = -K_1 E\alpha(\Delta T) \qquad 15.166$$

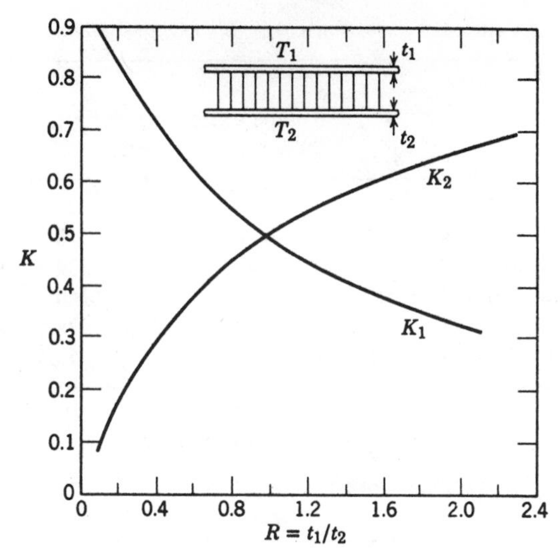

Fig. 15.29 Thermal stresses in a sandwich structure. [After Saelman (51), courtesy of the Rogers Publishing Co.]

and

$$\sigma_2 = \frac{E\alpha(\Delta T)}{1 + t_2/t_1} = K_2 E\alpha(\Delta T) \qquad 15.167$$

where

$$(\Delta T) = T_1 - T_2 \qquad (T_1 \text{ assumed} > T_2)$$

$$K_1 = \frac{1}{1+R}; \qquad K_2 = \frac{1}{1+1/R}$$

$R = t_1/t_2$
t = temperature
T = thickness of facing
α = coefficient of thermal expansion

The plot in Fig. 15.29 facilitates solution of the above stress equations. This problem has also been considered in some detail by Kuraniski (37).

Piping flexibility

The following example is presented to illustrate some of the calculations involved in analyzing the stresses in process piping operating at elevated temperature. The system is shown in Fig. 7.32*B*; a length of pipe containing a loop connects two pieces of equipment. The connection at *A* is to a spherical shell head and at *G* the connection is to a cylinder. Both the shell and pipe stresses will be examined.

Although most piping problems are solved by various computer programs, this example utilizes the theory outlined in Chapter 7. From Eq. 7.24, considering only bending.

$$U = \int_0^H \frac{M_{AC}{}^2\,dy}{2EI} + \int_0^\pi \frac{M_{CE}{}^2 R\,d\theta}{2(1/K)EI} + \int_0^L \frac{M_{EG}{}^2\,dy}{2EI}$$

where K, a factor defining the flexibility of a curved pipe was given in Chapter 3 as

$$K = 1 - \frac{9}{10 + 12(tR/a^2)^2} = \frac{1 + 12\lambda^2}{10 + 12\lambda^2}$$

and

$$\lambda = tR/a^2$$

R is the radius of curvature of the curved section, t is the wall thickness of the pipe, and a is the midwall radius.

At location A, the rotation is zero; therefore, from Eq. 7.55,

$$0 = \partial U/\partial M_A$$

or

$$0 = \int_0^H \frac{M_{AC}}{EI}\frac{\partial M_{AC}}{\partial M_A}\,dy + \int_0^\pi \frac{M_{CE}}{\left(\frac{1}{K}\right)EI}\frac{\partial M_{CE}}{\partial M_A}R\,d\theta + \int_0^L \frac{M_{EG}}{EI}\frac{\partial M_{EG}}{\partial M_A}\,dy$$

But

$$M_{AC} = M_A - F_x y$$
$$M_{CE} = M_A - F_x(H + R\sin\theta) + F_y R(1 - \cos\theta)$$
$$M_{EG} = M_A - F_x(H - y) + 2F_y R$$

$$\frac{\partial M_{AC}}{\partial M_A} = \frac{\partial M_{CE}}{\partial M_A} = \frac{\partial M_{EG}}{\partial M_A} = 1$$

From the foregoing relationships,

$$M_A = \frac{F_x\left(\frac{H^2}{2} + \frac{\pi HR}{1/K} + \frac{2R^2}{1/K} + HL - \frac{L^2}{2}\right) - F_y\left(\frac{\pi R^2}{1/K} + 2RL\right)}{H + \frac{\pi R}{1/K} + L}$$

Similarly, from Eq. 7.53,

$$\delta_x = \partial U/\partial F_x \quad \text{and} \quad \delta_y = \partial U/\partial F_y$$

Therefore

$$\delta_y = \frac{1}{EI}\left[M_A\left(\frac{\pi R^2}{1/K} + 2LR\right) - F_x\left(\frac{\pi HR^2}{1/K} + \frac{2R^3}{1/K} + 2HLR - L^2R\right)\right.$$
$$\left. + F_y\left(\frac{3\pi R^3}{2/K} + 4R^2L\right)\right]$$

and

$$\delta_x = \frac{1}{EI}\left[-M_A\left(\frac{H^2}{2} + \frac{\pi RH}{1/K} + HL + \frac{2R^2}{1/K} - \frac{L^2}{2}\right) + F_x\left(\frac{H^3}{3} + \frac{\pi RH^2}{1/K} + \frac{4R^2H}{1/K} + \frac{\pi R^3}{2/K} + H^2L - HL^2 + \frac{L^3}{3}\right) - F_y\left(\frac{\pi HR^2}{1/K} + \frac{2R^3}{1/K} + 2RHL - RL^2\right)\right]$$

The expansions δ_x and δ_y are the deformations that the piping system has to absorb. Consider the following conditions:

$H = 100$ ft

$L = 40$ ft

$R = 10$ ft

Pipe O.D. is 10.75 in. (stainless steel)
Pipe wall thickness is 0.25 in.
Moment of inertia of pipe is 113.7 in.^{4}
Operating temperature is 270°F
Modulus of steel at 270°F is 26.89×10^6 psi
Modulus of steel at 70°F is 27.40×10^6 psi
Coefficient of thermal expansion is 9.4×10^{-6} in./in./°F
K is 1.535 (value obtained from pipe code, $1.65a^2/tR$). From formula in this example K is 1.59)
Radius of spherical and cylindrical shells is 50 in.
Wall thickness of shells is 1.0 in.

Using the foregoing data and equations,

$$\delta_x = x\alpha\Delta T = (9.4 \times 10^{-6})(200)(20)(12) = 0.452 \text{ in.}$$
$$\delta_y = y\alpha\Delta T = (9.4 \times 10^{-6})(200)(60)(12) = 1.355 \text{ in.}$$
$$EI = (27.4 \times 10^6)(113.7) = 3.1 \times 10^6 \text{ lb-in.}^2$$
$$(F_x)_A = 32 \text{ lb}$$
$$(F_y)_A = -223 \text{ lb}$$
$$M_A = 8897 \text{ in.-lb}$$
$$(F_x)_G = -32 \text{ lb}$$
$$(F_y)_G = 223 \text{ lb}$$
$$M_G = -39{,}423 \text{ in.-lb}$$

The pipe stress can now be computed. For the curved portion, as shown in Chapter 3, the stress is greater than for straight pipe and a stress intensification factor β was defined. The pipe code now defines this factor as

$$\beta = \frac{0.9}{(tR/a^2)^{2/3}}$$

which for this example becomes 1.0 since a lesser value is never used. Therefore, for each section, applying the bending stress formula

$$\sigma = M/Z$$

it is found that the maximum stress occurs at location G and is 1863 psi, which is well below the allowable stress for the pipe.

It is now necessary to examine the effect of the piping reactions on the vessels to which the pipe is attached. In the example it was assumed that the pipe was rigidly attached to the vessels and that the vessels themselves do not distort as a result of temperature. Computation of shell reactions is complicated, and for exact results recourse to formulas published by Bijlaard (4B) is required. In the Bijlaard analysis the shell stresses due to piping reactions are given by the equation

$$\sigma = \frac{N}{t} \pm \frac{6M}{t^2}$$

where N is a membrane force and M is a bending moment. For a variety of cases Bijlaard provides plots enabling the quick determination of the quantities in the foregoing equation. However, because of the number of directions and reactions involved all of Bijlaard's plots cannot be reproduced here, so that it is convenient to use an approximate theory. Such a theory, based on the theory of beams on elastic foundations (see Chapter 3), has been developed by the Kellogg Co. (76). The Kellogg formula, after rearrangement to collect terms, is as follows:

$$\sigma = \frac{0.372}{tR_n}\left(\frac{R}{t}\right)^{1/2}\left(\frac{M}{R_n} + 0.75P\right)$$

where σ is the shell stress for spherical or cylindrical shells, R_n is the mean radius of the pipe connection, t is the shell thickness, R is the inside radius of the shell, M is the applied bending moment (for cylindrical shells the moment has to be in the longitudinal direction in order to apply the Kellogg formula), and P is the radial thrust. If P is directed toward the shell, the resulting shell stress is negative. Thus, utilizing the data given in the example and the Kellogg formula,

$$\sigma_{\text{sphere}} = \frac{0.372}{5.25}\left(\frac{50}{1.0}\right)^{1/2}\left[\frac{8897}{5.25} + 0.75(223)\right]$$
$$= 933 \text{ psi} \quad \text{(Compression)}$$

$$\sigma_{\text{cy linder}} = \frac{0.372}{5.25}\left(\frac{50}{1.0}\right)^{1/2}\left[\frac{39{,}423}{5.25} + 0.75(223)\right]$$
$$= 3845 \text{ psi} \quad \text{(Tension)}$$

In the foregoing example stress concentration effects at the juncture of pipe and vessels have not been included.

15-3 TIME-DEPENDENT THERMOMECHANICS

Thermal shock

The subject of thermal shock and thermal shock parameters, together with descriptions of test apparatus and methods of evaluation, has been reviewed by Manson (43). Some of the highlights of the analytical procedure will be considered here. If, in any solid body, the surface is suddenly subjected to a change in temperature ΔT, the surface is stressed an amount $(\Delta T \alpha E)/(1 - \nu)$. Thermal shock stresses are greater than those obtained under normal temperature gradients because the gradients are much steeper. Brittle materials often fail in the first cycle of a thermal shock test, whereas ductile materials usually stand up for several cycles. In any event, failure is eventually due to thermal fatigue. Shock tests are conveniently performed on flat plates, and for this reason many of the analyses of this phenomenon are based on such specimens.

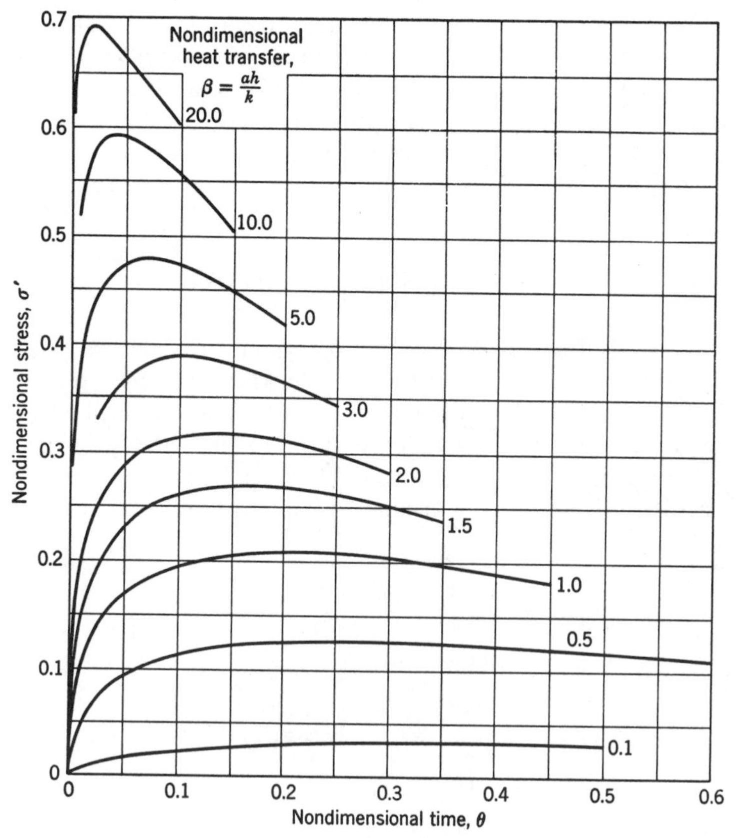

Fig. 15.30 Stress versus time as a function of Biot's modulus. [After Manson (43), courtesy of the Penton Publishing Co.]

Temperature distribution. First, it is necessary to know what the temperature distribution is at a time T after the surrounding temperature has been changed; knowing this the stresses are calculated using the equations in the theory of elasticity. For simplicity in the present discussion it is assumed that the stress at any point in the thickness of a plate can be described by the relation

$$\sigma' = \frac{T_{avg} - T}{T_0} \tag{15.168}$$

where σ', the ratio of the stress developed to the stress that would be developed with no dilation, is

$$\sigma' = \frac{\sigma(1 - \nu)}{E\alpha T_0} \tag{15.169}$$

Surfaces stresses are obtained by determining the average temperature and how the surface temperature varies with time. The results plotted in Fig. 15.30 show the essentials of the entire solution of surface stress in the flat plate problem. In addition to σ', the *reduced stress*, there are two additional important variables, β and θ, which are defined as

$$\beta = \text{Biot's modulus} = ah/k \tag{15.170}$$

and

$$\theta = \text{nondimensional time} = kt/\rho c a^2 \tag{15.171}$$

where a = plate half-thickness

h = heat transfer coefficient (amount of heat transferred from a unit area of the plate surface per unit temperature difference between the plate and its surroundings)

k = thermal conductivity

t = time

ρ = density

c = specific heat

Stress equations. The maximum nondimensional stress at the surface for each value of β, from Fig. 15.30, is plotted versus β in Fig. 15.31, which is adequate for most practical situations. Use can now be made of the approximate formulas to correlate the maximum stress developed in a material with the physical and mechanical properties of the material.

Small values of Biot's modulus. In many cases Biot's modulus is low and for this case (43),

$$T_0 = \left(\frac{3.25k\sigma_{max}}{E\alpha}\right)\left(\frac{1 - \nu}{ah}\right) \tag{15.172}$$

At failure it is assumed that σ_{max} is equal to the breaking stress in bending σ_b, so that

$$(T_0)_{max} = \left(\frac{3.25k\sigma_b}{E\alpha}\right)\left(\frac{1 - \nu}{ah}\right) \tag{15.173}$$

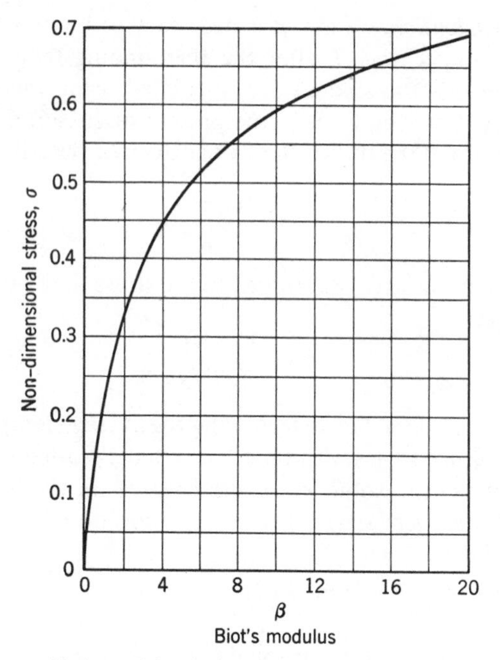

Fig. 15.31 Stress versus Biot's modulus. [After Manson, (43), courtesy of the Penton Publishing Co.]

where T_0 = the initial uniform temperature of the plate above ambient temperature

$(T_0)_{\max}$ = the maximum shock temperature

$k\sigma_b/E\alpha$ = the thermal shock parameter

Equation 15.173 gives a numerical measure of the shock temperature that will cause failure and thus provides a means for rating materials (see Table 15.7).

Large values of Biot's modulus. For large values of Biot's modulus,

$$(T_0)_{\max} = \frac{\sigma_b(1-\nu)}{E\alpha} \tag{15.174}$$

which results when $(\sigma')_{\max}$ approaches unity, corresponding to a large a value, a large h value, or a small k. If a is large, the surface can be brought to the temperature of the surroundings before bulk changes occur in the body and contraction is eliminated. If h is large, the same is true, and if k is small, only the surface layers can realize the imposed shock, and restraint is induced. These factors emphasize the critical nature of the test conditions, particularly of the necessity to make tests on specimens geometrically similar in size and shape to the structure being analyzed. Depending on test conditions, the results can be completely reversed and false evaluations made.

*Table 15.7 Correlation of Material Properties with Resistance to Fracture by Thermal Shock**

Material (by order of merit)	Thermal Shock Cycles before Failure for Upper Temperature (°C)				Coefficient of Thermal Expansion, in./in./°C ($\times 10^{-6}$)	Modulus Elasticity at 1000°C (psi) ($\times 10^{7}$)	Ultimate Strength at 1000°C (psi)	Thermal Shock Parameter
	1000	1100	1200	1300				
80 TiC, 20 Co	25	25	25	25	3.04	6	35,000	25,000
TiC	25	25	25	17	2.53	6	17,000	15,000
BeO	25	3			2.83	4.3	6,000	3,000
$ZrSiO_4$	1				1.39	2.4	9,000	2,000
MgO	1/2				4.26	1.2	3,000	900
94 ZrO_2, 6 CaO	0				3.06	2.5	7,000	700

* Data compiled by Manson (43) from tests of Bobrowsky ("The Applicability of Ceramics and Ceramals as Turbine Blade Materials for the Newer Aircraft Power Plants," *Transactions of the American Society of Mechanical Engineers*, **71,** No. 6, 621–629, August 1949). See also Reference 9 and *Missiles and Rockets*, **10** (18), 35, April 30 (1962) for discussion of thermantic structures which combine the high thermal shock resistance of porous ceramics and the structural reliability of high alloy metals.

Time-dependent rigidity

In a beam, for example, rigidity can be increased simply by increasing the moment of inertia. Another important physical parameter, however, influencing rigidity is the modulus, and this varies with temperature as illustrated in Fig. 15.23. Data reported in Fig. 15.23 imply relatively short times of loading so that Hooke's law can be applied. If loads are imposed for long lengths of time, the data would only be indicative. Better results are obtained by using *corrected* figures provided by *isochronous* plots, an example of which is shown in Fig. 15.28. Thus the modulus for aluminum at about 180°C is indicated to be approximately 9.5×10^5 psi from Fig. 15.23. However, from Fig. 15.28 the same material at this temperature for 1000 hr has a modulus of 5.7×10^5 psi, or a drop in modulus of 40%.

Analysis of creep behavior

Creep refers to the continuous deformation of a material as a function of time. Basically, creep behavior of a material should be analyzed in terms of *true stress* and *true deformation* (see Chapter 1); however, for many practical applications when deformations exceeding a percent or so are not tolerated and when design is based on a low limiting creep deformation, use can be made of the conventional condition of constant load on a part rather than a constant stress. It must be kept in mind, however, that this compromise is made in the interest

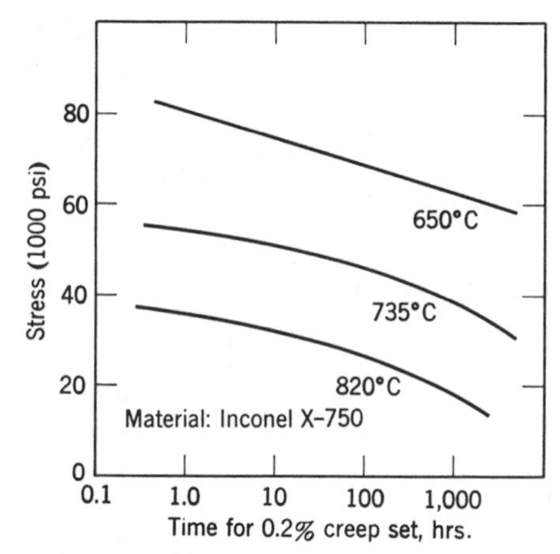

Fig. 15.32 Creep stress-deformation data. [After Goldin (22), courtesy of The Institute of Aeronautical Sciences.]

of practicality and that figures calculated are indications or trends and not exact representations of the behavior of the material. On the other hand, in large turbines, the creep involved over a period of years is considered undesirable (sometimes even dangerous) at deformations that do not materially deviate from a constant stress condition. Often 0.20% creep is allowed and design curves such as shown in Fig. 15.32 are used; naturally, because of the variation in modulus among different materials, 0.20% deformation due to creep results in different total deformations for, for example, steel and aluminum, since steel has three times the modulus of aluminum.

The mechanism of creep behavior, like that of all other physical processes, is constantly under examination. As time goes on and better instruments and techniques are made available to probe the micro- and atomic structures of materials, more lucid explanations of why materials behave as they do can be expected. Early attempts to rationalize creep behavior depicted a race between yielding of the material and strain hardening caused by the yielding at temperatures below the lowest temperature of recrystallization. At some particular stress, the applied load balanced the resistance to flow and creep stopped. However, if the temperature was raised above the lowest temperature for recrystallization, strain hardening was impeded and flow could occur at very low stress levels. These observed effects were still not explained, however, until the advent of the *theory of dislocations*. This subject is treated completely in Finnie and Heller's treatise on the creep of engineering materials (16).

Creep in tension. A typical plot of tension creep data is shown in Fig. 15.33. When a specimen is stressed, it first behaves elastically, analogous to ordinary stress-strain behavior; this provides an initial deformation before creep actually starts. The first stage of creep is usually characterized by a decreasing creep rate, the second stage by a steady-state rate, and the third stage by an increasing creep rate. Figure 15.34 illustrates diagrammatically the type of plot constructed

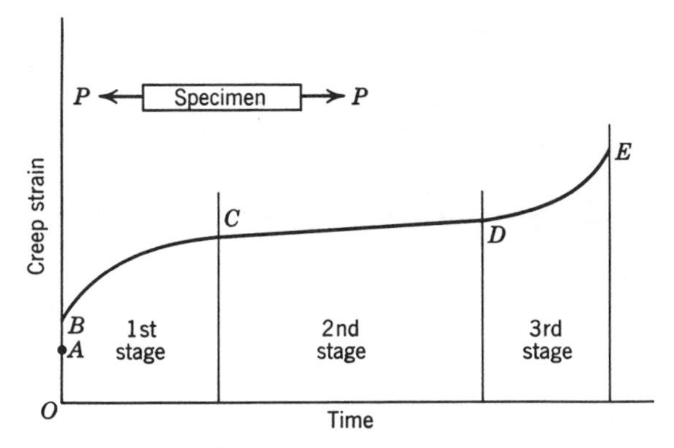

Fig. 15.33 Typical creep curve.

for tension creep experimental purposes. The tests are performed at a series of increasing stresses and corresponding creep rates are obtained, as illustrated. The portion of the creep curve most commonly used in design work is the steady-state or second stage, although the initial deformation is of great importance when dealing with short-life items like missiles. The steady-state stage of creep usually plots linear on log coordinates where the slope defines a creep rate or deformation per unit time—its use will be demonstrated later. Creep behavior of materials is sensitive not only to stress and time but also to atmosphere, physical properties of the material, past strain history of the material, etc.; therefore, for reliable design, complete creep curves should be available characterizing the material to be used. Extrapolations of short-time data to long times are not always reliable.

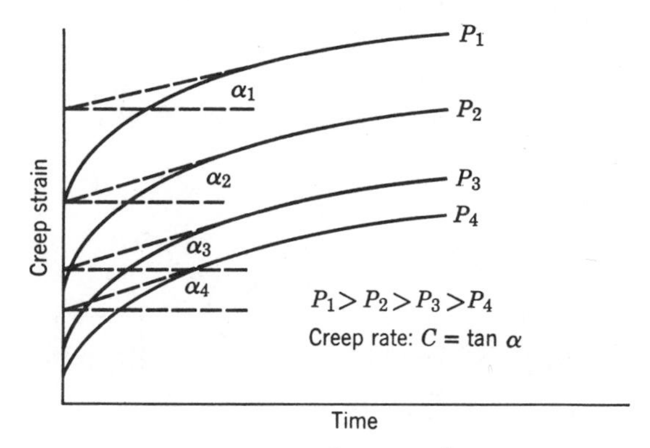

Fig. 15.34 Creep rate from second stage creep.

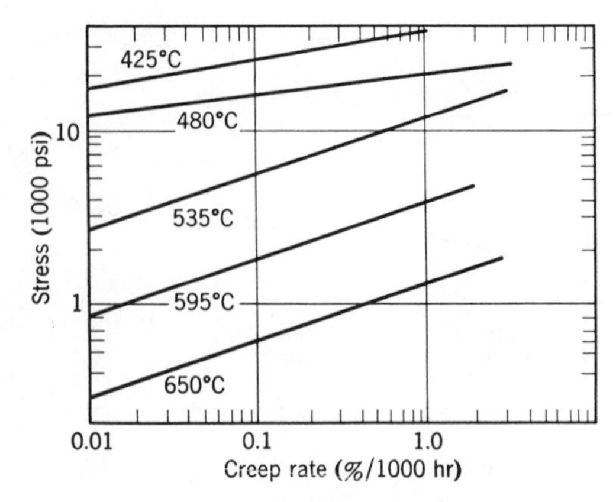

Fig. 15.35 Creep data for "killed" carbon steel. [From data through courtesy of the Timken Roller Bearing Co.]

Consideration will now be given to a mathematical theory of creep that will provide a description of the creep behavior of a material subjected to multiaxial loads on the basis of tension creep test data. The basis of this theory is the following relationship which has ample support from numerous experiments reported in the literature.

$$C = B\sigma^n \qquad 15.175$$

where C = creep rate in tension

σ = applied tensile stress

n = slope of line on log-log coordinates

B = stress intercept at log creep rate 1.0

Data illustrating this theory are shown in Fig. 15.35.

Creep in bending. The creep law expressed by Eq. 15.175 can be applied to problems in bending. Consider, for example, a beam of symmetrical cross section subjected to pure bending (Fig. 15.36). It is assumed that the bending

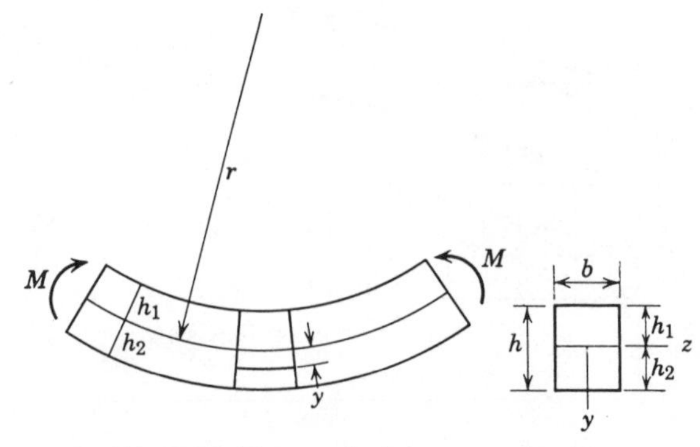

Fig. 15.36 Rectangular beam in pure bending.

forces act in the plane of symmetry, that plane sections remain plane, and that the properties are essentially the same in tension and compression. Under these conditions the unit elongation or compression of any fiber of the beam will be

$$\varepsilon = y/r \tag{15.176}$$

where y is the distance from the neutral axis to the fiber and r is the radius of curvature of the deflection curve. From Eq. 15.175,

$$C = B\sigma^n = \varepsilon/t \tag{15.177}$$

where t is the time. Combining Eqs. 15.176 and 15.177 gives

$$\varepsilon = tB\sigma^n = y/r \tag{15.178}$$

and solving for the stress gives

$$\sigma = (y/rtB)^{1/n} \tag{15.179}$$

The largest stresses occur at the outer fibers; thus from Eq. 15.179

$$(\sigma_{max})_{tension} = (h_1/rbt)^{1/n} \tag{15.180}$$

and

$$(\sigma_{max})_{compression} = (h_2/rBt)^{1/n} \tag{15.181}$$

At any distance y from the neutral axis

$$(\sigma)_{tension} = (\sigma_{max})_{tension}(y/h_1)^{1/n} \tag{15.182}$$

and

$$(\sigma)_{compression} = (\sigma_{max})_{compression}(y/h_2)^{1/n} \tag{15.183}$$

For equilibrium the tensile and compression forces must balance; therefore

$$(\sigma_{max})_{tension} \int_0^{h_1} \left(\frac{y}{h_1}\right)^{1/n} dA = (\sigma_{max})_{compression} \int_0^{h_2} \left(\frac{y}{h_2}\right)^{1/n} dA \tag{15.184}$$

from which the ratio h_1/h_2 can be computed when the width of the beam is known. If $h_1 = h_2$, the maximum tension and compressive stresses are equal and

$$\sigma_{max} = (h/2rBt)^{1/n} \tag{15.185}$$

and

$$\sigma = \sigma_{max}(2y/h)^{1/n} \tag{15.186}$$

Since the bending moment M is equal to the internal moment caused by the forces distributed over the cross section,

$$M = 2\sigma_{max} \int_0^{h/2} (2y/h)^{1/n} y\, dA \tag{15.187}$$

from which σ_{max} can be computed. For a rectangular section of width b,

$$M = 2b\sigma_{max} \int_0^{h/2} (2y/h)^{1/n} y\, dy \tag{15.188}$$

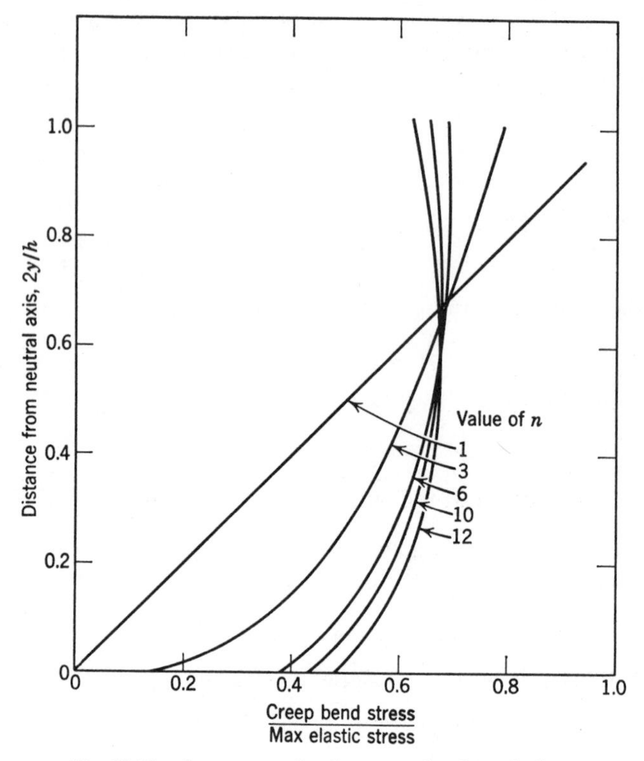

Fig. 15.37 Creep stress distributions in bending of a beam.

from which

$$\sigma_{max} = \frac{(Mh/2I)(2n+1)}{3n} \tag{15.189}$$

where I is the moment of inertia. The stress on any fiber is then

$$\sigma = \frac{(Mh/2I)(2n+1)}{3n}\left(\frac{2y}{h}\right)^{1/n} \tag{15.190}$$

The stress distributions for values of n of 1, 3, 6, 10, and 12 are shown in Fig. 15.37. The elastic stress distribution is shown for the case $n = 1$. It is clear from these distributions that for larger values of n the stress is only slightly affected by variation in n. The value of n comes from test data and Eq. 15.175.

The curvature is obtained from Eqs. 15.180 and 15.181. If the neutral and symmetry axes coincide, then

$$1/r = (2Bt/h)(\sigma_{max})^n \tag{15.191}$$

where B is obtained from Eq. 15.177. The corresponding deflection equation for the rectangular section is

$$d^2y/dx^2 = \pm(2Bt/h)(\sigma_{max})^n$$

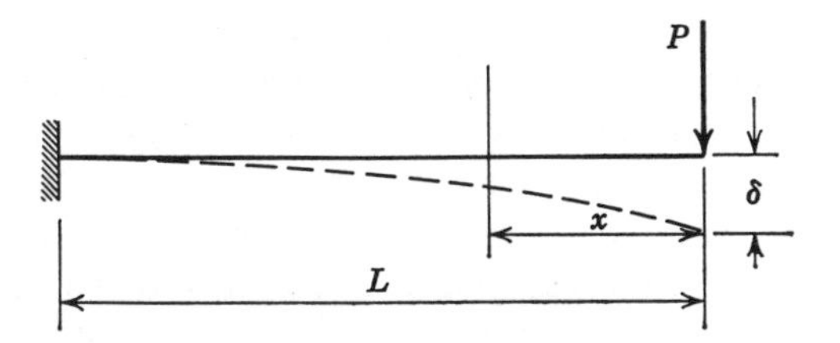

Fig. 15.38 Cantilever beam with end load.

or

$$\frac{d^2y}{dx^2} = \pm\left(\frac{2M^nBt}{h}\right)\left(\frac{h}{2I}\right)^n\left(\frac{2n+1}{3n}\right)^n \qquad 15.192$$

Equation 15.192 was derived on the basis of *pure bending*, but if it is assumed that the shear deformations involved are small, then the equation can be applied to any condition of bending. Thus

$$M^n = (d^2y/dx^2)(\phi/t) \qquad 15.193$$

where

$$\phi = 1/B\{(h/2)^{2n+1}[bn/(2n+1)]^n\}$$

As an example of the use of Eq. 15.193 consider the cantilever beam shown in Fig. 15.38. On any section x the moment is Px; therefore, using this condition in Eq. 15.193,

$$\delta = \frac{tP^n}{\phi}\left[\frac{x^{n+2}}{(n+1)(n+2)} - \frac{xL^{n+1}}{n+1} + \frac{L^{n+2}}{n+2}\right] \qquad 15.194$$

and

$$\delta_{\max} = \frac{tP^n}{\phi}\left(\frac{L^{n+2}}{n+2}\right) \qquad 15.195$$

Creep stresses and deflections of columns and beam columns have been treated by Lin (40, 41).

The elastic analog. Hoff (27) introduced a method of creep analysis that is particularly useful since it can be applied to statically indeterminate structures. In the literature this method has become known as the *elastic analog*. It is usually assumed in this theory that the creep strains are of the order of a percent or so; thus primary creep is disregarded and the analysis is based on the steady-state phase which for a variety of materials follows the law

$$\varepsilon = (\sigma/K)^n \qquad 15.196$$

Since the strains are assumed to be large, elastic deformations are negligible, and the limiting state of stress and strain approached as the creep strain becomes large compared to the elastic strain can be determined from the nonlinear stress-strain rate law

$$d\varepsilon/dt = (\sigma/K)^n \qquad 15.197$$

which is analogous to Eq. 15.196. For creep, the analogous elastic problem exists when the structure is loaded in the same manner, has geometric displacement boundary conditions everywhere proportional to the creep rate boundary conditions of the creep problem, and has an elastic stress-strain law given by Eq. 15.196. Under these circumstances the stresses in the *elastic structure* are numerically equal to the corresponding creep rates. Illustration of the theory by a model is fairly simple. Consider an assembly of bars of material in steady-state creep following Eq. 15.197; the total deformation for any bar is

$$\varepsilon = \int_0^t \frac{d\varepsilon}{dt}\, dt = \left(\frac{d\varepsilon}{dt}\right) t \qquad 15.198$$

Equation 15.198 indicates that time t acts to change the scale of deformation without affecting the relative deformation of different bars. Thus strain compatibility is maintained with the passage of time, and the elastic material following Eq. 15.196 presents a deformation pattern identical to that produced by creep. The essential advantage in using the principle of the *elastic analog* is that a multitude of known elastic solutions to problems can be utilized in obtaining solutions to plasticity problems. In particular, methods of elasticity can be used such as energy methods for statically indeterminate situations, and approximate solutions can be obtained when exact solutions are either very difficult or impossible to obtain.

The technique embodies two essential analytical tools, the nonlinear stress-strain relationship given by Eq. 15.196, which is qualitatively similar to the nonlinear creep relationship of Eq. 15.197, and the complementary energy defined in Chapter 7 as

$$u = \int_0^\sigma \varepsilon\, d\sigma \qquad 15.199$$

which is the complement of the elastic strain energy

$$u' = \int_0^\varepsilon \sigma\, d\varepsilon \qquad 15.200$$

A physical picture of these energies is shown in Chapter 7 and also in Fig. 15.39; u is the area bounded by the stress-strain curve and the stress axis, whereas u' is the area bounded by the curve and the strain axis. The sum of u and u' always equals the product of final stress and strain. Application of the theory will now be demonstrated.

Consider the statically indeterminate frame in Fig. 15.40; it consists of two deformable members 01 and 02 of equal cross-sectional area and a third rigid member 03. All joints are pin-connected and W is the only load applied. The steady-state creep forces and the rate of vertical deflection of the frame at the point of application of the load W will be determined.

If X is the force in the member 02, then by statics

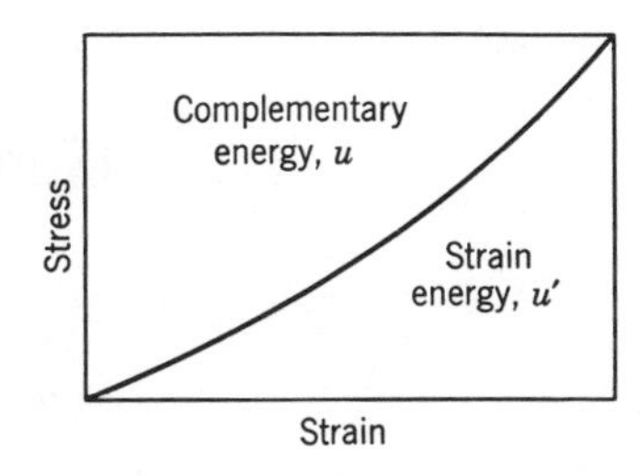

Fig. 15.39 Nonlinear elastic stress-strain relationship (see also Fig. 7.13).

$$F_{01} = \sqrt{5}W/2 - \sqrt{10}X/4 \qquad 15.201$$

$$F_{02} = X \qquad 15.202$$

$$F_{03} = -W/2 - \sqrt{2}X/4 \qquad 15.203$$

In ordinary elastic loading the unknown force X is easily found through application of the principle of minimum potential energy (see Chapter 7) which states,

$$\partial U'/\partial X = 0 = (\partial/\partial X)(F^2\, dL/2EA) \qquad 15.204$$

where U' is the total strain energy of the system, dL the differential length of the members, and A the cross-sectional area. Similarly, the vertical deflection δ is easily found since

$$\delta = \partial U'/\partial W \qquad 15.205$$

Now, if members 01 and 02 are subject to creep, the creep rate is assumed to follow the law given by Eq. 15.197, which is analogous to Eq. 15.196. Substitution of Eq. 15.197 into Eq. 15.199 gives an expression for the complementary energy,

$$u = [K/(n+1)](\sigma/K)^{n+1} \qquad 15.206$$

and the steady-state creep forces are found by solving for X and referring to Eqs. 15.201 to 15.203. The force X is found by noting that

$$\partial U/\partial X = 0 \qquad 15.207$$

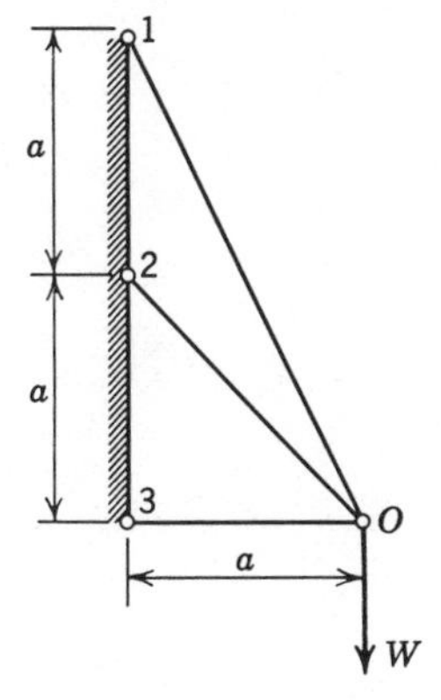

Fig. 15.40 Statically indeterminate frame.

where U, the total complementary energy, is defined as

$$U = \int_v u \, dv \qquad 15.208$$

where v is the volume of the system. From these relationships

$$X = \frac{\sqrt{5}\,W}{2[(\frac{4}{5})^{1/n} + \sqrt{10}/2]} \qquad 15.209$$

The vertical deflection rate at load W, by analogy to Eq. 15.205 becomes

$$\delta = \partial U / \partial W \qquad 15.210$$

from which

$$\delta = \frac{\sqrt{5}\,W}{AK[2(\frac{4}{5})^{1/n} + \sqrt{10}/2]^n} \qquad 15.211$$

The same procedure can also be applied to bending problems (68).* In this case, Eq. 15.196 is replaced by a moment-curvature law, which for a beam of rectangular cross section and height h is

$$h/r = (M/\beta)^n \qquad 15.212$$

where r = the radius of curvature

M = the bending moment

$\beta = Kch^2[2n/(2n+1)]$

n, K are as previously defined.

The complementary energy for this case is

$$u = \int_0^M \frac{1}{r} dM \qquad 15.213$$

Now consider the case of a center-loaded beam as shown in Fig. 15.41. From Eq. 15.213

$$u = \int_0^M \frac{dM}{r} = \frac{M^{n+1}}{(n+1)(h)(\beta)^n} \qquad 15.214$$

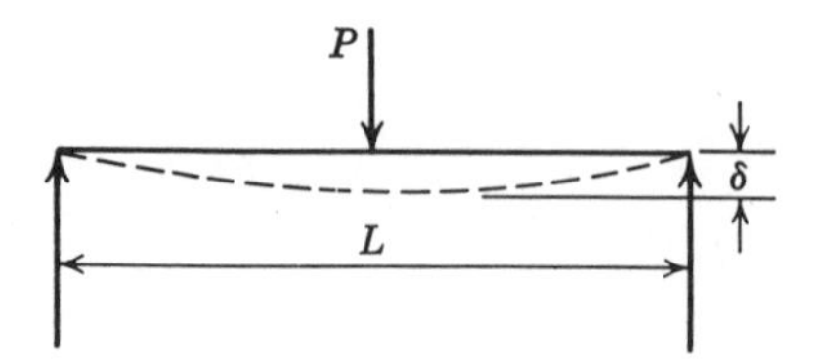

Fig. 15.41 Beam with center load.

*Courtesy of The Brooklyn Polytechnic Institute.

The total complementary energy is twice that of half of the beam

$$U = 2\int_0^a u\,dx \qquad 15.215$$

or

$$U = \frac{2a^{n+2}}{(n+1)(n+2)(h)(\beta)^n}\left(\frac{P^{n+1}}{2^{n+1}}\right) \qquad 15.216$$

The deflection under the load P is

$$\delta = \partial U/\partial P = (Pa/2\beta)^n(a^2/h)/(n+2) \qquad 15.217$$

Analysis of multiaxial creep. Many parts such as pipes, pressure vessels, and turbine disks are subjected continuously to loads that result in creep deformation. Such deformations have to be allowed for in design calculations; therefore it is necessary to have a convenient method for handling such calculations. A method will now be outlined for calculating creep deformations that take place under conditions of combined stresses on the basis of data obtained from simple tensile creep tests. The general theory for combined stresses is treated in Chapter 3.

The three principal plastic strains are given by equations analogous to those for the three principal elastic strains as follows (see Eqs. 6.19 to 6.21):

$$\varepsilon_1 = \phi[\sigma_1 - \tfrac{1}{2}(\sigma_2 + \sigma_3)] \qquad 15.218$$

$$\varepsilon_2 = \phi[\sigma_2 - \tfrac{1}{2}(\sigma_1 + \sigma_3)] \qquad 15.219$$

$$\varepsilon_3 = \phi[\sigma_3 - \tfrac{1}{2}(\sigma_1 + \sigma_2)] \qquad 15.220$$

where ϕ is a *plasticity modulus* and Poisson's ratio is $\frac{1}{2}$ because of the assumption of constancy of volume. As shown in Chapter 6, these equations can be combined into a single equation called the *equivalent stress*, which is directly related to the ordinary tension stress as follows (Eq. 6.27):

$$\sigma_0 = (\sigma_1{}^2 + \sigma_2{}^2 + \sigma_3{}^2 - \sigma_1\sigma_2 - \sigma_2\sigma_3 - \sigma_1\sigma_3)^{1/2} \qquad 15.221$$

Thus, for example, for calculating the yield pressure of a cylinder under internal pressure, the terms σ_1, etc., in Eq. 15.221 would be replaced by the elastic equations for the hoop, radial, and longitudinal stresses at the bore of the cylinder and σ_0 would be the yield strength of the material in tension. For the creep problem a conversion of Eqs. 15.218 to 15.220 to creep rate equations is accomplished by dividing through by t, thus

$$C_1 = \varepsilon_1/t = (\phi/t)[\sigma_1 - \tfrac{1}{2}(\sigma_2 + \sigma_3)] \qquad 15.222$$

$$C_2 = \varepsilon_2/t = (\phi/t)[\sigma_2 - \tfrac{1}{2}(\sigma_1 + \sigma_3)] \qquad 15.223$$

$$C_3 = \varepsilon_3/t = (\phi/t)[\sigma_3 - \tfrac{1}{2}(\sigma_1 + \sigma_2)] \qquad 15.224$$

For simple tension, all stresses are zero except σ_1; therefore, from the preceding equations

$$\sigma_0 = \sigma = \sigma_1 = Ct/\phi \qquad 15.225$$

However, from Eq. 15.175, for tension creep,

$$C = B\sigma^n \tag{15.175}$$

Therefore, if $\theta = \phi/t$, from Eqs. 15.175 and 15.225

$$\theta = B\sigma^{n-1} \tag{15.226}$$

Equations 15.222 to 15.224 can now be rewritten using Eqs. 15.225 and 15.226

$$C_1 = B\sigma_0{}^{n-1}[\sigma_1 - \tfrac{1}{2}(\sigma_2 + \sigma_3)] \tag{15.227}$$

$$C_2 = B\sigma_0{}^{n-1}[\sigma_2 - \tfrac{1}{2}(\sigma_1 + \sigma_3)] \tag{15.228}$$

$$C_3 = B\sigma_0{}^{n-1}[\sigma_3 - \tfrac{1}{2}(\sigma_1 + \sigma_2)] \tag{15.229}$$

In terms of the stress ratios $\sigma_2/\sigma_1 = \alpha$ and $\sigma_3/\sigma_1 = \beta$, these equations become

$$C_1 = B\sigma_1{}^n(\alpha^2 + \beta^2 - \alpha\beta - \alpha - \beta + 1)^{(n-1)/2}(1 - \alpha/2 - \beta/2) \tag{15.230}$$

$$C_2 = B\sigma_1{}^n(\alpha^2 + \beta^2 - \alpha\beta - \alpha - \beta + 1)^{(n-1)/2}(\alpha - \beta/2 - \tfrac{1}{2}) \tag{15.231}$$

$$C_3 = B\sigma_1{}^n(\alpha^2 + \beta^2 - \alpha\beta - \alpha - \beta + 1)^{(n-1)/2}(\beta - \alpha/2 - \tfrac{1}{2}) \tag{15.232}$$

These equations completely define the principal creep rates, in terms of the principal creep stresses and the experimental tension creep parameters B and n. Thus on the basis of simple tension creep test data, a log-log plot can be constructed from which values of B and n may be determined, and then by the use of Eqs. 15.227 to 15.232 creep rates for any state of stress can be predicted.

For two-dimensional creep, the principal stresses are evaluated in the same manner as principal static stresses (Chapter 3), so that in terms of component creep stresses

$$\frac{\sigma_1}{\sigma_2} = \frac{\sigma_x + \sigma_y}{2} \pm \sqrt{[(\sigma_x - \sigma_y)/2]^2 + \tau_{xy}{}^2} \tag{15.233}$$

where σ_x and σ_y are the component normal stresses and τ_{xy} is the component shear stress. For Example, for a thin-walled tube subjected to torsion plus internal pressure (Fig. 15.42),

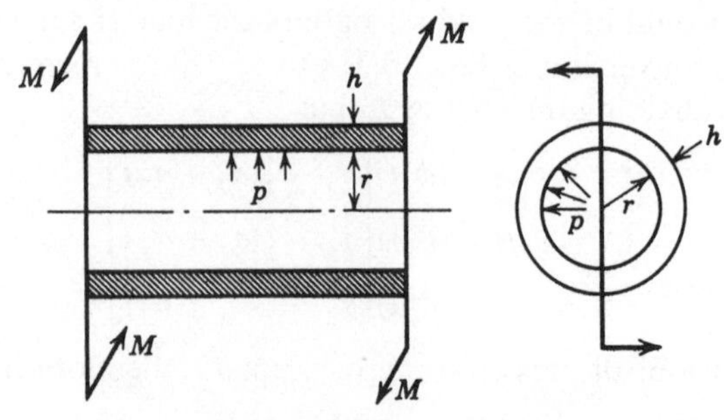

Fig. 15.42 Thin-walled tube subjected to torsion and internal pressure.

σ_x = longitudinal stress due to pressure = $pr/2h$

σ_y = hoop stress due to pressure = pr/h

τ_{xy} = shear stress due to torsion = $M/2\pi rh$

Then from Eqs. 15.230 and 15.231,

$$\text{hoop creep rate} = C_1 = B\sigma_1{}^n(\alpha^2 - \alpha + 1)^{(n-1)/2}(1 - \alpha/2) \qquad 15.234$$

$$\text{axial creep rate} = C_2 = B\sigma_1{}^n(\alpha^2 - \alpha + 1)^{(n-1)/2}(\alpha - \tfrac{1}{2}) \qquad 15.235$$

Putting the component stresses into Eqs. 15.234 and 15.235,

$$C_1 = \frac{B\sigma_1{}^n}{2}\left\{\frac{\sigma_x{}^2 + \sigma_y{}^2 - \sigma_x\sigma_y + 3\tau_{xy}{}^2}{\sigma_x{}^2/2 + \sigma_y{}^2/2 + \tau_{xy}{}^2 + (\sigma_x + \sigma_y)\sqrt{[(\sigma_x - \sigma_y)/2]^2 + \tau_{xy}{}^2}}\right\}^{(n-1)/2}$$
$$\times \left\{\frac{[(\sigma_x + \sigma_y)/2] + 3\sqrt{[(\sigma_x - \sigma_y)/2]^2 + \tau_{xy}{}^2}}{(\sigma_x + \sigma_y)/2 + \sqrt{[(\sigma_x - \sigma_y)/2]^2 + \tau_{xy}{}^2}}\right\} \qquad 15.236$$

and

$$C_2 = \frac{B\sigma_1{}^n}{2}\left\{\frac{\sigma_x{}^2 + \sigma_y{}^2 - \sigma_x\sigma_y + 3\tau_{xy}{}^2}{\sigma_x{}^2/2 + \sigma_y{}^2/2 + \tau_{xy}{}^2 + (\sigma_x + \sigma_y)\sqrt{[(\sigma_x - \sigma_y)/2]^2 + \tau_{xy}{}^2}}\right\}^{(n-1)/2}$$
$$\times \left\{\frac{[(\sigma_x + \sigma_y)/2] - 3\sqrt{[(\sigma_x - \sigma_y)/2]^2 + \tau_{xy}{}^2}}{(\sigma_x + \sigma_y)/2 + \sqrt{[(\sigma_x - \sigma_y)/2]^2 + \tau_{xy}{}^2}}\right\} \qquad 15.237$$

Equations 15.236 and 15.237 represent a general case since τ_{xy} is involved. If *only internal pressure* acts, then $\tau_{xy} = 0, \sigma_x = \sigma_y/2$, and

$$C_1 = B\sigma_1{}^n(\tfrac{3}{4})^{(n+1)/2} \qquad 15.238$$

$$C_2 = \frac{B\sigma_1{}^n}{8}(\tfrac{3}{4})^{(n-1)/2} \qquad 15.239$$

If *only torque* is applied, then $\sigma_x = \sigma_y = 0$ and Eqs. 15.236 and 15.237 reduce to

$$C_1 = (\tfrac{3}{2})B(\sigma_1)^n(3)^{(n-1)/2} \qquad 15.240$$

and

$$C_2 = (-\tfrac{3}{2})B(\sigma_1)^n(3)^{(n-1)/2} \qquad 15.241$$

Elastic analog applied to multiaxial creep. The elastic analog method given earlier in this section can also be applied to combined stress. Consider, for example, creep deformation in a thin-walled tube subjected to internal pressure and end loading.

For this combined stress problem a *generalized* elastic stress is used in place of Eq. 15.221; thus for the two-dimensional case,

$$(\sigma_0)_{\text{elastic}} = \sqrt{\sigma_1{}^2 + \sigma_2{}^2 - 2\nu\sigma_1\sigma_2} \qquad 15.242$$

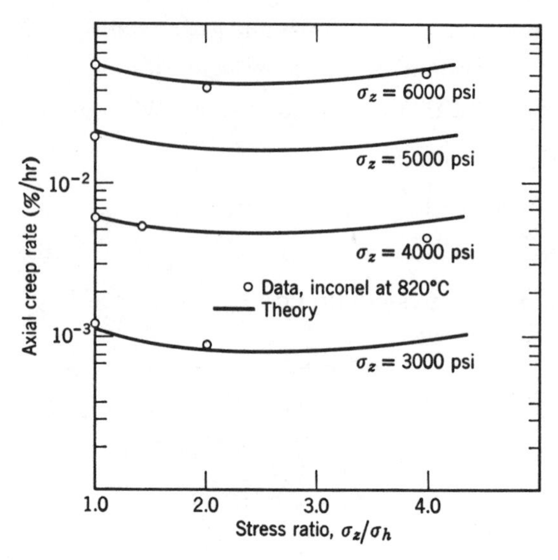

Fig. 15.43 Application of elastic analog theory to combined stress creep rates. [Calculated from data of Kennedy, Harms, and Douglas (34).]

The complementary energy is

$$u = [K/(n+1)](1/K)^{n+1}[\sqrt{\sigma_1{}^2 + \sigma_2{}^2 - 2\nu\sigma_1\sigma_2}]^{n+1} \qquad 15.243$$

Then if the axial creep is to be calculated and it is assumed that the axial creep is in the maximum principal stress direction (that is, the end load governs),

$$\varepsilon_1 = \partial u/\partial\sigma_1 \qquad 15.244$$

and combination of Eqs. 15.243 and 15.244 gives the result

$$\varepsilon_1 = (1/K)^n(\sigma_1 - \nu\sigma_2)[\sigma_1{}^2 + \sigma_2{}^2 - 2\nu\sigma_1\sigma_2]^{(n-1)/2} \qquad 15.245$$

where ε_1 is now the *axial creep rate.* In the elastic case, $n = 1$, $K =$ the modulus of elasticity E, and Eq. 15.245 reduces to the familiar elastic solution for *strain,*

$$\varepsilon_1 = (1/E)(\sigma_1 - \nu\sigma_2) \qquad 15.246$$

Equation 15.245 gives good results as shown in Fig. 15.43, where it was used to calculate the creep rate curves compared with data on Inconel tubes tested at 820°C at the Oak Ridge National Laboratory (34).

Multiaxial creep under nonuniform stress distributions

In the previous section the discussion was concerned with multiaxial creep in which the stress distribution was uniform such as in thin-walled tubes under various conditions of loading. Consideration will now be given to cases where the distribution of stress is not uniform but varies throughout the part.

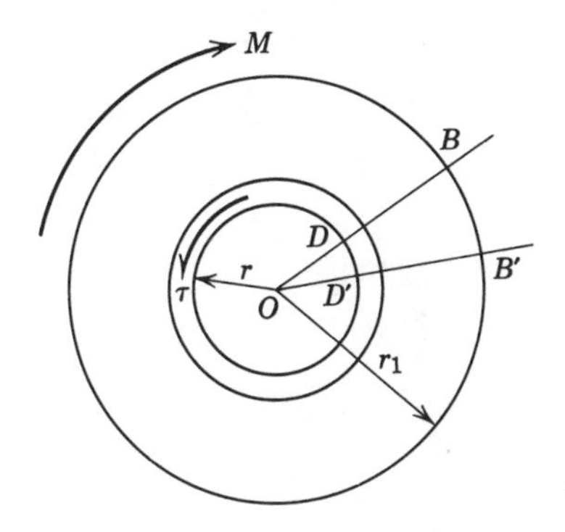

Fig. 15.44 Torsion of a circular rod.

Torsion of a solid bar. Consider the bar of circular cross section shown in Fig. 15.44. It is assumed that there is no longitudinal distortion of the cross section, that plane sections remain plane, and that torsional creep rates (C_s, C_s') are proportional to strain. Because of the torque M, a section OB moves to location OB', and

$$DD'/BB' = r/r_1 = C_s/C_s' \qquad 15.247$$

A torsional creep law similar to Eq. 15.75 for tension will be used, that is,

$$C_s = A\tau^m \qquad 15.248$$

and

$$C_s' = A\tau_1{}^m \qquad 15.249$$

From the preceding three equations

$$\tau = (r/r_1)^{1/m}\tau_1 \qquad 15.250$$

where τ is the shear stress at any radius r in the cross section. For equilibrium it is required that

$$M = \int_0^{r_1} \tau_r(2\pi r\, dr) \qquad 15.251$$

or, using Eq. 15.250

$$M = \int_0^{r_1} \left(\frac{r}{r_1}\right)^{1/m} \tau_1 r(2\pi r\, dr) \qquad 15.252$$

which reduces to

$$M = \frac{2\pi m\tau_1(r_1)^3}{3m+1} \qquad 15.253$$

Again, using Eq. 15.250

$$M = \frac{2\pi m(r_1)^3\tau(r/r_1)^{-1/m}}{3m+1} \qquad 15.254$$

from which it can be determined that at any section r

$$\tau = Mr_1(3m+1)(r/r_1)^{1/m}/(4mr_1 J) \qquad 15.255$$

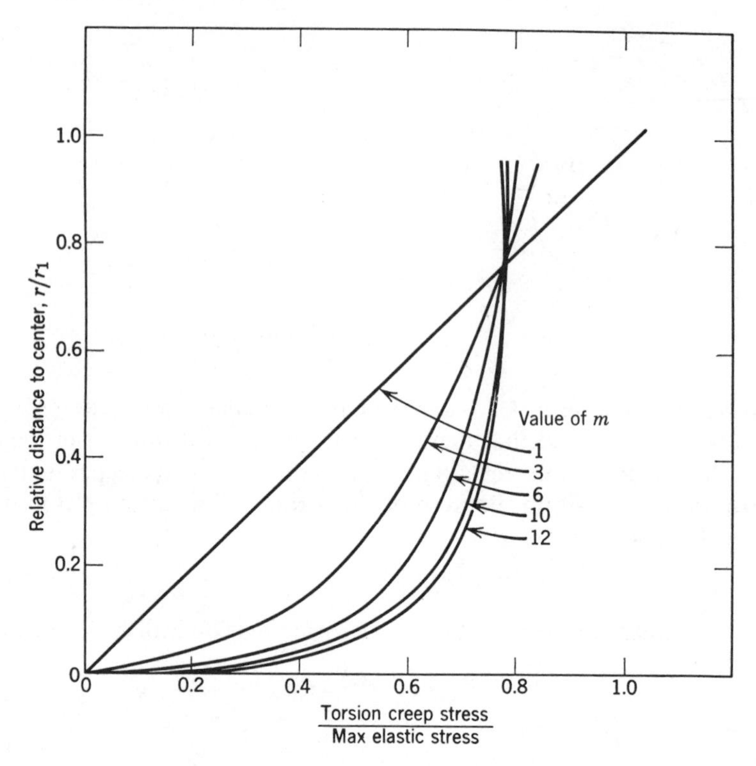

Fig. 15.45 Creep stress distribution in torsion.

where J is the polar moment of inertia $\pi(r_1)^4/2$. The stress distribution for various values of m are shown in Fig. 15.45. This plot is similar to that of Fig. 15.37 for bending. When $m = 1$, Eq. 15.255 reduces to the elastic solution

$$\tau = Mr/J \tag{15.256}$$

Equation 15. 255 defines values of shear stress; thus for any particular material characterized by parameter m, shaft sizes can be computed corresponding to a given allowable shear stress τ, which is a function of time. The results can also be expressed in terms of creep rates, which, for the present case, can be expressed as $r\beta/t$, where β is the angle of twist and t is time. Referring again to Fig. 15.44, the creep rate for the extreme fiber is

$$C_s' = r_1 \beta/t \tag{15.257}$$

Then from Eqs. 15.248 and 15.249

$$\tau_1{}^m = C_s'/A = r_1 \beta/tA \tag{15.258}$$

Finally, using Eqs. 15.253 and 15.258,

$$M = \frac{2\pi m(r_1)^3(r_1 \beta/tA)^{1/m}}{3m+1} \tag{15.259}$$

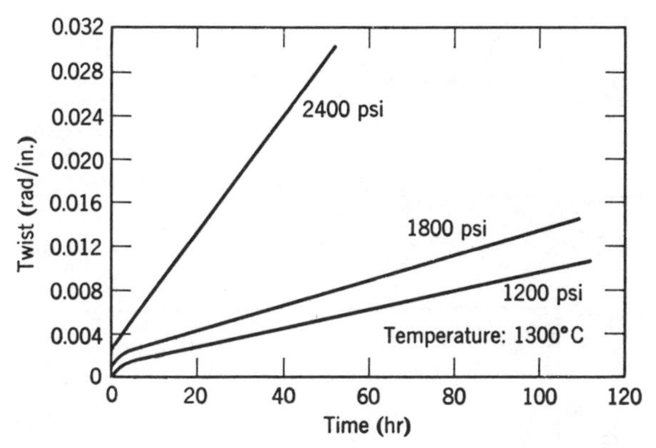

Fig. 15.46 Torsional creep of magnesia. [After Wygant (69), courtesy of the American Ceramic Society.)

from which

$$\beta = At(3m + 1)(Mr_1/J)^m/(4mr_1) \qquad 15.260$$

where β is the allowable angle of twist for a given time t. The creep behavior of solid and hollow circular sections subjected to various combinations of bending, torque, and axial load has been considered by Johnson et al. (32).

At elevated temperature creep also occurs in ceramics as illustrated by data in Fig. 15.46. These plots for a ceramic are reminiscent of those obtained for most metals, which suggest that at least some of the design procedures known to work for metals can also be applied to ceramic materials.

Heavy-walled cylinder under internal pressure at uniform temperature. Various methods have been proposed for calculating creep in heavy-walled cylinders; one procedure is that due to Siebel and Schwaigerer (53), which will be illustrated by performing some design calculations on a closed-end cylinder of Type 347 stainless steel whose life can be limited to 1000 hr of operation. The method can, of course, be used for different conditions and materials.

The basis of the method is in allowing a certain amount of creep to occur before the vessel is considered unusable. If, for example, the allowable bore creep is 1%, then analyses of safe pressures can be made starting with 1% bore expansion as a basis. Since it is assumed that the volume remains constant in the plastic range of strain, the following relationship can be used which relates the bore strain to any other hoop strain in the cylinder wall; thus, referring to Fig. 15.47,

$$\varepsilon_a = \varepsilon_r (R_1)^2 \qquad 15.261$$

Using Eq. 15.261 and the condition that $\varepsilon_a = 1\%$, Fig. 15.48 can be prepared. The safe maximum internal pressure for 1000 hr of operation at elevated temperature will now be calculated for a cylinder having a diameter ratio of 2.76. Creep data for 1000 hr operation at various temperatures are given in Fig. 15.49 for uniaxial loading. With Figs. 15.48 and 15.49 available, Fig. 15.50

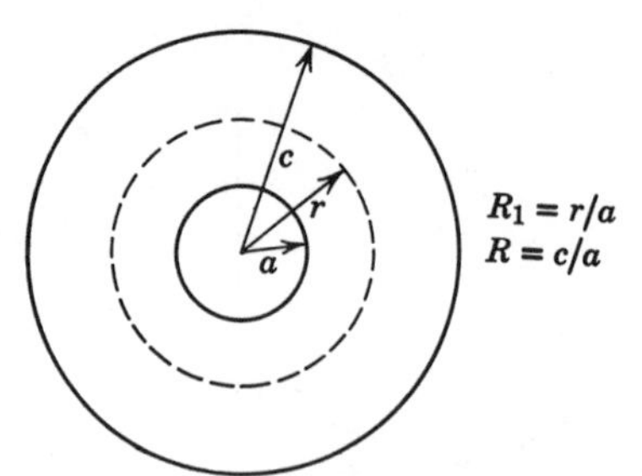

Fig. 15.47 Geometry of cylinder.

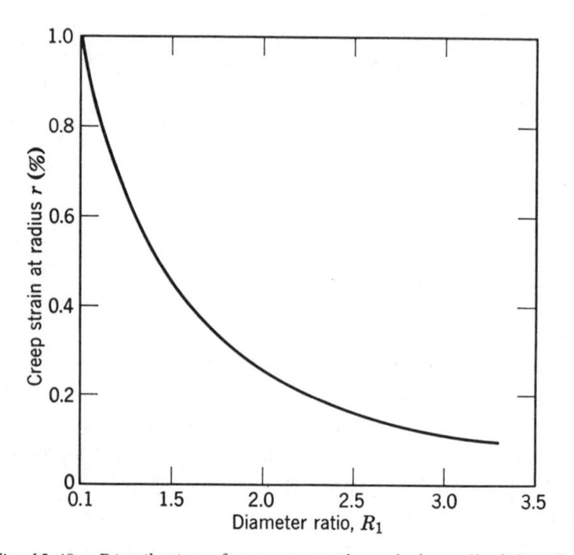

Fig. 15.48 Distribution of creep strain through the wall of the tube.

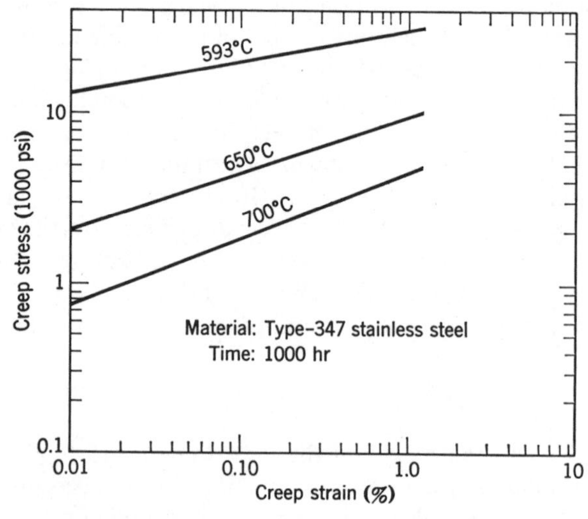

Fig. 15.49 Creep stress-creep strain curves.

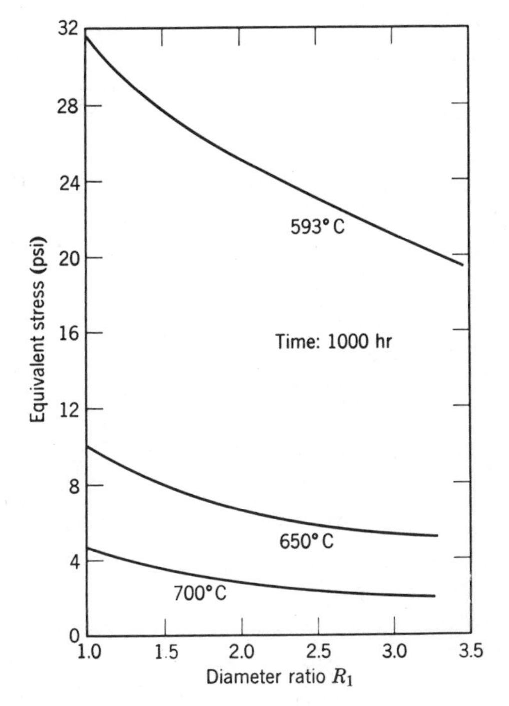

Fig. 15.50 Distribution of equivalent stress through the cylinder wall.

is obtained by superposition; this latter figure shows, for diameter ratios R_1, the equivalent uniaxial creep stress. For example, for 650°C, 0.10% creep strain represents a stress of about 5000 psi (Fig. 15.49); from Fig. 15.48 a strain of 0.10% corresponds to a diameter ratio R_1 of 3.25; thus on Fig. 15.50 for 650°C, a point is plotted at 5000 psi and $R_1 = 3.25$. The ordinate, *equivalent stress*, is that given by Eq. 15.221. Figure 15.50 furnishes the basic information from which to determine permissible internal pressures for the cylinder in question.

For equilibrium in a cylinder it is required that

$$p = \int_a^c \frac{(\sigma_h - \sigma_r)\, d\sigma_r}{r} \tag{15.262}$$

in which (see Chapter 6) it is assumed that

$$\sigma_h - \sigma_r = 2\sigma_0/\sqrt{3} \tag{15.263}$$

Equation 15.263 expresses the shear as an equivalent uniaxial value (the Von Mises condition). Thus, using Eqs. 15.262 and 15.263 and integrating for various diameter ratios R_1, the stress values as given by Fig. 15.50 are used to obtain the curves plotted in Fig. 15.51, which show the permissible internal pressures for cylinders to operate at a maximum of 1% bore strain for 1000 hr.

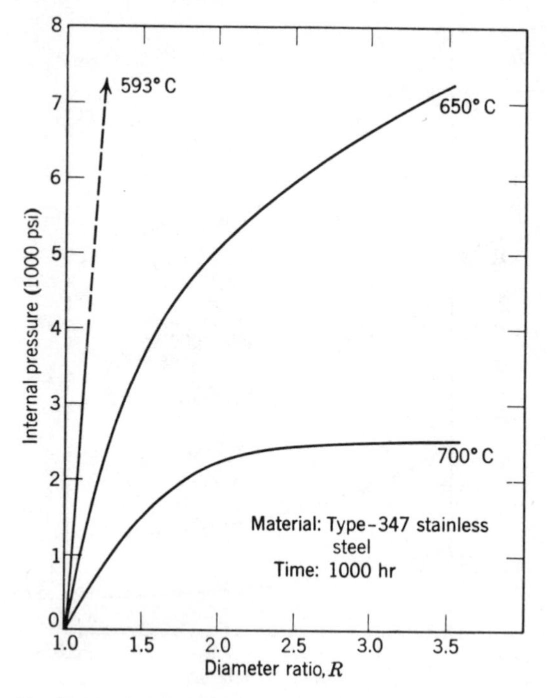

Fig. 15.51 Permissible internal pressure for 1% bore creep.

For example, for a diameter ratio of 2.76, Fig. 15.51 shows that the maximum permissible internal pressure at 650°C is 6250 psi.

The stress distribution is obtained from Eqs. 15.221, 15.262, and 15.263 with the following additions. The equilibrium condition expressed by Eq. 15.262 is rewritten in terms of stresses alone; thus for equilibrium

$$\sigma_h - \sigma_r = r\, d\sigma_r/dr \qquad 15.264$$

By definition, the hoop and radial strains are as follows:

$$\varepsilon_h = u/r \qquad 15.265$$

and

$$\varepsilon_r = du/dr \qquad 15.266$$

where u is the radial deformation. The creep rate is obtained by differentiating Eq. 15.265 and dividing by the time t; this operation plus substituting ε_r for du/dr obtained in differentiating Eq. 15.265 gives the result

$$(d/dr)C_h = (1/r)(C_r - C_h) \qquad 15.267$$

where C_h and C_r are creep rates in the hoop and radial directions respectively. These values can now be used in Eqs. 15.227 to 15.229. Assuming that the cylinder is in the fully plastic condition C_3, the longitudinal creep rate can be considered vanishingly small; therefore

$$\sigma_z = (\sigma_h + \sigma_r)/2 \tag{15.268}$$

and Eq. 15.221 can be rewritten as

$$\sigma_0 = (\sqrt{3}/2)(\sigma_h - \sigma_r) \tag{15.269}$$

Using this new value for σ_0 and Eqs. 15.227 and 15.228 with σ_z defined by Eq. 15.268

$$C_h = B(\sigma_h - \sigma_r)^n(\sqrt{3}/2)^{n+1} \tag{15.270}$$

and

$$C_r = B(\sigma_r - \sigma_h)^n(\sqrt{3}/2)^{n+1} \tag{15.271}$$

Now, substituting Eqs. 15.270 and 15.271 into Eq. 15.267 gives

$$(d/dr)(\sigma_h - \sigma_r)^n = (-2/r)(\sigma_h - \sigma_r)^n \tag{15.272}$$

which on integrating gives

$$(\sigma_h - \sigma_r)^n = N_1/r^2 \tag{15.273}$$

where N_1 is a constant of integration. The value of $(\sigma_h - \sigma_r)$ from Eq. 15.273 now substituted into Eq. 15.264, gives, after integrating,

$$\sigma_r = \frac{-n(N_1)^{1/n}(r)^{-2/n}}{2} + N_2 \tag{15.274}$$

where N_2 is another constant of integration. The radial stress equals $-p$ at the bore and is zero at the outside diameter; therefore the boundary conditions are established and the stresses are as follows:

$$\sigma_h = \frac{p}{n}\left[\frac{n + (2-n)(c/r)^{2/n}}{R^{2/n} - 1}\right] \tag{15.275}$$

$$\sigma_r = p\left[\frac{1 - (c/r)^{2/n}}{R^{2/n} - 1}\right] \tag{15.276}$$

$$\sigma_z = \frac{p}{R^{2/n} - 1}\left[1 + \left(\frac{1-n}{n}\right)\left(\frac{c}{r}\right)^{2/n}\right] \tag{15.277}$$

These three equations define the stress distribution for steady-state creep and are valid only as long as steady-state creep persists. Typical stress distributions are plotted in Fig. 15.52 for a cylinder having a diameter ratio of 2.0, subjected to 12,000 psi internal pressure. The actual creep rates in the hoop and radial directions can be determined from Eqs. 15.270, 15.271, and the three stress equations. Considerable caution should be exercised when applying creep rate data in design problems since very small differences in the stresses assumed to be acting can result in very large differences in calculated creep rates. The reason for this is that the creep curve flattens out in the direction of the abscissa.

In the preceding example, if there is a temperature differential through the

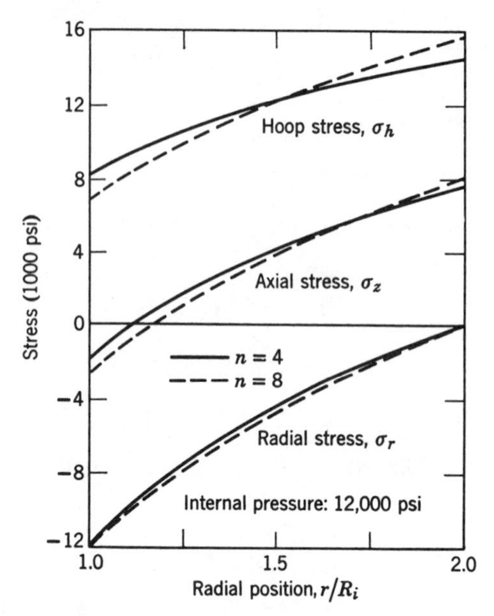

Fig. 15.52 Creep stress distributions for a heavy-walled cylinder of diameter ratio $R = 2$. [Courtesy of N. C. Dahl (12).]

wall, with the heat flowing towards the bore, the temperature distribution under steady-state conditions is given by

$$t = t_0 + \frac{HB}{k}\left[\frac{\log r}{\sqrt{a+b}}\right] \tag{15.278}$$

and the stresses given by Eqs. 15.275 to 15.277 can be used by substituting for n the value

$$\frac{n}{1 + Hb/2kB} \tag{15.279}$$

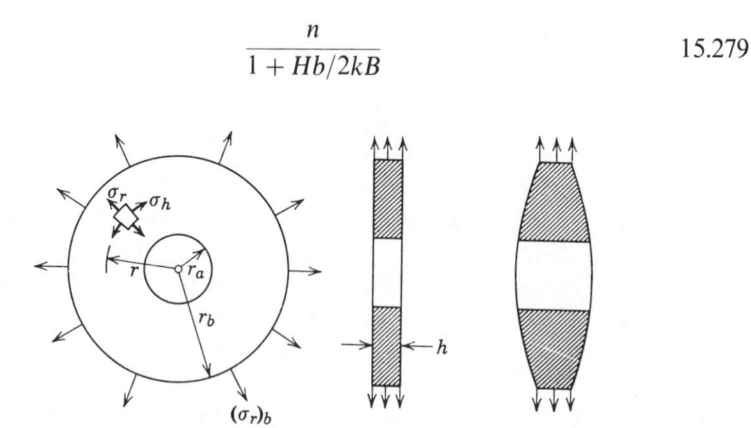

Fig. 15.53 Geometry of rotating disks. [After Wahl (65), courtesy of The American Society of Mechanical Engineers.]

where H = heat transfer rate at outside surface
k = thermal conductivity
t_0 = mean temperature

Rotating disks. Considerable work on this problem has been done by Wahl and his associates (65) and by Bailey (3). Wahl, for example, considered disks with both constant and variable thickness as shown in Fig. 15.53. The disk is subjected to a radial stress $(\sigma_r)_b$ at the outside edge; the disk is thin so that axial stresses can be considered negligible, and it is assumed that the hoop stress generated is always larger than the radial stress. It is also assumed that the temperature is uniform throughout the disk, a steady state of stress exists, and that the following creep relations are valid:

$$C_h = B\sigma_h{}^n \qquad 15.280$$

$$C_r = 0 \qquad 15.281$$

$$C_z = -B\sigma_h{}^n \qquad 15.282$$

The equilibrium conditions to be satisfied are as follows:

disk of constant thickness $\quad (d/dr)(r\sigma_r) - \sigma_h + \gamma\omega^2 r^2/g = 0 \quad$ 15.283

disk of variable thickness $\quad (d/dr)(rh\sigma_r) - h\sigma_h + \gamma\omega^2 r^2 h/g = 0 \quad$ 15.284

For the disk of constant thickness the following equations express the hoop and radial stresses:

$$\sigma_h = \frac{[(n-1)/n][\gamma\omega^2(b^3 - a^3)/(3g) + b(\sigma_r)_b]}{(b^{(n-1)/n} - a^{(n-1)/n})r^{1/n}} \qquad 15.285$$

and

$$\sigma_r = \frac{1}{r}\left[\frac{\gamma\omega^2(b^3 - a^3)}{3g} + b(\sigma_r)_b\right] \times \left[\frac{r^{(n-1)/n} - a^{(n-1)/n}}{b^{(n-1)/n} - a^{(n-1)/n}}\right] - \left[\frac{\gamma\omega^2(r^3 - a^3)}{3gr}\right] \qquad 15.286$$

where γ = density of material
ω = angular velocity
g = gravitational acceleration

A typical stress distribution is shown in Fig. 15.54, where

$$(\sigma_h)_{avg} = \gamma\omega^2(b^3 - a^3)/[3g(b-a)] + b(\sigma_r)_b/(b-a) \qquad 15.287$$

For the disk of variable thickness, $h = cr^m$

$$\sigma_h = \frac{\alpha[cb^{m+1}(\sigma_r)_b + \gamma\omega^2 I_0/g]F}{c(b^\alpha - a^\alpha)r^{1/n}} \qquad 15.288$$

and

$$\sigma_r = A\left[\frac{[cb^{m+1}(\sigma_r)_b + \gamma\omega^2 I_0/g](r^\alpha - a^\alpha)F}{(b^\alpha - a^\alpha)} - \frac{\gamma\omega^2 I}{g}\right] \qquad 15.289$$

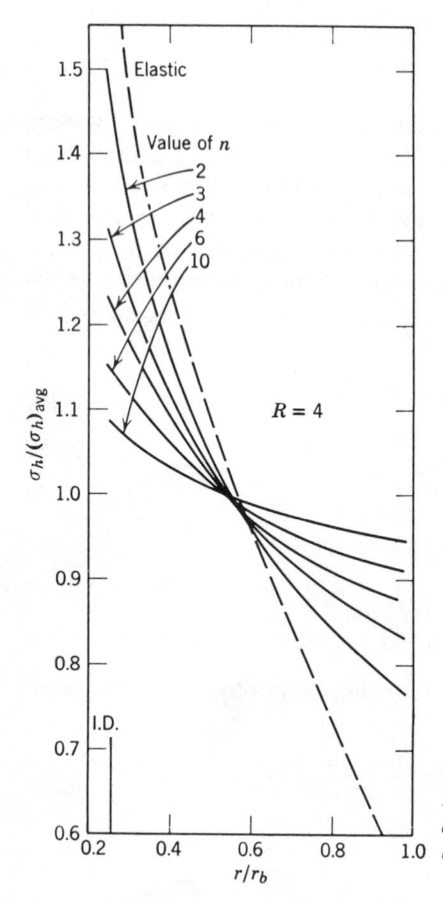

Fig. 15.54 Hoop-stress distribution for rotating disk of constant thickness. [After Wahl, (65), courtesy of The American Society of Mechanical Engineers.]

where $A = 1/(cr^{m+1})$

$\alpha = (mn + n - 1)/n$

$I = c(r^{3+m} - a^{3+m})/(3 + m)$

$I_0 = c(b^{3+m} - a^{3+m})/(3 + m)$

A typical stress distribution for a variable thickness disk is shown in Fig. 15.55. The disk in this case corresponds to that shown in Fig. 15.53, the thickness at the outside diameter being about one-half that at the bore. Thus $m = -0.5$ and $\alpha = (0.5 - 1)/n$. The value of the average hoop stress used on the ordinate is given by

$$(\sigma_h)_{avg} = cb^{m+1}(\sigma_r)_b/F + \gamma\omega^2 I_0/gF \qquad 15.290$$

where

$$F = \frac{c(b^{m+1} - a^{m+1})}{m + 1}$$

Special cases of constant thickness disks in which the temperature varied in the radial direction were also considered by Wahl (65).

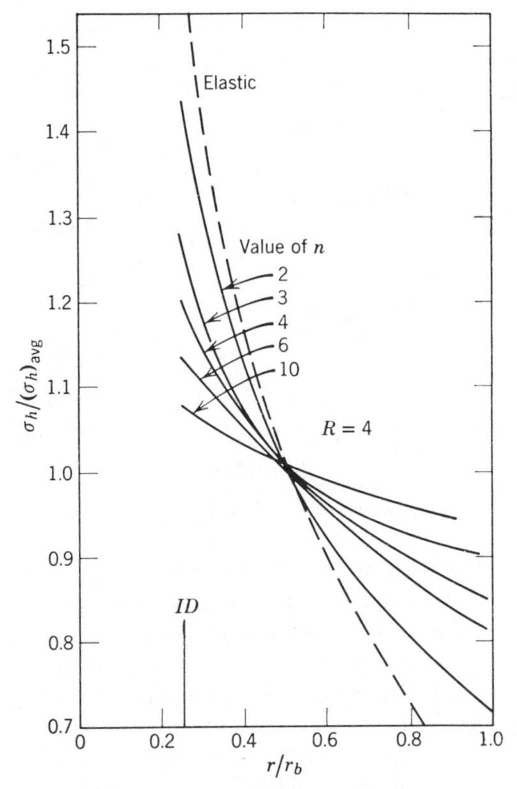

Fig. 15.55 Hoop-stress distribution for rotating disk of variable thickness. [After Wahl, (65), courtesy of The American Society of Mechanical Engineers.]

Stress rupture

Stress rupture is defined as the steadily applied stress that will cause fracture to take place in a material in a given time. Usually, the lower the stress the longer will be the time to fracture. By its very nature, stress rupture is closely associated with creep behavior—it may even be thought of as the terminus of the creep curve. Therefore in any problem involving creep, stress rupture has to be considered also since, under certain circumstances, it may be the governing factor in the design. This is particularly true for low stresses for materials (especially polymers, as discussed in Chapter 5) that exhibit brittle failure under these conditions. In other words, although creep may not be a problem in a particular case, it is always possible that stress-rupture failure could occur.

Some stress-rupture data have been obtained on tubes subjected to internal pressure, on torsion specimens, and on diaphragms, but the bulk of data have been obtained from tension and bend tests. Presentation of stress-rupture data is generally done by plotting applied stress versus time to rupture on log or semi-log coordinates as shown in Fig. 15.56. Note that the plots consist of

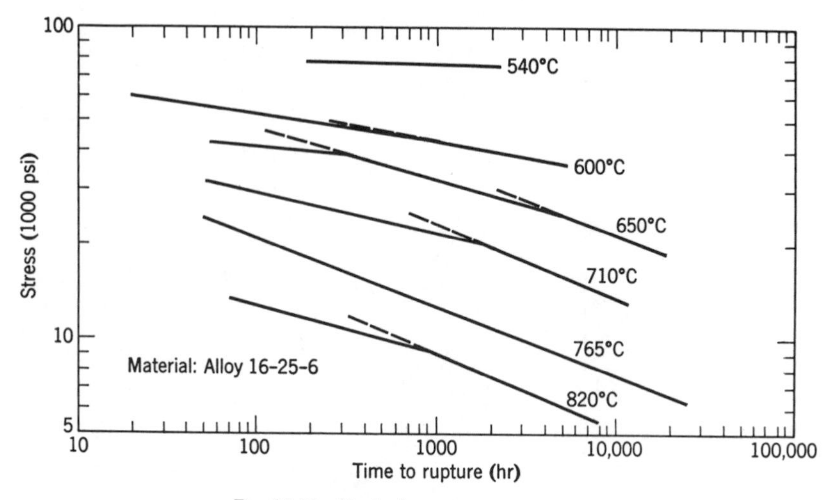

Fig. 15.56 Typical stress-rupture curves.

several lines of different slope for a given condition. This is usually observed, for most materials, particularly steels, undergo phase or other microstructural changes when heated and stressed for a period of time. The exact cause of the slope change is not known. On the basis of such plots design stress values can be obtained simply by picking a stress to avoid failure in a time considered practical for the life of the part. In addition to the usual stress-rupture curve it is also very helpful to have available data such as shown in Fig. 15.57, which shows deformation at rupture as well as the stress and time involved (11).

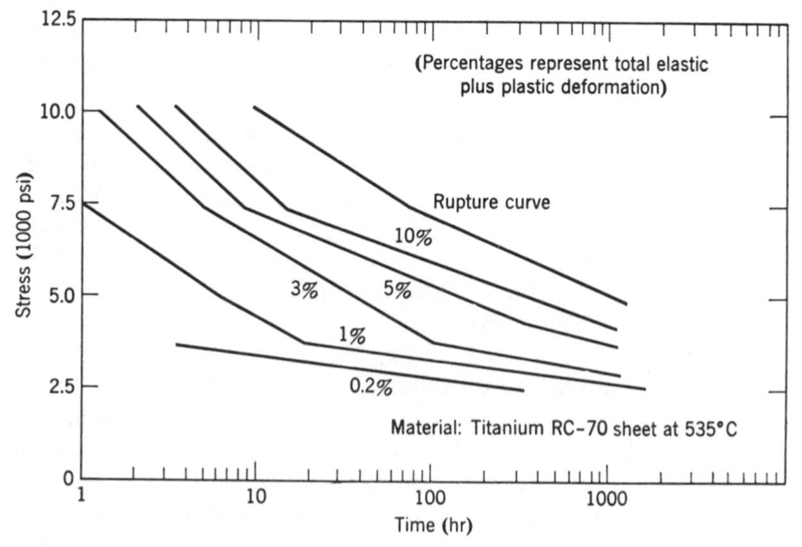

Fig. 15.57 Typical high temperature design data. [After Cross (11), courtesy of the McGraw-Hill Publishing Co.]

For tension applications such as beams the procedure mentioned is fairly reliable. Some designers use the stress-rupture plot and calculate a stress of about $\frac{1}{2}$ to $\frac{2}{3}$ the allowable, thus building a factor of safety into the design (22). When large factors of safety cannot be tolerated (some aircraft applications, for example), design may be based on the yield stress with due allowance for buckling. For most cases when testing cannot simulate the field condition, past experience is the best guide (36).

Various complications arise to make the assessment of stress rupture very difficult. Often stresses vary and the part may be at a high stress for a short time followed by a longer period at a lower stress (or vice versa). Frequently, in a situation like this, it is assumed that fractional life for one set of conditions is independent of other fractions of the life under different conditions (50). This implies the very questionable existence of an equation of state. In other words, if a structure spends a fraction F_1 of its total life at stress σ_1 and temperature t_1, then it later operates at σ_2 and t_2 for another fraction F_2, rupture is assumed to occur when $F_1 + F_2 + \ldots$, equals unity or

$$\sum_{n=1}^{n} (F_n)\sigma_n t_n = 1.0 \qquad 15.291$$

One of the main difficulties with stress-rupture analytics is in the extrapolation of short-time data to long times, sometimes over several decades of time on a log scale. In general, the safest procedure is to obtain data for the expected life of the part, but this is usually impossible except for times up to perhaps two years or so. If direct extrapolation is considered too unreliable, a theory has to be used. An appraisal of several theories has been presented by Goldhoff (21); for purposes of illustration here, one due to Larson and Miller (39) will be described. The following description of Larson's theory is based on a lecture presented at the Massachusetts Institute of Technology by Shank (52).

The Larson-Miller analysis concerning the prediction of stress-rupture behavior was first applied to creep, which was assumed to be a rate process. The extension of the theory to include stress rupture involves the assumption that the time to fracture depends on a summation of creep rates to rupture so that time to fracture is inverse to rate, or at constant stress (see chapter 5),

$$2.3RT(C + \log t) = Q \qquad 15.292$$

which comes from the rate equation

$$1/t = Ae^{-Q/RT} \qquad 15.293$$

where $C = \log A$, t is time to rupture, R is the gas constant, Q the activation energy, and T the absolute temperature. Although Eq. 15.292 states that C is a constant for constant stress, it is implied that C is insensitive to both stress and material. If C is truly a constant, then Eq. 15.292 is linear for log rupture time plotted against $1/T$ (See Fig. 15.58). In this figure the lines represent different stress levels. Various points making up or falling on these lines represent different temperatures and rupture times for the stresses. The intercept on the log

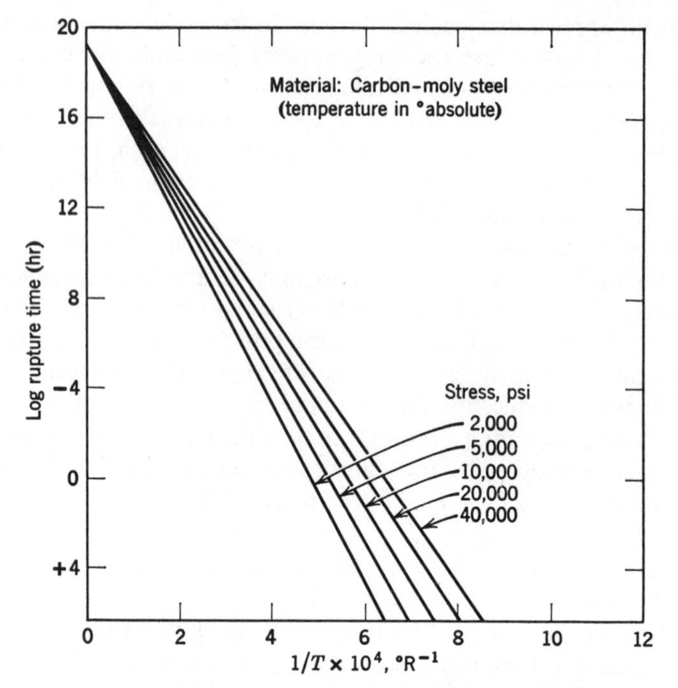

Fig. 15.58 Log rupture time versus reciprocal of R. [Data of Larson and Miller (39), compiled and included here through courtesy of M. E. Shank (52).]

time axis, when $1/T = 0$ is C; for this case C has a value of about 20.

If the quantity $T(C + \log t)$ is taken as a parameter independent of stress, then the following is implied. For a given stress, if a particular temperature T_1 and time to rupture t_1 are known, and C has been determined, any corresponding set of T_n and t_n can be determined for rupture or

$$T_1(20 + \log t_1) = T_n(20 + \log t_n) \tag{15.294}$$

Using this concept, the following combinations, for example, should have equivalent rupture stresses:

$$10^4 \text{ hr at } 550°\text{C} = 25 \text{ hr at } 650°\text{C}$$
$$10^3 \text{ hr at } 650°\text{C} = 12 \text{ hr at } 735°\text{C}$$
$$10^3 \text{ hr at } 150°\text{C} = 2.2 \text{ hr at } 205°\text{C}$$

This is supported by Larson's data as follows:

Steel	Stress for 10,000 hr Rupture at 550°C	Stress for 25 hr Rupture at 650°C
Plain carbon	7,000	8,000
C-Mo	12,000	12,000
2Cr-Mo	15,000	14,000
5Cr-Mo	17,000	13,000

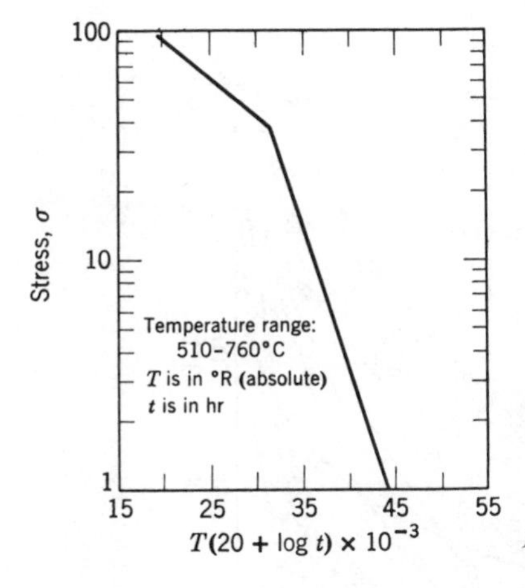

Fig. 15.59 Master rupture curve for C-Mo steel (52).

Data such as these for a variety of stress levels for one material can therefore be plotted on a *master curve* as a function of the parameter $T(20 + \log t)$ as shown in Fig. 15.59; this curve consists of a multitude of data points obtained at temperatures from 510 to 760°C.

In actual practice, for a wide range of materials, plots such as given in Fig. 15.58 do not always converge at a single point, indicating that C may not be a single-valued function. In practice, the value of C is usually picked by trial and error to give minimum scatter of data.

Combined stresses. Another problem of stress-rupture analysis is combined stress under stress-rupture conditions. For example, many cases of failure are known where tubes under internal pressure have spontaneously cracked with little or no evidence of ductility. Thus the possibility of a stress-rupture failure must be carefully considered in tube design since this type of failure could well be the limiting design consideration. In order to determine the stress-rupture condition substitute maximum values of the tensile stresses in Eqs. 15.275 to 15.277 into Eq. 15.221, where σ_0 in this case is the uniaxial stress for rupture of tension specimens. The maximum tensile stresses occur at the outside surface; therefore from Eqs. 15.275 and 15.277 with $r = b$ and substituting into Eq. 15.221, an expression is obtained for allowable pressure as a function of geometry and n, thus,

$$p = (n\sigma_0/\sqrt{3})(R^{2/n} - 1) \qquad 15.295$$

Equation 15.295 can now be used to calculate pressures for stress rupture.

Short-time creep and rapid heating

The design of components for relatively short life (like missiles) requires considerations quite different from those applied to long-time creep. The situation does have one convenient feature, however, in that tests to determine parameters can be extended over the entire expected life of the part. For very short lives

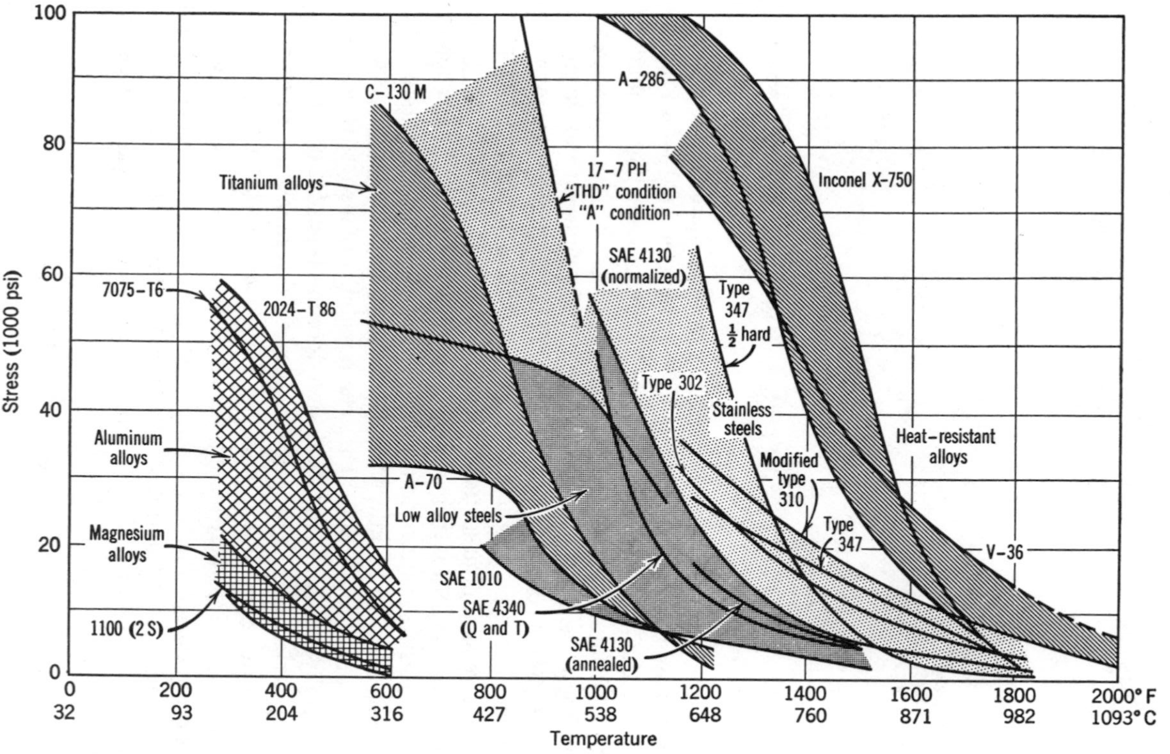

Fig. 15.60 Curves of stress versus temperature for 3% total deformation in 10 minutes. [After Van Echo (67), courtesy of The American Society for Metals.]

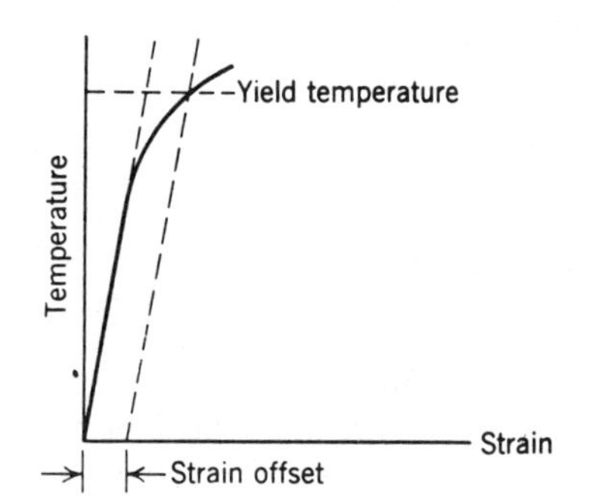

Fig. 15.61 Temperature-strain curve. [After Smith and Robinson (54), courtesy of The American Society for Metals.]

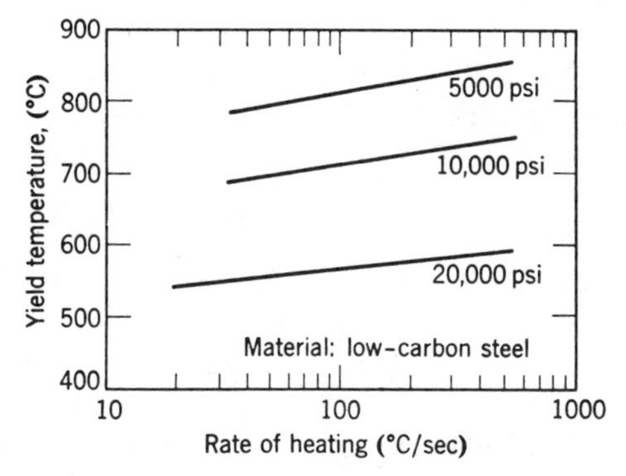

Fig. 15.62 Effect of heating rate on yield temperature. [After Smith and Robinson (54), courtesy of The American Society for Metals.]

(less than a few minutes) strain-rate effects become very important so that ordinary short-time tests do not always provide the correct values of stress and strain required for design. For example, some data by Van Echo (67) are shown in Fig. 15.60 for several alloys. The criterion here is the stress required to produce 3% deformation in 10 min. In longer life tests the stresses would probably all be lower because of creep effects.

Rate of heating is also an important consideration. For example, if temperature is continuously increased while a material is stressed, a temperature strain curve is obtained (Fig. 15.61) which resembles the usual stress-strain curve (71). Here a yield temperature for a given load and heating rate is obtained. An example of a specific effect of heating rate on *yield temperature* is shown in Fig. 15.62 for low-carbon steel (54).

Stress relaxation

If a material is strained to some deformation Δ and then this deformation held constant, the stress required to maintain Δ will diminish as some function of time. This phenomenon is called *stress relaxation* and it has practical implications.

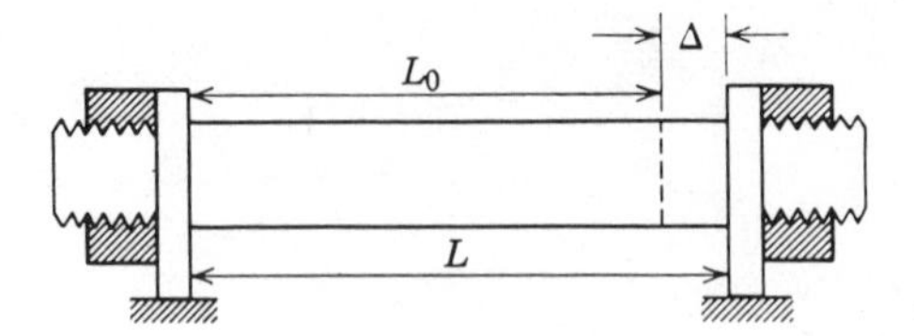

Fig. 15.63 Stress-relaxation specimen.

One of the most common manifestations of stress relaxation is the gradual loosening of joints in flanged connections after the initial pull-up. Consider, for example, the bolt shown in Fig. 15.63. This bolt of original length L_0 is strained an amount Δ and held at this deformation. If the initial stress is σ_0, then the elastic strain produced is

$$\varepsilon = \sigma_0/E \qquad 15.296$$

Now, as time passes the stress diminishes to some value σ because of creep. However, since the bar has a fixed deformation Δ,

$$\sigma_0/E = \sigma/E + \varepsilon' \qquad 15.297$$

where ε' is the strain due to creep. Assuming now the creep law given by Eq. 15.175 applies,

$$(d/dt)(\varepsilon') = B\sigma^n \qquad 15.298$$

Differentiating Eq. 15.297 with respect to time gives

$$(d/dt)(\varepsilon') + (1/E)(d\sigma/dt) = 0 \qquad 15.299$$

or

$$d\varepsilon'/dt = (-1/E)(d\sigma/dt) \qquad 15.300$$

and combining Eqs. 15.298 and 15.300 gives

$$dt = -d\sigma/B\sigma^n E \qquad 15.301$$

which on integrating within the limits of σ and σ_0 gives the result

$$t = \left[\frac{1}{(n-1)(BE)}\right]\left(\frac{1}{\sigma^{n-1}}\right)\left[1 - \left(\frac{\sigma}{\sigma_0}\right)^{n-1}\right] \qquad 15.302$$

where t is the time required for the initial stress σ_0 to decrease to σ.

Equation 15.302 gives a general relationship describing the stress relaxation process. However, there are practical difficulties which should be noted and allowed for in a specific case. First, the relaxation characteristics are not necessarily identical for all initial σ_0 values; usually the higher the initial σ_0 value the more rapid the immediate relaxation will be. The B and n values also differ. Therefore, depending on the problem, the actual conditions should be reproduced as faithfully as possible in the test and B and n values selected by using incremental portions of the creep curve. Another complication is that the material in the assembly holding the bar to a deformation Δ also strains and relaxes and this contributes to the over-all effect. This latter condition has been examined by Marin (44), who gives the following equation (similar to Eq. 15.302) for a practical case of a bolted flange connection:

$$t = \left(\frac{\phi}{BE}\right)\left(\frac{1}{n-1}\right)\left[\frac{1}{\sigma^{n-1}}\left(\frac{1-\sigma^{n-1}}{\sigma_0^{n-1}}\right)\right] \qquad 15.303$$

where $\phi = (1+b)/(1+a)$

a = total elastic deformation of flange at bolt circle divided by total elastic deformation of bolt.

b = total creep of flange at bolt circle divided by total creep of bolt.

In order to maintain a high stress after a given time, the additional elasticity, represented by b, can be increased by using long bolts with collars; this requires larger creep deformation in the bolts before a lower stress level is reached. Bolted joints are also discussed by Waters and by Wesstrom and Bergh; see (74).

Thermal fatigue

The analysis of the fatigue behavior of materials is discouragingly difficult even at room temperature; consequently the problem at elevated temperature is even less amenable to calculation. In a very rough way, the same analyses used for room temperature fatigue discussed in Chapter 12 can be used. Thus the method would be to conduct fatigue tests at elevated temperatures and use the data from these tests in the design formulas. At elevated temperature the most severe condition appears to be complete reversal of load, and design curves such as shown in Fig. 15.64 are commonly used. An example has already been shown in Fig. 15.1 of the effect of temperature on the fatigue characteristics of a turbine-blading alloy. Langer (38) has proposed methods for the application of fatigue data where thermal stresses are also involved, and Tapsell (57) has analyzed creep due to fluctuating loads.

When a part is continuously subjected to thermal fluctuations that give rise to internal stresses, failures frequently occur much the same as in externally

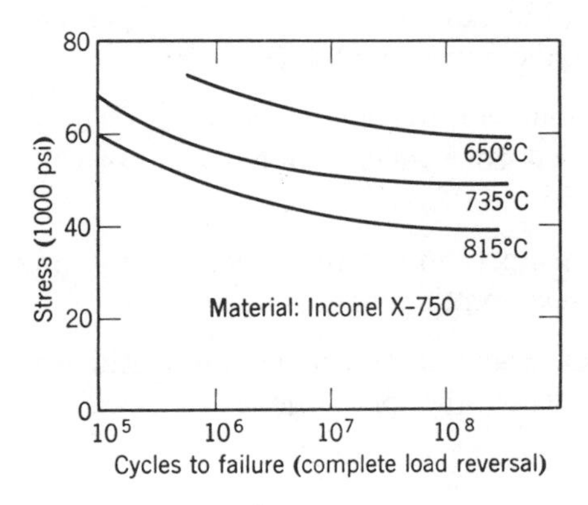

Fig. 15.64 High temperature fatigue data. [After Goldin (22), courtesy of The Institute of Aeronautical Sciences.]

Fig. 15.65 Thermal fatigue cracks in a carburized tray. [*After Avery (2), courtesy of The American Society for Metals.*]

applied loads. Thermal fatigue has also been called *fire cracking* and is considered one of the principal causes of deterioration of many heat-resistant materials. A photograph showing the result of thermal fatigue is shown in Fig. 15.65. This problem has been studied at some length by Avery (2), who concludes that, based on experimental evidence, the cause of thermal fatigue is plastic flow induced by expansion and contraction during heating and cooling.

If the thermal stresses generated are within the elastic limit of the material, then fatigue curves resembling those prepared from observations of external loading effects can be drawn. However, although there is a similarity between thermal and mechanical fatigue there are these important differences:

1. In thermal fatigue plastic deformation tends to concentrate in the regions of highest temperature, thus creating a weak zone which then has to absorb all deformation.

2. In mechanical fatigue at constant temperature there is alternation from tension to compression that tends to induce some strengthening.

3. Variations in temperature in a material can lead to nonhomogeneous structural effects and thus cause anisotropy of other properties.

4. Strain rates are usually different.

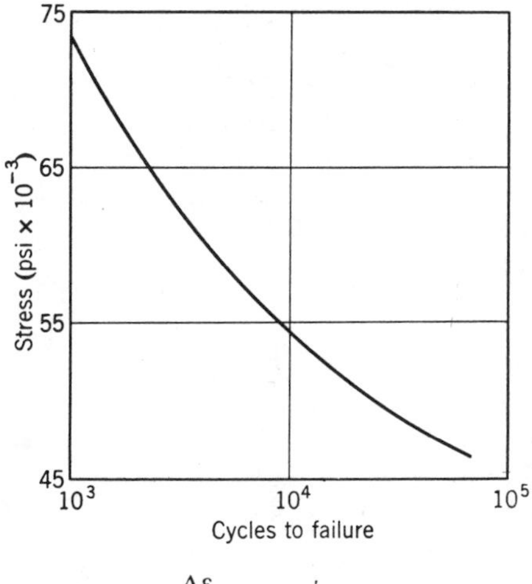

Fig. 15.66 Thermal cycling behavior of type 347 stainless steel. [After Manson (43) courtesy of The Penton Publishing Co.]

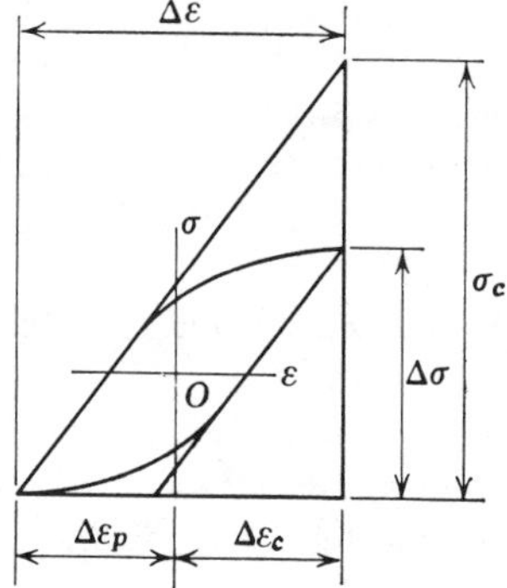

Fig. 15.67 Basic thermal stress fatigue relationship.

Some data illustrating the similarity of thermal and mechanical fatigue are shown in Fig. 15.66

The application of thermal fatigue data to design has been discussed by Coffin (9), who has proposed the following general analysis. Assume that the plastic straining in thermal fatigue is critical; therefore the value of this strain has to be determined. From load-unload curves such as shown in Fig. 15.67,

$$\Delta\varepsilon_p = \Delta\varepsilon - \Delta\sigma/E \qquad 15.304$$

where $\Delta\varepsilon_p$ = plastic strain range

$\Delta\varepsilon$ = total strain range

$\Delta\sigma/E$ = the elastic component

In order to determine the plastic strain assume completely elastic action and convert stresses to elastic strains; then the plastic strain change is

$$\Delta\varepsilon_p = \Delta\varepsilon' - \Delta\sigma/E \qquad 15.305$$

Table 15.8. *Values of C for Various Steels*

Material and Condition	Value of C for Eq. 15.306
347 Stainless (cycled 200°C)	0.78
347 Stainless (cycled 350°C)	0.74
347 Stainless (cycled 500°C)	0.59
347 Stainless (cycled 600°C)	0.56
13% Cr Steel (cycled 20°C)	0.64
Cr-Mo Steel (cycled 20°C)	0.61
304 Stainless (cycled 300°C)	0.75
Cr-Mo Steel (cycled 300°C)	0.57
304 Stainless (cycled 500°C)	0.63
Cr-Mo Steel (cycled 500°C)	0.43

where $\Delta\varepsilon_p$ is the plastic strain change, $\Delta\varepsilon'$ is the fictitious elastic strain calculated by assuming that the material was elastic, and $\Delta\sigma$ is the stress change. The value of $\Delta\varepsilon_p$ is now used in the following equation which fits experimental data:

$$\Delta\varepsilon_p = (C)/\sqrt{N} \qquad 15.306$$

where N is the number of cycles to failure. Values for C are listed in Table 15.8. Coffin gives the following example:*

The temperature in the exhaust port of a diesel engine cycles from 375°C *at idle to* 650°C *at full load. Stress analysis shows that a thermal stress of* 120,000 *psi is developed in the* 13*Cr steel port liner. Estimate the approximate thermal fatigue life of the liner.*

SOLUTION

1. Elastic thermal stress $= \sigma_c = 120{,}000$ psi.
2. Equivalent total strain range $= \Delta\varepsilon = 120{,}000/(30 \times 10^6) = 0.004$ in./in.
3. Assuming that yield is not reached, the elastic strain range $\Delta\varepsilon_c = (2)(30{,}000)/(30 \times 10^6) = 0.002$ in./in.
4. Plastic strain range $= \Delta\varepsilon_p = \Delta\varepsilon - \Delta\varepsilon_c = 0.002$ in./in.
5. Cycles to failure from Eq. 15.306,

$$\sqrt{N}\Delta\varepsilon_p = C = 0.57 \qquad \text{(see Table 15.8)}$$

or

$$N = 81{,}000 \text{ cycles}$$

The problem of thermal fatigue has also been considered in some detail by Manson (43).

A problem associated with thermal fatigue is the situation where cyclic thermal stresses are superposed on some steady stress state in a structure (46). For example, in a vessel under sustained internal pressure, the addition of cyclic

*Courtesy of The McGraw-Hill Publishing Co.

thermal stresses can lead to progressive expansion of the vessel if the stresses are high enough. This expansion caused by thermal cycling has been called the *thermal ratchet effect*; the following brief description is based on the analysis provided by Miller in (46).

Miller's equation for thermal growth is

$$y/\varepsilon_y = 2\{m - 2[m(1-n)]^{1/2}\}(1-\nu)$$

where y is the thermal growth in a pressurized cylinder, ε_y is the yield strain, and

$m = \sigma_t/\sigma_y$

$n = \sigma_p/\sigma_y$

σ_t = thermal stress (dependent on type of thermal gradient)

σ_p = pressure stress

σ_y = yield strength of the material

In the cylinder problem it is desired to have no thermal growth; therefore y is zero and the growth equation becomes

$$0 = m - 2[m(1-n)]$$

From this condition the limits on thermal stress to prevent cyclic growth assuming a linear thermal law are

$$\sigma_t = \sigma_y^2/\sigma_p \qquad \text{for} \qquad 0 < \sigma_p/\sigma_y < 0.50$$

and

$$\sigma_t = 4\sigma_y(1 - \sigma_p/\sigma_y) \qquad \text{for} \qquad 0.5 < \sigma_p/\sigma_y < 1.0$$

Similar computation can be made for a logarithmic thermal gradient.

REFERENCES

1. L. H. Abraham, *Structural Design of Missiles and Spacecraft*, McGraw-Hill Book Co., New York, N.Y. (1962).
2. H. S. Avery, *Met. Prog.*, **76** (2), 67–70 (August 1959).
3. R. W. Bailey, *Proc. Inst. Mech. Eng. (London)*, **131**, 131–344 (1935).
4. (A) P. P. Bijlaard, and R. J. Dohrmann, *Trans. Am. Soc. Mech. Eng. (J. of Eng. for Ind.)*, **83** (4), Series B, 467–475 (November 1961). See also a later paper, 62-WA-226.
   (B) P. P. Bijlaard, *Weld. J. Res. Suppl.*, **33**(12). 615s–623s (December 1954); **34**(12). 608s–617s, (December 1955); *Weld. Res. Council Bull.* No. 50, Welding Research Council, New York, N.Y. (May 1959).
5. D. R. Bland, *J. Mech. Phys. Solids*, **4**, 209–229 (1956)
6. R. V. Blaser, *Proc. Soc. for Exp. Stress Anal.*, **15** (2), 131–142 (1958).
7. B. A. Boley and J. H. Weiner, *Theory of Thermal Stress*, John Wiley and Sons, Inc., New York, N.Y. (1960).
8. C. C. Chang and W. H. Chu, *J. Appl. Mech.*, **21** (2), 101–108 (June 1954).
9. L. F. Coffin, *Prod. Eng.*, **28** (6), 175–179 (June 1957).
10. W. E. Cooper, *Proc. Soc. Exp. Stress Anal.*, **15** (2), 143–147 (1958).
11. H. C. Cross, *Prod. Eng.*, **24** (10), 182–187 (October 1953).

12. N. C. Dahl, Lecture: *Stress Analysis in the Presence of Creep*, Massachusetts Institute of Technology, Cambridge, Mass. (June 1958).
13. R. J. Diefendorf, and E. R. Stover, *Met. Prog.*, **81** (5), 103–108 (May 1962).
14. J. E. Dorn, Ed., *Mechanical Behavior of Materials at Elevated Temperature*, McGraw-Hill Book Co., New York, N.Y. (1961).
15. T. F. Durham, R. M. McClintock, and R. P. Reed, *Cryogenic Materials Data Handbook*, P. B. Report 171809, U.S. Department of Commerce (National Bureau of Standards), Washington, D.C.
16. I. Finnie and W. R. Heller, *Creep of Engineering Materials*, McGraw-Hill Book Co., New York, N.Y. (1959).
17. R. G. Frank and W. F. Zimmerman, *Materials for Rockets and Missiles*, The Macmillan Co., New York, N.Y. (1959).
18. A. M. Freudenthal, "Thermal Stress Analysis and Mechanical Design," in *Nuclear Engineering*, C. F. Bonilla, Ed., McGraw-Hill Book Co., New York, N. Y. (1957). See also I. I. Gol'denblat and N.A. Nikolaenko, *Calculation of Thermal Stresses in Nuclear Reactors*, Consultants Bureau, New York, N.Y. (1964).
19. B. E. Gatewood, *Q. Appl. Math.*, **6** (1), 84–87 (April 1948).
20. B. E. Gatewood, *Thermal Stresses*, McGraw-Hill Book Co., New York, N.Y. (1957).
21. R. M. Goldhoff, *Mat. Des. Eng.*, **49** (4), 93–97 (April 1959). See also Paper 58-A-121, *Trans. Am. Soc. Mech. Eng.*
22. R. Goldin, *Aeron. Eng. Rev.*, **16** (12), 36–41 (December 1957).
23. J. N. Goodier, "Thermal Stresses," in *Design Data and Methods*, 74–77 Am. Soc. Mech. Eng., New York, N.Y. (1953).
24. M. L. Gossard, P. Seide, and W. M. Roberts, *Thermal Bucking of Plates*, Tech. Note No. 1771, NACA, Washington, D.C. (August 1952).
25. R. W. Guard, *Prod. Eng.*, **27** (10), 160–174 (October 1956).
26. J. J. Harwood, *Mat. Methods*, **44** (6), 84–89 (December 1956).
27. N. J. Hoff, *Q. Appl. Math.*, **12** (1), 49–55 (April 1954).
28. N. J. Hoff, *J. Aeron. Soc. (London)*, **61** (536), 756–774 (November 1957).
29. G. Horvay, *Proc. First National Congress of Applied Mechanics*, 247–257, Am. Soc. Mech. Eng. (1952).
30. A. Hurlich, and J. F. Watson, *Met. Prog.*, **79** (4), 65–72 (April 1961).
31. D. J. Johns, *J. Aeron. Space Sci.*, **25**(8), 524–525 (August 1958).
32. A. E. Johnson, J. Henderson, and B. Khan, *Int. J. Mech. Sci.*, **4**, 195–203 (1962).
33. W. Johnson and P. B. Mellor, *Inst. J. Mech. Sci.*, **4**, 147–158 (1962).
34. C. R. Kennedy, W. O. Harms, and D. A. Douglas, *Trans. Am. Soc. Mech. Eng. (J. Basic Eng.)*, **81** (4), Series D, 599–609 (December 1959).
35. E. P. Kingsbury and E. Rabinowicz, *Trans. Am. Soc. Mech. Eng. (J. Basic Eng.)*, **81** (2), Series D, 118–122 (June 1959).
36. W. J. Knapp and F. R. Shanley, *Aeron./Space Eng.*, **17** (12), 34–38 (December 1958).
37. M. Kuraniski, Paper No. 26, "The Behavior of Sandwich Structures Involving Stress, Temperature and Time Dependent Factors," in *Non-Homogeneity in Elasticity and Plasticity*, 323–338, Pergamon Press, New York (1959).
38. B. F. Langer, *Weld. J. Res. Suppl.*, **37** (9), 411s–417s (September 1958).
39. F. R. Larson and J. Miller, *Trans. Am. Soc. Mech. Eng.*, **74** (5), 765–775 (July 1952).
40. T. H. Lin, *J. Appl. Mech.*, **23** (2), 214–218 (June 1956).
41. T. H. Lin, *J. Appl. Mech.*, **25** (1), 75–78 (March 1958).
42. R. R. Maccary, *Chem. Eng.*, **67** (24), 131–136 (November 28, 1960).

43. S. S. Manson, *Mach. Des.*, **30** (6), 114–120 (June 12, 1958); **30** (9), 126–133 (September 4, 1958); also, *Behavior of Materials Under Conditions of Thermal Stress*, National Advisory Committee for Aeronautics (NACA), Report 1170, Washington, D.C. (1954).

44. J. Marin, *Mechanical Properties of Materials and Design*, McGraw-Hill Book Co., New York, N.Y. (1942).

45. R. M. McClintock and H. P. Gibbons, *Mechanical Properties of Structural Materials at Low Temperatures*, U.S. Department of Commerce (National Bureau of Standards), Washington, D.C. (June 1, 1960).

46. D. R. Miller, *Trans. Am. Soc. Mech. Eng. (J. Basic Eng.)*, Series D, **81** (2), 190–196 (June 1959).

47. F. I. Ordway, Ed., *Advances in Space Science*, Vol. 2; see F. L. Bagby. "Materials in Space," 143–213, Academic Press, New York, N.Y. (1960).

48. M. W. Riley, *Mat. Des. Eng.*, **47** (6), 100–104 (June 1958).

49. R. J. Roark, and W. C. Young, *Formulas for Stress and Strain*, 5th ed. McGraw-Hill Book Co., New York, N.Y. (1975).

50. E. L. Robinson, *Trans. Am. Soc. Mech. Eng.*, **74** (5), 777–781 (July 1952); **60** (3), 253–259 (1938).

51. B. Saelman, *Des. News*, **16** (7), 68 (March 27, 1961).

52. M. E. Shank, Lecture: *Utilization of Creep Data for Engineering Design Purposes*, Massachusetts Institute of Technology, Cambridge, Mass. (June 1958).

53. E. Siebel and S. Schwaigerer, *Brennstoff-Wärme-Kraft*, **3** (5), 141–143 (May 1951).

54. W. K. Smith and A. T. Robinson, "Strength of Metals Undergoing Rapid Heating," in *Short-Time High Temperature Testing*, 5–35, Am. Soc. for Metals, Novelty, Oh. (1958).

55. J. Starczewski, *British Chem. Eng.*, **8** (3), 177–179 (March 1963).

56. G. F. Sulfrian, *Chem. Process.*, **24** (7), Part 2, 31–34 (July 1961).

57. H. J. Tapsell, P. G. Forrest, and G. R. Tremain, *Engineering*, **170** (4413), 189–191 (August 25, 1950).

58. A. S. Thompson, *J. Aero. Sci.*, **19** (7), 476–480 (July 1952).

59. S. Timoshenko, *J. Optical Soc. of America*, **11** (3), 233–255 (September 1925).

60. S. Timoshenko and J. M. Gere, *Theory of Elastic Stability*, 2nd ed., McGraw-Hill Book Co., New York, N.Y. (1961).

61. S. Timoshenko, *Strength of Materials*, 3rd ed., Part 1, D. Van Nostrand Co., Princeton, N.J. (1955).

62. S. Timoshenko, *Strength of Materials*, Part 2, 3rd ed., D. Van Nostrand Co., Princeton, N.J. (1956).

63. S. Timoshenko and S. Woinowsky-Kreiger, *Theory of Plates and Shells*, 2nd ed., McGraw-Hill Book Co., New York, N.Y. (1959).

64. T. Wah, *J. Franklin Inst.*, **272** (1), 45–60 (July 1961).

65. A. M. Wahl, *J. Appl. Mech.*, **24** (2), 299–305 (June 1957).

66. N. A. Weil, *Trans. New York Acad. Sci.*, **24** (3), 291–305 (January 1962).

67. J. A. Van Echo, "Short-Time Creep of Structural Sheet Metals," in *Short-Time High Temperature Testing*, 58–91, Am. Soc. for Metals, Novelty, Oh. (1958).

68. B. Venkatraman, *Solution of Some Problems in Steady Creep*, Report PIBAL-402, Polytechnic Institute of Brooklyn, Brooklyn, N.Y. (July 1957).

69. J. F. Wygant, *J. Am. Ceram. Soc.*, **34** (12), 374–380 (December 1951).

70. *American Institute of Physics Handbook*, McGraw-Hill Book Co., New York, N.Y. (1972).

71. *Short-Time High Temperature Testing*, Am. Soc. for Metals, Novelty, Oh. (1958).

72. *Steel*, 70–73 (October 3, 1960); also *Missiles and Rockets*, **9** (25), 32–33 (December 18, 1961).

73. *ASTM Symposium on Effect of Temperature on the Brittle Behavior of Metals with Particular Reference to Low Temperature*, ASTM Special Tech. Pub. No. 158 (1954).

74. *Pressure Vessel and Piping Design*, Am. Soc. for Mech. Eng., New York, N.Y. (1960).

75. *Piping Design and Engineering*, The Grinnell Co., Providence, R.I. (1951).

76. *Piping Design*, 2nd ed., John Wiley and Sons, New York, N.Y. (for M. W. Kellogg Co.) (1956).

77. *Catalog TRU-1*, Metals Controls Corporation, Attleboro, Mass.

78. *Temperature and Thermal Stress Distribution in Some Structural Elements Heated at a Constant Rate*, NACA Tech. Note No. 4306, by W. A. Brooks, Washington, D.C. (August 1958).

79. *Symposium on Stress-Strain-Time-Temperature Relationships in Materials*, ASTM Special Publication 325, Am. Soc. for Testing and Materials, Philadelphia, Pa. (1962).

80. "Mechanical Design for Low Temperature Piping and Vessels, a special report by V. L. Surdi, D. Romaine, L. P. Zick, and M. B. Clapp, *Hydrocarbon Process, Pet. Refiner*, **43** (6), 115–132 (June 1964).

81. H. L. Martin et al., *Effects of Low Temperature on the Mechanical Properties of Structural Metals*, NASA SP-5012 (01), National Aeronautics and Space Administration, Washington, D.C. (1968).

PROBLEMS

15.1 A 10-in. standard schedule 40 steel pipe, 1500 ft long, is subjected to a temperature variation of 300°F. Assume the pipe to be fixed at both ends, and find the size of an expansion loop for this application.

15.2 A 1-in. diameter copper and a 2-in. diameter aluminum rod are held between fixed supports (Fig. P 15.2). The temperature varies linearly from the left side to the right side by 150°F. What are the stresses in the members? If they were free to expand, what would the new lengths of the rods be?

15.3 A long, thick wall pipe is subjected to an internal pressure of 2000 psi at a temperature of 480°C. If material in the pipe is that of Fig. 15.35, find the maximum radial and tangential stresses in the wall and the corresponding tangential strain rate. How much will the pipe increase in diameter after 10,000 hr?

15.4 Select a material that will allow a 6-in. inside diameter boiler cylinder to operate at 5 kpsi internal pressure at 1400° F for five years without rupture.

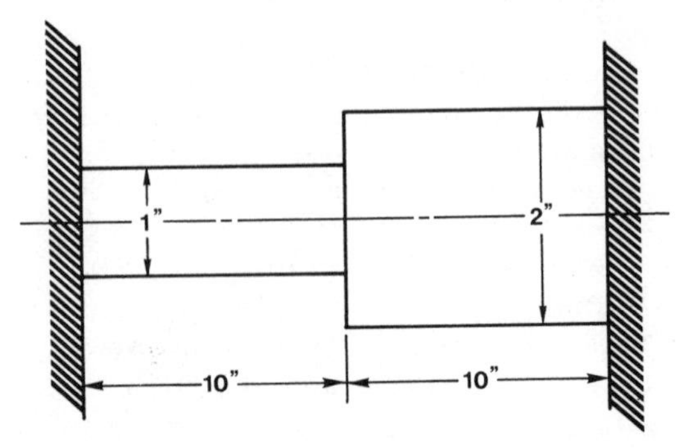

Fig. P15.2

15.5 In Problem 15.2, if the temperature is 150°F higher on the bottom of the rod and the temperature variation is linear from top to bottom, find the maximum stresses in the rods. If the rods were released on the left end, find the tip deflection and slope.

15.6 A 3-in. diameter steel ball at room temperature slips through a ring with zero clearance. If the ball is heated in oil to 400°F and set on the ring at room temperature, find how long the ball would sit on the ring before it falls through.

15.7 Two proposed designs are shown (Figs. P15.7*A* and P15.7*B*). Each has a

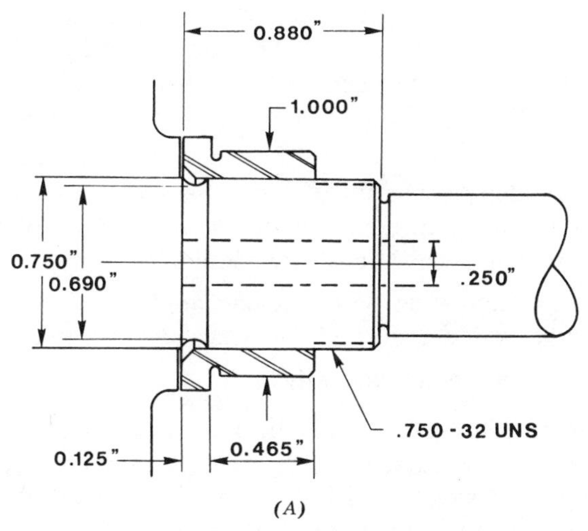

(*A*)

Fig. P15.7A

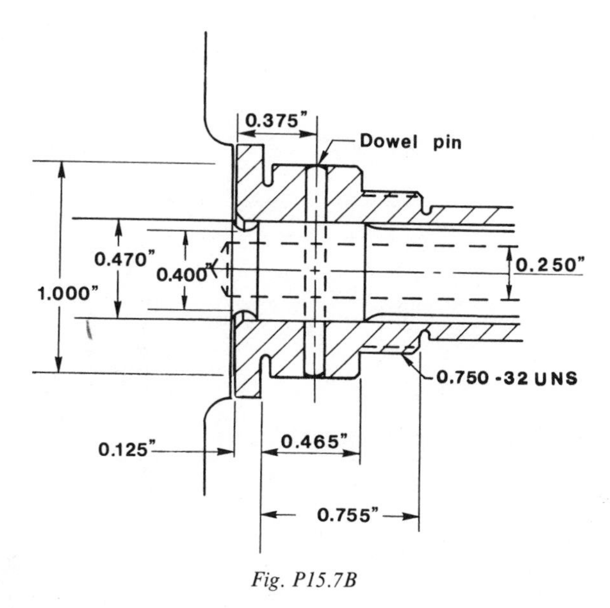

Fig. P15.7B

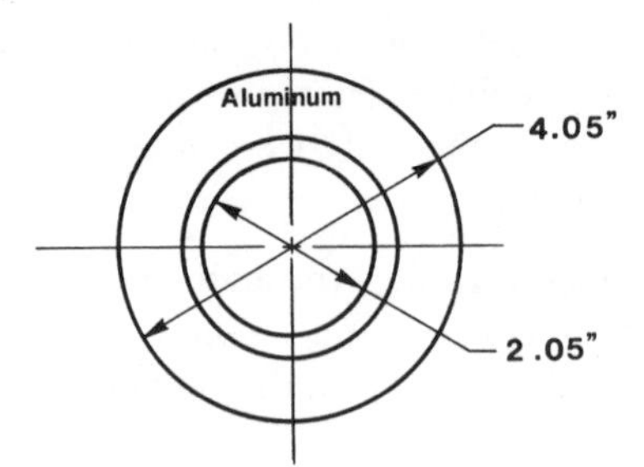

Fig. P15.8

thermal environment of $-70°F$ to $135°F$ and has preloaded steel bearings on the 1.00-in. diameter surfaces. They have 0.0002 in. clearance before the bearings make surface contact. The shafts are aluminum (6061-T6) with α of 13.9×10^{-6} in./in.°F and the collars are stainless steel (416) with α of 5.5×10^{-6} in./in.°F. Find:

a. The thermal expansion for both designs and the press fit required to maintain contact throughout the thermal range. What effect does the 0.250-in. hole have on the two designs?

b. Find the effects of an 85-lb preload applied by the 0.750–32 UNS threads on both designs and the critical areas of stress.

c. Which design is better and why?

15.8 An aluminum housing that holds the bearings in Problem 15.7 requires a (416) stainless steel insert to prevent 0.0002 in. thermal expansion and seizing of the bearing (Fig. P15.8). The bearing width is 0.500 in. Find the thickness of the steel insert and the press fit to maintain surface contact.

appendix A

CENTER OF GRAVITY—CENTROIDS

The center of gravity, or centroid, of any geometrical figure such as a line, curve, area, or volume is that point through which a single force must act to support the body. It is independent of the choice of axes. The mathematical development of centroids is given in texts on calculus and applied mathematics and will not be repeated here. In the following paragraphs the final results will be given with examples of application to specific problems.

A-1 CENTROID OF A LINE

The centroid of any straight line is at its mid-length; for a curved line the following analysis is required. Consider Fig. A1, showing a curve with respect to cartesian coordinates x and y. The length of the line s is given by

$$s = \int ds = \int_a^b [\sqrt{1 + (dy/dx)^2}]\, dx \tag{A1}$$

The centroid is then

$$\bar{x} = \frac{\int x\, ds}{s} = \frac{\int x[\sqrt{1 + (dy/dx)^2}]\, dx}{s} \tag{A2}$$

and

$$\bar{y} = \frac{\int y\, ds}{s} = \frac{\int y[\sqrt{1 + (dx/dy)^2}]\, dy}{s} \tag{A3}$$

Example. Centroid of a section of a circular arc, Fig. A2.

$$\bar{x} = \frac{\int x\, ds}{s} = \frac{\int_\theta^0 x[\sqrt{1 + (dy/dx)^2}]\, dx}{R\theta} \tag{A4}$$

However,

$$R^2 = x^2 + y^2,$$

so that

$$0 = 2x\, dx + 2y\, dy$$

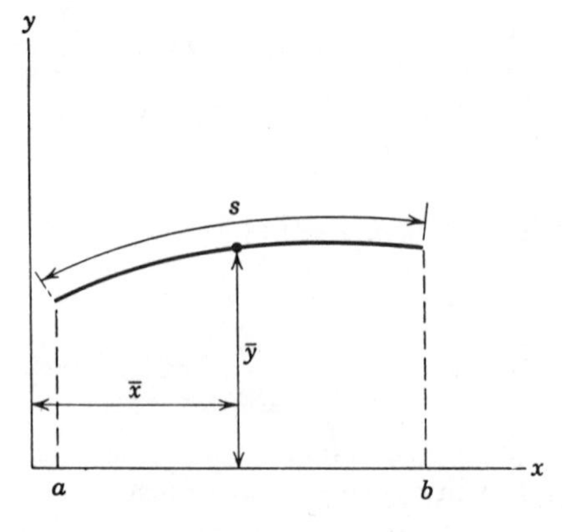

Fig. A1 General line in Cartesian coordinates.

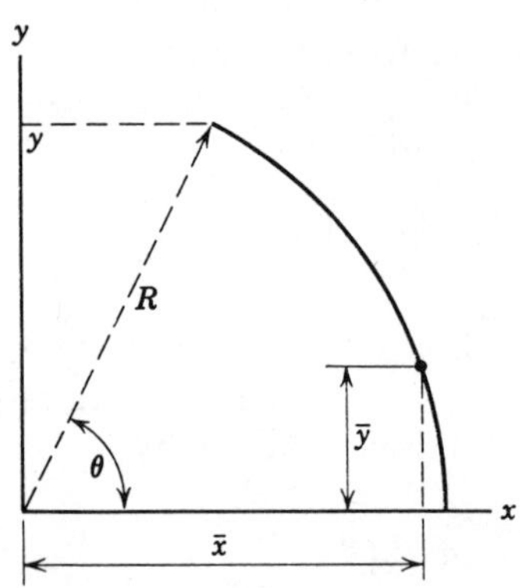

Fig. A2. Section of a circular arc.

from which

$$dy/dx = -x/y \tag{A5}$$

Therefore, using Eq. A5,

$$\bar{x} = \frac{\int_{\theta}^{0} x(\sqrt{1 + x^2/y^2})dx}{R\theta} \tag{A6}$$

or

$$\bar{x} = \frac{\int_{\theta}^{0} \frac{x}{y} R\,dx}{R\theta} = \frac{\int_{\theta}^{0} \frac{xR\,dx}{\sqrt{R^2 - x^2}}}{R\theta} = \frac{-R^3}{R\theta} \int_{\theta}^{0} \frac{\sin\theta\cos\theta\,d\theta}{\sqrt{R^2 - R^2\cos^2\theta}}$$

from which

$$\bar{x} = \frac{R\sin\theta}{\theta} \tag{A7}$$

Similarly,

$$\bar{y} = \frac{\int_{0}^{\theta} y(\sqrt{1 + (dx/dy)^2})\,dy}{R\theta} = \frac{R}{\theta}(1 - \cos\theta) \tag{A8}$$

A-2 CENTROID OF AN AREA

The centroid of a plane area is given by

$$\bar{x} = \frac{\int x\,dA}{A} \qquad \text{A9}$$

and

$$\bar{y} = \frac{\int y\,dA}{A} \qquad \text{A10}$$

For the triangular area shown in Fig. A3, for example,

$$\bar{y} = \frac{\int y\,dA}{A} = \frac{\int_0^h yx\,dy}{bh/2} = \int_0^h \frac{y[b(h-y)]\,dy}{bh^2/2} \qquad \text{A11}$$

from which

$$\bar{y} = h/3 \qquad \text{A12}$$

As another example consider the area enclosed by the arc of a circle of radius

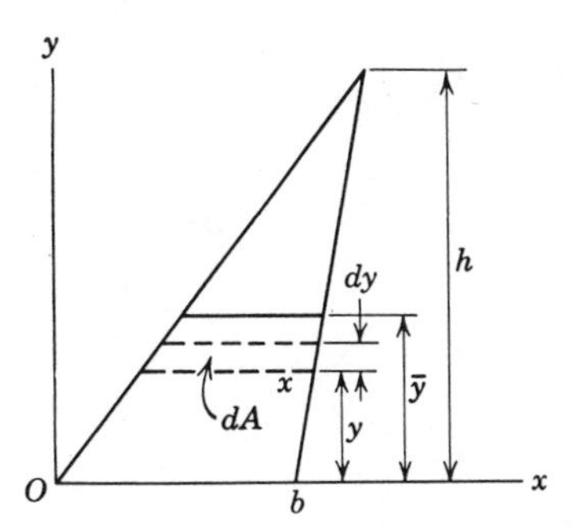

Fig. A3 General triangle.

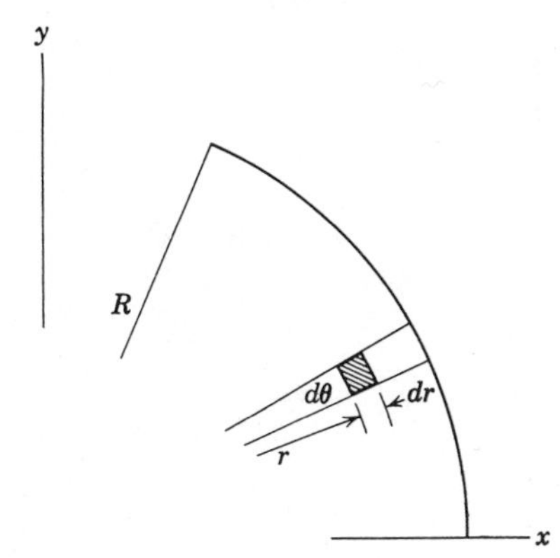

Fig. A4 Area enclosed by arc of a circle.

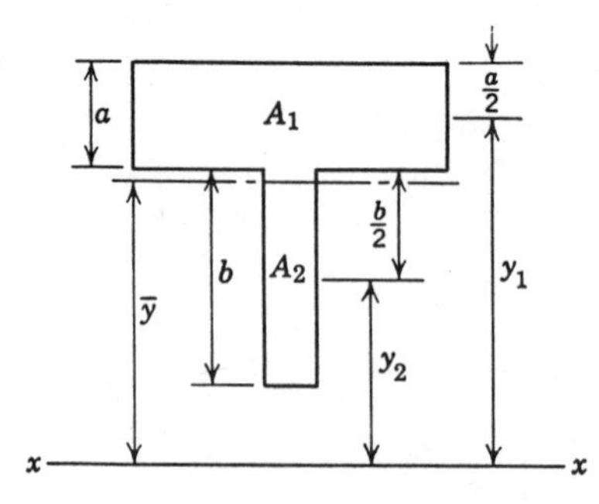

Fig. A5 Symmetrical structural section.

R, Fig. A4.

$$\bar{x} = \frac{\iint x\,dA}{A} = \frac{\int_0^{\theta}\int_0^{R} r\cos\theta(r\,d\theta\,dr)}{\dfrac{\pi R^2}{(2\pi/\theta)}} \qquad \text{A13}$$

from which

$$\bar{x} = \frac{2R\sin\theta}{3\theta} \qquad \text{A14}$$

If $\theta = 60°$, for example,

$$(\bar{x})_{60°} = \frac{2R\sqrt{3}}{(2)(\pi/3)(3)} = \frac{R\sqrt{3}}{\pi} \qquad \text{A15}$$

Consider the section shown in Fig. A5. In this case,

$$\bar{y} = \frac{\int y\,dA}{A} = \frac{A_1 y_1 + A_2 y_2}{A_1 + A_2} \qquad \text{A16}$$

A-3 CENTROID OF A SOLID

The centroid of any solid may be found from the relationship

$$\bar{x} = \frac{\iiint x\,dv}{v} \qquad \text{A17}$$

For example, consider the parallelopiped in Fig. A6; for this case

$$\bar{x} = \frac{\int_0^{a}\int_0^{b}\int_0^{c} x\,dx\,dy\,dz}{abc} = \frac{a}{2} \qquad \text{A18}$$

Similarly,

$$\bar{y} = b/2 \qquad \text{A19}$$

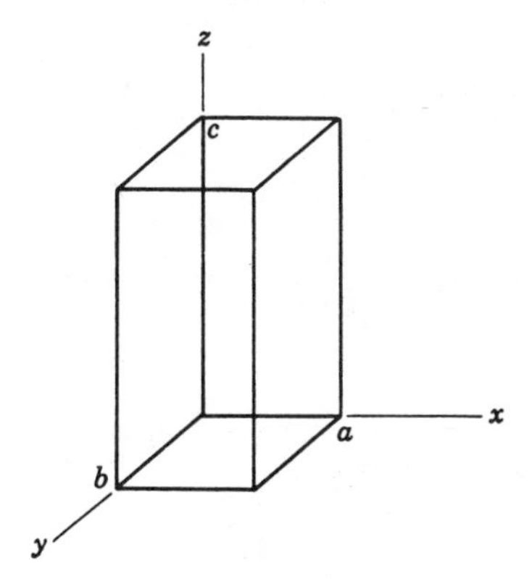

Fig. A6 Solid parallelopiped.

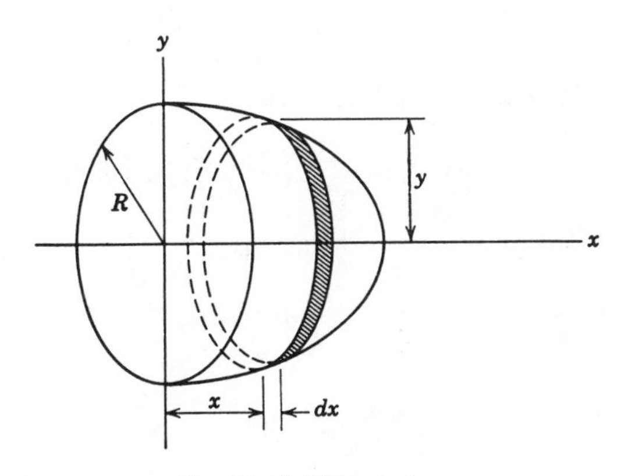

Fig. A7 Solid hemisphere.

and

$$\bar{z} = c/2 \tag{A20}$$

Triple integration is not always necessary. For example, consider the hemisphere shown in Fig. A7; here

$$\bar{x} = \frac{\iiint x\pi y^2\,dx}{\frac{1}{2}(\frac{4}{3}\pi R^3)} = \frac{6\int_0^R \pi R(R^2 - x^2)\,dx}{\pi R^3} = \frac{3R}{8} \tag{A21}$$

For the tetrahedron shown in Fig. A8, bounded by the planes $x/a + y/b + z/c = 1, x = 0,\ y = 0,$ and $z = 0$, the line ab is given by

$$x/a + y/b = 1 \tag{A22}$$

so that

$$\bar{x} = \frac{\int_0^b \int_0^{a(1-y/b)} x\,dv}{v} = \frac{\int_0^b \int_0^{a(1-y/b)} xc\left(1 - \frac{x}{a} - \frac{y}{b}\right)dx\,dy}{abc/6}$$

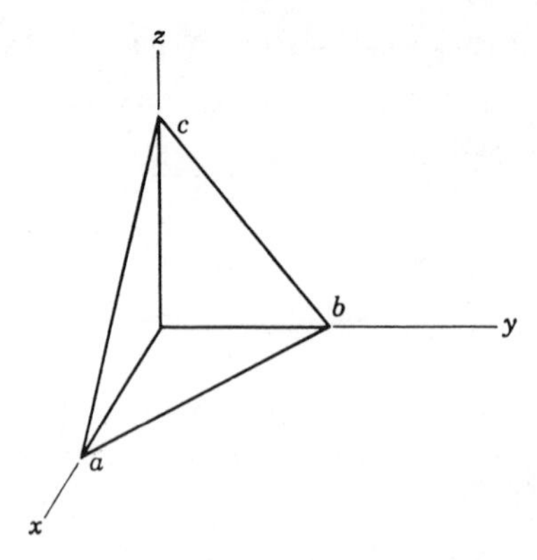

Fig. A8 Solid tetrahedron.

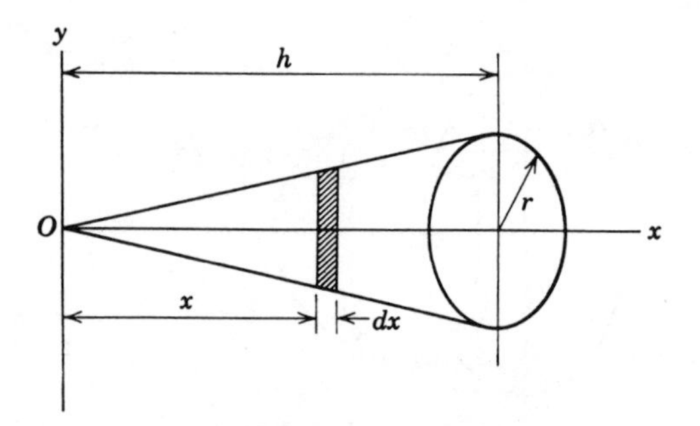

Fig. A9 Solid right circular cone.

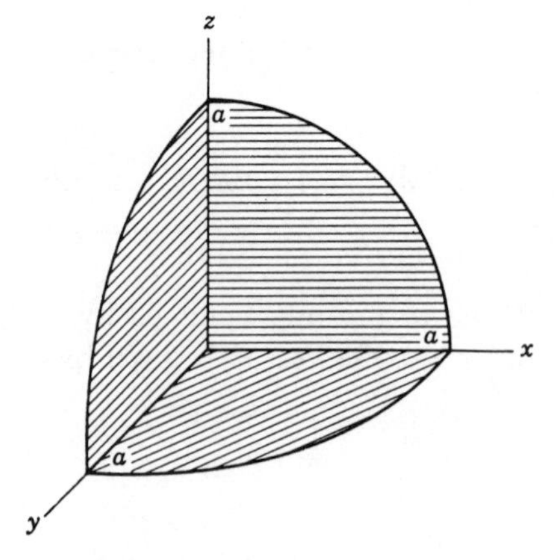

Fig. A10 Solid octant of a sphere.

or

$$\bar{x} = a/4 \qquad \text{A23}$$

Similarly, $\bar{y} = b/4$ and $\bar{z} = c/4$. Consider the cone shown in Fig. A9; for this solid of revolution

$$\bar{x} = \int_0^h \frac{x\,dv}{v} = \int_0^h \frac{x\pi y^2\,dx}{\pi r^2 h/3} = 3\int_0^h \frac{\pi x^2 r^2 (x\,dx)}{h^2(\pi r^2 h)}$$

or

$$\bar{x} = 3h/4 \qquad \text{A24}$$

As another example consider the octant of a sphere shown in Fig. A10. This solid is bounded by the three coordinate planes and the spherical surface $x^2 + y^2 + z^2 = a^2$. Here, as in some others, it is convenient to use either cylindrical or spherical coordinates, which (see Fig. A11) have the following relationship with cartesian coordinates.

Cartesian System	Cylindrical Coordinates	Spherical Coordinates
x	$\rho \cos\theta$	$\rho \sin\phi \cos\theta$
y	$\rho \sin\theta$	$\rho \sin\phi \sin\theta$
z	z	$\rho \cos\phi$
r	ρ	$\rho \sin\phi$

where θ is the longitude angle, ϕ the colatitude angle,

$$\rho = \sqrt{r^2 + z^2}$$

and

$$\cos\phi = \frac{z}{\sqrt{r^2 + z^2}}$$

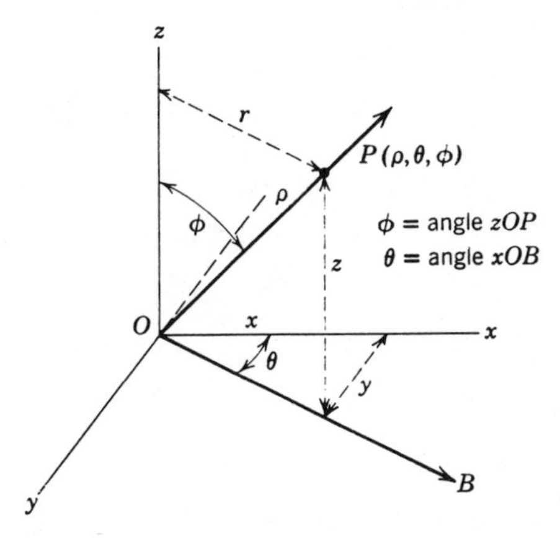

Fig. A11 Geometric relationship between Cartesian, Cylindrical, and Spherical coordinates.

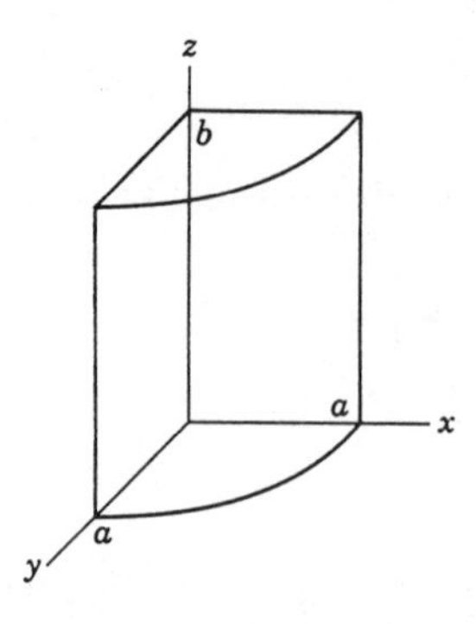

Fig. A12 Solid quarter of a right circular cylinder.

Thus in spherical coordinates for Fig. A10

$$\bar{x} = \frac{\iiint x\,dv}{v} = \frac{\int_0^{\pi/2}\int_0^{\pi/2}\int_0^{a} (\rho \sin\phi \cos\theta)(\rho^2 \sin\phi\, d\rho\, d\phi\, d\theta)}{\pi a^3/6}$$

or

$$\bar{x} = 3a/8 \qquad \text{A25}$$

As a final example, consider the cylindrical body in Fig. A12 bounded by the surface $x^2 + y^2 = a^2$ and the planes $z = 0$ and b. Using cylindrical coordinates

$$\bar{x} = \frac{\iiint x\,dv}{v} = \frac{\int_0^{a}\int_0^{\pi/2}\int_0^{b} (\rho \cos\theta)\rho\, d\rho\, d\theta\, dz}{\pi a^2 b/4}$$

or

$$\bar{x} = 4a/3\pi \qquad \text{A26}$$

A general compilation of results is given by E. W. Jenkins, *Product Engineering*, **34** (6), 70–77, (*March* 18, 1963).

appendix B

MOMENT OF INERTIA

Throughout the field of mechanics certain integral expressions frequently appear. For example, in many problems the following integrals

$$\int x^2\,dA; \quad \int y^2\,dA; \quad \int r^2\,dA \tag{B1}$$

representing moments of inertia appear. These expressions refer to plane areas; for solids of density γ or mass M the moment of inertia is expressed as

$$I = \int r^2\,dM \tag{B2}$$

Another integral frequently appearing is that for the *product of inertia*

$$\int xy\,dA \qquad \text{or} \qquad \int xy\,dM \tag{B3}$$

In the following paragraphs methods for determining these integrals for various geometrical cases will be given.

B-1 PLANE AREAS

In determining moments of inertia of plane areas the moment is ordinarily determined with respect to an axis in the plane. For example, for the rectangular cross section shown in Fig. B1, the moment of inertia with respect to the symmetry axis x is

$$I_x = 2\int_0^{h/2} z^2\,dA = 2\int_0^{h/2} z^2(b\,dz) = \frac{bh^3}{12} \tag{B4}$$

Correspondingly, the moment of inertia with respect to the symmetry axis z is

$$I_z = hb^3/12 \tag{B5}$$

For a triangle, Fig. B2, the moment of inertia with respect to the base is

$$I_x = \int_0^h z^2\,dA = \int_0^h z^2\left(\frac{h-z}{h}\right)b\,dz = \frac{bh^3}{12} \tag{B6}$$

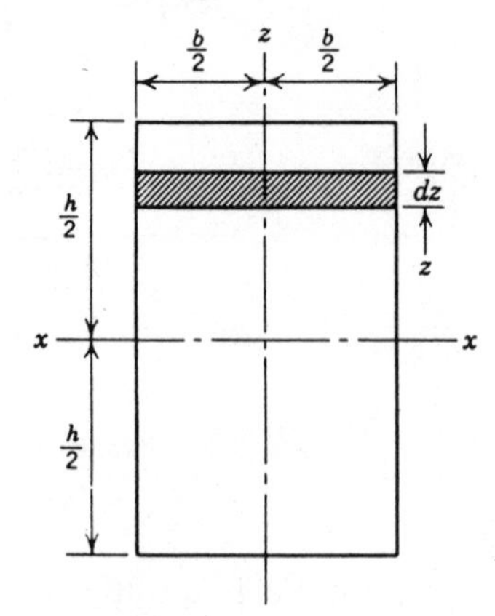

Fig. B1 Element of rectangular cross section.

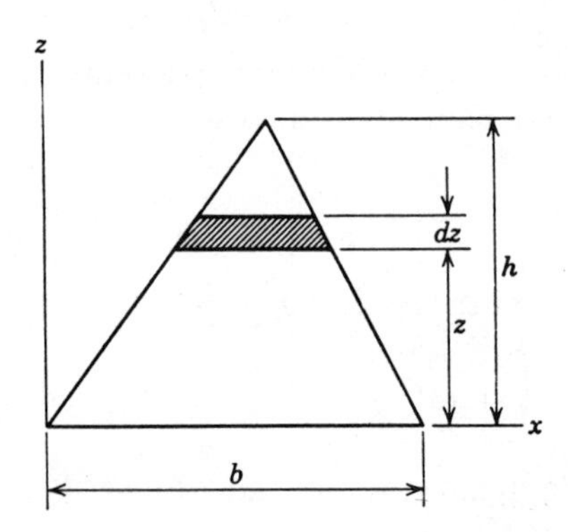

Fig. B2 Element of triangular cross section.

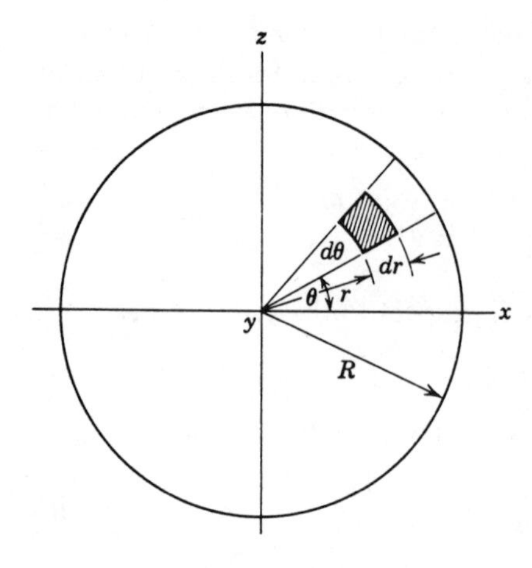

Fig. B3 Element of circular cross section.

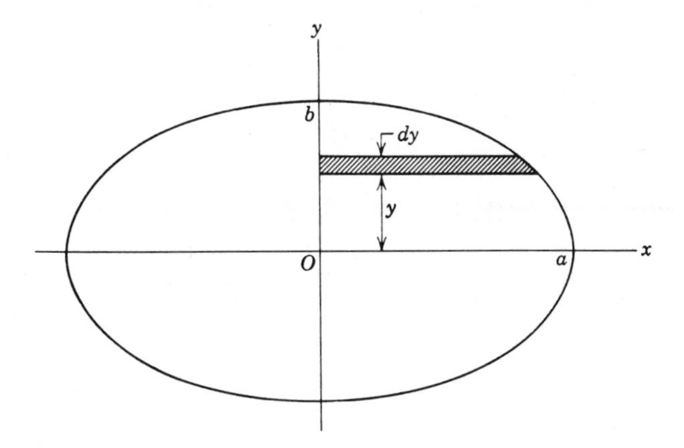

Fig. B4 Element of elliptic cross section.

For a circle, Fig. B3, the moment of inertia with respect to the diameter is

$$I_x = \int z^2\, dA = \int_0^R \int_0^{2\pi} (r^2 \sin^2 \theta) r\, d\theta\, dr = \frac{\pi R^4}{4} \qquad \text{B7}$$

When the axis of the body is normal to the plane of the cross section, a quantity called the *polar moment of inertia* is calculated. This quantity is used in torsional problems, for example. For an area of circular cross section as shown in Fig. B3, the polar moment of inertia with respect to the y axis is

$$J = \int r^2\, dA = \int_0^R \int_0^{2\pi} (r^2 r\, d\theta)\, dr = \frac{\pi R^4}{2} \qquad \text{B8}$$

From Eqs. B7 and B8 it is evident that $J = \int r^2\, dA = \int (x^2 + z^2)\, dA = I_x + I_z$, that is, the polar moment of inertia is equal to the sum of the plane moments of inertia about the symmetry axes.

Consider now the ellipse shown in Fig. B4 whose equation is

$$x^2/a^2 + y^2/b^2 = 1$$

Thus

$$I_x = \int y^2\, dA = 4 \int_0^b y^2 x\, dy = \frac{4a}{b} \int_0^b (y^2 \sqrt{b^2 - y^2}) dy$$

or

$$I_x = \frac{4a}{b} \int_0^b (y^2 \sqrt{b^2 - y^2}) dy$$

from which

$$I_x = \pi a b^3/4 \qquad \text{B9}$$

Similarly,

$$I_y = \pi b a^3/4 \qquad \text{B10}$$

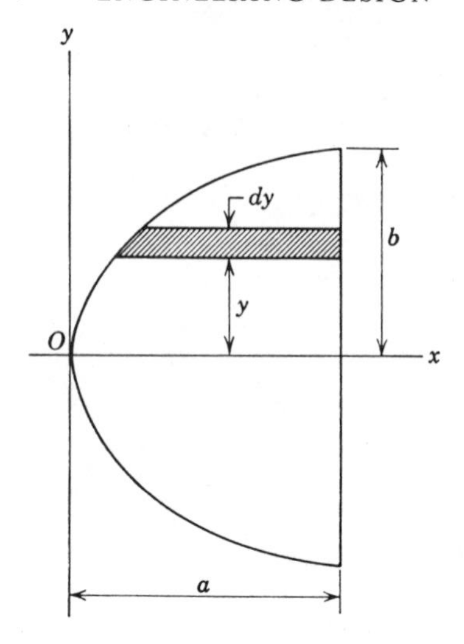

Fig. B5 Element of parabolic cross section.

and the polar moment of inertia is

$$J = I_x + I_y = (\pi ab/4)(b^2 + a^2) \qquad \text{B11}$$

For the parabolic segment shown in Fig. B5 whose equation is

$$y^2 = 4kx$$

$$I_x = \int y^2 \, dA = 2\int_0^b y^2(a - x)\, dy$$

From the above and Fig. B5, $b^2 = 4ka$, from which

$$k = b^2/4a$$

so that

$$I_x = 2\int_0^b y^2\left(a - \frac{4y^2 a}{4b^2}\right)dy = \frac{4ab^3}{15} \qquad \text{B12}$$

Similarly,

$$I_y = 4a^3b/7 \qquad \text{B13}$$

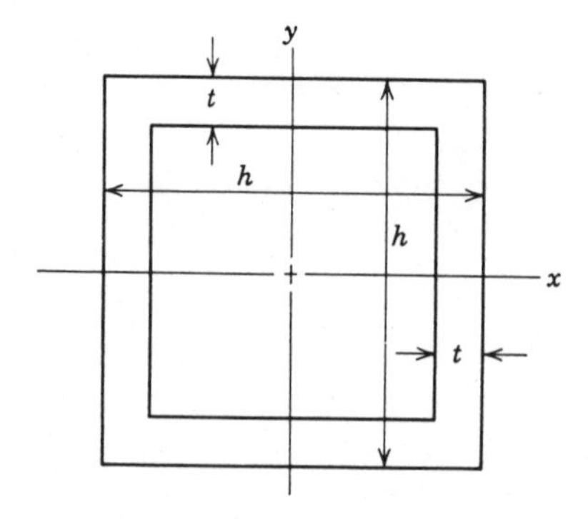

Fig. B6 Element of symmetrical square cross section.

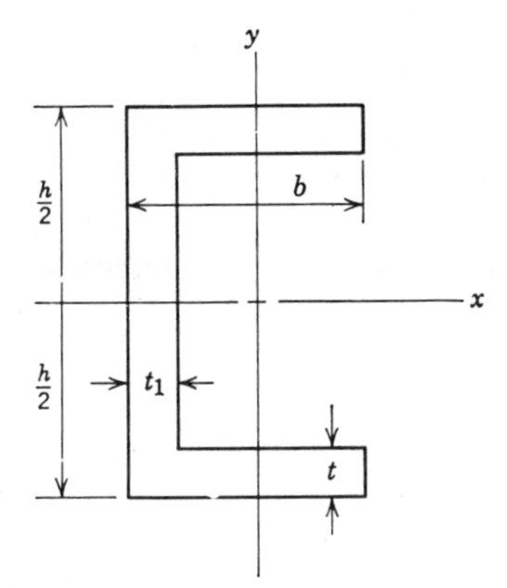

Fig. B7 Symmetrical structural cross section.

In many cases the cross section consists of a variety of shapes, and, for these, the moment of inertia may be determined by considering the individual inertias of the parts making up the cross section. For example, consider the section shown in Fig. B6. From Eq. B5,

$$I_x = \tfrac{1}{12}bh^3 = \tfrac{1}{12}h(h^3) - \tfrac{1}{12}(h-2t)(h-2t)^3$$

or

$$I_x = h^4/12 - (h-2t)^4/12 \qquad \text{B14}$$

For the section shown in Fig. B7,

$$I_x = \frac{bh^3}{12} - \frac{(b-t_1)(h-2t)^3}{12} \qquad \text{B15}$$

Another type of structural shape that requires considerable computation in the derivation of the moment of inertia is the corrugated shape. For approximate results a procedure has been developed where both the neutral axis and the moment of inertia can be deduced for a wide range of corrugated-type cross sections.* In this method the equations are as follows:

$$\bar{y} = \frac{K_1 t(2 - K_3 - K_2 K_3)}{2} \qquad \text{B16}$$

and

$$I = (Lt^3/12)(K_1{}^2 A + 1) = Lt^3 B/12 \qquad \text{B17}$$

where $\bar{y}$ is the distance from the base line to the neutral axis (see Fig. B8), A, K_2, and K_3 are parameters obtained from Fig. B9, and the other symbols are defined by Fig. B8. The symbol B is a magnification factor indicating, with respect to other corrugations, how much stiffer this particular one is. The symbol K_1 is the ratio d/t.

*E. A. Phillips, *Product Engineering*, **31** (37), 75–77 (September 12, 1960). Included here through courtesy of The McGraw-Hill Publishing Co.

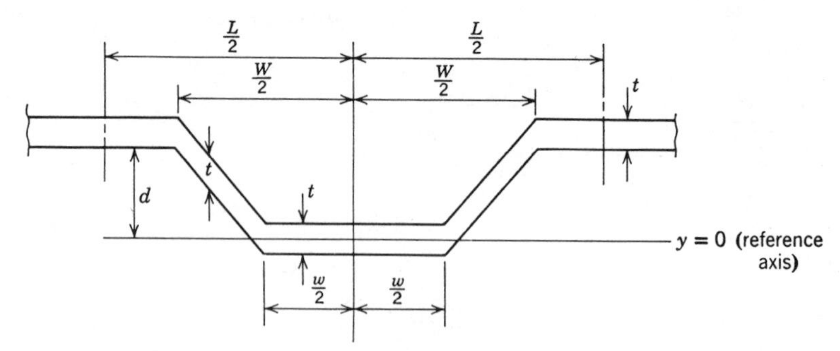

Fig. B8 Geometry of a pressed corrugation.

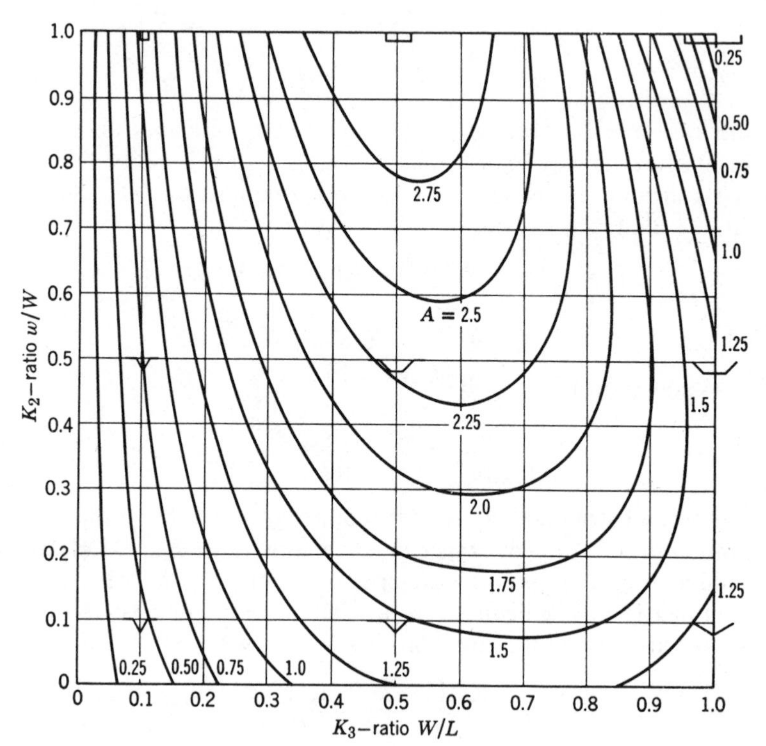

Fig. B9 Parameter chart for calculation of corrugation inertia.

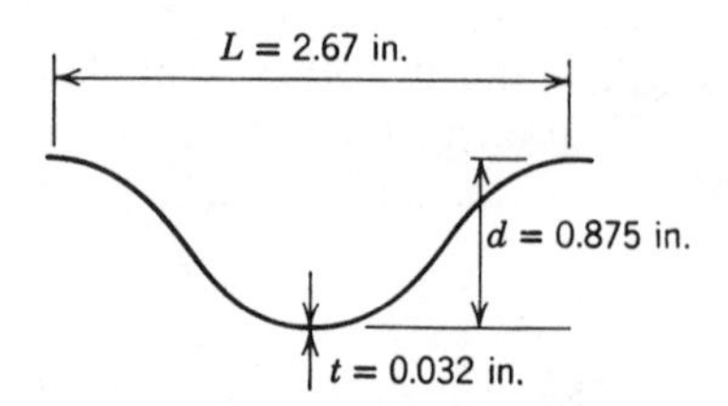

Fig. B10 Constant thickness curved web.

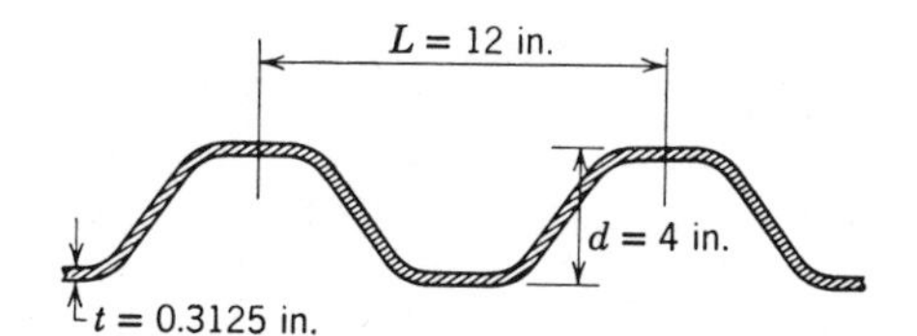

Fig. B11 Constant thickness corrugation with straight webs.

Consider Fig. B10, where

$$W = L/2 + d = 2.67/2 + 0.875 = 2.21 \text{ in.}$$
$$w = L/2 - d = 2.67/2 - 0.875 = 0.56 \text{ in.}$$
$$K_1 = d/t = 0.875/0.032 = 27.4$$
$$K_2 = w/W = 0.56/2.21 = 0.25$$
$$K_3 = W/L = 2.21/2.67 = 0.83$$

A (from Fig. B9) $= 1.75$

$$B = (K_1{}^2 A + 1) = (27.4)^2(1.75) + 1 = 1310$$

and

$$I = \frac{Lt^3 B}{12} = \frac{2.67(0.032)^3(1310)}{12} = 0.0095 \text{ in.}^4 \text{ per corrugation}$$

Finally,

$$I = \frac{(0.0095)(12)}{2.67} = 0.043 \text{ in.}^4 \text{ per foot of width.}$$

As another example consider the section shown in Fig. B11. By measurement $W = 8.8$ and $w = 3.4$; then

$$K_1 = d/t = 4.0/0.3125 = 12.8$$
$$K_2 = w/W = 3.4/8.8 = 0.386$$
$$K_3 = W/L = 8.8/12 = 0.73$$

A (from Fig. B9) $= 2.1$

$$B = (K_1{}^2 A + 1) = (12.8)^2(2.1) + 1 = 345$$

and

$$I = \frac{Lt^3 B}{12} = \frac{12(0.3125)^3(345)}{12} = 10.5 \text{ in.}^4 \text{ per corrugation.}$$

Transfer of axes

If the moment of inertia is known for an area with respect to its centroid, then the moment of inertia with respect to any parallel axis can be computed from the *parallel axis theorem*

$$I = I_c + Ad^2 \qquad \text{B18}$$

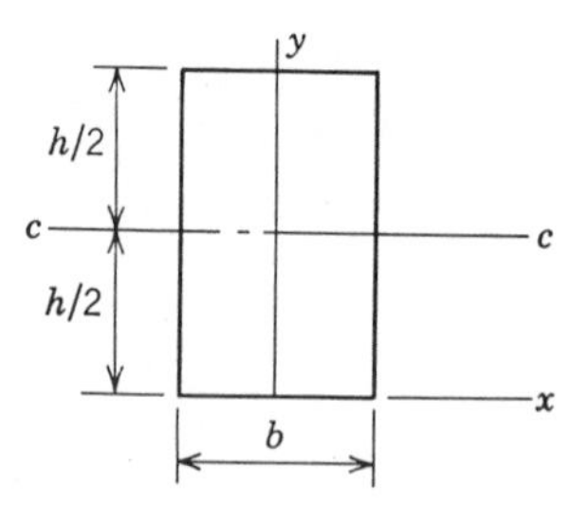

Fig. B12 Element of rectangular cross section.

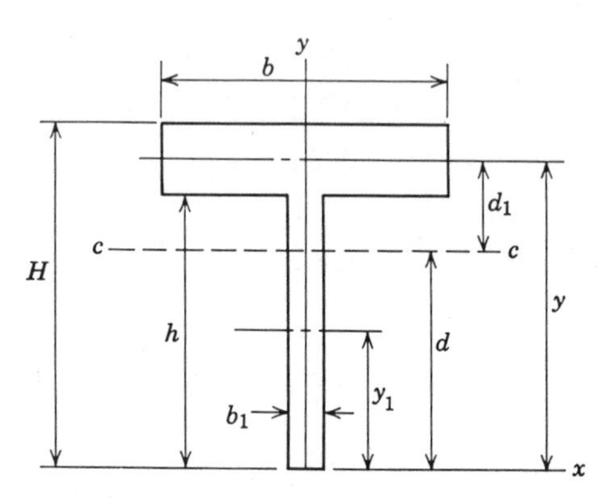

Fig. B13 Symmetrically shaped structural element.

where I_c is the certroidal moment of inertia, A the cross-sectional area, and d the distance from the centroidal axis to the axis in question. For example, for the section shown in Fig. B12

$$I_x = I_c + Ad^2 = \tfrac{1}{12}bh^3 + bh(h/2)^2 = bh^3/3 \qquad \text{B19}$$

For the section shown in Fig. B13 the value of I_c is found by first using Eq. A9 to locate the centroid

$$\bar{y} = \frac{\int y\,dA}{A} = \frac{(H-h)(b)y + hb_1y_1}{(H-h)(b) + b_1h} \qquad \text{B20}$$

where y is the distance from x to the centroid of the area $(H-h)\,(b)$ and y_1 is the distance $h/2$. Having Eq. B20,

$$I_c = [(b/12)(H-h)^3 + (H-h)(b)(d_1)^2] + [(b_1h^3/12) + b_1h(d-y_1)^2] \qquad \text{B21}$$

Similarly,

$$I_x = I_c + [(H-h)(b) + b_1h]d^2 \qquad \text{B22}$$

Product of inertia

The product of inertia I_{xy} appeared in the section dealing with unsymmetrical bending in Chapter 2; with respect to any pair of Cartesian axes it is the limit of

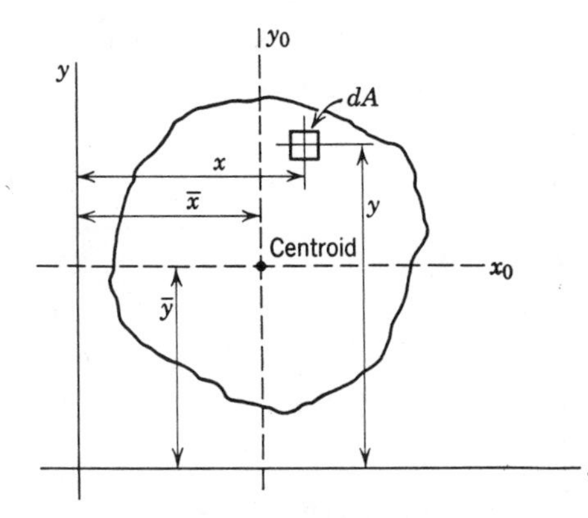

Fig. B14 Axis relationship for product of inertia of an area.

the sum of the products obtained by multiplying the areas of the element into which the whole area may be considered to be divided by the coordinates to these axes. Physically, the product of inertia has no significance, but it is used extensively in unsymmetrical analyses for the determination of limiting values of I_x and I_y in locating centers of pressure on irregular plane areas, and in the solution of indeterminate structural frames and bents. Mathematically, the product of inertia is defined as

$$I_{xy} = \int xy\, dA \tag{B23}$$

Note that the dimensions of I_{xy} are in.^{4} and also, since both x and y can vary, it may be positive, negative, or even zero. The product of inertia is zero if either axis of the plane area is an axis of symmetry; thus a *principal axis* may be defined as an axis through the centroid of the area for which the product of inertia is zero. If the product of inertia is known for one set of Cartesian axes intersecting at the centroid, the product of inertia with respect to two parallel axes may be determined by the parallel axis theorem. Referring to Fig. B14,

$$I_{xy} = I_{x_0y_0} + \bar{x}\bar{y}\, dA \tag{B24}$$

If the axes are rotated as shown in Fig. B15, the moments of inertia and

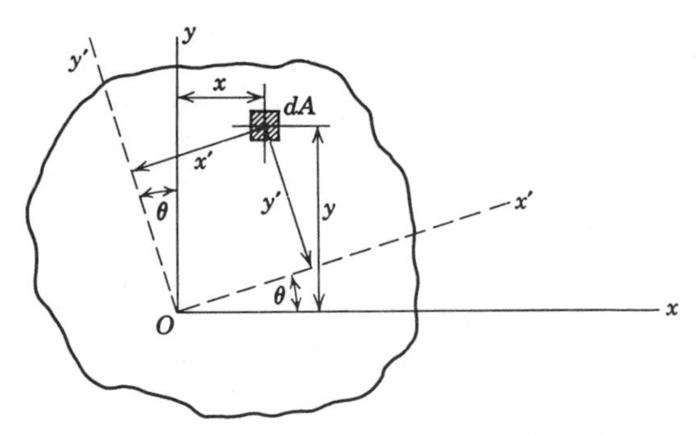

Fig. B15 Rotation of axes for product of inertia.

product of inertia with respect to the new axes can be determined. When the new xy axes are rotated an angle θ with respect to the original axes, the coordinates of a point x^1 and y^1 are as follows:

$$x^1 = x\cos\theta + y\sin\theta \qquad \text{B25}$$

and

$$y^1 = y\cos\theta - x\sin\theta \qquad \text{B26}$$

Then from Eqs. B1

$$I_{x^1} = \int (y^1)^2\, dA = \int (y\cos\theta - x\sin\theta)^2\, dA \qquad \text{B27}$$

or

$$I_{x^1} = I_x\cos^2\theta + I_y\sin^2\theta - 2I_{xy}\sin\theta\cos\theta \qquad \text{B28}$$

similarly,

$$I_{y^1} = I_x\sin^2\theta + I_y\cos^2\theta + 2I_{xy}\sin\theta\cos\theta \qquad \text{B29}$$

In order to evaluate minimum and maximum values of I_x and I_y it is convenient to rearrange Eqs. B28 and B29. From trigonometry,

$$\cos^2\theta = \frac{1+\cos 2\theta}{2}; \qquad \sin^2\theta = \frac{1-\cos 2\theta}{2}; \qquad \sin 2\theta = 2\sin\theta\cos\theta \qquad \text{B30}$$

Therefore,

$$I_{x^1} = \frac{I_x + I_y}{2} + \frac{I_x - I_y}{2}\cos 2\theta - I_{xy}\sin 2\theta \qquad \text{B31}$$

and

$$I_{y^1} = \frac{I_x + I_y}{2} + \frac{I_x - I_y}{2}\cos 2\theta + I_{xy}\sin 2\theta \qquad \text{B32}$$

Similarly,

$$I_{x^1y^1} = \int x^1 y^1\, dA = \int (x\cos\theta + y\sin\theta)(y\cos\theta - x\sin\theta)\, dA$$

or

$$I_{x^1y^1} = \frac{I_x - I_y}{2}\sin 2\theta + I_{xy}\cos 2\theta \qquad \text{B33}$$

Considering now all possible values of θ it can be shown that there is a set of axes for which the moments of inertia I_{x^1} and I_{y^1} are maximum and minimum, which are called the *principal axes*. From Eq. B31, for example, by differentiating with respect to θ and setting the result equal to zero the maximum or minimum value can be found; thus,

$$\frac{dI_{x^1}}{d\theta} = -2\left(\frac{I_x - I_y}{2}\right)\sin 2\theta - 2I_{xy}\cos 2\theta = 0$$

or

$$\tan 2\theta_m = \frac{2I_{xy}}{I_y - I_x} \qquad \text{B34}$$

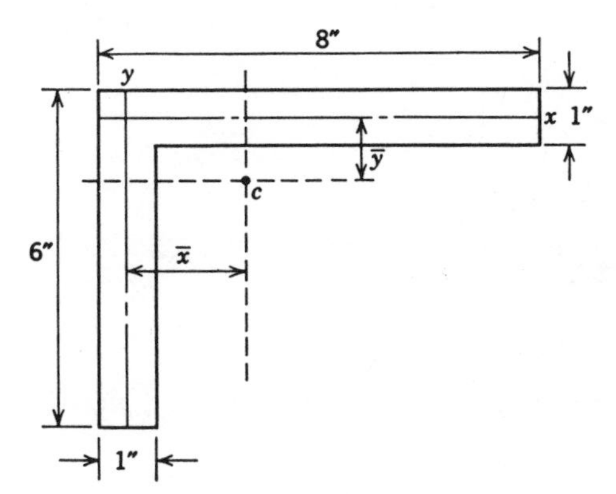

Fig. B16 Nonsymmetrical structural shape.

where θ_m is the angle for minimum or maximum I_{x^1}. Substituting the value of θ_m from Eq. B34 into Eq. B31 then gives

$$I_{\text{max}} = \frac{I_x + I_y}{2} + \sqrt{[(I_x - I_y)/2]^2 + (I_{xy})^2} \qquad \text{B35}$$

and

$$I_{\text{min}} = \frac{I_x + I_y}{2} - \sqrt{[(I_x - I_y)/2]^2 + (I_{xy})^2} \qquad \text{B36}$$

Equation B34 defines the direction of the principal axes, and Eqs. B35 and B36 define the values of the moments of inertia about these axes. If θ_m from Eq. B34 is placed in Eq. B33, it is seen that $I_{x^1y^1} = 0$. Note that Eqs. B35 and B36 are similar in form to the equations for principal stress (3.18 and 3.19); thus Mohr's circle method can be utilized in determining I_{max} and I_{min} by simply substituting for σ_x, σ_y, and τ_{xy} the values of I_{x^1}, I_{y^1}, and $I_{x^1y^1}$.

As an example the principal moments of inertia will be determined for the angle shown in Fig. B16. First it is necessary to locate the centroid C; from Eqs. A9 and A10,

$$\bar{x} = \frac{(5)(1)(0) + (8)(3.5)}{(5)(1) + (8)(1)} = 2.15 \text{ in.}$$

and

$$\bar{y} = \frac{(7)(1)(0) + (6)(1)(2.5)}{(5)(1) + (8)(1)} = 1.15 \text{ in.}$$

Then

$$I_{x^1} = \left[\left(\frac{1}{12}\right)(8)(1)^3 + (8)(1.15)^2 + \left(\frac{1}{12}\right)(1)(5)^3 + (5)(1.85)^2\right] = 38.63 \text{ in.}^4$$

$$I_{y^1} = \left[\left(\frac{1}{12}\right)(6)(1)^3 + (6)(2.15)^2 + \left(\frac{1}{12}\right)(1)(7)^3 + (7)(1.85)^2\right] = 78.10 \text{ in.}^4$$

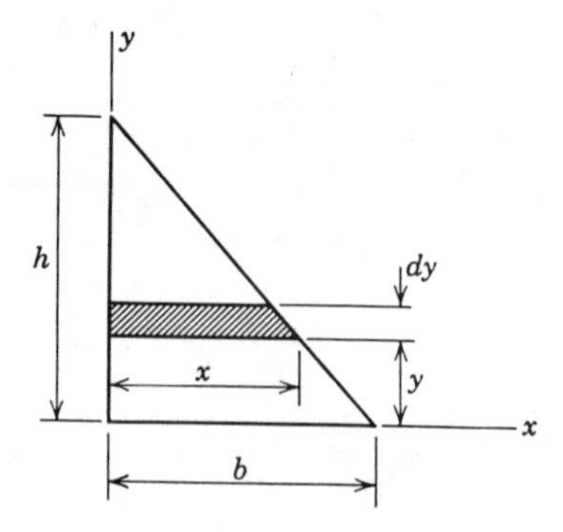

Fig. B17 Element of triangular section.

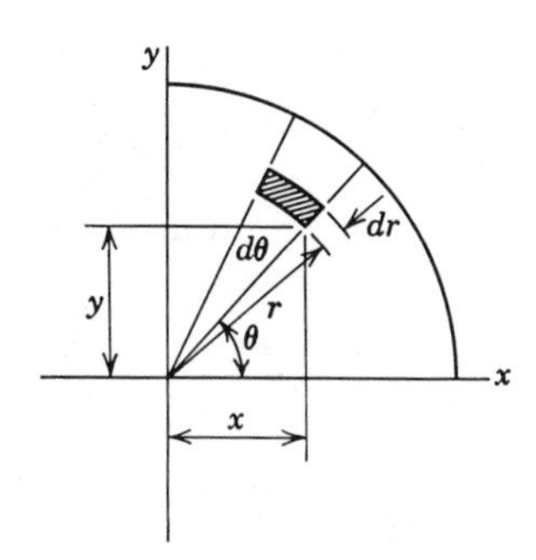

Fig. B18 Element of quadrant of a circle.

and

$$I_{x^1y^1} = 0 + 13(2.15)(1.15) = 32.2 \text{ in.}^4$$

From Eq. B34,

$$\tan 2\theta_m = \frac{(2)(32.2)}{78.10 - 38.63} = 1.63$$

or $\theta_m = 29°15'$ and the principal moments of inertia are calculated from Eqs. B35 and B36.

Consider the triangle in Fig. B17. For this case

$$I_{xy} = \int xy\,dA = \int_0^h \int_0^{[(h-y)/h]b} xy\,dy\,dx \qquad \text{B37}$$

or

$$I_{xy} = b^2h^2/24 \qquad \text{B38}$$

For a quadrant of a circle (Fig. B18),

$$I_{xy} = \int xy\,dA = \int_0^R \int_0^{\pi/2} (r \cos\theta)(r \sin\theta)(r)\,d\theta\,dr = \frac{R^4}{8} \qquad \text{B39}$$

As an illustration of the use of Mohr's circle method consider the rectangular

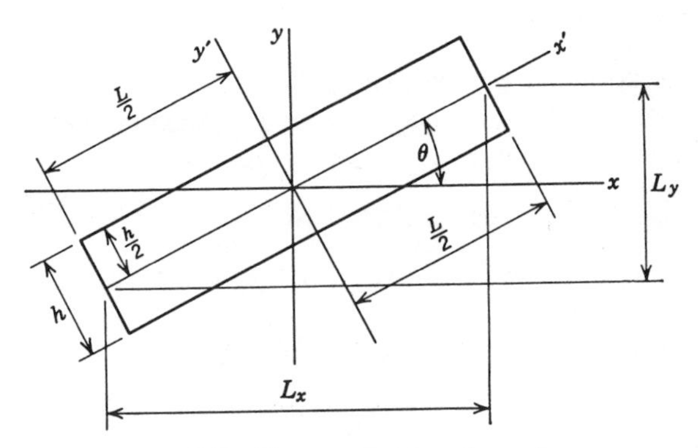

Fig. B19 Element of rectangular section.

element in Fig. B19. Mohr's circle for this case is shown in Fig. B20.

$$I_{\max} = \tfrac{1}{12} hL^3 \qquad \text{B40}$$

and

$$I_{\min} = \tfrac{1}{12} Lh^3 \qquad \text{B41}$$

With these values the center of the circle can be located giving definition for $(I_x + I_y)/2$ and $(I_y - I_x)/2$. From these quantities the distance OA is

$$OA = \sqrt{[(I_y - I_x)/2] + I_{xy}{}^2} \qquad \text{B42}$$

and by further trigonometric computation,

$$I_x = (hL)(L_y)^2/12 \qquad \text{B43}$$

$$I_y = (hL)(L_x)^2/12 \qquad \text{B44}$$

and

$$I_{xy} = (hL)(L_x)(L_y)/12 \qquad \text{B45}$$

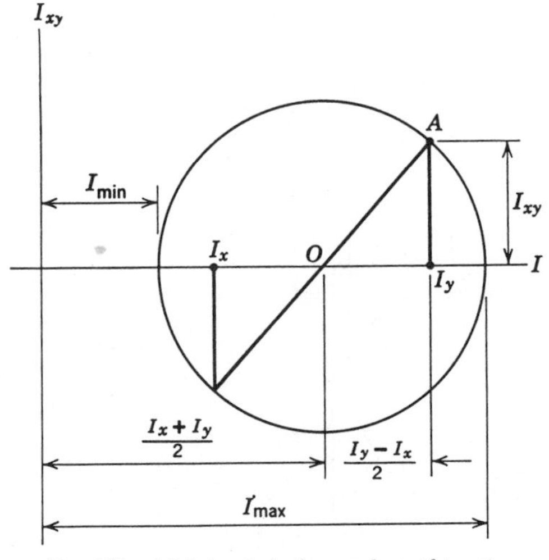

Fig. B20 Mohr's circle for product of inertia.

Radius of gyration

The radius of gyration is a distance such that, if the area of a section were concentrated at a point at that distance from an axis, the moment of inertia would be the same as that obtained by considering the entire area with respect to the axis. Thus, from Eqs. B1

$$I_x = \int x^2\, dA = r_x{}^2 A; \qquad I_y = \int y^2\, dA = r_y{}^2 A \qquad \text{B46}$$

where r is the radius of gyration. It is thus evident that

$$r = \sqrt{I/A} \qquad \text{B47}$$

As with moments of inertia the parallel axis theorem can also be used to define radii of gyration with respect to axes parallel to the reference axis; thus

$$r^2 = r_0{}^2 + d^2 \qquad \text{B48}$$

where r_0 is the radius of gyration with respect to a reference axis distant d from the new axis and parallel to it. The radii of gyration corresponding to the principal axes are the *principal radii of gyration.* In order to determine a radius of gyration or moment of inertia with respect to any other axes x^1, y^1, the *ellipse of inertia* may be used. In Fig. B21, for example, x and y are the principal axes and x^1, y^1 are orthogonal axes inclined θ degrees to axes x-y. The general equation of the ellipse is

$$x^2/a^2 + y^2/b^2 = 1 \qquad \text{B49}$$

and from Fig. B21 it can be seen that

$$x = r_y{}^1 \cos\theta \qquad \text{B50}$$

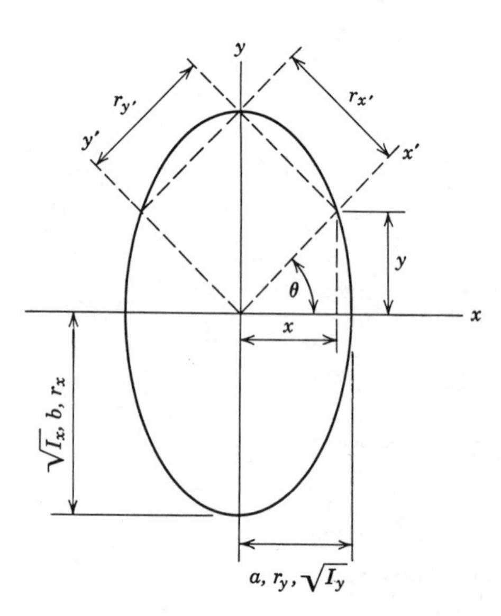

Fig. B21 Ellipse of inertia.

and

$$y = r_{y}{}^{1} \sin\theta \qquad \text{B51}$$

Substituting Eqs. B50 and B51 into Eq. B49 then gives the result

$$a^2 \sin^2\theta + b^2 \cos^2\theta = a^2b^2/(r_{y^1})^2 \qquad \text{B52}$$

However, the product of inertia with respect to principal axes is zero; therefore Eqs. B28 and B29 become

$$I_{x^1} = I_x \cos^2\theta + I_y \sin^2\theta \qquad \text{B53}$$

and

$$I_{y^1} = I_y \cos^2\theta + I_x \sin^2\theta \qquad \text{B54}$$

It can thus be seen from an examination of Eqs. B52 to B54 that

$$I_x = b^2 \quad \text{and} \quad I_y = a^2 \qquad \text{B55}$$

so that

$$I_{x^1} = a^2b^2/(r_{y^1})^2 = I_xI_y/(r_{y^1})^2 \qquad \text{B56}$$

Similarly,

$$I_{y^1} = I_xI_y/(r_{x^1})^2 \qquad \text{B57}$$

where r are the radii of gyration.

B-2 SOLIDS

Consider the solid cylinder shown in Fig. B22. If the density is γ (mass per unit volume), then from Eq. B2

$$I_z = \int r^2\,dM = \int_0^R \int_0^{2\pi} \gamma r^2 Lr\,(d\theta\,dr) = \frac{\pi\gamma L R^4}{2} \qquad \text{B58}$$

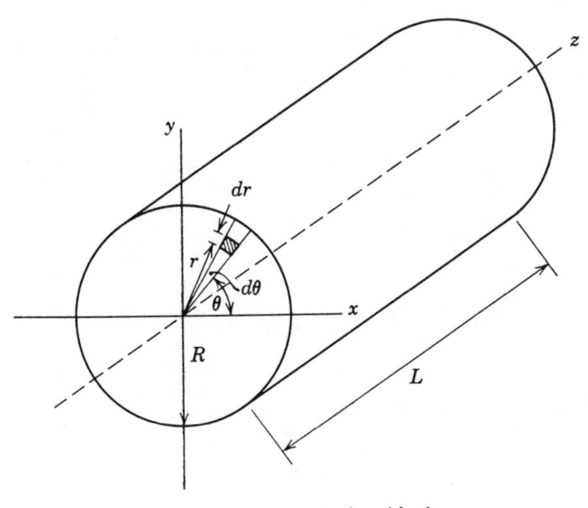

Fig. B22 Solid cylindrical body.

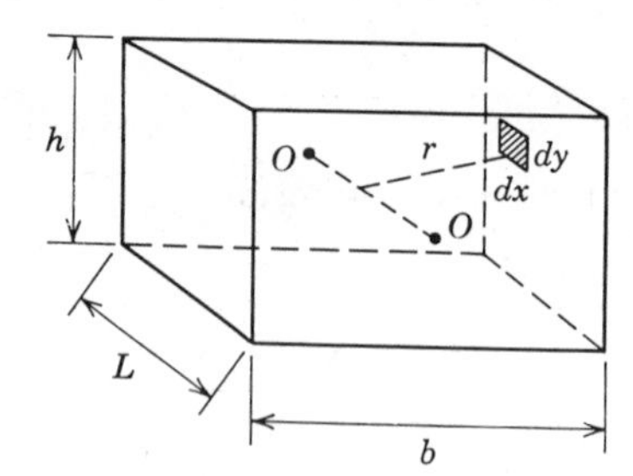

Fig. B23 Solid parallelopiped.

For the parallelopiped shown in Fig. B23 the moment of inertia with respect to axis 0–0 parallel to side L is

$$I_0 = \int r^2\, dM = \int r^2 \gamma L\, dx\, dy$$

or

$$I_0 = 4\gamma L \int_0^{b/2} \int_0^{h/2} (x^2 + y^2)\, dx\, dy = \frac{\gamma L}{12}(b^3 h + h^3 b) \qquad \text{B59}$$

For an axis along the L side,

$$I_L = \int r^2\, dM = \int \gamma r^2 L\, dx\, dy = \gamma L \int_0^b \int_0^h (x^2 + y^2)\, dx\, dy$$

from which

$$I_L = (\gamma L/3)(b^3 h + h^3 b) \qquad \text{B60}$$

For very thin plates the h^3 term may be neglected in Eq. B60 so that

$$I_L = \gamma L h b^3/3 \qquad \text{B61}$$

As an example of the use of the triple integral consider the solid shown in Fig. B24 which is bounded by the cylindrical surface $x^2 + y^2 = R^2$ and the

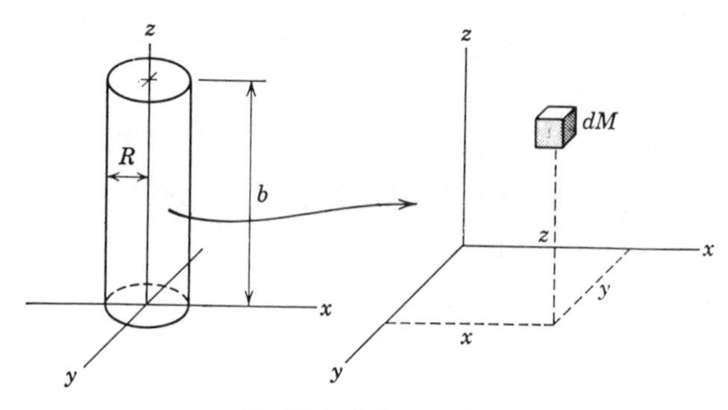

Fig. B24 Solid cylinder.

planes $z = 0$ and $z = b$. The moment of inertia about the x-axis is

$$I_x = \int r^2\, dM = \int (z^2 + y^2)\gamma\, dz\, dy\, dx$$

or

$$I_x = 4 \int_0^R \int_0^{\sqrt{R^2 - x^2}} \int_0^b (z^2 + y^2)\, dz\, dy\, dx$$

from which

$$I_x = \frac{\gamma \pi R^2 b}{12}(3R^2 + 4b^2) \qquad \text{B62}$$

appendix C

LARGE ELASTIC DEFORMATIONS

The analysis of large elastic deformations is beyond the scope of this book. However, for reference, the results for three particular cases are presented. For further information see the references cited.

C-1 CENTILEVER BEAM

In the elementary theory of bending of beams the square of the first derivative in the curvature equation (Eq. 2.77) is neglected. In addition, no provision is made for the shortening of the moment arm as the deflection is increased. Consider the beam in Fig. C1. From the elementary theory (see Chapter 2)

$$\text{end deflection} = \delta = PL^3/3EI \qquad \text{C1}$$

$$\text{maximum bending moment} = M = PL \qquad \text{C2}$$

$$\text{bend stress} = \sigma = M/Z \qquad \text{C3}$$

For example, letting

$$E = 30 \times 10^6 \text{ psi}$$

$$\delta = 2 \text{ in.}$$

$$L = 3 \text{ in.}$$

$$h = 0.030 \text{ in.}$$

$$b = 0.40 \text{ in.}$$

$$\nu = 0.30$$

and

$$I = 9 \times 10^{-7} \text{ in.}^4$$

the elementary theory would give

$$\text{deflection} = 2 = \frac{P(3)^3}{3(30 \times 10^6)(9 \times 10^{-7})} = \frac{P}{3}$$

or

$$P = 6 \text{ lb}$$

$$\text{moment} = (6)(3) = 18 \text{ in.-lb}$$

$$\text{stress} = \frac{(18)(0.015)}{(9 \times 10^{-7})} = 300{,}000 \text{ psi}$$

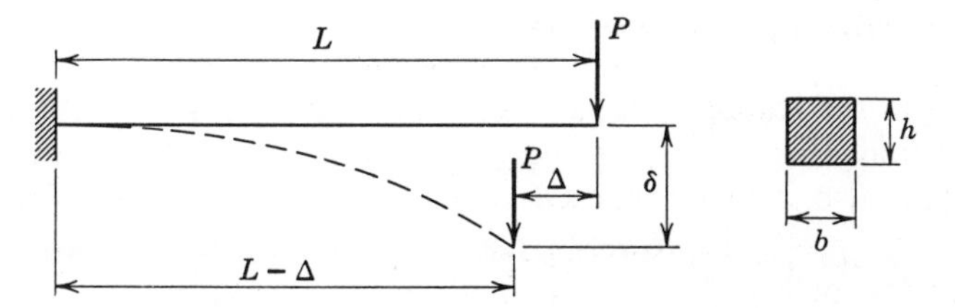

Fig. C1 Large elastic deflection of a cantilever beam.

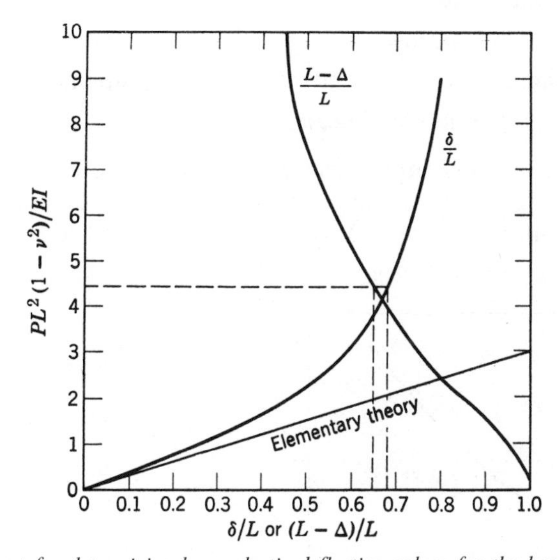

Fig. C2 Chart for determining large elastic deflection values for the beam in Fig. C1.

By the exact method, use is made of Fig. C2. For this case $\delta/L = \frac{2}{3} = 0.67$ which corresponds to the ordinate 4.5; therefore

$$\frac{PL^2(1 - \nu^2)}{EI} = 4.5,$$

from which

$$P = 14.85 \text{ lb}$$

Again, from Fig. C2,

$$(L - \Delta)/L = 0.64$$

Then, $L - \Delta - 0.64(3) = 1.92$, or a shortening of the moment arm of

$$\Delta = L - 1.92 = 3 - 1.92 = 1.08 \text{ in.}$$

Then, the *moment* is $M = (14.85)(1.92) = 25.95$ in.-lb and the *stress* is $\sigma = [25.95(0.015)/(9 \times 10^{-7})] = 432{,}000$ psi

C-2 THIN CIRCULAR PLATES

In thin plates the standard equations assume zero stress at midthickness; in large deflections this is not true and the *diaphragm stress* assists in supporting the load and also increases the effective stiffness of the plate. The following equation developed by Isakower (6) has been found useful for analyzing various types of plates for both large and small deflections,

$$pR^4/t^4 E = Fy/t \qquad \text{C4.}$$

where p = pressure, psi

R = plate radius, in.

t = plate thickness, in.

F = diaphragm factor (dependent on edge conditions)

y = center deflection of plate, in.

Figures C3 to C5 supply supplementary information to be used with Eq. C4. For example, to find the maximum deflection of a circular steel plate 0.25 in. thick with a 20 in. radius under 5 psi pressure with fixed edges, use is made of

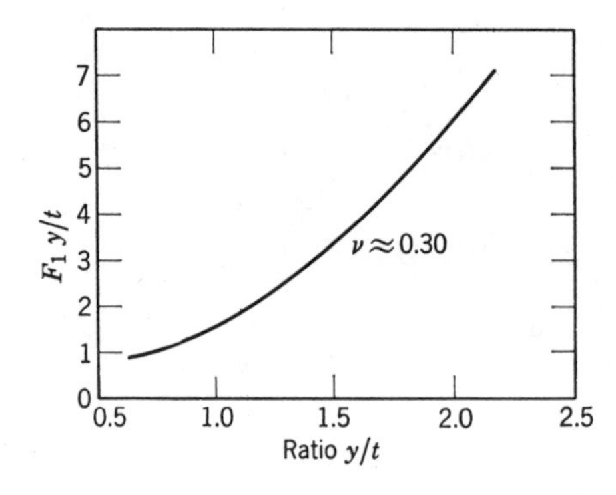

Fig. C3 Design chart for large elastic deflections of simply supported circular plate.

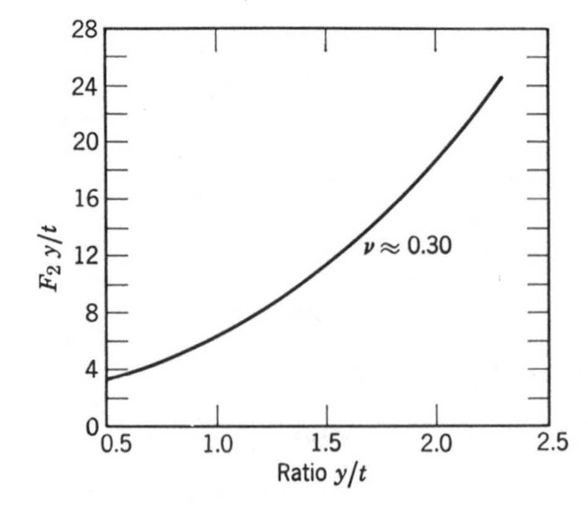

Fig. C4 Design chart for large elastic deflections of circular plates with edge restrained in vertical plane.

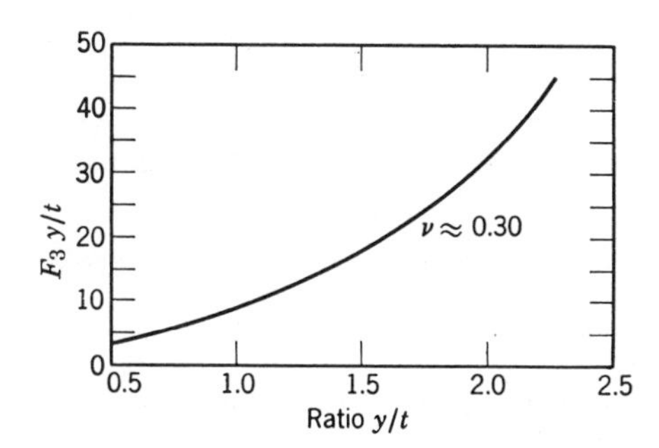

Fig. C5 Design chart for large elastic deflections of circular plate with fixed edge.

Fig. C5. From Eq. C4,

$$\frac{F_3 y}{t} = \frac{pR^4}{t^4 E} = \frac{(5)(20)^4}{(0.25)^4(30 \times 10^6)} = 6.84$$

which, from the figure gives $y/t = 0.87$. Therefore $y = 0.87(0.25) = 0.218$ in.

C-3 THIN RINGS

Consider the thin ring shown in Fig. C6 loaded along a diameter by a load $2P$. The small deformation theory for this structure was given in Chapter 7. As the load is applied the portions at A and B, Fig. C7, flatten and when the curvature is zero (9), the bending moment is

$$M = EI/r \qquad \text{C5}$$

and

$$P = 1.393(EI/r^2) \qquad \text{C6}$$

$$x = 1.198r \qquad \text{C7}$$

and

$$y = 0.718r \qquad \text{C8}$$

If the ring is flattened between two rigid plates (Fig. C8), $1 - \phi$ is defined as the

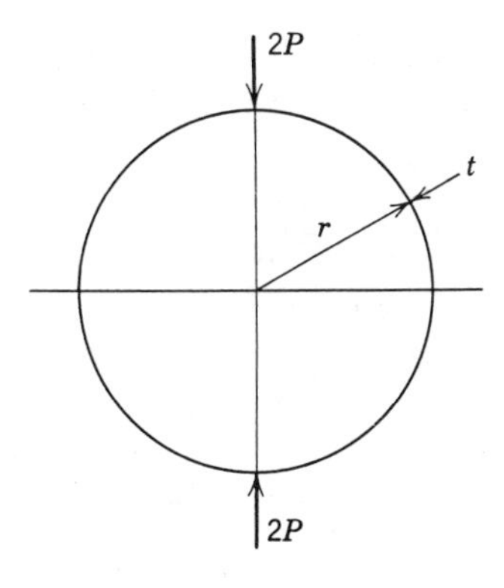

Fig. C6 Thin ring loaded on the diameter.

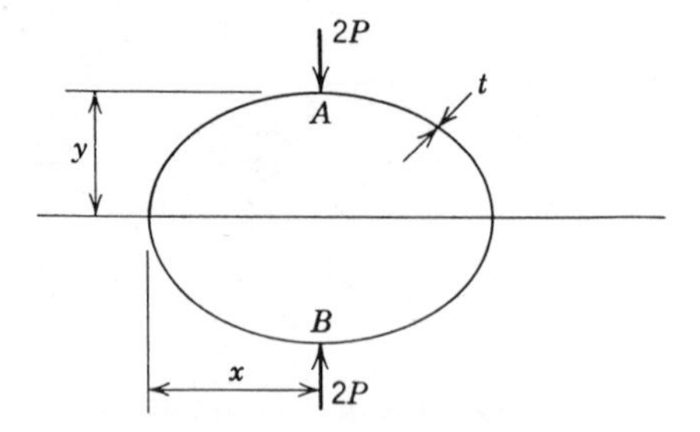

Fig. C7 Flattening of thin ring under diametral loading.

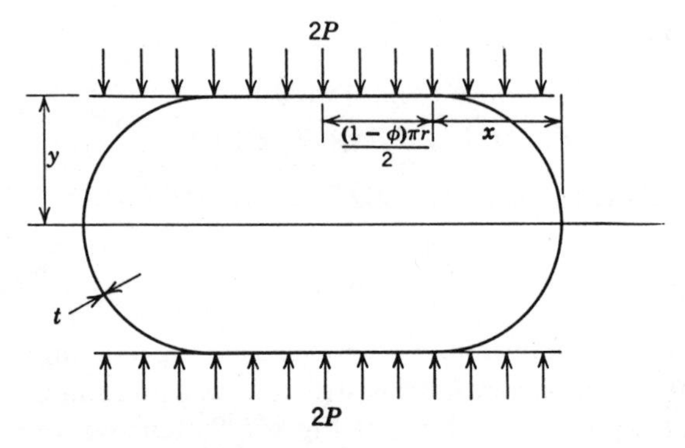

Fig. C8 Thin ring flattened between two rigid plates.

fraction of "flat circumference" and the "free" circumference is thus $2\pi r\phi$; then

$$P = \frac{1.393EI}{\phi^2 r^2} \tag{C9}$$

$$x = 1.198r\phi \tag{C10}$$

and

$$y = 0.718r\phi \tag{C11}$$

REFERENCES

1. H. M. Berger, *J. Appl. Mech.*, **77** (4), 465–472 (December 1955).
2. R. A. Beth, and C. P. Wells, *J. Appl. Phys.*, **22** (6), 742–746 (June 1951).
3. K. E. Bisshopp, and D. C. Drucker, *Q. Appl. Math.*, **3** (3), 272–275 (October 1945).
4. H. D. Conway, *Phil. Mag.*, **38** (Series 7), 905–911 (December 1947).
5. A. E. Green, and J. E. Adkins, *Large Elastic Deformations*, Clarendon Press, Oxford (1971).
6. R. I. Isakower, *Prod. Eng.*, **32** (17), 81–83 (April 24, 1961).
7. E. H. Mansfield, and P. W. Kleeman, *Aircraft Eng. (London)*, **27** (4), 102–108 (April 1955).
8. F. D. Murnaghan, *Finite Deformation of an Elastic Solid*, John Wiley and Sons, Inc. New York, N.Y. (1951).

9. H. H. Pan, *Australian J. Appl. Sci.*, **12** (4), 446–52 (1961).

10. R. S. Rivlin, "Some Topics in Finite Elasticity," in *Structural Mechanics*, 169–198, Pergamon Press, New York (1960).

11. B. Saelman, *J. Franklin Inst.*, **257** (2), 125–132 (February 1954).

12. A. E. Seames, and H. D. Conway, *J. Appl. Mech.*, **79** (2), 289–294 (June 1957).

13. S. Way, *Trans. Am. Soc. Mech. Eng.*, **56**, 627–36, Paper APM-56-12 (1934).

14. M. L. Williams, *J. Appl. Mech.*, **77** (4), 458–464 (December 1955).

appendix D

JOINTS AND CONNECTIONS

D-1 RIVETED CONNECTIONS

In the elementary analysis of riveted connections such factors as bending and tensile stresses in the rivets, friction between parts, and residual stresses are neglected. Shear is assumed to be uniform; tensile stress in the next section is also assumed uniform. Finally, the bearing stress (shear load divided by projected area of hole) for each rivet is assumed to be uniform. For allowable values of stress to be used reference is made to various codes and works dealing with structural members (3, 4, 7, 15, 18). For a discussion of the effect of bearing ratio on the static strength of riveted joints see (6).

Analysis of lap joint

Consider the triple-riveted lap joint shown in Fig. D1. Each rivet takes $F/3$ of the load; therefore the shear stress in the rivets is

$$\tau = \frac{F}{3A_r} = \frac{4(20{,}000)}{3\pi(1)} = 8500 \text{ psi}$$

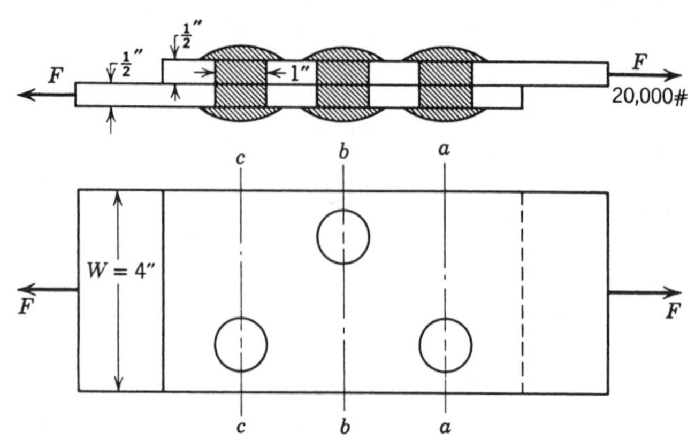

Fig. D1 Triple riveted lap joint.

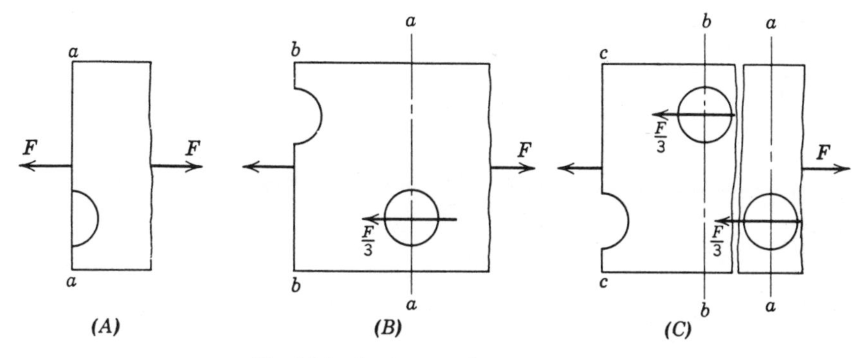

Fig. D2A Section a-a of joint in Fig. D1.
Fig. D2B Section b-b of joint in Fig. D1.
Fig. D2C Section c-c of joint in Fig. D1.

where A_r is the cross-sectional area of a rivet. The bearing stress is

$$\sigma_b = \frac{F}{3A_b} = \frac{20{,}000}{(3)(1)(0.5)} = 13{,}350 \text{ psi}$$

Starting with section a–a and assuming it to carry the full load, the tensile stress in the next section b–b, assuming uniform load distribution is (Fig. D2A)

$$(\sigma_t)_{a-a} = \frac{F}{(A_t)_{a-a}} = \frac{20{,}000}{(4-1)(0.5)} = 13{,}350 \text{ psi}$$

For section b–b, since section a–a carries one-third of the load (Fig. D2B),

$$(\sigma_t)_{b-b} = \frac{\frac{2}{3}(20{,}000)}{(4-1)(0.5)} = 8900 \text{ psi}$$

For section c–c, since a–a and b–b each carry one-third of the load (Fig. D2C),

$$(\sigma_t)_{c-c} = \frac{\frac{1}{3}(20{,}000)}{(4-1)(0.5)} = 4450 \text{ psi}$$

The over-all tensile stress is

$$\sigma = \frac{F}{A} = \frac{20{,}000}{4(0.5)} = 10{,}000 \text{ psi}$$

Suppose this type of joint is used to fasten a 50 in. diameter boiler together along a longitudinal section as shown in Fig D3. From Chapter 3, for a thin-walled cylinder under internal pressure the hoop stress is given as

$$\sigma_h = pD/2t$$

so that based on the repeating section w,

$$F = \sigma w t = (pD/2t)(w)(t) \qquad \text{D1}$$

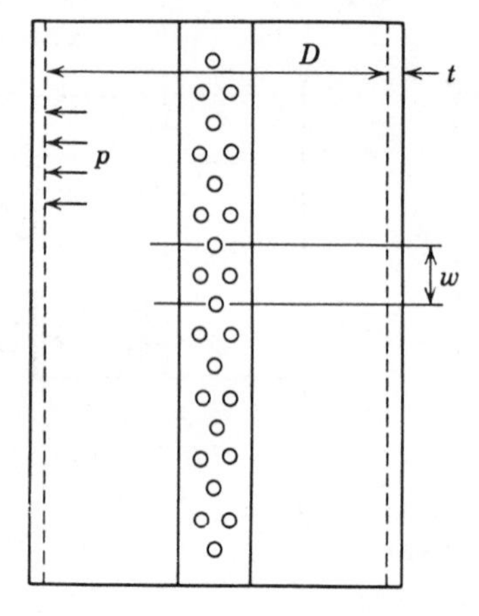

Fig. D3 Longitudinal riveted seam in a boiler.

or

$$20{,}000 = \frac{p(50)}{2(0.5)}(4)(0.5) = 100p$$

from which the allowable internal pressure is found to be

$$p = 200 \text{ psi}$$

Often allowable stress values are given and it is required to compute the allowable pressure corresponding to these allowable stress values. In the foregoing example let

$$\text{allowable shear stress} = \tau_w = 10{,}000 \text{ psi}$$
$$\text{allowable tensile stress} = (\sigma_t)_w = 15{,}000 \text{ psi}$$
$$\text{allowable bearing stress} = (\sigma_b)_w = 20{,}000 \text{ psi}$$

Using these allowable stress values the corresponding allowable force values are as follows:

$$(F_w)_{\text{shear}} = 3A_\tau(10{,}000) = \frac{30{,}000(\pi)(1)}{4} = 23{,}550 \text{ lb}$$

$$(F_w)_{\text{tension in }(a-a)} = A_t(15{,}000) = (4-1)(0.5)(15{,}000) = 22{,}500 \text{ lb}$$
$$(F_w)_{\text{tension in }(b-b)} = (\tfrac{3}{2})(4-1)(0.5)(15{,}000) = 33{,}750 \text{ lb}$$
$$(F_w)_{\text{tension in }(c-c)} = (3)(4-1)(0.5)(15{,}000) = 67{,}500 \text{ lb}$$
$$(F_w)_{\text{gross}} = (4)(0.5)(15{,}000) = 30{,}000 \text{ lb}$$
$$(F_w)_{\text{bearing}} = (3)(1)(0.5)(20{,}000) = 30{,}000 \text{ lb}$$

From these computations it is seen that the limiting value of F_w is for tension in section a–a; therefore the allowable pressure based on this value is

$$p_w = \frac{2F_w}{wD} = \frac{(2)(22{,}500)}{(4)(50)} = 225 \text{ psi}$$

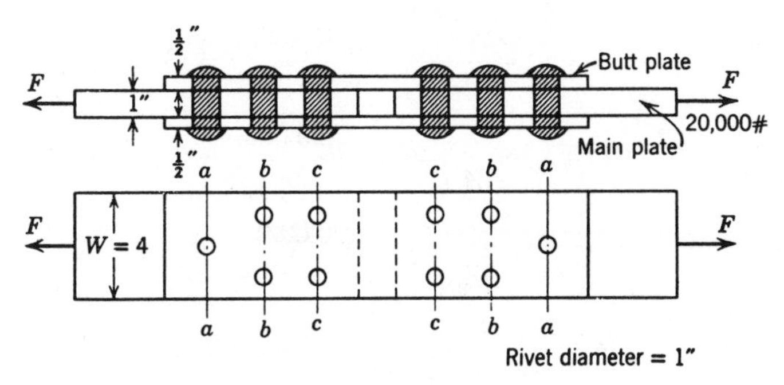

Fig. D4 Riveted butt joint.

The efficiency of this joint is

$$\text{efficiency} = \frac{22{,}500(100)}{(\sigma_w)wt} = \frac{2{,}250{,}000}{(15{,}000)(4)(0.5)} = 75\%$$

Analysis of butt joint

Consider the butt joint in Fig. D4. In this case each rivet takes one-tenth of the load and is subjected to double shear. The shear stress in the rivets is

$$\tau = \frac{F}{10A_\tau} = \frac{4(20{,}000)}{10\pi} = 2547 \text{ psi}$$

The bearing stress due to the main plate is

$$(\sigma_b)_{\text{main plate}} = \frac{F}{5A_b} = \frac{20{,}000}{(5)(1)(1)} = 4000 \text{ psi}$$

The tensile stress in section a–a of the main plate is

$$(\sigma_t)_{a-a} = \frac{20{,}000}{(A_t)_{a-a}} = \frac{20{,}000}{(4-1)(1)} = 6667 \text{ psi}$$

$$(\sigma_t)_{b-b} = \frac{F}{(A_t)_{b-b}} = \frac{\frac{4}{5}(20{,}000)}{(4-2)(1)} = 8000 \text{ psi}$$

$$(\sigma_t)_{c-c} = \frac{F}{(A_t)_{c-c}} = \frac{\frac{2}{5}(20{,}000)}{(4-2)(1)} = 4000 \text{ psi}$$

$$(\sigma_t)_{\text{gross}} = \frac{F}{A} = \frac{20{,}000}{(4)(1)} = 5000 \text{ psi}$$

The butt plate stresses are as follows:

$$(\sigma_b)_{\text{butt plate}} = \frac{F}{A_b} = \frac{20{,}000}{(10)(1)(0.5)} = 4000 \text{ psi}$$

$$(\sigma_t)_{a-a} = \frac{F}{(A_t)_{a-a}} = \frac{\frac{1}{10}(20{,}000)}{(4-1)(0.5)} = 1330 \text{ psi}$$

$$(\sigma_t)_{b-b} = \frac{F}{(A_t)_{b-b}} = \frac{\frac{3}{10}(20{,}000)}{(4-2)(0.5)} = 6000 \text{ psi}$$

$$(\sigma_t)_{c-c} = \frac{F}{(A_t)_{c-c}} = \frac{\frac{1}{2}(20{,}000)}{(4-2)(0.5)} = 10{,}000 \text{ psi}$$

$$(\sigma_t)_{\text{gross}} = \frac{F}{A} = \frac{\frac{5}{10}(20{,}000)}{(4)(0.5)} = 5000 \text{ psi}$$

The efficiency of this joint is then

$$\text{efficiency} = \frac{(2{,}547)(100)}{5{,}000} = 51\%$$

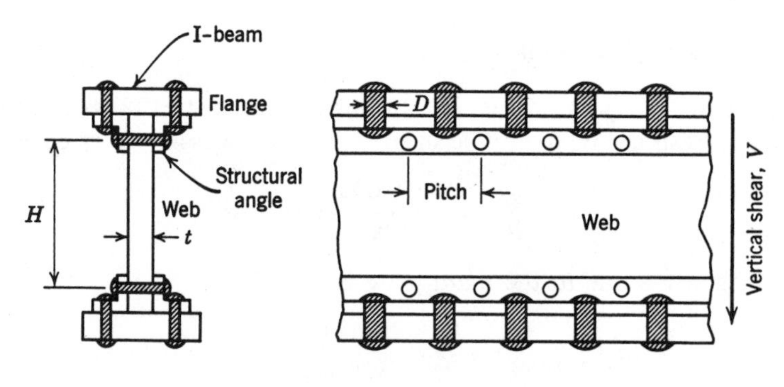

(A) Riveted plate girder

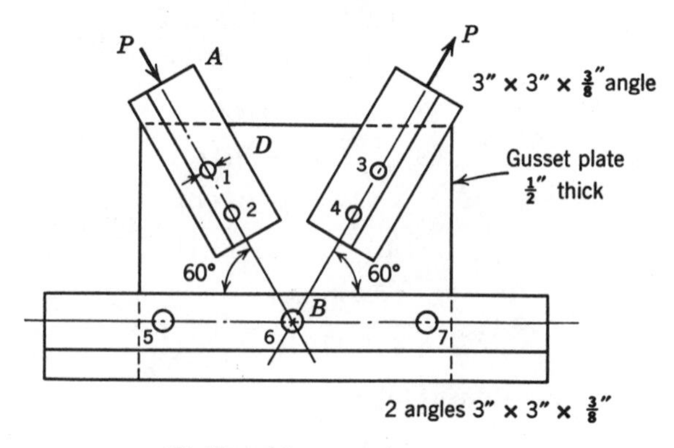

(B) Riveted truss member

Fig. D5 Examples of riveted structural members.

Plate girder

Structural I-beams are frequently strengthened by angles riveted in place as shown in Fig. D5*A*; in such a section it is assumed (as discussed in Chapter 3) that the shear load is carried by the web. Thus for the structure in Fig. D5*A* the average shear stress in the web is

$$\tau = V/ht \qquad \text{D2}$$

which acts (for equilibrium) both horizontally and vertically. If the allowable shearing force for a rivet is F_τ, then, from this equation

$$F_\tau = \tau pt = (V/ht)(pt) = Vp/h \qquad \text{D3}$$

from which the required pitch distance is

$$p = hF_\tau/V \qquad \text{D4}$$

In this last equation the value of F_τ has to be determined; if the allowable shearing stress is τ_w, then

$$F_\tau = \tau_w(2\pi D^2/4) \qquad \text{D5}$$

or, alternately, for an allowable bearing stress of $(\sigma_b)_w$,

$$F_\tau = (\sigma_b)_w tD \qquad \text{D6}$$

and the smaller value of F_τ governs.

Simple truss member

A simple truss member with axial loads is shown in Fig. D5*B*. In this structure each rivet is subjected to double shear; thus for rivets 1, 2, 3, and 4

$$\tau = \frac{P}{4A_\tau} = \frac{P}{4(\pi D^2/4)} = \frac{P}{\pi D^2} \qquad \text{D7}$$

The bearing stress is the greater of the following:

$$(\sigma_b)_{\text{angle}} = \frac{P}{(A_b)_{\text{angle}}} = \frac{P}{4D(\frac{3}{8})} \qquad \text{D8}$$

or

$$(\sigma_b)_{\text{gusset}} = \frac{P}{(A_b)_{\text{gusset}}} = \frac{P}{2D(\frac{1}{2})} \qquad \text{D9}$$

For rivets 5, 6, and 7

$$\tau = \frac{F}{6A_\tau} = \frac{2P\cos 60^\circ}{6(\pi D^2/4)} \qquad \text{D10}$$

and the bearing stress for the lower chord is the maximum of

$$(\sigma_b)_{\text{angle}} = \frac{F}{(A_b)_{\text{angle}}} = \frac{2P\cos 60^\circ}{(6)(D)(\frac{3}{8})}$$

or

$$(\sigma_b)_{\text{gusset}} = \frac{F}{(A_b)_{\text{gusset}}} = \frac{2P \cos 60^\circ}{(3)(D)(\frac{1}{2})}$$

Eccentric rivet groups

Various structural members are required to transmit bending moments; several types will be mentioned here.

Case I. Horizontal group. For the connection shown in Fig. D6*A* four rivets are symmetrically placed about an axis and are required to resist the eccentric load P. It is seen in Fig. D6*B* that each rivet takes $P/4$ vertically and that the load P generates a moment PL at the center of resistance O. This moment PL is resisted by a moment generated by the rivet forces and moment arms to O; thus, for equilibrium, letting the total rivet forces be F_1 and F_2 it is seen that

$$M = PL = 2F_1x_1 + 2F_2x_2 \tag{D11}$$

It now remains to establish a relationship between F_1 and F_2 and this is done by noting that the shear stress is proportional to the distance from the centroid, thus

$$\tau_1/\tau_2 = x_1/x_2 \tag{D12}$$

However $F_1 = \tau_1 A_1$ and $F_2 = \tau_2 A_2$; therefore, since the rivets are the same

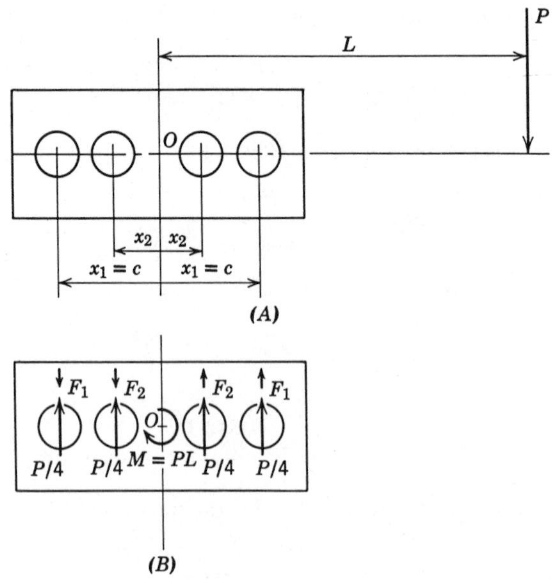

Fig. D6 Horizontal eccentric rivet group.

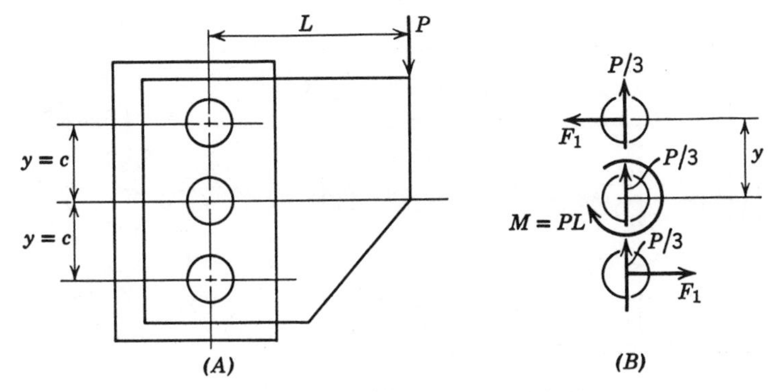

Fig. D7 Vertical eccentric rivet group.

size, $A_1 = A_2$ and

$$F_1/F_2 = x_1/x_2$$

or

$$F_2 = F_1(x_2/x_1)$$

Using this value of F_2 it is then found that

$$M = PL = 2(F_1x_1 + F_2x_2) = (2F_1/x_1)(x_1^{\,2} + x_2^{\,2}) \qquad \text{D13}$$

where, now, the maximum force F_1 is

$$F_1 = \frac{P}{4} + \frac{Mx_1}{2(x_1^{\,2} + x_2^{\,2})} \qquad \text{D14}$$

Case II. Vertical group. For the connection shown in Fig. D7*A*, three rivets are symmetrically located with respect to an eccentric load *P*. Here each rivet (Fig. D7*B*) takes $P/3$ of the load and a moment PL is generated at the center of resistance. This moment is balanced by a moment generated by the rivet forces and moment arms. In this case the resisting forces F_1 act horizontally; thus

$$M = PL = 2F_1y \qquad \text{D15}$$

and the resultant rivet force R_1 is

$$R_1 = \sqrt{(P/3)^2 + (M/2y)^2} \qquad \text{D16}$$

Case III. Simple compound group. A simple eccentric rivet group consisting of four symmetrically located rivets is shown in Fig. D8*A*. Each rivet (Fig. D8*B*) takes $P/4$ of the load and a moment FL is generated at the center of resistance. This moment is resisted by the moment generated by the rivet forces and moment arms. Here

$$M = FL = 4F_1r \qquad \text{D17}$$

from which

$$F_1 = FL/4r \qquad \text{D18}$$

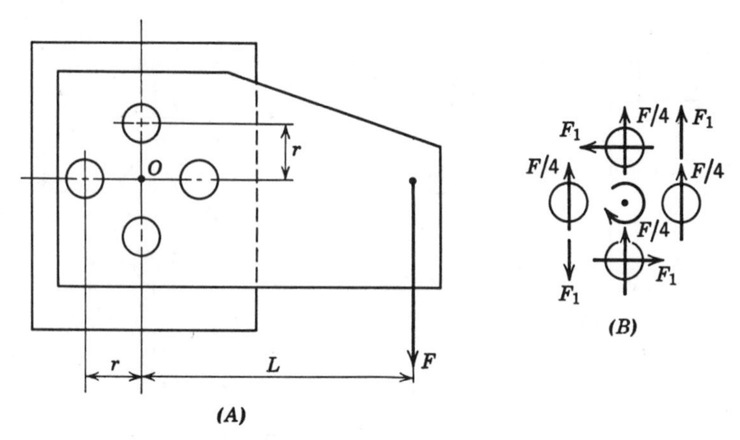

Fig. D8 Simple compound eccentric rivet group.

The maximum resultant load is for the rivet nearest the applied load or

$$R_1 = F_1 + F/4 = (F/4r)(r + L) \tag{D19}$$

Case IV. Simple compound group. Another type of simple compound rivet grouping is shown in Fig. D9*A*, in which each rivet takes $P/6$ of the load (Fig. D9*B*). The center of resistance moment PL is resisted by the rivet moment as follows:

$$M = PL = 4F_1c + 2F_2b \tag{D20}$$

Again, by proportional shear loading from the centroid

$$F_2 = F_1(b/c) \tag{D21}$$

so that from

$$M = PL = 4F_1c + 2F_2b$$

$$F_1 = \frac{PL}{4c + 2(b^2/c)} \tag{D22}$$

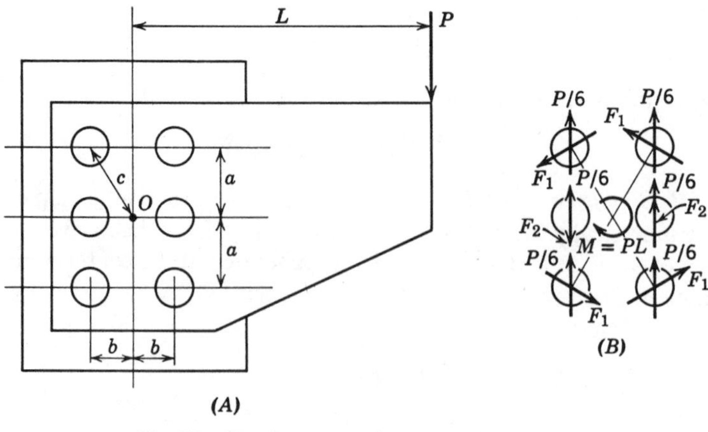

Fig. D9 Simple compound eccentric rivet group.

and

$$F_2 = \frac{PL}{4(c^2/b) + 2b} \tag{D23}$$

These forces plus the $P/6$ forces then give resultant forces. The maximum resultant force is for the top rivet closest to the load

$$R = \sqrt{F_1^{\ 2} + (P/6)^2 + 2F_1(P/6)\cos\theta} \tag{D24}$$

where

$$\theta = \tan^{-1}(a/b) \tag{D25}$$

Strength of connections

The actual strength of a riveted connection depends on several things such as thicknesses of parts, strength of parts, friction, and state of stress. Such items as tensile, bearing, and shear strength have already been mentioned. Another important item is rivet *tear-out*, that is, a situation where the force applied to the structure results in tearing out a section that bears on the rivet. This has been discussed by Shanley (15) and by Melcon and Hoblit (10) in connection with the design of lugs and shear pins. As an approximation the load per rivet for tear out is given by

$$P/\text{rivet} = 2Lt\tau_u \tag{D26}$$

where t is plate thickness, τ_u is the ultimate shear strength of the material, and L is a distance defined in Fig. D10. In general the strength of riveted connections can be improved by the devices shown in Fig. D11; in Fig. D11A for example, the addition of two rivets at the edge of a lap joint increases the joint strength by absorbing bending stresses otherwise imparted to the interior rivets which are supposed to resist only a shear load. In Fig. D11B, the addition of *dummy* rivets results in lowering the stress concentration associated with the interior rivets (see Chapter 11). Finally, in Fig. D11C the geometry includes a tapered butt plate that helps distribute the load more evenly, taking excess load off the first rivets.

In many instances the rivets holding two members together are staggered as

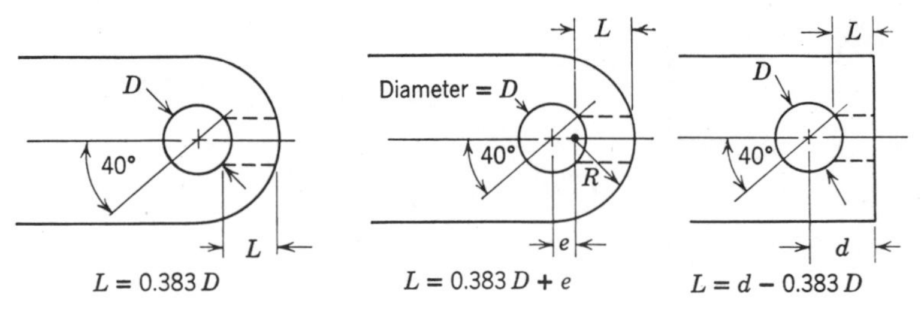

Fig. D10 Rivet tear-out geometry.

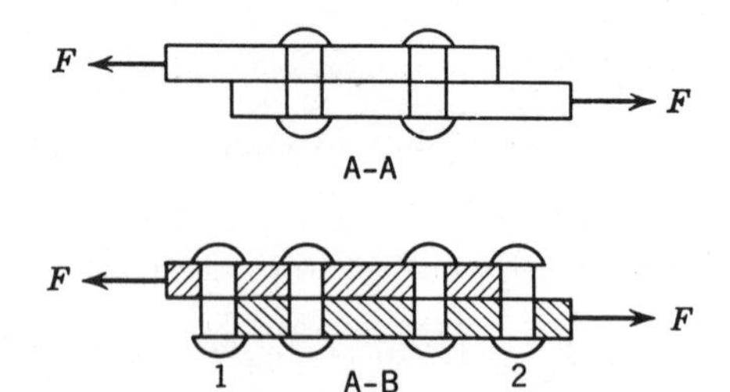

Same as A-A but with rivets 1 and 2 added to absorb bending load

(A)

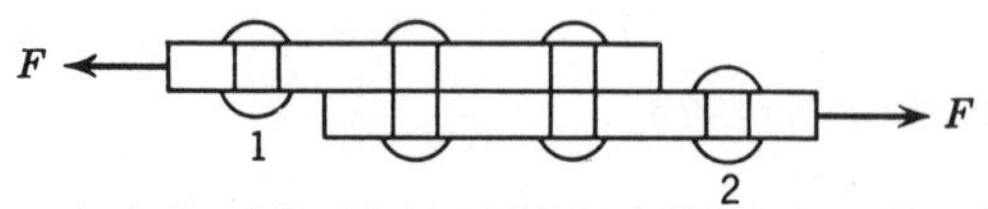

Dummy rivets 1 and 2 added to share load with end-working rivets (see Chapter 11 on Stress Concentration)

(B)

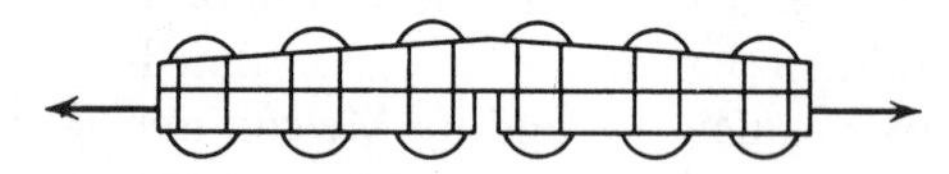

Tapered butt plate to take excessive load off first row of rivets

(C)

Fig. D11 Some devices for strengthening riveted connections.

shown in Fig. D12, and failure may take place by a rupture following the path *abcd*. For such cases the net area is often assumed to be

$$A_{\text{net}} = A_{ae} - Dt - Dt(1 - L^2/4HD) \qquad \text{D27}$$

where A_{ae} is the area across section $a - e$. For plates of uniform thickness t, the preceding equation is divided by t to give the net width W_{net}, thus,

$$W_{\text{net}} = \overline{ae} - 2D + L^2/4H \qquad \text{D28}$$

For repeating sections (additional rivets) this procedure is continued in suc-

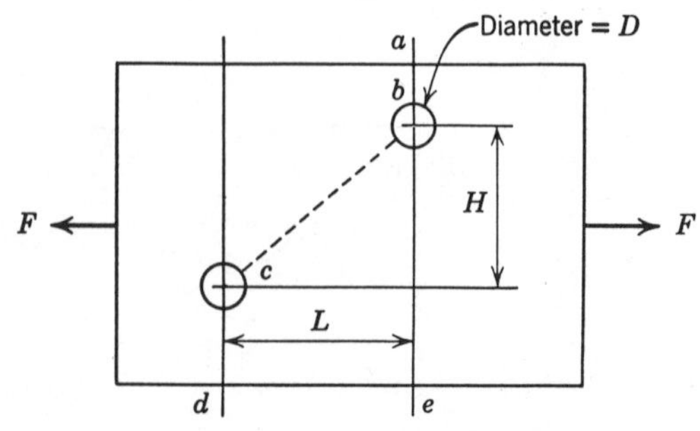

Fig. D12 Staggered rivet connection.

cession for each hole (3) so that for any section such as shown in Fig. D12 the net width is determined by deducting from the over-all width the sum of the diameters of all the rivet holes and adding the quantity $L^2/4H$ for each repeating section.

D-2 WELDED CONNECTIONS

The two main classifications of welded connections are butt and fillet, and within each type there are many variations. For these details reference is made to the *Welding Handbook* (12) and to pages 517 to 526 [(36), Chapter 3]. For present purposes only two simple types of welded connections will be considered. A simple butt weld is shown in Fig. D13*A*; if the allowable tensile stress value is σ_w, then the allowable load P is

$$P = (\sigma_w) b t_b \qquad \text{D29}$$

For the double fillet weld shown in Fig. D13*B* the critical distance is the weld throat; for equal weld legs the throat dimension is

$$t_{\text{throat}} = t_f = t \sin 45^\circ = 0.707t \qquad \text{D30}$$

The allowable load P for this case is based on the working stress in shear τ_w so that the allowable load P for this case is

$$P = (\tau_w)(0.707)(t)(2b) \qquad \text{D31}$$

For the fillet weld connection shown in Fig. D13*C* the allowable load is

$$P = (\tau_w)(0.707)(t)(L_1 + L_2) \qquad \text{D32}$$

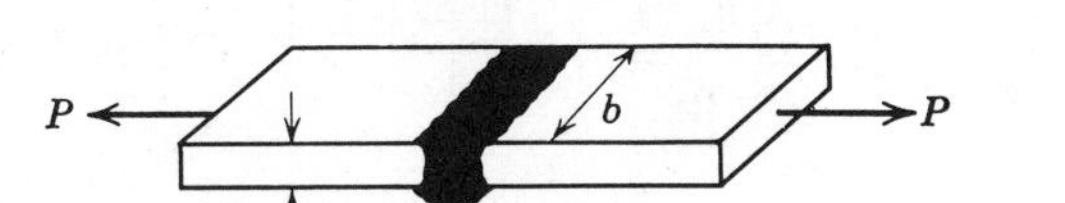

(*A*) One type of butt weld

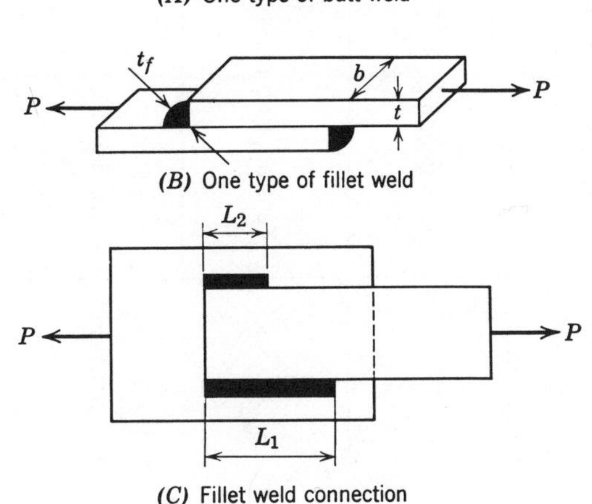

(*B*) One type of fillet weld

(*C*) Fillet weld connection

Fig. D13 Some types of welded joint connections.

Table D1 Summary of Equations for Analysis of Various Adhesive Scarf Joints

Type	Loading	Geometry	Shear Stress	Normal Stress
Flat Scarf	Tension, Compression	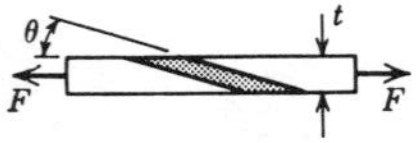 F = force/unit width	$\tau = \frac{F}{t} \sin\theta \cos\theta$	$\sigma = \frac{F}{t} \sin^2\theta$
	Pure Bending	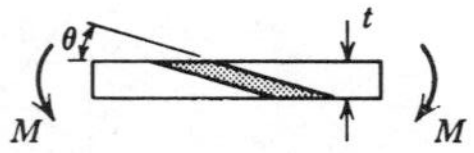 M = moment/unit width	$\tau = \frac{6M}{t} \sin\theta \cos\theta$	$\sigma = \frac{6M}{t} \sin^2\theta$
Tubular Scarf	Tension, Compression	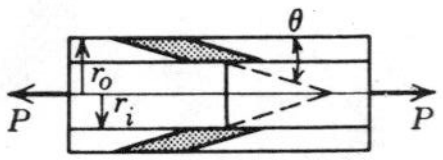 P = axial force	$\tau = \frac{P}{2\pi r_o t} \sin\theta \cos\theta$	$\sigma = \frac{P}{2\pi r_o t} \sin^2\theta$
	Pure Bending	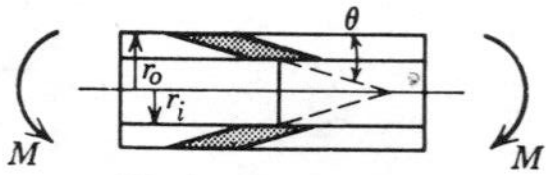 M = bending moment	$\tau = \frac{2M(r_o + r_i)}{\pi(r_o^4 - r_i^4)} \sin\theta \cos\theta$	$\sigma = \frac{2M(r_o + r_i)}{\pi(r_o^4 - r_i^4)} \sin^2\theta$
	Pure Torsion	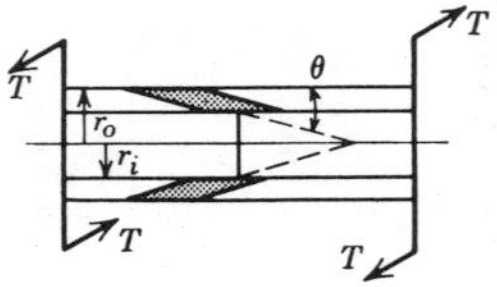 T = torque	$\tau = \frac{2T \sin\theta}{\pi(r_o + r_i)^2}$	$\sigma = 0$

D-3 ADHESIVE AND BONDED JOINTS

The analysis of lap, butt, scarf, flat-offset lap, double flat lap, landed, and tubular adhesive bonded joints has been given by Perry (11); a few of his equations useful in analyzing various scarf adhesive joints are given in Table D1.* Structural aircraft joint design, particularly adhesive bonded compression and shear panels, has been discussed by De Bruyne (2) and Lunsford (9). The analysis of the strength of these joints is quite complex and too lengthy to include here. However, for illustration, a simple adhesive lap joint will be considered. The analysis follows that suggested by Hahn and Fouser (5).

Consider the joint shown in Fig. D14; the joint is symmetrical and made of an adhesive plus two adherends of the same material and geometry. When the load F is applied, the adherends bend cylindrically and the deflection can be expressed by the plate equation in Chapter 3 (Eq. 3.310),

$$d^2y/dx^2 = -M/D \qquad \text{D33}$$

where M is a moment and

$$D = \frac{Et^3}{12(1-\nu^2)} \qquad \text{D34}$$

The stress distribution obtained by Hahn and Fouser for this case is

$$\sigma = (F/bt)(1 + C_1/C_2) \qquad \text{D35}$$

where b = adherend width

$$C_1 = 3[e^{K_1 d} + e^{-K_1 d}](1+q)e^{-K_2 x_2} \qquad \text{D36}$$

$$C_2 = (\sqrt{(2+q)^3} + 1)e^{K_1 d} - (\sqrt{(2+q)^3} - 1)e^{-K_1 d} \qquad \text{D37}$$

$$K_1 = \left(\sqrt{\frac{F(12)(1-\nu^2)t^2}{bE(2t+h_1)^3}}\right)(1/t) \qquad \text{D38}$$

$$K_2 = \left(\sqrt{\frac{F(12)(1-\nu^2)}{btE}}\right)(1/t) \qquad \text{D39}$$

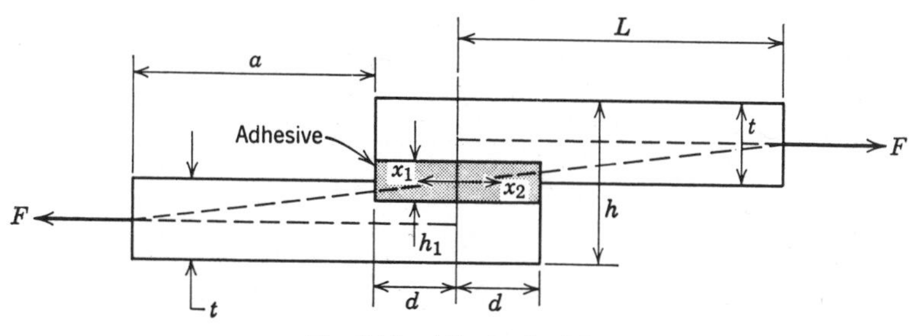

Fig. D14 Adhesive lap joint.

*Courtesy of The McGraw-Hill Publishing Co.

and

$$q = h_1/t \tag{D40}$$

An analysis for different materials and geometries is outlined in (5).

D-4 BOLTED CONNECTIONS

Two problems considered here are (a) selection of a bolt to take the applied load and (b) the torque required for pull-up.

Bolt strength

Various codes govern bolt calculations; here only an approximate method is given for finding the general size of bolt required—code calculations, where applicable, would have to be made also.

Consider the connection shown in Fig. D15. In this illustration D_p is the pitch, and D_0 is the major diameter. For such an assembly the load that can be carried in tension in approximately

$$F_T = (\pi/4)(D_p - 0.65/n)^2 \sigma_w \tag{D41}$$

where σ_w is the working stress of the material and n is the number of threads per inch. If it is assumed that the working shear stress is half the tensile yield strength, then the allowable shear load would be

$$F_\tau = \frac{\pi D_p}{4} H \sigma_w \tag{D42}$$

For satisfactory performance of bolts loaded in tension the tensile stress in the bolt as a result of tightening and external load should not exceed the allowable stress value for the material. Furthermore, the tightening load should be at least equal to the external load to insure tightness of the assembly. For situations involving cyclic loading the mean load and load range have to be properly adjusted to reduce the chances of a fatigue failure.

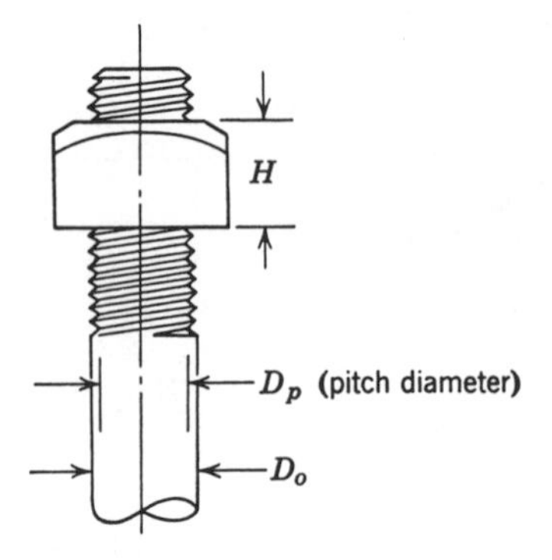

Fig. D15 Bolt and nut connection.

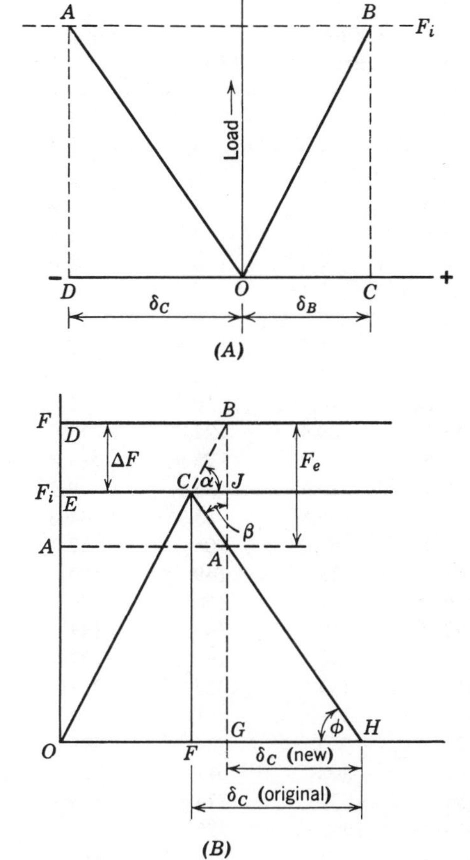

Fig. D16 Graphical representation of bolt load and deformation.

Static loading. When a bolt is torqued up, the initial axial load due to the torque is approximately*

$$F_i = T/0.2D_o \qquad \text{D43}$$

As shown in Fig. D16*A* the preload force F_i causes a deformation $+\delta_B$ in the bolt and $-\delta_C$ in the bolted material; the *spring rate* for the bolt is thus

$$K_B = \frac{F_i}{\delta_B} \qquad \text{D44}$$

and for the bolted material

$$K_C = \frac{F_i}{\delta_C} \qquad \text{D45}$$

*The factor 0.2 is an average value for plain bolts. For various plated bolts see (16).

Table D2 Maximum Torque Values for Bolts (in.-lb)*

Nominal Size	Structural Steel	Austenitic Stainless Steel	Brass	Silicon Bronze	Monel	Aluminum 2024-T4
2–56	2.2	2.5	2.0	2.3	2.5	1.4
2–64	2.7	3.0	2.5	2.8	3.1	1.7
3–48	3.5	3.9	3.2	3.6	4.0	2.1
3–56	4.0	4.4	3.6	4.1	4.5	2.4
4–40	4.7	5.2	4.3	4.8	5.2	2.9
4–48	5.9	6.6	5.4	6.1	6.7	3.6
5–40	6.9	7.7	6.3	7.1	7.8	4.2
5–44	8.5	9.4	7.7	8.7	9.6	5.1
6–32	8.7	9.6	7.9	8.9	9.8	5.3
6–40	10.9	12.1	9.9	11.2	12.3	6.6
8–32	17.8	19.8	16.2	18.4	20.2	10.8
8–36	19.8	22.0	18.0	20.4	22.4	12.0
10–24	20.8	22.8	18.6	21.2	25.9	13.8
10–32	29.7	31.7	25.9	29.3	34.9	19.2
$\frac{1}{4}$–20	65.0	75.2	61.5	68.8	85.3	45.6
$\frac{1}{4}$–28	90.0	94.0	77.0	87.0	106	57
$\frac{5}{16}$–18	129	132	107	123	149	80
$\frac{5}{16}$–24	139	142	116	131	160	86
$\frac{3}{8}$–16	212	236	192	219	266	143
$\frac{3}{8}$–24	232	259	212	240	294	157
$\frac{7}{16}$–14	338	376	317	349	427	228
$\frac{7}{16}$–20	361	400	327	371	451	242
$\frac{1}{2}$–13	465	517	422	480	584	313
$\frac{1}{2}$–20	487	541	443	502	613	328
$\frac{9}{16}$–12	613	682	558	632	774	413
$\frac{9}{16}$–18	668	752	615	697	855	656
$\frac{5}{8}$–11	1,000	1,110	907	1,030	1,330	715
$\frac{5}{8}$–18	1,140	1,244	1,016	1,154	1,482	798
$\frac{3}{4}$–10	1,259	1,530	1,249	1,416	1,832	980
$\frac{3}{4}$–16	1,230	1,490	1,220	1,382	1,790	958
$\frac{7}{8}$–9	1,919	2,328	1,905	2,140	2,775	1,495
$\frac{7}{8}$–14	1,911	2,318	1,895	2,130	2,755	1,490
1–8	2,832	3,440	2,815	3,185	4,130	2,205
1–14	2,562	3,110	2,545	2,885	3,730	1,995
$1\frac{1}{8}$–7	4,080	4,950	4,050	4,580	5,990	3,180
$1\frac{1}{8}$–12	3,860	4,400	3,810	4,330	5,650	3,020
$1\frac{1}{4}$–7	5,175	6,260	5,140	5,820	7.530	4,030
$1\frac{1}{4}$–12	4,750	5,750	4,730	5,380	6,900	3,700
$1\frac{1}{2}$–6	8,600	10,640	8,750	9,870	12,750	6,850
$1\frac{1}{2}$–12	6,950	8,420	6,900	7,800	10,100	5,400

* Table values are approximate and are torques to bring material just below yielding. Allowable values for design are corrected by various factors of safety depending on the code used.

When the bolted material consists of several materials, then

$$K_C = \frac{1}{1/K_{C_1} + 1/K_{C_2} + \ldots + 1/K_{C_n}} \qquad \text{D46}$$

The force F_i in the initial tightening is usually obtained by torquing a nut on the bolt; maximum torque values for several materials and bolt sizes are given in Table D2. In many instances, however, there is an externally applied load in addition to the preload, and this changes the values of allowable external load considerably. As shown in Fig. D16*B*, where the δ_C in Fig. D16*A* has been transferred to the plus side of the diagram, a change in load ΔF has the effect of increasing the bolt strain from *OF* to *OG*. However, the additional load has the effect of reducing the load on the bolted material so that the original compression *FH* is now reduced to *GH*; effectively this lowers the initial load F_i from *EE* to *AA*, or resulting in an *effective applied load* F_e equal to *AB*. By the geometry indicated in Fig. D16*B*

$$\frac{\Delta F}{\overline{CJ}} = \tan \alpha = K_B \qquad \text{D47}$$

and

$$\frac{\overline{CJ}}{F_e - \Delta F} = \tan \beta = \tan (90 - \phi) = \frac{1}{\tan \phi} = \frac{1}{K_C} \qquad \text{D48}$$

from which

$$\Delta F = \frac{(F_e)K_B}{K_B + K_C} \qquad \text{D49}$$

The total axial load is then $F_i + \Delta F$ or

$$F = F_i + \frac{F_e}{(1 + K_C/K_B)} \qquad \text{D50}$$

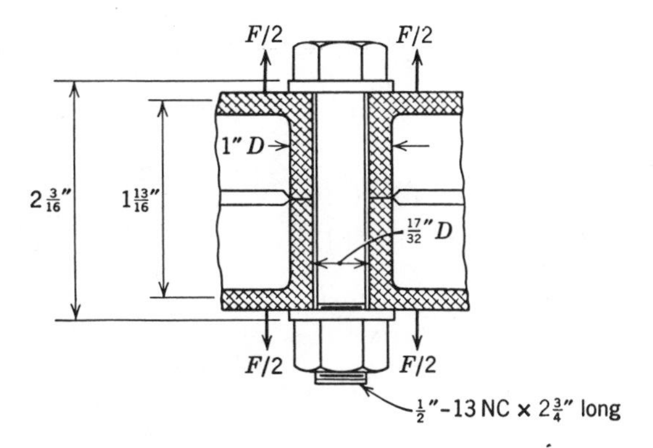

Fig. D17 Bolted assembly

The ratio $K_C/K_B = K$ is frequently referred to as the *stiffness constant* of the assembly. Often the K value is difficult to obtain for the bolted material, and experimental means must be used. For the bolt, or for the bolted material, if the geometry is simple,

$$K = F/\delta = AE/L \qquad \text{D51}$$

where A is the cross-sectional area and L is the length. For example, consider the assembly shown in Fig. D17. Two aluminum parts are bolted together with a steel bolt, and it is desired to determine the allowable external tension load and the required torque to draw up the nut. The effect of washers, and so on, is neglected, and a factor of 1.35 is allowed for the effect of shear stress induced by the torquing-up. The allowable tensile stress for the bolt is thus

$$\sigma_t = \left[1.35F_i + \frac{F_e}{(1 + K_C/K_B)}\right](1/A_r) \qquad \text{D52}$$

where A_r is the root area of the bolt. In order to fulfill the basic requirements of the joint

$$\sigma_t \leq \text{allowable stress for material}$$

and

$$F_i \geq \text{external load}$$

Therefore in the limiting case $F_i = F_e$, $K_C/K_B = 0$, and

$$\sigma_t = \frac{2.35F}{A_r} \qquad \text{D53}$$

If the allowable tensile stress for the material is 12,500 psi, then the allowable load F becomes

$$F = \frac{(12{,}500)(0.1257)}{2.35} = 665 \text{ lb}$$

The corresponding torque would then be

$$T = (665)(0.2)(0.531) = 76 \text{ in.-lb}$$

These values would be used when the constants cannot be determined. For this case, however,

$$K_C = \frac{E_C A_C}{L_C} = \frac{(10 \times 10^6)(0.785 - 0.2222)}{1.8125} = 3.11 \times 10^6$$

and

$$K_B = \frac{E_B A_B}{L_B} = \frac{(30 \times 10^6)(0.196)}{2.1875} = 2.69 \times 10^6$$

so that

$$K_C/K_B = 1.15$$

Then from Eq. D-52,

$$12{,}500 = F\left(1.35 + \frac{1}{2.15}\right)\left(\frac{1}{0.1257}\right)$$

from which

$$F = 860 \text{ lb}$$

Cyclic loading. For cyclic loading where fatigue failure would govern the criteria expressed by the modified Goodman's law (Soderberg's law) given by Eqs. 12.13 and 12.14 would be used. Thus for cyclic loading

$$\sigma_{\max} = \sigma_e + \sigma_m(1 - \sigma_e/\sigma_y) \qquad \text{D54}$$

$$\sigma_{\min} = -\sigma_e + \sigma_m(1 + \sigma_e/\sigma_y) \qquad \text{D55}$$

$$\sigma_m = \frac{\sigma_{\max} + \sigma_{\min}}{2} \qquad \text{D56}$$

and, for steel, as an approximation

$$\sigma_e = \sigma_u/2 \qquad \text{D57}$$

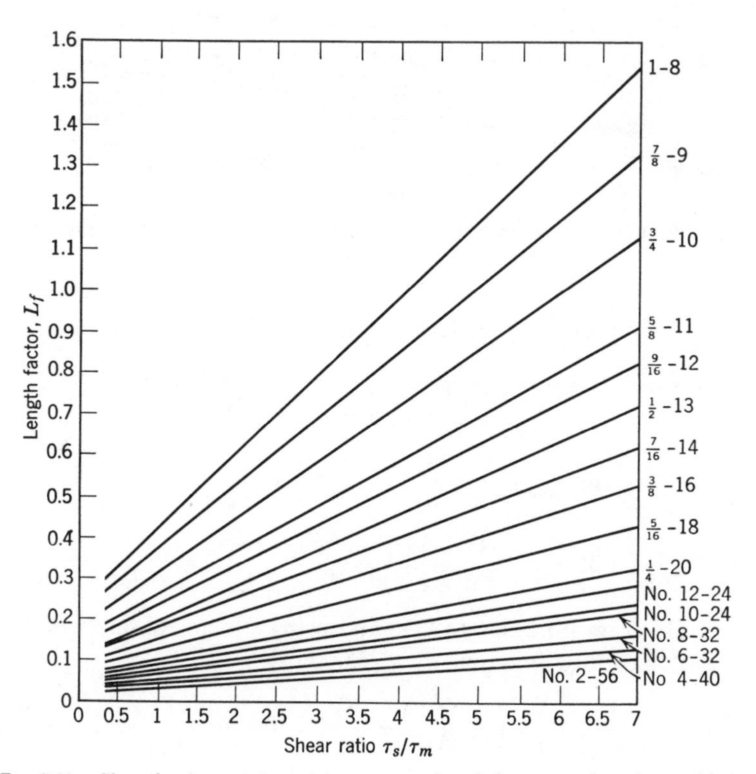

Fig. D18 Chart for determining minimum screw length for coarse thread tapped holes.

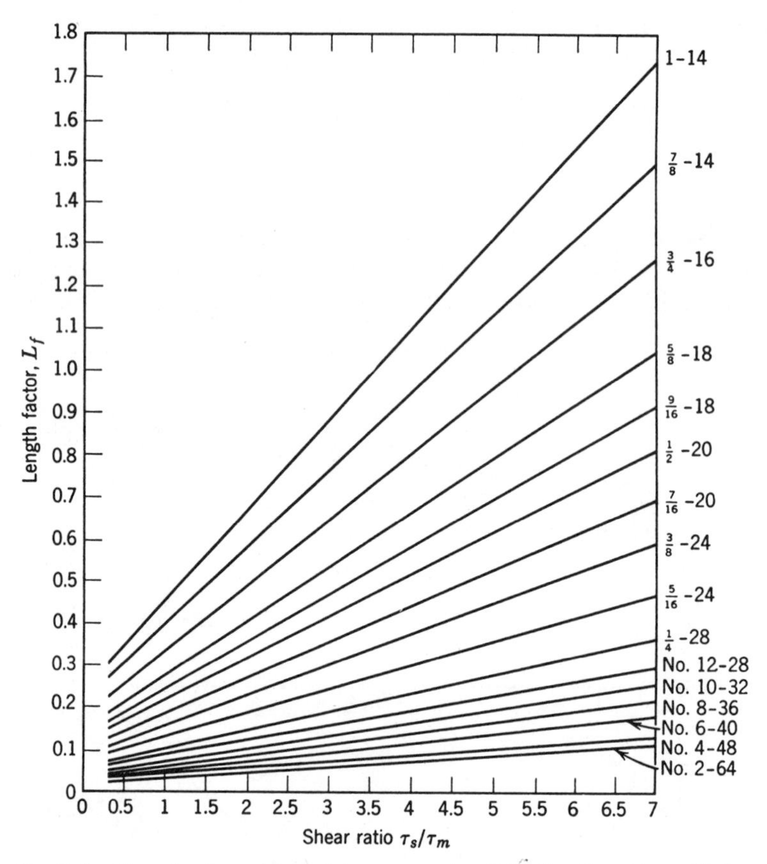

Fig. D19 Chart for determining minimum screw length for fine thread tapped holes.

Therefore, for example, for the situation in Fig. D17

$$\sigma_{\max} = F\left(1.35 + \frac{1}{1 + K_C/K_B}\right)\left(\frac{1}{A_r}\right) \tag{D58}$$

$$\sigma_{\min} = 1.35\frac{F}{A_r} \tag{D59}$$

$$\sigma_m = \frac{F}{2A_r}\left(1.35 + \frac{1}{1 + K_C/K_B}\right) + \frac{1.35F}{2A_r} = \frac{F}{2A_r}\left(2.70 + \frac{1}{1 + K_C/K_B}\right) \tag{D60}$$

Consequently, from Eqs. D54 to D56 and using a factor of safety of 2, as previously, with $\sigma_u = 80{,}000$ psi and $\sigma_y = 25{,}000$ psi

$$\frac{F}{A_r}\left(1.35 + \frac{1}{1 + K_C/K_B}\right) = \left(\frac{\sigma_u}{2}\right)\left(\frac{1}{2}\right) + \frac{F}{2A_r}\left(2.70 + \frac{1}{1 + K_C/K_B}\right)\left[1 - \frac{\sigma_u}{\sigma_y}\left(\frac{2}{2}\right)\right] \tag{D61}$$

from which the allowable cyclic load is

$$F = 432 \text{ lb}$$

This load is 50% less than the allowable for static loading. The corresponding torque, from Eq. D43, is

$$T = F(0.2)(D_0) = 432(0.2)(0.531) = 46 \text{ in.-lb}$$

Screw length for tapped holes. In general, the minimum length of thread is twice the bolt diameter plus $\frac{1}{4}$ in. for bolt length up to 6 in. and twice the bolt diameter plus $\frac{1}{2}$ in. for bolt lengths over 6 in. For design convenience the charts presented in Figs. D18 and D19 may be used (13).* The minimum length is given as

$$\text{minimum length} = \frac{\sigma_w L_f}{\tau_s} \qquad \text{D62}$$

where σ_w = maximum permissible tensile stress of screw material

L_f = length factor, from chart

τ_s = maximum permissible shear stress of screw material

On the abscissa of the charts τ_m is the permissible shear stress of the tapped material.

REFERENCES

1. H. D. Cobleigh, Ed., *ASME Screw Thread Manual*, American Society of Mechanical Engineers, New York, N.Y. (1953).
2. N. A. De Bruyne, *J. Appl. Poly. Sci.*, **6** (20), 122–129 (1962).
3. E. H. Gaylord and C. N. Gaylord, *Design of Steel Structures*, 2nd ed. McGraw-Hill Book Co., New York, N.Y.(1972).
4. L. E. Grinter, *Design of Modern Steel Structures*, 2nd ed. The MacMillan Co., New York, N.Y. (1960).
5. K. F. Hahn and D. F. Fouser, *J. Appl. Polym. Sci.*, **6** (20), 145–149 (1962).
6. J. Jones, *Fasteners*, **12** (2, 3), 10–13 (1957).
7. J. E. Lothers, *Advanced Design in Structural Steel*, Prentice-Hall, Englewood Cliffs, N.J. (1960).
8. J. L. Lubkin and E. Reissner, *Trans. Am. Soc. Mech. Eng.*, **78** (Paper No. 55-SA-59), 1213–1221 (August 1956).
9. L. R. Lunsford, *J. Appl. Polym. Sci.*, **6** (20), 130–135 (1962).
10. M. A. Melcon and F. M. Hoblit, *Prod. Eng.*, **24** (6), 160–170 (June 1953).
11. H. A. Perry, *Prod. Eng.*, **29** (27), 64–67 (July 7, 1958).
12. A. L. Phillips, Ed., *Welding Handbook*, 4th ed., American Welding Society, New York, N.Y. (1961).
13. G. Riske, *Prod. Eng.*, **30** (40), 69–71 (September 28, 1959).

*Courtesy of The McGraw-Hill Publishing Co.

14. B. E. Rossi, *Welding Engineering*, McGraw-Hill Book Co., New York, N.Y. (1954).

15. F. R. Shanley, *Basic Structures*, John Wiley and Sons, Inc. New York, N.Y. (1947).

16. V., Stimpel, *Prod. Eng.*, **31** (50), 73–75 (December 1960).

17. W. C. Wake, *Research*, **15** (5), 183–190 (May 1962).

18. *Steel Construction*, 7th ed., American Institute of Steel Construction, New York, N.Y. (1973).

AUTHOR INDEX

SUBJECT INDEX